Agricultural Waste

This book provides a comprehensive overview of agricultural waste and its utilization in various industrial technologies for sustainable management. It covers a wide range of topics, including the characterization of agricultural waste, its environmental impacts, and its utilization in various industrial sectors. This book also discusses the latest technological advancements in the conversion of agricultural waste into value-added products and the opportunities and challenges associated with their implementation.

Key Features:

- Discusses the environmental impacts of agricultural waste and the latest technological advancements in converting waste into value-added products
- Gives an overview of the use of agricultural waste in various industrial applications including biocomposite, pulp and paper, and bioenergy industries
- Considers economic, social, and environmental aspects of agricultural waste management
- Offers practical solutions for the effective utilization of agricultural waste toward sustainable development

Offering a multidisciplinary approach, this book serves as a valuable resource for researchers, academics, policymakers, and engineers interested in the sustainable management of agricultural waste.

R.A. Ilyas is Senior Lecturer in the Department of Chemical Engineering, Faculty of Chemical and Energy Engineering, and Deputy Director of the Centre for Advanced Composite Materials (CACM), Universiti Teknologi Malaysia.

M.I.H.M. Tahir earned his PhD in Chemical Engineering from the Universiti Teknologi Malaysia. His current research focuses on the thermochemical properties of municipal solid waste and agro-waste in Malaysia.

Muhammad Asyraf Muhammad Rizal is Senior Lecturer in the Department of Applied Mechanics and Design, Faculty of Mechanical Engineering, and Research Fellow at the Centre for Advanced Composite Materials (CACM), Universiti Teknologi Malaysia.

S.M. Sapuan is Professor of Composite Materials in the Department of Mechanical and Manufacturing Engineering, Universiti Putra Malaysia.

M.S.N. Atikah earned her MSc in Chemical Engineering from the Universiti Putra Malaysia. Her fields of specialization are microalgae, gasification, thermochemical conversion, biomass conversion, and microalgal biomass.

Agricultural Waste

Utilization and Industrial Technologies toward Sustainable Management

R.A. Ilyas, M.I.H.M. Tahir,
Muhammad Asyraf Muhammad Rizal,
S.M. Sapuan, and M.S.N. Atikah

CRC Press
Taylor & Francis Group
Boca Raton London New York

CRC Press is an imprint of the
Taylor & Francis Group, an **informa** business

Designed cover image: Shutterstock

First edition published 2026
by CRC Press
2385 NW Executive Center Drive, Suite 320, Boca Raton FL 33431

and by CRC Press
4 Park Square, Milton Park, Abingdon, Oxon, OX14 4RN

CRC Press is an imprint of Taylor & Francis Group, LLC

ISBN: 978-1-032-75054-5 (hbk)
ISBN: 978-1-032-75056-9 (pbk)
ISBN: 978-1-003-47220-9 (ebk)

DOI: 10.1201/9781003472209

Typeset in Times
by codeMantra

Contents

SECTION I Introduction

SECTION II Palm-Based Plant Waste

SECTION V Stalk-Based Plant Waste

SECTION VI Leaf-Based Plant Waste

SECTION VII Fruit-Based Plant Waste

SECTION VIII Tuber-Based Plant Waste

SECTION IX Challenges and Future Perspectives

Author Biographies

R.A. Ilyas is a senior lecturer in the Faculty of Chemical and Energy Engineering, Universiti Teknologi Malaysia, Malaysia. He is also a Fellow of International Association of Advanced Materials (IAAM), Sweden, Fellow of International Society for Development and Sustainability (ISDS), Japan, a member of Royal Society of Chemistry, UK and Institute of Chemical Engineers (IChemE), UK, Chair of Science Outreach for Young Scientists Network—Academy of Sciences Malaysia (YSN-ASM) 2023. He received his Diploma in Forestry, Bachelor's Degree (BSc) in Chemical Engineering, and Ph.D. degree in the field of Biocomposite Technology & Design at Universiti Putra Malaysia, Malaysia. He was the recipient of the MVP Doctor of Philosophy Gold Medal Award UPM 2019, for Best Ph.D. Thesis and Top Student Award, INTROP, UPM. He was awarded with National Young Scientist Award 2024 (Engineering) by Ministry of Science, Technology and Innovation (MOSTI), Malaysia, Outstanding Reviewer by Carbohydrate Polymers, Elsevier United Kingdom, Top Cited Article 2020–2021 Journal Polymer Composite, Wiley, 2022, and Best Paper Award at various International Conferences. R.A. Ilyas also was listed and awarded among World's Top 2% Scientist (Career-Long Achievement) Year 2022 and 2023, World's Top 2% Scientist (Subject-Wise) Citation Impact during the Single Calendar Year 2019–2023 by Stanford University, US, PERINTIS Publication Award 2021 and 2022 by Persatuan Saintis Muslim Malaysia, Emerging Scholar Award by Automotive and Autonomous Systems 2021, Belgium, Young Scientists Network— Academy of Sciences Malaysia (YSN-ASM) 2021, UTM Young Research Award 2021, UTM Publication Award 2021&2023, and UTM Highly Cited Researcher Award 2021. In 2021, he won Gold Award and Special Award (Kreso Glavac (The Republic of Croatia) at the Malaysia Technology Expo (MTE2022), Gold Award dan Special Award at International Borneo Innovation, Exhibition & Competition 2022 (IBIEC2022), and a Gold Award at New Academia Learning Innovation (NALI2022). He was awarded with Best Scientific Book Award from COMSTECH, Organization of Islamic Cooperation (OIC), Pakistan, and ModTech, Romania. His main research interests are (1) Polymer Engineering (Biodegradable Polymers, Biopolymers, Polymer composites, Polymer-gels) and (2) Material Engineering (Natural fiber reinforced polymer composites, Biocomposites, Cellulose materials, Nano-composites). To date, he has authored or co-authored more than 500 publications on green materials-related subjects.

M.I.H.M. Tahir is a chemical engineer and consultant specializing in environmental sustainability, solid waste management, and pollution control. He is a Professional Engineer (P.Eng – Chemical) registered with the Board of Engineers Malaysia and a Chartered Member of the Institution of Chemical Engineers (IChemE), UK. He holds a PhD in Chemical Engineering from Universiti Teknologi Malaysia (UTM), where he also completed his Master of Engineering. His undergraduate studies were completed through a joint program between the University of Sheffield and UniKL MICET. As a consultant, he has worked on various projects related to solid waste characterization, biomass utilization, thermal treatment systems, and waste-to-energy planning. His clients include the Ministry of Natural Resources and Environmental Sustainability (NRES), Port of Tanjung Pelepas (PTP), FGV R&D, and Nippon Steel Engineering, among others. He has also contributed to policy-related work such as the Renewable Energy Policy of Johor 2030. In addition to consultancy work, he is involved in research and technical writing. He has contributed to studies funded by Fundamental Research Grant Scheme (FRGS) and UTMPr grants, and has authored or co-authored publications in journals such as *Water*, *Journal of the Taiwan Institute of Chemical Engineers*, and *Chemical Engineering Transactions*. He is also a contributing author to a CRC Press book on agricultural waste utilization. His technical competencies include the use of tools such as ANSYS Fluent for computational fluid dynamics (CFD), artificial neural networks for modelling, and various statistical and engineering software. He has also participated in training sessions and public outreach activities on topics related to sustainability and environmental management.

Muhammad Asyraf Muhammad Rizal, also known as M.R.M. Asyraf, is currently a senior lecturer at the Department of Applied Mechanics and Design, Faculty of Mechanical Engineering, Universiti Teknologi Malaysia (UTM). He also serves as a Research Fellow at the Centre for Advanced Composite Materials (CACM), UTM. He is presently registered as a Professional Technologist and Graduate Technologist under the Malaysian Board of Technologists (MBOT). In addition, he holds the title of Graduate Engineer under the Board of Engineers Malaysia (BEM) and is a Corporate Member of the Persatuan Pembangunan Industri Enau Malaysia (PPIEM). Dr Asyraf obtained his Doctorate in Materials Engineering from Universiti Putra Malaysia (UPM), where he was awarded the IKM Research Prize in Polymer & Materials Science. For his undergraduate studies, he graduated with a Bachelor of Mechanical Engineering (First Class Honours), achieving a CGPA of 3.789. He was also the recipient of the Royal Education Award 2018 by the Conference of Rulers Malaysia, the Excellent Graduate Award 2019 by the Bank Rakyat Foundation, and the Excellent Academic Award by UPM Corporation. His main research interests include: (1) Design Development and Concurrent Engineering, (2) Characterization of Polymer Composites (3) Creep and Service Life Analysis and machine learning in composites. To date, he has authored and co-authored over 200 publications, including journal articles, book chapters, conference proceedings, bulletins, and books, primarily focusing on polymer composites and concurrent engineering. His current H-index is 43, with more than 6,900 citations, as recorded in the Scopus database. In addition, he has delivered more than 15 presentations at both local and international conferences and seminars, including engagements as keynote and invited speaker. He has published 1 authored and 1 edited book under Springer and CRC Press in the area of composite technology. His scientific roles include serving as an academic, associate editor, and editorial board member for several international journals, as well as guest editor for special issues in reputable publications. He has secured several prestigious research grants, including the Fundamental Research Grant Scheme (FRGS) from the Ministry of Higher Education, Malaysia, as well as industrial grants from Meiban Micro Sdn. Bhd. And UTM Research University Grants, amounting to a total value of RM191,158.50. He also contributes to various other grants, including industrial collaborations with Johor Port and SD Guthrie. He has also supervised more than five postgraduate students, including one Master's graduate and four ongoing PhD candidates.

S.M. Sapuan is Professor (Grade A) of composite materials, Department of Mechanical and Manufacturing Engineering, Universiti Putra Malaysia (UPM). He earned BEng in Mechanical Engineering, University of Newcastle, Australia, MSc in Engineering Design, Loughborough University, UK, and PhD in Material Engineering, De Montfort University, UK. He is Professional Engineer, and Fellow of the World Academy of Sciences, Society of Automotive Engineers International, International Artificial Intelligence Industry Alliance, Academy of Sciences Malaysia, International Society for Development and Sustainability, Plastic & Rubber Institute Malaysia, Malaysian Scientific Association, International Biographical Association, and Institute of Material Malaysia. He is Honorary Member of Asian Polymer Association and Founding Chairman and Honorary Member of Society of Sugar Palm Development and Industry, Malaysia. He is an Honorary

Adjunct Professor at National Textile University, Pakistan, and a visiting professor at UNITEN, Malaysia and BRIN, Indonesia. He is Editor in Chief of Journal of Natural Fibre Polymer Composites, co-editor-in-chief of Functional Composites and Structures, and editorial board member of 30 journals. He has produced over 2000 publications including 950 journal papers, 70 books, and 260 chapters in books. He has delivered 82 plenary and keynote lectures and 150 invited lectures. He organized 33 journal special issues as guest editor, reviewed 1500 journal papers and has 8 patents. He successfully supervised 100 PhD and 75 MSc students and 17 postdoctoral researchers. His h-index is 127 with 58,000 citations (Google Scholar). He received ISESCO Science Award, Khwarizimi International Award, National Book Award, Endeavour Research Promotion Award, Citation of Excellence Award, Emerald, UK, Frontier Researcher, Malaysia's Research Star Award 2025 from Clarivate, Publons Peer Review Award, USA, Professor of Eminence Award and Top Research Scientists' Malaysia Award. He was listed in World Top 2% Scientists, Stanford University, USA. He received SAE Subir Chowdhury Medal of Quality Leadership, International Society of Bionic Engineering Outstanding Contribution Award, China, The World Academy of Sciences (TWAS) Award in Engineering Sciences, Materials Science Leader Award by Research.com, IET Malaysia Leadership Award, William Johnson International Gold Medal, Life Time Achievement KnowTex Award and COMSTECH Best Scientific Book Award.

M.S.N. Atikah is a Malaysian researcher and chemical engineer affiliated with Universiti Putra Malaysia (UPM), where she specializes in sustainable materials, biopolymer composites, and nanocellulose-based technologies. She earned her Master of Science in Chemical Engineering with distinction (CGPA 4.00) and her Bachelor's Degree in Chemical Engineering, both from UPM. Her research contributions span over 44 indexed publications, with a total of more than 2,100 citations and an h-index of 16 (Scopus, May 2025). She has co-authored high-impact journal articles, including reviews on nanocellulose-starch biocomposites and sugar palm fiber-based materials. Her work is frequently cited in the field of sustainable engineering and green materials. She has presented her research in numerous international conferences and contributed to the development of eco-friendly materials for industrial applications. Her affiliations include membership with the Board of Engineers Malaysia (BEM) as a Graduate Engineer in the field of Chemical Engineering. She is known for her collaborative work with prominent researchers and her role in advancing biodegradable and renewable materials research in Southeast Asia.

Section I

Introduction

1 Agricultural Wastes Management

1.1 OVERVIEW OF AGRICULTURAL WASTE MANAGEMENT

Agricultural waste encompasses a wide range of materials generated from farming operations, including crop residues (Ilyas et al., 2019; Low et al., 2024), livestock manure (Koul et al., 2022), and by-products from agro-industrial processes (Mahardika et al., 2024). This waste can vary significantly in composition and volume, depending on the type of agriculture practiced and the scale of production. Crop residues, for instance, include stalks, leaves, and husks left over after harvest (Pulingam et al., 2022), while livestock operations produce manure and bedding materials (Saidu et al., 2020). Properly categorizing and understanding the scope of agricultural waste is essential for implementing effective management strategies that can minimize its negative impacts on the environment and enhance agricultural productivity.

Sustainable waste management is critical in addressing the challenges posed by agricultural waste. Figure 1.1 shows agricultural solid waste that can be valorized for industrial production, plant growth, animal feed production, soil improvement, and biosorbents. Bio-oil is representative of the leading industrial production directions. Through the production of biosorbents, organic carbon can be extracted and used for soil improvement. This facilitates the improvement of plant growth and the improvement of animal feed (Peng et al., 2023). By adopting practices such as composting, recycling, and utilizing waste-to-energy technologies, farmers can convert waste into

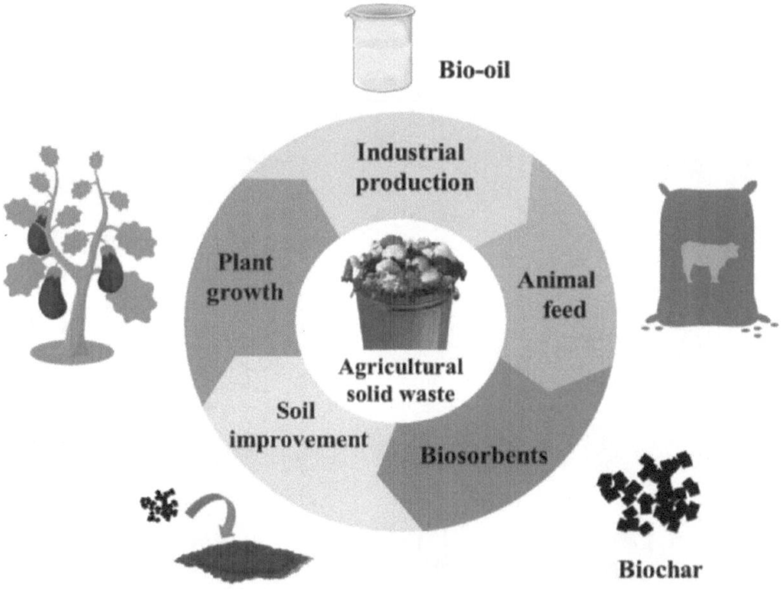

FIGURE 1.1 Application directions for agricultural solid waste are summarized as industrial production, plant growth, soil improvement, animal feed, and biosorbents (Peng et al., 2023).

DOI: 10.1201/9781003472209-2

valuable resources rather than letting it accumulate as pollution (Awalludin et al., 2015; Najafi et al., 2021; Wu et al., 2017). Sustainable management not only helps reduce waste but also improves soil fertility and reduces the need for chemical fertilizers. Moreover, engaging in sustainable practices can lead to economic benefits, as farmers can lower disposal costs and generate additional income through the sale of by-products. The holistic approach to managing agricultural waste fosters a circular economy (CE) within the agricultural sector, ensuring long-term sustainability.

The global impact of agricultural waste on the environment cannot be overstated. Improper disposal and management of agricultural waste contribute to significant environmental issues, including soil degradation, water pollution, and greenhouse gas (GHG) emissions. For instance, the decomposition of organic waste in landfills generates methane, a potent GHG that exacerbates climate change (Wang et al., 2024). Additionally, runoff from agricultural sites can carry excess nutrients into waterways, leading to eutrophication and harmful algal blooms (Feng et al., 2024). As the world is facing serious problems like climate change and losing natural resources, it is now more urgent than ever to find better ways to manage agricultural waste. By handling this waste more sustainably, we can reduce harm to the environment. This challenge needs everyone to work together not only in local communities but also on a national and global scale.

1.2 TYPES OF AGRICULTURAL WASTE

Agricultural waste is a broad category that encompasses various by-products generated from farming activities. Understanding the different types of agricultural waste is crucial for developing effective management strategies and promoting sustainability in the agricultural sector. The main types include crop residues (Shinde et al., 2022), process residues (Ruiz et al., 2013), animal waste (Maji et al., 2020), agro-industrial by-products (Onoja et al., 2019), forestry residues (Gupta et al., 2022),

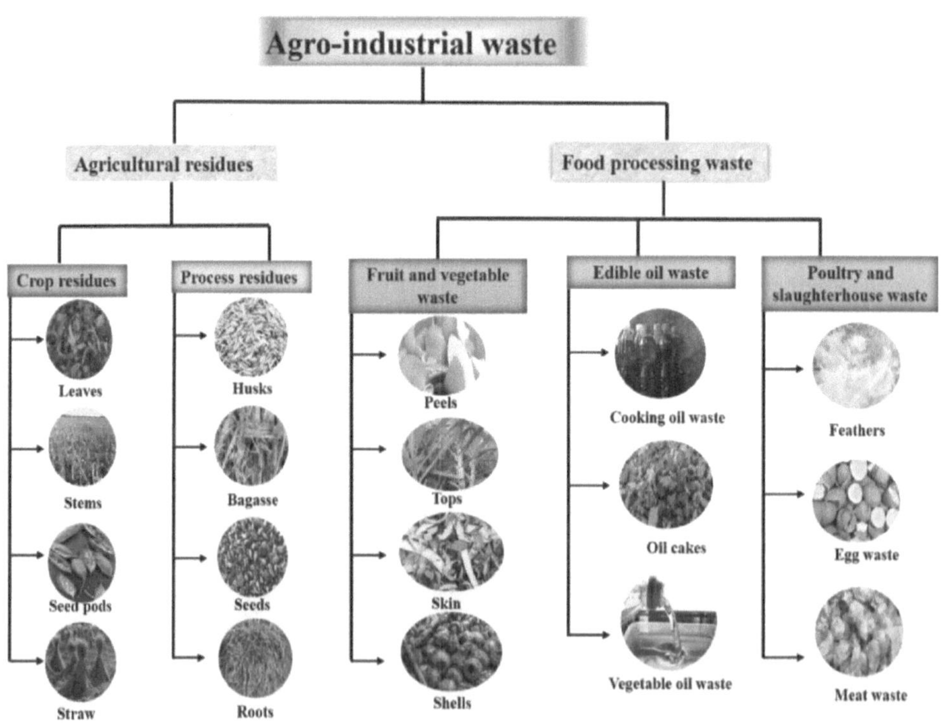

FIGURE 1.2 Classification of different types of agro-industrial food waste (Sharma et al., 2022).

and food waste (Capanoglu et al., 2022) from agricultural production (Figure 1.2). Each type has unique characteristics and implications for environmental health and resource management.

Crop residues are the remnants left in fields after harvesting. This includes stalks, leaves, and seed pods, which can account for a significant portion of agricultural waste (Shinde et al., 2022). While often considered a nuisance, crop residues can be valuable resources if managed properly. They can be incorporated back into the soil to improve fertility, used as feed for livestock, or converted into bioenergy through processes like anaerobic digestion (Dhull et al., 2023; Kadam et al., 2024). However, burning crop residues is a common practice in many regions, leading to air pollution and health issues, highlighting the need for sustainable disposal methods (Prateep Na Talang et al., 2024).

Animal waste, generated from livestock operations, represents another significant category of agricultural waste. This includes manure, bedding, and other by-products from animal husbandry (Meganathan et al., 2024). Proper management of animal waste is essential to prevent environmental contamination, as it can lead to nutrient runoff and water pollution. On the positive side, when treated and managed correctly, animal waste can be transformed into valuable fertilizers or biogas, contributing to a CE in agriculture (Hoyos-Sebá et al., 2024). The challenge lies in balancing waste management practices with animal husbandry's economic viability.

Agro-industrial by-products are derived from food processing and manufacturing activities. This category includes items like pulp, shells, and seeds, often produced in large quantities (Awoh et al., 2023; Shamsuddin et al., 2021). These by-products can be seen as waste, but they also hold potential for various applications, such as creating animal feed, compost, or bio-based materials. By finding innovative ways to utilize these by-products, the agricultural industry can reduce waste and improve overall resource efficiency, promoting a more sustainable food system.

Forestry residues, consisting of branches, leaves, and other by-products from logging and timber processing, contribute to agricultural waste in regions where forestry and agriculture intersect (Braghiroli & Passarini, 2020; Siol et al., 2023). These residues are often left on-site, leading to potential waste management challenges. However, like other types of agricultural waste, forestry residues can be repurposed for bioenergy, composting, or erosion control. Sustainable practices that incorporate these materials can enhance soil health and promote biodiversity, demonstrating the interconnectedness of agricultural and forestry practices.

Lastly, food waste from agricultural production includes unsold produce, spoiled goods, and other organic materials that do not make it to consumers (Perdana et al., 2023). This type of waste is particularly challenging as it not only represents a loss of resources but also contributes significantly to GHG emissions when disposed of in landfills. Strategies to mitigate food waste include donation programs, composting, and converting waste into bioenergy. By addressing food waste, the agricultural sector can improve efficiency and sustainability while contributing to food security and environmental health.

1.3 ENVIRONMENTAL IMPACTS OF AGRICULTURAL WASTE

The environmental impacts of agricultural waste are multifaceted and can have significant consequences for air, soil, and water quality. Improper disposal and management of agricultural residues, particularly through burning, can lead to air pollution, releasing harmful particulate matter and GHGs into the atmosphere (Figure 1.3) (Hamda et al., 2023). Additionally, the runoff of nutrients and chemicals from animal waste and agro-industrial by-products can contaminate nearby water sources, leading to eutrophication (Karaouzas, 2016). This process depletes oxygen in water bodies, causing fish kills and disrupting aquatic ecosystems (Vantarakis, 2021). Soil quality also suffers as excess nutrients and pathogens from waste accumulate, degrading soil health and its capacity to support plant life.

Greenhouse gas emissions are another critical concern linked to agricultural waste. Decomposing organic matter, particularly when it occurs anaerobically, releases methane, a potent GHG that contributes significantly to climate change. Manure management practices, such as lagoons and storage

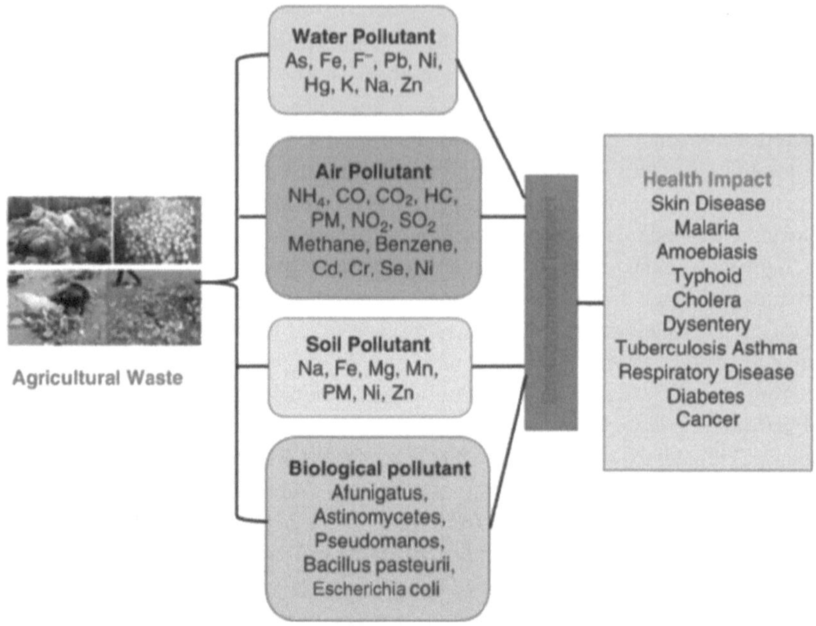

FIGURE 1.3 Agricultural waste, its disposal, and related environmental issues (Hamda et al., 2023).

tanks, can intensify methane emissions if not properly managed (Dalby et al., 2021). Moreover, the energy-intensive processes involved in waste treatment can further add to the carbon footprint of agricultural operations. Addressing these emissions is vital for reducing agriculture's overall impact on climate change and fostering a more sustainable agricultural system.

The impact of agricultural waste extends beyond pollution and emissions; it also poses threats to biodiversity and ecosystems. The over-application of fertilizers and pesticides, often resulting from improper waste management, can lead to habitat degradation and the decline of sensitive species (Narayanan et al., 2022). Furthermore, nutrient runoff can alter the balance of aquatic ecosystems, favoring certain species over others and reducing overall biodiversity. Healthy ecosystems are essential for agricultural resilience, as they provide critical services like pollination and pest control. Therefore, adopting sustainable waste management practices is crucial for protecting biodiversity and ensuring the long-term health of both agricultural and natural ecosystems.

1.4 TECHNOLOGICAL ADVANCEMENTS IN AGRICULTURAL WASTE MANAGEMENT

Technological advancements in managing agricultural waste are bringing important changes in how agricultural by-products are seen and used, changing waste into resources valuable for various industries. One of the most promising areas is in converting agricultural waste into biofuels (Awogbemi & Kallon, 2022). Through processes such as anaerobic digestion and fermentation, organic materials like crop residues and animal manure can be transformed into renewable energy sources like biogas and bioethanol. This shift helps reduce not only dependence on fossil fuels but also GHG emissions, which supports a more sustainable energy system. Such technology provides an alternative source of energy (Figure 1.4) and also helps farmers manage waste in a way that is profitable and effective.

In addition to biofuels, agricultural waste is increasingly used in the production of bioplastics and biocomposites (Mujtaba et al., 2023). Recent advancements in polymer science have made it possible to extract biopolymers from plant materials, which can then be processed into biodegradable

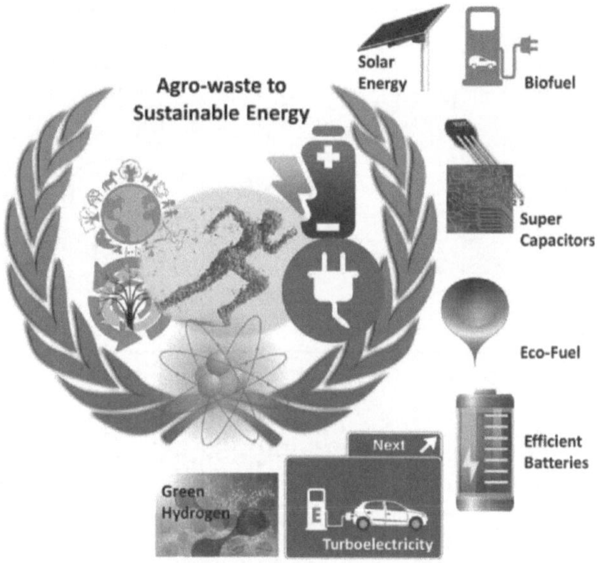

FIGURE 1.4 Agro-waste to sustainable energy (Sonu et al., 2023).

plastics (Nithya et al., 2024). This development helps address the serious problem of plastic pollution by offering a sustainable alternative that comes from renewable resources. By including agricultural waste in the production of bioplastics, industries can reduce their dependence on petroleum-based materials, lowering their carbon footprint, and creating new market opportunities for both farmers and manufacturers.

The development of sustainable biopolymer-based composites from agricultural waste biomass represents a promising approach to addressing today's environmental and economic challenges (Ibrahim et al., 2019). These innovative composites offer the potential to replace conventional materials in a variety of applications, including marine, furniture (Mazani et al., 2019), automotive, packaging (Halimatul et al., 2019; Jumaidin et al., 2019a,b; Nazrin et al., 2021), construction, structure (Alias et al., 2021), fire retardant (Suriani et al., 2021), household sectors, and additive manufacturing (Azlin et al., 2022), while also enhancing mechanical properties and biodegradability (Figure 1.5) (Asyraf et al., 2022; Nurazzi et al., 2021; Preethikaharshini et al., 2022). Biopolymers such as polylactic acid, polyhydroxyalkanoates, thermoplastic starch, and cellulose can be sourced from agricultural waste and, when combined with natural fibers through conventional fabrication methods, result in green biocomposites. Using agricultural waste biomass as a feedstock not only reduces waste and pollution but also provides a renewable and cost-effective source of raw materials (Faiz Norrrahim et al., 2021). Despite these advantages, further research is essential to optimize the properties of these biocomposites, improve their durability and performance, and scale up production to meet the increasing demand for sustainable materials (Asyraf et al., 2021). As research into sustainable materials progresses, breakthroughs in biocomposite modification and advanced technologies are expected to overcome current limitations, opening up numerous new applications (Jumaidin et al., 2021; Syafri et al., 2022). Thus, developing sustainable biopolymer-based composites from agricultural waste biomass presents an attractive solution for creating lightweight, eco-friendly materials with broad potential across multiple sectors.

Agricultural waste offers a valuable and sustainable source for isolating nanocellulose, transforming waste into high-performance materials for environmental applications (Norfarhana et al., 2022; Norfarhana et al., 2024a). This approach not only reduces waste but also aligns with global efforts to create sustainable solutions for pressing environmental issues (Imraan et al., 2023). One of the critical challenges today is water pollution, caused by contaminants like dyes, organic and inorganic

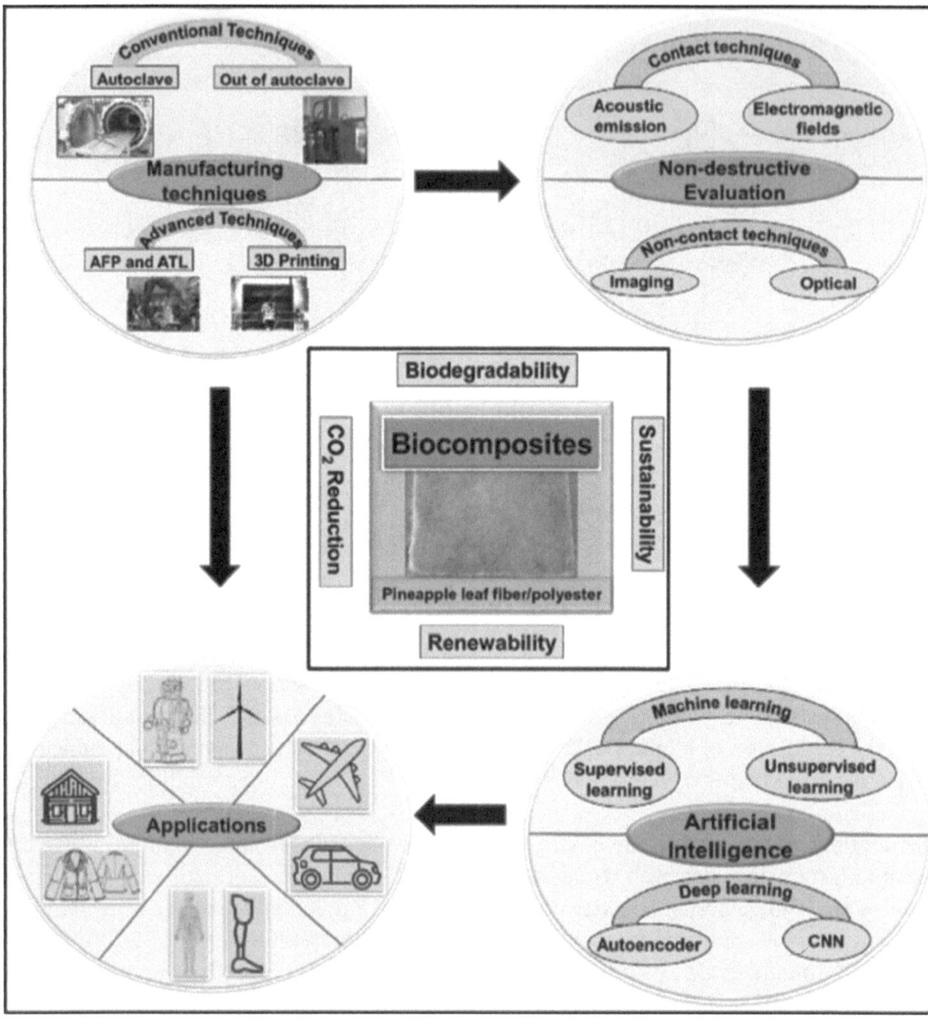

FIGURE 1.5 Agricultural waste for biocomposite applications (Preethikaharshini et al., 2022).

solvents, heavy metals, and oils, which pose significant risks to ecosystems and human health. Addressing this requires cost-effective, efficient, and eco-friendly water remediation strategies. Recently, green-adsorptive membranes have shown promise for effective water treatment, offering simple operation, high separation efficiency, low energy demands, and affordability. Nanocellulose, isolated from agricultural residues, has become a focus in membrane filtration research due to its exceptional properties, including a large specific surface area, anti-fouling behavior, high thermal resistance, mechanical strength, biodegradability, and biocompatibility (Norfarhana et al., 2024b). The high surface area of nanocellulose also allows for functionalization through abundant hydroxyl groups, enhancing its adsorptive capabilities. Studies show that modified nanocellulose is highly effective in removing heavy metals, separating oily substances, and extracting dyes from water. Thus, nanocellulose-based membranes, sourced sustainably from agricultural waste, offer a promising route for advanced water purification solutions.

Another major advancement in managing agricultural waste is through waste-to-energy technologies. These technologies can turn waste materials into usable energy through methods like incineration, gasification, and pyrolysis (Sharma, 2024). By making use of the energy in agricultural by-products, such as forestry residues and food waste, these systems can produce electricity or heat.

This offers the combined advantage of reducing waste and creating energy. By putting waste-to-energy systems in place, agricultural operations can lower their overall waste footprint, support local energy needs, and move closer to a CE.

Composting and producing organic fertilizers have also improved significantly due to technology (Noor et al., 2024). To prevent environmental degradation, swift and efficient agricultural waste management is essential. Aerobic digestion, or composting, is a prominent technique that transforms agricultural waste into a stable, sterilized organic fertilizer, thereby reducing the need for chemical fertilizers. However, conventional composting methods often face limitations, including the need for extensive land, lengthy fermentation periods, significant malodor emissions, low product quality, and limited demand for the resulting compost. Additionally, traditional composting can lead to substantial losses of organic nitrogen and carbon. This comprehensive chapter investigates sustainable and adaptable strategies to enhance the efficiency of agricultural waste composting. Noor et al.'s (2024) findings reveal that compost additives significantly influence the composting process by altering variables such as temperature, pH, and moisture content. By modifying the compost mixture, additives can effectively reduce gas emissions and limit the mobility of mineral ions. Specifically, the use of additives offers several advantages: (1) they enhance the physicochemical composition of the compost mixture; (2) accelerate the breakdown of organic material and stimulate microbial activity; (3) lower GHG and ammonia (NH_3) emissions, which reduces nitrogen (N) loss; and (4) help retain nutrients in the compost, promoting soil fertility, maturity, and reduced phytotoxicity. Modern composting techniques use controlled aeration, temperature control, and microbial inoculation to speed up decomposition and increase nutrient availability. These advancements turn agricultural waste into high-quality organic fertilizers that improve soil health and fertility. The use of compost in agriculture allows farmers to reduce their dependence on synthetic fertilizers, which are often linked to environmental harm (Srivastav et al., 2023). As awareness of organic farming's environmental benefits grow, the development of efficient composting technologies has become essential to achieving sustainable agricultural waste management. While synthetic fertilizers have played a transformative role in increasing food production, their excessive use has led to environmental concerns, particularly regarding water systems. Despite global nitrogen (N), phosphorus (P), and potassium (K) fertilizer usage rising from 64.9, 25.9, and 18.2 kg/ha in 2000 to 85.5, 33.2, and 20.4 kg/ha, respectively, these levels remain relatively low. However, the overuse of inorganic fertilizers has resulted in water contamination, harming ecosystems and introducing toxins into the food chain. This accumulation of harmful chemicals poses significant health risks, including severe illnesses such as cancer (Rashmi et al., 2020). For instance, nitrate concentrations in drinking water are limited to 45–50 mg/L in many countries to prevent adverse health impacts (Srivastav et al., 2023). To balance agricultural productivity with environmental and public health, there is a growing need to shift from chemical fertilizers to sustainable biofertilizers. Composts produced from various waste sources, such as poultry and dairy farms, have proven to be nutrient-rich, containing high levels of N, P, and K. For example, compost from dairy farm waste may contain approximately 45,100 mg/kg of nitrogen, 7,300 mg/kg of phosphorus, and 9,100 mg/kg of potassium. To make biofertilizers viable in agriculture, it is essential to raise awareness among farmers about sustainable agricultural practices, encouraging them to adopt environmentally friendly management approaches.

Additionally, agricultural waste is now being used to create products such as biochar, bio-bricks, and bio-coal (Arora et al., 2023). Biochar, which is made through pyrolysis, provides benefits like carbon storage and improved soil health (Singh Yadav et al., 2023). Bio-bricks and bio-coal present eco-friendly alternatives in construction and energy industries, supporting the movement toward environmentally sustainable materials (Koul et al., 2022). Progress in biotechnology also enables the production of biohydrogen, biomethane, and bioethanol from agricultural residues, resulting in renewable fuels that produce low emissions. Biohydrogen, for example, is attracting interest as a clean fuel for hydrogen fuel cells, while biomethane can be used for heating and transport, both helping to lower carbon emissions and supporting renewable energy (Srivastava et al., 2020).

Producing natural acids and bioelectricity from agricultural waste also shows the progress of technology in this field. Natural acids produced through microbial fermentation can replace synthetic chemicals in various industries. Bioelectricity, generated from microbial fuel cells (MFCs) that use organic materials, offers a new approach to renewable energy, producing electricity from agricultural waste and helping increase energy independence in rural areas (Pandit et al., 2021). Agro-industrial wastewater and waste residues are primarily rich in organic compounds, including cellulose, hemicellulose, and lignin, and they possess a relatively high energy potential. An efficient agro-waste treatment system offers multiple benefits, including energy recovery and waste stabilization. In response to the environmental impacts of fossil fuel consumption, there has been a renewed emphasis on utilizing renewable resources. Globally, efforts are focused on recovering energy from agricultural waste, to bolster energy security. To achieve this, various technologies and recovery methods have been developed, one of which is the MFC. MFCs have proven transformative for renewable energy production, contributing substantially to bioenergy generation (Pandit et al., 2021). The development of MFC technology enables agricultural waste to be used as a feedstock by diverse microbial communities within the anode compartment of the fuel cell. Due to the complex structure and crystallinity of agricultural biomass, biodegradation can be challenging; however, the high moisture content of such waste assists in overcoming this limitation. Agricultural and fruit wastewater resources support bacterial activity, allowing microbes to break down solid biomass effectively (Dwivedi et al., 2022). Additionally, it has been shown that different agricultural waste types and wastewater containing industrial by-products can enhance bioenergy production in MFCs through biochemical and thermochemical conversion routes. Several studies demonstrate the application of diverse substrates from different agro-industries in MFCs, showing promising results for simultaneous wastewater treatment and energy recovery. An appealing aspect of MFC technology is its dual function in wastewater management and bioenergy production, offering pollutant removal, environmental safety, and energy generation. This technology boasts several advantages, such as low-input energy costs and minimal residual biosolid production, making it an efficient bioenergy source. Research has further shown that concentrated wastewater from agro-waste enhances bioenergy production by fostering microbial consortia that drive oxidation–reduction reactions in MFCs using various anode and cathode materials.

Effective pretreatment methods have also been developed to address challenges posed by agro-waste with varying particle sizes and calorific values. There is a pressing need to promote sustainable environmental practices within agriculture, along with standardized agro-waste management, to minimize waste volumes entering the environment and open new avenues for bioenergy production. These innovations demonstrate a significant shift in how agricultural waste is managed, where waste is changed into a wide range of valuable products and energy sources. This development not only tackles the environmental challenges related to waste but also provides economic opportunities and supports sustainability across different sectors, marking an important step toward a circular and eco-friendly economy.

1.5 ECONOMIC AND SOCIAL ASPECTS OF AGRICULTURAL WASTE MANAGEMENT

Agricultural waste management offers substantial economic benefits through the process of waste valorization, which involves converting waste materials into usable products such as compost, bioenergy, or animal feed. By adopting these innovative practices, farmers can not only reduce costs associated with waste disposal but also create new revenue streams. This dual approach enhances overall profitability, allowing agricultural enterprises to invest in sustainable practices and technologies. Moreover, as waste valorization becomes more prevalent, it can stimulate local economies by promoting the development of related industries and creating jobs in waste processing and management.

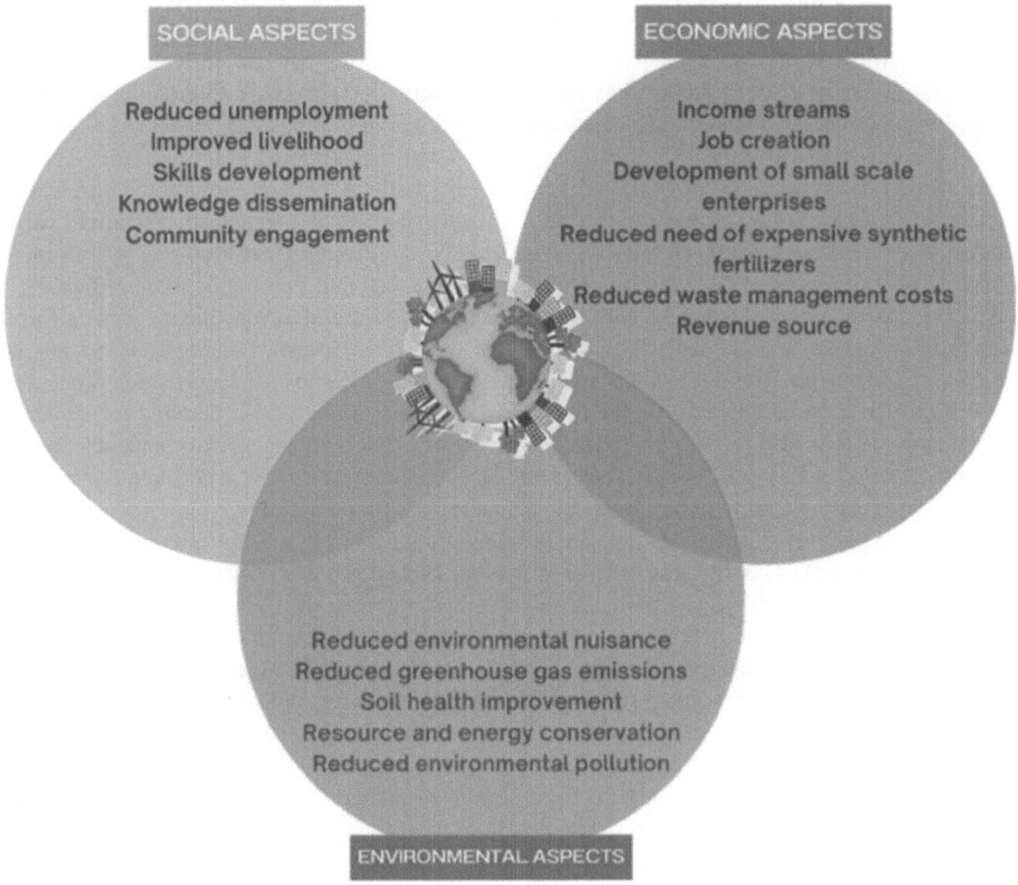

FIGURE 1.6 Social, economic, and environmental aspects of sustainability (Sheer et al., 2024).

Beyond the economic advantages, agricultural waste management has significant social implications for rural and agricultural communities (Figure 1.6) (Sheer et al., 2024). When farmers collaborate on waste management initiatives, they foster stronger community ties and promote knowledge sharing. This collective effort can lead to a sense of empowerment, as communities work together to address environmental challenges and improve local livelihoods. Additionally, effective waste management can enhance public health by reducing pollution and promoting a cleaner environment, contributing to the overall well-being of rural populations. Such social cohesion is crucial for building resilience in these communities, enabling them to adapt to changing economic and environmental conditions.

The success of agricultural waste management initiatives largely depends on the establishment of supportive policy and regulatory frameworks. Effective policies can provide the necessary incentives for farmers to adopt sustainable waste management practices, ensuring compliance with environmental standards and encouraging innovation (Sinha et al., 2021). Regulations that promote best practices in waste management can help mitigate potential environmental impacts while maximizing economic benefits. By integrating waste management into broader agricultural policies, governments can foster a more sustainable agricultural sector that not only benefits individual farmers but also enhances the well-being of rural communities and contributes to environmental conservation.

1.6 CHALLENGES IN AGRICULTURAL WASTE MANAGEMENT

Agricultural waste management faces significant challenges, primarily due to technological and infrastructural barriers. Many rural areas lack the necessary infrastructure to effectively collect, process, and recycle agricultural waste. Inadequate facilities for composting, anaerobic digestion, or recycling can hinder farmers' ability to implement waste management practices. Additionally, the availability of advanced technologies is often limited, particularly in developing regions, where farmers may not have access to the equipment or knowledge needed to transform waste into valuable resources (Budiman, 2021; Gontard et al., 2018). These technological gaps can lead to inefficiencies and increase the risk of environmental contamination from improperly managed waste.

Economic and market challenges further complicate agricultural waste management efforts. Farmers often face financial constraints that prevent them from investing in waste management systems or technologies. The initial costs associated with establishing waste valorization processes can be prohibitive, especially for smallholder farmers. Furthermore, market demand for recycled products may be low, limiting the financial incentives for farmers to engage in sustainable waste practices. Without a clear economic benefit, many farmers may be reluctant to change established practices, perpetuating a cycle of waste mismanagement and lost potential.

Policy and regulatory hurdles also play a critical role in shaping agricultural waste management outcomes. Inconsistent or poorly enforced regulations can create confusion and uncertainty for farmers, deterring them from adopting sustainable practices. Additionally, the lack of comprehensive policies that promote waste management and valorization can lead to missed opportunities for economic development and environmental protection. Governments may need to streamline regulations and provide clearer guidance on best practices to encourage farmers to participate actively in waste management initiatives. Without supportive policies, the potential for positive change remains stunted.

Finally, sustainability and environmental considerations are central to the discourse on agricultural waste management. While the goal is to minimize waste and reduce environmental impacts, many waste management practices can inadvertently cause harm if not implemented correctly. For example, improper composting techniques can lead to GHG emissions, while the disposal of certain waste products may contaminate soil and water sources. It is essential for waste management strategies to prioritize sustainability, ensuring that they not only address immediate waste issues but also contribute to long-term environmental health. Balancing economic, social, and environmental goals is vital for creating effective agricultural waste management systems that support both farmers and the ecosystems they rely on.

1.7 OPPORTUNITIES AND FUTURE TRENDS

The concept of a CE in agriculture presents a transformative opportunity for managing agricultural waste effectively (Velasco-Muñoz et al., 2021). The agricultural sector's unique characteristics including the perishable nature of products, seasonality, and dependency on natural ecosystems necessitate a tailored CE framework. Yet, current CE research often overlooks agriculture's specific needs, resulting in a lack of standardized frameworks and indicators. This study adopts the CE framework specifically for agriculture, proposing a definition, principles, and preliminary indicators to measure circularity, with an emphasis on sustainability. Existing indicators for CE in agriculture are limited, as they are primarily adopted from linear economy models and fail to capture the full complexity of agricultural processes. New, comprehensive indicators are essential to accurately assess agricultural circularity across environmental, economic, and social dimensions. These indicators should account for the sector's diverse activities, facilitate meso- and macro-level analyses, and evaluate circularity strategies such as narrowing, slowing, closing, and regenerating material loops. Adopting circular models in agriculture requires transforming supply and consumption patterns, favoring local product marketing, and promoting business models that enable nutrient recycling. Policies should support waste management, incentivize circular practices, and incorporate education and financial incentives. Shared infrastructure ownership can further enhance circularity (Velasco-Muñoz et al., 2021). By reimagining waste as a resource, farmers can adopt practices that minimize waste generation and maximize the reuse of materials. This shift not only promotes sustainability but also enhances economic resilience by creating closed-loop systems where organic waste is composted to enrich the soil, while by-products can be utilized in various agricultural processes. Emphasizing the CE encourages a holistic approach to agriculture that benefits both the environment and farmers' bottom lines, paving the way for a more sustainable and productive agricultural sector.

Innovations in waste utilization are emerging as key drivers in improving agricultural waste management. New technologies and processes, such as anaerobic digestion and biochar production, offer farmers efficient ways to convert waste into valuable products like biogas and soil amendments (Zhao et al., 2021). These innovations not only help reduce the volume of waste but also provide alternative energy sources and enhance soil fertility. As research and development continue to advance, the agricultural sector is likely to see increased adoption of these cutting-edge practices, leading to greater resource efficiency and sustainability.

The integration of digital and smart technologies plays a crucial role in enhancing agricultural waste management. Precision farming tools, such as IoT devices and data analytics, can help farmers monitor waste generation and optimize its utilization (Dhanaraju et al., 2022; Elijah et al., 2018; Farooqui et al., 2024). For instance, smart sensors can track the composition of waste, allowing for tailored treatment methods that maximize recovery and minimize environmental impact. Additionally, digital platforms can facilitate better coordination among farmers, processors, and consumers, streamlining the waste management process and fostering transparency in agricultural practices. As technology continues to evolve, its application in waste management is set to revolutionize how the agricultural sector approaches sustainability.

Collaborations between academia, industry, and policymakers are essential to harness the full potential of agricultural waste management. By fostering partnerships that bring together research institutions, private companies, and government agencies, stakeholders can share knowledge, resources, and best practices to drive innovation. These collaborations can also help shape effective policies that support sustainable waste management initiatives and promote the adoption of CE principles. By working together, these entities can create a robust framework for agricultural waste management that not only addresses current challenges but also paves the way for a sustainable and resilient agricultural future.

1.8 CONCLUSION

Agricultural waste management plays a crucial role in achieving sustainability by converting waste into renewable energy, bio-based materials, and nutrient-rich compost. This transformation not only reduces environmental impacts but also supports CE goals. Despite promising advancements in waste-to-energy and composting technologies, challenges such as limited infrastructure, economic barriers, and the absence of agriculture-specific CE indicators remain. For agriculture to fully integrate circular practices, it is essential to develop comprehensive, context-specific indicators and policies that support waste management at all scales. Future research should focus on improving CE frameworks, optimizing biocomposites, and enhancing stakeholder awareness. Ultimately, by promoting sustainable waste management practices, the agricultural sector can reduce its ecological footprint and contribute to a resilient, eco-friendly economy.

REFERENCES

Alias, A. H., Norizan, M. N., Sabaruddin, F. A., Asyraf, M. R. M., Norrrahim, M. N. F., Ilyas, A. R., Kuzmin, A. M., Rayung, M., Shazleen, S. S., Nazrin, A., Sherwani, S. F. K., Harussani, M. M., Atikah, M. S. N., Ishak, M. R., Sapuan, S. M., & Khalina, A. (2021). Hybridization of MMT/lignocellulosic fiber reinforced polymer nanocomposites for structural applications: A review. *Coatings*, *11*(11), 1355. https://doi.org/10.3390/coatings11111355

Arora, J., Ramawat, K. G., & Mérillon, J.-M. (2023). Disposal of agricultural waste and its effects on the environment, production of useful metabolites and energy: Potential and challenges (pp. 3–20). https://doi.org/10.1007/978-981-19-8774-8_1

Asyraf, M. R. M., Ishak, M. R., Norrrahim, M. N. F., Nurazzi, N. M., Shazleen, S. S., Ilyas, R. A., Rafidah, M., & Razman, M. R. (2021). Recent advances of thermal properties of sugar palm lignocellulosic fibre reinforced polymer composites. *International Journal of Biological Macromolecules*, *193*, 1587–1599. https://doi.org/10.1016/j.ijbiomac.2021.10.221

Asyraf, M. R. M., Syamsir, A., Zahari, N. M., Supian, A. B. M., Ishak, M. R., Sapuan, S. M., Sharma, S., Rashedi, A., Razman, M. R., Zakaria, S. Z. S., Ilyas, R. A., & Rashid, M. Z. A. (2022). Product development of natural fibre-composites for various applications: Design for sustainability. *Polymers*, *14*(5), 920. https://doi.org/10.3390/polym14050920

Awalludin, M. F., Sulaiman, O., Hashim, R., & Nadhari, W. N. A. W. (2015). An overview of the oil palm industry in Malaysia and its waste utilization through thermochemical conversion, specifically via liquefaction. *Renewable and Sustainable Energy Reviews, 50*, 1469–1484. Elsevier Ltd. https://doi.org/10.1016/j.rser.2015.05.085

Awogbemi, O., & Kallon, D. V. Von. (2022). Valorization of agricultural wastes for biofuel applications. *Heliyon*, *8*(10), e11117. https://doi.org/10.1016/j.heliyon.2022.e11117

Awoh, E. T., Kiplagat, J., Kimutai, S. K., & Mecha, A. C. (2023). Current trends in palm oil waste management: A comparative review of Cameroon and Malaysia. *Heliyon*, *9*(11), e21410. https://doi.org/10.1016/j.heliyon.2023.e21410

Azlin, M. N. M., Ilyas, R. A., Zuhri, M. Y. M., Sapuan, S. M., Harussani, M. M., Sharma, S., Nordin, A. H., Nurazzi, N. M., & Afiqah, A. N. (2022). 3D printing and shaping polymers, composites, and nanocomposites: A review. *Polymers*, *14*(1), 180. https://doi.org/10.3390/polym14010180

Braghiroli, F. L., & Passarini, L. (2020). Valorization of biomass residues from forest operations and wood manufacturing presents a wide range of sustainable and innovative possibilities. *Current Forestry Reports*, *6*(2), 172–183. https://doi.org/10.1007/s40725-020-00112-9

Budiman, I. (2021). The complexity of barriers to biogas digester dissemination in Indonesia: Challenges for agriculture waste management. *Journal of Material Cycles and Waste Management*, *23*(5), 1918–1929. https://doi.org/10.1007/s10163-021-01263-y

Capanoglu, E., Nemli, E., & Tomas-Barberan, F. (2022). Novel approaches in the valorization of agricultural wastes and their applications. *Journal of Agricultural and Food Chemistry*, *70*(23), 6787–6804. https://doi.org/10.1021/acs.jafc.1c07104

Dalby, F. R., Hafner, S. D., Petersen, S. O., VanderZaag, A. C., Habtewold, J., Dunfield, K., Chantigny, M. H., & Sommer, S. G. (2021). Understanding methane emission from stored animal manure: A review to guide model development. *Journal of Environmental Quality*, *50*(4), 817–835. https://doi.org/10.1002/jeq2.20252

Dhanaraju, M., Chenniappan, P., Ramalingam, K., Pazhanivelan, S., & Kaliaperumal, R. (2022). Smart farming: Internet of Things (IoT)-based sustainable agriculture. *Agriculture*, *12*(10), 1745. https://doi.org/10.3390/agriculture12101745

Dhull, P., Lohchab, R. K., Kumar, S., Kumari, M., Shaloo, & Bhankhar, A. K. (2023). Anaerobic digestion: Advance techniques for enhanced biomethane/biogas production as a source of renewable energy. *BioEnergy Research*, *17*(2), 1228–1249. https://doi.org/10.1007/s12155-023-10621-7

Dwivedi, S., Tanveer, A., Yadav, S., Anand, G., & Yadav, D. (2022). Agro-wastes for cost effective production of industrially important microbial enzymes. In *Microbial Biotechnology* (pp. 435–460). Wiley. https://doi.org/10.1002/9781119834489.ch23

Elijah, O., Rahman, T. A., Orikumhi, I., Leow, C. Y., & Hindia, M. H. D. N. (2018). An overview of Internet of Things (IoT) and data analytics in agriculture: Benefits and challenges. *IEEE Internet of Things Journal*, *5*(5), 3758–3773. https://doi.org/10.1109/JIOT.2018.2844296

Faiz Norrrahim, M. N., Ilyas, R. A., Nurazzi, N. M., Asmal Rani, M. S., Nur Atikah, M. S., & Shazleen, S. S. (2021). Chemical pretreatment of lignocellulosic biomass for the production of bioproducts: An overview. *Applied Science and Engineering Progress*, 1–18. https://doi.org/10.14416/j.asep.2021.07.004

Farooqui, N. A., Haleem, Mohd., Khan, W., & Ishrat, M. (2024). Precision agriculture and predictive analytics. In *Intelligent Techniques for Predictive Data Analytics* (pp. 171–188). Wiley. https://doi.org/10.1002/9781394227990.ch9

Feng, L., Wang, Y., Hou, X., Qin, B., Kuster, T., Qu, F., Chen, N., Paerl, H. W., & Zheng, C. (2024). Harmful algal blooms in inland waters. *Nature Reviews Earth & Environment*, *5*(9), 631–644. https://doi.org/10.1038/s43017-024-00578-2

Gontard, N., Sonesson, U., Birkved, M., Majone, M., Bolzonella, D., Celli, A., Angellier-Coussy, H., Jang, G.-W., Verniquet, A., Broeze, J., Schaer, B., Batista, A. P., & Sebok, A. (2018). A research challenge vision regarding management of agricultural waste in a circular bio-based economy. *Critical Reviews in Environmental Science and Technology*, *48*(6), 614–654. https://doi.org/10.1080/10643389.2018.1471957

Gupta, J., Kumari, M., Mishra, A., Swati, Akram, M., & Thakur, I. S. (2022). Agro-forestry waste management—A review. *Chemosphere*, *287*, 132321. https://doi.org/10.1016/j.chemosphere.2021.132321

Halimatul, M. J., Sapuan, S. M., Jawaid, M., Ishak, M. R., & Ilyas, R. A. (2019). Water absorption and water solubility properties of sago starch biopolymer composite films filled with sugar palm particles. *Polimery*, *64*(9), 27–35. https://doi.org/10.14314/polimery.2019.9.4

Hamda, A. S., Abo, L. D., Jayakumar, M., Selvakumar, K. V., Periyasamy, S., & Emana, A. N. (2023). Introduction: Growth of agricultural waste, its disposal, and related environmental issues. In *Agricultural Waste to Value-Added Products* (pp. 1–19). Springer Nature Singapore. https://doi.org/10.1007/978-981-99-4472-9_1

Hoyos-Sebá, J. J., Arias, N. P., Salcedo-Mendoza, J., & Aristizábal-Marulanda, V. (2024). Animal manure in the context of renewable energy and value-added products: A review. *Chemical Engineering and Processing: Process Intensification*, *196*, 109660. https://doi.org/10.1016/j.cep.2023.109660

Ibrahim, M., Sapuan, S., Zainudin, E. S., Zuhri, M. Y. M., & Edhirej, A. (2019). Corn (maize)—Its fibers, polymers, composites, and applications: A review. *Biodegradable Composites*, 13–36. https://doi.org/10.1515/9783110603699

Ilyas, R. A., Sapuan, S. M., Ibrahim, R., Abral, H., Ishak, M. R., Zainudin, E. S., Atikah, M. S. N., Mohd Nurazzi, N., Atiqah, A., Ansari, M. N. M., Syafri, E., Asrofi, M., Sari, N. H., & Jumaidin, R. (2019). Effect of sugar palm nanofibrillated celluloseconcentrations on morphological, mechanical andphysical properties of biodegradable films basedon agro-waste sugar palm (Arenga pinnata (Wurmb.) Merr) starch. *Journal of Materials Research and Technology*. https://doi.org/10.1016/j.jmrt.2019.08.028

Imraan, M., Ilyas, R. A., Norfarhana, A. S., Bangar, S. P., Knight, V. F., & Norrahim, M. N. F. (2023). Sugar palm (Arenga pinnata) fibers: New emerging natural fibre and its relevant properties, treatments and potential applications. *Journal of Materials Research and Technology*, *24*, 4551–4572. https://doi.org/10.1016/j.jmrt.2023.04.056

Jumaidin, R., Adam, N. W., Ilyas, R. A., Hussin, M. S. F., Taha, M. M., Mansor, M. R., Azlan, U. A. A., & Yob, M. S. (2019a). Water transport and physical properties of sugarcane bagasse fibre reinforced thermoplastic potato starch biocomposite. *Journal of Advanced Research in Fluid Mechanics and Thermal Sciences*, *61*(2), 273–281.

Jumaidin, R., Diah, N. A., Ilyas, R. A., Alamjuri, R. H., & Yusof, F. A. M. (2021). Processing and characterisation of banana leaf fibre reinforced thermoplastic cassava starch composites. *Polymers*, *13*(9), 1420. https://doi.org/10.3390/polym13091420

Jumaidin, R., Saidi, Z. A. S., Ilyas, R. A., Ahmad, M. N., Wahid, M. K., Yaakob, M. Y., Maidin, N. A., Rahman, M. H. A., & Osman, M. H. (2019b). Characteristics of cogon grass fibre reinforced thermoplastic cassava starch biocomposite: Water absorption and physical properties. *Journal of Advanced Research in Fluid Mechanics and Thermal Sciences*, *62*(1), 43–52.

Kadam, R., Jo, S., Lee, J., Khanthong, K., Jang, H., & Park, J. (2024). A review on the anaerobic co-digestion of livestock manures in the context of sustainable waste management. *Energies*, *17*(3), 546. https://doi.org/10.3390/en17030546

Karaouzas, I. (2016). Agro-industrial wastewater pollution in Greek river ecosystems (pp. 169–204). https://doi.org/10.1007/698_2016_453

Koul, B., Yakoob, M., & Shah, M. P. (2022). Agricultural waste management strategies for environmental sustainability. *Environmental Research*, *206*, 112285. https://doi.org/10.1016/j.envres.2021.112285

Low, L. Q., Ilyas, R. A., Jalil, R., Hawanis, H. S. N., Ibrahim, R., Azriena, H. A. A., Zuhri, M. Y. M., Ainun, Z. M. A., Fatriasari, W., Hastuti, N., & Amelia, D. (2024). Physical and mechanical properties enhancement of beaten oil palm trunk pulp and paper by optimizing starch addition: Towards sustainable packaging solutions. *Industrial Crops and Products*, *221*, 119232. https://doi.org/10.1016/j.indcrop.2024.119232

Mahardika, M., Zakiyah, A., Ulfa, S. M., Ilyas, R. A., Hassan, M. Z., Amelia, D., Knight, V. F., & Norrahim, M. N. F. (2024). Recent developments in Oil Palm Empty Fruit Bunch (OPEFB) fiber composite. *Journal of Natural Fibers*, *21*(1). https://doi.org/10.1080/15440478.2024.2309915

Maji, S., Dwivedi, D. H., Singh, N., Kishor, S., & Gond, M. (2020). Agricultural waste: Its impact on environment and management approaches (pp. 329–351). https://doi.org/10.1007/978-981-15-1390-9_15

Mazani, N., Sapuan, S. M., Sanyang, M. L., Atiqah, A., & Ilyas, R. A. (2019). Design and fabrication of a shoe shelf from kenaf fiber reinforced unsaturated polyester composites. In H. Ariffin, S. M. Sapuan, & M. A. Hassan (Eds.), *Lignocellulose for Future Bioeconomy* (1st ed., Issue 2000, pp. 315–332). Elsevier. https://doi.org/10.1016/B978-0-12-816354-2.00017-7

Meganathan, A., Shekar, K., Minhas, A. P., Jaganathan, M. K., Thiruvengadam, R., & Ramya Devi, K. T. (2024). Utilization of animal waste to produce fertilizer, biogas, bio-oils, and other value-added products. In *From Waste to Wealth* (pp. 1063–1087). Springer Nature Singapore. https://doi.org/10.1007/978-981-99-7552-5_47

Mujtaba, M., Fernandes Fraceto, L., Fazeli, M., Mukherjee, S., Savassa, S. M., Araujo de Medeiros, G., do Espírito Santo Pereira, A., Mancini, S. D., Lipponen, J., & Vilaplana, F. (2023). Lignocellulosic biomass from agricultural waste to the circular economy: A review with focus on biofuels, biocomposites and bioplastics. *Journal of Cleaner Production*, *402*, 136815. https://doi.org/10.1016/j.jclepro.2023.136815

Najafi, E., Castro, E., & Karimi, K. (2021). Biorefining for olive wastes management and efficient bioenergy production. *Energy Conversion and Management*, *244*, 114467. https://doi.org/10.1016/j.enconman.2021.114467

Narayanan, M., Kandasamy, S., He, Z., & Kumarasamy, S. (2022). Ecological impacts of pesticides on soil and water ecosystems and its natural degradation process. In *Pesticides in the Natural Environment* (pp. 23–49). Elsevier. https://doi.org/10.1016/B978-0-323-90489-6.00002-1

Nazrin, A., Sapuan, S. M., Zuhri, M. Y. M., Tawakkal, I. S. M. A., & Ilyas, R. A. (2021). Water barrier and mechanical properties of sugar palm crystalline nanocellulose reinforced thermoplastic sugar palm starch (TPS)/poly(lactic acid) (PLA) blend bionanocomposites. *Nanotechnology Reviews*, *10*(1), 431–442. https://doi.org/10.1515/ntrev-2021-0033

Nithya, R., Mohanrasu, K., & Arun, A. (2024). Agro-based bioplastic production and its application. In *Biodegradable Polymers, Blends and Biocomposites* (pp. 143–164). CRC Press. https://doi.org/10.1201/9781003304142-6

Noor, R. S., Shah, A. N., Tahir, M. B., Umair, M., Nawaz, M., Ali, A., Ercisli, S., Abdelsalam, N. R., Ali, H. M., Yang, S. H., Ullah, S., & Assiri, M. A. (2024). Recent trends and advances in additive-mediated composting technology for agricultural waste resources: A comprehensive review. *ACS Omega*, *9*(8), 8632–8653. https://doi.org/10.1021/acsomega.3c06516

Norfarhana, A. S., Ilyas, R. A., & Ngadi, N. (2022). A review of nanocellulose adsorptive membrane as multifunctional wastewater treatment. *Carbohydrate Polymers*, *291*, 119563. https://doi.org/10.1016/j.carbpol.2022.119563

Norfarhana, A. S., Ilyas, R. A., Ngadi, N., Othman, M. H. D., Misenan, M. S. M., & Norrrahim, M. N. F. (2024a). Revolutionizing lignocellulosic biomass: A review of harnessing the power of ionic liquids for sustainable utilization and extraction. *International Journal of Biological Macromolecules*, *256*, 128256. https://doi.org/10.1016/j.ijbiomac.2023.128256

Norfarhana, A. S., Khoo, P. S., Ilyas, R. A., Ab Hamid, N. H., Aisyah, H. A., Norrrahim, M. N. F., Knight, V. F., Rani, M. S. A., Septevani, A. A., Syafri, E., & Annamalai, P. K. (2024b). Exploring of cellulose nanocrystals from lignocellulosic sources as a powerful adsorbent for wastewater remediation. *Journal of Polymers and the Environment*. https://doi.org/10.1007/s10924-024-03227-3

Nurazzi, N. M., Harussani, M. M., Aisyah, H. A., Ilyas, R. A., Norrrahim, M. N. F., Khalina, A., & Abdullah, N. (2021). Treatments of natural fiber as reinforcement in polymer composites—A short review. *Functional Composites and Structures*, *3*(2), 024002. https://doi.org/10.1088/2631-6331/abff36

Onoja, E., Chandren, S., Abdul Razak, F. I., Mahat, N. A., & Wahab, R. A. (2019). Oil palm (Elaeis guineensis) biomass in Malaysia: The present and future prospects. *Waste and Biomass Valorization*, *10*(8), 2099–2117. Springer Netherlands. https://doi.org/10.1007/s12649-018-0258-1

Pandit, S., Savla, N., Sonawane, J. M., Sani, A. M., Gupta, P. K., Mathuriya, A. S., Rai, A. K., Jadhav, D. A., Jung, S. P., & Prasad, R. (2021). Agricultural waste and wastewater as feedstock for bioelectricity generation using microbial fuel cells: Recent advances. *Fermentation*, *7*(3), 169. https://doi.org/10.3390/fermentation7030169

Peng, X., Jiang, Y., Chen, Z., Osman, A. I., Farghali, M., Rooney, D. W., & Yap, P.-S. (2023). Recycling municipal, agricultural and industrial waste into energy, fertilizers, food and construction materials, and economic feasibility: A review. *Environmental Chemistry Letters*, *21*(2), 765–801. https://doi.org/10.1007/s10311-022-01551-5

Perdana, T., Kusnandar, K., Perdana, H. H., & Hermiatin, F. R. (2023). Circular supply chain governance for sustainable fresh agricultural products: Minimizing food loss and utilizing agricultural waste. *Sustainable Production and Consumption*, *41*, 391–403. https://doi.org/10.1016/j.spc.2023.09.001

Prateep Na Talang, R., Na Sorn, W., Polruang, S., & Sirivithayapakorn, S. (2024). Alternative crop residue management practices to mitigate the environmental and economic impacts of open burning of agricultural residues. *Scientific Reports*, *14*(1), 14372. https://doi.org/10.1038/s41598-024-65389-3

Preethikaharshini, J., Naresh, K., Rajeshkumar, G., Arumugaprabu, V., Khan, M. A., & Khan, K. A. (2022). Review of advanced techniques for manufacturing biocomposites: Non-destructive evaluation and artificial intelligence-assisted modeling. *Journal of Materials Science*, *57*(34), 16091–16146. https://doi.org/10.1007/s10853-022-07558-1

Pulingam, T., Lakshmanan, M., Chuah, J.-A., Surendran, A., Zainab-L, I., Foroozandeh, P., Uke, A., Kosugi, A., & Sudesh, K. (2022). Oil palm trunk waste: Environmental impacts and management strategies. *Industrial Crops and Products*, *189*, 115827. https://doi.org/10.1016/j.indcrop.2022.115827

Rashmi, I., Roy, T., Kartika, K. S., Pal, R., Coumar, V., Kala, S., & Shinoji, K. C. (2020). Organic and inorganic fertilizer contaminants in agriculture: Impact on soil and water resources. In *Contaminants in Agriculture* (pp. 3–41). Springer International Publishing. https://doi.org/10.1007/978-3-030-41552-5_1

Ruiz, H. A., Rodríguez-Jasso, R. M., Fernandes, B. D., Vicente, A. A., & Teixeira, J. A. (2013). Hydrothermal processing, as an alternative for upgrading agriculture residues and marine biomass according to the biorefinery concept: A review. *Renewable and Sustainable Energy Reviews*, *21*, 35–51. https://doi.org/10.1016/j.rser.2012.11.069

Saidu, M., Yuzir, A., Salim, M. R., Richard, A., & Afiz, B. (2020). Effect of operating parameter on the anaerobic digestion oil palm mesocarp fibre with cattle manure for biogas production. *IOP Conference Series: Earth and Environmental Science*, *476*, 012085. https://doi.org/10.1088/1755-1315/476/1/012085

Shamsuddin, R., Singh, G., Kok, H. Y., Hakimi Rosli, M., Dawi Cahyono, N. A., Lam, M. K., Lim, J. W., & Low, A. (2021). Palm oil industry—Processes, by-product treatment and value addition (pp. 121–143). https://doi.org/10.1007/978-3-030-61837-7_8

Sharma, R. (2024). Agro-industrial waste to energy—Sustainable management. *Sustainable Materials and Technologies*, *41*, e01117. https://doi.org/10.1016/j.susmat.2024.e01117

Sharma, V., Tsai, M.-L., Nargotra, P., Chen, C.-W., Kuo, C.-H., Sun, P.-P., & Dong, C.-D. (2022). Agro-industrial food waste as a low-cost substrate for sustainable production of industrial enzymes: A critical review. *Catalysts*, *12*(11), 1373. https://doi.org/10.3390/catal12111373

Sheer, A., Fahad Sardar, M., Younas, F., Zhu, P., Noreen, S., Mehmood, T., Ur Rahman Farooqi, Z., Fatima, S., & Guo, W. (2024). Trends and social aspects in the management and conversion of agricultural residues into valuable resources: A comprehensive approach to counter environmental degradation, food security, and climate change. *Bioresource Technology*, *394*, 130258. https://doi.org/10.1016/j.biortech.2023.130258

Shinde, R., Shahi, D. K., Mahapatra, P., Singh, C. S., Naik, S. K., Thombare, N., & Singh, A. K. (2022). Management of crop residues with special reference to the on-farm utilization methods: A review. *Industrial Crops and Products*, *181*, 114772. https://doi.org/10.1016/j.indcrop.2022.114772

Singh Yadav, S. P., Bhandari, S., Bhatta, D., Poudel, A., Bhattarai, S., Yadav, P., Ghimire, N., Paudel, P., Paudel, P., Shrestha, J., & Oli, B. (2023). Biochar application: A sustainable approach to improve soil health. *Journal of Agriculture and Food Research*, *11*, 100498. https://doi.org/10.1016/j.jafr.2023.100498

Sinha, A. K., Rakesh, S., Mitra, B., Roy, N., Sahoo, S., Saha, B. N., Dutta, S., & Bhattacharya, P. M. (2021). Agricultural waste management policies and programme for environment and nutritional security. In *Input Use Efficiency for Food and Environmental Security* (pp. 627–664). Springer Nature Singapore. https://doi.org/10.1007/978-981-16-5199-1_21

Siol, C., Thrän, D., & Majer, S. (2023). Utilizing residual biomasses from agriculture and forestry: Different approaches to set system boundaries in environmental and economic life-cycle assessments. *Biomass and Bioenergy*, *174*, 106839. https://doi.org/10.1016/j.biombioe.2023.106839

Sonu, Rani, G. M., Pathania, D., Abhimanyu, Umapathi, R., Rustagi, S., Huh, Y. S., Gupta, V. K., Kaushik, A., & Chaudhary, V. (2023). Agro-waste to sustainable energy: A green strategy of converting agricultural waste to nano-enabled energy applications. *Science of the Total Environment*, *875*, 162667. https://doi.org/10.1016/j.scitotenv.2023.162667

Srivastav, A. L., Patel, N., Rani, L., Kumar, P., Dutt, I., Maddodi, B. S., & Chaudhary, V. K. (2023). Sustainable options for fertilizer management in agriculture to prevent water contamination: A review. *Environment, Development and Sustainability*, *26*(4), 8303–8327. https://doi.org/10.1007/s10668-023-03117-z

Srivastava, R. K., Shetti, N. P., Reddy, K. R., & Aminabhavi, T. M. (2020). Biofuels, biodiesel and biohydrogen production using bioprocesses. A review. *Environmental Chemistry Letters*, *18*(4), 1049–1072. https://doi.org/10.1007/s10311-020-00999-7

Suriani, M. J., Radzi, F. S. M., Ilyas, R. A., Petrů, M., Sapuan, S. M., & Ruzaidi, C. M. (2021). Flammability, tensile, and morphological properties of oil palm empty fruit bunches fiber/pet yarn-reinforced epoxy fire retardant hybrid polymer composites. *Polymers*, *13*(8), 1282. https://doi.org/10.3390/polym13081282

Syafri, E., Jamaluddin, Sari, N. H., Mahardika, M., Amanda, P., & Ilyas, R. A. (2022). Isolation and characterization of cellulose nanofibers from Agave gigantea by chemical-mechanical treatment. *International Journal of Biological Macromolecules*, *200*(December 2021), 25–33. https://doi.org/10.1016/j.ijbiomac.2021.12.111

Vantarakis, A. (2021). Eutrophication and public health. In *Chemical Lake Restoration* (pp. 23–47). Springer International Publishing. https://doi.org/10.1007/978-3-030-76380-0_2

Velasco-Muñoz, J. F., Mendoza, J. M. F., Aznar-Sánchez, J. A., & Gallego-Schmid, A. (2021). Circular economy implementation in the agricultural sector: Definition, strategies and indicators. *Resources, Conservation and Recycling*, *170*, 105618. https://doi.org/10.1016/j.resconrec.2021.105618

Wang, Y., Fang, M., Lou, Z., He, H., Guo, Y., Pi, X., Wang, Y., Yin, K., & Fei, X. (2024). Methane emissions from landfills differentially underestimated worldwide. *Nature Sustainability*, *7*(4), 496–507. https://doi.org/10.1038/s41893-024-01307-9

Wu, Q., Qiang, T. C., Zeng, G., Zhang, H., Huang, Y., & Wang, Y. (2017). Sustainable and renewable energy from biomass wastes in palm oil industry: A case study in Malaysia. *International Journal of Hydrogen Energy*, *42*(37), 23871–23877. https://doi.org/10.1016/j.ijhydene.2017.03.147

Zhao, W., Yang, H., He, S., Zhao, Q., & Wei, L. (2021). A review of biochar in anaerobic digestion to improve biogas production: Performances, mechanisms and economic assessments. *Bioresource Technology*, *341*, 125797. https://doi.org/10.1016/j.biortech.2021.125797

Section II

Palm-Based Plant Waste

2 Oil Palm Waste Management

2.1 INTRODUCTION

In terms of global trade, palm oil is the most popular vegetable oil. With increased consumer wealth and population growth, demand will only continue to rise. Such economic effects on population growth are expected to impact the demand for palm oil over the next 10 years. Palm oil is very popular throughout most of Asia due to its affordability and reasonable freight costs from Southeast Asian sources. The oil palm segmentation process involves dividing and categorizing the industry according to various factors, such as geographical regions, production methods, end products, and market segments. These versatile trees bear fruit year-round, yielding oil-rich palm kernels that are extracted to produce palm oil, which finds its way into a wide range of consumer products, including cooking oils, snacks, cosmetics, and biofuels. Figure 2.1 shows the global palm oil marketed products.

Palm oil is one of the most valuable agricultural crops in the world, which makes palm oil biomass wastes the most suitable agricultural waste to be used in waste-to-energy projects in accordance with the Environmental Quality Act of 1974 (EQA74). Because of the significance of the exponentially rising energy demands throughout the past centuries, various ideal power alternatives have been researched as prospective replacements for traditional fossil fuels (Sadeghi et al., 2014). Every kind of organic material that has the potential to be employed in the production of energy is referred to as "biomass feedstock." It has been stated that this biomass feedstock is "theoretically carbon neutral." This is due to the fact that plants that are converted into biomass feedstock absorb

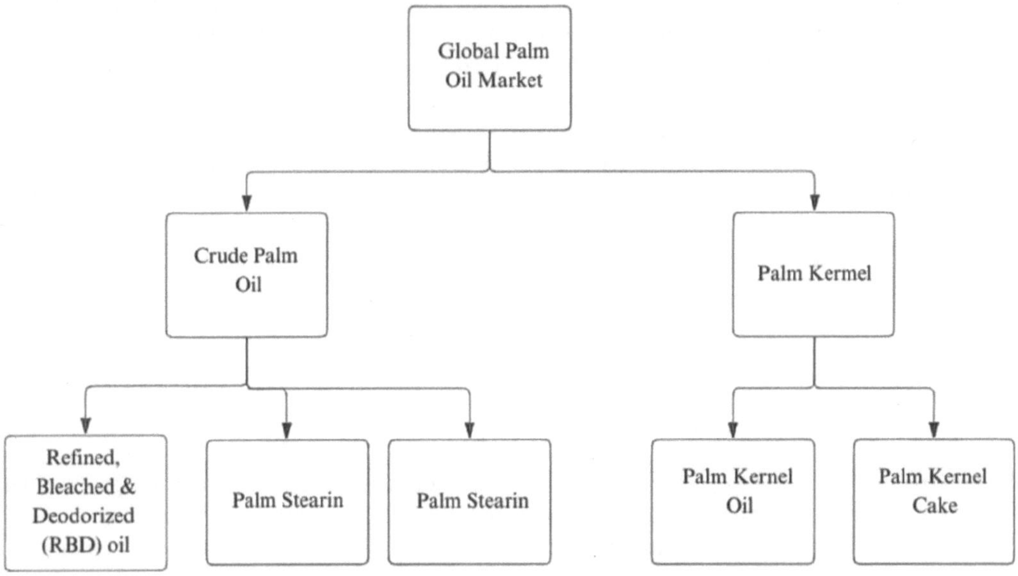

FIGURE 2.1 Segmentation by product.

DOI: 10.1201/9781003472209-4

carbon dioxide during the growth phase, and, when they are incinerated, they release the same quantity of carbon into the environment (Sadeghi et al., 2014).

Palm oil has potential environmental and social benefits, making it a controversial crop. Oil palm expansion has a number of negative environmental effects, including the destruction of natural ecosystems, the decline of woody biomass, and the draining of peatlands (Meijaard et al., 2020). When fire is used, this conversion often results in lower biodiversity and water quality as well as increased greenhouse gas (GHG) emissions. Oil palm development by large international and domestic companies is frequently associated with social issues such as land disputes and exploitation of workers, social inequality, and reductions in village well-being. Oil palm is a valuable crop in producing nations that promote economic growth in areas with few other alternatives for agricultural development. The adoption of oil palm by smallholder farmers results in significant benefits for their livelihoods on average (Meijaard et al., 2020).

This chapter intends to investigate industrial practices in the oil palm sector and its waste generation, including the impact of its operation from an environmental point of view. Subsequently, this chapter further analyzes the properties of reusable waste and then reviews all the existing methods or processes of recovery in making valuable products.

2.2 OIL PALM BIOMASS FROM NATURAL FIBER

Biomass is organic waste derived from animals and plants, such as sewage and municipal waste. In fact, biomass is also considered organic matter composed of carbon, hydrogen, oxygen, and nitrogen (Onoja et al., 2019). With 17.7 million tons of oil produced in 2008, Malaysia was the second-largest producer and exporter of oil palm in the world, accounting for 41% of global production (Abdul Aziz et al., 2011). Since Malaysia produces a significant amount of biomass waste, which offers a tremendous opportunity for this biomass to be commercialized in a variety of biomass-based goods, Malaysia is recognized as a significant participant in the palm oil sector.

Lignocellulosic biomass from oil palm describes the by-products of the oil palm industry's replanting, pruning, and milling processes (Asyraf et al., 2022a; Mahardika et al., 2024; Abdullah & Sulaiman, 2013a; Jafri et al., 2021). In fact, it represents a potential alternative energy source that can meet the rising demand for energy while reducing reliance on fossil fuels (Terry et al., 2021). About 75% of the solid waste comprises oil palm fronds (OPFs) and oil palm trunks (OPTs), readily available in plantations. The other 25% of the solid waste generated by oil palm consists of empty fruit bunch (EFB), mesocarp fibers (MFs), and palm kernel shells (PKSs) (Onoja et al., 2019). On the contrary, when palm oil is extracted by the wet process in the decanter, liquid waste from the oil palm is created. This liquid waste is called palm oil mill effluent (POME), cooling water, and sterile waste (Abdullah & Sulaiman, 2013a). Figure 2.2 shows the oil palm biomass waste mentioned in this section.

Some oil palm biomass waste components can be used in their natural state. Oil palm biomass has been used in many sectors due to the abundance of the resource, and many more studies are being done, particularly in the renewable energy sector. Table 2.1 briefly summarizes the various end products made directly or indirectly from oil palm biomass (Onoja et al., 2019).

2.3 ENVIRONMENTAL IMPACT FROM OIL PALM BIOMASS

Martin (2018) overviewed the environmental consequences associated with palm oil production. Palm oil is widely used globally due to its economic advantages. However, the increasing demand for palm oil has led to potential environmental problems, as land is cleared for palm oil plantations, resulting in deforestation. This deforestation poses significant threats to biodiversity, air quality, social well-being, and GHG emissions. Peat drainage, necessary for palm oil cultivation, further contributes to the decline of GHG emissions and water quality.

FIGURE 2.2 Natural fibers generated by oil palm. (a) Oil palm frond (OPF), (b) oil palm trunk (OPT), (c) empty fruit bunch (EFB), (d) mesocarp fibers (MFs), (e) palm kernel shells (PKSs), and (f) palm oil mill effluent (POME) (Onoja et al., 2019).

Addressing these environmental concerns, the European Union (EU) has taken action by revising its biofuels policy to phase out the use of palm oil-based biodiesel by 2030. This step reflects the EU's commitment to reducing the environmental impact of palm oil production, deforestation, and biodiversity loss, as emphasized in Russell's report. Ensuring sustainable production practices is vital as palm oil consumption continues to rise. According to the IUCN Red List of threatened species, expansion of oil palm cultivation significantly contributes to deforestation, particularly in Southeast Asia, destroying important natural habitats like tropical rainforests known for their rich biodiversity. This habitat loss results in declines in biodiversity, affecting various species, including endangered ones like orangutans, gibbons, and tigers, as they lose their homes. Compared to natural forests, oil palm plantations have lower species diversity and abundance, as their simplified and fragmented landscapes provide limited habitat for many species. Additionally, the introduction and spread of non-native species linked to oil palm cultivation can negatively impact the native biodiversity of the

TABLE 2.1
Oil Palm Biomass and Its Usage in the Industry

Biomass	Utilization in Industry
Oil palm frond (OPF)	• Paper making • Production of bindless boards • Production of syngas • Roughage for ruminant animals • Production of bio-oil and biochar
Oil palm trunk (OPT)	• Paper making • Production of plywood • Production of laminated veneer lumber • Production of particle board panels • Production of bio-oil and biochar
Empty fruit bunch (EFB)	• Paper making • Production of syngas or hydrogen • Production of fertilizer • Production of medium-density fiber boards and black boards • Production of briquettes • Production of biocomposite • Production of biochar • Production of activated carbon • Electricity generation • Production of hydrogen
Mesocarp fibers (MFs)	• Briquettes • For erosion control • For making mattress cushion • For soil stabilization • For horticulture and landscaping • For ceramic and brick production • For paper production • For acoustic control • As fertilizer • For animal feed • As filler in thermoplastic and thermosetting composite • Production of bio-oil
Palm kernel shells (PKSs)	• Syngas production • Briquettes production • For producing PKS or chitosan composite for the removal of heavy metals from wastewater • Production of carbon molecular sieve • Production of bio-oil
Palm oil mill effluent (POME)	• Biogas synthesis • For producing fertilizer • Syngas synthesis

area (Meijaard et al., 2020). Deforestation rates have been notably high in Asia and South America, while Mesoamerica and Africa have experienced lower rates. Figure 2.3 shows the vulnerable forest with the highest forest diversity. Southeast Asia accounts for about 68% of the oil palm cultivation area, but other regions also witness rapid expansion in this sector (Vijay et al., 2016).

In recent decades, oil palm has risen to prominence as the leading global supplier of vegetable oil, making up approximately 30% of overall vegetable oil consumption. This has transformed oil palm

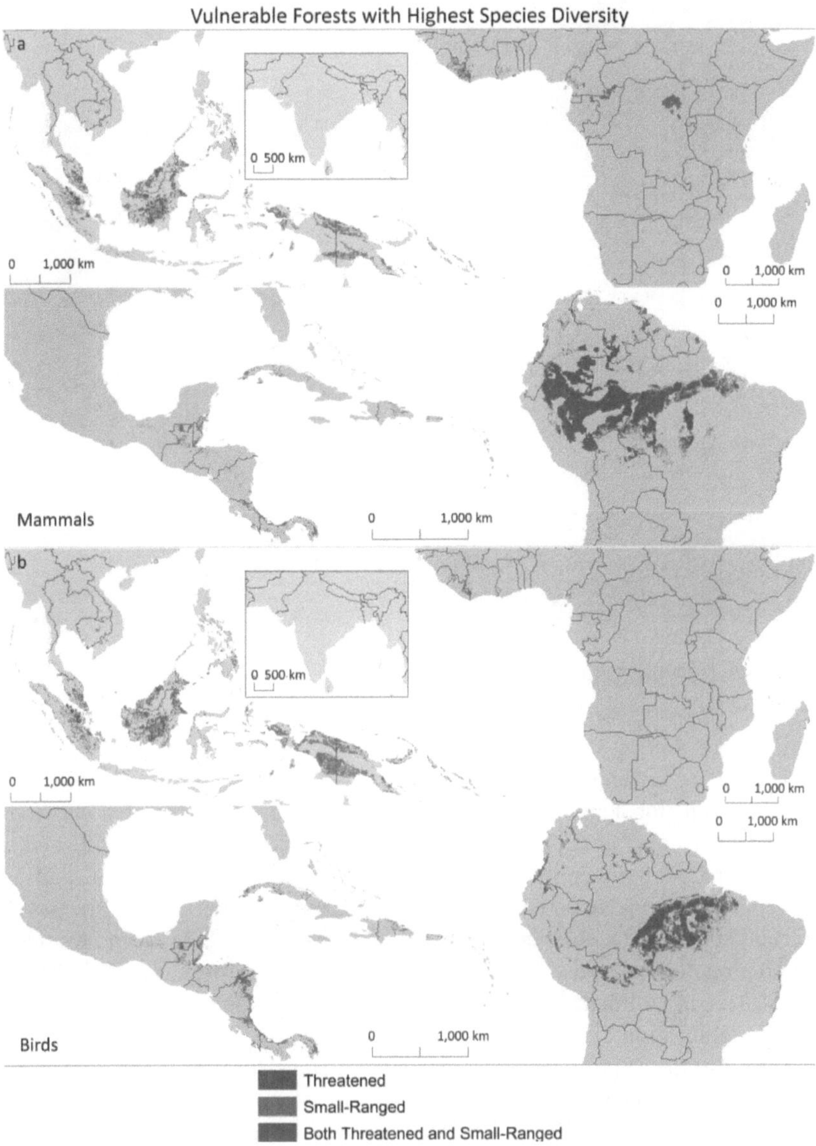

FIGURE 2.3 Vulnerable forest with the highest forest diversity: (a) mammals and (b) birds (Vijay et al., 2016).

production into a major industry in tropical regions (Lam et al., 2019). However, when oil palm is cultivated on peatlands, drainage systems are put in place to lower the water table, resulting in detrimental effects on peatland function. Formerly acting as carbon sinks, peatlands undergo a disturbing shift and start emitting GHGs like carbon dioxide (CO_2) and nitrous oxide (N_2O) (Lam et al., 2019; Page et al., 2011; Miettinen et al., 2016; Girkin et al., 2022). The increased CO_2 emissions occur due to accelerated decomposition of organic matter caused by higher oxygen levels resulting from peatland drainage (Miettinen et al., 2016), while the higher N_2O emissions are mainly attributed to nitrogen fertilization (Jauhiainen et al., 2012).

Studies have indicated that CO_2 emissions from oil palm plantations are one to two times higher than those from other major crops, particularly in Southeast Asia, where Malaysia is the second-largest global oil palm producer. The expansion of oil palm plantations has been rapid, extending even into deeper peatland areas, including forest zones (Cooper et al., 2020). However, specific data

FIGURE 2.4 Haze from the palm oil industry (WWF, 2022).

on GHG emissions resulting from this expansion and overall land-use changes in the peninsular region of Malaysia are lacking. Moreover, existing research on the impact of oil palm production has mainly focused on global and national scales, disregarding the high level of uncertainty in emissions at local levels (Cooper et al., 2020; Tonks et al., 2017; Charters et al., 2019). According to the World Wildlife Fund, oil palm cultivation contributes to air pollution by using fire for land clearing during site preparation, which can release smoke and haze, particularly in Southeast Asia. This can have detrimental effects on local air quality, posing health risks to nearby communities and contributing to overall air pollution levels. The smoke and haze from burning can contain harmful particles and pollutants that can be inhaled by humans and animals, leading to respiratory issues and other health problems (Kauffman et al., 2016). In addition to the immediate health risks, the release of smoke and haze into the atmosphere can contribute to smog formation and worsen air quality in surrounding areas, as shown in Figure 2.4.

According to Camara et al. (2019), the transformation of natural habitats for oil palm cultivation, which includes clearing land and the application of fertilizers and pesticides, can result in a decline in water quality. This decline in water quality takes place as sediments, nutrient runoff, and chemical pollutants from the cultivation activities enter adjacent water sources, adversely affecting aquatic ecosystems and the livelihoods of dependent communities (Mustafa & Hayder, 2021). Land clearing for oil palm plantations can cause erosion and sedimentation, as exposed soil is more susceptible to erosion during heavy rainfall, leading to increased sedimentation in rivers and streams. Moreover, the application of fertilizers and pesticides in oil palm cultivation can result in nutrient runoff, where excess nutrients such as nitrogen and phosphorus are carried off by rainwater into water bodies, causing eutrophication and disrupting the balance of aquatic ecosystems. Additionally, the use of pesticides can result in chemical pollution in water bodies, which can harm aquatic life and have cascading effects on the overall ecosystem health. This water quality degradation can have severe consequences for local communities that rely on these water bodies for drinking water, fishing, and other livelihood activities. Consequently, the effects of oil palm cultivation on water quality underscore the necessity for sustainable and conscientious practices to alleviate adverse impacts on aquatic ecosystems and the communities reliant on them

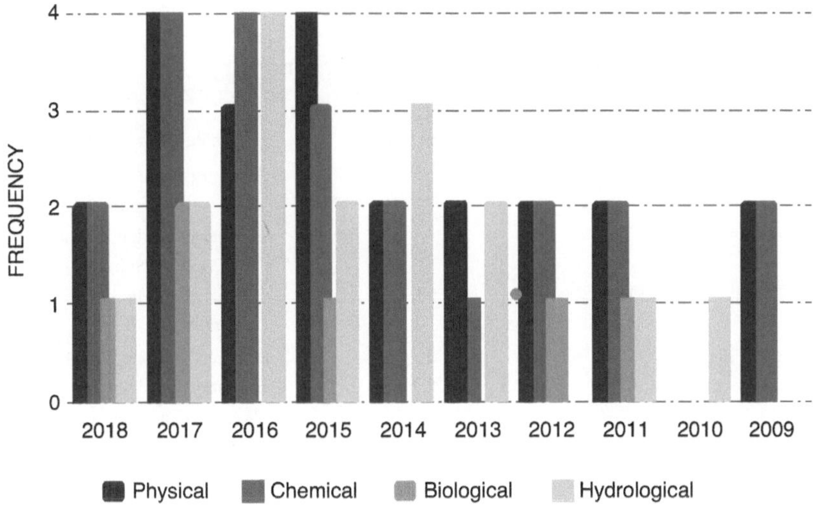

FIGURE 2.5 The frequency of water quality parameters utilized in various studies conducted in Malaysia (Camara et al., 2019).

(Dungani et al., 2018). The correlation analysis revealed a notable positive association between oil palm agriculture activities and the physical and chemical attributes of water quality (Kauffman et al., 2016). Notably, studies in Malaysia have predominantly focused on the physical parameters using the water quality index (DOE-WQI), as depicted in Figure 2.5.

The expansion of oil palm plantations, especially by major multinational and national companies, has raised concerns about its social impacts (Sakai et al., 2022). These concerns encompass issues like land grabbing, conflicts, labor exploitation, social inequality, and a decline in the well-being of local communities. These social problems can have significant consequences for both local communities and indigenous peoples residing in areas where oil palm is produced. While oil palm contributes to economic development in regions with limited agricultural options and can improve livelihoods for small farmers (Sakai et al., 2022), it also negatively impacts the environment and local communities, which cannot be ignored (Onoja et al., 2019). Addressing and mitigating these impacts is crucial and can be achieved through sustainable and responsible oil palm cultivation practices. Implementing certification schemes, better land-use planning, improved management practices, and engaging with local communities and stakeholders are essential to achieving this goal.

2.4 AVAILABILITY OF OIL PALM AGRO-WASTE

Researchers have extensively studied the quantity of agro-waste generated in the oil palm industry across various regions. The industry produces a substantial amount of waste during the production and processing of oil palm (Yaro et al., 2022), which can be put to use in multiple ways, including bioenergy production, animal feed, and enhancing soil quality (Kashyap et al., 2021). Below, we provide additional information on how agro-waste is utilized in the oil palm industry, and Table 2.2 presents the availability of agro-waste in this sector (Kaniapan et al., 2021).

In the palm oil industry, oil palm residues pose a significant concern, as about 90% of the material remains as waste biomass while only 10% is oil generated. The waste includes various forms like pressed fruit fibers,oil palm empty fruit bunch (OPEFB), oil palm kernel shell (OPKS), and POME, and it continues to accumulate as the trees reach their optimal lifespan through the OPT and OPF stages (Kaniapan et al., 2021). These palm oil wastes hold potential as sustainable industrial materials for various sectors, including eco-friendly asphalt pavement, as traditional asphalt

TABLE 2.2
The Availability of Agro-Waste in the Oil Palm Industry

Waste Type	Annual Production (million tons)	Potential Uses
Oil palm frond (OPF)	26	Bioenergy production, animal feed, soil amendment
Oil palm trunk (OPT)	4	Bioenergy production, animal feed, soil amendment
Empty fruit bunches (EFBs)	50	Bioenergy production, composting
Palm kernel shell (PKS)	3.3	Bioenergy production
Mesocarp fibers (MFs)	2.8	Bioenergy production
Palm oil mill effluent (POME)	23.7	Bioenergy production, fertilizer

is a significant contributor to air pollution and is a non-renewable resource. Replacing some of the asphalt in a pavement with waste palm oil wastes makes it possible to create a more sustainable and environmentally friendly pavement (Yaro et al., 2022).

Agro-waste contains valuable bioactive compounds that are beneficial for food purposes. These compounds offer essential health benefits, such as polyphenolic compounds, vitamins, minerals, fatty acids, anthocyanins, and pigments. Extracting these bioactive compounds makes it possible to fortify foods and create novel products that are both functional and nutritious (Bala et al., 2023). Using agro-industrial waste as raw materials can reduce production costs and decrease environmental pollution. Moreover, biofuels play a vital role as a sustainable alternative to fossil fuels. Research has demonstrated successful biofuel production from various agro-industrial residues, including rice straw, sweet potato waste, sawdust, corn stalks, sugarcane bagasse, and sugar beet waste (Murphy et al., 2021).

Using agro-waste in the oil palm industry has great potential for different applications, making the sector more sustainable and feasible (Van Dexter & Boopathy, 2020). Turning agro-waste into valuable bioproducts and bioactive compounds can be a significant move toward sustainable development in agriculture. However, the industry still faces challenges such as deforestation and haze creation, adversely affecting tropical landscapes and ecosystems (Sadh et al., 2018). Therefore, developing sustainable energy technologies and creating value-added bioproducts from agro-waste is essential to address these challenges.

2.5 PROPERTIES OF OIL PALM AGRO-WASTE FIBERS

2.5.1 Physical Properties

With the various kinds of recyclable agro-waste generated from the oil palm industry, which could be reutilized as a valuable product, physical properties are crucial aspects to be studied for further manufacturing. Researchers have investigated the physical properties of each type of agro-waste to improve their process pathway, according to the respective significance of their desired products. Referring to the collection of data obtained from the literature in Table 2.3, MF has the highest density among all the oil palm biomass, while OPT has the lowest density, with density only ranging from 300 to 600 kg/m^3 between all the layers of the trunk. However, according to Table 2.4, oil palm EFB has the lowest moisture content, and OPT has the highest moisture content among the various biomass. The trend in both Tables 2.3 and 2.4 indicates that biomass density is inversely proportional to moisture content.

TABLE 2.3
Density of Oil Palm Biomass

Biomass Type	Density Value (kg/m³)	References
OPF	430	Wahab et al. (2021)
OPT	• 300–350 (inner) • 300–400 (middle) • 500–600 (outer)	Bakar et al. (2013)
EFB	795.3	Ibrahim et al. (2014)
MF	1,260	Then et al. (2014)
PKS	740	Okoli and Okoli (2020)
POME	876	Ali and Tay (2013)

TABLE 2.4
Moisture Content of Oil Palm Biomass

Biomass Type	Moisture Content (%)	References
OPF	12	Guangul et al. (2013)
OPT	• 82.4 (inner) • 75.9 (middle) • 67.7 (outer)	Kosugi et al. (2010)
EFB	4.68	Abdullah and Sulaiman (2013b)
MF	10.1	Faizal et al. (2018)
PKS	2.8–30	Nizamuddin et al. (2015)

2.5.2 Chemical Composition

Multiple researchers have conducted characterization studies to determine the chemical composition of oil palm biomass, involving both ultimate and proximate analyses. The ultimate analysis performed on oil palm biomass has revealed the existence of various elements, including carbon (C), hydrogen (H), nitrogen (N), sulfur (S), and oxygen (O). Table 2.5 shows the list of chemical elements found in the six-oil palm biomass.

Based on the data reported by the researchers, the percentage of carbon in MFs was the highest compared to other biomass types. For oxygen content, it is shown that the OPT has the highest value. Since all the oil palm biomass was considered to have high carbon and low sulfur and nitrogen contents, they are suitable feed for torrefaction (Abdul Aziz et al., 2011). Oil generally makes up about 10% of the entire tree, leaving the remaining 90% as biomass. Fresh fruit contains only 21% palm oil; the rest is 14–15% dietary fiber, 6–7% palm kernel, 6–7% pericarp, and 23% biomass. Additionally, oil palm biomass typically has a higher oxygen content (40–50 wt%) and lower nitrogen (1 wt%) and sulfur content (0.2 wt%) than fossil fuels (Terry et al., 2021).

On the contrary, the main components of oil palm biomass are hemicellulose, cellulose, and lignin (Abdul Aziz et al., 2011; Norfarhana et al., 2024a, b). Cellulose, which has a molecular weight of approximately 100,000, is a polymer consisting of linear chains of glucopyranose units connected in an α configuration through 1,4 linkages (Imraan et al., 2023; Low et al., 2024; Norrrahim et al., 2022a). Hemicellulose is a complex mixture of various polysaccharides, including mannose, glucose, xylose, arabinose, methylglucoronic acid, and galaturonic acid. It has an average molecular weight of around 30,000 and is a cell wall component. Lignin, also present in the cell wall,

TABLE 2.5
Percentage of Elements in Oil Palm Biomass via Ultimate Analysis

	Percentage of Element (%)					
Biomass Type	**C**	**H**	**N**	**S**	**O**	**References**
OPF	44.58	4.53	0.71	0.07	48.80	Guangul et al. (2012)
OPT	42.72	5.61	0.44	—	51.25	Abnisa et al. (2011)
EFB	45.00	6.40	0.25	1.06	47.30	Abnisa et al. (2013)
MF	51.52	5.45	1.89	0.23	40.91	Samiran et al. (2015)
PKS	49.74	5.32	0.08	0.16	44.86	Idris et al. (2012)
POME	50.01	15.78	1.99	—	46.90	Loh (2017)

TABLE 2.6
Lignocellulosic Content of Oil Palm Biomass

	Lignocellulosic (%)			
Biomass Type	**Cellulose**	**Hemicellulose**	**Lignin**	**References**
OPF	40.01	30.78	29.50	Onoja et al. (2019)
EFB	23.70	21.60	29.20	
PKS	27.70	21.60	44.00	
OPT	34.50	31.80	30.00	Terry et al. (2021)
MF	34.5	31.8	25.7	Abdul Aziz et al. (2011)

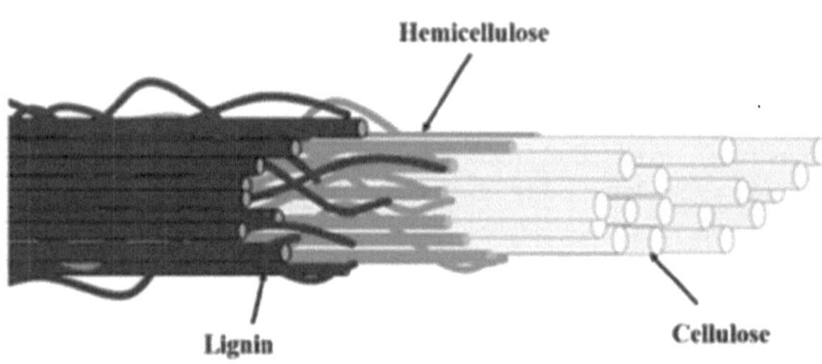

FIGURE 2.6 The overall complex structure of lignocellulosic fiber (Norrrahim et al., 2022b).

is a mononuclear aromatic polymer. Because hemicellulose and lignin are located closely together in the cell wall, they can combine to form a complex known as lignocellulose (Kelly-Yong et al., 2007). Table 2.6 shows the lignocellulosic content of oil palm biomass, while Figure 2.6 shows the structure of lignocellulosic fiber.

2.5.3 Mechanical Properties

EFB fibers are special fibers made up of multiple cells with a central part called lacuna. Their surface is porous, and this unique structure is essential for creating strong connections with the material used to make composites (Abdul Khalil et al., 2008; Nurazzi et al., 2021). The fibers are usually shaped as polygons with a compact bundle in the middle, and thick layers of cells protect them. These thick cell walls add strength, enabling the fibers to resist forces squeezing them from the sides (Bhat et al., 2010). OPF fibers are composed of vascular bundles of different sizes, typically containing thin-walled parenchyma tissues. These bundles consist of components, such as round gloves, vessels, fibers, phloem, and parenchyma tissue, with clear distinctions between the xylem and phloem tissues. Additionally, the phloem is further divided into two separate parts within each bundle. Research indicates that EFB fibers have a lower density, tensile strength, and young modulus compared to OPF fibers. However, EFB fibers have a higher elongation at break, making them more flexible (Keong, 2005).

Various methods can enhance the mechanical properties of oil palm solid waste. Chemical treatment improves the strength and stiffness of the fibers, while heat treatment enhances their thermal stability (Asyraf et al., 2021; Norrrahim et al., 2021; Syafri et al., 2022). The mechanical properties of oil palm solid waste play a crucial role in different applications. First, it can reinforce polymer composites like plastics and rubber, increasing their strength, stiffness, and toughness. Second, oil palm solid waste is valuable for producing biofuels like bioethanol and biodiesel, offering a renewable energy source. Lastly, the waste acts as an effective biosorbent to remove pollutants from water and wastewater, improving water quality. These applications demonstrate the versatility and potential benefits of using oil palm solid waste in sustainable and eco-friendly solutions.

Research in the area of mechanical properties of oil palm solid waste shows promise, as highlighted by Kaniapan et al. (2021). With further development, oil palm solid waste can be utilized to produce a wide range of products for various applications. The mechanical properties, as presented in Table 2.7, are crucial for reinforcing biomass in polymer composites.

2.5.4 Morphological Properties

The morphology properties of oil palm fiber were examined using scanning electron microscopy (SEM), providing valuable insights into its behavior and adhesion with the matrix. From Figure 2.7, SEM micrographs of the oil palm fiber composite samples with varying fiber loadings (0, 3, 5, and 7 wt%) were analyzed. Notably, the micrographs demonstrated a positive fiber/matrix adhesion, as no fiber pullouts or gaps were observed at lower fiber loadings (Sreekala et al., 1997). However, at higher fiber loadings (5 and 7 wt%), some pores and fiber pullouts were evident, indicating an increase in the number of pores and fractured fibers with higher fiber content (vol%). This observation aligns with previous research, where an increased natural fiber content in composite materials led to more gaps and fiber pullouts (Hussain et al., 2021; Chen et al., 2017; Norul Izani et al., 2013).

Furthermore, Figure 2.7d displayed a notable presence of a large pore and clear fractured fiber on the surface, attributed to inadequate fiber/matrix bonding and the use of higher fiber loading (7 wt%). Moreover, the appearance of large pores was linked to the degradation of oil palm fibers during the heating process, occurring within the temperature range of 240–250°C. To enhance the thermal stability of the fibers, researchers recommended chemically removing a certain proportion of

TABLE 2.7
Mechanical Properties of Oil Palm Solid Waste (EFB, OPF, PKS, and OPT)

Properties	EFB	OPF	PKS	OPT	References
Density (g/cm^3)	0.25–0.35	0.35–0.45	0.75–0.95	0.45–0.65	Aljuboori (2013)
Tensile strength (MPa)	0.9–2.4	1.4–2.8	8–12	20–30	
Young modulus (GPa)	1.5–3.0	2.0–3.5	5.0–6.5	10–12	Nasri et al. (2021)
Elongation at break (%)	3–6	4–8	1–3	1–2	

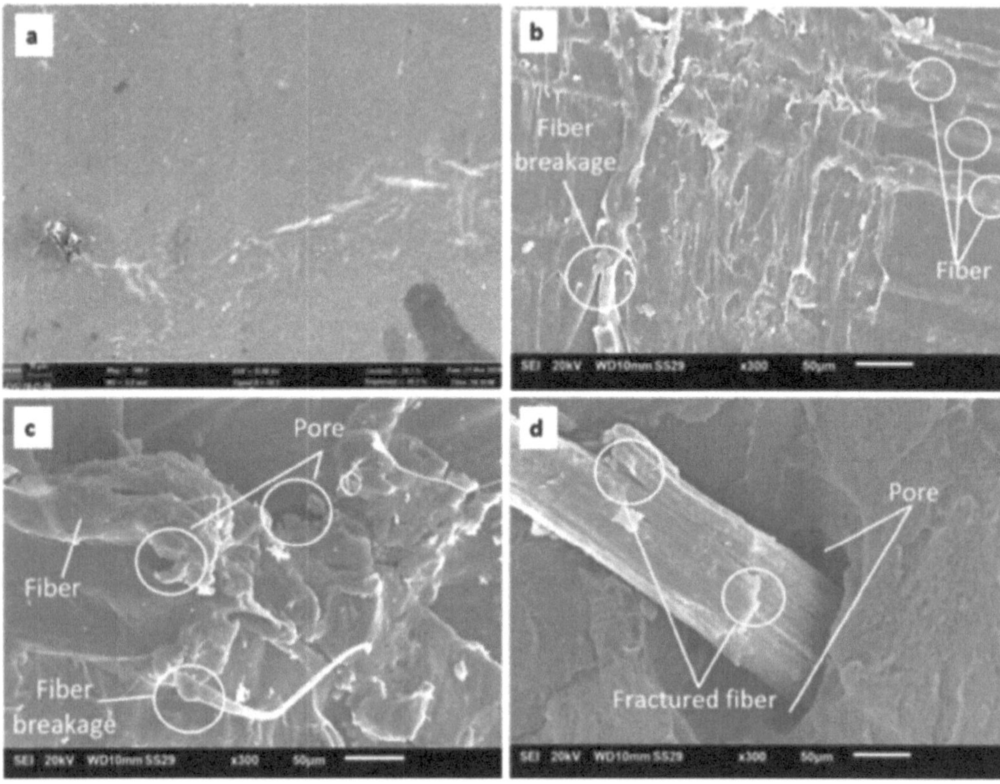

FIGURE 2.7 The surfaces of the oil palm fiber with different fiber loadings: (a) 0 wt%, (b) 3 wt%, (c) 5 wt%, and (d) 7 wt%. The scale bar in the inset represents 50 μm (Sreekala et al., 1997).

hemicellulose and lignin elements from the fibers. This process aimed to address natural fiber deterioration, a significant challenge encountered during the manufacturing of natural fiber composites, including curing, extrusion, and injection molding processes (Ahmad et al., 2023, 2021; Azlin et al., 2022). The SEM observations of oil palm fibers provided valuable insights into the fiber's adhesion with the matrix at varying loadings, as well as the impact of fiber content on the occurrence of pores and fractured fibers. Understanding these morphological properties is crucial for developing and improving natural fiber composites for various applications (Alias et al., 2021; Asyraf et al., 2022b; Mazani et al., 2019; Suriani et al., 2021).

2.5.5 Thermal Properties

Currently, little research has been done on the thermal properties of the general oil palm biomass itself. The combustion rate, specific heat capacity, and the heat released obtained from a study during gasification in the furnace are presented in Table 2.8 (Nyakuma et al., 2013). The variety of values for its properties may be due to factors such as the particle size, reaction temperature, heating rate, and the organic and inorganic composition of the fuels.

The general thermal conductivity for oil palm ranges from 0.055 and 0.091 W/m·K. In addition, one article mentions the thermal conductivity of biocomposite OPT. The OPT was classified into three categories of fineness: large, medium, and small. It was then boiled and pretreated in hot water and sodium hydroxide (NaOH) with the mixture of ramie fiber and tapioca starch at a ratio of 50:20:30. The study found that the pretreated OPT resulted in better thermal stability compared to those without the treatment. The results of thermal conductivity testing are summarized in Table 2.9. It was concluded that the larger the OPT particle size, the lesser thermal conductivity it has due to the density of the lower composite (Mawardi et al., 2022).

2.5.6 Calorific Value

All palm oil mills in Malaysia use fiber and shell as boiler fuel, which generates steam for the milling process and for other mill activities. By combusting biomass, we have more effectively disposed of the fuel than maximized its energy potential (Subramaniam et al., 2004). A group of researchers from the Malaysian Palm Oil Board have collected the collection data for calorific value for the various biomass from the palm oil industry, as displayed in Table 2.10, which can be used by interested parties regarding their potential energy (Subramaniam et al., 2004). Based on Table 2.10, of all the biomass generated from the oil palm industry, PKS has the highest calorific value while OPF has

TABLE 2.8
Thermal Properties for Some of the Biomass Types (Nyakuma et al., 2013)

Biomass Type	Combustion Rate ($\times 10^{-8}$ kg/s)	Heat Released, Q (J/g)	Specific Heat, c (J/kg K)
EFB	3.83	2,472.74	2,832.46
PKS	3.96	2,718.29	3,113.74
OPF	4.19	2,820.90	3,231.28

TABLE 2.9
Thermal Conductivity of Hybrid OPT (Mawardi et al., 2022)

Hybrid Composite	OPT Particle Size	Thermal Conductivity, k (W/m·K)
C1	Large	0.0723
C2	Medium	0.0793
C3	Small	0.1344

TABLE 2.10
Calorific Value of Oil Palm Biomass via ASTM 2015 Method (Subramaniam et al., 2004)

Biomass Type	Average Calorific Value (kJ/kg)	Calorific Range (kJ/kg)
OPF	15,719	15,400–15,950
OPT	17,471	17,000–17,800
EBF	18,795	18,000–19,920
MF	18,795	18,000–19,580
PKS	20,093	19,500–20,750
POME	16,992	16,100–17,650

the lowest. With the different amounts of generation annually on agro-waste relative to the biomass type, the optimization of the manufacture of green products from this recyclable waste could be figured out accordingly with this collection of data.

2.6 OIL PALM AGRO-WASTE MANAGEMENT

2.6.1 Compost

Composting should be used urgently in the modern day to transform organic wastes into resources since landfill space is becoming increasingly expensive and scarce. Composting results in the development of chemically stabilized composted materials by breaking down most of the biodegradable organic compounds and converting some of the remaining organic material into compounds that resemble humic acid (Supriatna et al., 2022). Compost that has partially decomposed or is unstable when applied to agriculture causes nitrogen to become immobilized and reduces the oxygen content surrounding root systems as a result of the quick activation of bacteria (Supriatna et al., 2022).

The best technique for a particular composting application depends on a number of variables, including location, operation size, the effects of occasional odors, capital availability, feedstock type, environmental requirements, labor availability, and commercial goals (Supriatna et al., 2022). Simple techniques like turned windrows and passively aerated static piles (ASPs) are commonly used in composting facilities. These techniques are widely used since they involve little site modification, engineering work, and financial commitment. ASP composting is another straightforward, popular technique with many uses (Supriatna et al., 2022). EFBs are usually composted to create organic fertilizers. In another study, researchers mentioned a long period of composting, ranging from 60 to 80 days. During the period of nitrogen degradation, N content inside the composting pile prevents nutrient imbalance in the soil. One of the researchers stated the product of the compost is worth MYR 1.78/ton of chemical fertilizer (Singh et al., 2010).

2.6.2 Pulp and Paper

The use of agricultural biomass in various industries has gained increasing attention as a sustainable approach to waste management and resource utilization. Among the available biomass resources, oil palm biomass presents significant potential, particularly in the pulp and paper industry. With vast amounts of oil palm waste generated annually, researchers have explored its potential for repurposing into valuable products. Numerous studies have demonstrated the effectiveness of incorporating oil palm biomass, such as EFBs, into paper production to enhance both environmental sustainability and material performance. According to Singh et al. (2013), utilizing oil palm biomass in the pulp and paper industry brings several benefits for waste management and environmental sustainability. In a study conducted by Rafidah et al. (2017), they blended oil palm EFB fibers with kenaf fibers to produce paper. The results demonstrated a significant improvement in the paper's mechanical properties, making it more durable and of higher quality. This highlights the potential of using oil palm biomass as a valuable raw material for top-notch paper production.

In the study by Rafidah et al. (2017), EFB fibers were collected from Sabutek Sdn. Bhd., cleaned and cut before pulping. Kenaf fibers from the National Kenaf and Tobacco Board were processed into chips. Both types of fibers were separately pulped using different methods. After pulping, the pulps were washed and screened to obtain cleaned and unbleached pulps. Bleaching with chlorine dioxide and peroxide-enhanced extraction were used for both EFB and kenaf pulps, followed by beating and blending at various proportions. The blended pulps were then used to produce paper sheets following standard procedures. Based on TAPPI standards, the paper sheets were tested for characteristics like opacity, smoothness, tensile strength, and tear strength. Additionally, a SEM was utilized to observe the morphology of the produced pulp and paper sheets. Figure 2.8 presents a pie chart illustrating the percentage of pulp and paper production from different sources, with oil palm biomass occupying a more significant share over time.

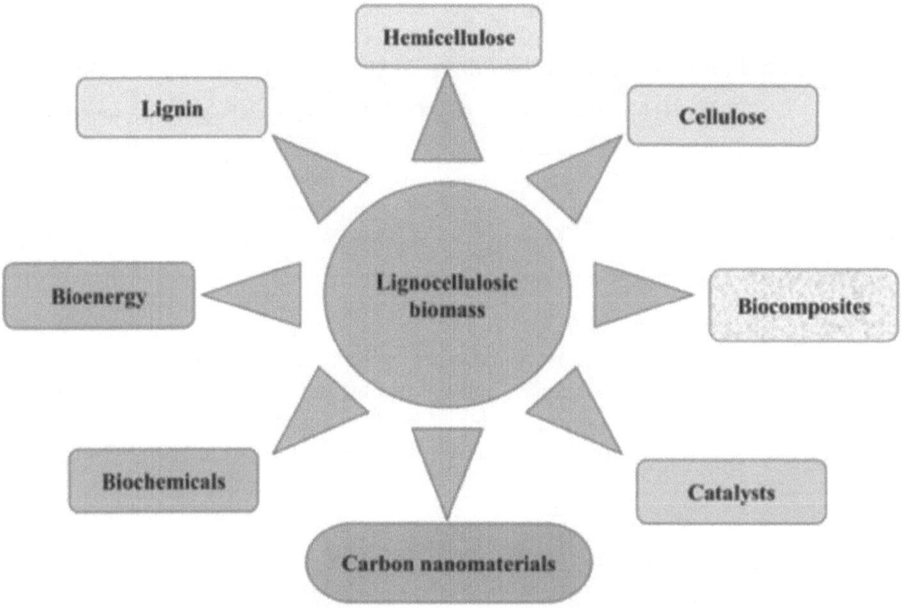

FIGURE 2.8 Potential of oil palm biomass to replace traditional wood sources in the pulp and paper industry (Pillai et al., 2018).

Furthermore, researchers have explored the potential of using oil palm leaves as a valuable resource in the papermaking industry. Despite being considered waste, these leaves contain fibrous properties that can be utilized for pulp and paper production (Singh et al., 2013). For instance, Norlin et al. (2023) investigated the feasibility of employing palm leaves as a raw material for papermaking. Fresh green leaves were cleaned to remove impurities before undergoing chemical pulping. The leaves were then cut into smaller pieces, ground, and soaked in sodium hydroxide (NaOH) solution for delignification at 80°C. Various treatment times were tested to determine the effectiveness of lignin removal for fiber production. After delignification, the resulting fibers were bleached using hydrogen peroxide (H_2O_2) and thoroughly washed. The fibers were shaped using a mold and deckle, and the produced paper was dried in an oven until the pulp was dehydrated. By utilizing oil palm leaves as a raw material, the paper industry can reduce reliance on traditional wood sources and promote sustainable resource management (Soloi & Hou, 2019).

Additionally, the utilization of oil palm fiber soda pulp to enhance recycled paper has been investigated in various studies (Wanrosli et al., 2005). The process involves collecting oil palm fibers from EFB and preparing them by removing impurities. These fibers undergo soda pulping and are cooked with sodium hydroxide and sodium sulfide solution to break down lignin and separate the fibers. After pulping, the fibers are washed and may undergo bleaching for improved brightness. The oil palm fiber soda pulp is then blended with recycled paper pulp, with as little as 10% of beaten virgin pulp sufficient to strengthen the recycled paper. The blended pulp is transformed into a pulp slurry and processed on a paper machine through forming, pressing, drying, and finishing stages to produce the final paper product.

This process offers a promising approach to upgrading recycled paper using sustainable oil palm fibers. By incorporating oil palm fiber soda pulp into the production of recycled paper, its strength can effectively be restored, ensuring that the recycled paper retains its original quality and performance (Wanrosli et al., 2005). This demonstrates how oil palm biomass can contribute to the circular economy by enhancing the value and usability of recycled materials. Table 2.11 summarizes the properties of oil palm biomass and its potential use in the pulp and paper industry. Utilizing oil

TABLE 2.11
Properties of Oil Palm Biomass and Its Potential Use in Pulp and Paper Industry

Property	Oil Palm EFB	Oil Palm Leaves	Oil Palm Fiber Soda Pulp	References
Fiber length	Long	Long	Long	Singh et al. (2013)
Fiber strength	Strong	Strong	Strong	Soloi and Hou (2019)
Cellulose content	High	High	High	Wanrosli et al. (2005)
Potential use	Papermaking	Papermaking	Upgrading recycled paper	Wanrosli et al. (2005)

palm biomass in the pulp and paper industry and other applications presents many opportunities for waste management, resource diversification, and environmental sustainability (Kelly-Yong et al., 2007). By tapping into the potential of oil palm residues, we can improve the quality and value of paper products while reducing waste, lessening environmental pollution, and contributing to a more sustainable future (Singh et al., 2013).

2.6.3 Biocomposite Materials

Wood substitutes can be derived from cellulosic raw materials such as OPTs and OPFs, which can be used to develop biocomposite products (Suhaily et al., 2012). It is also possible that wood products made with OPT may have an impact on the furniture industry. Since OPT is similar in quality to rubberwood (RW) and other solid wood species, it is being researched and processed as a substitute for wood furniture. Additionally, OPT can help furniture sectors dependent solely on natural resources, which can impact a whole country's environment. Its superior physical and mechanical properties, thermal stability, and high resistance to pests like termites make OPT lumber an ideal RW substitute (Suhaily et al., 2012). These characteristics are important for furniture production because they guarantee the furniture's quality, strength, resilience, and longevity, helping users use the items for years to come.

With the thermal properties described, agro-waste from the oil palm industry is suitable as an insulation material for heating equipment. In a study by Erwinsyah and Richter (2008) on manufacturing thermal insulation material made from oil palm EFB fibers, a variety of characteristics of cut EFB fibers were investigated, including fiber length distribution, density, moisture, and oil content, as well as the optimization process and production of insulation material. The results showed that fabricated EFB fiber material contained fibers, calyx, and other components with various fiber lengths ranging between 45 and 55 mm. According to the cut EFB fibers, moisture content averaged 11.36%, and density was 40.5 kg/m^3. The study concluded that the method of fabrication of EFB as an insulation material is practically acceptable (Erwinsyah & Richter, 2008).

The minimal environmental impact of these applications and the increased innovation and technology have encouraged their efforts to improve their value. Oil palm wastes may also be used to make fruit bunch fibers, similar to conventional methods that use OPTs and fronds. In terms of sustainability, the oil palm tree is superior to other wood substitutes because every component is used. Fiber reinforcement, on the contrary, enhances the strength and stiffness of the composites (Dungani et al., 2018; Halimatul et al., 2019; Jumaidin et al., 2019a,b, 2021; Nazrin et al., 2021).

2.6.4 Bio-Coal

Biomass and coal both contain the same fundamental components but in different ratios. As a result, both materials exhibit distinct combustion behavior. Compared to coal, biomass has a nearly four-fold increase in oxygen content, a decrease in sulfur and nitrogen content, a greater amount of volatile matter, and a higher level of reactivity. With better ignitability and combustibility than coal, it burns more quickly. An efficient way to create coal briquette formulations that yield a higher-quality fuel is to use biomass from agricultural waste (Sulaiman et al., 2011).

Briquetting is a densification process that uses pressure on materials to obtain a compact, durable, and high-quality fuel. The briquette process includes drying, grinding, sieving, compacting, cooling, and packing as necessary. The process aids in lowering handling, transportation, and storage costs. Increased bulk density and decreased moisture content can both increase the volumetric calorific value of the briquette. Another study has found that the waste generated from 90 million tons of crude palm oil can generate 1.19×10^8 GJ of electricity by co-firing at a presumed efficiency of 80% (Sulaiman et al., 2011).

In a study on the physical properties of biomass fuel briquette from oil palm residues (Ho et al., 2017), a collection of PKSs was dried for complete removal of palm oil and pulverized and then sieved to British Standard at 350, 250, and 150 μm with Octagon digital sieving machine. MFs and PKSs were mixed by manual stirrer in the following ratios: 90 to 10, 80 to 20, and 70 to 30, respectively. The amount of binder employed, which made up 30% of the aggregates' total mass, was 500 mL of water. The mixtures are then compacted. Once the piston was set for its removal, the briquette was delicately evacuated from the mold. The remaining moisture and palm oil were then allowed to dry on the briquette samples over the course of a week in the sun. These samples were then dried and burned to determine their moisture, ash, and volatile content, which is the result shown in Table 2.12.

In the process of making briquettes, moisture content percentage is a crucial factor. Briquette 350 μm (90:10) aggregate had the highest percentage of moisture, followed by 350 μm (80:20) and 250 μm (90:10). Given that the briquettes were sun-dried and that the PKS particle size is 350 m, briquettes with smaller grain particles tend to dry up more quickly, whereas those with bigger grain particles have higher moisture percentages since it took longer for them to dry up. Comparatively speaking, the aggregate 70:30 briquette series has the least moisture. In this regard, it can be deduced that particle size is one crucial characteristic when considering the dryness factor of any briquette (Ho et al., 2017). According to the result in Table 2.12, with an increase in MFs and in a manner that is directly proportional to the grain particles, the percentage ash content rises. Finer grain particles should ignite and burn more rapidly because of the higher combustion rate. The fact that MFs exhibits favorable burning properties, and largely uniform combustion may be the cause of this. They are anticipated to burn and convert to ash faster than PKS and ignite quickly (Ho et al., 2017). Table 2.12 shows the proportion of volatile matter in solid fuels with different grain sizes and mixing ratios. Series 150 m solid fuels have the most significant proportion, followed

TABLE 2.12
Moisture, Ash, and Volatile Matter Contents of Different Mixtures of Biomass Fuel (Ho et al., 2017)

Biomass Fuel (Solid)	Grain Sizes (μm)	Moisture (%)	Ash (%)	Volatile Matter (%)
90 PKS:10 MF	350	9.9	1.5	69.8
80 PKS:20 MF	350	9	1.7	70
70 PKS:30 MF	350	7.1	1.9	71.1
90 PKS:10 MF	250	8.5	3	71
80 PKS:20 MF	250	7	3.19	71.2
70 PKS:30 MF	250	6.5	3.27	72.1
90 PKS:10 MF	150	8	3.5	72.1
80 PKS:20 MF	150	6	3.61	72.5
70 PKS:30 MF	150	6.3	3.7	72.8

TABLE 2.13
Higher Heating Values HHV of Different Mixtures of Biomass Fuel (Ho et al., 2017)

Biomass Fuel (Solid)	Grain Sizes (μm)	HHV (kJ/kg)
90 PKS:10 MF	350	17.7653
80 PKS:20 MF	350	17.5822
70 PKS:30 MF	350	18.1063
90 PKS:10 MF	250	17.237
80 PKS:20 MF	250	17.6557
70 PKS:30 MF	250	17.6257
90 PKS:10 MF	150	17.0121
80 PKS:20 MF	150	17.6005
70 PKS:30 MF	150	17.4026

by series 250 and 350 m. The difference in volatile matter composition might be caused by incorrect aggregate mixing during manufacture, which could trap inflammable gases like nitrogen and carbon dioxide. The present study's findings demonstrate that briquettes can ignite readily with or without assistance since the percentage of volatile stuff reflects how quickly any fuel sample would ignite (Ho et al., 2017).

The study also tested the oil palm biomass samples to determine their calorific value. The Parikh formula was used to determine the heat energy released by the sample through combustion or higher heating value, HHV. The calculated HHV for the fuel samples is displayed in Table 2.13 (Ho et al., 2017). In the results, the HHV of the fuel samples from briquettes is shown at various mixing ratios. The largest calorific value for briquettes is 350 mm (70:30), followed by 350 mm (90:10) and 250 mm (80:20), respectively. Briquettes measuring 150 mm (90:10) and 250 mm (90:10) have the lowest heating values among biomass solid fuels. It is clear that the improved HHV of series 350 m is most likely due to the grain-like PKS particles and greater amounts of MF, which affect its moisture content. In another study, the HHV value is explained to be inversely proportional to the moisture content of the biomass fuel.

2.6.5 Biochar

Biochar, a solid carbon-based material, can be produced from oil palm waste through a process called pyrolysis, where the waste is heated without much oxygen (Konda et al., 2012). Various oil palm waste materials, like EFBs and OPFs, can be used as feedstocks for biochar production (Enaime & Lübken, 2021). This approach offers multiple advantages and applications. First, it helps manage waste in the oil palm industry by utilizing residues that would otherwise be discarded or cause pollution (Lee & Ofori-Boateng, 2013). Turning waste into biochar stabilizes carbon content, reducing GHG emissions and environmental impact (Enaime & Lübken, 2021).

Biochar from oil palm waste can be used for carbon dioxide capture, aiding in mitigating climate change by storing carbon emissions (Khedulkar et al., 2023). It can also improve wastewater treatment by adsorbing contaminants and pollutants due to its porous structure (Khedulkar et al., 2023). Applying biochar to soil in agriculture enhances soil health, water retention, and nutrient availability, leading to better plant growth and productivity. Moreover, biochar acts as a long-term carbon sink, helping to reduce carbon dioxide levels (Yadav et al., 2023). The production of biochar from oil palm waste aligns with a circular bioeconomy, transforming waste into a valuable resource, promoting resource efficiency, and supporting environmental sustainability in the oil palm industry (Yaashikaa et al., 2020). Table 2.14 summarizes some of the key benefits.

TABLE 2.14
Application of Biochar Derived from Oil Palm Waste

Benefit	Description	References
Sustainable waste management	Biochar can be produced from oil palm waste, which is a major environmental problem in many parts of the world. By converting this waste into a valuable product, biochar can help to reduce pollution and conserve resources.	Enaime and Lübken (2021)
Reduced greenhouse gas emissions	Biochar is a carbon-rich substance that can capture carbon dioxide from the air. This helps combat climate change and lessens the impact of global warming.	Khedulkar et al. (2023)
Improved wastewater quality	Biochar can be used to treat wastewater by adsorbing contaminants and pollutants. This can help to improve the quality of wastewater and reduce the environmental impact of wastewater treatment.	Khedulkar et al. (2023)
Enhanced soil health	Adding biochar to the soil enhances soil structure, retains water better, and makes nutrients more available. This leads to improved plant growth, higher productivity, and a reduction in soil erosion.	Yadav et al. (2023)
Circular bioeconomy	Making biochar fits into the idea of a circular bioeconomy, which is about reducing waste and making the most of resources. By turning waste into a valuable product like biochar, we can contribute to a more sustainable and circular economy.	Yaashikaa et al. (2020)

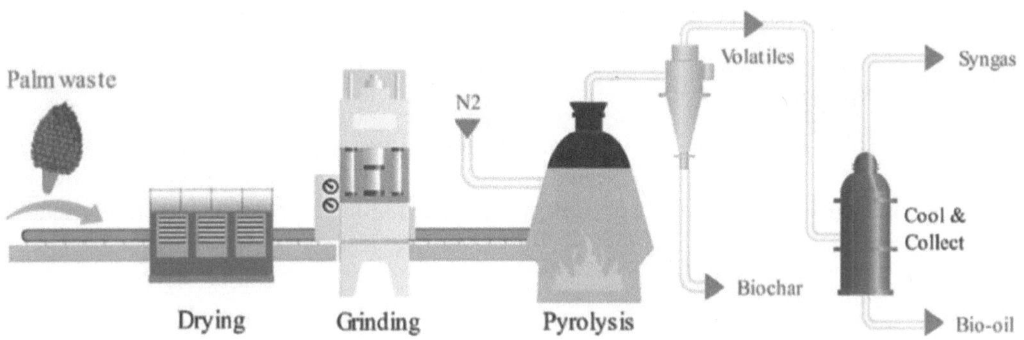

FIGURE 2.9 Simplified process flow of oil palm waste pyrolysis (Parthasarathy et al., 2023).

According to Parthasarathy et al. (2023), making biochar from oil palm waste is a hopeful new technology that can bring many environmental and economic advantages. While more research is required to fully grasp its potential, this technology could play a crucial role in promoting sustainable agriculture and waste management in the future. Figure 2.9 illustrates the process of converting oil palm waste into biochar using pyrolysis. Transforming oil palm waste into biochar using pyrolysis has numerous advantages. It helps manage waste sustainably, lowers GHG emissions, improves wastewater, and boosts soil health. Additionally, it supports a circular bioeconomy. By utilizing oil palm waste for biochar, the oil palm industry can reduce its environmental impact and promote a greener future.

2.6.6 Bio-Bricks

Another potential avenue for managing oil palm waste is through the production of bio-bricks. While the specific connection between oil palm waste and bio-bricks may not be explicitly stated, exploring the potential relationship and benefits is possible. Bio-bricks are environmentally friendly alternatives to traditional bricks, typically produced using organic or agricultural waste materials. In the context of oil palm waste management, it is conceivable that specific components or by-products of the waste, such as fiber or biomass residues, could be utilized as feedstock in the production of bio-bricks (Kolli et al., 2023). Incorporating oil palm waste into bio-bricks would contribute to the sustainable management of the waste while reducing reliance on conventional building materials. Moreover, oil palm waste can be used as a renewable resource to create bioenergy sources like biodiesel and biofuels. This provides a sustainable energy option and helps reduce waste using the oil palm waste for energy production (Mahlia et al., 2019). The energy generated from oil palm waste can also support bio-brick manufacturing, where renewable materials are used to make bricks, fostering a connection between waste management and sustainable construction practices. Additionally, oil palm biomass holds promise as a renewable resource for producing renewable materials, aligning with bio-bricks. Bio-bricks are typically made from renewable sources, and oil palm biomass has been recognized as a valuable option for creating various renewable materials (Hassan et al., 2019). By redirecting oil palm waste to produce renewable materials, including bio-bricks, we contribute to the circular economy and encourage sustainable resource management. Table 2.15 summarizes the potential benefits of using oil palm waste for bio-brick production.

Meanwhile, Figure 2.10 shows the potential process for producing bio-bricks from oil palm waste. The figure illustrates three key steps in making bio-bricks from oil palm waste. The first step is preparing the waste, which includes shredding it into small pieces and getting rid of any impurities. The second step is to form the oil palm waste into bricks. This can be done using a variety of methods, such as compression molding or extrusion. The third step is to dry the bricks. This can be done naturally or using an oven. Once the bricks are dry, they can be used to construct buildings or other structures (Khurshid et al., 2023).

2.6.7 Biohydrogen

Biohydrogen is an eco-friendly form of renewable energy, making it a promising alternative to address the decline in fossil oil reserves and minimize the environmental effects caused by emissions, with great demand in today's world (Pasaribu et al., 2020). Based on the literature, due to the

TABLE 2.15
The Potential Benefits of Using Oil Palm Waste to Produce Bio-Bricks

Benefit	Description	References
Reduced environmental impact	Making bio-bricks from oil palm waste can significantly reduce the environmental impact of the oil palm industry. Bio-bricks help minimize waste generation and decrease reliance on conventional building materials that harm the environment.	Kolli et al. (2023)
Increased sustainability	Using oil palm waste for bio-bricks enhances the sustainability of the oil palm industry by establishing a circular economy, where waste is reused and recycled instead of being discarded.	Mahlia et al. (2019)
Economic benefits	Moreover, producing bio-bricks from oil palm waste creates economic opportunities by generating jobs and income within the oil palm industry.	Hassan et al. (2019)

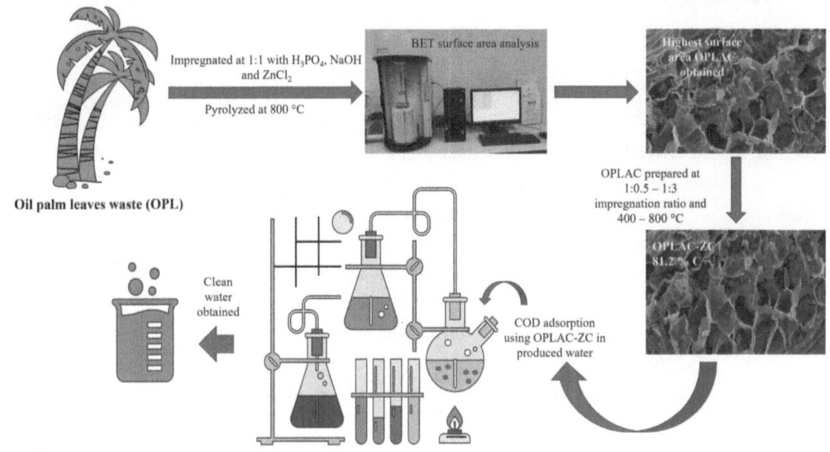

FIGURE 2.10 The potential process for producing bio-bricks from oil palm waste (Khurshid et al., 2023).

high organic content in POME, a lot of research has been conducted on the possibility of POME being converted into renewable sources, including biohydrogen (Prautama & Ardyansyah, 2018).

Several researchers studied the dark fermentation of POME to produce biohydrogen. Dark fermentation is one of the anaerobic fermentation processes applied for biohydrogen production (Rosman et al., 2020). One research utilized dark fermentation using *Clostridium butyricum*. In their study, POME underwent several procedures, including pretreatment, bacteria cultivation, dark fermentation, and analysis of biohydrogen products in gas chromatography. As a result of their research, biohydrogen was successfully produced from a concentration of sodium hydroxide (NaOH) 0.3%, which is 2,824.2 mg H_2/L POME. The best fermentation retention time for producing the highest levels of biohydrogen is 32 h at a 0.3% NaOH concentration (Pasaribu et al., 2020).

Apart from POME, EFB has been utilized as a main fermentation feedstock for biohydrogen production. In the study, the EFB underwent both diluted acetic acid (DAA) and ammonia fiber expansion (AFEX) pretreatment to compare the results before undergoing enzymatic hydrolysis. At the end of the study, it was concluded that the total amount of biohydrogen produced from AFEX EFB hydrolysate was approximately 45% greater than the amount produced in DAA EFB molasses fermentation. When AFEX EFB hydrolysate was fermented with *Enterobacter* sp. *KBH 6958*, the yield of biohydrogen was measured to be 1.55 mol/mol, while the yield from DAA EFB molasses was 1.19 mol/mol (Luthfi et al., 2019).

2.6.8 Biomethane

Biomethane, a renewable energy source, can be generated by breaking down organic waste, like POME (Golden Agri-Resources, 2020; Arthur & Glover, 2012). POME is wastewater produced during the processing of fresh fruit bunches into crude palm oil at palm oil mills (Golden Agri-Resources, 2020). If POME is not appropriately treated in open ponds, it decomposes organically and releases GHGs like methane and carbon dioxide (Golden Agri-Resources, 2020) (Table 2.16).

Research by Khurshid et al. (2023) studied continuous biogas production from pH-adjusted POME using two reactors: reactor R1 with oil palm ash and reactor R2 with biogas effluent recycling, as shown in Figure 2.11. The process involves three main steps. First, the POME is prepared by removing contaminants like solids and heavy metals. Second, the POME undergoes anaerobic digestion, breaking down organic matter without oxygen. Lastly, the produced biogas is collected and purified using filters and separators, making it suitable for generating electricity, heat, or powering vehicles.

TABLE 2.16
Potential Benefits of Biomethane Production from POME

Benefit	Description	References
Reduced greenhouse gas emissions	Biomethane can help reduce greenhouse gas emissions by capturing and utilizing methane from POME.	Golden Agri-Resources (2020)
Energy self-sufficiency	Biomethane can be used to generate renewable energy to power mills and biogas plants.	Arthur and Glover (2012)
Sustainable and circular approach	Biomethane production can help ensure a more sustainable and circular approach to palm oil production by repurposing waste biomass and utilizing effluent as fertilizer.	Seekao et al. (2021)

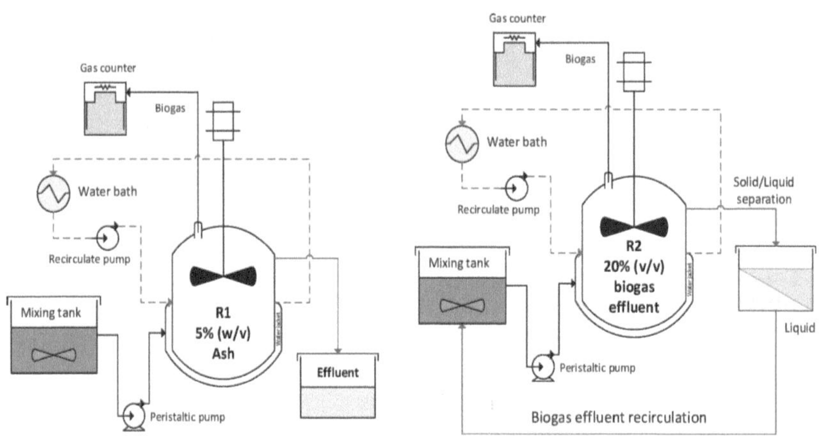

FIGURE 2.11 Straightforward schematic diagram illustrating the continuous biogas production process using pH-adjusted palm oil mill effluent (POME) with oil palm ash in reactor R1 and biogas effluent recycling in reactor R2 (Khurshid et al., 2023).

Implementing biogas plants to harness the waste from palm oil mills effectively mitigates GHG emissions. By capturing and utilizing the biogas produced from POME, both the mills and the biogas plants can be powered, leading to a reduction in emissions. This approach provides a sustainable energy source for the mills and helps minimize the environmental impact of POME (Golden Agri-Resources, 2020). The remaining effluent after biogas production can be utilized as fertilizer in palm oil plantations, closing the nutrient loop and enhancing soil fertility (Golden Agri-Resources, 2020). This practice supports sustainable agriculture using nutrient-rich effluent to benefit the plantation ecosystem. Co-digestion, which involves combining POME with other organic waste streams, such as chicken manure and crude glycerol, can improve the overall biogas production and address waste management challenges associated with POME (Seekao et al., 2021). This synergistic approach maximizes the energy potential of various waste sources and provides an efficient solution to waste management issues. Moreover, leftover plant materials from the palm oil industry, like EFBs, OPFs, and trunks, can be reused to create eco-friendly energy, like biogas (Dickinson et al., 2019). This way, the palm oil industry can help produce renewable energy and efficiently handle waste materials.

The palm oil industry can achieve multiple benefits by harnessing the potential of biomethane and implementing these waste management strategies. These include reducing GHG emissions, promoting energy self-sufficiency by generating renewable energy, and adopting a more sustainable and circular approach to palm oil production by repurposing waste biomass and utilizing effluent as fertilizer. These strategies minimize waste and maximize resource efficiency, enabling the industry to move toward a greener and more environmentally conscious approach while contributing to a sustainable future.

2.6.9 Bioethanol

Bioethanol is a substitute fuel derived from alcohol, obtained through the fermentation of organic substances rich in starch and sugar content (Mahlia et al., 2019). It has been suggested that wasted crops, such as oil palm waste, can serve as an alternative feedstock for biofuel production, replacing traditional starch crops and mitigating conflicts with human food consumption (Wardani et al., 2021). There are three types of palm oil biomass waste that can be converted into bioethanol, which are OPT, EFB, and OPF (Mahlia et al., 2019).

The expanding palm oil industry generates a significant amount of palm oil residue, including OPT waste, which holds substantial potential as a renewable energy source. OPT consists of approximately 70% sugar and 30% cellulosic residue, making it an auspicious material for utilization as feedstock in second-generation ethanol production (Wardani et al., 2021). Both EFB and OPT are classified as cellulosic materials, holding significant potential for conversion into fermentable sugars. However, the main challenge lies in their more challenging processing requirements. Pretreatments such as mechanical, chemical, and steam methods are necessary to break down the complex structure of lignocellulosic materials. Additionally, enzymes are employed to convert cellulose into sugars. This overall process increases the cost of sugar production and necessitates an appropriate wastewater treatment system due to the use of chemicals in the process (Mahlia et al., 2019).

In another research, the researchers concluded that OPF is the simplest and most inexpensive oil palm biomass to be processed as bioethanol. OPF juice is recognized for its rich content of various simple sugars such as glucose, fructose, and sucrose (Gloria et al., 2020). In 2014, Lee and Halim conducted research demonstrating that OPF juice contains a significant amount of sugar, specifically approximately 647.76 g/L of total free sugars, including glucose, fructose, and sucrose. This high sugar content makes OPF juice an excellent candidate for bioethanol production, as it can be extracted and subjected to fermentation. Utilizing OPF juice as a fermentation substrate offers several advantages, including its renewability, cost-effectiveness, and abundant availability (Gloria et al., 2020).

2.6.10 Biobutanol

Biobutanol is expected to play a crucial role in the next generation of biofuel. It is a derivation of butanol that is obtained through fermentation and is considered an exceptional type of biofuel. It can be mixed into regular gasoline in a similar manner to ethanol, but it offers numerous advantages, including a greater amount of energy per unit and lower vapor pressure, making it easier to store and transport. It is also incompatible with water, has improved compatibility with gasoline and diesel fuel, and can be utilized in conventional internal combustion engines without any alterations (Asri et al., 2019).

One research study reported the potential of converting OPTs for biobutanol production. The sap extracted from the cut OPT is a promising option as it does not require additional processing to facilitate fermentation. The sap is obtained by squeezing the trunk with acetone and ethanol by *Clostridium* spp. to produce biobutanol. The fiber wastes of the trunk after sap squeezing were hydrolyzed to fermentable sugars by acid treatment before their use as a substrate for butanol production. The researchers concluded that further study using detoxification methods should be done to improve the production of biobutanol from OPT (Komonkiat & Cheirsilp, 2013).

Previous studies showed that biobutanol can be biosynthesized from OPFs (Asri et al., 2019). This is due to the high content of glucose in the OPF juice. Moreover, the extracted juice contains a high concentration of minerals and essential nutrients that are crucial for the fermentation process in most microorganisms (Aliyu et al., 2015). However, this study is quite new, yet there is much research about it. Based on the readings, one study performed a laboratory-scale experiment on converting OPF to biobutanol by *Clostridium acetobutylicum ATCC 824* with gas stripping technology to reduce toxicity. It was concluded that gas stripping not only prevented the biobutanol toxicity but also increased the fermentation rate and the biobutanol yield (Asri et al., 2019).

2.6.11 Natural Acids

Based on readings, OPT from oil palm biomass is an excellent lignocellulosic feedstock for bioconversion of succinic acids due to its abundance and low cost. Succinic acids are usually used as a source of food additives, pharmaceuticals, and biodegradable plastics. To produce succinic acids, OPT needs to undergo acid pretreatment. In the study by Bukhari et al. (2020), among the organic acids tested, citric acid demonstrated superior performance compared to oxalic acid and formic acid. The utilization of OPT hydrolysates containing both glucose and xylose by *Actinobacillus succinogenes 130Z* showcases promising prospects to produce bio-based succinic acid. However, further research is needed to enhance the process of succinic acid fermentation using organic acids for the pretreatment of OPT (Harun et al., 2022). In addition to that research, there were other studies on the production of succinic acid from different types of biomasses, which are OPF juice, EFB, and OPT sap. The details of the research are shown in Table 2.17.

2.6.12 Bioelectricity

Bio-electrochemical systems are based on microbial interactions with solid electron acceptors/donors and include microbial fuel cells (MFCs), microbial electrolysis cells, and microbial desalination cells. In the process of producing electricity, microorganisms oxidize organic substances, and the electrons produced are then transported to the electrode (Ge et al., 2015).

In the context of wastewater treatment, MFCs can generate power while simultaneously removing pollutants. The substrate for MFCs has previously been made from a variety of wastewaters, including household, agricultural, and many more. However, the MFCs' removal capacities were constrained, and effective wastewater treatment required an integrated system incorporating additional technologies (Cheng et al., 2010). Integrated two-stage MFCs and IBAFs are studied by Cheng et al. (2010) to treat POME and remove partial ammonia and major organic pollutants associated with simultaneous electricity generation. Using both the up-flow anaerobic sludge blanket and MFC stage, over 90% of the total COD can be removed at current densities of 187 and 273 mA/m^2,

TABLE 2.17
Research for Succinic Acid Production from Various Types of Oil Palm Biomass

Type of Oil Palm Biomass	Pretreatment Process	Yield (g/g)	References
OPF juice	Alkaline pretreatment, enzymatic hydrolysis	0.66	Luthfi et al. (2019)
EFB	Dilute acid-microwave-alkaline pretreatment, enzymatic hydrolysis	0.47	Akhtar and Idris (2017)
OPT sap	—	0.36	Harun et al. (2022)

the maximum power density of MFC1 is 41.8 mW/m², 800 kW, and MFC2 is 44.6 mW/m², 400 kW. By scaling up this design, the implementation of energy recovery could also achieve pollution treatment.

2.7 POTENTIAL APPLICATIONS OF OIL PALM AGRO-WASTE

As mentioned in the previous subtopic, oil palm biomass has many potential applications. In addition to all the input mentioned above, some research mentions the current development or future application of oil palm waste that can be approached. Figure 2.12 shows the implementation of the application mentioned in the paragraphs below leading toward the sustainable approach that contributes to the circular bioeconomy.

2.7.1 Application of POME in Wastewater Treatment

One journal mentioned the application of oil palm waste, specifically the wastewater treatment of POME. Due to their low cost and ease of use, over 85% of mills in Malaysia still use traditional wastewater treatment techniques such as waste stabilization ponds. However, some mills use more advanced methods, such as POME treatment with microorganisms, membrane treatment, bioreactors, and oil trapping systems. The Department of Environment (DOE) Malaysia has established discharge standards that must be met for the treated POME, where compliance with these standards is necessary. The primary contributors to carbon emissions are the incineration process, which involves burning biomass residues and the treatment of POME. According to that research, the average GHG emission associated with producing 1 ton of crude palm oil is 1,100 kg CO_2-eq. This figure considers factors such as boiler operations, biogas production, mill activities, transportation, and electricity from the grid. Biogas generated from POME accounts for over 90% of these emissions. Therefore, by implementing an appropriate biogas capture system, it is possible to reduce a total of 80% of the 1,100 kg CO_2-eq GHG emissions (Cheah et al., 2023).

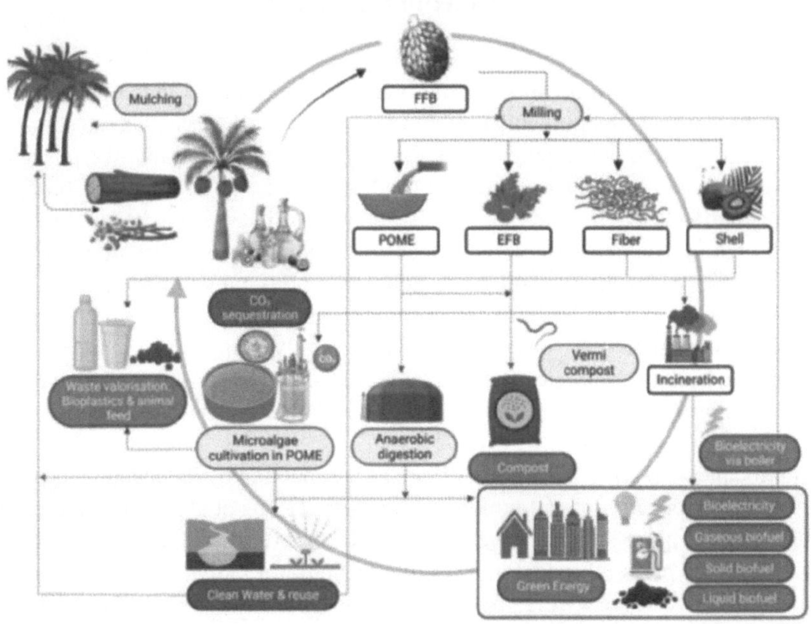

FIGURE 2.12 The potential of oil palm biomass that aligns with the concept of circular bioeconomy (Cheah et al., 2023).

2.7.2 Microalgal-Based Treatment for POME

Moreover, microalgae can be used for POME phycoremediation and carbon biosequestration. The industrial plant for treating POME cultivates microalgae species as part of the tertiary treatment process. This cultivation serves the purpose of remedying the POME. Microalgae species also engage in photosynthesis, capturing and converting carbon dioxide (CO_2). Microalgae biomass can be processed into valuable products such as biodiesel, bioethanol, and biopharmaceuticals. Microalgae have the potential to remove nitrogen, phosphorus, and hazardous components from various types of wastewater altogether, resulting in biomass production. This treatment is more affordable, cost-effective, and eco-friendly than traditional wastewater treatment methods. This method has been applied in the United States, as it produces 60% of the world's total biofuel. The United States collaborated with Asian and European countries to expand the bioenergy industry using microalgae (Cheah et al., 2023).

2.7.3 Waste Valorization of Oil Palm Biomass

In addition, the potential of oil palm waste is waste valorization. Waste valorization is the process of turning waste into useful goods. The majority of Malaysian palm oil mills follow a zero-waste strategy by creating high-quality composts from POME and biomass waste leftovers, which are then sold in the market to generate profit. Global Green Synergy Sdn. Bhd., one of the businesses that used this technique, claimed to be selling organic compost that adhered to the standards of the U.S. Composting Council, the EU, and the Australian standard for composting. Additionally, palm oil wastes may be used as a raw material for the manufacturing of bioplastics, and lignocellulosic biomass can be used to make polysaccharide-based bioplastics. Additionally, the biomass from palm oil may be utilized to make biogas and as an inexpensive adsorbent to clean up oil spills. However, further investment from business entities is required to realize the full potential of waste valorization at a commercial scale (Cheah et al., 2023; Norfarhana et al., 2022).

2.8 CHALLENGES ASSOCIATED WITH PROCESSING AND COMMERCIALIZATION

Although an abundance of palm oil waste could be utilized in making green products promising energy and nutrient recovery, the implementation of this idea is still lacking due to some challenges obstructing its execution and development. Kong et al. (2014) explained that the physicochemical nature of oil palm biomass remains a significant hindrance to the development of biochar production despite the abundance of oil palm biomass that may be utilized as feedstock. Optimization of the processing is required along with enhancement of equipment to assure product quality. Due to water intake during plant growth, oil palm biomass retains moisture to a significant extent; as a result, moving wet biomass over greater distances is more expensive and energy-inefficient. Because they were not created for easy adoption in a highly competitive market like the most recent ultra-supercritical technologies and conventional coal, biomass-based power generation technologies cannot provide a competitive advantage over other conventional proven technologies (Kong et al., 2014).

Another significant obstacle to the growth of biochar production by independent producers is the lack of long-term contracts with oil palm biomass providers (Kong et al., 2014). Neither the owners nor the producers of oil palm biomass have developed a comprehensive technological or commercial supply chain. In the palm oil industry, it is quite difficult for biomass end users and producers to get reliable feedstock contracts. It is possible to make agreements for short-term deliveries, but making long-term plans to guarantee supply at a fair price is challenging. These manners could be persistent as the awareness and acceptance of this technology by the public is still deficient (Kong et al., 2014).

In the view of biofuel production, for biofuels to compete with conventional fossil fuels, technological advancement in this field is crucial (Lim & Lee, 2012). In general, technology can be used throughout the planting, processing, and end-use phases of the biofuel supply chain. The palm oil output (measured in terms of fresh fruit bunch) and overall plantation cost are the two most crucial factors to take into account in palm oil plantations for biodiesel. The need for transport fuels will continue to rise due to population growth and rapid economic expansion. Due to the amount of land set aside for palm oil crops, the supply of biodiesel will thus be severely constrained compared to the demand (Lim & Lee, 2012).

The cost differential between biofuel blends and conventional transportation fuels is one of the primary challenges to their widespread use in transportation (Lim & Lee, 2012). Because it significantly affects so many aspects of the economy, the growing cost of transport fuels might lead to high inflation and rising living expenses. Even if the technology for producing biofuels is well-established, their commercialization is frequently postponed and, in some cases, totally abandoned because of the high cost of manufacturing. Furthermore, several nations have been implementing their fuel subsidy schemes so that people in these nations can buy them for less money. The economic gap between conventional fossil fuels and biofuels as transportation fuels will widen further as a result of subsidies for transportation fuels (Lim & Lee, 2012).

2.9 CONCLUSIONS

In conclusion, oil palm waste management is a pivotal aspect of sustainable practices within the oil palm industry. As this sector continues to grow to meet global demands for palm oil and its by-products, responsible management of waste becomes paramount. The proper management of waste not only minimizes environmental impacts but also presents significant economic opportunities through the use of waste for value-added products and renewable energy sources. Embracing innovative waste management techniques, such as composting, biogas generation, and biochar production, can effectively reduce GHG emissions, combat deforestation, and protect biodiversity. Additionally, supporting and incentivizing research and investments in waste management technologies will foster a more environmentally conscious oil palm industry. This will ensure that it strikes a harmonious balance between meeting economic needs and preserving our planet's precious natural resources. Collaborative efforts between governments, industries, and stakeholders are crucial to drive positive change and steer the oil palm industry toward a more sustainable and responsible future. Further research on energy, material, and nutrient potential recovery in the oil palm industry, along with a better awareness program, should be highly endorsed to optimize the current management of oil palm biomass in Malaysia, satisfying the pillars of sustainability, which are environment, economic, and social.

ACKNOWLEDGMENT

This chapter was written with Faiqah Batrisyia binti Usof, Nor Husna Binti Omar, and Nuh Farhan Bin Tajul Azwa.

REFERENCES

Abdul Aziz, M., Uemura, Y., & Sabil, K. (2011). Characterization of oil palm biomass as feed for torrefaction process. In *2011 National Postgraduate Conference—Energy and Sustainability: Exploring the Innovative Minds, NPC 2011*. https://doi.org/10.1109/NatPC.2011.6136260

Abdul Khalil, H. P. S., Siti Alwani, M., Ridzuan, R., Kamarudin, H., & Khairul, A. (2008). Chemical composition, morphological characteristics, and cell wall structure of Malaysian oil palm fibers. *Polymer-Plastics Technology and Engineering*, *47*(3), 273–280. https://doi.org/10.1080/03602550701866840

Abdullah, N., & Sulaiman, F. (2013a). The oil palm wastes in Malaysia. In M. D. Matovic (Ed.), *Biomass Now* (p. Ch. 3), 1–26. IntechOpen. https://doi.org/10.5772/55302

Abdullah, N., & Sulaiman, F. (2013b). The properties of the washed empty fruit bunches of oil palm. *Journal of Physical Science*, *24*(2), 117–137.

Abnisa, F., Arami Niya, A., Daud, W., Sahu, J., & Noor, I. (2013). Utilization of oil palm tree residues to produce bio-oil and bio-char via pyrolysis. *Energy Conversion and Management*, *76*, 1073–1082. https://doi.org/10.1016/j.enconman.2013.08.038

Abnisa, F., Daud, W. M. A. W., Husin, W. N. W., & Sahu, J. N. (2011). Utilization possibilities of palm shell as a source of biomass energy in Malaysia by producing bio-oil in pyrolysis process. *Biomass and Bioenergy*, *35*(5), 1863–1872. https://doi.org/10.1016/j.biombioe.2011.01.033

Ahmad, I., Sohail, M., Hameed, M., Fatima, S., Ahmad, M. S. A., Ahmad, F., Mehmood, A., Basharat, S., Asghar, A., Shah, S. M. R., & Ahmad, K. S. (2023). Morpho-anatomical determinants of yield potential in Olea europaea L. cultivars belonging to diversified origin grown in semi-arid environments. *PLoS One*, *18*(6), e0286736.

Ahmad, M. N., Ishak, M. R., Taha, M. M., Mustapha, F., & Leman, Z. (2021). Rheological and morphological properties of oil palm fiber-reinforced thermoplastic composites for fused deposition modeling (FDM). *Polymers*, *13*(21), 3739.

Akhtar, J., & Idris, A. (2017). Oil palm empty fruit bunches a promising substrate for succinic acid production via simultaneous saccharification and fermentation. *Renewable Energy*, *114*, 917–923. https://doi.org/10.1016/j.renene.2017.07.113

Ali, E. N., & Tay, C. I. (2013). Characterization of biodiesel produced from palm oil via base catalyzed transesterification. *Procedia Engineering*, *53*, 7–12. https://doi.org/10.1016/j.proeng.2013.02.002

Alias, A. H., Norizan, M. N., Sabaruddin, F. A., Asyraf, M. R. M., Norrrahim, M. N. F., Ilyas, A. R., Kuzmin, A. M., Rayung, M., Shazleen, S. S., Nazrin, A., Sherwani, S. F. K., Harussani, M. M., Atikah, M. S. N., Ishak, M. R., Sapuan, S. M., & Khalina, A. (2021). Hybridization of MMT/lignocellulosic fiber reinforced polymer nanocomposites for structural applications: A review. *Coatings*, *11*(11), 1355. https://doi.org/10.3390/coatings11111355

Aliyu, A. S., Aziz, A. A., Yahya, A., & Lattiff, Z. A. (2015). Potential of oil palm frond liquid extract and fiber as feedstock for bio-butanol production. *Jurnal Teknologi*, *74*, 63–67. https://doi.org/10.11113/jt.v74.4835

Aljuboori, A. H. R. (2013). Oil palm biomass residue in Malaysia: Availability and sustainability. *International Journal of Biomass and Renewables*, *2*(1), 13–18.

Arthur, R., & Glover, K. (2012). Biomethane potential of the POME generated in the palm oil industry in Ghana from 2002 to 2009. *Bioresource Technology*, *111*, 155–160.

Asri, F. M., Masngut, N., & Zahari, M. A. K. M. (2019). Biobutanol production from oil palm frond juice in 2L stirred tank bioreactor with in situ gas stripping recovery. *IOP Conference Series: Materials Science and Engineering*, *702*(1), 012004. https://doi.org/10.1088/1757-899X/702/1/012004

Asyraf, M. R. M., Ishak, M. R., Norrrahim, M. N. F., Nurazzi, N. M., Shazleen, S. S., Ilyas, R. A., Rafidah, M., & Razman, M. R. (2021). Recent advances of thermal properties of sugar palm lignocellulosic fibre reinforced polymer composites. *International Journal of Biological Macromolecules*, *193*, 1587–1599. https://doi.org/10.1016/j.ijbiomac.2021.10.221

Asyraf, M. R. M., Ishak, M. R., Syamsir, A., Nurazzi, N. M., Sabaruddin, F. A., Shazleen, S. S., Norrrahim, M. N. F., Rafidah, M., Ilyas, R. A., Rashid, M. Z. A., & Razman, M. R. (2022a). Mechanical properties of oil palm fibre-reinforced polymer composites: A review. *Journal of Materials Research and Technology*, *17*, 33–65. https://doi.org/10.1016/j.jmrt.2021.12.122

Asyraf, M. R. M., Syamsir, A., Zahari, N. M., Supian, A. B. M., Ishak, M. R., Sapuan, S. M., Sharma, S., Rashedi, A., Razman, M. R., Zakaria, S. Z. S., Ilyas, R. A., & Rashid, M. Z. A. (2022b). Product development of natural fibre-composites for various applications: Design for sustainability. *Polymers*, *14*(5), 920. https://doi.org/10.3390/polym14050920

Azlin, M. N. M., Ilyas, R. A., Zuhri, M. Y. M., Sapuan, S. M., Harussani, M. M., Sharma, S., Nordin, A. H., Nurazzi, N. M., & Afiqah, A. N. (2022). 3D printing and shaping polymers, composites, and nanocomposites: A review. *Polymers*, *14*(1), 180. https://doi.org/10.3390/polym14010180

Bakar, B. F. A., Tahir, P. M., Karimi, A., Bakar, E. S., Uyup, M. K. A., & Yong Choo, A. C. (2013). Evaluations of some physical properties for oil palm as alternative biomass resources. *Wood Material Science & Engineering*, *8*(2), 119–128. https://doi.org/10.1080/17480272.2012.701666

Bala, M., Bansal, S. K., & Fatima, F. (2023). Nanotechnology: A boon for agriculture. *Materials Today: Proceedings*, *73*, 267–270. https://doi.org/10.1016/j.matpr.2022.09.498

Bhat, R., Rai, R. V. and Karim, A. A. (2010). Mycotoxins in food and feed: Present status and future concerns. *Comprehensive Reviews in Food Science and Food Safety* 9, 57–81.

Bukhari, N. A., Jahim, J. M., Loh, S. K., Nasrin, A. B., Harun, S., & Abdul, P. M. (2020). Organic acid pretreatment of oil palm trunk biomass for succinic acid production. *Waste and Biomass Valorization*, *11*, 5549–5559.

Camara, M., Jamil, N. R., & Abdullah, A. F. B. (2019). Impact of land uses on water quality in Malaysia: A review. *Ecological Processes*, *8*(1). https://doi.org/10.1186/s13717-019-0164-x

Charters, L. J., Aplin, P., Marston, C. G., Padfield, R., Rengasamy, N., Bin Dahalan, M. P., & Evers, S. (2019). Peat swamp forest conservation withstands pervasive land conversion to oil palm plantation in North Selangor, Malaysia. *International Journal of Remote Sensing*, *40*(19), 7409–7438. https://doi.org/10.1080/01431161.2019.1574996

Chen, L., Tian, Y., Zhang, Z., Tong, Q., Sun, B., Rashed, M. M., & Jin, Z. (2017). Effect of pullulan on the digestible, crystalline and morphological characteristics of rice starch. *Food Hydrocolloids*, *63*, 383–390.

Cheng, J., Zhu, X., Ni, J., & Borthwick, A. (2010). Palm oil mill effluent treatment using a two-stage microbial fuel cells system integrated with immobilized biological aerated filters. *Bioresource Technology*, *101*(8), 2729–2734. https://doi.org/10.1016/j.biortech.2009.12.017

Cheah, W. Y., Siti-Dina, R. P., Leng, S. T. K., Er, A. C., & Show, P. L. (2023). Circular bioeconomy in palm oil industry: Current practices and future perspectives. *Environmental Technology & Innovation, 30*, 103050. pages 1-12. https://doi.org/10.1016/j.eti.2023.103050

Cooper, H. V., Evers, S., Aplin, P., Crout, N., Dahalan, M. P. B., & Sjogersten, S. (2020). Greenhouse gas emissions resulting from conversion of peat swamp forest to oil palm plantation. *Nature Communications*, *11*(1), 407.

Dickinson, E., Harrison, M., Parker, M., Dickinson, M., Donarski, J., Charlton, A., … & Wilson, J. (2019). From waste to food: Optimising the breakdown of oil palm waste to provide substrate for insects farmed as animal feed. *PLoS One*, *14*(11), e0224771.

Dungani, R., Aditiawati, P., Aprilia, S., Yuniarti, K., Karliati, T., Suwandhi, I., & Sumardi, I. (2018). Biomaterial from oil palm waste: Properties, characterization and applications. In *Palm Oil*. https://doi.org/10.5772/intechopen.76412

Enaime, G., & Lübken, M. (2021). Agricultural waste-based biochar for agronomic applications. *Applied Sciences*, *11*(19), 8914.

Erwinsyah, & Richter, C. (2008). Thermal insulation material made from oil palm empty fruit bunch fibers. *BIOTROPIA – The Southeast Asian Journal of Tropical Biology*, *14*(1), 32–50. https://doi.org/10.11598/btb.2007.14.1.23

Faizal, H. M., Shamsuddin, H. S., Heiree, M. H. M., Hanaffi, M. F. M. A., Abdul Rahman, M. R., Rahman, M. M., & Latiff, Z. A. (2018). Torrefaction of densified mesocarp fiber and palm kernel shell. *Renewable Energy*, *122*, 419–428. https://doi.org/10.1016/j.renene.2018.01.118

Ge, Z., Wu, L., Zhang, F., & He, Z. (2015). Energy extraction from a large-scale microbial fuel cell system treating municipal wastewater. *Journal of Power Sources*, *297*, 260–264. https://doi.org/10.1016/j.jpowsour.2015.07.105

Girkin, N. T., Cooper, H. V., Ledger, M. J., O'Reilly, P., Thornton, S. A., Åkesson, C. M., … Roucoux, K. H. (2022). Tropical peatlands in the Anthropocene: The present and the future. *Anthropocene*, *40*, 100354. https://doi.org/10.1016/j.ancene.2022.100354

Gloria, N., Legowo, E., & Kartawiria, I. (2020). Bioethanol Production from Oil Palm Fronds Biomass Waste Review. https://doi.org/10.1145/3429789.3429854

Golden Agri-Resources. (2020, December 23). Palm Supply Chain Traceability & Transformation. Retrieved 15 February 2024, from https://www.goldenagri.com.sg/sustainability/responsible-sourcing/palm-supply-chain-traceability-and-transformation/

Guangul, F. M., Sulaiman, S. A., Moni, M. N. Z., Atnaw, S. M., & Konda, R. E. (2013). Determination of the equilibrium moisture content of oil palm fronds feedstock for gasification process. *Asian Journal of Scientific Research*, *6*, 360–366.

Guangul, F. M., Sulaiman, S. A., & Ramli, A. (2012). Gasifier selection, design and gasification of oil palm fronds with preheated and unheated gasifying air. *Bioresource Technology*, *126*, 224–232. https://doi.org/10.1016/j.biortech.2012.09.018

Halimatul, M. J., Sapuan, S. M., Jawaid, M., Ishak, M. R., & Ilyas, R. A. (2019). Water absorption and water solubility properties of sago starch biopolymer composite films filled with sugar palm particles. *Polimery*, *64*(9), 27–35. https://doi.org/10.14314/polimery.2019.9.4

Harun, S., Luthfi, A. A. I., Abdul, P. M., Bukhari, N. A., & Jahim, J. M. (2022). Chapter 11- Oil palm biomass zero-waste conversion to bio-succinic acid. In S. Yusup & N. A. Rashidi (Eds.), *Value-Chain of Biofuels* (pp. 249–275). Elsevier. https://doi.org/10.1016/B978-0-12-824388-6.00011-7

Hassan, M. A., Farid, M. A. A., Shirai, Y., Ariffin, H., Othman, M. R., Samsudin, M. H., & Hasan, M. Y. (2019). Oil palm biomass biorefinery for sustainable production of renewable materials. *Biotechnology Journal*, *14*(6), 1800394.

Ho, M., Odusote, J., & Adeleke, A. (2017). Physical properties of biomass fuel briquette from oil palm residues. *Journal of Applied Sciences and Environmental Management*, *21*, 777–782. https://doi.org/10.4314/jasem.v21i4.19

Hussain, M., Levacher, D., Leblanc, N., Zmamou, H., Maigre, I. D., & Razakamanantsoa, A. (2021). Influence of palm oil fibers length variation on mechanical properties of reinforced crude bricks. In *Proceedings of the 4th International Conference on Bio-Based Building Materials ICBBM,* Barcelona, Spain. (pp. 628–633).

Ibrahim, S. H., Sia, W. K., Baharun, A., Nawi, M. N. M., & Affandi, R. (2014). Thermal performance of oil palm fiber and paper pulp as the insulation materials. *Journal of Civil Engineering, Science and Technology*, *5*(2). https://doi.org/10.33736/jcest.135.2014

Idris, S., Abd Rahman, N., & Ismail, K. (2012). Combustion characteristics of Malaysian oil palm biomass, sub-bituminous coal and their respective blends via thermogravimetric analysis (TGA). *Bioresource Technology*, *123*, 581–591. https://doi.org/10.1016/j.biortech.2012.07.065

Imraan, M., Ilyas, R. A., Norfarhana, A. S., Bangar, S. P., Knight, V. F., & Norrrahim, M. N. F. (2023). Sugar palm (Arenga pinnata) fibers: New emerging natural fibre and its relevant properties, treatments and potential applications. *Journal of Materials Research and Technology*, *24*, 4551–4572. https://doi.org/10.1016/j.jmrt.2023.04.056

Jafri, N. H. S., Jimat, D. N., Azmin, N. F. M., Sulaiman, S., & Nor, Y. A. (2021). The potential of biomass waste in Malaysian palm oil industry: A case study of Boustead Plantation Berhad. *IOP Conference Series: Materials Science and Engineering*, *1192*(1), 012028. https://doi.org/10.1088/1757-899X/1192/1/012028

Jauhiainen, J., Hooijer, A., & Page, S. E. (2012). Carbon dioxide emissions from an Acacia plantation on peatland in Sumatra, Indonesia. *Biogeosciences*, *9*(2), 617–630. https://doi.org/10.5194/bg-9-617-2012

Jumaidin, R., Adam, N. W., Ilyas, R. A., Hussin, M. S. F., Taha, M. M., Mansor, M. R., Azlan, U. A. A., & Yob, M. S. (2019a). Water transport and physical properties of sugarcane bagasse fibre reinforced thermoplastic potato starch biocomposite. *Journal of Advanced Research in Fluid Mechanics and Thermal Sciences*, *61*(2), 273–281.

Jumaidin, R., Diah, N. A., Ilyas, R. A., Alamjuri, R. H., & Yusof, F. A. M. (2021). Processing and characterisation of banana leaf fibre reinforced thermoplastic cassava starch composites. *Polymers*, *13*(9), 1420. https://doi.org/10.3390/polym13091420

Jumaidin, R., Saidi, Z. A. S., Ilyas, R. A., Ahmad, M. N., Wahid, M. K., Yaakob, M. Y., Maidin, N. A., Rahman, M. H. A., & Osman, M. H. (2019b). Characteristics of cogon grass fibre reinforced thermoplastic cassava starch biocomposite: Water absorption and physical properties. *Journal of Advanced Research in Fluid Mechanics and Thermal Sciences, 62*(1), 43–52.

Kaniapan, S., Hassan, S., Ya, H., Patma Nesan, K., & Azeem, M. (2021). The utilisation of palm oil and oil palm residues and the related challenges as a sustainable alternative in biofuel, bioenergy, and transportation sector: A review. *Sustainability*, *13*(6), 3110. https://doi.org/10.3390/su13063110

Kashyap, M., Samadhiya, K., Ghosh, A., Anand, V., Lee, H., Sawamoto, N., … Bala, K. (2021). Synthesis, characterization and application of intracellular Ag/AgCl nanohybrids biosynthesized in Scenedesmus sp. as neutral lipid inducer and antibacterial agent. *Environmental Research*, *201*, 111499. https://doi.org/10.1016/j.envres.2021.111499

Kauffman, J. B., Arifanti, V. B., Basuki, I., Kurnianto, S., Novita, N., Murdiyarso, D., … & Warren, M. W. (2016). Protocols for the Measurement, Monitoring, and Reporting of Structure, Biomass, Carbon Stocks and Greenhouse Gas Emissions in Tropical Peat Swamp Forests. Working Paper 221, Center for International Forestry Research, Bogor.

Kelly-Yong, T. L., Lee, K. T., Mohamed, A. R., & Bhatia, S. (2007). Potential of hydrogen from oil palm biomass as a source of renewable energy worldwide. *Energy Policy*, *35*(11), 5692–5701. https://doi.org/10.1016/j.enpol.2007.06.017

Keong, C. Y. (2005). Recovering renewable energy from palm oil waste and biogas. *Energy Sources*, *27*(7), 589–596. https://doi.org/10.1080/00908310490448406

Khedulkar, A. P., Pandit, B., & Doong, R. A. (2023). Agricultural waste to real worth biochar as a sustainable material for supercapacitor. *Science of the Total Environment*, *869*, 161441.

Khurshid, H., Mustafa, M. R. U., & Kilic, Z. (2023). Removal of naphthalene, fluorene and phenanthrene by recyclable oil palm leaves' waste activated carbon supported nano zerovalent iron (N-OPLAC) composite in wastewater. *Water*, *15*(5), 967.

Kolli, M., Mugada, K. K., Kumar, A., & Rakesh, S. G. (2023). Next-generation waste residue composite materials. In M. Jawaid, M. Thariq, & M. S. Hossain (Eds.), Waste Residue Composites (Vol. 16, pp. 1–24). Springer Nature, Singapore.

Komonkiat, I., & Cheirsilp, B. (2013). Felled oil palm trunk as a renewable source for biobutanol production by Clostridium spp. *Bioresource Technology*, *146*, 200–207. https://doi.org/10.1016/j.biortech.2013.07.067

Konda, R. E., Sulaiman, S. A., & Ariwahjoed, B. (2012). Syngas production from gasification of oil palm fronds with an updraft gasifier. *Journal of Applied Sciences*, *12*, 2555–2561. https://doi.org/10.3923/jas.2012.2555.2561

Kong, S.-H., Loh, S.-K., Bachmann, R. T., Rahim, S. A., & Salimon, J. (2014). Biochar from oil palm biomass: A review of its potential and challenges. *Renewable and Sustainable Energy Reviews*, *39*, 729–739. https://doi.org/10.1016/j.rser.2014.07.107

Kosugi, A., Tanaka, R., Magara, K., Murata, Y., Arai, T., Sulaiman, O., Hashim, R., Hamid, Z. A. A., Yahya, M. K. A., Yusof, M. N. M., Ibrahim, W. A., & Mori, Y. (2010). Ethanol and lactic acid production using sap squeezed from old oil palm trunks felled for replanting. *Journal of Bioscience and Bioengineering*, *110*(3), 322–325. https://doi.org/10.1016/j.jbiosc.2010.03.001

Lam, M., Chen, C.-Y., Li, Z., Martin, A. R., Bryois, J., Ma, X., … Huang, H. (2019). Comparative genetic architectures of schizophrenia in East Asian and European populations. *Nature Genetics*, *51*(12), 1670–1678. https://doi.org/10.1038/s41588-019-0512-x

Lee, K. T., & Ofori-Boateng, C. (2013). Environmental sustainability assessment of biofuel production from oil palm biomass. In K. T. Lee & C. Ofori-Boateng (Eds.), *Sustainability of Biofuel Production from Oil Palm Biomass* (pp. 149–187). Springer, Singapore.

Lim, S., & Lee, K. T. (2012). Implementation of biofuels in Malaysian transportation sector towards sustainable development: A case study of international cooperation between Malaysia and Japan. *Renewable and Sustainable Energy Reviews*, *16*(4), 1790–1800. https://doi.org/10.1016/j.rser.2012.01.010

Loh, S. K. (2017). The potential of the Malaysian oil palm biomass as a renewable energy source. *Energy Conversion and Management*, *141*, 285–298. https://doi.org/10.1016/j.enconman.2016.08.081

Low, L. Q., Ilyas, R. A., Jalil, R., Hawanis, H. S. N., Ibrahim, R., Azriena, H. A. A., Zuhri, M. Y. M., Ainun, Z. M. A., Fatriasari, W., Hastuti, N., & Amelia, D. (2024). Physical and mechanical properties enhancement of beaten oil palm trunk pulp and paper by optimizing starch addition: Towards sustainable packaging solutions. *Industrial Crops and Products*, *221*, 119232. https://doi.org/10.1016/j.indcrop.2024.119232

Luthfi, A. A. I., Tan, J. P., Harun, S., Manaf, S., & Jamaliah, M. (2019). Homogeneous solid dispersion (HSD) system for rapid and stable production of succinic acid from lignocellulosic hydrolysate. *Bioprocess and Biosystems Engineering*, *42*. https://doi.org/10.1007/s00449-018-2019-8

Mahardika, M., Zakiyah, A., Ulfa, S. M., Ilyas, R. A., Hassan, M. Z., Amelia, D., Knight, V. F., & Norrrahim, M. N. F. (2024). Recent developments in Oil Palm Empty Fruit Bunch (OPEFB) fiber composite. *Journal of Natural Fibers*, *21*(1). https://doi.org/10.1080/15440478.2024.2309915

Mahlia, T. M. I., Ismail, N., Hossain, N., Silitonga, A. S., & Shamsuddin, A. H. (2019). Palm oil and its wastes as bioenergy sources: A comprehensive review. *Environmental Science and Pollution Research*, *26*(15), 14849–14866. https://doi.org/10.1007/s11356-019-04563-x

Martin, R. (2018). At a Glance—Palm oil: Economic and Environmental Impacts EN. European Union Agency for Asylum, Malta. Retrieved 13 Feb 2024, from https://policycommons.net/artifacts/3367645/at-a-glance-palm-oil/4166309/

Mawardi, I., Aprilia, S., Faisal, M., & Rizal, S. (2022). Investigation of thermal conductivity and physical properties of oil palm trunks/ramie fiber reinforced biopolymer hybrid composites as building bio-insulation. *Materials Today: Proceedings*, *60*, 373–377. https://doi.org/10.1016/j.matpr.2022.01.249

Mazani, N., Sapuan, S. M., Sanyang, M. L., Atiqah, A., & Ilyas, R. A. (2019). Design and fabrication of a shoe shelf from kenaf fiber reinforced unsaturated polyester composites. In H. Ariffin, S. M. Sapuan, & M. A. Hassan (Eds.), *Lignocellulose for Future Bioeconomy* (1st ed., Issue 2000, pp. 315–332). Elsevier. https://doi.org/10.1016/B978-0-12-816354-2.00017-7

Meijaard, E., Brooks, T., Carlson, K., Slade, E., Garcia-Ulloa, J., Gaveau, D., Lee, J. S. H., Santika, T., Juffe-Bignoli, D., Struebig, M., Wich, S., Ancrenaz, M., Koh, L., Zamira, N., Abrams, J., Prins, H., Sendashonga, C., Murdiyarso, D., Furumo, P., & Sheil, D. (2020). The environmental impacts of palm oil in context. *Nature Plants*, *6*, 1418. https://doi.org/10.1038/s41477-020-00813-w

Miettinen, J., Shi, C., & Liew, S. C. (2016). Land cover distribution in the peatlands of Peninsular Malaysia, Sumatra and Borneo in 2015 with changes since 1990. *Global Ecology and Conservation*, *6*, 67–78. https://doi.org/10.1016/j.gecco.2016.02.004

Murphy, D. J., Goggin, K., & Paterson, R. R. M. (2021). Oil palm in the 2020s and beyond: Challenges and solutions. *CABI Agriculture and Bioscience*, *2*(1), 39. https://doi.org/10.1186/s43170-021-00058-3

Mustafa, H. M., & Hayder, G. (2021). Recent studies on applications of aquatic weed plants in phytoremediation of wastewater: A review article. *Ain Shams Engineering Journal*, *12*(1), 355–365. https://doi.org/10.1016/j.asej.2020.05.009

Nasri, N. N., Nawi, N. M., Baharuddin, A. S., & Lazim, S. M. (2021). Determination of tensile properties for twisted fiber bundles of oil palm empty fruit bunch at different diameters. *Basrah Journal of Agricultural Sciences*, *34*, 149–156.

Nazrin, A., Sapuan, S. M., Zuhri, M. Y. M., Tawakkal, I. S. M. A., & Ilyas, R. A. (2021). Water barrier and mechanical properties of sugar palm crystalline nanocellulose reinforced thermoplastic sugar palm starch (TPS)/poly(lactic acid) (PLA) blend bionanocomposites. *Nanotechnology Reviews*, *10*(1), 431–442. https://doi.org/10.1515/ntrev-2021-0033

Nizamuddin, S., Shrestha, S., Athar, S., Ali, B. S., & Siddiqui, M. A. (2015). A critical analysis on palm shell from oil palm industry as a feedstock for solid char production. *Reviews in Chemical Engineering*, *32*. https://doi.org/10.1515/revce-2015-0062

Norfarhana, A. S., Ilyas, R. A., & Ngadi, N. (2022). A review of nanocellulose adsorptive membrane as multifunctional wastewater treatment. *Carbohydrate Polymers*, *291*, 119563. https://doi.org/10.1016/j.carbpol.2022.119563

Norfarhana, A. S., Ilyas, R. A., Ngadi, N., Othman, M. H. D., Misenan, M. S. M., & Norrrahim, M. N. F. (2024a). Revolutionizing lignocellulosic biomass: A review of harnessing the power of ionic liquids for sustainable utilization and extraction. *International Journal of Biological Macromolecules*, *256*, 128256. https://doi.org/10.1016/j.ijbiomac.2023.128256

Norfarhana, A. S., Khoo, P. S., Ilyas, R. A., Ab Hamid, N. H., Aisyah, H. A., Norrrahim, M. N. F., Knight, V. F., Rani, M. S. A., Septevani, A. A., Syafri, E., & Annamalai, P. K. (2024b). Exploring of cellulose nanocrystals from lignocellulosic sources as a powerful adsorbent for wastewater remediation. *Journal of Polymers and the Environment*. https://doi.org/10.1007/s10924-024-03227-3

Norlin, B., An, S., Granfeldt, T., Krapohl, D., Lai, B., Rahman, H., Zeeshan, F., & Engstrand, P. (2023). Visualisation of sulphur on single fiber level for pulping industry. *Journal of Instrumentation: An IOP and SISSA Journal*, *18*(01), C01012. https://doi.org/10.1088/1748-0221/18/01/c01012

Norrahim, M. N. F., Farid, M. A. A., Jenol, M. A., Nurazzi, N. M., Hakimi, M. I., Shazleen, S. S., & Ilyas, R. A. (2022a). Economic insights into the production of cellulose nanofibrils from oil palm biomass. In S. M. Sapuan, M. N. F. Norrrahim, R. A. Ilyas, & C. Soutis (Eds.), *Industrial Applications of Nanocellulose and Its Nanocomposites* (pp. 39–48). Woodhead Publishing, Cambridge, United Kingdom.

Norrahim, M. N. F., Farid, M. A. A., Lawal, A. A., Yasim-Anuar, T. A. T., Samsudin, M. H., & Zulkifli, A. A. (2022b). Emerging technologies for value-added use of oil palm biomass. *Environmental Science: Advances*, *1*(3), 259–275. https://doi.org/10.1039/D2VA00029F

Norrahim, M. N. F., Ilyas, R. A., Nurazzi, N. M., Rani, M. S. A., Atikah, M. S. N., & Shazleen, S. S. (2021). Chemical pretreatment of lignocellulosic biomass for the production of bioproducts: An overview. *Applied Science and Engineering Progress*, 1–18. https://doi.org/10.14416/j.asep.2021.07.004

Norul Izani, M. A., & Paridah, M. T. (2013). Effects of different treatments of oil palm empty fruit bunch (OPEFB) on residual oil content and medium density fiberboard (MDF) performance. Paper presented at the International Conference on Agriculture and Biotechnology (ICABT 2013), Kuala Lumpur, Malaysia.

Nurazzi, N. M., Harussani, M. M., Aisyah, H. A., Ilyas, R. A., Norrrahim, M. N. F., Khalina, A., & Abdullah, N. (2021). Treatments of natural fiber as reinforcement in polymer composites—A short review. *Functional Composites and Structures*, *3*(2), 024002. https://doi.org/10.1088/2631-6331/abff36

Nyakuma, B., Johari, A., & Ahmad, A. (2013). Thermochemical analysis of palm oil wastes as fuel for biomass gasification. *Jurnal Teknologi*, *62*. https://doi.org/10.11113/jt.v62.1891

Okoli, I. C., & Okoli, L. (2020, November 9). Oil Palm Tree Wastes 6: Uses of the Palm Kernel Shell. *Research Tropica*. Retrieved July 24, 2023.

Onoja, E., Chandren, S., Razak, F., Mahat, N., & Wahab, R. (2019). Oil palm (Elaeis guineensis) biomass in Malaysia: The present and future prospects. *Waste and Biomass Valorization*, *10*. https://doi.org/10.1007/s12649-018-0258-1

Page, S. E., Rieley, J. O., & Banks, C. J. (2011). Global and regional importance of the tropical peatland carbon pool. *Global Change Biology*, *17*(2), 798–818. https://doi.org/10.1111/j.1365-2486.2010.02279.x

Parthasarathy, V., Senthil, A., Nagarajan, D., & Sundaresan, B. (2023). Palm fiber-based epoxy composites. In M. Jawaid, M. S. Hossain, & N. Saba (Eds.), Epoxy-Based Biocomposites (pp. 151–162). CRC Press, Boca Raton, FL, USA.

Pasaribu, M., Purba, F. S. P., Sitompul, E. Y. R., & Nainggolan, E. A. (2020). Biohydrogen production from palm oil mill effluent by dark fermentation using clostridium butyricum. *IOP Conference Series: Materials Science and Engineering*, *885*, 012049. https://doi.org/10.1088/1757-899X/885/1/012049

Pillai, H. U. K., Anilkumar, N., Achuthankutty, C. T., Mendes, C. R., Sabu, P., Jayalakshmi, K. V., … Menon, N. R. (2018). Planktonic food web structure at SSTF and PF in the Indian sector of the Southern Ocean during austral summer 2011. *Polar Research*, *37*(1), 1495545. https://doi.org/10.1080/17518369.2018.1495545

Prautama, C., & Ardyansyah, F. A. (2018). Production of biohydrogen as a future energy solution from palm oil mill effluent by mixed culture in endemic area of Pakenjeng, Garut, Indonesia. *SSRN Electronic Journal*. https://dx.doi.org/10.2139/ssrn.3192535

Rafidah, D., Ainun, Z. M. A., Hazwani, H. A., Rushdan, I., Luqman, C. A., Sharmiza, A., Paridah, M. T., & Jalaluddin, H. (2017). Characterisation of pulp and paper manufactured from oil palm empty fruit bunches and kenaf fibers. *Pertanika Journal of Tropical Agricultural Science*, *40*(3), 449–458.

Rosman, F. S., Yusoff, M. Z. M., Zakaria, M. R., Maeda, T., & Hassan, M. A. (2020). Dark fermentative biohydrogen production from palm oil mill effluent: Operation factors and future progress of biohydrogen energy. *Pertanika Journal of Science and Technology*, *28*. https://doi.org/10.47836/pjst.28.s2.19

Sadeghi, O., Fazeli, A., Bakhtiarinejad, M., & Sidik, N. A. C. (2014). An overview of waste-to-energy in Malaysia. *Applied Mechanics and Materials*, *695*, 792–796.

Sadh, P. K., Duhan, S., & Duhan, J. S. (2018). Agro-industrial wastes and their utilization using solid state fermentation: A review. *Bioresources and Bioprocessing*, *5*(1). https://doi.org/10.1186/s40643-017-0187-z

Sakai, S., Otsuka, T., Matsuda, S., Sakairi, Y., Uchida, R., Sugahara, K., Kano, A., & Yang, D. (2022). Subnanomolar sensitive stable isotopic determination in CO_2 by tunable infrared laser absorption spectroscopy. *Analytical Chemistry*, *94*(17), 6446–6450. https://doi.org/10.1021/acs.analchem.1c05458

Samiran, N. A., Jaafar, M. N. M., Chong, C. T., & Jo-Han, N. (2015). A review of palm oil biomass as a feedstock for syngas fuel technology. *Jurnal Teknologi*, *72*. https://doi.org/10.11113/jt.v72.3932

Seekao, N., Sangsri, S., Rakmak, N., Dechapanya, W., & Siripatana, C. (2021). Co-digestion of palm oil mill effluent with chicken manure and crude glycerol: Biochemical methane potential by monod kinetics. *Heliyon*, *7*(2), e06204.

Singh, P., Sulaiman, O., Hashim, R., Peng, L. C., & Singh, R. P. (2013). Using biomass residues from oil palm industry as a raw material for pulp and paper industry: Potential benefits and threat to the environment. *Environment, Development and Sustainability*, *15*, 367–383.

Singh, R. P., Ibrahim, M. H., Esa, N., & Iliyana, M. S. (2010). Composting of waste from palm oil mill: A sustainable waste management practice. *Reviews in Environmental Science and Bio/Technology*, *9*, 331–344. https://doi.org/10.1007/s11157-010-9199-2

Soloi, S., & Hou, E. K. Z. (2019, November). The potential of oil palm leaf fiber in paper-making industry. *Journal of Physics: Conference Series*, *1358*(1), 012005.

Sreekala, M. S., Kumaran, M. G., & Thomas, S. (1997). Oil palm fibers: Morphology, chemical composition, surface modification, and mechanical properties. *Journal of Applied Polymer Science*, *66*(5), 821–835.

Subramaniam, V., Chin, C. M., & Ngan M. A. (2004). Energy database of oil palm. *Palm Oil Engineering Bulletin*, *21–22*, 15–16.

Suhaily, S. S., Jawaid, M., Khalil, H. P. S. A., Mohamed, A. R., & Ibrahim, F. (2012). A review of oil palm biocomposites for furniture design and applications: Potential and challenges. *BioResources*, *7*(3), 4400–4423.

Sulaiman, F., Abdullah, N., Gerhauser, H., & Shariff, A. (2011). An outlook of Malaysian energy, oil palm industry and its utilization of wastes as useful resources. *Biomass and Bioenergy*, *35*(9), 3775–3786. https://doi.org/10.1016/j.biombioe.2011.06.018

Supriatna, J., Setiawati, M. R., Sudirja, R., Suherman, C., & Bonneau, X. (2022). Composting for a more sustainable palm oil waste management: A systematic literature review. *The Scientific World Journal*, 2022, 1–20. https://doi.org/10.1155/2022/5073059

Suriani, M. J., Radzi, F. S. M., Ilyas, R. A., Petrů, M., Sapuan, S. M., & Ruzaidi, C. M. (2021). Flammability, tensile, and morphological properties of oil palm empty fruit bunches fiber/pet yarn-reinforced epoxy fire retardant hybrid polymer composites. *Polymers*, *13*(8), 1282. https://doi.org/10.3390/polym13081282

Syafri, E., Jamaluddin, Sari, N. H., Mahardika, M., Amanda, P., & Ilyas, R. A. (2022). Isolation and characterization of cellulose nanofibers from Agave gigantea by chemical-mechanical treatment. *International Journal of Biological Macromolecules*, *200*(December 2021), 25–33. https://doi.org/10.1016/j.ijbiomac.2021.12.111

Terry, L. M., Li, C., Chew, J. J., Aqsha, A., How, B. S., Loy, A. C. M., Chin, B. L. F., Khaerudini, D. S., Hameed, N., Guan, G., & Sunarso, J. (2021). Bio-oil production from pyrolysis of oil palm biomass and the upgrading technologies: A review. *Carbon Resources Conversion*, *4*, 239–250. https://doi.org/10.1016/j.crcon.2021.10.002

Then, Y. Y., Ibrahim, N. A., Zainuddin, N., Ariffin, H., Yunus, W. M. Z. W., & Chieng, B. W. (2014). The influence of green surface modification of oil palm mesocarp fiber by superheated steam on the mechanical properties and dimensional stability of oil palm mesocarp fiber/poly(butylene succinate) biocomposite. *International Journal of Molecular Sciences*, *15*(9), 15344–15357. https://doi.org/10.3390/ijms150915344

Tonks, A. J., Aplin, P., Beriro, D. J., Cooper, H., Evers, S., Vane, C. H., & Sjögersten, S. (2017). Impacts of conversion of tropical peat swamp forest to oil palm plantation on peat organic chemistry, physical properties and carbon stocks. *Geoderma*, *289*, 36–45. https://doi.org/10.1016/j.geoderma.2016.11.018

Van Dexter, S., & Boopathy, R. (2020). Analysis of termite microbiome and biodegradation of various phenolic compounds by a bacterium isolated from the termite gut in Louisiana, USA. In *Valorisation of Agro-Industrial Residues – Volume I: Biological Approaches* (pp. 163–184). https://doi.org/10.1007/978-3-030-39137-9_8

Vijay, V., Pimm, S. L., Jenkins, C. N., & Smith, S. J. (2016). The impacts of oil palm on recent deforestation and biodiversity loss. *PloS One*, *11*(7), e0159668. https://doi.org/10.1371/journal.pone.0159668

Wahab, R., Rasat, M. S. M., Fauzi, N. M., Sulaiman, M. S., Samsi, H. W., Mokhtar, N., Ghani, R. S. M., & Razak, M. H. (2021). Processing and properties of oil palm fronds composite boards from Elaeis guineensis. In M. Jawaid, M. Asim, & N. Saba (Eds.), *Elaeis guineensis: Bioenergy, Biomaterials, and Sustainability* (pp. 259–285). Springer, Singapore.

Wanrosli, W. D., Zainuddin, Z., & Roslan, S. (2005). Upgrading of recycled paper with oil palm fiber soda pulp. *Industrial Crops and Products*, *21*(3), 325–329.

Wardani, A. K., Sutrisno, A., Faida, T. N., Yustina, R. D., & Murdiyatmo, U. (2021). Ethanol production from oil palm trunk: A combined strategy using an effective pretreatment and simultaneous saccharification and cofermentation. *International Journal of Microbiology*, 2021, 2509443. https://doi.org/10.1155/2021/2509443

Yaashikaa, P. R., Kumar, P. S., Varjani, S., & Saravanan, A. (2020). A critical review on the biochar production techniques, characterization, stability and applications for circular bioeconomy. *Biotechnology Reports*, *28*, e00570.

Yadav, N. K., Singh, S. K., Patel, A. B., Meitei, M. M., Meena, D. K., Yadav, M. K., ... & Choudhary, B. K. (2023). Biochar production methods vis-a-vis aquaculture applications: A strategy for sustainable paradigm. In A. Dey & R. Kesavamoorthy (Eds.), *Advances in Resting-State Functional MRI* (pp. 537–559). Springer, Singapore.

Yaro, N. S. A., Sutanto, M. H., Habib, N. Z., Napiah, M., Usman, A., Jagaba, A. H., & Al-Sabaeei, A. M. (2022). Application and circular economy prospects of palm oil waste for eco-friendly asphalt pavement industry: A review. *Journal of Road Engineering*, *2*(4), 309–331. https://doi.org/10.1016/j.jreng.2022.10.001

3 Date Palm Waste Management

3.1 INTRODUCTION

Date palm (*Phoenix dactylifera*) is a majestic and resilient tree that has been revered and cultivated for thousands of years (Al-Yousef et al., 2018). Known as the "tree of life" in many cultures, the date palm holds a special place in human history and the natural world (Zaid & de Wet, 2002). With its tall stature, graceful fronds, and succulent fruit clusters, this remarkable tree is not only a symbol of beauty but also a vital source of sustenance and economic prosperity in arid regions (Food and Agriculture Organization of the United Nations [FAO], 2021). Along with the camel, the date palm was responsible for opening the vast desert territories for human activity and the development of oasis agro-ecosystems. The oasis represents the climax of the rigorous management of scarce water and land resources in alliance with the date palm (Jaradat, 2011). The date palm's distinctive crown of feathery leaves can reach impressive heights, creating an iconic silhouette against the vast desert landscapes where it thrives (FAO, 2021). Its deep-rooted resilience allows it to withstand harsh conditions, including scorching temperatures, sandy soils, and limited water availability, making it an invaluable resource in arid and semiarid regions (Al-Shammari et al., 2014).

One of the date palm's most treasured gifts is its delicious and nutritious fruit. Date clusters hang in abundant clusters beneath the fronds, ripening into a rich caramel hue as they soak up the sun's warmth (Abdul Rahiman et al., 2012). Renowned for their natural sweetness and chewy texture, dates have been savored as a delectable treat since ancient times (Al-Farsi & Lee, 2008). Beyond their delightful taste, dates are packed with essential nutrients, including fiber, potassium, and various vitamins. They are a valuable dietary component in regions where other food sources may be scarce (Al-Shahib & Marshall, 2003). Beyond their culinary allure, date palms have played a crucial role in sustaining livelihoods and promoting cultural heritage. Date cultivation has been practiced for generations, with communities passing down their expertise in nurturing and harvesting these trees (Chao et al., 2010). The labor-intensive process of hand-pollination, pruning, and tending to the date palm groves has become a cherished tradition, fostering a deep connection between people and the land (Sabalpara et al., 2019). The economic significance of date palms extends to the trade of date products, including dried dates, date syrup, and even date-based cosmetics, creating opportunities for local industries and fostering economic growth (Zaid & de Wet, 2002).

In conclusion, the date palm is not just a majestic tree but a symbol of resilience, nourishment, and cultural heritage (Al-Shahib & Marshall, 2003). From its towering presence in arid landscapes to its succulent fruit bursting with flavor and nutrients, the date palm has captivated the hearts and palates of people around the world. As we continue to appreciate the date palm's remarkable qualities and harness its potential, we honor its profound role in sustaining life and preserving centuries-old traditions (FAO, 2021). The objectives of this study are to investigate the physical, chemical, mechanical, morphological, and thermal properties of date palm waste, to study the environmental impacts of date palm waste, and to identify the advanced potential uses of date palm waste.

DOI: 10.1201/9781003472209-5

3.2 DATE PALM WASTE

Date palm waste refers to the residual material generated during the cultivation, harvesting, and processing of dates, which are the fruits of the date palm tree (*P. dactylifera*). Date palm trees are primarily grown in arid regions of the world, such as the Middle East, North Africa, and parts of Asia, where they are an essential crop with significant cultural and economic importance (Ahmad et al., 2012a; Zaid & de Wet, 2002). The date palm waste comprises various components, including the leaves, stems, fruit pits known as seeds, and fibrous matter (Sabalpara et al., 2019). These by-products are often generated in large quantities due to the extensive cultivation of date palm trees and the high demand for dates. While dates are a highly valued fruit and widely consumed, the waste generated during their production presents both challenges and opportunities (Abdul Rahiman et al., 2012). One of the key challenges associated with date palm waste is its disposal. Improper waste management can lead to environmental pollution and pose a threat to human and animal health. However, there is a growing recognition of the potential value and diverse applications of date palm waste, leading to increased efforts in developing sustainable solutions for its utilization (Al-Yousef et al., 2018).

Date palm waste can be transformed into various useful products through different processes. For instance, fibrous matter can be utilized as a source of biomass for energy production, either through direct combustion or by converting it into biofuels (Ahmad et al., 2012a). The fruit pits can be processed to obtain oil, which has potential applications in the food, cosmetics, and pharmaceutical industries (Al-Farsi & Lee, 2008). The leaves and stems can also be used in manufacturing furniture, handicrafts, and building materials (Chao et al., 2010). Furthermore, date palm waste can serve as a valuable resource for agricultural purposes. It can be composted to produce organic fertilizers, which enhance soil fertility and promote sustainable agriculture practices (Al-Shahib & Marshall, 2003). The waste can also be utilized as animal feed, contributing to the livestock industry (Sabalpara et al., 2019). Efforts are being made to explore innovative technologies and techniques to maximize the value extracted from date palm waste while minimizing its environmental impact. By adopting sustainable waste management practices and developing efficient conversion processes, the utilization of date palm waste can contribute to resource conservation, renewable energy production, and the development of a circular economy (Ahmad et al., 2012a; Al-Shammari et al., 2014).

Hence, date palm waste, a by-product of date palm cultivation and processing, holds significant potential for utilization in various industries and applications. Its conversion into valuable products and integration into sustainable waste management systems can contribute to environmental sustainability, economic growth, and resource efficiency in prevalent regions of date palm cultivation.

3.3 ENVIRONMENTAL IMPACT FROM DATE PALM WASTE

Date palm waste can have both positive and negative environmental impacts. On the positive side, the utilization of date palm waste can contribute to waste reduction and promote resource efficiency. Finding beneficial uses for the waste materials, such as composting or converting them into bioenergy, reduces the amount of waste sent to landfills and mitigates potential environmental pollution and greenhouse gas emissions (Mallaki & Fatehi, 2014). However, if date palm waste is not managed properly, it can pose environmental challenges. One concern is the potential for air pollution when waste materials, such as pruned leaves and trunks, are burned or left to decay. Improper burning of palm waste can release pollutants into the atmosphere, contributing to air pollution and adversely affecting air quality (Mallaki & Fatehi, 2014).

Another potential environmental impact is the disposal of this waste in water bodies. If date palm waste is discarded in or near water sources, it can degrade water quality and harm aquatic ecosystems. The decomposition of organic matter in water bodies can lead to oxygen depletion and the growth of harmful algae, negatively impacting aquatic plants, animals, and overall biodiversity (Alshabanat, 2019). Furthermore, improper disposal of date palm waste can result in the proliferation of pests and diseases. Waste materials left unattended can serve as breeding grounds for insects and pathogens, potentially affecting the health and productivity of other plants in the vicinity (Giwa et al., 2019).

Proper waste management practices are crucial to minimize the environmental impact of date palm waste. This includes implementing sustainable waste treatment methods such as composting, anaerobic digestion, or controlled burning that meet environmental regulations. Additionally, promoting awareness and education among date palm growers and processors regarding the importance of responsible waste management can help mitigate the potential environmental risks associated with date palm waste (Alshabanat, 2019).

3.4 AVAILABILITY OF DATE PALM WASTE

Date palm is a primary agricultural production throughout the world's dry and semiarid regions, particularly in the Middle East and North Africa (MENA). Around 120 million date palm trees worldwide produce several million tons of dates each year in addition to secondary products such as stems, coir, palm midribs, leaves, and fronds (Zafar, 2016). Zafar (2016) also mentioned that more than 84 million date palm trees are grown in the Arab world, with the majority in Iran, Egypt, Iraq, Algeria, Morocco, Saudi Arabia, the United Arab Emirates, and Tunisia.

With 1.47 million tons of dates produced annually in 2012, nearly one-fifth of the world's total production has made Egypt the world's largest date pioneer. More than 23 million date palm trees may be found in Saudi Arabia, generating roughly 1 million tons of dates yearly (Zafar, 2016). A date palm has an average economic life of 40–50 years, although some can be productive for up to 150 years (Chao & Krueger, 2007). Date palm demands specific care over its lifespan. The branch should be cut, and dead or broken leaves should be taken off regularly. Just one date palm will make about 20 kg of trash every year. Some estimates have reported that Saudi Arabia makes more than 200,000 tons of date palm waste yearly (Zafar, 2016).

The date palm may be replenished in a very short amount of time, making it a renewable natural resource (Ali & Alabdulkarem, 2017). Date palms need 4–8 years after planting to bear fruit and 7–10 years to provide sustainable yields for commercial harvest (Zafar, 2016). Consequently, this production of date will also generate enormous amounts of agricultural waste, including dry leaves, stems, pits, and seeds. Local farmers said they generally gather and burn this waste. Small amounts of rubbish are recycled as animal feed by being shredded and mixed with other biowastes, or they can degrade naturally and turn into fertilizer (Ali & Alabdulkarem, 2017).

Date palm biomass is mainly composed of cellulose, hemicelluloses, and lignin. Date palm also has a high concentration of volatile solids and a low moisture content. Because of these features, date palm biomass is a good waste-to-energy resource in the MENA area. The date palm tree schematic shows several waste components and locations in Figure 3.1. A variety of thermal and biochemical processes are available to transform the energy held in date palm biomass into usable forms of energy. Due to its low moisture content, date palm waste is well-suited to thermochemical conversion processes such as combustion, gasification, and pyrolysis. On the contrary, the high volatile solids content of date palm biomass suggests good biogas potential in anaerobic digestion facilities, perhaps by co-digestion with sewage sludge, animal wastes, and/or food wastes (Zafar, 2016). The cellulose component of date palm waste may be processed into biofuel (bioethanol). Thus, the quantity of date palm trees in the Gulf Cooperation Council, particularly in Saudi Arabia, might catalyze the development of the region's biomass and biofuels sector.

3.5 PROPERTIES OF DATE PALM WASTE FIBER

3.5.1 Physical Properties

Date palm physical properties have been researched numerous times in different types of research, as they are crucial for the ultimate applications (Al-Kaabi et al., 2005). Various industrial applications necessitate specific fiber properties, including density, diameter, cost, availability, aspect ratio,

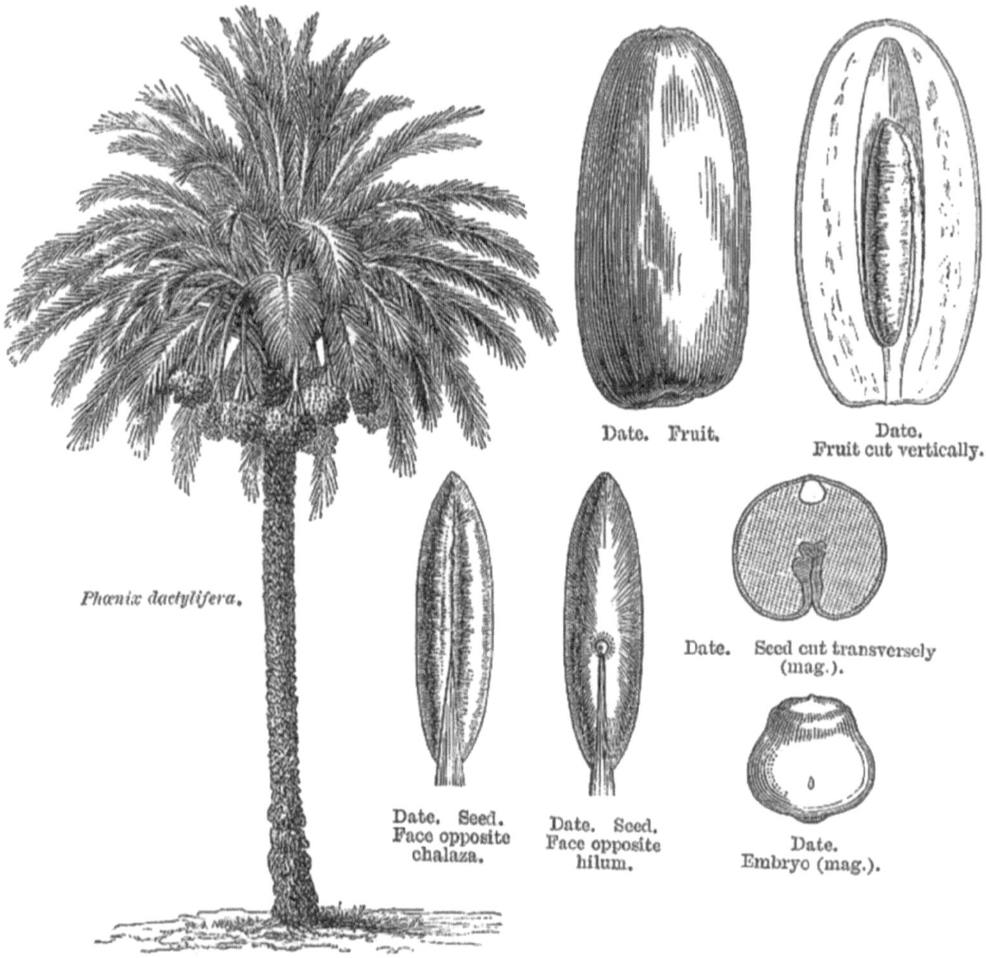

FIGURE 3.1 Representation of the date palm structure, illustrating components of morphological features and specialized offshoot structures (Watson and Dallwitz, 1992).

TABLE 3.1
Date Palm Physical Properties

Fiber Type	Date Palm
Density (g/cm^3)	0.9–1.2
Length (mm)	20–250
Diameter (μm)	100–1,000
Specific modulus (approximate)	7
Thermal conductivity (W/m·k)	0.083

thermal conductivity, and fiber length. Density, one of the essential physical properties, can play a significant role in the application of natural fibers (Al-Oqla & Sapuan, 2014). As demonstrated in Table 3.1, the date palm has a lower density but longer fiber. In addition, the diameter of the date palm is bigger, causing it to have an intermediate aspect ratio value, which could be a useful characteristic for various applications (Jonoobi et al., 2019).

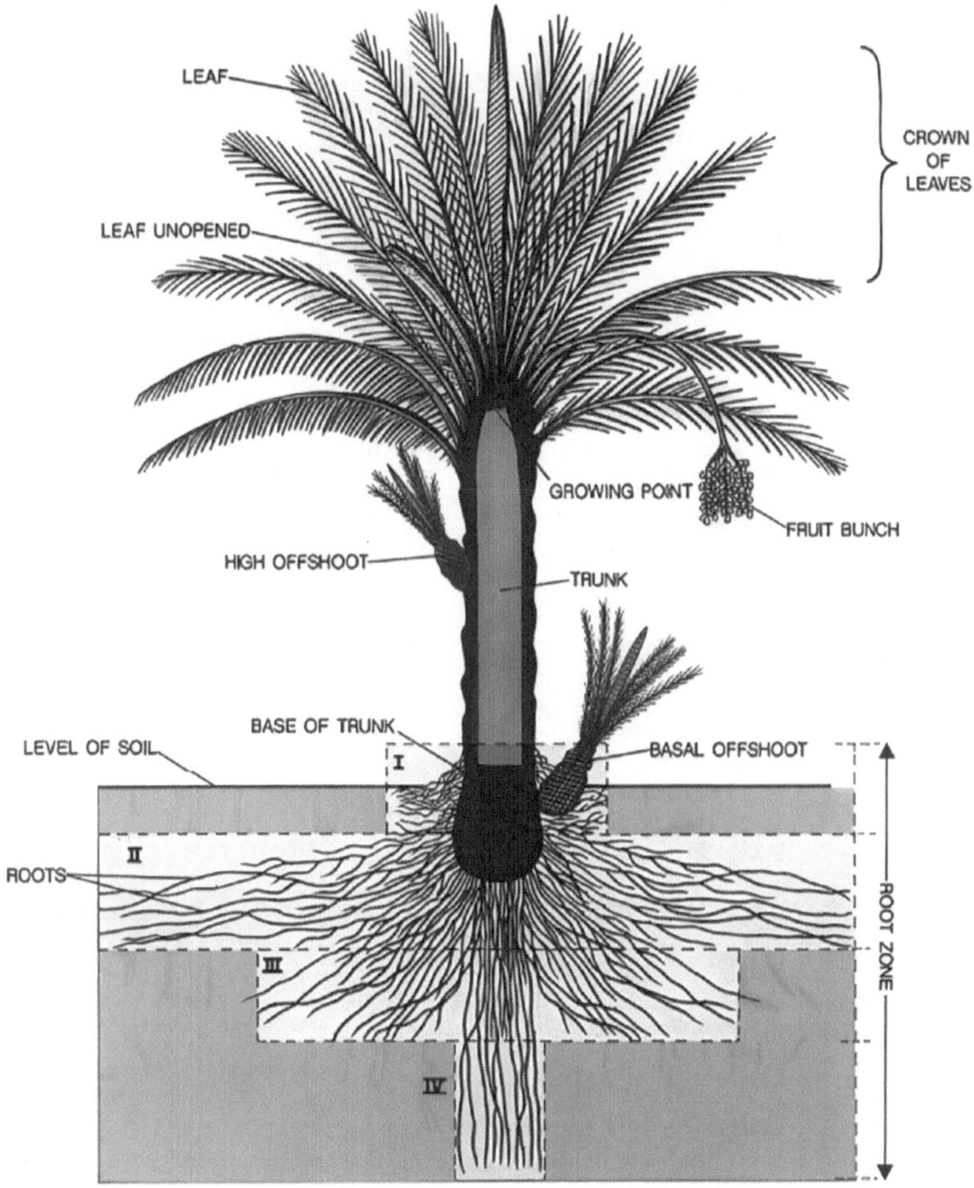

FIGURE 3.2 Date palm tree schematic (Chao & Krueger, 2007).

3.5.2 Chemical Composition

Date palm is usually made from fiber that can be studied from the aspect of chemical properties (Al-Oqla et al., 2014). Chemical composition studies are important to distinguish between crucial characteristics such as recyclability, degradability, and resistance (Azwa et al., 2013). The natural fiber structure is shown in Figure 3.2. Date palm is predominantly composed of cellulose encased in lignin. By separating the leaf into two distinct parts (leaflet and rachis), as shown in Figure 3.3, Table 3.2 lists the chemical composition of the leaflet and rachis (Figure 3.4).

Considering how much cellulose is in the cell wall and how it helps keep the cells together, the date palm has a lot of promise. The amount of cellulose and lignin in a date palm leaf shows how excellent and competitive date palm fibers are compared to coir, hemp, and sisal. Since

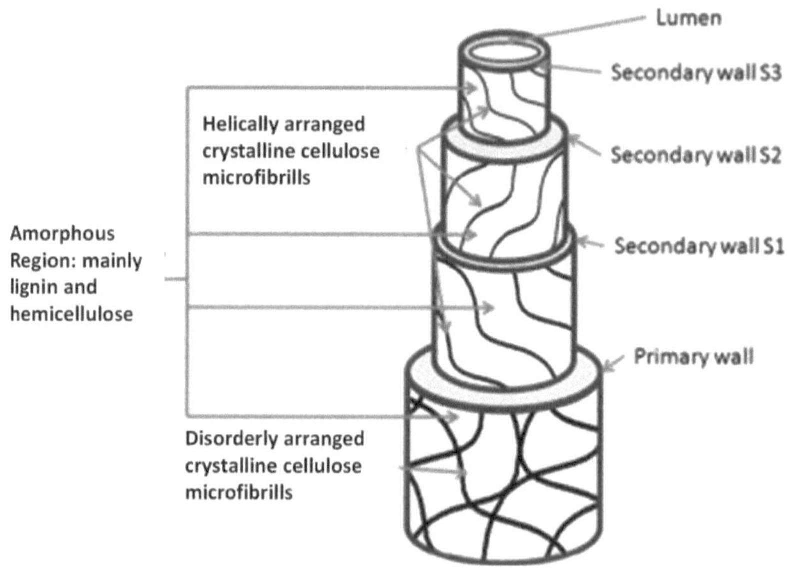

FIGURE 3.3 Natural fiber structure (Jonoobi et al., 2019).

TABLE 3.2
The Average Percentage of Leaflet and Rachis Weight for Chemical Structure

Constituents (%)	Cellulose	Lignin	Hemicellulose	Extractive	Ash	References
Leaflet	40.21	32.2	12.8	4.25	10.54	Mirmehdi et al. (2014)
	54.75	15.3	20	8.2	1.75	Sbiai et al. (2010)
	47.14	36.73	16.13	32.86		Nasser et al. (2016)
	34.87	14.03	19.84			
Rachis	38.26	22.53	28.17	5.08	5.96	Mirmehdi et al. (2014)
	45.16	26.68	28.16	17.45		Nasser et al. (2016)
	40.40	12.49	33.08			

there is less date palm fiber than hemp or sisal, it soaks up less water. On the contrary, date palm fibers contain more cellulose than lignin, making them appropriate for automotive applications. When searching for alternative uses and products for waste materials, it may be advantageous to consider the chemical composition of date palm waste. The lower heating value (LHV) of date palm fronds from various studies is presented in Table 3.3. The preponderance is desert-based. Comparing the LHV of date palm leaves to that of bituminous coal and wood revealed the LHV of date palm leaves is approximately half that of bituminous coal and comparable to that of wood (Al-Oqla & Sapuan, 2014).

3.5.3 Mechanical Properties

Mechanical qualities can be affected and determined by a wide variety of other variables and parameters, such as cellular size, flaws, structure, chemical composition, and microfibrillar angle.

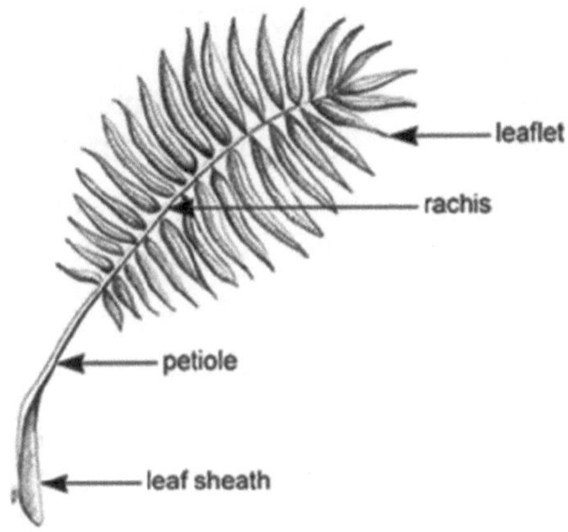

FIGURE 3.4 Date palm rachis and leaflet (Jonoobi et al., 2019).

TABLE 3.3
Characteristics of Date Palm Leaf

Source	Part	Moisture (%)	Volatile Matter (%)	Fixed Carbon (%)	Lower Heating Value (MJ/kg)	Ash (%)	References
Wood	—	20	82	17	18	1	McKendry (2002)
Bituminous coal	—	11	35	45	34	9	McKendry (2002)
Date palm	Leaflet	5	78.1	5.2	17.9	11.7	Sait et al. (2012)
		7.1	68	9.7	19	15.2	El may et al. (2012)
		8.50	72.28	7.64	—	11.58	Bensidhom et al. (2018)
	Rachis	17.7	55.3	7.8	10.9	19.2	Sait et al. (2012)
		12.1	73.6	8.3	15.2	6	El may et al. (2012)
		7.27	78.11	9.12		5.50	Bensidhom et al. (2018)
	Leaf (rachis and leaflet)	17.6	70.59	22.30	—	7.11	Usman et al. (2015)
		6.2	69.9	20.9	—	2.9	Jouiad et al. (2015)
		5	78.1	5.2	—	11.7	Bassyouni et al. (2014)
		5.3	52.1	6.1	—	20.2	Hussain et al. (2014)

The microfibrillar angle, the angle between microfibrils and the fiber axis, determines the mechanical characteristics of fibrils (Jonoobi et al., 2019). The fiber's stiffness and strength are improved due to the decreased angle. More potent mechanical properties are often related to longer cell length, higher cellulose content, higher degrees of cellulose polymerization, and lower microfibrillar angle (John & Thomas, 2008; Norfarhana et al., 2022, 2024a,b). Additionally, when cellulose content and cell wall thickness rise, important mechanical characteristics like Young's modulus and tensile strength tend to increase (Methacanon et al., 2010) (Table 3.4).

TABLE 3.4
Mechanical Properties of Date Palm Fibers (Jonoobi et al., 2019)

Mechanical Properties	Value
Tensile strength (MPa)	58–203
Young's modulus (GPa)	2–7.5
Elongation at break (%)	5–10

3.5.4 Thermal Properties

To determine the date palm waste thermal properties, samples were prepared with a size of (10×10×4) cm, attached with a bonding substance made by diluting polystyrene waste in petrol, and compressed for 15 days. The obtained material average density was determined to be equal to 377.38±4.03 kg/m^3, corresponding to 0.0738±0.0006 W/m·K for thermal conductivity (Khoudja et al., 2021). These results are obtained within the range identified by a number of researchers. In this aspect, Oushabi et al. (2017a) determined that the thermal conductivity of a sample consisting of date palm waste mixed with urea-formaldehyde resin before compression by heating was 0.0572±0.0017 W/m·K. Ali and Alabdulkarem (2017) found values from 0.0475 to 0.0697 W/m·K for samples made of cornstarch-bound date palm fiber. In a separate investigation, Agoudjil et al. (2011) found date palm residue samples had 0.08 W/m thermal conductivity. Masri et al. (2018) analyzed a composite made from ground date palm leaves and found a thermal conductivity range of 0.112–0.159 W/m·K.

3.5.5 Calorific Value

Thermogravimetric data were utilized to characterize biomass samples and the relativity of the kinetics of the reaction from thermal degradation. When put together with the account of transport processes, the kinetics of biomass burning could give designers and optimizers of thermochemical systems a lot of useful information (Sait et al., 2012). Several methods for calculating kinetic parameters can be found in the current scientific literature (Guo & Lua, 2001; White et al., 2011). The thermogravimetric and derivative thermogravimetric results of date palm biomass in nitrogen and air environments indicated the peak temperature associated with the disintegration of date palm biomass compositions consisting of cellulose, hemicelluloses, and lignin in both environments. Since the seed, leaf, and leaf stem biomass from a date palm all have different peak temperatures during pyrolysis and combustion, the kinetic parameters for each biomass component were calculated separately for the pyrolysis and combustion phases. The thermal degradation aspects and kinetic parameters, activation energy and pre-exponential factor of date biomass, were determined for the distinct peaks (Sait et al., 2012). Table 3.5 shows the kinetic parameters for the component of date palm waste.

TABLE 3.5
Kinetic Parameters for the Components of Date Palm Waste (Sait et al., 2012)

Biomass	Activation Energy, *E* (kJ/kmol)		Frequency Factor, *A* (s^{-1})	
	In N_2	In Air	In N_2	In Air
Leaf	25.85	29.00	3,930	11,434
Stem	12.03	13.45	413	1,012
Seed	20.24	27.15	1,646	9,139

TABLE 3.6
Kinetic Parameters for Date Palm Biomass in Atmosphere (Sait et al., 2012)

Biomass	Activation Energy, *E* (kJ/mol)	Reaction Order (*n*)	Frequency Factor, *A* (s^{-1})	R^2
		Peak 1		
Leaf	22.1	1.9	1,192	0.93
Stem	9.7	1.9	207	0.81
Seed	12.7	2.0	178	0.81
		Peak 2		
Leaf	43.6	1.00	191,634	0.99
Stem	32.8	1.00	26,326	0.99
Seed	30.7	0.90	13,485	0.95

During the thermochemical conversion of biomass, the chemical and physical processes highly interacted at the particle and reaction environment levels. Studies of the kinetics of a reaction involving lignocellulosic biomass materials are highly predictive. Since a comprehensive kinetic analysis of date palm waste in the air and environment with nitrogen has not yet been conducted, the scientific literature cannot validate current kinetic findings. Table 3.6 summarizes the development of thermal degradation in the first reaction zones of seed, leaf, and leaf stem of pyrolysis for peaks 1 and 2. Based on the kinetic diagrams, it was determined that the degradation reactions of date biomass depend on the Arrhenius equation. Different date palm biomasses have different chemical reaction regimes during heat degradation, which can be observed in the varying activation energy (*E*). The activation energy for the leaf, leaf stem, and seed of the first apex were 22, 9, and 12.7 kJ/mol, respectively. The second peak for leaf, leaf stem, and seed activation energies were 43.6, 32.8, and 30.7 kJ/mol, respectively. The second peak zone had higher kinetic characteristics for all date biomasses. The Arrhenius equation showed that activation energy is proportional to the pre-exponential component.

Table 3.7 displays the kinetic parameters for the leaf, leaf stem, and seeds that correspond to peaks 1, 2, and 3 in the atmospheric environment. In the N_2 atmosphere, the activation energy of date palm biomass ranged from 9.7 to 43.6 kJ/mol, whereas, in air, it ranged from 9.04 to 30.95 kJ/mol. The seed material had the highest activation energy value, indicating its high-temperature sensitivity in air. As a result of the leaf stem's high moisture content, a substantial quantity of energy may be required to evaporate the moisture from the biomass. In addition, kinetics suggested that the thermal degradation of date palm biomass in an oxygen atmosphere may be affected by thermal and mass transfer limitations. In accordance with the findings of Munir et al. (2009), activation energy in the air was found to be greater than in the nitrogen atmosphere.

3.6 DATE PALM WASTE MANAGEMENT

Date palm waste is a naturally occurring raw resource with enormous potential. However, due to several variables like tree and harvesting conditions, natural raw materials exhibit significant property differences (Fąiad et al., 2022). Well-known qualities of the material are essential for industrial production, especially in the mechanical sectors. Due to the controllability of the production parameters, the qualities of synthetic raw materials are well-defined, in contrast to those of natural materials (Faiad et al., 2022). However, this rule does not apply to natural materials like dates and palm leaves because they are recycled garbage. Date palm waste properties can be controlled through pre-treatment procedures. The attributes can be defined and modified, with additional desirable qualities added or undesirable ones removed.

TABLE 3.7
Kinetic Parameters for Date Palm Biomass in Air Atmosphere (Sait et al., 2012)

Biomass	Activation Energy, *E* (kJ/mol)	Reaction Order (*n*)	Frequency Factor, *A* (s^{-1})	R^2
		Peak 1		
Leaf	21.94	1.0	1,469	0.94
Stem	9.04	1.0	101	0.92
Seed	14.94	0.9	136	0.822
		Peak 2		
Leaf	24.46	1.0	4,557	0.98
Stem	20.53	1.0	3,613	0.98
Seed	30.95	1.0	12,177	0.95
		Peak 3		
Leaf	—	—	—	—
Stem	10.26	1.0	482	0.90
Seed	28.02	1.0	9,666	0.98

3.6.1 Physical and Chemical Treatments

To control the properties of palm fiber, it is necessary to manage its length, aspect ratio, surface bonding, and contents of volatile matter, extractive, water, and ash (inorganic chemical waste products comprising calcium and potassium) (Faiad et al., 2022; Nurazzi et al., 2021; Syafri et al., 2022). Cutting, shredding, grinding, and heating are used to regulate the majority of these qualities (Faiad et al., 2022). Surface bonding is managed by other physical procedures like surface fibrillation and electric discharge (Hakeem et al., 2014). Ash and extractive contents, surface bonding, cellulose, hemicelluloses, and lignin concentrations are all within the bounds of palm fiber (Faiad et al., 2022). However, chemical treatments mostly involve washing/soaking in water/organic compounds to reduce surface bonding, extractive content, and ash (Faiad et al., 2022). The chemical and physical processes researchers employed in the publications under evaluation are listed in Table 3.8.

3.6.1.1 Pre-Treatment Methods

Pre-treatment can be prepared based on the needed raw material (fiber or powder) and the desired end product (active carbon, biochar, composite, etc.). As a result, two types of pre-treatments were applied to date palm waste: fiber and powder pre-treatments (Faiad et al., 2022). The waste was first sorted into the following categories: leaflet, seed, rachis, and fruit bunch, as the first stage in the preparation of either fiber or powder (Faiad et al., 2022).

3.6.1.2 Fiber Pre-Treatment

The truck, leaf, and mesh fiber that surrounds the trunk of a date palm all contain fiber. The primary sources are the fruit bunch and the leaf, yearly waste items taken from the tree. Before starting the treatment, the majority that studied the date palm fibers chopped the fibers (Kyaw et al., 2021). The aspect ratio lowers as a result of the operation, which also reduces the fiber's overall mechanical characteristics to a few centimeters. Additionally, because the length and diameter of the fibers produced by methods like shredding are not well-defined, they are unsuitable for technical applications, particularly in composites where the reinforcement's length and diameter greatly influence the qualities of the composite (Faiad et al., 2022).

A fiber extraction procedure should be utilized to obtain the complete length of the date palm fiber as well as the regulated length and diameter (Faiad et al., 2022). Processes for removing natural

TABLE 3.8
Physical and Chemical Treatments of Date Palm Waste Natural Fibers

Treatment	Process Applied	References
Physical	Sizing: cutting, shredding, grinding	Kyaw et al. (2021)
	Drying	Puligundla et al. (2016)
	Surface fibrillation	Hakeem et al. (2014)
	Electric discharge	Hakeem et al. (2014)
	Steam treatment	Hakeem et al. (2014)
Chemical	Sodium hydroxide treatment	Hakeem et al. (2014)
	Acetylation	Hakeem et al. (2014)
	Graft copolymerization	Hakeem et al. (2014)
	Permanganate treatment	Hakeem et al. (2014)
	Silane treatment	Li et al. (2007)
	Benzoylation treatment	Sanjay et al. (2018)
	Acrylation and acrylonitrile grafting	Sanjay et al. (2018)
	Maleated coupling gents	Belgacem et al. (2020)
	Peroxide treatment	Belgacem et al. (2020)

fibers are divided into the following categories (Imraan et al., 2023). For instance, the biological extraction method employs a retting process where microorganisms are used to dissolve the bonds between the fibers (Faiad et al., 2022). Water and dew rettings are two different types of retting. Such a procedure takes a while and may take longer than 20 days. It also impacts the fiber's chemical and mechanical characteristics (Faiad et al., 2022). However, the most cost-effective extraction technique is biological extraction. Other methods, such as chemical extraction, involve immersing the fiber source in an organic chemical like NaOH. As the concentration, duration, and temperature of such an extraction method significantly affect the fiber quality, they vary depending on the fiber source and the desired characteristics. Alawar et al. (2009) investigated the effects of treatment with sodium hydroxide and hydrochloric acid at different concentrations on the fibers of the date palm. Moreover, the fastest form of extraction is mechanical, which employs a device called a decorticator (Tarabi et al., 2016). Unfortunately, no research has been found on the impact of date palm fibers' mechanical extraction. Due to the rapidity of the extraction process and the absence of chemicals or microorganisms that may change the fiber's characteristics, decortication is a preferred extraction technique. This makes chemical or physical treatment quicker, simpler, and more effective.

The wasted leaf, bunch, and trunk of a date palm are potential sources of fiber. Cleaning is necessary to obtain the maximum fiber quality because such waste is subjected to soil and weather conditions. Hegazy et al. (2015) investigated how washing affected the reinforced composite qualities as well as the mechanical properties of date palm fiber. The ash and certain extracts can be removed by washing with water or chemicals (Faiad et al., 2022). Hence, drying is an essential pre-treatment procedure for reducing the fiber's weight and density by removing the extractives and water. The sun (or the open air) or an oven can both be used for drying. Although solar drying is more cost-effective, oven drying is recommended because of its controllability and quick process.

3.6.2 Compost

Soil organic matter is one of the most essential soil ingredients owing to its direct and indirect impacts on plant growth, which produces a conducive environment for plant root growth and soil bacteria (Raja et al., 2021). Through their impact on the physical, chemical, and biological aspects of the soil, such organic materials help plants flourish by preserving soil structure, improving water holding capacity, delivering metabolic energy, and supplying nutrients.

Organic fertilizers are carbon-based substances that enhance plant quality, productivity, and growth (Ghouili et al., 2023). Compost has the highest concentration of plant nutrients of all organic fertilizers. Increased soil organic matter, soil nutrient availability, water usage efficiency, and soil characteristics from compost application can all lead to higher crop yields (Asaye et al., 2022). Several researchers showed that, with adequate composting, organic leftovers such as animal manures, sewage sludge, and plant wastes may be utilized as a growth medium with extremely good outcomes and favorable effects on plant growth (Raja et al., 2021). It is interesting to note that organic fertilizers, such as compost, are sources of organic matter and nutrients as well as microbial population stimulation and enhancement of the physical, chemical, and biological characteristics of the soil, resulting in sustainable agricultural output (Ho et al., 2022).

The full destruction of over 15 million date palm trees left behind millions of abandoned trees with bare trunks, enormous amounts of collected debris, and yearly seasonal pruning of the palm trees. A distinctive aspect of sustainable agriculture is its reduced reliance on chemical fertilizers and recycling of agricultural waste to preserve and/or enhance soil quality (Raja et al., 2021). To lessen the effects of its buildup and/or prevent environmental contamination, agro-waste products can be applied as organic rowing culture, especially after thorough composting.

Waste from date palms, including fronds, trunks, and fruit bunches, may be utilized to fertilize and enhance the soil through compost. Date palm trash may be a useful source of organic material for making compost rich in nutrients. Composting is a natural process of degrading organic materials. According to several studies, applying compost enhances microbial biomass and activity, nutrient absorption by plants, and disease incidence, all of which promote plant development. The activation of numerous genes involved in the absorption and transport of nitrate, ammonium, and phosphorus in the leaves and roots of plants may be responsible for the increase in macronutrient intake (Ghouili et al., 2022).

Composting date palm waste not only recycles organic material but also lessens the amount of garbage that ends up in landfills and gives agricultural practices a sustainable supply of nutrients. As the need for adequate organic substrate has increased, research has been concentrated on finding locally accessible, inexpensive (low cost), and sufficiently physicochemical materials. A previous study by Ghouili et al. (2022) found the application of date palm waste as compost. It stated that the biological interactions between soil microorganisms and plant roots, which significantly impact plant development and health, are well recognized. At the cultivation, booting, and ripening phases of barley development, Ghouili et al. (2022) showed that using date palm waste compost greatly boosted soil microbial richness and variety. Compost amendment also greatly impacted the makeup and organization of soil microbial populations. Beneficial microbes, including *Chaetomium*, *Actinobacteriota*, *Talaromyces*, and *Mortierella*, increased with compost treatment, whereas detrimental germs like *Alternaria*, *Aspergillus*, and *Neocosmospora* were decreased. Additionally, the application of compost changed the trophic modes, metabolic processes, and bacterial and fungal community activities in the soil. Overall, the findings of Ghouili et al.'s (2022) study showed that compost can support advantageous microorganisms, increasing the availability of vital nutrients to barley plants, while also producing substances necessary for plant growth and suppressing pathogens, improving crop productivity and disease resistance.

Another finding of date palm waste as compost was found by Raja et al. (2021). The study was done to assess the efficacy of compost manufacture utilizing finely ground date palm waste and residues to compare its performance as a replacement for commercial peat moss in potting media. This date palm waste is made up of cellulose, hemicellulose, lignin, and others that are used in various biological activities (Raja et al., 2021). The use of this waste in the composting process is crucial from an industrial, agricultural, and environmental perspective.

Numerous studies on date palm natural composites have been examined to demonstrate the value of date palm fibers as a reinforcement. The following items have been created and tested utilizing laboratory tools, making time a constraint on industrial-level production. For instance, Haque et al. (2009) looked into the mechanical characteristics of palm fiber used to strengthen a thermoplastic polymer matrix made of polypropylene. To increase the surface adhesion with the matrix, they employed full-length fibers that had undergone chemical treatment with a benzene diazonium salt. Raw, untreated fibers were employed as a benchmark. Tensile strength significantly decreased, although flexural and impact strengths significantly improved up to 30wt% fiber. In all their tests, adding more fiber weight to the matrix had no impact. Due to higher matrix adherence and decreased wettability, chemical treatment of date palm fibers considerably enhanced their mechanical properties compared to untreated fibers. In addition, the mechanical characteristics of date palm fiber in an epoxy matrix were studied by Saba et al. (2019). After cleaning and drying the fiber in the oven, they used a laboratory grinding machine to turn it into a 0.8–1mm powder. The composite's tensile strength, impact resistance, and plasticity were significantly improved by including date palm fiber as reinforcement. They discovered that the ideal weight percentage for the reinforcement was 50wt%.

3.6.3 Pulp and Paper

Wood consumption and pulp demand increased due to the significant growth in pulp output over the last few years, reaching 180 million tons of pulp in 2008 (Marrakchi et al., 2011). Therefore, new rapidly expanding non-wood sources of lignocellulosic fibers must be discovered to maintain the forest supply, meet the growing demand for pulps, particularly in the papermaking sectors, and produce innovative materials like those based on cellulose fibers (Marrakchi et al., 2011).

According to the Tunisian Ministry of Agriculture, date palms are one of the most widely grown plants in Tunisia and cover more than 4 million hectares (Marrakchi et al., 2011). This crop generates enormous quantities of rachises, either burned to recover energy or left to biodegrade and fertilize. The date palm has not been thoroughly investigated for use in papermaking. Still, some researchers suggested that Egyptian date palm leaves might be a useful source of cellulose fibers. In contrast, Khristova et al. (2006) investigated Sudanese date palm leaves and rachises for papermaking applications.

Even though the raw pulps of Tunisian date palm rachis contained a high amount of polysaccharides, according to a prior work by Marrakchi et al. (2011), it resulted in paper sheets with strong mechanical qualities. This study, a natural progression from previous research, aimed to investigate the features of refined date palm rachis pulp and its pulps and papers (Marrakchi et al., 2011).

Other studies by Ghosh and Nag (2009) discussed how date palm leaf is an ideal raw material for creating pulp and paper of various grades because of its high alpha cellulose (50–55%), hemicellulose (26–30%), and ultimate fiber length (1.25–2.50mm). High levels of hemicellulose in the pulp were discovered to be beneficial for beating processes and the development of links between the fibers in paper sheets (Ghosh & Nag, 2009). The paper sheet's tensile, bursting, and rip strengths, as well as folding endurance, may all be achieved without using natural or synthetic polymers or additives, according to Ghosh and Nag (2009). Figure 3.5 shows the process flow chart for making pulp and paper using date palm leaf as the raw material.

3.6.4 Biochar

Biochar is a solid substance that may be integrated into soils for agronomic or environmental management (Usman et al., 2015). It is made by thermally converting unstable carbon-enriched materials into stable carbon-enriched charred materials (Lehmann et al., 2011). A successful method to reduce greenhouse gas emissions has been said to be the integration of biochar into the soil (Lehmann & Joseph, 2009). Other studies by Yuan et al. (2011) also documented the use of biochar to improve the characteristics and fertility of the soil and clean up polluted soil. According to several other studies, biochars can be utilized as an alternative to traditional sorbents to remove various kinds of organic and inorganic pollutants from aqueous solutions (Usman et al., 2015). According to another study by Tayade et al. (2012), biochar is recommended for usage in severely

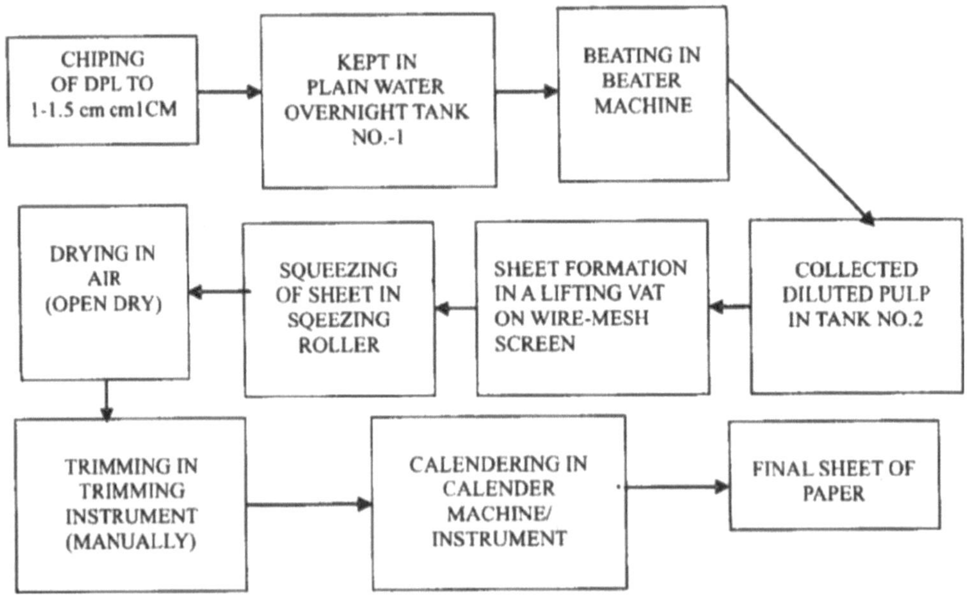

FIGURE 3.5 Mechanical pulping process flow chart (Ghosh & Nag, 2009).

polluted areas to reduce CO_2 emissions and successfully lowered the concentration of ambient CO_2 when mixed into soil.

Biochar is generally recognized as a low-cost, renewable, and long-lasting biomaterial. Through a process known as pyrolysis, date palm waste is transformed into biochar, a carbon-rich solid that resembles charcoal. It is formed by heating biomass, such as agricultural waste, without the presence of oxygen. It may be used in a variety of ways to improve agriculture, the environment, and carbon sequestration. At much lower temperatures, biochar may be produced from a range of waste materials, including sewage, food waste, and agricultural waste (Salem et al., 2021). It is speculated that date palm leaf waste may be used as a resource for biochar that is both affordable and effective at absorbing carbon dioxide, which helps with the management of CO_2 capture and storage. As a result, the research by Salem et al. (2021) revealed that the date palm leaf waste had excellent properties for biochar pyrolyzed at 500°C for 5 h. However, it is reasonable to say that raw date palm leaf waste biochar has the potential to effectively adsorb 20% of carbon dioxide based on the study of data mentioned for CO_2 adsorption (Salem et al., 2021).

Additionally, biochar can accelerate both the rates of nitrogen reduction and the retention of nitrogen in soil. Studies showed that adding biochar also reduced the rate of net nitrification, nitrate (NO_3) and ammonium (NH^{4+}) leachings, ammonia (NH_3) volatilization, and nitrogen immobilization (Sizirici et al., 2021). Biochar is frequently utilized to trap carbon, lower greenhouse gas emissions, and enhance soil quality. Residues from agricultural crops, forestry, wood, industrial solid, municipal, and animal manures are among the diverse biomasses used in biochar production (Burezq & Davidson, 2023).

There are several thermal, physiochemical, and biochemical processes that may be used to use date palm biomass sustainably. Although date palm residue recycling is not a new technological advancement, it is still regarded as the most significant biological activity within the environmental system that attempts to maintain the environmental balance. The ecosystem may be restored by using biological processes carried out by microorganisms to enrich the soil by transferring agricultural leftovers to soil enrichment. Biochar has the potential to reduce problems with date palm residue clearing due to its low cost and ability to turn date palm residues into financial rewards (Burezq & Davidson, 2023).

3.6.5 Bioethanol

One of the earliest domesticated plants, the date palm is native to the Arabian Peninsula and has been an essential part of human everyday existence for 7,000 years (Sallon et al., 2020). Pakistan produces 0.47 million tons of dates annually, placing it sixth in the world behind Egypt, Iran, Algeria, Saudi Arabia, and Iraq in terms of date palm output (Ahmad et al., 2012a). Although Pakistan is the world's leading date producer, a substantial amount of the fruit is lost because of monsoon rains and their poor market value, which causes the farmers to suffer a significant financial loss (Ahmad et al., 2012a).

To increase economic output, date palm waste must be converted into valuable items. Boulal et al. (2016) produced ethanol using solar energy and date palm waste. The outcomes showed that ethanol was produced from date palm waste at a rate of 250 mL/kg. The primary biological mechanism in the environmental structure that aims to achieve environmental balance is thought to be the waste recycling process for date fruit (Ahmad et al., 2012a). It is acknowledged that dates are the main source of natural antioxidants, such as carotenoids, phenolics, tocopherols, flavonoids, and ascorbic acid (Ahmad et al., 2012a). These antioxidants can appear in a variety of amounts and forms depending on the genotype of the date palm and the mechanics of the postharvest processing.

Dates are typically whole foods that degrade naturally. Their fermentable sugars promote anaerobic fermentation, which is then utilized to produce bioethanol by distillation. Due to the high sugar content (60–70%), its sap is extremely nutritious and is considered excellent for microbial fermentation (Banaruee & Amirizad, 2017). Sixteen percent of sugar in fermented molasses results in alcohol concentrations of up to 5.6–6% (Fadel et al., 2013). Numerous bacteria, including *Zymomonas mobilis* and *Saccharomyces cerevisiae*, can be employed in fermentation operations to create fructose and bioethanol (Ahmad et al., 2012a). In general, using mutations of these bacteria reduces fructose intake (Yao et al., 2014). Lignocelluloses, starches, and sugars are three major groups of feedstocks utilized in bioethanol manufacturing (Ahmad et al., 2012a).

As a safe, regenerative, environmentally beneficial, and recyclable alternative to fossil fuels, ethanol derived from waste production provides benefits (Iqbal et al., 2017). Recently, the production of bioethanol in the petrol industry has surpassed 20 million gallons. Agriculture by-product ethanol is renewable and generates bioethanol on a big scale, which is preferable to petrol (Saini et al., 2014). The study by Ahmad et al. (2012a) demonstrated that *S. cerevisiae* was preferable for efficient ethanol synthesis. The findings showed that yeast consumed more sugar from 36 to 72 h and depleted to zero afterward. Since there was more sugar intake in the 25% inoculum, there was also higher ethanol generation as a result of higher sugar consumption. The Fourier transform infrared study findings verified the presence of the C_2H_5OH ethanol group (Ahmad et al., 2012a). The heterogeneous chemical composition of the nearby contents of date palm fruits was robotized by near-infrared spectroscopy.

Another finding by Boulal et al. (2016) was the utilization of date palm waste as bioethanol using solar energy. Over 236,807 tons, or 30% of the date palm fruits produced in Algeria each year, are wasted during picking, storing, and selling (Boulal et al., 2016). This enormous biomass may be gasified to produce biogas that can be converted into usable products like bioethanol, biodiesel, and petrol (Boulal et al., 2016). By chemically synthesizing or anaerobically fermenting biomass, bioethanol is becoming the primary biofuel generated and is significant for industrial growth, investment, and consumption. It is more environmentally friendly, reasonably priced, and cleaner than other gases (Boulal et al., 2016).

Bioenergy, which includes bioethanol, biodiesel, and other chemicals, is technically possible and has economically emerged as the future answer to energy and environmental problems. Bioethanol has a stronger burning effect than other biogases and a lower environmental impact (Raghavi et al., 2016). The United States, Brazil, EU, and China produced 50.3%, 25.5%, 4.5%, and 2% million gallons of ethanol in 2014, respectively, according to the Renewable Fuels Association (Quintero et al., 2012).

3.7 POTENTIAL APPLICATIONS

3.7.1 Composites of Date Palm Fibers

Primarily polymer-based composite materials are usually natural fiber composites (NFCs) that use natural fibers derived from kenaf, jute, ramie, sisal, coir, and pineapple leaf, among other sources (Faiad et al., 2022; Alias et al., 2021; Mazani et al., 2019; Suriani et al., 2021). However, researchers have only recently begun to investigate the possibility of using biomass refuse as a source of natural fiber. Natural fiber has attracted a great deal of attention since fiberglass has very high advantages over conventional materials for various applications (Asyraf et al., 2022; Halimatul et al., 2019; Jumaidin et al., 2019a,b, 2021). Natural fibers are substantially less expensive and more accessible than conventional low-density fibers; they have excellent mechanical and thermal properties (Asyraf et al., 2021; Nazrin et al., 2021). Additionally, natural fibers are biodegradable (Saheb & Jog, 1999). However, their hydrophobic polymer matrices issues, a large variation in mechanical components, lack of moisture resistance, low disintegration temperatures (Sgriccia et al., 2008), and processing complexities are significant obstacles to their use in industrial investments. Since there are large numbers of date palm trees, with date palm fiber being highly accessible in the Middle East, researchers, primarily in the Gulf countries, have focused their research on date palm fibers. Numerous studies of NFC are reviewed to demonstrate the usefulness of date palm fiber as reinforcements. Laboratory apparatus have played important roles in preparing and testing the products, posing a time constraint for industrial production (Faiad et al., 2022).

3.7.1.1 Composites of Polypropylene

Haque et al. (2009) investigated the mechanical components of date palm fiber as a source for a polypropylene standard made of thermoplastic polymer. Fibers were mixed up with benzene diazonium salt to enhance surface adhesion with the matrix. In comparison, unprocessed fibers were utilized. Up to 30 wt% fiber resulted in a precipitous deplete in tensile strength but relatively increments in flexural and impact strengths. In all of their experiments, the weight of fiber in the matrix also gave no reaction. Date palm waste undergoing chemical treatment made the fibers relatively enhanced mechanical components compared to untreated fibers due to improved matrix adhesion and decreased wettability (Faiad et al., 2022).

3.7.1.2 Composites of Epoxy

Saba et al. (2019) examined the mechanical components of date palm fiber embedded in an epoxy standard matrix. The fiber underwent drying with the oven before being ground using laboratory grinding equipment to reduce the fiber length to 0.8–1 mm. The benefits of date palm fiber as a source relatively increased the composite's tensile and impact strengths. However, in their morphological images (Figure 3.6), there were air bubbles present in the epoxy standard matrix that compromised the mechanical components, in contrast to the pure epoxy resin, in which air bubbles were absent. It is recommended to find a superior manufacturing process to improve the mechanical components of such a composite (Faiad et al., 2022). Scanning electron microscopy (SEM) can be used to analyze the surface properties and distribution of the infill within the range (Saba et al., 2019). Figures 3.7–3.9 depict SEM micrographs of 40%, 50%, and 60% DPF/epoxy composites. Figure 3.7 is a micrograph of pure epoxy resin displaying brittle plasticity, a vitreous and smooth exterior, and numerous fractures resembling streams (Saba et al., 2016a).

Remarkably, undulating fractures indicated the material's high propensity for tensile fracturing, rupture, and propagation with significantly lower energy consumption. Other researchers have observed identical SEM images for purified epoxy resin (Saba et al., 2016a, 2017). At first observation, the SEM morphology (Figures 3.7–3.9) of all DPF/epoxy composites appeared identical but also distinct from pure epoxy resin. As shown in Figures 3.7–3.9, the addition of DPF caused the epoxy surface to become more angular and irregular, indicating a significant reduction in breaking point and ductility. Figure 3.6 demonstrates that the 40% DPF filler incline marginally coarsened

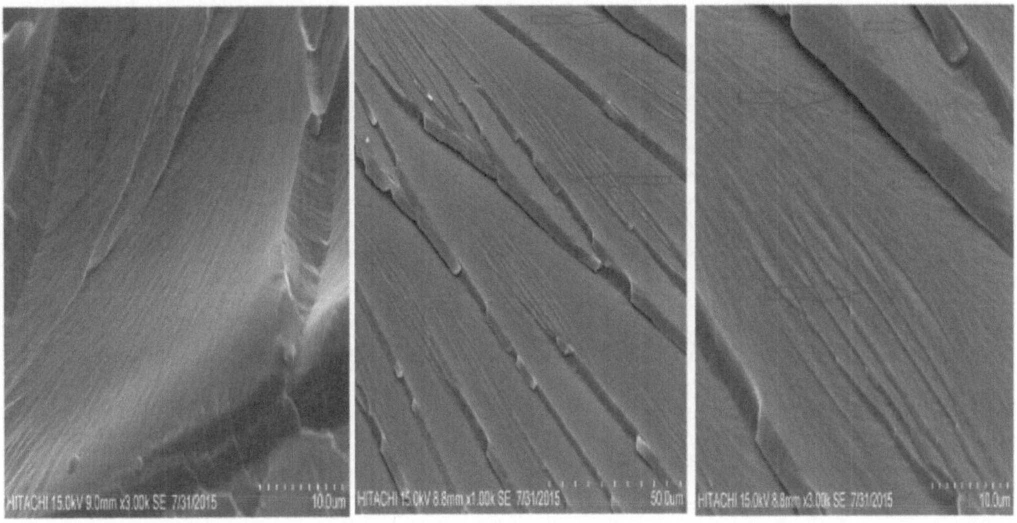

FIGURE 3.6 Tensile fractured images of epoxy resin from SEM (Saba et al., 2019).

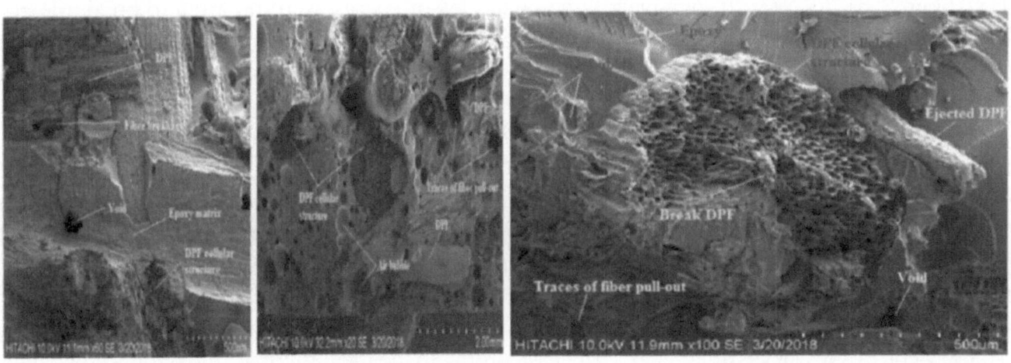

FIGURE 3.7 Tensile fractured sample of 40% DPF/epoxy composites from SEM (Saba et al., 2019).

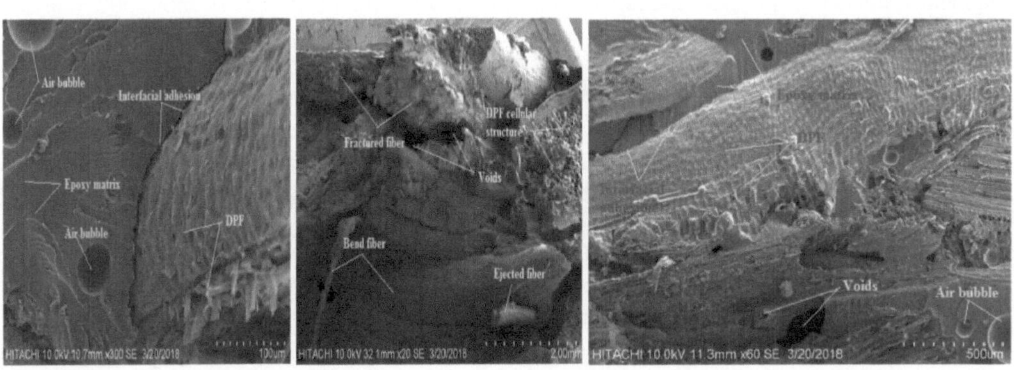

FIGURE 3.8 Tensile fractured sample of 50% DPF/epoxy composites (Saba et al., 2019).

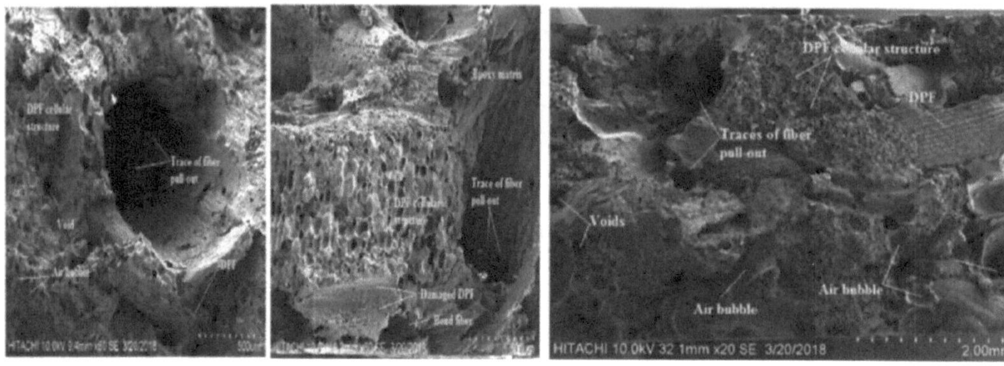

FIGURE 3.9 Tensile fractured sample of 60% DPF/epoxy composites (Saba et al., 2019).

and roughened the surface compared with pure epoxy resin. As the DPF filler impeded the fracture angle and crack initiation mechanisms, it slowed the premature failure of the examined composites (Saba et al., 2019). Extension of 50% DPF infill to the epoxy (Figure 3.8) resulted in a more abrasive and coarser surface compared to 40% loading (Figure 3.7) but significantly less than 60% DPF/epoxy composites (Figure 3.9). In addition, the SEM micrograph of 50% DPF/epoxy composites revealed significantly less chunking, aggregation, fiber surface, and void formation than the remainder (Figure 3.8). As a result, relatively large quantities of energy were required to break the 50% DPF/epoxy composites because these fillers impeded the crack injection path and redirected it onto a highly convoluted and curved path. These factors enhanced the resistance to pre-breaking and, ultimately, the tensile properties (Saba et al., 2019).

For 60% DPF/epoxy composites, however, the situation was quite different because of the buildup of DPF infill and microvoids within the range due to insufficient epoxy matrix for effective wettability of DPF. All of these factors contributed to the beginning and spread of fractures that led to the failure of composites when tensile stress was applied, as shown in Figure 3.9. These factors were governed by the poor diffusion and minimal adhesion of incompatible DPF infill, which reduced the value of mechanical properties. The statements were also true when alkali-treated date palm fiber was reinforced in polyurethane standard (Oushabi et al., 2017a). At high-filler loadings of glass fiber, nanoclay, or epoxy composites (Jeyakumar et al., 2017), according to identical findings, the presence of cavities and aggregates in the matrix led to the premature rupture of pure epoxy resin. Intriguingly, reasonable arguments were also presented for date palm fiber extracted from the mesh portion of the palm at various loadings (20, 50, and 70 wt%) and reinforced in biodegradable thermoplastic starch (Saleh et al., 2017). Due to excellent wettability, 50% fiber loading produced a comparable SEM image with more fragmented fibers than pullouts and higher matrix-fiber adhesion. Fiber content above 50% increased the applied load that caused fatigue damage in stressed composites (Saba et al., 2019).

3.7.1.3 Date Palm Used for Oriented Strand Board

Date palm stalks collected in Al-Kharj were used to effectively manufacture oriented strand board (OSB), an engineering wood-based panel that can be used widely. OSB with superior mechanical and dimensional stability were manufactured. However, the adhesion and compression processes could be improved to produce superior results. This application has tremendous economic potential since it is easier to refine and produce, and there is a high demand for OSB (Hegazy et al., 2015).

3.7.1.4 Medium-Density Fiberboard (MDF)

Hosseinkhani et al. (2014) invented medium-density fiberboard (MDF) from date palm stalks with excellent mechanical components. To evaluate the factors that affect MDF quality, 15 specimens of MDF board were manufactured using two types of resin, three pressing levels, and two resin loadings of 10 and 12 wt% resin. They demonstrated that the mechanical properties and wettability resistance of MDF produced from date palm fronds exceeded the standard for MDF.

3.7.1.5 Date Palm Used for Particle Board

In Saudi Arabia and Algeria, Amirou et al. (2013) and Hegazy et al. (2015) investigated the invention of particle boards using the date palm trunk and rachis. Amirou et al. (2013) demonstrated that the material is highly compatible with EN particleboard specifications for such applications. Date palm rachis was utilized in a study by Hegazy and Ahmed (2015) that looked at how particle size, panel density, hot water extraction, and date palm cultivar affected the mechanical characteristics of particleboard. For their low-cost particleboard, Ferrández-García et al. (2018) also employed palm rachis, comparing those from date palms, canary palms, and Washingtonia palms. Particleboard made from date palm waste has been the subject of a number of studies, and the researchers who looked into the process came to a range of different findings on the best manufacturing parameters to use.

3.7.1.6 Gypsum Board Reinforcement

Gypsum board reinforcement refers to the process of enhancing the structural integrity and durability of gypsum boards, also known as drywall or plasterboard. Reinforcement is typically achieved by incorporating various materials within the gypsum board's core or applying additional layers of reinforcing elements. Braiek et al. (2017) studied the thermophysical effect of date palm fiber-reinforced gypsum board. One common method involves embedding fiberglass mesh or paper tape within the joints between gypsum board panels. This reinforcement technique helps to prevent cracks and joint separation, ensuring a more robust and stable surface. Steel or aluminum studs may also be used as framing elements to support and strengthen gypsum board installations, particularly in load-bearing applications. According to Al-Rifaie and Al-Niami (2016), the reinforcement contributed to the overall strength and resilience of gypsum boards, making them more resistant to impacts, vibrations, and other mechanical stresses.

3.7.2 Biochar

Biochar is charcoal produced by the pyrolysis of biomass, and it has been used as a soil amendment for carbon sequestration and soil health benefits in the absence of oxygen. It is feasible to make plant nutrients accessible to the soil and improve soil structure by converting plant biomass into biochar that can be mixed with the soil. Biochar is a product of either slow or rapid biomass pyrolysis, although slower biomass pyrolysis produces more biochar. Loss of 20–80 t/ha of carbon (C) results in degraded soil and water qualities and decreased biomass production (Ahmad et al., 2019). Through a process known as pyrolysis, date palm waste such as palm fronds, tree trunks, and date pits can be converted into biochar. Biochar derived from date palm residue has numerous beneficial properties for soil enhancement and carbon sequestration. It has a highly porous structure that improves soil water retention and nutrient availability, thereby fostering plant growth and reducing irrigation needs. In addition, biochar functions as a long-term carbon absorber, mitigating climate change by sequestering atmospheric carbon dioxide. Biochar from date palm residue is a sustainable and environmentally responsible option for soil management, as its application to agricultural fields increases soil fertility and reduces greenhouse gas emissions (Ahmad et al., 2019).

3.7.2.1 Converting Date Palm Waste into Biochar

Utilizing concentrated solar thermal energy, the pyrolysis of palm waste is economical and environmentally beneficial. The procedure is depicted in Figure 3.10. The process includes waste collection and cleansing, drying, volume reduction, charging the solar heating unit, and storage (Giwa et al., 2019). Water washing reduced the amount of dust and ash-forming components or inorganic minerals present in bulk flow biomass. The cleaned biomass was dried in a rotary dryer using hot air for 6h.

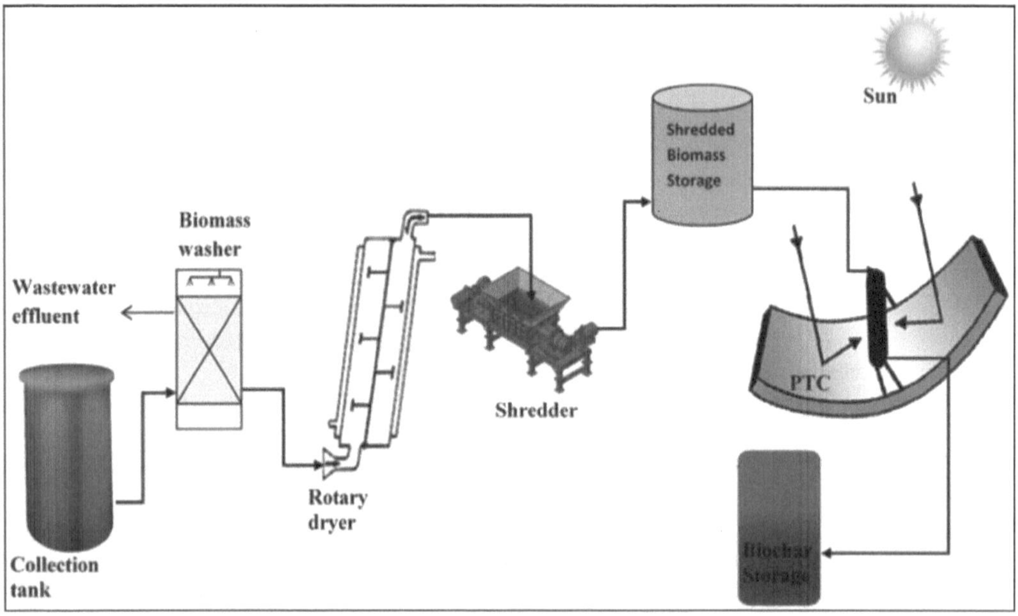

FIGURE 3.10 Converting date palm waste into biochar (Faiad et al., 2022).

Because the consistency of the result depends on the particle size of the biomass fed into the pyrolyzer, the dried biomass was shredded and milled. Belt conveyors then transported the milled, dried particles into the concentrated solar energy pyrolyzer receiver or stirred reactor. Depending on the size and scope of the operation, the solar pyrolyzer can be manually- or batch-filled. Improving biochar stability was the focus of research conducted by Ahmad et al. (2019), who found that impregnating biomass with silica or zeolite according to the method described by Yao et al. (2014) was successful.

3.7.3 Activated Carbon

Activated carbon, also referred to as carbonaceous material with 72–90% carbon, is amorphous and has a variety of other names (Haimour & Emeish, 2006). Activated carbon derived from date palm waste is a valuable adsorbent material with various applications. Date palm waste, including palm fronds, tree trunks, and date pits, can be utilized to produce activated carbon through a series of processing steps. The carbonization process involves heating the biomass in the absence of oxygen to convert it into char (Haimour & Emeish, 2006). Subsequently, the char undergoes activation, which involves treatment with steam or chemicals to create a highly porous structure with a large surface area. The resulting activated carbon has excellent adsorption properties, effectively trapping and removing impurities, contaminants, and pollutants from liquids and gases (Ayinla et al., 2019). Final activated carbon properties depend heavily on precursor properties, precursor pre-treatment, and the activation procedure. The production factor is determined based on the activated carbon's applications and requisite properties. Activated carbon derived from date palm waste presents a sustainable and eco-friendly solution for waste utilization while offering effective purification and filtration (Faiad et al., 2022).

3.7.3.1 Active Carbon Production Pre-Treatment

The production of activated carbon from date palm waste involves several pre-treatment stages to prepare the biomass for carbonization and activation processes. The suggested pre-treatment stages for date palm waste are (1) isolation, (2) washing, (3) drying, and (4) sizing. Typically, date palm waste is not segregated during the extraction process. Due to the significant disparities in their properties, leaflets and rachises must be separated. Date palm fruits cannot use the precursor for activated carbon; only the seeds have been utilized by scientists. The date seeds are the primary by-product of date processing industries (Faiad et al., 2022).

After collection, date palm waste is usually left on the soil, and subsequently, leaching and absorption of soil minerals happen, diminishing the waste's calorific value. Washing has two phases: rinsing and soaking with water, removing particles that are not chemically pulverized to waste. In the second phase, a weak acid is introduced to dissolve chemically bound pollutants and unwanted particulates, such as ash. Some extractives, such as fat, can be dissolved by washing with an acidic solution. Cleansing is required to guarantee that the activated carbon is effective since ash particles have a major impact on the quality of the activated carbon. By making the date palm waste a raw material rather than waste, washing processes can be made much simpler and cheaper. This is accomplished by accumulating fronds directly in the warehouse rather than leaving them on the soil due to climate conditions (Ahmad et al., 2012b).

The objective of the drying process is to reduce moisture content, as excessive moisture can hinder efficient carbonization. Pradeep et al. (2016) investigated the possibility of using microwaves to dry lignocellulosic biomass. This is followed by a process called pyrolysis, where the biomass is heated in the absence of oxygen to convert it into char. The pyrolysis temperature and residence time are carefully controlled to ensure optimal carbonization. Oven drying is preferred due to its controllability, simplicity of operation, dependable heating efficacy, and quicker process. The size of the precursor has a direct impact on the size of the final activated carbon particles. The precursor is initially crushed and sieved to eliminate particle size variability (Faiad et al., 2022). These pre-treatment stages are essential to produce high-quality active carbon from date palm waste, ensuring optimal adsorption properties and performance.

3.7.3.2 Activation Methods

Physical and chemical activations are the two most common types of activation. The active carbon's characteristics are altered by each technique. The term "thermal activation" or physical activation refers to this two-step pyrolytic process, which entails carbonization and activation. First, water, organic acid, and volatile chemicals are removed from the preparation by a process called carbonization, similar to char creation. In addition, carbonization is the first step in the development of surfaces and voids. The second step, activation, involves reacting carbonized char with steam, gas, or a mixture of the two to further increase the surface area established in the first step and to generate more pores. The vast majority of scientists advocate for employing carbon dioxide as an activation gas (Ayinla et al., 2019).

Chemical activation is a "wet oxidation" technique wherein the precursor is carbonized and activated with a chemical addition at temperatures lower than those required by physical activation. Unlike its chemical counterpart, initiating physical activation takes more time and effort (Ayinla et al., 2019).

3.7.3.3 Application of Date Palm-Activated Carbon

There is an abundance of uses for activated carbon. This material has several potential applications, including electrode material for supercapacitors, food decolonization and deodorization, gas adsorption, water purification, and gas masks. Kyaw et al. (2021) produced highly efficient activated carbon by employing chemical activation. The resulting activated carbon exhibited exceptional performance as multipurpose electrodes in storing deionization systems used for water desalination. Their innovative system successfully removed salt ions, decomposed harmful substances, and eliminated dye molecules, showcasing its effectiveness in water treatment. Haimour and Emeish (2006) achieved activated carbon production through physical activation, although the results did not meet the specifications outlined by ASTM. On the contrary, Vohra (2020) successfully remediated gases polluted with ammonia using a chemically activated date pit. It finds applications in diverse industries such as water treatment, air purification, industrial processes, and even in the production of filters and masks. Activated carbon derived from date palm waste presents a sustainable and eco-friendly solution for waste utilization while offering effective purification and filtration.

3.7.4 Water Depolluting and Desalination

Approximately 70% of the earth's surface is enveloped by water, yet a mere 2.5% of this vast expanse consists of freshwater. However, it is worth noting that only 20% of the available freshwater is liquid (El-Dessouky & Ettouney, 2002). Freshwater is used for more than just drinking and washing; it also has many industrial and other uses. Due to this sort of use, water becomes contaminated with harmful chemicals and metals. Industries are discharging dirty water straight into rivers and seas since desalination processes are expensive and need high-priced components. The use of agricultural residue as inexpensive absorbents for organic and inorganic pollution is a topic of intense interest among researchers. Date palm fiber and seed debris have been shown in multiple tests to be a good absorber for a wide variety of contaminants. The majority of the referenced research has been put into practice in Gulf nations due to the high demand for water desalination and the widespread availability of date palm waste (Faiad et al., 2022).

3.7.4.1 Date Palm Waste Preparation as Absorbent

Biochar made from discarded date palms was used in the study. Almost all of them used date seed as their primary ingredient, processing it through a biochar transformation before either grinding it into powder or activating it into activated carbon (Al-Ghamdi et al., 2013). Date palm fiber was harvested from the trunk, cleaned, dried, and then milled into a powder by Amin et al. (2017).

3.7.4.2 Water Desalination

Research has demonstrated that date palm residue is an efficient adsorbent owing to its favorable chemical characteristics and significant potential as a precursor for biochar and active carbon (Faiad et al., 2022). The following uses for date palm waste as a desalination adsorbent have been researched and validated by researchers, such as heavy metal removals. Heavy metals such as zinc, cadmium (Al-Ghamdi et al., 2013), copper (Al-Ghouti et al., 2010), lead (Mahdi et al., 2018), boron (Al.Haddabi et al., 2016), and arsenic (Amin et al., 2017), among others, have contaminated natural water resources as a result of the Industrial Revolution. The World Health Organization established guidelines for the maximum quantity of each heavy metal that can be safely consumed in drinking water. Following this directive, a large number of researchers examined the ability of date palm waste to absorb heavy metals. Due to its advantageous surface functional group, they concentrated primarily on the date seed (Al-Ghouti et al., 2010).

Other than that, it can be used as a dye. Dyes are used in a wide variety of products, for example, textiles, paper, rubber, plastics, leather, cosmetics, drugs, food, and paint. Dye leftovers can be found in the effluent of several businesses. Aside from nonionic (disperse), cationic (basic), and anionic (direct, acidic, and reactive), dyes also come in a third form called anionic (basic) (Ahmad et al., 2012a). Each study isolated a single dye because of the wide variety of dyes' chemical makeup. However, they proved that date seeds could take in particles of color. Banat et al. (2003) looked at methylene blue dye adsorption onto raw, activated carbon from date pits. Using unprocessed date seed, Mahmoodi et al. (2010) studied the elimination of acid green 25, acid black 26, and acid blue 7 from raw date stones.

Moreover, date palm waste has a high potential for the removal of phenolic pollution. When the hydroxyl (OH) group is attached to an aromatic ring carbon atom, the organic molecule is called a phenol. Monohydroxybenzene (C_6H_5OH), known as benzenol or carbolic acid, is the simplest member of the phenol family and goes by the term phenol. Sources of phenolic pollution are both artificial and natural. The breakdown of dead plants and animals (organic matter) in the water and the production of phenolic compounds by microbes and aquatic plants are natural contributors to water pollution. There are many uses for phenol in industry and agriculture (Okasha & Ibrahim,

2010). Phenol is present in the effluent in pesticides, paint, and solvents in the pharmaceutical, petroleum, and petrochemical industries. Phenol is a very hazardous compound that may enter the human body by drinking water or even by direct skin contact at high concentrations, resulting in burns and damage to the kidney, muscles, heart, liver, gastrointestinal tract, and muscles (Anku et al., 2017). In scientific research, it has been demonstrated that raw date seed powder and activated date pit carbon efficiently eliminate phenolic contaminants from water. Banat et al. (2004) found that activated carbon from date pits had an adsorption capacity that was 16 times higher than that of unprocessed date pits.

3.7.4.3 Removal of Pesticides

As a result of agricultural activities, pesticides have evolved into one of the most harmful water contaminants due to their mobility and propensity to accumulate in climbing food chains. Bakouri et al. (2010) investigated the effectiveness of using treated date stones to eliminate the drin pesticides (dieldrin, endrin, and aldrin). Maximum effectiveness in terms of waste elimination was 93%. Ashour (2008) looked at how well-activated carbon derived from date pits may soak up the weed-killing pesticide pendimethalin. High pendimethalin adsorption capabilities were shown for date pit-activated carbon by Ashour (2008).

3.7.5 Other Potential Applications

Date palm waste, a by-product of the date palm industry, has gained attention as a versatile and sustainable resource for numerous innovative applications across various sectors. One remarkable use is in the field of medical diagnostics, where porous nano-sheets derived from date palm waste have been developed for the simultaneous detection of uric acid and dopamine. According to Ahammad et al. (2019), these nano-sheets can help diagnose neurological conditions such as Parkinson's and Alzheimer's diseases, offering a sensitive and selective platform for capturing target molecules in the presence of interfering species. Beyond medical applications, date palm waste has proven effective in the construction and acoustic industries. Taban et al. (2021) demonstrated how processed date palm fibers can be transformed into sound-absorbing materials. These fibers, with their intricate structure of air pockets and fibers, dissipate sound energy and minimize noise pollution, making them suitable for eco-friendly insulation and soundproofing in homes and industrial settings. In addition to these high-tech applications, date palm waste offers significant benefits to the food and fuel industries. Putra et al. (2020) successfully extracted fructose from date fruit residue, providing a natural sweetener alternative for food and beverage products. Similarly, Azam and Ahmad (2019) explored the conversion of date palm cellulose into glucose and lactic acid, which can be utilized in fermentation processes. Moreover, date palm waste shows great potential as a renewable energy source. Researchers like Galiwango et al. (2021) and Kamil et al. (2019) have demonstrated how date palm waste can be processed into biogas and biodiesel, contributing to sustainable fuel alternatives that help reduce reliance on fossil fuels.

3.7.5.1 Porous Nano-Sheet

Porous nano-sheets derived from date palm waste offer a promising platform for the simultaneous detection of uric acid and dopamine. Dopamine and uric acid levels in humans were detected by Ahammad et al. (2019) despite the presence of confounding species. Diagnosing neurological illnesses like Parkinson's and Alzheimer's requires this sort of detection. By functionalizing the nano-sheets with specific receptors or enzymes, they can selectively capture and detect uric acid and dopamine molecules.

3.7.5.2 Sound Absorption

After undergoing the right processing and treatment, date palm waste may be converted into acoustic panels or insulation materials that effectively attenuate sound waves. Date palm fiber was used by Taban et al. (2021) and found that sound energy was dissipated and trapped inside the waste material's complex web of fibers and air pockets, which lessened reverberation and dampened noise transmission. As a result, date palm trash may be used as a green and long-lasting option for soundproofing in homes, businesses, and factories. Date palm waste may be effectively reused for sound absorption, which helps with waste management, makes for a more pleasant listening environment, and lessens the negative effects of noise pollution.

3.7.5.3 Fructose Sweetener

Putra et al. (2020) recovered sweetener fructose from date fruit residue using selective fermentation. Fructose's strong flavor and solubility make it a popular ingredient in the food and drink industries. Using date fruit waste to extract fructose is a frugal option due to the nature of waste. The first step is to remove the pits from the dates so that the kernels within may be retrieved. Enzymatic hydrolysis is then employed to convert the complex carbohydrates in the kernels into simpler sugars, such as fructose. Fructose is refined and purified by subjecting it to a series of subsequent purification and filtering procedures. Not only is date palm waste being put to good use by being converted into fructose sweeteners, but it also provides customers with a natural and healthier option to conventional sweeteners.

3.7.5.4 Lactic Acid and Glucose

Similar to Putra et al. (2020), Azam and Ahmad (2019) suggested that it is feasible to produce glucose from waste cellulosic material derived from date palms with the aid of cellulose enzymes produced by specific microorganisms. Using lactic acid-producing microbes, the cellulosic material derived from date palms can be fermented into lactic acid, similar to the method described previously.

3.7.5.5 Biofuel Source

Since the petroleum quantity is finite, researchers and engineers have suggested biofuels as petroleum-based fuel substitutes (Faiad et al., 2022). Lattieff (2016) converted date palm fruit waste into fuel-ready biogas via anaerobic digestion. He proved that waste from processing date fruits could be used to produce biogas. Galiwango et al. (2021) utilized direct catalytic depolymerization of lignocellulose from date palm waste to create liquid fuel. Date fruit seeds were used in the creation of biodiesel by Kamil et al. (2019). DSO biodiesel was created using a Model of Extraction System named "BUCHI Model B-811." After the extraction process was completed, they used a single-cylinder CI diesel engine to put the biodiesel/diesel fuel blends through their paces. According to the results, their mixture produced lower CO_2, CO, and HC levels but higher NO_x levels. By simulating pyrolysis, gasification, and fermentation, date palm waste was studied for its potential as a biofuel by Martis et al. (2020) in the United Arab Emirates. Based on their findings, pyrolysis and gasification are preferable to fermentation from a financial standpoint.

Date palm waste has emerged as a versatile resource, offering sustainable applications across various industries (Table 3.9). In the field of NFCs, it has been utilized to produce composites of polypropylene, epoxy, and OSBs, among others, as shown in studies by Belgacem et al. (2020) and Yusoff et al. (2010). Additionally, date palm waste has been used to create MDF and particle boards (Hegazy & Ahmed, 2015). Its application as a reinforcement in gypsum boards has also been explored (Braiek et al., 2017). Beyond construction materials, date palm waste serves as a biochar source for fertilizers (Hegazy & Ahmed, 2015) and as activated carbon for gas absorption and water treatment (Vohra, 2020; Ahmad et al., 2012a). It even functions as an electrode material for supercapacitors (Haimour & Emeish, 2006). In water desalination and depollution, date palm waste shows efficiency in filtering heavy metals, removing dyes, and tackling phenolic pollution and pesticides (Mahdi et al., 2018; Banat et al., 2003). In the food industry, fructose sweeteners, glucose, and lactic acid have been successfully extracted from date palm waste (Putra et al., 2020; Azam & Ahmad, 2019). Additionally, its potential as a biofuel source has been proven through the production of biogas and liquid fuels (Lattieff, 2016; Galiwango et al., 2021). Other innovative applications

TABLE 3.9
Summary of Date Palm Waste Applications

Application	Products	References
Natural fiber composites (NFCs)	Composites of polypropylene	Belgacem et al. (2020)
		Haque et al. (2009)
	Composites of epoxy	Yusoff et al. (2010)
		Saba et al. (2019)
	Oriented strand board	Hegazy and Ahmed (2015)
	Medium-density fiberboard (MDF)	Hosseinkhani et al. (2014)
	Particle board from date palm	Hegazy and Ahmed (2015)
	Gypsum board reinforcement	Braiek et al. (2017)
Biochar	Fertilizer	Hegazy and Ahmed (2015)
Activated carbon	Gas absorber	Vohra (2020)
	Water treatment	Ahmad et al. (2012b)
	Electrode material of supercapacitor	Haimour and Emeish (2006)
Water desalination and depolluting	Heavy metal filters	Mahdi et al. (2018)
	Dyes removal	Banat et al. (2003)
		Mahmoodi et al. (2010)
	Phenolic pollution removal	Banat et al. (2004)
	Pesticides removal	Bakouri et al. (2010)
Food industry's raw material	Fructose sweetener	Putra et al. (2020)
	Lactic acid and glucose	Azam and Ahmad (2019)
Biofuel source	Biogas	Lattieff (2016)
	Liquid fuel	Galiwango et al. (2021)
Other applications	Porous nano-sheet	Ahammad et al. (2019)
	Sound absorber	Taban et al. (2021)

include porous nano-sheets for medical diagnostics (Ahammad et al., 2019) and sound-absorbing materials for acoustic applications (Taban et al., 2021), making date palm waste a valuable resource for sustainable solutions across diverse sectors.

3.8 CHALLENGES ASSOCIATED WITH PROCESSING AND COMMERCIALIZING

Processing and commercializing date palm waste can pose several challenges. Some of the key challenges include collecting and handling, waste composition and variability, moisture content, limited processing technologies, lack of awareness and infrastructure, economic viability, and environmental consideration. During the harvesting of date palms and the subsequent processing of the dates, a significant amount of trash is produced. The collection and management of this garbage might provide a number of logistical challenges. The garbage is frequently bulky and heavy, and it is necessary to have the appropriate machinery and infrastructure in place for it to be collected and transported efficiently (Al-Farsi et al., 2005). In addition, waste from date palms consists of several distinct components, including fronds, trunks, pits, and fibers; each has its unique features and chemical makeup. It may be challenging to design uniform processing processes because the content of trash might vary greatly. It is necessary to have a detailed characterization and an understanding of the waste's composition to find appropriate commercial applications (Belgacem et al., 2021). Moreover, waste from date palms tends to have a high moisture content, which makes it more susceptible to deterioration and rotting. To stop the growth of microorganisms, stop rotting, and prevent the loss of important compounds in the waste, proper drying and preservation processes need to be used (Oladzad et al., 2021). On the contrary, it can be difficult to create processing technologies that are both successful and economical when used to waste from date palms. There is a demand for creative and environmentally responsible processes to turn waste into goods with additional value. There is a possibility that the traditional methods of processing are neither adequate nor economically viable for large-scale commercialization (Chandrasekaran & Bahkali, 2013).

The exploitation of waste from date palms is still in a relatively undeveloped stage, and there may be a lack of information regarding the various applications and advantages of this waste. In addition, there may be an absence of infrastructure and facilities that have been created expressly for processing waste from date palms, which could make its commercialization more difficult (Castillo et al., 2023). Furthermore, a substantial obstacle is presented by the question of whether or not the processing and commercialization of waste from date palms is economically viable. It is necessary to do an accurate cost-benefit analysis, considering the costs involved in collecting, processing, and market demand for the products obtained. Creating a sustainable business model and searching for markets for the products that can be extracted from the waste might be tough (Oladzad et al., 2021). Another challenge is processing and selling waste from date palms, which requires strict adherence to all applicable legislations and consideration of the surrounding environment. It is necessary to consider the legislation governing waste management, the requirements for sustainability, and the assessments of the effects on the environment. It can be difficult and time-consuming to ensure that the methods of processing and the goods obtained from those techniques meet all of the relevant regulatory criteria (Anli et al., 2020).

Addressing these challenges requires a collaborative effort involving researchers, industry stakeholders, government bodies, and local communities. Innovations in processing technologies, research on product development, raising awareness, and creating supportive policies and infrastructure can contribute to the successful processing and commercialization of date palm waste.

3.9 CONCLUSION

In conclusion, the management of date palm waste presents an important opportunity for sustainable use and a range of possible applications. Date palm agriculture produces significant amounts of waste that may be successfully handled and turned into useful goods, minimizing negative environmental effects and increasing resource efficiency. Waste from date palms, including the leaves, fronds, and other biomass leftovers, may be effectively collected and processed for a variety of uses, including biomass, bioethanol, pulp and papermaking, compost, biochar, and the production of natural acid. Overall, managing date palm waste offers the chance to transform a large agricultural by-product into useful resources. This can aid in the development of sustainable agriculture, the restoration of the environment, and the mitigation of climate change while minimizing waste and maximizing resource efficiency by using the potential of date palm waste through the creation of biochar and its diverse uses. The agricultural industry may gain economic, environmental, and social benefits from date palm waste by putting in place effective waste management procedures and investigating these possible uses. To maximize the value and use of date palm waste and to support a circular economy and sustainable agricultural practices, further research, development, and process optimization must be done.

ACKNOWLEDGMENT

This chapter was written with Muhammad Hassanuddin Bin Zainul 'Abidin, Kavinesh A.L. Ganesan, and Seri Afra Indahni binti A. Razak.

REFERENCES

Abdul Rahiman, B., Jaleel, C. A., Abdurahman, I. M., & Aziz, M. A. (2012). The date palm (*Phoenix dactylifera* L.): Overview of biology, uses, and cultivation. *Journal of Medicinal Plants Research*, *6*(39), 5113–5120.

Agoudjil, B., Benchabane, A., Boudenne, A., Ibos, L., & Fois, M. (2011). Renewable materials to reduce building heat loss: Characterization of date palm wood. *Energy and Buildings*, *43*(2), 491–497. https://doi.org/10.1016/j.enbuild.2010.10.014

Ahammad, A. J. S., Odhikari, N., Shah, S. S., Hasan, Md. M., Islam, T., Pal, P. R., Ahmed Qasem, M. A., & Aziz, Md. A. (2019). Porous tal palm carbon nanosheets: Preparation, characterization and application for the simultaneous determination of dopamine and uric acid. *Nanoscale Advances*, *1*(2), 613–626. https://doi.org/10.1039/C8NA00090E

Ahmad, M., Ahmad, M., Usman, A. R. A., Al-Faraj, A. S., Abduljabbar, A., Ok, Y. S., & Al-Wabel, M. I. (2019). Date palm waste-derived biochar composites with silica and zeolite: Synthesis, characterization and implication for carbon stability and recalcitrant potential. *Environmental Geochemistry and Health*, *41*(4), 1687–1704. https://doi.org/10.1007/s10653-017-9947-0

Ahmad, M., Yaseen, M., Rafique, S., & Khan, R. A. (2012a). An overview of date palm biomass for sustainable energy and bioproducts: Perspectives and challenges. *Renewable and Sustainable Energy Reviews*, *16*(2), 1384–1399.

Ahmad, T., Danish, M., Rafatullah, M., Ghazali, A., Sulaiman, O., Hashim, R., & Ibrahim, M. N. M. (2012b). The use of date palm as a potential adsorbent for wastewater treatment: A review. *Environmental Science and Pollution Research*, *19*(5), 1464–1484. https://doi.org/10.1007/s11356-011-0709-8

Alawar, A., Hamed, A. M., & Al-Kaabi, K. (2009). Characterization of treated date palm tree fiber as composite reinforcement. *Composites Part B: Engineering*, *40*(7), 601–606. https://doi.org/10.1016/j.compositesb.2009.04.018

Al-Farsi, M., Alasalvar, C., Morris, A., Baron, M., & Shahidi, F. (2005). Compositional and sensory characteristics of three native sun-dried date (*Phoenix dactylifera* L.) varieties grown in Oman. *Journal of Agricultural and Food Chemistry*, *53*(19), 7586–7591.

Al-Farsi, M., & Lee, C. (2008). Nutritional and functional properties of dates: A review. *Critical Reviews in Food Science and Nutrition*, *48*(10), 877–887.

Al-Ghamdi, A., Altaher, H., & Omar, W. (2013). Application of date palm trunk fibers as adsorbents for removal of Cd+2 ions from aqueous solutions. *Journal of Water Reuse and Desalination*, *3*(1), 47–54. https://doi.org/10.2166/wrd.2013.031

Al-Ghouti, M. A., Li, J., Salamh, Y., Al-Laqtah, N., Walker, G., & Ahmad, M. N. M. (2010). Adsorption mechanisms of removing heavy metals and dyes from aqueous solution using date pits solid adsorbent. *Journal of Hazardous Materials*, *176*(1), 510–520. https://doi.org/10.1016/j.jhazmat.2009.11.059

Al.Haddabi, M., Ahmed, M., Al.Jebri, Z., Vuthaluru, H., Znad, H., & Al.Kindi, M. (2016). Boron removal from seawater using date palm (*Phoenix dactylifera*) seed ash. *Desalination and Water Treatment*, *57*(11), 5130–5137. https://doi.org/10.1080/19443994.2014.1000385

Ali, M. E., & Alabdulkarem, A. (2017). On thermal characteristics and microstructure of a new insulation material extracted from date palm trees surface fibers. *Construction and Building Materials*, *138*, 276–284. https://doi.org/10.1016/j.conbuildmat.2017.02.012

Alias, A. H., Norizan, M. N., Sabaruddin, F. A., Asyraf, M. R. M., Norrahim, M. N. F., Ilyas, A. R., Kuzmin, A. M., Rayung, M., Shazleen, S. S., Nazrin, A., Sherwani, S. F. K., Harussani, M. M., Atikah, M. S. N., Ishak, M. R., Sapuan, S. M., & Khalina, A. (2021). Hybridization of MMT/lignocellulosic fiber reinforced polymer nanocomposites for structural applications: A review. *Coatings*, *11*(11), 1355. https://doi.org/10.3390/coatings11111355

Al-Kaabi, K., Al-Khanbashi, A., & Hammami, A. (2005). Date palm fibers as polymeric matrix reinforcement: DPF/polyester composite properties. *Polymer Composites*, *26*(5), 604–613. https://doi.org/10.1002/pc.20130

Al-Oqla, F. M., Alothman, O. Y., Jawaid, M., Sapuan, S. M., & Es-Saheb, M. H. (2014). Processing and properties of date palm fibers and its composites. In K. R. Hakeem, M. Jawaid, & U. Rashid (Eds.), *Biomass and Bioenergy: Processing and Properties* (pp. 1–25). Springer International Publishing. https://doi.org/10.1007/978-3-319-07641-6_1

Al-Oqla, F. M., & Sapuan, S. M. (2014). Natural fiber reinforced polymer composites in industrial applications: Feasibility of date palm fibers for sustainable automotive industry. *Journal of Cleaner Production*, *66*, 347–354. https://doi.org/10.1016/j.jclepro.2013.10.050

Al-Rifaie, W. N., & Al-Niami, M. (2016). Mechanical performance of date palm fiber-reinforced gypsums. *Innovative Infrastructure Solutions*, *1*(1), 18. https://doi.org/10.1007/s41062-016-0022-y

Alshabanat, M. (2019). Morphological, thermal, and biodegradation properties of LLDPE/treated date palm waste composite buried in a soil environment. *Journal of Saudi Chemical Society*, *23*(3), 355–364. https://doi.org/10.1016/J.JSCS.2018.08.008

Al-Shahib, W., & Marshall, R. J. (2003). The fruit of the date palm: Its possible use as the best food for the future? *International Journal of Food Sciences and Nutrition*, *54*(4), 247–259.

Al-Shammari, K. I., Al-Hosni, K., & Al-Othman, A. (2014). Cultivation and characterization of date palm. *Journal of Date Palm Research*, *33*(1), 1–24.

Al-Yousef, H. M., Al-Fawaz, S., Al-Abbasi, F. A., Al-Qurainy, F., & Daffalla, H. M. (2018). Cultivation of date palm (*Phoenix dactylifera* L.): A review. *Journal of Food and Nutrition Research*, *6*(10), 645–651.

Amin, M. T., Alazba, A. A., & Amin, M. N. (2017). Absorption behaviours of copper, lead, and arsenic in aqueous solution using date palm fibers and orange peel: Kinetics and thermodynamics. *Polish Journal of Environmental Studies*, *26*(2), 543–557. https://doi.org/10.15244/pjoes/66963

Amirou, S., Zerizer, A., Pizzi, A., Haddadou, I., & Zhou, X. (2013). Particleboards production from date palm biomass. *European Journal of Wood and Wood Products*, *71*(6), 717–723. https://doi.org/10.1007/s00107-013-0730-3

Anku, W. W., Mamo, M. A., & Govender, P. P. (2017). Phenolic compounds in water: Sources, reactivity, toxicity and treatment methods. In M. Soto-Hernández, M. Palma-Tenango, & M. del R. García-Mateos (Eds.), *Phenolic Compounds: Natural Sources, Importance and Applications* (pp. 419–443). London, UK: IntechOpen.

Anli, M., Baslam, M., Tahiri, A., Raklami, A., Symanczik, S., Boutasknit, A., Ait-El-Mokhtar, M., Ben-Laouane, R., Toubali, S., & Ait Rahou, Y. (2020). Biofertilizers as strategies to improve photosynthetic apparatus, growth, and drought stress tolerance in the date palm. *Frontiers in Plant Science*, *11*, 516818.

Asaye, Z., Kim, D.-G., Yimer, F., Prost, K., Obsa, O., Tadesse, M., Gebrehiwot, M., & Brüggemann, N. (2022). Effects of combined application of compost and mineral fertilizer on soil carbon and nutrient content, yield, and agronomic nitrogen use efficiency in maize-potato cropping systems in southern Ethiopia. *Land*, *11*(6), 784. https://doi.org/10.3390/land11060784

Ashour, S. S. (2008). Adsorption of herbicide (pendimethalin) onto activated carbons developed from date pits. *Egyptian Journal of Chemistry*, *51*, 55–67.

Asyraf, M. R. M., Ishak, M. R., Norrrahim, M. N. F., Nurazzi, N. M., Shazleen, S. S., Ilyas, R. A., Rafidah, M., & Razman, M. R. (2021). Recent advances of thermal properties of sugar palm lignocellulosic fibre reinforced polymer composites. *International Journal of Biological Macromolecules*, *193*, 1587–1599. https://doi.org/10.1016/j.ijbiomac.2021.10.221

Asyraf, M. R. M., Syamsir, A., Zahari, N. M., Supian, A. B. M., Ishak, M. R., Sapuan, S. M., Sharma, S., Rashedi, A., Razman, M. R., Zakaria, S. Z. S., Ilyas, R. A., & Rashid, M. Z. A. (2022). Product development of natural fibre-composites for various applications: Design for sustainability. *Polymers*, *14*(5), 920. https://doi.org/10.3390/polym14050920

Ayinla, R. T., Dennis, J. O., Zaid, H. M., Sanusi, Y. K., Usman, F., & Adebayo, L. L. (2019). A review of technical advances of recent palm bio-waste conversion to activated carbon for energy storage. *Journal of Cleaner Production*, *229*, 1427–1442. https://doi.org/10.1016/j.jclepro.2019.04.116

Azam, M. T., & Ahmad, A. (2019). Date palm waste: An efficient source for production of glucose and lactic acid. In Mu. Naushad and Eric Lichtfouse (Eds.), *Sustainable Agriculture Reviews 34: Date Palm for Food, Medicine and the Environment* (pp. 155–178). Cham, Switzerland: Springer International Publishing.

Azwa, Z. N., Yousif, B. F., Manalo, A. C., & Karunasena, W. (2013). A review on the degradability of polymeric composites based on natural fibers. *Materials & Design*, *47*, 424–442. https://doi.org/10.1016/j.matdes.2012.11.025

Bakouri, H. El, Morillo, J., Usero, J., Vanderlinden, E., & Vidal, H. (2010). Effectiveness of acid-treated agricultural stones used in biopurification systems to avoid pesticide contamination of water resources caused by direct losses: Part I. Equilibrium experiments and kinetics. *Bioresource Technology*, *101*(14), 5084–5091. https://doi.org/10.1016/j.biortech.2010.01.131

Banaruee, H., & Amirizad, S. A. (2017). Internationalization of date industry through small industries development. *European Journal of Management and Marketing Studies*, *2*(3), 46–60.

Banat, F., Al-Asheh, S., & Al-Makhadmeh, L. (2003). Evaluation of the use of raw and activated date pits as potential adsorbents for dye containing waters. *Process Biochemistry*, *39*(2), 193–202. https://doi.org/10.1016/S0032-9592(03)00065-7

Banat, F., Al-Asheh, S., & Al-Makhadmeh, L. (2004). Utilization of raw and activated date pits for the removal of phenol from aqueous solutions. *Chemical Engineering & Technology*, *27*(1), 80–86. https://doi.org/10.1002/ceat.200401868

Bassyouni, M., ul Hasan, S. W., Abdel-Aziz, M. H., Abdel-hamid, S. M.-S., Naveed, S., Hussain, A., & Ani, F. N. (2014). Date palm waste gasification in downdraft gasifier and simulation using ASPEN HYSYS. *Energy Conversion and Management*, *88*, 693–699. https://doi.org/10.1016/j.enconman.2014.08.061

Belgacem, C., Serra-Parareda, F., Tarrés, Q., Mutjé, P., Delgado-Aguilar, M., & Boufi, S. (2021). Valorization of date palm waste for plastic reinforcement: Macro and micromechanics of flexural strength. *Polymers*, *13*(11), 1751.

Belgacem, C., Tarres, Q., Espinach, F. X., Mutjé, P., Boufi, S., & Delgado-Aguilar, M. (2020). High-yield lignocellulosic fibers from date palm biomass as reinforcement in polypropylene composites: Effect of fiber treatment on composite properties. *Polymers*, *12*(6), 1423. https://doi.org/10.3390/polym12061423

Bensidhom, G., Ben Hassen-Trabelsi, A., Alper, K., Sghairoun, M., Zaafouri, K., & Trabelsi, I. (2018). Pyrolysis of date palm waste in a fixed-bed reactor: Characterization of pyrolytic products. *Bioresource Technology*, *247*, 363–369. https://doi.org/10.1016/j.biortech.2017.09.066

Boulal, A., Kihal, M., Khelifi, C., & Benali, B. (2016). Bioethanol production from date palm fruit waste fermentation using solar energy. *African Journal of Biotechnology*, *15*(30), 1621–1627.

Braiek, A., Karkri, M., Adili, A., Ibos, L., & Ben Nasrallah, S. (2017). Estimation of the thermophysical properties of date palm fibers/gypsum composite for use as insulating materials in building. *Energy and Buildings*, *140*, 268–279. https://doi.org/10.1016/j.enbuild.2017.02.001

Burezq, H., & Davidson, M. K. (2023). Biochar from date palm (*Phoenix dactylifera* L.) residues—A critical review. *Arabian Journal of Geosciences*, *16*(2), 101.

Castillo, A. B., Cortes, D. J. D., Sorino, C. F., Soriño, C. K. P., El-Naas, M. H., & Ahmed, T. (2023). Bioethanol production from waste and nonsalable date palm (*Phoenix dactylifera* L.) fruits: Potentials and challenges. *Sustainability*, *15*(4), 2937.

Chandrasekaran, M., & Bahkali, A. H. (2013). Valorization of date palm (*Phoenix dactylifera*) fruit processing by-products and wastes using bioprocess technology—Review. *Saudi Journal of Biological Sciences*, *20*(2), 105–120.

Chao, C. T., & Krueger, R. R. (2007). The date palm (*Phoenix Dactylifera* L.): Overview of biology, uses, and cultivation. *HortScience*, *42*(5), 1077–1082. https://doi.org/10.21273/HORTSCI.42.5.1077.

Chao, C. T., Krueger, R. R., & Brecht, J. K. (2010). Date palm. In E. M. Yahia (Ed.), *Postharvest Biology and Technology of Tropical and Subtropical Fruits* (pp. 347–368). Oxford, UK: Woodhead Publishing Limited.

El may, Y., Jeguirim, M., Dorge, S., Trouvé, G., & Said, R. (2012). Study on the thermal behavior of different date palm residues: Characterization and devolatilization kinetics under inert and oxidative atmospheres. *Energy*, *44*(1), 702–709. https://doi.org/10.1016/j.energy.2012.05.022

El-Dessouky, H. T., & Ettouney, H. M. (2002). *Fundamentals of Salt Water Desalination*. Amsterdam, Netherlands: Elsevier.

Fadel, M., Keera, A. A., Mouafi, F. E., & Kahil, T. (2013). High level ethanol from sugar cane molasses by a new thermotolerant *Saccharomyces cerevisiae* strain in industrial scale. *Biotechnology Research International*, 2013, 1–6. https://doi.org/10.1155/2013/253286

Faiad, A., Alsmari, M., Ahmed, M. M. Z., Bouazizi, M. L., Alzahrani, B., & Alrobei, H. (2022). Date palm tree waste recycling: Treatment and processing for potential engineering applications. *Sustainability (Switzerland)*, *14*(3), 1134. https://doi.org/10.3390/su14031134

FAO (2021). Special Report: 2021 FAO Crop and Food Supply Assessment Mission to the Syrian Arab Republic—December 2021. https://doi.org/10.4060/cb8039en

Ferrández-García, C.-E., Ferrández-García, A., Ferrández-Villena, M., Hidalgo-Cordero, J. F., García-Ortuño, T., & Ferrández-García, M.-T. (2018). Physical and mechanical properties of particleboard made from palm tree prunings. *Forests*, *9*(12), 755. https://doi.org/10.3390/f9120755

Galiwango, E., Al-Marzuoqi, A. H., Khaleel, A. A., & Abu-Omar, M. M. (2021). Catalytic depolymerization of date palm waste to valuable C5–C12 compounds. *Catalysts*, *11*(3), 371. https://doi.org/10.3390/catal11030371

Ghosh, S. K., & Nag, D. (2009). Chemical free hand made paper from date-palm (*Phoenix Dactylifera* 4) leaves. *IPPTA Journal*, *21*(2), 95–97.

Ghouili, E., Abid, G., Hogue, R., Jeanne, T., D'Astous-Pagé, J., Sassi, K., Hidri, Y., M'Hamed, H. C., Somenahally, A., Xue, Q., Jebara, M., Nefissi Ouertani, R., Riahi, J., de Oliveira, A. C., & Muhovski, Y. (2023). Date palm waste compost application increases soil microbial community diversity in a cropping barley (Hordeum vulgare L.) field. *Biology*, *12*(4), 546. https://doi.org/10.3390/biology12040546

Ghouili, E., Hidri, Y., M'Hamed, H. C., Somenahally, A., Xue, Q., Znaïdi, I. E. A., Jebara, M., Ouertani, R. N., Muhovski, Y., & Riahi, J. (2022). Date palm waste compost promotes plant growth and nutrient transporter genes expression in barley (Hordeum vulgare L.). *South African Journal of Botany*, *149*, 247–257.

Giwa, A., Yusuf, A., Ajumobi, O., & Dzidzienyo, P. (2019). Pyrolysis of date palm waste to biochar using concentrated solar thermal energy: Economic and sustainability implications. *Waste Management*, *93*, 14–22. https://doi.org/10.1016/j.wasman.2019.05.022

Guo, J., & Lua, A. C. (2001). Kinetic study on pyrolytic process of oil-palm solid waste using two-step consecutive reaction model. *Biomass and Bioenergy*, *20*(3), 223–233. https://doi.org/10.1016/S0961-9534(00)00080-5

Haimour, N. M., & Emeish, S. (2006). Utilization of date stones for production of activated carbon using phosphoric acid. *Waste Management*, *26*(6), 651–660. https://doi.org/10.1016/j.wasman.2005.08.004

Hakeem, K. R., Jawaid, M., & Rashid, U. (Eds.). (2014). *Biomass and Bioenergy*. https://doi.org/10.1007/978-3-319-07578-5

Halimatul, M. J., Sapuan, S. M., Jawaid, M., Ishak, M. R., & Ilyas, R. A. (2019). Water absorption and water solubility properties of sago starch biopolymer composite films filled with sugar palm particles. *Polimery*, *64*(9), 27–35. https://doi.org/10.14314/polimery.2019.9.4

Haque, Md. M., Hasan, M., Islam, Md. S., & Ali, Md. E. (2009). Physico-mechanical properties of chemically treated palm and coir fiber reinforced polypropylene composites. *Bioresource Technology*, *100*(20), 4903–4906. https://doi.org/10.1016/j.biortech.2009.04.072

Hegazy, S. S., & Ahmed, K. (2015). Effect of date palm cultivar, particle size, panel density and hot water extraction on particleboards manufactured from date palm fronds. *Agriculture*, *5*(2), 267–285. https://doi.org/10.3390/agriculture5020267

Hegazy, S., Ahmed, K., & Hiziroglu, S. (2015). Oriented strand board production from water-treated date palm fronds. *BioResources*, *10*(1), 448–456.

Ho, T. T., Tra, V. T., Le, T. H., Nguyen, N.-K.-Q., Tran, C.-S., Nguyen, P.-T., Vo, T.-D.-H., Thai, V.-N., & Bui, X.-T. (2022). Compost to improve sustainable soil cultivation and crop productivity. *Case Studies in Chemical and Environmental Engineering*, *6*, 100211. https://doi.org/10.1016/j.cscee.2022.100211

Hosseinkhani, H., Euring, M., & Kharazipour, A. (2014). Utilization of date palm (*Phoenix dactylifera* L.) pruning residues as raw material for MDF manufacturing. *Journal of Materials Science Research*, *4*(1). https://doi.org/10.5539/jmsr.v4n1p46

Hussain, A., Farooq, A., Bassyouni, M. I., Sait, H. H., El-Wafa, M. A., Hasan, S. W., & Ani, F. N. (2014). Pyrolysis of Saudi Arabian date palm waste: A viable option for converting waste into wealth. *Life Science Journal*. https://hdl.handle.net/10754/337219

Imraan, M., Ilyas, R. A., Norfarhana, A. S., Bangar, S. P., Knight, V. F., & Norrrahim, M. N. F. (2023). Sugar palm (Arenga pinnata) fibers: New emerging natural fibre and its relevant properties, treatments and potential applications. *Journal of Materials Research and Technology*, *24*, 4551–4572. https://doi.org/10.1016/j.jmrt.2023.04.056

Iqbal, M., Khera, R. A., Hussain, T., Sadia, H., Abbas, M., Nazir, A., Arshad, M., & Younas, U. (2017). Cytotoxicity and bioactivity evaluation of pygmy date palm extracts. *Current Science Perspectives*, *3*(2), 106.

Jaradat, A. A., (2011). Biodiversity of date palm. In *Encyclopedia of Life Support Systems (EOLSS): Land Use, Land Cover and Soil Sciences*, Developed under the Auspices the UNESCO, Eolss Publishers, Oxford, UK, https://www.eolss.net, Retrieved June 25, 2011.

Jeyakumar, R., Sampath, P. S., Ramamoorthi, R., & Ramakrishnan, T. (2017). Structural, morphological and mechanical behaviour of glass fiber reinforced epoxy nanoclay composites. *The International Journal of Advanced Manufacturing Technology*, *93*(1), 527–535. https://doi.org/10.1007/s00170-017-0565-x

John, M. J., & Thomas, S. (2008). Biofibers and biocomposites. *Carbohydrate Polymers*, *71*(3), 343–364. https://doi.org/10.1016/j.carbpol.2007.05.040

Jonoobi, M., Shafie, M., Shirmohammadli, Y., Ashori, A., Zarea-Hosseinabadi, H., & Mekonnen, T. (2019). A review on date palm tree: Properties, characterization and its potential applications. *Journal of Renewable Materials*, *7*(11), 1055–1075. https://doi.org/10.32604/jrm.2019.08188

Jouiad, M., Al-Nofeli, N., Khalifa, N., Benyettou, F., & Yousef, L. F. (2015). Characteristics of slow pyrolysis biochars produced from Rhodes grass and fronds of edible date palm. *Journal of Analytical and Applied Pyrolysis*, *111*, 183–190. https://doi.org/10.1016/j.jaap.2014.10.024

Jumaidin, R., Adam, N. W., Ilyas, R. A., Hussin, M. S. F., Taha, M. M., Mansor, M. R., Azlan, U. A. A., & Yob, M. S. (2019a). Water transport and physical properties of sugarcane bagasse fibre reinforced thermoplastic potato starch biocomposite. *Journal of Advanced Research in Fluid Mechanics and Thermal Sciences*, *61*(2), 273–281.

Jumaidin, R., Diah, N. A., Ilyas, R. A., Alamjuri, R. H., & Yusof, F. A. M. (2021). Processing and characterisation of banana leaf fibre reinforced thermoplastic cassava starch composites. *Polymers*, *13*(9), 1420. https://doi.org/10.3390/polym13091420

Jumaidin, R., Saidi, Z. A. S., Ilyas, R. A., Ahmad, M. N., Wahid, M. K., Yaakob, M. Y., Maidin, N. A., Rahman, M. H. A., & Osman, M. H. (2019b). Characteristics of cogon grass fibre reinforced thermoplastic cassava starch biocomposite: Water absorption and physical properties. *Journal of Advanced Research in Fluid Mechanics and Thermal Sciences, 62*(1), 43–52.

Kamil, M., Ramadan, K., Olabi, A. G., Ghenai, C., Inayat, A., & Rajab, M. H. (2019). Desert palm date seeds as a biodiesel feedstock: Extraction, characterization, and engine testing. *Energies*, *12*(16), 3147.

Khoudja, D., Taallah, B., Izemmouren, O., Aggoun, S., Herihiri, O., & Guettala, A. (2021). Mechanical and thermophysical properties of raw earth bricks incorporating date palm waste. *Construction and Building Materials, 270*. https://doi.org/10.1016/j.conbuildmat.2020.121824

Khristova, P., Kordsachia, O., Patt, R., Karar, I., & Khider, T. (2006). Environmentally friendly pulping and bleaching of bagasse. *Industrial Crops and Products, 23*(2), 131–139.

Kyaw, H. H., Al-Mashaikhi, S. M., Myint, M. T. Z., Al-Harthi, S., El-Shafey, E.-S. I., & Al-Abri, M. (2021). Activated carbon derived from the date palm leaflets as multifunctional electrodes in capacitive deionization system. *Chemical Engineering and Processing: Process Intensification, 161*, 108311. https://doi.org/10.1016/j.cep.2021.108311

Lattieff, F. A. (2016). A study of biogas production from date palm fruit wastes. *Journal of Cleaner Production, 139*, 1191–1195. https://doi.org/10.1016/j.jclepro.2016.08.139

Lehmann, J. & Joseph, S. (2009). *Biochar for Environmental Management: Science and Technology*. Earthscan, London, pp. 405.

Lehmann, J., Rillig, M. C., Thies, J., Masiello, C. A., Hockaday, W. C., & Crowley, D. (2011). Biochar effects on soil biota—A review. *Soil Biology and Biochemistry, 43*(9), 1812–1836. doi:10.1016/j.soilbio.2011.04.022

Li, X., Tabil, L. G., & Panigrahi, S. (2007). Chemical treatments of natural fiber for use in natural fiber-reinforced composites: A review. *Journal of Polymers and the Environment, 15*(1), 25–33. https://doi.org/10.1007/s10924-006-0042-3

Mahdi, Z., Yu, Q. J., & El Hanandeh, A. (2018). Removal of lead(II) from aqueous solution using date seed-derived biochar: Batch and column studies. *Applied Water Science, 8*(6), 181. https://doi.org/10.1007/s13201-018-0829-0

Mahmoodi, N. M., Hayati, B., & Arami, M. (2010). Textile dye removal from single and ternary systems using date stones: Kinetic, isotherm, and thermodynamic studies. *Journal of Chemical & Engineering Data, 55*(11), 4638–4649. https://doi.org/10.1021/je1002384

Mallaki, M., & Fatehi, R. (2014). Design of a biomass power plant for burning date palm waste to cogenerate electricity and distilled water. *Renewable Energy, 63*, 286–291. https://doi.org/10.1016/j.renene.2013.09.036

Marrakchi, Z., Khiari, R., Oueslati, H., Mauret, E., & Mhenni, F. (2011). Pulping and papermaking properties of Tunisian Alfa stems (Stipa tenacissima)—Effects of refining process. *Industrial Crops and Products, 34*(3), 1572–1582.

Martis, R., Al-Othman, A., Tawalbeh, M., & Alkasrawi, M. (2020). Energy and economic analysis of date palm biomass feedstock for biofuel production in UAE: Pyrolysis, gasification and fermentation. *Energies, 13*(22). https://doi.org/10.3390/en13225877

Masri, T., Ounis, H., Sedira, L., Kaci, A., & Benchabane, A. (2018). Characterization of new composite material based on date palm leaflets and expanded polystyrene wastes. *Construction and Building Materials, 164*, 410–418. https://doi.org/10.1016/j.conbuildmat.2017.12.197

Mazani, N., Sapuan, S. M., Sanyang, M. L., Atiqah, A., & Ilyas, R. A. (2019). Design and fabrication of a shoe shelf from kenaf fiber reinforced unsaturated polyester composites. In H. Ariffin, S. M. Sapuan, & M. A. Hassan (Eds.), *Lignocellulose for Future Bioeconomy* (1st ed., Issue 2000, pp. 315–332). Elsevier. https://doi.org/10.1016/B978-0-12-816354-2.00017-7

McKendry, P. (2002). Energy production from biomass (part 1): Overview of biomass. *Bioresource Technology, 83*(1), 37–46. https://doi.org/10.1016/S0960-8524(01)00118-3

Methacanon, P., Weerawatsophon, U., Sumransin, N., Prahsarn, C., & Bergado, D. T. (2010). Properties and potential application of the selected natural fibers as limited life geotextiles. *Carbohydrate Polymers, 82*(4), 1090–1096. https://doi.org/10.1016/j.carbpol.2010.06.036

Mirmehdi, S. M., Zeinaly, F., & Dabbagh, F. (2014). Date palm wood flour as filler of linear low-density polyethylene. *Composites Part B: Engineering, 56*, 137–141. https://doi.org/10.1016/j.compositesb.2013.08.008

Munir, S., Daood, S. S., Nimmo, W., Cunliffe, A. M., & Gibbs, B. M. (2009). Thermal analysis and devolatilization kinetics of cotton stalk, sugar cane bagasse and shea meal under nitrogen and air atmospheres. *Bioresource Technology, 100*(3), 1413–1418. https://doi.org/10.1016/j.biortech.2008.07.065

Nasser, R. A., Salem, M. Z. M., Hiziroglu, S., Al-Mefarrej, H. A., Mohareb, A. S., Alam, M., & Aref, I. M. (2016). Chemical analysis of different parts of date palm (*Phoenix dactylifera* L.) using ultimate, proximate and thermo-gravimetric techniques for energy production. *Energies, 9*(5). https://doi.org/10.3390/en9050374

Nazrin, A., Sapuan, S. M., Zuhri, M. Y. M., Tawakkal, I. S. M. A., & Ilyas, R. A. (2021). Water barrier and mechanical properties of sugar palm crystalline nanocellulose reinforced thermoplastic sugar palm starch (TPS)/poly(lactic acid) (PLA) blend bionanocomposites. *Nanotechnology Reviews, 10*(1), 431–442. https://doi.org/10.1515/ntrev-2021-0033

Norfarhana, A. S., Ilyas, R. A., & Ngadi, N. (2022). A review of nanocellulose adsorptive membrane as multifunctional wastewater treatment. *Carbohydrate Polymers, 291*, 119563. https://doi.org/10.1016/j.carbpol.2022.119563

Norfarhana, A. S., Ilyas, R. A., Ngadi, N., Othman, M. H. D., Misenan, M. S. M., & Norrrahim, M. N. F. (2024a). Revolutionizing lignocellulosic biomass: A review of harnessing the power of ionic liquids for sustainable utilization and extraction. *International Journal of Biological Macromolecules, 256*, 128256. https://doi.org/10.1016/j.ijbiomac.2023.128256

Norfarhana, A. S., Khoo, P. S., Ilyas, R. A., Ab Hamid, N. H., Aisyah, H. A., Norrrahim, M. N. F., Knight, V. F., Rani, M. S. A., Septevani, A. A., Syafri, E., & Annamalai, P. K. (2024b). Exploring of cellulose nanocrystals from lignocellulosic sources as a powerful adsorbent for wastewater remediation. *Journal of Polymers and the Environment*. https://doi.org/10.1007/s10924-024-03227-3

Nurazzi, N. M., Harussani, M. M., Aisyah, H. A., Ilyas, R. A., Norrrahim, M. N. F., Khalina, A., & Abdullah, N. (2021). Treatments of natural fiber as reinforcement in polymer composites—A short review. *Functional Composites and Structures, 3*(2), 024002. https://doi.org/10.1088/2631-6331/abff36

Okasha, A. Y., & Ibrahim, H. G. (2010). Phenol removal from aqueous systems by sorption of using some local waste materials. *EJEAFChe, 9*, 796–807.

Oladzad, S., Fallah, N., Mahboubi, A., Afsham, N., & Taherzadeh, M. J. (2021). Date fruit processing waste and approaches to its valorization: A review. *Bioresource Technology, 340*, 125625.

Oushabi, A., Sair, S., Abboud, Y., Tanane, O., & Bouari, A. El. (2017a). An experimental investigation on morphological, mechanical and thermal properties of date palm particles reinforced polyurethane composites as new ecological insulating materials in building. *Case Studies in Construction Materials, 7*, 128–137. https://doi.org/10.1016/j.cscm.2017.06.002

Oushabi, A., Sair, S., Oudrhiri Hassani, F., Abboud, Y., Tanane, O., & El Bouari, A. (2017b). The effect of alkali treatment on mechanical, morphological and thermal properties of date palm fibers (DPFs): Study of the interface of DPF–Polyurethane composite. *South African Journal of Chemical Engineering, 23*, 116–123. https://doi.org/10.1016/j.sajce.2017.04.005

Pradeep, P., Sang-Eun, O., & Chulkyoon, M. (2016). Microwave-assisted pre-treatment technologies for the conversion of lignocellulosic biomass to sugars and ethanol: A review. *Carbon Letters, 17*(1), 1–10. https://doi.org/10.5714/CL.2016.17.1.001

Puligundla, P., Oh, S.-E., & Mok, C. (2016). Microwave-assisted pre-treatment technologies for the conversion of lignocellulosic biomass to sugars and ethanol: A review. *Carbon Letters, 17*(1), 1–10. https://doi.org/10.5714/cl.2016.17.1.001

Putra, M. D., Abasaeed, A. E., & Al-Zahrani, S. M. (2020). Prospective production of fructose and single cell protein from date palm waste. *Electronic Journal of Biotechnology, 48*, 46–52. https://doi.org/10.1016/j.ejbt.2020.09.007

Quintero, J. A., Cardona, C. A., Felix, E., Moncada, J., Sánchez, Ó. J., & Gutiérrez, L. F. (2012). Techno-economic analysis of bioethanol production in Africa: Tanzania case. *Energy, 48*(1), 442–454.

Raghavi, S., Sindhu, R., Binod, P., Gnansounou, E., & Pandey, A. (2016). Development of a novel sequential pre-treatment strategy for the production of bioethanol from sugarcane trash. *Bioresource Technology, 199*, 202–210.

Raja, A. M., Khalaf, N. H., & Alkubaisy, S. A. (2021). Utilization of date palm waste compost as substitute for peat moss. *IOP Conference Series: Earth and Environmental Science, 904*(1), 012041.

Saba, N., Alothman, O. Y., Almutairi, Z., Jawaid, M., & Ghori, W. (2019). Date palm reinforced epoxy composites: Tensile, impact and morphological properties. *Journal of Materials Research and Technology, 8*(5), 3959–3969. https://doi.org/10.1016/j.jmrt.2019.07.004

Saba, N., Mohammad, F., Pervaiz, M., Jawaid, M., Alothman, O. Y., & Sain, M. (2017). Mechanical, morphological and structural properties of cellulose nanofibers reinforced epoxy composites. *International Journal of Biological Macromolecules, 97*, 190–200. https://doi.org/10.1016/j.ijbiomac.2017.01.029

Saba, N., Paridah, M. T., Abdan, K., & Ibrahim, N. A. (2016a). Effect of oil palm nano filler on mechanical and morphological properties of kenaf reinforced epoxy composites. *Construction and Building Materials, 123*, 15–26. https://doi.org/10.1016/j.conbuildmat.2016.06.131

Saba, N., Tahir, P. M., Abdan, K., & Ibrahim, N. A. (2016b). Fabrication of epoxy nanocomposites from oil palm nano filler: Mechanical and morphological properties. *BioResources, 11*(3), 7721–7736. https://doi.org/10.15376/biores.11.3.7721-7736

Sabalpara, A. N., Kharadi, V. B., Patel, V. J., Patel, H. G., & Patel, N. V. (2019). A review on *Phoenix dactylifera* Linn. (Date Palm): Cultivation, uses and medicinal properties. *Journal of Drug Delivery and Therapeutics, 9*(4), 413–420.

Saheb, D. N., & Jog, J. P. (1999). Natural fiber polymer composites: A review. *Advances in Polymer Technology*, *18*(4), 351–363. https://doi.org/10.1002/(SICI)1098-2329(199924)18:4<351::AID-ADV6>3.0.CO;2-X

Saini, J. K., Saini, R., & Tewari, L. (2014). Lignocellulosic agriculture wastes as biomass feedstocks for second-generation bioethanol production: Concepts and recent developments. *3 Biotech*, *5*(4), 337–353. https://doi.org/10.1007/s13205-014-0246-5

Sait, H. H., Hussain, A., Salema, A. A., & Ani, F. N. (2012). Pyrolysis and combustion kinetics of date palm biomass using thermogravimetric analysis. *Bioresource Technology*, *118*, 382–389. https://doi.org/10.1016/j.biortech.2012.04.081

Saleh, M. A., Al Haron, M. H., Saleh, A. A., & Farag, M. (2017). Fatigue behavior and life prediction of biodegradable composites of starch reinforced with date palm fibers. *International Journal of Fatigue*, *103*, 216–222. https://doi.org/10.1016/j.ijfatigue.2017.06.005

Salem, I. B., Saleh, M. B., Iqbal, J., El Gamal, M., & Hameed, S. (2021). Date palm waste pyrolysis into biochar for carbon dioxide adsorption. *Energy Reports*, *7*, 152–159.

Sallon, S., Cherif, E., Chabrillange, N., Solowey, E., Gros-Balthazard, M., Ivorra, S., Terral, J.-F., Egli, M., & Aberlenc, F. (2020). Origins and insights into the historic Judean date palm based on genetic analysis of germinated ancient seeds and morphometric studies. *Science Advances*, *6*(6), eaax0384.

Sanjay, M. R., Madhu, P., Jawaid, M., Senthamaraikannan, P., Senthil, S., & Pradeep, S. (2018). Characterization and properties of natural fiber polymer composites: A comprehensive review. *Journal of Cleaner Production*, *172*, 566–581. https://doi.org/10.1016/j.jclepro.2017.10.101

Sbiai, A., Maazouz, A., Fleury, E., Sautereau, H., Kaddami, H., & Einstein, A. (2010). Oxidized palm fiber composites. *BioResources*, *5*(2), 672–689.

Sgriccia, N., Hawley, M. C., & Misra, M. (2008). Characterization of natural fiber surfaces and natural fiber composites. *Composites Part A: Applied Science and Manufacturing*, *39*(10), 1632–1637. https://doi.org/10.1016/j.compositesa.2008.07.007

Sizirici, B., Fseha, Y. H., Yildiz, I., Delclos, T., & Khaleel, A. (2021). The effect of pyrolysis temperature and feedstock on date palm waste derived biochar to remove single and multi-metals in aqueous solutions. *Sustainable Environment Research*, *31*, 1–16.

Suriani, M. J., Radzi, F. S. M., Ilyas, R. A., Petrů, M., Sapuan, S. M., & Ruzaidi, C. M. (2021). Flammability, tensile, and morphological properties of oil palm empty fruit bunches fiber/pet yarn-reinforced epoxy fire retardant hybrid polymer composites. *Polymers*, *13*(8), 1282. https://doi.org/10.3390/polym13081282

Syafri, E., Jamaluddin, Sari, N. H., Mahardika, M., Amanda, P., & Ilyas, R. A. (2022). Isolation and characterization of cellulose nanofibers from Agave gigantea by chemical-mechanical treatment. *International Journal of Biological Macromolecules*, *200*(December 2021), 25–33. https://doi.org/10.1016/j.ijbiomac.2021.12.111

Taban, E., Amininasab, S., Soltani, P., Berardi, U., Abdi, D. D., & Samaei, S. E. (2021). Use of date palm waste fibers as sound absorption material. *Journal of Building Engineering*, *41*, 102752. https://doi.org/10.1016/j.jobe.2021.102752

Tarabi, N., Mousazadeh, H., Jafari, A., & Taghizadeh-Tameh, J. (2016). Evaluation of properties of bast fiber extracted from Calotropis (Millkweed) by a new decorticator machine and manual methods. *Industrial Crops and Products*, *83*, 545–550. https://doi.org/10.1016/j.indcrop.2015.12.071

Tayade, P. R., Sapkal, V. S., Sapkal, R. S., Deshmukh, S. K., Rode, C. V, Shinde, V. M., & Kanade, G. S. (2012). A method to minimize the global warming and environmental pollution. *Journal of Environmental Science and Engineering*, *54*(2), 287–293.

Usman, A. R. A., Abduljabbar, A., Vithanage, M., Ok, Y. S., Ahmad, M., Ahmad, M., Elfaki, J., Abdulazeem, S. S., & Al-Wabel, M. I. (2015). Biochar production from date palm waste: Charring temperature induced changes in composition and surface chemistry. *Journal of Analytical and Applied Pyrolysis*, *115*, 392–400. https://doi.org/10.1016/j.jaap.2015.08.016

Vohra, M. (2020). Treatment of gaseous ammonia emissions using date palm pits based granular activated carbon. *International Journal of Environmental Research and Public Health*, *17*(5). https://doi.org/10.3390/ijerph17051519

Watson, L. & Dallwitz, M. J. (1992 onwards). The Families of Flowering Plants: Descriptions, Illustrations, Identification, and Information Retrieval. Version: 1st June 2007. delta-intkey.com.

White, J. E., Catallo, W. J., & Legendre, B. L. (2011). Biomass pyrolysis kinetics: A comparative critical review with relevant agricultural residue case studies. *Journal of Analytical and Applied Pyrolysis*, *91*(1), 1–33. https://doi.org/10.1016/j.jaap.2011.01.004

Yao, Y., Gao, B., Fang, J., Zhang, M., Chen, H., Zhou, Y., Creamer, A. E., Sun, Y., & Yang, L. (2014). Characterization and environmental applications of clay–biochar composites. *Chemical Engineering Journal*, *242*, 136–143. https://doi.org/10.1016/j.cej.2013.12.062

Yuan, J.-H., Xu, R.-K., & Zhang, H. (2011). The forms of alkalis in the biochar produced from crop residues at different temperatures. *Bioresource Technology*, *102*(3), 3488–3497. doi:10.1016/j.biortech.2010.11.018

Yusoff, M. Z. M., Salit, M. S., Ismail, N., & Wirawan, R. (2010). Mechanical properties of short random oil palm fiber reinforced epoxy composites. *Sains Malaysiana*, *39*(1), 87–92.

Zafar, S. (2016). Biomass Potential of Date Palm Wastes. Agriculture, Biomass Energy, Middle East, Renewable Energy, Waste Management. Retrieved from https://www.ecomena.org/biomass-date-palm-wastes/

Zaid, A., & de Wet, P. F. (2002). Date Palm Cultivation. Food and Agriculture Organization of the United Nations. Retrieved from https://www.fao.org/4/y4360e/y4360e00.htm

4 Sago Waste Management

4.1 INTRODUCTION

Sago, as shown in Figure 4.1, a term from Javanese, refers to the starch-containing pith of the *Metroxylon sagu* plant. The word "metro" signifies the parenchyma of the pith, while "xylon" refers to the xylem (Singhal et al., 2008). The sago palm stands out as a superior starch crop, surpassing cassava, rice, and corn, with its impressive starch production capacity ranging from 20 to 25 tons per hectare per year (Lai et al., 2013). Within Malaysia, the Mukah division in Sarawak boasts the most extensive sago palm plantations, making it a prominent global exporter of sago starch. Annually, Sarawak exports over 40,000 tons of sago starch to various countries, including Peninsular Malaysia (Lai et al., 2013).

Nevertheless, as worldwide demand increases, the export volume is anticipated to escalate, resulting in higher waste generation. The primary types of waste generated by the sago industry encompass sago hampas, bark, and wastewater. These waste materials predominantly consist of

FIGURE 4.1 Sago palms (*Metroxylon sagu*) in Sarawak, Malaysia (Singhal et al., 2008).

DOI: 10.1201/9781003472209-6

starch and lignocellulosic compounds, namely, lignin, cellulose, and hemicellulose. Despite their high value, sago wastes have been largely overlooked and underutilized as products with enhanced value. Furthermore, improper disposal practices, such as discharging untreated waste into nearby rivers currently employed for these waste materials, pose detrimental effects on the environment.

Sago pith waste (SPW), as shown in Figure 4.2, locally known as sago hampas, is a fibrous residue abundantly containing sago starch. It is a by-product of the sago starch extraction process, remaining after most starch has been extracted from the rasped pith of the sago palm (Singhal et al., 2008). SPW consists of a significant proportion (60–80 wt%) of sago starch granules entrapped within a matrix of lignocellulosic fibers. Manufacturers have observed that approximately 1 ton of SPW is generated for every ton of sago starch produced (on a dry weight basis). Regrettably, manufacturers have not effectively utilized SPW and often combined it with wastewater, ultimately discharging it into rivers. According to statistics from the Malaysia Statistics Department, Malaysia produced 52,000 tons of sago starch in 2011, a 64% increase compared to 2001 (Lai et al., 2013).

Consequently, a similar quantity of SPW was discarded into rivers. This prevailing practice has led to two significant consequences. First, it wastes valuable raw material, as a substantial amount of starch is dumped into the rivers. The estimated quantity of starch disposed of as SPW in Sarawak alone accounted for almost half of the total annual imports of starch in Malaysia (40,000 tons) (Mohd et al., 2001). Second, it severely compromises water quality, threatening aquatic life. The microbiological degradation of the waste consumes dissolved oxygen in the water, leaving insufficient oxygen levels to sustain higher life forms.

Considering the actual starch content present in SPW, it is posited that this waste material can be effectively plasticized to form a biodegradable composite without the need for synthetic plastics as binders. In this context, the plasticized starch is the matrix that securely holds the reinforcing fibers. This unique characteristic sets SPW apart from other fibrous wastes. Consequently, the resultant plasticized SPW is believed to hold significant potential for producing various consumer products through different processing techniques such as compression, injection, and thermoforming. Previous research has explored similar approaches using cassava bagasse, a fibrous waste material derived from cassava starch extraction facilities. Cassava bagasse, characterized by its predominant starch content (40–60%) and fibrous composition (15–50%), was successfully plasticized and processed into natural composites (de Morais Teixeira et al., 2005). The same raw material was used to produce a composite resembling cardboard, wherein kraft paper was introduced as long fibers to enhance the final product's mechanical strength and water resistance properties (Matsui et al., 2004).

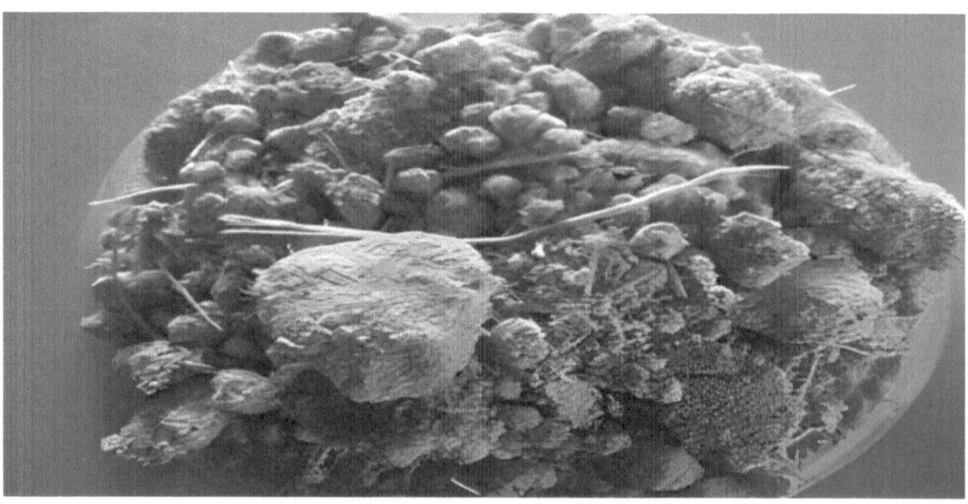

FIGURE 4.2 Sago pith waste (Lai et al., 2013).

If not effectively managed, this waste can have adverse environmental effects. Improper disposal of SPW, such as pouring it into rivers, can pollute water and harm aquatic ecosystems. The increased chemical oxygen demand (COD) and biological oxygen demand (BOD) caused by SPW decomposition can deplete oxygen levels in water bodies, resulting in the mortality of aquatic organisms. Moreover, the physical presence of SPW can impede water flow, negatively influencing the health of river ecosystems. To mitigate these environmental impacts and protect water resources, it is essential to identify sustainable solutions for managing SPW (Wahi et al., 2022).

Traditionally, both humans and animals ingest sago starch extracted from palms. However, approximately 50% of the sago palm is wasted during processing. This presents an opportunity to investigate potential applications for sago waste, such as converting it into products with added value, minimizing waste generation during sago processing, and promoting the sustainable use of sago palm resources. Appropriate waste management strategies and utilization of sago waste could reduce environmental impact and encourage sustainable sago palm cultivation and processing practices (Wahi et al., 2022). The improper processing and disposal of SPW, such as dumping it into rivers, is a significant problem related to sago waste management. This can result in environmental pollution, negatively impacting aquatic life due to elevated COD and BOD levels. Developing sustainable solutions for managing SPW to mitigate these ecological effects is essential. Implementing proper waste management practices, such as decomposition or anaerobic digestion, can facilitate the conversion of SPW into valuable products while minimizing its environmental impact. In addition, educating sago processing industries and local communities about the significance of responsible waste management practices can play a crucial role in reducing SPW-caused pollution (Wahi et al., 2022).

Researchers have recently investigated novel methods for transforming sago waste into products with added value. Extracting cellulose fibers from SPW, a highly abundant natural polymer, is an intriguing application. These cellulose fibers can be converted into cellulose nanofibrils (CNFs) and then aerogels composed of CNF. Unique characteristics of CNF aerogels include high porosity, large specific surface areas, and biodegradability. These characteristics render them eco-friendly and versatile materials with applications in various fields. CNF aerogels can be used for waste adsorption, oil/water separation, thermal insulation, and biomedical materials and as carriers for metal nanoparticles/metal oxides, among other applications. The use of SPW for the production of CNF aerogels not only reduces waste generation during sago processing but contributes to the development of sustainable and environmentally benign materials with various applications (Beh et al., 2020). Notably, although the extraction of cellulose fibers from SPW offers significant potential, there may be obstacles and constraints associated with their implementation.

This chapter highlights the importance of sago waste, particularly SPW, as an underutilized resource with significant potential for sustainable applications. The primary purpose of this chapter is to examine the environmental and economic issues arising from improper SPW disposal and to explore its potential as a valuable material for creating biodegradable products, biofuels, and other high-value chemicals. Sago waste, with its rich composition of starch and lignocellulosic fibers, presents a promising opportunity for transforming industrial by-products into eco-friendly alternatives. By emphasizing innovative uses of SPW, such as producing biodegradable composites and extracting cellulose fibers for various applications, this chapter aims to promote sustainable waste management practices in the sago industry and encourage the shift toward a circular economy, ultimately reducing environmental pollution and resource wastage.

4.2 BIOMASS FROM NATURAL FIBERS

Over centuries, biomass has served as a significant fuel source and holds immense importance as an energy provider, particularly in countries like India. In global rankings, India is the third-largest biomass energy producer following China and the United States (Jain & Sharma, 2011b). In recent years, biomass energy has garnered increased attention due to its potential to mitigate greenhouse gas emissions and offer renewable energy alternatives (Jain & Sharma, 2010). As the impending challenge of an insufficient energy supply looms, bioenergy emerges as a viable solution (Garg, 2012). Bioenergy, categorized as a form of renewable energy, is generated from biomass, encompassing materials that were biologically alive not long ago (Jain & Sharma, 2011d). Biomass contains organic matter that once existed in the environment. Bioenergy can be converted into various valuable forms, such as vehicle fuel, heat, and power.

Biomass encompasses any organic matter, ranging from wood and cereals to seaweed and animal waste, capable of being converted into usable energy. It is one of the earliest energy sources utilized by humans, second only to the sun. For millennia, burning wood has been a prevalent means of heating homes and cooking meals (Jain, 2019). Notable biomass sources include municipal solid waste, animal waste, forestry residues, farm waste, industrial waste, agricultural residue, and sewage sludge (Goyal et al., 2013; Jain & Sharma, 2011a,c). Figure 4.3 depicts the general classification of biomass into three main categories: crop-derived biomass, biomass derived from natural vegetation, and biomass derived from organic waste and residues. The category of organic waste and residues can be further subdivided into various subcategories, including floating plant waste, animal waste, forest residue, agricultural residue, municipal waste (both solid and liquid), and industrial waste. These classifications provide a comprehensive overview of the diverse sources from which biomass can be obtained. Crop-derived biomass encompasses biomass obtained from cultivated crops, while biomass derived from natural vegetation refers to biomass obtained from wild or naturally occurring plant sources. On the contrary, organic waste and residues include a wide range of biomass sources derived from various waste materials, such as plant waste, animal waste, and by-products generated from industrial and agricultural processes. This systematic categorization aids in understanding the different origins and types of biomass, facilitating effective utilization and management strategies for sustainable energy production and waste management.

Biomass encompasses a wide range of components with varying chemical compositions. Plant-derived leftovers, for example, consist mainly of cellulose, hemicellulose, and lignin, albeit in different proportions (Faiz Norrahim et al., 2021; Norfarhana et al., 2022, 2024a, b; Syafri et al., 2022). Conversely, cow dung is abundant in proteins, while grains primarily consist of starch. These distinct

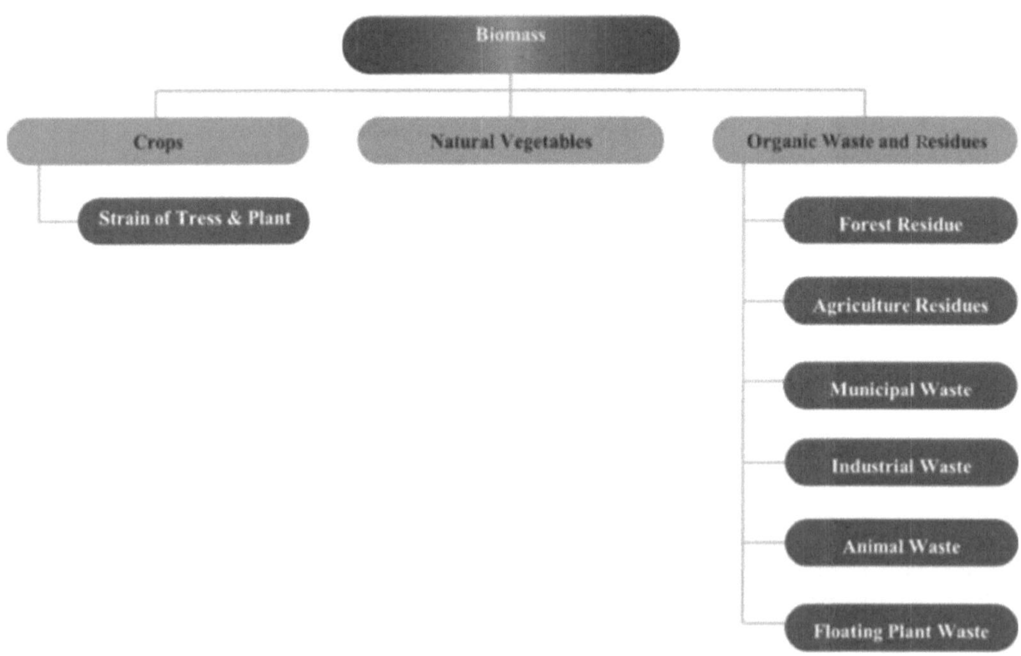

FIGURE 4.3 Classification of biomass (Chauhan & Pratap Singh, 2023).

chemical structures give rise to diverse chemical characteristics. Cellulose, a complex carbohydrate known as a polysaccharide, is a linear polymer composed of up to 10,000 monomeric units of D-glucose interconnected by β-1,4 glycosidic linkages. Hemicelluloses, another type of non-cellulosic polysaccharide, predominantly exhibit β-1,4 glycosidic links in their structure. Lignin, the second-most abundant organic component in plants, is a heterogeneous aromatic polymer formed during plant cell wall biosynthesis in conjunction with carbohydrates. Lignin comprises a family of complex organic polymers and serves as a vital structural component in the support tissues of vascular plants and certain algae. Cellulose, hemicellulose, and lignin possess significant applications in various industries, including biomedicine and chemicals. Notably, lignin can be utilized for biofuel production.

Cellulose, the primary constituent of mature plant fibers, plays a crucial role in determining the strength and purity of fiber biomass. With cellulose accounting for over 90% of the total biomass, it dramatically influences fibers' mechanical properties, rigidity, and durability. Higher cellulose content generally results in superior mechanical properties and durability, while compounds like callose, hemicellulose, or lignin can negatively impact fiber strength and quality. Researchers have been focused on developing techniques and strategies to modify and optimize cellulose content and quality in fiber biomass. Various approaches, such as chemical treatments, enzymatic processes, and genetic engineering, have been explored to enhance cellulose content, structure, and properties. These modifications aim to improve fiber strength, functionality, and durability, making them suitable for applications in textiles, composites, and building materials (Hu et al., 2022).

Biomass-based materials sourced from plants and agricultural waste are gaining attention for their abundant supply, cost-effectiveness, and favorable mechanical properties. These materials, with their natural fibers and structures, offer versatility in various applications. By modifying the chemical groups, biomass-based materials can be functionalized to enhance specific properties like strength, durability, and water resistance, making them suitable for packaging, textiles, composites, and building materials. Their use provides advantages over conventional materials by reducing environmental impact and promoting resource efficiency through renewable resources and agricultural waste utilization. Additionally, biomass-based materials offer cost savings through local sourcing and utilizing low-cost or waste materials. Researchers are focused on optimizing cellulose content, structure, and properties in fiber biomass, as cellulose is the critical component affecting strength and purity. Techniques such as chemical treatments, enzymatic processes, and genetic engineering are being explored to enhance fiber strength, functionality, and durability for textiles, composites, and building materials applications. Developing biomass-based materials as sustainable alternatives is gaining momentum in various industries (Ye et al., 2023).

4.3 ENVIRONMENTAL EFFECT FROM BIOMASS

The heavy reliance on fossil fuels in current energy production has resulted in notable environmental pollution and the emission of greenhouse gases (Sharma et al., 2020). Finding alternative and renewable fuels with minimal emissions is of paramount importance in tackling these challenges (Chia et al., 2018). Lignocellulosic biomass, which consists of cellulose, hemicellulose, and lignin, presents a sustainable and plentiful source of energy that enables cleaner and more environmentally friendly power generation (Hoang et al., 2021a). The pre-treatment of biomass is essential in addressing the natural resistance of lignocellulosic materials, enhancing their susceptibility to chemical and enzymatic reactions for efficient breakdown (Nurazzi et al., 2021). Various pretreatment techniques, including physical, hydrothermolysis, chemical catalysts, physicochemical, and biological approaches, can be utilized to optimize the productivity and yield of biomass conversion. Integration of multiple processes helps overcome technical challenges and improves the efficiency of the final product. Acid-based pre-treatment is commonly favored due to its effectiveness in breaking down lignocellulosic structures. However, this method has drawbacks, such as the generation of inhibitory compounds and the discharge of acidic effluents, which can have adverse effects on the environment and human health (Hoang et al., 2021b). Figure 4.4 provides a visual

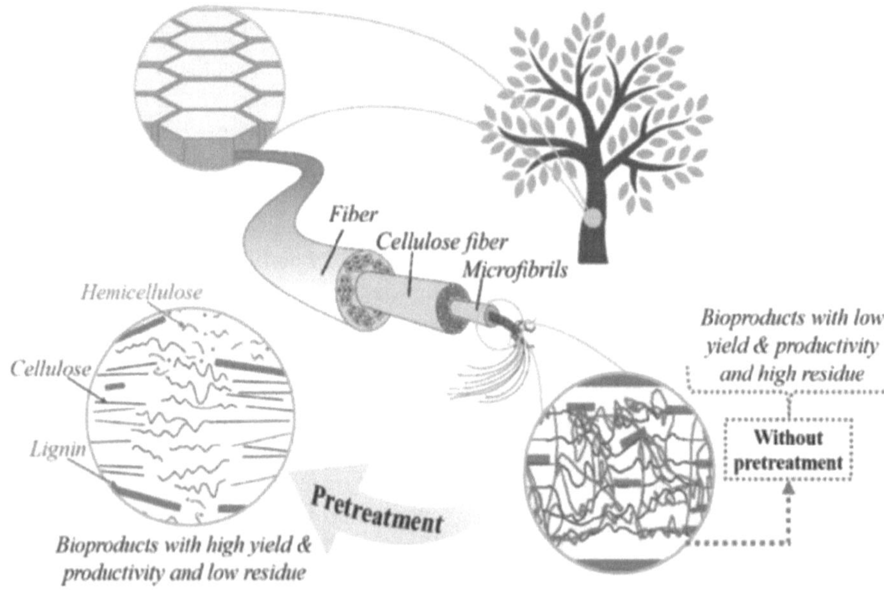

FIGURE 4.4 Differences in the pre-treatment effects of lignocellulosic biomass on bioproducts (Hoang, et al., 2021a).

representation of the fundamental distinctions between biomass that has undergone pre-treatment and biomass that has not. It highlights how these differences influence the production process of biofuels, specifically in terms of yield, productivity, and residue.

Biomass, as a renewable energy source, offers significant potential for reducing greenhouse gas emissions. When derived sustainably, biomass provides a sustainable solution for waste management and contributes to mitigating climate change. It has a neutral CO_2 conversion, balancing the carbon dioxide emitted during combustion with the carbon dioxide being absorbed during biomass growth. This minimizes the release of carbon dioxide into the atmosphere, helping to regulate global temperatures. Furthermore, biomass can be utilized to produce various valuable products, including biofuels, sorbents, fertilizers, construction materials, and minerals. This versatile resource not only contributes to resource conservation by reducing reliance on non-renewable fuels but also has the ability to reduce harmful emissions and capture and store toxic components. Overall, biomass offers environmental benefits through its potential to reduce greenhouse gas emissions, promote waste management, and provide a sustainable source of valuable products (Vassilev et al., 2015).

In conclusion, biomass exhibits varied compositions that are influenced by variables such as biomass resources, plant species, and development conditions. It offers numerous benefits as a renewable energy source, notably its CO_2-neutral conversion that does not contribute to global warming. Biomass can produce various valuable products while reducing the demand for fossil fuels and mitigating harmful emissions. Utilizing biomass as a versatile and abundant resource, we can contribute to developing sustainable energy solutions and preserving the environment.

4.4 AVAILABILITY OF SAGO WASTE

The starchy palm pith is called "sago" in Javanese. The term "Metro" stands for "Xylon," which refers to the plant's parenchyma of pith and is referred to scientifically as *M. sagu*. With an annual yield of 20–25 tons of starch per hectare, the sago palm is recognized for its remarkable production potential. This surpasses the amount of starch produced by other crops like cassava, rice, or maize

(Amin et al., 2019). The state of Sarawak in Malaysia, which exports more than 40,000 tons of sago starch annually to several Asian nations, relies heavily on the sago starch business as a significant source of income. These numbers are anticipated to rise since starch output and export value have steadily increased from 15.0% to 20.0% per year (Amin et al., 2019). A significant amount of residual solid debris, including bark, is produced during starch extraction from the sago palm trunk. Unfortunately, this material is frequently burned or dumped into adjacent waterways for disposal. As a result, the sago starch sector has complex waste management problems that could result in environmental contamination and health dangers. Addressing these waste management challenges and the rising demand for sago starch has become crucial (Amin et al., 2019).

For comparison, for every ton of sago flour produced, around 0.75 tons of sago bark (SB) waste is created. This trash provides a sizable prospective supply of biowaste that might be successfully utilized, given the anticipated yearly production of 32,000 tons of SB in Sarawak alone (Wahi et al., 2022). In conclusion, "sago" primarily refers to the starch-containing palm pith known technically as *M. sago*. With rising starch production and export volumes, the sago starch sector in Sarawak contributes significantly to the local economy. But handling sago waste, such as bark and hampas, has emerged as a critical environmental concern. We may prevent ecological contamination, lower health hazards, and effectively use sago waste as a valuable resource by considering sustainable options for sago waste management.

4.4.1 Physical Properties of Sago Waste

Sago waste exhibits distinct physical properties that aid in its identification and processing in industrial applications. It is easily recognizable with a color ranging from white to light brown, and its fibrous texture, attributed to leftover starch, provides tactile cues. Table 4.1 lists the physical properties of sago waste. The moisture content of sago waste, typically between 60% and 80%, influences its handling and processing characteristics. Understanding and controlling the moisture content is crucial for determining the suitable treatment and utilization of sago waste in various industries. Overall, the physical properties, including color, texture, and moisture content, play a significant role in the identification and industrial processing of sago waste (Tenriawaru et al., 2018).

4.4.2 Chemical Properties of Sago Waste

The chemical properties of substances, determined by their atomic and molecular structures, play a crucial role in understanding their behavior and reactions. Table 4.2 lists the chemical properties of sago waste. Sago waste, a complex mixture mainly consisting of starch, lignin, cellulose, and hemicellulose, exhibits varying chemical compositions depending on processing techniques and production

TABLE 4.1
Physical Properties of Sago Waste (Tenriawaru et al., 2018)

Physical Properties	Value
Specific density	0.45
Water adsorption	45%
Bulk density	1270 kg/m^3
Moisture content	33.74%
Acid resistance	No discoloration
Alkaline resistance	No discoloration
Compression strength	64.09 N/mm^2

TABLE 4.2
Chemical Properties of Sago Waste (Amin et al., 2019)

Composition (%)	Sago Hampas	Sago Bark
Starch	49.5	–
Cellulose	20.2	44.1
Hemicellulose	7.9	21.1
Lignin	3.3	23.3

factors. The presence of cellulose, alongside starch and lignin, enables sago waste to be utilized in various industries. Understanding the chemical properties of sago waste is essential for exploring its industrial applications and developing sustainable strategies for its use. This knowledge can lead to the extraction of valuable components like cellulose for biofuels and biodegradable materials. Additionally, comprehending the chemical properties of sago waste contributes to efficient waste management practices and promotes the sustainable utilization of sago palm resources (Amin et al., 2019).

4.4.3 Morphological Properties of Sago Waste

The morphological properties of sago residue, derived from sago starch production, are key in determining its potential applications. Factors such as the origin, age, and processing methods of the sago palm influence its unique morphology. Sago residue is primarily composed of lignocellulosic materials, including cellulose, hemicellulose, and lignin. Cellulose fibers provide structural integrity and rigidity, with their lengths, widths, and orientations influencing the overall texture and mechanical properties of sago waste. Hemicellulose surrounds cellulose fibers, offering flexibility and adhesion, while lignin fills spaces between fibers, enhancing support and rigidity. Understanding the morphological properties of sago residue is essential for determining its suitability for applications such as composite materials and paper production. The fiber characteristics also affect mechanical strength, permeability, and absorbency. Additionally, the morphological properties influence the behavior of sago waste during processing and manufacturing processes, impacting flow properties and packing density. Optimizing these properties can contribute to sustainable and value-added applications, enhancing the economic and environmental viability of sago waste (Khairunnisa et al., 2021). Table 4.3 provides the morphological properties of sago waste. Overall, the morphological properties of sago waste can influence its potential applications as a feedstock for energy production, as well as its potential use in other applications such as adsorption, insulation, and rheology modifiers (Khairunnisa et al., 2021).

4.4.4 Thermal Properties of Sago Waste

Thermal properties are those properties of a material which is related to its conductivity of heat. In other words, these are the properties that are exhibited by a material when the heat is passed through it. Table 4.4 lists the thermal properties of sago waste. These properties are important for understanding how materials transfer, store, and dissipate heat and can influence their performance in a wide range of applications. Therefore, the thermal properties of sago waste are important to consider when evaluating its potential uses and applications.

In addition, specific heat capacity is an important thermal property that is the amount of heat energy required to raise the temperature of a substance by a certain amount. Moreover, the composition, moisture content, and density of sago waste can all affect its specific heat capacity. Organic materials, on average, have lower specific heat capacity than metals and inorganic compounds. Therefore,

TABLE 4.3
Morphological Properties of Sago Waste (Khairunnisa et al., 2021)

Morphological Properties	Description
Particle size	Sago waste was composed of irregularly shaped particles ranging in size from a few microns to several millimeters. The particle size can affect the bulk density, flowability, and packing behavior of the material.
Porosity	Sago waste is highly porous, with pore sizes ranging from a few nanometers to several micrometers. This property can influence the material's water absorption capacity, gas permeability, and thermal insulation properties.
Surface area	Sago waste has a relatively large surface area due to its irregular particle shape and high porosity. This can enhance its adsorption properties and make it useful in applications such as water treatment and pollution control.
Chemical composition	Sago waste is primarily composed of cellulose, hemicellulose, and lignin. The relative proportions of these components can vary depending on the processing method used to extract sago starch. This can affect the material's chemical and physical properties, including its reactivity, strength, and durability.
Density	The density of sago waste is relatively low due to its porous and fibrous structure, which can influence its thermal and insulation properties.

TABLE 4.4
Thermal Properties of Sago Waste Thermal Properties of Sago Waste (Siruru et al., 2022)

Thermal Properties	Description
Thermal conductivity	The thermal conductivity of sago waste is relatively low, ranging from 0.07 to 0.14 W/mK, which makes it a good insulator.
Ash content	Sago waste has a high ash content, ranging from 12% to 18%, which can affect its combustion properties.
Ignition temperature	The ignition temperature of sago waste is relatively high, ranging from 300°C to 350°C, which can make it difficult to ignite.
Pyrolysis	Sago waste can undergo pyrolysis, a process that involves heating the material in the absence of oxygen to produce biochar, bio-oil, and syngas.

there is no exact value for the specific heat capacity of sago waste (Nurmalasari et al., 2022). On the contrary, thermal stability refers to the capacity of a material to resist high temperatures without substantial deterioration or structural changes. Overall, sago waste has the potential to be used as a feedstock for energy production through combustion or pyrolysis. Its low thermal conductivity also makes it a good insulator, which could have potential applications in building materials. However, the high ash content and difficulty in igniting may limit its use in certain applications (Siruru et al., 2022).

4.4.5 Caloric Calorific Value of Sago Waste

"Caloric" and "calorific" are terms used interchangeably to refer to the energy content of a substance, typically food or fuel. The caloric or calorific value represents the amount of energy that can be obtained from a given amount of the substance through combustion or metabolism. Sago waste's caloric or calorific value is determined by several parameters, including moisture content, chemical makeup, and processing methods. However, sago waste has an average calorific content of 3,800–4,500 kcal/

kg (Sahupala, 2022). Despite having a lower caloric value than fossil fuels, sago waste can be used as a biomass fuel for heat and power generation. Sago waste can be burned in boilers or furnaces to produce steam, which can then be used for electricity generation or other industrial processes (Siruru et al., 2022). The heat generated by the combustion of sago waste can also be used for drying or in thermal applications. In addition, sago waste may be transformed into biofuels by a variety of techniques, including pyrolysis and fermentation. These techniques can convert the organic components of sago waste into liquid biofuels or biogas, which can be utilized as transportation fuels or as a source of heat and energy. It is crucial to remember that the moisture concentration of sago waste might impact its caloric value. Higher moisture content decreases waste material energy content because energy is required to evaporate excess water during combustion. To maximize the caloric value of sago waste, suitable drying or moisture reduction processes are frequently used. Sago waste is a by-product of the sago business, which extracts starch from the sago palm tree. The waste consists primarily of fibrous material, such as bark, wood chips, and leaves, which are often burned to create heat or fed to animals.

It is crucial to note that the calorific value of sago waste varies depending on the kind and grade of waste, as well as the techniques employed for collecting, storage, and processing. As a result, it is advised that waste be properly analyzed to verify its actual calorific content before using it as fuel or feed (Hammado & Utomo, 2018). In addition, the caloric value is significant because it gives vital information about a substance's energy content, allowing for efficient energy generation, fuel selection, waste management, carbon footprint calculations, and economic analyses. We may make educated decisions to optimize energy production, limit environmental consequences, and encourage sustainable energy practices by knowing and utilizing caloric value.

4.5 AGRO-WASTE MANAGEMENT

Agro-waste management refers to the processes and strategies used to handle and dispose of agricultural waste in an environmentally sustainable manner (El-Ramady et al., 2022). Agricultural waste includes organic matter such as crop residues, animal manure, and food processing waste, which can pose a significant environmental hazard if not managed properly (El-Ramady et al., 2022). Effective agro-waste management aims to reduce the negative impact of agricultural waste on the environment and human health, while also generating economic benefits by using waste products for other purposes such as energy production, composting, and animal feed. Some common agro-waste management practices include composting, anaerobic digestion, biomass conversion, and recycling. These methods can help reduce greenhouse gas emissions, improve soil health, and prevent water pollution caused by agricultural waste.

Sago waste, also known as tapioca waste, is the residue left over from the processing of sago starch, which is extracted from the trunk of the sago palm. During sago starch extraction, a large amount of sago tree bark would be separated, and wastewater mixed with fiber waste would be discharged daily. Therefore, the management and processing of the by-products (sago waste) is a key factor in sustainable sago agroindustry (Rasyid et al., 2020). There are a few methods of agro-waste management such as compost, pulp and paper, and biocomposite materials (Matsagar & Wu, 2022).

4.5.1 Compost

Composting is a natural process that turns organic waste into compost, which is a nutrient-rich soil amendment. Sago trash, which is a by-product of the production of sago starch from the trunk of the sago palm, may be composted successfully to provide high-quality organic fertilizer. Sago waste has various properties that make it appropriate for composting. To begin with, it is high in organic matter, particularly cellulose, hemicellulose, lignin, and other plant-derived substances. These elements contribute a useful carbon source to the composting process. Second, the fibrous and spongy nature of sago waste allows for air movement within the compost pile. Adequate aeration is critical for sustaining the activity of aerobic bacteria that decompose. Finally, sago waste includes moisture, which is necessary for the compost pile's appropriate moisture content. Sago waste should be shredded or cut into smaller pieces to assist decomposition before beginning the composting process. To achieve a balanced carbon-to-nitrogen ratio, it is advantageous to combine the waste with nitrogen-rich resources such as green plants or manure. This ratio is critical for microbial activity and organic matter decomposition.

Microorganisms such as bacteria, fungi, and actinomycetes break down the organic elements in sago waste during composting. A range of biological and chemical reactions occur throughout the breakdown process. Heat is created as microorganisms consume the carbon-rich components of sago waste, resulting in a rise in temperature within the compost pile. Proper composting process management, such as turning the pile regularly, ensures that the temperature stays within the optimal range of 50–65°C to encourage effective decomposition and destroy possible pathogens and weed seeds. Composting sago waste has various advantages. In the beginning, it aids in the diversion of organic waste from landfills, hence lowering environmental strain and greenhouse gas emissions. Sago waste may be converted into a profitable resource rather than being wasted or burned. Second, the compost made from sago waste is a high-nutrient soil supplement. It improves soil structure, increases water retention capacity, and provides plants with critical nutrients. Compost's organic content contributes to soil fertility, microbial activity, and soil erosion reduction.

Composting also aids in the stabilization of sago waste. The volume and weight of the garbage decrease as it decomposes. The resultant compost is easier to handle, transport, and apply to agricultural areas or gardens as a consequence. In addition, composting sago waste helps promote circular economy practices and sustainable waste management. Composting minimizes reliance on synthetic fertilizers and encourages the use of organic, locally accessible inputs by completing the loop and using trash as a useful resource. Composting sago waste is a helpful and environmentally friendly practice. Sago waste is appropriate for composting due to its high organic matter content, fibrous texture, and moisture level. The resultant compost is rich in nutrients, increases soil fertility, and promotes sustainable agriculture.

4.5.2 Pulp and Paper

The pulp and paper business consumes a lot of natural resources, especially wood fibers. Alternative raw materials, on the contrary, can be investigated to lessen reliance on traditional wood-based sources. Sago waste is an example of a by-product of the manufacturing of sago starch from the trunk of the sago palm, which has the potential and option for use in the pulp and paper sector (Figure 4.5) (Rasyid et al., 2020). Sago waste has properties that make it useful for pulp and paper manufacture. It is high in cellulose, hemicellulose, and lignin, all of which are important components of plant fibers. Because of the fibrous structure of sago waste, cellulose fibers may be extracted and utilized as a raw material in paper manufacture. Then, sago waste is plentiful and easily accessible in areas where sago palm growth is common, making it a potentially sustainable resource for the business. Various pulping procedures may be used to obtain fibers from sago waste. To separate the fibers from the waste material, mechanical pulping procedures such as grinding, or refining

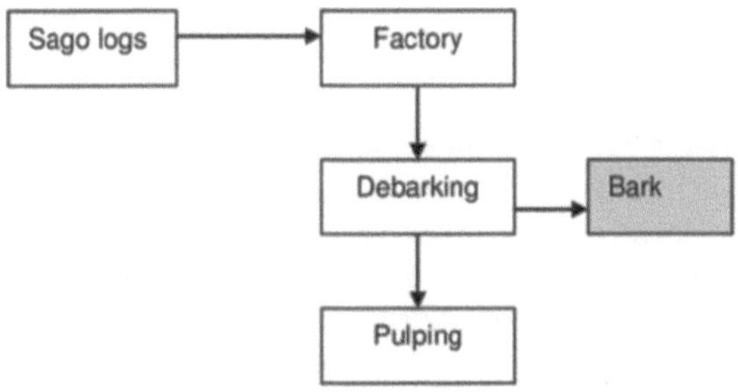

FIGURE 4.5 Flow paper production from sago waste (Awg-Adeni et al., 2010).

can be utilized. To dissolve the lignin and separate the fibers, chemical pulping procedures such as alkaline or kraft pulping can be used. The pulping technique used is determined by the desired properties of the finished paper product as well as the unique processing requirements.

There are various advantages to using sago waste fibers in the manufacturing of pulp and paper. For starters, it diversifies the raw material base, lowering reliance on wood supplies. This can help with sustainable resource management and the environmental impact of deforestation. The industry may help forest and biodiversity conservation by integrating sago waste fibers. Sago waste fibers can also improve the quality of paper. They help to boost tensile strength, which is important for applications that require durability and tearing resistance. The fibers can increase paper opacity, allowing for higher print quality and less ink show-through. Next, the fibrous nature of sago waste helps promote drainage in the papermaking process, resulting in improved sheet formation and uniformity.

Sago waste can be used as a filler or additive in papermaking in addition to fiber production. Sago waste's high cellulose content and fibrous structure make it ideal for improving paper qualities such as brightness, opacity, and ink absorption. The waste may be ground into a fine powder and used in the papermaking process, minimizing the requirement for chemical additives and improving the finished product's sustainability. Furthermore, sago waste may be utilized to make specialized papers. Sago waste's specific features, such as texture, color, and absorbency, may be used to make papers with distinguishing characteristics. These specialty papers have a wide range of uses, including food packaging, cosmetics, and crafts. The use of sago waste in the pulp and paper sector helps to promote sustainable waste management practices. The industry decreases trash output and supports circular economy concepts by transforming a waste item into a useful resource. On the contrary, using sago waste fibers or additives in paper production might assist decrease the industry's carbon footprint by decreasing the consumption of energy-intensive wood pulping operations.

Sago waste holds significant potential for use in the pulp and paper industry. Its fibrous nature, cellulose content, and unique properties make it suitable for fiber production and as a filler or additive in papermaking. By utilizing sago waste, the industry can reduce its reliance on traditional wood fibers, enhance paper properties, and promote sustainable practices. Incorporating sago waste in the pulp and paper production process contributes to waste reduction, resource conservation, and the development of environmentally friendly paper products.

4.5.3 Biocomposite Materials

Biocomposite materials, which combine a matrix material with natural fibers, have gained significant attention as sustainable alternatives to conventional synthetic composites (Halimatul et al., 2019; Jumaidin et al., 2019a,b, 2021; Nazrin et al., 2021). Sago waste, the by-product generated during the processing of sago starch from the trunk of the sago palm (*M. sagu*), offers great potential as a natural fiber reinforcement in biocomposite materials. Here, we explore the characteristics and applications of sago waste in biocomposites. The beneficial qualities of sago waste make it appropriate for biocomposite applications. Mechanical strength and stiffness are provided by the fibrous structure of sago waste, which is made of cellulose, hemicellulose, and lignin. The waste material is also low in density, making it suitable for lightweight applications. Then, sago waste is plentiful, renewable, and biodegradable, providing environmental benefits over traditional synthetic fibers. Sago waste fibers are often mixed with a matrix material, such as biodegradable polymers like polylactic acid, polyhydroxyalkanoates, or starch-based polymers, to make biocomposites. The matrix material connects the fibers, ensuring structural integrity and improving overall mechanical qualities.

Several stages are involved in the production of sago waste biocomposites. The sago waste fibers must first be removed and treated. Fibers can be obtained mechanically, such as grinding or refining, or chemically, by removing non-fibrous components. Once the fibers have been separated, they are combined with the matrix material, which is often in the form of pellets or granules, and then processed using processes such as extrusion, compression molding, or injection molding to create the final biocomposite product. The use of sago waste fibers in biocomposite products has various advantages. For starters, the fibers improve the composite's mechanical qualities, such as tensile strength, flexural strength, and impact resistance. Sago waste biocomposites are thus suited for a variety of applications requiring structural stability, such as automobile components, construction materials, and consumer items.

The applications for sago waste biocomposites are numerous. They may be used to make packaging materials such as trays, containers, and disposable cutlery. Biocomposites may also be used in the building industry to make panels, boards, and structural components (Alias et al., 2021; Asyraf et al., 2022; Azlin et al., 2022; Mazani et al., 2019; Suriani et al., 2021). Sago waste biocomposites are also used in the car sector for interior elements including door panels and dashboard components. Moreover, sago waste biocomposites may be used to make consumer goods including furniture, toys, and household items. Sago waste fibers' natural and distinctive aesthetics might add value to these items, appealing to eco-conscious consumers.

Sago waste, specifically the residue and pulp, shows promise in producing composites with enhanced mechanical properties. The strength and stiffness of sago fibers make them suitable for reinforcing composites. By incorporating sago waste into composite materials, not only can the value of the by-product be maximized, but it also contributes to resource conservation and waste reduction. The improved mechanical properties of sago-based composites make them suitable for applications in the construction, packaging, automotive, and furniture industries (Supu & Jaya, 2018). From Table 4.5, the mechanical behavior of a composite is significantly influenced by the quality of bonding at the fiber–matrix interface and the interconnectivity of phases within the composite. Additionally, the composite's particulate size and concentration have an effect. In general, larger particles and higher concentrations result in stronger adhesion and a greater Young's modulus. On the contrary, the presence of voids can reduce the composite's mechanical strength, and the particle size influences the occurrence of voids. The weight fraction and particulate size are crucial factors that affect the composite material's overall mechanical properties. The chemical and mechanical bonding between particles and matrix is crucial for stress transfer and determining the composite's mechanical behavior. The matrix content has a direct influence on the mechanical properties of sago waste particleboard, with an increase in matrix content resulting in a decrease in Young's modulus (Chiang et al., 2015).

As shown in Table 4.5, the tensile strengths for 0.6 mm and 1.18 mm particulates decreased consistently with a reduction in weight fraction. The weight fraction had a significant effect on tensile

TABLE 4.5
Mechanical Properties of Particleboard (Chiang et al., 2015)

Particleboard	Weight Fraction of Sago (wt%)	Young Modulus (MPa)	Tensile Strength (MPa)	Impact Strength (J/m^2)
0.60 mm	90	1,181.75	5.23	22,112.83
	85	911.3	4.04	31,737.04
	80	883.71	3.91	46,554.13
	75	87.10	3.87	49,214.15
	70	766.7	3.39	51,181.56
1.18 mm	90	2,492.20	9.98	23,648.56
	85	2,203.05	8.82	28,574.16
	80	1,913.8	7.66	29,482.03
	75	989.75	3.96	39,862.92
	70	826.12	3.31	31,131.06
2.00 mm	90	1,141.10	4.17	22,985.67
	85	1,266.15	4.62	38,767.18
	80	1,998.50	7.30	36,032.39
	75	1,869.95	6.83	27,232.82
	70	1,741.40	6.36	20,416.52

performance, with 90 wt% of 1.18 mm particles exhibiting the greatest strength (Chiang et al., 2015). The interfacial adhesion between particles and matrix substantially affected tensile strength and stress transfer. Increased matrix weakened the material due to poor adhesion, while decreased particle fraction led to debonding and load transfer issues. Excessive matrix and moisture content reduced bonding and tensile strength. A homogeneous dispersion and enhanced performance required strong chemical bonds between the particles and matrix. Sago granules introduced a new interfacial region that influences stress transfer, while networked urea formaldehyde (UF) particleboard dispersion improved tensile strength.

A variety of factors influence the mechanical properties of sago/UF particleboard. The particle loading, fiber–matrix composition, morphology, dispersion, and adhesion all influence Young's modulus, which assesses stiffness. Reducing the particleboard's weight fraction increases Young's modulus as a result of increased space between particles and diminished fiber–matrix bonding. Compared to other particle diameters, 1.18 mm particles have a greater Young's modulus due to improved fiber–matrix adhesion and compaction, whereas excessive matrix content decreases modulus (Chiang et al., 2015). The adhesion between particles and the matrix, hydrophilicity, and weight fraction all affect tensile strength. Increasing matrix content diminishes tensile strength, whereas decreasing particle weight fraction impedes stress transfer. The addition of sago granules to the UF matrix affects stress transfer and tensile strength. Impact resistance is dependent on particle size and weight fraction, with an excessive matrix content decreasing impact resistance. Optimal weight fractions varied depending on particle size, with 0.6 mm particles exhibiting greater impact resistance (Chiang et al., 2015). Due to poor interfacial adhesion and high particle content, mechanical properties degrade beyond the optimal weight fraction. Important roles are played by particle size, chemical composition (cellulose, hemicellulose, lignin), and natural fiber characteristics in determining the mechanical properties and impact resistance of particleboard (Chiang et al., 2015).

Sago waste fibers hold immense potential as a reinforcement material in biocomposites. Their fibrous structure, low density, and sustainable attributes make them suitable for various applications. Incorporating sago waste fibers in biocomposite materials can enhance mechanical properties, promote sustainability, and reduce the environmental impact of conventional composites. The utilization of sago waste in biocomposites contributes to waste valorization, resource efficiency, and the development of environmentally friendly materials for a wide range of industries.

4.5.4 Bio-Coal

Bio-coal, also known as torrefied biomass or black pellets, is a type of solid fuel that is produced by heating biomass to high temperatures in the absence of oxygen. This process, known as torrefaction, results in a product that has a similar appearance, energy content, and combustion characteristics as coal. Sago trash, a by-product of the processing of sago starch from the trunk of the sago palm, has the potential to be used as a feedstock in the manufacturing of bio-coal (Figure 4.6). This renewable energy source has a number of advantages and applications (Siruru et al., 2022). The biomass used to produce bio-coal can come from a variety of sources, including wood chips, agricultural waste, and other organic materials. During the torrefaction process, the biomass is heated to temperatures between 200°C and 300°C, causing it to release moisture and volatile compounds. The remaining solid material is then densified into pellets or briquettes. Bio-coal has several advantages over traditional coal, including lower greenhouse gas emissions, reduced ash content, and higher energy density. It can be used in existing coal-fired power plants without requiring significant modifications, making it a potential low-carbon alternative to coal (Siruru et al., 2022).

In addition to its use in power generation, bio-coal is being explored for use in other applications, such as industrial heating and cooking. It can also be used as a replacement for charcoal in the production of iron and steel, which could help reduce carbon emissions in the steel industry. However, the production of bio-coal still requires energy and can result in emissions of greenhouse gases, particularly if the biomass used is not sustainably sourced. As such, the sustainability and

FIGURE 4.6 Bio-coal (Cheng et al., 2020).

environmental impact of bio-coal production needs to be carefully considered (Siruru et al., 2022). In addition, bio-coal has superior fuel characteristics compared to raw biomass. It has a lower moisture content, which improves combustion efficiency and minimizes greenhouse gas emissions. Torrefaction also increases the carbon-to-oxygen ratio in bio-coal, resulting in fewer volatile organic compound emissions and a higher heating value. Bio-coal has also improved in grindability and hydrophobicity, making it more appropriate for co-firing with coal in current coal-fired power plants. Bio-coal has physical qualities comparable to conventional coal, enabling simpler integration and use in existing infrastructure without needing large changes (Siruru et al., 2022).

Bio-coal made from sago waste has a wide range of uses. It may be used directly in place of coal in power generation, reducing greenhouse gas emissions and reliance on fossil fuels. Bio-coal's high energy density and enhanced fuel qualities make it a desirable option for businesses that require heat and steam, such as cement, steel, and paper. Furthermore, bio-coal may be utilized as a renewable and sustainable fuel source for domestic heating and cooking, particularly in locations where power or clean cooking fuels are scarce. Bio-coal combustion emits less emissions than typical biomass fuels, contributing to enhanced indoor air quality and lowering health risks linked with indoor air pollution (Siruru et al., 2022).

Hence, sago waste has the potential to be used as a feedstock in the manufacturing of bio-coal. The torrefaction process converts sago waste into a solid fuel with increased energy density and fuel characteristics. The use of bio-coal has various advantages, including increased energy content, lower emissions, and greater grindability. Sago waste bio-coal may be utilized in a variety of applications, including power generation, industrial operations, and home heating. We can contribute to renewable energy development, decrease environmental effects, and promote sustainable energy solutions by utilizing the energy potential of sago waste through bio-coal manufacturing.

4.5.5 Biochar

Biochar is a carbon-rich solid substance made through pyrolysis, which involves heating biomass in the absence of oxygen. Sago waste has the potential to be used as a feedstock for biochar synthesis (Figure 4.7) (Rambli et al., 2019). The process is typically carried out in a specialized pyrolysis kiln or retort. In addition, sago waste has various properties that make it ideal for biochar synthesis. It has a high concentration of lignocellulosic components, including cellulose, hemicellulose, and lignin, which contribute to its carbon content (Rambli et al., 2019). After that, sago waste is abundant, particularly in areas where sago palm agriculture is prominent, making it a sustainable and abundant feedstock for biochar synthesis. To create biochar from sago waste, the biomass is roasted

FIGURE 4.7 Biochar (Lehmann, 2007).

to high temperatures (usually between 300°C and 700°C) in an oxygen-limited environment. The organic components of sago waste are degraded using heat, eliminating volatile compounds and leaving behind biochar, a stable carbon-rich material.

Biochar generated from sago waste offers a variety of benefits and applications. For starters, it may be used as a soil supplement in agriculture. When added to soil, biochar improves soil structure, water retention, and nutrient availability. Its porous structure encourages the growth of beneficial microorganisms, which boosts soil fertility and microbial activity. Biochar is also a long-term carbon sink, storing carbon in the soil and reducing greenhouse gas emissions. Biochar has the potential to be used for energy as well. It may be used to produce renewable energy through combustion or gasification processes. Biochar burning can generate heat or electricity, providing a renewable energy source (Rambli et al., 2018). Biochar can also be utilized as a feedstock for the manufacture of biofuels or as a precursor for the manufacturing of activated carbon (Rambli et al., 2018).

Biochar production can be carried out at various scales, from small-scale systems for home use to large-scale industrial systems. The properties of the resulting biochar can be influenced by various factors, including the type of feedstock, the pyrolysis temperature and duration, and the post-treatment of the biochar. It is important to note that biochar production should be carried out using sustainable and environmentally friendly practices to ensure that the process does not have negative impacts on the environment or local communities (Rambli et al., 2019). Furthermore, sago waste biochar may be utilized in waste management and odor control. Because of its adsorption capabilities, it is useful in absorbing and decreasing odorous substances from organic waste streams or wastewater treatment plants. When utilized as a soil cover or liner material, biochar can also assist minimize leachate and odor emissions from landfills (Rambli et al., 2018).

Sago waste has a lot of potential as a feedstock for making biochar. Pyrolysis can produce a carbon-rich substance with many advantages from the thermal conversion of sago waste. In agriculture, biochar generated from sago waste can improve soil fertility, water retention, and nutrient availability. It can also help with carbon sequestration, environmental cleanup, and the development of renewable energy. Sago waste-derived biochar has a wide range of uses, including waste management and odor control. We can encourage sustainable agriculture, environmental remediation, and waste management practices by using the potential of sago waste for biochar production.

4.5.6 Bio-Bricks

The sago waste has a lot of potential for use in making bio-bricks, among other things. In place of conventional clay bricks, there are building materials called bio-bricks that are manufactured from biomass (Figure 4.8). Bio-bricks are building materials made from natural agro-waste, which have a lower carbon footprint compared to standard building materials and are cost-effective to produce (Gaurav et al., 2022). They are made from a variety of organic materials, including sawdust,

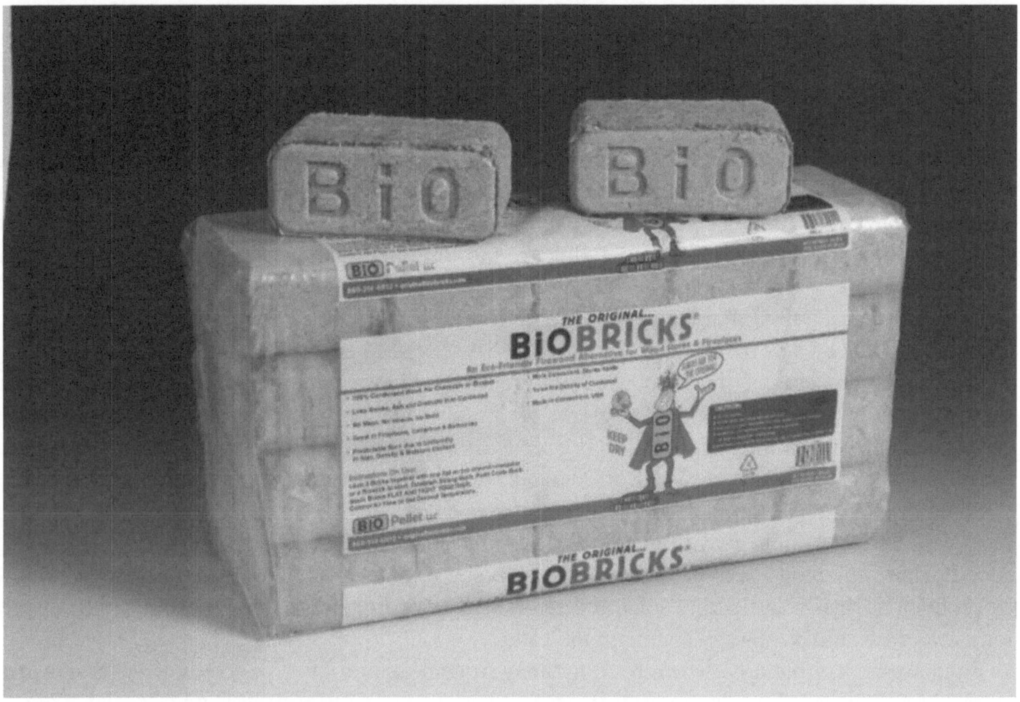

FIGURE 4.8 Bio-bricks (Shetty et al., 2008).

wood chips, and agricultural residues (Hadi Izaan et al., 2022). The properties of the bio-bricks, such as their density and heating value, can be influenced by various factors, including the type of feedstock, the compression pressure, and the size and shape of the bio-bricks. Bio-bricks are a sustainable alternative to traditional fossil fuels, as they are made from renewable biomass sources and emit fewer pollutants when burned. They can be used for a variety of heating applications, including residential heating, commercial heating, and industrial processes. It is important to note that bio-bricks production should be carried out using sustainable and environmentally friendly practices to ensure that the process does not have negative impacts on the environment or local communities. Additionally, proper handling and storage of bio-bricks is important to prevent fires and other safety hazards.

Sago waste bio-bricks offer several environmental benefits. By utilizing this waste material, they reduce the amount of sago waste sent to landfills or discarded into aquatic environments, thereby preventing potential contamination and promoting more sustainable waste management practices. (Hadi Izaan et al., 2022). Then, the creation of bio-bricks contributes to the conservation of natural resources by lowering the demand for clay and sand, both of which are often utilized in traditional brick making (Hadi Izaan et al., 2022). These bio-bricks are also thermally efficient. Sago waste's cellulose composition functions as insulation, offering superior heat resistance than clay bricks. This property qualifies them for the construction of energy-efficient buildings that demand less energy for heating and cooling. Additionally, sago waste bio-bricks may be constructed to absorb more water, assisting in rainfall management. This characteristic enables the bricks to gently absorb and release water, lowering the danger of floods and boosting water conservation. Sago waste bio-bricks are a more environmentally friendly and sustainable alternative to regular clay bricks. Their manufacture aids in the reduction of waste, conservation of natural resources, and reduction of carbon emissions. Sago waste bio-bricks have the potential to become a viable option for eco-friendly construction practices, supporting a greener and more sustainable future with additional study and development.

4.5.7 Biohydrogen

Biohydrogen production from sago waste is a potential sustainable energy generation method. Sago waste, which is generated during the processing of sago palm plants, is a rich source of organic materials that may be transformed into biohydrogen via a biological process known as dark fermentation (Sinha & Pandey, 2011). In this section, we will focus on the biohydrogen generation potential of sago waste and its relevance in the renewable energy environment (Jenol et al., 2013). Figure 4.9 shows the flow production of biohydrogen from sago waste. Sago waste contains a variety of organic substances like starch, cellulose, hemicellulose, and lignin. Through fermentation, microbes may break down these complex chemical molecules into simpler ones such as sugars. Dark fermentation is a form of fermentation that happens in the absence of light and oxygen, allowing hydrogen gas (H_2) to be produced as a by-product (Jenol et al., 2013).

A two-step process is commonly used to harness the biohydrogen potential of sago waste. The sago waste is processed in the first phase to break down the complex polysaccharides and lignocellulosic components into fermentable sugars. Milling, acid or alkaline hydrolysis, or enzymatic hydrolysis are examples of physical or chemical pre-treatment operations. The idea is to reduce complex organic stuff into simple sugars that microbes can readily absorb. Then, in the fermentation tank, a consortium of hydrogen-producing microbes, such as *Clostridium* species, is introduced. These microbes use the carbohydrates in sago waste to create hydrogen gas, as well as by-products such as volatile fatty acids and alcohols. The hydrogen gas may be caught and collected for use as a sustainable energy source in the future.

Sago waste has plenty of benefits for biohydrogen generation. For example, it is a widely available and plentiful waste material in areas where sago palm farming is widespread. We may decrease environmental contamination related to disposal by using it. Second, sago waste is abundant in starch, which may readily be turned into sugars for fermentation. Starch is a particularly promising substrate for biohydrogen production because it produces more hydrogen gas than other biomass feedstocks. In addition, sago waste provides a carbon-neutral or even carbon-negative method for hydrogen generation. When hydrogen produced by biohydrogen methods is used as a fuel, it emits only water vapor, resulting in no net carbon dioxide emissions (Jenol et al., 2013). This helps reduce greenhouse gas emissions and address climate change issues. However, there are several issues that must be solved in the generation of sago waste biohydrogen. The pretreatment procedure for converting complex carbs into fermentable sugars may be time-consuming and expensive. Furthermore, getting high hydrogen yields requires maintaining appropriate fermentation parameters such as temperature, pH, and substrate concentration (Jenol et al., 2013).

Sago waste has a high potential for biohydrogen generation by dark fermentation. It provides an abundant and renewable fuel for microbial activities that convert to hydrogen gas (Sinha & Pandey, 2011).

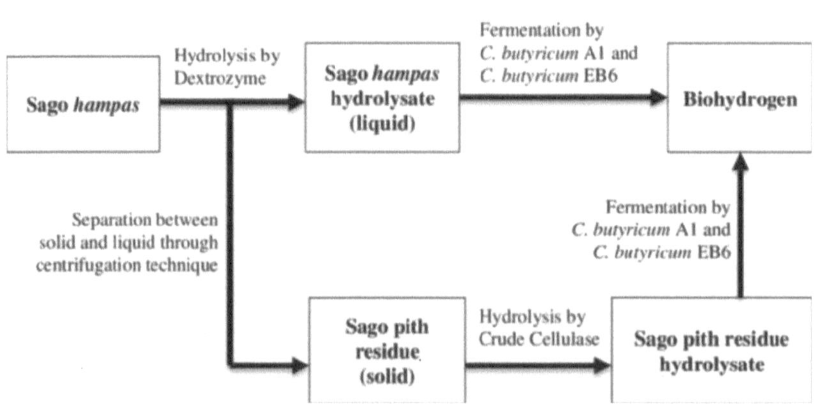

FIGURE 4.9 Flow production of biohydrogen from sago waste (Jenol et al., 2013).

Sago waste biohydrogen is an ecologically benign alternative to fossil fuels that can help accelerate the transition to a more sustainable energy future. Continued research and development in this sector is required to optimize the process and make large-scale biohydrogen generation from sago waste commercially viable.

4.5.8 Biomethane

Anaerobic digestion is a versatile process that holds great potential for producing biomethane, a renewable energy source. Sago waste, the residual material left after extracting sago starch, is particularly suitable for biomethane production due to its high organic content. During anaerobic digestion, microorganisms break down the organic compounds in sago waste, resulting in biogas rich in methane and carbon dioxide (Shi & Xu, 2023). Biomethane derived from sago waste has multiple valuable applications, such as a renewable transportation fuel and a source of heat and energy for various industries and residential uses. By utilizing sago waste for biomethane production, we can efficiently manage waste and contribute to a low-carbon economy, reducing greenhouse gas emissions and promoting renewable energy generation. In conclusion, anaerobic digestion of sago waste offers a promising solution for producing sustainable biomethane, contributing to a greener and more sustainable future (Shi & Xu, 2023).

4.5.9 Bioethanol

Developing renewable and sustainable fuels is crucial due to the scarcity of fossil fuels and environmental pollution (Balat, 2007). Bioethanol derived from biomass and bioenergy crops has emerged as a viable alternative to gasoline (Demirbas, 2011). While current large-scale bioethanol production primarily relies on sugar-containing substances and starch grains, their use may not be ideal due to their value as food resources (Balat et al., 2008). Lignocellulosic biomass, which includes agricultural, forestry, and industrial waste, holds significant potential for economically viable bioethanol production worldwide (Limayem & Ricke, 2012). Starchy lignocellulosic biomass, such as waste from starch processing factories and food and beverage industries, also shows promise as a feedstock for large-scale production in tropical regions.

SPW from the sago palm is a promising feedstock for bioethanol production due to its abundance and high content of starch, cellulose, hemicellulose, and lignin. Various methods, including enzymatic hydrolysis, simultaneous fermentation, and dilute acid hydrolysis, have been explored to efficiently convert the biomass into fermentable sugars and ethanol. Researchers have focused on non-edible sources like lignocellulosic biomass, which does not compete with food production, to address concerns about food shortages and fuel conflicts. Studies on sago hampas, primarily composed of lignocellulosic materials, have shown impressive results in ethanol production through enzymatic hydrolysis cycles and simultaneous co-saccharification and fermentation (Sc-SF) techniques (Amin et al., 2019). Another method, sequential saccharification and simultaneous fermentation, was proposed to achieve high ethanol yields from sago hampas. Dilute acid hydrolysis has also proven to be effective, producing a substantial number of fermentable sugars and achieving a higher yield compared to traditional enzymatic techniques. The use of sago hampas and lignocellulosic biomass as feedstocks for bioethanol production offers a sustainable and renewable energy solution. These alternative biomass resources can alleviate pressure on food crops and reduce environmental impacts. Additionally, bioethanol from sago hampas can be utilized as a renewable fuel, contributing to a greener and more sustainable energy sector (Amin et al., 2019). Overall, the exploration of alternative feedstocks and innovative techniques has shown promising results in maximizing ethanol yields and optimizing the bioethanol production process from sago hampas and lignocellulosic biomass. This progress signifies a significant step toward a more sustainable and diversified renewable energy future (Amin et al., 2019).

4.5.10 Biobutanol

The biotechnological production of biofuels, particularly biobutanol and acetone–butanol–ethanol (ABE), from sago hampas has gained significant attention. *Clostridium acetobutylicum* ATCC 824 can efficiently produce ABE from fermentable sugars derived from sago hampas enzymatic hydrolysis, making it a promising biofuel substitute. Additionally, microwave pyrolysis has been explored as a method to convert sago hampas into bio-oils with a high calorific value, making it suitable for use as a liquid fuel. This process not only offers an alternative energy source but also contributes to the sustainable management of agricultural waste (Amin et al., 2019).

Sago hampas fermentation with *Clostridium butyricum* A1 has demonstrated higher biohydrogen production yields compared to synthetic glucose, showcasing its potential as a renewable resource for clean energy production. Furthermore, sago hampas can be co-fired with other biomass residues to produce energy, and techno-economic analysis indicates its viability as a feedstock for combined heat and power systems, leading to reduced carbon emissions. The various applications of sago hampas in biofuel production highlight its potential as a valuable renewable energy resource. Utilizing sago hampas not only provides environmental benefits but also contributes to waste management and the development of a circular economy. Continued research and development in optimizing conversion processes and scaling up biofuel production from sago hampas will be essential to fully realize its potential as a sustainable energy source (Amin et al., 2019).

4.5.11 Natural Acid

Sago residue is rich in complex carbohydrates and can be utilized as a plentiful source for microbial fermentation to produce organic acids. This fermentation process involves introducing bacteria or yeast to the sago waste, which then metabolizes the carbohydrates to form organic acids. The production of organic acids from sago waste offers a sustainable approach to obtaining valuable compounds that have diverse industrial applications, such as acetic acid and citric acid. This not only contributes to waste valorization but also presents an opportunity to create economically valuable products (Loh et al., 1999). Table 4.6 lists all types of organic acids that can be obtained naturally. The fermentation process of sago waste produces some of these organic acids.

Immobilized-cell reactors (ICRs) have shown promise in achieving efficient organic acid production. These reactors confine microorganisms within a matrix or support material, allowing for repeated use and enhanced productivity. Studies have demonstrated that ICRs can convert a significant portion of glucose and xylose present in sago residue into organic acids within a short timeframe, making them a promising approach for sustainable organic acid production. Utilizing sago waste for organic acid production aligns with circular economy principles by transforming waste materials into valuable chemical compounds. This approach reduces dependency on petroleum-derived feedstocks and provides a more sustainable alternative to conventional organic acid production methods, which often rely on fossil fuels (Loh et al., 1999).

4.5.12 Bioelectricity

Bioelectricity produced by sago waste provides great promise as a renewable energy source. Sago waste, a by-product of sago processing, is high in organic materials, primarily cellulose and hemicellulose. This organic matter can be used to generate power by anaerobic digestion or burning, providing a sustainable and environmentally beneficial alternative to fossil fuels (Jenol et al., 2019). Anaerobic digestion is a method of producing bioelectricity from sago waste. The waste is deposited in a controlled environment where microorganisms break down the organic content and produce biogas as a by-product (Pandit et al., 2021). Biogas is mostly composed of methane and carbon dioxide, and it can be used to power a generator. This method not only provides renewable electricity but also yields digest, a valuable by-product that may be utilized as a nutrient-rich fertilizer.

TABLE 4.6
Type of Organic Acid (Loh et al., 1999)

Acid	Chemical Name	p*Ka*
Formic	Formic acid	3.75
Propionic	Acetic acid	4.76
Butyric	2-Propanoic acid	4.88
Lactic	Butanoic acid	3.83
Sorbic	2-Hydroxypropanoic acid	4.76
Fumaric	2,4-Hexandienoic acid	3.02
Malic	Hydroxybutanedioic acid	3.40
Tartaric	2,3-Dihydroxy-butanedioic acid	2.93
Citric	2-Hydroxy-1,2,3-propanetricarboxylic acid	3.13
Benzoic	Benzenecarboxylic acid	4.19

Another method is to directly burn sago trash to generate electricity. This process includes burning waste in a boiler to generate heat, which is then utilized to generate steam. Steam powers a turbine, which is linked to a generator, transforming mechanical energy into electrical energy. This procedure is similar to regular power plants in that it uses renewable biomass instead of fossil fuels (Jenol et al., 2019). There are many benefits to producing bioelectricity from sago waste. First and foremost, it reduces greenhouse gas emissions and aids in the fight against climate change (Wan et al., 2016). The release of methane, a powerful greenhouse gas, from decomposing sago waste can be reduced by using a waste product for energy production. Next, the production of bioelectricity decreases dependency on fossil fuels, improving energy security and decreasing the environmental problems associated with their extraction and burning (Wan et al., 2016).

4.6 POTENTIAL APPLICATION

Sago is made from the starch-rich pith of the tropical sago palm, a popular staple meal in many regions of the world. However, the manufacture of sago generates substantial volumes of waste, which might pose environmental concerns if not properly controlled. Recent research and innovation, fortunately, have shown possible applications for sago waste, presenting the opportunity for sustainable utilization and waste reduction (Rasyid et al., 2020). One interesting application for sago waste is bioenergy production. Sago waste includes a high concentration of cellulose and hemicellulose, which can be turned into biofuels like ethanol or biogas using biochemical or thermochemical processes (Rasyid et al., 2020). We may minimize our dependency on fossil fuels and cut greenhouse gas emissions by utilizing the energy component of sago waste, thus contributing to the fight against climate change. Sago waste can also be used as a feedstock for the synthesis of high-value biochemicals. Sago waste carbohydrates can be fermented to yield a variety of compounds, including organic acids, enzymes, and bioplastics (Rasyid et al., 2020). These chemicals have a wide range of industrial applications, including the production of food additives and pharmaceuticals, as well as the development of biodegradable packaging materials. We can establish a circular economy model that maximizes resource efficiency and minimizes trash output by extracting valuable biochemicals from sago waste.

Additionally, sago waste has the potential to be a nutrient-rich organic fertilizer. The waste contains significant levels of nitrogen, phosphorus, and potassium, all of which are necessary for plant growth (Rasyid et al., 2020). We can improve soil fertility, minimize the demand for chemical fertilizers, and encourage sustainable farming practices by composting or processing sago waste into organic fertilizers. Not only does this increase crop yields, but it also helps to avoid fertilizer runoff and subsequent water pollution. Another interesting application is using sago waste as a substrate for mushroom culture (Senghie et al., 2021). Because of its high carbohydrate content, sago waste makes an excellent medium for mushroom cultivation. We can develop edible and therapeutic mushrooms utilizing sago waste as a substrate, creating extra income prospects for farmers and supporting sustainable agriculture (Senghie et al., 2021).

The potential applications for sago waste are wide and varied. Sago waste offers several potentials for sustainable utilization, ranging from bioenergy production and biochemical synthesis to organic fertilizer and mushroom culture. We can contribute to the development of a more sustainable and circular economy by utilizing these potentials, which will not only reduce environmental pollution but will also produce economic value. Researchers, policymakers, and stakeholders must work together to encourage further study and implementation of these applications, assuring the effective utilization of sago waste and realizing its full potential.

4.7 CHALLENGES ASSOCIATED WITH PROCESSING AND COMMERCIALIZING

Even though sago waste has enormous potential for a variety of applications, numerous hurdles must be overcome to properly process and commercialize this waste product (Rasyid et al., 2020). These obstacles vary from scientific and logistical issues to economic and regulatory considerations, all of which necessitate careful consideration for the successful utilization of sago waste. The variation in the composition and qualities of sago waste is a serious challenge. The content of sago waste can vary based on a number of factors, including the sago palm species used, the processing method used, and the geographical region of production (Rasyid et al., 2020). This unpredictability makes standardization and consistent quality control during processing difficult. Innovative methods and procedures are required to properly treat and convert sago waste into valuable goods. Furthermore, the processing of sago waste requires the use of specialized equipment and infrastructure (Rasyid et al., 2020). Often, complicated and energy-intensive methods such as enzymatic hydrolysis, fermentation, or anaerobic digestion are used to convert sago waste into biofuels, biochemicals, or fertilizers. Creating the requisite processing facilities and purchasing the necessary equipment can be costly, especially for small-scale producers or regions with limited resources. To overcome these obstacles and enable broad processing and commercialization of sago waste, infrastructure development, technical competence, and financial backing are required.

Another challenge is the affordability and market demand for sago waste-based products (Rasyid et al., 2020). While sago waste can be turned into a variety of valuable commodities, their economic viability must be carefully assessed. Consider factors like as competition from current alternatives, market price changes, and manufacturing scalability. Moreover, establishing market demand for sago waste-based products may necessitate marketing initiatives and consumer education to highlight the environmental and social benefits of these alternatives (Rasyid et al., 2020). Furthermore, constitutional frameworks and policies play an important influence in shaping the commercialization of sago waste. A region's waste management legislation, bioenergy strategies, and agricultural guidelines can either help or hinder the processing and utilization of sago waste. Ensuring a favorable policy climate that supports sustainable waste utilization, provides incentives for innovation, and facilitates market access is critical for overcoming regulatory barriers and supporting commercialization initiatives. Developing sustainable and effective supply networks for sago waste can be challenging. Sago palm cultivation and processing are often concentrated in specific places that are geographically remote from prospective end consumers or processing facilities. Having a consistent supply of sago waste, minimizing transportation costs, and properly managing logistics are all important considerations in developing profitable supply chains. Collaboration among many stakeholders, including farmers, processors, and end users, is required to build well-organized and sustainable sago waste supply chains.

4.8 CONCLUSION

In conclusion, sago waste provides numerous prospects for sustainable utilization and waste reduction. This waste product can be transformed into important commodities such as biofuels, biochemicals, organic fertilizers, and substrates for mushroom growing using novel processing processes. These uses help create a circular economy by lowering dependency on fossil fuels, encouraging sustainable agriculture, and reducing environmental pollution. However, the processing and commercialization of sago waste are not without difficulties. The requirement for specialized infrastructure, cost-effectiveness, regulatory considerations, and the establishment of efficient supply chains are among the challenges that must be overcome. Collaboration among researchers, industry participants, policymakers, and financial institutions is critical to addressing these difficulties and creating a conducive ecosystem for sago waste utilization. Despite the constraints, the potential benefits of utilizing sago waste are significant. By seizing the opportunities presented by this waste product, we may not only reduce environmental impacts but also generate economic value, improve resource efficiency, and promote long-term growth. To fully realize the potential of sago waste and pave the road for a more sustainable future, we must continue to engage in research, innovation, and collaboration.

ACKNOWLEDGMENT

This chapter was written with Wong Poh Soon and Muhammad Ainul Naim Bin Mohd Asri.

REFERENCES

Alias, A. H., Norizan, M. N., Sabaruddin, F. A., Asyraf, M. R. M., Norrrahim, M. N. F., Ilyas, A. R., Kuzmin, A. M., Rayung, M., Shazleen, S. S., Nazrin, A., Sherwani, S. F. K., Harussani, M. M., Atikah, M. S. N., Ishak, M. R., Sapuan, S. M., & Khalina, A. (2021). Hybridization of MMT/lignocellulosic fiber reinforced polymer nanocomposites for structural applications: A review. *Coatings*, *11*(11), 1355. https://doi.org/10.3390/coatings11111355

Amin, N., Sabli, N., Izhar, S., & Yoshida, H. (2019). Science & Technology Sago wastes and its applications. *Pertanika Journal of Science & Technology*, *27*(4), 1841–1862.

Asyraf, M. R. M., Syamsir, A., Zahari, N. M., Supian, A. B. M., Ishak, M. R., Sapuan, S. M., Sharma, S., Rashedi, A., Razman, M. R., Zakaria, S. Z. S., Ilyas, R. A., & Rashid, M. Z. A. (2022). Product development of natural fibre-composites for various applications: Design for sustainability. *Polymers*, *14*(5), 920. https://doi.org/10.3390/polym14050920

Awg-Adeni, D., Abd-Aziz, S., Bujang, K., & Hassan, M. (2010). Bioconversion of sago residue into value added products. *African Journal of Biotechnology*, *9*(14), 2016–2021. https://www.ajol.info/index.php/ajb/article/view/78406

Azlin, M. N. M., Ilyas, R. A., Zuhri, M. Y. M., Sapuan, S. M., Harussani, M. M., Sharma, S., Nordin, A. H., Nurazzi, N. M., & Afiqah, A. N. (2022). 3D printing and shaping polymers, composites, and nanocomposites: A review. *Polymers*, *14*(1), 180. https://doi.org/10.3390/polym14010180

Balat, M. (2007). Global bio-fuel processing and production trends. *Energy Exploration & Exploitation*, *25(3)*, 195–218.

Balat, M., Balat, H., & Öz, C. (2008). Progress in bioethanol processing. *Progress in Energy and Combustion Science*, *34*(5), 551–573. https://doi.org/10.1016/j.pecs.2007.11.001

Beh, J. H., Lim, T. H., Lew, J. H., & Lai, J. C. (2020). Cellulose nanofibril-based aerogel derived from sago pith waste and its application on methylene blue removal. *International Journal of Biological Macromolecules*, *160*, 836–845. https://doi.org/10.1016/j.ijbiomac.2020.05.227

Chauhan, K., & Pratap Singh, V. (2023). Prospect of biomass to bioenergy in India: An overview. *Materials Today: Proceedings*. https://doi.org/10.1016/j.matpr.2023.01.419

Cheng, B. H., Huang, B. C., Zhang, R., Chen, Y. L., Jiang, S. F., Lu, Y., Zhang, X. S., Jiang, H., & Yu, H. Q. (2020). Bio-coal: A renewable and massively producible fuel from lignocellulosic biomass. *Science Advances*, *6*(1), 0748. https://doi.org/10.1126/sciadv.aay0748

Chia, S. R., Ong, H. C., Chew, K. W., Show, P. L., Phang, S. M., Ling, T. C., Nagarajan, D., Lee, D. J., & Chang, J. S. (2018). Sustainable approaches for algae utilisation in bioenergy production. *Renewable Energy*, *129*, 838–852. https://doi.org/10.1016/j.renene.2017.04.001

Chiang, T. C., Hamdan, S., & Shahrilosman, M. (2015). Mechanical properties of sago/urea formaldehyde particleboard affected by the weight fraction of sago. *MATEC Web of Conferences*, *27*. https://doi.org/10.1051/matecconf/20152701001

de Morais Teixeira, E., Da Róz, A. L., de Carvalho, A. J. F., & da Silva Curvelo, A. A. (2005). Preparation and characterisation of thermoplastic starches from cassava starch, cassava root and cassava bagasse. *Macromolecular Symposia*, *229*(1), 266–275. https://doi.org/10.1002/masy.200551133

Demirbas, A. (2011). Competitive liquid biofuels from biomass. *Applied Energy*, *88*(1), 17–28. https://doi.org/10.1016/j.apenergy.2010.07.016

El-Ramady, H., Brevik, E. C., Bayoumi, Y., Shalaby, T. A., El-Mahrouk, M. E., Taha, N., Elbasiouny, H., Elbehiry, F., Amer, M., Abdalla, N., Prokisch, J., Solberg, S. Ø., & Ling, W. (2022). An overview of agro-waste management in light of the water-energy-waste nexus. *Sustainability*, *14*, 15717. https://doi.org/10.3390/su142315717

Faiz Norrrahim, M. N., Ilyas, R. A., Nurazzi, N. M., Asmal Rani, M. S., Nur Atikah, M. S., & Shazleen, S. S. (2021). Chemical pretreatment of lignocellulosic biomass for the production of bioproducts: An overview. *Applied Science and Engineering Progress*, 1–18. https://doi.org/10.14416/j.asep.2021.07.004

Garg, P. (2012). Energy scenario and vision 2020 in India. *Journal of Sustainable Energy & Environment*, *3*(1), 7–17.

Gaurav, R., Vaibhav, Y., Gaurav, P., Dhiraj, D., & Mane, S. (2022). Bio- bricks. *International Journal of Advanced Research in Science, Communication and Technology*, 562–565. https://doi.org/10.48175/ijarsct-5604

Goyal, P., Sharma, M. P., & Jain, S. (2013). Optimization of transesterification of Jatropha curcas oil to biodiesel using response surface methodology and its adulteration with kerosene. *Journal of Materials and Environmental Science*, *4*(2), 277–284.

Hadi Izaan, I., Suraya Hani, A., Norhayati, A. W., Mohamad Hairi, O., Zalipah, J., Noor Azlina, A. H., Norhafizah, S., & Wimala, M. (2022). Preliminary study of sago fine waste as a sand replacement material for cement brick. *IOP Conference Series: Earth and Environmental Science*. https://doi.org/10.1088/1755-1315/1022/1/012052

Halimatul, M. J., Sapuan, S. M., Jawaid, M., Ishak, M. R., & Ilyas, R. A. (2019). Water absorption and water solubility properties of sago starch biopolymer composite films filled with sugar palm particles. *Polimery*, *64*(9), 27–35. https://doi.org/10.14314/polimery.2019.9.4

Hammado, N., & Utomo, S. (2018). Physicochemical characteristic of sago hampas and sago wastewater in Luwu Regency. *E3S Web of Conferences*, *73*. https://doi.org/10.1051/e3sconf/20187307007

Hoang, A. T., Nizetic, S., Ong, H. C., Chong, C. T., Atabani, A. E., & Pham, V. V. (2021a). Acid-based lignocellulosic biomass biorefinery for bioenergy production: Advantages, application constraints, and perspectives. *Journal of Environmental Management*, *296*. https://doi.org/10.1016/j.jenvman.2021.113194

Hoang, A. T., Nižetić, S., Ong, H. C., Mofijur, M., Ahmed, S. F., Ashok, B., Bui, V. T. V., & Chau, M. Q. (2021b). Insight into the recent advances of microwave pretreatment technologies for the conversion of lignocellulosic biomass into sustainable biofuel. *Chemosphere*, *281*. https://doi.org/10.1016/j.chemosphere.2021.130878

Hu, W., Gao, M., Xu, B., Wang, S., Wang, Y., & Zhou, Z. (2022). Co-occurring elevated temperature and drought stresses during cotton fiber thickening stage inhibit fiber biomass accumulation and cellulose synthesis. *Industrial Crops and Products*, *187*, 115348. https://doi.org/10.1016/J.INDCROP.2022.115348

Jain, S. (2019). The current and future perspectives of biofuels. In *Biomass, Biopolymer-Based Materials, and Bioenergy: Construction, Biomedical, and Other Industrial Applications* (pp. 495–517). https://doi.org/10.1016/B978-0-08-102426-3.00021-7

Jain, S., & Sharma, M. P. (2010). Prospects of biodiesel from Jatropha in India: A review. *Renewable and Sustainable Energy Reviews*, *14*(2), 763–771. https://doi.org/10.1016/j.rser.2009.10.005

Jain, S., & Sharma, M. P. (2011a). Study of oxidation stability of Jatropha curcas biodiesel/diesel blends. *International Journal of Energy and Environment*. *2*(3). www.IJEE.IEEFoundation.org

Jain, S., & Sharma, M. P. (2011b). Long term storage stability of Jatropha curcas biodiesel. *Energy*, *36*(8), 5409–5415. https://doi.org/10.1016/j.energy.2011.06.055

Jain, S., & Sharma, M. P. (2011c). Oxidation and thermal behavior of Jatropha curcas biodiesel influenced by antioxidants and metal contaminants. *International Journal of Engineering, Science and Technology*, *3*(4). www.ijest-ng.com

Jain, S., & Sharma, M. P. (2011d). Power generation from MSW of Haridwar city: A feasibility study. *Renewable and Sustainable Energy Reviews*, *15*(1), 69–90. https://doi.org/10.1016/j.rser.2010.09.007

Jenol, M. A., Ibrahim, M. F., Kamal Bahrin, E., Kim, S. W., & Abd-Aziz, S. (2019). Direct bioelectricity generation from sago hampas by Clostridium beijerinckii SR1 using microbial fuel cell. *Molecules*, *24*(13), 2397. https://doi.org/10.3390/molecules24132397

Jenol, M. A., Ibrahim, M. F., Lai Yee, P., Md Salleh, M., & Abd-Aziz, S. (2013). Sago biomass as a sustainable source for biohydrogen production by Clostridium butyricum A1. *BioResources*, *9*(1). https://doi.org/10.15376/biores.9.1.1007-1026

Jumaidin, R., Adam, N. W., Ilyas, R. A., Hussin, M. S. F., Taha, M. M., Mansor, M. R., Azlan, U. A. A., & Yob, M. S. (2019a). Water transport and physical properties of sugarcane bagasse fibre reinforced thermoplastic potato starch biocomposite. *Journal of Advanced Research in Fluid Mechanics and Thermal Sciences*, *61*(2), 273–281.

Jumaidin, R., Diah, N. A., Ilyas, R. A., Alamjuri, R. H., & Yusof, F. A. M. (2021). Processing and characterisation of banana leaf fibre reinforced thermoplastic cassava starch composites. *Polymers*, *13*(9), 1420. https://doi.org/10.3390/polym13091420

Jumaidin, R., Saidi, Z. A. S., Ilyas, R. A., Ahmad, M. N., Wahid, M. K., Yaakob, M. Y., Maidin, N. A., Rahman, M. H. A., & Osman, M. H. (2019b). Characteristics of cogon grass fibre reinforced thermoplastic cassava starch biocomposite: Water absorption and physical properties. *Journal of Advanced Research in Fluid Mechanics and Thermal Sciences, 62*(1), 43–52.

Khairunnisa, N., Fathi, M., Abd, S. M., Abdullah, A., Fhaizal, M., Bukhori, M., Wahi, R., & Zailani, M. A. (2021). Physicochemical properties of sago bark biochar and its potential as plant growth media (Sifat Fizikokimia Bio-Arang Sisa Kulit Sagu dan Potensinya Sebagai Media Pertumbuhan Tanaman). *Malaysian Journal of Analytical Sciences*, *25*, 622–636.

Lai, J. C., Rahman, W. A. W. A., & Toh, W. Y. (2013). Characterisation of sago pith waste and its composites. *Industrial Crops and Products*, *45*, 319–326. https://doi.org/10.1016/j.indcrop.2012.12.046

Lehmann, J. (2007). Bio-energy in the black. *Frontiers in Ecology and the Environment*, *5*(7), 381–387. https://doi.org/10.1890/1540-9295(2007)5[381:bitb]2.0.co;2

Limayem, A., & Ricke, S. C. (2012). Lignocellulosic biomass for bioethanol production: Current perspectives, potential issues and future prospects. *Progress in Energy and Combustion Science*, *38*(4), 449–467. https://doi.org/10.1016/j.pecs.2012.03.002

Loh, C. W., Fakhru'l-Razi, A., Hassan, M. A., & Karim, M. I. A. (1999). Production of organic acids from kitchen wastes. *Artificial Cells, Blood Substitutes, and Immobilization Biotechnology*, *27*(5–6), 455–459. https://doi.org/10.3109/10731199909117719

Matsagar, B. M., & Wu, K. C.-W. (2022). Agricultural waste-derived biochar for environmental management. In *Biochar in Agriculture for Achieving Sustainable Development Goals* (pp. 3–13). https://doi.org/10.1016/B978-0-323-85343-9.00026-4

Matsui, K. N., Larotonda, F. D. S., Paes, S. S., Luiz, D. B., Pires, A. T. N., & Laurindo, J. B. (2004). Cassava bagasse-Kraft paper composites: Analysis of influence of impregnation with starch acetate on tensile strength and water absorption properties. *Carbohydrate Polymers*, *55*(3), 237–243. https://doi.org/10.1016/J.CARBPOL.2003.07.007

Mazani, N., Sapuan, S. M., Sanyang, M. L., Atiqah, A., & Ilyas, R. A. (2019). Design and fabrication of a shoe shelf from kenaf fiber reinforced unsaturated polyester composites. In H. Ariffin, S. M. Sapuan, & M. A. Hassan (Eds.), *Lignocellulose for Future Bioeconomy* (1st ed., Issue 2000, pp. 315–332). Elsevier. https://doi.org/10.1016/B978-0-12-816354-2.00017-7

Mohd, A. M. Dos, Islam, Md. N., & Noor, B. Mohd. (2001). Enzymic extraction of native starch from sago (Metroxylon sagu) waste residue. *Starch - Stärke*, *53*(12), 639–643. https://doi.org/10.1002/1521-379X(200112)53:12<639::AID-STAR639>3.0.CO;2-2

Nazrin, A., Sapuan, S. M., Zuhri, M. Y. M., Tawakkal, I. S. M. A., & Ilyas, R. A. (2021). Water barrier and mechanical properties of sugar palm crystalline nanocellulose reinforced thermoplastic sugar palm starch (TPS)/poly(lactic acid) (PLA) blend bionanocomposites. *Nanotechnology Reviews*, *10*(1), 431–442. https://doi.org/10.1515/ntrev-2021-0033

Norfarhana, A. S., Ilyas, R. A., & Ngadi, N. (2022). A review of nanocellulose adsorptive membrane as multifunctional wastewater treatment. *Carbohydrate Polymers*, *291*, 119563. https://doi.org/10.1016/j.carbpol.2022.119563

Norfarhana, A. S., Ilyas, R. A., Ngadi, N., Othman, M. H. D., Misenan, M. S. M., & Norrrahim, M. N. F. (2024a). Revolutionizing lignocellulosic biomass: A review of harnessing the power of ionic liquids for sustainable utilization and extraction. *International Journal of Biological Macromolecules*, *256*, 128256. https://doi.org/10.1016/j.ijbiomac.2023.128256

Norfarhana, A. S., Khoo, P. S., Ilyas, R. A., Ab Hamid, N. H., Aisyah, H. A., Norrahim, M. N. F., Knight, V. F., Rani, M. S. A., Septevani, A. A., Syafri, E., & Annamalai, P. K. (2024b). Exploring of cellulose nanocrystals from lignocellulosic sources as a powerful adsorbent for wastewater remediation. *Journal of Polymers and the Environment*. https://doi.org/10.1007/s10924-024-03227-3

Nurazzi, N. M., Harussani, M. M., Aisyah, H. A., Ilyas, R. A., Norrahim, M. N. F., Khalina, A., & Abdullah, N. (2021). Treatments of natural fiber as reinforcement in polymer composites—A short review. *Functional Composites and Structures*, *3*(2), 024002. https://doi.org/10.1088/2631-6331/abff36

Nurmalasari, Lestari, S. J., Syamsuddin, M., Sukarti, & Alam, M. N. (2022). Thermal stability analysis of cellulose from sago fiber waste (Metroxylon sago). *EduChemia (Jurnal Kimia dan Pendidikan)*, *7*(2), 182. https://doi.org/10.30870/educhemia.v7i2.13491

Pandit, S., Savla, N., Sonawane, J. M., Sani, A. M., Gupta, P. K., Mathuriya, A. S., Rai, A. K., Jadhav, D. A., Jung, S. P., & Prasad, R. (2021). Agricultural waste and wastewater as feedstock for bioelectricity generation using microbial fuel cells: Recent advances. *Fermentation*, *7*(3), 169. https://doi.org/10.3390/fermentation7030169

Rambli, J., Azlina, W., Ab, W., Ghani, K., Amran, M., Salleh, M., & Khezri, R. (2019). Sago-derived biochar. *BioResources*, *14*(1), 1928–1940.

Rasyid, T. H., Kusumawaty, Y., & Hadi, S. (2020). The utilization of sago waste: Prospect and challenges. *IOP Conference Series: Earth and Environmental Science*, *415*(1), 012023. https://doi.org/10.1088/1755-1315/415/1/012023

Sahupala, P. (2022). The utilization of sago due waste as an alternative fuel. *Engineering and Technology Journal*, *07*(09). https://doi.org/10.47191/etj/v7i9.11

Senghie, H., Bolhassan, M. H., & Awg-Adeni, D. S. (2021). The effects of sago (Metroxylon sagu) bark and frond waste as substrates on the growth and yield of grey oyster mushrooms (Pleurotus sajor-caju). *Pertanika Journal of Tropical Agricultural Science*. https://doi.org/10.47836/pjtas.44.2.04

Sharma, S., Kundu, A., Basu, S., Shetti, N. P., & Aminabhavi, T. M. (2020). Sustainable environmental management and related biofuel technologies. *Journal of Environmental Management*, *273*. https://doi.org/10.1016/j.jenvman.2020.111096

Shetty, R. P., Endy, D., & Knight, T. F. (2008). Engineering BioBrick vectors from BioBrick parts. *Journal of Biological Engineering*, *2*(1). https://doi.org/10.1186/1754-1611-2-5

Shi, Y., & Xu, J. (2023). A multi-objective approach to kitchen waste and excess sludge co-digestion for biomethane production with anaerobic digestion. *Energy*, *262*, 125243. https://doi.org/10.1016/J.ENERGY.2022.125243

Singhal, R. S., Kennedy, J. F., Gopalakrishnan, S. M., Kaczmarek, A., Knill, C. J., & Akmar, P. F. (2008). Industrial production, processing, and utilization of sago palm-derived products. *Carbohydrate Polymers*, *72*(1), 1–20. https://doi.org/10.1016/j.carbpol.2007.07.043

Sinha, P., & Pandey, A. (2011). An evaluative report and challenges for fermentative biohydrogen production. *International Journal of Hydrogen Energy*, *36*(13), 7460–7478. https://doi.org/10.1016/j.ijhydene.2011.03.077

Siruru, H., Syafii, W., Wistara, I. N. J., Pari, G., & Budiman, I. (2022). Properties of sago waste charcoal using hydrothermal and pyrolysis carbonization. *Biomass Conversion and Biorefinery*, *12*, 5543–5554. https://doi.org/10.1007/s13399-020-00983-9

Supu, I., & Jaya, I. (2018). Synthesis and compression strength properties of composite based on sago pulp fiber waste. *IOP Conference Series: Earth and Environmental Science*, 187(1). https://doi.org/10.1088/1755-1315/187/1/012005

Suriani, M. J., Radzi, F. S. M., Ilyas, R. A., Petrů, M., Sapuan, S. M., & Ruzaidi, C. M. (2021). Flammability, tensile, and morphological properties of oil palm empty fruit bunches fiber/pet yarn-reinforced epoxy fire retardant hybrid polymer composites. *Polymers*, *13*(8), 1282. https://doi.org/10.3390/polym13081282

Syafri, E., Jamaluddin, Sari, N. H., Mahardika, M., Amanda, P., & Ilyas, R. A. (2022). Isolation and characterization of cellulose nanofibers from Agave gigantea by chemical-mechanical treatment. *International Journal of Biological Macromolecules*, *200*(December 2021), 25–33. https://doi.org/10.1016/j.ijbiomac.2021.12.111

Tenriawaru, E., Supu, I., & Cambaba, S. (2018). Physical properties of sago bark. *IOP Conference Series: Earth and Environmental Science*, *187*, 012007. https://doi.org/10.1088/1755-1315/187/1/012007

Vassilev, S. V., Vassileva, C. G., & Vassilev, V. S. (2015). Advantages and disadvantages of composition and properties of biomass in comparison with coal: An overview. *Fuel*, *158*, 330–350. https://doi.org/10.1016/J.FUEL.2015.05.050

Wahi, R., Zuhaidi, N. F. Q., Yahaya, S. A., Lam, S. S., Imran-Shaukat, M., Aziz, S. M. A., Ngaini, Z., & Mohamad Pauzan, A. S. (2022). Engineered microwave biochar from sago bark waste for heavy metals adsorption. *Materials Today: Proceedings*, *57*, 1403–1414. https://doi.org/10.1016/j.matpr.2022.03.270

Wan, Y. K., Sadhukhan, J., & Ng, D. K. S. (2016). Techno-economic evaluations for feasibility of sago-based biorefinery, Part 2: Integrated bioethanol production and energy systems. *Chemical Engineering Research and Design*, *107*, 102–116. https://doi.org/10.1016/j.cherd.2015.09.017

Ye, X., Chi, R., Wu, Z., Chen, J., Lv, Y., Lin, C., Liu, Y., & Luo, W. (2023). A biomass fiber adsorbent grafted with phosphate/amidoxime for efficient extraction of uranium from seawater by synergistic effect. *Journal of Environmental Management*, *337*, 117658. https://doi.org/10.1016/J.JENVMAN.2023.117658

5 Sugar Palm Waste Management

5.1 INTRODUCTION

Humanity now faces serious problems related to climate change, worldwide warming, forest destruction, non-biodegradable waste, water pollution, and air pollution. Finding solutions that guarantee the maintenance of sustainable forests and the environment is what motivates scientists and engineers. Synthetic fibers are now frequently found in high-quality commercial items, particularly in the composites sector. Natural fibers, as an alternative, are now attracting more attention due to the detrimental effects synthetic fibers have on the environment and human health (Reddy & Yang, 2015). A large quantity of agro-industrial waste or lignocellulosic biomass can be produced from natural fibers as well. This lignocellulosic biomass can be transformed into a variety of bioproducts, including biocomposite, biofuels, biopolysaccharides, bioadsorbents, and more (Sanyang et al., 2016; Norfarhana et al., 2022, 2024a,b; Nurazzi et al., 2021b; Syafri et al., 2022).

Natural forests are home to the sugar palm, a member of the Palmae family of plants. It has a reputation for advancing quickly and maturing in just 10 years (Mogea et al., 1991). The sugar palm is recognized by more than 150 different local names in Indonesia, but it is more frequently called enau or kabung in Malaysia (Heyne, 1950). The sap of the sugar palm, which is gathered and used as a sweetener and in the production of numerous foods and beverages, is the sugar palm's main claim to fame. Making cuts in the tree trunk and letting the sap drop into a container is how the sap is collected. The sap can subsequently be reduced by boiling to create sugar, which is frequently used in Southeast Asian pastries and beverages. Other applications for the sugar palm include building, weaving, and the production of biofuel (Olsson et al., 2019).

Sugar palm, also known as *Arenga pinnata*, is a species of palm tree that is native to Southeast Asia. It is a multi-stemmed palm tree with a trunk diameter of up to 50 cm and a height that can reach up to 30 m. A range of meals and beverages, goods made of wood, biofibers, biopolymers, and biocomposites can all be made from this versatile palm species. By fermenting the sap of the sugar palm, it is now employed as a source of renewable energy in the form of bioethanol. Although the sugar palm can yield a variety of goods, the three most common are palm sugar, fruits, and fibers (Ishak et al., 2013a). According to Siregar (2005), the sugar palm is renowned for its versatility and has the ability to produce useful goods from practically every section of the tree. Besides palm sap, the sugar palm may be utilized to generate a number of other important goods such as palm neera, freshly squeezed juices, conventional glucose blocks, toddy, crystalline and brown sugar, vinegar (Redhead, 1989), starch from the trunk, seawater-resistant fiber, edible heart (Mogea et al., 1991), bioethanol (Lay, 2009), fruits, leaves for roofing, brooms, matting, baskets, cigarette papers (Davis and Johnson, 1987), and cattle feeds (Sath et al., 2008).

In the tropical regions of South and Southeast Asia, this plant is frequently found in hot, humid climates. Asia's most economically viable plant, the sugar palm, is found throughout the Indo-Malay Archipelago, which includes countries in Southeast Asia like the Philippines, Indonesia, and Malaysia (Sanyang et al., 2016; Ilyas et al., 2017). Along the rivers in rural Malaysia, sugar palm trees can be found. Their plantation growth encompasses approximately 892 hectares (Sahari et al., 2012b). The sugar palm is a physically tall palm tree with long, angular leaves and ornamental trunk fibers. The efficient cultivation of sugar palm trees has lately undergone a new method. In Tomohon,

 DOI: 10.1201/9781003472209-7

Indonesia, the first sugar palm factory ever built around the world was completed. Sugar palm fibers (SPFs) will be widely available as by-products as the sugar palm plantation expands. In addition to having the most uses, sugar palm is a multipurpose palm species.

Currently, the SPFs are left to waste with no purpose for the industry since large amounts of SPF have been dubbed to be abundant. If this ongoing problem is not addressed, there may be additional storage capacity issues at some impacted places in the future. Many different natural fiber classes are currently being modified and used in a variety of sectors. Through the use of several methods, SPF can be converted into cellulose for use in a variety of sectors, including the automotive, packaging, and other industries involved. This chapter expects to discuss the potential and management of sugar palm wastes in the agro-industry sector to reduce the use of any non-green techniques. These efforts could lead to waste reduction, hazardous waste disposal, and energy savings.

5.2 SUGAR PALM FIBERS

5.2.1 Sugar Palm Waste from Natural Fiber

The by-products or residue produced during the manufacturing of SPF are referred to as sugar palm trash from natural fibers. These fibers are obtained from sugar palm trees and are frequently used to create a variety of goods, such as furniture, baskets, and mats. Some waste elements are produced throughout the production process, including the tree's outer bark, leaves, and small branches. These waste materials can be further processed to yield natural fibers that are used to create goods like ropes, twines, and nets. Since it is a renewable resource and may be utilized to make products with little influence on the environment, the sugar palm waste from natural fibers is a sustainable and environmentally beneficial material (Ishak et al., 2012).

Natural fibers can be categorized into three types: plant, animal, and mineral fibers, as pointed out by Mathivanan et al. (2015). Among these, natural plant fibers are most commonly employed as reinforcement materials in biocomposites. The classification of plant fibers is based on the specific plant or plant part they are extracted from, as indicated by Lilholt and Lawther (2000). Nowadays, natural fibers are widely used as reinforcement in various applications, including composite production. SPF, in particular, exhibits superior durability, tensile strength, and resistance to moisture and heat compared to coir fiber, according to Huzaifah et al. (2017).

Natural fibers have several advantages over synthetic fibers. The benefits of natural fibers frequently outweigh their disadvantages. Natural fibers have a low density and are readily available. Additionally, it diminishes rapidly and permits simple machining with minimal tool wear. In spite of their poor durability in heat, reliability, and resilience to the environment, natural fibers can replace glass fiber and other synthetic fibers in outdoor applications provided certain technological challenges can be overcome. Natural fiber composites are currently the topic of intense study and development due to their excellent shape retention, availability, renewability, cost-effectiveness, and environmental friendliness (Garkhail et al., 2000).

Based on Mukhtar et al. (2016), abaca, kenaf, flax, hemp, coir, jute, and sisal are among the most popular and widely used natural fibers. SPF is widely recognized for its traditional multipurpose uses in a variety of foods and drinks in rural communities. A multipurpose palm species, the sugar palm (*A. pinnata*) produces a wide range of goods, including food, drinks, timber, biofibers, biopolymers, and biocomposites. Through the recent development of bioethanol from sugar palm sap, it has also been investigated as a sustainable energy source. Fruits, fibers, and palm sugar are the three primary products made from sugar palms. Due to their woven shape, these fibers are naturally strong, seawater resistant, and simple to process. The trunk of the tree is covered in fibers from top to bottom, as shown in Figure 5.1.

SPF chemical structure varies according to plant parts, plant years of age, conditions for growth, soil quality, climate impact, methods of testing, and plant height. So Ishak et al. (2013b) examined

FIGURE 5.1 The tree's trunk encircled by sugar palm fiber from top to bottom (Afzaluddin et al., 2019).

FIGURE 5.2 Sugar palm tree (Huzaifah et al., 2017).

the chemical. SPF composition from a single sugar palm tree at heights of 1, 3, 5, 7, 9, 11, 13, and 15 m. They discovered that SPF from the bottom of the sugar palm tree (1–3 m) contained poorer fibers with the lowest amounts of lignin, hemicellulose, and cellulose. This observation was made in light of the aging of sugar palm fronds and fibers discovered close to the tree's base. Figure 5.2 shows the parts of the sugar palm tree such as the fronds, trunk, and ijuk.

According to Imraan et al. (2023), natural fiber made from plants mostly consists of cellulose, hemicellulose, and lignin. Cellulose is responsible for the fiber's mechanical toughness. The cellulose-hemicellulose network, which is the primary structural element of fiber cells, is made up of cellulose microfibrils that are hydrogen-linked to hemicellulose molecules, which act as a cementing matrix between them. By stiffening cellulose-hemicellulose and acting as a coupling agent, lignin gives plants their rigidity. SPF provides better tensile strength, durability, and resistance to heat and moisture than coir fiber (Huzaifah et al., 2017). Figure 5.3 shows the structure of natural fibers made from plants.

Lignocellulose fibers, also referred to as natural fibers, are made from cellulose or lignocellulose, protein, and minerals. To make products or act as an ingredient for strengthening, these fibers can be crocheted, matted, or braided. They are typically spun into rope, thread, or filament. According to the variety of their sources, natural fibers are divided into those with plant, animal, and mineral origins, as shown in Figure 5.4 (Bogataj et al., 2019; Sanyang et al., 2016). Some of the most famous natural fibers that are mostly sourced from plants include hemp, abaca, flax, cotton, jute, coir, ramie, and sisal. The term "fiber" in this context refers to a multicellular structure. Van der Waals forces, hydrogen bonds, and mechanical contacts between hemicellulose, lignin, and cellulose microfibril domains are used to arrange and connect these cells.

Since the majority of natural fibers come from waste from agriculture that is primarily incinerated for fuel, they have an adverse effect on the environment. Thirty-three percent of the world's ethanol is created from crop residue, according to Yevich and Logan (2003) estimation of biofuel

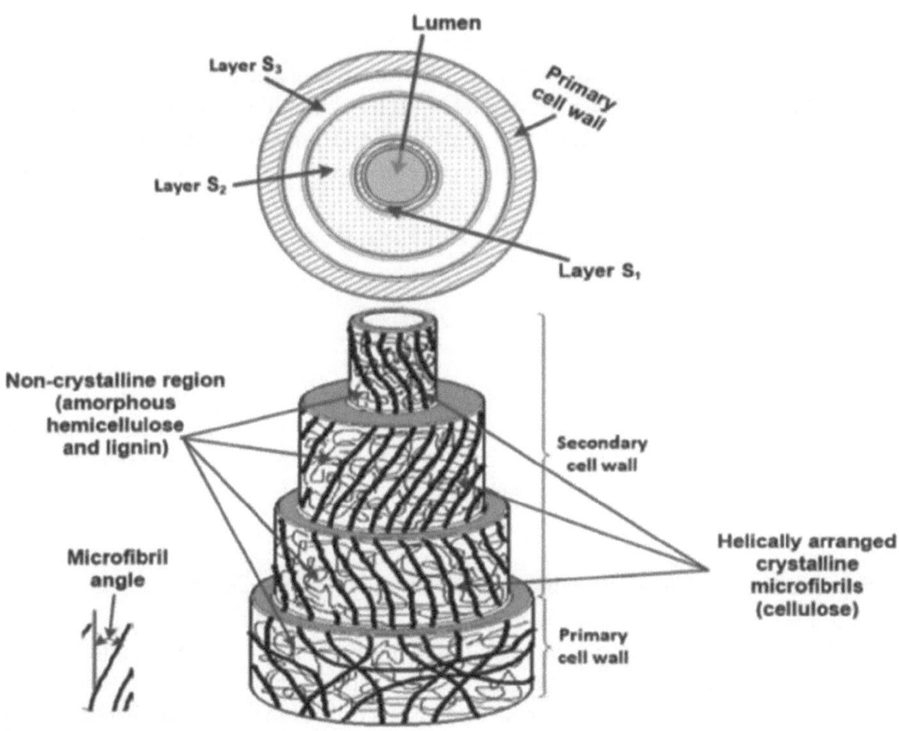

FIGURE 5.3 Natural fiber's structure (Kalia et al., 2009; Rong et al., 2001).

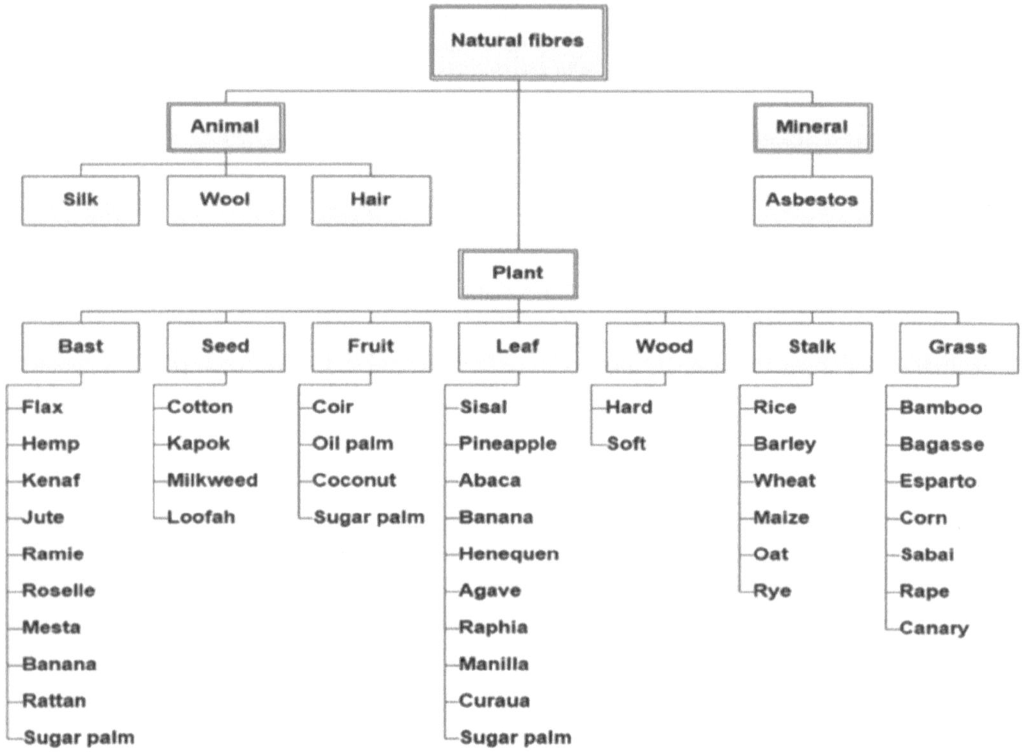

FIGURE 5.4 Classification of natural fibers (Jawaid & Abdul Khalil, 2011; Sanyang et al., 2016).

usage and field burning of waste from agriculture. Agricultural waste is produced in plenty and at a low cost, in the billions of tons. The majority of this waste might be utilized in polymeric composites as fibers and fillers. Natural fibers are available in large quantities in many nations, which surely supports their usage in a variety of industrial applications. Given how natural fibers are categorized, SPF, which is more recent than other natural fibers, can occupy a variety of positions. This explains why the frond, bunch, or trunk of the sugar palm plant may contain fiber (Sahari et al., 2012c; Ishak et al., 2012). Therefore, under the classification of natural fibers as described in Figure 5.4, SPF can appear under fruit, bast, and leaf.

5.2.2 Environmental Impact of Sugar Palm Waste

The usage of sugar palm waste may have both beneficial and detrimental effects on the environment. Positively, using sugar palm waste as a source of bioenergy can assist in lowering the usage of fossil fuels, which can then help lower greenhouse gas emissions and lessen climate change. To promote healthier soil and reduce the demand for synthetic fertilizers, sugar palm waste can also be utilized as a fertilizer or soil amendment. In the developing "green" economy focused on energy efficiency, natural fibers will be crucial. The carbon emissions caused by the burning of plastics will be decreased by using biodegradable materials in polymer products. Negatively, incorrect sugar palm waste management can cause the ecosystem to deteriorate. For instance, if the garbage is burned or allowed to decay outdoors, dangerous contaminants may be released into the air, causing air pollution and respiratory issues. The garbage can also poison water sources and harm aquatic habitats if it is dumped in bodies of water. Therefore, it's critical to make sure that waste from sugar palms is correctly managed and disposed of in a way that respects the environment, such as by using it to produce renewable energy or turning it into goods with additional value like biochar or bioplastics.

The use of natural fibers with polymer matrix allows for the production of inexpensive polymeric-reinforced composites and helps to reduce habitat destruction. The addition of SPF to polymer matrices further aids in addressing environmental issues brought on by the disposal of conventional composites that are not biodegradable. Due to the continued massive solid waste disposal, landfill areas are drastically diminishing. The creation of composites using SPF reinforcement was greatly influenced by these environmental effects. According to Sanyang et al. (2016), it is interesting to note that sugar palm tree fibers, which once could be dumped as agricultural waste, are now used to create biocomposite materials. For sugar palm producers, this has created a fresh opportunity to turn waste into income. Agro-waste materials are also considered for use in economically viable goods; otherwise, they would be disposed of. Additionally, SPF is a great replacement for commonly used non-renewable fibers, just as other natural fibers. The main causes of natural fibers' low cost are their availability and capacity to be renewed.

The usage of synthetic fibers is not sustainable given the rapid depletion of world petroleum resources. A cost-saving and environmental protection measure is the switch to fibers that are more environmentally friendly. By using renewable fibers, the "cradle-to-cradle" idea of consumer goods is replaced with the "cradle-to-grave" idea. These natural fiber-reinforced composites spontaneously break down into water and carbon dioxide (CO_2) at the end of their useful lifetimes. The non-renewable and non-biodegradable nature of their synthetic fiber cousins eventually results in pernicious solid waste management issues (Sanyang et al., 2016). A worrisome worry is spreading around the world that anthropogenic CO_2 emissions are causing climate change. Excessive CO_2 emission is responsible for over 70% of the growing greenhouse effect, which has caused global warming and had disastrous environmental repercussions like more frequent storms, floods, and droughts. Therefore, it is crucial to implement greener technology and use environmentally friendly materials to satisfy the global demand for CO_2 reduction. SPFs are therefore considered to be renewable resources, and using them could assist in lowering atmospheric CO_2 emissions (Sanyang et al., 2016).

5.2.3 Availability of Sugar Palm Waste

The natural habitat of sugar palm waste includes forested areas although they are seldom far from populated areas, ravines, streams, slopes, and semi-cultivated areas. Unlike secondary forests, which are typically located close to human settlements, primary woods are rarely home to sugar palm waste. From sea level to 1,400 m elevation, it thrives in hot, humid regions. Bangladesh, Cambodia, Brunei, India, Indonesia, Malaysia, Papua New Guinea, Myanmar, Sri Lanka, Singapore, Philippines, Thailand, and Vietnam are among the nations where the sugar palm is indigenous. It belongs to the most varied and useful tree species in human culture (Hazrati et al., 2019). In Malaysia's rural areas, sugar palm trees are abundant beside rivers and bushes. For

FIGURE 5.5 Geographical location of sugar palm availability (Imraan et al., 2023).

instance, the sugar palm is common and grows wild in numerous locations. As shown in Figure 5.5, around the size of 800 hectares of sugar palm trees have been planted by Kebun Rimau Sdn. Bhd. in Tawau (Sabah, West Malaysia), while 50 hectares of sugar palm trees have been planted in Benta and Pahang. However, this species has a much lower area of plantation as compared to the other palm species as the oil palm and coconut (Sanyang et al., 2016).

5.2.4 Properties of Sugar Palm Waste Fibers

5.2.4.1 Physical Properties

Sugar palm waste is brown in color which is collected from the sugar palm tree based on the section wanted ranging from the top to the bottom. The waste is very rough and rigid physically. It is long and has a very small diameter measurement for the circumference. Figure 5.6 shows the sugar palm waste that was collected from the middle section of the sugar palm tree (Asyraf et al., 2021). The sugar palm is generally a very large, tall palm with a single stem that ranges from 10 to 20 m tall and 65 cm in diameter. The trunk usually exists in long black fibers and ripped leaf roots. Starch can also be kept in the trunk. The fiber's properties are usually discussed physically based on the diameter, density, and moisture content. Table 5.1 lists the physical properties of the SPF and cellulose (Adawiyah et al., 2013).

One of the most important considerations when assessing novel natural ingredients as prospective fillers, especially for polymer composites, is their physical properties, such as moisture content. It is therefore an important factor for identifying the properties and ultimate uses of the waste, commonly known as fibers (Suriani et al., 2021a). Figure 5.6 shows the extraction of fiber, also well known as waste from a sugar palm tree. Fibers generally have a high moisture content which

FIGURE 5.6 Sugar palm tree alongside extracted naturally woven fibers (Ilyas et al., 2017; Shahroze et al., 2018).

TABLE 5.1
Typical Diameter, Density, and Moisture Content of Sugar Palm Fiber and Cellulose (Sahari, 2012)

Sample	Diameter (mm)	Density (g/cm³)	Moisture Content (wt%)
Sugar palm fiber	212.01	1.5	8.36
Sugar palm cellulose	8.81–11.84	1.36–1.28	4.19–3.83

TABLE 5.2
Chemical Compositions of the Sugar Palm Waste on Multiple Sections of the Tree (Ishak et al., 2013c)

Composition	Sugar Palm Frond	Sugar Palm Bunch	Ijuk	Sugar Palm Trunk
Cellulose (%)	66.49	61.76	52.29	40.56
Helocellulose (%)	81.22	71.78	65.62	61.1
Lignin (%)	18.89	23.48	31.52	46.44
Ash (%)	3.05	3.38	4.03	2.38
Moisture (%)	2.74	2.7	7.4	1.45
Extraction (%)	2.46	2.24	4.39	6.3

may have an adverse effect on the durability of consideration toward polymer composites in the specification of the development of porosity, tensile strength, and dimensions. This is so because the chemical structure of cellulose contains hydroxyl groups that are typically prone to water. As a result, the fibers that underwent cellulose extraction produced an abundance of hydrophilic ionic groups on their surfaces and consequently indirectly increased moisture absorption. Raw fibers (%) and sugar palm cellulose (%) both have an increase in moisture content (Fitriana et al., 2020).

5.2.4.2 Chemical Compositions

The chemical makeup of SPF varies according to the fiber's type and purpose. The properties of each fiber are affected by the characteristics of each component. Between plants and within the same plant, there are large variations in chemical makeup. For sugar palm waste, also known as fiber, overall cell walls are primarily made of cellulose and hemicellulose, two sugar-based polymers (Hassan & Wagner, 2016). In addition, different sections of the tree display different compositions of the sugar palm waste (Huzaifah et al., 2017). Table 5.2 lists the chemical compositions of sugar palm waste in different sections of the tree.

The composition of sugar palm frond contains the highest percentage of cellulose, being that it is one of the most valuable materials to be used in many industries through further treatment and extraction processes. Plants and multiple parts of the same plant have quite different chemical compositions. The bulk of SPFs are waxes, with cellulose (-cellulose), pectin, lignin, and hemicellulose serving as a compatibility between cellulose and lignin (Asyraf et al., 2021). The sugar palm tree's fronds contain the highest concentration of cellulose, and over 90% of its total fiber content is contained in *ijuk* fiber. Additionally, in comparison to the height of the sugar palm tree, the chemical composition of the fibers varies. Ishak et al. (2013b) found that the cellulose, hemicellulose, and lignin content of sugar palm trees increased as their height grew. However, it was anticipated that their contents would decline as the tree became older, which was connected to the fiber juvenile stage.

5.2.4.3 Mechanical Properties

This cellulosic fiber has been widely employed in a range of applications due to its durability and strength, notably home items (Ishak et al., 2012). In its country of origin, the SPF is also known

TABLE 5.3
Mechanical Properties for Multiple Parts of Sugar Palm Tree (Sahari et al., 2012a)

Fibers	Sugar Palm Frond	Sugar Palm Bunch	Sugar Palm Trunk	ijuk
Tensile strength (MPa)	4,241.4	365.1	198.3	276.6
Tensile modulus (GPa)	10.4	8.6	3.1	5.9
Elongation at break (%)	9.8	12.5	29.7	22.3

as fiber and is a black material that is frequently used to create ropes, brushes, roofing, brooms, and other handicrafts (Asyraf et al., 2021). One of the main benefits of SPF over coir fiber is that it is resilient and has a long lifespan because it is unaffected by moisture and heat (Shahroze et al., 2019). Table 5.3 lists some of the most important characteristics of SPFs, including the mechanical characteristics of the fiber's major components. As reinforcing materials, polymeric polymers and SPF have been used. SPF is also employed in composite materials because of its low cost, low density, high toughness, ease of separation, high biodegradability, and improved energy recovery. Unlike synthetic fibers like glass, SPF gives composites strength and stiffness, while also reducing the risk of fiber fractures (Izwan et al., 2020).

The mechanical properties of SPF can be related more to the biocomposites. As SPF loading increased, the tensile strength and modulus of SPF biocomposites exhibited an upward trend, whereas the addition of SPF caused the elongation to decrease from 8.03% to 3.32%. The significant increase in tensile strength that occurred as SPF increased suggested that the cellulose fiber played a significant role in the establishment of a strong link between SPS and SPF (Ammar et al., 2018). The scanning electron microscopic analyses of some samples of the broken tensile samples established justification for the statement that the tensile modulus increased as the fiber loading increased. Since the SPF biocomposite has a low-stiffness matrix and high-stiffness fibers, its stiffness can be increased by adding up the weight percentage of the fibers. While fiber loading increased, elongation showed a trend toward decreasing (Afzaluddin et al., 2019). The hydrophilic properties of SPF, which lead to excellent compatibility, may be the cause of the hybrid composites' better tensile strength than those of their individual components. The improvement in mechanical properties of the produced composites is frequently linked to the combination of appropriate materials (Jumaidin et al., 2017).

5.2.4.4 Morphological Properties

The chemical makeup of SPF differs according to the type and character of the fiber. All cell walls are mostly made of a combination of hemicellulose and cellulose. The characteristics of each composition affect those of each fiber. Plants and diverse portions of the same plant have quite different chemical makeup. The bulk of SPFs are composed of waxes, and cellulose (-cellulose), pectin, lignin, and hemicellulose act as a matrix between lignin and cellulose (Talib et al., 2011). Figure 5.7 shows the morphological micrographs of the raw SPF.

5.2.4.5 Thermal Properties

The thermal deterioration of SPF is studied in this section. Figures 5.8 and 5.9 indicate, respectively, the effects of increasing temperature on SPF mass and derivative mass loss rate. Shortly said, 1 and 3–15 m represent the two biggest disparities between the behaviors of the fibers as determined by thermogravimetric analysis (TGA) and derivative thermogravimetry (DTG).

The behavior of fibers and composites can be significantly influenced by thermal characteristics, as demonstrated by prior research. Natural fibers' poor temperature stability is a well-known barrier to their usage as reinforcement in biocomposites (Latif et al., 2019). The thermal stability and

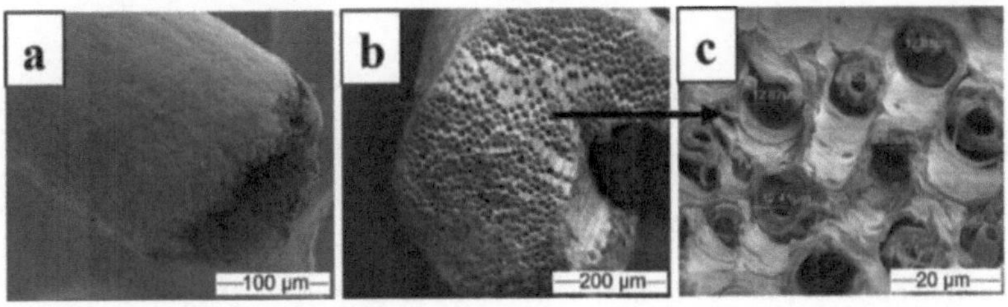

FIGURE 5.7 FESEM photography of the raw sugar palm fiber (SPF): (a) longitudinal section, (b) cross-section, and (c) primary and secondary cell walls (Ilyas et al., 2017).

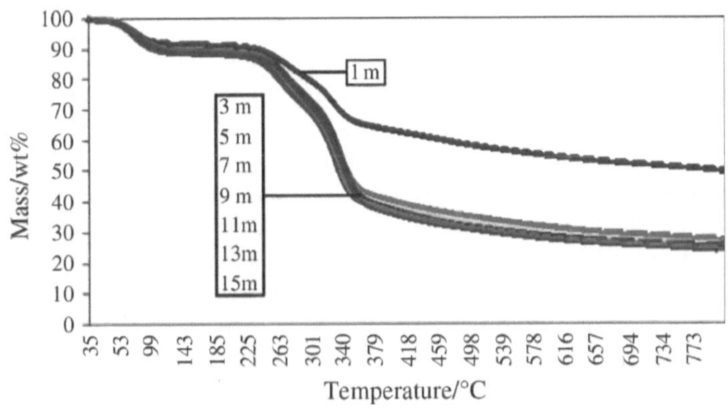

FIGURE 5.8 Thermogravimetric analysis (TGA) of sugar palm fibers collected from multiple heights of sugar palm tree (Ishak et al., 2012).

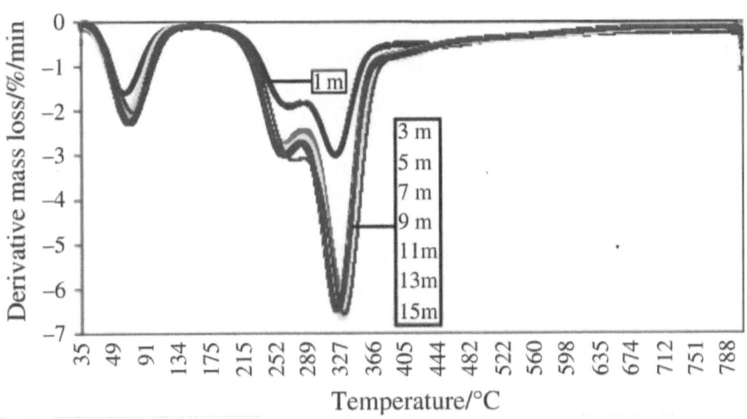

FIGURE 5.9 Derivative thermogravimetry (DTG) of sugar palm fibers collected from multiple heights of a sugar palm tree (Ishak et al., 2012).

thermal decomposition of polymeric systems were investigated using TGA and DTG curves. Each step of heat deterioration is identified by a distinct peak in the DTG curves and a corresponding weight loss on the TGA graph. To maximize a polymer's processing potential, statistics on degradation and decomposition mode during heat exposure must be gathered (Yang et al., 2007).

According to a study by Ilyas et al. (2017), there are four stages in the thermal decomposition process: moisture evaporation (45°C–123°C), hemicellulose decomposition (220°C–300°C),

TABLE 5.4
Onset Temperature, Degradation Temperature of the Maximum Weight Loss Rate, Weight Loss, and Char Yield for the Untreated and Treated Sugar Palm Waste (Ilyas et al., 2017)

	Water Evaporation			First Thermal Degradation			Second Thermal Degradation			Char Yield
Sample	T_{onset} (°C)	T_{max} (°C)	W_L (%)	T_{onset} (°C)	T_{max} (°C)	W_L (%)	T_{onset} (°C)	T_{max} (°C)	W_L (%)	W_L (%)
SPF	41.7	106.8	10.4	210.6	281	15.1	308.1	345.5	43.8	30.7
Treated SPF	43.5	101.2	8.6	207.9	345.1	73.7	—	—	—	17.7

cellulose decomposition (300°C–400°C), lignin decomposition (160°C–900°C), leaving the ash as a residue. The ability of fiber to tolerate the temperature applied during the production process or application of the product was often assessed using these temperature ranges. Table 5.4 provides the water evaporation and degradation level for each stage for untreated and treated sugar palm waste. This lack of volatile extractive movement was caused by water moving from inside the fiber to the major surface during heating, vaporizing the water on the fiber surface. Only losses of water in the cell lumen and cell wall, as well as volatile extractives in the fibers, occurred because of the low-temperature heating process. At 106.78°C, the moisture in the SPF had completely evaporated as was seen (Yang et al., 2007).

5.2.4.6 Caloric Calorific Value

For the design of useful equipment, the precise energy needed to create briquettes is crucial. Sanjay et al. (2016) have discovered values for barley straw that fall between 5 and 25 MJ t1. The specific energy required to compress cotton stalks in circular dies is 7.2 MJ t^1. According to the density of the biomass residue, densified biomass residue (DBR) in circular dies produces a specific energy value between 5 and 10 MJ t^1 The SPF characteristics are excellent as they can survive soaking as well as good resistance to mechanical disintegration. The estimated gross calorific value is 16.4 MJ/kg (Husain et al., 2002).

5.2.5 Sugar Palm Waste Management

5.2.5.1 Compost

There are many methods of waste management. One of them is composting. Microorganisms degrade and transform complex degradable materials into organic and inorganic wastes during the aerobic process of composting. The by-products contain "humic-like" chemicals, which set them apart from those found in native soil, coals, and peats. Through the composting process, various degradable wastes can be transformed into items that can be used effectively and safely as biofertilizers and soil additives (Ayilara et al., 2020). All materials that may be spontaneously transformed into eco-friendly chemicals by microbes are biodegradable (Ahmad et al., 2007). Among these appropriate materials, agricultural wastes play a significant role in composting because they include a sufficient amount of organic material (Ahmad et al., 2007; Khalil et al., 2008). Composting has also made use of a number of biomasses, such as wheat straw, rice straw, coffee husks, vegetables, and olive tree pruning (Vakili et al., 2015). Considering that discarded leaves, fronds, and trunk fibers remain as waste after harvesting the sugar palm fruits, they can be composted to produce organic fertilizer. They can be transformed into useful, manageable, and eco-friendly products through composting.

According to Larasati et al. (2017), as the material from this compost is a mixture of compost from sugar palm solid waste, manure, and microbial starter, it is extremely beneficial to use.

TABLE 5.5
Results of the Chemical Analysis of Sugar Palm Bagasse Compost (Sarjiyah & Handoko, 2022)

Criteria	Result
C-Organic	11.98%
N total	1.64%
C/N ratio	7.28

Compost made from sugar palm solid waste is incredibly rich in organic matter and nutrients. Meanwhile, Sarjiyah and Handoko (2022) found that making compost is a possible substitute for correctly handling bagasse. The high fiber content and C/N ratio of sugar palm bagasse, however, slow down the composting process. To increase N content, decrease C/N ratio content, and accelerate composting, blended ingredients are necessary. *Gliricidia* leaves are one possible mixture component. Table 5.5 lists the characteristics of the compost created by combining *Gliricidia* leaves with sugar palm bagasse.

5.2.5.2 Pulp and Paper

Another method to make use of these wastes generated from SPFs is making pulp and paper. The pulp and paper industry has been growing significantly on a global scale. Raw materials for pulp and paper have therefore been in great demand. Wood, non-wood, and recycled waste paper are the three raw ingredients needed to manufacture paper (Abd El-Sayed et al., 2020). Non-wood resources offer a variety of benefits over producing pulp and paper from wood, including simplicity of pulping, superior bleached pulp, and good sources for specialist types of paper. There are a variety of non-wood plant fibers that are useful as fiber resources in places like China, India, and Egypt where there are not enough trees to go around. These non-wood plant fibers are becoming more and more significant in these places. According to Mohamed et al. (2021), the potential of a material to be used in pulp and papermaking activities is checked through the chemical and physical properties of the materials. As summarized in Table 5.2, every raw material tested positive for cellulose content greater than 40%. The holocellulose percentage for all raw materials, however, is shown to be greater than 60%. Except for the trunk of the sugar palm, the lignin percentage of all raw materials is less than 40%. These showed that the sugar palm ijuk, bunch, and frond could all be employed in the pulping and papermaking processes.

5.2.5.3 Biochar

Another type of waste management includes thermal conversion such as pyrolysis, gasification, and combustion. Two of the thermochemical biomass conversion techniques that are frequently studied are pyrolysis and gasification. With a maximum temperature of 700°C, pyrolysis is the thermal breakdown of biomass that takes place in the absence of oxygen at operating temperatures between 350°C and 550°C. The pyrolysis process converts organic molecules into mixes of solid, liquid, and gas (Lee et al., 2019). Environmentally responsible biomass waste utilization and chemical fuel conversion are now regarded as one of the most successful methods for creating renewable energy. Gasification differs from pyrolysis in that it can generate fuel gas that can be burned to generate heat. The pyrolysis process, in contrast, yields pyrolysis oil (py-oil) or bio-oil, a liquid fuel that can take the place of fuel oil in the generation of electricity or in the use of static heating (Lee et al., 2019). Figure 5.10 shows different types of pyrolysis processes.

Over the years, biomass has generated a lot of scholarly discussion since it may be converted into products with significant added value, like biofuel and charcoal (Ogunlalu et al., 2021). Thus, sugar palm waste, including fronds, empty fruit bunches, and fibers, can be converted into heat,

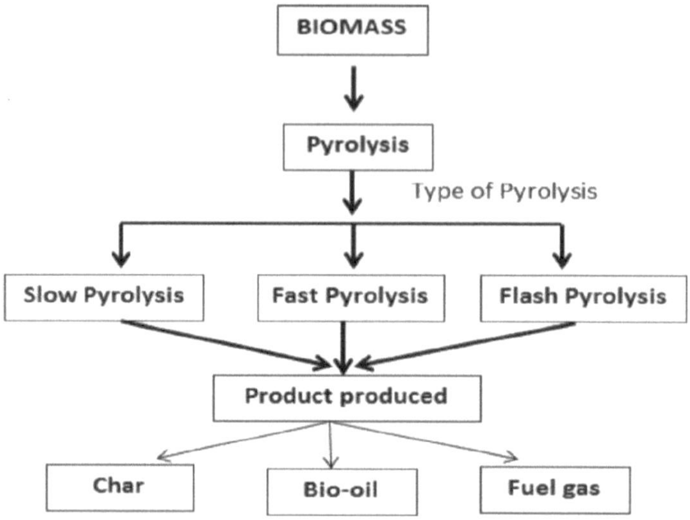

FIGURE 5.10 Different types of pyrolysis process (Lee et al., 2019).

TABLE 5.6
Thermochemical Methods and Process Conditions (Iwuozor et al., 2022)

Methods	Temperature (°C)	Residence Time	Biochar Yield (%)
Pyrolysis	400–550 (slow)	40 min (slow)	31–34 (slow)
	400–800 (fast)	Hour—day (fast)	47 (fast)
Hydrothermal carbonization	180–230	13–40 h	78
Gasification	700–900	10–30 min	10–13
Torrefaction	160–300	2 h	55–95

electricity, or biofuels. These methods with different processing settings are crucial for detecting the physical and chemical properties of the biochar, such as surface area, microporosity, and hydrophobicity, according to studies on biochar made from sugarcane bagasse (Moradi-Choghamarani et al., 2019b). Along with the physical characteristics of the biochar, the process parameters also significantly affect its organo-chemical characteristics (Moradi-Choghamarani et al., 2019a). The numerous thermochemical conversion methods are summarized in Table 5.6, along with the conditions that govern them.

5.2.5.4 Bioethanol

In terms of organic chemistry, bioethanol (C_2H_5OH), also known as ethyl alcohol, is a form of alcohol that has recently gained popularity as a potential petrol substitute for roads and as a renewable bioenergy source. It is also a clear, colorless liquid that degrades quickly (Hossain & Jalil, 2017). Typically, bioethanol is made by fermenting sucrose or simple sugars from different types of biomass, whether they come from feedstock or sources other than feedstock (Gnansounou & Dauriat, 2005). Currently, a different solution to the energy, economic, and environmental problems the world is experiencing is the synthesis of bioethanol from cellulosic and lignocellulosic materials, notably wastes (Srivastava et al., 2007). Consequently, research on the production of bioethanol from plant-based waste biomass is currently necessary to comprehend environmental and energy concerns.

Sugarcane waste biomass, such as sugarcane bagasse and sugarcane molasses, are among the top options for lignocellulosic biomass in the production of bioethanol due to their wide availability in the categories of forest and industrial residual waste biomass. As sugar palm waste is rich in lignocellulosic materials as provided in Table 5.2, they have rising potential in bioethanol production. According to Alisjahbana and Busch (2017), an average of 3.5 L of ethanol can be produced for every 10 L of sugar palm sap. In light of this, 73,710 L of ethanol may be produced by 1 hectare of palm trees. The potential for palm sap biofuel can reach 2 million kiloliters per year if output is enough. As a result, biofuel has a lot of promise and has to be developed (Ansar et al., 2021).

An extremely well-known, long-established, and traditional natural metabolic method for converting lignocellulosic biomass to bioethanol is fermentation, in which an organism changes complex carbohydrates into simple sugar and sugar into an alcohol or an acid. Yeast, bacteria, or enzymes are used experimentally in this fermentation process. Because yeast fermentation is a technique that is effective, efficient, and simple to utilize, it has been the focus of this study. Compared to other fermenting processes, yeast fermentation has fewer setbacks (Hossain & Jalil, 2017).

5.2.5.5 Biocomposite

The ability of sugar palms to create biopolymers like starch is another alluring prospect. Starch can be found in the sugar palm stem's inside. It has served as a source of raw materials for the manufacture of starch and glue. It can be used as "green" material and is harvested from mature palm tree trunks. In the presence of a plasticizer at a high temperature, such as water, glycerol, or sorbitol, starch will behave as a biopolymer (Sahari et al., 2012c). Using the starch from the sugar palm tree trunk, a biodegradable polymer can be made, which can then be reinforced with natural fibers to produce eco-friendly composites. In addition to being widely accessible, affordable, and renewable, this combination is advantageous (especially in tropical nations like the Philippines, Cambodia, and Malaysia). Their future in the field of biocomposite materials is therefore bright (Sanyang et al., 2016).

The overall technical features of natural fiber are greatly influenced by the characteristics and proportions of each chemical component. The cellulose percentage of SPF, which mostly contributes to strength, is higher than that of coir and kenaf but still falls within the range of abaca, flax, and jute when compared to other natural fibers. The possibility of using SPF as reinforcement in natural fiber composites is supported by this convincing data (Mukhtar et al., 2016).

The SPF has been studied extensively as a polymer composite reinforcing material. Bachtiar et al. (2010) studied the effect of alkaline treatment on the flexural properties of SPF-reinforced epoxy composites. The fibers were treated using sodium hydroxide (NaOH) with 0.25 and 0.5 M concentration solution for 1, 4, and 8 h soaking time. The purpose of treating fibers with alkali was to enhance the interfacial bonding between matrix and fiber surfaces. The maximum flexural strength occurred at 0.25 M NaOH solution with 1 h of soaking time, i.e. 96.71 MPa, improving by 24.41% from untreated fiber composite. However, the maximum flexural modulus took place at 0.5 M NaOH solution with 4 hours soaking time, i.e. 6,948 MPa, improving by 148% from untreated composite.

Rashid et al. (2017) prepared SPF composites made of phenolic resin. The SPF was subjected to either a 30-day treatment in seawater or a 4-h treatment in a 0.5% alkaline solution. The composite samples were created using a hot press machine, and the composites contained 30% (vol.) SPF in powdered form. On the mechanical (flexural, impact, and compressive), thermal, and morphological properties of the composites, the effects of the fiber treatments were examined. Compared to the untreated composite, the SPF treatments significantly enhanced the composites' mechanical qualities. The flexural and impact strength of the composites were increased the greatest by the alkaline treatment. The compressive strength was improved the most by the seawater treatment, though. Morphological investigations showed that the surface treatments enhanced the connection between the fibers and the matrix. The thermal stability of the composites was only marginally impacted by the alkaline and seawater treatments of the SPF, according to the thermal degradation investigation. As a result, SPF works well as a substitute natural fiber for strengthening biocomposites (Rashid et al., 2017). Table 5.7 summarizes other studies on SPF-reinforced polymer composites

TABLE 5.7
SPF-Reinforced Polymer Composites

Matrix	Treatment	Process	Findings
Thermo-set	Alkali treatment	Hand lay-up	Tensile strength=41.88 MPa, tensile modulus=3,780 GPa
Phenolic	Alkali treatment	Hot press	Tensile strength=32.5 MPa, tensile modulus=0.263 GPa
Polypropylene	Silane treatment	Mixing Brabender	Tensile strength=23 MPa, tensile modulus=1,096 GPa
Polypropylene	Silane treatment	Mixing Brabender	Tensile strength=173.44 MPa, tensile modulus=10.07 GPa
Unsaturated polyester	Alkali treatment	Sheet molding	Thermal degradation properties (first phase=80–130c), (second phase=250c), (third phase=250–370c), (fourth phase=360–400)
Epoxy	Thermo-set	Hand lay-up	Tensile strength=22.7 MPa, tensile modulus=3.62 GPa
Thermoplastic polyurethane	Alkali treatment (6 wt% NaOH)	Hot press	Thermal degradation properties (first phase=146 nc), (fourth phase=480 c)
Unsaturated polyester	Seawater	Hand lay-up	Tensile strength=18.33 MPa, tensile modulus=4.374 GPa
Polyurethane	Combine	Extruding	Tensile strength=18.42 MPa, tensile modulus=1.307 GPa
Polyurethane	Combine	Mixing Brabender	Tensile strength=142.09 MPa, tensile modulus=7.75 GPa

In addition, the introduction of hybrid composites, which contain two or more fibers in one matrix, is considered a way to improve the properties of natural fiber-reinforced polymer composites. Combining one natural fiber with another natural fiber/synthetic fiber in one matrix improves the thermal and mechanical properties of the composite (Nurazzi et al., 2021a). This has demonstrated that hybrid composites are more reliable in a variety of applications and are also more environmentally friendly. Natural–natural fibers, natural–synthetic fibers, natural fiber with carbonaceous materials, and natural fibers with metal can all be used to hybridize natural fiber-based reinforced polymer composites (Suriani et al., 2021a).

Safri et al. (2019) determined how adding glass fiber and benzoylating SPF composites affect their dynamic mechanical, impact, and post-impact properties. The hybrid composites were created by hand-laying up various volume fractions of glass fiber and SPF. In comparison to EP/UTSPF composites, EP/30TSPF/70GF composites had the largest loss and storage modulus, according to the results of dynamic mechanical analysis. The experimental findings for the low-velocity impact tests indicate that the impact response and impact energy had a very good relationship. The photos from the ultrasonic C-scan reveal that as the impact energy rises, the damaged area expands. The impact damage had a clear relationship with the impact response, according to the compression after impact investigations, and the EP/30TSPF/70GF composite's compressive strength dropped as the impact energy increased. Overall, it can be said that the sugar palm composites' dynamic mechanical impact and post-impact properties are improved by the benzoylation process and the addition of glass fiber (Safri et al., 2019).

Atiqah et al. (2019) prepared an SPF-based hybrid composite to determine how different treatments, such as 6% alkaline (TNSP), 2% silane (TSSP), and 6% alkaline-2% silane (TNSSP), affected the physical and thermal properties of composites made of sugar palm, glass, and thermoplastic polyurethane. The combined alkaline-silane treated hybrid composites (TNSSP) had the lowest density, thickness swelling, and water absorption when compared to other composites when considering physical attributes. Compared to untreated SPF-based composites, treated composites were shown to have better thermal stability. Overall, the hybrid composites of processed sugar palm, glass, and thermoplastic polyurethane are suited for the production of automobiles (Atiqah et al., 2019).

The impact of microwave treatment on the tensile characteristics of thermoplastic polyurethane composites reinforced with treated SPF and 6% NaOH was studied by Mohammed et al. (2018). First, a 6% alkali solution was used to cure the SPFs. The alkali-treated SPF was then subjected to microwave treatment. In the microwave treatment, three different temperatures (i.e., 70°C, 80°C, and 90°C) were used. The polyurethane resin and SPFs were combined, and the hot press and extruder equipment were used to create the composites. Following the ASTM D-638 standard, the tensile properties (i.e., tensile strength, tensile modulus, and elongation at break) were investigated. With a microwave temperature of 70°C and a 6% alkali pre-treatment, the highest tensile strength of 18.42 MPa was recorded. Consequently, the temperature is 70°C (Mohammed et al., 2018).

A study by Mukhtar et al. (2019) examined how sodium bicarbonate treatment affects the physical and mechanical characteristics of hybrid and nonhybrid laminate composites made of glass fiber-reinforced polypropylene and sugar palm. The results will be contrasted with the standard alkali therapy. Utilizing a combination of the hot compression procedure and film stacking, laminate composites were created. Prior to manufacturing, sodium bicarbonate and 4 and 10 wt% of alkali, respectively, were applied to the naturally woven SPF in the mat. All of the laminate composites were tested for tensile strength, flexibility, and impact resistance as well as for water absorption and morphological analysis. Both sodium bicarbonate and alkaline treatments increased the tensile strength of the hybrid and nonhybrid composites. The alkaline-treated SPF composite (L03), which showed the highest value of 61.75 MPa, and the sodium bicarbonate-treated SPF composite (L04), which recorded 58.76 MPa compared to 53.01 MPa for the untreated SPF composite, both showed a rise that was more pronounced (L02). The flexural strength followed the same pattern. Overall, alkaline treatment outperformed sodium bicarbonate treatment in terms of performance (Mukhtar et al., 2019). Table 5.8 summarizes several other studies on SPF-reinforced hybrid composites.

5.2.5.6 Bioelectricity

Sugar palm bioelectricity refers to the generation of electrical energy using the sugar palm plant (*A. pinnata*) as a renewable biomass feedstock. It involves converting the energy stored in sugar palm biomass into electricity through various methods. Bioelectricity can be generated by burning lignocellulose feedstock, which is obtained from biomass sources such as agricultural waste and by-products, plantation forests, sawdust, and native forests (Lee et al., 2019).

The same processes that are used to manufacture other types of bioenergy, such as thermochemical conversion, can also be employed to make bioelectricity. The combustion process is one method for producing bioelectricity. In a chemical reaction known as combustion, biomass, and oxygen are mixed at high temperatures to produce carbon dioxide, water, and heat. The fuel's chemical energy is stored during combustion and released during combustion as heat, light, radiation, and other forms of energy. Biomass is converted during combustion into char and volatiles, which are then combined with oxygen to produce heat. To generate energy, the steam produced by the combustion process will run the steam turbine and produce bioelectricity (Lee et al., 2019). Figure 5.11 shows several types of thermochemical conversion that could lead to fuels and power.

For the production of renewable energy in Malaysia, a large portion of biomass from palm oil has the potential to replace coal. This palm oil biomass can reportedly generate 5,000 MW of energy at a 40% operating efficiency. Malaysia's yearly dependency on coal may be replaced by this substantial electricity potential. However, the potential has not been realized because of issues with technological stability, financial limits, and other issues with government policy. To achieve the main benefits of renewable energy and carbon neutrality, a coordinated decarbonization effort needs to be integrated into a framework. The creation of biofuel pellets, a palm oil production strategy, and

TABLE 5.8
SPF-Reinforced Hybrid Composites

Hybrids Fiber	Matrix	Process	Mechanical Findings	References
Woven glass	Unsaturated polyester	Compression molding		
Chopped glass fiber	Unsaturated polyester	Hand lay-up	Flexural strength = 17.5 MPa, tensile strength = 21.5 MPa, impact strength = 20.6 kJ/m^2	Afzaluddin et al. (2019)
Glass fiber	Thermoplastic polyurethane	Melt compounding		Nik Baihaqi et al. (2021)
Carbon fiber	Epoxy	Hand lay-up		Safri et al. (2020)
Glass fiber	Epoxy	Hand lay-up		Safri et al. (2019)
Seaweed fiber	Thermoplastic sugar palm starch/agar	Hydraulic thermo-process		Olivier-Bourbigou et al. (2010)
Roselle short fiber	Polyurethane composites	Melt mixing and hot compression	Flexural strength = 10.19 MPa, tensile strength = 13.6 MPa, impact strength = 29.4 kJ/m^2	Radzi et al. (2018)
Cornhusk	Cornstarch	Solution casting		Ibrahim et al. (2020)
Glass fiber	Unsaturated polyester	Sheet molding		Edhirej et al. (2017)
Cassava fiber	Cassava starch	Casting		Bachtiar et al. (2014)
Kenaf fiber	Polypropylene	Compression molding		Edhirej et al. (2020)
Cornstalk fiber	Cornstarch	Solution casting		Siregar et al. (2021)
Ramie fiber	Epoxy	Compression molding		Razali et al. (2018)
Roselle fiber	Vinyl ester	Hand lay-up		Mukhtar et al. (2019)
Glass fiber	Polypropylene	Film stacking and hot compression		Davoodi et al. (2010)
Seaweed (long fiber)	SP starch/agar	Melt mixing and hot compression	Flexural strength = 31.25 MPa, tensile strength = 17.74 MPa, impact strength = 6.0 kJ/m^2	Jumaidin et al. (2017)
Ramie (woven fiber)	Epoxy		Flexural strength = 80.70 MPa, tensile strength = 52.66 MPa	Siregar et al. (2021)

a biogas plant for the production of electricity and steam are a few of the suggested measures to decarbonize the energy sector (Zamri et al., 2022). As the sugar palm waste is available throughout the year and is abundant, these wastes can be utilized to produce bioelectricity.

5.2.6 Potential Applications of Sugar Palm Waste

Sugar palm waste has various potential industrial applications due to its rich chemical composition, which includes cellulose, hemicellulose, and lignin. Some potential industrial applications of sugar palm waste are listed below.

5.2.6.1 Automotive Industry

A modern car's total body weight is made up of 100–150 kg of polymer materials. The primary drivers of this increase in demand for biobased polymers and composites in automotive applications are the following: (1) lighter cars have better fuel economy; (2) lower cars have lower carbon dioxide emissions; (3) less reliance on non-renewable materials; and (4) more flexible designs have simpler assembly and disassembly. Government regulations that support consumer preferences and

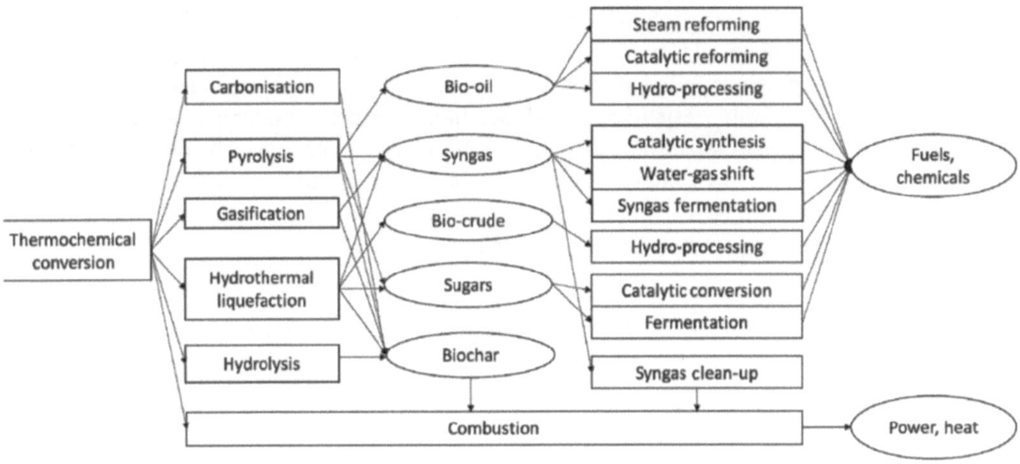

FIGURE 5.11 Options for producing fuels, chemicals, power, and heat via thermochemistry (Zamri et al., 2022).

environmental preservation are another factor in the rise in the usage of biobased components by the auto industry. The European Commission published Directive 2000/53/EG in 2005, which stipulated that 85% of a vehicle's weight must be recyclable. By 2015, this ratio is expected to increase to 95% (Leman et al., 2008).

In exterior and interior components for semi- and non-structural applications, plant fibers currently meet performance standards for elongation, ultimate breaking force, impact strength, flexibility, flammability, fogging characteristics, acoustic absorption, odor, dimensional stability, suitability for processing dwell time and temperature, crash resistance, and water absorption. Natural fiber composites are already being used by a number of well-known manufacturers, including Volkswagen-Audi, Daimler-Chrysler, and Opel-GM, in components of passenger cars such as the rear parcel shelf, door trim panels, and seat squabs (Magurno, 1999).

5.2.6.2 Packaging Industries

According to Jawaid and Abdul Khalil (2011), natural fibers can be divided into three groups: those with an animal origin, those with a mineral origin, and those with a plant origin (cellulose). In composites for food packaging, natural fibers like cellulose and lignocellulose are frequently used (Halimatul et al., 2019; Jumaidin et al., 2019a,b, 2021; Nazrin et al., 2021; Suriani et al., 2021b). Utilizing a biodegradable polymer matrix has been found to have various benefits, including low cost, low density with good features, and non-harmful processing, in investigations of natural fibers. As a result, the filler and matrix will affect how biodegradable the composite for food packaging is. Water availability affects biodegradation processes by promoting microbial attack and matrix hydrolysis. Natural fibers are a harsher substrate for bacterial growth because of their capacity to absorb water. The rate of biodegradation is affected by temperature, humidity, and the total number of bacteria present because these elements hasten the process (Ludueña et al., 2012).

The oxygen penetration into the packaging material that comes into touch with the food must be considered while creating food packaging, among other factors (Cichello, 2015). In their investigation into ways to improve food packaging's water vapor permeability (Berthet et al., 2015) discovered that composite materials are suitable for items that breathe, such as fruits. It has been revealed that wheat straw fibers speed up the evaporation of water, making the composite suitable for use in food products. A lot of focus has recently been placed on the packaging materials cellophane, cellulose acetate, polylactic acid (PLA), and starch. While products like cheese and coffee can be packaged with cellophane, PLA is frequently combined with other biodegradable polymers to enhance its packaging-friendly features. Furthermore, cellulose acetate is rarely employed in product packaging due to its higher cost compared to poly(butylene adipate-co-terephthalate) (PBAT), polycaprolactone (PCL), and polybutylene succinate (Shahroze et al., 2018).

Throughout this study, starch–polymer blends have been suggested as a viable choice for sustainable packaging. A green material formed from plants called starch is composed of amylose and amylopectin. A branching polymer called amylopectin can improve a material's mechanical properties, processing capacity, and water resistance. Amylopectin is a branching polymer, whereas amylose is a linear polymer. The starch-based nanocomposite film has also proven to be extremely durable and has the potential to be environmentally friendly plastic bags and food packaging. According to research by Hazrati et al. (2019), nanofiller can boost mechanical, gas, and water vapor barrier properties while also using less material during production.

5.2.7 Challenges Associated with Processing and Commercializing

SPFs have a number of benefits, but they also have limitations that still limit their use in the composites sector. SPFs are unable to offer a physical property pattern that is consistent and uniform. They generally differ from one harvesting season to another or even from one plant to another in terms of their features (Faruk et al., 2012). SPFs differ in their qualities depending on four factors: (1) the plant's growth environment (i.e., the amount of rain and the soil), (2) its age, (3) the portion of the plant from which the fibers are removed, and (4) the method used to harvest and prepare the fiber (Sanyang et al., 2016).

Natural fibers' high moisture sensitivity has severely limited their ability to be used in the production of durable composite for outdoor applications (such as exterior car parts), and SPF is no exception. SPFs' low microbial resistance and susceptibility to rotting are caused by their hydrophilic nature (Koronis et al., 2013). Consideration should be given to this issue, particularly during composite processing, long-term storage, and shipping. Because cellulose in SPFs is hygroscopic, it can absorb water from the environment and swell. Their swelling alters the composite's dimensional stability as well as its mechanical and physical properties (Burgueño et al., 2005).

Their poor fiber–matrix adhesion is one of the most difficult issues facing the application of natural fibers. Natural fibers are hydrophilic, but most polymeric matrices (such as thermoplastics and thermosets) are hydrophobic ("non-polar" in nature). The compatibility between the two is therefore low. For the matrix to smoothly transfer the load to the stiff fibers through shear stresses, the interface between SPFs and the matrix needs to be bonded strongly (Ghelli & Minak, 2011). Therefore, the mechanical properties of SPF-reinforced polymer composites are significantly decreased by poor fiber–matrix adhesion. Under these circumstances, the composite's full potential cannot be realized, and it may be subject to environmental attacks that severely reduce its mechanical properties and cause delamination. Chemical "coupling" or "compatabilizing" chemicals can improve the affinity and adhesion between natural fibers and thermoplastic matrices during manufacture (Koronis et al., 2013).

Natural fibers' poor thermal stability is another barrier to their widespread use in composites (Rashid et al., 2016). Below 200°C, natural fibers can withstand processing temperatures; above that point, they begin to deteriorate and shrink. When fibers are exposed to high temperatures, depolymerization, hydrolysis, oxidation, dehydration, decarboxylation, and recrystallization take place, which alter the fibers' physical and/or chemical structures (Azwa et al., 2013). Limiting the processing temperature, pressure, and time range is necessary to prevent these processing problems (Villa et al., 2005).

5.3 CONCLUSIONS

In conclusion, managing sugar palm waste is essential for both economic growth and environmental sustainability. There are many opportunities for the leaves, fronds, trunks, and fibers produced by sugar palm trees to be used in productive and sustainable ways. The use of ecologically friendly techniques like composting and mulching is a crucial component of sugar palm waste management. Composting the waste allows for the return of beneficial nutrients to the soil, increasing its fertility and obviating the need for artificial fertilizers. Mulching can also assist in maintaining soil moisture, inhibit weed growth, and enhance overall soil health. The creation of biofuels and bioenergy from sugar palm waste is a significant additional use. Sugar palm waste is a great source of cellulose and lignocellulosic material for ethanol, biogas, and solid biomass fuel generation. This not only lessens reliance on fossil fuels but also gives rural communities the chance to make money and increase their level of energy independence. Additionally, there is a growing trend toward using sugar palm waste in a variety of sectors. The fibers and trunks can be converted into plywood, medium-density fiberboard, and particleboard, which are sustainable substitutes for conventional timber products. The leaves and fronds can be used to make paper, packaging, and handicrafts, promoting regional craftsmanship and generating extra income. In summary, efficient sugar palm waste treatment and application can support the circular economy, rural development, and environmental sustainability. We can increase the value of sugar palm waste, lessen its negative effects on the environment, and generate economic opportunities while promoting a greener and more sustainable future by putting forward creative and sustainable practices.

ACKNOWLEDGEMENT

This chapter was written with Muhammad Imraan Khamarul Arifin, Ng Kai En, and Kee Nadhrah Izzatul Iman Uzair.

REFERENCES

Abd El-Sayed, E. S., El-Sakhawy, M., & El-Sakhawy, M. A. M. (2020). Non-wood fibers as raw material for pulp and paper industry. *Nordic Pulp and Paper Research Journal*, *35*(2), 215–230. https://doi.org/10.1515/npprj-2019-0064

Adawiyah, D. R., Sasaki, T., & Kohyama, K. (2013). Characterization of arenga starch in comparison with sago starch. *Carbohydrate Polymers*, *92*(2), 2306–2313. https://doi.org/10.1016/j.carbpol.2012.12.014

Afzaluddin, A., Jawaid, M., Salit, M. S., & Ishak, M. R. (2019). Physical and mechanical properties of sugar palm/glass fiber reinforced thermoplastic polyurethane hybrid composites. *Journal of Materials Research and Technology*, *8*(1), 950–959. https://doi.org/10.1016/j.jmrt.2018.04.024

Ahmad, R., Jilani, G., Arshad, M., Zahir, Z. A., & Khalid, A. (2007). Bio-conversion of organic wastes for their recycling in agriculture: An overview of perspectives and prospects. *Annals of Microbiology*, *57*(4), 471–479.

Alisjahbana, A. S., & Busch, J. M. (2017). Forestry, forest fires, and climate change in Indonesia. *Bulletin of Indonesian Economic Studies*, *53*(2), 111–136. https://doi.org/10.1080/00074918.2017.1365404

Ammar, I. M., Huzaifah, M. R. M., Sapuan, S. M., Ishak, M. R., & Leman, Z. B. (2018). Development of sugar palm fiber reinforced vinyl ester composites. In *Natural Fibre Reinforced Vinyl Ester and Vinyl Polymer Composites*. Elsevier Ltd. https://doi.org/10.1016/b978-0-08-102160-6.00011-1

Ansar, Nazaruddin, Azis, A. D., & Fudholi, A. (2021). Enhancement of bioethanol production from palm sap (Arenga pinnata (Wurmb) Merr) through optimization of Saccharomyces cerevisiae as an inoculum. *Journal of Materials Research and Technology*, *14*, 548–554. https://doi.org/10.1016/j.jmrt.2021.06.085

Asyraf, M. R. M., Ishak, M. R., Norrrahim, M. N. F., Nurazzi, N. M., Shazleen, S. S., Ilyas, R. A., Rafidah, M., & Razman, M. R. (2021). Recent advances of thermal properties of sugar palm lignocellulosic fibre reinforced polymer composites. *International Journal of Biological Macromolecules*, *193*(PB), 1587–1599. https://doi.org/10.1016/j.ijbiomac.2021.10.221

Atiqah, A., Jawaid, M., Sapuan, S. M., Ishak, M. R., Ansari, M. N. M., & Ilyas, R. A. (2019). Physical and thermal properties of treated sugar palm/glass fibre reinforced thermoplastic polyurethane hybrid composites. *Journal of Materials Research and Technology*, *8*(5), 3726–3732. https://doi.org/10.1016/j.jmrt.2019.06.032

Ayilara, M. S., Olanrewaju, O. S., Babalola, O. O., & Odeyemi, O. (2020). Waste management through composting: Challenges and potentials. *Sustainability (Switzerland)*, *12*(11). https://doi.org/10.3390/su12114456

Azwa, Z. N., Yousif, B. F., Manalo, A. C., & Karunasena, W. (2013). A review on the degradability of polymeric composites based on natural fibres. *Materials and Design*, *47*, 424–442. https://doi.org/10.1016/j.matdes.2012.11.025

Bachtiar, D., Sapuan, S. M., & Hamdan, M. M. (2010). Flexural properties of alkaline treated sugar palm fibre reinforced epoxy composites. *International Journal of Automotive and Mechanical Engineering*, *1*(1), 79–90. https://doi.org/10.15282/ijame.1.2010.7.0007

Bachtiar, D., Siregar, J. P., bin Sulaiman, A. S., & bin Mat Rejab, M. R. (2014). Tensiles properties of hybrid sugar palm/kenaf fibre reinforced polypropylene composites. *Applied Mechanics and Materials, 695*, 155–158.

Berthet, M. A., Angellier-Coussy, H., Machado, D., Hilliou, L., Staebler, A., Vicente, A., & Gontard, N. (2015). Exploring the potentialities of using lignocellulosic fibres derived from three food by-products as constituents of biocomposites for food packaging. *Industrial Crops and Products*, *69*, 110–122. https://doi.org/10.1016/j.indcrop.2015.01.028

Bogataj, V., Fajs, P., Peñalva, C., Omahen, M., Čop, M., & Henttonen, A. (2019). Utilization of recycled polypropylene, cellulose and newsprint fibres for production of green composites. *Detritus*, *7*(September), 36–43. https://doi.org/10.31025/2611-4135/2019.13857

Burgueño, R., Quagliata, M. J., Mohanty, A. K., Mehta, G., Drzal, L. T., & Misra, M. (2005). Hybrid biofiber-based composites for structural cellular plates. *Composites Part A: Applied Science and Manufacturing*, *36*(5), 581–593. https://doi.org/10.1016/j.compositesa.2004.08.004

Cichello, S. A. (2015). Oxygen absorbers in food preservation: A review. *Journal of Food Science and Technology*, *52*(4), 1889–1895. https://doi.org/10.1007/s13197-014-1265-2

Davis, T. A., & Johnson, D. V. (1987). Current utilization and further development of the palmyra palm (Borassus flabellifer L., Arecaceae) in Tamil Nadu State, India. *Economic Botany*, *41*(2), 247–266.

Davoodi, M. M., Sapuan, S. M., Ahmad, D., Ali, A., Khalina, A., & Jonoobi, M. (2010). Mechanical properties of hybrid kenaf/glass reinforced epoxy composite for passenger car bumper beam. *Materials & Design*, *31*(10), 4927–4932. https://doi.org/10.1016/j.matdes.2010.05.021

Edhirej, A., Sapuan, S. M., Jawaid, M., & Zahari, N. I. (2017). Cassava/sugar palm fiber reinforced cassava starch hybrid composites: Physical, thermal and structural properties. *International Journal of Biological Macromolecules*, *101*, 75–83. https://doi.org/10.1016/j.ijbiomac.2017.03.045

Edhirej, A., Sapuan, S. M., Jawaid, M., Zahari, N. I. (2020). Processing and characterization of cornstalk/sugar palm fiber reinforced cornstarch biopolymer hybrid composites. In Advanced Processing, Properties, and Applications of Starch and Other Bio-Based Polymers, Mohammad Jawaid, Mohammad Thariq Hameed Sultan, and Naheed Saba (Eds.), (pp. 35–46). Amsterdam, Netherlands: Elsevier.

Faruk, O., Bledzki, A. K., Fink, H. P., & Sain, M. (2012). Biocomposites reinforced with natural fibers: 2000–2010. *Progress in Polymer Science*, *37*(11), 1552–1596. https://doi.org/10.1016/j.progpolymsci.2012.04.003

Fitriana, N. E., Suwanto, A., Jatmiko, T. H., Mursiti, S., & Prasetyo, D. J. (2020). Cellulose extraction from sugar palm (Arenga pinnata) fibre by alkaline and peroxide treatments. *IOP Conference Series: Earth and Environmental Science*, 462(1). https://doi.org/10.1088/1755-1315/462/1/012053

Garkhail, S. K., Heijenrath, R. W. H., & Peijs, T. (2000). Mechanical properties of natural-fibre-mat-reinforced thermoplastics based on flax fibres and polypropylene. *Applied Composite Materials*, *7*(5–6), 351–372. https://doi.org/10.1023/A:1026590124038

Ghelli, D., & Minak, G. (2011). Low velocity impact and compression after impact tests on thin carbon/epoxy laminates. *Composites Part B: Engineering*, *42*(7), 2067–2079. https://doi.org/10.1016/j.compositesb.2011.04.017

Gnansounou, E., & Dauriat, A. (2005). Ethanol fuel from biomass: A review. *Journal of Scientific & Industrial Research, 64,* 809–821.

Halimatul, M. J., Sapuan, S. M., Jawaid, M., Ishak, M. R., & Ilyas, R. A. (2019). Water absorption and water solubility properties of sago starch biopolymer composite films filled with sugar palm particles. *Polimery*, *64*(9), 27–35. https://doi.org/10.14314/polimery.2019.9.4

Hassan, M. M., & Wagner, M. H. (2016). Surface modification of natural fibers for reinforced polymer composities: A critical review. *Reviews on Adhesion and Adhesives*, 4, 1–46.

Hazrati, K. Z., Sapuan, S. M., & Ilyas, R. A. (2019). Biobased food packaging using natural fibre: A review. *Prosiding Seminar Enau Kebangsaan 2019* (pp. 140–142), April. Kuala Pilah, Negeri Sembilan, Malaysia.

Heyne, K. (1950). *De Nuttige planten van Indonesie [The Useful Plants in Indonesia]*. NV Uitgeverij W. van Hoeve, Bandung, Indonesia.

Hossain, N., & Jalil, R. (2017). Sugar and bioethanol production from oil palm trunk (OPT). *Asia Pacific Journal of Energy and Environment*, *4*(1). https://doi.org/10.18034/apjee.v4i1.237

Husain, Z., Zainac, Z., & Abdullah, Z. (2002). Briquetting of palm fibre and shell from the processing of palm nuts to palm oil. *Biomass and Bioenergy*, *22*, 505–509.

Huzaifah, M. R. M., Sapuan, S. M., Leman, Z., Ishak, M. R., & Maleque, M. A. (2017). A review of sugar palm (Arenga pinnata): Application, fibre characterisation and composites. *Multidiscipline Modeling in Materials and Structures*, *13*(4), 678–698. https://doi.org/10.1108/MMMS-12-2016-0064

Ibrahim, M. I. J., Sapuan, S. M., Zainudin, E. S., & Zuhri, M. Y. M. (2020). Preparation and characterization of cornhusk/sugar palm fiber reinforced Cornstarch-based hybrid composites. *Journal of Materials Research and Technology*, *9*(1), 200–211. https://doi.org/10.1016/j.jmrt.2019.10.045

Ilyas, R. A., Sapuan, S. M., Ishak, M. R., & Zainudin, E. S. (2017). Effect of delignification on the physical, thermal, chemical, and structural properties of sugar palm fibre. *BioResources*, *12*(4), 8734–8754. https://doi.org/10.15376/biores.12.4.8734-8754

Imraan, M., Ilyas, R. A., Norfarhana, A. S., Bangar, S. P., Knight, V. F., & Norrrahim, M. N. F. (2023). Sugar palm (Arenga pinnata) fibers: New emerging natural fibre and its relevant properties, treatments and potential applications. *Journal of Materials Research and Technology*, *24*, 4551–4572. https://doi.org/10.1016/j.jmrt.2023.04.056

Ishak, M. R., Leman, Z., Salit, M. S., Rahman, M. Z. A., Anwar Uyup, M. K., & Akhtar, R. (2013a). IFSS, TG, FT-IR spectra of impregnated sugar palm (Arenga pinnata) fibres and mechanical properties of their composites. *Journal of Thermal Analysis and Calorimetry*, *111*(2), 1375–1383. https://doi.org/10.1007/s10973-012-2457-5

Ishak, M. R., Leman, Z., Sapuan, S. M., Rahman, M. Z. A., & Anwar, U. M. K. (2013b). Chemical composition and FT-IR spectra of sugar palm (Arenga pinnata) fibers obtained from different heights. *Journal of Natural Fibers*, *10*(2), 83–97. https://doi.org/10.1080/15440478.2012.733517

Ishak, M. R., Sapuan, S. M., Leman, Z., Rahman, M. Z. A., & Anwar, U. M. K. (2012). Characterization of sugar palm (Arenga pinnata) fibres tensile and thermal properties. *Journal of Thermal Analysis and Calorimetry*, *109*(2), 981–989. https://doi.org/10.1007/s10973-011-1785-1

Ishak, M. R., Sapuan, S. M., Leman, Z., Rahman, M. Z. A., Anwar, U. M. K., & Siregar, J. P. (2013c). Sugar palm (Arenga pinnata): Its fibres, polymers and composites. *Carbohydrate Polymers*, *91*(2), 699–710. https://doi.org/10.1016/j.carbpol.2012.07.073

Iwuozor, K. O., Emenike, E. C., Ighalo, J. O., Omoarukhe, F. O., Omuku, P. E., & Adeniyi, A. G. (2022). A review on the thermochemical conversion of sugarcane bagasse into biochar. *Cleaner Materials*, *6*. Elsevier Ltd. https://doi.org/10.1016/j.clema.2022.100162

Izwan, S. M., Sapuan, S. M., Zuhri, M. Y. M., & Mohamed, A. R. (2020). Effects of benzoyl treatment on NaOH treated sugar palm fiber: Tensile, thermal, and morphological properties. *Journal of Materials Research and Technology*, *9*(3), 5805–5814. https://doi.org/10.1016/j.jmrt.2020.03.105

Jawaid, M., & Abdul Khalil, H. P. S. (2011). Cellulosic/synthetic fibre reinforced polymer hybrid composites: A review. *Carbohydrate Polymers*, *86*(1), 1–18. https://doi.org/10.1016/j.carbpol.2011.04.043

Jumaidin, R., Adam, N. W., Ilyas, R. A., Hussin, M. S. F., Taha, M. M., Mansor, M. R., Azlan, U. A. A., & Yob, M. S. (2019a). Water transport and physical properties of sugarcane bagasse fibre reinforced thermoplastic potato starch biocomposite. *Journal of Advanced Research in Fluid Mechanics and Thermal Sciences*, *61*(2), 273–281.

Jumaidin, R., Diah, N. A., Ilyas, R. A., Alamjuri, R. H., & Yusof, F. A. M. (2021). Processing and characterisation of banana leaf fibre reinforced thermoplastic cassava starch composites. *Polymers*, *13*(9), 1420. https://doi.org/10.3390/polym13091420

Jumaidin, R., Saidi, Z. A. S., Ilyas, R. A., Ahmad, M. N., Wahid, M. K., Yaakob, M. Y., Maidin, N. A., Rahman, M. H. A., & Osman, M. H. (2019b). Characteristics of cogon grass fibre reinforced thermoplastic cassava starch biocomposite: Water absorption and physical properties. *Journal of Advanced Research in Fluid Mechanics and Thermal Sciences*, *62*(1), 43–52.

Jumaidin, R., Sapuan, S. M., Jawaid, M., Ishak, M. R., & Sahari, J. (2017). Thermal, mechanical, and physical properties of seaweed/sugar palm fibre reinforced thermoplastic sugar palm Starch/Agar hybrid composites. *International Journal of Biological Macromolecules*, *97*, 606–615. https://doi.org/10.1016/j.ijbiomac.2017.01.079

Kalia, S., Kaith, B. S., & Kaur, I. (2009). Pretreatments of natural fibers and their application as reinforcing material in polymer composites—A review. *Polymer Engineering and Science*, *49*(7), 1253–1272. https://doi.org/10.1002/pen.21328

Khalil, A., Domeizel, M., & Prudent, P. (2008). Monitoring of green waste composting process based on redox potential. *Bioresource Technology*, *99*(14), 6037–6045. https://doi.org/10.1016/j.biortech.2007.11.043

Koronis, G., Silva, A., & Fontul, M. (2013). Green composites: A review of adequate materials for automotive applications. *Composites Part B: Engineering*, *44*(1), 120–127. https://doi.org/10.1016/j.compositesb.2012.07.004

Larasati, N. D., Budiyanto, G., & Widyastuti, T. (2017). Application of cow rumen liquid in palm sugar waste compost for cultivating sweet corn in coastal sandy soil of Samas Beach Bantul. *Planta Tropika: Journal of Agro Science*, *5*(2). https://doi.org/10.18196/pt.2017.069.96-105

Latif, R., Wakeel, S., Khan, N. Z., Noor Siddiquee, A., Lal Verma, S., & Akhtar Khan, Z. (2019). Surface treatments of plant fibers and their effects on mechanical properties of fiber-reinforced composites: A review. *Journal of Reinforced Plastics and Composites*, *38*(1), 15–30. https://doi.org/10.1177/0731684418802022

Lay, A. (2009). Rekayasa teknologi alat pengolahan bioetanol dari nira aren. *Bulletin Palma*, *37*(2009, December), 100–114.

Lee, S. Y., Sankaran, R., Chew, K. W., Tan, C. H., Krishnamoorthy, R., Chu, D.-T., & Show, P.-L. (2019). Waste to bioenergy: A review on the recent conversion technologies. *BMC Energy*, *1*(1). https://doi.org/10.1186/s42500-019-0004-7

Leman, Z., Sapuan, S. M., Saifol, A. M., Maleque, M. A., & Ahmad, M. M. H. M. (2008). Moisture absorption behavior of sugar palm fiber reinforced epoxy composites. *Materials and Design*, *29*(8), 1666–1670. https://doi.org/10.1016/j.matdes.2007.11.004

Lilholt, H., & Lawther, J. M. (2000). Natural organic fibers. In *Comprehensive Composite Materials* (pp. 303–325). Elsevier. https://doi.org/10.1016/b0-08-042993-9/00048-6

Ludueña, L., Vázquez, A., & Alvarez, V. (2012). Effect of lignocellulosic filler type and content on the behavior of polycaprolactone based eco-composites for packaging applications. *Carbohydrate Polymers*, *87*(1), 411–421. https://doi.org/10.1016/j.carbpol.2011.07.064

Magurno, A. (1999). Vegetable fibres in automotive interior components. *Die Angewandte Makromolekulare Chemie*, *272*, 99–107. https://doi.org/10.1002/(SICI)1522-9505(19991201)272:1<99::AID-APMC99>3.0.CO;2-C

Mathivanan, D., Parlaungan Siregar, J., Ruzaimi Mat Rejab, M., Bachtiar, D., & Cionita, T. (2015). The mechanical properties of alkaline treated pineapple leaf fibre to reinforce tapioca based bioplastic resin composite. *Materials Science Forum*, 882, 66–70

Mogea, M. J., Seibert, B., & Smits, W. (1991). Multipurpose palms: The sugar palm (Arenga pinnata (Wurmb) Merr.). *Agroforestry Systems*, *13(2),* 111–129.

Mohamed, A. Z., Ainun, Z. M. A., Norzita, Y., Latifah, J., Amran, S., Ilyas, R. A., & Huzaifah, M. R. M. (2021). Pulp and papermaking potential of sugar palm. *Nanocellulose reinforced Biopolymer View Project International Conference on Sugar Palm and Allied Fibre Polymer Composites 2021*. https://www.researchgate.net/publication/357186710

Mohammed, A. A., Bachtiar, D., Rejab, M. R. M., & Siregar, J. P. (2018). Effect of microwave treatment on tensile properties of sugar palm fibre reinforced thermoplastic polyurethane composites. *Defence Technology*, *14*(4), 287–290. https://doi.org/10.1016/j.dt.2018.05.008

Moradi-Choghamarani, F., Moosavi, A. A., & Baghernejad, M. (2019a). Determining organo-chemical composition of sugarcane bagasse-derived biochar as a function of pyrolysis temperature using proximate and Fourier transform infrared analyses. *Journal of Thermal Analysis and Calorimetry*, *138*(1), 331–342. https://doi.org/10.1007/s10973-019-08186-9

Moradi-Choghamarani, F., Moosavi, A. A., Sepaskhah, A. R., & Baghernejad, M. (2019b). Physico-hydraulic properties of sugarcane bagasse-derived biochar: The role of pyrolysis temperature. *Cellulose*, *26*(12), 7125–7143. https://doi.org/10.1007/s10570-019-02607-6

Mukhtar, I., Leman, Z., Ishak, M. R., & Zainudin, E. S. (2016). Sugar palm fibre and its composites: A review of recent developments. *BioResources*, *11*(4), 10756–10782.

Mukhtar, I., Leman, Z., Zainudin, E. S., & Ishak, M. R. (2019). Hybrid and nonhybrid laminate composites of sugar palm and glass fibre-reinforced polypropylene: Effect of alkali and sodium bicarbonate treatments. *International Journal of Polymer Science*, *2019*. https://doi.org/10.1155/2019/1230592

Nazrin, A., Sapuan, S. M., Zuhri, M. Y. M., Tawakkal, I. S. M. A., & Ilyas, R. A. (2021). Water barrier and mechanical properties of sugar palm crystalline nanocellulose reinforced thermoplastic sugar palm starch (TPS)/poly(lactic acid) (PLA) blend bionanocomposites. *Nanotechnology Reviews*, *10*(1), 431–442. https://doi.org/10.1515/ntrev-2021-0033

Nik Baihaqi, N. M. Z., Khalina, A., Mohd Nurazzi, N., Aisyah, H. A., Sapuan, S. M., & Ilyas, R. A. (2021). Effect of fiber content and their hybridization on bending and torsional strength of hybrid epoxy composites reinforced with carbon and sugar palm fibers. *Polimery*, *66*(1), 36–43. https://doi.org/10.14314/polimery.2021.1.5

Norfarhana, A. S., Ilyas, R. A., & Ngadi, N. (2022). A review of nanocellulose adsorptive membrane as multifunctional wastewater treatment. *Carbohydrate Polymers*, *291*, 119563. https://doi.org/10.1016/j.carbpol.2022.119563

Norfarhana, A. S., Ilyas, R. A., Ngadi, N., Othman, M. H. D., Misenan, M. S. M., & Norrahim, M. N. F. (2024a). Revolutionizing lignocellulosic biomass: A review of harnessing the power of ionic liquids for sustainable utilization and extraction. *International Journal of Biological Macromolecules*, *256*, 128256. https://doi.org/10.1016/j.ijbiomac.2023.128256

Norfarhana, A. S., Khoo, P. S., Ilyas, R. A., Ab Hamid, N. H., Aisyah, H. A., Norrahim, M. N. F., Knight, V. F., Rani, M. S. A., Septevani, A. A., Syafri, E., & Annamalai, P. K. (2024b). Exploring of cellulose nanocrystals from lignocellulosic sources as a powerful adsorbent for wastewater remediation. *Journal of Polymers and the Environment*. https://doi.org/10.1007/s10924-024-03227-3

Nurazzi, N. M., Asyraf, M. R. M., Fatimah Athiyah, S., Shazleen, S. S., Rafiqah, S. A., Harussani, M. M., Kamarudin, S. H., Razman, M. R., Rahmah, M., Zainudin, E. S., Ilyas, R. A., Aisyah, H. A., Norrahim, M. N. F., Abdullah, N., Sapuan, S. M., & Khalina, A. (2021a). A review on mechanical performance of hybrid natural fiber polymer composites for structural applications. *Polymers*, *13*(13), 2170. https://doi.org/10.3390/polym13132170

Nurazzi, N. M., Harussani, M. M., Aisyah, H. A., Ilyas, R. A., Norrahim, M. N. F., Khalina, A., & Abdullah, N. (2021b). Treatments of natural fiber as reinforcement in polymer composites—A short review. *Functional Composites and Structures*, *3*(2), 024002. https://doi.org/10.1088/2631-6331/abff36

Ogunlalu, O., Oyekunle, I. P., Iwuozor, K. O., Aderibigbe, A. D., & Emenike, E. C. (2021). Trends in the mitigation of heavy metal ions from aqueous solutions using unmodified and chemically-modified agricultural waste adsorbents. *Current Research in Green and Sustainable Chemistry*, *4*. https://doi.org/10.1016/j.crgsc.2021.100188

Olivier-Bourbigou, H., Magna, L., & Morvan, D. (2010). Ionic liquids and catalysis: Recent progress from knowledge to applications. *Applied Catalysis A: General*, *373*(1–2), 1–56. https://doi.org/10.1016/j.apcata.2009.10.008

Olsson, J., Novy, V., Nielsen, F., Wallberg, O., & Galbe, M. (2019). Sequential fractionation of the lignocellulosic components in hardwood based on steam explosion and hydrotropic extraction. *Biotechnology for Biofuels*, *12*(1), 1–12. https://doi.org/10.1186/s13068-018-1346-y

Radzi, A. M., Sapuan, S. M., Jawaid, M., & Mansor, M. R. (2018). Mechanical performance of roselle/sugar palm fiber hybrid reinforced polyurethane composites. *BioResources*, *13*(3), 6238–6249. https://doi.org/10.15376/biores.13.3.6238-6249

Rashid, B., Leman, Z., Jawaid, M., Ghazali, M. J., & Ishak, M. R. (2016). Physicochemical and thermal properties of lignocellulosic fiber from sugar palm fibers: Effect of treatment. *Cellulose*, *23*(5), 2905–2916. https://doi.org/10.1007/s10570-016-1005-z

Rashid, B., Leman, Z., Jawaid, M., Ghazali, M. J., & Ishak, M. R. (2017). Influence of treatments on the mechanical and thermal properties of sugar palm fibre reinforced phenolic. *BioResources*, *12*, 1447–1462.

Razali, N., Sapuan, S. M., & Razali, N. (2018). Mechanical properties and morphological analysis of roselle/sugar palm fiber reinforced vinyl ester hybrid composites. In Vinyl Ester-Based Biocomposites, Mohammad Jawaid, Mohamed Thariq Hameed Sultan, and Naheed Saba (Eds.), (pp. 169–180). Boca Raton, Florida, USA: CRC Press (Taylor & Francis Group).

Reddy, N., Yang, Y. (2015). *Innovative Biofibers from Renewable Resources* (p. 1e454). https://doi.org/10.1007/978-3-662-45136-6.

Redhead, J. (1989). Utilization of Tropical Foods: Trees. FAO Food and Nutrition Paper No 47: 3. FAO, Rome, p. 52.

Rong, M. Z., Zhang, M. Q., Liu, Y., Yang, G. C., & Zeng, H. M. (2001). The effect of fiber treatment on the mechanical properties of unidirectional sisal-reinforced epoxy composites. *Composites Science and Technology*, *61*(10), 1437–1447. https://doi.org/10.1016/s0266-3538(01)00046-x

Safri, S. N. A., Sultan, M. T. H., Jawaid, M., & Majid, M. S. A. (2019). Analysis of dynamic mechanical, low-velocity impact and compression after impact behaviour of benzoyl treated sugar palm/glass/epoxy composites. *Composite Structures*, *226*(July), 111308. https://doi.org/10.1016/j.compstruct.2019.111308

Safri, S. N. A., Sultan, M. T. H., & Shah, A. U. M. (2020). Characterization of benzoyl treated sugar palm/glass fibre hybrid composites. *Journal of Materials Research and Technology*, *9*(5), 11563–11573. https://doi.org/10.1016/j.jmrt.2020.08.057

Sahari, J., Sapuan, S. M., Ismarrubie, Z. N., Rahman, M. Z. A. (2012a). Physical and chemical properties of different morphological parts of the sugar palm fibres. *Fibres & Textiles in Eastern Europe*, *20*, 21–24.

Sahari, J., Sapuan, S. M., Zainudin, E. S., & Maleque, M. A. (2012b). A new approach to use arenga pinnata as sustainable biopolymer: Effects of plasticizers on physical properties. *Procedia Chemistry*, *4*, 254–259. https://doi.org/10.1016/j.proche.2012.06.035

Sahari, J., Sapuan, S. M., Zainudin, E. S., & Maleque, M. A. (2012c). Sugar palm tree: A versatile plant and novel source for biofibres, biomatrices, and biocomposites. *Polymers from Renewable Resources*, *3*(2), 61–78. https://doi.org/10.1177/204124791200300203

Sanjay, M. R., Arpitha, G. R., Naik, L. L., Gopalakrishna, K., & Yogesha, B. (2016). Applications of natural fibers and its composites: An overview. *Natural Resources*, *07*(03), 108–114. https://doi.org/10.4236/nr.2016.73011

Sanyang, M. L., Sapuan, S. M., Jawaid, M., Ishak, M. R., & Sahari, J. (2016). Recent developments in sugar palm (Arenga pinnata) based biocomposites and their potential industrial applications: A review. *Renewable and Sustainable Energy Reviews*, *54*, 533–549. https://doi.org/10.1016/j.rser.2015.10.037

Sarjiyah, & Handoko, N. A. (2022). The balance composition of urea and sugar palm bagasse compost and its effects on growth and yield of soybean (Glycine Max L.). *IOP Conference Series: Earth and Environmental Science*, *985*(1). https://doi.org/10.1088/1755-1315/985/1/012059

Sath, K., Borin, K., Preston, T. R. (2008). Survey on feed utilization for cattle production in Takeo province. *Livestock Research for Rural Development*, *20*(Suppl.), 2008.

Sahari, **J.,** Sapuan, S. M**.,** Ismarrubie**,** Z. N**., &** Rahman**,** M. Z. A**.** (2012). Physical and Chemical Properties of Different Morphological Parts of Sugar Palm Fibres. Fibres & Textiles in Eastern Europe, *20*(2), 23–26.

Shahroze, R. M., Ishak, M. R., Sapuan, M., Leman, Z., Asim, M., & Chandrasekar, M. (2018). Sugar palm nanocomposites. *BioResources*, *13*(4), 7430–7444.

Shahroze, R. M., Muthukumar, C., Krishnasamy, S., & Senthilmuthukumar, T. (2019). A review on the various fibre treatment techniques used for the fibre surface. *Prosiding Seminar Enau Kebangsaan 2019* (pp. 48–52), April. Kuala Pilah, Negeri Sembilan, Malaysia.

Siregar, J. P. (2005). Tensile and flexural properties of Arenga Pinnata filament (Ijuk filament) reinforced epoxy composites. Master's thesis, Universiti Putra Malaysia, Serdang, Malaysia.

Siregar, J. P., Zalinawati, M., Cionita, T., Rejab, M. R. M., Mawarnie, I., Jaafar, J., & Hamdan, M. H. M. (2021). Mechanical properties of hybrid sugar palm/ramie fibre reinforced epoxy composites. *Materials Today: Proceedings*, *46*, 1729–1734. https://doi.org/10.1016/j.matpr.2020.07.565

Srivastava, A. K., Agrawal, P., & Rahiman, A. (2007). Delignification of rice husk and production of bioethanol. *International Journal of Innovative Research in Science, Engineering and Technology (An ISO (Vol. 3297)*, *3*(3), 10187–10194. www.ijirset.com

Suriani, M. J., Ilyas, R. A., Zuhri, M. Y. M., Khalina, A., Sultan, M. T. H., Sapuan, S. M., Ruzaidi, C. M., Wan, F. N., Zulkifli, F., Harussani, M. M., Azman, M. A., Radzi, F. S. M., & Sharma, S. (2021a). Critical review of natural fiber reinforced hybrid composites: Processing, properties, applications and cost. *Polymers*, *13*(20), 3514. https://doi.org/10.3390/polym13203514

Suriani, M. J., Radzi, F. S. M., Ilyas, R. A., Petrů, M., Sapuan, S. M., & Ruzaidi, C. M. (2021b). Flammability, tensile, and morphological properties of oil palm empty fruit bunches fiber/pet yarn-reinforced epoxy fire retardant hybrid polymer composites. *Polymers*, *13*(8), 1282. https://doi.org/10.3390/polym13081282

Syafri, E., Jamaluddin, Sari, N. H., Mahardika, M., Amanda, P., & Ilyas, R. A. (2022). Isolation and characterization of cellulose nanofibers from Agave gigantea by chemical-mechanical treatment. *International Journal of Biological Macromolecules*, *200*(December 2021), 25–33. https://doi.org/10.1016/j.ijbiomac.2021.12.111

Talib, R. A., Tawakkal, I. S. M. A., & Abdan, K. (2011). The influence of mercerised kenaf fibres reinforced polylactic acid composites on dynamic mechanical analysis. *Key Engineering Materials*, *471–472*, 815–820.https://doi.org/10.4028/www.scientific.net/KEM.471-472.815

Vakili, M., Rafatullah, M., Ibrahim, M. H., Salamatinia, B., Gholami, Z., & Zwain, H. M. (2015). A review on composting of oil palm biomass. *Environment, Development and Sustainability*, *17*(4), 691–709. https://doi.org/10.1007/s10668-014-9581-2

Villa, S., Garriga, R., Pérez, P., Gracia, M., González, J. A., García De La Fuente, I., & Cobos, J. C. (2005). Thermodynamics of mixtures with strongly negative deviations from Raoult's law: Part 9. Vapor-liquid equilibria for the system 1-propanol+di-n-propylamine at six temperatures between 293.15 and 318.15 K. *Fluid Phase Equilibria*, *231*(2), 211–220. https://doi.org/10.1016/j.fluid.2005.01.013

Yang, H., Yan, R., Chen, H., Lee, D. H., & Zheng, C. (2007). Characteristics of hemicellulose, cellulose and lignin pyrolysis. Fuel, 86(12–13), 1781–1788. https://doi.org/10.1016/j.fuel.2006.12.013

Yevich, R., & Logan, J. A. (2003). An assessment of biofuel use and burning of agricultural waste in the developing world. *Global Biogeochemical Cycles*, *17*(4). https://doi.org/10.1029/2002gb001952

Zamri, M. F. M. A., Milano, J., Shamsuddin, A. H., Roslan, M. E. M., Salleh, S. F., Rahman, A. A., Bahru, R., Fattah, I. M. R., & Mahlia, T. M. I. (2022). An overview of palm oil biomass for power generation sector decarbonization in Malaysia: Progress, challenges, and prospects. *Wiley Interdisciplinary Reviews: Energy and Environment*, *11*(4). https://doi.org/10.1002/wene.437

6 Coconut Waste Management

6.1 INTRODUCTION

Due to the rising population and agricultural production, agricultural waste management (AWM) is becoming increasingly important (Son, 2021). Improper agricultural waste (AW) disposal produces greenhouse gases, endangering both the natural and human environments. AWM can help prevent the underutilization of livestock manure and the reckless burning of crop residues and contribute to clean, safe, and sustainable agriculture (Koul et al., 2021). Bioeconomic strategies based on AWM can convert AW into valuable resources, provide job opportunities for youth, reduce GHG pollution and reliance on fossil fuels, and improve soil aeration and water retention. AW includes residues from crops, livestock and poultry, agro-industrial, abattoir, and aquaculture. Using bioeconomic strategies to manage AW can minimize waste while producing value-added products. Compost, bio-coal, biochar, bio-bricks, biohydrogen, biomethane, bioethanol, and biobutanol can all be made from lignocellulose, a major component of agricultural biomass. Greater public and agricultural community's understanding of the benefits of biological and biotechnological management of AW would eliminate preconceived notions and fearmongering.

According to Saeed et al. (2023), coconut waste is a major source of pollution in the environment due to the large amount produced and improper management. Coconut industries contribute significantly to the economy, with high import and export values, and coconuts are produced in 92 countries around the world. Coconut industries' waste biomass, peels, shells, seeds, and other parts of solid agro-waste are potential sources of monomeric sugars, polyphenols, and extractive proteins that can be used to make enzymes, biofuels, catalysts, and biochar (Varghese et al., 2023). To address the problem of coconut waste, it is critical to investigate its potential as a valuable resource for a variety of applications. Using coconut waste can help the environment and the economy by opening up new business opportunities (Zakaria, 2022). With the growing global demand for environmentally friendly and sustainable solutions, the utilization of coconut waste could be a promising area for further research and development.

Coconut waste is a major polluter in Malaysia, with coconut shells being the most prominent, producing approximately 3.18 million tons of waste annually (Nadzri et al., 2020). On the contrary, coconut waste can be used in various applications, including shell, husk, coir, and mesocarp. Coconut shells have excellent properties, are inexpensive and easily renewable, and have a high tensile strength (Somashekhar et al., 2018). Coconut waste has a massive surface area per unit volume and a network of sub-microscopic pores where foreign substances are absorbed. Coconut waste can be used for various purposes, including as a capacitor and electrode material, to improve electrical and mechanical properties, and as a good source of carbon particles with a high specific modulus (Divyashree et al., 2016). Coconut is one of Malaysia's most important trees, with a per capita consumption of 17.0 kg per year (Omar et al., 2022).

Hence, there are a few ways to manage coconut waste in Malaysia, such as recycling and reusing. Coconut shells can be used as an alternative fuel source for industrial boilers, while coconut husks can be processed into coir fiber, which can be used to make mats, ropes, and brushes, among other things (Lebedeva et al., 2022). Coconut mesocarp can be fed to livestock or composted to make organic fertilizer (Global Organics, 2016). Coconut waste can be diverted

DOI: 10.1201/9781003472209-8

from landfills by recycling and reusing it, reducing the amount of waste sent for disposal and conserving natural resources. The Malaysian government has implemented initiatives to promote sustainable coconut waste management, such as offering incentives to businesses that use coconut biomass as a renewable energy source and developing awareness campaigns to encourage recycling and reuse (Isa et al., 2021).

Overall, proper coconut waste management in Malaysia can lead to environmental and economic benefits such as reduced pollution, conservation of natural resources, and new revenue streams from recycling and reusing coconut waste. Coconut waste can also be used in various industrial applications, promoting proper waste management. This chapter aims to learn about proper AWM, with a particular emphasis on coconut trash, and to investigate the possibilities for industrial applications derived from coconut waste. The project's goal is to address the vital issue of waste management in the agricultural sector, with a special emphasis on coconuts, and to identify feasible methods for the usage of coconut waste in diverse industrial contexts.

Agro-waste management, particularly the management of coconut waste, is essential for a number of reasons. For starters, it helps to reduce environmental pollution caused by solid waste accumulation, as coconut waste is a significant contributor to solid waste in many countries (Sadh et al., 2018). Second, converting waste into valuable products can contribute to developing a circular economy where waste is viewed as a useful resource rather than a problem (Kalkanis et al., 2022). Third, it can provide new economic opportunities for communities through the sale of coconut-based products (Zainol et al., 2023).

Additionally, coconut waste has the potential to yield valuable products such as biofuels, enzymes, polyphenols, and extractive proteins (Bolivar-Telleria et al., 2018). These products have a wide range of industrial applications, including energy production, food processing, and biomedicine. Furthermore, using coconut waste as a raw material can help reduce reliance on nonrenewable resources like fossil fuels and contribute to developing a more sustainable economy (Clauser et al., 2021). Finally, proper agro-waste management practices, including coconut waste management, can help the United Nations achieve the Sustainable Development Goals, particularly concerning environmental sustainability, economic growth, and poverty reduction.

6.2 NATURAL FIBERS FROM BIOMASS

Every year, a massive amount of biomass, about 140 gigatons, is discarded as waste, causing various environmental issues (Jayaprakash et al., 2022). As the world's population grows, so does the demand for agricultural products and biomass, resulting in increased disposal of biomass residues and their application in a variety of fields. The pattern of discarded biomass disposal or utilization has shifted from traditional methods to biogas or biofuel production and other value-added bioproducts. However, because of the techno-economics of production, the potential of recycling options for producing natural fibers has been overlooked. Fibers are long, continuous filaments in high demand, with more than 70% of fibers used in textile production by 2025. Asia, particularly China, is the world leader in fiber production. Converting potential agricultural biomass into fibers can alleviate the demand for synthetic polymer production that relies on fossil fuels (Mujtaba et al., 2023). This goal can be met with integrated biorefinery and bioprocessing expenditures that generate multiple bioproducts.

6.2.1 Extraction of Fibers from Agro-Waste

According to Balla et al. (2019), natural fibers are unmodified biodegradable polymers that are continuous long- or short-chain filaments derived from plants, animals, insects, and minerals. Plant-based natural fibers have a low density and tensile properties comparable to glass or carbon fibers. In some cases, fibers are extracted from these sources using retting, mechanical processing methods, or a combination of both methods. Retting is the controlled removal of gummy materials that hold fiber bundles together (Sisti et al., 2017). There are various types of retting depending on the process:

FIGURE 6.1 Water retting (Nair et al., 2013).

a. **Water Retting**: Plant materials are soaked under controlled conditions and removed by natural microbial action. This process produces uniform fiber quality but is less appealing due to high water requirements, foul odors, and high costs (Figure 6.1).
b. **Dew Retting**: Plant materials with pectinaceous properties are left in a field for natural air decomposition. This method is simple but results in inconsistent fiber quality and properties (Chabbert et al., 2020).
c. **Enzymatic Retting**: Under controlled conditions, specific enzymes such as pectinase and xylanase are used to remove pectinaceous substances. This process yields low-strength fibers, which is a significant disadvantage.
d. **Chemical Retting**: Chemicals such as sodium hydroxide, sodium benzoate, or hydrogen peroxide are used to remove pectinaceous substances, resulting in long, clean fibers with natural color and tensile strength. One disadvantage is the cost of processing.
e. **Mechanical Processing or Mechanical Retting**: This process takes only a few minutes and involves decortication followed by cleaning, resulting in a large number of short fibers. Decortication is accomplished through the use of ball mills, hammer mills, or crushing rollers, and the materials produced are sorted, screened, and size-separated for cleaning and use (Lee et al., 2020).

Moreover, there has been a growing interest in using cellulosic agricultural biomass to produce fibers via bioprocessing in recent years (Norrrahim et al., 2021). These agricultural residues, which include bagasse, coir, banana, and rice husk, are classified as secondary sources of fiber and fall into two categories: field residues left after harvest and processed residues generated during the main product's extraction (Vitrone et al., 2021). The extraction process is designed to discard non-cellulosic components such as pectin, lignin, and hemicellulose, with the chemical composition of the agricultural residue determining the extraction method used. Figure 6.2 shows the diverse hybrid natural/synthetic reinforced fiber polymer matrix applications.

In accordance with Jayaprakash et al. (2022), cellulosic fibers can be extracted from agricultural residues using one of four methods: chemical, mechanical, combined, or biological (Norfarhana et al., 2022, 2024a,b; Syafri et al., 2022). Non-cellulosic materials are removed using acids such as sulfuric or hydrochloric acid or hydrogen peroxide. Mechanical methods break down cell walls and isolate fibers using homogenization, ultrasonication, and steam explosion techniques. Combined chemical and mechanical methods are often more efficient, whereas biological methods, which use enzymes to extract the cellulose, require careful optimization and selection of biological agents.

FIGURE 6.2 Diverse hybrid natural/synthetic reinforced fiber polymer matrix applications. (a) Sugar palm fiber, (b) pineapple fiber, (c) hemp fiber, (d) corn fiber, (e) jute fiber, (f) flax fiber, (g) coir fiber, (h) kenaf fiber, and (i) EFB fiber (Birniwa et al, 2023).

Each method has advantages and disadvantages, with chemical and mechanical methods being more time-efficient and biological methods being less expensive. On the contrary, biological methods necessitate more process engineering and optimization, whereas chemical methods may produce harmful by-products. Meanwhile, combined methods strike a balance between cost and efficiency.

6.2.2 Coconut Waste as Source of Natural Fiber

Coconut fiber is a thick natural fiber derived from the husk of coconuts. It is primarily grown in tropical countries such as India, Sri Lanka, Indonesia, the Philippines, and Malaysia (Go et al., 2019). Coir fiber, a type of coconut fiber, is a lightweight and strong fiber that is highly valued for its distinct properties and availability. Coconut fiber has a higher lignin content, lower cellulose and hemicellulose contents, and a high microfibrillar angle compared to other natural fibers (Adeniyi et al., 2019). These properties give it several valuable characteristics, including resilience, strength, wear resistance, and high elongation at break. Coir fiber is used to make a variety of products, including ropes, mats, mattresses, and brushes, as well as in the upholstery industry, agriculture, construction, and other industries (Girijappa et al., 2019).

Coir coconut fiber is a type of natural cellulose fiber derived from coconut husk. It is the fibrous material found between the outer coat and the hard internal shell of a coconut. Coconut farming typically produces coconut oil, milk, water, and other products such as desiccated coconut. The coconut husk is normally discarded as waste but can be converted into sustainable products (Hasan et al., 2021). Coir coconut fiber is extracted from the husks using traditional retting or mechanical and biological methods. The action of microorganisms and moisture to dissolve or rot cellulose tissues and pectin away from bast-fiber bundles is known as retting. The retted pulp is traditionally beaten with a wooden mallet to separate the fibers from the pith and outer shell (Nair et al., 2013). In recent years, the fibers have been separated by hand or in a rotating drum with spikes on the inside. Coir coconut fibers can be white or brown, depending on the age or maturity of the coconuts at the time of harvest (Stelte et al., 2022).

White fibers from immature or green coconuts are harvested on the plant after 6 to 12 months. These fibers are less strong than brown fibers but are smoother and finer (Kiron, 2021). White fibers can be spun into yarn, woven into mats, or twisted into rope or twine. Fully ripened and mature coconuts yield brown fibers (Textile Value Chain, 2021). The coconuts are harvested when the nutritious layer around the seed is ready to be processed into copra and desiccated coconut. It produces coarse but strong fibers with high abrasion resistance. Freshwater is retted with ripe coconuts. Brown and white coir fibers range from 10–30 cm long (Ali et al., 2022). Long fibers that are at least 20 cm long are referred to as bristle fiber, while shorter fibers are known as mattress fiber. Bristle fiber accounts for roughly one-third of the coir coconut fiber produced by each coconut. Shorter fibers are finer in texture and are typically spun into yarn. The residue after fiber extraction is known as coir pith (Cowley, 2021).

Coir coconut fiber has numerous advantages. First, it is environmentally friendly and biodegradable. It is derived from a renewable source, and the fibers or fabrics are completely biodegradable (Kao-Walter et al., 2011). It is tough, long-lasting, water-resistant, and the only fiber that is saltwater resistant. The fibers of coconut husk dry quickly. It absorbs moisture 50% quicker than polyester (Liné Cowley, 2019). It eliminates odors naturally without additional treatments or chemicals (James & Yadav, 2021). Even after several washes, coir coconut fibers absorb harmful ultraviolet rays, providing UV protection. However, there is one drawback to using coconut coir fiber. Concerns have been raised about the amount of water used in retting and extracting coir coconut fiber from the husks. Traditional retting causes water pollution by leaving many organic pollutants, such as pectin, tannin, fat, pectosan, toxic polyphenols, and various bacteria, including *Salmonella* (Tondo, 2022).

After all, new uses for coir coconut fiber are being discovered and promoted as technology advances and production increases. Coir is becoming a popular material for geotextiles. Geotextiles are permeable fabrics that, when combined with soil, can filter, separate, protect, reinforce, or drain. Natural fibers and geotextiles will become more popular as people abandon non-biodegradable synthetics.

6.2.3 Application of Natural Fibers

Plants, animals, and minerals are the sources of natural fibers. They have been used for a variety of purposes for centuries and remain popular today. For instance, the textile industry employs natural fibers such as cotton, linen, wool, and silk, which are widely used to make clothing. These fibers are preferred because they are comfortable, breathable, and long-lasting (Dhir, 2022). Moreover, natural fibers such as jute, sisal, and coir are used to make rugs, carpets, and mats. Cotton and linen are popular for bed sheets, pillowcases, and curtains. Wool is used in the production of blankets and upholstery (Cohen, 2022). In the pulp and paper section, paper is made from natural fibers such as bamboo, hemp, and flax. These fibers are preferred because of their strength and durability (Kamarudin et al., 2022). Meanwhile, ropes and twines are commonly made from natural fibers such as sisal, hemp, and jute. Because of their strength and flexibility, these fibers are preferred. These natural fibers have been used to manufacture construction materials such as roofing tiles, wall panels, and insulation. These fibers are preferred for their sustainability and environmental friendliness (Dhaliwal, 2020). Other than that, jute and cotton are used to make eco-friendly packaging materials like bags and sacks. These fibers are favored because of their strength and biodegradability (Girijappa et al., 2023). Lastly, cotton and silk are used to produce medical textiles such as bandages and dressings. These fibers are suitable because they are comfortable and breathable (Saber & El-Aziz, 2021).

According to Girijappa et al. (2019), natural fibers are increasingly used in the automotive and aerospace industries to manufacture interior components, such as those found in Mercedes-Benz vehicles. They are also used in insulation, construction, furniture, and clothing. Natural fibers also show promise in architecture, as they can be used for sunscreens, cladding, walling, and flooring (Steffens et al., 2017). Furthermore, natural fibers are used to manufacture exterior auto body components and are experiencing broad growth with promising opportunities in the automotive and construction industries.

Europe is expected to remain the largest market for natural fiber-reinforced composites (Khalid et al., 2021). For automotive applications, bast fibers such as hemp, kenaf, and flax are preferred, whereas wood plastic composite is the material of choice for construction industries. While natural fiber composites are relatively new in the electrical, electronics, and sporting industries, they have the potential to capture a significant market share in the future (Raj et al., 2020). Natural fiber-reinforced polymer composites will grow in new potential areas as material performance improves. Natural fibers are versatile and have numerous applications in various industries. They are preferred because of their environmental friendliness, sustainability, and durability.

6.3 ENVIRONMENTAL IMPACT FROM BIOMASS

Although biomass is an appealing renewable fuel option with low sulfur emissions, its use as an energy source can have negative environmental consequences. The primary concern is competition for agricultural land used to produce food and fiber (Kanianska, 2016). Other environmental effects of biomass feedstock production and the use of forest and agricultural residues for energy include soil disturbance, nutrient depletion, and impaired water quality (Trimble et al., 1984). The magnitude of these impacts varies by location and necessitates regional assessments. Processes for converting biomass into fuel may generate air pollutants, solid waste, and wastewater, all of which can have negative environmental consequences. Particulates, hydrogen sulfide, carbon monoxide, and polycyclic organic matter can be produced during the thermochemical conversion of biomass to fuels (Teh et al., 2021). There are also environmental concerns regarding the generation and disposal of solid waste and wastewater. However, with careful planning and conservation practices, the use of appropriate environmental control technology, and the utilization of any by-products, the environmental impacts of biomass production and conversion can be reduced (Ferronato & Torretta, 2019).

In addition, biomass energy is becoming more popular due to perceived environmental benefits, but it may have serious health consequences. Biomass refers to organic materials that are burned to generate energy, such as wood, AW, and charcoal (McFarland, 2019). The combustion of biomass emits harmful pollutants such as fine particulate matter, carbon monoxide, and nitrogen oxides, which can have a negative impact on the health of those who live nearby. Particulate matter, which consists of tiny particles less than 2.5 μm in diameter, has been linked to respiratory and cardiovascular diseases, including lung cancer, heart attacks, and strokes (Awopeju, 2021). Burning of biomass emits a large amount of these fine particles into the air, which can cause serious health problems for those exposed to it. According to studies, people who live near biomass-burning facilities have higher rates of hospitalization for respiratory illnesses and heart attacks than those who live further away (Harvard T.H. Chan School of Public Health, 2022).

Using biomass for energy can have economic and social consequences for communities. Biomass production necessitates a large amount of land, which can result in deforestation, soil erosion, and biodiversity loss. This can harm local communities that rely on forests for a living (Ayompe et al., 2021). Furthermore, using biomass for energy can lead to unequal access to energy, as people living in areas with high levels of biomass production may have better access to energy than those living in areas with low levels of biomass production (Benti et al., 2021). As a result, while biomass can be an environmentally friendly energy source, it is critical to consider its potential health, economic, and social impacts when making decisions about its use.

Some of the environmental disadvantages of bioenergy can be minimized by more sustainable forest management and careful selection of the type of biomass produced for fuel and how it is harvested (Rashidi et al., 2022). Research and technology advancements, as well as policy reform, can help ensure that future bioenergy investments are more environmentally friendly.

6.4 AVAILABILITY OF AGRO-WASTE

Agro-waste is a significant and readily available resource for many countries. Agro-waste is produced during the cultivation and processing of agricultural crops such as rice, wheat, and sugarcane and can include residues such as straw, husks, stalks, and bagasse (Maraveas, 2020). The crop type, production methods, and degree of processing determine the amount of agro-waste produced. Agro-waste is abundant in many developing countries but is frequently underutilized or wasted. This waste, however, could be converted into valuable resources such as biofuels, fertilizer, and animal feed with the right technologies and investment (Neh, 2020). Therefore, agro-waste has a high potential to be a significant resource for sustainable development, particularly in rural areas where agriculture is a major economic activity. However, agro-waste availability and utilization are affected by various factors, including government policies, infrastructure investment, and technological advancements. These factors must be considered when evaluating the resource potential of agro-waste (Adejumo & Adebiyi, 2021). Global coconut production has been continuously expanding over the years as of 2021. Global coconut production is expected to reach 63.7 million metric tons in 2021, as depicted in Figure 6.3.

Coconut waste is abundant as a result of the widespread production and consumption of coconuts. Countries such as Indonesia, the Philippines, India, Sri Lanka, Thailand, and Brazil contribute significantly to the worldwide coconut industry, resulting in significant amounts of coconut waste (James & Yadav, 2021). Coconut trees are commonly planted in these areas, and coconuts are widely used for food, oil production, and a variety of other applications. As a result, the residues produced by these operations increase the availability of coconut waste. Coconut waste is commonly available and accessible in places with coconut farms due to the great demand for coconut goods and the extensive cultivation of coconut plants (Beveridge et al., 2022). The quantity of coconut trash opens up possibilities for its use in various industries, from agriculture and horticulture to energy production and waste management.

6.5 PROPERTIES OF AGRO-WASTE FIBERS

Due to their economic and cultural importance worldwide, AW fibers are widely used for building materials, decorative products, and raw materials (Souza, 2023). These fibers are environmentally friendly, low cost, long lasting, and strong, making them appealing for composites (Karimah et al., 2021a). AW is a major global environmental issue that must be addressed. As a result, researchers

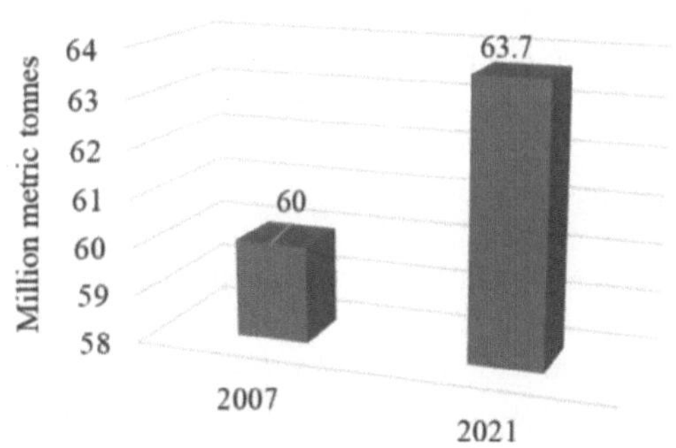

FIGURE 6.3 Global coconut production in 2007 and 2021.

are investigating the potential of AW fibers as a substitute for synthetic fibers to create safe and environmentally friendly products. AW is a promising renewable energy source that can be obtained from different industries, including oil palm plantations, rice husks, rice straw, sugarcane, pineapple, banana, and coconut (ASEAN Briefing, 2021). These industries generate a lot of biomass, mostly made up of natural fibers. However, only about 10% of these fibers are used as alternative raw materials in many industries (Asyraf et al., 2022), including biocomposites, automotive components, furniture (Mazani et al., 2019), biomedical products (Dungani et al., 2016), fire retardant (Suriani et al., 2021) and packaging (Alias et al., 2021; Halimatul et al., 2019; Jumaidin et al., 2019a,b, 2021; Nazrin et al., 2021). As a result of their properties, low cost, and environmental sustainability, AW fibers have a significant potential for increased use in various industries. This can lead to global environmental conservation while creating new economic opportunities in the agricultural sector.

6.5.1 Physical Properties

Natural fibers, also known as plant fibers, are derived from the stems, leaves, seeds, and roots of plants. Their excellent physical properties have made them useful in a variety of applications (Textileblog, 2023). Natural fibers' key physical properties include length, diameter, density, moisture content, and color (Widyasanti et al., 2020). Firstly, natural fiber length and diameter vary depending on the plant species. Cotton fibers, for example, are short and thin, whereas jute fibers are long and thick. Fiber length and diameter influence mechanical properties such as tensile strength and stiffness (Wong et al., 2008). Second, the density of natural fibers varies according to plant species. Natural fibers are generally less dense than synthetic fibers, making them lightweight and easy to handle (Karimah et al., 2021b).

Table 6.1 summarizes the physical features of coconut fiber. The physical properties (diameter, density, and weight gain by water absorption of a material) of coir fiber composites are heavily influenced by the properties of their constituent materials and their interaction. Coir fibers have a density of 1.1–1.5 g/cm^3, Young's modulus of 2–8 GPa, and tensile strength of 105–593 MPa. Water absorption ranges from 10% to 180%, with elongation at break ranging from 15% to 51%. These properties can vary depending on the fiber's origin, pre-treatment, and extraction methods (Adeniyi et al., 2019).

6.5.2 Chemical Composition

Plant fibers are primarily composed of cellulose (60–80%), hemicellulose, amorphous lignin, and moisture. Natural fibers also contain pectin, wax, protein, tannins, inorganic salt, and water-soluble compounds, which vary depending on the climatic conditions of the plant. Cellulose is a primary component of the cell wall, whereas hemicellulose has a random unshaped structure, and lignin acts as a cellulose fiber bonding agent.

According to Kumar et al. (2021), a single fiber extracted from plants such as bast, leaves, and grasses is made up of many fiber divisions and can be considered a hollow composite. Figure 6.4 shows the structure of a single fiber cell The lumen, secondary wall (S3, S2, and S1), and primary wall (from inside to outside) are the layers in a single fiber. The primary layer accumulates cellulose and hemicellulose, while the secondary wall contains helically arranged crystalline cellulose microfibers that provide mechanical strength to the fiber. The middle lamella, also known as the outer layer, contains lignin and pectin, which hold the fiber bundle together and form the final lignocellulosic fiber structure.

Coir fibers are primarily composed of lignin, hemicellulose, and cellulose. According to the research, the average chemical composition of coir fibers ranges from 32% to 50% cellulose, 0.15–15% hemicellulose, 30–46% lignin, and about 3–4% pectin. This emphasizes that cellulose and lignin are the two most abundant constituents of plant fibers. The variations in coir fiber properties can be attributed to the source of the coconut plant from which the fibers are extracted or the method of extraction used (Adeniyi et al., 2019).

TABLE 6.1
Summary of Physical Properties for Coconut Fibers (Widyasanti et al., 2020)

Physical Properties	Explanation
Length and diameter	Coconut fibers are often coarse in texture, dense, and lengthy. Coconut fibers' lengthy and dense nature contributes to their strength and durability.
Color	Natural coconut fibers range in color from light brown to golden brown, depending on the maturity of the coconut husk. Factors such as processing procedures and the presence of contaminants can affect the color.
Texture	The texture of coconut fibers is rough and fibrous, with a natural curliness. The roughness adds friction and improves interlocking, making them excellent for applications requiring mechanical strength, such as ropes or matting.
Moisture content	Coconut fibers have a low moisture content and are resistant to deterioration and microbiological development. Coir fibers' minimal moisture absorption guarantees that they keep their strength and stability even when exposed to dampness.
Density	Coconut fibers have a low density, ranging between 0.1 and 0.3 g/cm^3. Because of their low density, they are lightweight, which is helpful in applications where weight is a factor.
Thermal conductivity	Thermal insulation is provided by coir fibers. They have a low thermal conductivity. Thus, they are effective at reducing heat transfer. As a result, coconut fibers are excellent for building insulation or as liners in horticulture applications to protect roots from high-temperature variations.
Water absorption and retention	Coconut fibers have a great capacity for water absorption. They can absorb up to their weight in water, and the presence of natural lignin and cellulose in the fibers allows them to retain water for long periods of time. Because of this feature, coir fibers are used in horticulture as a growing medium for plants or in erosion control applications to stabilize soil and preserve moisture.
Biodegradability	Coconut fibers are biodegradable and eco-friendly. They degrade naturally over time without harming the environment, making them a more environmentally friendly alternative to synthetic fibers.

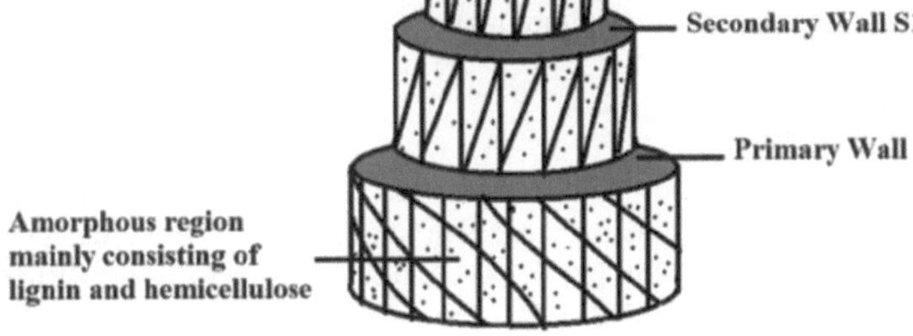

FIGURE 6.4 Structure of single fiber cell (Kumar et al., 2021).

6.5.3 Mechanical Properties

The spiral angle between the fiber axis and the microfiber determines the mechanical properties of the fiber, such as strength and modulus, with a smaller spiral angle suggesting greater fiber strength and modulus. Lignin and pectin both have lower mechanical strength and interface characteristics between the matrix and the fiber (Kumar et al., 2021) (Table 6.2).

TABLE 6.2
Summary of Mechanical Properties for Coconut Fibers (Mendoza-Quiroga et al., 2022)

Mechanical Properties	Explanation
Tensile strength	Coir fibers have high tensile strength, which means they can sustain pulling or stretching forces without breaking. Coir fiber has tensile strengths ranging from 200 to 600 MPa. Because of their strength, coir fibers are appropriate for applications requiring high tensile load resistance, such as ropes, twines, and geotextiles.
Flexural strength	Flexural strength is a material's capacity to resist bending or deformation under an applied force. Coir fibers offer a moderate flexural strength, making them excellent for applications requiring shape retention and resistance to bending or warping, such as mats, upholstery, or composite materials.
Impact resistance	Coir fibers offer high impact resistance, allowing them to absorb and transmit impact energy without breaking or fracturing. This feature qualifies coir fibers for applications requiring impact resistance, such as geotextiles for erosion control or padding in packaging materials.
Abrasion resistance	Coir fibers have high abrasion resistance, allowing them to survive rubbing, scraping, or friction without deterioration. Because of this feature, coir fibers are appropriate for applications requiring durabilities, such as doormats, carpets, or erosion control mats.
Cohesion and interlocking	Coir fibers have a natural curliness and rough texture, contributing to their cohesion and interlocking ability. This feature increases their strength and stability, making them appropriate for mechanical interlocking applications, such as ropes, nets, or soil erosion control blankets.
Moisture resistance	Coir fibers have an inherent resistance to moisture, allowing them to preserve mechanical qualities even when damp or humid conditions exist. Because of their moisture resistance, coir fibers are excellent for outdoor use or products requiring moisture resistance, such as erosion control products or horticulture substrates.

6.5.4 Morphological Properties

Coconut fibers have particular morphological features that contribute to their wide range of applications. Table 6.3 shows the summary of the morphological properties of coconut fibers. The coir fibers derived from coconut waste have a significant length and strength, making them robust and strong (Yan, 2016). The morphological features of coconut waste make it useful in a variety of industries. Coconut waste's morphological qualities contribute to its wide applications, encouraging sustainable practices in a variety of disciplines and businesses.

6.5.5 Thermal Properties

When heat is applied to a solid body, a liquid, or a gas, thermal properties are associated with a material-dependent reaction. This reaction can result in a rise in temperature, a phase transition, a change in length or volume, the beginning of a chemical reaction, or a change of another physical or chemical property (Buck & Rudtsch, 2011). Heat insulation materials are essential in automobiles, airplanes, electrical devices, industry, and daily life.

Researchers are always looking for thermal insulation materials that are strong, durable, and environmentally safe. They are manufactured from renewable resources like composites from AW, such as coconut waste. Table 6.4 shows the thermal conductivity of selected materials in coconut waste (Ismail et al., 2021). Due to its ability to maintain insulation and increase residential dwellings' energy efficiency, coconut fiber was proven to be an effective insulator. The outcomes also indicate that coconut fiber successfully enhanced the physical models' envelope's energy efficiency (Iwaro & Mwasha, 2019). Table 6.5 shows the summary of the thermal properties of coconut fibers. The thermal conductivity of coconut waste materials is listed in Table 6.4.

TABLE 6.3
Summary of Morphological Properties for Coconut Fibers (Fu et al., 2023)

Morphological Properties	Explanation
Fiber length	Coir fibers are generally long and coarse. Individual fiber can range in length from a few to 30cm or more, depending on the type and processing methods used. The long coir fibers add to their strength, making them appropriate for applications requiring high tensile strength.
Fiber diameter	Coir fibers have bigger diameters than other natural fibers ranging from 10 to 30 μm, contributing to their strength and longevity. The increased diameter also improves their resistance to mechanical loads.
Fiber surface texture	The surface roughness of coir fibers is rough and fibrous. The fibers' surface can be uneven or somewhat twisted, giving them a natural curliness. This surface pattern adds friction and improves interlocking, making them excellent for mechanical strength and stability applications, such as ropes, matting, or geotextiles.
Cross-sectional shape	Coir fibers' cross-sectional shape can range from circular to oval or irregular. The mechanical properties of the fibers, such as flexibility and bending resistance, are influenced by the form of the cross-section.
Surface morphology	Coir fibers' surfaces may contain abnormalities, nodes, or irregularly distributed tiny ridges. These surface characteristics can help the fibers interlock and improve their cohesiveness when employed in applications such as matting or composites.
Microstructure	Coir fibers are composed of cellulose fibrils embedded in a lignin matrix and have a hierarchical microstructure. The cellulose fibrils provide structural strength and rigidity, while the lignin functions as a binding agent, holding the fibrils together. This microstructure influences coir fibers mechanical qualities such as tensile strength and flexibility.

TABLE 6.4
Thermal Conductivity of Selected Materials in Coconut Waste (Ismail et al., 2021)

Material	k (W/m·K)
Coconut shell	0.030–0.125
Coconut fiber, tapioca, and PVA	0.105
Coconut fiber	0.048–0.049
Coconut fiber and epoxy resin	0.140
Coconut shell and epoxy resin	0.143

6.5.6 Calorific Value

The energy content of the samples is determined by their calorific value. It is an essential factor in the processes of thermochemical conversion. Coconut shell has a calorific value of 19.4 MJ/kg and a fuel value index of 4,441. The higher heating value, lower ash value, and high density of the feedstock material may be the reasons for the high fuel energy value index of coconut shell biomass (Ahmad et al., 2022).

TABLE 6.5
Summary of Thermal Properties for Coconut Fibers (Mahmud et al., 2023; Hassan et al., 2020; Ichim et al., 2022)

Thermal Properties	Explanation
Thermal conductivity	Coir fibers have a low thermal conductivity, which means they can withstand heat transmission. This feature makes coir fibers useful for thermal insulation applications since they can aid in reducing heat movement between different environments. Coir fiber, for example, can be used as insulation in buildings, packaging, and other products requiring temperature management.
Heat resistance	Coir fibers are heat resistant and can sustain moderate temperatures without substantial degradation or damage. They have a greater ignition temperature than other natural fibers, which makes them less prone to burning or charring. However, like most natural fibers, coir fibers are still organic components and can be flammable under extreme heat settings.
Moisture and humidity	Coir fibers can absorb and hold moisture, which might affect their thermal qualities. Coir fibers can retain water and modify their thermal behavior when wet or exposed to humid conditions. Moisture can have an impact on their thermal conductivity and heat resistance.
Thermal stability	When subjected to temperature changes, coir fibers remain relatively stable. When exposed to temperature swings, they do not easily shrink, expand, or undergo substantial changes in shape or structure. Because of their thermal stability, coir fibers are useful for applications requiring dimensional stability, such as composites or building materials.
Thermal decomposition	Coir fibers thermally decompose when exposed to high temperatures. This decomposition process involves the breakdown of the organic elements that make up the fiber structure, such as cellulose and lignin. Coir fibers generally decompose at temperatures higher than 200°C (392°F). Fiber density, moisture content, and heating conditions can all influence the rate and amount of decomposition.
Thermal resistance	Coir fibers are resistant to thermal deterioration. They can preserve structural integrity and mechanical qualities at high temperatures for a limited time. Because of their thermal resistance, coir fibers can be utilized in applications requiring moderate heat exposure, such as thermal insulation, soundproofing, and fire-resistant materials.

6.6 AGRO-WASTE MANAGEMENT

6.6.1 Compost

The ideal starting material for compost is coconut. It has various benefits, including being more clean, efficient, and sustainable. Coconut waste degrades into compost in approximately 6 months. As a result of the slow rot of the hard shell and husk, this takes longer than most raw organic material. Determining the amount of feedstock needed in a specific area and space depends on the ultimate compost made from the waste. Unlike microorganisms composting alone, earthworm activity and the surface area for microbial action are increased when coconut husk and livestock manure are blended and mixed. This resulted in a faster rate of decomposition. Coconut husk compost promotes the blooming of plants, fruits, vegetables, and flowers (Swarnam et al., 2016). One alternate method for converting coconut waste into nutrient-rich compost that is beneficial to plants and soil while reducing their adverse environmental effects is vermicomposting. Although earthworms are the key catalysts of the process by modifying the substrate and changing biological activity, microorganisms are ultimately responsible for the metabolic breakdown of organic materials (Swarnam et al. 2016; Ismail et al. 2021).

6.6.2 Pulp and Paper

Coconut fiber could take the place of wood in the production of paper. Coconut coir has a high enough content of cellulose fiber to be employed in pulp production. Moreover, it can be used to create composite materials that combine HVS pulp and paper pulp made from coir fibers. About 5–7% of the world's total paper and pulp production is derived from non-wood sources (Ward et al., 2008). This raw material is more environmentally friendly, easier to recycle, and more biodegradable. Moreover, waste paper pulp mixed with fiber pulp with the highest alpha-cellulose content is used as a raw material in producing composite paper. In addition, the properties of composite paper, such as strong pull, the ability to absorb ink, writing ability, and endurance, have been tested. Compared to wood, non-wood resources offer the advantage of quickly producing excellent pulp and fibers for specialized papers (Fahmy et al., 2017; Gonzalo et al., 2017).

This kind of paper is made through a very complex procedure. They need to be cooked, chopped, and then formed into sheets after some coconut husk pieces have been divided into little pieces. After that, soft paper is made in the form of sheets in the vat, with fabric placed between every sheet. A sizable sheet of dark brown coconut paper will result from the drying process (Othman et al., 2013).

6.6.3 Biocomposite Materials

Utilizing AW to create biocomposites is one of the environmentally friendly waste management strategies. Due to their environmental benefits, sustainability, and performance characteristics that are comparable to those of their synthetic counterparts, biocomposites are considered environmentally friendly materials. Biocomposite materials can reduce fuel consumption and conserve energy in applications like automobiles because of their low weight and excellent strength (Das et al., 2022). The coconut shells are organic wastes that are frequently used as solid fuel. When burned, they produce smoke, which causes air pollution. However, zero-emission applications can use powdered coconut shell particles. Figure 6.5 shows the preparation of coconut shell powder (Yulianto et al., 2017).

A biocomposite is a composite material with natural fiber reinforcement and resin as the matrix. Myceliated coco pith biocomposite waste treatment for insulator products is a very effective and sustainable approach. This is because of the absence of chemicals in the composite formation process. Instead, *Ganoderma lucidum lingzhi*, a wood fungus, uses cocopith as a substrate. *G. lucidum lingzhi* is maintained by the growth conditions, allowing the mycelium of the fungus to attach to each space between the cocopith particles and form a solid or novel material. The advantages of

FIGURE 6.5 Coconut shell powder preparation (Das et al., 2022).

cocopith biocomposite as an insulator are that these biocomposite characteristics can be mechanically conditioned in accordance with industry needs, based on tensile strength and pressure test, which can reach maximum flexural stress, where strong press value and strong biocomposite pull are directly proportional to the density value of the biocomposite itself (Mardijanti et al., 2021). Figure 6.5 shows the coconut shell powder preparation.

6.6.4 Bio-Coal

After being consumed with its juice and copra, tender coconut is typically disposed of as waste biomass along with its husk and shell. Tender coconut waste (husks and shells) should be disposed of properly to prevent poor sanitation, air pollution, and drain blockages that encourage the growth of mosquitoes and other hazardous microorganisms (Obeng et al., 2020). One of the practical and appealing alternatives to using this waste as a replacement for fossil coal and facilitating its thermal applications, particularly in the home and industrial sectors, is bio-coal (Ghosal et al., 2016). As the name implies, bio-coal is a form of renewable biomass used to make fossil coal. Because it can be handled and burned in the same manner as fossil coal and has physical characteristics similar to coal, it is regarded as a coal substitute (Agar & Wihersaari, 2012).

Bio-coal is generated from readily available, inexpensive, and readily available coal and AW such as coconut waste. Due to the bulkiness of the husk, these residues have a low heating value per unit volume and significant shipping and storage expenses. In the existing coal-fired power plants, boilers for heat and steam, food processing industries, fuels for gasifiers, domestic cooking, and water heating, bio-coal is attractive because of its high energy content and coal-like properties. Carbonization and densification of wasted tender coconut are easy and inexpensive approaches to producing high-density, energy-concentrated briquette char (bio-coal) in situ. Most volatile substances are eliminated during carbonization, which transforms biomass into char, a carbon-rich solid product with a higher calorific value, enhancing the substantial contribution of its burning properties (Tumuluru et al., 2011). The steps to prepare bio-coals are presented in Figure 6.6 (Ghosal & Roy, 2021).

6.6.5 Biochar

Another treatment option for coconut wastes is a conversion to a carbon-rich biochar. Biochar is typically produced by combusting biomass fuel, such as wood and AW, to generate sustainable energy. To create coconut shell and coconut husk biochar, pyrolysis, gasification, and self-sustained carbonization are used (Adorna et al., 2020). The properties of biochar made from coconut husk, coconut shell, and activated biochar from coconut shells make these carbon-rich materials potentially beneficial for industrial and environmental uses. The global coconut industry can use coconut shell and husk biomass as potential resources for their particular industrial development and environmental treatment since the practicality of biochar and activated biochar from coconut waste biomass produced using different production technologies and activation processes are perceived (Ajien et al., 2023). Figure 6.7 shows biochar production from coconut husk.

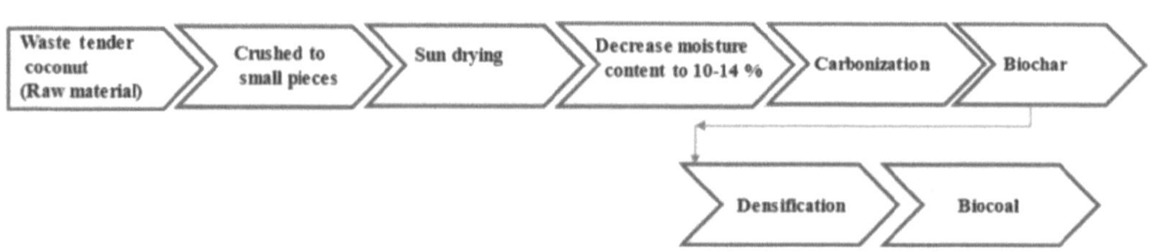

FIGURE 6.6 Steps to prepare bio-coal from waste tender coconut.

FIGURE 6.7 Biochar production from coconut husk (Prabhu et al., 2022).

6.6.6 Bio-Brick

Since bio-brick is a new material, little study has been done to assess its end-of-life applications and disposal. The absorbed carbon dioxide is trapped in the substance since it cannot be decomposed (Prétot et al., 2014). Burning crop residue can be supplemented with bio-brick, an agro-waste-based, sustainable, affordable, and financially successful alternative (Rautray et al., 2021). These bio-bricks can be used as an excellent source of sound and heat insulation in walls because they have a reduced thermal conductivity of about 0.27 W/m·k. While having various advantages, bio-bricks cannot be utilized directly to construct load-bearing structures. In low-cost housing, they can be used with wooden or metal structural frames. Research is ongoing to increase the ability of bio-bricks to support loads (Koul et al., 2022). Additionally, Kimeng et al. noted that, due to their inadequate compressive strength, concrete panels made utilizing groundnut shells in place of aggregate could not be used in heavy structural applications. The AW was first cut to the required size before being combined by hand or a mixer with a lime-based slurry and water. The substance was put into molds and firmly packed with a wooden dowel to create compact bricks. Then the sides of the molds were opened and left to dry for 15–20 days after being cured for 24 hours.

First, plants absorb atmospheric carbon dioxide throughout their growth and store it in their cellulose. Second, the calcium hydroxide absorbs carbon dioxide during the carbonation process to create the more stable calcium bicarbonate or calcium carbonate. Hempcrete is, therefore, a particularly sustainable building material because the amount of carbon dioxide created during production is lower than the carbon dioxide retained in the block. It may be a practical way to reduce the pollution brought on by burning AWs while also meeting the rising demand for building supplies (Arrigoni et al., 2017). Figure 6.8 shows the bio-brick from coconut husk.

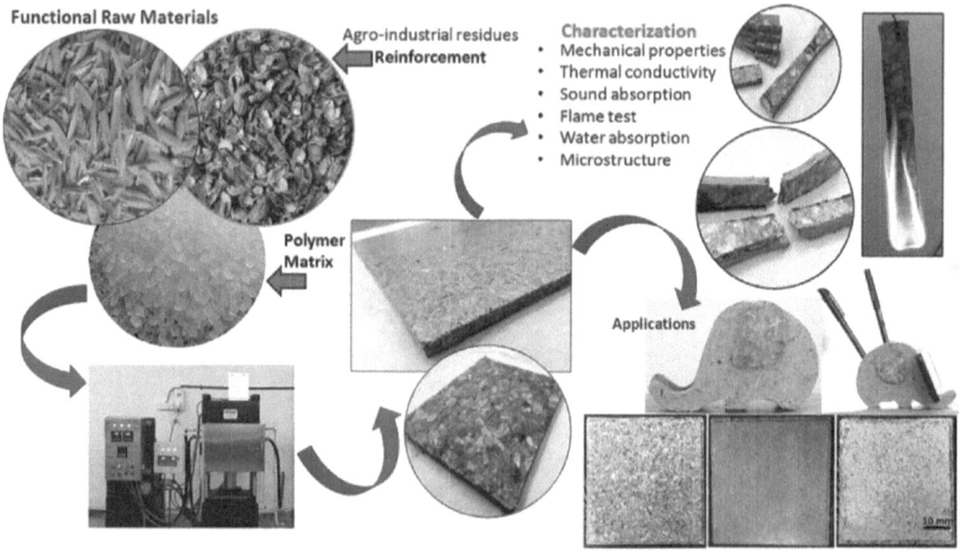

FIGURE 6.8 Bio-brick from coconut husk (Maraveas, 2020).

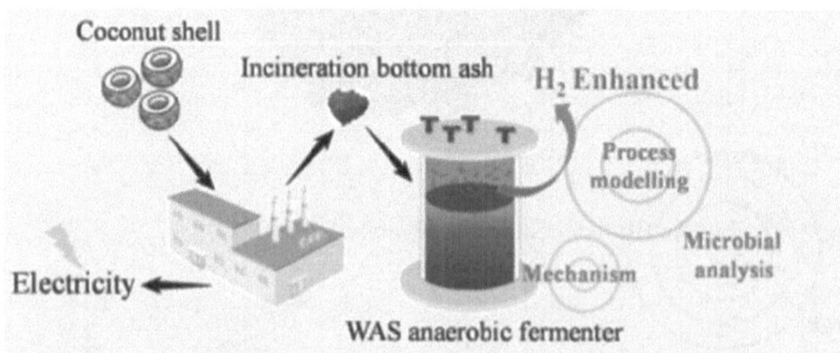

FIGURE 6.9 Biohydrogen generated from coconut shell (Foo and Hameed, 2010).

6.6.7 Biohydrogen

A significant and sustainable alternative energy source to fossil fuels is biohydrogen. Figure 6.9 shows the biohydrogen generated from coconut shells. To alleviate the burden of limited resources and encourage the use of ecologically friendly technology, biohydrogen serves as a beneficial energy source (Kim et al., 2011). As hydrogen releases a significant amount of energy (286 kJ/mol) and only produces water when it burns, it has been viewed as a potential alternative energy source. Co-pyrolysis is a suitable method for producing hydrogen from food waste and coconut fiber, allowing for the synthesis of products with a high added value. Bottom ash from burning coconut shells is rich in a number of inorganic minerals, including K, Ca, and Na, that are necessary for the growth and metabolism of microorganisms. It is a mediator to improve waste-activated sludge anaerobic fermentation for microbial hydrogen production (Wei et al., 2021).

6.6.8 Biomethane

Biomethane is a flexible and convenient fuel that may be used everywhere without requiring any adjustments to equipment designed to function on natural gas (Urban, 2013). The methane content within biogas directly influences its energy potency, typically ranging from 50% to 80%,

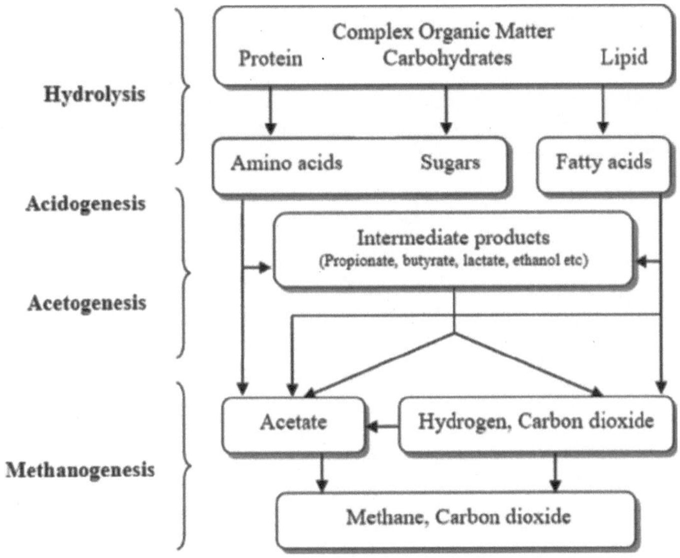

FIGURE 6.10 Pathways of the anaerobic digestion of organic substrate.

accompanied by varying proportions of carbon dioxide, nitrogen dioxide, oxygen dioxide, and 0.1–1% of other gases. Coconut copra emerges as a promising renewable energy source, demonstrating potential as a local, sustainable, and cost-effective alternative to fossil fuels. Through the decomposition of organic matter or substrates, anaerobic digestion triggers a series of unique metabolic transformations. This intricate process unfolds across three distinct stages: hydrolysis, acidogenesis, and acetcogenesis phases wherein acids are produced, and, finally, methanogenesis (White, 2011). Figure 6.10 shows the pathways of the anaerobic digestion of organic substrate.

6.6.9 Bioethanol

Bioethanol can be used as a fuel, chemical feedstock, and solvent in a variety of sectors. Because of its availability, coconut waste can be a substitute feedstock for biofuel. Therefore, there are three major steps required to produce bioethanol from coconut waste: biomass pre-treatment, hydrolysis, and fermentation (Bolivar-Telleria et al., 2018). Figure 6.11 shows the production of bioethanol and biodiesel using coconut waste. After the coconut waste was obtained, it was dried at 50°C until its weight was consistent. Then, 5 g of coconut waste was added to 100 mL of a 50% sodium hydroxide (NaOH) solution to perform the alkaline pre-treatment. The mixture was then autoclaved for 40 min at a temperature of 121°C and a pressure of 1 atm. The mixture was then filtered to separate the solid from the liquid phase. The amount of sugar in the liquid was determined according to Miller's method (Miller, 1959), which involved measuring total reducing sugar (TRS) following hydrolysis with sulfuric acid (H_2SO_4) solution and neutralization with NaOH. After being rinsed with distilled water, the solid was dried until it achieved a consistent weight and became neutral. After that, enzymatic hydrolysis of the pre-treated coconut waste was performed.

The pre-treated solid waste was then added to a solution consisting of a citrate buffer, distilled water, and enzymes to perform enzyme hydrolysis. The hydrolysis reaction was carried out for 72 hours in an orbital shaker incubator (Sangkharak, 2011). Filtration was used to separate the liquid and solid components. The liquid phase was evaluated for its TRS level before being subjected to a further fermentation of ethanol. The enzymatic conversion was calculated as the percentage of monosaccharides produced in relation to the amount of hydrolyzed solids. The overall percentage yield, which shows how much sugar is produced during hydrolysis relative to the amount in natural

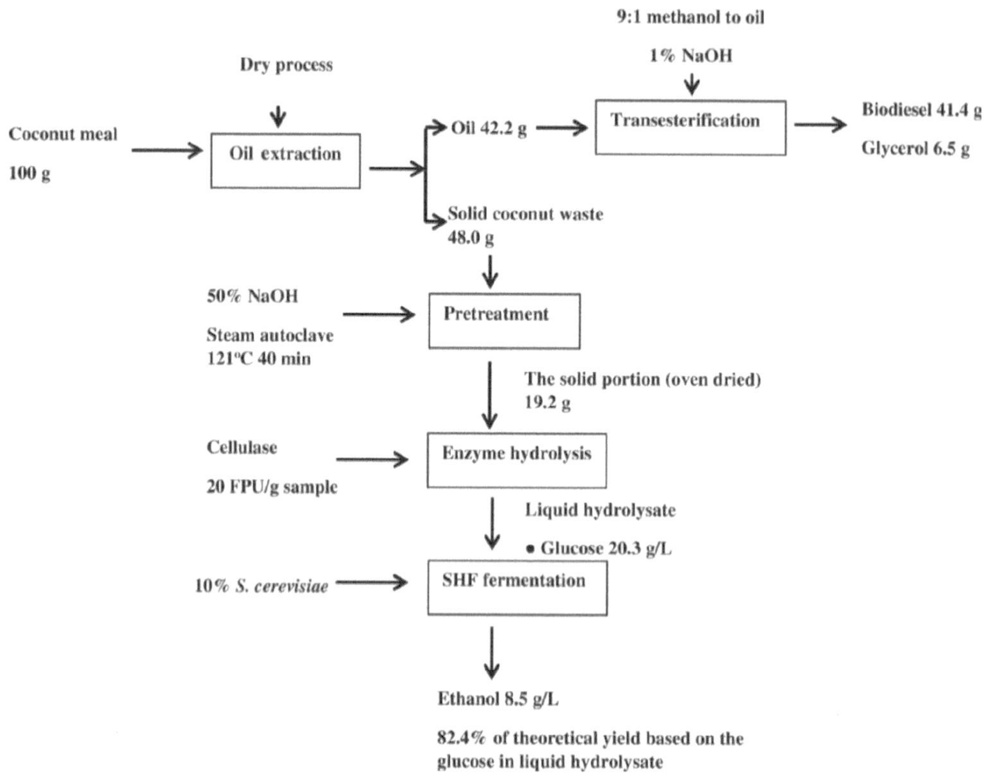

FIGURE 6.11 The production of bioethanol and biodiesel using coconut waste (Sangkharak et al., 2019).

coconut, was calculated using enzyme conversion and the alkaline pre-treatment yield (Cabral et al., 2016). Lastly, the liquid underwent fermentation to produce ethanol.

6.6.10 Biobutanol

Like the process used for bioethanol production, coconut waste can generate butanol through three primary stages: biomass pre-treatment, hydrolysis, and fermentation. Microorganisms like *Clostridium acetobutylicum* are employed during fermentation to convert sugars into butanol. Figure 6.12 shows the production of biobutanol from food waste. Upon obtaining the coconut waste, it was dried at 50°C until a consistent weight was achieved. Subsequently, 5 g of the waste was mixed with a 50% sodium hydroxide (NaOH) solution and subjected to alkaline pre-treatment, followed by autoclaving at 121°C and 1 atm pressure for 40 min. The resulting mixture underwent filtration to separate solids from liquids. TRS in the liquid was quantified using Miller's method, involving sulfuric acid (H_2SO_4) hydrolysis and subsequent neutralization with NaOH. The solid waste was neutralized with distilled water and subjected to enzymatic hydrolysis (Miller, 1959).

The pre-treated solid waste was then combined with a solution comprising citrate buffer, distilled water, and enzymes for enzyme hydrolysis conducted for 72 hours in an orbital shaker incubator (Sangkharak, 2011). Filtration separated the liquid and solid components, with the TRS level in the liquid assessed before proceeding to ethanol fermentation. The enzymatic conversion was determined as the percentage of monosaccharides produced relative to the hydrolyzed solids. The overall percentage yield, indicative of sugar production during hydrolysis relative to natural coconut,

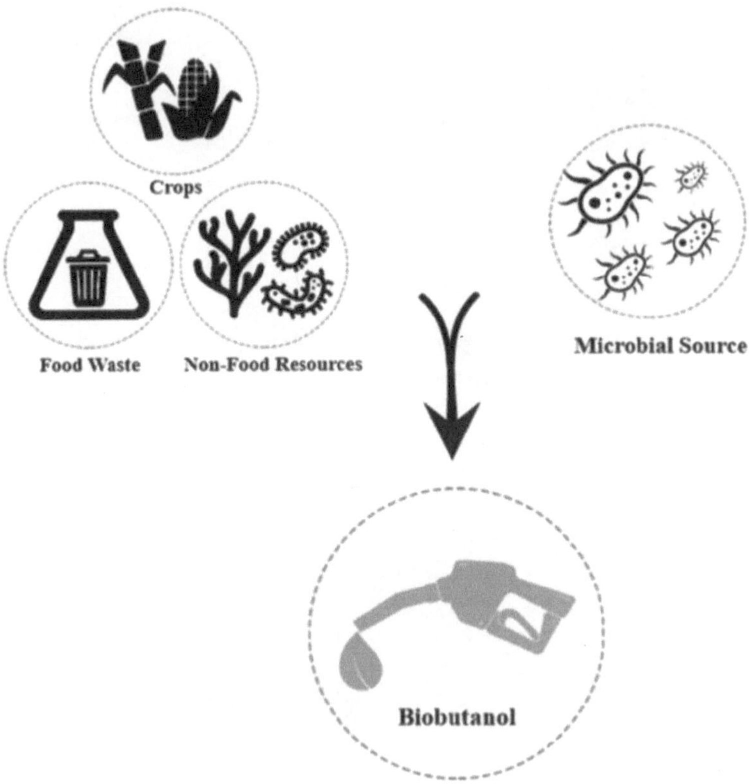

FIGURE 6.12 Production of biobutanol from food waste.

was calculated considering enzyme conversion and alkaline pre-treatment yield. Finally, the liquid underwent fermentation to yield ethanol (Cabral et al., 2016).

6.6.11 Natural Acids

In the tropics and subtropics, coconut husk is a plentiful biomass waste rich in cellulose and lignin. Although many concerns have been raised about using coconut waste, its cellulose and lignin components structure makes for a highly efficient and selective conversion of coconut waste into chemicals (Ding et al., 2018). It was initially suggested that coconut waste be easily converted into high-selectivity formic and acetic organic acids under mild hydrothermal conditions. It was discovered that a straightforward acidic pre-treatment was effective for producing formic and acetic acids in large yields. The maximum yield can reach up to 50.2% of formic and acetic acids using biomass like coconut waste (Ding et al., 2018).

Using concentrated or diluted acids, simple acidic pre-treatment is a technique for breaking down the stiff structure of lignocellulosic materials. Diluted sulfuric acid, which has been used to pre-treat a range of biomass types, including coconut husks, is most frequently utilized. Furfural is traditionally produced using diluted sulfuric acid. The research discusses hydrothermal methods to produce acetic and formic acids with a simple pre-treatment (Figure 6.13) (Ding et al., 2018). Acid pre-treatment is a popular and efficient technique for getting high sugar yields from lignocellulosic biomass. Coconut waste was prepared with dilute acid and then autoclaved (Bolivar-Telleria et al., 2018). Effective pre-treatment of lignin-rich wastes from coconuts was investigated to utilize coconut wastes for fuel, chemicals, and materials. In general, lignocellulosic materials can be broken down using simple acidic pre-treatment to provide value-added products like natural acids.

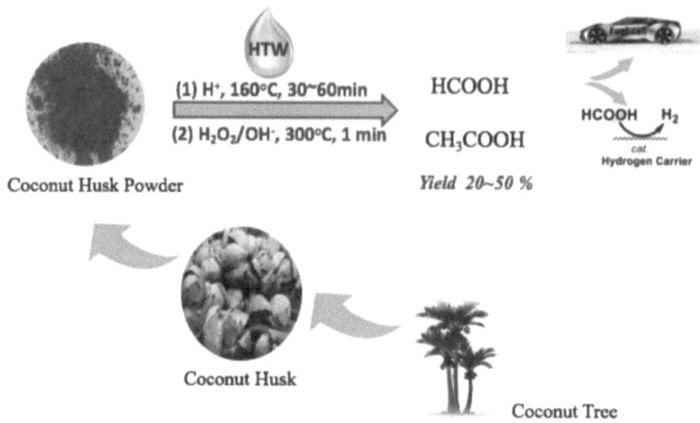

FIGURE 6.13 Process to produce acid from coconut husk (Ding et al., 2018).

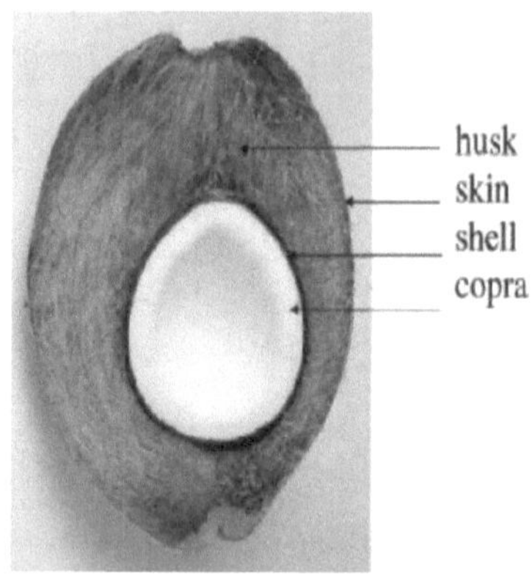

FIGURE 6.14 Coconut fruit parts (Bradley et al., 2006).

6.6.12 Bioelectricity

Various parts of the coconut can produce fuel, such as the husk, the shell, and the oil in the copra (Bradley et al., 2006). Figure 6.14 shows the coconut fruit parts. Coconut waste can be used to produce energy in three ways. First, the oil from the coconut can be obtained and turned into biodiesel, which can be utilized to generate electricity by powering the diesel generator. The other choice is to burn the coconut oil in a modified diesel generator (Bradley et al., 2006). Second, coconut waste can be converted to combustible gases, which can help generate heat and electricity through the use of gas turbines (Bradley et al., 2006). Third, charcoal and combustible gases can be produced from the coconut shell and used for heating or cooking (Bradley et al., 2006).

There are two alternatives to turning coconut oil into energy. Coconut oil can first be chemically processed into biodiesel using the same processes as other vegetable oils. Methanol is a reactant in the procedure, known as transesterification, and lye is a catalyst (Bradley et al., 2006). Coconut biodiesel production equipment is affordable to buy and easy to use. The biodiesel produced from coconut waste performs incredibly well in a diesel generator, emitting just about one-third of the emissions of normal diesel fuel made from petroleum. The diesel engine does not need to be altered to operate the biodiesel. About \$0.20/L more is spent on processing, but selling the glycerin should more than cover this increased expense.

Directly burning coconut oil in a diesel generator is a different option. The diesel engine must be modified to accomplish this. Coconut oil has a high viscosity, which makes it hard to handle by the diesel engine fuel injection system in the diesel engine. However, at a temperature of 70°C, the viscosity is quite similar to the regular diesel, allowing for direct combustion in a diesel engine. Therefore, the choices are to alter the fuel to produce coconut biodiesel or to alter the diesel engine so that the fuel can be preheated before use. Utilizing a diesel generator, energy from coconut oil, coconut oil biodiesel, and regular diesel can be converted into electricity with relative conversion efficiencies of 19.7%, 20.3%, and 22.4%, respectively (Bradley et al., 2006).

6.7 POTENTIAL APPLICATIONS

The utilization of coconut waste presents a compelling opportunity for sustainable resource management and economic diversification in various industries. By repurposing coconut shells, husks, fibers, and water, a multitude of valuable products can be derived, ranging from fuel and activated carbon to animal feed, fertilizers, construction materials, and even cosmetics. This discussion underscores the importance of adopting innovative approaches to maximize the potential of coconut waste, thereby reducing environmental impact and fostering economic growth in coconut-producing regions.

One of the potential applications of coconut waste in the energy sector is fuel. Coconut shells can be used as a source of biomass fuel, which can be burned to generate electricity or heat since they have a high calorific value (Ahmad et al., 2022). In addition, activated carbon can be produced from this waste since it can be applied with high-temperature steam activation. It has well-developed micropores, strong adsorption, high hardness, and low ash content and is environmentally friendly (Ahmad et al., 2022). It is utilized extensively in various liquid and gas phase adsorption, gold recovery, water filtration, food and beverage adsorption, and other applications.

In the agricultural sector, coconut waste can be used as animal feed, particularly for ruminant animals, such as cattle and goats. Copra meal is another name for coconut meal. It is a by-product of the process that turns coconut meat, or copra, into coconut oil. This coconut meal is valuable for livestock, poultry, and other grass-fed animals. Essentially, it is used as a protein supplement, on its own or in combination with other protein sources. The main benefit of utilizing coconut meal as animal feed is that it is more readily available and affordable than other options. The nutritional potential of coconut meal for feeding animals is tremendous. It is rich in protein but of low quality, has high fiber content, has oil residue as an energy source, and has certain vitamins or minerals. These nutrients are very beneficial for developing, improving, and maintaining animal health (Wahyuni et al., 2018). Coconut husks and fibers can be processed into a soil conditioner and fertilizer known as coir, which is rich in potassium and other nutrients. For coconut trees grown on an organic medium like deep peat, coconut husk ash proved to be a very effective fertilizer. In particular, it offers the minerals K and Cl, which are crucial for coconut (Bonneau et al., 2010).

For construction industries, coconut fibers can be used to make a range of building materials, including rope, mats, and insulation. About 70% of the coconut's flesh and 30% of its fibers are found in the husk. Lignin, a material that makes up ligneous cell walls, is found in significant quantities in the flesh. That works well as an adhesive for the coconut fibers when melted. To make coconut boards, the husk of the coconut is separated. Small fragments of the husk are treated before being hot-pressed into boards. The board material (known as ecoco boards) produced from coconut husk is on par with or better than medium-density fiberboard in terms of its qualities, such as durability and fire safety (Hazlegreaves, 2018).

In the food industry, coconut water can be processed into a range of beverages, including coconut water, milk, and cream. Coconut oil can also be extracted from coconut for cooking and to be used in food products. Coconut oil and coconut water are commonly used in cosmetics and personal care products, including soaps, shampoos, and moisturizers. For the skin and hair, these substances are well-known for their moisturizing, nourishing, and soothing benefits (Pandiselvam et al., 2019). With further research and innovation, more uses for coconut waste will likely be discovered in the future.

6.8 CHALLENGES ASSOCIATED WITH PROCESSING AND COMMERCIALIZING

The effective management and commercialization of coconut waste pose significant challenges across multiple fronts, encompassing technological limitations, economic feasibility, market demand dynamics, environmental impacts, and supply chain coordination. Technologically, the lack of specialized machinery and expertise tailored for coconut waste processing impedes efficient resource recovery and waste reduction efforts. Investing in research and development initiatives aimed at innovating processing techniques is crucial to overcoming these hurdles and maximizing the value extracted from coconut by-products.

Economic viability stands as another formidable barrier, with businesses facing substantial upfront costs associated with infrastructure, machinery, and processing methods. Moreover, the unpredictable market demand for coconut waste-derived products further complicates financial planning and investment decisions. Governments and financial institutions play a pivotal role in fostering a conducive environment for investment by offering incentives such as grants, subsidies, and low-interest loans; encouraging sustainable practices; and stimulating market growth.

Market promotion and consumer awareness are pivotal in driving demand for coconut waste-derived products. However, low consumer awareness and misconceptions about these products hinder their market penetration and expansion. Efforts to educate consumers and businesses about the benefits of coconut waste-derived items can bolster market acceptance and foster sustainable consumption patterns. Moreover, creating regulatory frameworks and standards, alongside promoting environmentally friendly processing methods and circular economy principles, is essential for mitigating environmental pollution and ensuring sustainable waste management practices.

Effective supply chain coordination and stakeholder engagement are imperative for efficient coconut waste management. Given the complexity of the coconut supply chain involving growers, processors, decision-makers, and consumers, concerted efforts and shared objectives are vital. Engaging stakeholders through collaborative platforms, knowledge-sharing initiatives, and inclusive decision-making processes can facilitate comprehensive waste management strategies and promote long-term sustainability. By addressing these multifaceted challenges and embracing innovative solutions, the coconut industry can transform waste into valuable resources, fostering circular economy principles and sustainable development paradigms.

6.9 CONCLUSIONS

In conclusion, the management of coconut waste emerges as a multifaceted challenge, necessitating a nuanced examination of its economic, environmental, and societal implications. The interconnected nature of these variables underscores the complexity of the problem, suggesting that a systematic solution requires comprehensive consideration of each component. Effectively processing and commercializing coconut by-products presents a significant hurdle, with complexities ranging from requiring specialized technology and substantial investments to addressing consumer reluctance and misconceptions. However, amid these challenges lies immense profitability and sustainable resource utilization potential. Innovative approaches can transform coconut waste into valuable resources, from biofuel derived from husks to activated carbon from shells, fostering employment opportunities, and promoting sustainable growth across industries.

A holistic approach to coconut waste management necessitates collaboration and coordination among diverse stakeholders, encompassing coconut farmers, processors, policymakers, and community members. By engaging in partnership, comprehensive waste management plans can be developed, considering every aspect of the coconut value chain from cultivation to disposal. Such an approach mandates creating and adopting innovative, sustainable techniques tailored to ecological benefits, energy efficiency, and commercial viability. Additionally, educating stakeholders about the value and opportunities associated with coconut waste by-products is crucial for generating demand and establishing markets.

Achieving economically feasible, socially acceptable, and environmentally sustainable coconut waste management requires dedicated support and assistance for local communities. Empowering communities with capacity-building initiatives, resources, and technical support fosters active participation and ensures solutions are tailored to their unique needs. In essence, navigating the complexities of coconut waste management demands a concerted effort to develop holistic solutions, integrating collaboration, innovation, awareness, and community support. Through such endeavors, coconut waste management can evolve into a catalyst for sustainable development, realizing economic prosperity, social equity, and environmental stewardship in coconut-producing regions and beyond.

ACKNOWLEDGMENT

This chapter was written with Chong Zhi Xuen, Muhammad Ezzat Syameer Syameer Suib, and Nurul Zahirah Muhamad Zakri.

REFERENCES

Adejumo, I. O., & Adebiyi, O. A. (2021). Agricultural solid wastes: Causes, effects, and effective management. IntechOpen eBooks. https://doi.org/10.5772/intechopen.93601

Adeniyi, A. G., Onifade, D. V., Ighalo, J. O., & Adeoye, S. A. (2019). A review of coir fiber reinforced polymer composites. *Composites Part B-Engineering*, *176*, 107305. https://doi.org/10.1016/j.compositesb.2019.107305

Adorna, J., Borines, M., & Dang, V. D., & Doong, R.-A. (2020). Coconut shell derived activated biochar–manganese dioxide nanocomposites for high performance capacitive deionization. *Desalination*, *492*, 114602.

Agar, D., & Wihersaari, M. (2012). Bio-coal, torrefied lignocellulosic resources—Key properties for its use in co-firing with fossil coal—Their status. *Biomass and Bioenergy*, *44*, 107–111. https://doi.org/10.1016/j.biombioe.2012.05.004

Ahmad, R. K., Anwar Sulaiman, S., Yusup, S., Sham Dol, S., Inayat, M., & Aminu Umar, H. (2022). Exploring the potential of coconut shell biomass for charcoal production. *Ain Shams Engineering Journal, 13*(1), 101499. https://doi.org/10.1016/j.asej.2021.05.013

Ajien, A., Idris, J., Md Sofwan, N., Husen, R., & Seli, H. (2023). Coconut shell and husk biochar: A review of production and activation technology, economic, financial aspect and application. *Waste Management & Research, 41*(1), 37–51. https://doi.org/10.1177/0734242x221127167

Ali, B., Hawreen, A., Kahla, N. B., Amir, M. T., Azab, M., & Raza, A. (2022). A critical review on the utilization of coir (coconut fiber) in cementitious materials. *Construction and Building Materials*, *351*, 128957. https://doi.org/10.1016/j.conbuildmat.2022.128957

Alias, A. H., Norizan, M. N., Sabaruddin, F. A., Asyraf, M. R. M., Norrahim, M. N. F., Ilyas, A. R., Kuzmin, A. M., Rayung, M., Shazleen, S. S., Nazrin, A., Sherwani, S. F. K., Harussani, M. M., Atikah, M. S. N., Ishak, M. R., Sapuan, S. M., & Khalina, A. (2021). Hybridization of MMT/lignocellulosic fiber reinforced polymer nanocomposites for structural applications: A review. *Coatings*, *11*(11), 1355. https://doi.org/10.3390/coatings11111355

Arrigoni, A., Pelosato, R., Melià, P., Ruggieri, G., Sabbadini, S., & Dotelli, G. (2017). Life cycle assessment of natural building materials: The role of carbonation, mixture components and transport in the environmental impacts of hempcrete blocks. *Journal of Cleaner Production*, *149*, 1051–1061. https://doi.org/10.1016/j.jclepro.2017.02.161

ASEAN Briefing (2021, January 8). Biomass industry in the Philippines. *ASEAN Business News*. https://www.aseanbriefing.com/news/biomass-industry-philippines/

Asyraf, M. R. M., Syamsir, A., Zahari, N. M., Supian, A. B. M., Ishak, M. R., Sapuan, S. M., Sharma, S., Rashedi, A., Razman, M. R., Zakaria, S. Z. S., Ilyas, R. A., & Rashid, M. Z. A. (2022). Product development of natural fibre-composites for various applications: Design for sustainability. *Polymers*, *14*(5), 920. https://doi.org/10.3390/polym14050920

Awopeju, O. F. (2021). Health effect of biomass fuel smoke. IntechOpen eBooks. https://doi.org/10.5772/intechopen.94611

Ayompe, L. M., Schaafsma, M. R., & Egoh, B. N. (2021). Towards sustainable palm oil production: The positive and negative impacts on ecosystem services and human wellbeing. *Journal of Cleaner Production*, *278*, 123914. https://doi.org/10.1016/j.jclepro.2020.123914

Balla, V. K., Kate, K. H., Satyavolu, J., Singh, P., & Tadimeti, J. G. D. (2019). Additive manufacturing of natural fiber reinforced polymer composites: Processing and prospects. *Composites Part B-Engineering*, *174*, 106956. https://doi.org/10.1016/j.compositesb.2019.106956

Benti, N. E., Gurmesa, G. S., Argaw, T., Aneseyee, A. B., Gunta, S., Kassahun, G. B., Aga, G. S., & Asfaw, A. A. (2021). The current status, challenges and prospects of using biomass energy in Ethiopia. *Biotechnology for Biofuels*, *14*(1). https://doi.org/10.1186/s13068-021-02060-3

Beveridge, F. C., Kalaipandian, S., Yang, C., & Adkins, S. W. (2022). Fruit biology of coconut (Cocos nucifera L.). *Plants*, *11*(23), 3293. https://doi.org/10.3390/plants11233293

Birniwa, A. H., Abdullahi, S. S., Ali, M., Mohammad, R. E. A., Jagaba, A. H., Amran, M., Avudaiappan, S., Maureira-Carsalade, N., & Flores, E. I. S. (2023). Recent trends in treatment and fabrication of plant-based fiber-reinforced epoxy composite: A review. *Journal of Composites Science*, *7*(3), 120. https://doi.org/10.3390/jcs7030120

Bolivar-Telleria, M., Turbay, C., Favarato, L., Carneiro, T., de Biasi, R. S., Fernandes, A. A., Santos, A. M., & Fernandes, P. M. (2018). Second-generation bioethanol from coconut husk. *BioMed Research International*, 2018, 1–20. https://doi.org/10.1155/2018/4916497

Bonneau, X., Haryanto, I., Karsiwan, T. (2010). Coconut husk ash as a fertilizer for coconut palms on peat. *Experimental Agriculture*, *46*(3), 401–414. https://doi.org/10.1017/s0014479710000025

Bradley, W. L., Poel, J., & Huang, H. (2006). Cocos nucifera: An abundant renewable source of energy. *Proceedings of the International Conference on Renewable Energy for Developing Countries-2006.* (pp. 1–5). Washington, D.C., USA.

Buck, W., & Rudtsch, S. (2011). Thermal properties. In H. Czichos, T. Saito, & L. Smith (Eds.), *Springer Handbook of Metrology and Testing* (pp. 453–483). Berlin, Heidelberg: Springer.

Cabral, M. M., Abud, A. K., Silva, C. E., & Almeida, R. M. (2016). Bioethanol production from coconut husk fiber. *Ciência Rural*, *46*(10), 1872–1877. https://doi.org/10.1590/0103-8478cr20151331

Chabbert, B., Padovani, J., Djemiel, C., Ossemond, J., Lemaitre, A., Yoshinaga, A., Hawkins, S., Grec, S., Beaugrand, J., & Kurek, B. (2020). Multimodal assessment of flax dew retting and its functional impact on fibers and natural fiber composites. *Industrial Crops and Products*, *148*, 112255. https://doi.org/10.1016/j.indcrop.2020.112255

Clauser, N. M., González, G., Mendieta, C. M., Kruyeniski, J., Area, M. C., & Vallejos, M. E. (2021). Biomass waste as sustainable raw material for energy and fuels. *Sustainability*, *13*(2), 794. https://doi.org/10.3390/su13020794

Cohen, A. (2022). Textile tips: Seagrass, coir, jute, and sisal. Natural fiber elegance. *FiberSeal Northeast.* https://www.fibersealnortheast.com/blog/textile-tips-what-is-sisal

Cowley, L. (2021). Coir—The natural fiber from coconut husk. *Eco World.* Retrieved from https://ecoworldonline.com/coir-the-natural-fiber-from-coconut-husk/

Das, O., Babu, K., Shanmugam, V., Sykam, K., Tebyetekerwa, M., Neisiany, R. E., Ramakrishna, S. (2022). Natural and industrial wastes for sustainable and renewable polymer composites. *Renewable and Sustainable Energy Reviews*, *158*, 112054. https://doi.org/10.1016/j.rser.2021.112054

Dhaliwal, J. (2020). Natural fibers: Applications. IntechOpen eBooks. https://doi.org/10.5772/intechopen.86884

Dhir, Y. J. (2022). Natural fibers: The sustainable alternatives for textile and non-textile applications. IntechOpen eBooks. https://doi.org/10.5772/intechopen.106393

Ding, K., Le, Y., Yao, G., Ma, Z., Jin, B., Wang, J. J., & Jin, F. (2018). A rapid and efficient hydrothermal conversion of coconut husk into formic acid and acetic acid. *Process Biochemistry*, *68*, 131–135. https://doi.org/10.1016/j.procbio.2018.02.021

Divyashree, A., Manaf, S. A. B. A., Yallappa, S., Chaitra, K., Kathyayini, N., & Hegde, G. (2016). Low cost, high performance supercapacitor electrode using coconut wastes: Eco-friendly approach. *Journal of Energy Chemistry*, *25*(5), 880–887. https://doi.org/10.1016/j.jechem.2016.08.002

Dungani, R., Karina, M., Subyakto, S., Sulaeman, A., Hermawan, D., & Hadiyane, A. (2016). Agricultural waste fibers towards sustainability and advanced utilization: A review. *Asian Journal of Plant Sciences*, *15*(1–2), 42–55. https://doi.org/10.3923/ajps.2016.42.55

Fahmy, Y., Fahmy, T. Y. A., Mobarak, F., El-Sakhawy, M., & Fadl, M. (2017) Agricultural residues (wastes) for manufacture of paper, board, and miscellaneous products: Background overview and future prospects. *International Journal of Chemtech Research, 10*(2), 424–448.

Ferronato, N., & Torretta, V. (2019). Waste mismanagement in developing countries: A review of global issues. *International Journal of Environmental Research and Public Health, 16*(6), 1060. https://doi.org/10.3390/ijerph16061060

Foo, K. Y., & Hameed, B. H. (2010). Insights into the modeling of adsorption isotherm systems. *Chemical Engineering Journal, 156*(1), 2–10.

Fu, S., Wu, H., Zhu, K., Zhao, Z., & Liang, Z. (2023). The unique morphology of coconut petiole fibers facilitates the fabrication of plant composites with high impact performance. *Polymers, 15*(9), 2200. https://doi.org/10.3390/polym15092200

Ghosal, M. K., Mahapatra, N., Sahoo, N., & Rout, P. K. (2016). A study on preparation and heating value of bio-coal from green coconut shell for domestic cooking fuel. *Ecology, Environment and Conservation Paper, 22*, (Sept. Suppl.), 91–97.

Ghosal, M. K., & Roy, K. (2021). Physical characteristics studies of biocoal from waste tender coconut. *Biological Forum: An International Journal, 13*(3a): 691–695.

Girijappa, Y. G. T., Nagaraju, S. B., Puttegowda, M., Verma, A., Rangappa, S. M., & Siengchin, S. (2023). Biopolymer-based composites: An eco-friendly alternative from agricultural waste biomass. *Journal of Composites Science, 7*(6), 242. https://doi.org/10.3390/jcs7060242

Girijappa, Y. G. T., Sanjay, M. R., & Siengchin, S. (2019). Natural fibers as sustainable and renewable resource for development of eco-friendly composites: A comprehensive review. *Frontiers in Materials, 6*. https://doi.org/10.3389/fmats.2019.00226

Global Organics. (2016, July 11). Sustainable coconut—Let's count the ways. *Global Organics News & Views*. https://www.global-organics.com/product-category/coconut/

Go, A. W., Conag, A. T., Igdon, R. M. B., Toledo, A. S., & Malila, J. S. (2019). Potentials of agricultural and agro-industrial crop residues for the displacement of fossil fuels: A Philippine context. *Energy Strategy Reviews, 23*, 100–113. https://doi.org/10.1016/j.esr.2018.12.010

Gonzalo, A., Bimbela, F., Sánchez, J. L., Labidi, J., Marín, F., & Arauzo, J. (2017). Evaluation of different agricultural residues as raw materials for pulp and paper production using a semichemical process. *Journal of Cleaner Production, 156*, 184–193.

Halimatul, M. J., Sapuan, S. M., Jawaid, M., Ishak, M. R., & Ilyas, R. A. (2019). Water absorption and water solubility properties of sago starch biopolymer composite films filled with sugar palm particles. *Polimery, 64*(9), 27–35. https://doi.org/10.14314/polimery.2019.9.4

Harvard, T. H. Chan School of Public Health. (2022, November 24). Health consequences of using biomass for energy. *News*. https://www.hsph.harvard.edu/news/hsph-in-the-news/health-consequences-of-using-biomass-for-energy/

Hasan, K. M. A., Horvath, P., Bak, M., & Alpár, T. (2021). A state-of-the-art review on coir fiber-reinforced biocomposites. *RSC Advances, 11*(18), 10548–10571. https://doi.org/10.1039/d1ra00231g

Hassan, T., Jamshaid, H., Mishra, R., Khan, M. Q., Petru, M., Novák, J., Chotěborský, R., & Hromasová, M. (2020). Acoustic, mechanical and thermal properties of green composites reinforced with natural fibers waste. *Polymers, 12*(3), 654. https://doi.org/10.3390/polym12030654

Hazlegreaves, S. (2018, September 20). Cocoboards: Affordable building material made from coconut husks. *Open Access Government*. https://www.openaccessgovernment.org/affordable-building-material/52270/

Ichim, M., Stelea, L., Filip, I., Lisa, G., & Muresan, E. I. (2022). Thermal and mechanical characterization of coir fiber–reinforced polypropylene biocomposites. *Crystals, 12*(9), 1249. https://doi.org/10.3390/cryst12091249

Isa, N. M., Sivapathy, A., & Kamarruddin, N. F. (2021). Malaysia on the way to sustainable development: Circular economy and green technologies (pp. 91–115). Emerald Publishing Limited eBooks. https://doi.org/10.1108/978-1-80043-806-420211009

Ismail, I., Hatta, Z. N., Mursal, Jalil, Z., & Fadzullah, S. H. S. M. (2021). Thermal conductivity of coconut shell particle epoxy resin composite. *Journal of Physics: Conference Series, 1816*(1), 012031. https://doi.org/10.1088/1742-6596/1816/1/012031

Iwaro, J., & Mwasha, A. (2019). Effects of using coconut fiber–insulated masonry walls to achieve energy efficiency and thermal comfort in residential dwellings. *Journal of Architectural Engineering, 25*(1), 04019001. https://doi.org/10.1061/(ASCE)AE.1943-5568.0000341

James, A., & Yadav, D. (2021). Valorization of coconut waste for facile treatment of contaminated water: A comprehensive review (2010–2021). *Environmental Technology and Innovation, 24*, 102075. https://doi.org/10.1016/j.eti.2021.102075

Jayaprakash, K., Osama, A., Rajagopal, R., Goyette, B., & Karthikeyan, O. P. (2022). Agriculture waste biomass repurposed into natural fibers: A circular bioeconomy perspective. *Bioengineering*, *9*(7), 296. https://doi.org/10.3390/bioengineering9070296

Jumaidin, R., Adam, N. W., Ilyas, R. A., Hussin, M. S. F., Taha, M. M., Mansor, M. R., Azlan, U. A. A., & Yob, M. S. (2019a). Water transport and physical properties of sugarcane bagasse fibre reinforced thermoplastic potato starch biocomposite. *Journal of Advanced Research in Fluid Mechanics and Thermal Sciences*, *61*(2), 273–281.

Jumaidin, R., Diah, N. A., Ilyas, R. A., Alamjuri, R. H., & Yusof, F. A. M. (2021). Processing and characterisation of banana leaf fibre reinforced thermoplastic cassava starch composites. *Polymers*, *13*(9), 1420. https://doi.org/10.3390/polym13091420

Jumaidin, R., Saidi, Z. A. S., Ilyas, R. A., Ahmad, M. N., Wahid, M. K., Yaakob, M. Y., Maidin, N. A., Rahman, M. H. A., & Osman, M. H. (2019b). Characteristics of cogon grass fibre reinforced thermoplastic cassava starch biocomposite: Water absorption and physical properties. *Journal of Advanced Research in Fluid Mechanics and Thermal Sciences*, *62*(1), 43–52.

Kalkanis, K., Alexakis, D., Kyriakis, E., Kiskira, K., Llanes, J. L., Themelis, N. J., & Psomopoulos, C. S. (2022). Transforming waste to wealth, achieving circular economy. *Circular Economy and Sustainability*, *2*(4), 1541–1559. https://doi.org/10.1007/s43615-022-00225-2

Kamarudin, S. H., Basri, M. S. M., Rayung, M., Abu, F., Ahmad, S., Norizan, M. N., Osman, S., Sarifuddin, N., Desa, M. S. Z. M., Abdullah, U. H., Tawakkal, I. S. M. A., & Abdullah, L. C. (2022). A review on Natural Fiber Reinforced Polymer Composites (NFRPC) for sustainable industrial applications. *Polymers*, *14*(17), 3698. https://doi.org/10.3390/polym14173698

Kanianska, R. (2016). Agriculture and its impact on land-use, environment, and ecosystem services. InTech eBooks. https://doi.org/10.5772/63719

Kao-Walter, S., Mfoumou, E., & Ndikontar, M. K. (2011). Mechanical properties and life-cycle sustainability aspects of natural fiber. *Advanced Materials Research*, *347–353*, 1887–1893. https://doi.org/10.4028/www.scientific.net/amr.347-353.1887

Karimah, A., Ridho, M., Munawar, S. S., Adi, D. S., Ismadi, Damayanti, R., Subiyanto, B., Fatriasari, W., & Fudholi, A. (2021b). A review on natural fibers for development of eco-friendly bio-composite: Characteristics, and utilizations. *Journal of Materials Research and Technology*, *13*, 2442–2458. https://doi.org/10.1016/j.jmrt.2021.06.014

Karimah, A., Ridho, M. R., Munawar, S. S., Ismadi, Amin, Y., Damayanti, R., Lubis, M. Z., Wulandari, A. P., Nurindah, Iswanto, A. H., Fudholi, A., Asrofi, M., Saedah, E., Sari, N. H., Pratama, B. I., Fatriasari, W., Nawawi, D. S., Sanjay, M. R., & Siengchin, S. (2021a). A comprehensive review on natural fibers: Technological and socio-economical aspects. *Polymers*, *13*(24), 4280. https://doi.org/10.3390/polym13244280

Khalid, M., Rashid, A. A., Arif, Z. U., Ahmed, W., Arshad, H., & Zaidi, A. A. (2021). Natural fiber reinforced composites: Sustainable materials for emerging applications. *Results in Engineering*, *11*, 100263. https://doi.org/10.1016/j.rineng.2021.100263

Kim, M. S., Kim, D. H., & Lee, J. K. (2011). 3.11- Biohydrogen. In M. Moo-Young (Ed.), *Comprehensive Biotechnology* (Second Edition) (pp. 115–125). Burlington: Academic Press.

Kiron, M. I. (2021). Coconut or coir fiber: Properties, manufacturing process and applications. *Textile Learner*. https://textilelearner.net/coconut-or-coir-fiber-properties-manufacturing/

Koul, B., Sharma, K., & Shah, M. P. (2021) Phycoremediation: A sustainable alternative in wastewater treatment (WWT) regime. *Environmental Technology & Innovation*, *25*, 102040.

Koul, B., Yakoob, M., & Shah, M. P. (2022). Agricultural waste management strategies for environmental sustainability. *Environmental Research, 206*, 112285. https://doi.org/10.1016/j.envres.2021.112285

Kumar, S., Prasad, L., Patel, V. K., Kumar, V., Kumar, A., Yadav, A. K., & Winczek, J. (2021). Physical and mechanical properties of natural leaf fiber-reinforced epoxy polyester composites. *Polymers*, *13*(9), 1369. https://doi.org/10.3390/polym13091369

Lebedeva, D., Hijmans, S., Mathew, A. P., Subbotina, E., & Samec, J. S. M. (2022). Waste-to-fuel approach: Valorization of lignin from coconut coir pith. *ACS Agricultural Science & Technology*, *2*(2), 349–358. https://doi.org/10.1021/acsagscitech.1c00248

Lee, C. H., Abdan, K., Lee, S. H., & Liu, M. (2020). A comprehensive review on BAST fiber retting process for optimal performance in fiber-reinforced polymer composites. *Advances in Materials Science and Engineering*, 2020, 1–27. https://doi.org/10.1155/2020/6074063

Liné Cowley (2019). *Coir – The Natural Fiber from Coconut Husk*. Eco World. Retrieved from https://ecoworldonline. com/coir-the-natural-fiber-from-coconut-husk/.

Mahmud, M. A., Abir, N., Anannya, F. R., Khan, A. N., Rahman, A. N. M. M., & Jamine, N. (2023). Coir fiber as thermal insulator and its performance as reinforcing material in biocomposite production. *Heliyon*, *9*(5), e15597. https://doi.org/10.1016/j.heliyon.2023.e15597

Maraveas, C. (2020). "Production of sustainable construction materials using agro-wastes." *Materials*, *13*(2), 262.

Mardijanti, D. S., Megantara, E. N., Bahtiar, A., & Sunardi, S. (2021). Turning the cocopith waste into myceliated biocomposite to make an insulator. *International Journal of Biomaterials,* 2021, 6630657. https://doi.org/10.1155/2021/6630657

Mazani, N., Sapuan, S. M., Sanyang, M. L., Atiqah, A., & Ilyas, R. A. (2019). Design and fabrication of a shoe shelf from kenaf fiber reinforced unsaturated polyester composites. In H. Ariffin, S. M. Sapuan, & M. A. Hassan (Eds.), *Lignocellulose for Future Bioeconomy* (1st ed., Issue 2000, pp. 315–332). Elsevier. https://doi.org/10.1016/B978-0-12-816354-2.00017-7

McFarland, K. (2019). Biomass advantages and disadvantages. *SynTech Bioenergy*. https://www.syntechbioenergy.com/blog/biomass-advantages-disadvantages

Mendoza-Quiroga, R., Mercado-Caruso, N., Fabregas-Villegas, J., Domínguez, J. M., & Chimento, J. (2022). Effect of the morphology measuring methods of coconut fiber on the determination of mechanical tensile properties. *Journal of Natural Fibers*, *19*(15), 11818–11830. https://doi.org/10.1080/15440478.2022.2044962

Miller, G. L. (1959). Use of dinitrosalicylic acid reagent for determination of reducing sugar. *Analytical Chemistry*, *31*(3), 426–428. https://doi.org/10.1021/ac60147a030

Mujtaba, M., Fraceto, L. F., Fazeli, M., Mukherjee, S., Savassa, S. M., de Medeiros, G. A., Pereira, A. D. E. S., Mancini, S. D., Lipponen, J., & Vilaplana, F. (2023). Lignocellulosic biomass from agricultural waste to the circular economy: A review with focus on biofuels, biocomposites and bioplastics. *Journal of Cleaner Production*, *402*, 136815. https://doi.org/10.1016/j.jclepro.2023.136815

Nadzri, S. N. A., Sultan, M. T. H., Shah, A. U. M., Safri, S. N. A., Talib, A. R. A., Sapuan, S., & Basri, A. A. (2020). A comprehensive review of coconut shell powder composites: Preparation, processing, and characterization. *Journal of Thermoplastic Composite Materials*, *35*(12), 2641–2664. https://doi.org/10.1177/0892705720930808

Nair, G. R., Singh, A., Zimniewska, M., & Raghavan, V. (2013). Comparative evaluation of physical and structural properties of water retted and non-retted flax fibers. *Fibers*, *1*(3), 59–69. https://doi.org/10.3390/fib1030059

Nazrin, A., Sapuan, S. M., Zuhri, M. Y. M., Tawakkal, I. S. M. A., & Ilyas, R. A. (2021). Water barrier and mechanical properties of sugar palm crystalline nanocellulose reinforced thermoplastic sugar palm starch (TPS)/poly(lactic acid) (PLA) blend bionanocomposites. *Nanotechnology Reviews*, *10*(1), 431–442. https://doi.org/10.1515/ntrev-2021-0033

Neh, A. (2020). Agricultural Waste Management System [AWMS] in Malaysian. *Open Access Journal of Waste Management & Xenobiotics*, *3*(2), 1–2. https://doi.org/10.23880/oajwx-16000140

Norfarhana, A. S., Ilyas, R. A., & Ngadi, N. (2022). A review of nanocellulose adsorptive membrane as multifunctional wastewater treatment. *Carbohydrate Polymers*, *291*, 119563. https://doi.org/10.1016/j.carbpol.2022.119563

Norfarhana, A. S., Ilyas, R. A., Ngadi, N., Othman, M. H. D., Misenan, M. S. M., & Norrrahim, M. N. F. (2024a). Revolutionizing lignocellulosic biomass: A review of harnessing the power of ionic liquids for sustainable utilization and extraction. *International Journal of Biological Macromolecules*, *256*, 128256. https://doi.org/10.1016/j.ijbiomac.2023.128256

Norfarhana, A. S., Khoo, P. S., Ilyas, R. A., Ab Hamid, N. H., Aisyah, H. A., Norrrahim, M. N. F., Knight, V. F., Rani, M. S. A., Septevani, A. A., Syafri, E., & Annamalai, P. K. (2024b). Exploring of cellulose nanocrystals from lignocellulosic sources as a powerful adsorbent for wastewater remediation. *Journal of Polymers and the Environment.* https://doi.org/10.1007/s10924-024-03227-3

Norrrahim, M. N. F., Ilyas, R. A., Nurazzi, N. M., Asmal Rani, M. S., Nur Atikah, M. S., & Shazleen, S. S. (2021). Chemical pretreatment of lignocellulosic biomass for the production of bioproducts: An overview. *Applied Science and Engineering Progress*, 1–18. https://doi.org/10.14416/j.asep.2021.07.004

Obeng, G. Y., Amoah, D. Y., Opoku, R., Sekyere, C., Adjei, E. A., & Mensah, E. (2020). Coconut wastes as bioresource for sustainable energy: Quantifying wastes, calorific values and emissions in Ghana. *Energies*, *13*(9), 2178. https://doi.org/10.3390/en13092178

Omar, H., Malek, N. S. A., Nurfazianawatie, M. Z., Rosman, N., Bunyamin, I., Abdullah, S., Khusaimi, Z., Rusop, M., & Asli, N. A. (2022). A review of synthesis graphene oxide from natural carbon based coconut waste by Hummer's method. *Materials Today: Proceedings*, *75*, 188–192. https://doi.org/10.1016/j.matpr.2022.11.427

Othman, M. H., Main, N. M., Kunchi Mon, S. Z., & Zaleha, M. (2013). Development of paper using coir fibers as a packaging product. *Asian Journal of Scientific Research, 6*, 207–216. https://doi.org/10.3923/ajsr.2013.207.216

Pandiselvam, R., Manikantan, M. R., Beegum, S., & Mathew, A. C. (2019). Virgin coconut oil infused healthy cosmetics. *Indian Coconut Journal, 13*(9), 30–32. https://www.researchgate.net/publication/336232293_Virgin_Coconut_Oil_infused_healthy_cosmetics

Prabhu, P., et al. (2022). Mechanical, tribology, dielectric, thermal conductivity, and water absorption behaviour of Caryota urens woven fiber-reinforced coconut husk biochar toughened wood-plastic composite. *Biomass Conversion and Biorefinery*, 1–8.

Prétot, S., Collet, F., & Garnier, C. (2014). Life cycle assessment of a hemp concrete wall: Impact of thickness and coating. *Building and Environment, 72*, 223–231.

Raj, S. S., Dhas, J. E. R., & Jesuthanam, C. P. (2020). Challenges on machining characteristics of natural fiber-reinforced composites—A review. *Journal of Reinforced Plastics and Composites, 40*(1–2), 41–69. https://doi.org/10.1177/0731684420940773

Rashidi, N. A., Chai, Y. X., & Yusup, S. (2022). Biomass energy in Malaysia: Current scenario, policies, and implementation challenges. *Bioenergy Research, 15*(3), 1371–1386. https://doi.org/10.1007/s12155-022-10392-7

Rautray, P., Roy, A., Deepak, J. M., & Eisenbart, B. (2021) Bio-bricks: Circular economy and new products. *8th International Conference on Research into Design (ICoRD2021)*, IDC School of Design IIT, Bombay

Saber, D., & El-Aziz, K. A. (2021). Advanced materials used in wearable health care devices and medical textiles in the battle against coronavirus (COVID-19): A review. *Journal of Industrial Textiles, 51*(1_suppl), 246S–271S. https://doi.org/10.1177/15280837211041771

Sadh, P. K., Duhan, S., & Duhan, J. S. (2018). Agro-industrial wastes and their utilization using solid state fermentation: A review. *Bioresources and Bioprocessing, 5*(1). https://doi.org/10.1186/s40643-017-0187-z

Saeed, M., Mohammad, A., Singh, P., Lal, B., Suliman, M., Alshahrani, M. Y., & Sharma, M. (2023). Coconut waste valorization to produce biochar catalyst and its application in cellulose-degrading enzymes production via SSF. *International Journal of Biological Macromolecules*, 240, *124382*. https://doi.org/10.1016/j.ijbiomac.2023.124382

Sangkharak, K. (2011). Optimization of enzymatic hydrolysis for ethanol production by simultaneous saccharification and fermentation of wastepaper. *Waste Management & Research: The Journal for a Sustainable Circular Economy, 29*(11), 1134–1144. https://doi.org/10.1177/0734242x10387656

Sangkharak, K., Chookhun, K., Numreung, J., & Prasertsan, P. (2019). Utilization of coconut meal, a waste product of milk processing, as a novel substrate for biodiesel and bioethanol production. *Biomass Conversion and Biorefinery, 10*(3), 651–662. https://doi.org/10.1007/s13399-019-00456-8

Sisti, L., Totaro, G., Vannini, M., & Celli, A. (2017). Retting process as a pre-treatment of natural fibers for the development of polymer composites. In *Springer Series on Polymer and Composite Materials* (pp. 97–135). Springer International Publishing. https://doi.org/10.1007/978-3-319-68696-7_2

Somashekhar, T. M., Naik, P., Nayak, V., Mallikappa, & Rahul, S. (2018). Study of mechanical properties of coconut shell powder and tamarind shell powder reinforced with epoxy composites. *IOP Conference Series: Materials Science and Engineering, 376*, 012105. https://doi.org/10.1088/1757-899x/376/1/012105

Son, N. H. (2021, June 29). The current status of agricultural wastes and residuals management and recycling in Vietnam. *FFTC Agricultural Policy Platform (FFTC-AP)*. https://ap.fftc.org.tw/article/2786

Souza, E. (2023). Agro-waste design: Husks, bagasse and straw transformed into efficient building materials. *ArchDaily*. https://www.archdaily.com/976034/agro-waste-design-husks-bagasse-and-straw-transformed-into-efficient-building-materials

Steffens, F., Steffens, H., & Oliveira, F. R. (2017). Applications of natural fibers on architecture. *Procedia Engineering, 200*, 317–324. https://doi.org/10.1016/j.proeng.2017.07.045

Stelte, W., Reddy, N., Barsberg, S., & Sanadi, A. R. (2022). Coir from coconut processing waste as a raw material for applications beyond traditional uses. *Bioresources, 18*(1). https://doi.org/10.15376/biores.18.1.stelte

Suriani, M. J., Radzi, F. S. M., Ilyas, R. A., Petrů, M., Sapuan, S. M., & Ruzaidi, C. M. (2021). Flammability, tensile, and morphological properties of oil palm empty fruit bunches fiber/pet yarn-reinforced epoxy fire retardant hybrid polymer composites. *Polymers, 13*(8), 1282. https://doi.org/10.3390/polym13081282

Swarnam, T. P., Velmurugan, A., Pandey, S. K., & Dam Roy, S. (2016). Enhancing nutrient recovery and compost maturity of coconut husk by vermicomposting technology. *Bioresource Technology, 207*, 76–84. doi:https://doi.org/10.1016/j.biortech.2016.01.046

Syafri, E., Jamaluddin, Sari, N. H., Mahardika, M., Amanda, P., & Ilyas, R. A. (2022). Isolation and characterization of cellulose nanofibers from Agave gigantea by chemical-mechanical treatment. *International Journal of Biological Macromolecules, 200*(December 2021), 25–33. https://doi.org/10.1016/j.ijbiomac.2021.12.111

Teh, J. S., Teoh, Y. H., How, H. G., Le, T. D., Jason, Y. J. J., Tho, N. H., & Loo, D. L. (2021). The potential of sustainable biomass producer gas as a waste-to-energy alternative in Malaysia. *Sustainability*, *13*(7), 3877. https://doi.org/10.3390/su13073877

Textile Value Chain (2021, April 29). Clothing from coconut fibers. Textile Magazine, Textile News, Apparel News, Fashion News. https://textilevaluechain.in/in-depth-analysis/articles/clothing-from-coconut-fibres

Textileblog. (2023). Properties, classification and application of plant fibers. *Textile Blog*. https://www.textileblog.com/plant-fibres-properties-types-and-uses/

Tondo, M. (2022). Coconut husk, biomass from coconut: Fibers are known as coir. *Lampoon Magazine*. https://www.lampoonmagazine.com/article/2020/12/27/coconut-husk-fibers/

Trimble, J., Van Hook, R. I., & Folger, A. G. (1984). Biomass for energy: The environmental issues. *Biomass*, *6*(1–2), 3–13. https://doi.org/10.1016/0144-4565(84)90003-9

Tumuluru, J. S., Sokhansanj, S., Hess, J. R., Wright, C. T., & Boardman, R. D. (2011). REVIEW: A review on biomass torrefaction process and product properties for energy applications. *Industrial Biotechnology*, *7*(5), 384–401. https://doi.org/10.1089/ind.2011.7.384

Urban, W. (2013). 16- Biomethane injection into natural gas networks. In A. Wellinger, J. Murphy, & D. Baxter (Eds.), *The Biogas Handbook* (pp. 378–403). Sawston, Cambridge, UK: Woodhead Publishing.

Varghese, S. A., Pulikkalparambil, H., Promhuad, K., Srisa, A., Laorenza, Y., Jarupan, L., Nampitch, T., Chonhenchob, V., & Harnkarnsujarit, N. (2023). Renovation of agro-waste for sustainable food packaging: A review. *Polymers*, *15*(3), 648. https://doi.org/10.3390/polym15030648

Vitrone, F., Ramos, D., Ferrando, F., & Salvadó, J. (2021). Binderless fiberboards for sustainable construction. Materials, production methods and applications. *Journal of Building Engineering*, *44*, 102625. https://doi.org/10.1016/j.jobe.2021.102625

Wahyuni, T. H., Ginting, N., Yunilas, Hasnudi, Mirwandono, E., Siregar, G. A., Sinaga, I. G., & Sembiring, I. (2018). The utilization of coconut waste fermentated by Aspergillus niger and Saccharomyces cerevisiae on meat quality of weaning males rex rabbit. *IOP Conference Series*, *122*, 012129. https://doi.org/10.1088/1755-1315/122/1/012129

Ward, A. J., Hobbs, P. J., Holliman, P. J., & Jones, D. L. (2008). Optimization of the anaerobic digestion of agricultural resources. *Bioresource Technology*, *99*(17), 7928–7940. https://doi.org/10.1016/j.biortech.2008.02.044

Wei, W., Shi, X., Wu, L., & Ni, B.-J. (2021). Insights into coconut shell incineration bottom ash mediated microbial hydrogen production from waste activated sludge. *Journal of Cleaner Production, 322*, 129157. doi:https://doi.org/10.1016/j.jclepro.2021.129157

White, J. (2011). Biogas generation potential of coconut copra in the anaerobic digestion process. Thesis. Department of Civil Engineering, University of Canterbury.

Widyasanti, A., Napitupulu, L. O. B., & Thoriq, A. (2020). Physical and mechanical properties of natural fiber from Sansevieria trifasciata and Agave sisalana. *IOP Conference Series*, *462*(1), 012032. https://doi.org/10.1088/1755-1315/462/1/012032

Wong, W. S., Baji, A., & Leng, S. (2008). Effect of fiber diameter on tensile properties of electrospun poly(ε-caprolactone). *Polymer*, *49*(21), 4713–4722. https://doi.org/10.1016/j.polymer.2008.08.022

Yan, Y. (2016). Developments in fibers for technical nonwovens. In *Advances in Technical Nonwovens* (pp. 19–96). Elsevier. https://doi.org/10.1016/b978-0-08-100575-0.00002-4

Yulianto, M., Agustina, S. E., Hartulistiyoso, E., Nelwan, L. O., & Nurlela. (2017). Study of temperature characterization of agricultural waste in the development of stove for combine heat power. *AIP Conference Proceedings*, *1826*(1), 020031. https://doi.org/10.1063/1.4979247

Zainol, F. A., Arumugam, N., Daud, W. N. W., Suhaimi, N. A. M., Ishola, B. D., Ishak, A. Z., & Afthanorhan, A. (2023). Coconut value chain analysis: A systematic review. *Agriculture*, *13*(7), 1379. https://doi.org/10.3390/agriculture13071379

Zakaria, M. H. (2022). Business potential of coconut-based products in the global markets. *FFTC Agricultural Policy Platform (FFTC-AP)*. https://ap.fftc.org.tw/article/3046

Section III

Seed-Based Plant Waste

7 Soybean Waste Management

7.1 INTRODUCTION

Soybean, a highly versatile and widely cultivated legume crop, has gained global significance due to its high protein content and oil extraction capabilities (Islam et al., 2022). However, the processing of soybeans results in the generation of substantial amounts of waste materials, including soybean hulls, soybean meal, and soybean oil sludge, which pose significant environmental challenges if not effectively managed (Finkenstadt et al., 2019). The improper disposal of soybean waste can contribute to water and soil pollution, release of greenhouse gases, and other environmental issues, necessitating sustainable waste management strategies and innovative industrial applications to mitigate these impacts.

In recent years, researchers and industries have been actively exploring ways to utilize soybean waste as a valuable resource, rather than a burden, through diverse industrial applications ranging from animal feed and food ingredient production to biofuel, bioplastics, and functional materials (Singh et al., 2022). For instance, soybean hulls, which are the fibrous outer covering of soybeans, have been utilized as a source of dietary fiber, animal feed, or a feedstock for the production of biofuels through fermentation or thermochemical conversion processes (Finkenstadt et al., 2019). Soybean meal, which is the protein-rich by-product of soybean oil extraction, has been used as an ingredient in animal feed and can also be further processed to extract bioactive compounds for functional food or pharmaceutical applications . Additionally, soybean oil sludge, a by-product of soybean oil refining, has been converted into biodiesel and bio-based lubricants or used as a raw material for the production of bioplastics (De Oliveira et al., 2018).

The potential industrial applications of soybean waste offer several benefits, including reduced waste generation, decreased environmental impact, and the development of sustainable and value-added products. However, the successful implementation of these applications requires overcoming challenges such as economic viability, scalability, and technological feasibility (Karim et al., 2025). Therefore, a comprehensive understanding of the current research and industrial applications of soybean waste management is crucial to unlock the full potential of soybean waste as a valuable resource in various industrial sectors.

7.2 BIOMASS FROM NATURAL FIBERS

Biomass from fiber is a promising renewable energy source that has gained significant attention due to its potential to replace fossil fuels and reduce greenhouse gas emissions. Fiber-rich organic materials, such as agricultural residues, forestry residues, and industrial residues, can be utilized as biomass feedstocks for various applications. This chapter aims to provide an overview of biomass from fiber, including its properties, production methods, and potential applications.

Biomass from fiber exhibits unique properties that make it suitable for a wide range of applications (Norrahim et al., 2021; Norfarhana et al., 2022, 2024a,b; Nurazzi et al., 2021; Syafri et al., 2022). One of the main properties of biomass from fiber is its high cellulose content. Cellulose is a complex carbohydrate that is the main component of plant cell walls and can be converted into biofuels or other value-added products through various chemical, thermal, or biological processes.

DOI: 10.1201/9781003472209-10

The high cellulose content of biomass from fiber makes it a promising feedstock for bioenergy production, including bioethanol, biogas, and bio-oil (Vidal, 2018).

In addition to cellulose, biomass from fiber contains other valuable components, such as hemicellulose and lignin. Hemicellulose is a polysaccharide that can be hydrolyzed into sugars and fermented into biofuels, while lignin is a complex polymer that can be used as a source of renewable chemicals or as a fuel. The presence of hemicellulose and lignin in biomass from fiber adds to its potential for value-added applications, such as biorefinery processes, where multiple products can be produced from a single biomass feedstock (Chen et al., 2019).

Biomass from fiber can be produced using various methods, depending on the type of fiber and the desired properties of the final product. One common production method involves pretreating fiber-rich organic materials to remove impurities, such as lignin and hemicellulose, and increase the accessibility of cellulose. Pre-treatment methods can include mechanical methods, such as milling or grinding; chemical methods, such as acid or alkali treatments; or biological methods, such as enzymatic treatments (Ragauskas et al., 2014).

After pre-treatment, the fiber can be further processed to convert cellulose into bioenergy or other value-added products. Cellulose can be hydrolyzed into sugars through enzymatic or acid-catalyzed processes, and the resulting sugars can be fermented into bioethanol or other biofuels. Alternatively, cellulose can be thermochemically converted into bio-oil or syngas through processes such as pyrolysis or gasification. The production methods of biomass from fiber can be optimized by controlling factors such as temperature, pressure, reaction time, and catalysts, to achieve the desired properties of the biomass and the target product (Chen et al., 2019).

Biomass from fiber has a wide range of potential applications in various sectors, including energy, materials, and chemicals. One of the main applications of biomass from fiber is in bioenergy production. Cellulose-rich biomass can be converted into biofuels, such as bioethanol or biogas, which can be used as renewable alternatives to fossil fuels in transportation or electricity generation. The thermochemical conversion of cellulose into bio-oil or syngas can also provide renewable sources of heat, power, or chemicals (Vidal, 2018).

In addition to bioenergy, biomass from fiber can be used for the production of a wide range of materials and chemicals. Cellulose, hemicellulose, and lignin in biomass from fiber can be utilized to produce bioplastics, biocomposites, and bio-based chemicals. For example, cellulose can be processed into cellulose acetate, a biodegradable and renewable material used in films, coatings, and packaging. Hemicellulose can be used as a source of renewable chemicals, such as furfural or xylitol, which have applications in pharmaceuticals, food additives, and other industrial processes. Lignin can be converted into bio-based resins, adhesives, and antioxidants (Chen et al., 2019).

Furthermore, biomass from fiber can find applications in the agriculture and environmental sectors. For instance, the use of biomass-derived biochar, a carbon-rich material produced through pyrolysis of biomass, as a soil amendment can enhance soil fertility, carbon sequestration, and water retention in agricultural fields. Biomass from fiber can also be used in bioremediation processes to remove pollutants from soil and water, as well as in wastewater treatment to remove contaminants and generate biogas for energy production (Ragauskas et al., 2014).

Biomass from fiber is a promising renewable energy source with a wide range of potential applications. Its high cellulose content, along with hemicellulose and lignin, makes it suitable for bioenergy production and the production of various materials and chemicals. Production methods such as pre-treatment and enzymatic or thermochemical conversion can be optimized to achieve the desired properties of the biomass and the target product. Biomass from fiber can be utilized in various sectors, including energy, materials, chemicals, agriculture, and environmental applications, contributing to the development of a sustainable and circular bio-based economy.

7.3 ENVIRONMENTAL IMPACT FROM BIOMASS

Biomass has gained attention as a renewable energy source due to its potential to reduce greenhouse gas emissions and mitigate climate change. However, like any form of energy production, biomass utilization also has environmental impacts that need to be considered. One of the key advantages of biomass utilization is its potential to reduce greenhouse gas emissions. Biomass is a carbon-neutral energy source because the carbon released during combustion is offset by the carbon absorbed by plants during their growth (Paavola, 2019). This makes biomass an attractive alternative to fossil fuels, which release carbon that has been stored for millions of years in the form of fossilized organic matter. Biomass can also replace other carbon-intensive sources of energy, such as coal or natural gas, in power generation and heat production, resulting in reduced emissions of greenhouse gases, including carbon dioxide (CO_2), methane (CH_4), and nitrous oxide (N_2O) (Cherubini et al., 2019). Additionally, biomass can be used to produce biofuels, such as bioethanol and biodiesel, which have lower carbon emissions compared to fossil fuels. Biofuels produced from biomass can help reduce the dependence on fossil fuels and contribute to mitigating climate change. For example, bioethanol can be blended with gasoline, reducing greenhouse gas emissions from transportation. Biodiesel, on the contrary, can be used as a substitute for diesel fuel, reducing emissions of particulate matter, sulfur dioxide, and other air pollutants (Behera et al., 2019).

Moreover, biomass utilization can have positive impacts on waste management and land use. Biomass can be derived from various sources, including agricultural residues, forest residues, and food waste, which can help divert organic waste from landfills, reducing methane emissions from decomposition (Abhilash et al., 2019). Additionally, biomass production can be integrated with sustainable land management practices, such as agroforestry and afforestation, which can help restore degraded lands, enhance biodiversity, and promote soil carbon sequestration (Fargione et al., 2018). However, biomass utilization also has environmental disadvantages that need to be addressed. One of the main concerns is the potential for negative impacts on biodiversity and ecosystem services. Biomass production, particularly large-scale monoculture plantations, can lead to the conversion of natural habitats, loss of biodiversity, and disruption of ecosystem services, such as water purification, soil erosion control, and pollination (Haberl et al., 2018). The use of agrochemicals, such as fertilizers and pesticides, in biomass production can also have negative effects on soil and water quality and can potentially contribute to eutrophication and contamination of aquatic ecosystems (Paavola, 2019).

Another concern is the potential for competition with food production. Biomass production, particularly bioenergy crops, can compete with food production for land, water, and other resources, leading to land use changes, food price increases, and food security issues (Searchinger et al., 2018). Additionally, the sustainability of biomass production depends on the management practices employed, including the use of fertilizers, irrigation, and land use planning, which can have significant environmental implications. Unmanaged or poorly managed biomass production can result in land degradation, water depletion, and other negative environmental impacts (Cherubini et al., 2019). In conclusion, biomass utilization has both advantages and disadvantages in terms of its environmental impact. On the one hand, biomass has the potential to reduce greenhouse gas emissions, promote waste management, and contribute to sustainable land use practices. On the other hand, biomass utilization can also have negative impacts on biodiversity, ecosystem services, and food production. To ensure the sustainable utilization of biomass, careful planning, monitoring, and management practices are essential. This includes the adoption of sustainable biomass production methods, such as agroforestry and afforestation, as well as the use of best management practices to minimize negative environmental impacts, such as integrated pest management and nutrient management strategies (Behera et al., 2019). Additionally, the development of policies and regulations that promote sustainable biomass production and protect biodiversity, ecosystem services, and food security is crucial.

7.4 AVAILABILITY OF SOYBEAN WASTE

Soybeans are a widely cultivated crop with various applications, and their processing generates significant amounts of by-products and waste materials. This introduction aims to provide an overview of the availability of soybean waste and its potential for valorization and utilization. Soybeans are primarily grown for their oil and protein content, and they are one of the major agricultural crops worldwide. According to the International Institute for Sustainable Development (Voora et al.,

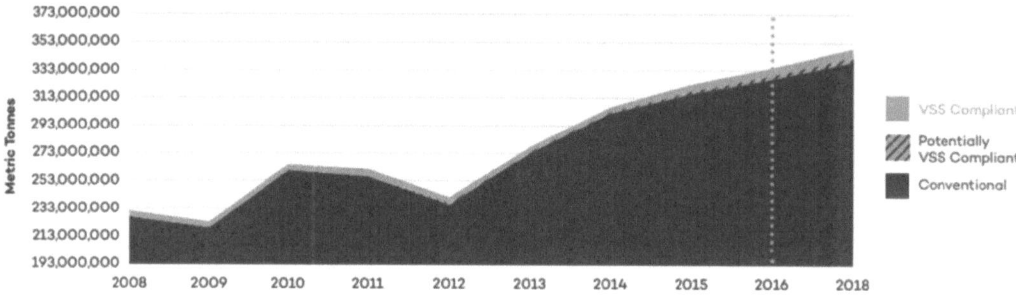

FIGURE 7.1 Global soybean production (Voora et al., 2020).

2020), the production of soybeans has been increasing, with Brazil overtaking the United States as the leading soybean-producing country. Figure 7.1 shows global soybean production. In 2021/2022, Brazil produced approximately 129.5 million metric tons of soybeans, highlighting its significance in global soybeans. Soybean waste has gained attention as a potential feedstock for bioenergy and bioproducts due to its availability, low cost, and renewable nature. Soybean husks and shells, also known as soybean hulls, are the outer protective coverings of soybeans and are rich in cellulose and hemicellulose, making them suitable for bioenergy production through various conversion technologies such as combustion, gasification, and pyrolysis (Abdullah et al., 2018). Soybean meal residue, which is the by-product of soybean oil extraction, is rich in protein and can be utilized to produce animal feed, fertilizer, and bioplastics (Nanda et al., 2017). The utilization of soybean waste as a biomass resource can reduce the environmental burden associated with waste disposal and contribute to the diversification of the bioenergy and bioproducts feedstock portfolio.

The availability of soybean waste as a biomass resource for bioenergy and bioproducts presents several sustainability implications. On the positive side, the utilization of soybean waste can reduce the reliance on fossil fuels, mitigate greenhouse gas emissions, and contribute to the circular economy by converting waste into valuable products (Nanda et al., 2017). Soybean waste can also provide additional income streams for farmers and create employment opportunities in the bioenergy and bioproducts sectors, thus contributing to rural development and economic growth (Abdullah et al., 2018). Moreover, the utilization of soybean waste can potentially reduce the pressure on land use for biomass production, as it is a by-product of the existing soybean processing industry and does not require additional land for cultivation (Santos et al., 2019).

However, the utilization of soybean waste as a biomass resource also raises sustainability concerns. The increased demand for soybean waste for bioenergy and bioproduct production may compete with its other uses, such as animal feed or soil amendment, which can impact the availability and affordability of these products (Nanda et al., 2017). Furthermore, the utilization of soybean waste should be accompanied by proper management practices to avoid negative environmental impacts, such as air emissions from combustion, soil and water pollution from improper waste disposal, and potential land use changes (Abdullah et al., 2018).

It is essential to carefully assess and monitor the sustainability aspects of soybean waste utilization to ensure that it contributes to the overall sustainability goals and does not compromise other environmental or social values. Soybean waste represents a significant biomass resource for bioenergy and bioproduct production, offering potential economic, environmental, and social benefits. However, its utilization should be approached with careful consideration of sustainability implications, including potential impacts on food and feed availability, land use, and environmental quality. Proper management practices, regulatory frameworks, and stakeholder engagement are crucial to ensure the sustainable utilization of soybean waste as a biomass resource. Further research and development efforts are also needed to optimize conversion technologies, improve product yields, and assess the life cycle environmental and social impacts of soybean waste utilization for bioenergy and bioproduct production. Integrating sustainability principles into the planning, implementation, and monitoring of soybean waste utilization can contribute to a more sustainable and circular bioeconomy.

7.5 PROPERTIES OF SOYBEAN WASTE FIBERS

Soybean waste fibers, derived from soybean plant residues, have gained significant attention as a renewable resource with potential applications in various industries, including bioenergy, bioproducts, and composites. These fibers possess certain properties that play a crucial role in determining their suitability and performance in different applications. The chemical composition of soybean waste fibers is an important factor to consider. Soybean plant residues, such as soybean straw, husks, and shells, contain cellulose, hemicellulose, and lignin, which are the main components of plant fibers. The cellulose content of soybean waste fibers ranges from 30% to 45%, while the hemicellulose and lignin content can vary depending on the specific part of the soybean plant and the processing methods used (Makeri et al., 2016). The mechanical strength, thermal stability, and chemical reactivity of soybean waste fibers are influenced by these chemical constituents, making them suitable for various applications (Makeri et al., 2016). For instance, the presence of lignin in soybean waste fibers contributes to their high thermal stability, making them suitable for high-temperature applications such as bioenergy production through thermochemical processes like pyrolysis and gasification (Makeri et al., 2016). In addition to their chemical composition, the physical properties of soybean waste fibers play a crucial role in determining their performance. Fiber length, diameter, and aspect ratio are important physical characteristics that affect their behavior in different applications (Makeri et al., 2016). Longer fibers with a higher aspect ratio are preferred in composites and bioproducts due to their ability to provide better reinforcement and mechanical properties (Makeri et al., 2016). The diameter of soybean waste fibers can vary depending on the specific part of the soybean plant and the processing methods used. The surface properties of soybean waste fibers, such as surface area, roughness, and hydrophobic/hydrophilic nature, also influence their interactions with other materials and their suitability for different applications (Makeri et al., 2016). These properties affect the adhesion and compatibility of soybean waste fibers with matrices in composite materials.

Furthermore, the environmental properties of soybean waste fibers are important to consider. These fibers are derived from agricultural residues, making them a renewable and sustainable resource that contributes to waste reduction, resource conservation, and environmental sustainability (Makeri et al., 2016). The production process of soybean waste fibers typically involves low energy input and generates minimal environmental pollution, further enhancing their environmental friendliness (Makeri et al., 2016). To maximize the utilization of soybean waste fibers as a sustainable and renewable resource, it is crucial to understand and optimize their properties for specific applications (Makeri et al., 2016). Ongoing research and development efforts are needed to further enhance the properties of soybean waste fibers and promote their sustainable utilization in bioenergy, bioproducts, and composites.

7.5.1 Physical Properties

Soybean waste fibers, derived from the residues generated during soybean processing, have garnered significant attention as a promising renewable and sustainable material for diverse applications. These fibers possess distinct physical properties that make them suitable for reinforcement in composite materials. It is crucial to understand the physical properties of soybean waste fibers to optimize their utilization in various applications. In Table 7.1, a summary of the physical properties of soybean waste fibers is presented. The table provides essential information such as fiber length, diameter, aspect ratio, surface area, surface roughness, and hydrophobic/hydrophilic nature. These properties play a vital role in influencing the performance of soybean waste fibers in composites and other bioproducts (Kibar & Öztürk, 2008).

7.5.2 Chemical Composition

Understanding the elemental and ultimate composition of soybeans is essential for evaluating their nutritional and functional properties. Soybean is a widely used and highly nutritious legume that

TABLE 7.1
Physical Properties of Soybean (Reinprecht et al., 2015)

Physical Properties	Description
Fiber length	Varies depending on the specific part of the soybean plant and the processing methods used
Fiber diameter	Varies depending on the specific part of the soybean plant and the processing methods used
Aspect ratio	Longer fibers with a higher aspect ratio are typically preferred for composites and bioproducts
Surface area	Determined by the fiber length, diameter, and surface roughness
Surface roughness	Influences fiber–matrix interactions and composite properties
Hydrophobic/hydrophilic nature	Affects the fiber's compatibility with different matrices and applications

TABLE 7.2
Chemical Composition of Soybean (Perkins, 1995)

Element	Content (%)
Carbon (C)	40–50
Hydrogen (H)	6–7
Nitrogen (N)	4–5
Oxygen (O)	38–45
Sulfur (S)	0.2–0.4
Ash	4–6
Moisture	10–12

is rich in macronutrients such as carbon (C), hydrogen (H_2), nitrogen (N_2), and oxygen (O_2), as well as micronutrients such as sulfur (S). Additionally, soybeans contain ash and moisture, which play important roles in their overall composition. Table 7.2 provides an overview of the approximate ranges of elemental and ultimate analysis of soybeans, based on available literature (Perkins, 1995).

7.5.3 Mechanical Properties

The mechanical properties of soybean waste fibers are important factors influencing their potential use as reinforcement in composite materials. These fibers, derived from soybean processing residues, provide a renewable and sustainable alternative to traditional reinforcements. Understanding the mechanical properties of soybean waste fibers is critical for optimizing performance and ensuring successful integration into various composite systems. Table 7.3 summarizes the mechanical properties of soybean waste fibers in this context. The table includes important parameters for evaluating the mechanical performance of soybean waste fibers in composite materials, such as tensile strength, modulus, and elongation at break (Kibar & Öztürk, 2008).

TABLE 7.3
Mechanical Properties of Soybean (Kibar & Öztürk, 2008)

Fiber Type	Tensile Strength (MPa)	Tensile Modulus (GPa)	Elongation at Break (%)
Soybean hull fiber	40–120	1.5–5.2	1–4
Soybean stem fiber	60–180	3.2–9.5	1–5
Soybean leaf fiber	30–100	1.8–4.5	1–4
Soybean straw fiber	20–80	0.8–3.5	1–3

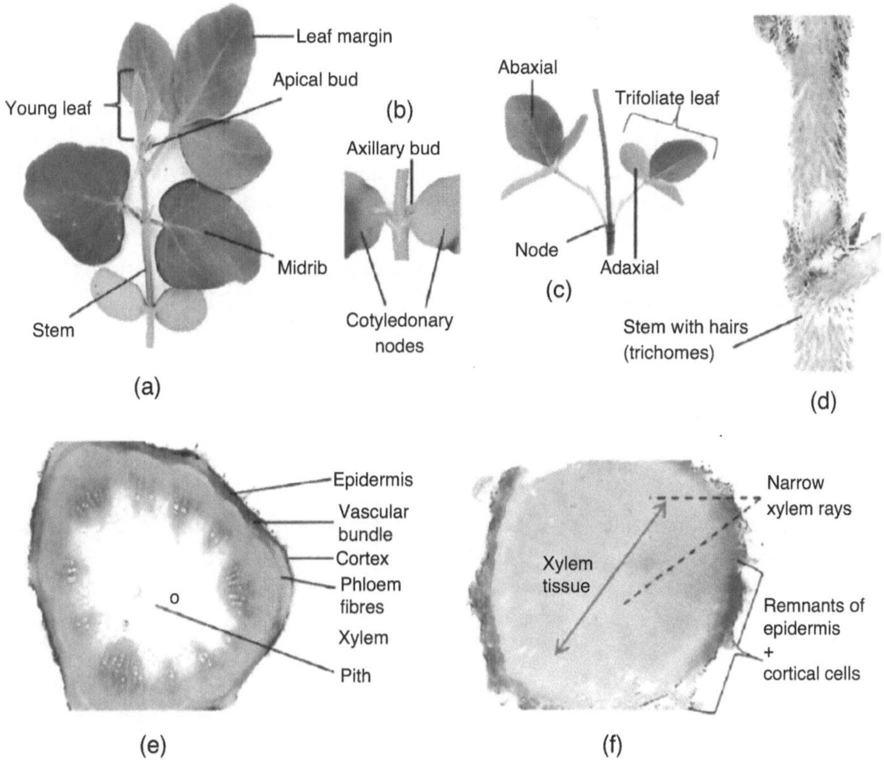

FIGURE 7.2 Morphology and anatomy of the soybean plant. (a) The vegetative first trifoliate (V1) stage. (b) Cotyledons and axillary buds at the axil. (c) Trifoliate leaves with adaxial–abaxial leaf surfaces. (d) A soybean stem with epidermal hairs. (e) A cross-section of a soybean stem micrograph. (f) A cross-section of a soybean root micrograph (Mangena, 2018).

7.5.4 Morphological Properties

Understanding the intricate features and characteristics of crops is of utmost importance for achieving optimal yield and resource management. Among the diverse array of crops, soybean (*Glycine max*) holds a significant place due to its wide-ranging applications as a food source, feedstock, and industrial raw material. Figure 7.2 shows the morphology and anatomy of the soybean plant. Exploring the morphological properties of soybean plants allows us to explore the fascinating world of their physical structures, shedding light on their growth patterns, reproductive traits, and overall phenotypic development. Seed morphology encompasses both the external and internal features, nutritional value, and chemical composition of soybean seeds (Sun & Yuan, 2022). This comprehensive exploration not only aids in understanding the domestication process and regulation of soybean seeds but also paves the way for targeted breeding efforts to enhance their desirable traits.

Moreover, characterizing the morphological variability and diversity among soybean plants has been a subject of keen interest for researchers. By analyzing a wide range of soybean germplasm, scientists have unraveled the immense potential for developing agronomically superior soybean varieties with enhanced biochemical traits (Shilpashree et al., 2021). This endeavor contributes to the development of vegetable-type soybean varieties, catering to the increasing demand for these versatile legumes. The morphological properties of soybean plants have implications beyond their structural characteristics. They also influence plant responses to environmental factors and cultivation practices, such as intercropping and plant density (Ningsih et al., 2019). Exploring the adaptive morphological and physiological responses of soybean plants can provide valuable insights into optimizing cultivation techniques and maximizing agricultural productivity.

7.5.5 Thermal Properties

Thermal properties are crucial characteristics of materials, including soybeans, as they govern their behavior when exposed to heat. Understanding the thermal properties of soybeans is vital for various applications, such as food processing, feed production, and bio-based materials development. Table 7.4 provides an overview of the thermal properties of soybeans, including melting point, thermal conductivity, heat capacity, thermogravimetric analysis, differential scanning calorimetry, thermal stability, and decomposition temperature (Zhang et al., 2015). These properties may vary depending on factors such as the soybean variety, processing methods, and storage conditions. The information presented in this table serves as a valuable reference for researchers, engineers, and other stakeholders interested in utilizing soybeans in various thermal applications.

7.5.6 Caloric Calorific Value

The calorific value of soybean, also known as the energy content, is an important parameter that determines the amount of heat energy released when soybean is burned or used as a fuel. It is a crucial factor in evaluating the potential of soybeans as a renewable energy source. Calorific value is typically expressed in units of energy per unit mass, such as megajoules per kilogram (MJ/kg)

TABLE 7.4
Thermal Properties of Soybean (Zhang et al., 2015)

Property	Description
Melting point	Typically ranges from 150°C to 250°C, depending on the soybean variety
Thermal conductivity	Generally low, ranging from 0.03 to 0.15 W/m·K
Heat capacity	Varies between 1.4 and 2.4 J/g·K
Thermogravimetric analysis (TGA)	Usually exhibits multiple weight loss stages, indicating complex decomposition patterns
Differential scanning calorimetry (DSC)	Typically shows endothermic peaks associated with melting, crystallization, and other thermal transitions
Thermal stability	Resistant to thermal degradation at temperatures below the melting point
Decomposition temperature	Typically, above 300°C, indicating good thermal stability

or kilocalories per kilogram (kcal/kg). The calorific value of soybean can vary depending on factors such as the soybean variety, moisture content, processing methods, and storage conditions. For example, research studies have reported the calorific value of soybean to range from approximately 16–20 MJ/kg (Zhang et al., 2015). The energy content of soybean is largely attributed to its composition, which includes carbohydrates, proteins, fats, and fiber. The high protein and oil content of soybean contributes to its relatively high calorific value compared to other biomass feedstocks.

7.6 AGRO-WASTE MANAGEMENT

7.6.1 Compost

Horticultural waste has been co-composed of a variety of materials, including chicken feces, market waste, and waste from slaughterhouses to speed compost stabilization and enhance product quality. Fruit peels, soybean leftovers, and food waste all contain high levels of nutrients but are frequently burned before disposal. Utilizing the lost nutrients and enhancing composting are both possible benefits of co-composting with these materials.

As a by-product of processing soybeans, soybean wastes are made up of hulls, stems, and seeds that are not fit for human consumption. Because of their high carbon-to-nitrogen (C/N) ratio, soybean residues can be used in composting as a source of nitrogen and phosphorus as well as a bulking agent. Studies have demonstrated that, when combined with other organic waste, such as leaves and sawdust, soybean residues can be a useful co-composting material (Choy et al., 2015). Due to the high nutrient content of soybean residues, co-composting with them has been found to produce a compost with high levels of nitrogen, phosphorus, and potassium.

Soybean waste not only is a source of nutrients but also helps in the composting process. It has been discovered to produce compost at high temperatures, which is crucial for killing pathogens and breaking down organic matter. High temperatures also cause the total carbon to degrade more quickly and lower the C/N ratio, which makes the final product more stable (Choy et al., 2015). Generally speaking, soybean waste is a valuable co-composting component that can accelerate composting and produce high-quality organic fertilizer.

7.6.2 Biocomposite Materials

Biocomposites are sustainable materials created by combining natural fibers, with a polymer matrix, which can be either synthetic or bio-based. These natural fibers serve as reinforcement, providing strength and rigidity, while the matrix binds the fibers together, distributing stress and helping the material maintain its shape. Unlike traditional composites, which often rely on petroleum-based materials, biocomposites are more eco-friendly, renewable, and often biodegradable (Azlin et al., 2022; Nurazzi et al., 2021). This makes them increasingly popular in applications focused on sustainability and reducing environmental impact (Alias et al., 2021). Applications of biocomposites span various industries due to their balance of mechanical performance and eco-friendliness (Alias et al., 2021; Asyraf et al., 2022; Halimatul et al., 2019; Jumaidin et al., 2019a,b, 2021; Nazrin et al., 2021; Suriani et al., 2021). In the automotive sector, biocomposites are used for interior panels, dashboards, and other non-structural components to reduce vehicle weight and enhance sustainability. The construction industry also employs biocomposites for applications like decking, wall panels, and insulation materials, benefiting from their strength, thermal properties, and lower environmental impact. Additionally, biocomposites are becoming common in packaging, consumer goods, and furniture, where they offer a sustainable alternative to plastics. Biomedical applications are also emerging, with biocomposites being used in prosthetics, implants, and scaffolding for tissue engineering due to their compatibility with human tissues and biodegradability.

Soy proteins, which are readily available as by-products or wastes from the agricultural and horticultural industries, are used as the main ingredient to make biocomposite material. The processing conditions of protein-based biocomposites affect their final properties, and the most popular processing methods in the plastics industry are extrusion and injection molding. But, when proteins are extruded or injected, they are subjected to mechanical shear, high temperatures, and high pressures, which can cause denaturation and degradation of the proteins (Garrido et al., 2016). To lessen these associations, enhance processability, and ultimately improve the properties of biocomposite materials, plasticizers can be used. This research is unique in that it uses extrusion and injection techniques to create biocomposites based on soy proteins, expanding the market for these materials' potential uses.

Using soybean waste as a filler material in biocomposites can have a double benefit. Wastes can be disposed of in a more effective way, and value-added products can be obtained at lower cost, bringing both environmental and economic benefits to the development of more sustainable composites (Garrido et al., 2016). Overall, the use of soybean waste in the development of biocomposites shows great potential for creating sustainable and environmentally friendly alternatives to traditional composites.

7.6.3 Biochar

Soybean residue (SR) as a feedstock has great potential in the production of biofuels and carbon materials, which can have a significant impact on waste management and sustainable development. SR is a by-product of the extraction of soybean oil and the production of soymilk and tofu. It has a high nutritional value and is a great source of vegetable protein for animal feed. Studies suggest that SR, given its high reserves globally, is a suitable feedstock for the production of biofuels and chemicals due to its carbonaceous nature. The process used to turn SR into biofuels and carbon materials is pyrolysis (Sun et al., 2021). Pyrolysis is the thermal breakdown of organic material without the presence of oxygen, producing biochar, gas, and oil. It is possible to dry waste soybeans before feeding them into a pyrolysis reactor. The material is exposed to a controlled environment with little oxygen as the reactor is heated to a specific temperature. The type of biochar produced depends on the temperature and length of the heating process.

The volatile organic compounds in the soybean waste are evaporated during the pyrolysis process and collected as bio-oil, which can then be processed into a variety of chemicals and fuels (Sun et al., 2021). Biochar, a black carbon-rich substance that can be used as a soil amendment as well as in several industrial and environmental applications, makes up the remaining solid portion. Due to its high nitrogen and ash content, biochar made from waste soybeans has a distinct composition. The nitrogen in soy waste can enrich the biochar with valuable nutrients, making it a potent fertilizer (Sun et al., 2021). Furthermore, the high ash content of soybean waste can result in biochar with high alkalinity, which is advantageous for regulating the pH of the soil. Overall, the production of biochar from soybean waste is a valuable and affordable way to reduce agricultural waste and create a product with a wide range of potential uses.

7.6.4 Biodegradable Materials

Soybean, a widely cultivated crop, is grown worldwide and mainly consumed in East Asia in various forms, both as unfermented foods such as cooked soybeans, tofu, and tofu skin, and as fermented foods such as miso, soy sauce, and natto. Although soybeans contain about 35% protein and 20% fat by weight, they are mainly used to extract oil. After removing the oil, the remaining defatted soybean contains a considerable amount of protein. Although some defatted soybeans are used for human and animal consumption, a large proportion is disposed of as industrial waste worldwide. Previous studies have investigated the use of defatted soybeans in composites with cellulose, glycerol, polyacrylamide, and graphene. However, the processes for producing these composites with defatted soybeans are complex. Therefore, a simpler process for the production of bioplastics that is consistent with green chemistry principles is desirable.

In a study conducted by Yamada et al. (2020), on the preparation of bioplastic using soy protein, the soy protein (200 mg) was pressed into 13-mm diameter pellets using a hand press at a pressure of about 600 MPa for 15 min (Yamada et al., 2020). These pellets were then soaked for 24 hours at room temperature in HCHO solutions with concentrations ranging from 0% to 37% (Yamada et al., 2020). After the soaked pellets were removed from the HCHO solutions, they were washed with ultrapure water, air-dried at room temperature for 12 hours, and finally heated at 80°C for 2 h (Yamada et al., 2020). In this study, it was reported that the highest bending strength of the bioplastic achieved 35 MPa, which was the same as that of commercial polyethylene (Yamada et al., 2020). Hence, the soybean protein-based bioplastic shows promising results as a biodegradable material suitable for applications like agricultural materials, industrial components, and disposable products.

7.6.5 Biohydrogen

As global petroleum reserves decrease over the years, the demand for alternative energy sources has increased significantly. Biohydrogen is proposed as a clean and renewable energy source that can be produced by utilizing biomass, specifically from tempeh and tofu processing. Through a procedure known as fermentation, microorganisms convert the organic compounds in the waste into hydrogen gas in the biohydrogen production from soybean waste. The fermentation process can be fed with the waste products from the production of tofu and tempeh, which are primarily made of carbohydrates. The waste materials are put in a sealed container with a culture of microorganisms, typically a combination of bacteria and/or fungi, during fermentation. Anaerobic respiration, a process used by microorganisms to break down carbohydrates, results in the release of hydrogen gas as a by-product (Rengga et al., 2017). The hydrogen gas can then be gathered and cleaned up to be used as fuel.

The fact that soybean waste is a renewable resource with plenty of supply and easy accessibility is one benefit of using it to produce biohydrogen (Rengga et al., 2017). Additionally, the use of soybean waste in the production of biohydrogen can lessen waste and environmental pollution. The composition of the waste materials, the kind of microorganisms used, and the circumstances in which the fermentation process is carried out are just a few of the variables that can influence how effectively biohydrogen is produced from soybean waste. The process can be made more efficient overall and the yield of hydrogen gas can be increased by optimizing these factors.

7.6.6 Bioethanol

Soybean waste, such as okara, has the potential to be used as a feedstock for bioethanol production. Okara is a by-product of the manufacturing of soymilk and tofu and is a rich source of carbohydrates, including ethanol-producing fermentable sugars like glucose, galactose, and mannose (Choi et al., 2015). In addition, pectin, a complex polymer made of galacturonic acid bound by various monosaccharides like arabinose and rhamnose, is present in okara.

Okara's high moisture content and the presence of lignocellulosic materials that may obstruct the fermentation process present one of the difficulties in using okara for the production (Choi et al., 2015) of bioethanol. The conversion of okara into bioethanol has thus been improved through the investigation of various pre-treatment techniques. The lignocellulosic content of okara can be effectively reduced through autoclave pre-treatment, which increases the material's accessibility to the enzymes that convert the carbohydrates into fermentable sugars.

Enzymatic hydrolysis is used to transform the carbohydrates into fermentable sugars after pre-treatment. Cellulases, hemicelluloses, and pectinases are a few of the specific enzymes that are used in this process to convert complex carbohydrates into simple sugars. After being hydrolyzed, the sugars are then fermented by bacteria or yeast to create ethanol. In addition to offering an alternative energy source, the production of bioethanol from soybean waste aids in the utilization of agricultural waste streams that would otherwise be dumped in landfills, creating environmental issues. The competition between crops grown for food and those grown for fuel is lessened and the cost of raw materials for the production of bioethanol is reduced when soybean waste is used.

7.6.7 Biobutanol

Biobutanol is said to be the best alternative solution for biofuel, whereas its properties are almost the same as those of gasoline and slightly stronger than those of ethanol (Ezeji et al., 2007). The process includes fermentation of diluted sulfuric acid-treated soybean star hydrolysate, which would help the production of biobutanol (Qureshi & Blaschek, 2006). In the process, fermentation takes place where bacteria convert the sugar into solvents. The conversion process of sugars involves ABE fermentation into acetone, butanol, and ethanol with a ratio of 3:2:1 (Qureshi et al., 2018).

The cost of production of butanol can be saved alternatively with the use of a fermentation substrate (Formanek et al., 1997). Soybean straw, due to its lignocellulosic nature, presents a promising and abundant renewable resource. To utilize it for biofuel production, the straw is first ground and blended into fine particles using a mechanical grinder. The hydrolysate preparation is carried out at room temperature in a 150 mL screw-capped container. Specifically, 10 g of soybean straw is mixed thoroughly with 10 mL of sulfuric acid (H_2SO_4). After the reaction, the mixture is allowed to cool to room temperature, and the pH is adjusted to 6.8 using 2 M sodium hydroxide (NaOH). This pre-treatment facilitates the breakdown of complex carbohydrates. Analytical results indicate that sulfuric acid-pretreated soybean stalks yield a maximum sugar concentration of 39 g/L through hydrolysis. Subsequently, a substantial amount of biobutanol is generated through fermentation, confirming the effectiveness of the acid pre-treatment process.

7.6.8 Natural Acids

Nowadays, vegetable oils are generally used for deep-frying (Romano et al., 2021) due to the rich presence of polyunsaturated fatty acids, which promote deterioration during frying. When the oil is exposed to higher temperatures (Romano et al., 2021) in the frying process, various reactions lead to the production of free fatty acids, aldehydes, peroxides (Hou et al., 2021), and trans-fatty acids (Günal-Köroğlu et al., 2019). Furthermore, with a global consumption value of 59.48 million tons, soybean is the most popular vegetable oil used for deep-frying (Statista, 2022). Acid value (AV) in the raw soybean oil increased dramatically during the frying process, rising from 0.29 0.00 mg/g (day 0) to 4.52 0.04 mg/g (day 5). After two frying cycles, raw soybean oil had an acid value of 1.99 0.04 mg/g; however, from the third to the fifth frying cycle, the AV nearly doubled, causing the raw oil to significantly degrade (Li et al., 2021). The fatty acid of the soybean could also be determined through the process of frying, which inhibits the oxidation degree of the oil. Unsaturated fatty acids (USFAs) typically undergo oxidation processes (Li et al., 2021) that break or saturate their double bonds (Manzoor et al., 2022), increasing the content of saturated fatty acids (SFAs) at the expense of USFAs. Additionally, USFAs go through cis/trans isomerization events, which increase the trans-fatty acid content (Szabo et al., 2022).

These reactions resulted in an increase of 53% in SFAs when the soybean oil is used for frying, with a notable increase in oleic and linoleic acids (Romano et al., 2021). It was found that the composition of fatty acids inside a soybean contains palmitic acid (11.71%), oleic acid (23.62%), stearic acid (4.14%), linoleic acid (51.38%), and arachidic acid (0.49%). Thus, the percentage of SFAs found in soybean oil fatty acids was 16.34% and that of USFAs was around 83.66% (Abdo et al., 2023).

7.6.9 Bioelectricity

The electrical potentials of several soybean plant areas were measured in an electronically shielded chamber using a JSD-731 Electro-Physiological Instrument. A sample fixer identified the area of a plant that needed to be measured. The spacing between the stimulation and induction probes could be adjusted. Stainless steel or beryllium probes were utilized. The output resistance was set to 1.1 kOhm, and the voltage was set at 5.0 with a range of 2–10 V. The amplification factor was 100+3%, the whole period cycle was 0.05 s, and the amplifier's input resistance was 9.4 million omegas.

Soybean electrical resistance can also be identified using a 57-7 Meridian Instrument, which was developed by Beijing Aviation University, as researched by Hou et al. (1994). The equipment is able to identify the frequency and potential of the surfaces of the plant tissue through impulse scanning of direct current. It identified the values of the average current recorded using a micro-ammeter through a circuit connection and converted it to resistance for adjustments. In the experiments to identify the electrical resistance, its voltage was set to 80–120 V with a frequency of 500 Hz.

On the upper and bottom surfaces of both young and old leaves, the bioelectrical potentials of the major vein, minor vein, and mesophyll of soybean were measured. The major vein, the minor veins, and the mesophyll all had mean potentials of −1.28, −2.30, and −9.25 mV, respectively. The findings indicate that the main vein's potential was significantly higher than that of the tiny veins and mesophyll ($p = 0.0022$ and $p = 0.0064$, respectively). The main veins of a leaf seem to have the greatest potential of the three leaf components, followed by the minor veins and then the mesophyll.

7.7 POTENTIAL APPLICATIONS

Soybeans are currently the most important food crop in the world, especially in Europe, where they have a wide range of applications, especially in the form of cooking oil. Additionally, they are processed into soy protein, milk, tofu, and oil. The two most widely grown soybean kinds are yellow and black. In China and Japan, they are additionally known as edamame. In addition to being high in protein, fiber, and a variety of other vital vitamins and minerals, soybeans are low in fat and calories. Several medical diseases and symptoms can benefit from soybean consumption. Soybeans are eaten in a variety of ways, including roasted, salted, and boiled. Other than that, they are frequently substituted for meat. They are a great supplement source with various health advantages, especially for vegans who are unable to eat meat such as poultry due to health and religious purposes.

Soybean consumption can also reduce the signs of aging like wrinkles, dark spots, fine lines, and skin discoloration. Estrogen tends to decrease in females as they get older, which is a crucial hormone for preserving skin suppleness; it is medically believed that soybean consumption could provide the necessities, as researched by Linda Funk in one of her journals—"Research Studies Show Soy's Benefits for Skin Health," 2019. The ability to release phytoestrogen in soybeans encourages the body to manufacture more estrogen. Moreover, another advantage of soy waste is that most of it will end up in livestock feed, where the hay or straw of the residue from the soy would turn into food for animals to consume.

Other than that, soy grit waste is also advantageous as a by-product. To make soy grits, whole soybeans must be washed, toasted, and defatted. Their wild flavor is enhanced by toasting. Soy grits are a "healthier" alternative to white and yellow maize grits since they are high in protein and fiber (Tadimalla, 2023). Soy products can be used to add protein and texture to various dishes, especially starchy foods such as cereals and bread. Soy grits, in particular, are also valuable as filler ingredients and are widely used in fish and livestock feed formulations. Additionally, instead of being solely processed into meal, soybean waste can be utilized for oil extraction. Soybean oil, derived from the seeds, has diverse applications. In the culinary field, it is commonly used for cooking due to its versatility. In industrial applications, soybean oil serves as a base for products such as printing ink and oil-based paints, showcasing its multifunctional potential beyond food use.

Lisboa et al. (2020) studied cement composites incorporating alkali-treated soybean pods and eucalyptus wood particles. The results showed a 26% increase in modulus of rupture (MOR) and a 22% increase in modulus of elasticity (MOE) compared to control composites without the fillers. These improvements in mechanical performance highlight the effectiveness of soybean agro-waste as a reinforcing filler, supporting its use in value-added sustainable construction materials. Its utilization offers a cost-effective solution, as soybean waste is inexpensive and readily available, while also contributing to waste reduction and sustainability by diverting biomass from landfills. Moreover, the use of soybean as a filler may provide improved tensile strength, attributed to its favorable morphological characteristics. Table 7.5 highlights the diverse applications of soybean, ranging from human consumption to its industrial potential as a value-added material.

Another by-product that was generated from the processing of soil oil was soy hull. It is the fibrous outer layer of soybeans, which consists of 5% of the soybean weight. However, soy hull is a useful source of ruminant feed since it has a high concentration of cellulose (29–52%), hemicellulose (10–25%), structured protein (11%–15%), pectin (4%–8%), and lignin (1%–4%). Thus, soy hull is rarely used in human food (Liu & Li, 2017) However, a sizable fraction of soy hull is utilized to make dietary fiber, ethanol, microfibril, bio-oil, herbal medicines, and peptides. Soy hulls have also been found to be an effective source for wastewater treatment (Weidner & Grant, 1994).

Furthermore, soy oil processing leads to lecithin by-products. These substances are usually used widely in the food and pharmaceutical industry due to their emulsifying properties. Many food products, including salad dressing, sweets, margarine, chocolate, baking, and frying goods, benefited from lecithin's ability to stabilize fat and improve texture. Lecithin decreases blood cholesterol and low-density lipoprotein levels and improves cardiovascular health (Mourad et al., 2010).

Another by-product from the effects of soy oil processing is soymeal or soy grit and defatted soy flakes. Soy flakes and grit are rich sources of calcium, fiber, and protein (40–54%). These leftovers are employed in the production of plant-based protein, baked goods, bread, meat alternatives, and animal feed. Fully extracted soymeal has a high protein content and a low lipid content; moreover, ejected soymeal, on the contrary, has a high fat content (Banaszkiewicz, 2011). Thus, adding value to soymeal could contribute to the cheaper manufacture of plant-based proteins. Soymeal is mostly utilized as animal feed (98%) as soy protein concentrate, soy protein isolates, textured soy flour, metabolized energy concentrate, and crude protein concentrate. The remaining 2% is used for other purposes (Purdum, 2014). Table 7.6 provides details of the bioactive component of each 100 g defatted soymeal.

TABLE 7.5
Alternative Usage of Soybean and Soybean Waste

Materials	Usage	Human Consumption	References
Soymeal	Plant-based protein powder	Pharmaceuticals, biodiesel, bioplastics, soap, and cosmetics	Humphreys (2019)
Isoflavones, saponins, lecithin, phytic acids	Fermented soy products	Animal feed, plant protein concentrate	Ahmad et al. (2014)
Soy oil	Omega-3 fatty acids, polyunsaturated fatty acids	Pharmaceuticals (reduce risk of cardiovascular disease)	Phan et al. (2009)
Soybean	Swine feed, poultry feed, dairy cattle feed, pet feed, bee feed, calf milk replacers, beef cattle feed, fish food	Animal feed	Colletti et al. (2020)
	Phytosterols, emulsifying agents, protease inhibitors, antioxidants, sterols, protein supplements	Medicinal products	
	Ethanol, soap and detergent, cosmetics, stabilizing agents, Anti-corrosion agents, lubricants	Other industrial applications	

TABLE 7.6
Bioactive Component and Anti-Nutritional Profile of 100 g Defatted Soymeal

Bioactive and Anti-Nutritional Factors	Amount	Health Impact	References
Protease inhibitor	4–8 mg/g	It binds to the active site of the protease enzyme, which acts as a precursor for protein digestion.	Mesén-Porras et al. (2020)
Trypsin inhibitor	21.1–31.1 mg/g	It inhibits the activity of the digestive enzymes trypsin and chymotrypsin.	Aderibigbe et al. (2020)
Lectins	50–200 mg/g	They have the capability to agglomerate blood cells together and damage the intestinal wall, which influences the absorption of nutrients.	Fasina et al. (2003)
Saponins	0.6%	They inhibit cholesterol absorption. In addition, they show anti-inflammatory, anti-carcinogenic, cardioprotective, and antimicrobial properties.	Tarade et al. (2006)

7.8 CHALLENGES ASSOCIATED WITH PROCESSING AND COMMERCIALIZING

The main issue with commercializing agro-waste in its potential industrial application is the leading of environmental pollution. It comes mainly from the residual herbicides, pesticides, fertilizers, and packaging of what is left in production. Organic waste in urban settings such as products that were not sold can have negative effects on the environment, such as unpleasant odor, which would knowingly attract vermin to the area. Furthermore, the groundwater surrounding the neighborhood will have an impact on these irregularities by affecting their tap water for daily consumption. This will lead to diseases such as cholera if it is not controlled.

The main challenge in commercializing agro-waste for industrial applications lies in its potential to contribute to environmental pollution. This pollution primarily stems from residual herbicides, pesticides, fertilizers, and packaging materials left over from agricultural production. In urban areas, organic waste such as unsold perishable products can deteriorate and cause several negative environmental effects, including foul odors that attract pests and vermin. Additionally, the leaching of contaminants from unmanaged waste can affect nearby groundwater sources, posing risks to the safety of tap water used for daily consumption. If not properly managed, such contamination may lead to serious public health issues, including waterborne diseases such as cholera.

The soybean pressing cost is also one of the challenges that each company should be aware of. To commercialize soybeans better, plenty of steps need to be taken to manage the process properly, including the consumption of energy required. The amount of steam used during soybean crushing is a crucial key performance indicator for assessing an organization's management effectiveness and the level of technological and equipment advancement. Numerous new and expanded press oil facilities have been constructed as a result of the oil processing industry's rapid development. Different levels of technology and equipment exist in the enterprises, resulting in the variance in steam consumption per unit due to factors including capital, scale, and factory construction time. About 45–50% of the overall cost of crushing soybeans is attributable to the cost of the steam. High steam consumption will increase factory costs and unit production costs, decreasing the competitive edge of items on the market.

Even while the recent decline in the high rates of soil erosion associated with the production of soybeans is encouraging, the rate is still much greater than what is sustainable. Even if modern methods like conservation tillage have helped to prevent erosion, soybeans are still grown in "highly erodible" areas. In many large-scale soybean farms, soil compaction is an issue given how heavily mechanized soy farming is. Ecosystems that are vulnerable, including rainforests and savannahs, are put under stress as soybean farming spreads throughout South America and other regions. The fast spread of soybean farming into central Brazil's natural ecosystem worries the World Wildfire Fund, in particular. In some instances, the growth of soybean plantations has resulted in the displacement of smallholders who produce crops for subsistence. Several greenhouse gases produced by the soy sector contribute to global warming. Major sources of emissions for tropical countries like Brazil, Argentina, and Paraguay are due to deforestation and land conversion. According to the Brazilian government, carbon dioxide emissions from Cerrado conversion are equivalent to more than half of all emissions produced in the UK in 2009.

7.9 CONCLUSIONS

As a conclusion, soybean agro-waste management is an important aspect that should be looked into as the waste produced from the production of soybean foods can be utilized for other purposes as well. In a soybean manufacturing process, conducting market research for a better understanding of the market is needed as a business plan roadmap for a sustainable waste environment. Thus, with a proper medium of disposal or recycling soybean waste, one can generate income and conserve natural resources. It is important to promote the three Rs, reuse, reduce, and recycle, to ensure the sources being used to its good values without any wastage. This is to ensure it matches with the United Nations Sustainable Goals, affordable and clean energy, and responsible consumption and production for the benefit of future generations. It will take a concerted effort from many stakeholders to overcome these obstacles. Soybean waste producers should prioritize waste reduction and adopt sustainable practices such as composting, anaerobic digestion, or bioenergy production, at both industrial and individual levels. Additionally, educating customers about the harm that wasted soybeans cause to the environment and encouraging prudent consumption can help mitigate the effects. By considering soybean waste as a useful resource rather than garbage, one can accomplish the sustainability of natural resources without producing any undesired waste and instead contribute to the creation of greater benefits.

ACKNOWLEDGMENT

This chapter was written with Ebrahim Abdo Mohammed Gaber, Muhammad Syafiq Adly Mohd Rafiz, and Tang Zhi Ying.

REFERENCES

Abdo, E. M., Shaltout, O. E., & Mansour, H. M. (2023, January). Natural antioxidants from agro-wastes enhanced the oxidative stability of soybean oil during deep-frying. *LWT*, *173*, 114321. https://doi.org/10.1016/j.lwt.2022.114321

Abdullah, N., Sulaiman, F., Gerhauser, H., & Kaltschmitt, M. (2018). Techno-economic and environmental assessment of bioenergy production from soybean waste: A review. *Renewable and Sustainable Energy Reviews*, *81*(Part 1), 1252–1262. https://doi.org/10.1016/j.rser.2017.08.064

Abhilash, P. C., Pandey, V. C., & Srivastava, P. (2019). Biomass energy and environment: A sustainable development perspective. *Biomass Conversion and Biorefinery*, *9*(2), 259–267.

Aderibigbe, A., Cowieson, A. J., Sorbara, J. O., Pappenberger, G., & Adeola, O. (2020). Growth performance and amino acid digestibility responses of broiler chickens fed diets containing purified soybean trypsin inhibitor and supplemented with a monocomponent protease. *Poultry Science*, *99*(10), 5007–5017. https://doi.org/10.1016/j.psj.2020.06.051

Ahmad, A., Hayat, I., Arif, S., Masud, T., Khalid, N., & Ahmed, A. (2014). Mechanisms involved in the therapeutic effects of soybean (Glycine max). *International Journal of Food Properties*, *17*(6), 1332–1354. https://doi.org/10.1080/10942912.2012.714828

Alias, A. H., Norizan, M. N., Sabaruddin, F. A., Asyraf, M. R. M., Norrahim, M. N. F., Ilyas, A. R., Kuzmin, A. M., Rayung, M., Shazleen, S. S., Nazrin, A., Sherwani, S. F. K., Harussani, M. M., Atikah, M. S. N., Ishak, M. R., Sapuan, S. M., & Khalina, A. (2021). Hybridization of MMT/lignocellulosic fiber reinforced polymer nanocomposites for structural applications: A review. *Coatings*, *11*(11), 1355. https://doi.org/10.3390/coatings11111355

Asyraf, M. R. M., Syamsir, A., Zahari, N. M., Supian, A. B. M., Ishak, M. R., Sapuan, S. M., Sharma, S., Rashedi, A., Razman, M. R., Zakaria, S. Z. S., Ilyas, R. A., & Rashid, M. Z. A. (2022). Product development of natural fibre-composites for various applications: Design for sustainability. *Polymers*, *14*(5), 920. https://doi.org/10.3390/polym14050920

Azlin, M. N. M., Ilyas, R. A., Zuhri, M. Y. M., Sapuan, S. M., Harussani, M. M., Sharma, S., Nordin, A. H., Nurazzi, N. M., & Afiqah, A. N. (2022). 3D printing and shaping polymers, composites, and nanocomposites: A review. *Polymers*, *14*(1), 180. https://doi.org/10.3390/polym14010180

Banaszkiewicz, T. (2011). Nutritional value of soybean meal. In *Soybean and Nutrition*. IntechOpen. Retrieved November 8, 2021, https://www.intechopen.com/chapters/19972.

Behera, S. K., Sharma, R. K., & Mahapatra, D. M. (2019). Environmental impacts of biomass-based bioenergy production: A comprehensive review. *Renewable and Sustainable Energy Reviews*, *112*, 634–657.

Chen, L., Wang, W., & Wei, Q. (2019). Recent advances in biomass-derived materials for advanced applications. *Bioresources*, *14*(1), 1–3.

Cherubini, F., Peters, G. P., Berntsen, T., Strømman, A. H., & Hertwich, E. G. (2019). Assessing the environmental impact of biomass utilization: A review of life cycle assessments. *Renewable and Sustainable Energy Reviews*, *107*, 37–50.

Choi, I. S., Kim, Y. G., Jung, J. K., & Bae, H. J. (2015). Soybean waste (okara) as a valorization biomass for the bioethanol production. *Energy*, *93*, 1742–1747. https://doi.org/10.1016/j.energy.2015.09.093

Choy, S. Y., Wang, K., Qi, W., Wang, B., Chen, C. L., & Wang, J. Y. (2015). Co-composting of horticultural waste with fruit peels, food waste, and soybean residues. *Environmental Technology (United Kingdom)*, *36*(11), 1448–1456. https://doi.org/10.1080/09593330.2014.993728

Colletti, A., Attrovio, A., Boffa, L., Mantegna, S., & Cravotto, G. (2020). Valorisation of by-products from soybean (Glycine max (L.) merr.) processing. *Molecules*, *25*(9), 2129. https://doi.org/10.3390/molecules25092129

De Oliveira, L. S., da Silva, C. S., & Santos, V. L. (2018). Valorization of soybean oil deodorizer distillate for biodiesel production. *Renewable Energy*, *125*, 341–349. https://doi.org/10.1016/j.renene.2018.03.075

Ezeji, T. C., Qureshi, N., & Blaschek, H. P. (2007, June). Bioproduction of butanol from biomass: From genes to bioreactors. *Current Opinion in Biotechnology*, *18*(3), 220–227. https://doi.org/10.1016/j.copbio.2007.04.002

Fargione, J., Bassett, S., Boucher, T., Bridgham, S. D., Conant, R. T., Cook-Patton, S. C., Ellis, P. W., Falcucci, A., Fourqurean, J. W., Gopalakrishna, T., & Gu, H. (2018). Natural climate solutions for the United States. *Science Advances*, *4*(11), eaat1869.

Fasina, Y., Classen, H., Garlich, J., Swaisgood, H., & Clare, D. (2003). Investigating the possibility of monitoring lectin levels in commercial soybean meals intended for poultry feeding using steam-heated soybean meal as a model. *Poultry Science*, *82*(4), 648–656. https://doi.org/10.1093/ps/82.4.648

Finkenstadt, V. L., da Silva, E. M., & Visentainer, J. V. (2019). Soybean processing waste: Current and potential industrial applications. *Industrial Crops & Products*, *142*, 111838. https://doi.org/10.1016/j.indcrop.2019.111838

Formanek, J., Mackie, R., & Blaschek, H. P. (1997). Enhanced butanol production by Clostridium beijerinckii BA101 grown in semidefined P2 medium containing 6 percent maltodextrin or glucose. *Applied and Environmental Microbiology*, *63*(6), 2306–2310.

Funk, L. (2019, February 11). Research studies show soy's benefits for skin health. Retrieved 24 February 2024, from U.S. Soy website: https://ussoy.org/research-studies-show-soys-benefits-for-skin-health/

Garrido, T., Peñalba, M., De La Caba, K., & Guerrero, P. (2016). Injection-manufactured biocomposites from extruded soy protein with algae waste as a filler. *Composites Part B: Engineering*, *86*, 197–202. https://doi.org/10.1016/j.compositesb.2015.09.058

Günal-Köroğlu, D., Turan, S. Y., Kıralan, M., & Ramadan, M. F. (2019). Enhancement of sunflower oil stability during deep-frying using extracts from olive oil by-products and soy lecithin. *International Food Research Journal, 26*(4), 1269–1277.

Haberl, H., Marques, A., & Schmidt, J. (2018). Anthropogenic conversion of biomass: A natural and cultural history of biofuels. *Journal of Industrial Ecology, 22*(2), 229–241.

Halimatul, M. J., Sapuan, S. M., Jawaid, M., Ishak, M. R., & Ilyas, R. A. (2019). Water absorption and water solubility properties of sago starch biopolymer composite films filled with sugar palm particles. *Polimery, 64*(9), 27–35. https://doi.org/10.14314/polimery.2019.9.4

Hou, J., Zhou, X., Yu, T., Sop, R. T., Ma, J., Wang, M., Wu, Q., Zheng, X., & Jiang, Z. (2021). Ziziphi spinosae semen oil enhance the oxidative stability of soybean oil under frying conditions. *European Journal of Lipid Science and Technology, 123*(11), 2100060. https://doi.org/10.1002/ejlt.202100060

Hou, T. Z., Dawitof, M., Wang, J. Y., & Li, M. D. (1994). Experimental evidence of a plant meridian system: I. Bioelectricity and acupuncture effects on electrical resistance of the soybean (Glycine max). *The American Journal of Chinese Medicine, 22*(1), 1–10. https://doi.org/10.1142/S0192415X94000024

Humphreys, L. (2019). Factors poised to shape the future of the soybean meal market, February 24, 2019. *U.S. Soy*. https://ussoy.org/factors-poised-to-shape-the-future-of-the-soybean-meal-market/

Islam, M.S., Muhyidiyn, I., Islam, M.R., Hasan, M.K., Hafeez, A.G., Hosen, M.M., Saneoka, H., Ueda, A., Liu, L., Naz, M., & Barutçular, C., (2022). Soybean and sustainable agriculture for food security. In *Soybean-Recent Advances in Research and Applications*. IntechOpen. https://doi.org/10.5772/intechopen.104129

Jumaidin, R., Adam, N. W., Ilyas, R. A., Hussin, M. S. F., Taha, M. M., Mansor, M. R., Azlan, U. A. A., & Yob, M. S. (2019a). Water transport and physical properties of sugarcane bagasse fibre reinforced thermoplastic potato starch biocomposite. *Journal of Advanced Research in Fluid Mechanics and Thermal Sciences, 61*(2), 273–281.

Jumaidin, R., Diah, N. A., Ilyas, R. A., Alamjuri, R. H., & Yusof, F. A. M. (2021). Processing and characterisation of banana leaf fibre reinforced thermoplastic cassava starch composites. *Polymers, 13*(9), 1420. https://doi.org/10.3390/polym13091420

Jumaidin, R., Saidi, Z. A. S., Ilyas, R. A., Ahmad, M. N., Wahid, M. K., Yaakob, M. Y., Maidin, N. A., Rahman, M. H. A., & Osman, M. H. (2019b). Characteristics of cogon grass fibre reinforced thermoplastic cassava starch biocomposite: Water absorption and physical properties. *Journal of Advanced Research in Fluid Mechanics and Thermal Sciences, 62*(1), 43–52.

Karim, A., Osse, E. F., & Khalloufi, S. (2025). Innovative strategies for valorization of byproducts from soybean industry: A review on status, challenges, and sustainable approaches towards zero-waste processing systems. *Heliyon*, 11, e42118. https://doi.org/10.1016/j.heliyon.2025.e42118

Kibar, H., & Öztürk, T. (2008). Physical and mechanical properties of soybean. *International Agrophysics, 22*(3), 239–244.

Li, P., Yang, X., Lee, W. J., Huang, F., Wang, Y., & Li, Y. (2021). Comparison between synthetic and rosemary-based antioxidants for the deep frying of French fries in refined soybean oils evaluated by chemical and non-destructive rapid methods. *Food Chemistry, 335*, 127638. https://doi.org/10.1016/j.foodchem.2020.127638

Lisboa, F.J.N., Scatolino, M.V., de Paula Protásio, T., Júnior, J.B.G., Marconcini, J.M., & Mendes, L.M. (2020). Lignocellulosic materials for production of cement composites: Valorization of the alkali treated soybean pod and eucalyptus wood particles to obtain higher value-added products. *Waste and Biomass Valorization, 11*, pp.2235–2245. https://doi.org/10.1007/s12649-018-0488-2

Liu, H. M., & Li, H.-Y. (2017). Application and conversion of soybean hulls. In *Soybean: The Basis of Yield, Biomass and Productivity*. IntechOpen. Retrieved November 8, 2021, from

Makeri, M. U., Karim, R., Abdulkarim, M. S., Ghazali, H. M., Miskandar, M. S., & Muhammad, K. (2016). Comparative analysis of the physico-chemical, thermal, and oxidative properties of winged bean and soybean oils. *International Journal of Food Properties, 19*(12), 2769–2787.

Mangena, P. (2018). Water stress: Morphological and anatomical changes in soybean (Glycine max L.) plants. In *Plant, Abiotic Stress and Responses to Climate Change* (pp. 9–31). London, UK: IntechOpen.

Manzoor, S., Masoodi, F. A., Rashid, R., & Dar, M. M. (2022). Improving oxidative stability of soyabean oil by apple pomace extract during deep frying of French fries. *Food Bioscience, 49*, 101874. https://doi.org/10.1016/j.fbio.2022.101874

Mesén-Porras, E. A., Dahdouh-Cabia, S., Jimenez-Quiros, C., Mora-Castro, R., Rodríguez, C., & Pinto-Tomás, A. A. (2020). Soybean protease inhibitors increase Bacillus thuringiensis subs. israelensis toxicity against Hypothenemus hampei. *Agronomía Mesoamericana*, 31(2), 461–478. https://doi.org/10.15517/am.v31i2.36573

Mourad, A. M., de Carvalho Pincinato, E., Mazzola, P. G., Sabha, M., & Moriel, P. (2010). Influence of soy lecithin administration on hypercholesterolemia. *Cholesterol*, 824813. https://doi.org/10.1155/2010/824813.

Nanda, S., Ghosh, J., & Sanadi, A. R. (2017). Soybean residues: A sustainable biomass resource for the production of bioenergy and bioproducts. *Industrial Crops and Products*, *95*, 465–474. https://doi.org/10.1016/j.indcrop.2016.09.024

Nazrin, A., Sapuan, S. M., Zuhri, M. Y. M., Tawakkal, I. S. M. A., & Ilyas, R. A. (2021). Water barrier and mechanical properties of sugar palm crystalline nanocellulose reinforced thermoplastic sugar palm starch (TPS)/poly(lactic acid) (PLA) blend bionanocomposites. *Nanotechnology Reviews*, *10*(1), 431–442. https://doi.org/10.1515/ntrev-2021-0033

Ningsih, F., Zubaidah, S., & Kuswantoro, H. (2019, May). Diverse morphological characteristics of soybean (Glycine max L. Merill) pods and seeds germplasm. *IOP Conference Series: Earth and Environmental Science*, *276*(1), 012014.

Norfarhana, A. S., Ilyas, R. A., & Ngadi, N. (2022). A review of nanocellulose adsorptive membrane as multifunctional wastewater treatment. *Carbohydrate Polymers*, *291*, 119563. https://doi.org/10.1016/j.carbpol.2022.119563

Norfarhana, A. S., Ilyas, R. A., Ngadi, N., Othman, M. H. D., Misenan, M. S. M., & Norrrahim, M. N. F. (2024a). Revolutionizing lignocellulosic biomass: A review of harnessing the power of ionic liquids for sustainable utilization and extraction. *International Journal of Biological Macromolecules*, *256*, 128256. https://doi.org/10.1016/j.ijbiomac.2023.128256

Norfarhana, A. S., Khoo, P. S., Ilyas, R. A., Ab Hamid, N. H., Aisyah, H. A., Norrrahim, M. N. F., Knight, V. F., Rani, M. S. A., Septevani, A. A., Syafri, E., & Annamalai, P. K. (2024b). Exploring of cellulose nanocrystals from lignocellulosic sources as a powerful adsorbent for wastewater remediation. *Journal of Polymers and the Environment*. https://doi.org/10.1007/s10924-024-03227-3

Norrrahim, M. N. F., Ilyas, R. A., Nurazzi, N. M., Asmal Rani, M. S., Nur Atikah, M. S., & Shazleen, S. S. (2021). Chemical pretreatment of lignocellulosic biomass for the production of bioproducts: An overview. *Applied Science and Engineering Progress*, 1–18. https://doi.org/10.14416/j.asep.2021.07.004

Nurazzi, N. M., Harussani, M. M., Aisyah, H. A., Ilyas, R. A., Norrrahim, M. N. F., Khalina, A., & Abdullah, N. (2021). Treatments of natural fiber as reinforcement in polymer composites—A short review. *Functional Composites and Structures*, *3*(2), 024002. https://doi.org/10.1088/2631-6331/abff36

Paavola, J. (2019). A review of the environmental impacts of biomass energy production. *Renewable and Sustainable Energy Reviews*, *105*, 1–15.

Perkins, E. G. (1995). Composition of soybeans and soybean products. In D. R. Erickson (Ed.), *Practical Handbook of Soybean Processing and Utilization* (pp. 9–28). Champaign, Illinois: AOCS Press.

Phan, L., Brown, H., White, J., Hodgson, A., & Jessop, P. G. (2009). Soybean oil extraction and separation using switchable or expanded solvents. *Green Chemistry*, *11*(1), 53–59. https://doi.org/10.1039/B810423A

Purdum, S. E. (2014, April 1). Soybean meal research update. *Feed Strategy*. https://www. feedstrategy.com/animal-nutrition-formulation/soybean-meal-research-update/

Qureshi, N., & Blaschek, H. P. (2006). Butanol production from agricultural biomass. In Shetty, K., Paliyath, G., Pometto, A., Levin, R. E. (Eds.) *Food Biotechnology* (pp. 525–549). Boca Raton, FL: Taylor & Francis.

Qureshi, N., Harry-O'Kuru, R. E., Liu, S., & Saha, B. (2018). Yellow top (Physaria fendleri) presscake: A novel substrate for butanol production and reduction in environmental pollution. *Biotechnology Progress*, *35*(3), e2767. https://doi.org/10.1002/btpr.2767.

Ragauskas, A. J., Beckham, G. T., Biddy, M. J., Chandra, R., Chen, F., Davis, M. F., Davison, B. H., Dixon, R. A., Gilna, P., Keller, M., Langan, P., Naskar, A. K., Saddler, J. N., Tschaplinski, T. J., Tuskan, G. A., & Wyman, C. E. (2014). Lignin valorization: Improving lignin processing in the biorefinery. *Science*, *344*(6185), 1246843. https://doi.org/10.1126/science.1246843

Reinprecht, Y., Arif, M., Simon, L. C., & Pauls, K. P. (2015). Genome regions associated with functional performance of soybean stem fibers in polypropylene thermoplastic composites. *Plos One*, *10*(7), e0130371.

Rengga, W. D. P., Wati, D. S., Siregar, R. Y., Wulandari, A. R., Lestari, A. A., & Chafidz, A. (2017). Bio-hydrogen production from tempeh and tofu processing wastes via fermentation process using microbial consortium: A mini-review. *AIP Conference Proceedings*, *1818*. https://doi.org/10.1063/1.4976910

Romano, R., Filosa, G., Pizzolongo, F., Durazzo, A., Lucarini, M., Severino, P., Souta, E. B., & Santini, A. (2021). Oxidative stability of high oleic sunflower oil during deep-frying process of purple potato Purple Majesty. *Heliyon*, *7*(3), e06294.

Santos, E. M., Freitas, T. M. D., Mendes, T. D., Aguiar, V. M., & Freitas, L. A. P. (2019). Soybean straw as a potential feedstock for bioenergy production: A review. *Renewable and Sustainable Energy Reviews*, *110*, 357–372. https://doi.org/10.1016/j.rser.2019.04.006

Searchinger, T., Wirsenius, S., Beringer, T., Dumas, P., & Hillier, J. (2018). Assessing the efficiency of changes in land use for mitigating climate change. *Nature*, *564*(7735), 249–253.

Shilpashree, N., Devi, S. N., Manjunathagowda, D. C., Muddappa, A., Abdelmohsen, S. A. M., Tamam, N., Elansary, H. O., El-Abedin, T. K. Z., Abdelbacki, A. M. M., & Janhavi, V. (2021). Morphological characterization, variability and diversity among vegetable soybean (Glycine max L.) genotypes. *Plants*, *10*(4), 671.

Singh, P., & Krishnaswamy, K. (2022). Sustainable zero-waste processing system for soybeans and soy by-product valorization. *Trends in Food Science & Technology*, *128*, 331–344.

Statista. (2022). Consumption of vegetable oils worldwide from 2013/14 to 2020/2021, by oil type. *Statista*. Retrieved 12 February, 2022 from https://www.statista.com/statistics/263937/vegetable-oils-global-consumption/

Sun, L. & Yuan, Z. (2022). Seed morphology of soybean. https://doi.org/10.1016/bs.abr.2022.03.004.

Sun, Y., Li, C., Zhang, S., Li, Q., Gholizadeh, M., Wang, Y., Hu, S., Xiang, J., & Hu, X. (2021). Pyrolysis of soybean residue: Understanding characteristics of the products. *Renewable Energy*, *174*, 487–500. https://doi.org/10.1016/j.renene.2021.04.063

Suriani, M. J., Radzi, F. S. M., Ilyas, R. A., Petrů, M., Sapuan, S. M., & Ruzaidi, C. M. (2021). Flammability, tensile, and morphological properties of oil palm empty fruit bunches fiber/pet yarn-reinforced epoxy fire retardant hybrid polymer composites. *Polymers*, *13*(8), 1282. https://doi.org/10.3390/polym13081282

Syafri, E., Jamaluddin, Sari, N. H., Mahardika, M., Amanda, P., & Ilyas, R. A. (2022). Isolation and characterization of cellulose nanofibers from Agave gigantea by chemical-mechanical treatment. *International Journal of Biological Macromolecules*, *200*(December 2021), 25–33. https://doi.org/10.1016/j.ijbiomac.2021.12.111

Szabo, Z., Marosvölgyi, T., Szabo, E., Koczka, V., Verzar, Z., Figler, M., & Decsi, T. (2022). Effects of repeated heating on fatty acid composition of plant-based cooking oils. *Foods*, *11*(2). https://doi.org/10.3390/foods11020192

Tadimalla, R. T. (2023). Are grits healthy for you. *Stylecraze*. https://www.stylecraze.com/articles/health-benefits-of-grits/

Tarade, K. M., Singhal, R. S., Jayram, R. V., & Pandit, A. B. (2006). Kinetics of degradation of saponins in soybean flour (Glycine max.) during food processing. *Journal of Food Engineering*, *76*(3), 440–445. https://doi.org/10.1016/j.jfoodeng.2005.05.044

Vidal, B. C. (2018). *Biomass as a Sustainable Energy Source for the Future: Fundamentals, Methods and Current Issues*. Amsterdam, Netherlands: Elsevier.

Vitali, I. (2011). Soya and the Cerrado: Brazil's forgotten jewel. World Wildlife Foundation Report. https://assets.wwf.org.uk/downloads/soya_and_the_cerrado.pdf

Voora, V., Larrea, C., & Bermudez, S. (2020). Global market report: Soybeans. Winnipeg, Manitoba: International Institute for Sustainable Development. https://www.iisd.org/publications/report/global-market-report-soybeans

Weidner, S. J., & Grant, R. J. (1994). Soyhulls as a replacement for forage fiber in diets for lactating dairy cows. *Journal of Dairy Science*, *77*(2), 513–521. https://doi.org/10.3168/jds.S0022-0302(94)76979-4

Yamada, M., Morimitsu, S., Hosono, E., & Yamada, T. (2020). Preparation of bioplastic using soy protein. *International Journal of Biological Macromolecules*, *149*, 1077–1083. https://doi.org/10.1016/j.ijbiomac.2020.02.025

Zhang, Y., Wang, X., Wu, T., & Xu, X. (2015). Effects of drying methods on the energy content and combustion characteristics of soybean. *Energy*, *91*, 606–613. https://doi.org/10.1016/j.energy.2015.08.023

8 Coffee Waste Management

8.1 INTRODUCTION

Considering its potential environmental and economic effects, efficient agricultural waste management has been a growingly relevant problem in recent years. Agricultural waste improperly disposed of can lead to resource depletion, greenhouse gas (GHG) emissions, and soil and water contamination (Chávez et al., 2019). A significant source of agricultural waste is the processing phase of coffee production, which generates vast quantities of waste material during all stages of production, as shown in Figure 8.1. However, these waste materials have industrial applications that could assist with the creation of a sustainable coffee industry (McNutt and He, 2019). In this study, we will look into the possible industrial applications of coffee residue, concentrating on the various approaches to handling and utilizing these waste materials. By proposing workable and persistent solutions for managing coffee waste, this study could assist the coffee industry's transition to a circular economy.

With an estimated 2.25 billion cups of coffee daily, coffee is among the most popular drinks globally (Miskon et al., 2023). There are several phases involved in the production of coffee, including harvesting, processing, roasting, and packaging. During the processing phase, the pulp, parchment, and shell are removed from the coffee cherries to obtain the coffee beans. This procedure generates a substantial quantity of waste, posing a significant challenge for coffee producers around the globe. Several potential industrial applications exist for the organic matter, nutrients, and bioactive compounds in coffee processing by-products (Arya et al., 2022). However, misguided coffee residue disposal can lead to soil and water pollution, GHG emissions, and environmental deterioration (Sasikumar et al., 2020, Suganya and Senthil Kumar, 2018). Therefore, developing a circular economy in the coffee sector depends on finding sustainable ways to handle coffee waste.

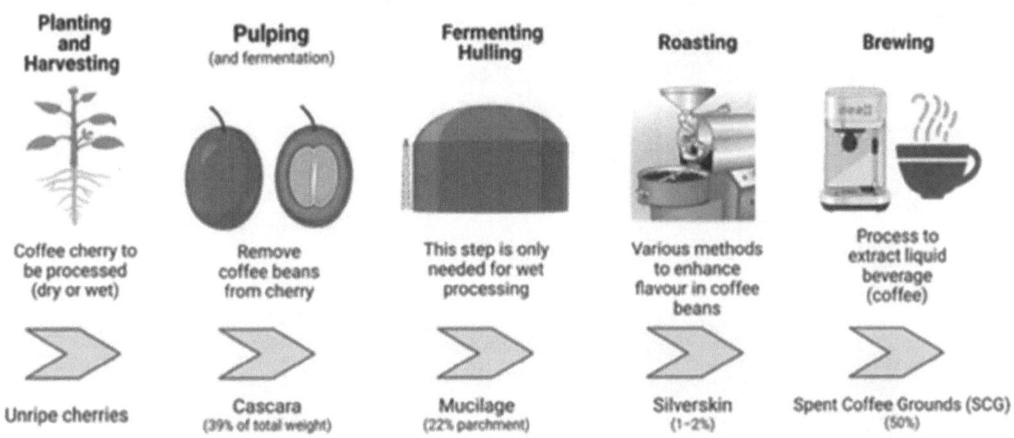

FIGURE 8.1 The different stages of coffee production with generated waste (Bevilacqua et al., 2023).

DOI: 10.1201/9781003472209-11

Composting, animal feed, biofuels, and the extraction of bioactive substances are just a few possible industrial uses for used coffee waste (Franca and Oliveira, 2022). Coffee waste can produce compost, a natural fertilizer that can enhance soil fertility and plant growth. Nitrogen, phosphate, and potassium are abundant in coffee waste and are vital nutrients for plant development. Compost made from coffee grounds strengthens soil, enhances crop yields, and lessens the demand for artificial fertilizers (Kasongo et al., 2011). Coffee waste may be used to create biofuels such as biodiesel and bioethanol. Due to its high oil content, coffee pulp is a promising feedstock for biodiesel production (da Rosa Pin et al., 2020). Biofuels may be produced from coffee waste, which can help reduce dependability on fossil fuels and contribute to developing a sustainable system.

During the processing phase, the coffee industry generates significant quantities of waste, including pulp, parchment, and residue, which poses a considerable challenge for coffee producers worldwide. GHG emissions, soil and water pollution, and improper coffee residue disposal can all occur. Additionally, coffee waste is a valuable resource with a variety of possible industrial uses, including decomposition, animal feed, biofuels, mushroom culture, and the extraction of medicinal compounds. However, improper management and utilization of coffee residue presents a problem for the coffee industry, resulting in resource loss and environmental degradation. For the coffee industry to establish a circular economy, it is essential to discover sustainable solutions for managing coffee residue.

This chapter review the current management practices of coffee waste and their environmental impact, to study the possible economic and environmental advantages associated with recycling coffee waste for commercial use, to identify the challenges and barriers to the adoption of sustainable coffee waste management practices, and lastly to develop recommendations for the effective management and utilization of coffee waste for sustainable coffee production.

8.2 AGRO-WASTE

8.2.1 Biomass from Natural Fibers

Natural fibers originated from plants, animals, or minerals and have been utilized for centuries in textiles, building materials, and paper production, among other applications (Dhir, 2022). Natural fibers

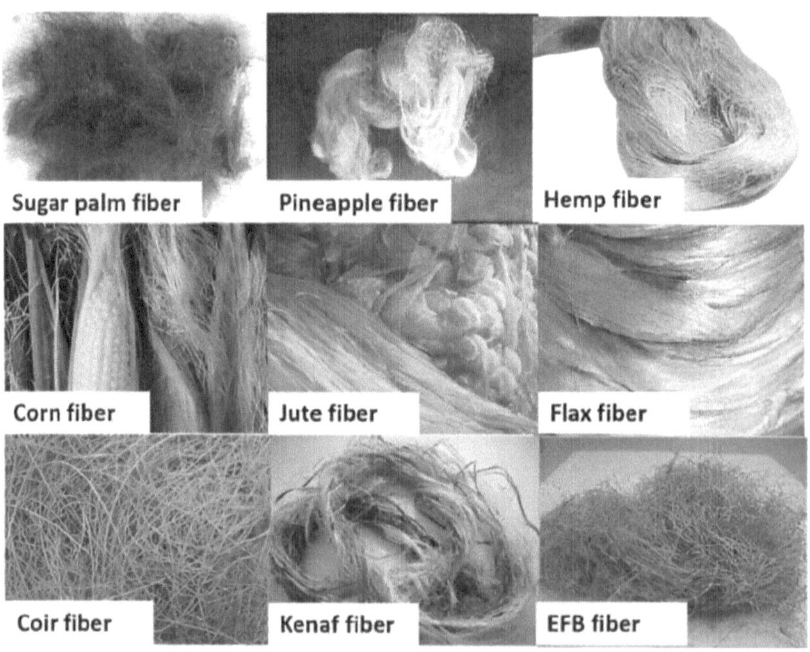

FIGURE 8.2 The different types of natural fibers (Nurazzi, 2021).

are again receiving attention as a potential source of biomass for several industrial applications due to the growing demand for environmentally friendly and sustainable products (Pegoretti et al., 2019).

Natural fibers are ubiquitous and renewable resources that can be derived from a variety of plant sources, such as jute, hemp, flax, cotton, and others, as shown in Figure 8.2. The presence of cellulose, hemicellulose, lignin, and other organic compounds in these fibers makes them an optimal source of biomass (Daria et al., 2020). The organic material created by living or recently extinct creatures is known as biomass, and it may be utilized to create biofuels, electricity, and other goods with added value. Natural fibers provide a number of benefits over fossil fuels like coal, oil, and natural gas as a source of biomass (Ross, 2019). Burning natural fibers produces far fewer GHG emissions and other impurities since they are renewable and biodegradable. Additionally, natural fibers may be grown sustainably and environmentally benign, assisting in the creation of a circular economy (Molino, 2019).

Potential industrial applications for natural fibers include energy production, biofuels, paper production, and construction materials. The direct combustion or gasification of natural fibers can produce energy. In direct combustion, natural fibers are consumed to produce heat that can be used to generate electricity or warm buildings. Gasification is the process of converting natural fibers into a gas that can be used to generate electricity or produce biofuels (Sikarwar et al., 2017, Osman, 2021). Natural fibers can be used as feedstock for biofuels, such as bioethanol and biodiesel production. Natural fibers' cellulose and hemicellulose can be converted to carbohydrates, which are then fermented into bioethanol. Oil extracted from natural fibers can be converted into biodiesel (Tran, 2019). Utilizing natural fibers, paper and other paper products are produced. High cellulose content makes natural fibers like jute and hemp suitable for paper production. Utilizing natural fibers in paper manufacturing reduces the demand for virgin wood pulp, conserving natural resources (Pegoretti et al., 2019).

Taking advantage of natural fibers as a biomass source presents several obstacles and opportunities. The availability and purity of natural fibers are the most significant obstacles. The production of natural fibers is affected by environmental factors such as climate, soil conditions, and parasites. In addition, the quality of natural fibers can vary considerably based on their origin, affecting their suitability for particular applications (Korol et al., 2020). Additionally, using natural fibers as a source of biomass necessitates substantial infrastructure and apparatus investments, which can impede adoption.

Regarding these challenges, there are a number of options for using natural fibers as a source of biomass. By minimizing waste and fostering resource efficiency, the growth of a sustainable biomass sector based on natural fibers might help create a circular economy. Furthermore, using natural fibers as a biomass resource can lessen dependency on fossil fuels and help mitigate climate change by reducing GHG emissions (Popp et al., 2014). Additionally, by opening up new markets for natural fibers and job possibilities, the utilization of natural fibers as a biomass source may assist farmers and rural communities economically.

8.2.2 Environmental Impact from Biomass

Biomass is a renewable energy source that has garnered popularity as a substitute for fossil fuels due to its ability to reduce glasshouse gas emissions and promote sustainable development (Kumar and Rathore, 2023). However, biomass usage can have environmental consequences that must be considered. The derived carriers of the resulting energy are shown in Figure 8.3. These carriers consume biomass and fix CO_2 through photosynthesis to transform solar energy into chemical energy.

Biomass production requires extensive land use, which can result in land use change and habitat loss. The transformation of natural ecosystems, such as forests and grasslands, into agricultural land for biomass production can reduce biodiversity and ecosystem services (Englund, 2019, Lawler, 2014). In addition, the intensification of agriculture for biomass production can contribute to the use of pesticides and fertilizers, which can negatively affect the condition of soil and water (McLennon et al., 2021). The loss of natural ecosystems as a result of a shift in land use for biomass production can have substantial effects on biodiversity. The loss of natural ecosystems can result in the extinction of numerous species and the disruption of ecological processes. In addition, intensification of

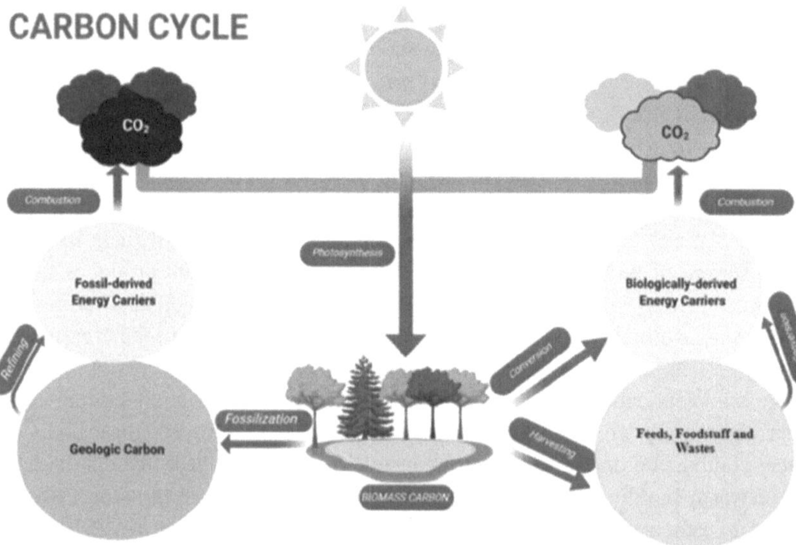

FIGURE 8.3 Carbon cycle showing the origin of energy sources in biomass (Ahmed, 2021).

agriculture for biomass production can result in the use of genetically modified crops and the loss of crop and livestock genetic diversity (Semenchuk, 2022).

Biomass production requires substantial water consumption, which can have an effect on water availability and quality. The intensification of cultivation for biomass production can increase irrigation, deplete groundwater reserves, and increase the risk of soil salinization (Yadav et al., 2022). In addition, using fertilizers and pesticides can lead to water contamination, which can harm aquatic environments and human health (Opoku-Kwanowaa et al., 2020). Air quality may be impacted by biomass production and use. When biomass is burned for energy, particulate matter, carbon and nitrogen monoxides, and oxygenated hydrocarbons can be released (Dhahak et al., 2019), leading to air pollution and respiratory illnesses (Bălă et al., 2021). Additionally, using fertilizers and pesticides can cause the release of GHGs such as nitrous oxide and methane when biomass is produced (Ashitha et al., 2021, Heimpel et al., 2013).

Several strategies can be employed to mitigate the potential environmental impacts of biomass production and utilization. One approach is utilizing biomass from sustainable sources such as agricultural residues, refuse timber, and energy products cultivated on degraded land. This strategy can reduce the impact on natural ecosystems and biodiversity while simultaneously promoting sustainable development. Another tactic is to deploy effective biomass conversion technologies, including gasification and pyrolysis, which can lower emissions of GHGs and other pollutants (Antar et al., 2021). Additionally, using biomass in combination with carbon capture and storage technology can produce negative emissions, aiding in mitigating climate change.

Significant environmental impacts can result from the production and supply of biomass, including land use change, biodiversity loss, water consumption, and air pollution. However, the potential environmental impacts can be mitigated through sustainable biomass production practices and efficient biomass conversion technologies. Additional research and investment are needed to create and promote sustainable biomass production and usage methods that help create a circular economy while minimizing the negative effects on natural ecosystems and fostering sustainable development.

8.2.3 Availability of Agro-Waste

Coffee is one of the most popular drinks consumed worldwide. Around 3.5 billion cups of coffee are consumed daily worldwide (Blinová et al., 2017). It is cultivated in more than 70 nations and

TABLE 8.1
Total Production of Coffee by All Exporting Countries in 2016 (International Coffee Organization, 2017)

Country	Weight (kg)
Arabica	95,204
Africa	16,353
Asia and Oceania	43,110
Mexico and Central America	17,740
South America	74,420

accounts for the annual production of more than 16 billion pounds of beans, according to Table 8.1. That is a lot of beans, and as they are only utilized once before being thrown away, a significant amount of waste is also created. Coffee was traditionally developed as a colonial cash crop. An estimated 11 million hectares of the world's farmland are dedicated to coffee cultivation (Blinová et al., 2017).

Given the amount of coffee consumed globally and the amount of residue created during the extraction of the beverage from coffee powder with hot water, it may be estimated that tons of waste coffee from cafeterias and domestic manufacturing are produced (Tokimoto et al., 2005). Although the precise values of the quantities eliminated are unknown, they can be calculated using industry-generated figures. This sector of the economy consumes about 50% of the world's coffee in the production of instant coffee, generating about 6 million tons of coffee waste annually (Tokimoto et al., 2005). Hence, the amount of coffee waste generated from cafeterias and domestic production must be large. Although some potential uses for these residues have been suggested, the majority are still unutilized and may be dumped into the environment, harming human and environmental health.

8.3 PROPERTIES OF AGRO-WASTE FIBERS

8.3.1 Physical Properties

The moisture content of spent coffee grounds (SCGs) is 6.0 g/100 g (Aguilar-Raymundo et al., 2019). The flavor of the coffee can range from mild and delicious to burned, depending on how it has been roasted. The majority of coffee will smell somewhat nutty and slightly caramelized. A stale cup of coffee will have a musty, cardboard-like aroma (Espresso & Coffee Guide, 2023).

According to a prior study, the density of biocomposite made from SCGs ranged from 0.605 to 0.789 g/cm^3. The porosity of biocomposite samples was found to be in the range of 12–14%. A thermogravimetric measurement revealed that the biocomposite from SCGs maintained its weight up to 250°C. After 250°C, the sample quickly decomposed. When the height was 11 mm, the SCGs took fewer than 6 days to dry in the open air (Ismail et al., 2023). More than 96% of SCGs were between 100 and 500 µm in particle size (Kang et al., 2017).

8.3.2 Chemical Properties

SCGs contain significant amounts of organic materials, including fatty acids, lignin, cellulose, hemicellulose, and other polysaccharides, which can be utilized to produce goods with added value (Campos-Vega et al., 2015). Table 8.2 provides the chemical composition of SCGs. The SCGs came

TABLE 8.2
Chemical Composition of SCGs (Ballesteros et al., 2014)

Chemical Components	Composition (g/100 g Dry Material)
Cellulose (glucose)	12.40±0.79
Hemicellulose	39.10±1.94
• Arabinose	• 3.60±0.52
• Mannose	• 19.07±0.85
• Galactose	• 16.43±1.66
• Xylose	• Not detected
Lignin	23.90±1.70
• Insoluble	• 17.59±1.56
• Soluble	• 6.31±0.37
Fat	2.29±0.30
Ashes	1.30±0.10
Protein	17.44±0.10
Nitrogen	2.79±0.10
Carbon/nitrogen (C/N ratio)	16.9±0.10
Total dietary fiber	60.46±2.10
• Insoluble	• 50.78±1.58
• Soluble	• 9.68±2.70

from NovaDelta Comércio e Industria de Cafés, S.A. (Campo Maior, Portugal) and were made from Arabica and Robusta coffee varietals and blends (Ballesteros et al., 2014).

8.3.3 Mechanical Properties

According to research, some mechanical qualities only slightly diminish when SCGs are added to epoxy resin, while other mechanical properties are markedly increased. The study also revealed that the epoxy resin composite made from SCGs had high mechanical characteristics and was fire resistant (Nguyen et al., 2021). Due to their high carbon content (>58%) (Pujol et al., 2013), SCGs have been mentioned as a viable biomaterial for flame retardant fillers (Neo, 2019). This carbon concentration may cause a dense char layer to form during combustion, protecting against fire. When coated with SCGs, epoxy resin can be utilized as a matrix for fiber-reinforced composites or as a coating that resists flames. Due to its high compressibility and microscopic layered structure, which provided a sizable specific surface area that facilitated the absorption of transmitted sound, SCG panels could be a useful addition to prefabricated noise barriers, according to some studies that focused on the acoustic properties of SCGs panels (Yun et al., 2020, Berardi and Iannace, 2015). Previous research showed that SCGs can be a starting point for creating biopolymers and composites. The tensile strength, Young's modulus, and elongation at break of polyhydroxybutyrate (PHB) films manufactured from SCGs were 16.0 MPa, 1.0 GPa, and 1.3%, respectively (Cruz et al., 2014). SCGs can be said to be the best material for composites (Espresso & Coffee Guide, 2023, Lee et al., 2023).

8.3.4 Morphological Properties

The scanning electron microscopy (SEM) images of Figures 8.4 and 8.5 show a morphological surface of thin sheets of material that looked like sawdust. The N_2 adsorption/desorption isotherms

FIGURE 8.4 Spent coffee grounds (SCGs) (Jagdale et al., 2019).

FIGURE 8.5 Micrographs by scanning electron microscopy (SEM) of SCGs. Magnification of (a) 200-fold and (b) 2,000-fold (Lachheb et al., 2019).

showed that the porosity of SCGs was relatively high. However, very little N_2 was absorbed by SCGs, indicating that their mesoporosity was underdeveloped (Lachheb et al., 2019). Since microporous materials exhibited N_2 adsorption/desorption isotherms with a propensity to create a plateau at low relative pressure (Mussatto et al., 2009), which was not confirmed in the current analysis for any of the cases, it was also confirmed that there were no micropores in the sample. BET surface area (SBET) studies showed that SCGs had reasonably high SBET (4.3 m^2/g), while the SBET was very low for the sample, and micropores were not discovered. These results led to similar conclusions. The total volume of pores was also very close to the residue (0.004 cm^3/g) (Lachheb et al., 2019). Previous research suggested adding SCGs to room-temperature cure thermosets to enhance their qualities because of the porous nature of the grounds. The thermosets improved with SCGS and had higher glass transition temperatures and rubbery plateau modulus (Mussatto et al., 2009). A prior study assessed the possible use of mixing SCGs with clay to create ceramic bricks. Ceramic samples were molded using clay pastes that contained 5%, 10%, 15%, and 20% SCGs, respectively. The outcomes demonstrated that adding SCGs could boost the perceived porosity and water absorption (Sena da Fonseca et al., 2014).

8.3.5 Thermal Properties

The thermal conductivity of the SCG biocomposite ranged from 0.038 to 0.054 W/m·K, demonstrating that it is perfectly suitable for thermal insulation (Ismail et al., 2023). SCGs were added to the building wall plaster to increase thermal conductivity (Lachheb et al., 2019). A previous study found that coffee husks were added to a polyethylene matrix to increase the material's modulus and thermal properties and improve building thermal insulation materials (Huang et al., 2018). They discovered that as SCG concentration increased, thermal conductivity significantly decreased. When 17% of SCGs was added to burned clay bricks, thermal conductivity was lowered by 50%, while adding 6% by weight of SCGs to plaster composites reduced it from 0.5 to 0.31 W/(mK) (Muñoz Velasco et al., 2016). In a prior study, clay bricks were mixed with waste materials, including SCGs and olive mill residue, and the outcomes appeared favorable. Thermal conductivity increased by 19%, and compressive strength was comparable to the value of the reference bricks (Eliche-Quesada et al., 2011).

8.3.6 Calorific Value

SCGs have a high calorific value (over 20 MJ/kg) and a high carbon content (above 50%) (Coletti, 2017). According to other studies, SCGs used as fuel have a lower heating value of roughly 18.8 MJ/kg (4,500 kcal/kg with a water content of 10%) (Kang et al., 2017). The material's characteristics are improved via torrefaction, which increases its calorific value to 32 MJ/kg (Jeníček et al., 2022). Torrefaction, a method of roasting coffee beans, was first developed in Portugal, Spain, and France. The extended shelf life and flavor retention that this method of coating the beans in sugar was thought to offer are benefits that all coffee connoisseurs may appreciate (MacDonnell, 2022). SCGs are still useful even after brewing. SCGs are still a possible source of energy because they comprise significant levels of sugars, lipids, antioxidants, and other valuable substances (McNutt and He, 2019).

8.4 AGRO-WASTE MANAGEMENT

8.4.1 Introduction to Agro-Waste Management

Given the amount of coffee consumed globally and the amount of residue created during the extraction of the beverage from coffee powder with hot water, it may be estimated that tons of waste coffee from cafeterias and domestic manufacturing are produced (Fernandes, 2017). All previous waste management plans ignored the coffee waste's valuable organic matter content and concentrated on its disposal or reuse (Banu, 2021). Currently, the methods used for effective coffee waste management are built around the idea of biorefineries, which support circular economies and sustainable development. The use and recycling of such wastes depend on the characteristics and makeup of each component (Mirón-Mérida et al., 2021). Applications for coffee waste include animal feeding, enzyme production, composting, the creation of biogas and biochars, and the production of bioethanol. The possible applications of coffee waste are discussed in the next section.

8.4.2 Compost

Coffee waste is rich in plant-derived nutrients and biodegradable organic components such as fatty acids, lignin, cellulose, hemicellulose, and other polysaccharides that can be used in composting along with other organic wastes (Pujol et al., 2013). The organic materials undergo spontaneous biological decomposition during composting, mostly in an aerobic atmosphere. According to Dadi et al. (2019), the two main coffee by-products, husk and pulp, can be composted successfully on their own or in combination with other degradable fractions of municipal solid waste in varying ratios to produce very mature compost that is stabilized and sanitized to ensure its maximum benefit for agriculture. The reason is that coffee waste compost contributes to iron and potassium, the macronutrients that help plants grow. Moreover, it is important to thoroughly understand the chemistry of the local soil before making and applying compost samples to the soil (Dadi et al., 2019).

For instance, if the soil's pH is acidic, the finished compost should be somewhat basic; otherwise, it could impact the nearby soil and waste time, effort, and resources. Additionally, the compost must be sufficiently rich in potassium if the soil lacks certain nutrients, like potassium. From the experiment, coffee waste was discovered to enhance lettuces' dry weight with a relatively higher degree of mineralization, which may have nutritional advantages (Cervera-Mata et al., 2018).

8.4.3 Pulp and Paper

Coffee wastes from farms, such as coffee husk and coffee pulp, end up in landfills (Rangarajan and Tharian, 2019). However, there are ways to utilize coffee waste, such as using it as a raw material for pulp and paper production (Blinová et al., 2017). The major coffee by-products are coffee husks, pulps, silverskin, and SCGs. Coffee husk, or the skin of the coffee bean, is frequently a leftover by-product abandoned by coffee planters after the pulping process in coffee production. When combined with cotton and linen rags, coffee husk creates stunning, distinctive textured paper (Haeng and Chul, 2019).

There are two methods to make coffee paper. The first step involves cleaning, cooking, and adding the coffee pulp to the cotton fibers. Figure 8.6 shows the notebook made of coffee waste. The paper becomes colored when the coffee husk combines with the white cotton (VIVATI Group, 2019). The pulp fragments can be grounds to your desired fineness. The coffee pulp is ingested into the sheets, turning them into a cohesive piece of paper. This lovely material is used to create bags, gift wraps, and packaging (Triantafillopoulos and Koukoulas, 2020). The second procedure involves cleaning and drying the dried husk (Yoon, 2016). The husk is then lowered into the vats and raised with cotton rags. The husk and several cotton colors are blended in this instance. Once the paper has dried, the husk lends a new texture. Some husk appears to fall out of the paper, which is a nice treat for anyone who loves sustainability (Kueh, 2021).

FIGURE 8.6 Coffee waste notebook (VIVATI Group, 2019).

Sustainability is considered when selecting paper. Even though trees have a very low cellulose content, they are used and cut down to make paper. Additionally, the pulp and paper produced using this process can be used to create goods that have the ability to eliminate odor and stop topical skin issues like decubitus ulcers, allergic responses, and eczema (Sun, 2014).

8.4.4 Biocomposite Materials

Coffee waste can be utilized as biocomposite materials due to its unique chemical and physical properties. Biocomposite materials are made up of two or more materials, one of which is derived from natural sources and can be used to create a wide range of products, including construction materials, furniture, and packaging (Mohanty et al., 2018). One of the main components of coffee waste is cellulose, a natural polymer that can be used to create biocomposites (Nguyen et al., 2021). Cellulose is extracted from the coffee waste and mixed with a binding agent, such as a natural resin or bio-based polymer (Leow et al., 2021). This mixture is then molded or shaped into the desired form using techniques such as compression molding or extrusion (Yiga et al., 2020). Another component of coffee waste is lignin, which is a natural polymer that can also be used as a binding agent. Lignin can be extracted from coffee waste and mixed with other natural fibers, such as hemp or flax, to create biocomposite materials (Yiga et al., 2020).

Thus, coffee waste can be used to create biodegradable packaging materials. For example, coffee waste is mixed with biodegradable polymers such as polylactic acid to create a strong and durable packaging material that can be easily composted after use (Nguyen and Nguyen, 2021). Coffee waste has great potential as a source of biocomposite materials due to its abundance, low cost, and unique properties (Nguyen and Nguyen, 2021).

Furthermore, coffee waste-based composite adsorbents have been created and tested for the removal of contaminants from aqueous solutions. The encapsulation of coffee wastes in polysaccharides like calcium alginate and chitosan has been studied in certain studies in this regard (Leow et al., 2021). For example, SCGs were encapsulated using calcium alginate to boost Ni, Cd, and Cu adsorption capability. In comparison to calcium alginate beads alone, the results demonstrated high adsorption capacities and quick adsorption rates (Cerino-Córdova et al., 2020).

8.4.5 Bio-Coal

Coffee waste can be converted into a useful and sustainable energy source in the form of bio-coal through several methods, such as hydrothermal carbonization (HTC), pyrolysis, and torrefaction (Kim and Oh, 2021). The HTC process includes heating coffee waste in the presence of water under high pressure to create a solid comparable to coal in terms of its energy content and combustion properties (Massaya et al., 2021). Afterward, this bio-coal can be utilized in a variety of businesses as a green fuel supply. One such technique is pyrolysis, which entails heating the coffee waste without oxygen to create a solid material with properties similar to coal in terms of energy content and burning. Pyrolysis can be carried out at various temperatures and residence durations to produce bio-coal with diverse qualities (Vuppaladadiyam, 2023). Torrefaction is a different technique, including heating the SCGs at a lower temperature than standard pyrolysis (Lee et al., 2023). The coffee waste is subjected to this procedure to eliminate moisture and volatile components, creating a solid with enhanced fuel characteristics (Zhang, 2018). By replacing traditional fossil fuels with bio-coal, it helps to lower GHG emissions and slow down climate change. Furthermore, bio-coal has a high energy density, which makes it a competitive substitute for conventional coal in industrial applications (Sundqvist et al., 2021).

8.4.6 Biochar

The solid waste product left over after biomasses are thermally cracked in an oxygen-free environment is known as biochar (Bartoli et al., 2020). Biochar is a carbonaceous source with a high added value that can be used in a variety of industries, including the creation of composite materials, catalysts, and composites. These applications for biochar may result in long-lasting and affordable solutions for a range of environmental and energy-related problems (Cha, 2016).

SCGs are favored for the adsorption of gaseous molecules due to the unique structural characteristics of their porous surface. In the current study by Jagdale et al. (2019), the possibility and affordability of SCGs as a source of carbon for biochar have been discussed. Biochar made from SCGs was studied as a potential sensing component for humidity sensors to replace the recent integration of carbon nanomaterials into electrodes, which will cause toxicity issues to the environment (Lee et al., 2021). From the result, the sensor responded when the relative humidity was above 20%, and both the response and recovery times were relatively quick (Jagdale et al., 2019). Coffee grounds biochar displays n-type semiconductor characteristics with an exponential trend as the humidity level rises (Lee et al., 2021). Thus, it can be an essential source for creating carbon compounds for sensor applications.

8.4.7 Bio-Bricks

Bricks are conventional building materials and currently meet significant challenges in industrial research, such as lowering production costs, reducing energy consumption, and enhancing the physical qualities and durability of the finished products (Coletti, 2017). Much research has been conducted to find novel materials created to solve environmental concerns and society's welfare in response to the growing demand for sustainable production (Yoo et al., 2019). The use of waste products from industry is one potential strategy for achieving these objectives and satisfying these various requirements. According to Eliche-Quesada et al. (2011), mixing 3 wt% of SCGs with raw clay has shown good mechanical, physical, and thermal properties as a result of the bio-bricks produced. Figure 8.7 shows the bio-bricks made from coffee waste (Moussa, 2022). More SCGs were able to exhibit lower densities and, consequently, higher porosity, demonstrating a higher insulating capacity and an appropriate mechanical resistance (Saberian et al., 2021).

FIGURE 8.7 Bio-bricks made from coffee waste (Moussa, 2022).

8.4.8 Biohydrogen

Coffee waste is rich in organic material, specifically cellulose and lignin, which microbes can break down to make hydrogen gas (Vítěz et al., 2016). Coffee waste must first undergo pre-treatment to culture the bacteria to decompose the complex compounds and make them more palatable to microorganisms for biohydrogen production (Elmously et al., 2019). Through a technique known as anaerobic fermentation (Dzulkarnain et al., 2022), coffee waste can be used as a feedstock for biohydrogen production. The process takes place in the absence of light and oxygen and involves the microbial breakdown of organic materials to create organic acids, hydrogen gas (H_2), and carbon dioxide (CO_2). Through the process, by-products such as organic acids, alcohols, and other gases are also created (Dzulkarnain et al., 2022). These by-products can be employed in various industrial processes, including synthesizing chemicals and biofuels.

Overall, utilizing coffee waste as a feedstock for biohydrogen production through anaerobic fermentation can provide an eco-friendly solution for managing coffee waste while producing a valuable source of renewable energy. However, there are some limitations to using coffee waste in biohydrogen production (Banu, 2021). Due to its high moisture content, coffee waste can be challenging to handle and store. Its moisture concentration can also dilute the organic matter, lowering the possibility of producing biohydrogen (Kamaraj et al., 2020). The content of coffee waste varies depending on factors such as coffee kind, roasting procedure, and disposal method. It may be challenging to optimize the biohydrogen generation process, resulting in low-yield production (Saratale et al., 2013). Figure 8.8 shows the process of hydrogen production from coffee waste. Hydrothermal pre-treatment and co-digestion are involved in hydrogen production (Montoya et al., 2020).

8.4.9 Biomethane

The waste from coffee production can be used as a feedstock to create biomethane, a renewable natural gas that can be used as a sustainable energy source. Anaerobic digestion is the process by which organic matter is broken down by microbes without the presence of oxygen (Luz et al., 2017). Coffee waste must first be gathered and transferred to an anaerobic digestion facility to be converted into biomethane. An anaerobic digester at the facility receives waste after it has been combined with water (Luz et al., 2017). The organic material in the waste stream is broken down by microorganisms inside the digester, creating biogas as a by-product. Methane and carbon dioxide make up the

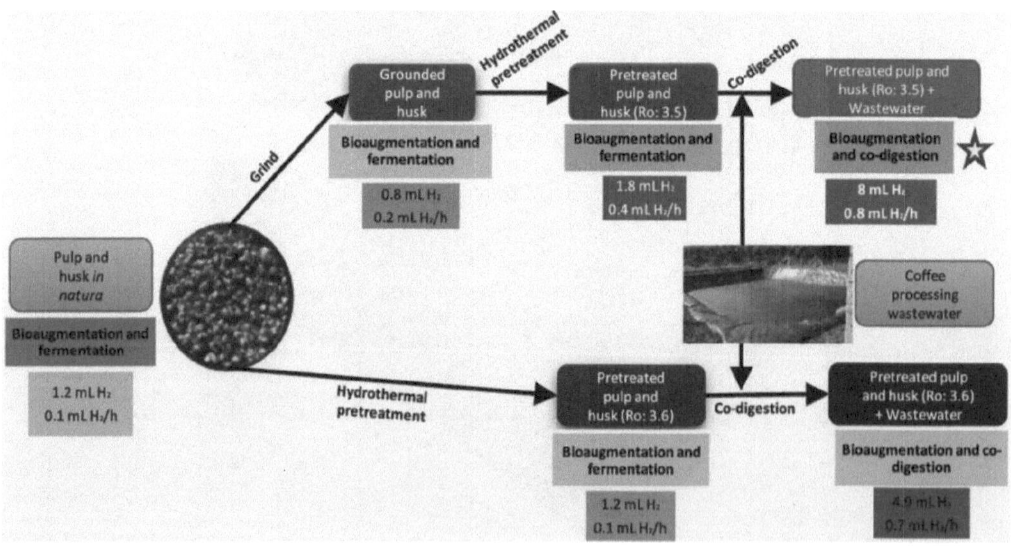

FIGURE 8.8 Hydrogen production from coffee waste (Montoya et al., 2020).

majority of the biogas made from coffee waste, with trace amounts of other gases (Marchetti and Vasmara, 2018). The carbon dioxide and other impurities are taken out of the biogas during purification to create biomethane, leaving a highly pure methane stream (Golmakani et al., 2022).

There are several methods to purify the biomethane gas from the biogas stream: pressure swing adsorption, membrane separation, chemical absorption, and cryogenic separation (Lee, 2019). Using a selectively permeable membrane, which lets methane pass through while trapping contaminants, membrane separation entails passing the biogas through the membrane (Luz et al., 2018). The cleaned methane is then collected after the pollutants have been released. The generated biomethane can either be pumped into the natural gas grid or used as a renewable transportation fuel, replacing fossil fuels and cutting GHG emissions (Browne, 2014).

8.4.10 Bioethanol

Bioethanol is a cutting-edge method that can lessen agricultural waste pollution while being environmentally benign. It burns more thoroughly than petrol because it contains a lot of oxygen. Based on the research done by Woldesenbet et al. (2016), Harsono et al. (2015), and Nguyen and Nguyen (2021), using the waste from coffee processing to make bioethanol, which offers an alternative energy source from waste biomass and addresses both the environmental waste disposal issue and the human health issue. It is a possible replacement feedstock for developing second-generation bioethanol made from agricultural waste (Nguyen and Nguyen, 2021). The processes of coffee waste to bioethanol production are pre-treatment, enzymatic hydrolysis, fermentation, color removal, and pervaporation. After being processed with ethanol at a high temperature (Gigliobianco, 2020), the coffee waste was hydrolyzed with enzymes to create sugars. In the experiment conducted by Woldesenbet et al. (2016), the sample was hydrolyzed by 0.4 M H_2SO_4 for 1 h at 100°C, resulting in the highest production of ethanol (78%), which was then fermented for 24 h at an initial pH of 4.5. Figure 8.9 shows some applications of coffee waste, which include biofuel, biogas, bio-oil, and bioethanol production (Battista et al., 2020).

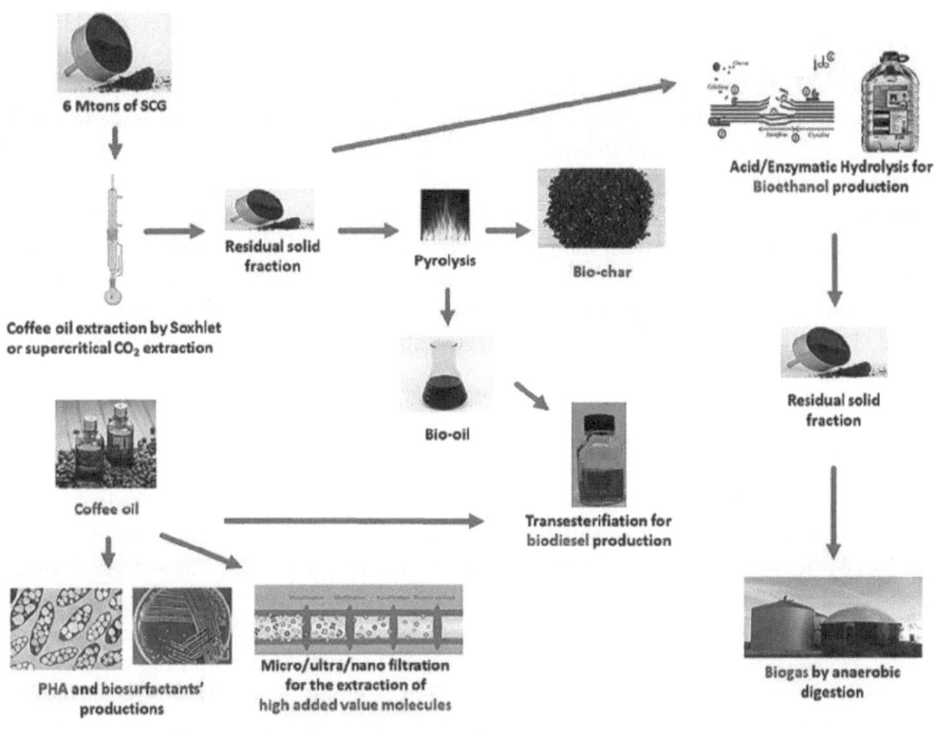

FIGURE 8.9 Bioethanol production from coffee waste (Battista et al., 2020).

8.4.11 Biobutanol

Because of its rich content of carbohydrates, such as cellulose and hemicellulose, which specific bacteria can ferment, coffee waste represents a valuable biomass source for producing biobutanol (Procentese et al., 2018, Hijosa-Valsero et al., 2018). The process of converting coffee waste into biobutanol encompasses several stages: pre-treatment, fermentation, separation, and purification.

The purpose of the pre-treatment is to convert the complex carbohydrates in the coffee waste into simpler sugars that can be fermented easily. The waste first undergoes pre-treatment via enzymatic hydrolysis and acid hydrolysis (Sarghini, 2021). In fermentation, microorganisms like *Clostridium acetobutylicum* or *Clostridium beijerinckii* are used to break down the pre-treated coffee waste. These bacteria use a mechanism known as acetone–butanol–ethanol fermentation to transform the carbohydrates into biobutanol (López-Linares et al., 2021). The biobutanol must be removed from the fermentation broth after fermentation. Different techniques, including distillation or solvent extraction, are capable of helping accomplish this. Purification is necessary to raise the quality of the biobutanol and remove any leftover contaminants. Adsorption or molecular sieves are two methods that can be used to achieve this (Ibrahim et al., 2022).

8.4.12 Natural Acids

Coffee waste can be used as an alternative source of natural acids, which have several manufacturing purposes and commercial uses. Quinic, citric, malic, and acetic acids are only a few of the organic acids found in coffee (Suárez-Quiroz et al., 2014). Coffee waste can be utilized to extract these organic acids, which can then be utilized in a variety of ways (Haile, 2014). To produce natural acids from coffee waste involves a drying process, an acid extraction method such as Soxhlet or solid–liquid extraction (Acevedo, 2013), followed by filtration to remove the impurities and purification using crystallization to obtain high purity of natural acids. The desired function approach was used to try to maximize extraction efficiency (Acevedo, 2013). The method used in this case also included a reliability assessment of the total desirability function to find a more reliable optimal zone, which included all feasible solutions adhering to the specified restrictions with a probability of less than 95% (Frackowiak and Béguin, 2001).

The natural acids extracted from coffee waste can be applied to food and beverage industries as a flavor enhancer, acidulant, and preservative (Nam and Xiang, 2019). Furthermore, due to its high nitrogen content and the organic acids it contains, coffee waste can be utilized as a natural fertilizer to increase soil acidity and nutrient availability (Nam and Xiang, 2019). This alternative solution helps to improve the agricultural management and pollution control system. For the cosmetics industry, coffee waste extracts can be used in cosmetics and personal care products due to their natural acid content, which can improve skin health and prevent aging. In addition, natural acids from coffee waste also contribute to the pharmaceutical industry due to their medicinal properties (Prihadi and Maimulyanti, 2021).

8.4.13 Bioelectricity

Coffee waste can be utilized as a source of bioelectricity, which involves the generation of electricity through the use of microorganisms that can produce electricity from organic matter (El Achkar et al., 2018). Coffee waste contains a high amount of organic matter, such as carbohydrates, proteins, and lipids, which microorganisms can break down to produce electrons that can be harnessed to generate electricity (McNutt and He, 2019). There are several methods for using coffee waste to generate bioelectricity. For instance, microbial fuel cells are technological innovations that use microbes to transform organic material into power (Garcia-Freites et al., 2020). By consuming the organic material in the coffee waste, the bacteria in microbial fuel cells produce electrons that an electrode may collect to provide power. Moreover, anaerobic digestion is one of the methods that

TABLE 8.3
Result of Columbian Coffee Case Study Gasification to Bioenergy Production (Garcia-Freites et al., 2020)

Types of Gas	Producer Gas Parameters		
	Simulation Data (mole %)	Experimental Data (mole %)	Difference
Hydrogen	22.3%	19.53%	2.77
Carbon monoxide	18.8%	16.32%	2.48
Carbon dioxide	13.8%	13.77%	0.03
Methane	1.2%	3.42%	2.22
Nitrogen	43.4%	46.69%	3.09
$LHV_{d.b}$	4.7 MJ/kg	4 MJ/kg	0.7
Gas yield	2.52 kg gas/kg biomass	2.84 kg gas/kg biomass	0.32

includes breaking down organic material without the presence of oxygen to produce biogas, which can be utilized to create energy. Anaerobic digestion of coffee waste can produce biogas, which can be utilized to create energy (Garcia-Freites et al., 2020).

In addition, coffee waste can be gasified to produce syngas through gasification, which can be used to generate electricity (Haochun et al., 2016). Coffee waste is a valuable solid fuel for thermochemical conversion, such as gasification, due to its high lignocellulosic content, low calorific value of 17.5–18 MJ/kg, and low ash content. Employing process modeling and the Colombian coffee industry as a case study, this study assesses the viability of using these residues in small-scale downdraft gasifiers connected to internal combustion engines for the generation of power and low-grade heat (Haochun et al., 2016). The producer gas's characteristics (5.6 MJ/Nm^3) and the gasifier's capabilities imply that this gas might be used to generate electricity. When the system's low-grade heat is recovered for use outside the system, such as in the drying process for coffee, a cogeneration system efficiency of 45.6% would be possible (Garcia-Freites et al., 2020). Table 8.3 shows the Columbian coffee case study results on bioenergy production. The results showed that the gas yield from coffee waste can be up to 2.5 kg per kg of biomass (Garcia-Freites et al., 2020).

8.5 POTENTIAL APPLICATIONS

Some potential applications of coffee husk in the future have not yet been discussed in the previous chapter. Coffee husks may be utilized for mushroom cultivation to promote the production of agriculture and improve the quality of the mushroom (Lindsay, 2020). The primary component of the substrate mixture for mushroom development can be coffee husk, which offers a sustainable and easily accessible medium for growing mushrooms. It is frequently mixed with other substances, such as straw, sawdust, or gypsum, to enhance the substrate's nutrient composition and physical attributes (Carrasco-Cabrera et al., 2019). The ability of coffee husk to retain moisture is advantageous for mushroom growing. For the growth of mushroom mycelium and subsequent fruiting, it may hold water while allowing for sufficient aeration, generating the ideal moisture environment.

Furthermore, the coffee husk is a source of nourishment for the mycelium of mushrooms because it includes lignin, sugars, and other organic materials (Adam, 2023). Figure 8.10 shows mushroom cultivation using coffee waste. The mushroom's vegetative component, the mycelium, colonizes and decomposes the coffee husk, transforming it into the fruiting body. After being sprayed with mushroom spawn, the mycelium of the selected mushroom species, the coffee husk substrate is placed

FIGURE 8.10 Mushroom cultivation by coffee waste (Elmously et al., 2019).

in a controlled environment with ideal temperature, humidity, and lighting conditions for fruiting (Jae Sul, 2017). For optimal fruiting, several mushroom species have particular needs. For the development of mushrooms, coffee husk offers a sustainable and affordable substrate solution. A by-product that would otherwise be thrown away can be supplied locally from coffee manufacturing facilities, decreasing waste. Additionally, the coffee husk's physical attributes, such as its nutrient content and moisture retention, are advantageous for mushroom culture (Li, 2014).

In addition, the coffee husk is a sustainable substitute for conventional building materials in the construction industry (Saberian et al., 2021). Coffee husk can be processed and mixed with a binding substance like resin or adhesive to make particleboards or fiberboards. To create panels, which can be used for a variety of purposes in construction, including wall partitions, furniture, and flooring, the coffee husk fibers or particles are compressed and glued together (La Scalia et al., 2021). It also can be used as insulation material (Lachheb et al., 2019). It features built-in thermal insulation, which helps to control temperature and save energy. Insulation made of coffee husks can be used on floors, roofs, and walls to increase energy efficiency and lessen heat transfer. Moreover, the coffee husk can be utilized in construction as a decorative component. It can be used to create panels or tiles, giving surfaces inside or outside a distinctive texture and appearance (Kua, 2017). Coffee husk-based decorative elements can be used to decorate facades, ceilings, or walls to make rooms look beautiful. Due to their excellent moisture resistance properties, they are suitable for use in spaces with greater moisture levels, like bathrooms or kitchens, because they can endure moisture and humidity when processed and treated properly (Moussa, 2022). Figure 8.11 shows the effect of adhesive content on the moisture content of coffee waste fiberboard. The result showed the higher the adhesive content, the higher the moisture content of the coffee waste (Rachtanapun et al., 2012).

Other than that, interest in finding a replacement for the term "waste" for used coffee is always rising. Given their toxicity to various plants, microbes, and aquatic species, the high concentration of polyphenols, caffeine, and tannins in this residue may cause soil and water pollution. However, because they can stop or delay the oxidation of the substrate, these same antioxidants, particularly caffeine and chlorogenic acid, find widespread use in the food, cosmetic, and pharmaceutical industries (Garcia and Kim, 2021).

The use of substances recovered from SCGs in active packaging has promise they allow the sensory and nutritional qualities of food to be preserved while prolonging the product's shelf life (Garcia and Kim, 2021). Antioxidants were extracted from SCGs using a high-pressure and temperature extraction technique. The obtained extract created zein-based films intended for active packaging applications. Solvent casting and electrospinning were used to develop SCG extract biopolymer films (Tsang et al., 2014). The morphology and antiradical potency of the released active ingredients were used to categorize the films. Therefore, coffee waste is a potential ingredient for biopackaging active ingredients (Tsang et al., 2014). Coffee waste food packaging is safe and provides a healthy guarantee to consumers due to its food-grade characteristics. The young generation,

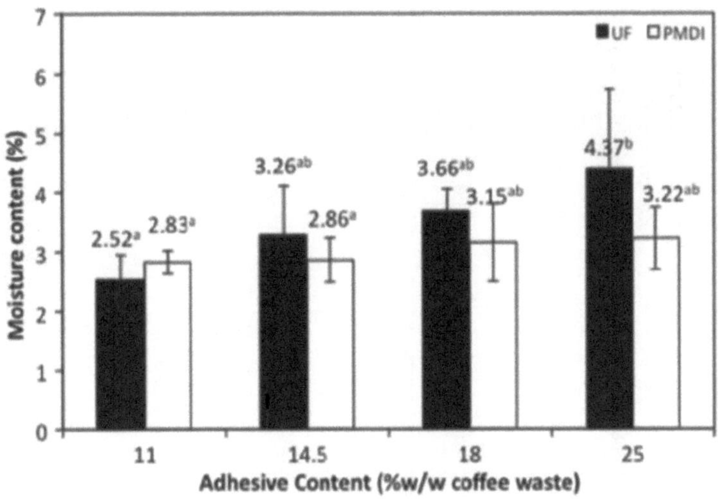

FIGURE 8.11 The effect of adhesive content on the moisture content of coffee waste fiberboard (Rachtanapun et al., 2012).

who care more about natural ingredients, has also become one of the potential customers of coffee waste-based cosmetic products (Lu, 2019).

Finding alternate uses for SCGs has been the focus of numerous recent studies. Some of these applications are listed in Table 8.4 and Figure 8.12 (de Otálora et al., 2020). Agriculture generates a lot of waste, so the first step in finding a replacement to ensure that it is not thrown away without fully utilizing its benefits is to concentrate on applying conversion to biofuels like bio-oil, biogas, biodiesel, bioethanol, fertilizers, and animal supplements. A recent study identified the potential use of SCGs at lower doses as a functional ingredient in the diet of sheep (de Otálora et al., 2020). As shown in Figure 8.12, there are numerous potential applications for SCGs that may need further research. In the chemical industry, SCGs can be used in cosmetics, surfactants, and even as foaming agents. In terms of material applications, flame retardant, thermal insulation, photothermal, and electromagnetic materials are some of the alternatives, as SCGs have been found to be excellent thermal insulators. Some other applications are in the food sector, in which SCGs can be used as food ingredients and food products and even used in food packaging as biopolymers and fibers.

8.6 CHALLENGES

When it comes to managing agricultural waste, utilizing coffee fiber for mushroom cultivation and construction purposes shows significant potential. Coffee fiber, a by-product of the coffee industry, offers a sustainable and readily available material for growing fungi and serves as an alternative to traditional building materials (Alemu et al., 2022). However, several hurdles need to be addressed for successful processing and commercialization. This section examines various challenges linked to coffee residue, including sourcing and consistency, pre-treatment and processing, standardization and quality assurance, technical knowledge and expertise, market acceptance and consumer awareness, and economic feasibility. Overcoming these challenges is crucial to unlocking the potential of coffee fiber for mushroom cultivation and construction uses. This section provides an overview of the obstacles related to the processing and commercialization of coffee residue for potential applications, as outlined in Table 8.5. By surmounting these barriers, the utilization of coffee fiber in industrial applications can contribute significantly to efficient agricultural waste management.

Employing coffee residue for mushroom cultivation and construction applications presents a number of challenges that must be overcome for commercialization. Obtaining a consistent and

TABLE 8.4
Recent Proposals for Alternative Uses of Spent Coffee Grounds (SCGS)

Starting Material	Technology	Product/Application	References
	Energy		
Wood sawdust and SCGS	Pallet pressing	Solid fuel pellets	Nosek et al. (2020)
SCF and larch sawdust or spruce shavings	High-pressure hydraulic briquetting press	Solid fuel briquettes	Brunerová et al. (2019)
SCGS	Pyrolysis	Biochar, carbon cloth electrode, electricity generation, and storage	Andrade et al. (2020)
SCGS and reduced graphene oxide	Vacuum-assisted impregnation	Composite phase change material and solar energy storage	Hu et al. (2021)
	Chemicals		
SCGS extract and *Xanthophyllomyces dendrorhous*	Drying, sterilization, and water extraction	Astaxanthin, cosmetics, supplements, and food	Hirono-Hara et al. (2022)
SCGS extract and *Millerozyma farinosa*	Drying, sterilization, and water extraction	Glutathione/medicine, food supplements, and cosmetics	Hirono-Hara et al. (2021)
SCGS	Defatting + drying + hot water extraction + incubation with ammonium sulfate + centrifugation + ultra-filtration + vacuum drying	Surfactants	Ishwarya and Nisha (2022)
	Adsorbents		
SCGS	Mixture with KOH and carbonization	Biochar/ammonia removal from water	Nguyen et al. (2021)
SCGS	Torrefaction	Biochar/removal of diesel mixed in water	Lee et al. (2021)
SCGS	No treatment	Biosorbent/heavy metal (Cd) removal from aqueous solution	Kim and Kim (2020)
	Materials		
SCGS and epoxy resin	Mixing and curing	Flame retardant in polymer	Vahabi et al. (2021)
SCGS and polylactide (PLA) powder	Decolorization, micro- and nano-processing, and extrusion	Polylactic acid composite for 3D-printing	Li et al. (2021)
	Food Products and Ingredients		
SCGS	Acid extraction followed by precipitation	Bioactive peptides with antihypertensive and antioxidant potentials	Ribeiro et al. (2021)
SCGS	Microwave-assisted extraction of antioxidants, fermentation (*Saccharomyces cerevisiae*) and distillation	Fermented and distilled alcoholic beverages	Machado et al. (2018)
SCGS	Hydrolysis followed by fermentation with wine yeasts	Alcoholic beverages	Liu et al. (2021)

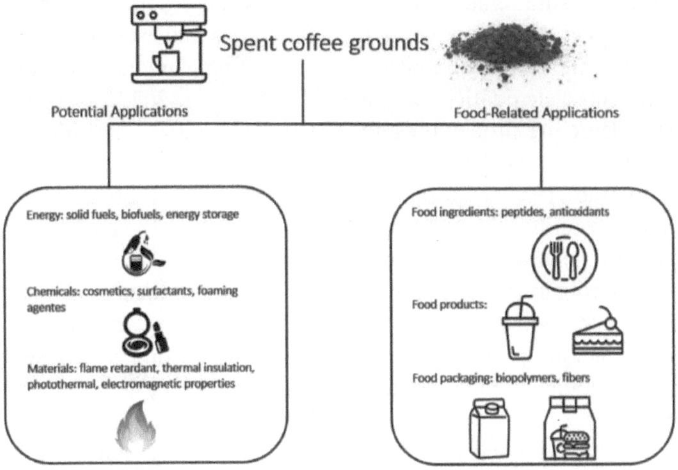

FIGURE 8.12 Summarized illustrative overview of potential uses for SCGs (Franca) (de Otálora et al., 2020).

TABLE 8.5
Summary of the Challenges Associated with Processing and Commercializing

Challenges	Description
Sourcing and consistency	Ensuring a reliable and consistent supply of coffee residue for large-scale production, coordinating with coffee manufacturing facilities
Pre-processing and treatment	Developing efficient and cost-effective techniques for pre-processing and treating coffee fiber to remove impurities and contaminants
Standardization and quality control	Establishing quality control measures and guidelines for coffee hull processing, handling, and storage to ensure a dependable and uniform substrate
Market acceptance and consumer awareness	Promoting market acceptance and creating consumer awareness about the benefits and applications of coffee residue for commercialization
Economic viability	Assessing the economic viability of coffee residue refining and commercialization by considering sourcing costs, treatment costs, market demand, and pricing

(Alemu et al., 2022, Addisie and Tebarek, 2022, Skorupa et al., 2022, Bager et al., 2022, Awasthi, 2020, Tan et al., 2019, Martinez-Saez, 2014, Gómez et al., 2003, Giovannucci and Koekoek, 2013, Mata et al., 2018, Atabani, 2019).

dependable supply of coffee residue can be a logistical challenge due to regional and volumetric variations in coffee production (Addisie and Tebarek, 2022). In addition, thorough pre-treatment and processing of coffee residue are required, including eliminating impurities and attaining a uniform particle size (Skorupa et al., 2022). It is essential to develop efficient and cost-effective pre-treatment and treatment techniques.

One of the most difficult aspects of refining and commercializing coffee residue is ensuring a reliable and consistent supply. The availability of coffee residue may vary by location and season, making it difficult to establish a reliable source for large-scale production. This challenge can be mitigated by coordinating with coffee manufacturing facilities and establishing long-term partnerships. Additionally, coffee hull quality can differ, necessitating cautious selection and quality control measures to ensure substrate quality consistency (Bager et al., 2022, Awasthi, 2020).

Coffee fiber must undergo pre-processing and treatment to remove impurities such as coffee residue, grime, and foreign substances. This procedure involves cleansing, dehydrating, and potentially sterilizing the shell to eradicate contaminants that could inhibit mushroom growth or diminish the quality of building materials (Tan et al., 2019). Developing efficient and cost-effective pre-processing techniques is essential for the widespread use of coffee shells.

Controlling the standardization and quality of coffee fiber products is essential for mushroom cultivation and construction applications. Establishing quality control measures and guidelines for coffee hull processing, handling, and storage is necessary to ensure a dependable and uniform substrate. This includes monitoring the substrate's moisture content, nutrient content, and overall performance (Martinez-Saez, 2014, Gómez et al., 2003). Implementing quality control protocols can assist in preserving product integrity and conforming to industry standards.

Promoting market acceptance and creating consumer awareness about the advantages and applications of coffee residue is crucial for its commercialization. The demand for coffee residue can be increased by educating potential users, such as mushroom producers, construction companies, architects, and consumers, about its sustainable aspects and benefits (Giovannucci and Koekoek, 2013). Marketing initiatives, product demonstrations, and collaborations with industry stakeholders can assist in establishing credibility and familiarity with coffee shell products.

The economic viability of coffee residue refining and commercialization are essential factors to consider. Costs associated with sourcing, pre-treatment, transportation, and treatment, as well as market demand and pricing, can influence the viability of large-scale utilization (Mata et al., 2018, Atabani, 2019). To establish a sustainable business model, it is essential to develop cost-effective processing methods, investigate potential revenue sources, and evaluate the overall economic viability.

8.7 CONCLUSION

SCGs present a huge challenge due to their environmental implications, which demand attention and practical solutions, considering the significant amount of daily waste. Coffee production produces a considerable amount of SCGs, one of the most extensively consumed beverages in the world, which frequently end up in landfills and cause resource and environmental waste. Additionally, the enormous volume of SCGs produced by households, coffee shops, and industrial-scale coffee production presents difficulties in managing, storing, and processing. However, recent studies have revealed that SCGs have potential applications in the food business as a sustainable replacement for catalysts, cosmetics, composites, building materials, and food additives. To lessen its negative effects on the environment and to promote sustainability, it is essential to research and put into practice solutions for the valorization of SCGs. However, to achieve quality and consistency, challenges must also be addressed. Challenges include the need for a regular and reliable supply of coffee residue, pre-treatment and a few other processes to produce particles of a uniform size, and technical expertise when using coffee. Market acceptance and consumer awareness, the economic viability of coffee residue refining, commercialization, and controlling the standardization and quality of coffee can be overcome by improving education and awareness to the public as the first step.

To address these challenges, concerted efforts from various stakeholders are essential. Coffee producers, at both the industrial and individual levels, should prioritize waste reduction and implement sustainable practices such as composting, anaerobic digestion, or bioenergy production. Ultimately, raising awareness among consumers about the environmental impact of coffee grounds waste and promoting responsible consumption practices can contribute to mitigating the issue. By viewing SCGs not as mere waste but as valuable resources, we can unlock their potential for creating sustainable solutions and reducing the overall environmental footprint of coffee production. Through collective action, we can transform the waste of SCGs into a valuable opportunity for environmental stewardship and the transition toward a more circular economy.

ACKNOWLEDGMENT

This chapter was written with Sim Jia Yi, Maged Abdullah Abdo Shehari, and Nadzrul Amin Bin Jamaluddin.

REFERENCES

F. Acevedo et al., "Spent coffee grounds as a renewable source of bioactive compounds," *Journal of Biobased Materials and Bioenergy*, vol. 7, no. 3, pp. 420–428, Jul. 2013, doi: 10.1166/jbmb.2013.1369.

S. Adam, "Growing Mushrooms in Coffee Grounds," *GroCycle*, 2023.

G. Addisie and L. Tebarek, "Upgrading opportunities and challenges for small coffee producers in Sidama region of Ethiopia," *International Journal of Rural Management*, Feb. 2022, doi: 10.1177/09730052221080884.

V. G. Aguilar-Raymundo, R. Sánchez-Páez, A. L. Gutiérrez-Salomón, and J. A. Barajas-Ramírez, "Spent coffee grounds cookies: Sensory and texture characteristics, proximate composition, antioxidant activity, and total phenolic content," *Journal of Food Processing and Preservation*, no. 43, no. 12, 2019, doi: 10.1111/jfpp.14223.

I. Ahmed et al., "Socio-economic and environmental impacts of biomass valorisation: A strategic drive for sustainable bioeconomy," *Sustainability*, vol. 13, no. 8, p. 4200, Apr. 2021, doi: 10.3390/SU13084200.

D. Alemu, M. Tafesse, and A. K. Mondal, "Mycelium-based composite: The future sustainable biomaterial," *International Journal of Biomaterials*, 2022, doi: 10.1155/2022/8401528.

T. S. Andrade, J. Vakros, D. Mantzavinos, and P. Lianos, "Biochar obtained by carbonization of spent coffee grounds and its application in the construction of an energy storage device," *Chemical Engineering Journal Advances*, vol. 4, p. 100061, 2020, doi: 10.1016/j.ceja.2020.100061.

M. Antar, D. Lyu, M. Nazari, A. Shah, X. Zhou, and D. L. Smith, "Biomass for a sustainable bioeconomy: An overview of world biomass production and utilization," *Renewable and Sustainable Energy Reviews*, 2021, doi: 10.1016/j.rser.2020.110691.

S. S. Arya, R. Venkatram, P. R. More, and P. Vijayan, "The wastes of coffee bean processing for utilization in food: A review," *Journal of Food Science and Technology*, vol. 59, no. 2, pp. 429–444, Feb. 01, 2022, doi: 10.1007/s13197-021-05032-5.

A. Ashitha, K. R. Rakhimol, and J. Mathew, "Fate of the conventional fertilizers in environment," In *Controlled Release Fertilizers for Sustainable Agriculture*, pp. 25–39, Jan. 2021, doi: 10.1016/B978-0-12-819555-0.00002-9.

A. E. Atabani et al., "Valorization of spent coffee grounds into biofuels and value-added products: Pathway towards integrated bio-refinery," *Fuel*, vol. 254, p. 115640, Oct. 2019, doi: 10.1016/J.FUEL.2019.115640.

M. K. Awasthi et al., "Refining biomass residues for sustainable energy and bio-products: An assessment of technology, its importance, and strategic applications in circular bio-economy," *Renewable and Sustainable Energy Reviews*, vol. 127, p. 109876, Jul. 2020, doi: 10.1016/J.RSER.2020.109876.

S. L. Bager, B. Düdder, F. Henglein, J. M. Hébert, and H. Wu, "Event-based supply chain network modeling: Blockchain for good coffee," *Frontiers in Blockchain*, vol. 5, p. 5, Jun. 2022, doi: 10.3389/FBLOC.2022.846783.

G. -P. Bălă, R.-M. Râjnoveanu, E. Tudorache, R. Motişan, and C. Oancea, "Air pollution exposure—The (in) visible risk factor for respiratory diseases," *Environmental Science and Pollution Research*, 2021, doi: 10.1007/s11356-021-13208-x.

F. Ballesteros, J. A. Teixeira, and S. I. Mussatto, "Chemical, functional, and structural properties of spent coffee grounds and coffee silverskin," *Food and Bioprocess Technology*, vol. 7, no. 12, pp. 3493–3503, 2014, doi: 10.1007/s11947-014-1349-z.

J. R. Banu et al., "A critical review on limitations and enhancement strategies associated with biohydrogen production," *International Journal of Hydrogen Energy*, vol. 46, no. 31, pp. 16565–16590, May 2021, doi: 10.1016/j.ijhydene.2021.01.075.

M. Bartoli, M. Giorcelli, P. Jagdale, M. Rovere, and A. Tagliaferro, "A review of non-soil biochar applications," *Materials*, vol. 13, no. 2, p. 261, Jan. 2020, doi: 10.3390/ma13020261.

F. Battista, E. M. Barampouti, S. Mai, P. Pavan, D. Malamis, K. Moustakas, and M. Loizidou, "Added-value molecules recovery and biofuels production from spent coffee grounds," *Renewable & Sustainable Energy Reviews*, 131, 110007, 2020, doi: 10.1016/j.rser.2020.110007.

U. Berardi and G. Iannace, "Acoustic characterization of natural fibers for sound absorption applications," *Building and Environment*, vol. 94, pp. 840–852, 2015, doi: 10.1016/j.buildenv.2015.05.029.

E. Bevilacqua, V. Cruzat, I. Singh, R. B. Rose'Meyer, S. K. Panchal, and L. Brown, "The potential of spent coffee grounds in functional food development," *Nutrients,* vol. 15, no. 4, p. 994, Feb. 2023, doi: 10.3390/NU15040994.

L. Blinová, M. Sirotiak, A. Bartošová, and M. Soldán, "Review: Utilization of waste from coffee production," *Research Papers Faculty of Materials Science and Technology Slovak University of Technology*, vol. 25, no. 40, pp. 91–101, Jun. 2017, doi: 10.1515/rput-2017-0011.

J. D. Browne, "Biomethane Production from Food Waste and Organic Residues," PhD Thesis, University College Cork, p. 209, 2014.

A. Brunerová, H. Roubík, M. Brožek, A. Haryanto, U. Hasanudin, D. A. Iryani, and D. Herák, "Valorization of bio-briquette fuel by using spent coffee grounds as an external additive," *Energies*, vol. 13, no. 1, p. 54, 2019, doi: 10.3390/en13010054.

R. Campos-Vega, G. Loarca-Piña, H. A. Vergara-Castañeda, and B. D. Oomah, "Spent coffee grounds: A review on current research and future prospects," *Trends in Food Science & Technology*, vol. 45, no. 1, pp. 24–36, 2015, doi: 10.1016/j.tifs.2015.04.012.

C. P. Carrasco-Cabrera, T. L. Bell, and M. A., Kertesz, "Caffeine metabolism during cultivation of oyster mushroom (Pleurotus ostreatus) with spent coffee grounds," *Applied Microbiology and Biotechnology*, vol. 103, no. 14, pp. 5831–5841, 2019, doi: 10.1007/S00253-019-09883-Z.

F. J. Cerino-Córdova, N. E. Dávila-Guzmán, A. M. García León, J. J. Salazar-Rabago, and E. Soto-Regalado, "Revalorization of coffee waste," In *Coffee: Production and Research*, 2020, IntechOpen, doi: 10.5772/intechopen.92303.

A. Cervera-Mata, S. Pastoriza, J. Á. Rufián-Henares, J. Párraga, J. M. Martín-García, and G. Delgado, "Impact of spent coffee grounds as organic amendment on soil fertility and lettuce growth in two Mediterranean agricultural soils," *Archives of Agronomy and Soil Science*, vol. 64, no. 6, pp. 790–804, May 2018, doi: 10.1080/03650340.2017.1387651.

J. S. Cha et al., "Production and utilization of biochar: A review," *Journal of Industrial and Engineering Chemistry*, vol. 40, pp. 1–15, Aug. 2016, doi: 10.1016/j.jiec.2016.06.002.

R. P. Chávez, E. C. C. Pizarro, and Y. L. Galiano, "Landfill leachate treatment using activated carbon obtained from coffee waste," *Engenharia Sanitaria e Ambiental*, vol. 24, no. 4, pp. 833–842, Jul. 2019, doi: 10.1590/s1413-41522019178655.

C. Coletti, "Bricktech: Assessment for the Use of Waste in the Brick Production. Petrophysical Characterization of New Mix Designs and Optimization of the Firing Conditions," 2017. https://hdl.handle.net/10481/46426

M. V. Cruz, A. Paiva, P. Lisboa, F. Freitas, V. D. Alves, P. Simões, S. Barreiros, and M. A. M. Reis, "Production of polyhydroxyalkanoates from spent coffee grounds oil obtained by supercritical fluid extraction technology," *Bioresource Technology*, vol. 157, pp. 360–363, 2014, doi: 10.1016/j.biortech.2014.02.013.

B. V. da Rosa Pin et al., "Energetic use of biogas from the anaerobic digestion of coffee wastewater in southern Minas Gerais, Brazil," *Renewable Energy*, vol. 146, pp. 2084–2094, Feb. 2020, doi: 10.1016/j.renene.2019.08.004.

D. Dadi, G. Daba, A. Beyene, P. Luis, and B. Van der Bruggen, "Composting and co-composting of coffee husk and pulp with source-separated municipal solid waste: A breakthrough in valorization of coffee waste," *International Journal of Recycling of Organic Waste in Agriculture*, vol. 8, no. 3, pp. 263–277, Sep. 2019, doi: 10.1007/s40093-019-0256-8.

M. Daria, L. Krzysztof, and M. Jakub, "Characteristics of biodegradable textiles used in environmental engineering: A comprehensive review," *Journal of Cleaner Production*, vol. 268, Sep. 2020, doi: 10.1016/J.JCLEPRO.2020.122129.

X. D. de Otálora, R. Ruiz, I. Goiri, J. Rey, R. Atxaerandio, D. San Martin, M. Orive, B. Iñarra, J. Żufia, J. Urkiza, and A. García-Rodríguez, "Valorisation of spent coffee grounds as functional feed ingredient improves productive performance of Latxa dairy ewes. *Animal Feed Science and Technology*, vol. 264, p. 114461, 2020, doi: 10.1016/j.anifeedsci.2020.114461.

A. Dhahak, R. Bounaceur, C. Le Dreff-Lorimier, G. Schmidt, G. Trouve, and F. Battin-Leclerc, "Development of a detailed kinetic model for the combustion of biomass," *Fuel*, vol. 242, pp. 756–774, Apr. 2019, doi: 10.1016/J.FUEL.2019.01.093.

Y. J. Dhir, "Natural fibers: The sustainable alternatives for textile and non-textile applications," *Natural Fiber,* Aug. 2022, doi: 10.5772/INTECHOPEN.106393.

E. L. N. Dzulkarnain, J. O. Audu, W. R. Z. Wan Dagang, and M. F. Abdul-Wahab, "Microbiomes of biohydrogen production from dark fermentation of industrial wastes: Current trends, advanced tools and future outlook," *Bioresour Bioprocess*, vol. 9, no. 1, p. 16, Dec. 2022, doi: 10.1186/s40643-022-00504-8.

J. H. El Achkar, A. Baydoun, D. Salameh, N. Louka, Z. Hobaika, and R. G., Maroun, "Can Coffee Grounds Be Considered as a Potential for Green Energy Production," 2018, doi: 10.1109/REDEC.2018.8598105.

D. Eliche-Quesada, C. Martínez-García, M. L. Martínez-Cartas, M. T. Cotes-Palomino, L. Pérez-Villarejo, N. Cruz-Pérez, and F. A. Corpas-Iglesias, "The use of different forms of waste in the manufacture of ceramic bricks," *Applied Clay Science*, vol. 52, no. 3, pp. 270–276, 2011, doi: 10.1016/j.clay.2011.03.003.

D. Eliche-Quesada, L. Pérez-Villarejo, F. J. Iglesias-Godino, C. Martínez-García, and F. A. Corpas-Iglesias, "Incorporation of coffee grounds into clay brick production," *Advances in Applied Ceramics*, vol. 110, no. 4, pp. 225–232, May 2011, doi: 10.1179/1743676111Y.0000000006.

M. Elmously, N. Jäger, A. Apfelbacher, R. Daschner, and A. Hornung, "Thermo-catalytic reforming of spent coffee grounds," *Bioresources and Bioprocessing*, vol. 6, no. 1, pp. 1–12, 2019, doi: 10.1186/S40643-019-0281-5.

O. Englund et al., "Beneficial land use change: Strategic expansion of new biomass plantations can reduce environmental impacts from EU agriculture," *Global Environmental Change*, 2019, doi: 10.1016/j.gloenvcha.2019.101990.

A. S. Fernandes et al., "Impacts of discarded coffee waste on human and environmental health," *Ecotoxicology and Environmental Safety*, vol. 141, pp. 30–36, Jul. 2017, doi: 10.1016/j.ecoenv.2017.03.011.

E. Frackowiak and F. Béguin, "Carbon materials for the electrochemical storage of energy in capacitors," *Carbon* N Y, vol. 39, no. 6, pp. 937–950, May 2001, doi: 10.1016/S0008-6223(00)00183-4.

A. S. Franca and L. S. Oliveira, "Potential uses of spent coffee grounds in the food industry," *Foods*, vol. 11, no. 14, Jul. 01, 2022, doi: 10.3390/foods11142064.

C. V. Garcia and Y. T. Kim, "Spent coffee grounds and coffee silverskin as potential materials for packaging: A review," *Journal of Polymers and the Environment*, vol. 29, no. 8, pp. 2372–2384, 2021, doi: 10.1007/S10924-021-02067-9.

S. Garcia-Freites, A. Welfle, A. Lea-Langton, P. Gilbert, and P. Thornley, "The potential of coffee stems gasification to provide bioenergy for coffee farms: A case study in the Colombian coffee sector," *Biomass Convers Biorefin*, vol. 10, no. 4, pp. 1137–1152, Dec. 2020, doi: 10.1007/s13399-019-00480-8.

M. R. Gigliobianco et al., "Optimization of the extraction from spent coffee grounds using the desirability approach," *Antioxidants*, vol. 9, no. 5, p. 370, Apr. 2020, doi: 10.3390/antiox9050370.

D. Giovannucci and F. J. Koekoek, "The state of sustainable coffee: A study of twelve major markets," In *The Top 50 Sustainability Books*, pp. 52–55, Apr. 2013.

A. Golmakani, S. A. Nabavi, B. Wadi, and V. Manovic, "Advances, challenges, and perspectives of biogas cleaning, upgrading, and utilization," *Fuel*, vol. 317, p. 123085, Jun. 2022, doi: 10.1016/j.fuel.2021.123085.

M. Gómez, F. Ronda, C. A. Blanco, P. A. Caballero, and A. Apesteguía, "Effect of dietary fiber on dough rheology and bread quality," *European Food Research and Technology*, vol. 216, no. 1, pp. 51–56, 2003, doi: 10.1007/S00217-002-0632-9.

C. J. Haeng and H. Y. Chul, "Eco-Friendly Coating Composition for Food Packing Paper," 2019.

M. Haile, "Integrated volarization of spent coffee grounds to biofuels," *Biofuel Research Journal*, vol. 1, no. 2, pp. 65–69, 2014, https://www.biofueljournal.com/article_5548_c803c72f8473e7d18d9013726ea9a611.pdf

Z. Haochun, J. Liang, M. Zhuang, S. Lei, M. Ningning, L. Yongqiang, and Z. Zhiyu, "Method for Extracting Biodiesel Based on Waste Coffee Grounds," 2016.

S. S. Harsono, Salahuddin, M. Fauzi, G. S. Purwono, D. Soemarno, and Kissinger, "Second generation bioethanol from Arabica coffee waste processing at smallholder plantation in Ijen plateau region of East Java," *Procedia Chemistry*, vol. 14, pp. 408–413, 2015, doi: 10.1016/j.proche.2015.03.055.

G. E. Heimpel, Y. Yang, J. D. Hill, and D. W. Ragsdale, "Environmental consequences of invasive species: Greenhouse gas emissions of insecticide use and the role of biological control in reducing emissions," *PLoS One*, vol. 8, no. 8, p. 72293, 2013, doi: 10.1371/journal.pone.0072293.

M. Hijosa-Valsero, J. Garita-Cambronero, A. I. Paniagua-García, and R. Díez-Antolínez, "Biobutanol production from coffee silverskin," *Microbial Cell Factories*, vol. 17, no. 1, p. 154, Dec. 2018, doi: 10.1186/s12934-018-1002-z.

Y. Hirono-Hara, A. Kato, Y. Mizutani, H. Kikukawa, and K. Y. Hara, "Effect of spent coffee grounds extract on astaxanthin production by Xanthophyllomyces dendrorhous," *Bioresource Technology Reports*, vol. 17, p. 100953, 2022, doi: 10.1016/j.biteb.2022.100953.

Y. Hirono-Hara, Y. Mizutani, K. Murofushi, K. Iwahara, S. Sakuragawa, H. Kikukawa, and K. Y. Hara, "Glutathione fermentation by Millerozyma farinosa using spent coffee grounds extract and seawater," *Bioresource Technology Reports*, vol. 15, p. 100777, 2021, doi: 10.1016/j.biteb.2021.100777.

X. Hu, H. Huang, Y. Hu, X. Lu, and Y. Qin, "Novel bio-based composite phase change materials with reduced graphene oxide-functionalized spent coffee grounds for efficient solar-to-thermal energy storage," *Solar Energy Materials and Solar Cells*, vol. 219, p. 110790, 2021, doi: 10.1016/j.solmat.2020.110790.

L. Huang, B. Mu, X. Yi, S. Li, and Q. Wang, "Sustainable use of coffee husks for reinforcing polyethylene composites," *Journal of Polymers and the Environment*, vol. 26, no. 1, pp. 48–58, 2018, doi: 10.1007/s10924-016-0917-x.

M. F. Ibrahim, N. N. Talib, N. H. Alias, I. N. A. Bakar, S. Abd Aziz, and P. L. Yee, "Utilization of agricultural biomass for bio-butanol production," pp. 235–248, 2022, doi: 10.1016/B978-0-12-824388-6.00019-1.

S. P. Ishwarya and P. Nisha, "Insights into the composition, structure-function relationship, and molecular organization of surfactants from spent coffee grounds," *Food Hydrocolloids*, vol. 124, p. 107204, 2022, doi: 10.1016/j.foodhyd.2021.107204.

I. Ismail, Q. Aini, C. R. Maulida, M. Mursal, Z. Jalil, and S. H. S. M. Fadzullah, "Thermal properties of spent coffee grounds biocomposite using epoxy resin matrix," *AIP Conference Proceedings*, vol. 2613, 020009, 2023, doi: 10.1063/5.0119827.

O. Jae Sul, "Mushroom Cultivation Temperature and Humidity Automatic Control System Using Coffee Grounds," 2017.

P. Jagdale, D. Ziegler, M. Rovere, J. Tulliani, and A. Tagliaferro, "Waste coffee grounds biochar: A material for humidity sensors," *Sensors*, vol. 19, no. 4, p. 801, Feb. 2019, doi: 10.3390/s19040801.

L. Jeníček, B. Tunklová, J. Malaťák, M. Neškudla, and J. Velebil, "Use of spent coffee grounds as an alternative fuel and possible soil amendment," *Materials*, vol. 15, no. 19, p. 6722, 2022, doi: 10.3390/ma15196722.

M. Kamaraj, K. K. Ramachandran, and J. Aravind, "Biohydrogen production from waste materials: Benefits and challenges," *International Journal of Environmental Science and Technology*, vol. 17, no. 1, pp. 559–576, 2020, doi: 10.1007/S13762-019-02577-Z.

S. B. Kang, H. Y. Oh, J. J. Kim, and K. S. Choi, "Characteristics of spent coffee grounds as a fuel and combustion test in a small boiler (6.5kW)," *Renewable Energy*, vol. 113, pp. 1208–1214, 2017, doi: 10.1016/j.renene.2017.06.092.

R. K. Kasongo, A. Verdoodt, P. Kanyankagote, G. Baert, and E. Van Ranst, "Coffee waste as an alternative fertilizer with soil improving properties for sandy soils in humid tropical environments," *Soil Use and Management*, vol. 27, no. 1, pp. 94–102, Mar. 2011, doi: 10.1111/j.1475-2743.2010.00315.x.

H. -J. Kim and S.-C. Oh, "Hydrothermal carbonization of spent coffee grounds," *Applied Sciences*, vol. 11, no. 14, p. 6542, 2021, doi: 10.3390/APP11146542.

M. -S. Kim and J.-G. Kim, "Adsorption characteristics of spent coffee grounds as an alternative adsorbent for cadmium in solution," *Environments*, vol. 7, no. 4, p. 24, 2020, doi: 10.3390/environments7040024.

J. Korol, A. Hejna, D. Burchart-Korol, and J. Wachowicz, "Comparative analysis of carbon, ecological, and water footprints of polypropylene-based composites filled with cotton, jute and kenaf fibers," *Materials*, vol. 13, no. 16, p. 3541, Aug. 2020, doi: 10.3390/MA13163541.

T. A. Kua, "Application of Spent Coffee Grounds as a Road Subgrade Construction Material," 2017.

A. B. H. Kueh, "Spent grounds coffee—Awaking the sustainability prospects," *Environmental and Toxicology Management*, vol. 1, no. 1, pp. 1–6, 2021, https://doi.org/10.33086/etm.v1i1.2016.

S. Kumar and K. Rathore, "Renewable energy for sustainable development goal of clean and affordable energy," *International Journal of Materials, Manufacturing and Sustainable Technologies (IJMMST)*, vol. 2, pp. 1–15, 2023, doi: 10.56896/IJMMST.2023.2.1.001.

G. La Scalia, M. Saeli, P. P. Miglietta, and R. Micale, "Coffee biowaste valorization within circular economy: An evaluation method of spent coffee grounds potentials for mortar production," *International Journal of Life Cycle Assessment*, vol. 26, no. 9, pp. 1805–1815, 2021, doi: 10.1007/S11367-021-01968-0.

A. Lachheb, A. Allouhi, M. El Marhoune, R. Saadani, T. Kousksou, A. Jamil, M. Rahmoune, and O. Oussouaddi, "Thermal insulation improvement in construction materials by adding spent coffee grounds: An experimental and simulation study," *Journal of Cleaner Production*, vol. 209, pp. 1411–1419, 2019, doi: 10.1016/j.jclepro.2018.11.140.

J. J. Lawler et al., "Projected land-use change impacts on ecosystem services in the United States," *Proceedings of the National Academy of Sciences U. S. A.*, vol. 111, no. 20, pp. 7492–7497, May 2014, doi: 10.1073/PNAS.1405557111.

K. T. Lee, C. L. Cheng, D. S. Lee, W. H. Chen, D. V. N. Vo, L. Ding, and S. S. Lam, "Spent coffee grounds biochar from torrefaction as a potential adsorbent for spilled diesel oil recovery and as an alternative fuel," *Energy*, 239, 122467, 2021, doi: 10.1016/J.ENERGY.2021.122467.

K. T. Lee, J. T. Du, W. H. Chen, A. T. Ubando, and K. T. Lee, "Green additive to upgrade biochar from spent coffee grounds by torrefaction for pollution mitigation," *Environmental Pollution*, vol. 285, p. 117244, 2021, doi: 10.1016/J.ENVPOL.2021.117244.

K. -T. Lee, Y.-T. Shih, S. Rajendran, Y.-K. Park, and W.-H. Chen, "Spent coffee grounds torrefaction for waste remediation and valorization," *Environmental Pollution*, vol. 324, p. 121330, 2023, doi: 10.1016/j.envpol.2023.121330.

M. Lee et al., "Sequential production of lignin, fatty acid methyl esters and biogas from spent coffee grounds via an integrated physicochemical and biological process," *Energies (Basel)*, vol. 12, no. 12, p. 2360, Jun. 2019, doi: 10.3390/en12122360.

Y. Leow, P. Y. M. Yew, P. L. Chee, X. J. Loh, and D. Kai, "Recycling of spent coffee grounds for useful extracts and green composites," *RSC Advances*, vol. 11, no. 5, pp. 2682–2692, 2021, doi: 10.1039/D0RA09379C.

J. Li, "Mushroom Culture Medium Prepared from Coffee Grounds," 2014.

S. Li, C. Shi, S. Sun, H. Chan, H. Lu, A. Nilghaz, J. Tian, and R. Cao, "From brown to colored: Polylactic acid composite with micro/nano-structured white spent coffee grounds for three-dimensional printing," *International Journal of Biological Macromolecules*, vol. 174, pp. 300–308, 2021, doi: 10.1016/j.ijbiomac.2021.01.176.

C. Lindsay, "How to Grow Mushrooms in Coffee Grounds," *Modern Farmer Media*, Sep. 13, 2020.

Y. Liu, Y. Lu, and S. Q. Liu, "The potential of spent coffee grounds hydrolysates fermented with Torulaspora delbrueckii and Pichia kluyveri for developing an alcoholic beverage: The yeasts growth and chemical compounds modulation by yeast extracts," *Current Research in Food Science*, vol. 4, pp. 489–498, 2021, doi: 10.1016/j.crfs.2021.07.004.

J. C. López-Linares, M. T. García-Cubero, M. Coca, and S. Lucas, "Efficient biobutanol production by acetone-butanol-ethanol fermentation from spent coffee grounds with microwave assisted dilute sulfuric acid pre-treatment," *Bioresource Technology*, vol. 320, p. 124348, 2021, doi: 10.1016/J.BIORTECH.2020.124348.

H. Lu, "Coffee Grounds-Containing Plastic Composite Material and Preparation Method Thereof," 2019.

F. C. Luz, S. Cordiner, A. Manni, V. Mulone, and V. Rocco, "Anaerobic digestion of coffee grounds soluble fraction at laboratory scale: Evaluation of the biomethane potential," *Applied Energy*, vol. 207, pp. 166–175, 2017, doi: 10.1016/J.APENERGY.2017.06.042.

F. C. Luz, M. Volpe, L. Fiori, A. Manni, S. Cordiner, V. Mulone, and V. Rocco, "Spent coffee enhanced biomethane potential via an integrated hydrothermal carbonization-anaerobic digestion process," *Bioresource Technology*, vol. 256, pp. 102–109, 2018, doi: 10.1016/J.BIORTECH.2018.02.021.

K. MacDonnell, "What Is Coffee Torrefaction? The Interesting Answer! Coffee Affection," October 28, 2022. https://coffeeaffection.com/what-is-coffee-torrefaction/

E. Machado, S. Mussatto, J. Teixeira, M. Vilanova, and J. Oliveira, "Increasing the sustainability of the coffee agro-industry: Spent coffee grounds as a source of new beverages," *Beverages*, vol. 4, no. 4, p. 105, 2018, doi: 10.3390/beverages4040105.

R. Marchetti and C. Vasmara, "Spent coffee grounds from coffee vending machines as feedstock for biogas production," *Environmental Engineering and Management Journal*, vol. 17, no. 10, pp. 2401–2408, 2018, doi: 10.30638/EEMJ.2018.238.

N. Martinez-Saez et al., "A novel antioxidant beverage for body weight control based on coffee silverskin," *Food Chemistry*, vol. 150, pp. 227–234, May 2014, doi: 10.1016/J.FOODCHEM.2013.10.100.

J. Massaya, G. Pickens, B. Mills-Lamptey, and C. J. Chuck, "Enhanced hydrothermal carbonization of spent coffee grounds for the efficient production of solid fuel with lower nitrogen content," *Energy & Fuels*, vol. 35, no. 11, pp. 9462–9473, Jun. 2021, doi: 10.1021/acs.energyfuels.1c00870.

T. M. Mata, A. A. Martins, and N. S. Caetano, "Bio-refinery approach for spent coffee grounds valorization," *Bioresource Technology*, vol. 247, pp. 1077–1084, Jan. 2018, doi: 10.1016/J.BIORTECH.2017.09.106.

E. McLennon, B. Dari, G. Jha, D. Sihi, and V. Kankarla, "Regenerative agriculture and integrative permaculture for sustainable and technology driven global food production and security," *Agronomy Journal*, vol. 113, no. 6, pp. 4541–4559, Nov. 2021, doi: 10.1002/AGJ2.20814.

J. McNutt and Q. (Sophia) He, "Spent coffee grounds: A review on current utilization," *Journal of Industrial and Engineering Chemistry*, vol. 71, pp. 78–88, Mar. 25, 2019, doi: 10.1016/j.jiec.2018.11.054.

V. A. Mirón-Mérida, B. E. Barragán-Huerta, and P. Gutiérrez-Macías, "Coffee waste: A source of valuable technologies for sustainable development," In *Valorization of Agri-Food Wastes and By-Products*, pp. 173–198, 2021, Elsevier, doi: 10.1016/B978-0-12-824044-1.00009-X.

M. T. Miskon, M. H. F. Rahiman, and M. N. Taib, "Assessing the effectiveness of fuzzy-PID controller under temperature setpoint changes during coffee roasting process," *International Journal of Emerging Technology and Advanced Engineering*, vol. 9001, no. 03, 2023, doi: 10.46338/ijetae0323_02.

A. K. Mohanty, S. Vivekanandhan, J.-M. Pin, and M. Misra, "Composites from renewable and sustainable resources: Challenges and innovations," *Science (1979)*, vol. 362, no. 6414, pp. 536–542, Nov. 2018, doi: 10.1126/science.aat9072.

A. Molino et al., "Power production by biomass gasification technologies," In *Current Trends and Future Developments on (Bio-) Membranes: Renewable Energy Integrated with Membrane Operations*, pp. 293–318, Jan. 2019, doi: 10.1016/B978-0-12-813545-7.00012-X.

A. C. V. Montoya, R. C. da Silva Mazareli, E. L. Silva, and M. B. A. Varesche, "Improving the hydrogen production from coffee waste through hydrothermal pre-treatment, co-digestion and microbial consortium bioaugmentation," *Biomass Bioenergy*, vol. 137, p. 105551, Jun. 2020, doi: 10.1016/j.biombioe.2020.105551.

T. Moussa et al., "Spent coffee grounds as building material for non-load-bearing structures," *Materials*, vol. 15, no. 5, p. 1689, Feb. 2022, doi: 10.3390/ma15051689.

P. Muñoz Velasco, M. A. Mendívil, M. P. Morales, and L. Muñoz, "Eco-fired clay bricks made by adding spent coffee grounds: A sustainable way to improve buildings insulation," *Materials and Structures*, vol. 49, no. 1–2, pp. 641–650, 2016, doi: 10.1617/s11527-015-0525-6.

S. Mussatto, M. Fernandes, G. Rocha, J. J. M. Órfão, J. Teixeira, and I. C. Roberto, "Production, characterization and application of activated carbon from Brewer's spent grain lignin," *Bioresource Technology*, vol. 101, pp. 2450–2457, 2009, doi: 10.1016/j.biortech.2009.11.025.

C. Nam and C. Xiang, "Natural dyeing application of used coffee grounds as a potential resource," *International Journal of Fashion Design, Technology and Education*, vol. 12, no. 3, pp. 335–345, 2019, doi: 10.1080/17543266.2019.1640291.

D. M. Nguyen, V. T. Nhung, T. C. Le Do, C. N. Ha-Thuc, and P. Perré, "Effective synergistic effect of treatment and modification on spent coffee grounds for sustainable biobased composites," *Waste and Biomass Valorization*, 1–10, 2021, doi: 10.1007/S12649-021-01566-Z.

T. A. Nguyen and Q. T. Nguyen, "Hybrid biocomposites based on used coffee grounds and epoxy resin: Mechanical properties and fire resistance," *International Journal of Chemical Engineering*, vol. 2021, pp. 1–12, 2021, doi: 10.1155/2021/1919344.

V. -T. Nguyen, T.-D.-H. Vo, T. Tran, T.-N. Nguyen, T.-N.-C. Le, X.-T. Bui, and L.-G. Bach, "Biochar derived from the spent coffee grounds for ammonium adsorption from aqueous solution," *Case Studies in Chemical and Environmental Engineering*, vol. 4, p. 100141, 2021, doi: 10.1016/j.cscee.2021.100141.

R. Nosek, M. M. Tun, and D. Juchelkova, "Energy utilization of spent coffee grounds in the form of pellets," *Energies*, vol. 13, no. 5, p. 1235, 2020, doi: 10.3390/en13051235.

N. M. Nurazzi et al., "A review on natural fiber reinforced polymer composite for bullet proof and ballistic applications," *Polymers*, vol. 13, no. 4, p. 646, Feb. 2021, doi: 10.3390/POLYM13040646.

Y. Opoku-Kwanowaa, R. K. Furaha, L. Yan, and D. Wei, "Effects of planting field on groundswater and surface water pollution in China," *Clean (Weinh)*, vol. 48, no. 5–6, p. 1900452, Jun. 2020, doi: 10.1002/CLEN.201900452.

A. I. Osman et al., "Conversion of biomass to biofuels and life cycle assessment: A review," *Environmental Chemistry Letters*, vol. 19, pp. 4075–4118, 2021, doi: 10.1007/s10311-021-01273-0.

A. Pegoretti, R. Cercená, S. Siengchin, Y. G. Thyavihalli Girijappa, S. M. Rangappa, and J. Parameswaranpillai, "Natural fibers as sustainable and renewable resource for development of eco-friendly composites: A comprehensive review," *Frontiers in Materials*, vol. 6, p. 226, 2019, doi: 10.3389/fmats.2019.00226.

J. Popp, Z. Lakner, M. Harangi-Rákos, and M. Fári, "The effect of bioenergy expansion: Food, energy, and environment," *Renewable and Sustainable Energy Reviews*, vol. 32, pp. 559–578, Apr. 2014, doi: 10.1016/J.RSER.2014.01.056.

A. R. Prihadi and A. Maimulyanti, "Chemical compounds of coffee grounds and spent coffee grounds for pharmaceutical products," *Pharmaceutical and Biomedical Sciences Journal (PBSJ)*, vol. 2, no. 2, pp. 49–52, 2021.

A. Procentese, F. Raganati, L. Navarini, G. Olivieri, M. E. Russo, and A. Marzoccchella, "Coffee silverskin as a renewable resource to produce butanol and isopropanol," *Chemical Engineering Transactions*, vol. 64, pp. 139–144, 2018, doi: 10.3303/CET1864024.

D. Pujol, C. Liu, J. Gominho, M. À. Olivella, N. Fiol, I. Villaescusa, and H. Pereira, "The chemical composition of exhausted coffee waste," *Industrial Crops and Products*, vol. 50, pp. 423–429, 2013, doi: 10.1016/j.indcrop.2013.07.056.

P. Rachtanapun, T. Sattayarak, and N. Ketsamak, "Correlation of density and properties of particleboard from coffee waste with urea–formaldehyde and polymeric methylene diphenyl diisocyanates," *Journal of Composite Materials*, vol. 46, no. 15, pp. 1839–1850, Jul. 2012, doi: 10.1177/0021998311426624.

P. Rangarajan and J. A. Tharian, "Coffee waste management—An overview," *International Journal of Current Research*, 2019. https://www.researchgate.net/publication/330825000

E. Ribeiro, T. de S. Rocha, and S. H. Prudencio, "Potential of green and roasted coffee beans and spent coffee grounds to provide bioactive peptides," *Food Chemistry*, vol. 348, p. 129061, 2021, doi: 10.1016/j.foodchem.2021.129061.

J. R. H. Ross, "CO2 and energy," Elsevier, 2019, doi: 10.1016/B978-0-444-63474-0.00009-6.

M. Saberian, J. Li, A. Donnoli, E. Bonderenko, P. Oliva, B. Gill, S. Lockrey, and R. Siddique, "Recycling of spent coffee grounds in construction materials: A review," *Journal of Cleaner Production*, vol. 289, p. 125837, 2021, doi: 10.1016/J.JCLEPRO.2021.125837.

G. D. Saratale, R. G., Saratale, and J. S. Chang, "Biohydrogen from renewable resources," pp. 185–221, 2013, doi: 10.1016/B978-0-444-59555-3.00009-X.

F. Sarghini et al., "Acid hydrolysis of spent coffee grounds: Effects on possible prebiotic activity of oligosaccharides," *Chemical and Biological Technologies in Agriculture*, vol. 8, no. 1, p. 67, Dec. 2021, doi: 10.1186/s40538-021-00262-3.

K. S. K. Sasikumar, N. Saravanan, M. Sambathkumar, and R. Guekndran, "Acoustic characterization of farm residues for sound absorption applications," *Materials Today: Proceedings*, pp. 2917–2922, 2020, doi: 10.1016/j.matpr.2020.02.884.

P. Semenchuk et al., "Relative effects of land conversion and land-use intensity on terrestrial vertebrate diversity," *Nature Communications*, 2022, doi: 10.1038/s41467-022-28245-4.

B. Sena da Fonseca, A. Vilão, C. Galhano, and J. A. R. Simão, "Reusing coffee waste in manufacture of ceramics for construction," *Advances in Applied Ceramics*, vol. 113, no. 3, pp. 159–166, 2014, doi: 10.1179/1743676113Y.0000000131.

V. S. Sikarwar, M. Zhao, P. S. Fennell, N. Shah, and E. J. Anthony, "Progress in biofuel production from gasification," *Progress in Energy and Combustion Science*, vol. 61, pp. 189–248, 2017, doi: 10.1016/J.PECS.2017.04.001.

A. Skorupa, M. Worwąg, and M. Kowalczyk, "Coffee industry and ways of using by-products as bioadsorbents for removal of pollutants," *Water*, vol. 15, no. 1, p. 112, Dec. 2022, doi: 10.3390/W15010112.

M. L. Suárez-Quiroz, A. A. Campos, G. V. Alfaro, O. Gonzalez-Rios, P. Villeneuve, and M.-C. Figueroa-Espinoza, "Isolation of green coffee chlorogenic acids using activated carbon," *Journal of Food Composition and Analysis*, vol. 33, no. 1, pp. 55–58, 2014, doi: 10.1016/J.JFCA.2013.10.005.

S. Suganya and P. Senthil Kumar, "Influence of ultrasonic waves on preparation of active carbon from coffee waste for the reclamation of effluents containing Cr(VI) ions," *Journal of Industrial and Engineering Chemistry*, vol. 60, pp. 418–430, Apr. 2018, doi: 10.1016/j.jiec.2017.11.029.

W. K. Sun, "Method for Manufacturing Functional Pulp and Paper by Using Coffee Grounds," Korean Patent (KR Patent No. WO2014081093A1), May 30, 2014.

L. Sundqvist Ökvist and M. Lundgren, "Experiences of bio-coal applications in the blast furnace process—Opportunities and limitations, "*Minerals*, vol. 11, no. 8, p. 863, 2021, doi: 10.3390/MIN11080863.

J. Y. Tan, P. J. Ker, K. Y. Lau, M. A. Hannan, and S. C. H. Tang, "Applications of photonics in agriculture sector: A review," *Molecules*, vol. 24, no. 10, p. 2025, May 2019, doi: 10.3390/MOLECULES24102025.

T. Tokimoto, N. Kawasaki, T. Nakamura, J. Akutagawa, and S. Tanada, "Removal of lead ions in drinking water by coffee grounds as vegetable biomass," *Journal of Colloid and Interface Science*, vol. 281, no. 1, pp. 56–61, 2005, doi: 10.1016/j.jcis.2004.08.083.

T. T. A. Tran et al., "Bioethanol production from lignocellulosic biomass," *Xiandai Huagong/Modern Chemical Industry*, vol. 31, no. SUPPL. 2, pp. 40–44, Jun. 2019

N. Triantafillopoulos and A. A. Koukoulas, "The future of single-use paper coffee cups: Current progress and outlook," *Bioresources*, vol. 15, no. 3, pp. 7260–7287, 2020.

K. C. Tsang, J. C. Heijdenrijk, and J. D. Glaser, "Packaging Filled with Roast and Grounds Coffee Beans," 2014.

H. Vahabi, M. Jouyandeh, T. Parpaite, M. R. Saeb, and S. Ramakrishna, "Coffee wastes as sustainable flame retardants for polymer materials," *Coatings*, vol. 11, no. 9, p. 1021, 2021, doi: 10.3390/coatings11091021.

T. Vítěz, T. Koutný, M. Šotnar, and J. Chovanec, "On the spent coffee grounds biogas production," *Acta Universitatis Agriculturae et Silviculturae Mendelianae Brunensis*, vol. 64, no. 4, pp. 1279–1282, 2016, doi: 10.11118/ACTAUN201664041279.

A. K. Vuppaladadiyam et al., "Bio-oil and biochar from the pyrolytic conversion of biomass: A current and future perspective on the trade-off between economic, environmental, and technical indicators," *Science of the Total Environment*, vol. 857, p. 159155, Jan. 2023, doi: 10.1016/j.scitotenv.2022.159155.

A. G. Woldesenbet, B. Woldeyes, and B. S. Chandravanshi, “Bio-ethanol production from wet coffee processing waste in Ethiopia,” *SpringerPlus*, vol. 5, no. 1, p. 1903, Dec. 2016, doi: 10.1186/s40064-016-3600-8.

M. Yadav, B. B. Vashisht, S. K. Jalota, A. Kumar, and D. Kumar, “Sustainable water management practices for intensified agriculture,” pp. 131–161, 2022, doi: 10.1007/978-3-031-12059-6_8.

V. A. Yiga, S. Pagel, M. Lubwama, S. Epple, P. W. Olupot, and C. Bonten, “Development of fiber-reinforced polypropylene with NaOH pre-treated rice and coffee husks as fillers: Mechanical and thermal properties,” *Journal of Thermoplastic Composite Materials*, vol. 33, no. 9, pp. 1269–1291, Sep. 2020, doi: 10.1177/0892705718823255.

J. Yoo, S. J. Chang, S. Wi, and S. Kim, “Spent coffee grounds as supporting materials to produce bio-composite PCM with natural waxes,” *Chemosphere*, vol. 235, pp. 626–635, 2019, doi: 10.1016/J.CHEMOSPHERE.2019.06.195.

B. Y. Yun, H. M. Cho, Y. U. Kim, S. C. Lee, U. Berardi, and S. Kim, “Circular reutilization of coffee waste for sound absorbing panels: A perspective on material recycling,” *Environmental Research*, vol. 184, p. 109281, 2020, doi: 10.1016/j.envres.2020.109281.

P. Zhang, “Application study of Bio-FGD based on environmental safety during the coal combustion,” *IOP Conference Series: Earth and Environmental Science*, vol. 154, no. 1, p. 012018, 2018, doi: 10.1088/1755-1315/154/1/012018.

H. Yoon, "Paper Coating Material Having Environment-Friendly, Water-Proof and Oil-Proof Properties, and Method of Manufacturing the Same.," Korean Patent (KR Patent No. US20180094165A1), 2016.

9 Cotton Waste Management

9.1 INTRODUCTION

The textile industry is known as the world's second-largest polluting industry, according to its substantial greenhouse emissions to the environment (Tausif et al., 2018). In the textile industry, several waste products, such as fiber, silver, yarn, and woven, are produced at various stages by various textile processes (Isci & Demirer, 2007). The production and consumption of cotton waste textiles are rising steadily. In general, technical and economic considerations are needed for the accounting of cotton waste for trash recycling (Ertaş et al., 2010). Moreover, environmental studies must take into account these two concerns and offer suggestions for ways to get manufacturers to treat cotton waste (Esteve-Turrillas & de la Guardia, 2017). The lack of post-consumer waste disposal facilities is a problem in many developing nations, raising major environmental issues. Using textile waste, such as cotton waste, will help with trash disposal issues and add value to the waste. As a result, recycling and turning trash into resources is crucial to reduce pollution (Hamawand et al., 2016).

Cotton is the most widely used natural fiber in the world produced from the processed covering of seeds of the *Gossypium* plant, which is specific to Indian territory as well as tropical areas of Africa and America. Various types of fabrics can be produced when weaving cotton yarn with yarns from other fibers. However, this process generates waste known as "cotton textile waste." This phrase refers to processed textile waste, which can include fibers, threads, or pieces of fabric in various sizes (Haule et al., 2016). The format of cotton waste varies depending on the stage of transformation and the company being evaluated, such as the spinning, weaving, or textile lab industries (Petrucci et al., 2015). Furthermore, it is feasible to prevent landfills from becoming overflowing with textile waste and provide a sustainable alternative to the manufacture of clothing and other items from virgin cotton by collecting waste textiles, particularly those components utilized in numerous industrial processes, such as cotton (Shirvanimoghaddam et al., 2019).

Cotton waste refers to the by-products and residual materials generated during cotton production, processing, or use (Maciel et al., 2019). Cotton, a natural fiber obtained from the seed pods of cotton plants, is widely used in the textile industry to manufacture various products like clothing, towels, and bed sheets (Thyavihalli Girijappa et al., 2019). Different forms of waste are produced throughout the cotton production and processing cycle. First, during the ginning process, cotton fibers are separated from the seeds, producing lint waste. Lint waste comprises short fibers, cotton dust, and small cotton fragments. Cotton seeds are not used for fiber production and are considered waste material (Paletto et al., 2019). However, they have economic value and find applications in oil extraction, animal feed, and fertilizer production. Next, fabric scraps are another form of cotton waste generated during the cutting and sewing stages of textile manufacturing (Senthil Kumar & Yaashikaa, 2019). If originating from cotton fabric, these scraps can be classified as cotton waste. Additionally, post-consumer waste is generated when cotton products, such as clothing, towels, or bedding, become worn out or damaged after use (Fahim et al., 2022).

Effectively managing cotton waste is crucial for environmental sustainability and resource efficiency. Various methods are employed to handle cotton waste. Recycling is one such method, where cotton waste is mechanically shredded and reprocessed into new fibers that can be used to create

DOI: 10.1201/9781003472209-12

new cotton-based products (Patel, 2017). Composting is another approach, where cotton waste and other organic materials are composted to produce nutrient-rich soil amendments (Halimi et al., 2007). This allows for the organic decomposition of cotton waste, returning valuable nutrients to the soil. Cotton waste also finds application in various industrial sectors (Hamawand et al., 2016). It can be used to produce insulation materials, upholstery stuffing, and paper manufacturing, among others. The textile industry and sustainability-conscious individuals are actively working to reduce cotton waste through efficient production methods, promoting recycling initiatives, and embracing a circular economy approach (Mahamude et al., 2021).

In summary, cotton waste represents the leftover materials that are generated during the production, processing, and use of cotton. Managing and repurposing cotton waste effectively is essential for promoting sustainability, minimizing environmental impact, and maximizing the value extracted from cotton fibers (Ütebay et al., 2019). The aim of this chapter is to explore the sustainable management of cotton waste, highlighting the potential for recycling and valorization of cotton by-products. Through detailed analyses of cotton waste properties, including chemical, physical, mechanical, and thermal aspects, this chapter provides insights into how these materials can be repurposed for various applications, such as geotechnical uses, wastewater treatment, and the production of advanced materials. By investigating current research, methods, and standards in cotton waste fiber utilization, this chapter aims to offer practical solutions to reduce environmental pollution while enhancing resource efficiency in the textile industry.

9.2 BIOMASS FROM NATURAL FIBERS

Every nation in the world produces natural fibers. It is used to make a variety of items, including textiles, ropes, brushes, carpets, mattresses, mats, paper, board materials, etc. The rise of the chemical fiber sector is additionally supported by the development of renewable, biodegradable biomass fiber (Wang et al., 2020). Currently, China has seen significant growth in the biomass fiber sector. Over 200 million tons of biomass fibers are produced there. With such huge biomass fiber resources, this business has a lot of room to grow (Saidur et al., 2011).

Biomass obtained from natural fibers comprises the organic material derived from plant-based fibers, which can be used as a renewable energy source or raw material for various industrial applications (Bhuiyan et al., 2023). Natural fibers, including cotton, hemp, flax, jute, bamboo, and sisal, contain cellulose, hemicellulose, and lignin, making them suitable for biomass utilization (Tamburini et al., 2023; Norfarhana et al., 2022, 2024a,b; Nurazzi et al., 2021). The applications of biomass derived from natural fibers are diverse (Alias et al., 2021; Azlin et al., 2022; Mazani et al., 2019). It can be utilized as a renewable energy source by burning it for heat and electricity generation or converting it into biofuels like bioethanol and biodiesel (Cambero & Sowlati, 2014). Natural fiber biomass can also be incorporated into polymer matrices to create bio composites, offering sustainable alternatives for automotive, construction (Asyraf et al., 2022), fire retardant (Suriani et al., 2021), and packaging materials (Zafar et al., 2021; Halimatul et al., 2019; Jumaidin et al., 2019a, b, 2021; Nazrin et al., 2021). In the pulp and paper industry, natural fiber biomass, mainly wood fibers, is commonly used to produce pulp for paper and cardboard manufacturing (Tausif et al., 2018). Additionally, natural fibers can be directly used in textile production to create fabrics for clothing, upholstery, and other textile applications (Shogren et al., 2019).

Biomass from natural fibers is also beneficial in horticulture and agriculture, such as organic mulch or compost, improving soil fertility and moisture retention, controlling weeds, and serving as animal bedding (Crini et al., 2020). Moreover, it can be a valuable feedstock for producing bio-based chemicals, including bioplastics, solvents, and platform chemicals. Using biomass from natural fibers offers numerous advantages, such as reduced greenhouse gas emissions and decreased dependence on fossil fuels. However, responsible sourcing and processing of natural fibers are essential to mitigate potential negative environmental impacts (Koul et al., 2022; Norrrahim et al., 2021).

9.3 ENVIRONMENTAL IMPACT FROM BIOMASS

The environmental impact of biomass can vary depending on factors such as the type of biomass, sourcing methods, production processes, and specific applications. While biomass is often considered a renewable energy source with potential sustainability benefits, it is important to carefully assess its environmental implications (Abbasi & Abbasi, 2010). One significant concern is the emission of carbon dioxide when biomass is burned. If biomass comes from sustainably managed sources and the carbon dioxide released during combustion is offset by carbon absorption from new plant growth, it can be considered carbon-neutral (Paletto et al., 2019). However, if unsustainable practices or deforestation are involved, biomass combustion can contribute to a net increase in carbon emissions and exacerbate climate change (Senthil Kumar & Yaashikaa, 2019). Around 220 million metric tons of carbon dioxide are emitted annually through cotton production, significantly known as biomass (Fahim et al., 2022). Synthetic fertilizers, which release nitrous oxide into the environment, are heavily used to grow regular cotton. A powerful greenhouse gas, nitrous oxide is 310 times more potent than carbon dioxide (Esteve-Turrillas & de la Guardia, 2017). Deforestation and extensive pesticide use are two more unsustainable practices that increase greenhouse gas emissions.

Huge amounts of water can be consumed to yield cotton. About 69% of the water footprint of producing fiber for the textile industry is accounted for by cotton agriculture (Delate et al., 2020). The Aral Sea in Central Asia had a catastrophe that reduced its surface water supplies to less than 10% of its former volume (Guiding_2017, n.d.). The leading cause of this was cotton producers' use of seawater for irrigation. One of the worst climate disasters that humanity has ever experienced was the Aral Sea calamity (Abbasi & Abbasi, 2010). About 97% of the Indus River's wastewater was used to irrigate cotton fields due to cotton's need for water. Typically, cotton farms are monocultural. Additionally, soil quality deteriorates, just like in monocultural systems maintained for years (Eyupoglu, 2019). Over the past 70 years, the area used to grow cotton has mainly remained stable. Exhaustion has prompted growth in new directions (Sandin & Peters, 2018). Deforestation and the loss of wildlife habitat are direct effects of the development. Due to the lack of tree cover in those fields, topsoil erosion is also a serious concern. A harmful degree of sediment builds up for aquatic life can result from erosions. According to research, invertebrates and fish populations in streams and rivers can decrease due to heavy sedimentation (Zabaniotou & Andreou, 2010). Additionally, it may raise the water's temperature and lower its oxygen content. The use of chemicals in cotton farming also degrades the soil. The natural soil fertility is harmed, and the soil's capacity to filter water and sequester carbon is compromised.

Consumers are drawn to cotton that is vividly colored or immaculately white, but many are unaware of the environmental implications. Twenty percent of industrial water contamination is attributable to the fashion sector (Al Afif et al., 2020). That amount reflects the several chemical processes cotton undergoes, including dyeing, bleaching, and other processes (Tausif et al., 2018). Pesticides, hazardous colors, chlorine, and other chemicals required to turn the crop into fiber and garments can be found in run-off effluent from processing (Wang et al., 2020). These contaminants affect subsurface aquifers, rivers, lakes, wetlands, and other water systems. Nineteen pesticides, 12 of which can be directly linked to cotton growing, were identified in the rainwater in the Brazilian cotton region (Shogren et al., 2019). Humans can be impacted by pollution from cotton processing into textiles or clothing just as much as wildlife and the environment (Tausif et al., 2018).

9.4 AVAILABILITY OF COTTON WASTE

The availability of cotton waste exists due to its recycling process from most recycled cotton waste. Mechanical recycling is used to claim the majority of reclaimed cotton (Ütebay et al., 2019). Materials are first arranged according to color. After sorting, the fabrics are run through a machine, which shreds the fabric into yarn and then raw fiber. The fiber is severely strained throughout this challenging procedure. Fibers frequently break and tangle when being shredded (Yousef et al., 2019). After that, the unprocessed fiber is spun again into yarns for usage in other goods. Recycled fiber will never offer quality attributes that are on par with virgin fiber. In particular, the length of homogeneity and fiber length will be affected, which will restrict the end-use application (De Silva & Byrne, 2017).

The availability of cotton waste benefits the current generation due to less production of energy and dye usage through the process of recycling. Moreover, the availability of cotton is used to save energy, reduce emissions, and minimize fossil fuel usage (Haque et al., 2021). For decision-making on end-of-life options, a precise estimation of the quantity and quality of waste cotton available at various supply chain stages is necessary (Wang, 2010). Second, while some studies have examined the technical viability and environmental advantages of multiple uses for waste cotton, few have offered a thorough analysis and comparison of various applications from a life cycle viewpoint (Johnson et al., 2020).

Cotton waste is influenced by various factors, including the scale of cotton production, the methods used during processing, and the demand for cotton-based products (Eyupoglu, 2019). The availability of cotton waste is typically linked to different stages in the cotton supply chain, such as ginning, textile manufacturing, and post-consumer use (Johnson et al., 2020). Countries with significant cotton cultivation and textile industries, like China, India, the United States, and Pakistan, generally have a greater abundance of cotton waste (Ute et al., 2019). These countries possess large-scale cotton production and extensive textile sectors, leading to a higher volume of cotton waste generated.

The availability of cotton waste can also fluctuate depending on the specific type of waste (Eyupoglu, 2019). For instance, lint waste, comprising short fibers, cotton dust, and small fragments, is usually abundant during ginning (Guiding_2017, n.d.). Fabric scraps emerge during textile manufacturing, while post-consumer cotton waste becomes accessible as consumers dispose of or recycle cotton products (Koul et al., 2022). Moreover, the availability of cotton waste can be influenced by market demand, economic factors, and recycling initiatives (Rahman & Uddin, 2022). Growing awareness of sustainability and improved waste management practices in the textile industry have contributed to efforts to reduce waste, potentially impacting the quantity of available cotton waste (Lu et al., 2023). Overall, the availability of cotton waste is shaped by factors such as cotton production levels, the activity of the textile industry, and consumer behaviors. The type of cotton waste and regional considerations also contribute to determining its availability.

9.5 PROPERTIES OF COTTON WASTE FIBERS

9.5.1 Physical Properties

The values of moisture content, geometric mean diameter, and bulk density of ground cotton stalks are provided in Table 9.1. The particle size distribution obtained by sieve analysis is shown in Figure 9.1. Before the pelletizing operation, the moisture content of dried and ground raw material was approximately 10%. Three different moisture contents (10%, 15%, and 20%) were determined, where PC indicates pallet for cotton. The sieve analysis determined that the particles were mostly in groups 0.425–0.85 mm. Particles in these groups consist of 50% of the total material. Since cotton stalks are relatively woody and brittle, the amount of dust material smaller than 0.125 mm in

TABLE 9.1
Physical Properties of Ground Raw Materials (Yilmaz et al., 2020)

Raw Material	Bulk Density (kg/m³)	Geometric Mean Diameter (mm)	Moisture Range (%, wb)	Moisture Content of Raw Material after Moistening (%)	
Cotton stalk	230	0.443	10–11	PC1	10.70
			14–15	PC2	15.13
			20–21	PC3	20.37

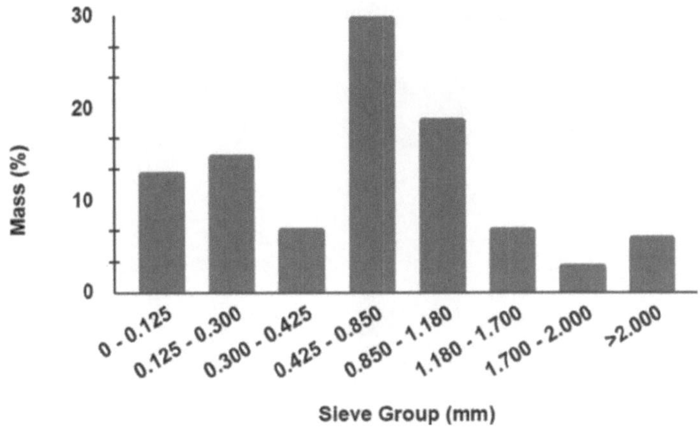

FIGURE 9.1 Particle size distribution of ground cotton stalk (Yilmaz et al., 2020).

TABLE 9.2
Properties and Structure of Cotton Fabric (Alomayri et al., 2014)

Properties	Value/Structure
Fabric thickness (mm)	0.41
Fabric geometry	Woven (plain weave)
Yarn nature	Bundle
Filament size (mm)	0.0413
Number of filaments in a bundle	24
Bundle diameter (mm)	0.23
Opening size (mm)	0.5
Fabric density (g/cm^3)	1.6
Tensile strength (MPa)	287–597

diameter is higher in the cotton stalk (Yilmaz et al., 2020). Table 9.2 lists the properties and structure of cotton fabric. The strength and tenacity of the bamboo/cotton mixed yarn were compared in different research to yarn made entirely of cotton, which had identical physical attributes. When bamboo content in the mix increased, it was discovered that each count of yarns had a decreasing characteristic. The blend's bamboo-to-cotton content determined the qualitative attributes (Latif et al., 2018).

9.5.2 Chemical Properties

These fibers' primary structural element is cellulose, and their secondary structural elements include hemicelluloses, lignin, pectin, lipids, and waxes. Different amounts of these chemicals depend on the fiber type, geographic region, growing practices, and first processing. The significant component of cotton fibers is cellulose, which makes up approximately 83–99%. Other minor components include lignin (up to 6%), hemicelluloses, and pectin (up to 5%). Figure 9.2 shows the chemical composition of cotton yarns: Cy with α-cellulose, 94%; hemicelluloses, 5%; and lignin, 6%. The chemical composition and the location of the chemical components in the fiber structure influence the type and amount of surface functional groups and availability (Vukcevic et al., 2023).

The chemical characteristics of cotton stalks compared to various annual plants, conifers, and deciduous wood, the primary fibrous raw materials are given in Table 9.3. In comparison to most

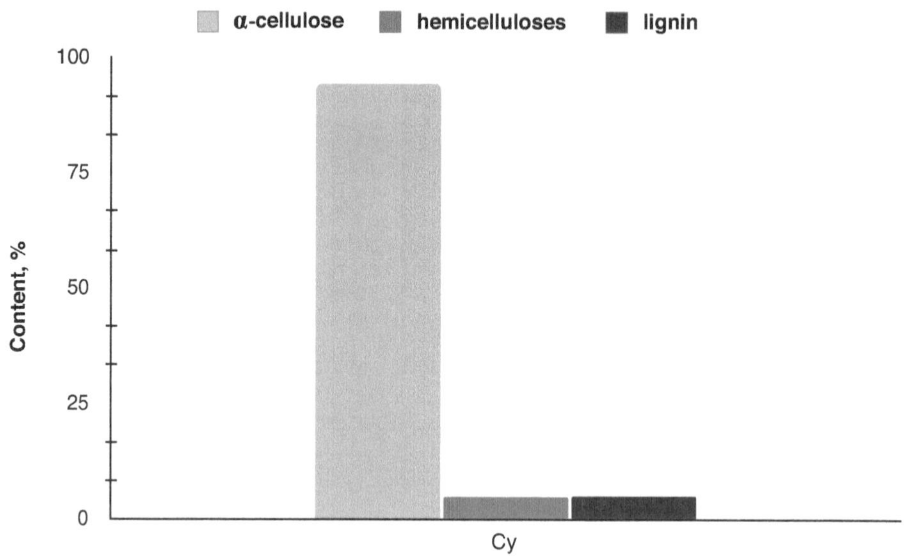

FIGURE 9.2 Chemical properties of flax and hemp fibers and cotton yarns (Vukcevic et al., 2023).

TABLE 9.3
Fiber Properties of Cotton Stalks and Comparison with Common Papermaking Fibers (Tutus et al., 2010)

Samples	Fiber Length (mm)	Fiber Width (μm)	Lumen Width (μm)	Cell Wall Thickness (μm)
Cotton stalks	0.81	24.98	16.75	4.12
Wheat straw	0.74	13.20	4.02	4.59
Rice straw	0.89	14.80	6.40	4.20
Canola stalks	1.17	23.02	12.50	5.26
Coniferous	3–7	32–43	15–30	13–17

annual plants and conifers, which had holocellulose contents between 70% and 81% on average, cotton stalks had an average holocellulose content of 75.6%, making them richer in cellulose (45.5%) than certain annual plants, softwoods, and hardwoods (35–45%). The lignin concentration of cotton stalks was found to be 18.2%, which is comparable to all softwoods (25–32%), including scoots pine, which was 26.0; it is, however, significantly lower than all hardwoods (21–25%), including 24.5% in bamboo, 17–26% in the same annual plants, and hardwoods. The solubility of cotton stalks with some annual plants was likewise comparable. Hardwoods and softwoods, however, are less soluble than cotton stalks and other annual plants. According to Table 9.3, all basic chemical properties and specific solubility values are within the usual range when compared to coniferous woods. Cotton stalks' higher rate of NaOH solubility than coniferous stalks may be caused by the presence of low-molecular-weight carbohydrates and other alkali-soluble substances (Tutus et al., 2010).

9.5.3 Mechanical Properties

Mechanical properties and fiber length are sensitive to extrusion parameters such as screw configuration and compounding temperature. Short fibers resulted in better mechanical properties than long fibers, and water absorption and volumetric swelling increased with fiber length. Increasing fiber length had beneficial effects on tensile, flexural modulus of elasticity, and toughness of wood-plastic

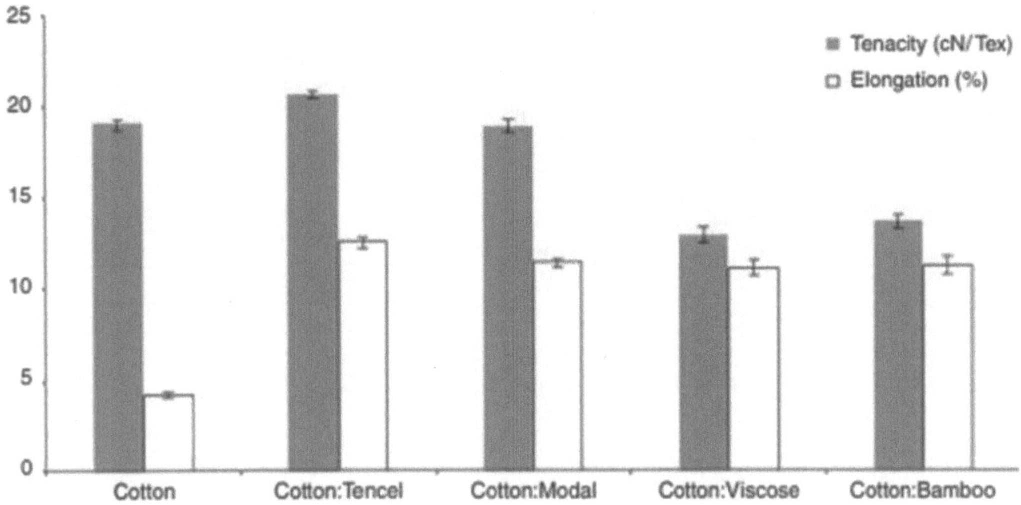

FIGURE 9.3 Tenacities and elongations at break percent of blended yarns (Latif et al., 2018).

composites (El Messiry et al., 2015). When the fabric weave is the same for all textiles, the yarn strength is the critical factor determining the tensile strength of the fabric. Figure 9.3 illustrates a comparison of the tensile strength of several textiles. Contrary to other blends and fabrics made entirely of cotton, cotton: Tencel fabric has better tensile strengths in the warp (30 kgf) and weft (28 kgf), with wise orientations. This is because the cotton: Tencel yarn has a greater tensile strength (20.6 cN/Tex). Cotton displays the minimum tensile strength in both directions (15 and 17 kgf). Due to cotton's lower tensile strength, viscose fabric is produced using viscose yarn (12.97 cN/Tex). The statistical analysis (one-way ANOVA) demonstrates that the tensile strength in the warp and weft directions for cotton, and its mixes has substantially different values since the *p*-value is less than 0.05 (Latif et al., 2018).

The fiber, yarn, and fabric properties determine the tear strength. The characteristics that determine the strength of yarn and, eventually, fabric, in the case of fiber, are strength and fineness. The variables determining strength include single yarn strength, elongation, and twist. In the case of cloth, the strength is influenced by the weave type and weight per square meter. This work maintained a steady yarn linear density and weave pattern. As illustrated in Figure 9.4, it was discovered that cotton: Tencel fabric exhibits the most tear strength (warp-wise 1,540 gf and weft-wise 1,560 gf), whereas cotton: viscose fabric exhibits the least strength (warp-wise 725 gf and weft-wise 685 gf) due to the weaker cotton: viscose yarn (12.97 cN/tex). Overall, 100% cotton fabric (warp-wise 1,100 gf and weft-wise 1,165 gf) provides less tear strength than cotton: Tencel fabric (warp-wise 1,540 gf and weft-wise 1,560 gf), but it does so at a level that is comparable to that of cotton: modal fabric (warp-wise 1,160 gf and weft-wise 1,165 gf). Additionally, cotton fabric has a higher tear strength than textiles made entirely of cotton and viscose or bamboo (both of which have warp and weft weights of 775 and 740 gf, respectively). Because the cotton: Tencel yarn (20.65 cN/Tex) is stronger, the fabric in this combination has the maximum tear strength. Given that the *p*-value is less than 0.05, the statistical analysis (one-way ANOVA) reveals that cotton and its mixes' rip strength in the warp and weft directions have statistically different values (Latif et al., 2018).

9.5.4 Morphological Properties

Figure 9.5 shows a Vizopan microscope examination of a macerated cotton stalk sample. Table 9.3 provides an overview of the cotton stalk's physical characteristics and comparisons with typical sources of fiber for papermaking. The findings demonstrate that the cotton stalks had short

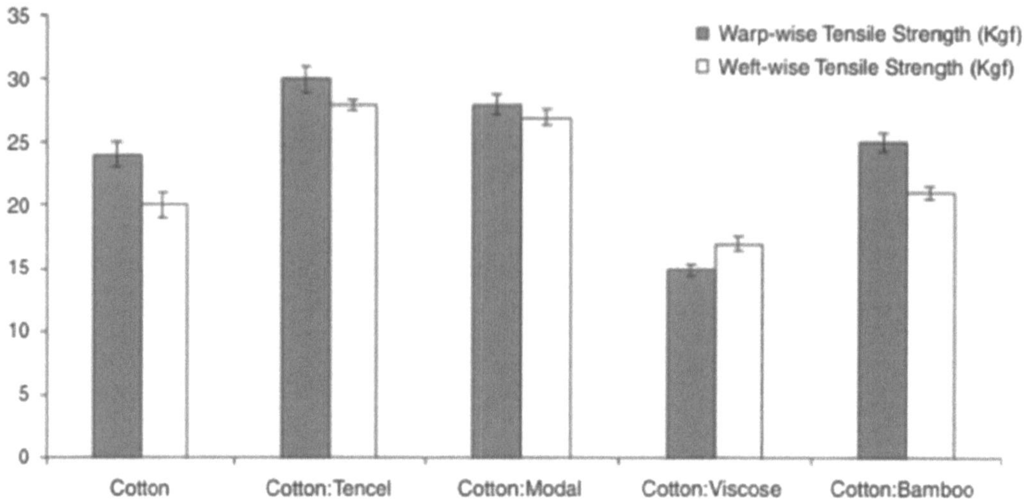

FIGURE 9.4 Tensile strength (warp-wise and weft-wise) of fabrics (Latif et al., 2018).

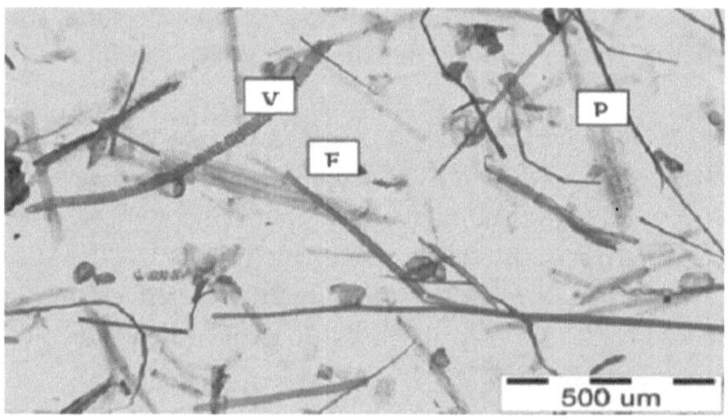

FIGURE 9.5 Macerated sample of cotton stalks stem (F, fibers; P, parenchyma cells; V, vessels) (Tutus et al., 2010).

fibers with a mean length of 0.81 mm. The fibers of cotton stalks are as long as the short fibers of non-wood plants like rice straw. Despite using the same hardwood fibers, they are longer than wheat straw. Cotton stalk fibers have fiber diameters and lumen widths similar to canola stalk fibers and are the same as softwood fibers. However, cotton stalk fibers have stronger cell walls than identical non-wood, hard, and softwood fibers. Despite being considered short fibers, the overall morphological characteristics of cotton stalk fibers are suitable for papermaking (Tutus et al., 2010).

9.5.5 Thermal Properties and Calorific Value

The cotton textiles' qualities for providing thermal comfort were assessed. Fabrics created from a 50/50 blend of traditional modal/viscose fiber with cotton fiber and 100% cotton fiber were compared to those made from a 50/50 combination of modal/viscose fiber in microfiber form with cotton fiber. The thermal conductivity values of the modal microfiber blended textiles were greater than those of the traditional modal fiber blended fabrics and lower than those of cotton fabrics. The modal microfiber mixed materials had the coolest sensation compared to the other two fabric types since they had the highest thermal absorptive properties (Gun & Tiber, 2011). Figure 9.6 illustrates the relationship between cotton stalk calorific values and moisture content.

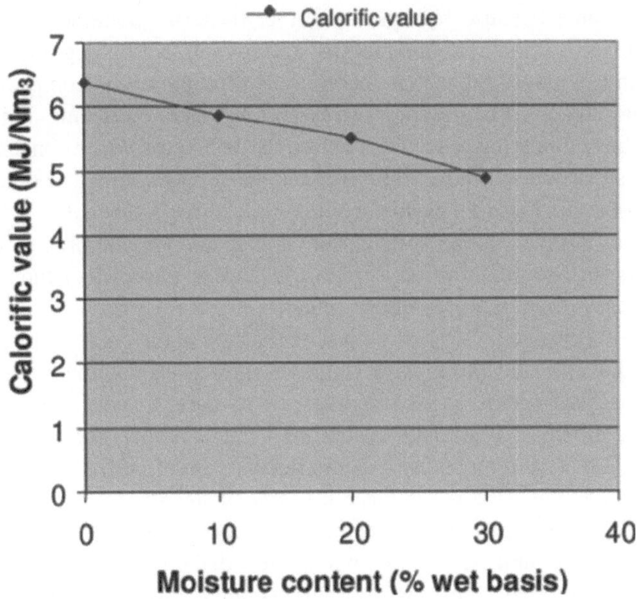

FIGURE 9.6 Calorific value against moisture content at 800°C for cotton stalks biomass (Koroneos & Lykidou, 2011).

As predicted, the calorific values fall as moisture content rises. Due to a drop in CO content and an increase in CO_2, the calorific value of the generated gas falls as moisture content rises (Koroneos & Lykidou, 2011).

9.6 AGRO-WASTE MANAGEMENT

Waste cotton management refers to managing and disposing of cotton waste generated during various stages of cotton production and consumption. This can include waste generated during the spinning, weaving, cutting, and sewing of cotton products and waste generated from the disposal of cotton products at the end of their useful life (Lu et al., 2023). The leftover cotton fiber from the cotton industry is where the waste cotton is derived from. Effective waste cotton management involves reducing waste generated, reusing or recycling waste where possible, and safely disposing of any remaining waste (Shen et al., 2004). There are several ways to manage the cotton waste from the factory in the existing industry, such as the textile industry and spinning mills (Ute et al., 2019). The most common ways to manage cotton waste are compost, pulp and paper, biocomposite materials, bio-coal, biochar, bio-bricks, biohydrogen, biomethane, bioethanol, biobutanol, and natural acids.

9.6.1 Compost

Composting is the ideal method of handling a variety of organic wastes because it yields stable organic matter that may be used to enrich the soil and create the substrate for plant growth (Tsatiris et al., 1998). Compost is being used more and more in organic and intensive farming. It is preferred to apply stabilized mature composts to agricultural soils to raise soil organic matter and nutrients (Sayara et al., 2020). Composting agricultural waste, particularly cotton waste, is an efficient and eco-friendly approach to handling organic waste and turning it into compost that is rich in nutrients.

Cotton is biodegradable because bacteria and other bacteria spontaneously break it down. It degrades relatively quickly, taking about 5 months, and organic cotton degrades even more rapidly (TOTP, 2021). Cotton is a natural fiber that decomposes throughout the composting process and is biodegradable (Silvia, 2021). To begin composting cotton waste, collecting and sorting the waste are the first steps in the composting process for cotton waste. Cotton waste consists of stalks, leaves, and other organic waste products produced throughout the supply chain. The cotton waste gathered is ensured to be free of impurities like plastic or synthetic materials. Then, the cotton waste is shredded into smaller pieces to make composting easier. The decomposition process can be sped up by increasing the waste's surface area through shredding. Smaller particles disintegrate more quickly, producing compost that is finer and of a higher caliber.

The proper carbon-to-nitrogen (C/N) ratio must be reached in the compost pile for composting to be successful (Saad et al., 2013). Cotton waste is a high-carbon (brown) material. Add nitrogen-rich (green) items, such as food scraps, grass clippings, or manure, to your compost pile to create a well-balanced mix. Typically, 25–30 parts carbon to 1 part nitrogen constitutes the optimal C/N ratio (Singh et al., 2022). The composting process depends on effective moisture management. It is crucial to keep the compost pile moist but not soggy. Water the pile as necessary to keep the moisture level constant, especially in dry weather. An excessively wet compost pile might result in anaerobic conditions and unpleasant odors, while a too-dry pile will slow decomposition (Singh et al., 2022). In addition, frequently turning the compost pile is essential for giving the microorganisms that cause decomposition access to oxygen. Cotton waste and other organic materials eventually decompose into compost, turning black, crumbly, and earthy smelling. The compost pile is prepared for curing after completely decomposing and resembles mature compost. The compost must sit for a few weeks to many months to cure. The compost stabilizes and matures throughout this time, producing a high-quality final product. Figure 9.7 shows the result for cotton waste mixed with composting materials (Singh et al., 2022).

Composting cotton waste is a sustainable and beneficial method for handling agricultural waste because it has several advantages and a variety of uses. Cotton waste composting helps to reduce the amount of agricultural waste produced during cotton production and processing (Silvia, 2021).

(a) (b)

FIGURE 9.7 (a) Blended cotton waste ready to be mixed, and (b) mixed cotton waste with composting materials (Singh et al., 2022).

As a result, less cotton waste ends up in landfills, which is better for the environment. Next, cotton waste is composted to provide a nutrient-rich soil supplement. It provides vital minerals for plant growth, including nitrogen (N), phosphorus (P), potassium (K), and micronutrients (Tsatiris et al., 1998). Compost is a slow-release nutrient supply when mixed into the soil, enabling healthy plant growth and greater agricultural harvests. Then, by promoting soil aggregation and aeration, compost strengthens soil structure (Tsatiris et al., 1998). It aids in releasing compacted soils, making it easier for roots to delve into them and improving water penetration and drainage. The oxygen and nutrients available to plant roots are also increased by improved soil structure. Most importantly, cotton waste composting reduces greenhouse gas emissions. Under aerobic composting conditions, organic matter decomposes without releasing methane (CH_4), a powerful greenhouse gas. Instead, carbon dioxide (CO_2) is produced (Kammann et al., 2012). Composting cotton waste provides several benefits and uses. It supports a circular economy, lowers greenhouse gas emissions, enhances soil health, and decreases waste. Composting cotton waste can support sustainable agriculture and lessen the harmful effects of the fashion industry on the environment.

9.6.2 Pulp and Paper

The manufacturing of paper and paper-based products from various raw materials, principally wood, but also recycled fibers like cotton waste and other sources of cellulose constitutes the large pulp and paper production industry. Cotton is a natural seed fiber from plants widely used to make yarn in the textile industry (Rahman & Uddin, 2022). The pulp and paper industries use cotton linters as an alternative supply of cellulosic raw materials since they are a valuable by-product of cotton processing (Sczostak, 2009). Converting cotton waste into pulp is known as cotton waste pulping. Meanwhile, paper made from cotton waste pulp is known as cotton paper.

Cotton waste pulping can start with the collection and sorting of cotton waste. Collect cotton waste materials, ensuring they contain nothing harmful like buttons, zippers, or synthetic fibers. Separate the cotton waste from other non-cotton items by sorting it. Then, to expand its surface area and make the pulping process easier, the cotton waste is shredded into smaller pieces. After that, the cotton is defibered chemically to separate the cotton fibers from other materials like dyes and finishing chemicals. The most common chemical ways to convert cotton waste into pulp are Kraft and Sulfite pulpings (Mboowa, 2021). In the Kraft pulping process, the waste cotton fibers are shredded and fried at a high temperature of 160–170°C in a solution of sodium hydroxide (NaOH) and sodium sulfide (Na_2S) (Gellerstedt, 2008). The lignin that holds the fibers together is broken down during this process, and the cellulose fibers are separated from the trash. As a result, pulp of exceptional quality is created. Meanwhile, sulfite pulping is a type of chemical pulping where the cooking liquid is a solution of sulfuric acid and bisulfite ions (Gellerstedt, 2008). This process can create less complicated pulps to bleach than Kraft pulps by using the acid solution to break down the lignin linkages between the wood fibers.

The process of creating paper starts once the cotton waste is transformed into pulp. After turning cotton waste into pulp, the pulp is washed and bleached to improve the fibers' brightness and whiteness, as illustrated in Figure 9.8. A slurry of water and cotton pulp is used to make paper. This slurry is evenly distributed onto a conveyor made of moving wire mesh, where the water drains away and is replaced by a thin layer of fibers (Bajpai, 2015). Next, large rollers squeeze the wet paper sheet to squeeze out more water, making the paper drier and more consistent (Bajpai, 2015). The remaining moisture is then dried out of the pressed pulp. A drying machine can be used to do this. Once the pulp is dry, it can be used to create paper for packing, filter, and magazine paper.

This recycling method encourages sustainable practices in the paper and textile industries while reducing waste and conserving natural resources. An efficient and environmentally responsible technique to utilize waste materials and promote a more circular economy is to convert cotton waste into paper. The process necessitates specialized facilities and equipment, and adherence to environmental standards provides responsible waste management and environmentally friendly paper

FIGURE 9.8 Example of white kraft paper produced (Jones, 2020).

production methods. Making paper and pulp from cotton waste is a promising first step toward a brighter future as more sectors adopt eco-consciousness.

9.6.3 Biocomposite Materials

A biocomposite material is a composite of two or more different constituent materials, at least one of which is naturally produced, that is combined to create a new material with better performance than the sum of its parts. Biocomposite materials can be made from cotton waste (Shi et al., 2023). Natural fibers, such as cotton waste, are combined with a biopolymer matrix, often produced from renewable sources, to create biocomposites, a class of materials (Shi et al., 2023). Cotton is a biodegradable natural fiber made from plants. When cotton waste is used in biocomposites as reinforcement, it increases the material's biodegradability, making it more environmentally friendly. Furthermore, cotton fibers work well with various biodegradable polymers, including those based on corn starch (Reza et al., 2023). This compatibility allows for adequate bonding between the cotton fibers and the polymer matrix, resulting in improved mechanical properties of the biocomposite (Reza et al., 2023). Figure 9.9 shows the biocomposite product from cotton waste.

Converting cotton waste into biocomposite materials involves a series of steps, including fiber extraction, preparation of the biopolymer matrix, mixing the fibers with the biopolymer, shaping the composite, and curing the material (Anandjiwala & Blouw, 2007). The cotton waste must first be clean and separate from any impurities and contaminants. The choice of an appropriate biopolymer matrix depends on the particular needs of the biocomposite and the uses for which it is intended. Polylactic acid (PLA), polyhydroxyalkanoates (PHAs), and starch-based polymers, among others, are often used as biopolymers in the manufacture of biocomposite materials (Elsawy et al., 2017). The biopolymer is subsequently made into pellets or a solution that may be mixed with cotton fibers.

Next, the cotton fibers are combined with the biopolymer matrix for compounding. Compounding can be done in many ways, such as melt mixing, extrusion, or solution casting. A homogeneous mixture is produced by the mixing procedure, which guarantees that the fibers are evenly distributed throughout the biopolymer matrix (Shaghaleh et al., 2018). Compression molding, injection molding, or extrusion are molding processes that can be used to shape the mixed biocomposite material into the desired shape (Mohanty et al., 2004). A typical technique is compression molding, where

FIGURE 9.9 Biocomposite materials from cotton waste in powder (Fernandes et al., 2021).

the mixture is put into a mold, and then heated and compressed to form it into a specific pattern or shape. Some formulations may need to cure or solidify after the biocomposite has been shaped to enhance the material's mechanical stability (Mohanty et al., 2004). Heat treatment, chemical cross-linking, or other curing procedures particular to the chosen biopolymer can all be used to achieve curing. The biocomposite material is tested and characterized for its mechanical, thermal, and biodegradation properties to meet the desired specifications. The biocomposite produced here can be applied to many sectors, such as packaging, building, automotive, and agriculture.

There are many advantages to converting cotton waste into biocomposite materials. Biocomposites from cotton waste promote the efficient use of renewable and abundant resources (Kocak et al., 2020). By repurposing cotton waste that would otherwise be discarded, the process reduces the need for new raw materials and minimizes the environmental impact associated with waste disposal. Besides that, it is also sustainable due to reduced carbon footprint (Kocak et al., 2020). Biocomposites from cotton waste have a lower carbon footprint than conventional materials from fossil fuels. Compared to manufacturing plastics derived from petroleum, biopolymers, like PLA or PHA, often produce less greenhouse gas emissions (Kocak et al., 2020). This is because the production of biopolymers involves the use of renewable resources and reduces the dependence on fossil fuels. Overall, turning cotton waste into biocomposite materials offers a variety of economic and environmental benefits, assisting in the shift to more environmentally friendly and durable material options. Utilizing biocomposites' potential allows us to promote resource efficiency and ethical waste management while lowering the environmental effect of numerous businesses.

9.6.4 Bio-Coal

Bio-coal is a solid biofuel produced through a thermal conversion process called torrefaction (Carlson & Ebben, 2016). It is from biomass materials like wood scraps, food waste, energy crops, and other organic elements. It is created from biomass using a pyrolysis and carbonization process under temperature and residence time controls (Cheng et al., 2020). The torrefaction process involves heating the biomass without oxygen at temperatures between 200°C and 300°C (Saadon et al., 2014).

A by-product of the processing of textile products, cotton waste, has the potential to be used as a feedstock in the production of bio-coal. The majority of fossil fuels, including coal, oil, and natural gas, have been rapidly used by the world's population (Wang et al., 2018). Energy depletion and global warming are the two main crises brought on by the widespread usage of fossil fuels. As a result, the 21st century has made the development of renewable energy sources and reducing carbon dioxide emissions a top goal (Waheed et al., 2022). Therefore, cotton waste is an excellent alternative to fossil fuels as raw materials for bio-coal production. This involves several stages of converting cotton waste into bio-coal.

Torrefaction is a method of thermal conversion used to turn cotton waste into bio-coal. Torrefaction, a controlled heating process, turns biomass, such as cotton waste, into a solid biofuel with characteristics similar to coal by removing moisture and volatile chemicals (Chen et al., 2021). The process begins with drying the cotton waste to reduce its moisture content. Next, torrefaction is used to treat wasted cotton that has been dried. Under anaerobic conditions, the procedure is usually conducted in a reactor between 200°C and 300°C (Negi et al., 2020). The biomass cannot burn or combust due to the reactor. Instead, it proceeds through a series of chemical processes that lead to the release of volatile chemicals and the creation of solid bio-coal. The bio-coal is cooled to an appropriate temperature following torrefaction before being handled or stored. The bio-coal's quality can be preserved with quick cooling. Figure 9.10 shows the bio-coal product from biomass after the torrefaction process.

The torrefaction process produces bio-coal, a solid carbon-rich material similar to coal. To suit the criteria, the bio-coal can undergo additional processing and refinement. The generated bio-coal can be used as a renewable and carbon-neutral fuel source (Sundqvist Ökvist & Lundgren, 2021). It can replace fossil coal in many industrial operations, including producing electricity and heating systems. For easier handling and storage, the bio-coal can be further processed into the necessary forms, such as pellets or briquettes. These forms can be employed in various applications and are simpler to transport.

The conversion of cotton waste into bio-coal through torrefaction offers numerous environmental and economic benefits, contributing to a more sustainable and circular economy (Sundqvist Ökvist & Lundgren, 2021). Creating bio-coal from cotton waste provides a sustainable way to utilize trash

FIGURE 9.10 Pelleted bio-coal from the torrefaction process (Crnogaća, 2017).

efficiently. Cotton waste is a readily available biomass resource made from a variety of sources, such as used garments, textiles, and other items made of cotton. The procedure lessens the environmental impact of trash disposal and supports a circular economy philosophy by converting this garbage into a valuable energy source (Carlson & Ebben, 2016). In addition, bio-coal produces less pollution and greenhouse gas emissions when burned than conventional coal. Chemicals that cause the release of hazardous gases during combustion, such as hemicellulose and certain lignin, are eliminated during the torrefaction process. As a result, bio-coal helps to reduce emissions of carbon dioxide (CO_2), sulfur dioxide (SO_2), nitrogen oxides (NO_x), and particulate matter, improving air quality and having a less negative impact on the environment (Carlson & Ebben, 2016).

Hence, cotton waste bio-coal offers significant social, economic, and environmental benefits. The use of cotton waste bio-coal helps to create a more sustainable and responsible energy future in a variety of ways, including lower emissions, sustainable resource utilization, and energy diversification. To combat climate change, preserve natural resources, and promote a cleaner and more resilient energy landscape for future generations, it is essential to embrace this renewable and environmentally friendly energy source.

9.6.5 Biochar

Biochar is a type of charcoal produced from biomass through a process called pyrolysis (Spears, 2018). The organic material is transformed into biochar during pyrolysis, a stable form of carbon that is difficult to escape into the atmosphere. Pyrolysis produces heat or energy that can be caught and used to provide clean energy. Regarding stabilizing carbon, biochar is far more effective and cleaner than other types of charcoal (Spears, 2018). The pyrolysis technique can be used to create biochar from cotton waste. Pyrolysis is heating organic materials, like cotton waste, without oxygen to produce biochar and other by-products like gases and bio-oil (Kar et al., 2022), as shown in Figure 9.11.

Pyrolysis starts with the dried cotton waste being loaded into a pyrolysis reactor. To prevent total combustion, the reactor is made to run in anaerobic oxygen-free conditions (Kan et al., 2016). The reactor can be in the shape of a screw reactor, rotary kiln, fixed-bed reactor, or fluidized bed reactor. The desired temperature range is reached in the pyrolysis reactor, which is normally between 300°C and 800°C (572°F and 1,472°F). The temperature is carefully controlled and maintained throughout the operation to obtain the intended product output and quality. The finished biochar product can be added to soil to boost nutrient retention, the condition of the soil, and fertility. Additionally, it can be applied in a variety of industrial settings or used as a renewable energy source.

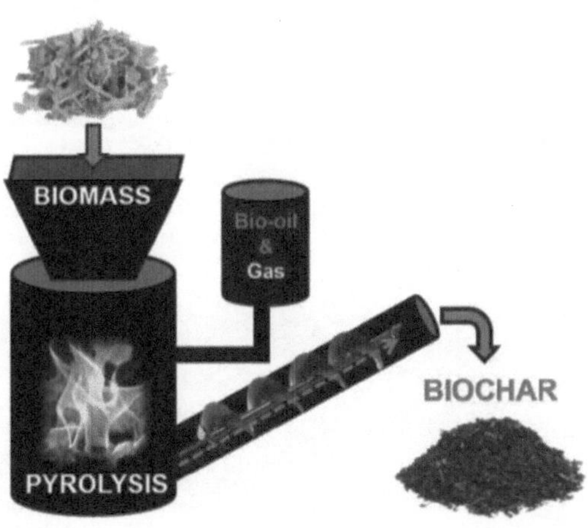

FIGURE 9.11 Biochar production from the pyrolysis process (Lévesque et al., 2021).

The benefits of using cotton waste-derived biochar make it a valuable resource with a wide range of uses. Carbon is efficiently sequestered in the soil by employing biochar made from cotton waste as a soil additive (Çay et al., 2020). Biochar's stable carbon content helps to store carbon in a way that resists decomposition, lowering carbon dioxide (CO_2) emissions and lessening the effects of greenhouse gases. In addition, biochar production offers a responsible and sustainable way to handle cotton waste (Çay et al., 2020). Cotton waste is transformed into biochar, a valuable and practical product that lowers the quantity of trash transported to landfills or incineration plants. Less combustion occurs; thus, less carbon gas is emitted into the atmosphere (Mowrer et al., 2022).

Overall, the pyrolysis process for biochar production involves heating cotton waste at high temperatures in the absence of oxygen, which breaks down the organic components of the waste and converts them into biochar. The process can be optimized by controlling the temperature, heating ramp, and cooling rate and pre-processing the biomass. The produced biochar can be used for soil amendment, environmental remediation, and bioenergy production.

9.6.6 Bio-Bricks

Bio-bricks, also known as biomass briquettes, are solid biofuels made from compressed organic materials, as shown in Figure 9.12. Agro-waste, like cotton, has potential applications as an alternative to produce bio-bricks (Rautray et al., 2019). These bio-bricks can be adjusted to fit the local market and building construction types based on the availability of cotton waste resources. Using cotton waste as a construction material can reduce the use of natural resources and energy consumption (Rautray et al., 2019).

The process of converting cotton waste into bio-bricks begins with collecting discarded cotton materials, including clothing, textiles, and other cotton-based products. Then, the waste is sorted to ensure it is free from contaminants. Combine the cotton waste with a binding substance to produce a homogeneous mixture. Clay, cement, or lime are often used as binding agents. The bricks' strength comes from the binding chemical, which also aids in keeping the cotton waste particles together (Teklehaimanot et al., 2021). The pre-processed cotton waste and the chosen binder should be combined in a container for mixing. The ratio of cotton waste to binder will vary based on the specific binder's characteristics and the desired strength and properties of the final bio-bricks (Johnson et al., 2020). Usually, a 1:3 ratio by weight is applied when using chopped husk and lime slurry as binding agents. The ingredients are mixed thoroughly until evenly blended (Çay et al., 2020). After that, the molds are filled with the blended cotton waste and binder in the desired shapes and sizes. Depending on how the bio-bricks will be used, the molds may be rectangular, square, or specially created. The mixture is crushed during molding to guarantee a uniform density and get rid of air voids.

FIGURE 9.12 Bio-bricks used in industry (Rautray et al., 2021).

Next, dry the molded bricks to remove excess moisture. This can be done by air-drying the bricks in a well-ventilated area or using a drying chamber. Proper drying is important to prevent cracking or deformation of the bricks. Conduct quality inspections to make sure the bio-bricks fulfill the required criteria for strength, appearance, and performance once they have dried completely (Poornima et al., 2021). The bio-bricks are put through quality control to make sure they are appropriate for their intended use in building applications. In sustainable building designs, bio-bricks can be used as load-bearing walls, non-load-bearing walls, partitions, and visual features (Poornima et al., 2021).

The unique properties of bio-bricks make them suitable for various applications, contributing to sustainable building practices. Bio-bricks have a low-carbon footprint since they are created from organic materials like agro-waste (Rautray et al., 2021). They can be utilized as sustainable building materials to lessen the negative effects of construction on the environment. Furthermore, construction projects can use bio-bricks as a cost-effective solution. Since they can be made from materials that are close by, transportation expenses and carbon emissions are minimized (Rautray et al., 2021). Bio-bricks can also be used as decorative elements. Bio-bricks can be employed as decorative elements in architectural projects because of their aesthetic appeal and natural texture, bringing a unique touch to building facades or interior spaces (Rautray et al., 2021).

There are many benefits to turning cotton waste into bio-bricks, including waste reduction, a smaller carbon impact, and the encouragement of a circular economy. This method lessens reliance on non-renewable resources, promotes responsible waste management, and lessens the environmental effects of conventional brick manufacture by using leftover cotton components. Additionally, using bio-bricks encourages the use of sustainable building techniques, making them a desirable option for architects, builders, and anyone who cares about the environment.

9.6.7 Biohydrogen

Biohydrogen is defined as hydrogen gas (H_2) created by living things or specific microbes that go through biological processes. It is regarded as a type of renewable energy since it may be created from materials that are renewable (Bhatia et al., 2021). The splitting of water molecules by biological processes, the conversion of biomass, and solar thermal splitting of water are three sustainable sources of biohydrogen, which is now regarded as the best biofuel among all others because it can provide all energy demands.

Utilizing agricultural waste or leftovers as a feedstock, biohydrogen production from agro-waste uses biological processes to produce hydrogen gas. Agro-waste is a term that refers to a variety of organic waste products from agriculture, including cotton waste, straw, husks, shells, and other by-products. Cotton could be seen as the potential to produce hydrogen (Sołowski et al., 2020b). Cotton is a naturally occurring polymer primarily made of cellulose (Binczarski et al., 2022). A method known as dark fermentation can turn cotton waste into biohydrogen. With no light or oxygen present, microbes degrade organic material in this process, generating hydrogen gas as a by-product (Sołowski et al., 2020a). This method involves several steps in converting cotton waste into biohydrogen.

First, textile manufacturers can collect cotton waste, such as cotton yarn fibers, and prepare it for hydrolysis. Then, acid is used to hydrolyze cotton waste, converting cellulose and hemicellulose into simple sugars. It has been demonstrated that the cotton waste hydrolysates provide a reliable supply of glucose for fermentation. The hydrolysates made from cotton waste are used as a substrate for dark fermentation. During fermentation, microbes convert glucose into organic acids like acetic acid, butyric acid, and hydrogen gas (Sotelo-Navarro & Poggi-Varaldo, 2021). The process can be more efficient by altering the C/N ratio, initial total solids, and temperature (Sotelo-Navarro & Poggi-Varaldo, 2021).

The fermentation process produces carbon dioxide (CO_2), methane (CH_4), and other gases in addition to hydrogen. Biogas is the name given to the mixture of gases that is produced. Despite being a greenhouse gas, methane produced during dark fermentation can be useful if it is used up instead of being released into the atmosphere (Bhatia et al., 2021). Depending on the particular circumstances, the selected microbes, and the makeup of the cotton waste, the dark fermentation process might last for a variety of lengths of time. The process normally takes a few days to finish. Several factors, including temperature, pH level, substrate concentration, and the presence of additional co-substrates, affect the fermentation's success (Mao et al., 2015).

After the fermentation process finishes, the gas produced is collected and separated to gain hydrogen from the gas mixture. Hydrogen can be separated and purified from biogas using a number of methods, including membrane separation, pressure swing adsorption (PSA), and water scrubbing. High-purity hydrogen gas may be isolated using these techniques, making it appropriate for use in fuel cells, hydrogen-powered cars, and combustion-based electricity generation, among other things.

Using biohydrogen as an energy source offers numerous advantages, making it an attractive and promising option for sustainable energy production. An agricultural residue that may otherwise be wasted is repurposed through the generation of biohydrogen from cotton waste. By using cotton waste as a feedstock for biohydrogen, we support environmentally friendly waste management techniques and lessen the load that trash disposal has on the environment. In addition, biohydrogen produced from cotton waste is a green energy source. A practically carbon-neutral energy cycle is created when the carbon dioxide (CO_2) emitted during the burning of biohydrogen is equal to the CO_2 absorbed by cotton plants throughout their growth (Cao et al., 2022). By doing so, greenhouse gas emissions are reduced, and climate change is fought. Other than that, using cotton waste for biohydrogen production supports the circular economy concept. By converting waste materials back into valuable energy resources, we promote a more sustainable and resource-efficient system.

Overall, producing energy using biohydrogen from cotton waste is sustainable and environmentally responsible. By using this agricultural waste as a source of clean energy, we cut waste and greenhouse gas emissions, contributing to a future where energy is more dependable and sustainable. However, there are some challenges associated with biohydrogen production. For instance, it could be expensive to solve technological and engineering issues. Additionally, numerous established industrial techniques like steam reforming of natural gas compete with biohydrogen, at least for commercial purposes.

9.6.8 Biomethane

Biomethane, also known as renewable natural gas, is a promising and environmentally friendly biofuel that is gaining attention as a sustainable alternative to conventional natural gas. Biogas is transformed into biomethane through purification, giving it characteristics similar to those of natural gas.

Anaerobic digestion is a technique that can turn cotton waste into biomethane. Anaerobic bacteria are used to break down the organic content in cotton waste in the absence of oxygen to produce biomethane (Twizerimana et al., 2021). Biogas is created as a result of this process, largely made up of methane (CH_4) and carbon dioxide (CO_2) with traces of other gases. Raw biogas obtained from anaerobic digestion contains contaminants, including CO_2 and hydrogen sulfide (H_2S), making it less clean than finished biomethane (Twizerimana et al., 2021). The biogas goes through a biogas upgrading procedure to make it appropriate for usage as biomethane. Several methods can be used to remove contaminants and raise the methane concentration, including membrane separation, PSA, and water washing. The upgraded biogas is then separated from the purified biomethane. Now that it is methane-rich and of excellent quality, the biomethane can be used as a clean, sustainable energy source.

The use of biomethane as a renewable energy source has a number of significant advantages, making it a desirable and long-term choice for the generation of clean energy. Biomethane can be used to replace natural gas, a significant part of oil, and coal, as it is identical to the methane or natural gas produced by fossil fuels (Brynolf et al., 2014). Carbon dioxide and a few other greenhouse gases are released when biomethane is burned for energy production, heating, or fuel, but the amount of carbon dioxide in the atmosphere is not increased. This is because if organic matter, which serves as the raw material for the production of biomethane, were simply allowed to spontaneously degrade, the same quantity of carbon dioxide would be released into the sky (Brynolf et al., 2014).

The energy source biomethane is regarded as carbon-neutral. A carbon footprint that is practically neutral is created when the carbon dioxide released when burning or using biomethane is balanced by the CO_2 absorbed by the biomass during its growth. Because fossil fuels release more CO_2 from deep carbon reservoirs into the atmosphere, biomethane is a cleaner substitute for them due to this property. Then, changing the raw material for the production of biomethane into cotton waste could promote the reduction of greenhouse gas emissions (Bagi et al., 2017). Capturing and utilizing methane as biomethane has a major positive influence on the environment because it is a potent greenhouse gas with a far larger potential for global warming than carbon dioxide.

Energy storage and grid balancing may benefit greatly from using biomethane. The power system can be stabilized using biomethane as a renewable energy source since it can be stored and used during high energy demand or limited supply of other renewable energy sources, such as solar and wind (Bagi et al., 2017). In general, producing biomethane from cotton waste is an environmentally responsible and sustainable method of producing energy. Cotton waste-derived biomethane has the potential to significantly contribute to the worldwide transition to a low-carbon economy as biomethane technology continues to progress.

9.6.9 Bioethanol

In this era of increasing concerns about climate change, energy security, and environmental sustainability, bioethanol is seen as one of the alternatives that can be used as a replacement for fuel. Suppose we continue to use non-renewable energy such as petrol and diesel. In that case, this action will lead to the problem of running out of energy and have a negative impact on the environment. Bioethanol is produced from plant-based feedstocks through a process called bioalcohol fermentation. A renewable, non-toxic, biodegradable resource like bioethanol helps fight climate change, lower greenhouse gas emissions, lessen reliance on crude oil, clean up the air, and add jobs to the agricultural and ethanol production sectors (Stein, 2023). Bioethanol can be produced from cotton waste. Several studies have revealed that various pre-treatment procedures may efficiently convert cotton waste biomass into bioethanol (Ranjithkumar et al., 2017). Cotton clothing and discarded cotton products have also been noted as potential feedstocks for the manufacturing of bioethanol (Nikolić et al., 2017).

Cotton waste can be converted into bioethanol through a process called bioethanol production or bio alcohol fermentation. There are several steps involved in turning cotton waste into bioethanol, including pre-treatment, hydrolysis, fermentation, and distillation, as shown in Figure 9.13. Pre-treatment is required to convert the complex structure of cellulose and hemicellulose in cotton waste into simple sugars that can be quickly fermented into ethanol (Mahalakshmi et al., 2011). Several techniques, including acid hydrolysis, steam explosion, and enzymatic hydrolysis, can be used for pre-treatment (Ranjithkumar et al., 2017). After pre-treatment, the cellulose and hemicellulose are hydrolyzed using cellulase and other enzymes to become simple sugars (Saini et al., 2014). The hydrolyzed sugars are fermented by yeast or bacteria to create ethanol. Finally, distillation is used to extract the ethanol from the fermentation broth.

As we know, bioethanol offers several benefits in the field of environmental conservation. Bioethanol is a renewable biofuel made from plant-based feedstocks. Burning bioethanol produces carbon dioxide (CO_2), although it is balanced by the CO_2 absorbed by plants during growth, unlike burning conventional gasoline (gasoline). As a result, a closed carbon cycle is created, making bioethanol almost carbon-neutral. We can dramatically lower greenhouse gas

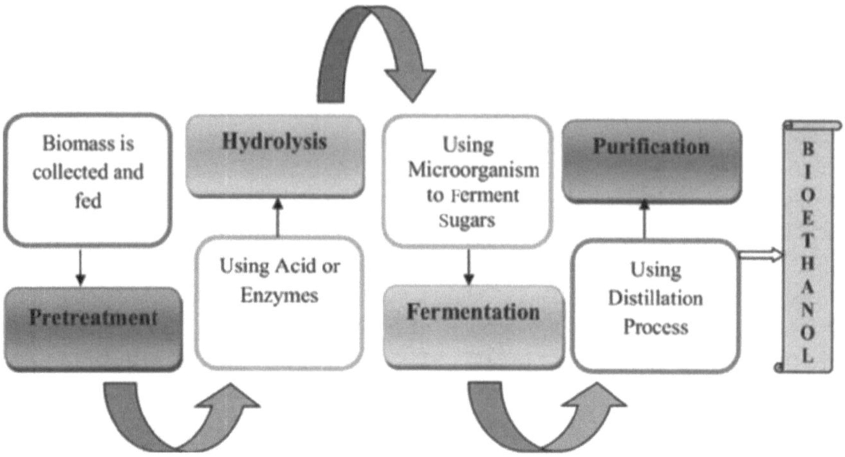

FIGURE 9.13 Process involved in bioethanol production (Devi et al., 2022).

emissions by using bioethanol, which will lessen the effects of climate change and its effects on the ecosystem (Stein, 2023).

The utilization of bioethanol as fuel also can improve air quality. Compared to petrol, bioethanol combustion emits less hazardous pollutants like nitrogen oxides (NO_x) and particulates (Lu, 2011). These contaminants play a role in smog formation, air pollution, and respiratory issues. We can enhance air quality and lessen the harmful effects of transportation on the environment and human health by employing bioethanol in cars and other uses. Furthermore, using sustainable plant-based resources to produce bioethanol requires less dependence on non-renewable fossil fuels like crude oil (Bagi et al., 2017). Using fewer fossil fuels helps maintain delicate ecosystems, reduce environmental damage from oil production, and preserve limited natural resources.

In conclusion, the economic viability of producing bioethanol from cotton waste depends on several variables, including the accessibility of cotton waste as a feedstock, the effectiveness of the conversion process, and the market demand for bioethanol. However, bioethanol synthesis from cotton waste represents a possible route for waste valorization and renewable energy generation as the globe progresses toward cleaner and more sustainable energy sources.

9.6.10 Biobutanol

A four-carbon alcohol called biobutanol is created when biomass is fermented. A type of renewable biofuel member of the alcohol family is biobutanol, sometimes referred to as butanol. It is created when specific microbes ferment organic materials or biomass. Its characteristics are similar to those of petrol. Even some gasoline-powered vehicles that haven't been upgraded can use biobutanol. Biobutanol represents a realistic and appropriate substitute for gasoline and gasoline–ethanol blend fuels since it can lower carbon emissions by 85% compared to gasoline.

The most common method to produce biobutanol from feedstock, such as cotton waste, is fermentation. The process is similar to bioethanol production but uses different substrates. Cellulosic materials made from cotton waste are a plentiful source of biomass worldwide. These facts encouraged an in-depth study to develop a process for effectively converting lignocellulosic materials into sugar monomers for subsequent fermentation to ethanol (Kabir et al., 2020). Other biomasses such as starches, lignocelluloses, and citrus waste need to be pre-treated to make sugars available for the following fermentation stage, in contrast to simple sugars, which can be turned to ethanol directly (Nikolić et al., 2017). Fermentable sugars and other substances are released during pre-treatment and hydrolysis. The most efficient way to release simple sugars from cellulosic materials to make compounds like ethanol, glucose, and furfural is through hydrolysis utilizing the right enzymes/substrate or chemically using acids (Mahalakshmi et al., 2011).

There are various benefits to using renewable biofuels like biobutanol rather than traditional fossil fuels. One of them is that biobutanol has a higher energy density than other biofuels like bioethanol (Zafar, 2022). It is a more energy-efficient fuel with increased mileage and engine performance due to its higher energy content. Thus, biobutanol has a much lower loss of fuel economy (Christine & Gable, 2019).

The compatibility of biobutanol with existing petrol engines and infrastructure is one of its key benefits. Without requiring significant changes to cars or fuelling facilities, it can be used as a straight substitute for petrol or combined at different ratios (Christine & Gable, 2019). This trait makes switching to cleaner fuels easier without substantial expenditures on new infrastructure. In addition, biobutanol also can be flexible in fuel blends (Pennington, 2013). At different concentrations, biobutanol is easily blended with petrol or gasoline. The switch to renewable fuels can be facilitated by this flexibility, which allows for various fuel mixtures. Switching to greener fuels can be made smoothly by substituting biobutanol for petrol or blending it with higher proportions (Pennington, 2013). This feature facilitates a more rapid transition to renewable biofuels by greatly simplifying the adoption process.

In conclusion, it is vital to keep in mind that commercial-scale synthesis of biobutanol from cotton waste is currently in the research and development stage and confronts several difficulties. These difficulties include maximizing fermentation productivity, ensuring the procedure is economical, and out-competing other biofuels like bioethanol and biodiesel. However, as biotechnology and biofuel research continue to improve, producing biobutanol from cotton waste and other biomass sources may become economically feasible, offering a potential and sustainable alternative to traditional fossil fuels.

9.6.11 Natural Acids

Natural acids, also known as organic acids, are a wide family of chemical compounds found naturally in various living things, including plants, animals, and microorganisms. In general, they are weak acids that do not fully dissociate in water. Several classes of natural acids, such as carboxylic, phenolic, and amino acids, can be divided depending on their chemical composition. Table 9.4 provides the common type of organic acid and source from agro-waste (Yusuf, 2017).

Cotton waste is capable of being fermented to create a variety of natural acids, primarily acetic acid and lactic acid. These typical acids, acetic and lactic, are produced during fermentation, which uses microbes to break down the organic stuff in the cotton waste (Clark, 2013). Certain bacteria are selected for the manufacture of lactic acid and acetic acid because they have the capacity to effectively ferment the sugars in cotton waste into these organic acids. Acetogenic bacteria create acetic acid, whereas lactic acid is produced by lactic acid bacteria. Once the feedstock is prepared, it is subjected to fermentation. In separate fermenters, the cotton waste is inoculated with lactic acid bacteria for lactic acid production and acetogenic bacteria for acetic acid production (Sharma et al., 2020).

Next for the fermentation process, through a series of enzymatic processes, lactic acid bacteria, including *Lactobacillus* and *Streptococcus* species, metabolize the sugars in the cotton waste. Along with water and carbon dioxide, glucose and other carbohydrates are transformed into lactic acid throughout this process (Sołowski et al., 2020b). Typically, the fermentation process takes place in an anaerobic (oxygen-free) setting. The sugars in cotton waste are metabolized by acetogenic bacteria, including *Acetobacter* and *Gluconobacter* species, to produce acetic acid (Sołowski et al., 2020a). The sugars are initially transformed into acetic acid and carbon dioxide. Anaerobic conditions are also necessary for the fermentation process. After the fermentation duration, the broth should be separated from the biomass and bacteria. Separation methods like

TABLE 9.4
Production of Various Organic Acids from Agro-Industrial Wastes by Solid-State Fermentation Process (Yusuf, 2017)

Agro-Industrial	Microbial Isolate Wastes	Products Obtained
Kitchen wastes	Self-inoculated media/isolates	Short-chain organic acids (lactic, acetic, propionic, and butyric acids)
Outer cover of gallo seeds	*Rhizopus oryzae*	Gallic acid
Tea wastes with sugarcane molasses	*Aspergillus niger*	Gluconic acid
Wheat kernels	*Aspergillus oryzae*	Oxalic acid
Carrot-processing waste	*Rhizopus oryzae*	Lactic acid
Sweet sorghum	*Lactobacillus paracasei*	Lactic acid
Sugarcane bagasse	*Rhizopus oryzae*	Lactic acid
Cassava bagasse and sugarcane bagasse	*Lactobacillus delbrueckii*	Lactic acid
Pineapple wastes	*Aspergillus foetidus*	Citric acid
Apple pomace, potato starch residues, coffee husk, orange peel, corncob, sugarcane bagasse, wheat bran, rice bran, pineapple waste, mixed fruit waste, beet molasses, date syrup, wood hemicellulose, rice hulls, cassava fibrous residue, palm and olive oil residues, etc.	*Arthrobacter paraffinens, Bacillus licheniformis, Corynebacterium* sp., *Aspergillus niger, Aspergillus aculeatus, Aspergillus carbonarius*	Citric acid

filtration or centrifugation are used to distinguish between the solid and liquid phases (Pleissner et al., 2016). Concentrating the liquid component to increase the amount of lactic acid or acetic acid. Higher-purity lactic acid or acetic acid is created by purifying the organic acids to get rid of impurities and undesired by-products. After fermentation, broth should be separated from biomass and bacteria (Bagi et al., 2017).

There are several benefits to using natural acids made from cotton waste in a variety of industries and applications. Waste valorization is one of the main benefits of employing natural acids derived from cotton waste. Cotton waste produced during agricultural or textile production can be transformed into useful compounds like lactic acid and acetic acid. This procedure contributes to a more sustainable approach to waste management by promoting a circular economy and lowering the environmental impact of garbage disposal. Lactic and acetic acids, natural acids from cotton waste, are important in the food and beverage industries (Chakravarty & Gaur, 2017). Lactic acid preserves and enhances the flavor of foods like dairy, bread, and fermented goods. Acetic acid is frequently used in cooking, pickling, and salad dressings in the form of vinegar. Natural acids are incredibly versatile substances with a wide range of uses in numerous industries. In the food, beverage, pharmaceutical, and cosmetic industries, lactic acid is used as a preservative, flavor enhancer, pH regulator, and moisturizing agent (Ruiz-Ruiz et al., 2018). Acetic acid is a vital component of vinegar and is used in cleaning products, chemical synthesis, and food preservation.

An environmentally responsible and sustainable method is fermenting cotton waste into lactic and acetic acids to gain natural ones. It generates essential organic acids with a variety of industrial applications, minimizes waste disposal, and valorizes agricultural waste. This biotechnological procedure has the potential to improve efficiency and reach a wider audience as research and technology evolve, paving the way for a future that is greener and more sustainable.

9.7 POTENTIAL APPLICATIONS

9.7.1 Textile Recycling

Cotton waste is converted into new textile materials through textile recycling, minimizing the negative environmental effects of the textile industry and fostering a circular economy (Leal Filho et al., 2019). Gathering and sorting cotton waste, which might include fabric remnants, used clothing, and other abandoned cotton-based products, is usually the first step in recycling (Piribauer & Bartl, 2019). Recycling and treating cotton waste makes it possible to create new textile materials. Cotton waste fibers can be shredded and spun again to produce yarns and fabrics that can be used to make new apparel, home textiles, and industrial textiles (Halimi et al., 2008).

9.7.2 Nonwoven Products

Cotton scraps can be used to make nonwoven products. Nonwoven products made from cotton have diverse applications and advantages in different industries. These engineered fabrics are not created through traditional weaving or knitting methods but by bonding or interlocking cotton fibers (Luitel et al., 2015). Cotton nonwovens are used in various sectors, such as wipes, medical and hygiene products, filtration systems, insulation materials, and industrial applications (Venkataraman et al., 2023). For instance, cotton nonwoven wipes are absorbent, gentle on the skin, and biodegradable, making them suitable for personal care and cleaning. In the medical field, cotton nonwovens are utilized in surgical drapes, medical gowns, and hygiene products due to their softness, breathability, and hypoallergenic properties (Husain et al., 2019). Cotton nonwovens also serve well in filtration applications, providing effective air and liquid filtration in HVAC systems, water filtration, and automotive filters. Furthermore, they are used in insulation materials for their thermal and acoustic properties. Cotton nonwovens are also employed in industrial cleaning, spill absorption, and as reinforcement materials. The use of cotton nonwovens offers environmental benefits as cotton is a renewable and biodegradable fiber. With their versatility and advantageous qualities, cotton nonwovens are a valuable choice for various applications across industries (Mourad et al., 2016).

9.7.3 Paper Industry

Cotton waste fibers can be used in the production of paper. Cotton fibers are ideal for making high-quality papers, including stationery, currency notes, and specialized papers, since they are strong, lengthy, and absorbent (Rajput et al., 2012). The paper industry makes use of cotton fibers as a valuable resource in producing high-quality papers. Cotton fibers, renowned for their strength and durability, are utilized in specialty papers, stationery, and currency notes (Gopal et al., 2018). The process begins with collecting and preparing cotton waste or fibers, ensuring their purity and suitability for paper production. The fibers are then transformed into a pulp by mixing them with water, refining the mixture to enhance bonding and paper characteristics (Pandey & Shaikh, 1987). Additives may be introduced to achieve desired features. The refined pulp is formed into sheets or rolls through sheet formation, pressing, and drying (Ali et al., 1993). Cotton-based papers offer durability, strength, and a smooth texture. They exhibit excellent printing properties, brightness, and archival qualities, making them suitable for fine stationery, legal documents, and archival records (Halimi et al., 2008). The utilization of cotton fibers by the paper industry diversifies the paper range available and results in exceptional quality papers (Rahman & Uddin, 2022).

9.7.4 Automotive Industry

Cotton is utilized in various ways within the automotive industry to enhance vehicle comfort, functionality, and sustainability. Cotton fibers and cotton-based materials are incorporated into different automotive applications (Mueller & Krobjilowski, 2003). One important use is in the production of automotive interiors, including seat covers, upholstery, and interior linings (Bhat & Parikh, 2010). Cotton's natural softness, breathability, and comfort make it a preferred choice for creating a pleasant and cozy environment inside vehicles. Cotton is also used in the manufacturing of insulation materials for automotive applications, providing sound absorption and thermal insulation to reduce noise and enhance cabin comfort (Islam et al., 2020). Additionally, cotton waste can be repurposed to create composite materials used in automotive components, offering lightweight yet durable solutions. Cotton-based materials are also employed in air and cabin filters, improving air quality and enhancing the performance of HVAC systems (He et al., 2021). The use of cotton in the automotive industry provides various benefits, including its renewable and biodegradable nature, comfort, and aesthetic appeal. It also supports sustainability by reducing reliance on synthetic materials. Overall, cotton's utilization in automotive applications demonstrates its versatility and contribution to creating comfortable, sustainable vehicles (Yetisen et al., 2016).

9.7.5 Construction Materials

Cotton waste fibers can be combined with other materials to make environmentally friendly building supplies. Cotton fibers and cotton-based materials have diverse applications in the construction industry, providing numerous benefits and supporting sustainable building practices (Madurwar et al., 2013). Cotton is found in various construction materials and processes. One common application is insulation, where recycled cotton fibers derived from textile waste or post-consumer cotton products are used (Binici et al., 2012). Cotton insulation offers thermal properties, contributing to building energy efficiency and temperature regulation. It is easy to handle and non-irritating and possesses sound absorption capabilities, making it a viable alternative to conventional insulation materials. Cotton-based materials are also utilized in acoustic panels, effectively absorbing sound waves and improving the acoustic performance of spaces (Binici & Aksogan, 2015). These panels are employed in theaters, studios, offices, and other environments that require noise control. Furthermore, cotton fibers can be incorporated into concrete mixes to enhance their strength and impact resistance (Rajput et al., 2012). Concrete, a composite material made by combining cotton fibers and cement, reinforces structures such as beams, columns, and slabs, enhancing their durability and structural integrity (Sadrolodabaee et al., 2021).

Cotton fabrics and textiles find application in interior finishes, including wall coverings, curtains, and upholstery, adding both visual appeal and comfort to build environments. Flame-retardant finishes can be applied to cotton fabrics for applications where fire safety is a concern (Teklehaimanot et al., 2021). The use of cotton in construction materials provides numerous advantages. It is a renewable and biodegradable resource, promoting sustainable construction practices (Algin & Turgut, 2008). Additionally, cotton offers excellent thermal and acoustic insulation properties, contributing to energy savings and indoor comfort. Moreover, the utilization of cotton-based materials supports recycling efforts and reduces landfill waste, promoting a circular economy (Rajput et al., 2012).

9.7.6 Animal Bedding

Cotton waste can be used to make bedding for animals and pets. Because its absorbent qualities offer comfort and moisture absorption, it is appropriate for animal bedding (Uvarov et al., 2020). Cotton is commonly employed as animal bedding in agricultural and livestock settings, offering numerous advantages such as comfort, cleanliness, and absorbency for the animals (Sharma-Shivappa & Chen, 2008). Cotton bedding is widely used for horses due to its softness, comfort, and reduced likelihood of causing skin irritation or allergies. It effectively absorbs moisture and urine, keeping the bedding dry and minimizing odors. Furthermore, cotton provides cushioning and insulation, creating a cozy resting surface for the animals (Grimes et al., 2007).

9.7.7 Mushroom Cultivation

As a substrate, mushrooms can be grown on cotton waste, such as cottonseed hulls. Cotton waste, including cottonseed hulls and bolls, can be employed as a substrate for cultivating mushrooms (Baysal & Peker, 2001). Mushrooms require a nutrient-rich medium for growth, and cotton waste serves as a viable alternative in mushroom cultivation (Weon et al., 2005). A balanced and nourishing substrate can be created by processing and combining cotton waste with materials like straw, wood chips, or sawdust. This substrate provides the necessary elements, such as carbon, nitrogen, and moisture, for the mushroom mycelium to thrive and develop (Kwon et al., 2007). Various mushroom species, such as oyster mushrooms and shiitake mushrooms, have been successfully cultivated using cotton waste-based substrates. These mushrooms are renowned for their culinary value and are commercially grown on a large scale (Sardar et al., 2022).

The cultivation process involves sterilization or pasteurization of the substrate to eliminate competing organisms. Mushroom spawn is then introduced, allowing the mycelium to colonize and expand throughout the cotton waste (Adebayo et al., 2009). Finally, the mushrooms mature and are harvested under controlled environmental conditions, including temperature, humidity, and light (Rajapakse, n.d.). Utilizing cotton waste in mushroom cultivation brings numerous advantages. It offers a sustainable solution for repurposing cotton waste, reducing environmental impact, and promoting circular economy principles. Moreover, cotton waste substrates can be cost-effective, particularly when sourced locally from textile industries or cotton-growing regions (Kim et al., 2017).

9.7.8 Biofuel Production

Cotton waste is a useful feedstock for the creation of biofuels. Through procedures like fermentation or anaerobic digestion, cotton waste biomass can be transformed into biofuels like bioethanol or biogas (Patel, 2017). Cotton can be utilized as a renewable energy source in the production of biofuels (Venkataraman et al., 2023). Biofuels, derived from organic matter, can be generated using cotton biomass as a feedstock (Hamawand et al., 2016). Cotton waste, including cottonseed hulls and stalks, can be processed to extract the cellulose content. Rich in sugars, this cellulose can be converted into biofuels through various biochemical or thermochemical methods (Akhmedov et al., 2017).

9.7.9 Composting

To make nutrient-rich compost, cotton waste can be composted with other organic materials (Rajput et al., 2012). Cotton waste decomposition in composting systems helps soil enrichment and has uses in agriculture, horticulture, and gardening. Cotton waste can be effectively utilized in composting, making a valuable contribution to the management of organic waste (Tsatiris et al., 2020). When cotton waste, such as fabric scraps or cottonseed hulls, is composted, it naturally decomposes alongside other organic materials. Composting creates an environment that encourages the breakdown of organic matter by microorganisms, resulting in the formation of nutrient-rich compost (Aishwariya & Amsamani, 2012).

Composting cotton waste offers numerous benefits. It helps balance the carbon-to-nitrogen ratio in compost, improves its structure and aeration, and enhances moisture retention and airflow (Islam et al., 2020). As cotton waste decomposes, it releases valuable nutrients and organic matter into the compost, which can then be used to enrich soil and support plant growth in gardens, agricultural fields, and landscaping projects (Hamawand et al., 2016). By composting cotton waste, we promote environmentally friendly waste management. It reduces the amount of waste sent to landfills, thereby minimizing methane emissions and alleviating the strain on landfill capacity (Paredes et al., 1996). Instead, cotton waste can be recycled and returned to the soil as a valuable resource through composting.

9.8 CHALLENGES ASSOCIATED WITH PROCESSING AND COMMERCIALIZING

Processing and commercializing cotton waste present various challenges that need to be addressed for effective implementation. Cotton is a widely cultivated crop globally, and its production and processing generate substantial amounts of cotton waste, such as cotton stalks, cottonseed shells, and other by-products. Although using this waste to create bioproducts can have many positive environmental and economic effects, many challenges must be overcome. There are several challenges associated with processing and commercializing cotton waste.

The most common factor is the contamination of cotton recycling. Cotton waste can contain contaminants such as pesticides, herbicides, and other chemicals. This can make it difficult to process and can pose a risk to human health and the environment. Recycled cotton is more likely to be contaminated by other fibers, which could lower the end product's quality and make it more challenging to market commercially (Wojnowska-Baryła et al., 2022). Cotton waste from the production of textiles could contain leftover dyes, finishing agents, or other impurities that would impact the color and purity of recovered cotton (Ruuth et al., 2022). On-textile components from abandoned garments or fabrics, including buttons, zippers, or metal, can contaminate recycled cotton and other factors (Ruuth et al., 2022).

Inconsistent supply chain management of cotton waste is also one of the difficulties happening in the industry. Setting up an effective supply chain for recycling used cotton is essential. Collecting and sorting cotton waste from various sources, including textile mills and manufacturing facilities, is required (Johnson et al., 2020). The supply paths for wasted cotton are not well understood. The supply chain of waste cotton between different businesses and locations can be complex and very variable. For decision-making on end-of-life options, a precise estimation of the quantity and quality of waste cotton available at various supply chain stages is necessary (Johnson et al., 2020). To identify locations appropriate for recycling and reuse, it will be helpful to have a thorough awareness of the kind, quantity, and quality of waste cotton alone throughout the supply chain (Johnson et al., 2020). Figure 9.14 shows an estimation of the amount of cotton waste produced throughout the worldwide supply chain for the production of clothing and cotton.

Then, cotton can be sorted and processed in various forms, including lint, yarn, and fabric scraps. Sorting and processing these materials can be labor-intensive and expensive (Beall, 2020). Cost is labor-intensive because the process takes a long time. Post-consumer waste is especially challenging to sort through due to its numerous color blends and fabric hues (Celep et al., 2023). Cotton waste also affects the end-use application. Many various low-grade items, including insulation, mop heads, rags, and stuffing, can be made from recycled cotton. This makes customers or users unable to trust and have the confidence to use the materials from cotton waste. In addition, impacts

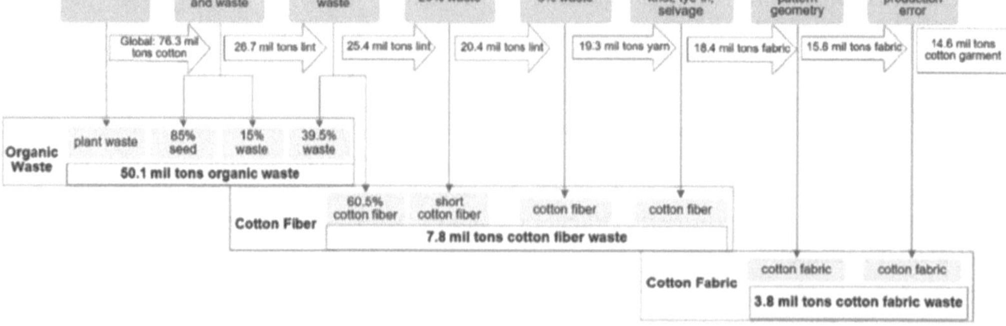

FIGURE 9.14 Annual fiber waste generated during global cotton garment production (Johnson et al., 2020).

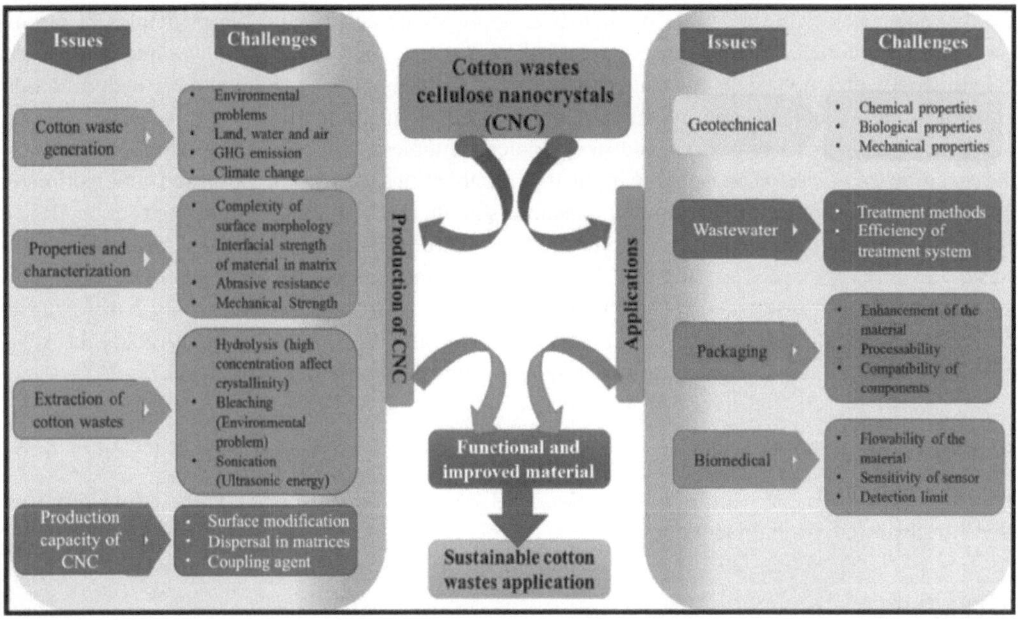

FIGURE 9.15 Issue challenges of extraction methods to produce cellulose nanocrystals (CNCs) (Rizal et al. (2021).

on recycled cotton's fiber length and homogeneity could limit its ability to be used for specific end uses (Celep et al., 2023). Thus, it is necessary to mix recycled cotton with other fibers to improve its properties, such as length, uniformity, and resistance. The quality of cotton waste can also be reduced due to the quality of recycled cotton fiber derived from cotton waste will never be identical to that of the original fiber, which may restrict the possibility of high-value end-use applications (Celep et al., 2023).

According to Rizal et al. (2021), the main problem when using natural fibers as reinforcement in a composite related to the manufacturing of cellulose nanocrystals (CNCs) from waste cotton is the hydrophilic behavior of the fibers and the hydrophobicity of the matrix. Research is still being done extensively on the problem of enhancing the reinforced matrix interfaces in composites utilizing cotton fiber as reinforcement. The application may be constrained by the difficulty of dispersing the reinforcement in the polymer matrix without causing the biopolymer to degrade. Figure 9.15 illustrates the issue of extraction methods for the production of CNCs (Rizal et al., 2021). Despite these limitations, creating textiles from recycled cotton is an environmentally friendly process that lowers waste, saves resources, and does not use toxic materials. The method significantly saves water, energy, and carbon dioxide emissions.

9.9 CONCLUSIONS

In the sector of industrial applications, the management of cotton waste presents a complex yet promising landscape. This approach involves transforming discarded cotton materials into valuable resources across diverse industries. Through innovative techniques and processes, cotton waste is repurposed, creating a sustainable cycle that reduces environmental impact while maximizing resource efficiency. From textiles to construction materials, biofuels, and beyond, cotton waste demonstrates the potential to revolutionize traditional waste management practices. By adopting sustainable practices such as recycling, upcycling, and repurposing, cotton waste alchemy not only minimizes landfill burden but also contributes to the conservation of natural resources. Moreover, the management of cotton waste for industrial applications aligns with the principles of circular economy, where materials are reused, recycled, and regenerated to create new value chains. This holistic approach emphasizes the importance of collaboration among stakeholders, including manufacturers, policymakers, and consumers, to realize the full potential of cotton waste. Through strategic investments in research and development, coupled with robust regulatory frameworks, the integration of cotton waste into industrial applications can be streamlined and optimized. Furthermore, raising awareness among consumers about the benefits of utilizing products derived from cotton waste market demand and promotes a culture of sustainability. In conclusion, cotton waste alchemy represents a paradigm shift in waste management practices, where discarded cotton materials are transformed into valuable assets for industrial applications. By harnessing the principles of sustainability, innovation, and collaboration, cotton waste alchemy holds the key to creating a more sustainable and resource-efficient future for industries worldwide.

ACKNOWLEDGMENT

This chapter was written with Agathianprakash A.L. Asokan, Muhammad Hariz Abdul Hamid, and Nur Ain Syamilah Mohamad Noor.

REFERENCES

Abbasi, T., & Abbasi, S. A. (2010). Biomass energy and the environmental impacts associated with its production and utilization. *Renewable and Sustainable Energy Reviews*, *14*(3), 919–937.

Adebayo, G. J., Omolara, B. N., & Toyin, A. E. (2009). Evaluation of yield of oyster mushroom (Pleurotus pulmonarius) grown on cotton waste and cassava peel. *African Journal Biotechnology*, *8*, 215–218.

Aishwariya, S., & Amsamani, S. (2012). Evaluating the efficacy of compost evolved from bio-managing cotton textile waste. *Journal of Environmental Research and Development*, *6*(4), 941–952.

Akhmedov, S. K., Krepl, V., Muntean, A. B., & Ivanova, T. (2017). Research on solid biofuels from cotton waste biomass–alternative for Tajikistan's energy sector development. *Agronomy Research*, *15*(5), 1846–1855.

Al Afif, R., Anayah, S. S., & Pfeifer, C. (2020). Batch pyrolysis of cotton stalks for evaluation of biochar energy potential. *Renewable Energy*, *147*, 2250–2258.

Algin, H. M., & Turgut, P. (2008). Cotton and limestone powder wastes as brick material. *Construction and Building Materials*, *22*(6), 1074–1080.

Ali, F., Sarma, T. C., & Saikia, C. N. (1993). Pulp and paper from certain fast-growing plant species. *Bioresource Technology*, *45*(1), 65–67.

Alias, A. H., Norizan, M. N., Sabaruddin, F. A., Asyraf, M. R. M., Norrrahim, M. N. F., Ilyas, A. R., Kuzmin, A. M., Rayung, M., Shazleen, S. S., Nazrin, A., Sherwani, S. F. K., Harussani, M. M., Atikah, M. S. N., Ishak, M. R., Sapuan, S. M., & Khalina, A. (2021). Hybridization of MMT/lignocellulosic fiber reinforced polymer nanocomposites for structural applications: A review. *Coatings*, *11*(11), 1355. https://doi.org/10.3390/coatings11111355

Alomayri, T., Assaedi, H., Shaikh, F. U. A., & Low, I. M. (2014). Effect of water absorption on the mechanical properties of cotton fabric-reinforced geopolymer composites. *Journal of Asian Ceramic Societies*, *2*(3), 223–230. https://doi.org/10.1016/j.jascer.2014.05.005

Anandjiwala, R. D., & Blouw, S. (2007). Composites from bast fibers-prospects and potential in the changing market environment. *Journal of Natural Fibers*, *4*(2), 91–109. https://doi.org/10.1300/j395v04n02_07

Asyraf, M. R. M., Syamsir, A., Zahari, N. M., Supian, A. B. M., Ishak, M. R., Sapuan, S. M., Sharma, S., Rashedi, A., Razman, M. R., Zakaria, S. Z. S., Ilyas, R. A., & Rashid, M. Z. A. (2022). Product development of natural fibre-composites for various applications: Design for sustainability. *Polymers*, *14*(5), 920. https://doi.org/10.3390/polym14050920

Azlin, M. N. M., Ilyas, R. A., Zuhri, M. Y. M., Sapuan, S. M., Harussani, M. M., Sharma, S., Nordin, A. H., Nurazzi, N. M., & Afiqah, A. N. (2022). 3D printing and shaping polymers, composites, and nanocomposites: A review. *Polymers*, *14*(1), 180. https://doi.org/10.3390/polym14010180

Bagi, Z., Ács, N., Böjti, T., Kakuk, B., Rákhely, G., Strang, O., Szuhaj, M., Wirth, R., & Kovács, K. L. (2017). Biomethane: The energy storage, platform chemical and greenhouse gas mitigation target. *Anaerobe*, *46*, 13–22. https://doi.org/10.1016/j.anaerobe.2017.03.001

Bajpai, P. (2015). Basic overview of pulp and paper manufacturing process. In *Green Chemistry and Sustainability in Pulp and Paper Industry* (pp. 11–39). https://doi.org/10.1007/978-3-319-18744-0_2

Baysal, E., & Peker, H. (2001). An alternate to waste paper recycling: Mushroom cultivation. *Teknoloji, 3–4*, 9–13.

Beall, A. (2020, July 13). Why Clothes Are so Hard to Recycle. *BBC*. https://www.bbc.com/future/article/20200710-why-clothes-are-so-hard-to-recycle

Bhat, G., & Parikh, D. (2010). Biodegradable materials for nonwovens. https://doi.org/10.1533/9781845699741.1.46

Bhatia, S. K., Jagtap, S. S., Bedekar, A. A., Bhatia, R. K., Rajendran, K., Pugazhendhi, A., Rao, C. V., Atabani, A. E., Kumar, G., & Yang, Y.-H. (2021). Renewable biohydrogen production from lignocellulosic biomass using fermentation and integration of systems with other energy generation technologies. *Science of the Total Environment, 765*, 144429. https://doi.org/10.1016/j.scitotenv.2020.144429

Bhuiyan, M. A. R., Ali, A., Mohebbullah, Md., Hossain, M. F., Khan, A. N., & Wang, L. (2023). Recycling of cotton apparel waste and its utilization as a thermal insulation layer in high performance clothing. *Fashion and Textiles, 10*(1), 22. https://doi.org/10.1186/s40691-023-00342-y

Binczarski, M. J., Malinowska, J. Z., Berlowska, J., Cieciura-Wloch, W., Borowski, S., Cieslak, M., Puchowicz, D., & Witonska, I. A. (2022). Concept for the use of cotton waste hydrolysates in fermentation media for biofuel production. *Energies, 15*(8), 2856. https://doi.org/10.3390/en15082856

Binici, H., & Aksogan, O. (2015). Engineering properties of insulation material made with cotton waste and fly ash. *Journal of Material Cycles and Waste Management, 17*(1), 157–162. https://doi.org/10.1007/s10163-013-0218-6

Binici, H., Gemci, R., Kucukonder, A., & Solak, H. H. (2012). Investigating sound insulation, thermal conductivity and radioactivity of chipboards produced with cotton waste, fly ash and barite. *Construction and Building Materials, 30*, 826–832. https://doi.org/10.1016/j.conbuildmat.2011.12.064

Brynolf, S., Fridell, E., & Andersson, K. (2014). Environmental assessment of marine fuels: Liquefied natural gas, liquefied biogas, methanol and bio-methanol. *Journal of Cleaner Production, 74*, 86–95. https://doi.org/10.1016/j.jclepro.2014.03.052

Cambero, C., & Sowlati, T. (2014). Assessment and optimization of forest biomass supply chains from economic, social and environmental perspectives—A review of literature. *Renewable and Sustainable Energy Reviews, 36*, 62–73. https://doi.org/10.1016/j.rser.2014.04.041

Cao, T. N.-D., Mukhtar, H., Yu, C.-P., Bui, X.-T., & Pan, S.-Y. (2022). Agricultural waste-derived biochar in microbial fuel cells towards a carbon-negative circular economy. *Renewable and Sustainable Energy Reviews, 170*, 112965. https://doi.org/10.1016/j.rser.2022.112965

Carlson, C., & Ebben, A. (2016). Biocoal from Biomass: An Emerging Opportunity for Renewable Energy. https://feeco.com/biocoal-biomass-emerging-opportunity-renewable-energy/

Çay, A., Yanık, J., Akduman, Ç., Duman, G., & Ertaş, H. (2020). Application of textile waste derived biochars onto cotton fabric for improved performance and functional properties. *Journal of Cleaner Production, 251*, 119664. https://doi.org/10.1016/j.jclepro.2019.119664

Celep, G., Tetik, G. D., & Yilmaz, F. (2023). Limitations of textile recycling: The reason behind the development of alternative sustainable fibers. In *Next-Generation Textiles*. https://doi.org/10.5772/intechopen.105118

Chakravarty, K., & Gaur, S. (2017). Lactic acid bacteria: Applications in food and health. *Indo Global Journal of Pharmaceutical Sciences, 07*(01). https://doi.org/10.35652/igjps.2017.44

Chen, W.-H., Lin, B.-J., Lin, Y.-Y., Chu, Y.-S., Ubando, A. T., Show, P. L., Ong, H. C., Chang, J.-S., Ho, S.-H., Culaba, A. B., Pétrissans, A., & Pétrissans, M. (2021). Progress in biomass torrefaction: Principles, applications and challenges. *Progress in Energy and Combustion Science, 82*, 100887. https://doi.org/10.1016/j.pecs.2020.100887

Cheng, B.-H., Huang, B.-C., Zhang, R., Chen, Y.-L., Jiang, S.-F., Lu, Y., Zhang, X.-S., Jiang, H., & Yu, H.-Q. (2020). Bio-coal: A renewable and massively producible fuel from lignocellulosic biomass. *Science Advances, 6*(1), eaay0748. https://doi.org/10.1126/sciadv.aay0748

Christine, & Gable, S. (2019, January 31). How Biobutanol Can Be Used as a Motor Fuel. *Treehugger*. https://www.treehugger.com/pros-cons-biobutanol-85645

Clark, J. (2013, October 3). Organic Acids. *Chemistry LibreTexts*. https://chem.libretexts.org/Bookshelves/Organic_Chemistry/Supplemental_Modules_(Organic_Chemistry)/Fundamentals/Organic_Acids_and_Bases/Organic_Acids

Crini, G., Lichtfouse, E., Chanet, G., & Morin-Crini, N. (2020). Applications of hemp in textiles, paper industry, insulation and building materials, horticulture, animal nutrition, food and beverages, nutraceuticals, cosmetics and hygiene, medicine, agrochemistry, energy production and environment: A review. *Environmental Chemistry Letters, 18*(5), 1451–1476. https://doi.org/10.1007/s10311-020-01029-2

Crnogaća, B. R. (2017). Torrefaction as a process for biomass conversion into bio-coal. *Tehnika*, *72*(3), 323–327. https://doi.org/10.5937/tehnika1703323c

De Silva, R., & Byrne, N. (2017). Utilization of cotton waste for regenerated cellulose fibers: Influence of degree of polymerization on mechanical properties. *Carbohydrate Polymers*, *174*, 89–94. https://doi.org/10.1016/j.carbpol.2017.06.042

Delate, K., Heller, B., & Shade, J. (2020). Organic cotton production may alleviate the environmental impacts of intensive conventional cotton production. In *Renewable Agriculture and Food Systems*. Cambridge University Press. https://doi.org/10.1017/S1742170520000356

Devi, A., Bajar, S., Kour, H., Kothari, R., Pant, D., & Singh, A. (2022). Lignocellulosic biomass valorization for bioethanol production: A circular bioeconomy approach. *BioEnergy Research*. https://doi.org/10.1007/s12155-022-10401-9

El Messiry, M., El Deeb, R., & El-Tarfawy, S. (2015). Mechanical and Water Absorption Properties of Micro-Cellulose Obtained from Cotton Fiber Waste Reinforced Polyvinyl Alcohol (PVA) Composites. Conference paper.

Elsawy, M. A., Kim, K.-H., Park, J.-W., & Deep, A. (2017). Hydrolytic degradation of polylactic acid (PLA) and its composites. *Renewable and Sustainable Energy Reviews*, *79*, 1346–1352. https://doi.org/10.1016/j.rser.2017.05.143

Ertaş, M., Acemioĝlu, B., Alma, M. H., & Usta, M. (2010). Removal of methylene blue from aqueous solution using cotton stalk, cotton waste and cotton dust. *Journal of Hazardous Materials*, *183*(1–3), 421–427. https://doi.org/10.1016/j.jhazmat.2010.07.041

Esteve-Turrillas, F. A., & de la Guardia, M. (2017). Environmental impact of recover cotton in textile industry. *Resources, Conservation and Recycling*, *116*, 107–115. https://doi.org/10.1016/j.resconrec.2016.09.034

Eyupoglu, S. (2019). Organic cotton and environmental impacts. In *Textile Science and Clothing Technology* (pp. 157–176). Springer Singapore. https://doi.org/10.1007/978-981-10-8782-0_8

Fahim, I. S., Chbib, H., Elhakim, Y., Shabrawy, S., & Azouz, M. (2022). Organic binding materials in cotton waste. In *NILES 2022–4th Novel Intelligent and Leading Emerging Sciences Conference, Proceedings* (pp. 243–247). https://doi.org/10.1109/NILES56402.2022.9942390

Fernandes, P. R. B., Contin, B., Siqueira, M. U., Ruschel-Soares, R., & Baruque-Ramos, J. (2021). Biocomposites from cotton denim waste for footwear components. *Materials Circular Economy*, *3*(1). https://doi.org/10.1007/s42824-021-00045-z

Gellerstedt, G. (2008, August). Moodle USP: e-Disciplinas. Edisciplinas.usp.br. https://edisciplinas.usp.br/pluginfile.php/5597689/mod_resource/content/1/Pulping%20Chemistry%20and%20Technology%20-Vol%202-%20De%20Gruyter.pdf#page=101

Gopal, P. M., Sivaram, N. M., & Barik, D. (2018). Paper industry wastes and energy generation from wastes. In *Energy from Toxic Organic Waste for Heat and Power Generation* (pp. 83–97). Elsevier. https://doi.org/10.1016/B978-0-08-102528-4.00007-9

Grimes, J. L., Carter, T. A., Gernat, A. E., & Godwin, J. L. (2007). A novel bedding material made from cotton waste, gypsum, and old newsprint for rearing Turkeys. *Journal of Applied Poultry Research*, *16*(4), 598–604. https://doi.org/10.3382/japr.2007-00034

Gun, A. D., & Tiber, B. (2011). Color, color fastness and abrasion properties of 50/50 bamboo/cotton blended plain knitted fabrics in three different stitch lengths. *Textile Research Journal*, *81*(18), 1903–1915.

Halimatul, M. J., Sapuan, S. M., Jawaid, M., Ishak, M. R., & Ilyas, R. A. (2019). Water absorption and water solubility properties of sago starch biopolymer composite films filled with sugar palm particles. *Polimery*, *64*(9), 27–35. https://doi.org/10.14314/polimery.2019.9.4

Halimi, M. T., Hassen, M. B., & Sakli, F. (2008). Cotton waste recycling: Quantitative and qualitative assessment. *Resources, Conservation and Recycling*, *52*(5), 785–791.

Halimi, T. M., Hassen, B. M., Azzouz, B., & Sakli, F. (2007). Effect of cotton waste and spinning parameters on rotor yarn quality. *Journal of the Textile Institute*, *98*(5), 437–442. https://doi.org/10.1080/00405000701547649

Hamawand, I., Sandell, G., Pittaway, P., Chakrabarty, S., Yusaf, T., Chen, G., Seneweera, S., Al-Lwayzy, S., Bennett, J., & Hopf, J. (2016). Bioenergy from cotton industry wastes: A review and potential. *Renewable and Sustainable Energy Reviews*, *66*, 435–448. https://doi.org/10.1016/j.rser.2016.08.033

Haque, A. N., Remadevi, R., & Naebe, M. (2021). A review on cotton gin trash: Sustainable commodity for material fabrication. *Journal of Cleaner Production*, *281*, 125300. https://doi.org/10.1016/j.jclepro.2020.125300

Haule, L. V., Carr, C. M., & Rigout, M. (2016). Preparation and physical properties of regenerated cellulose fibers from cotton waste garments. *Journal of Cleaner Production*, *112*, 4445–4451. https://doi.org/10.1016/j.jclepro.2015.08.086

He, X., Su, D., Cai, W., Pehlken, A., Zhang, G., Wang, A., & Xiao, J. (2021). Influence of material selection and product design on automotive vehicle recyclability. *Sustainability (Switzerland)*, *13*(6). https://doi.org/10.3390/su13063407

Husain, S. N. H., Rashid, A. H. A., Shaari, M. F., Hassan, N. N. M., Ahmad, M. R., & Rashid, R. A. (2019, November). Mechanical properties evaluation of nonwoven industrial cotton waste produced by needle punching method. *IOP Conference Series: Materials Science and Engineering*, *670*(1), 012035.

Isci, A., & Demirer, G. N. (2007). Biogas production potential from cotton wastes. *Renewable Energy*, *32*(5), 750–757. https://doi.org/10.1016/j.renene.2006.03.018

Islam, G. M. N., Ali, A., & Collie, S. (2020). Textile sensors for wearable applications: A comprehensive review. *Cellulose*, *27*(11), 6103–6131. https://doi.org/10.1007/s10570-020-03215-5

Johnson, S., Echeverria, D., Venditti, R., Jameel, H., & Yao, Y. (2020). Supply chain of waste cotton recycling and reuse: A review. *AATCC Journal of Research*, *7*(1_suppl), 19–31. https://doi.org/10.14504/ajr.7.s1.3

Jones, M. (2020, August 13). 6 Types of Kraft Paper and Their Best Business Uses. www.oren-intl.com. https://www.oren-intl.com/blog/bid/359426/6-types-of-kraft-paper-and-their-best-business-uses

Jumaidin, R., Adam, N. W., Ilyas, R. A., Hussin, M. S. F., Taha, M. M., Mansor, M. R., Azlan, U. A. A., & Yob, M. S. (2019a). Water transport and physical properties of sugarcane bagasse fibre reinforced thermoplastic potato starch biocomposite. *Journal of Advanced Research in Fluid Mechanics and Thermal Sciences*, *61*(2), 273–281.

Jumaidin, R., Diah, N. A., Ilyas, R. A., Alamjuri, R. H., & Yusof, F. A. M. (2021). Processing and characterisation of banana leaf fibre reinforced thermoplastic cassava starch composites. *Polymers*, *13*(9), 1420. https://doi.org/10.3390/polym13091420

Jumaidin, R., Saidi, Z. A. S., Ilyas, R. A., Ahmad, M. N., Wahid, M. K., Yaakob, M. Y., Maidin, N. A., Rahman, M. H. A., & Osman, M. H. (2019b). Characteristics of cogon grass fibre reinforced thermoplastic cassava starch biocomposite: Water absorption and physical properties. *Journal of Advanced Research in Fluid Mechanics and Thermal Sciences, 62*(1), 43–52.

Kabir, K., Batagarawa, S. M., Muhammad, I., Nazifi, A., Haruna, I., Ahmed, H. I., Bamanga, R., & Bugaje, I. (2020). Bioethanol production from waste cotton lint using Saccharomyces cerevisiae. *Journal of Applied Sciences*, *20*(2), 58–66. https://doi.org/10.3923/jas.2020.58.66

Kammann, C., Ratering, S., Eckhard, C., & Müller, C. (2012). Biochar and hydrochar effects on greenhouse gas (carbon dioxide, nitrous oxide, and methane) fluxes from soils. *Journal of Environment Quality*, *41*(4), 1052. https://doi.org/10.2134/jeq2011.0132

Kan, T., Strezov, V., & Evans, T. J. (2016). Lignocellulosic biomass pyrolysis: A review of product properties and effects of pyrolysis parameters. *Renewable and Sustainable Energy Reviews*, *57*, 1126–1140. https://doi.org/10.1016/j.rser.2015.12.185

Kar, S., Santra, B., Kumar, S., Ghosh, S., & Majumdar, S. (2022). Sustainable conversion of textile industry cotton waste into P-dopped biochar for removal of dyes from textile effluent and valorisation of spent biochar into soil conditioner towards circular economy. *Environmental Pollution*, *312*, 120056. https://doi.org/10.1016/j.envpol.2022.120056

Kim, S. Y., Lee, S. Y., Weon, H. Y., Sang, M. K., & Song, J. (2017). Complete genome sequence of Bacillus velezensis M75, a biocontrol agent against fungal plant pathogens, isolated from cotton waste. *Journal of Biotechnology*, *241*, 112–115. https://doi.org/10.1016/j.jbiotec.2016.11.023

Kocak, E. D., Taşdemir, M., & Gul, C. (2020). Sustainable green-based composites from renewable resources in textile: Industrial cotton wastes. https://doi.org/10.1007/978-3-030-38013-7_3

Koroneos, C., & Lykidou, S. (2011). Equilibrium modeling for a dwndraft biomass gasifier for cotton stalks biomass in comparison with experimental data. *Journal of Chemical Engineering and Materials Science*, *2*(4), 61–68.

Koul, B., Yakoob, M., & Shah, M. P. (2022). Agricultural waste management strategies for environmental sustainability. *Environmental Research*, *206*. https://doi.org/10.1016/j.envres.2021.112285

Kwon, S. W., Lee, S. Y., Kim, B. Y., Weon, H. Y., Kim, J. B., Go, S. J., & Lee, G. B. (2007). Bacillus niabensis sp. nov., isolated from cotton-waste composts for mushroom cultivation. *International Journal of Systematic and Evolutionary Microbiology*, *57*(8), 1909–1913. https://doi.org/10.1099/ijs.0.64178-0

Latif, W., Basit, A., Ali, Z., & Ahmad Baig, S. (2018). The mechanical and comfort properties of cotton and regenerated fibers blended woven fabrics. *International Journal of Clothing Science and Technology*, *30*(1), 112–121. https://doi.org/10.1108/IJCST-07-2017-0101

Leal Filho, W., Ellams, D., Han, S., Tyler, D., Boiten, V. J., Paco, A., Moora, H., & Balogun, A. L. (2019). A review of the socio-economic advantages of textile recycling. *Journal of Cleaner Production*, *218*, 10–20. https://doi.org/10.1016/j.jclepro.2019.01.210

Lévesque, V., Oelbermann, M., & Ziadi, N. (2021). Biochar in temperate soils: Opportunities and challenges. *Canadian Journal of Soil Science*, 1–26. https://doi.org/10.1139/cjss-2021-0047

Lu, J. I. (2011). Environmental Effects of Vehicle Exhausts, Global and Local Effects: A Comparison between Gasoline and Diesel (Master's thesis, Halmstad University, Sweden). Retrieved from https://www.diva-portal.org/smash/get/diva2:427347/FULLTEXT01.pdf

Lu, L., Fan, W., Meng, X., Xue, L., Ge, S., Wang, C., Foong, S. Y., Tan, C. S. Y., Sonne, C., Aghbashlo, M., Tabatabaei, M., & Lam, S. S. (2023). Current recycling strategies and high-value utilization of waste cotton. *Science of the Total Environment*, *856*. https://doi.org/10.1016/j.scitotenv.2022.158798

Luitel, K. P., Hudson, D., & Ethridge, D. (2015). Economics and marketing: Evaluating cotton utilization in nonwoven textiles. *The Journal of Cotton Science*, 19, 298–306.

Maciel, M. M. Á. D., Benini, K. C. C. de C., Voorwald, H. J. C., & Cioffi, M. O. H. (2019). Obtainment and characterization of nanocellulose from an unwoven industrial textile cotton waste: Effect of acid hydrolysis conditions. *International Journal of Biological Macromolecules*, *126*, 496–506. https://doi.org/10.1016/j.ijbiomac.2018.12.202

Madurwar, M. V., Ralegaonkar, R. V., & Mandavgane, S. A. (2013). Application of agro-waste for sustainable construction materials: A review. *Construction and Building Materials*, *38*, 872–878. https://doi.org/10.1016/j.conbuildmat.2012.09.011

Mahalakshmi, M., Angayarkanni, J., Rajendran, R., & Rajesh, R. (2011). Bioconversion of cotton waste from textile mills to bioethanol by microbial saccharification and fermentation. *Annals of Biological Research*, *2*(3), 380–388.

Mahamude, A. S. F., Harun, W. S. W., Kadirgama, K., Ramasamy, D., & Farhana, K. (2021). Cotton waste research follows the effect of pre-treatment and the observation of physical appearance. *IOP Conference Series: Materials Science and Engineering*, *1078*(1), 012014. https://doi.org/10.1088/1757-899x/1078/1/012014

Mammino, L. Editor, (2020). *Biomass Burning in Sub-Saharan Africa Chemical Issues and Action Outreach. A Year of Young Elements*, *42*(3), 31.

Mao, C., Feng, Y., Wang, X., & Ren, G. (2015). Review on research achievements of biogas from anaerobic digestion. *Renewable and Sustainable Energy Reviews*, *45*, 540–555. https://doi.org/10.1016/j.rser.2015.02.032

Mazani, N., Sapuan, S. M., Sanyang, M. L., Atiqah, A., & Ilyas, R. A. (2019). Design and fabrication of a shoe shelf from kenaf fiber reinforced unsaturated polyester composites. In H. Ariffin, S. M. Sapuan, & M. A. Hassan (Eds.), *Lignocellulose for Future Bioeconomy* (1st ed., Issue 2000, pp. 315–332). Elsevier. https://doi.org/10.1016/B978-0-12-816354-2.00017-7

Mboowa, D. (2021). A review of the traditional pulping methods and the recent improvements in the pulping processes. *Biomass Conversion and Biorefinery*. https://doi.org/10.1007/s13399-020-01243-6

Mohanty, A. K., Wibowo, A., Misra, M., & Drzal, L. T. (2004). Effect of process engineering on the performance of natural fiber reinforced cellulose acetate biocomposites. *Composites Part A: Applied Science and Manufacturing*, *35*(3), 363–370. https://doi.org/10.1016/j.compositesa.2003.09.015

Mourad, A. H. I., Ayad, O. G., Rahman, A., Hilal-Alnaqbi, A., & Abu-Jdayil, B. I. (2016, July). Experimental investigation of Kevlar KM2Plus nano-reinforced laminated composite thermo-mechanical properties. In *Pressure Vessels and Piping Conference* (Vol. 50428, Paper no. V06AT06A030). American Society of Mechanical Engineers. Vancouver, British Columbia, Canada.

Mowrer, J., Dembele, G., Strzelec, A., & Thapa, B. (2022). Cotton waste-derived biochar and compost improve early growth and AMF association in cotton. *Agrosystems, Geosciences & Environment*, *5*(3), e20298.

Mueller, D. H., & Krobjilowski, A. (2003). New discovery in the properties of composites reinforced with natural fibers. *Journal of Industrial Textiles*, *33*(2), 111–129. https://doi.org/10.1177/152808303039248

Nazrin, A., Sapuan, S. M., Zuhri, M. Y. M., Tawakkal, I. S. M. A., & Ilyas, R. A. (2021). Water barrier and mechanical properties of sugar palm crystalline nanocellulose reinforced thermoplastic sugar palm starch (TPS)/poly(lactic acid) (PLA) blend bionanocomposites. *Nanotechnology Reviews*, *10*(1), 431–442. https://doi.org/10.1515/ntrev-2021-0033

Negi, S., Jaswal, G., Dass, K., Mazumder, K., Elumalai, S., & Roy, J. K. (2020). Torrefaction: A sustainable method for transforming of agri-wastes to high energy density solids (bio-coal). *Reviews in Environmental Science and Bio/Technology*, *19*(2), 463–488. https://doi.org/10.1007/s11157-020-09532-2

Nikolić, S., Lazić, V., Veljović, Đ., & Mojović, L. (2017). Production of bioethanol from pre-treated cotton fabrics and waste cotton materials. *Carbohydrate Polymers*, *164*, 136–144. https://doi.org/10.1016/j.carbpol.2017.01.090

Norfarhana, A. S., Ilyas, R. A., & Ngadi, N. (2022). A review of nanocellulose adsorptive membrane as multifunctional wastewater treatment. *Carbohydrate Polymers*, *291*, 119563. https://doi.org/10.1016/j.carbpol.2022.119563

Norfarhana, A. S., Ilyas, R. A., Ngadi, N., Othman, M. H. D., Misenan, M. S. M., & Norrrahim, M. N. F. (2024a). Revolutionizing lignocellulosic biomass: A review of harnessing the power of ionic liquids for sustainable utilization and extraction. *International Journal of Biological Macromolecules*, *256*, 128256. https://doi.org/10.1016/j.ijbiomac.2023.128256

Norfarhana, A. S., Khoo, P. S., Ilyas, R. A., Ab Hamid, N. H., Aisyah, H. A., Norrrahim, M. N. F., Knight, V. F., Rani, M. S. A., Septevani, A. A., Syafri, E., & Annamalai, P. K. (2024b). Exploring of cellulose nanocrystals from lignocellulosic sources as a powerful adsorbent for wastewater remediation. *Journal of Polymers and the Environment*. https://doi.org/10.1007/s10924-024-03227-3

Norrrahim, M. N. F., Ilyas, R. A., Nurazzi, N. M., Rani, M. S. A., Atikah, M. S. N., & Shazleen, S. S. (2021). Chemical pretreatment of lignocellulosic biomass for the production of bioproducts: An overview. *Applied Science and Engineering Progress*, 1–18. https://doi.org/10.14416/j.asep.2021.07.004

Nurazzi, N. M., Harussani, M. M., Aisyah, H. A., Ilyas, R. A., Norrrahim, M. N. F., Khalina, A., & Abdullah, N. (2021). Treatments of natural fiber as reinforcement in polymer composita short review. *Functional Composites and Structures*, *3*(2), 024002. https://doi.org/10.1088/2631-6331/abff36

Paletto, A., Bernardi, S., Pieratti, E., Teston, F., & Romagnoli, M. (2019). Assessment of environmental impact of biomass power plants to increase the social acceptance of renewable energy technologies. *Heliyon*, *5*(7). https://doi.org/10.1016/j.heliyon.2019.e02070

Pandey, S. N., & Shaikh, A. J. (1987). Utilisation of cotton plant stalk for production of pulp and paper. *Biological Wastes*, *21*(1), 63–70.

Paredes, C., Bernal, M. P., Cegarra, J., Roig, A., & Navarro, A. F. (1996). Nitrogen Tranformation During the Composting of Different Organic Wastes. In *Progress in Nitrogen Cycling Studies: Proceedings of the 8th Nitrogen Workshop held at the University of Ghent, 5–8 September, 1994* (pp. 121–125). Springer Netherlands.

Patel, V. R. (2017). Cost-effective sequential biogas and bioethanol production from the cotton stem waste. *Process Safety and Environmental Protection*, *111*, 335–345. https://doi.org/10.1016/j.psep.2017.07.019

Pennington, D. (2013, February 28). Biobutanol: A Renewable Fuel for the Future. *MSU Extension*. https://www.canr.msu.edu/news/biobutanol_a_renewable_fuel_for_the_future

Petrucci, R., Nisini, E., Puglia, D., Sarasini, F., Rallini, M., Santulli, C., Minak, G., & Kenny, J. M. (2015). Tensile and fatigue characterisation of textile cotton waste/polypropylene laminates. *Composites Part B: Engineering*, *81*, 84–90. https://doi.org/10.1016/j.compositesb.2015.07.005

Piribauer, B., & Bartl, A. (2019). Textile recycling processes, state of the art and current developments: A mini review. *Waste Management & Research*, *37*(2), 112–119. https://doi.org/10.1177/0734242x18819277

Pleissner, D., Schneider, R., Venus, J., & Koch, T. (2016). Separation of lactic acid and recovery of salt-ions from fermentation broth. *Journal of Chemical Technology & Biotechnology*, *92*(3), 504–511. https://doi.org/10.1002/jctb.5023

Poornima, V., Venkatasubramani, R., Sreevidya, V., & Chandrasekar, P. (2021). Study on properties of bio-bricks. *Materials Today: Proceedings*. https://doi.org/10.1016/j.matpr.2021.08.315

Rahman, S., & Uddin, A. J. (2022). Unusable cotton spinning mill waste: A viable source of raw material in paper making. *Heliyon*, *8*(8). https://doi.org/10.1016/j.heliyon.2022.e10055

Rajapakse, P. (2011). New cultivation technology for paddy straw mushroom (Volvariella volvacea). In *Proceedings of the 7th International Conference on Mushroom Biology and Mushroom Products (ICMBMP7)* (pp. 446–451). Arcachon, France.

Rajput, D., Bhagade, S. S., Raut, S. P., Ralegaonkar, R. V., & Mandavgane, S. A. (2012). Reuse of cotton and recycle paper mill waste as building material. *Construction and Building Materials*, *34*, 470–475. https://doi.org/10.1016/j.conbuildmat.2012.02.035

Ranjithkumar, M., Ravikumar, R., Sankar, M. K., Kumar, M. N., & Thanabal, V. (2017). An effective conversion of cotton waste biomass to ethanol: A critical review on pretreatment processes. *Waste and Biomass Valorization*, *8*(1), 57–68. https://doi.org/10.1007/s12649-016-9563-8

Rautray, P., Roy, A., Mathew, D. J., & Eisenbart, B. (2019). Bio-brick—Development of sustainable and cost effective building material. *Proceedings of the Design Society: International Conference on Engineering Design*, *1*(1), 3171–3180. https://doi.org/10.1017/dsi.2019.324

Rautray, P., Roy, A., Mathew, D. J., & Eisenbart, B. (2021). *Bio-bricks: Circular Economy and New Products*. In H. S. S. R. S. (Ed.), Smart Innovation, Systems and Technologies (Vol. 221, pp. 845–857). Springer. https://doi.org/10.1007/978-981-16-0041-8_69

Reza, M. M., Begum, H. A., & Uddin, A. J. (2023). Potentiality of sustainable corn starch-based biocomposites reinforced with cotton filter waste of spinning mill. *Heliyon*, *9*(5), e15697. https://doi.org/10.1016/j.heliyon.2023.e15697

Rizal, S., H. P. S., A. K., Oyekanmi, A. A., Gideon, O. N., Abdullah, C. K., Yahya, E. B., Alfatah, T., Sabaruddin, F. A., & Rahman, A. A. (2021). Cotton wastes functionalized biomaterials from micro to nano: A cleaner approach for a sustainable environmental application. *Polymers*, *13*(7), 1006. https://doi.org/10.3390/polym13071006

Ruiz-Ruiz, F., Mancera-Andrade, E. I., Parra-Saldivar, R., Keshavarz, T., & Iqbal, H. M. N. (2018). Drug delivery and cosmeceutical applications of poly-lactic acid based novel constructs—A review. *Current Drug Metabolism*, *18*(10). https://doi.org/10.2174/1389200218666170919170335

Ruuth, E., Sanchis-Sebastiá, M., Larsson, P. T., Teleman, A., Jiménez-Quero, A., Delestig, S., Sahlberg, V., Salén, P., Sanchez Ortiz, M., Vadher, S., & Wallberg, O. (2022). Reclaiming the value of cotton waste textiles: A new improved method to recycle cotton waste textiles via acid hydrolysis. *Recycling*, *7*(4), 57. https://doi.org/10.3390/recycling7040057

Saad, N. F., Maâ min, N. N., Md Zain, S., Ahmad Basri, N. E., & Md Zaini, N. S. (2013). Composting of mixed yard and food wastes with effective microbes. *Jurnal Teknologi*, *65*(2). https://doi.org/10.11113/jt.v65.2196

Saadon, S., Uemura, Y., & Mansor, N. (2014). Torrefaction in the presence of oxygen and carbon dioxide: The effect on yield of oil palm kernel shell. *Procedia Chemistry*, *9*, 194–201. https://doi.org/10.1016/j.proche.2014.05.023

Sadrolodabaee, P., Claramunt, J., Ardanuy, M., & de la Fuente, A. (2021). Mechanical and durability characterization of a new textile waste micro-fiber reinforced cement composite for building applications. *Case Studies in Construction Materials*, *14*, e00492.

Safaya, G., Hoekstra, A. Y., & Savenije, H. H. G. (2017). *Guiding Farmers Towards Sustainable Cotton Production*. Water Footprint Network. Retrieved from https://waterfootprint.org/resources/Guiding_Farmers_Towards_Sustainable_Cotton_Prod_2017.pdf

Saidur, R., Abdelaziz, E. A., Demirbas, A., Hossain, M. S., & Mekhilef, S. (2011). A review on biomass as a fuel for boilers. *Renewable and Sustainable Energy Reviews*, *15*(5), 2262–2289. https://doi.org/10.1016/j.rser.2011.02.015

Saini, J. K., Saini, R., & Tewari, L. (2014). Lignocellulosic agriculture wastes as biomass feedstocks for second-generation bioethanol production: Concepts and recent developments. *3 Biotech*, *5*(4), 337–353. https://doi.org/10.1007/s13205-014-0246-5

Sandin, G., & Peters, G. M. (2018). Environmental impact of textile reuse and recycling—A review. *Journal of Cleaner Production*, *184*, 353–365. https://doi.org/10.1016/j.jclepro.2018.02.266

Sardar, H., Anjum, M. A., Hussain, S., Ali, S., Shaheen, M. R., Ahsan, M., Ejaz, S., Ahmad, K. S., Naz, S., & Shafique, M. (2022). Deciphering the role of moringa leaf powder as a supplement in the cotton waste substrate for the growth and nutrition of king oyster mushroom. *Scientia Horticulturae*, *293*. https://doi.org/10.1016/j.scienta.2021.110694

Sayara, T., Basheer-Salimia, R., Hawamde, F., & Sánchez, A. (2020). Recycling of organic wastes through composting: Process performance and compost application in agriculture. *Agronomy*, *10*(11), 1838. https://doi.org/10.3390/agronomy10111838

Sczostak, L. (2009). *Cotton Linters: An Alternative Cellulosic Raw Material*. Cellulose Chemistry and Technology, 43(9-10), 419–426. Retrieved from https://www.cellulosechemtechnol.ro/pdf/CCT9-10-2009/p.419-426.pdf

Senthil Kumar, P., & Yaashikaa, P. R. (2019). Organic cotton and its environmental impacts. In M. A. Gardetti & S. S. Muthu (Eds.), *Organic Cotton: Is it a Sustainable Solution?* (pp. 127–139). Springer Singapore. https://doi.org/10.1007/978-981-10-8782-0_7

Shaghaleh, H., Xu, X., & Wang, S. (2018). Current progress in production of biopolymeric materials based on cellulose, cellulose nanofibers, and cellulose derivatives. *RSC Advances*, *8*(2), 825–842.

Sharma, N., Sahota, P. P., & Singh, M. P. (2020). Organic acid production from agricultural waste (pp. 415–438). https://doi.org/10.1007/978-981-33-4347-4_17

Sharma-Shivappa, R. R., & Chen, Y. (2008). Conversion of cotton wastes to bioenergy and value-added products. *Transactions of the ASABE*, *51*(6), 2239–2246.

Shen, L. Y., Tam, V. W. Y., Tam, C. M., & Drew, D. (2004). Mapping approach for examining waste management on construction sites. *Journal of Construction Engineering and Management*, *130*(4), 472–481. https://doi.org/10.1061/(ASCE)0733-9364(2004)130:4(472)

Shi, Y., Jiang, J., Ye, H., Sheng, Y., Zhou, Y., Foong, S. Y., Sonne, C., Chong, W. W. F., Lam, S. S., Xie, Y., Li, J., & Ge, S. (2023). Transforming municipal cotton waste into a multilayer fiber biocomposite with high strength. *Environmental Research*, *218*, 114967. https://doi.org/10.1016/j.envres.2022.114967

Shirvanimoghaddam, K., Czech, B., Wiącek, A. E., Ćwikła-Bundyra, W., & Naebe, M. (2019). Sustainable carbon microtube derived from cotton waste for environmental applications. *Chemical Engineering Journal, 361*, 1605–1616. https://doi.org/10.1016/j.cej.2018.11.157

Shogren, R., Wood, D., Orts, W., & Glenn, G. (2019). Plant-based materials and transitioning to a circular economy. *Sustainable Production and Consumption, 19*, 194–215. https://doi.org/10.1016/j.spc.2019.04.007

Silvia. (2021, September 29). Composting Ideas: Can Cotton Be Composted? https://enviromom.com/can-cotton-be-composted/

Singh, V., Wyatt, J., Zoungrana, A., & Yuan, Q. (2022). Evaluation of vermicompost produced by using post-consumer cotton textile as carbon source. *Recycling, 7*(1), 10. https://doi.org/10.3390/recycling7010010

Sołowski, G., Konkol, I., & Cenian, A. (2020a). Methane and hydrogen production from cotton wastes in dark fermentation process under anaerobic and microaerobic conditions. In *Frontiers in Water-Energy-Nexus—Nature-Based Solutions, Advanced Technologies and Best Practices for Environmental Sustainability: Proceedings of the 2nd WaterEnergyNEXUS Conference* (pp. 285–287), November 2018, Salerno, Italy. Springer International Publishing.

Sołowski, G., Konkol, I., Shalaby, M., & Cenian, A. (2020b). Rapid hydrogen generation from cotton wastes by mean of dark fermentation. *SN Applied Sciences, 2*(8). https://doi.org/10.1007/s42452-020-03247-3

Sotelo-Navarro, P. X., & Poggi-Varaldo, H. M. (2021). Hydrogen from dark fermentation of the organic fraction of waste diapers: Optimization based on response surface experiments. *Frontiers in Energy Research, 9*. https://doi.org/10.3389/fenrg.2021.630212

Spears, S. (2018, May 16). What Is Biochar? *Regeneration International*. https://regenerationinternational.org/2018/05/16/what-is-biochar/

Stein, Z. (2023, July 19). Bioethanol | Definition, Production Process, Uses, & Benefits. www.carboncollective.co. https://www.carboncollective.co/sustainable-investing/bioethanol

Sundqvist Ökvist, L., & Lundgren, M. (2021). Experiences of bio-coal applications in the blast furnace process—Opportunities and limitations. *Minerals, 11*(8), 863.

Suriani, M. J., Radzi, F. S. M., Ilyas, R. A., Petrů, M., Sapuan, S. M., & Ruzaidi, C. M. (2021). Flammability, tensile, and morphological properties of oil palm empty fruit bunches fiber/pet yarn-reinforced epoxy fire retardant hybrid polymer composites. *Polymers, 13*(8), 1282. https://doi.org/10.3390/polym13081282

Tamburini, M., Pernetti, R., Anelli, M., Oddone, E., Morandi, A., Osuchowski, A., Villani, S., Montomoli, C., & Monti, M. C. (2023). Analysing the impact on health and environment from biogas production process and biomass combustion: A scoping review. *International Journal of Environmental Research and Public Health, 20*(7). https://doi.org/10.3390/ijerph20075305

Tausif, M., Jabbar, A., Naeem, M. S., Basit, A., Ahmad, F., & Cassidy, T. (2018). Cotton in the new millennium: Advances, economics, perceptions and problems. *Textile Progress, 50*(1), 1–66. https://doi.org/10.1080/00405167.2018.1528095

Teklehaimanot, M., Hailay, H., & Tesfaye, T. (2021). Manufacturing of ecofriendly bricks using microdust cotton waste. *Journal of Engineering*, 2021, 1–10. https://doi.org/10.1155/2021/8815965

Thyavihalli Girijappa, Y. G., Mavinkere Rangappa, S., Parameswaranpillai, J., & Siengchin, S. (2019). Natural fibers as sustainable and renewable resource for development of eco-friendly composites: A comprehensive review. *Frontiers in Materials, 6*. Retrieved from https://www.frontiersin.org/articles/10.3389/fmats.2019.00226

Tsatiris, V.D., Siminis, C.I., Kefaki, M.D., Manios, V.I., & Davaris, D. (1998). Composting Cotton Mill Waste. In *Proceedings of the World Cotton Research Conference-2* (pp. 1121–1125). Athens, Greece. https://www.icac.org/Content/EventDocuments/PdfFiles6d70c992_e5cc_4157_904f_4dd96caaa48b/Composting%20Cotton%20Mill%20Waste.pdf

Tutus, A., Ezici, A. C., & Ates, S. (2010). Chemical, morphological and anatomical properties and evaluation of cotton stalks (Gossypium hirsutum l.) in pulp industry. *Scientific Research and Essays, 5*(12), 1553–1560. https://www.academicjournals.org/SRE

Twizerimana, M., M'Arimi, M. M., Nganyi, E. O., Omara, T., Olomo, E., & Kawelamzenje, N. A. (2021). Anaerobic digestion of cotton yarn wastes for biogas production: Feasibility of using sawdust to control digester temperature at room temperature. *OALib, 08*(07), 1–15. https://doi.org/10.4236/oalib.1107654

Ute, T. B., Celik, P., & Uzumcu, M. B. (2019). Utilization of cotton spinning mill wastes in yarn production. In *Textile Industry and Environment*. https://doi.org/10.5772/intechopen.85127

Ütebay, B., Çelik, P., & Çay, A. (2019). Effects of cotton textile waste properties on recycled fiber quality. *Journal of Cleaner Production, 222*, 29–35. https://doi.org/10.1016/j.jclepro.2019.03.033

Uvarov, R.A., Briukhanov, A., Semenov, B., & Nazarova, A.V. (2020). Cattle barn bedding from recycled manure: Some veterinary, technological and economic aspects of application. *BIO Web of Conferences, 27*, 7.

Venkataraman, D., Shabani, E., & Park, J. H. (2023). Advancement of nonwoven fabrics in personal protective equipment. *Materials*, *16*(11). https://doi.org/10.3390/ma16113964

Vukcevic, M., Maletic, M., Pejic, B., Karić, N., Trivunac, K., & Perić-Grujić, A. (2023). Waste hemp and flax fibers and cotton and cotton/polyester yarns for removal of methylene blue from wastewater: Comparative study of adsorption properties. *Journal of the Serbian Chemical Society*, *88*(6), 669–683. doi:10.2298/JSC221213015V

Waheed, A., Naqvi, S. R., & Ali, I. (2022). Co-torrefaction progress of biomass residue/waste obtained for high-value bio-solid products. *Energies*, *15*(21), 8297. https://doi.org/10.3390/en15218297

Wang, T., Zhai, Y., Li, H., Zhu, Y., Li, S., Peng, C., Wang, B., Wang, Z., Xi, Y., Wang, S., & Li, C. (2018). Co-hydrothermal carbonization of food waste-woody biomass blend towards biofuel pellets production. *Bioresource Technology*, *267*, 371–377. https://doi.org/10.1016/j.biortech.2018.07.059

Wang, Y. (2010). Fiber and textile waste utilization. *Waste and Biomass Valorization*, *1*(1), 135–143. https://doi.org/10.1007/s12649-009-9005-y

Wang, Z., Bui, Q., Zhang, B., & Pham, T. L. H. (2020). Biomass energy production and its impacts on the ecological footprint: An investigation of the G7 countries. *Science of the Total Environment*, *743*. https://doi.org/10.1016/j.scitotenv.2020.140741

Weon, H. Y., Kim, B. Y., Kwon, S. W., Park, I. C., Cha, I. B., Tindall, B. J., Stackebrandt, E., Trüper, H. G., & Go, S. J. (2005). Leadbetterella byssophila gen. nov., sp. nov., isolated from cotton-waste composts for the cultivation of oyster mushroom. *International Journal of Systematic and Evolutionary Microbiology*, *55*(6), 2297–2302. https://doi.org/10.1099/ijs.0.63741-0

Wojnowska-Baryła, I., Bernat, K., & Zaborowska, M. (2022). Strategies of recovery and organic recycling used in textile waste management. *International Journal of Environmental Research and Public Health*, *19*(10), 5859. https://doi.org/10.3390/ijerph19105859

Yetisen, A. K., Qu, H., Manbachi, A., Butt, H., Dokmeci, M. R., Hinestroza, J. P., Skorobogatiy, M., Khademhosseini, A., & Yun, S. H. (2016). Nanotechnology in textiles. *ACS Nano*, *10*(3), 3042–3068. https://doi.org/10.1021/acsnano.5b08176

Yilmaz, H., Topakci, M., Karayel, D., & Çanakci, M. (2020). Comparison of the physical properties of cotton and sesame stalk pellets produced at different moisture contents and combustion of the finest pellets. *Energy Sources, Part A: Recovery, Utilization and Environmental Effects*. https://doi.org/10.1080/15567036.2020.1850931

Yousef, S., Tatariants, M., Tichonovas, M., Sarwar, Z., Jonuškienė, I., & Kliucininkas, L. (2019). A new strategy for using textile waste as a sustainable source of recovered cotton. *Resources, Conservation and Recycling*, *145*, 359–369. https://doi.org/10.1016/j.resconrec.2019.02.031

Yusuf, M. (2017). Agro-industrial waste materials and their recycled value-added applications: Review. https://doi.org/10.1007/978-3-319-48281-1_48-1.

Zabaniotou, A., & Andreou, K. (2010). Development of alternative energy sources for GHG emissions reduction in the textile industry by energy recovery from cotton ginning waste. *Journal of Cleaner Production*, *18*(8), 784–790.

Zafar, M. W., Sinha, A., Ahmed, Z., Qin, Q., & Zaidi, S. A. H. (2021). Effects of biomass energy consumption on environmental quality: The role of education and technology in Asia-Pacific Economic Cooperation countries. *Renewable and Sustainable Energy Reviews*, *142*. https://doi.org/10.1016/j.rser.2021.110868

Zafar, S. (2022, June 8). Advantages of Biobutanol. *BioEnergy Consult*. https://www.bioenergyconsult.com/tag/advantages-of-biobutanol/

Section IV

Bast-Based Plant Waste

10 Banana Waste Management

10.1 INTRODUCTION

Banana is a fruit that is grown in most countries especially countries that possess tropical climates. It is one of the most produced fruits, contributing to almost 16% of the global production of fruit (Wobiwo et al., 2017). It is also the fourth most significant food crop, after rice, wheat, and maize. Table 10.1 provides the world banana production in 2019 of several main producer countries of bananas. The banana is an evergreen herbaceous plant with high calcium, magnesium, and nitrogen assimilation emissions that belongs to the Zingiberales order, Musaceae family, and *Musa* genus. The banana is a fruit that varies in size, color, and firmness. Its green, fleshy skin turns yellow as it ages and turns brown when it is fully ripe (Mostafa, 2021). Cone-shaped fruit develops from the plant's top up to 15 m in length during 80–180 days. Bananas, which include more than 50 species and dozens of hybrids, have huge, underground stems that are rhizomatous from which their large leaves and strongly spiraled pods emerge, giving the appearance of a fake stem (pseudo-stem), as shown in Figure 10.1.

The banana is an herbaceous plant with a high nutritional value, which includes a lot of minerals like calcium, potassium, and the like. The assimilation emissions that belong to the Zingiberales order, Musaceae family, and *Musa* genus include emission in nitrogen (Mostafa, 2021). The banana is a fruit that varies in size, color, and firmness. Its green, fleshy skin turns yellow as it ages and turns brown when it is fully ripe. FAOSTAT (2018) estimates that 115.7 million tons of fruit are produced globally each year. With a global production share of 26.7% and an annual output of over 30.81 million tons from an area of 0.88 million hectares, India is the world's top banana producer. China, Indonesia, Brazil, Ecuador, and the Philippines are all major producers. India accounts only for 0.38% of global exports because most output is consumed locally (Ahmad & Danish, 2018).

Although bananas are a very popular fruit, about 60% of the biomass is wasted after harvest. According to statistics, approximately 114.08 million metric tons of trash from banana production results in environmental issues such as excessive greenhouse gas (GHG) emissions. These wastes have a high concentration of materials that are extremely important for the industrial world, like natural fibers, which can be processed by a series of specific unit operations, like bacterial fermentation and anaerobic degradation, to create bioplastics, organic fertilizers, and biofuels like ethanol,

TABLE 10.1
World Banana Production in 2019 (Alzate Acevedo et al., 2021)

Countries	Banana Production (Millions of Tons)
India	30.4
China	11.6
Indonesia	7.2
Brazil	6.8
Ecuador	6.5
Philippines	6

DOI: 10.1201/9781003472209-14

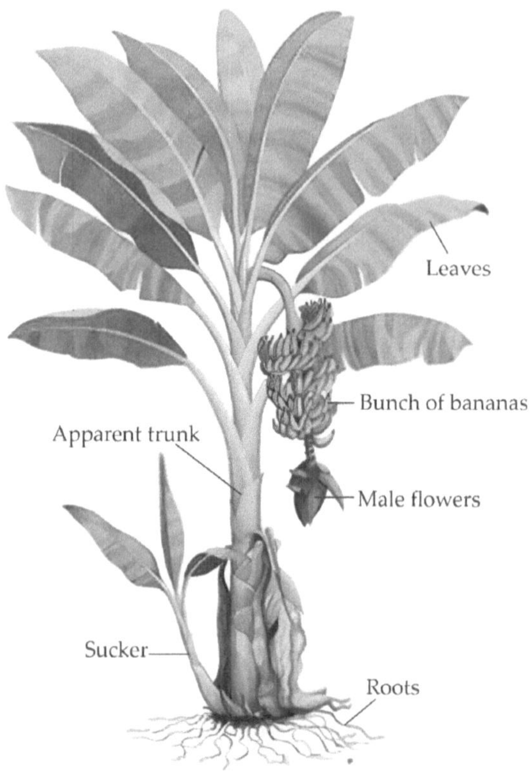

FIGURE 10.1 Parts of the banana plant.

biogas, and hydrogen (Alzate Acevedo et al., 2021). Thus, the chapter's focus is on the expanding utilization of banana waste in many fields. A thorough explanation of its features and several therapies is provided. In addition, the use of banana wastes in pulp and paper, biocomposites, biochar, bio-bricks, bioplastics, biohydrogen, biofertilizer, biomethane, bioethanol, biobutanol, and bioelectricity is also studied.

10.2 BANANA FIBER: A NATURAL AND SUSTAINABLE BIORESOURCES

Natural fibers of plant and animal origin have undergone extensive research over the past 20 years and are now taking precedence over synthetic counterparts due to their superior biochemical and strength capabilities, biodegradability, and renewability. Bananas, coir, sisal, jute, kenaf, and a variety of other plant-based natural fibers have been researched for industrial applications. Banana fiber is particularly interesting because it is one of the most consumed fruits worldwide and is grown in 130 countries, making banana waste available domestically. Annual banana production was 115.7 million tons in 2018. Fruit, fruit peel, flower buds, leaves, and pseudo-stem (banana trunk) from banana plants are utilized in a variety of industrial applications (Balda et al., 2021). The majority of banana waste biomass is pseudo-stem, which creates high-quality fiber with industrial applications in the production of sanitary pads, textiles, pulp and paper, food, and reinforced composite materials for use in aerospace, vehicles, and other composite products. Masks made of banana fiber have recently been developed by the Philippine Department of Science and Technology to reduce the use of single-use plastic. This chapter describes several fiber extraction techniques, banana fiber's biochemical and mechanical characteristics, and potential uses (Balda et al., 2021). Figure 10.2 shows a general flowchart of processing the banana pseudo-stem (BPS) via several steps for different applications.

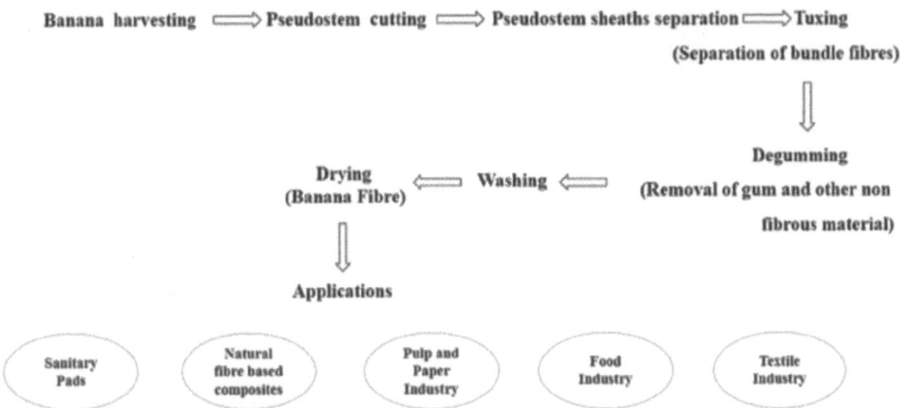

FIGURE 10.2 Banana fiber: a natural and sustainable bioresource for eco-friendliness (Balda et al., 2021).

10.2.1 Characteristics of Banana Fiber

Banana fiber, which originates from the plant's pseudo-stem, has a similar appearance to ramie and bamboo fibers. Pseudo-stems have a structure of 14–18 sheaths and produce fibers of different quality depending on their location. Coarse fiber (the outermost two to three sheaths), which breaks easily and is thus rejected; soft, glossy fiber (the middle sheaths); and extremely soft fiber (a few middle and inside sheaths except the innermost sheaths) are among these fibers. The pulpy substance found in the innermost sheaths is not processed to create fibers. Over 600 kg/ha of banana fiber can be generated using pseudo-stem (Sawarkar et al., 2022).

10.2.2 Fiber Recovery

Different reactors were given up at various points throughout digestion. About 0.5 mm mesh sieves were used to remove the fibers from the digestate, and the mixed fluid was put into the sieve after the bioreactors had been shaken. A freshwater jet was used to wash away the fibers that had been caught on the sieves until the filtrate was clear. After washing, the fibers were placed into a beaker for a second cleaning, where non-fibrous residues were removed by agitating the fibers suspended in new water, as shown in Figure 10.3. The procedure was done at least five times to obtain clean fibers. When fibers were extracted, they were either taken for qualitative and quantitative tests immediately or were stored at 20°C (Wobiwo et al., 2017).

FIGURE 10.3 Fiber recovery: (a) sieve topped digestate; (b) fibers in a beaker (Wobiwo et al., 2017).

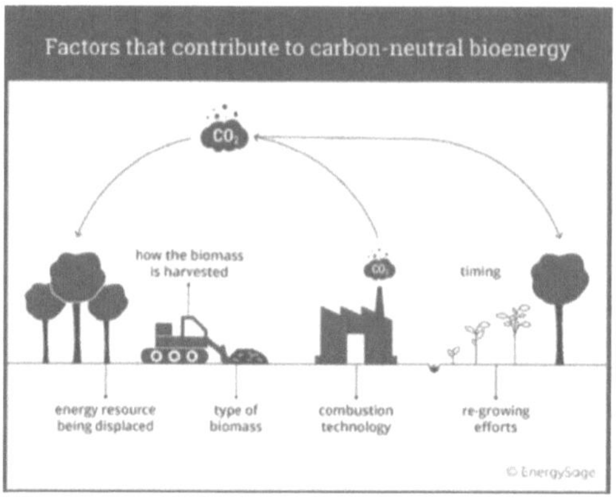

FIGURE 10.4 Factors that contribute to carbon-neutral bioenergy (Wobiwo et al., 2017).

10.3 ENVIRONMENTAL IMPACT OF BIOMASS

When people burn biomass for heat or energy, carbon dioxide is emitted into the atmosphere. However, during the photosynthetic process, biomass sources like trees and agricultural crops also trap carbon dioxide. While trees and other plants absorb as much carbon dioxide as is emitted when burning biomass, as shown in Figure 10.4, the carbon cycle is in equilibrium (Wobiwo et al., 2017). The carbon impact of bioenergy is dependent on the combustion method, how the biomass is gathered, any attempts at re-growth, the type of biomass used, timing, and the energy resource it is replacing, so it's not quite that straightforward (Balda et al., 2021).

Bioethanol can be made from banana waste due to its high amounts of cellulose (more than 30% of the total on a dry basis) and hemicelluloses (25% of the total) (Rojas-Flores et al., 2022). These wastes are produced in significant quantities during the banana growing and harvesting stages. This study proposes utilizing other organic acids (tartaric, oxalic, and citric) during the acid pre-treatment step in addition to sulfuric acid to limit the production of unwanted compounds and improve bioethanol synthesis. Instead, a solid waste flow from the manufacturing of bioethanol is managed in an anaerobic digestion (AD) facility to produce digestate, which may be used as a biofertilizer, and biogas, which can be turned into electricity (Rojas-Flores et al., 2022). Figure 10.5 illustrates the system bounds of biomass biorefinery in general.

10.4 AVAILABILITY OF BANANA WASTE

A banana peel is the outside protective skin of a regular yellow banana. Bananas are typically 15–23 cm in length and come in bunches of six to eight bananas. The medium-thick, unique yellow banana skin turns from deep green to yellow as it ripens and then progressively develops brown spots until it is entirely coated in brown. The skin encloses and protects the banana's semi-starchy, ivory-cream-colored flesh. Unripe banana peels are thicker and more fibrous; as they ripen, they thin and become sweeter (Balda et al., 2021).

Banana peels are available year-round and widely accessible. After harvest, over 60% of the biomass produced by bananas is squandered (Rojas-Flores et al., 2022). Globally, around 114.08 million metric tons of banana garbage are created, causing environmental difficulties such as excessive GHG emissions (Rojas-Flores et al., 2022). The fruit can be harvested 4–6 weeks after the fingers finish growing. When the small blooms at the end are dry and easy to rub off, and the bananas appear well-rounded between the ribs, they are ready to be picked. For bananas, it is recommended to remove the entire stalk (Rojas-Flores et al., 2022).

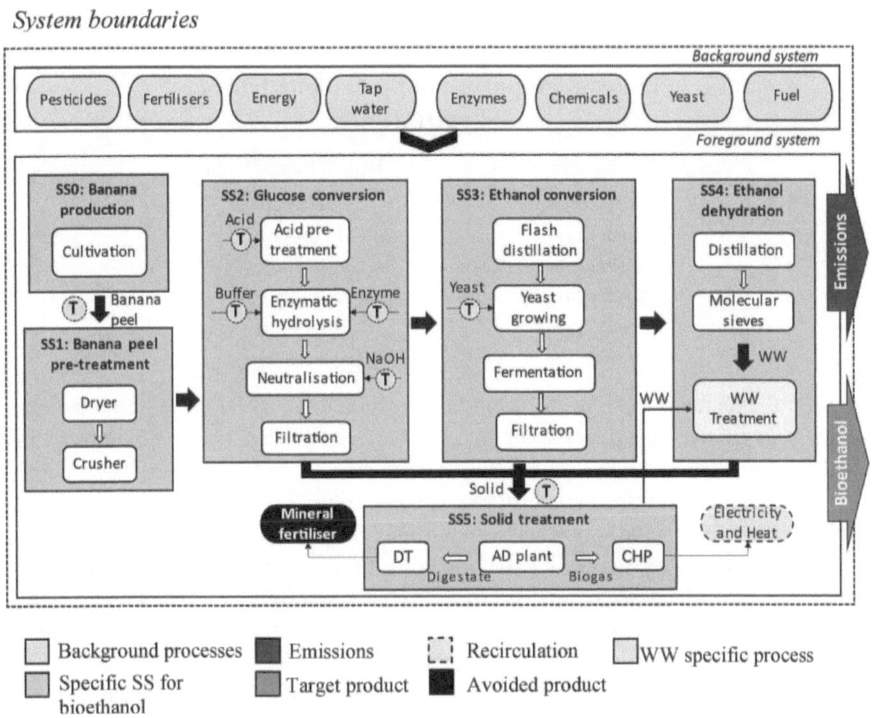

FIGURE 10.5 System bounds of the case study suggested as a biomass refinery approach. AD, anaerobic digestion; CHP, combined heat and power; DT, sewage treatment; SS, subsystem; T, transport; WW, wastewater (Rojas-Flores et al., 2022).

10.5 PROPERTIES OF BANANA WASTE FIBER

Fibers can be generally categorized into synthetic fiber and natural fiber. Synthetic fiber is defined as fibers that are made via chemical synthesis, especially from petrochemical raw material, while natural fibers are derived directly from living organisms. In recent years, due to the rising awareness of the global population toward sustainability, an alternative synthetic fiber, which is mostly non-biodegradable, is gradually being replaced by natural fiber. Natural fiber is more environmentally friendly due to its biodegradability, non-toxicity, and possession of a low carbon footprint (Akatwijuka et al., 2022). Natural fiber can be further divided into plant, animal, and mineral fibers. In the context of agro-waste, the plant fibers would be focused particularly on banana waste fibers as it is one of the crucial fruit crops worldwide. For banana waste, the main plant parts that are involved in the synthesis of the agro-waste fiber are the leaf and stalk (Akatwijuka et al., 2022). Therefore, to develop new technology, studies on various properties are crucial to researchers and technologists to gain deeper insights into the banana waste fiber, which would be useful in the development of new technologies or materials (Akatwijuka et al., 2022).

10.5.1 Physical Properties

The characterization in the context of the physical properties of the banana waste fiber includes the diameter of the single-cell fiber, the length of the single-cell fiber, aspect ratio, bulk density, color, and moisture content. Therefore, the range of value of these physical properties can be very wide. According to Bhatnagar et al. (2015), a typical range of some of the physical properties based on the single-cell fiber of banana waste fiber are given in Table 10.2 (Bhatnagar et al., 2015).

TABLE 10.2
Physical Properties of the Banana Fibers (Bhatnagar et al., 2015)

Physical Properties	Typical Range/Value
Diameter (μm)	80–250
Length (mm)	1,000–5,000
Moisture content (%)	60
Aspect ratio (L/d)	150
Density (kg/m^3)	950–750

However, for the diameter part, based on an investigation done by Pappu et al. on the research conducted by different scholars, the diameter of banana fiber is an average of 206.5 micrometers, while for the length of the fiber, the variation is an average of 90cm according to the pseudo-stem's length (Pappu et al., 2015). It was also found that the moisture content is approximately 20.23% compared to the 60% given in Table 10.2, while the density should vary from 1,300 to 1,400 kg/m^3 (Pappu et al., 2015). This discrepancy shows that it is very difficult to obtain uniform-quality banana fibers. This is because physical properties could be influenced by various factors such as harvesting time, plant part, origin, plant age, and extraction method (Akatwijuka et al., 2022). The color of the banana fiber is clear white.

The diameter of pseudo-stem fibers extracted mechanically using methods such as manual extraction or mechanical decortication is the highest. Meanwhile, fibers extracted chemically using methods such as alkali or acid treatment have lower diameters, and hybrid extraction techniques yield the lowest-diameter fibers. The diameter decreases regardless of methods because impurities adhered to the fiber's outer surfaces are excluded (Akatwijuka et al., 2022).

The density of banana fiber can be measured using a pycnometer, a device that measures the density of a substance by determining its volume and weight. First, a sample of the fiber is weighed, and its volume is measured by filling the pycnometer with a known amount of liquid and then immersing the fiber in the liquid. The volume of the displaced liquid is then measured, and the density of the fiber is calculated by dividing its weight by its volume (Rohen et al., 2017). It is also important to note that there is no one specific standard tailored for physical properties measurement.

10.5.2 Chemical Composition

Table 10.3 provides the chemical composition of banana plants according to a chapter done by Bhatnagar et al. (2015). It can be noticed from the table that cellulose should be the main component in the banana plant waste with a composition of at least 60%, followed by hemicellulose, lignin, and finally pectin or extractive. According to the data compiled by Pappu et al. (2015) from various studies and illustrated in Figure 10.6, the values and ratios of each component show slight differences compared to those in Table 10.3. However, the overall trend in component abundance remains similar, with cellulose identified as the primary component in banana plant fiber, which contributes to its white color (Venkateshwaran & Elayaperumal, 2010). It is important to note that the chemical composition of banana fiber is strongly influenced by the banana species and the environmental conditions in which it is cultivated.

10.5.3 Mechanical Properties

In general, the main mechanical properties that were being measured were tensile strength and tensile modulus (Young modulus). Other mechanical properties such as elongation, strain, and impact strength were also being studied. The typical range for the tensile strength and young modulus is approximately an average of 721.5 MPa and 29.5 GPa, respectively (Bhatnagar et al., 2015), while

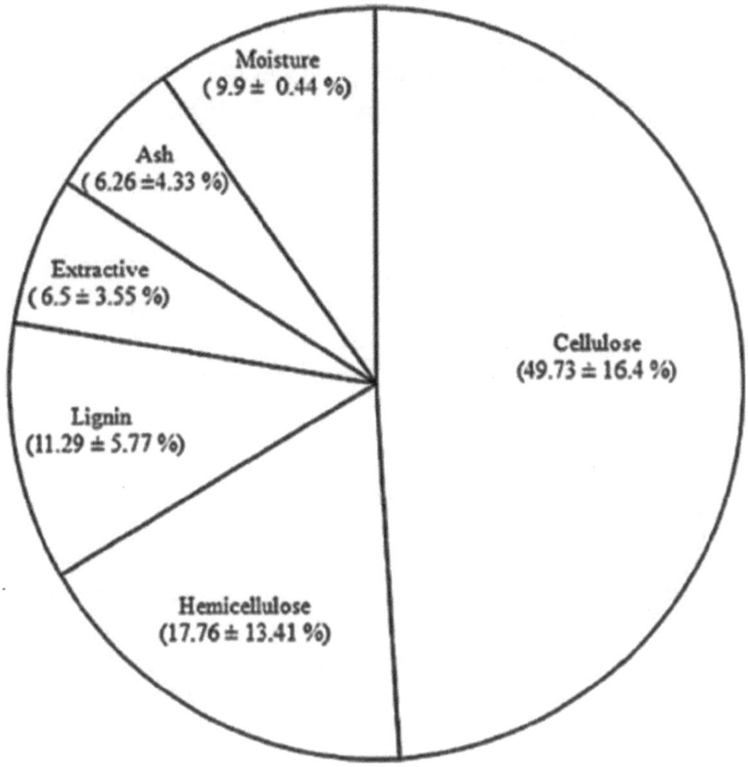

FIGURE 10.6 Biological macromolecules of banana fibers (Pappu et al., 2015).

TABLE 10.3
Chemical Constitution of Typical Banana Fibers (Bhatnagar et al., 2015)

Components	Content (%)
Cellulose	60–65
Hemicellulose	6–19
Lignin	5–10
Pectin	3–5
Ash	1–3
Extractive	3–6

from the investigation done by Pappu et al. (2015), the tensile strength is illustrated to be an average of 458±257 MPa and 17.14±10.72 GPa, respectively (Pappu et al., 2015). One of the crucial factors that affect the mechanical properties of the banana waste fiber is the chemical composition such as cellulose, hemicellulose, and particularly lignin, which affect the tensile and impact strength significantly. Therefore, the selection of treatment methods for banana waste fiber such as alkali treatment and manual extraction is very important to ensure that the banana waste fiber possesses the desired mechanical properties before further processing (Nadhari et al., 2019).

There are several methods being used to test the mechanical properties of banana fiber. To test the tensile strength, the typical equipment being used is the universal testing machine, referred to as the ASTM D638 standard, which is tailored for the testing of resin mechanical properties for banana fiber-reinforced composites (Ramesh et al., 2014). The typical mechanical parameters can be found on ASTM C1557-03. Table 10.4 provides the typical mechanical properties of banana waste fibers.

TABLE 10.4
Typical Mechanical Properties of Banana Wastes Fibers (Bhatnagar et al., 2015)

Mechanical Properties	Value
Tensile strength (MPa)	529–614
Young's modulus (GPa)	27–32
Specific tensile strength (MPa)	392–677
Specific Young's modulus (GPa)	20–24
Density (kg/m^3)	750–950
Failure strain (%)	1–3

10.5.4 Morphological Properties

From the atomic bonding perspective using the FTIR spectroscopy method, it was found that there is a strong, broader band of 3326, indicating the presence of hydroxyl group stretching, and for the untreated banana fiber, the presence of impurity was indicated by wavenumber of 2273. However, the impurities could be removed by proper treatment (Jagadeesh et al., 2015). Besides the hydroxyl group stretching, a more detailed spectrum includes an intense band at 2,900 cm^{-1} due to the C–H sigma bond stretching, 1,036 cm^{-1} due to the C–O bond stretching, 1,725 cm^{-1} due to the carbonyl group stretching, 1,150 cm^{-1} due to the ether group stretching, and 1,500–1,600 cm^{-1} due to the aromatic carbon–carbon bond skeletal vibration, as shown in Figure 10.7 (Milani et al., 2016).

Next, from the morphological characterization using scanning electron microscopy (SEM), in general, a single banana fiber has elliptical shape hoses in its cross-sectional structure (Jagadeesh et al., 2015). Therefore, it indicated that a bundle of banana fibers consists of multiple hollow sub-fibers, which possess a lumen structure with numerous interconnected air pores and cavities (Bakri et al., 2017). Figure 10.8 illustrates the structure of both treated and untreated banana fibers from several perspectives using SEM and an optical microscope; the different morphological description is also provided in the figure. For the analysis using X-ray powder diffraction, it was found that there is a presence of both crystalline and amorphous regions in banana fibers of various species. The trend of the diffractogram illustrated in Figure 10.9 showed the similarity of the banana fibers with other lignocellulose dominant fibers (Milani et al., 2016).

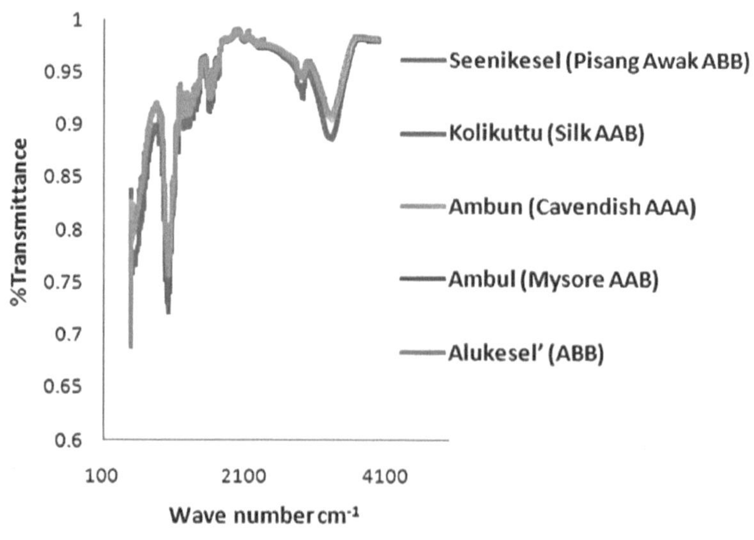

FIGURE 10.7 FTIR spectra of banana fiber of five various species in an experiment conducted by Milani et al. (2016).

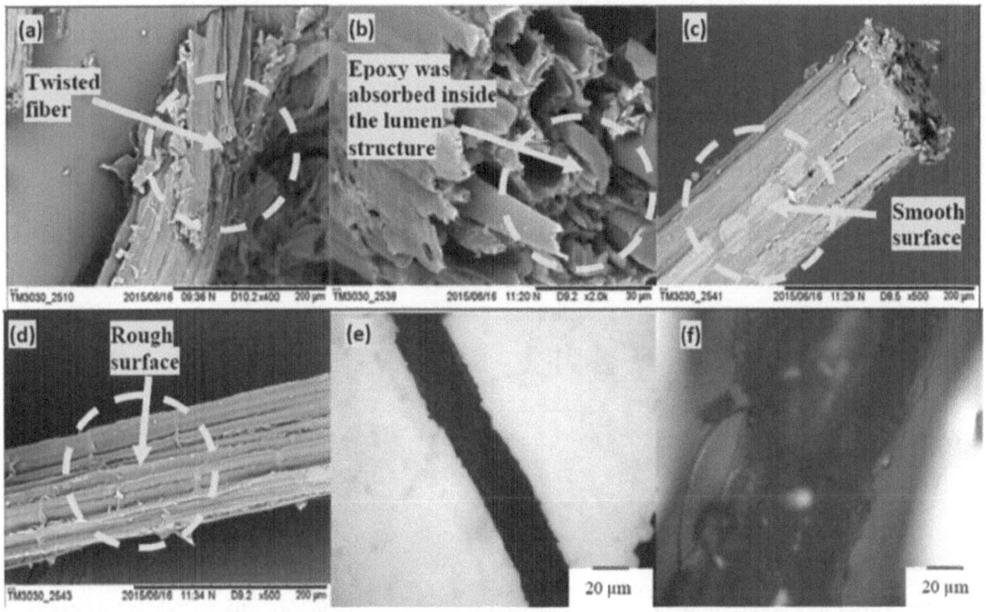

FIGURE 10.8 Image from SEM micrographs of banana fibers: (a) tangled banana fibers, (b) hollow lumen structure, (c) untreated banana fiber, and (d) treated banana fiber. Light/optical micrographs of banana fiber illustrating (e) untreated banana fiber and (f) treated banana fiber (Bakri, 2015).

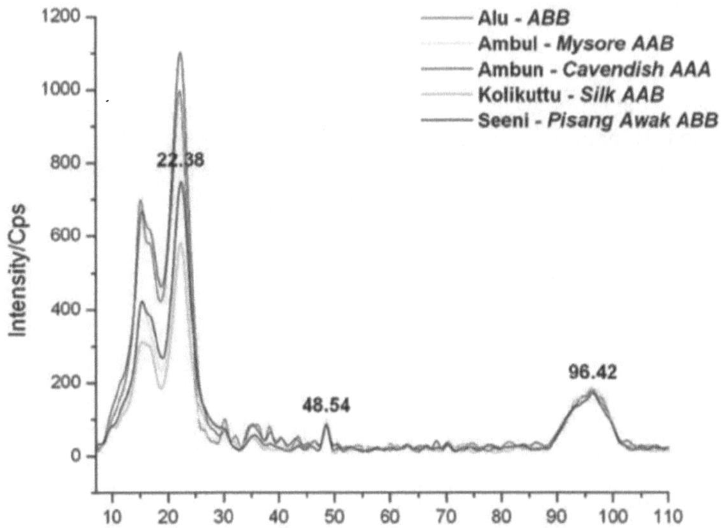

FIGURE 10.9 X-ray diffraction patterns of banana fibers of five types of banana species according to an experiment done by Milani et al. (2016).

10.5.5 Thermal Properties

The common criteria used to assess the thermal properties of the banana fibers include thermal stability, which is determined by the maximum temperature at which the fiber starts to decompose. Next, the thermal properties also covered thermal conductivity, heat capacity, and the like. The typical methods available to test those properties include differential scanning calorimetry (DSC) and thermogravimetric analysis (TGA). For the DSC testing, according to the research done by Milani et al. (2016), the thermal stability of the banana fiber ranged from 280 to 350 from the exothermic peaks.

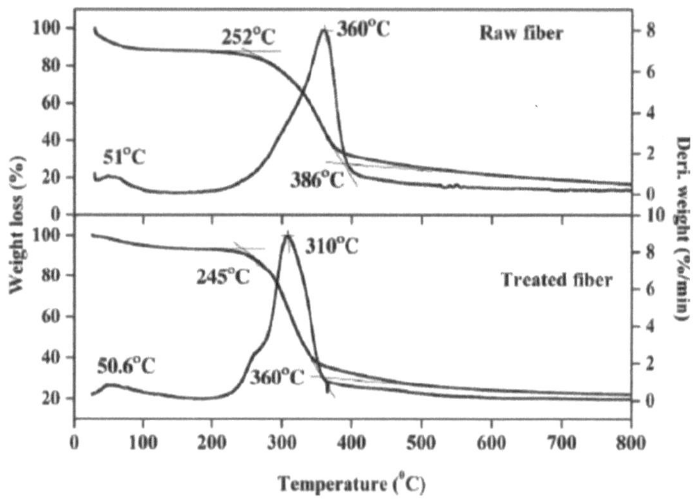

FIGURE 10.10 TG (Thermogravimetry)/DTG (Derivative Thermogravimetry) of raw and treated banana fibers (Ronald Aseer et al., 2013).

This is due to the degradation of the chemical composition of the banana fibers, particularly hemicellulose and α-cellulose due to subjection to high temperature. It is also important that the result is almost similar to the research done by Ronald Aseer et al. (2013), which possesses the first exothermic peak at a range of 280–320 cm^{-1}. The initial endothermic trough is due to the heat of vaporization of water, which had been attached to the fiber.

For the TGA and its differential analysis, based on the result obtained by Ronald Aseer illustrated in Figure 10.10 and tabulated in Table 10.5 in terms of the percentage of weight loss, the trend of degradation can be observed (Ronald Aseer et al., 2013). It can be shown that raw banana fiber undergoes three stages of thermal degradation. The first stage involves moisture loss at 51°C with 6% degradation (Ronald Aseer et al., 2013). The main stage occurs at 252–386°C with 60% degradation due to cellulose depolymerization and loss of hydroxyl groups, while the final stage occurs at 386–700°C due to the decomposition of lignin, which is difficult to degrade (Ronald Aseer et al., 2013).

10.5.6 Calorific Value

The calorific value is usually obtained by burning a mass of sample in the calorimeter. There are a variety of calorimeters that could be used to determine the calorific value of the agricultural waste fiber. However, the type of calorimeter that is commonly used in agricultural waste is the adiabatic bomb calorimeter at which the sample is first burnt in a vessel (bomb) under oxygen to ensure complete combustion (Bomb Calorimeter, 2015). A bucket of water adjacent to it will absorb the heat, and the change in temperature can be detected. The calorific value can therefore be determined using a simple equation of $q = -c\Delta T$ over the mass of the

TABLE 10.5
Mass Loss of Treated and Raw Banana Fibers (in Percent) at Different Temperatures (Ronald Aseer et al., 2013)

Sample	100°C	200°C	300°C	400°C	500°C	600°C	Char Content (%)
Raw fiber	10	12	24	66	75	79	14
Treated fiber	5	7	26	67	74	76	21

sample, where c is the heat capacity of the calorimeter and ΔT is the temperature change. The standard normally used is the ASTM D2015 testing method, which provides accurate measurement of calorific value. The heating value of banana waste is estimated to be 15.5–17.7 (MJ/kg) (Abdullah et al., 2013).

10.6 BANANA WASTE MANAGEMENT

10.6.1 Biomethane

Biomethane, or biogas, is a bioenergy derived from biomass decomposition or chemical conversion (Koul et al., 2022). Another potential environmentally acceptable low-carbon fuel gas alternative to conventional cookstoves that burn wood or charcoal is biomethane (Koul et al., 2022). The use of banana waste has the potential to generate biomethane as stated by the previous studies. Wobiwo et al. (2017) state that three varieties of banana peduncles, which are Grande Naine (GN), Pelipita (PPT), and CRPP969 (phytopathogen-resistant hybrid plantain), obtained from the African Research Centre of Bananas and Plantains (CARBAP) and Plantations Haut Penja, in biomethane and fiber recovery through the anaerobic digester. Natural lignocellulosic fibers are separated and recovered using AD systems. The outcomes showed that GN and CRBP969 were found to be more beneficial for the generation of methane compared to the recovery of fiber because of the lower initial fiber concentration and more digestible fibers. A total of 200–204 m^3/(ha·year) of methane is produced per annum from GN banana peduncles, which is degraded in less than 20 days (Wobiwo et al., 2017).

Additionally, under a 7.5% total solids (TS) condition, AD of fresh banana peel produced a maximum biomethane, a production rate of 5.31 mL/h. The experiment was carried out in a 0.5 L batch reactor at pH 7 under a mesophilic condition of 37°C (Pisutpaisal et al., 2014). Housagul et al. (2014) investigated the use of banana peel with varied total solids in the range of 2.5–10% w/v, and banana peel and leftover or pure glycerol are two examples of the wastes employed. Every treatment was set up twice in a 0.5 L batch reactor during the mesophilic fermentation. Every 12 and 24 h, respectively, samples of gas and liquid were taken. At 7.5% banana peel, the potential of maximum methane production (181 mL) results at 188 mL/g TS. When 7.5% banana peel was combined with 7.5 g/L pure glycerol (in the presence of buffer), maximum methane production (467 mL) resulted in chemical oxygen demand (COD)of 151 mL/g, while when 7.5% banana peel was combined with 7.5 g/L waste glycerol (in the presence of buffer) (Housagul et al., 2014).

According to Nagarajan et al. (2022), the synthesis of hydrogen as well as methane from banana peels during a two-phase anaerobic degradation process appears promising. In a batch reactor with mesophilic incubation, the two-phase sequential hydrogen and methane fermentation process, together with the one-phase methane fermentation process, were compared for performance and total energy recovery at different F/M (feedstock to microbial inoculum) ratios from 2.5 to 10. The production, rates, and possibility of biogas are impacted by F/M ratios. One-phase fermentation of methane showed its best performance at 5.0 F/M; 251.3 mL/g VS, 2.05 mL/h, and 352.9 mL were the potential, production rate, and yield of methane, respectively, at this condition (Nagarajan et al., 2022).

Haosagul et al. (2019) studied the wastes used including the concentration of waste glycerol (7.5–90 g/L) and banana peels with an average moisture content of 2.5–10% w/v to produce biomethane. The ratio of the co-substrate mixture was used in a 0.5 L batch reactor that ran for 120 h at 37°C and pH 7. It was discovered that waste glycerol (7.5 g/L) combined with banana peel (7.5%) produced CH_4 at the fastest rate. The greatest CH_4 yield was at 0.28 m^3/kg COD, and the highest CH_4 production was 652 mL. The use of banana peel and banana peduncles indirectly reduces the organic waste generated by the community (Haosagul et al., 2019).

10.6.2 Bioelectricity

Many efforts have been made by researchers to replace fossil fuels as an electricity source in human lifestyle, since it brings many disadvantages to the environment. Innovative renewable energy is used to generate electricity such as microbial fuel cells (MFCs) that generate sustainable bioelectricity from biomass using microorganisms. Carbon sources, for instance, cellulose, sucrose, glucose, fructose, and vitamins formed by lignocellulosic waste, from the waste of fruit peel have the energy potentiality for the production of bioelectricity in MFC (Verma & Mishra, 2023).

Banana peel research was conducted using dried banana peel and banana peel slurry to generate bioelectricity using *Saccharomyces cerevisiae* as a biocatalyst in MFC. The result shows that 2.2 ± 0.1 and 86.9 ± 0.4 mW/m^2 dried banana peel and banana peel slurry, respectively, for the maximum power output generated. Indigenous microbial consortia that are present in MFCs make the higher power output from the use of banana slurry-based MFCs. Not just that, 70–88% of the removal of COD was achieved for banana slurry and 18–44% of the removal of COD was achieved for dried banana peel powder in MCFs (Verma & Mishra, 2023).

The potential of banana peel waste (BPW) and orange peel waste (OPW), which were the two different organic substrates as a feedstock for the MFCs, is examined by Elviliana et al. (2018). Every experiment in this research used MFC with a single chamber, without any additional mediators or chemical pre-treatment. For ten days, measurements of electricity current and potential were taken. The VS values of OPW (87.06% total solid) and BPW (75.32% total solid) from the characterization research show that OPW includes more organic components than BPW. The findings also revealed that, albeit to slightly varying extents, both wastes produced electricity. Electric current and voltage ratings for BPW are, respectively, 0.492 V and 0.101 mA on average. Meanwhile, 0.563 V and 0.017 mA are the average values for electric current and voltage ratings for OPW, respectively (Elviliana et al., 2018).

Kalagbor et al. (2020) studied the employing single-chamber microbial fuel cell (SMFC) technology to transform these agricultural wastes into bioelectricity or green energy. In this study, 5, 10, 15, and 20 kg of fruit waste were used (Kalagbor et al., 2020). The fruits were tomatoes, bananas, and pineapples (fruit and peel). The results obtained show that the amount of substrate enhanced the amount of electricity produced. The maximum voltage output produced on day 1 from waste bananas was 3.1 V. The current readings result in considerably inversely proportional results to the voltage readings. The physiochemical metrics, including conductivity, pH, BOD, DO, and COD, agreed with those from related investigations. To sustain a healthy, clean, and pollution-free environment, it is possible to remove these agricultural wastes from the ecosystem by reducing this biomass through biodegradation utilizing SMFC technology (Kalagbor et al., 2020).

Maximum currents (3.71667 ± 0.05304 mA) and voltages (1.01 ± 0.017 V) along with an ideal pH and a maximum electrical conductivity of the substrate of 4.023 ± 0.064 and 182.333 ± 3.51 μS/cm, respectively, were accomplished using banana waste and inexpensive laboratory-scale technique. The peaks associated with phenolic compounds, alkanes, and alkenes are reduced in the FTIR spectra of the original and final substrate. The peak voltage recorded was 1,006.95 mV, and the greatest power density was $5{,}736.112 \pm 12.62$ mW/cm^2 at 6.501 A/cm^2 current density. The species *Pseudomonas aeruginosa* (100%), *Paenalcaligenes suwonensis* (99.09%), *Klebsiella oxytoca* (99.39%), and *Raoultella terrigena* (99.8%) were found to be the main sources of electricity for this kind of substrate (Rojas-Flores et al., 2022).

A study shows how banana waste can be utilized for power production in combination with an MFC and AD. An anaerobic batch bioreactor was used to determine the biochemical methane potential (BMP), maximum methane production (MMP), and methane production rate (MPR), during 64 days of observation. The MFC was used to investigate the polarization curve, resistance, maximum power density (MPD), maximum voltage, and current using the digestate produced by the AD. The AD produced 320.3 mL, 373.3 mL CH_4/g VS, and 18.6 mL CH_4/L day of MMP, BMP, and MPR, respectively. The maximum voltage produced by the MFC (286 mV), MPD (41.3 mW/m^2), resistance (580.99 Ω), and current was 0.0002867 A. The bioenergy produced was 13.38 kJ/g VS overall by both procedures (Rincón-Catalán et al., 2022).

This integrated system demonstrated a strong and promising application of banana waste to produce sustainable bioenergy. As a consequence, the banana agro-industry might benefit from using this feedstock to create a circular economy and create sustainable processes.

10.6.3 Bioethanol

In the conversion process to bioethanol, the starchy substance's sugar content is transformed into alcohol under regulated circumstances, emitting carbon dioxide. Some of the examples that can be used as raw materials for alcoholic fermentation are sugar crops, starchy crops, and cellulosic plant parts (Koul et al., 2022). Several studies have been done on producing bioethanol from the use of banana waste. *Aspergillus fumigatus* and *Aspergillus ellipticus* were used to release reducing sugar, which was used as a substrate by the yeast strain *S. cerevisiae* NCIM 3570 in the production of ethanol. The use of cellulolytic fungi and yeast in producing biofuels through the fermentation of BPS is one of the encouraging ways to overcome the high demand for crude oil in India (Ingale et al., 2014).

In Malaysia, the second-largest waste biomass is the banana stem (BS). Hossain et al. (2019) studied ethanol produced from BS waste using fermentation and hydrolysis techniques in the presence of yeast *S. cerevisiae.* The largest glucose yields were after enzymatic (5.614 g/L) and sulfuric acid hydrolysis (40.61 g/L). The greatest minimum glucose yield and minimum glucose yield was 62.23 g/L at 12 h and 0.69 g/L at 72 h, respectively, during fermentation. A distillation procedure was then used to recover 99.8% pure bioethanol. In comparison to diesel emissions, bioethanol's CO_2 emissions were 97,161 kg/y and its SO_2 emissions were 513 kg/y (Hossain et al., 2019). The process of making bioethanol and energy generation from the stems of bananas can be portrayed as the strategy toward achieving Sustainable Development Goals(Hossain et al., 2019).

Santiago et al. (2022) suggested using other organic acids such as oxalic, tartaric, and citric acids in addition to the traditional sulfuric acid during the pre-treatment of the·acid stage to minimize the creation of undesirable chemicals and enhance the synthesis of bioethanol. The environmental parameters of four methods for valorizing banana peel trash to make bioethanol are analyzed using the life cycle assessment approach. The method using citric acid has a poor environmental profile, as measured by the results per kilogram of bioethanol, because of the background activities involved in the production of acid (about 55% for major impact categories). In contrast, the best environmental profile is the oxalic acid-based method, with a reduction of between 20% and 35% in comparison to the citric acid method, depending on the effect category. Regardless of the method used, the primary hotspot in many subsystems is the energy requirements for production, which is primarily thermal energy (depending on the impact category, varying from 30% and 50%) (Santiago et al., 2022).

Prakash et al. (2018) examined the viability of converting banana peels into bioethanol utilizing a combination of depolymerizing enzymes. *Geobacillus stearothermophilus* HPA19, which turns out a cocktail of thermo-alkali-stable xylano-pectino-cellulolytic enzyme/s using wheat bran, was isolated from a natural resource. The ideal temperature and pH for xylanase (80°C, 9.0), filter paper cellulase (70°C, 8.0), and pectinase (80°C, 9.0). The cocktail enzymes exhibited stability at high pH (10.0) and temperature (80°C). While Na^+, Ca^{2+}, and K^+ boosted pectinase activity, others promoted the relative activity of xylanase and FPase, which are Ni^{2+} and Zn^{2+}. The cocktail's ability to saccharify banana peel was evaluated. The amount of enzyme used has a significant impact on minimizing the sugar obtained, which was 37.06 mg/mL using the one variable at a time technique. Additionally, saccharification was optimized using response surface methods, which resulted in a twofold increase in reducing sugar. The fermentation process produced the most ethanol with an efficiency value of 21.1 g/L and 76.5%, respectively, after 30 h (Prakash et al., 2018).

According to Guerrero et al. (2018), to find the ideal condition of enzymatic hydrolysis to produce the concentration of glucose between 75 and 100 g/L, a full factorial experimental design was adopted for bioethanol production. For the rachis the ideal enzymatic hydrolysis conditions were 17.6% and 16.0 (filter paper activity units) FPU/g of solid loading glucan and enzyme dose, respectively, while pseudo-stem ideal enzymatic hydrolysis conditions were 15.1% and 14.9 FPU/g of solid loading and glucan, respectively. The best configuration for rachis was simultaneous saccharification and fermentation (SSF 8 h) after an 8-h pre-hydrolysis to get the highest productivity of volumetric ethanol. In contrast, for pseudo-stem, the configuration of SSF led to increased ethanol output. Pseudo-stem produced 112 L/t of ethanol under ideal conditions and 103 L/t produced from rachis (Guerrero et al., 2018).

Legodi et al. (2021) studied various pre-treatment techniques for BPS, the origin of sugars that can be utilized to make bioethanol. Following pre-treatment with 5% (v/v) H_2SO_4, 3% (w/v) NaOH, and liquid hot water on BPS, the amounts of cellulose that were still available were 48%, 52%, and 25%, respectively. *Trichoderma harzianum* LMLBP07 13-5 crude enzyme from *Trichoderma longibrachiatum* LMLSAUL 14-1 released 3.8–21.8 g/L, and *T. longibrachiatum* LMLSAUL 14-1 released 5.4–43.5 g/L glucose to the biomass when used to saccharify pre-treated BPS at substrate loading of 10–15% and temperature of 50°C. Separate hydrolysis and fermentation (SHF) of alkaline-pre-treated BPS hydrolysate employing *S. cerevisiae* UL01 at 100 rpm and 30°C produced ethanol. The 17.6 g/L of ethanol was the highest-ever produced (Legodi et al., 2021).

Nwabanne and Aghadi (2018) investigated parameters that affect the hydrolysis and fermentation process that can increase yields of bioethanol production from banana peel. In an SHF method, *Aspergillus niger*, which was separated from infected cassava tubers, was employed to hydrolyze banana peel (BP). The hydrolytes were fermented using commercial *S. cerevisiae*. According to the proximate analysis, the main constituents of banana peel are cellulose, fiber, and protein were 65.5%, 15.4%, and 6.0%, respectively. Temperature, pH, time, and yield of 34°C, pH 6.5, 5 days, and 122 mg/mL, respectively, were the ideal circumstances to lower the sugar yields from the process of hydrolysis. While temperature, pH, and time frame of 34°C, pH 6.0, and 5 days, respectively, 8.1% volume weight produced the highest production of bioethanol from the fermentation process (Nwabanne & Aghadi, 2018).

John et al. (2020) investigated the use of cellulase made in vitro to convert used banana peels into bioethanol, which would lower the cost of making the fuel. Banana peel that had been dried and powdered was subjected to ultrasonication for 1 h at a power output of 750 W and 20 kHz of operating frequency with sulfuric acid (2% v/v). Following pre-treatment, lignin was reduced by 67.50%, while hemicellulose (45.6%) and cellulose (69.3%) were recovered. By using the cellulase enzyme that can be produced by *A. niger*, the pre-treated banana peel was hydrolyzed. The isolated crude cellulase enzyme's activity was calculated to be 1.712 FPU/mL. To assess the impact of 10–50 FPU enzyme loading and a 24–96 h incubation period on decreasing sugar, the pre-treatment banana peel was hydrolyzed at 50°C. Using 24 h of activated *S. cerevisiae*, hydrolysate was fermented, and the fermentation impact times ranging from 18 to 120 h were examined; hence at the 24th hour of fermentation, 4.24 g/L of ethanol was produced (John et al., 2020).

Palacios et al. (2021) investigated the application of pre-saccharification, simultaneous saccharification, and fermentation (PSSF) and SSF with high solid loading (up to 25% w/w) as different process configurations using two yeasts, *Kluyveromyces marxianus* thermotolerant strain derived from Agave honey and *S. cerevisiae* commercial strain for bioethanol production from banana peel. Similar ethanol generation for 35°C of *S. cerevisiae* and 41°C of *K. marxianus* was accomplished at 10–12 h on SSF with banana peel of 10% w/w. The PSSF procedure, utilizing a commercial *S. cerevisiae* strain (25% w/v) banana peel, produced the largest amount of ethanol (32.6 g/L) at 64 h, while *K. marxianus* did not produce any ethanol under the identical assay conditions. This study showed that banana peels are a promising feedstock for PSSF's production of ethanol (Palacios et al., 2021).

Suhag et al. (2020) investigated the use of banana leaf waste as a suitable feedstock for bioethanol manufacturing by the use of *A. niger* JD-11 crude cellulases for saccharification of untreated and pre-treated (dilute acid, alkaline, and steam) waste of banana leaf. At 45°C, 15 FPU/g enzyme loading, 2% substrate loading, and 0.15% polyethylene glycol 6,000 (w/v) as the surfactant, the crude cellulases produced the highest amount (524.83 mg/g) of reducing sugars from the dilute acid pre-treated banana leaf waste. With the help of *S. cerevisiae*, the enzymatic hydrolysate was fermented using an initial reducing sugar concentration (40 g/L) in the medium at 150 rpm, pH 5.5, and 30°C for 30 h. After 12 h of the hydrolysate fermentation from the waste of banana leaf, a maximum ethanol yield and productivity of 0.38 g/g sugar and 1.28 g/L/h, respectively, were achieved (Suhag et al., 2020).

10.6.4 Biochar

Char, which is also referred to as charcoal, is defined as residue formed when organic matter such as bone or wood is partially heated under pyrolysis or burnt in limited excess of air. The residue is usually a black porous solid that consists of structureless carbon. It can be used as fuel or added to compost. However, biochar is made from biomass that is primarily meant for soil application according to the Swiss organization Biochar Info (Guo et al., 2015).

Most of the components presented in the banana waste can be processed into biochar to a certain extent, but the most used component in biochar for banana waste is the banana peel and peduncle. In general, to prepare the banana waste biochar, the chosen component must be dried for a certain period of time before being fed to the muffle oven to be pyrolyzed under high temperatures. Figure 10.11 shows the flowchart in biochar preparation in general (Pradhan et al., 2020).

For the application of banana peel-driven biochar, the experimental result of soil amendment using banana peel biochar produced at the temperature of 350°C conducted by Sial et al. (2019) suggested that the amendment of the soil using banana peel biochar together with various ways of treatment can significantly reduce the nitrous oxide emission by (37.1–54.8%) and methane gas emission by 20% compared to the soil amendment using banana peel itself. This result shows that biochar made from the banana peel can significantly reduce GHG emissions besides improving the soil biochemical properties (Sial et al., 2019).

A study shows that the biochar produced from the banana peduncle of *Musa* species, which is being pyrolyzed at 300–500°C, possesses heavy metal removal properties. In the adsorption experiment conducted by Karim et al. (2015), under acidic conditions, the biochar could absorb a significant amount (about 27–41%) of the heavy metal chromium (V) ions in the stock solution prepared. The biochar derived from banana waste has the potential to improve the quality of soil significantly and its application in wastewater treatment.

10.6.5 Bio-Briquette

Briquettes are described by the Oxford Dictionary as compacted blocks of coal dust or other flammable material used as fuel. Typically, pulverized coal or other materials like charcoal,

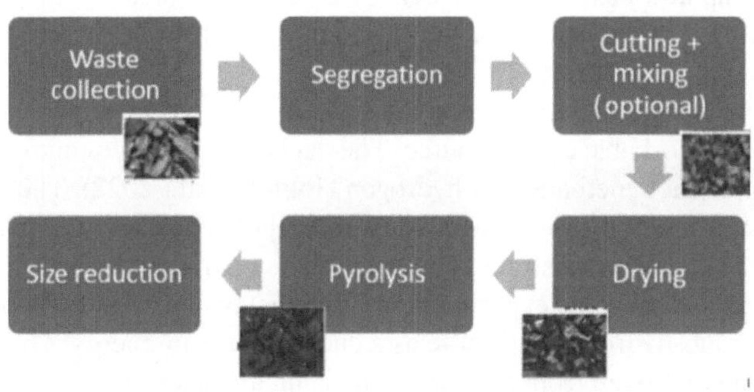

FIGURE 10.11 General steps in creating biochar (Pradhan et al., 2020).

sawdust, peat, or biomass are pressed together, either with or without a binder. Briquettes are frequently used as a source of fuel for cooking and heating, particularly in locations where conventional fuels like firewood or coal may be in short supply or expensive. They are frequently formed into compact, little blocks that are easy to handle and burn well. In the context of bio-briquette production, the most utilized banana waste is the banana peels, pseudo-stem (trunks), and the leaf (Faizal, 2017).

The bio-briquette made from the carbonized banana peels at the temperature of 300°C exhibited a calorific value of approximately 6,000 kcal/kg, which was higher than the standard calorific value of briquette of more than 5,000 kcal/kg (Faizal, 2017). Next, the experiment study also showed that the bio-briquette made solely from carbonized banana waste comprising different ratios of banana peels, pseudo-stem, and banana leaf met the ISO 17225 in terms of ash content and moisture content. The bio-briquette produced also has a maximum calorific value of 21.75 MJ/kg, which exceeds the minimum high heating value of ±17.5 MJ/kg, according to standard DIN 51731 (Ahmad et al., 2018). Therefore, it has been shown that there is a huge potential for the utilization of banana waste in bio-briquette production.

10.6.6 Bioplastic

A bioplastic can be defined as a polymer that has been made into a commercial product from a natural or renewable resource, making it biodegradable (Rudin & Choi, 2013). The demand for bioplastics increases due to the environmental problems arising from the use of petroleum-based plastic (synthetic plastic), which is non-biodegradable and thus accumulates significantly around the globe (Pongsuwan et al., 2022).

Research has shown that, due to banana waste's high starch content, especially in the banana peel part, the method of synthesis of bioplastic using the banana peel as a substrate can be established (Choudhary et al., 2022). Banana peels have been successfully used in several experiments for manufacturing biodegradable films. Using an alkaline leaching technique, starch recovered from BPW was used to create bioplastic. Research conducted on the development of banana peel/corn starch bioplastic film showed that a bioplastic produced using the banana peel starch as the main biopolymer and the corn starch as the co-biopolymer can have a maximum tensile strength of 34.72 N/m^2 at 4% corn starch concentration and an onset melting temperature of 54.41°C using hydrochloric acid and glycerol treatment (Sultan & Johari, 2017). This shows the feasibility of producing the commercial standard bioplastic via banana waste.

10.6.7 Bio-Hydrogen

A viable alternative method for producing fuel from inexpensive, renewable, ecologically benign, and non-polluting resources is biological hydrogen generation. Given the rising demand for hydrogen on a global scale, using organic waste as a substrate is a sustainable method of producing it. Using cheap and plentiful feedstocks is essential to keeping the process economically viable because the cost of the substrate plays a significant role in the economics of producing bio-hydrogen (Balda et al., 2021).

It would be feasible to see the utilization of banana organic waste as a foundation source to produce bio-hydrogen as a workable energy source. The peel of a banana contains a lot of carbohydrates and can be used to produce methane and hydrogen (Jokhio et al., 2022). The results of the current study showed that the creation of hydrogen is a very delicate process that is highly dependent on several variables, including the substrate's composition, temperature, the pH of the growth medium, and the quantity of inoculum. This experiment implied that banana waste might serve as a substrate for the production of bio-hydrogen and serve as a cheap source of energy. This study examined the viability of producing hydrogen from banana waste using an anaerobic fermentation technique, and significant amounts of hydrogen gas were produced as a result (Ahmad & Danish, 2018).

10.6.8 Biobutanol

The conversion of banana peel to ethanol or another renewable fuel is well established in the literature. An exciting alternative to more conventional biofuels like ethanol and methanol is butanol, commonly known as n-butanol. The advantage of butanol is that it has a longer carbon chain, which supports a higher heating value and less volatility, evaporative heat, and corrosivity, resulting in fewer ignition issues with engines. In addition, direct applications to diesel engines using pure butanol or butanol–diesel mixes are possible without modifying the architecture of the engines (Sawarkar et al., 2022).

Enzymatic hydrolysis, fermentation, and pre-treatment are all steps in the biobutanol production from banana peel. Pre-treatment of banana peel primarily aims to change the composition of the complex carbohydrates (hemicellulose and cellulose) and remove lignin (delignification). However, the composition of lignocellulosic biomass has a significant impact on pre-treatment effectiveness; as a result, optimizing that composition is a crucial step for a procedure to be effective. In low-lignin biomasses like banana peel, pre-treatment techniques combining physical (such as milling or grinding) and chemical (alkali/acid) treatment have been found to yield encouraging results (Mishra et al., 2020).

Petro-fuel may eventually be replaced with biobutanol since it is less caustic and volatile and has a lower affinity for moisture, further lowering the risk of groundwater pollution. (Koul et al., 2022). For delignification of banana peel biomass and maximum cellulose exposure, sulfuric acid (H_2SO_4) and sodium hydroxide (NaOH) pre-treatments were carried out using the response surface methodology technique. Compared to pre-treatment with NaOH, pre-treatment with H_2SO_4 exhibited increased cellulose exposure of 585.2 mg/g of biomass. Using an L16 Taguchi orthogonal array design, the sulfuric acid pre-treated banana peel (PBP) was then exposed to independent enzymatic hydrolysis and co-fermentation (SHCF). Maximum glucose was 677.3 mg/g of PBP, and xylose was 165.1 mg/g of PBP yield from SHCF. Furthermore, co-cultures of *S. cerevisiae* and *Pichia* sp. at an incubation time of 72 h resulted in greater butanol titers (7.4 g/L) from sugar loadings (20 g/L), which were 89.1% and 75.67% higher compared to the monocultures of *S cerevisiae* and *Pichia* sp. The higher sugar loadings, 90 g/L, used in scale-up tests resulted in decreased butanol concentrations (15.7 g/L) and overall butanol productivity (0.2 g/L/h). The total energy required to produce 1 g of fuel-grade butanol from reducing sugar loadings of 20 g/L shake flask was 138.1 kJ and 90 g/L fermenter was 85.7 kJ (Mishra et al., 2020).

Sivanarutselvi et al. (2019) studied the efficiency of alkali pre-treatment of BPS for anaerobic fermentation utilizing *Clostridium sporogenes* to produce biobutanol. When using 2% NaOH (w/v) and 30°C soaking temperature, the maximum delignification was 92%. Temperature increase caused an increase in delignification but also reduced cellulose recovery. From the pre-treated biomass, the best glucose production of 524 mg/g glucan was attained. When *C. sporogenes* fermented hydrolysate derived from alkali-soaking pre-treatment under anaerobic conditions, 10.12 g/L of butanol was produced, with a yield of 0.326 g/g of total sugar consumed, which was much greater than that with untreated biomass (Sivanarutselvi et al., 2019).

10.6.9 Natural Acids

Bananas are the most used fruit in Aceh, Malaysia. However, most people eat only a portion of the fruit, and the banana peels are thrown away as garbage. Aceh, Malaysia, is currently having trouble disposing of garbage from banana peels. It is well known that banana peels have higher concentrations of gallocatechin and catecholamine than the fruit. Water, protein, anthocyanin, cyanidin, carotenoids, sterols, fiber, lipid peroxidation inhibitors, vitamin B6, and others are also present in banana peels. This study's objective was to identify the solid-state fermentation process used by *A. niger* to produce citric acid from waste banana peels (Sawarkar et al., 2022).

10.6.10 Pulp and Paper

Due to a lack of available raw materials made from wood, paper and pulp production currently falls short of meeting the demands of a growing population, especially in developing nations. Deforestation has therefore started gradually in many regions of the world. In comparison to non-wood, wood resources offer the advantage of producing great pulp quickly and good fibers for specialized papers (Koul et al., 2022). Since the previous 10 years, numerous investigations on the pulping of various banana fiber components have been conducted (Ferdous et al., 2023).

Ferdous et al. (2023) studied that, pulp yields for the stem, peduncle, and leaf of bananas in the soda-anthraquinone (AQ) process, pulping at a 20% alkali charge, were 30.0%, 39.12%, and 34.55%, respectively, with a kappa value below 20. Meanwhile, the formic acid/peroxyformic pulping method produced more pulp than soda-AQ because of the sample's extremely high ash content that was maintained in the organic acid pulping process. To understand the pulping behavior, the alkaline nitrobenzene oxidation (ANO) of these fibers was studied. In our investigation, it was discovered that the leaf (midrib), stem, and peduncle of bananas all contain a sizable amount of cellulose. Lignin from the banana waste is made up of syringol and p-hydroxyphenyl propane units. The stem, leaf, and peduncle fibers from bananas can be thought of as longer fibers with a thinner cell wall, which leads to paper with a higher bonding strength. These fibers anatomically contain a lot of parenchyma cells, which produce fines during the pulping process and deliver the pulp with delayed drainage (Ferdous et al., 2023).

Pulping of banana leaf, banana peduncle, and pineapple was carried out by Fagbemigun et al. (Fagbemigun et al., 2016) using the method of soda pulping with 7% sodium hydroxide (NaOH) at a temperature of 120°C. According to the findings, short-fiber resources include the banana peduncle with an average fiber length of 1.27 mm, and the banana leaf with an average fiber length of 0.88 mm. The average fiber diameter was 15.81/11.56 m, while 9.79/3.84 m for the lumen width of the banana peduncle for the banana leaf. The Runkel ratios for the two samples were 0.37 and 1.69 for banana peduncle and banana leaf, respectively, with banana peduncle proving to be a good source of elastic fibers (Fagbemigun et al., 2016).

Rahman et al. (2014) assessed the pulp qualities of three banana plant parts by the Kraft process. Different banana plant (*Musa paradisica*) sections, including the leaf stalk, peduncle, and pseudo-stem, were chemically and morphologically evaluated to determine their suitability for chemical pulp. Every section was examined for extractives, the solubility of ash, sodium hydroxide, cellulose, lignin, and pentosan content, as well as for the width and length of its fibers. These segmented sections have comparatively high amounts of ash, alkali solubility with values of 7.6–19.1% and 32.4–51.4%, respectively, and water solubility, but low amounts of lignin with a value of 12.8–16.1%, moderate amounts of cellulose with a value of 31.6–43.3%. In terms of fiber length and width, the banana plant's morphological characteristics were discovered to be a medium fiber with a high slenderness ratio (Mote et al., 2023). The ideal pulping conditions of active alkali 16% for the leaf stalk, 22% for the peduncle, and 12% for the pseudo-stem at 150°C for 1 h resulted in pulp yields of roughly 30.5–40.5% on oven-dried raw material with 16.9–26.1 kappa numbers range (Rahman et al., 2014). Unbleached pulp's mechanical characteristics revealed that the pseudo-stem had a high tensile, rip, and burst index (Rahman et al., 2014).

Berhanu et al. (2018) analyzed the viability of employing sulfur-free ethanol-alkali pulping on false bananas (*Enset ventricosum*) fibers for the production of paper pulp. The impacts of four independent process factors (ethanol concentration, time, pulping temperature, and alkali concentration) on kappa number, yellowness, pulp yield, brightness, and whiteness of pulp were assessed using the approach of response surface with central composite experimental design. The ideal delignification conditions were projected to be alkaline concentration (15%) and ethanol concentration (46%) at 140°C with a 61 min of cooking period, resulting in the highest pulp yield and lowest kappa number. The pulp yield for the anticipated ideal pulping conditions was 69.9%, with 4.9 and 64.5% kappa number brightness, respectively. The viscosity and strength characteristics of the ethanol-alkali pulp produced under standard conditions were compared to Kraft pulp, demonstrating equivalent burst, tensile, and tear strengths (Berhanu et al., 2018).

According to Tripathi et al. (2019), the kraft and soda-AQ pulping processes were discovered to be the most suited BSs based on the produced pulp characteristics. The distinctive morphology of BS pulp and the high double fold and air resistance values of paper hand sheets made from it show that it is suitable for high-quality papers including envelope paper, banknote paper, barrier papers, and grease-proof papers. Rice straw and sarkanda pulps were superior to BS pulp in terms of bleachability, which was comparable to wheat straw pulp and inferior to sugarcane bagasse pulp. During the mill-scale testing, it was able to replace softwood pulp with BS pulp up to 75% in the pulp feed without having any negative effects on one of the best kraft paper grades' physical strength characteristics. Hence, BSs can be used for papermaking by substituting imported softwood pulp, according to the trial run (Tripathi et al., 2019).

Arafat et al. (2018) conducted a study of waste banana fiber (WBF) and extracted banana fiber (EBF) that satisfies the quality of handmade paper. To create handmade paper, the chemical and morphological characteristics of EBF waste were compared in this study. WBF was distinguished by having less cellulose, more lignin, and longer fiber length. With changing active alkali and boiling-temperature cooking times, EBF and WBF were pulped. With an 8% alkali charge and 120 min of cooking, WBF's pulp yield was 35.9%. The SR values or degrees of drainage resistance for EBF (65) and WBF (71) are in the undefeated state (Arafat et al., 2018).

Ramdhonee and Jeetah (2017) investigated the use of banana fiber for the production of wrapping paper. The ratios of banana fiber to wastepaper, sugarcane bagasse, and wastepaper were 20:80, 40:60, 60:40, and 80:20, respectively. About 12% NaOH and 4% Na_2S were used to chemically pulp banana and sugarcane bagasse for 1.5 h. The papers' average thickness was 0.263 mm, plus or minus 0.04 mm. The water absorption of 100% banana paper (2.15 s) and 100% bagasse paper (2.17 s) was 2.25 times greater than that of 100% wastepaper. The banana paper containing 20% wastepaper had the highest tensile index, which was 29.4 N m/g. The banana-bagasse mix ratio with the highest tensile index (16.04 N m/g) was 20% bagasse. The tensile index of 100% banana paper was 18.97 N m/g. The banana paper with a 20% wastepaper mix had the highest bursting index (2.00 kPa·m^2/g), while the highest bursting index for banana–bagasse mix ratios was illustrated by 20% bagasse addition (0.88 kPa·m^2/g). The burst index of 100% banana paper was 1.05 kPa·m^2/g. The abrasion resistance of 180 spins, as shown by 100% banana paper, was the highest, while the highest abrasion resistance was shown by 20% mix wastepaper (136 rotations) and 20% mix bagasse (120 rotations). As the pulp mix percentages with the lowest weight losses, wastepaper (20–40%) and bagasse (10–30%) with weight loss of 5.4–5.8% and 10.0–15.5%, respectively, were identified (Ramdhonee & Jeetah, 2017).

Utilizing the fibers from these sources to produce pulp and paper can aid in filling the gap left by the nation's lack of access to wood. It will offer inexpensive raw materials for cottage businesses, generating employment chances for young people and aspiring entrepreneurs. Additionally, it will contribute to the expansion of the food crop value chain, boost farmer revenue, and act as a productive waste management method.

10.6.11 Biocomposite

Biocomposite is a composite material made by employing a natural resource (often biological) as reinforcement and a polymer derived from natural or synthetic sources to make the matrix (Halimatul et al., 2019; Jumaidin et al., 2019a, b, 2021; Nazrin et al., 2021). Together, the reinforcement and matrix produce better qualities than either one of the natural components alone (Asyraf et al., 2022; Suriani et al., 2021). The matrix provides processability and safeguards the reinforcing material, while the reinforcement significantly contributes to stress transmission and strengthening of the biocomposite (Alias et al., 2021). Biocomposite materials come from natural sources, which appeals to industries in terms of cost and environmental responsibility (Manu et al., 2022). Composite materials' two main constituents are the matrix and the reinforcement. Fibers are the most typical reinforcements (Azlin et al., 2022; Mazani et al., 2019). Not just that, they offer the majority of the strength and stiffness. The matrix uses an effective load-transferring mechanism to bond the fibers together. Weight reduction, resistance to corrosion, and part-count reduction are the key driving forces for the use of composites. Because polymers and fibers are both employed as matrices that have a low density, composites are relatively light. More importantly, compared to most materials, nanofibers have better strength-to-weight, stiffness-to-weight ratios, and high surface area (Norfarhana et al., 2022, 2024a, b; Nurazzi et al., 2021; Syafri et al., 2022). The usage of wooden materials in buildings generally resulted in a 45% rise in deforestation rates and a 4.4% increase in consumption rates. Therefore, to create renewable biocomposites, different mixtures of industrial waste, including FA and groundnut ash, are utilized as natural plant fiber and filler materials, like jute fiber and banana fiber, as reinforcement (Rajendran & Nagarajan, 2022).

The waste from red banana peduncles as a precursor for the extraction of carboxylated cellulose nanocrystals (CCNCs) through acidified chlorination, alkalization, and citric acid hydrolysis, along with the evaluation of its reinforcing potential in polyvinyl alcohol. With good thermal resistivity against degradation up to 280°C, high crystallinity (89.7%), and 48 mV of negative zeta potential, the CCNCs have a rod-like form. Active packaging applications are possible with the bio-nanocomposite film's increased physical, functional, thermal, and biobreakdown features (Nagarajan et al., 2022).

Martins et al. (2022) investigated the biocomposites made from BPW at various contents of 0–20 wt% with an increment of 5, to increase the oil sorption capacity (system polyurethane foam [PU]/banana peel oil). According to the findings, the waste of banana peel was hydrogen-bonded to PU chains, boosting the density of the foams and reducing the size of the pore as the fiber concentration rose from 444 to 123 m. Banana peel fibers were added to PU to maintain the foams' hydrophobicity while reducing their crystallinity. The biocomposites demonstrated better thermal stability when compared to pure PU (Tonset from 247.6°C to 264°C). However, PU/BP20% showed superior sorption capacity because of an increase in the size of the pore, which directly influenced the oil sorption (Martins et al., 2022).

10.6.12 Biofertilizer

Sogani et al. (2023) investigated the application of banana peels as biological fertilizer applied to a black gram crop cultivated in a plant microbial fuel cell (PMFC). The germination rate of the black gram crop increased, as indicated by the findings, when the amount of extract of banana peel was increased. The black gram crop's germination rate from 12% of control without biofertilizer increased to 100% of 20 mL/L of biofertilizer after 7 days of sowing. The highest power density yet observed in a PMFC running on biofertilizer was 260 mW/m^2, which is more than the control (2.65 times). Within 60 working days, the addition of biofertilizer to the PMFC also resulted in higher biomass, which in turn increased the production of bioelectricity (Sogani et al., 2023).

According to Lim and Matu (2015), leftovers from fruits like bananas, papaya, pineapple, citrus orange, and watermelon were used to study economical approaches for making biofertilizers from agricultural waste. Biofertilizer was created using the solid-state fermentation process and then employed in vegetable plantations. The experiment's findings demonstrated that the plant samples that were being treated with biofertilizer made from papaya, watermelon, and banana wastes had favorable physical features. In this study, pH levels and the potassium content of the biofertilizers were studied. The pH value for the watermelon biofertilizer was determined to be the highest (5.15), while the potassium content of the banana biofertilizer was the greatest, at 3.93 g K/L with a pH value of 4.52 (Lim & Matu, 2015).

10.7 POTENTIAL APPLICATIONS

10.7.1 Potential Applications of Banana Waste

The banana is a great and economical food source for underdeveloped nations where the bulk of the population depends on bananas for nutrition. Banana plant components can be used to make numerous functional meals, including wine, alcohol, biogas, cow feed, and more. They can also act as a pesticide, antioxidant, and color absorber. The banana is more easily digestible and packed with nutrients than many other fruits. The banana fruit takes less time (105 min) to digest than the apple (210 min). In addition to being tasty, easy to peel, high in potassium and calcium, and low in sodium, bananas are also well-liked for their aroma and texture (Padam et al., 2014).

10.7.2 Banana Peel

Banana peel contains glucose (3%), crude protein (6–9%), crude fat (3.8–11%), and total dietary fiber (43.2–49.7%). It is also rich in pectin, essential amino acids (leucine, valine, phenylalanine, and threonine), and minerals (K, P, Ca, and Mg) (Mohapatra et al., 2010). All essential amino acids have contents that exceed FAO standards, except for lysine. Fruits mature by an increase in their soluble sugar content, a decrease in starch and hemicellulose content, and a slight rise in protein and fat content. An increase in soluble sugar concentration could be explained by endogenous enzymes that break down starch and hemicelluloses (Mohapatra et al., 2010).

Banana skins can also be used to extract banana oil (amyl acetate), which is used for flavoring meals. Also, banana peels are high in cellulose (7.6–9.6%), hemicelluloses (6.4–9.4%), lignin (6–12%), pectin (10–21%), and galactouronic acid. Banana peel pectin also contains glucose, galactose, arabinose, rhamnose, and xylose. Peels contained a greater quantity of micronutrients (Fe and Zn) than pulps. As a result, peels could be an excellent animal and poultry feed item. Banana peel can also be used to manufacture wine and ethanol, as a biogas substrate, and to extract pectin. Peel ash is a source of alkali that may be used to make soap as well as fertilizer for banana plants (Padam et al., 2014)

10.7.3 Removal of CO_2, NO, NO_x, and SO_2 Gases from Motorcycle Emission

Industrial progress has led the earth to a very alarming point of environmental collapse as a result of the unregulated increase of the usage of fossil fuels as the principal energy source during the past 50 years. As a result, the average global temperature will rise, changing the ecosystem. Gases including CO_2, CO, hydrocarbons, SO_2, and NO_x produce volatile organic compounds and particulate matter, and excessive petrol emissions from internal combustion engines, and industries have become one of the main reasons for environmental pollution. These chemicals are responsible for global warming, acid rain, photochemical smog, stratospheric ozone depletion, the greenhouse effect, and other environmental challenges (Viena et al., 2018).

The search for the most effective, reasonably priced, and environmentally friendly adsorbent from agricultural waste is still ongoing. In Aceh Province, Indonesia, banana production contributed 504,850 quintals, making it one of the major fruit commodities. The increasing output was accompanied by an increase in the amount of BPW. When bananas are harvested, only the meat is consumed, and the peels are dumped into the environment. Banana peels are no longer used for anything other than animal feed or fertilizer. Banana peel piles are effective as a raw material for producing natural activated carbon from biomass waste. These wastes might be utilized as a cheap bio-absorbent (Viena et al., 2018).

10.7.4 Banana Waste Utilization in Wastewater Treatment

Banana waste has been shown to be highly valuable because it includes a range of parts that can be used, such as banana fruit peels, trunks, pseudo-stems, leaves, and piths. Many agricultural wastes have been studied for use as adsorbents. These banana waste components have been extensively researched as adsorbents for a wide range of pollutants (Ahmad & Danish, 2018). Due to their widespread availability, lack of adequate use by farmers after fruit harvest, and considerable carbon compound content, banana wastes have caught the attention of researchers as useful raw material for adsorbents (Ahmad & Danish, 2018). Additionally, if banana tree waste is not properly managed, it may release GHGs when dumped in humid conditions, posing a significant environmental risk (Ahmad & Danish, 2018). Farmers typically threw banana tree trash in ponds and rivers, where it slowly decomposed, emitting methane and other gases that spread a foul aroma and harmed the local ecosystem. It makes sense to use banana trees and fruit waste as an excellent adsorbent material for a sustainable future (Ahmad & Danish, 2018).

10.8 CHALLENGES IN COMMERCIALIZATION AND PROCESSING

10.8.1 Emerging Techniques for Banana Processing

Typically, bananas are harvested when they are completely developed and green, which is usually during the pre-climacteric stage. Bananas that will be served locally are stored in ethylene-rich settings to speed the ripening process by adding various compounds such as ethephon, whereas green matured fruits that will be consumed as staple meals or specialized products are further cooked or processed (Maduwanthi & Marapana, 2019).

After cleaning, dewaxing, and pesticide treatment, the fruits are normally peeled by hand or mechanically. Since the pectin has not yet broken down and separated from the endocarp, green bananas are often peeled by hand. However, putting a ripe banana between two steel rolls to break the peel and separate it from the pulp is a simpler method for doing so (Maduwanthi & Marapana, 2019). Product quality is extremely important to consumers. The quality of a finished product is affected by numerous unit operations, including harvesting, handling, pre-processing treatments, processing settings, packing, and storage conditions. Many of the cutting-edge techniques being developed by researchers around the world are now in use in energy-efficient industrial production, with a focus on safety, stability, and quality (Maduwanthi & Marapana, 2019).

10.8.2 Challenges in Commercializing Banana Waste Biorefineries

Nowadays, in parallel to the Sustainability Development Goals especially on Goal 7: Affordable and Clean Energy, it has been suggested that biorefineries could be used to process biomass from fruit waste, including banana waste, into a variety of products, such as bioenergy, biochemicals, and biomaterials with several unit operations such as distillation and enzymatic hydrolysis. However, there are several challenges that need to be faced to commercialize, which include:

a. The viability from an economic perspective: in the past, bio-based products have cost more than their fossil-based counterparts. Furthermore, the initial capital is usually enormous, which subsequently makes the investment in these biorefineries a risky option. As a result, it can be difficult to get investors for innovative industrial-scale technology (Manhongo et al., 2022).
b. Despite the agro-industrial residues from banana plants are promising potential feedstock for biorefinery technologies due to their high growth rates and carbon neutrality, challenges regarding composition disparity in banana wastes must be addressed. The geographic location

of the banana plantation, plant species variety, and ripening stage are some of the main factors contributing to the variation in banana waste composition. Consequently, it is difficult to standardize the biorefinery processing of banana waste (Redondo-Gómez et al., 2020).

c. Although there are numerous methods for valorizing fruit processing waste, some fruit processing waste biorefinery technologies are still lab-scale and have a restricted range of products, product combinations, and process pathways. Therefore, it makes the forecast very difficult in the context of determining which mix of goods, raw materials, and technologies is sustainable (Redondo-Gómez et al., 2020). Next, current technologies need further improvement to get the most value out of banana waste, such as by attaining positive energy balances via integration with other biorefinery processes (Manhongo et al., 2022).

10.9 CONCLUSION

Some common properties in characterizing the banana waste fiber are being reviewed. The banana peel has mostly been utilized to generate bioplastics with high disintegration rates and biofuels like ethanol and diesel due to its high abundance of organic components rich in carbon, such as polysaccharides. The abundance of polysaccharides in banana waste fiber also enables it to be manufactured as paper and pulp. Banana plant leaves are used to make organic fertilizers, biodegradable containers and utensils, and packaging material. BPSs are utilized to make bioplastics with a quick biodegradation period due to their fibrous character. Banana waste-loss content, particularly cellulose and hemicellulose, is crucial for nanotechnological processes, particularly when creating green nanoparticles. Based on the observations, despite the banana waste is utilized in wastewater treatment and the production of bio-hydrogen, biochar, and the like, the main direction of agro-waste management of banana waste is more focused on the production of pulp and paper, bioethanol, bioelectricity, and biomethane. Therefore, the challenges of commercialization, which are related closely to setting up biorefinery for banana waste to produce biofuel and product, are being addressed.

ACKNOWLEDGMENT

This chapter was written with Goh Wee Hong, Nur Shahira Binti Jamil, and S.K. MD Amzad Hossani.

REFERENCES

Abdullah, N., Sulaiman, F., & Taib, R. M. (2013). Characterization of banana (Musa spp.) plantation wastes as a potential renewable energy source. *AIP Conference Proceedings*, *1528*. https://doi.org/10.1063/1.4803618

Ahmad, K. K., Sazali, K., & Kamarolzaman, A. A. (2018). Characterization of fuel briquettes from banana tree waste. *Materials Today: Proceedings*, *5*(10). https://doi.org/10.1016/j.matpr.2018.07.027

Ahmad, T., & Danish, M. (2018). Prospects of banana waste utilization in wastewater treatment: A review. *Journal of Environmental Management*, *206*, 330–348. https://doi.org/10.1016/j.jenvman.2017.10.061

Akatwijuka, O., Gepreel, M. A. H., Abdel-Mawgood, A., Yamamoto, M., Saito, Y., & Hassanin, A. H. (2022). Overview of banana cellulosic fibers: Agro-biomass potential, fiber extraction, properties, and sustainable applications. *Biomass Conversion and Biorefinery*. https://doi.org/10.1007/s13399-022-02819-0

Alias, A. H., Norizan, M. N., Sabaruddin, F. A., Asyraf, M. R. M., Norrrahim, M. N. F., Ilyas, A. R., Kuzmin, A. M., Rayung, M., Shazleen, S. S., Nazrin, A., Sherwani, S. F. K., Harussani, M. M., Atikah, M. S. N., Ishak, M. R., Sapuan, S. M., & Khalina, A. (2021). Hybridization of MMT/lignocellulosic fiber reinforced polymer nanocomposites for structural applications: A review. *Coatings*, *11*(11), 1355. https://doi.org/10.3390/coatings11111355

Alzate Acevedo, S., Díaz Carrillo, Á. J., Flórez-López, E., & Grande-Tovar, C. D. (2021). Recovery of banana waste-loss from production and processing: A contribution to a circular economy. *Molecules (Basel, Switzerland)*, *26*(17), 1–30. https://doi.org/10.3390/molecules26175282

Arafat, K., Nayeem, J., Quadery, A., Quaiyyum, M., & Jahan, M. S. (2018). Handmade paper from waste banana fibre. *Bangladesh Journal of Scientific and Industrial Research, 53*(2). https://doi.org/10.3329/bjsir.v53i2.36668

Asyraf, M. R. M., Syamsir, A., Zahari, N. M., Supian, A. B. M., Ishak, M. R., Sapuan, S. M., Sharma, S., Rashedi, A., Razman, M. R., Zakaria, S. Z. S., Ilyas, R. A., & Rashid, M. Z. A. (2022). Product development of natural fibre-composites for various applications: Design for sustainability. *Polymers, 14*(5), 920. https://doi.org/10.3390/polym14050920

Azlin, M. N. M., Ilyas, R. A., Zuhri, M. Y. M., Sapuan, S. M., Harussani, M. M., Sharma, S., Nordin, A. H., Nurazzi, N. M., & Afiqah, A. N. (2022). 3D printing and shaping polymers, composites, and nanocomposites: A review. *Polymers, 14*(1), 180. https://doi.org/10.3390/polym14010180

Bakri, M. K. B. (2015). Investigation of Acoustical and Mechanical Properties of Epoxy Based Natural Fibre Composites. https://www.researchgate.net/publication/303974294_Investigation_of_Acoustical_and_Mechanical_Properties_of_Epoxy_Based_Natural_Fibre_Composites

Bakri, M. K. B., Jayamani, E., & Hamdan, S. (2017). Processing and characterization of banana fiber/epoxy composites: Effect of alkaline treatment. *Materials Today: Proceedings, 4*(2). https://doi.org/10.1016/j.matpr.2017.02.167

Balda, S., Sharma, A., Capalash, N., & Sharma, P. (2021). Banana fibre: A natural and sustainable bioresource for eco-friendly applications. *Clean Technologies and Environmental Policy, 23*(5), 1389–1401. https://doi.org/10.1007/s10098-021-02041-y

Berhanu, H., Kiflie, Z., Neiva, D., Gominho, J., Feleke, S., Yimam, A., & Pereira, H. (2018). Optimization of ethanol-alkali delignification of false banana (Ensete ventricosum) fibers for pulp production using response surface methodology. *Industrial Crops and Products, 126*. https://doi.org/10.1016/j.indcrop.2018.08.093

Bhatnagar, R., Gupta, G., & Yadav, S. (2015). A review on composition and properties of banana fibers. *International Journal of Scientific & Engineering Research, 6*(5), 49–52.

Bomb Calorimeter. (2015). *Dictionary of Energy* (pp. 41–81). https://doi.org/10.1016/B978-0-08-096811-7.50002-0

Choudhary, P., Pathak, A., Kumar, P., Chetna, S., & Sharma, N. (2022). Commercial production of bioplastic from organic waste–derived biopolymers viz-a-viz waste treatment: A minireview. *Biomass Conversion and Biorefinery*. https://doi.org/10.1007/s13399-022-03145-1

Elviliana, T. O. S. L., Virginia, C., & Suhartini, S. (2018). Conversion banana and orange peel waste into electricity using microbial fuel cell. *IOP Conference Series: Earth and Environmental Science*, 209(1). https://doi.org/10.1088/1755-1315/209/1/012049

Fagbemigun, T., Fagbemi, O., Buhari, F., Mgbachiuzo, E., & Igwe, C. (2016). Fibre characteristics and strength properties of Nigerian pineapple leaf (Ananas cosmosus), banana peduncle and banana leaf (Musa sapientum)—Potential green resources for pulp and paper production. *Journal of Scientific Research & Reports, 12*, 1–13. https://doi.org/10.9734/JSRR/2016/29248

Faizal, M. (2017). Utilization biomass and coal mixture to produce alternative solid fuel for reducing emission of green house gas. *International Journal on Advanced Science, Engineering and Information Technology, 7*(3). https://doi.org/10.18517/ijaseit.7.3.2474

Ferdous, T., Quaiyyum, M. A., Jin, Y., Bashar, M. S., Yasin Arafat, K. M., & Jahan, M. S. (2023). Pulping and bleaching potential of banana pseudo stem, banana leaf and banana peduncle. *Biomass Conversion and Biorefinery, 13*(2). https://doi.org/10.1007/s13399-020-01219-6

Food and Agriculture Organization of the United Nations. FAOSTAT 2018. https://www.fao.org/faostat/en/#data/QC (accessed February 8, 2022).

Guerrero, A. B., Ballesteros, I., & Ballesteros, M. (2018). The potential of agricultural banana waste for bioethanol production. *Fuel, 213*, 176–185. https://doi.org/10.1016/j.fuel.2017.10.105

Guo, M., He, Z., & Uchimiya, S. M. (2015). Introduction to biochar as an agricultural and environmental amendment. In *Agricultural and Environmental Applications of Biochar: Advances and Barriers*. https://doi.org/10.2136/sssaspecpub63.2014.0034

Halimatul, M. J., Sapuan, S. M., Jawaid, M., Ishak, M. R., & Ilyas, R. A. (2019). Water absorption and water solubility properties of sago starch biopolymer composite films filled with sugar palm particles. *Polimery, 64*(9), 27–35. https://doi.org/10.14314/polimery.2019.9.4

Haosagul, S., Boonyawanich, S., & Pisutpaisal, N. (2019). Biomethane production from co-fermentation of agricultural wastes. *International Journal of Hydrogen Energy, 44*(11). https://doi.org/10.1016/j.ijhydene.2018.09.080

Hossain, N., Razali, A., Chowdhury, H., Ong, H. C., Shamsuddin, A. H., Silitonga, A. S., Mahlia, T. M. I., & Chowdhury, T. (2019). Experimental investigation, techno-economic analysis and environmental impact of bioethanol production from banana stem. *Energies, 12*, 3947. https://doi.org/10.3390/en12203947

Housagul, S., Sirisukpoka, U., Boonyawanich, S., & Pisutpaisal, N. (2014). Biomethane production from co-digestion of banana peel and waste glycerol. *Energy Procedia*, *61*. https://doi.org/10.1016/j.egypro.2014.12.113

Ingale, S., Joshi, S. J., & Gupte, A. (2014). Production of bioethanol using agricultural waste: Banana pseudo stem. *Brazilian Journal of Microbiology*, *45*(3). https://doi.org/10.1590/S1517-83822014000300018

Jagadeesh, D., Venkatachalam, R., & Nallakumarasamy, G. (2015). Characterisation of banana fiber—A review. *Journal of Environmental Nanotechnology*, *4*(2). https://doi.org/10.13074/jent.2015.06.152154

John, I., Yaragarla, P., & Appusamy, A. (2020). Production of Bioethanol from Banana Peel Using Isolated Cellulase from Aspergillus Niger (pp. 9–18). https://doi.org/10.1007/978-981-13-9213-9_2

Jokhio, I., Qureshi, S. S., Sahito, A. R., Qureshi, M. un N., Lashari, I. A., Mubarak, N. M., Tripathi, M., Al-Mohaimeed, A. M., & Elshikh, M. S. (2022). Enhanced bio-methane and bio-hydrogen production using banana plant waste and sewage sludge through anaerobic co-digestion. *International Journal of Hydrogen Energy*. https://doi.org/10.1016/j.ijhydene.2022.04.238

Jumaidin, R., Adam, N. W., Ilyas, R. A., Hussin, M. S. F., Taha, M. M., Mansor, M. R., Azlan, U. A. A., & Yob, M. S. (2019a). Water transport and physical properties of sugarcane bagasse fibre reinforced thermoplastic potato starch biocomposite. *Journal of Advanced Research in Fluid Mechanics and Thermal Sciences*, *61*(2), 273–281.

Jumaidin, R., Diah, N. A., Ilyas, R. A., Alamjuri, R. H., & Yusof, F. A. M. (2021). Processing and characterisation of banana leaf fibre reinforced thermoplastic cassava starch composites. *Polymers*, *13*(9), 1420. https://doi.org/10.3390/polym13091420

Jumaidin, R., Saidi, Z. A. S., Ilyas, R. A., Ahmad, M. N., Wahid, M. K., Yaakob, M. Y., Maidin, N. A., Rahman, M. H. A., & Osman, M. H. (2019b). Characteristics of cogon grass fibre reinforced thermoplastic cassava starch biocomposite: Water absorption and physical properties. *Journal of Advanced Research in Fluid Mechanics and Thermal Sciences, 62*(1), 43–52.

Kalagbor, I. A., Azunda, B. I., Igwe, B. C., & Akpan, B. J., (2020). Electricity generation from waste tomatoes, banana, pineapple fruits and peels using Single Chamber Microbial Fuel Cells (SMFC). *Journal of Waste Management & Xenobiotics*. https://doi.org/10.23880/oajwx-16000141

Karim, A., Kumar, M., Mohapatra, S., Panda, C., & Singh, A. (2015). Banana peduncle biochar: Characteristics and adsorption of hexavalent chromium from aqueous solution. *International Research Journal of Pure and Applied Chemistry*, *7*(1). https://doi.org/10.9734/irjpac/2015/16163

Koul, B., Yakoob, M., & Shah, M. P. (2022). Agricultural waste management strategies for environmental sustainability. *Environmental Research*, *206*. https://doi.org/10.1016/j.envres.2021.112285

Legodi, L. M., LaGrange, D. C., Jansen van Rensburg, E. L., & Ncube, I. (2021). Enzymatic hydrolysis and fermentation of banana pseudostem hydrolysate to produce bioethanol. *International Journal of Microbiology*, 2021, 1–14. https://doi.org/10.1155/2021/5543104

Lim, S.-F., & Matu, S. U. (2015). Utilization of agro-wastes to produce biofertilizer. *International Journal of Energy and Environmental Engineering*, *6*(1), 31–35. https://doi.org/10.1007/s40095-014-0147-8

Maduwanthi, S. D. T., & Marapana, R. A. U. J. (2019). Induced ripening agents and their effect on fruit quality of banana. *International Journal of Food Science*, *2019*. https://doi.org/10.1155/2019/2520179

Manhongo, T. T., Chimphango, A. F. A., Thornley, P., & Röder, M. (2022). Current status and opportunities for fruit processing waste biorefineries. *Renewable and Sustainable Energy Reviews*, *155*, 111823. https://doi.org/https://doi.org/10.1016/j.rser.2021.111823

Manu, T., Nazmi, A. R., Shahri, B., Emerson, N., & Huber, T. (2022). Biocomposites: A review of materials and perception. *Materials Today Communications*, *31*, 103308. https://doi.org/10.1016/j.mtcomm.2022.103308

Martins, L., Zanini, N., Pinheiro, L., & Mulinari, D. (2022). Valorization of banana peel waste used as filler in castor oil polyurethane foam for vegetal oil sorption. *Journal of Natural Fibers*, *19*(14), 7806–7817. https://doi.org/10.1080/15440478.2021.1958414

Mazani, N., Sapuan, S. M., Sanyang, M. L., Atiqah, A., & Ilyas, R. A. (2019). Design and fabrication of a shoe shelf from kenaf fiber reinforced unsaturated polyester composites. In H. Ariffin, S. M. Sapuan, & M. A. Hassan (Eds.), *Lignocellulose for Future Bioeconomy* (1st ed., Issue 2000, pp. 315–332). Elsevier. https://doi.org/10.1016/B978-0-12-816354-2.00017-7

Milani, M. D. Y., Samarawickrama, D. S., Dharmasiri, G. P. C. A., & Kottegoda, I. R. M. (2016). Study the structure, morphology, and thermal behavior of banana fiber and its charcoal derivative from selected banana varieties. *Journal of Natural Fibers*, *13*(3), 332–342. https://doi.org/10.1080/15440478.2015.1029195

Mishra, R. R., Samantaray, B., Behera, B. C., Pradhan, B. R., & Mohapatra, S. (2020). Process optimization for conversion of waste banana peels to biobutanol by a yeast co-culture fermentation system. *Renewable Energy*, *162*. https://doi.org/10.1016/j.renene.2020.08.045

Mohapatra, D., Mishra, S., & Sutar, N. (2010). Banana and its by-product utilisation: An overview. *Journal of Scientific and Industrial Research*, *69*(5), 323–329.

Mostafa, H. S. (2021). Banana plant as a source of valuable antimicrobial compounds and its current applications in the food sector. *Journal of Food Science*, *86*(9). https://doi.org/10.1111/1750-3841.15854

Mote, M. A., Gavade, M. A., Bandgar, M. S., & Narute, M. A. (2023). Handmade paper from MUSA accuminata (banana) stem. *International Journal of Advanced Research in Science, Communication and Technology*. https://doi.org/10.48175/ijarsct-7897

Nadhari, W. N. A. W., Danish, M., Nasir, M. S. R. M., & Geng, B. J. (2019). Mechanical properties and dimensional stability of particleboard fabricated from steam pre-treated banana trunk waste particles. *Journal of Building Engineering*, *26*. https://doi.org/10.1016/j.jobe.2019.100848

Nagarajan, K. J., Sanjay, M. R., Sathick Basha, K., Raghav, G. R., Ashok Kumar, R., Siengchin, S., Surya Rajan, B., Sabari Nath, P., & Khan, A. (2022). Extraction of cellulose nanocrystals from red banana peduncle agro-waste and application in environmentally friendly biocomposite film. *Polymer Composites*, *43*(8), 4942–4958. https://doi.org/10.1002/pc.26755

Nazrin, A., Sapuan, S. M., Zuhri, M. Y. M., Tawakkal, I. S. M. A., & Ilyas, R. A. (2021). Water barrier and mechanical properties of sugar palm crystalline nanocellulose reinforced thermoplastic sugar palm starch (TPS)/poly(lactic acid) (PLA) blend bionanocomposites. *Nanotechnology Reviews*, *10*(1), 431–442. https://doi.org/10.1515/ntrev-2021-0033

Norfarhana, A. S., Ilyas, R. A., & Ngadi, N. (2022). A review of nanocellulose adsorptive membrane as multifunctional wastewater treatment. *Carbohydrate Polymers*, *291*, 119563. https://doi.org/10.1016/j.carbpol.2022.119563

Norfarhana, A. S., Ilyas, R. A., Ngadi, N., Othman, M. H. D., Misenan, M. S. M., & Norrrahim, M. N. F. (2024a). Revolutionizing lignocellulosic biomass: A review of harnessing the power of ionic liquids for sustainable utilization and extraction. *International Journal of Biological Macromolecules*, *256*, 128256. https://doi.org/10.1016/j.ijbiomac.2023.128256

Norfarhana, A. S., Khoo, P. S., Ilyas, R. A., Ab Hamid, N. H., Aisyah, H. A., Norrrahim, M. N. F., Knight, V. F., Rani, M. S. A., Septevani, A. A., Syafri, E., & Annamalai, P. K. (2024b). Exploring of cellulose nanocrystals from lignocellulosic sources as a powerful adsorbent for wastewater remediation. *Journal of Polymers and the Environment*. https://doi.org/10.1007/s10924-024-03227-3

Nurazzi, N. M., Harussani, M. M., Aisyah, H. A., Ilyas, R. A., Norrrahim, M. N. F., Khalina, A., & Abdullah, N. (2021). Treatments of natural fiber as reinforcement in polymer composites—A short review. *Functional Composites and Structures*, *3*(2), 024002. https://doi.org/10.1088/2631-6331/abff36

Nwabanne, J., & Aghadi, C. (2018). Statistical modelling of enzymatic hydrolysis of banana peels for bioethanol production. *Current Journal of Applied Science and Technology*, *28*(4), 1–14. https://doi.org/10.9734/CJAST/2018/42906

Padam, B. S., Tin, H. S., Chye, F. Y., & Abdullah, M. I. (2014). Banana by-products: An under-utilized renewable food biomass with great potential. *Journal of Food Science and Technology*, *51*(12), 3527–3545. https://doi.org/10.1007/s13197-012-0861-2

Palacios, A. S., Ilyina, A., Ramos-González, R., Aguilar, C. N., Martínez-Hernández, J. L., Segura-Ceniceros, E. P., González, M. L. C., Aguilar, M., Ballesteros, M., Oliva, J. M., & Ruiz, H. A. (2021). Ethanol production from banana peels at high pretreated substrate loading: Comparison of two operational strategies. *Biomass Conversion and Biorefinery*, *11*(5), 1587–1596. https://doi.org/10.1007/s13399-019-00562-7

Pappu, A., Patil, V., Jain, S., Mahindrakar, A., Haque, R., & Thakur, V. K. (2015). Advances in industrial prospective of cellulosic macromolecules enriched banana biofibre resources: A review. *International Journal of Biological Macromolecules*, *79*. https://doi.org/10.1016/j.ijbiomac.2015.05.013

Pisutpaisal, N., Boonyawanich, S., & Saowaluck, H. (2014). Feasibility of biomethane production from banana peel. *Energy Procedia*, *50*. https://doi.org/10.1016/j.egypro.2014.06.096

Pongsuwan, C., Boonsuk, P., Sermwittayawong, D., Aiemcharoen, P., Mayakun, J., & Kaewtatip, K. (2022). Banana inflorescence waste fiber: An effective filler for starch-based bioplastics. *Industrial Crops and Products*, *180*, 114731. https://doi.org/10.1016/J.INDCROP.2022.114731

Pradhan, S., Abdelaal, A. H., Mroue, K., Al-Ansari, T., Mackey, H. R., & McKay, G. (2020). Biochar from vegetable wastes: Agro-environmental characterization. *Biochar*, *2*(4). https://doi.org/10.1007/s42773-020-00069-9

Prakash, H., Chauhan, P. S., General, T., & Sharma, A. K. (2018). Development of eco-friendly process for the production of bioethanol from banana peel using inhouse developed cocktail of thermo-alkali-stable depolymerizing enzymes. *Bioprocess and Biosystems Engineering*, *41*(7). https://doi.org/10.1007/s00449-018-1930-3

Rahman, M. M., Islam, T., Nayeem, J., & Jahan, S. (2014). Variation of chemical and morphological properties of different parts of banana plant (Musa paradisica) and their effects on pulping. *International Journal of Lignocellulosic Products*, *1*(12), 93–103.

Rajendran, M., & Nagarajan, C. K. (2022). Experimental investigation on bio-composite using jute and banana fiber as a potential substitute of solid wood based materials. *Journal of Natural Fibers*, *19*(12), 4557–4566. https://doi.org/10.1080/15440478.2020.1867943

Ramdhonee, A., & Jeetah, P. (2017). Production of wrapping paper from banana fibres. *Journal of Environmental Chemical Engineering*, *5*(5). https://doi.org/10.1016/j.jece.2017.08.011

Ramesh, M., Atreya, T. S. A., Aswin, U. S., Eashwar, H., & Deepa, C. (2014). Processing and mechanical property evaluation of banana fiber reinforced polymer composites. *Procedia Engineering*, *97*. https://doi.org/10.1016/j.proeng.2014.12.284

Redondo-Gómez, C., Quesada, M. R., Astúa, S. V., Zamora, J. P. M., Lopretti, M., & Vega-Baudrit, J. R. (2020). Biorefinery of biomass of agro-industrial banana waste to obtain high-value biopolymers. *Molecules*, *25*(17). https://doi.org/10.3390/molecules25173829

Rincón-Catalán, N. I., Cruz-Salomón, A., Sebastian, P. J., Pérez-Fabiel, S., Hernández-Cruz, M. del C., Sánchez-Albores, R. M., Hernández-Méndez, J. M. E., Domínguez-Espinosa, M. E., Esquinca-Avilés, H. A., Ríos-Valdovinos, E. I., & Nájera-Aguilar, H. A. (2022). Banana waste-to-energy valorization by microbial fuel cell coupled with anaerobic digestion. *Processes*, *10*(8). https://doi.org/10.3390/pr10081552

Rohen, L. A., Neves, A. C. C., de P. Mantovani, D., Carlos Maurício, F. V., da Silva Vieira, J., de A. Pontes, L., Margem, F. M., & Monteiro, S. (2017). Hemp fiber density using the pycnometry technique. In *Characterization of Minerals, Metals and Materials Series, Part F7* (pp. 423–428). https://doi.org/10.1007/978-3-319-51382-9_46

Rojas-Flores, S., De La Cruz-Noriega, M., Nazario-Naveda, R., Benites, S. M., Delfín-Narciso, D., Angelats-Silva, L., & Murga-Torres, E. (2022). Use of banana waste as a source for bioelectricity generation. *Processes*, *10*(5). https://doi.org/10.3390/pr10050942

Ronald Aseer, J., Sankaranarayanasamy, K., Jayabalan, P., Natarajan, R., & Priya Dasan, K. (2013). Morphological, physical, and thermal properties of chemically treated banana fiber. *Journal of Natural Fibers*, *10*(4). https://doi.org/10.1080/15440478.2013.824848

Rudin, A., & Choi, P. (2013). Biopolymers. In The Elements of Polymer Science & Engineering (pp. 521–535). https://doi.org/10.1016/B978-0-12-382178-2.00013-4

Santiago, B., Moreira, M. T., Feijoo, G., & González-García, S. (2022). Environmental comparison of banana waste valorisation strategies under a biorefinery approach. *Waste Management*, *142*. https://doi.org/10.1016/j.wasman.2022.02.005

Sawarkar, A. N., Kirti, N., Tagade, A., & Tekade, S. P. (2022). Bioethanol from various types of banana waste: A review. *Bioresource Technology Reports*, *18*(May), 101092. https://doi.org/10.1016/j.biteb.2022.101092

Sial, T. A., Khan, M. N., Lan, Z., Kumbhar, F., Ying, Z., Zhang, J., Sun, D., & Li, X. (2019). Contrasting effects of banana peels waste and its biochar on greenhouse gas emissions and soil biochemical properties. *Process Safety and Environmental Protection*, *122*, 366–377. https://doi.org/10.1016/J.PSEP.2018.10.030

Sivanarutselvi, S., Poornima, P., Muthukumar, K., & Velan, M. (2019). Studies on effect of alkali pretreatment of banana pseudostem for fermentable sugar production for biobutanol production. *Journal of Environmental Biology*, *40*(3). https://doi.org/10.22438/jeb/40/3/MRN-721

Sogani, M., Sonu, K., Syed, Z., & Rajvanshi, J. (2023). Preparation of biofertilizer blend from banana peels along with its application in agriculture and plant microbial fuel cell. *IOP Conference Series: Earth and Environmental Science*, 1151(1), 012034. https://doi.org/10.1088/1755-1315/1151/1/012034

Suhag, M., Kumar, A., & Singh, J. (2020). Saccharification and fermentation of pretreated banana leaf waste for ethanol production. *SN Applied Sciences*, *2*(8), 1448. https://doi.org/10.1007/s42452-020-03215-x

Sultan, N. F. K., & Johari, W. L. W. (2017). The development of banana peel/corn starch bioplastic film: A preliminary study. *Bioremediation Science and Technology Research*, *5*(1). https://doi.org/10.54987/bstr.v5i1.352

Suriani, M. J., Radzi, F. S. M., Ilyas, R. A., Petrů, M., Sapuan, S. M., & Ruzaidi, C. M. (2021). Flammability, tensile, and morphological properties of oil palm empty fruit bunches fiber/pet yarn-reinforced epoxy fire retardant hybrid polymer composites. *Polymers*, *13*(8), 1282. https://doi.org/10.3390/polym13081282

Syafri, E., Jamaluddin, Sari, N. H., Mahardika, M., Amanda, P., & Ilyas, R. A. (2022). Isolation and characterization of cellulose nanofibers from Agave gigantea by chemical-mechanical treatment. *International Journal of Biological Macromolecules*, *200*(December 2021), 25–33. https://doi.org/10.1016/j.ijbiomac.2021.12.111

Tripathi, S. K., Kant Bhardwaj, N., Chechi, S., & Varadhan, R. (2019). Suitability of banana stem pulp as replacement of softwood pulp for making superior grade unbleached paper from agro residue pulp. *Appita Journal*, *72*(3).

Venkateshwaran, N., & Elayaperumal, A. (2010). Banana fiber reinforced polymer composites—A review. *Journal of Reinforced Plastics and Composites*, *29*(15). https://doi.org/10.1177/0731684409360578

Verma, M., & Mishra, V. (2023). Bioelectricity generation by microbial degradation of banana peel waste biomass in a dual-chamber S. cerevisiae-based microbial fuel cell. *Biomass and Bioenergy*, *168*. https://doi.org/10.1016/j.biombioe.2022.106677

Viena, V., Elvitriana, & Wardani, S. (2018). Application of banana peels waste as adsorbents for the removal of CO2, NO, NOx, and SO2 gases from motorcycle emissions. *IOP Conference Series: Materials Science and Engineering*, *334*(1), 012037. https://doi.org/10.1088/1757-899X/334/1/012037

Wobiwo, F. A., Alleluya, V. K., Emaga, T. H., Boda, M., Fokou, E., Gillet, S., Deleu, M., & Gerin, P. A. (2017). Recovery of fibers and biomethane from banana peduncles biomass through anaerobic digestion. *Energy for Sustainable Development*, *37*, 60–65. https://doi.org/10.1016/j.esd.2017.01.005

Section V

Stalk-Based Plant Waste

11 Wheat Waste Management

11.1 INTRODUCTION

Wheat, a cereal grain, is a crucial food source all across the globe. Wheat is a grass that comes from the genus *Triticum*, and there are many distinct types of wheat that are cultivated in different parts of the world. Wheat is cultivated largely for its seeds, which are transformed into flour and utilized in the production of numerous staple meals such as cereals, bread, pasta, and pastries (Skendi & Papageorgiou, 2021). Grain is utilized not only to make beer and other alcoholic beverages but also as a vital source of nutrition for livestock (Dabija et al., 2021). Wheat is thought to have originated in the Fertile Crescent in the Middle East, where it has been grown for thousands of years. It is now an important crop in many nations, including the United States, China, India, Russia, and Canada, and is cultivated in virtually every country in the world. Wheat is an adaptable crop that thrives in both hot and dry places and cooler and rainy climates. It is also a high-yield crop, so it helps feed and support a lot of people and keeps a lot of farmers in business all around the world. The evolution of modern wheat is illustrated in Figure 11.1, which also shows examples of spikes and grain (Shewry, 2009).

Wheat waste consists of any wheat plant parts that are either not used in food production or thrown as a by-product (Sadh et al., 2018). This category includes the by-products of milling wheat into flour, such as the bran and germ, and the plant detritus, such as straw and chaff, that remain after the grain has been harvested. The vast amounts of agricultural and food processing waste that can be generated from wheat waste raise serious concerns for environmental sustainability.

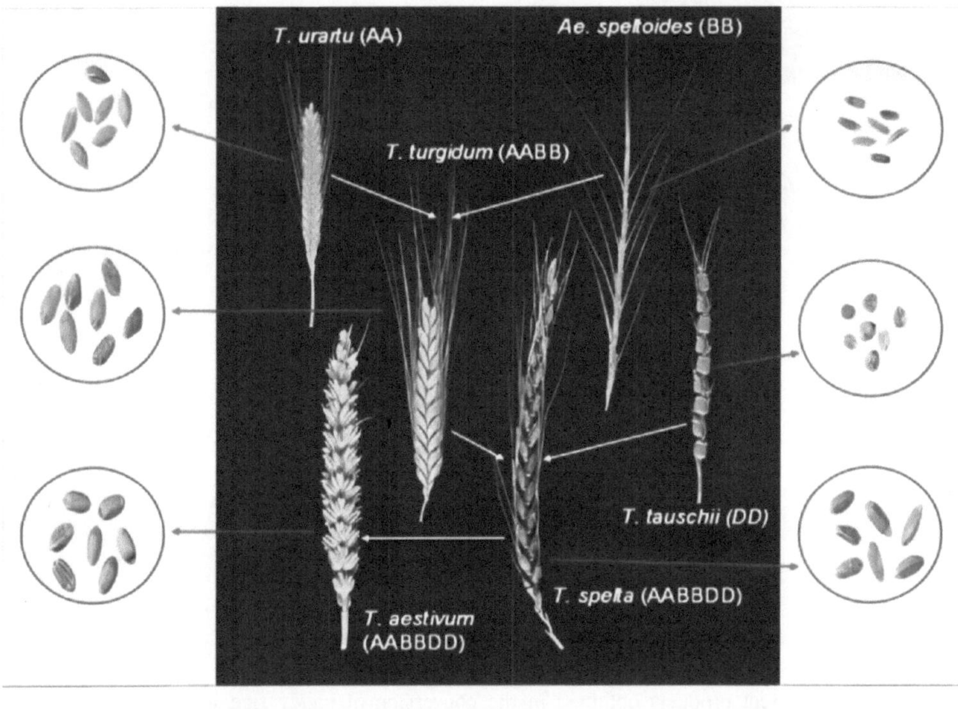

FIGURE 11.1 Examples of spikes and grain (Shewry, 2009).

DOI: 10.1201/9781003472209-16

However, it is also a resource that may be used for many other purposes. Animal feed is a major consumer of wheat by-products. Straw and other plant debris can be utilized to feed cattle, while bran and germ can be used to augment animal feed with a high concentration of protein (Irfan, 2022). Wheat hulls can also be burned to produce ethanol or utilized as a fuel source in other forms of energy production. The extraction of bioactive chemicals, enzymes, and fibers from wheat waste for use in applications as diverse as food, medicines, and personal care items is also an area of current research and development (Awad et al., 2021). Wheat waste is a problem for the agricultural and food industries, but it also offers chances to recycle valuable materials and cut down on waste (Skendi & Papageorgiou, 2021). Figure 11.2 illustrates the sequential stages of rice production and highlights the primary products and potential by-products (wastes) generated at each step.

11.2 BIOMASS FROM NATURAL FIBERS

Biomass is any organic material derived from living organisms that is burned for energy or put to some other industrial use. Everything from wood and agriculture wastes to municipal solid waste and algae can be classified as such. Biomass can be used to generate heat or electricity, or it can be processed further into biofuels, biogas, or charcoal to produce yet more energy (Semadeni, 2004). In contrast to finite fossil fuels, renewable biomass can be grown and replaced indefinitely, making it an ideal energy source. However, if it results in deforestation or other forms of land-use change, or if it is not managed sustainably, the use of biomass for energy can have severe environmental implications. Biomass can be produced from a wide range of plant and animal fibers, including cotton, hemp, flax, jute, kenaf, and silk. Biofuels, bioplastics, and other eco-friendly materials can all be made from these fibers (Sharma et al., 2023).

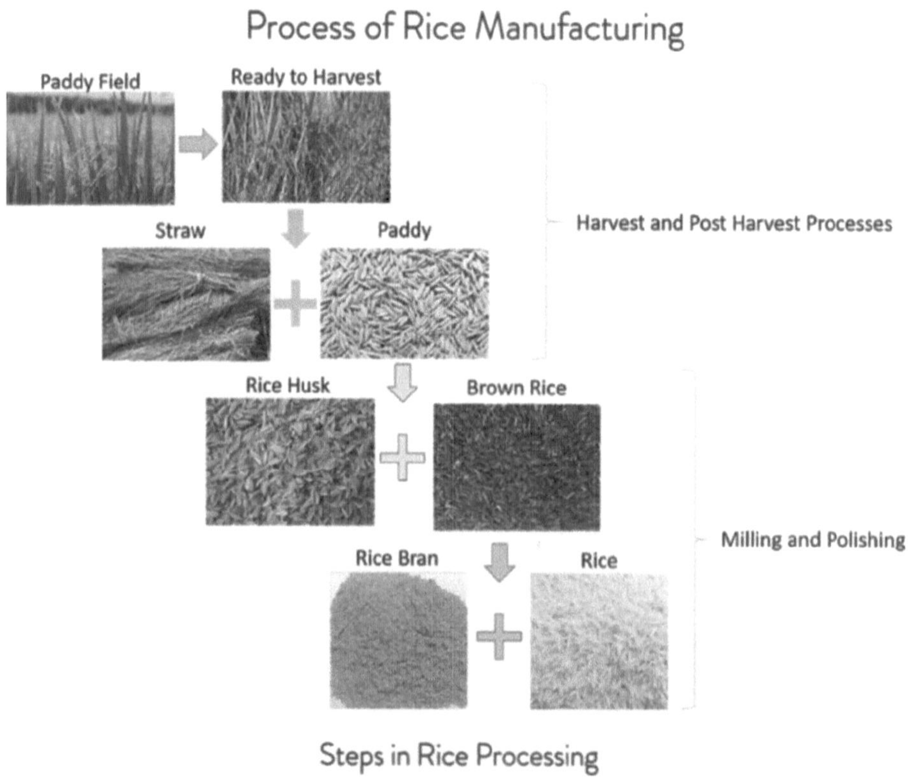

FIGURE 11.2 Different products obtained in the conversion of paddy rice with the proportion of the by-products.

Cellulosic ethanol, produced from plant fiber cellulose and hemicellulose, is one type of biofuel created from natural fibers. Since it does not pose a threat to food supplies and can be produced from a larger variety of feedstocks, this ethanol has the potential to be a more environmentally friendly option than corn-based ethanol (Sharma et al., 2023). Bioplastics, which are produced from renewable resources and are biodegradable or compostable, can also be created from natural fibers or to be reinforced with natural fiber (Halimatul et al., 2019; Jumaidinet al., 2019a,b, 2021; Nazrin et al., 2021). Cigarette filters, textiles, and packaging are all common places to find cellulose acetate, a bioplastic made from cellulose-based fibers (Yadav & Hakkarainen, 2021). Biomass-derived from natural fibers could serve as a pivotal input in the long-term viability of the energy and materials sectors (Alias et al., 2021; Asyraf et al., 2022; Azlin et al., 2022; Faiz Norrrahim et al., 2021; Mazani et al., 2019; Suriani et al., 2021). It is essential, however, that these fibers be cultivated and collected in an environmentally friendly manner.

11.3 ENVIRONMENTAL IMPACT FROM BIOMASS

Environmental impact refers to the effect that human activities have on the natural environment (Abdallah, 2017). It encompasses the changes and disruptions caused by various actions, processes, and developments that can have positive and negative environmental consequences. Human activities such as industrialization, urbanization, agriculture, transportation, and energy production contribute to environmental impact (Manisalidis et al., 2020). These activities often result in the depletion of natural resources, pollution of air, water, and soil, loss of biodiversity, habitat destruction, and climate change. While biomass can be a renewable source of energy, its use can also have environmental impacts that need to be carefully managed. Some of the potential environmental impacts of biomass are land usage, water usage, air pollution, soil quality, and biodiversity.

Land-use change from biomass refers to the conversion of natural land cover, such as forests or grasslands, into land dedicated to biomass production for energy or other reasons. Plants, crops, agricultural wastes, and forestry by-products are examples of organic resources that can be used as renewable energy sources or raw materials in various sectors. Land-use change for biomass production can have both beneficial and bad environmental consequences. On the positive side, biomass can be a sustainable alternative to fossil fuels, reducing greenhouse gas (GHG) emissions and mitigating climate change. Biomass energy systems can help replace the use of coal or oil, reducing carbon dioxide emissions and promoting a transition to renewable energy sources (Siles et al., 2023). However, converting natural ecosystems to biomass production can have unintended consequences. Deforestation or clearance of native vegetation for large-scale biomass plantations can result in biodiversity loss and habitat degradation. Natural habitat loss can alter ecosystems and endangered species' viability. Furthermore, increased biomass production may need fertilizers, pesticides, and irrigation, all of which can negatively influence water quality and availability. Excessive irrigation water use can decrease water supplies, especially in areas where water is already scarce. Moreover, monoculture plantations established for biomass production may lack the biodiversity and ecological functions of natural ecosystems. These monocultures often require the use of genetically modified organisms or highly productive plant varieties, which can have unintended consequences for biodiversity and ecosystem stability (Zilberman et al., 2018).

The use of water in biomass production can have serious environmental consequences, affecting both the quality and availability of water supplies. Biomass production, especially large-scale plantations, frequently necessitates enormous amounts of water for irrigation, processing, and biomass conversion. One of the most serious environmental issues is the loss of water resources. The high water demands of biomass crops can strain local water supplies, particularly in already water-stressed areas. Excessive water withdrawal for biomass production might restrict available water for other uses such as drinking water, agriculture, and ecosystem upkeep. This can exacerbate water scarcity issues, impacting both human populations and aquatic ecosystems (Laser et al., 2009). Furthermore, intensive irrigation for biomass crops can contribute to water pollution. Excess irrigation water, along with agrochemicals such as fertilizers and pesticides, can leach into water bodies, leading to nutrient runoff, eutrophication, and surface and groundwater contamination. These pollutants can degrade water quality, harm aquatic ecosystems, and pose risks to human health. The alteration of natural hydrological cycles due to land-use change for biomass production can also have detrimental environmental effects. Forests and wetlands, which naturally regulate water flow and recharge aquifers, are often cleared for biomass plantations. This disruption can result in changes to local water cycles, including altered rainfall patterns, reduced groundwater recharge, and increased surface runoff, leading to flooding or erosion (Inglezakis et al., 2016).

Air pollution caused by biomass combustion or biomass-related activities can have serious environmental consequences. When biomass, such as wood or crop leftovers, is burned for energy or other uses, it emits a variety of pollutants into the atmosphere. The emission of particulate matter (PM) is one of the key problems. Fine PMs, such as PM2.5 (particles with a diameter of 2.5 µm or less) and PM10 (particles with a diameter of 10 µm or less), can be produced by biomass burning. These particles can be inhaled deeply into the respiratory system, causing respiratory troubles, cardiovascular problems, and other negative health impacts in people. Additionally, PM can contribute to the formation of smog and reduce visibility (Kern, 2023). Biomass combustion also releases gaseous pollutants, such as carbon monoxide (CO), nitrogen oxides (NO_x), sulfur dioxide (SO_2), and volatile organic compounds (VOCs). These pollutants can have detrimental effects on human health and the environment. For example, CO is a toxic gas that can impair oxygen transport in the body, leading to respiratory and cardiovascular problems. NO_x and VOCs are precursors to the formation of ground-level ozone, a harmful pollutant that can cause respiratory issues and damage vegetation. Furthermore, biomass combustion can contribute to the release of GHGs such as carbon dioxide (CO_2) and methane (CH_4). While biomass is considered a renewable energy source, the combustion process still releases CO_2, contributing to global warming and climate change. Additionally, incomplete combustion of biomass can result in the emission of CH_4, a potent GHG with a greater warming potential than CO_2 over the short term (Lebel et al., 2022).

The impact of biomass production on soil quality is a complicated subject that varies depending on a variety of factors, such as biomass crop type, land management practices, and soil conditions. While biomass production has several benefits for soil health, such as organic matter input and nutrient cycling, it also has some problems and possible dangers to soil quality. One of the primary challenges is soil erosion and degradation. Large-scale land-use changes for biomass plantations can increase soil erosion, especially if suitable soil conservation measures are not adopted. Deforestation or the conversion of natural ecosystems can remove the protective cover of vegetation, making the soil more susceptible to erosion by wind and water. Soil erosion can result in nutrient loss, decreased soil fertility, and reduced water-holding capacity, negatively impacting agricultural productivity and ecosystem functioning (Maximillian et al., 2019). Additionally, intensive biomass production may require the use of fertilizers, pesticides, and irrigation, which can impact soil quality. The excessive use of fertilizers can lead to nutrient imbalances and soil acidification, affecting soil microbial communities and nutrient availability. The application of pesticides can have detrimental effects on soil organisms, disrupting soil biodiversity and ecosystem functioning. Moreover, improper irrigation practices can contribute to soil salinization or waterlogging, further compromising soil health. The choice of biomass crop and management practices can also influence soil quality. Some biomass crops, such as perennial grasses or legumes, can have positive effects on soil fertility, organic matter content, and soil structure (Edwards & Burney, 2005). These crops can contribute to increased carbon sequestration, enhance soil moisture retention, and promote beneficial soil microbial activity. On the contrary, if biomass crops are managed intensively, with practices such as monoculture, excessive tillage, or inadequate nutrient management, it can lead to soil degradation, compaction, and reduced biodiversity (Adetunji et al., 2020).

Depending on factors such as land-use practices, crop choices, and management tactics, biomass production can substantially impact biodiversity, both positively and negatively. While biomass crops may offer habitat and food for some species, the total impact on biodiversity is determined by the quantity and intensity of production, as well as the location and qualities of the area exploited. The conversion of natural ecosystems, such as forests or grasslands, into biomass plantations is one of the key challenges. This land-use change frequently results in habitat loss and fragmentation, displacing and disrupting native plant and animal species. Removing natural vegetation can reduce biodiversity, eliminating crucial habitats for various species, including threatened or endangered ones. Moreover, the loss of diverse ecosystems can negatively impact ecological interactions and disrupt ecosystem services (Duffy et al., 2016). However, if biomass production is planned and managed sustainably, it can contribute to the conservation of biodiversity. Certain biomass crops, such as perennial grasses or native species, can provide habitats and food sources for a range of organisms, including pollinators and birds (Jager & Kreig, 2018). Agroforestry systems that incorporate biomass crops with native trees can also support biodiversity by creating diverse and interconnected habitats. Proper land management practices such as maintaining buffer zones, protecting riparian areas, and preserving wildlife corridors can further enhance biodiversity conservation. Furthermore, the choice of biomass feedstock can influence the impact on biodiversity. Using agricultural residues or dedicated energy crops that do not compete with food production can help minimize the conversion of natural habitats (Gent et al., 2017). Additionally, selecting non-invasive species and utilizing diverse crop rotations can reduce the risk of introducing or spreading invasive species, which can have detrimental effects on native biodiversity.

To minimize these potential environmental impacts, it is important to use sustainable biomass production and management practices. This includes ensuring that biomass production does not result in the conversion of natural habitats, minimizing water use, implementing appropriate pollution control measures, and ensuring that biomass production is integrated with other land uses to maintain biodiversity and soil health (Rather et al., 2022).

11.4 AVAILABILITY OF WHEAT WASTE

Wheat waste, a by-product of wheat processing, encompasses materials like wheat bran, wheat germ, and wheat straw. It is obtained after harvesting wheat grains and has an annual global production of approximately 529 million tons (Tufail et al., 2018). On average, around 1.3–1.4 kg of wheat straw is yielded for every kg of grain produced (Memon et al., 2018), resulting in roughly 534 million tons of global wheat straw production. The availability of wheat waste depends on factors such as region, processing methods, and demand for wheat products.

The Food and Agriculture Organization of the United Nations (2020) reports that wheat is the second-most-produced cereal grain after maize, and its global trade surpasses that of all other crops combined. In 2020, the total global production of wheat reached 760 million tons. China, India, and Russia are the three largest individual wheat producers, contributing to about 41% of the world's total wheat production. Wheat is a vital crop globally, with its primary product being flour, along with by-products like bran and germ. These by-products are commonly used as animal feed or in the production of various food items, including breakfast cereals, baked goods, and pasta.

Apart from wheat bran and middlings, wheat straw represents another form of wheat waste with various potential applications. According to a study conducted by Paul et al. (2020) on the production of bio-oil from wheat straw, this crop is one of the most widely cultivated in the world, generating an annual production of 529 million tons of straw as its main by-product. Wheat straw finds diverse uses, such as animal bedding, a source of bioenergy, or a raw material for pulp and paper production. Wheat straw is highly regarded as an attractive, cost-effective, and promising feedstock for manufacturing different products (Memon et al., 2018). Within the EU, it stands as the most significant type of agricultural residue, accumulating approximately 144 million tons each year (Memon et al., 2018). Its applications range from fuel for heating, animal feed, and domestic animal bedding to being mixed into the soil or burned in open areas. It is considered a widely distributed, recyclable, dynamic, and productive source for producing biogas, bio-hydrogen in biorefineries, and bioethanol, promoting biomass utilization in an environmentally friendly setting. Due to its abundant availability, wheat straw should be utilized as a cost-effective resource for biofuel, bioethanol, and renewable energy. Proper disposal of wheat straw is of utmost importance, especially in countries where it is produced in significant quantities.

The demand for wheat waste and its by-products in the market can vary depending on several factors. In recent years, there has been an increasing demand for sustainable and environmentally friendly solutions, which has led to a growing interest in utilizing wheat waste. Various industries, including animal feed, bioenergy, bioplastics, and food ingredients, drive the demand for wheat waste products. In the animal feed sector, wheat bran and other wheat by-products are a valuable source of dietary fiber and nutrients for livestock. The demand for these products is influenced by factors such as the availability of alternative feed ingredients, regulations on animal nutrition, and consumer preferences for ethically sourced and sustainable feed.

In the bioenergy sector, agricultural residues like wheat straw play a crucial role as feedstocks for biofuel production, particularly in generating bioethanol and biogas. The demand for wheat waste as a bioenergy feedstock is heavily influenced by government policies that support renewable energy, the availability of other biomass sources, and the efficiency and cost-effectiveness of conversion technologies. Moreover, there is a growing demand for wheat waste in other industries as well. For instance, in the production of bioplastics and food ingredients, wheat by-products can be processed into valuable components like wheat protein isolates or dietary fibers, suitable for use in various food and beverage applications. This demand is driven by factors like increasing consumer preference for plant-based and sustainable products, the development of new food formulations, and the regulatory environment concerning food safety and labeling. The global market for wheat straw was valued at USD 643.6 million in 2021. It is expected to experience substantial growth with a compound annual growth rate of 9.50% during the forecast period of 2022–2029, ultimately reaching a value of USD 1,330.24 million by the end of 2029 (Data Bridge Market Research, 2023.). This data illustrates the increasing significance and potential of wheat waste in various industries and highlights its role in the development of sustainable and economically viable solutions.

Therefore, wheat waste is a readily available resource with significant potential for utilization in various industries. The by-products of wheat processing, such as straw, bran, husk, and chaff, offer sustainable waste management and value-addition opportunities. The availability of wheat waste is influenced by factors such as the global production of wheat, agricultural practices, and processing methods. As the demand for sustainable and eco-friendly solutions increases, there is a growing interest in utilizing wheat waste for applications such as animal feed, bioenergy production, bioplastics, and food ingredients. However, the specific availability of wheat waste may vary geographically and depend on regional agricultural practices and market dynamics. To assess the local availability of wheat waste, it is recommended to consider factors such as the volume of wheat production, waste management practices, and local industry requirements.

11.5 PROPERTIES OF WHEAT WASTE FIBERS

11.5.1 Physical Properties

The physical properties of wheat waste can vary depending on the specific form and composition, encompassing particle size, density, porosity, moisture content, angle of repose, flowability, and color. The particle size varies depending on the type of waste, such as bran or straw, with wheat bran particles ranging from 77 to 1,687 μm in size (Jacobs et al., 2015) and wheat straw particles measuring 0.38–0.69 mm in length (Zhang et al., 2012). The density of wheat waste depends on the type and moisture content; for example, wheat straw may have a density range of 97.52–177.23 kg/m^3. Additionally, the porosity of wheat waste is reported to be around 46.39–84.24% for wheat straw. Moisture content varies between 5.02% and 7.79% in wheat straw, depending on storage conditions and processing methods (Zhang et al., 2012). The angle of repose for wheat straw is approximately

23–28° (Suri & Singh, 2023). Furthermore, flowability is influenced by the type of waste and processing method, while the color of wheat waste can range from light yellow to dark brown, depending on the level of processing and exposure to heat and moisture.

11.5.2 Chemical Composition

Wheat waste encompasses several parts of the wheat plant, namely wheat bran, wheat germ, and wheat straw, which are left after processing. The chemical composition of wheat waste varies depending on the specific plant part and the processing method employed. Wheat bran, the outer layer of the wheat kernel, is rich in dietary fiber, protein, vitamins, and minerals. Wheat germ, the embryo of the wheat kernel, is also removed during milling and serves as a concentrated source of nutrients, including vitamins, minerals, and antioxidants. Wheat straw constitutes the stalks and stems of the wheat plant remaining after grain harvest, and it is notably high in cellulose and lignin, which form key components of plant cell walls. Wheat waste presents a valuable reservoir of nutrients and other compounds with diverse applications, including its utilization in animal feed, biofuels, and bioplastics.

The chemical composition of wheat waste, commonly known as wheat straw, exhibits variations influenced by factors such as soil type, climate, and the specific wheat variety. Sun et al. (2004) conducted an analysis that revealed the constituents and their respective proportions. Wheat straw predominantly consists of cellulose (46.2–49.2% of dry weight), providing its structural integrity, alongside hemicellulose (11.2–12.2% of dry weight). Contributing to the rigidity of wheat straw is lignin, a complex polymer present at 2.5–2.9% of dry weight. Additionally, wheat waste contains a small amount of protein (3–6.3% of dry weight) and inorganic minerals (ash) in quantities below 22.8% of dry weight. Furthermore, minor components like sugars, fats, and waxes may also be found. A comprehensive understanding of the chemical composition of wheat waste is vital for exploring its potential applications across various fields.

11.5.3 Mechanical Properties

Wheat waste refers to the residual materials generated during the processing of wheat, including straw, chaff, bran, and germ. The mechanical properties of wheat waste depend on various factors, such as the type of waste, its moisture content, and the direction and rate of the applied force. Some of the key mechanical properties of wheat waste include compressive strength, tensile strength, and flexural strength. The compressive strength of wheat waste is an important mechanical property that determines its ability to resist compressive loads. The compressive strength of wheat straw was found to range from 4.91 to 7.26 MPa depending on its moisture content and the direction of the applied force. The tensile strength of wheat waste determines its ability to resist pulling forces. Tensile strength, on the contrary, signifies the capacity of wheat waste to resist pulling forces. The tensile strength of wheat straw ranged from 21.2 to 31.2 MPa, depending on its moisture content and the direction of the applied force. Also, the flexural strength of wheat waste determines its ability to resist bending or flexure. Overall, the mechanical properties of wheat waste depend on various factors, such as the type of waste, its moisture content, and the direction and rate of the applied force. These properties can be useful in determining the potential applications of wheat waste in various industries, such as the development of biocomposites and construction materials.

11.5.4 Morphological Properties

Wheat waste refers to the by-products generated during the processing of wheat, such as straw, bran, and husk. These agricultural residues have garnered increasing attention due to their potential as a valuable resource for various applications. Their fibrous nature and complex composition characterize the morphology properties of wheat waste. Wheat straw, the most abundant form of wheat waste, possesses a hierarchical structure comprising cellulose, hemicellulose, and lignin. This fibrous morphology grants it exceptional mechanical properties, making it well-suited for various applications, such as reinforcement in composite materials or as a filler in biodegradable polymers (Pereira et al., 2017). Additionally, the bran and husk fractions of wheat waste display a more granular morphology characterized by an abundance of starch, proteins, fibers, and minerals (Al-Foudari et al., 2022). These properties render them valuable for diverse purposes, including animal feed, biofuel production, and the development of functional food ingredients. Furthermore, the morphology properties of wheat waste can be further improved through various processing techniques. For instance, physical or chemical treatments can modify the structure of wheat straw, leading to an increased surface area and enhanced reactivity. Such modifications enable the utilization of wheat waste in applications like water purification, the adsorption of heavy metals, and the removal of organic pollutants from wastewater (Xu et al., 2020). The versatility and potential of wheat waste underscore its significance in contributing to sustainable and innovative solutions across multiple industries.

Overall, wheat waste exhibits diverse morphology properties, including fibrous structures in wheat straw and granular compositions in bran and husk fractions. These properties make wheat waste a versatile resource with potential applications in a wide range of industries, including materials science, agriculture, and environmental engineering.

11.5.5 Thermal Properties and Calorific Value

Wheat waste is a by-product of wheat processing, including wheat bran, wheat straw, and wheat germ. These materials have been investigated for their thermal properties, which can be useful in designing thermal processes for their utilization. One study examined the thermal properties of wheat straw, including its thermal conductivity, specific heat, and thermal diffusivity. The results showed that the thermal conductivity of wheat straw increased with increasing temperature, while its specific heat decreased with increasing temperature. The thermal diffusivity of wheat straw was found to be relatively low compared to other agricultural wastes (Saad & Bellel, 2022).

Another study investigated the thermal properties of wheat bran, including its thermal conductivity, specific heat, and thermal diffusivity. The results showed that the thermal conductivity of wheat bran increased with increasing temperature while its specific heat and thermal diffusivity remained relatively constant. The study also found that wheat bran had a higher thermal conductivity than wheat straw (Gaydabrus et al., 2018). In addition, a study examined the thermal properties of wheat germ, including its thermal conductivity, specific heat, and thermal diffusivity. The results showed that the thermal conductivity of wheat germ increased with increasing temperature while its specific heat and thermal diffusivity remained relatively constant. The study also found that wheat germ had a lower thermal conductivity than wheat straw and wheat bran (Meriles et al., 2022). The calorific value of wheat waste may exhibit variability based on the particular type of waste and the measurement method employed. However, a research study conducted by Spirchez et al. (2019) revealed that wheat straw, one of the common forms of wheat waste, has a calorific value of 17.69 MJ/kg.

11.6 WHEAT WASTE MANAGEMENT

11.6.1 Compost

Composting wheat waste can be an effective way to reduce the environmental impact of this waste and produce a valuable soil amendment. Wheat waste, such as straw and chaff, can be composted with other organic materials to create a nutrient-rich soil amendment that can be used to improve soil fertility and structure. Composting is the process of decomposing organic matter by employing microorganisms like bacteria and fungi to create a long-lasting substance similar to humus.

Wheat straw is often composted with other organic materials like yard trash, food scraps, and manure for a proper carbon-to-nitrogen ratio. Composting requires a variety of microbes, all of which require oxygen, so the materials must be moistened and turned on a frequent basis (Xu et al., 2023).

Composting wheat waste provides several environmental and agricultural benefits. Wheat waste, including straw, chaff, and other plant wastes, can be composted successfully, providing a sustainable disposal solution while also generating important organic matter that can be restored to the land. One of the primary benefits of composting wheat waste is that it reduces GHG emissions. Wheat waste decomposes in landfills, releasing methane, a powerful GHG. Composting, on the contrary, promotes the breakdown of organic matter in an aerobic (oxygen-rich) environment, reducing methane emissions and helping to mitigate climate change (Liu et al., 2022).

Compost derived from wheat waste is rich in organic nutrients, such as nitrogen, phosphorus, and potassium. Applying compost to agricultural fields can recycle these nutrients and replenish soil fertility. This reduces the dependence on synthetic fertilizers, which can have negative environmental impacts, such as water pollution through runoff and GHG emissions during production. Compost enhances soil structure, improves water-holding capacity, and promotes beneficial microbial activity, leading to healthier and more productive soils (Harrison, 2008). Furthermore, compost application helps to control soil erosion. Wheat straw, a common by-product of wheat production, can be left on the soil surface as a protective mulch or incorporated into the topsoil during composting. This mulching effect helps reduce water runoff and soil erosion, preserving the integrity of the land and preventing sedimentation in water bodies (Croke, 2004).

11.6.2 Pulp and Paper

Paper and pulp wheat trash is the residue produced during the manufacturing of pulp and paper goods utilizing wheat as a raw material. Wheat straw, stalk, and other wheat wastes are frequently used in the pulp and paper industry to make fiber-based goods (Sain & Panthapulakkal, 2006). The non-fibrous components of the wheat plant, such as the straw or stalks, are separated and processed separately in the process of generating pulp and paper from wheat. In the context of pulp and paper manufacture, these non-fibrous leftovers are sometimes referred to as wheat waste. They are made up of lignocellulosic components leftover from the extraction of wheat fibers for papermaking.

Pulp and paper wheat waste can be recycled and used in various ways. One common application is as a source of raw material for the production of recycled paper products. The waste wheat fibers can be processed and combined with other recycled fibers to create new paper products, reducing the demand for virgin fiber resources and conserving natural resources (Kong et al., 2016). Pulp and paper wheat waste offers several benefits, both from an environmental and economic perspective. Utilizing wheat residues in the pulp and paper industry can transform these waste materials into valuable resources, contributing to sustainability and resource efficiency. One of the primary benefits of pulp and paper wheat waste is reducing the demand for virgin fiber resources. By incorporating wheat waste fibers into producing paper products, the industry reduces the need to harvest and process new trees. This helps conserve natural forests, preserve biodiversity, and reduce the environmental impact associated with deforestation and the conversion of forested lands.

Recycling pulp and paper wheat waste also contributes to trash reduction and the promotion of a circular economy. Waste materials are collected and reintroduced into the manufacturing process rather than being dumped or burnt. This reduces the amount of garbage delivered to landfills, saves landfill space, and reduces GHG emissions caused by waste breakdown. Additionally, using pulp and paper wheat waste as a biomass fuel source has both environmental and economic benefits (Jia et al., 2018). Waste materials can be used as a renewable and sustainable source of energy to power pulp and paper mill operations. By replacing fossil fuels with wheat waste biomass, the industry reduces its carbon footprint and dependency on non-renewable energy sources, contributing to GHG emission reductions and mitigating climate change (Wang et al., 2021).

From an economic perspective, pulp and paper wheat waste can provide cost savings for the industry. Recycling waste fibers reduces the need to acquire new raw materials, which can be expensive and resource-intensive. The industry can potentially lower production costs and improve overall profitability by utilizing wheat waste. Moreover, the utilization of pulp and paper wheat waste supports the development of a circular and sustainable bioeconomy. By maximizing the value of waste materials and integrating them back into the production cycle, the industry moves toward a more efficient and resource-conscious approach (Patwa et al., 2021).

However, there are constraints on using wheat waste in the papermaking process. Because wheat waste fibers are shorter than wood fibers, paper made from them may be less sturdy and unsuitable for some uses. Production costs may also rise due to the fact that using wheat waste may necessitate extra processing procedures like purifying the fibers and removing contaminants. While there are advantages to recycling wheat chaff for use in the paper industry, such advantages should be weighed against the potential negative environmental and economic effects (Cabalova et al., 2011).

11.6.3 Biocomposite Materials

Biocomposite materials derived from wheat waste can be a sustainable replacement for petroleum-based synthetic composites. Composite materials with enhanced mechanical qualities can be made by combining a natural fiber like wheat straw with a polymer matrix like a biodegradable plastic. Due to its high cellulose content and ready availability as an agricultural waste product, wheat straw can be employed as a reinforcing ingredient in biocomposites (Sánchez-Safont et al., 2018). To make a composite material that is strong, lightweight, and biodegradable, wheat straw fibers are often mixed with a biodegradable polymer matrix like polylactic acid or polyhydroxyalkanoates (Muthuraj et al., 2019).

Biocomposites made from wheat waste offer several benefits, combining the advantages of utilizing agricultural residues and the sustainability of composite materials. These innovative materials have gained attention in various industries due to their unique properties and environmental advantages. One of the key benefits of biocomposites made from wheat waste is their contribution to waste reduction and the circular economy. Wheat straw and other agricultural residues that would otherwise be discarded or burned can be repurposed as raw materials for biocomposite production. By utilizing these waste materials, biocomposites help divert agricultural residues from landfills and reduce environmental pollution, promoting a more sustainable and efficient use of resources (Karić et al., 2022).

Biocomposites made from wheat waste also offer significant environmental benefits compared to traditional petroleum-based materials. Wheat waste is a renewable resource, and using it as a composite material reduces the reliance on fossil fuels. This substitution decreases GHG emissions, mitigating climate change and contributing to a lower carbon footprint. Additionally, biocomposites are biodegradable or compostable, reducing their environmental impact at the end of their life cycle. Furthermore, biocomposites derived from wheat waste have desirable mechanical properties. The inherent strength and stiffness of wheat fibers contribute to the overall performance of the composite material. These biocomposites can exhibit excellent tensile strength, flexural strength, and impact resistance, making them suitable for a wide range of applications (Andrew & Dhakal, 2022).

Challenges, such as processing complexities and restrictions in their mechanical qualities, are linked with the usage of biocomposites generated from wheat waste, which is still relatively new. Biocomposite materials derived from wheat chaff waste have yet to reach their full potential in terms of performance and characteristics. However, this is expected to change as more research is conducted in the field (Mujtaba et al., 2023).

11.6.4 Bio-Coal

Bio-coal, sometimes called torrefied wheat straw or agrocoal, is a solid biofuel that can be used as a coal replacement in thermal power plants and other industrial settings. Bio-coal is made by heating wheat straw in the absence of oxygen at very high temperatures; this transforms the plant material into a carbon-rich product with qualities comparable to coal (Barskov et al., 2019). The resulting bio-coal has a higher energy density and improved combustion characteristics compared to the original wheat waste. Bio-coal derived from wheat waste offers several significant benefits as a renewable energy source and sustainable alternative to conventional coal (Akbari et al., 2021). This innovative form of solid fuel brings numerous advantages to the table, ranging from environmental sustainability to improved combustion properties.

One of the key advantages of bio-coal derived from wheat waste is its contribution to climate change mitigation. Bio-coal delivers a carbon-neutral or even carbon-negative energy solution using wheat straw and other agricultural leftovers as a feedstock. The carbon dioxide emitted during bio-coal combustion is compensated by the carbon absorbed by growing wheat crops, resulting in a closed carbon cycle. Compared to typical coal, this greatly reduces net GHG emissions, aiding the fight against global warming and minimizing climate impact. Another advantage of bio-coal from wheat waste is its better combustion characteristics. Wheat waste is converted into a more homogenous and energy-dense fuel by the torrefaction process. Because bio-coal has a larger energy content per unit weight, it is a more efficient fuel source. Its homogeneity and improved grindability make it easier to handle, transport, and co-fire with coal in existing power plants (Panahi et al., 2018). This adaptability enables a smooth transition to bio-coal without requiring major changes to the current infrastructure.

However, there are also significant difficulties in making and using bio-coal derived from wheat waste. Energy inputs for the torrefaction process can be high, and feedstock availability can vary widely. It is also possible that bio-coal is not cost-competitive with coal in some locations and would necessitate expensive upgrades to existing infrastructure before it could be used. However, further study and development are needed to perfect the manufacture and usage of this promising biofuel. Bio-coal generated from wheat waste can provide a sustainable alternative to coal in some applications (Guajardo et al., 2021).

11.6.5 Biochar

Biochar derived from wheat waste is a carbon-rich, porous, and stable form of charcoal that is produced through a process called pyrolysis. Wheat straw, husk, or other agricultural residues are subjected to high temperatures in an oxygen-limited environment, resulting in biomass conversion into biochar. The pyrolysis process involves heating the wheat waste to temperatures between 400°C and 700°C which triggers the decomposition of organic compounds. This leads to the release of volatile gases and the formation of biochar as a solid residue. The resulting biochar is a highly porous material with a large surface area, providing an ideal habitat for beneficial microorganisms and soil organisms (Tan et al., 2023). Biochar derived from wheat waste offers several significant benefits as a sustainable soil amendment and carbon sequestration tool. This carbon-rich material, produced through the pyrolysis of wheat straw or other agricultural residues, has transformative effects on soil health, agricultural productivity, and climate change mitigation (Oni et al., 2019).

One of the most important benefits of biochar made from wheat waste is its capacity to improve soil fertility and nutrient retention. The porous nature of biochar provides a home for beneficial bacteria and boosts soil water-holding capacity, allowing plants to absorb more nutrients. Biochar's high cation exchange capacity aids in retaining important nutrients like nitrogen, phosphorus, and potassium, minimizing nutrient leaching and boosting plant availability. This results in enhanced crop growth, higher yields, and more efficient fertilizer use, contributing to more sustainable agricultural practices. Biochar formed from wheat waste is important for carbon sequestration in addition to its impact on soil fertility. Biochar is a stable form of carbon that remains in the soil for an extended period, effectively storing carbon and reducing atmospheric GHG levels. By sequestering carbon from the atmosphere and incorporating it into the soil, biochar helps mitigate climate change and reduce the carbon footprint associated with agricultural activities (Lal, 2004).

However, there are also significant difficulties in making and using biochar from wheat waste. The pyrolysis process can be very expensive to run, both in terms of energy costs and the cost of the necessary machinery and facilities. Factors like soil type and climate can affect how biochar influences soil health and fertility. Biochar produced from wheat waste has potential as a sustainable agricultural practice with multiple positive outcomes for both the environment and crop production (Semida et al., 2019). The manufacture and application of biochar derived from wheat waste are likely to undergo further refinement as a result of ongoing research and development in this area.

11.6.6 Bio-Bricks

Wheat waste bio-bricks are a form of compressed biomass fuel that can be used as an alternative to firewood in home and business heating systems. Bio-bricks are consistent, solid bricks created from agricultural waste like wheat straw that can be burned in an indoor heating appliance (Biron, 2017). Wheat straw, husk, and other agricultural residues are transformed into compacted fuel blocks through a process that harnesses the potential of this abundant and often underutilized biomass. Bio-bricks from wheat waste offer several significant benefits as a renewable and sustainable fuel source. These compacted fuel blocks, derived from agricultural residues such as wheat straw or husks, have gained recognition for their positive impact on the environment, energy efficiency, and local economies.

One of the primary benefits of bio-bricks made from wheat waste is that they contribute to waste reduction and resource efficiency. Bio-bricks help to reduce agricultural waste and promote the circular economy by utilizing agricultural wastes that would otherwise be thrown or left unused. This decreases the environmental impact of waste disposal and maximizes resource utilization, making bio-bricks a sustainable energy choice. Furthermore, bio-bricks from wheat waste are considered renewable energy sources. Wheat straw and other agricultural residues are readily available on an annual basis, providing a continuous supply of feedstock for bio-brick production. This renewable aspect ensures a long-term and sustainable fuel solution, reducing reliance on non-renewable fossil fuels and contributing to a more sustainable energy system (Rahman & Wahid, 2021).

Bio-bricks also offer advantages in terms of energy efficiency and combustion performance. Their compact and uniform shape allows for a more controlled and efficient burn, resulting in higher heat output and longer burn times than traditional firewood. This increased energy efficiency translates into reduced fuel consumption and cost savings for consumers. Producing and utilizing bio-bricks derived from wheat waste has certain difficulties (Koul et al., 2022). Significant amounts of energy are needed in the production process, and in some locations, feedstock is scarce. It is also possible that a current heating setup will need to be adjusted, or a new cooker will need to use bio-bricks instead. Bio-bricks manufactured from wheat waste have significant potential as a renewable and eco-friendly heating option. The manufacturing and application of bio-bricks derived from agricultural waste are likely to benefit from ongoing research and development in this field (Maraveas, 2020).

11.6.7 Bio-Hydrogen

For more than a century, fossil fuels have been used. They have significantly harmed the environment and depleted scarce fuel supplies, which has prompted the hunt for eco-friendly alternative fuels. The development of clean and renewable energy supplies was taken into consideration globally as the demand for fossil fuels rises and environmental contamination worsens (Umar et al., 2021). Energy development is more promising thanks to hydrogen because it has several specialized, financial, and environmental benefits. Of all known fuels, it has the highest heating value per unit mass (142 kJ/g) (Soares et al., 2020). Handling hydrogen gas is safer than handling other known natural gases. The most important feature is that burning hydrogen produces just water as a reaction product, making it an environmentally friendly fuel. Because of these qualities, hydrogen is a strong candidate to replace fossil fuels (Sinha & Pandey, 2011). Due to the limited supply of fossil fuels and their detrimental environmental effects, developing carbon-free energy is necessary to meet current energy demands. Bio-hydrogen is a carbon-neutral energy source that is gaining popularity for its potential as a replacement as well as its dependability and endless supply in the expanding hydrogen economy. Thus, bio-hydrogen production through dark fermentation of organic waste, biomass, and wastewater is becoming increasingly popular these days.

Handling hydrogen gas is safer than handling other known natural gases. The most important feature is that burning hydrogen produces just water as a reaction product, making it an environmentally friendly fuel. Because of these qualities, hydrogen is a strong candidate to replace fossil fuels (Sinha & Pandey, 2011). Due to the limited supply of fossil fuels and their detrimental effects on the environment, developing carbon-free energy is necessary to meet current energy demands. Bio-hydrogen is a carbon-neutral energy source that is gaining popularity for its potential as a replacement, dependability, and endless supply in the expanding hydrogen economy. Thus, bio-hydrogen production through dark fermentation of organic waste, biomass, and wastewater is becoming increasingly popular these days. Handling hydrogen gas is safer than handling other known natural gases. The most important feature is that burning hydrogen produces just water as a reaction product, making it an environmentally friendly fuel. Because of these qualities, hydrogen is a strong candidate to replace fossil fuels (Sinha & Pandey, 2011). Due to the limited supply of fossil fuels and their detrimental effects on the environment, developing carbon-free energy is necessary to meet current energy demands.

Zhu et al. (2023) outlined the procedures for managing wheat trash to make bio-hydrogen. The collection and preparation of the wheat waste is the first phase. Wheat straw was gathered from past manufacturing locations. To remove extractives, the raw straw was processed through a superfine pulverizer to produce a 40–60 mesh powder, which was then filtered. The powder was preserved in an oven before being used as a feedstock to create bio-hydrogen. The wheat straw included 92.40% volatile solids (VS), 95.39% total solids, and 6.70% ash, respectively. The China General Microbiological Culture Collection Center provided the *Enterobacter aerogenes* ATCC13048, which was cultivated in an aerobic incubator shaker at 150 rpm and 37°C. The culture media (per liter) contains Luria–Bertani broth, NaCl, peptone, and yeast extract. The inoculum for the subsequent hydrogen fermentation was chosen to be *E. aerogenes* ATCC13048 in the logarithmic phase, and the inoculation process was carried out under sterile conditions.

Figure 11.3 depicts the process flow diagram for the integrated treatment of wheat straw, which underwent lyophilization, HTP, ultrasound-assisted alkali extraction, and dilute alkali cooking as pretreatments. The following is a description of the specific pretreatment practices and associated treatment circumstances: (1) A 2% (w/w) aqueous NaOH solution was poured over 4 g of straw. The mixture was initially seethed for 3 h at 100°C while being subjected to a 100 W ultrasonic cell disruptor for 1 h. (2) A 4 g sample of straw was lyophilized with distilled water and heated to 100°C for 3 h under the same concentration of aqueous NaOH solution. (3) A hydrothermal reaction involving 4 g of straw sample and 80 mL of distilled water was conducted for 1 h at 170°C. The insoluble leftovers were then carefully washed, filtered through a Buchner funnel, and freeze-dried. The resulting residues were further processed using the same alkali cooking setting as before—the previously stated integrated treatment procedures produced residues designated as RU, RL, and RHTP, respectively. As a control, the raw straw sample extracted directly under the same alkali cooking conditions was used; it was designated as RC. The leftovers from the pretreatment of wheat straw were then used for enzymatic saccharification. From Sigma-Aldrich (Shanghai, China), a cellulase (C2730) with an activity of 700 FPU/g was purchased. In more detail, 100 mL of sodium acetate buffer (pH 4.8) was added along with 3 g of pretreatment residues (RC, RU, RL, and RHTP), and cellulase with an activity of 20 FPU/g residues was then added. In a shaking incubator, the hydrolysis reaction temperature was maintained at 50°C for 72 h and shaken at 150 rpm. Also used as a control was raw wheat straw (RM). The enzymatic-saccharification mixes that were produced were employed to produce fermentative hydrogen.

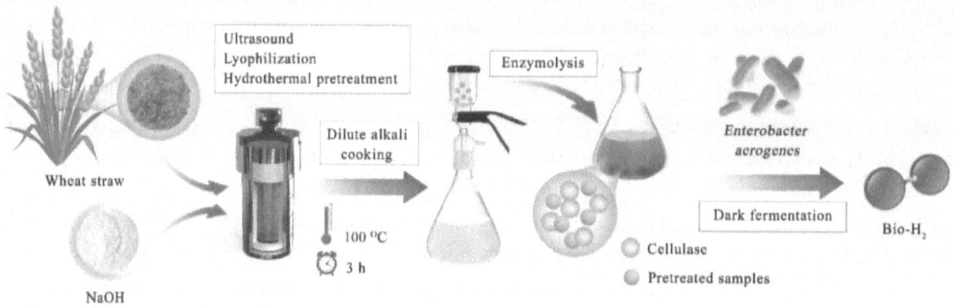

FIGURE 11.3 Process flow diagram for integrated treatment of wheat waste.

11.6.8 Biomethane

Due to its significant environmental, sociological, and economic advantages, methane production through anaerobic digestion (AD) of biomass has been viewed as an appealing and potential way to acquire renewable biofuel (Zhang et al., 2014). Organic waste is broken down by a variety of bacteria naturally during AD, which results in the production of biogas, which is mostly made up of methane (N_2) and carbon dioxide (CO_2) (Frigon & Guiot, 2010). China is a significant agricultural nation with abundant biomass resources. As an agricultural waste, wheat straw is seen as a plentiful renewable resource. According to Yang et al. (2010), China produced 109 million metric tons of wheat straw in 2007. An appealing technique for handling wheat straw is AD, which can reduce pollutants while also recovering energy (Solé-Bundó et al., 2017). Although the release of GHGs, such as CO_2 from fossil fuels, has become a global concern, over 88% of the world's energy still comes from these sources. The EU member states have decided to boost the share of renewable energy to 20% of total consumption by 2020 to address environmental concerns (EREC, 2008).

11.6.8.1 Biomethane Production through AD

The natural breakdown of organic compounds by bacteria is known as AD. According to Li et al. (2014), AD is a delicate multistage process that includes hydrolysis, acidogenesis, acetogenesis, and methanogenesis and relies on susceptible microorganisms to complete the digestion task. Organic in this context refers to being made of or derived from plants or animals. AD takes place in enclosed areas without air or oxygen. The letters AD can stand for either AD or the constructed system in which it occurs, also known as a digester. The technique employs wheat waste as a digester.

Four steps are involved in the microbial breakdown of organic material to produce biomethane: hydrolysis of complex organic polymers to soluble compounds; fermentation of the hydrolysis products into intermediate compounds; anaerobic oxidation of these intermediate products to produce acetate, H_2, and CO_2; and methane production by methanogenic Archaea. According to some theories, in the case of straw, the crystalline structure of the lignocelluloses prevents initial decomposition, making the hydrolysis of these insoluble chemicals the rate-limiting step (Lynd et al., 2002). To understand the mechanics of cellulose degradation and find strategies to speed up the process, many studies have looked at anaerobic cellulose-degrading bacteria and their enzymatic capacities. However, only a small number of these studies have looked at cellulose-degrading bacteria in biogas digesters; the majority of these investigations have been conducted on samples from soil and gut habitats (Ransom-Jones et al., 2012).

In general, anaerobic degradation of cellulose is performed by physiologically distinct taxa of microorganisms. Degradation occurs through the action of various cellulases, including endoglucanases, exoglucanases, β-glucosidases, and various enzymes that attack hemicellulose (Schwarz, 2001). Enzymes can be soluble or bound to cell-associated cellulosomes, such as in anaerobic bacteria. Glycoside hydrolases are classified into 120 families based on amino acid sequence according to the International Classification of Carbohydrate-Activating Enzymes (CAZy database). In anaerobic bacteria, cellulolytic glycoside hydrolases are mainly assigned to glycoside hydrolase families 5, 9, and 48 (Pereyra et al., 2010). Pereira et al. (2010) designed and validated degenerate consensus primers for amplifying the cel5 and cel48 families of glycoside hydrolase genes with the aim of enabling the characterization and quantification of cellulolytic bacteria in biogas digesters.

Here are the steps to produce biomethane from wheat: Harvest the wheat and separate the grain from the straw. Straw is the part of the plant that produces biomethane. The straw is then crushed into small pieces to increase surface area and placed in an anaerobic digester. Starting materials first undergo a separation and pretreatment process. After removing impurities, the material is shredded before being pumped into the buffer tank, allowing continuous feeding to the digester. Sanitation is the final step before digestion, which is performed at 70°C for 1 h to kill pathogenic microorganisms. In the fermenter, liquid manure is mixed with active microorganisms. AD of organic matter mainly consists of four processes: hydrolysis, acidogenesis, acetogenesis, and methanogenesis. An anaerobic microbial process produces biogas at a temperature of 37°C (mesophilic process). The gas produced is conditioned to remove H_2S and other unwanted contaminants before being processed into automotive fuel (>95 mol% CH_4). Water scrubbing technology, which uses water as a solvent, is used to treat biogas. According to this theory, the solubility of CH_4 in water is much lower than that of CO_2. The energy consumption of water scrubbing is about 0.45–0.90 kWh/Nm^3 of scrubbed gas and is mainly used for gas compression and water pumping (Li et al., 2017).

Producing biomethane from wheat straw offers a sustainable solution to both energy and waste management. Using organic materials to produce renewable natural gas can reduce our reliance on fossil fuels and decrease GHG emissions. With the right feedstock, monitoring, and optimization of the biomethane production process, we can produce a valuable resource that can be used as fuel, fertilizer, and more.

11.6.9 Bioethanol

Waste from the processing of wheat can be used to create bioethanol, a renewable fuel. Utilizing bioethanol can lessen our reliance on fossil fuels while lowering our net carbon dioxide emissions, the main GHG (Wyman, 1996). Bioethanol can be produced using wheat waste, including straw, chaff, and other agricultural wastes. However, because it uses food resources as raw materials, large-scale bioethanol production has come in for growing amounts of criticism. Brazil uses a lot of sugarcane in its bioethanol production, while the United States uses corn (Wheals et al., 1999). In Europe, other starchy grains like wheat and barley are primarily used (Galbe & Zacchi, 2002). Due to market rivalry, the usage of such sugar-rich raw materials raises food prices (Catic & Rujnic-Sokele, 2008). Therefore, lignocellulosic resources such as agricultural by-products, forest leftovers, industrial waste streams, or energy crops must serve as the foundation for future growth in the production of biofuel (Claassen et al., 1999). These feedstocks are more readily available and less expensive than those from food crops for the synthesis of second-generation (2G) bioethanol (Tan et al., 2008). Due to its widespread availability and low cost, wheat straw has the highest potential of all agricultural leftovers in Europe (Kim & Dale, 2004). Hemicellulose hydrolysis must be pretreated to increase the cellulose's accessibility to enzymes for effective utilization of lignocellulosic products. Steam explosion is one of the best pretreatments for lignocellulosic materials (McMillan, 1994). One of the most promising methods for commercializing this process is the combination of acid catalysis with steam explosion (Hu et al., 2008). According to certain studies, wheat straw impregnated with a modest amount of H_2SO_4 before being steam pretreated increases sugar output (Ballesteros et al., 2006).

Wheat (*Triticum aestivum* L.) is the most commonly grown crop in the world, grown in more than 115 countries under a wide variety of environmental conditions. Wheat yields have increased over the past 100 years, and in 2008, the world dry wheat annual production was estimated at over 650 Tg (Atwell, 2001). About 850 Tg of wheat residues are generated per year, assuming a residue/crop ratio of 1.3. Depending on the landowner's preference, the created straw may also be removed from the land, plowed back into the ground, or left in the field. Wheat straw incineration has been used for a very long time. The health impacts of smoke from burned fields have become a major worry in recent years, calling into question this practice (Kerstetter and Lyons, 2001). Large volumes of air pollutants like PM (PM10), CO, and NO_2 are produced while burning wheat straw. Therefore, finding different methods to get rid of extra wheat straw is very important and necessary. Complete removal of wheat residues can reduce soil organic matter and cause soil erosion.

The percentage of wheat straw that must be left in the field depends on the weather, crop rotation, existing soil fertility, slope, and cultivation method (Kerstetter and Lyons, 2001). According to Kim and Dale (2004), a soil coverage of 60% should be considered to prevent soil erosion, requiring approximately 1.7 Mg of wheat residue per hectare. The average global wheat yield is about 2.4 Mg, so about 430 Tg of wheat straw is available for the production of about 120 GL of bioethanol per year. According to Otero et al. (2007), this volume of ethanol may replace approximately 93(3) GL of gasoline after accounting for changes in volumetric energy content and octane number. According to several investigations, when engine parameters are correctly fine-tuned, ethanol-gasoline mixes result in better engine performance and decreased CO emissions (Bayraktar, 2005; Wu et al., 2004). Therefore, switching from burning wheat straw to producing ethanol from excess wheat straw looks to be a promising strategy to recover energy from waste biomass and partially reduce reliance on fossil fuels while also promoting GHG impacts and enhancing urban air quality.

Like all lignocellulosic biomass, wheat straw is made up mostly of cellulose, hemicellulose, and lignin, with minor amounts of soluble substrates (sometimes referred to as extracts) and ash (Norfarhana et al., 2022, 2024a,b; Nurazzi et al., 2021; Syafri et al., 2022). Depending on the variety of wheat, the soil, and the climate, there may be a little variation in the chemical makeup of wheat straws. According to Prasad et al. (2007), the percentages of cellulose, hemicellulose, and lignin in wheat straw are 33–40, 20–25, and 15–20 (wt%), respectively, so that neither water nor enzymes may flow through the structure and cellulose stalks are bunched and compactly packed (Laureano-Perez et al., 2005; Sun et al., 2004). Hemicellulose bridges lignin and cellulose fibers and is easily broken down by hemicellulase enzymes and diluted acids or bases.

Lignin–cellulose–xylan interactions significantly affect the digestibility of lignocellulosic materials because they are covalently bonded to cellulose and xylan, which is the main hemicellulose carbohydrate polymer in wheat straw (Laureano-Perez et al., 2005). Because of the intricate nature of the lignocellulosic matrix, at least four primary processes, including pretreatment, hydrolysis, fermentation, and distillation, are needed to produce ethanol from wheat straw. Unlike sucrose and starch, lignocellulosic biomass like wheat straw must be processed to make the cellulose accessible for effective enzymatic depolymerization (Figure 11.4).

11.6.10 Biobutanol

Wheat waste is one type of organic material that can be used to make biobutanol, a renewable fuel. Straw, chaff, and other agricultural waste from the production of wheat can be utilized as feedstock to make biobutanol. Biobutanol has a number of advantages compared to other biofuels like ethanol. It is less corrosive to engine and fuel system components, has a high energy content, and may be used in gasoline engines that are already in use without modification. Biobutanol is a versatile chemical that may also be employed as a solvent in numerous industrial processes.

According to Morone and Pandey (2014), renewable lignocellulosic biomass-based biofuels are anticipated to meet rising energy demands without causing GHG emissions to rise as they do with fossil fuels. Due to its high energy density (butanol has a liter energy density of 29.2 MJ, ethanol has a liter energy density of 19.6 MJ, and gasoline has a liter energy density of 32 MJ), butanol is one

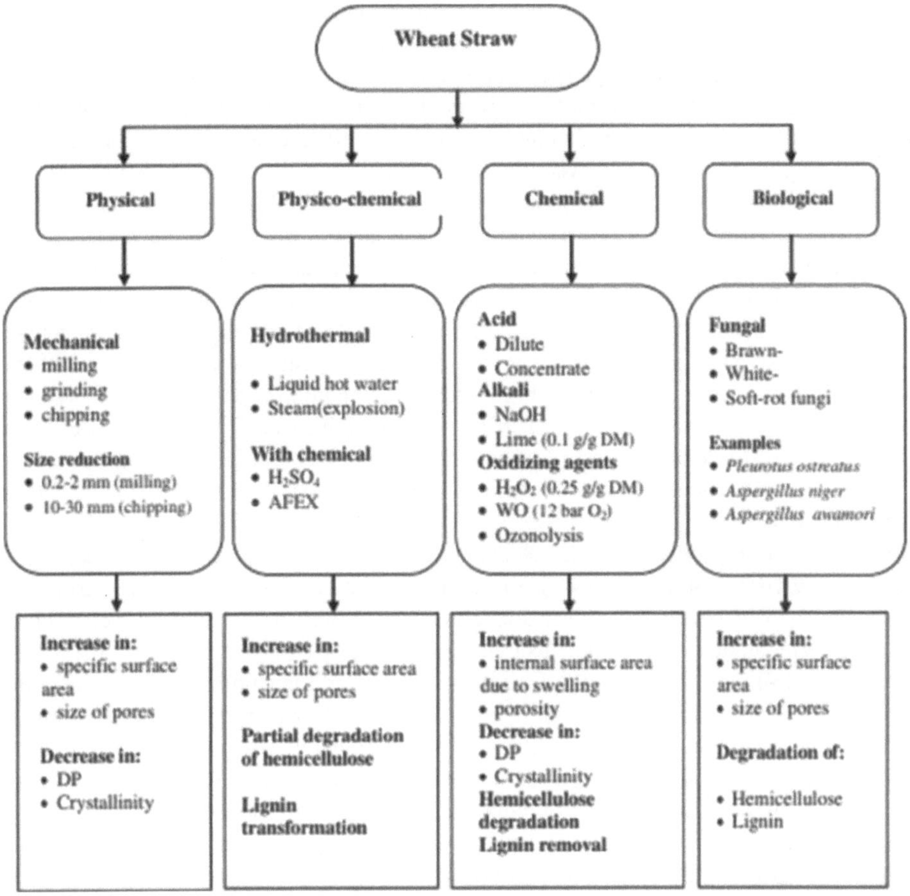

FIGURE 11.4 The most common pretreatment methods used on wheat straw and their possible effects (DP, degree of polymerization; WO, wet oxidation).

of the most promising biofuels (Cascone, 2008). The most common biomass on earth and a natural carbon source, cellulose has grown to be a crucial feedstock for manufacturing butanol in biorefinery processes (Leschine, 1995). The industry requires numerous steps to convert cellulose to biofuels, including pretreatment, enzymatic saccharification, detoxification, and fermentation. Therefore, it would be ideal to create a bioconversion system that does not require the aforementioned pretreatment stages and instead turns cellulosic biomass directly into biofuels (Olson et al., 2012). However, the resistance of cellulose and the absence of powerful microorganisms have hindered this endeavor. Finding novel microbes with distinctive genetic cassettes that can transform cellulosic resources into biofuels is a crucial first step in solving the problem (Demain et al., 2009). As previously mentioned, a number of solventogenic strains of the genus Clostridium have been used to produce butanol from starches like corn and cassava, as well as monosaccharides like glucose and xylose.

Butanol cannot be produced by these strains using polysaccharides like cellulose; only ethanol or hydrogen can be produced this way (Weber et al., 2010). Research has recently focused on developing microorganisms that can directly transform cellulose into biofuels in condensed bioprocesses (Lynd et al., 2016). For instance, natural cellulolytic microorganisms like *Caldicellulosiruptor bescii*, *Clostridium cellulolyticum*, and *Clostridium thermocellum* have had genes encoding for biosynthetic pathways of biofuels inserted into them. These organisms are capable of converting cellulose into ethanol and isobutanol at rates of 0.66 and 0.64 g/L, respectively (Higashide et al., 2011). By inserting a cellulosomal gene, *Clostridium acetobutylicum* underwent a similar endeavor.

However, the engineered strains were unable to utilize cellulose due to the complex structure and expression stability of functional minicellulosomes (Kovács et al., 2013).

Only a small number of wild-type species are now able to directly convert cellulosic biomass into large concentrations and yields of biofuels, particularly butanol. Therefore, the development of an integrated bioprocess technology for butanol production is still necessary. Remember that (1) product inhibition and carbon catabolite repression (Yu et al., 2015), as well as (2) the difficulty of downstream purification of butanol from other by-products like acetone and ethanol are obstacles to sustainable butanol synthesis. Therefore, it is desirable to find bacterial strains that can directly ferment cellulose and hemicellulose as major products of butanol to achieve and facilitate integrated bioprocessing.

Several crucial procedures are involved in producing biobutanol from discarded wheat. To prevent spoiling, wheat waste must first be gathered and stored in a dry atmosphere. Trash needs to be cleaned and treated to get rid of any potential toxins. Waste wheat is processed into small fragments using techniques like shredding and crushing. This fragmentation expands the waste's surface area, making extracting the sugars needed for further fermentation easier. The wheat waste is then given enzymes to start the saccharification process. These enzymes break down complex carbohydrates into simple sugars like glucose and fructose. The sugars are more readily available for fermentation after this enzymatic treatment.

The sugar solution produced by the saccharification process is coupled with bacteria, particularly *C. acetobutylicum*, in the following phase. Acetone–butanol–ethanol (ABE) fermentation is a process carried out by bacteria. Bacteria convert carbohydrates to ethanol, acetone, and biobutanol during ABE fermentation. Biobutanol, acetone, and ethanol are the end products of ABE fermentation. Distill the mixture to remove the biobutanol. Based on their boiling points, the components are separated using this technique. Distillation can be used to extract biobutanol from other by-products, allowing for purification. Biobutanol is further purified using purification processes such as fractional distillation, adsorption, and membrane separation. These techniques aid in the removal of contaminants, producing a product of biobutanol that has been refined.

Producing biobutanol from wheat waste offers a sustainable energy and waste management solution. By using agricultural residues to produce a versatile chemical that can be used as a fuel or solvent, we can reduce our reliance on fossil fuels and decrease GHG emissions. However, there are challenges to biobutanol production from wheat waste, such as feedstock availability, enzyme optimization, and bacterial contamination. With further research and development, biobutanol production from wheat waste can become a viable and sustainable source of renewable energy.

11.6.11 Natural Acids

The great renewable carbohydrates found in wheat straw include cellulose, hemicellulose, lignin, and others. Lactic acid is an example of an organic acid that is a vital chemical raw material that can be used to treat immunological and inflammatory conditions (Meesum et al., 2021). According to Basha et al. (2020), fumarate and silver together can speed up the healing of wounds. The manufacture of organic acids using wheat straw not only makes it possible to utilize biomass waste resources but also boosts the financial advantages of crops (Reza et al., 2015). Traditional techniques for creating organic acids include chemical synthesis, fermentation, and others (Eraky et al., 2021). These techniques have high conversion rates, but the processes are laborious, expensive, and time-consuming.

Gas and liquid substances generated in manufacturing are easily released into the environment and cause environmental pollution. Therefore, finding a low-emission, high-efficiency organic acid production method is the only way to achieve environmentally friendly organic acid production. Numerous studies have shown that choosing an appropriate pretreatment can disrupt lignin, hemicellulose, and microcrystalline cellulose crosslinks in straw, improve conversion efficiency, and achieve higher product profits (Valles et al., 2021). In addition to pretreatment, hydrothermal treatment is an environmentally friendly process. The heat treatment process is quick and easy compared to conventional fermentation techniques. The organic stuff inside the straw is encouraged to undergo pre-hydrolysis by hydrothermal treatment, which yields solids and liquids that are high in sugar.

Alkali treatment simultaneously breaks down cellulose hydrogen bonds; widens the gap between cellulose molecules; alters the composition, structure, and characteristics of fibers; increases moisture content; increases specific surface area and pore volume; and exposes free surface area (Karuna et al., 2022). These two factors make lignin and cellulose conversion easier. Comparing treated and untreated raw materials, alkaline hydrothermal treatment dramatically accelerates the complex saccharification rate (Romaní et al., 2016). However, nothing is known about how an alkaline hydrothermal environment can encourage crop straw to be converted to organic acids.

The alkaline hydrothermal treatment method was used in a study by Zhang et al. (2017) utilizing wheat straw as the raw material. The study looked into the best combinations of NaOH concentration, reaction temperature, and reaction time for turning straw biomass into organic acid. It lessens environmental pollution and enhances the ecological environment by converting low-value-added wheat straw into high-value-added organic acid. Additionally, it offers a fresh, environmentally friendly method for producing organic acids, achieves effective use of agricultural wastes like wheat straw, and uses waste straw biomass resources to significantly meet the rising material needs of people in society, resulting in positive economic, environmental, and social benefits.

Wheat waste, comprising straw, chaff, and other agricultural wastes, provides a different source of natural acids. Wheat waste is plentiful and easily accessible, making it a viable raw material for the synthesis of natural acids. The production of natural acids from wheat waste can also aid in waste reduction and the advancement of the circular economy. A microbial fermentation procedure is used to make natural acids from waste wheat by using microorganisms to break down waste carbohydrates. Fermentation methods must be properly tuned to provide the highest possible output of natural acids. Following fermentation, the natural acids are isolated and refined to produce a pure product with a wide range of applications.

There are a number of significant benefits to making natural acids from leftover wheat. First, wheat trash reduces our reliance on fossil fuels and aids in the fight against climate change by serving as a sustainable and renewable raw material supply. Utilizing wheat waste to create natural acids is in line with sustainable practices and encourages resource efficiency. Additionally, natural acids derived from wheat waste are extremely adaptable compounds that have a wide range of uses. They are used in a variety of industries, including pharmaceuticals, food preservation, and industrial operations. Natural acids are used in a variety of industries, which increases their significance and market demand (Rezaei et al., 2021). The creation of natural acids from wheat waste can also aid in waste reduction and the advancement of a circular economy. This procedure lessens waste formation and encourages the idea of material recycling and reuse by employing wheat waste as a raw material. This strategy boosts total resource efficiency and is consistent with sustainable practices. However, there are certain difficulties in producing natural acids from wheat waste. The improvement of the microbial fermentation process is a significant difficulty. The generation of natural acids must be maximized by careful fermentation process optimization. Given that various raw materials require various microbes, this can be a challenging undertaking that necessitates intensive research and development efforts.

Another challenge is the separation of natural acids from fermentation mixes. Because they are present in small quantities and have a tendency to mix with other chemicals, natural acids are difficult to separate and purify. Special techniques and procedures are required to successfully isolate and extract the necessary natural acids De Assis et al. (2015). Furthermore, producing natural acids from wheat waste would not be as economical as from more traditional sources like citrus fruits and sugar cane. Making use of wheat waste as a raw material could lead to pricey production. To keep costs competitive, a constant focus on technical advancement, process innovation, and economies of scale is necessary (Baska et al., 2021).

Producing natural acids from wheat waste offers a sustainable solution to both waste management and chemical production. By using agricultural residues to produce a versatile chemical that can be used in a variety of applications, we can reduce our reliance on fossil fuels and promote a circular economy. However, there are challenges to natural acid production from wheat waste, such as fermentation optimization, product separation, and cost-effectiveness. With further research and development, natural acid production from wheat waste can become a viable and sustainable source of natural acids.

11.6.12 Bioelectricity

Agricultural runoff is one of the biggest wastes since it contains a lot of carbon-containing substances that, if released untreated, might have an impact on incoming water (Yang et al., 2021). Because of the amount of energy and carbon sources they contain, agricultural residues are regarded as one of the most significant substrates. According to ElMekawy et al. (2015), their sugars are produced through weak acids or enzyme treatment. Wheat straw, which contains cellulose with an organic carbon concentration of around 34–40%, hemicellulose with an organic carbon content of roughly 21–26%, and lignin with an organic carbon content of roughly 11–23%, are examples of significant agricultural leftovers. All of these go through hydrolysis to create wheat straw hydrolysates, which in microbial fuel cells (MFCs) provide a significant quantity of electricity (Khan et al., 2012). Similar to this, when employing single-chamber MFCs, raw maize stalks are a substantial agricultural waste that also helps with power generation. Although efficient, this method had a substantially lower output yield than MFC, which used glucose as a substrate. Agro-industrial effluents and pre-harvest, harvest, and processing processes in agriculture produce agricultural wastes. Wastewater and agricultural/food waste are produced as a result of industrial and agricultural food processing activities. Agricultural wastes easily biodegrade because they are rich in organic matter conducive to microbial development and a wide variety of macro- and micronutrients. Large quantities of animal husbandry and agro-processing wastewater are among the many agricultural industrial wastewater with high concentrations of organic contaminants (Lin et al., 2009). However, these agricultural waste products and sewage can be seen as brand-new renewable energy sources that can be transformed into biochar-enabled products, such as fertilizers, MFCs, biofuels, biogas, and bioelectricity.

Farm wastes are first treated. Pretreatment procedures are typically thought to need the breakdown of agricultural wastes' intricate molecular structures into simpler monomers. As a result, it supports high output following conversion. The technological methods for digesting biomass can vary. These include physical techniques like crushing, thermal techniques like a steam explosion, biological techniques like enzymatic techniques, chemical techniques like utilizing acids, alkalis, etc., and combo therapies like thermochemical techniques. These techniques increase surface area while reducing operational expenses and facilitating easy access to enzymes for hydrolysis. For instance, by making the waste easily accessible to microbial populations and enzymes during hydrolysis, physical treatment of the waste increases the surface area. On the contrary, the thermochemical pretreatment method increases the heat and mass transfer rate. It promotes uniform temperature distribution within the agricultural waste particles, resulting in higher recovery efficiency of hydrolysis components during liquefaction. Similarly, thermochemical transformations involve two main methods.

11.7 POTENTIAL APPLICATIONS

Wheat waste holds significant potential for various applications beyond its use in bioelectricity production. One of these applications is animal feed, where wheat straw, chaff, and other agricultural residues can serve as a cost-effective alternative to traditional animal feed sources. Moreover, wheat waste can be utilized as a soil amendment, contributing to improved soil fertility and structure. With its high levels of organic matter and nutrients, wheat waste can enhance the nutrient content and water-holding capacity of the soil, promoting healthier plant growth (Adusumilli et al., 2014).

In addition, wheat straw and other agricultural residues can be employed as raw materials for the production of biodegradable products. These waste materials can be transformed into environmentally friendly alternatives such as packaging materials, disposable cutlery, and cups, reducing the reliance on non-biodegradable materials and supporting sustainable practices. Wheat waste also showcases potential in bioremediation applications. The microorganisms present in wheat waste have the capability to break down pollutants, making it valuable in cleaning up contaminated soil and water. This application aids in environmental restoration and reduces the impact of harmful substances (Zheng and Chi, 2012).

Composting represents another beneficial use of wheat waste. By composting wheat waste, a nutrient-rich soil amendment can be produced, enriching the soil with organic matter and vital nutrients. This contributes to improved soil health and plant growth while diverting waste from landfills. Furthermore, wheat waste can be utilized as a source of heat generation. Burning wheat waste can produce heat for industrial processes or be used to heat buildings. This offers a sustainable and renewable energy option while simultaneously reducing waste. Figure 11.5 presents a comprehensive schematic diagram illustrating the process of harnessing electricity from agricultural waste. In summary, wheat waste presents several potential applications. It can be used as animal feed, soil amendment, raw material for biodegradable products, bioremediation agent, compost for enriching soil, and a source of heat generation. Exploring these diverse applications contributes to waste reduction, resource efficiency, and sustainable practices across various industries.

11.8 CHALLENGES ASSOCIATED WITH PROCESSING AND COMMERCIALIZING

The processing and commercialization of wheat waste as a feedstock for various applications present several challenges that need to be addressed. One of the main challenges is the availability of wheat waste as a feedstock, which can be influenced by factors such as crop yields, harvesting practices, and transportation costs. The limited availability of wheat waste can lead to supply

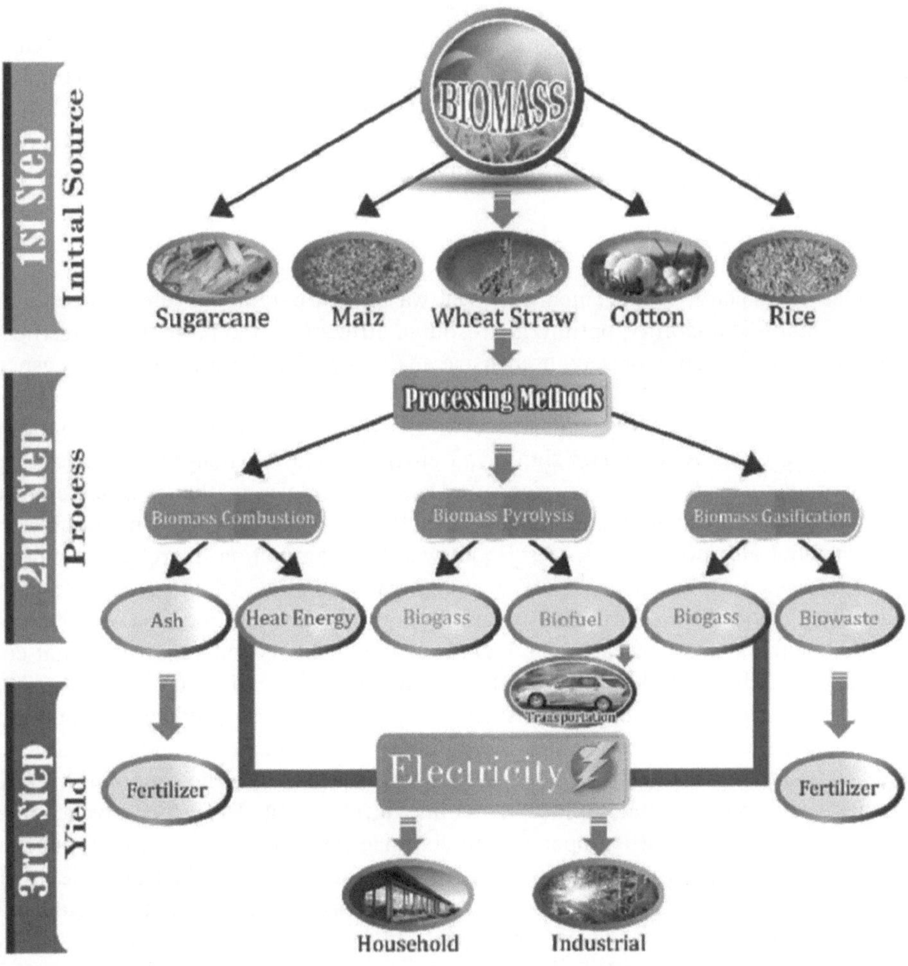

FIGURE 11.5 Schematic diagram to generate electricity using wheat straw.

fluctuations and impact the feasibility of processing and commercializing it. Another challenge lies in developing efficient and cost-effective processing technologies. Advanced technologies like enzymes and microorganisms are often required for the conversion of wheat waste, but they can be expensive and demand specialized expertise. Finding ways to optimize these technologies and reduce their costs is crucial for the successful commercialization of wheat waste as a feedstock (Arendt and Zannini, 2013).

Market demand also plays a significant role. Factors such as market competition, consumer preferences, and regulatory policies can influence the demand for wheat waste-derived products. If there is a lack of demand for these products, the commercialization of wheat waste becomes economically unfeasible. Therefore, understanding market dynamics and creating demand for wheat waste-derived products are important considerations. Infrastructure limitations pose another challenge. The lack of suitable storage, transportation, and processing facilities can hinder the commercialization of wheat waste-derived products. Developing the necessary infrastructure requires substantial investment, which can be particularly challenging in rural areas where wheat waste is typically generated. Economic plays a vital role in determining the viability of processing and commercializing wheat waste. The high capital costs associated with establishing and scaling up production facilities can be prohibitive, especially for small- and medium-sized enterprises. Finding cost-effective solutions and exploring funding opportunities are essential to overcome this challenge.

In conclusion, the challenges associated with processing and commercializing wheat waste as a feedstock require a collaborative effort from various stakeholders. Farmers, researchers, technology providers, and policymakers need to work together to develop innovative solutions. This includes advancing processing technologies, creating supportive policy frameworks, and establishing the necessary infrastructure. By addressing these challenges, we can unlock the full potential of wheat waste as a valuable resource, promote sustainability, and foster a circular economy.

11.9 CONCLUSIONS

In conclusion, the effective management of wheat waste is of paramount importance for ensuring environmental sustainability, optimizing resource utilization, and promoting a resilient and circular food system. Through our comprehensive analysis and exploration of various aspects of wheat waste management, we have gained valuable insights into the challenges, opportunities, and potential solutions associated with this critical issue. Our findings clearly demonstrated that mismanaging wheat waste poses significant environmental challenges. Improper disposal of wheat by-products leads to the emission of GHGs, particularly methane, which contributes to climate change. The burden on landfills due to inefficient waste management not only strains available space but also poses a risk of soil and water contamination, thereby threatening the integrity of ecosystems and human health. It is imperative that we adopt sustainable practices to mitigate these environmental impacts and minimize the ecological footprint of wheat waste. However, the significance of wheat waste management extends beyond environmental considerations. Our research has highlighted the substantial economic and social benefits associated with effective waste management practices. By valorizing wheat by-products through innovative technologies, such as biofuel production, animal feed formulation, and the extraction of high-value compounds, we can create new revenue streams, enhance the economic viability of the wheat industry, and contribute to local economic development. To further advance the field of wheat waste management, it is imperative to continue investing in research and innovation. Future studies should focus on evaluating the socio-economic impacts of sustainable practices, exploring emerging technologies for waste valorization, assessing the environmental performance of different waste management strategies, and monitoring the effectiveness of policy interventions. Continuous research and knowledge generation will enhance our understanding of this complex issue and support evidence-based decision-making. Therefore, effective wheat waste management requires a comprehensive and collaborative approach that spans the entire wheat value chain. By implementing sustainable practices, embracing technological advancements, and creating an enabling policy environment, we can transform wheat waste from a challenge into an opportunity. Let us seize this opportunity to build a more sustainable and resilient future where wheat waste is minimized, resources are optimized, and the ecological and socio-economic benefits are maximized.

ACKNOWLEDGMENT

This chapter was written with Lydia Hii Dai Jia, Nur Itqan Mohamad Ismail, and Mohamad Syahrul Naim Anuar.

REFERENCES

Abdallah, T. (Ed.). (2017). Environmental impacts. In: *Sustainable Mass Transit* (pp. 45–59). Elsevier. https://doi.org/10.1016/b978-0-12-811299-1.00004-6

Adetunji, A. T., Ncube, B., Mulidzi, R., & Lewu, F. B. (2020). Management impact and benefit of cover crops on soil quality: A review. *Soil and Tillage Research, 204*, 104717. https://doi.org/10.1016/j.still.2020.104717

Adusumilli, N., Lee, T., Rister, M. E., & Lacewell, R. D. (2014). The economics of mitigation of water pollution externalities from biomass production for energy. *Resources, 3*, 721–733.

Akbari, M., Oyedun, A. O., Gemechu, E., & Kumar, A. (2021). Comparative life cycle energy and greenhouse gas footprints of dry and wet torrefaction processes of various biomass feedstocks. *Journal of Environmental Chemical Engineering, 9*(4), 105415. https://doi.org/10.1016/j.jece.2021.105415

Al-Foudari, M., Sidhu, J. S., & Al-Hazza, A. (2022). Effect of psyllium husk and wheat mill bran fractions on the microstructure and mixograph characteristics of Arabic bread. *Saudi Journal of Biological Sciences, 29*(12), 103479–103479. https://doi.org/10.1016/j.sjbs.2022.103479

Alias, A. H., Norizan, M. N., Sabaruddin, F. A., Asyraf, M. R. M., Norrrahim, M. N. F., Ilyas, A. R., Kuzmin, A. M., Rayung, M., Shazleen, S. S., Nazrin, A., Sherwani, S. F. K., Harussani, M. M., Atikah, M. S. N., Ishak, M. R., Sapuan, S. M., & Khalina, A. (2021). Hybridization of MMT/lignocellulosic fiber reinforced polymer nanocomposites for structural applications: A review. *Coatings, 11*(11), 1355. https://doi.org/10.3390/coatings11111355

Andrew, J. J., & Dhakal, H. N. (2022). Sustainable biobased composites for advanced applications: Recent trends and future opportunities – A critical review. *Composites Part C: Open Access, 7*, 100220. https://doi.org/10.1016/j.jcomc.2021.100220

Arendt, E., & Zannini, E. (Eds.). (2013). Wheat and other Triticum grains. In: *Cereal Grains for the Food and Beverage Industries* (pp. 1–67e). Woodhead Publishing.

Asyraf, M. R. M., Syamsir, A., Zahari, N. M., Supian, A. B. M., Ishak, M. R., Sapuan, S. M., Sharma, S., Rashedi, A., Razman, M. R., Zakaria, S. Z. S., Ilyas, R. A., & Rashid, M. Z. A. (2022). Product development of natural fibre-composites for various applications: Design for sustainability. *Polymers, 14*(5), 920. https://doi.org/10.3390/polym14050920

Atwell, W. A. (2001). An overview of wheat development, cultivation, and production. Cereal Foods World, 46(2), 59–62.

Awad, A. M., Kumar, P., Ismail-Fitry, M. R., Jusoh, S., Ab Aziz, M. F., & Sazili, A. Q. (2021). Green extraction of bioactive compounds from plant biomass and their application in meat as natural antioxidant. *Antioxidants, 10*(9), 1465. https://doi.org/10.3390/antiox10091465

Azlin, M. N. M., Ilyas, R. A., Zuhri, M. Y. M., Sapuan, S. M., Harussani, M. M., Sharma, S., Nordin, A. H., Nurazzi, N. M., & Afiqah, A. N. (2022). 3D Printing and shaping polymers, composites, and nanocomposites: A review. *Polymers, 14*(1), 180. https://doi.org/10.3390/polym14010180

Ballesteros, I., Negro, M. J., Oliva, J. M., Cabanas, A., Manzanares, P., & Ballesteros, M. (2006). Ethanol production from steam-explosion pretreated wheat straw. *Applied Biochemistry and Biotechnology, 12*, 496–508.

Barskov, S., Zappi, M., Buchireddy, P., Dufreche, S., Guillory, J., Gang, D., Hernandez, R., Bajpai, R., Baudier, J., Cooper, R., & Sharp, R. (2019). Torrefaction of biomass: A review of production methods for biocoal from cultured and waste lignocellulosic feedstocks. *Renewable Energy, 142*, 624–642. https://doi.org/10.1016/j.renene.2019.04.068

Basha, S. I., Ghosh, S., Vinothkumar, K., Ramesh, B., Kumari, P. H. P., Mohan, K. V. M., & Sukumar, E. (2020). Fumaric acid incorporated Ag/agaragar hybrid hydrogel: A multifunctional avenue to tackle wound healing. *Materials Science & Engineering C, 111*, 110743.

Baska, S., Bollenbach, A., & Tsikas, D. (2021). GC-MS discrimination of citrulline from ornithine and homocitrulline from lysine by chemical derivatization: Evidence of formation of N5-carboxy-ornithine and N6-carboxy-lysine. *Molecules, 26*(8), 2301.

Bayraktar, H. (2005). Experimental and theoretical investigation of using gasoline–ethanol blends in spark-ignition engines. *Renewable Energy, 30*(11), 1733–1747.

Biron, M. (Ed.). (2017). Biobricks. In: *Industrial Applications of Renewable Plastics* (pp. 155–369). https://doi.org/10.1016/b978-0-323-48065-9.00005-4

Cabalova, I., Kacik, F., Geffert, A., & Kacikov, D. (2011). The effects of paper recycling and its environmental impact. In: Broniewicz, E. (Ed.), *Environmental Management in Practice*. IntechOpen. https://doi.org/10.5772/23110

Cascone, R. (2008). Biobutanol—A replacement for bioethanol? *Chemical Engineering Progress, 104*, S4–S9.

Catic, I., & Rujnic-Sokele, M. (2008). Agriculture products – Food for living being or for machinery. *Polimeri, 4*, 243–243.

Claassen, P. A. M., van Lier, J. B., Contreras, A. M. L., van Niel, E. W. J., Sijtsma, L., Stams, A. J. M., de Vries, S. S., & Weusthuis, R. A. (1999). Utilisation of biomass for the supply of energy carriers. *Applied Microbiology and Biotechnology, 6*, 741–755.

Croke, J. (2004). Hydrology | Soil erosion control. In: Burley, J. (Ed.), *Encyclopedia of Forest Sciences*, 387–397. Academic Press. https://doi.org/10.1016/b0-12-145160-7/00270-2

Dabija, A., Ciocan, M. E., Chetrariu, A., & Codină, G. G. (2021). Maize and sorghum as raw materials for brewing, a review. *Applied Sciences, 11*(7), 3139. https://doi.org/10.3390/app11073139

Data Bridge Market Research. (2023). Wheat straw market size, survey report, & industry analysis. www.databridgemarketresearch.com. Retrieved June 11, 2023, from https://www.databridgemarketresearch.com/reports/global-wheat-straw-market

De Assis Castro, R. C., & Roberto, I. C. (2015). Effect of nutrient supplementation on ethanol production in different strategies of saccharification and fermentation from acid pretreated rice straw. *Biomass and Bioenergy, 78*, 156–163.

Demain, A. L. (2009). Biosolutions to the energy problem. *Journal of Industrial Microbiology and Biotechnology, 36*, 319–332.

Duffy, J. E., Lefcheck, J. S., Stuart-Smith, R. D., Navarrete, S. A., & Edgar, G. J. (2016). Biodiversity enhances reef fish biomass and resistance to climate change. *Proceedings of the National Academy of Sciences, 113*(22), 6230–6235. https://doi.org/10.1073/pnas.1524465113

Edwards, L., & Burney, J. (2005). Cover crops. In: Hillel, D., & Hatfield, J. L. (Eds.), *Encyclopedia of Soils in the Environment* (pp. 311–318). Elsevier/Academic. https://doi.org/10.1016/b0-12-348530-4/00252-6

ElMekawy, A., Srikanth, S., Bajracharya, S., Hegab, H. M., Nigam, P. S., Singh, A., Mohan, S. V., & Pant, D. (2015). Food and agricultural wastes as substrates for bioelectrochemical system (BES): The synchronized recovery of sustainable energy and waste treatment. *Food Research International, 73*, 213–225.

Eraky, M., Jin, K., Zhang, Q. G., Zhang, Z. P., Ai, P., & Elsayed, M. (2021). Acidogenic biorefinery of rice straw for volatile fatty acids production via sequential two-stage fermentation: Effects of pre-treatments. *Environmental Technology & Innovation, 23*, 101686.

EREC. (2008). *Renewable Energy Technology Roadmap 20% by 2020*. European Renewable Energy Council, Brussels.

Faiz Norrrahim, M. N., Ilyas, R. A., Nurazzi, N. M., Asmal Rani, M. S., Nur Atikah, M. S., & Shazleen, S. S. (2021). Chemical pretreatment of lignocellulosic biomass for the production of bioproducts: An overview. *Applied Science and Engineering Progress, 14*, 1–18. https://doi.org/10.14416/j.asep.2021.07.004

FAO. (2020). FAOSTAT. www.fao.org. Retrieved from https://www.fao.org/faostat/en/#data/QCL/visualize

Food and Agriculture Organization of the United Nations. (2022). FAOSTAT 2018. https://www.fao.org/faostat/en/#data/QC (accessed February 8, 2022).

Frigon, J. C., & Guiot, S. R. (2010). Biomethane production from starch and lignocellulosic crops: A comparative review. *Biofuels, Bioproducts and Biorefining, 4*, 447–458. https://doi.org/10.1002/bbb.229

Galbe, M., & Zacchi, G. (2002). A review of the production of ethanol from softwood. *Applied Microbiology and Biotechnology, 6*, 618–628.

https://doi.org/10.1051/matecconf/201819401018

Gent, S., Twedt, M., Gerometta, C., & Almberg, E. (Eds.). (2017). Introduction to feedstocks. In: *Theoretical and Applied Aspects of Biomass Torrefaction* (pp. 17–39). Elsevier: Butterworth-Heinemann, Oxford. https://doi.org/10.1016/b978-0-12-809483-9.00002-6

Guajardo, D., Valdebenito, F., Díaz, J., Cifuentes, G., & Azócar, L. (2021). Carbonization of microalgae for bio-coal production as a solid biofuel similar to bituminous coal. *Fuel, 302*, 120963. https://doi.org/10.1016/j.fuel.2021.120963

Halimatul, M. J., Sapuan, S. M., Jawaid, M., Ishak, M. R., & Ilyas, R. A. (2019). Water absorption and water solubility properties of sago starch biopolymer composite films filled with sugar palm particles. *Polimery, 64*(9), 27–35. https://doi.org/10.14314/polimery.2019.9.4

Harrison, R. B. (2008). Composting and formation of humic substances. In: Jørgensen, S. E., & Fath, B. D. (Eds.), *Encyclopedia of Ecology* (pp. 713–719). Academic Press. https://doi.org/10.1016/b978-008045405-4.00262-7

Higashide, W., Li, Y., Yang, Y., & Liao, J. C. (2011). Metabolic engineering of *Clostridium cellulolyticum* for production of isobutanol from cellulose. *Applied and Environmental Microbiology, 77*, 2727–2733.

Hu, G., Heitmann, J. A., & Rojas, O. J. (2008). Feedstock pretreatment strategies for producing ethanol from wood, bark, and forest residues. *Bioresources, 1*, 270–294.

Inglezakis, V. J., Poulopoulos, S. G., Arkhangelsky, E., Zorpas, A. A., & Menegaki, A. N. (2016). Aquatic environment. In: Poulopoulos, S. G., & Inglezakis, V. J. (Eds.), *Environment and Development* (pp. 137–212). Elsevier. https://doi.org/10.1016/b978-0-444-62733-9.00003-4

Irfan Said, M. (2022). The role of the livestock farming industry in supporting the global agricultural industry. In: Asaduzzaman, M., & Afroz, M. (Eds.), *Agricultural Development in Asia - Potential Use of Nano-Materials and Nano-Technology*. IntechOpen. https://doi.org/10.5772/intechopen.97868

Jacobs, P. J., Hemdane, S., Dornez, E., Delcour, J. A., & Courtin, C. M. (2015). Study of hydration properties of wheat bran as a function of particle size. *Food Chemistry, 179*, 296–304. https://doi.org/10.1016/j.foodchem.2015.01.117

Jager, H. I., & Kreig, J. A. F. (2018). Designing landscapes for biomass production and wildlife. *Global Ecology and Conservation, 16*. https://doi.org/10.1016/j.gecco.2018.e00490

Jia, M., Sun, M., Li, X., Xu, X., Wang, Y., Shi, L., & Hu, H. (2018). A new era of straw-based pulping? Evidence from a carbon metabolism perspective. *Journal of Cleaner Production, 193*, 327–337. https://doi.org/10.1016/j.jclepro.2018.04.227

Jumaidin, R., Adam, N. W., Ilyas, R. A., Hussin, M. S. F., Taha, M. M., Mansor, M. R., Azlan, U. A. A., & Yob, M. S. (2019a). Water transport and physical properties of sugarcane bagasse fibre reinforced thermoplastic potato starch biocomposite. *Journal of Advanced Research in Fluid Mechanics and Thermal Sciences, 61*(2), 273–281.

Jumaidin, R., Diah, N. A., Ilyas, R. A., Alamjuri, R. H., & Yusof, F. A. M. (2021). Processing and characterisation of banana leaf fibre reinforced thermoplastic cassava starch composites. *Polymers, 13*(9), 1420. https://doi.org/10.3390/polym13091420

Jumaidin, R., Saidi, Z. A. S., Ilyas, R. A., Ahmad, M. N., Wahid, M. K., Yaakob, M. Y., Maidin, N. A., Rahman, M. H. A., & Osman, M. H. (2019b). Characteristics of cogon grass fibre reinforced thermoplastic cassava starch biocomposite: Water absorption and physical properties. *Journal of Advanced Research in Fluid Mechanics and Thermal Sciences 62, 62*(1), 43–52.

Karić, N., Maia, A. S., Teodorović, A., Atanasova, N., Langergraber, G., Crini, G., Ribeiro, A. R. L., & Đolić, M. (2022). Bio-waste valorisation: Agricultural wastes as biosorbents for removal of (in)organic pollutants in wastewater treatment. *Chemical Engineering Journal Advances, 9*, 100239. https://doi.org/10.1016/j.ceja.2021.100239

Karuna, N., Namwong, W., & Ruaysoongnern, N. (2022). Alkaline-extracted lignin from xylia xylocarpa and its antioxidant properties. *Journal of Sustainability Science and Management, 17*(4), 151–160. Penerbit UMT, Universiti Malaysia Terengganu. https://doi.org/10.46754/jssm.2022.4.012

Kern, M. (2023). Offshore oil and gas drilling: A comprehensive guide. OilPrice.com. https://oilprice.com/Energy/Crude-Oil/Offshore-Oil-And-Gas-Drilling-A-Comprehensive-Guide.amp.html?cv=1

Kerstetter, J. D., & Lyons, J. K. (2001). *Logging and Agricultural Residue Supply Curves for the Pacific Northwest*. Washington State University, Olympia, WA.

Khan, T. S., & Mubeen, U. (2012). Wheat straw: A pragmatic overview. *Current Research Journal of Biological Sciences, 4*, 673–675.

Kim, S., & Dale, B. E. (2004). Global potential bioethanol production from wasted crops and crop residues. *Biomass Bioenergy, 4*, 361–375.

Kong, L., Hasanbeigi, A., & Price, L. (2016). Assessment of emerging energy-efficiency technologies for the pulp and paper industry: A technical review. *Journal of Cleaner Production, 122*, 5–28. https://doi.org/10.1016/j.jclepro.2015.12.116

Koul, B., Yakoob, M., & Shah, M. P. (2022). Agricultural waste management strategies for environmental sustainability. *Environmental Research, 206*, 112285. https://doi.org/10.1016/j.envres.2021.112285

Kovács, K., Willson, B. J., Schwarz, K., Heap, J. T., Jackson, A., Bolam, D. N., Winzer, K., & Minton, N. P. (2013). Secretion and assembly of functional mini-cellulosomes from synthetic chromosomal operons in Clostridium acetobutylicum ATCC 824. *Biotechnology for Biofuels, 6*, 117.

Lal, R. (2004). Soil carbon sequestration to mitigate climate change. *Geoderma*, *123*(1–2), 1–22. https://doi.org/10.1016/j.geoderma.2004.01.032

Laser, M., Larson, E., Dale, B., Wang, M., Greene, N., & Lynd, L. R. (2009). Comparative analysis of efficiency, environmental impact, and process economics for mature biomass refining scenarios. *Biofuels, Bioproducts and Biorefining, 3*(2), 247–270. https://doi.org/10.1002/bbb.136

Laureano-Perez, L., Teymouri, F., Alizadeh, H., & Dale, B. E. (2005). Understanding factors that limit enzymatic hydrolysis of biomass: Characterization of pretreated corn stover. *Applied Biochemistry and Biotechnology*, *124*, 1081–1099.

Lebel, E. D., Finnegan, C. J., Ouyang, Z., & Jackson, R. B. (2022). Methane and NO_x emissions from natural gas stoves, cooktops, and ovens in residential homes. *Environmental Science & Technology*, *56*(4), 2529–2539. https://doi.org/10.1021/acs.est.1c04707

Leschine, S. B. (1995). Cellulose degradation in anaerobic environments. *Annual Review of Microbiology, 49*, 399–426.

Li, H., Mehmood, D., Thorin, E., & Yu, Z. (2017). Biomethane production via anaerobic digestion and biomass gasification. *Energy Procedia, 105*, 1172. https://doi.org/10.1016/j.egypro.2017.03.490

Li, J., Wei, L. Y., Duan, Q. W., Hu, G. Q., & Zhang, G. Z. (2014). Semi-continuous anaerobic co-digestion of dairy manure with three crop residues for biogas production. *Bioresource Technology, 156*, 307. https://doi.org/10.1016/j.biortech.2014.01.064

Lin, H., & Ying, Y. (2009). Theory and application of near infrared spectroscopy in assessment of fruit quality: A review. *Sensing and Instrumentation for Food Quality and Safety, 3*, 130–141.

Liu, Z., Wang, X., Li, S., Bai, Z., & Ma, L. (2022). Advanced composting technologies promotes environmental benefits and eco-efficiency: A life cycle assessment. *Bioresource Technology*, *346*, 126576. https://doi.org/10.1016/j.biortech.2021.126576

Lynd, L. R., Guss, A. M., Himmel, M. E., Beri, D., Herring, C., Holwerda, E. K., Murphy, S. J. L., Olson, D. G., Paye, J., Rydzak, T., Shao, X., Tian, L., & Worthen, R. (2016). Advances in consolidated bioprocessing using Clostridium thermocellum and Thermoanaerobacter saccharolyitcum. In: Wittmand, C., & Liao, J. C. (Eds.), *Industrial Biotechnology: Microorganisms* (pp. 365–394). Wiley-VCH Verlag GmbH & Co. KGaA.

Lynd, L. R., Weimer, P. J., van Zyl, W. H., & Pretorius, I. S. (2002). Microbial cellulose utilization: Fundamentals and biotechnology. *Microbiology and Molecular Biology Reviews, 66*, 506–577.

Manisalidis, I., Stavropoulou, E., Stavropoulos, A., & Bezirtzoglou, E. (2020). Environmental and health impacts of air pollution: A review. *Frontiers in Public Health*, *8*. https://doi.org/10.3389/fpubh.2020.00014

Maraveas, C. (2020). Production of sustainable construction materials using agro-wastes. *Materials*, *13*(2), 262. https://doi.org/10.3390/ma13020262

Maximillian, J., Brusseau, M. L., Glenn, E. P., & Matthias, A. D. (2019). Pollution and environmental perturbations in the global system. In: Brusseau, M. L., Pepper, I. L., & Gerba, C. P. (Eds.), *Environmental and Pollution Science* (pp. 457–476). Academic Press. https://doi.org/10.1016/b978-0-12-814719-1.00025-2

Mazani, N., Sapuan, S. M., Sanyang, M. L., Atiqah, A., & Ilyas, R. A. (2019). Design and fabrication of a shoe shelf from kenaf fiber reinforced unsaturated polyester composites. In: Ariffin, H., Sapuan, S. M., & Hassan, M. A. (Eds.), *Lignocellulose for Future Bioeconomy* (1st ed., Issue 2000, pp. 315–332). Elsevier, Amsterdam. https://doi.org/10.1016/B978-0-12-816354-2.00017-7

McMillan, J. D. (1994). Pretreatment of lignocellulosic biomass. In: Himmel, M. E., Baker, J. O., & Overend, R. P. (Eds.), *Enzymatic Conversion of Biomass for Fuels Production*. ACS Symposium Series 566 (pp. 292–324). Washington: American Chemical Society.

Meesum, S., Ananth, K. K., Brianna, C., Carole, A. O., & Hariharan, S. (2021). Lactic acid suppresses MRGPRX2 mediated mast cell responses. *Cellular Immunology, 368*, 104422.

Memon, S., Wahid, I., Khan, M., Tanoli, M., & Bimaganbetova, M. (2018). Environmentally friendly utilization of wheat straw ash in cement-based composites. *Sustainability, 10*(5), 1322. https://doi.org/10.3390/su10051322

Meriles, S. P., Curet, S., Boillereaux, L., Ribotta, P. D., & Penci, M. C. (2022). Thermo-physical properties of wheat germ: Heat and mass transfer during convective heating. *Journal of Food Process Engineering, 46*(1). https://doi.org/10.1111/jfpe.14190

Morone, A., & Pandey, R. A. (2014). Lignocellulosic biobutanol production: Gridlocks and potential remedies. *Renewable and Sustainable Energy Reviews, 37*, 21–35.

Mujtaba, M., Fernandes Fraceto, L., Fazeli, M., Mukherjee, S., Savassa, S. M., Araujo de Medeiros, G., do Espírito Santo Pereira, A., Mancini, S. D., Lipponen, J., & Vilaplana, F. (2023). Lignocellulosic biomass from agricultural waste to the circular economy: A review with focus on biofuels, biocomposites and bioplastics. *Journal of Cleaner Production, 402*, 136815. https://doi.org/10.1016/j.jclepro.2023.136815

Muthuraj, R., Lacoste, C., Lacroix, P., & Bergeret, A. (2019). Sustainable thermal insulation biocomposites from rice husk, wheat husk, wood fibers and textile waste fibers: Elaboration and performances evaluation. *Industrial Crops and Products, 135*, 238–245. https://doi.org/10.1016/j.indcrop.2019.04.053

Nazrin, A., Sapuan, S. M., Zuhri, M. Y. M., Tawakkal, I. S. M. A., & Ilyas, R. A. (2021). Water barrier and mechanical properties of sugar palm crystalline nanocellulose reinforced thermoplastic sugar palm starch (TPS)/poly(lactic acid) (PLA) blend bionanocomposites. *Nanotechnology Reviews, 10*(1), 431–442. https://doi.org/10.1515/ntrev-2021-0033

Norfarhana, A. S., Ilyas, R. A., & Ngadi, N. (2022). A review of nanocellulose adsorptive membrane as multifunctional wastewater treatment. *Carbohydrate Polymers, 291*, 119563. https://doi.org/10.1016/j.carbpol.2022.119563

Norfarhana, A. S., Ilyas, R. A., Ngadi, N., Othman, M. H. D., Misenan, M. S. M., & Norrrahim, M. N. F. (2024a). Revolutionizing lignocellulosic biomass: A review of harnessing the power of ionic liquids for sustainable utilization and extraction. *International Journal of Biological Macromolecules, 256*, 128256. https://doi.org/10.1016/j.ijbiomac.2023.128256

Norfarhana, A. S., Khoo, P. S., Ilyas, R. A., Ab Hamid, N. H., Aisyah, H. A., Norrrahim, M. N. F., Knight, V. F., Rani, M. S. A., Septevani, A. A., Syafri, E., & Annamalai, P. K. (2024b). Exploring of cellulose nanocrystals from lignocellulosic sources as a powerful adsorbent for wastewater remediation. *Journal of Polymers and the Environment, 32*, 1–31. https://doi.org/10.1007/s10924-024-03227-3

Nurazzi, N. M., Harussani, M. M., Aisyah, H. A., Ilyas, R. A., Norrrahim, M. N. F., Khalina, A., & Abdullah, N. (2021). Treatments of natural fiber as reinforcement in polymer composites—A short review. *Functional Composites and Structures, 3*(2), 024002. https://doi.org/10.1088/2631-6331/abff36

Olson, D. G., McBride, J. E., Shaw, A. J., & Lynd, L. R. (2012). Recent progress in consolidated bioprocessing. *Current Opinion in Biotechnology, 23*, 396–405.

Oni, B. A., Oziegbe, O., & Olawole, O. O. (2019). Significance of biochar application to the environment and economy. *Annals of Agricultural Sciences, 64*(2), 222–236. https://doi.org/10.1016/j.aoas.2019.12.006

Otero, J. M., Panagiotou, G., & Olsson, L. (2007). Fueling industrial biotechnology growth with bioethanol. *Advances in Biochemical Engineering/Biotechnology, 108*, 1–40.

Panahi, A., Tarakcioglu, M., Schiemann, M., Delichatsios, M., & Levendis, Y. A. (2018). On the particle sizing of torrefied biomass for co-firing with pulverized coal. *Combustion and Flame, 194*, 72–84. https://doi.org/10.1016/j.combustflame.2018.04.014

Patwa, N., Sivarajah, U., Seetharaman, A., Sarkar, S., Maiti, K., & Hingorani, K. (2021). Towards a circular economy: An emerging economies context. *Journal of Business Research, 122*, 725–735. https://doi.org/10.1016/j.jbusres.2020.05.015

Paul, A. S., Panwar, N. L., Salvi, B. L., Jain, S., & Sharma, D. (2020). Experimental investigation on the production of bio-oil from wheat straw. *Energy Sources, Part A: Recovery, Utilization, and Environmental Effects, 46*, 1–16. https://doi.org/10.1080/15567036.2020.1779416

Pereira, P. H. F., Waldron, K. W., Wilson, D. R., Cunha, A. P., Brito, E. S. de, Rodrigues, T. H. S., Rosa, M. F., & Azeredo, H. M. C. (2017). Wheat straw hemicelluloses added with cellulose nanocrystals and citric acid. Effect on film physical properties. *Carbohydrate Polymers, 164*, 317–324. https://doi.org/10.1016/j.carbpol.2017.02.019

Pereyra, L. P., Hiibel, S. R., Prieto Riquelme, M. V., Reardon, K. F., & Pruden, A. (2010). Detection and quantification of functional genes of cellulose-degrading, fermentative, and sulfate-reducing bacteria and methanogenic archaea. *Applied and Environmental Microbiology, 76*, 2192–2202.

Prasad, B., Carver, B. F., Stone, M. L., Babar, M. A., Raun, W. R., & Klatt, A. R. (2007). Potential use of spectral reflectance indices as a selection tool for grain yield in winter wheat under great plains conditions. *Crop Science, 47*(4), 1426–1440.

Rahman, M. N., & Wahid, M. A. (2021). Renewable-based zero-carbon fuels for the use of power generation: A case study in Malaysia supported by updated developments worldwide. *Energy Reports, 7*, 1986–2020. https://doi.org/10.1016/j.egyr.2021.04.005

Ransom-Jones, E., Jones, D., McCarthy, A., & McDonald, J. (2012) The fibrobacteres: an important phylum of cellulose-degrading bacteria. *Microbial Ecology, 63*, 267–281.

Rather, R. A., Wani, A. W., Mumtaz, S., Padder, S. A., Khan, A. H., Almohana, A. I., Almojil, S. F., Alam, S. S., & Baba, T. R. (2022). Bioenergy: A foundation to environmental sustainability in a changing global climate scenario. *Journal of King Saud University – Science, 34*(1), 101734. https://doi.org/10.1016/j.jksus.2021.101734

Reza, M. T., Rottler, E., Herklotz, L., & Wirth, B. (2015). Hydrothermal carbonization (HTC) of wheat straw: Influence of feedwater pH prepared by acetic acid and potassium hydroxide. *Bioresource Technology, 182*, 336.

Rezaei, M., Amiri, H., & Shafiei, M. (2021). Aqueous pretreatment of triticale straw for integrated production of hemicellulosic methane and cellulosic butanol. *Renewable Energy, 171*, 971.

Romaní, A., Tomaz, P. D., Garrote, G., Teixeira, J. A., & Domingues, L. (2016). Combined alkali and hydrothermal pretreatments for oat straw valorization within a biorefinery concept. *Bioresource Technology, 220*, 323.

Saad Azzem, L., & Bellel, N. (2022). Thermal and physico-chemical characteristics of plaster reinforced with wheat straw for use as insulating materials in building. *Buildings, 12*(8), 1119. https://doi.org/10.3390/buildings12081119

Sadh, P. K., Duhan, S., & Duhan, J. S. (2018). Agro-industrial wastes and their utilization using solid state fermentation: A review. *Bioresources and Bioprocessing, 5*(1). https://doi.org/10.1186/s40643-017-0187-z

Sain, M., & Panthapulakkal, S. (2006). Bioprocess preparation of wheat straw fibers and their characterization. *Industrial Crops and Products, 23*(1), 1–8. https://doi.org/10.1016/j.indcrop.2005.01.006

Sánchez-Safont, E. L., Aldureid, A., Lagarón, J. M., Gámez-Pérez, J., & Cabedo, L. (2018). Biocomposites of different lignocellulosic wastes for sustainable food packaging applications. *Composites Part B: Engineering, 145*, 215–225. https://doi.org/10.1016/j.compositesb.2018.03.037

Schwarz, W. (2001). The cellulosome and cellulose degradation by anaerobic bacteria. *Applied Microbiology and Biotechnology, 56*, 634–649.

Semadeni, M. (2004). Storage of energy, overview. In: Cleveland, C. J. (Ed.), *Encyclopedia of Energy* (pp. 719–738). Elsevier. https://doi.org/10.1016/b0-12-176480-x/00104-2

Semida, W. M., Beheiry, H. R., Sétamou, M., Simpson, C. R., Abd El-Mageed, T. A., Rady, M. M., & Nelson, S. D. (2019). Biochar implications for sustainable agriculture and environment: A review. *South African Journal of Botany, 127*, 333–347. https://doi.org/10.1016/j.sajb.2019.11.015

Sharma, S., Asolekar, S. R., Thakur, V. K., & Asokan, P. (2023). Valorization of cellulosic fiber derived from waste biomass of constructed wetland as a potential reinforcement in polymeric composites: A technological approach to achieve circular economy. *Journal of Environmental Management, 340*, 117850. https://doi.org/10.1016/j.jenvman.2023.117850

Shewry, P. R. (2009). Wheat. *Journal of Experimental Botany, 60*(6), 1537–1553. https://doi.org/10.1093/jxb/erp058

Siles, J. A., Vera, A., Díaz-López, M., García, C., van den Hoogen, J., Crowther, T. W., Eisenhauer, N., Guerra, C., Jones, A., Orgiazzi, A., Delgado-Baquerizo, M., & Bastida, F. (2023). Land-use- and climate-mediated variations in soil bacterial and fungal biomass across Europe and their driving factors. *Geoderma, 434*, 116474. https://doi.org/10.1016/j.geoderma.2023.116474

Sinha, P., & Pandey, A. (2011). An evaluative report and challenges for fermentative biohydrogen production. *International Journal of Hydrogen Energy, 36*, 7460–7478. https://doi.org/10.1016/j.ijhydene.2011.03.077

Skendi, A., & Papageorgiou, M. (2021). Introduction in wheat and breadmaking. In: Galanakis, C. M. (Ed.), *Trends in Wheat and Bread Making* (pp. 1–27). Academic Press, Amsterdam. https://doi.org/10.1016/b978-0-12-821048-2.00001-5

Soares, J. F., Confortin, T. C., Todero, I., Mayer, F. D., & Mazutti, M. A. (2020). Dark fermentative biohydrogen production from lignocellulosic biomass: Technological challenges and future prospects. *Renewable and Sustainable Energy Reviews, 117*, 109484. https://doi.org/10.1016/j.rser.2019.109484

Solé-Bundó, M., Eskicioglu, C., Garfí, M., Hélène, C., & Ivet, F. (2017). Anaerobic co-digestion of microalgal biomass and wheat straw with and without thermo-alkaline pretreatment. *Bioresource Technology, 237*, 89–98. https://doi.org/10.1016/j.biortech.2017.03.151

Spirchez, C., Lunguleasa, A., Ionescu, C., & Croitoru, C. (2019). Physical and calorific properties of wheat straw briquettes and pellets. *MATEC Web of Conferences, 290*, 11011. https://doi.org/10.1051/matecconf/201929011011

Sun, X.-F., Sun, R.-C., Su, Y., & Sun, J.-X. (2004). Comparative study of crude and purified cellulose from wheat straw. *Journal of Agricultural and Food Chemistry, 52*(4), 839–847. https://doi.org/10.1021/jf0349230

Suri, S., & Singh, A. (2023). Colored cereals: Beyond nutritional values. In: Bangar, S. P. & Kumar, M. (Eds.), *Functionality and Application of Colored Cereals* (pp. 47–72). Academic Press. https://doi.org/10.1016/b978-0-323-99733-1.00002-9

Suriani, M. J., Radzi, F. S. M., Ilyas, R. A., Petrů, M., Sapuan, S. M., & Ruzaidi, C. M. (2021). Flammability, tensile, and morphological properties of oil palm empty fruit bunches fiber/pet yarn-reinforced epoxy fire retardant hybrid polymer composites. *Polymers, 13*(8), 1282. https://doi.org/10.3390/polym13081282

Syafri, E., Jamaluddin, Sari, N. H., Mahardika, M., Amanda, P., & Ilyas, R. A. (2022). Isolation and characterization of cellulose nanofibers from Agave gigantea by chemical-mechanical treatment. *International Journal of Biological Macromolecules, 200*(December 2021), 25–33. https://doi.org/10.1016/j.ijbiomac.2021.12.111

Tan, K. T., Lee, K. T., & Mohamed, A. R. (2008). Role of energy policy in renewable energy accomplishment: The case of second-generation bioethanol. *Energy Policy, 9*, 3360–3365.

Tan, S., Zhou, G., Yang, Q., Ge, S., Liu, J., Cheng, Y. W., Yek, P. N., Wan Mahari, W. A., Kong, S. H., Chang, J.-S., Sonne, C., Chong, W. W., & Lam, S. S. (2023). Utilization of current pyrolysis technology to convert biomass and manure waste into biochar for soil remediation: A review. *Science of the Total Environment, 864*, 160990. https://doi.org/10.1016/j.scitotenv.2022.160990

Tufail, T., Saeed, F., Imran, M., Arshad, M. U., Anjum, F. M., Afzaal, M., Bader Ul Ain, H., Shahbaz, M., Gondal, T. A., & Hussain, S. (2018). Biochemical characterization of wheat straw cell wall with special reference to bioactive profile. *International Journal of Food Properties, 21*(1), 1303–1310. https://doi.org/10.1080/10942912.2018.1484759

Umar, M., Ji, X., Kirikkaleli, D., & Alola, A. A. (2021). The imperativeness of environmental quality in the United States transportation sector amidst biomass-fossil energy consumption and growth. *Journal of Cleaner Production, 285*, 124863. https://doi.org/10.1016/j.jclepro.2020.124863

Valles, A., Capilla, M., Álvarez-Hornos, F. J., García-Puchol, M., San-Valero, P., & Gabaldón, C. Optimization of alkali pretreatment to enhance rice straw conversion to butanol. *Biomass and Bioenergy, 150*, 106131, 2021.

Wang, F., Harindintwali, J. D., Yuan, Z., Wang, M., Wang, F., Li, S., Yin, Z., Huang, L., Fu, Y., Li, L., Chang, S. X., Zhang, L., Rinklebe, J., Yuan, Z., Zhu, Q., Xiang, L., Tsang, D. C. W., Xu, L., Jiang, X., … & Chen, J. M. (2021). Technologies and perspectives for achieving carbon neutrality. *The Innovation, 2*(4), 100180. https://doi.org/10.1016/j.xinn.2021.100180

Wheals, A. E., Basso, L. C., Alves, D. M. G., & Amorim, H. V. (1999). Fuel ethanol after 25 years. *Trends Biotechnol, 12*, 482–487.

Wu, C. W., Chen, R. H., Pu, J. Y., & Lin, T. H. (2004). The influence of air–fuel ratio on engine performance and pollutant emission of an SI engine using ethanol–gasoline-blended fuels. *Atmospheric Environment, 38*(40), 7093–7100.

Wyman, C. E. (Ed.). (1996). Ethanol production from lignocellulosic biomass: overview. In: *Handbook on Bioethanol: Production and Utilization* (pp. 1–18). Taylor & Francis, Washington.

Xu, C., Yuan, Q., Zhao, S., He, T., & Song, N. (2020). Effects of pretreatments on physical and chemical characteristics of wheat straw used as a maintenance-free compressed green roof substrate material. *Journal of Cleaner Production, 277*, 123381. https://doi.org/10.1016/j.jclepro.2020.123381

Xu, M., Sun, H., Chen, E., Yang, M., Wu, C., Sun, X., & Wang, Q. (2023). From waste to wealth: Innovations in organic solid waste composting. *Environmental Research, 229*, 115977. https://doi.org/10.1016/j.envres.2023.115977

Yadav, N., & Hakkarainen, M. (2021). Degradable or not? cellulose acetate as a model for complicated interplay between structure, environment and degradation. *Chemosphere, 265*, 128731. https://doi.org/10.1016/j.chemosphere.2020.128731

Yang, Y., Liew, R. K., Tamothran, A. M., Foong, S. Y., Yek, P. N. Y., Chia, P. W., Van Tran, T., Peng, W., & Lam, S. S. (2021). Gasification of refuse-derived fuel from municipal solid waste for energy production: A review. *Environmental Chemistry Letters, 19*, 2127–2140.

Yang, Y. L., Zhang, P. D., Zhang, W. L., Tian, Y. S., Zheng, Y. H., & Wang, L. S. (2010). Quantitative appraisal and potential analysis for primary biomass resources for energy utilization in China. *Renew. Sust. Energ. Rev. 14*, 3050–3058. https://doi.org/10.1016/j.rser.2010.07.054

Yu, L., Xu, M., Tang, I.-C., & Yang, S.-T. (2015). Metabolic engineering of Clostridium tyrobutyricum for n-butanol production through co-utilization of glucose and xylose. *Biotechnology and Bioengineering, 112*, 2134–2141.

Zhang, C., Su, H., Baeyens, J., & Tan, T. W. (2014). Reviewing the anaerobic digestion of food waste for biogas production. *Renewable and Sustainable Energy Reviews, 38*, 383–392. https://doi.org/10.1016/j.rser.2014.05.038

Zhang, N., Zhou, K., & Du, X. (2017). Application of fuzzy logic and fuzzy AHP to mineral prospectivity mapping of porphyry and hydrothermal vein copper deposits in the Dananhu-Tousuquan island arc, Xinjiang, NW China. *Journal of African Earth Sciences*, *128*, 84–96.

Zhang, Y., Ghaly, A. E., & Li, B. (2012). Physical properties of wheat straw varieties cultivated under different climatic and soil conditions in three continents. *American Journal of Engineering and Applied Sciences*, *5(2)*, 98–106. https://doi. org/10.3844/ajeassp.2012.98.106

Zheng, D., & Chi, F. Q. (2012). A comprehensive evaluation on strow return to soil in the sustainable development of agriculture. *Heilongjiang Agricultural Science, 1*, 133–138.

Zhu, J., Song, W., Chen, X., & Sun, S. (2023). Integrated process to produce biohydrogen from wheat straw by enzymatic saccharification and dark fermentation. *International Journal of Hydrogen Energy, 48*(30). ISSN 0360–3199. https://doi.org/10.1016/j.ijhydene.2022.05.056

Zilberman, D., Holland, T., & Trilnick, I. (2018). Agricultural gmos—What we know and where scientists disagree. *Sustainability*, *10*(5), 1514. https://doi.org/10.3390/su10051514

12 Sugarcane Waste Management

12.1 INTRODUCTION

Agricultural waste is rich in bioactive compounds. These by-products can be used as a secondary source for the production of a variety of items, including biogas, biofuel, mushrooms, and tempeh, which is employed as the main element in many research projects and commercial ventures. Utilizing agro-industrial wastes as a source of raw materials can reduce production costs and pollution levels in the environment. Agro-industrial wastes are transformed into biofuels, enzymes, vitamins, antioxidants, animal feed, antibiotics, and other substances by solid-state fermentation (SSF). A number of microorganisms are used through SSF techniques to manufacture these valuable products. As stated by Bos and Hamelinck (2014), SSF is therefore evaluated and discussed together with their effect on the development of value-added products.

Industries having an agricultural base produced the vast majority of residues each year. These residues could damage the environment and have a negative effect on both human and animal health if they are not properly disposed of. According to the study by Bos and Hamelinck (2014), the majority of agro-industrial wastes are not properly recycled and are instead burned, dumped, or accidentally landfilled. These untreated wastes raise a variety of greenhouse gases, which play a role in a number of problems related to climate change. Furthermore, the use of fossil fuels affects the emission of greenhouse gases. As a result, the development of alternative, cleaner, and renewable bioenergy supplies is now a global priority as noted by Okonko et al. (2009). These wastes pose a significant difficulty for disposal. For instance, the coffee and cereal industries both generated waste in the form of coffee pulp, along with a sizable amount of waste in the form of peels from the coffee and juice industries. Approximately 147.2 million metric tons of fiber sources have been discovered globally, as opposed to the anticipated 709.2 and 673.3 million metric tons of remaining wheat straw and rice straw from the 1990s, respectively, according to Rodríguez-Couto (2008). According to Graminha et al. (2008), these agro-industrial residues are classed as agro-industrial by-products because of their high nutritional potential and chemical make-up. As a result, they are drawing more attention to quality management.

Sugarcane or its scientific name *Saccharum officinarum*, as shown in Figure 12.1, is a tropical perennial grass in the Poaceae family, cultivated for its sweet juice, which is used to produce sugar and other related products. Yamane (2023) reported that sugarcane has a long history of cultivation and has been an important crop in various parts of the world for centuries. Originating in Southeast Asia, sugarcane was first domesticated in New Guinea around 6,000 years ago. From there, it spread to India, where it was extensively cultivated. It later made its way to China, the Middle East, and other regions through trade routes.

Sugarcane is a tall, robust plant that can grow up to 4–6 m in height. It has thick, fibrous stalks filled with sweet sap, which is the primary source of sugar. The plant has a complex root system that helps draw nutrients and moisture from the soil. It requires a warm and tropical climate with abundant sunlight and adequate rainfall for optimal growth as stated by Yamane (2023). Cultivating sugarcane involves planting stem cuttings, known as setts, into prepared fields. The plant takes about 9–24 months to mature, depending on the variety and growing conditions. Once mature, the stalks are harvested by cutting them close to the ground. The harvested sugarcane is then processed to extract the juice. The chapter elaborates on waste management of sugarcane waste to protect the environment by considerably reducing the emission of harmful elements, such

DOI: 10.1201/9781003472209-17

FIGURE 12.1 Sugarcane trees.

as greenhouse gases into the atmosphere, implementing efficient and comprehensive management practices for agricultural waste, playing a crucial role in promoting sustainability, and minimizing environmental effects. Additionally, by using effective waste management techniques, the risk of soil erosion and water contamination is significantly decreased, protecting the integrity of crucial ecosystems and maintaining the quality of natural resources necessary for a variety of ecological and human activities. Other than that, it is to conserve natural resources by utilizing agricultural waste in a sustainable manner. Agricultural waste exhibits tremendous potential in reducing the demand for virgin raw materials and acting as a forceful force in minimizing waste formation thanks to its ability to be recycled, reused, and turned into valuable goods. The efficient use of materials is maximized while the overall resource consumption is decreased through the strategic utilization of agricultural waste, whether through recycling procedures, repurposing initiatives, or transformative conversions, paving the way for a more sustainable and circular economy. By using these plentiful agricultural residues as a biomass feedstock, their organic matter can be effectively converted through a variety of processes, like combustion, gasification, or anaerobic digestion, unlocking their latent energy content and enabling their transformation into a clean and renewable source of heat and power, reducing reliance on fossil fuels and mitigating the negative environmental impacts associated with those fuels.

12.2 BIOMASS FROM NATURAL FIBERS

Biomass derived from natural fibers encompasses a diverse range of organic materials obtained from plant-based sources. These natural fibers hold immense potential for various applications, including energy generation, manufacturing, and the production of sustainable materials. Crop residues, such as straw, husks, stalks, and stems from agricultural crops like wheat, rice, corn, and cotton, form a significant portion of biomass from natural fibers. These residues can be harnessed for energy production through processes like combustion, gasification, or anaerobic digestion. Thyavihalli et al. (2019) stated that they can serve as valuable feedstocks for the production of bio-based materials, such as biofuels, bioplastics, and biocomposites, replacing fossil fuel-based products and reducing environmental impacts.

Furthermore, forest residues, including branches, bark, and tops resulting from timber harvesting or forest restoration activities, contribute to the biomass derived from natural fibers. These forest residues can be utilized as biomass fuel for heating or electricity generation, helping to meet energy demands in a sustainable manner. Additionally, they can be processed into wood pellets or used as raw materials for the production of bio-based products, such as paper, construction materials, and textiles.

Dedicated energy crops like miscanthus, switchgrass, and hemp are also grown for their potential biomass. These energy crops can be harvested yearly or on shorter rotations and have significant biomass yields. Their fibers can be used to create biomass pellets, briquettes, or densified biomass that can be burned or gasified to produce energy that is renewable. A renewable and sustainable alternative to fossil fuels, biomass made from natural fibers lowers greenhouse gas emissions and encourages a circular economy. Utilizing these natural fibers assists rural development, agricultural sustainability, the transition to a more ecologically friendly and resource-efficient future, as well as climate change mitigation, as notified by Thyavihalli et al. (2019).

12.3 ENVIRONMENTAL IMPACT FROM BIOMASS FROM NATURAL FIBERS

The study of Thyavihalli et al. (2019) shows that the utilization of biomass derived from natural fibers can have both positive and negative environmental consequences. On the positive side, biomass from natural fibers provides significant environmental advantages compared to fossil fuels. It is a renewable energy source that reduces greenhouse gas emissions and contributes to mitigating climate change. When managed appropriately, biomass production can also support sustainable land-use practices, including crop rotation, soil improvement, and erosion prevention.

Ungureanu et al. (2022) stated that there are potential environmental drawbacks associated with biomass from natural fibers. Large-scale cultivation of energy crops for biomass production can lead to land-use changes, such as deforestation or the conversion of natural habitats. These actions can result in biodiversity loss, disruption of ecosystems, and fragmentation of habitats. It is crucial to ensure that biomass production occurs on previously degraded or underutilized lands, minimizing negative impacts on natural ecosystems. Additionally, the intensive use of water, fertilizers, and pesticides in energy crop cultivation can have implications for water resources and quality. Efficient water management practices and the adoption of sustainable agricultural approaches are vital for minimizing these impacts. Furthermore, improper handling or disposal of biomass waste can contribute to air and water pollution, causing environmental degradation.

To mitigate the potential environmental impacts of biomass from natural fibers, it is important to adopt sustainable cultivation practices, prioritize the use of marginal lands, promote biodiversity conservation, and ensure responsible waste management. Moreover, utilizing efficient biomass conversion technologies and conducting life cycle assessments can help evaluate and minimize environmental impacts throughout the entire process of biomass production and utilization as clearly stated by Thyavihalli et al. (2019).

12.4 AVAILABILITY OF AGRO-WASTE

According to the study conducted by Rasul et al. (1999), the global production of sugarcane waste can differ each year due to factors like sugarcane production levels, harvesting methods, and processing technologies. Obtaining an exact figure for worldwide sugarcane waste production is challenging due to limited comprehensive data. However, based on an estimation from 2020, global sugarcane production was around 1.9 billion metric tons, and sugarcane waste, including bagasse, filter cake, and molasses, makes up a significant portion of the harvested crop.

Bagasse, which accounts for about 25–30% of the total sugarcane weight, along with filter cake and molasses generated during sugar extraction, contributes to the overall volume of sugarcane waste. It's important to note that the utilization and management of sugarcane waste can vary across regions, with some countries having established practices for converting it into valuable products such as bioenergy, animal feed, and organic fertilizers. To obtain more precise and up-to-date information on global sugarcane waste production, referring to agricultural and industry reports, statistical databases, or consulting organizations like the Food and Agriculture Organization or sugar industry associations is recommended, as stated by Rasul et al. (1999).

12.5 PROPERTIES OF AGRO-WASTE FIBERS

12.5.1 Physical Properties

Sugarcane bagasse is composed of three main components: the pith, fiber, and rind, each with varying proportions. These components differ from each other in terms of their shape and size. The pith particles in sugarcane bagasse have a spongy texture and are relatively uniform in shape. Their shape can be approximated as non-spherical (Pérez et al., 2019). This suggests that the pith particles are roughly spherical, with similar dimensions in all directions. In contrast, the fibers in sugarcane bagasse have a high length-to-width ratio. They are elongated and can be modeled as cylindrical structures. The sugarcane bagasse has a comparable lower density, which is only around 0.10 ± 0.02 g/cm^3 (Mendes et al., 2017).

The fibers are significantly longer than they are wide, giving them a slender and elongated appearance. The rind component of sugarcane bagasse consists of larger particles. These particles have a rectangular shape with a high length-to-width ratio. Unlike the spherical pith particles or the cylindrical fibers, the rind particles have a more elongated and flattened shape resembling rectangles. Interestingly, there are instances where fibers are located adjacent to the inner wall of the rind particles. This suggests that the fibers can be found in close proximity to the internal surface of the rind component. It's possible that these fibers are intertwined or embedded within the structure of the rind particles (Lee & Mariatti, 2008).

Understanding the composition and characteristics of the different elements in sugarcane bagasse is important for various applications. Having an understanding of the fibers' shape and size, along with the pith and rind, can significantly impact the processing techniques, mechanical characteristics, and potential applications of sugarcane bagasse in various industries, such as biofuel production, pulp and paper manufacturing, and composite materials (Mahmud et al., 2023).

12.5.2 Chemical Composition

Sugarcane bagasse is composed of various components, and its composition can vary. Typically, it contains approximately 40–50% cellulose and around 25–35% hemicellulose. The remaining portion consists of other substances such as lignin, wax, and ash. Cellulose is a major component of sugarcane bagasse, and it has a crystalline structure. The crystallinity of cellulose can range from 50% to 90%, depending on the source of the cellulose. This means that a varying proportion of the cellulose molecules are arranged in an ordered, crystalline manner. The crystalline structure contributes to the strength and rigidity of the material (Mahmud & Anannya, 2021).

Hemicellulose is a compound that has an amorphous structure. It is a complex carbohydrate that contains different sugar units, such as xylose, glucose, and others. Unlike cellulose, hemicellulose lacks a well-defined crystalline arrangement and instead has a more disordered, amorphous structure. Cellulose and hemicellulose are interconnected through lignin. Lignin acts as a glue or binder between the cellulose and hemicellulose components in the sugarcane bagasse. It helps enhance the rigidity and structural integrity of the material by providing a binding matrix between the fibers (Mariana et al., 2021).

The three different types of hydroxyl groups present in cellulose contribute to the formation of strong intermolecular forces and intramolecular bonding within the cellulose structure. These interactions further strengthen the overall material. Understanding the composition and properties of cellulose, hemicellulose, and lignin in sugarcane bagasse is important for various applications.

TABLE 12.1
The Chemical Composition of the Sugarcane Bagasse (Cao et al., 2006)

Name of the Content	Percentage (%)
Cellulose	26–47
Hemicellulose	19–33
Lignin	14–23
Ash	1–5

For example, knowledge of the cellulose content and structure can inform processes like biofuel production or cellulose-based material manufacturing. The presence of hemicellulose and lignin also impacts the potential uses and processing methods of sugarcane bagasse in industries such as papermaking or bio-based materials development. The chemical composition of the sugarcane bagasse has been concluded in Table 12.1.

12.5.3 Mechanical Properties

The tensile strength of sugarcane bagasse refers to its ability to resist breaking or deforming under tensile (pulling) forces. The normal tensile strength of sugarcane bagasse is relatively low, around 70.9 megapascals (Boontima et al., 2015). This means that sugarcane bagasse is prone to cracking or deforming when subjected to stress. However, the tensile strength of sugarcane bagasse can be improved through an alkali treatment. After an alkali treatment, the tensile strength of sugarcane bagasse can be enhanced by approximately 61% (Jalalah et al., 2022). This treatment modifies the structure of the bagasse, making it more resistant to breaking or deforming under tension.

In addition to tensile strength, the other types of mechanical properties of the sugarcane bagasse are important to consider. The low-impact resistance of sugarcane bagasse is due to its low toughness and poor resistance to shock loading. This means that it may not withstand sudden impacts or shocks very well and can fracture or fail under such conditions (Danso et al., 2015). Sugarcane bagasse also exhibits low compressive strength, which is the ability to resist being crushed or compacted. The compressive strength of sugarcane bagasse is typically in the range of 1–2 MPa, indicating that it can easily be crushed or compacted under compressive forces. On the contrary, sugarcane bagasse has a flexural strength of around 10–20 MPa. Flexural strength refers to its ability to withstand bending and twisting without breaking. This indicates that sugarcane bagasse can tolerate certain degrees of bending and twisting forces before it fractures.

Overall, the mechanical properties of sugarcane bagasse, including its tensile strength, impact resistance, compressive strength, and flexural strength, play a crucial role in determining its potential applications and limitations in various industries, such as construction, packaging, or composite materials.

12.5.4 Morphological Properties

Morphological properties refer to the physical structure and characteristics of a material, such as size, shape, surface texture, and pore structure. These properties play a critical role in determining the physical and mechanical properties of the material, as well as its potential applications in various industries (Thommes et al., 2015).

The average diameter of a sugarcane bagasse is around 0.49 mm, and the average length is around 9.13 mm. Figures 12.2 and 12.3 depict the longitudinal and cross-sectional views, respectively, of sugarcane bagasse fibers as captured by scanning electron microscopy. The fiber found in sugarcane bagasse exhibits several distinct physical characteristics. First, it is multicellular and

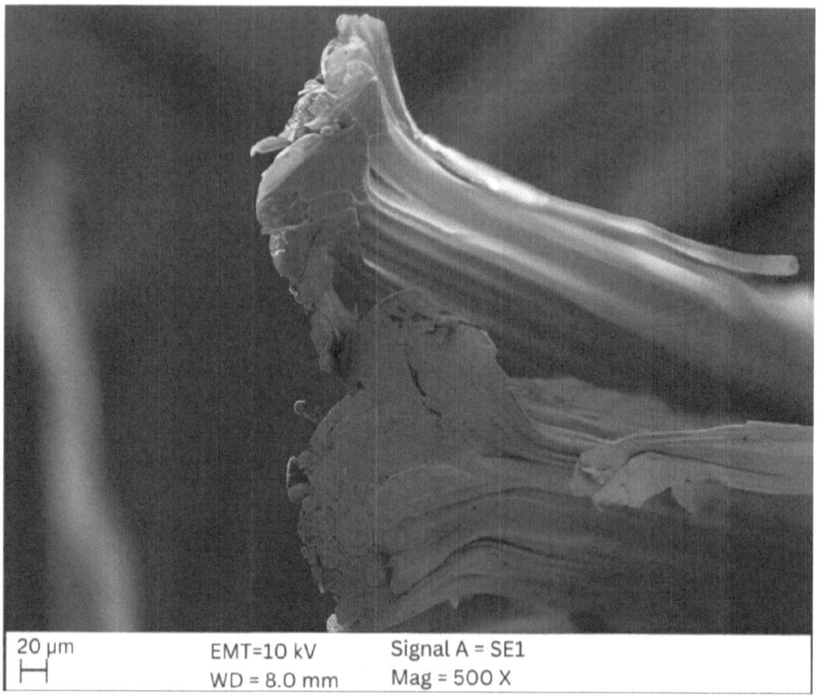

FIGURE 12.2 Longitudinal section of sugarcane bagasse fiber (Mehrzad et al., 2022).

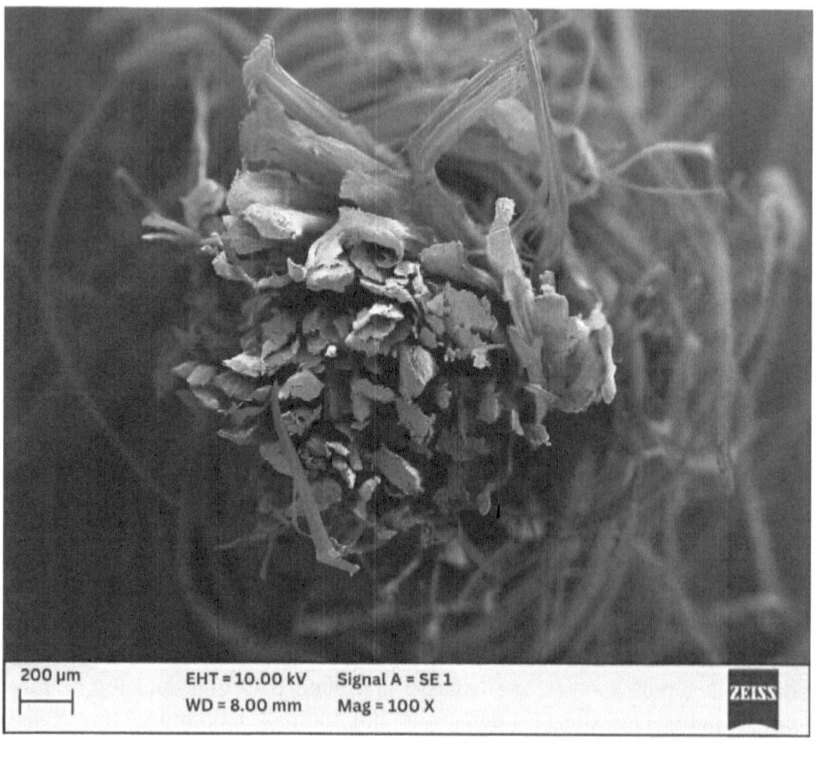

FIGURE 12.3 Cross-section of sugarcane bagasse fiber (Mehrzad et al., 2022).

lacks a standard form, displaying an irregular cross-sectional shape. The surface of the fiber appears flagellated, indicating a rough or irregular texture. Longitudinal striations are visible, suggesting a non-smooth surface (Mehrzad et al., 2022).

Notably, alkali treatment has a significant effect on the fibers. It effectively separates them from fibrous bundles and removes binding gummy substances and lignin. As a result, the fibers become well-separated and free of these substances. The fiber itself is solid, lacking any pores or hollow void spaces in its cross-section. This solid nature indicates a dense structure without empty spaces or air gaps (Mehrzad et al., 2022). Understanding these physical characteristics of sugarcane bagasse fibers is crucial for various applications. For instance, in the pulp and paper industry, knowledge of fiber properties is essential for assessing processability, strength, and overall performance (Mehrzad et al., 2022).

12.5.5 Thermal Properties

Thermal properties refer to the characteristics of a material or object in response to heat or temperature changes. These properties play a significant role in understanding how materials behave when it comes to heat transfer, storage, and release. This knowledge is crucial for evaluating the performance and suitability of materials for various applications (Morais et al., 2022). In the case of sugarcane bagasse, understanding its thermal properties is important for assessing its potential uses and applications. These properties can provide insights into how sugarcane bagasse interacts with heat and how efficiently it can be used as a thermal resource. The thermal properties of sugarcane bagasse are concluded in Table 12.2.

12.5.6 Caloric Calorific Value

The calorific value, commonly referred to as the heating value, gauges the energy content present in a fuel or combustible substance. It assesses the quantity of heat energy liberated when a particular amount of the material undergoes complete combustion. Such evaluations are usually carried out under controlled conditions, either using a bomb calorimeter to maintain constant volume or a flow calorimeter to maintain constant pressure (El-Ramady et al., 2022). The calorific value is expressed in units of energy per unit mass of the material being burned. In the case you mentioned, the calorific value of sugarcane bagasse is approximately 3,813.83 calories per gram (cal/g). This means that when 1 g of sugarcane bagasse is burned completely, it releases approximately 3,813.83 calories of heat energy (El-Ramady et al., 2022).

The high caloric value of sugarcane bagasse can be attributed to its high carbon content. Carbon-rich materials tend to have higher calorific values because carbon has a high energy content and releases significant heat energy when it undergoes combustion. The sugarcane waste, which is also a type of fibrous residue that is left after sugarcane stalks, is crushed to extract juice for sugar

TABLE 12.2
Thermal Properties of Sugarcane Bagasse Fiber

Thermal Properties	Description	References
Thermal conductivity	The thermal conductivity of the sugarcane bagasse is around 0.034 and 0.042 W/mK, making it more effective as a thermal insulator.	Morais et al. (2022)
Ignition temperature	250–263°C	Zanatta et al. (2016)
Pyrolysis	Occurs at 270–480°C with an average efficiency of 85.23%	Sosa-Arnao and Nebra (2009)
Specific heat capacity	Between 1,600 and 2,200 kcal/kg	Getu et al. (2021)

production and contains a considerable amount of carbon, making it a valuable source of energy. It can be used as a biofuel in various applications, such as electricity generation, or as a feedstock for producing heat or bio-based products (Ajala et al., 2021).

12.6 AGRO-WASTE MANAGEMENT

The process of managing waste produced by agricultural activities in an efficient and sustainable manner, known as agro-waste management, encompasses a broad range of strategies and techniques aimed at harnessing the economic potential inherent in agricultural waste. These strategies include but are not limited to composting, recycling, incineration, and the production of biogas (Gupta et al., 2022). In the upcoming subtopic, each of these approaches will be comprehensively elucidated, shedding light on their mechanisms and benefits. In 2022, a comprehensive study conducted by El-Ramady et al. (2022) sheds light on the pivotal and profound role that agro-waste management plays in the attainment of the Sustainable Development Goals (SDGs)—an interconnected framework established by the United Nations to address pressing global challenges. The study delves into the multifaceted benefits and potential positive impacts that can be harnessed through the adoption and effective implementation of agro-waste management practices, which have the capacity to significantly contribute to the realization of the SDGs encompassing a wide spectrum of environmental, social, and economic objectives (El-Ramady et al., 2022). By strategically embracing and integrating innovative agro-waste management approaches, humanity can proactively navigate toward a more sustainable future, where a harmonious balance between human activities and nature can be achieved. The integration of these practices can lead to profound transformations in various spheres, ranging from environmental conservation and ecological restoration to social welfare improvement and economic growth (Srivastav & Kumar, 2021).

Notably, the shift toward sustainable waste management alternatives offers a myriad of ecological advantages, bolstering biodiversity conservation, minimizing habitat destruction, and mitigating climate change impacts. Through the responsible disposal and recycling of agro-waste, significant reductions in greenhouse gas emissions can be achieved, paving the way for a greener and more resilient planet (Srivastav & Kumar, 2021). Furthermore, the economic viability of these agro-waste management strategies cannot be underestimated. By adopting cost-effective practices, societies can witness enhanced resource efficiency and reduced financial burdens, ultimately leading to improved livelihoods for communities reliant on agricultural activities (El-Ramady et al., 2022). This integration of sustainable waste management techniques can also foster green innovation and promote the emergence of a circular economy, where resources are conserved, reused, and reimagined for greater long-term prosperity. In comparison to conventional waste treatment methods, which often carry detrimental environmental implications and substantial economic costs, the incorporation of agro-waste management aligns with the principles of environmental friendliness and fosters a circular economic model, underlining its unparalleled potential to realize the SDGs in a holistic and inclusive manner (El-Ramady et al., 2022).

In conclusion, El-Ramady et al.'s study, conducted in 2022, underscores the urgency and importance of recognizing agro-waste management as a critical catalyst in steering our global community toward achieving the SDGs. By conscientiously embracing these sustainable practices and working collaboratively on a global scale, societies can collectively navigate the challenges that lie ahead, forging a path toward a more sustainable, equitable, and prosperous future for generations to come.

12.6.1 Compost

Composting agro-waste is a sustainable and economical process to manage organic waste, lower the emission of greenhouse gases, and encourage sustainable agriculture. As for the sugarcane waste, it can be effectively composted to provide a nutrient-rich fertilizer that can be applied to crops or gardens. The fibrous by-product left after the juice has been extracted from sugarcane stalks is known as bagasse or sugarcane waste (Ansari et al., 2021). Based on the study by Salman et al. (2023), the addition of sugarcane mud to compost can improve crop performance. This is due to the increase

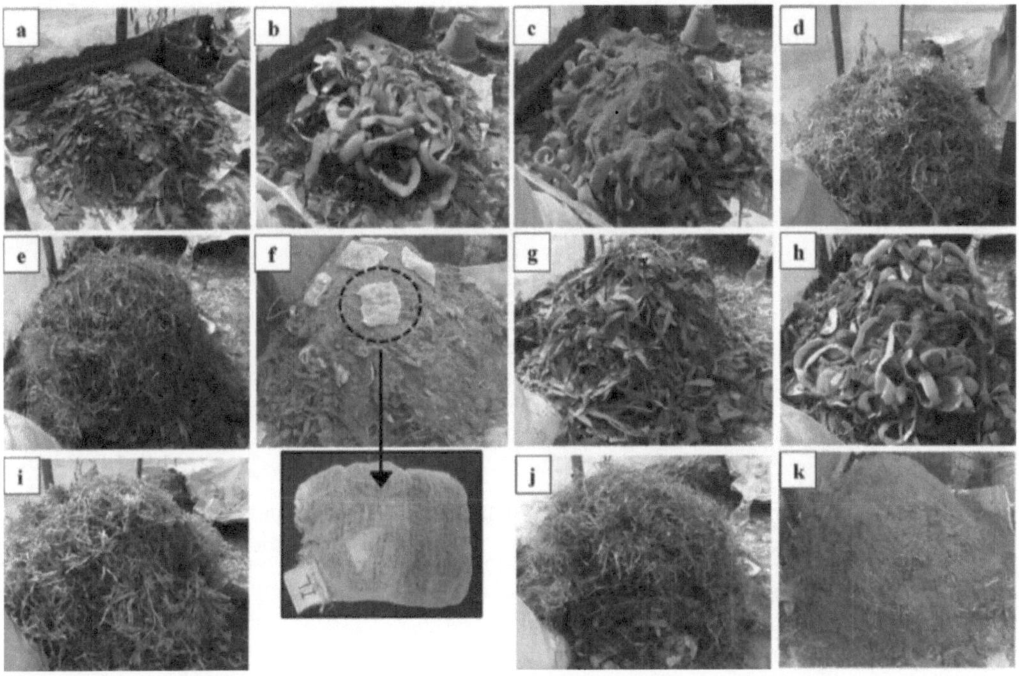

FIGURE 12.4 Formation of compost pile. Setting up of compost pile (from a–k) with different feedstocks while (f) shows the strategy used to set sugarcane bagasse (SCB) in place, in center of compost pile for subsequent analysis after composting (Ansari et al., 2021).

of nitrogen availability provided to the soil when the composts are mixed with sugarcane filter mud (Salman et al., 2023).

The method being used for composting the sugarcane waste is by using one of the common methods for composting systems—the Windrows composting system. This system was explained in the study by Ansari et al. (2021), as shown in Figure 12.4. The compost pile is set up as shown from **a to k** using different feedstocks, which are green leaves, soil, shredded paper, grass clippings, soil, and untreated sugarcane bagasse. Then, at **f** the sugarcane bagasse is added to the center of the compost pile. The layering process was made until a height of two meters was achieved. The compost can be applied as a nutrient-rich soil amendment in gardens, farms, and cultivating crops to enhance the health and fertility of the soil (Ansari et al., 2021).

12.6.2 Pulp and Paper

The management of sugarcane waste in the pulp and paper industry started by collecting the waste and transferring it to the pulp and paper mill. As sugarcane waste is solid waste; the waste is being sorted to ensure there are no impurities that could affect the next process. A study made by Rainey and Covey (2016) stated that the most crucial part of pulp and paper production using sugarcane bagasse is the pulping process (Rainey & Covey, 2016). The sorted and processed waste from sugarcane is next pulped, which breaks it down into fibers that can be used to produce paper. This can be accomplished by a variety of methods including mechanical pulping, chemical pulping, or a combination of the two. The bleaching process is applied to the pulp to gain the desired quality and color of the final product. The sequence of the bleaching process is given in Table 12.3. By undergoing this process, the impurities in the pulp can be removed, and the desired brightness and whiteness of the paper are achieved. The process is then continued with the papermaking process. Rainey and Covey (2016) state that there are two main types of paper formers converting pulp into paper goods

TABLE 12.3
Bleaching Sequence (Rainey & Covey, 2016)

Chemical	Description
Chlorination	Reaction with chlorine in an acidic medium
Alkaline extraction	Dissolution of reaction products with NaOH
Hypochlorite	Reaction with hypochlorite in an alkaline solution
Chlorine dioxide	Reaction with ClO_2 in an acidic medium
Peroxide	Reaction with peroxides in an alkaline medium
Oxygen	Reaction with elemental oxygen at pressure in an alkaline medium
Chelating agent	Function to remove transition metals, which would otherwise consume peroxide in the derived pulp
Ozone	Reaction with ozone

(Rainey & Covey, 2016): the conventional Fourdrinier formers, which have decreased as a percentage of the industry overall, and twinwire formers. The final product paper is produced by pressing and finishing the sheets during this process.

12.6.3 Biocomposite Materials

According to Manu et al. (2022), biocomposite materials are composite materials made by employing a natural resource, which is usually a biological reinforcement, and a polymer from a natural or synthetic environment to generate the matrix. Due to their inexpensive cost, natural fibers were investigated as candidates for reinforcing synthetic resin matrices (Manu et al., 2022). Currently, the use of sugarcane bagasse is being used widely in plastic material production due to its characteristics where it has a suitable value for natural fibers. Compared to the previous plastic material such as petroleum-based polymers, it is non-biodegradable, which makes it not sustainable for the environment. Hence, many researchers started to use biodegradable plastic using natural waste to overcome these problems.

The properties of the sugarcane bagasse consist of 45% of cellulose. Moreover, its bundled fiber structure contributes to the production of lightweight components, making it a viable alternative to petroleum-based polymers.(Mahmud et al., 2023). Based on the study made by Mahmud et al (2023), the composite material between starch and sugarcane cellulose can be used to produce biodegradable plastics (Mahmud et al., 2023). Table 12.4 compares the composite between different types of packaging materials. Sugarcane is used in this innovation by providing low-density composite material.

TABLE 12.4
Comparison of Composites between Different Types of Packaging Materials (Mahmud et al., 2023; Doroudiani & Kortschot, 2003; Arun & Hemalatha, 2017)

Properties Material	Density (g/cm^3)	Flexural Strength (MPa)	Impact Resistance (J/m)
Modified starch composite	0.61	13–22	10–14
Polystyrene foam	0.2–0.4	0.1–0.5	5–22
Cardboard	1.65	—	—

12.6.4 Bio-Coal

A solid fuel created through the torrefaction process, which involves heating biomass, including sugarcane waste, in the absence of oxygen, is known as bio-coal. However, the creation of bio-coal is not typically connected with sugarcane waste. Sugarcane waste contains fibrous residue left over after the juice is extracted. Hence, it can be utilized as a feedstock for torrefaction to create bio-coal (Wang et al., 2014).

The carbonizing temperature for sugarcane bagasse is 250–400°C to produce biochar. The yield of 65% biochar is then ground to produce a particle size of 300 μm. Then, the briquetting process was done at 2–8 MPa to produce bio-coal at 5 and 5 cm sizes (Wang et al., 2014). The use of bio-coal reduces the emission of carbon dioxide and carbon monoxide emission to 0.9% and 0.02%, respectively (Manyuchi et al., 2019). The process flow diagram is shown in Figure 12.5.

12.6.5 Biochar

It is stated that at least half of the bagasse produced can be used to produce heat and electricity for the sugar processing industry (Adeniyi et al., 2021). The management of sugarcane bagasse in the biochar industry is started by preparing the feedstock. The sugarcane waste is collected from the sources and washed using distilled water to remove the impurities, which are dust and dirt that attach to it. After that, the drying process of the wet sugarcane bagasse will be done for at least 2 days under the sun to remove the moisture of the combustion fuel.

The sugarcane bagasse conversion was then conducted in the reactor with a height of 48.5 cm. The temperature profile for the peak temperature and biochar yield being collected for this experiment study were calculated by Adeniyi et al. (2021). The result of the reactor performance is given in Table 12.5. It states that the biochar yield for sugarcane bagasse conversion is lower than the hybrid conversion due to the availability of plastic in the hybrid conversion. Compared to other types of feedstocks such as elephant grass, plantain fibers, oil palm fibers, orange peels, and orange albedo, sugarcane bagasse gives the highest specific surface area (Adeniyi et al., 2021).

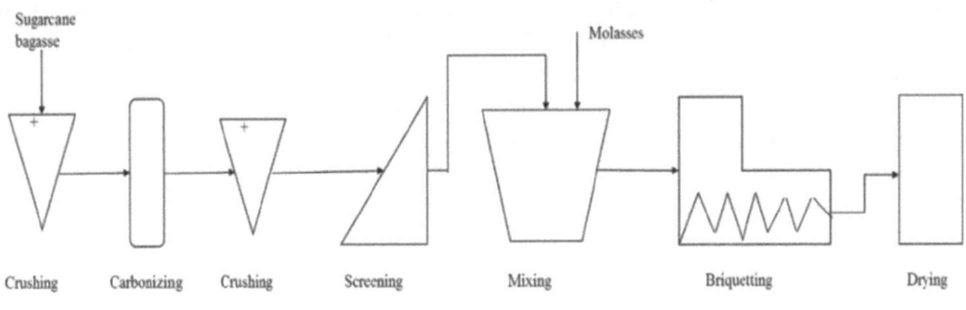

FIGURE 12.5 Process flow diagram for bio-coal production from bagasse (Wang et al., 2014).

TABLE 12.5
Reactor Performance (Pallavi et al., 2013)

Index	Sugarcane Bagasse Conversion	Hybrid Conversion
Process time	70 min	90 min
Peak temperature	349°C	250°C
Biochar yield	16.67 wt%	45.46%

Toda et al. (2005) conducted research on biodiesel synthesis, focusing on the utilization of sugar as a heterogeneous catalyst. The sugar catalyst underwent treatment through sulfonation of incompletely carbonized sucrose and D-glucose. Subsequently, numerous studies have been conducted, exploring the creation of heterogeneous solid catalysts from various precursors for the production of biodiesel through fatty acid esterification (Hidayat et al., 2021; Toda et al., 2005). Hence, the use of sugarcane bagasse can be used as a solid catalyst.

12.6.6 Bio-Bricks

The use of bagasse fibers derived from sugarcane waste as reinforcement in the fabrication of concrete composites has gained attention due to their potential to enhance mechanical properties and promote sustainability. This is due to the benefits of the properties of bagasse fiber whereby its origin is agricultural waste and it is of low cost; is biodegradable; has comparatively high tensile strength; high impact strength, and low weight; and it is easily available (Adeniyi et al., 2021). The process of development of sugarcane bagasse as bio-bricks is started by collecting the sugarcane waste from the sugar mill. Then, the impurities from the waste are removed, and the sugarcane bagasse are dried. The dry sugarcane bagasse is then chopped into fine pieces. To design strong bio-bricks made of sugarcane bagasse, lime, stone dust, and water should be added.

The chopped sugarcane bagasse, lime, water, and stone dust were mixed properly. The addition of the lime slurry needs to be at the calibrated ratio between other components as it is important to achieve the right composition for the bio-bricks. After that, the mixture is compacted into the cast properly. The bio-bricks that are already cast will be left for a month to be allowed to dry in the air. The study stated that, by using these bio-bricks, greenhouse gas emission is reduced, which is beneficial for the environment (Anitha & Senthilselvan, 2022).

In the realm of construction, the predominant material of choice has been fired clay bricks, which unfortunately contribute to a substantial release of greenhouse gases during their production (Koul et al., 2022; Kulkarni & Rao, 2016). However, a promising eco-friendly alternative known as bio-brick has emerged, derived from agro-waste materials, which offers numerous advantageous features such as carbon negativity, sustainability, cost-effectiveness, and profitability. This innovative bio-brick displays a significantly lower thermal conductivity, approximately 0.27 W/mk, making it an excellent option for providing both sound and heat insulation within walls (Koul et al., 2022; Rautray et al., 2021).

The development of bio-bricks has unveiled an impressive capability, wherein a single block can sequester an impressive 322.2 g of CO_2, further contributing to its environmental benefits (Koul et al., 2022). Despite the multitude of advantages that bio-bricks present, they currently lack the capacity to be directly utilized in the construction of load-bearing buildings. Nevertheless, they prove to be ideal for application in low-cost housing when combined with wood or metal structural frameworks, offering a viable and eco-conscious solution (Koul et al., 2022). As researchers continue to explore and enhance the load-bearing capacity of bio-bricks, there remains great potential for these innovative construction materials to play an increasingly significant role in environmentally conscious and sustainable building practices.

12.6.7 Biohydrogen

By using a sustainable and creative process called dark fermentation, sugarcane bagasse may be converted into biohydrogen, unlocking the resource potential to produce green energy. Anaerobic fermentation known as dark fermentation is crucial to this procedure. Collecting and preparing the bagasse is the first stage in creating biohydrogen from sugarcane waste. To make it less resistant to enzymatic hydrolysis, the fibrous by-product of sugar extraction, known as sugarcane bagasse, is pre-treated. The complicated structure of bagasse is broken down during this pre-treatment process by physical, chemical, or biological techniques, which makes the material less unpleasant for fermentation (Rautray et al., 2019).

The bagasse is first pre-treated before undergoing enzymatic hydrolysis. To convert the cellulose and hemicellulose found in bagasse into fermentable carbohydrates like glucose and xylose, enzymes are used (Bundhoo, 2019). By transforming the bagasse's carbohydrate content into soluble sugars, the hydrolysis procedure tries to unlock the potential energy held within it. Following hydrolysis, dark fermentation occurs on the bagasse. Specialized microorganisms, usually bacteria or archaea, are added to the hydrolysate during dark fermentation (Bundhoo, 2019). These bacteria have the capacity to ferment the hydrolysate's sugars, resulting in hydrogen gas as the main by-product. Lack of light and exogenous electron acceptors like oxygen or nitrate are two characteristics of dark fermentation (Bundhoo, 2019).

Biohydrogen gas was separated using gas stripping, membrane separation, or pressure swing adsorption. The recovered biohydrogen gas must be purified to ensure its quality and get rid of any last-minute contaminants, including carbon dioxide or sulfur compounds. For this, purification procedures like pressure swing adsorption or membrane-based methods are frequently used (Rautray et al., 2019). Figure 12.6 shows the schematic diagram of biohydrogen production from cellulose. Clean energy can be generated from the biohydrogen created by the dark fermentation of sugarcane bagasse. It can be processed further for use in fuel cells, where it can produce electricity with water as the only by-product, or it can be used directly to generate heat and power. In terms of lowering greenhouse gas emissions and dependence on fossil fuels, this eco-friendly energy source has promise (Zhu & Lin, 2016).

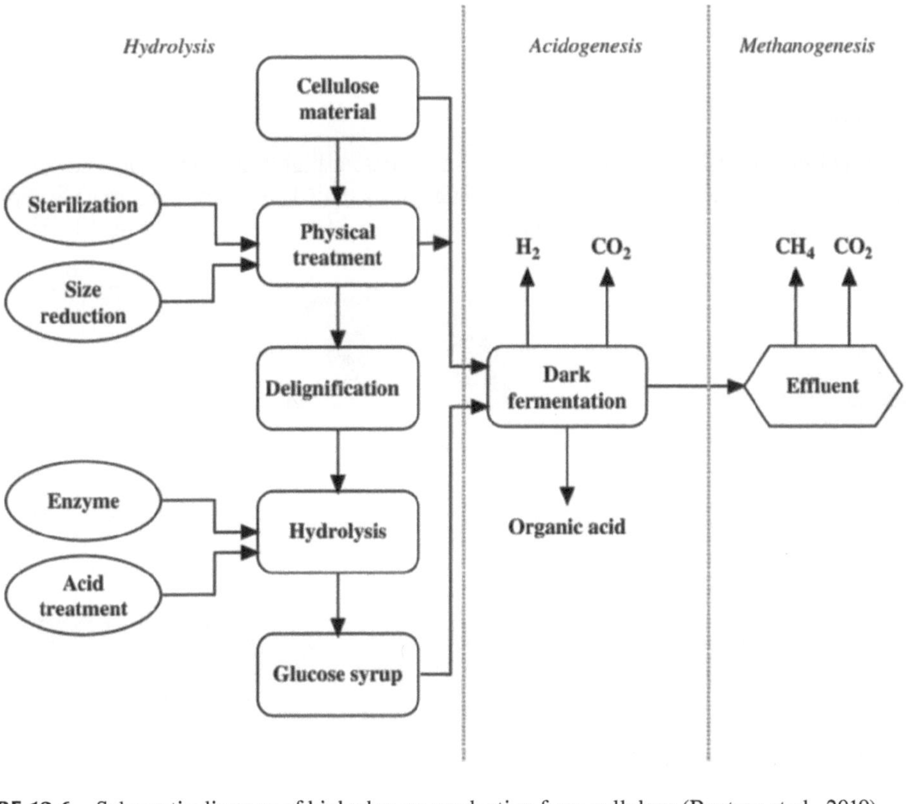

FIGURE 12.6 Schematic diagram of biohydrogen production from cellulose (Rautray et al., 2019).

12.6.8 Biomethane

Anaerobic digestion of sugarcane residues, such as bagasse or vinasse, can be used to produce biomethane (Ghaleb et al., 2021). A renewable natural gas that can be used as a sustainable energy source is biomethane. The following text explains the steps that are commonly included in producing biomethane from sugarcane waste. The first step is the feedstock preparation. Waste from sugarcane is gathered and prepared for anaerobic digestion, such as bagasse or vinasse, a liquid by-product of ethanol manufacturing (Ghaleb et al., 2021). To increase the surface area and improve the digestion of the biomass, the waste may be chopped or shredded. Then, anaerobic digestion continues. The processed sugarcane waste is introduced into an anaerobic digester, a sealed container, or a system where microorganisms decompose organic material in the absence of oxygen. Methane (CH_4) and carbon dioxide (CO_2), together with traces of other gasses, are the two main components of the biogas produced during anaerobic digestion of the complex organic molecules in sugarcane waste (Chong et al., 2009).

After biomethane is produced, it must be filtered to get rid of contaminants such as moisture, carbon dioxide, and hydrogen sulfide. This purification procedure, also known as biogas upgrading, can be carried out utilizing a variety of methods, such as membrane separation, pressure swing adsorption, or water scrubbing (Nadaleti & Lourenco, 2021). The objective is to increase the biomethane's methane content to the required levels for usage as a fuel for vehicles or injection into natural gas pipelines (Chong et al., 2009). Figure 12.7 shows the overview of the production process of biomethane and its application in the industry. There are many uses for the biomethane produced from sugarcane bagasse. It can be injected into the natural gas network and used for cooking, heating, or producing electricity. As an alternative, it can be compressed and utilized as a sustainable transportation fuel to power compressed natural gas-powered cars. Additionally, biomethane can be used in combined heat and power systems, which concurrently produce heat and electricity (Yaqoob et al., 2021).

12.6.9 Bioethanol

The bioethanol derived from sugarcane bagasse represents a highly advantageous substitute fuel for modern high-performance internal combustion engines. This is primarily due to its exceptional cellulose content and the abundant availability of bagasse, which is a vital and readily accessible

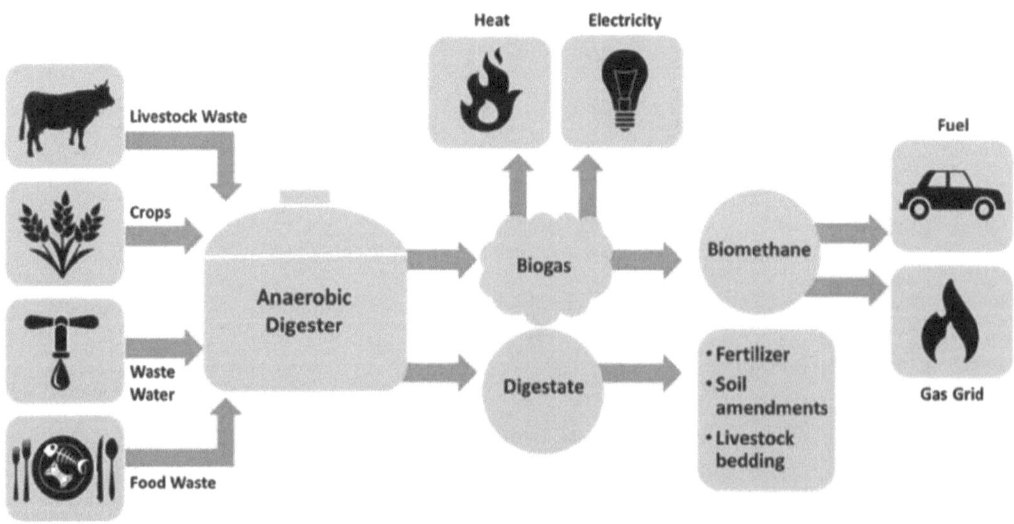

FIGURE 12.7 Process of the production of biomethane and its application (Tanigawa, 2017).

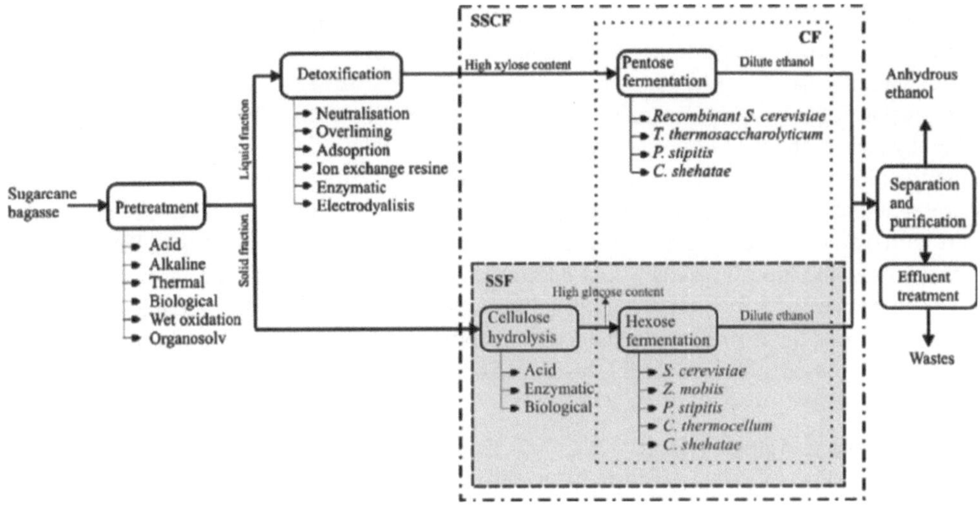

FIGURE 12.8 Process scheme of fuel ethanol production from sugarcane bagasse (Koul et al., 2022).

resource from the sugar industry. As a result, bagasse has emerged as a fundamental and essential feedstock for the efficient production of bioethanol (Akinkuade et al., 2022). The definition of bioethanol is created by turning any starchy substance sugar content into alcohol. Then, under carefully controlled conditions, carbon dioxide is released. The production of bioethanol starts with a pre-treatment process, where the sugarcane waste goes through acid hydrolysis to remove the lignocellulosic constituents to identify the simple sugar needed for yeast activity. The further process is called anaerobic fermentation according to the Emden–Meyerhoff pathway in which the enzymes will act as the catalyst. It is mentioned that sugarcane bagasse has the potential to produce 23.2–59.1 L/kg of bioethanol (Agarwal et al., 2022). Figure 12.8 shows the process scheme of fuel ethanol production from sugarcane bagasse.

Bioethanol, an eco-friendly energy source well-suited for ground transportation, is anticipated to reach a global ethanol production of 55% by 2026, with corn accounting for 65% and sugarcane contributing to 35% of the feedstocks (Song et al., 2020). The choice of feedstocks for bioethanol production varies based on regional factors, climate conditions, national priorities, and biomass availability in each country (Soleymani et al., 2021). Possible feedstock options include corn, sorghum, straw, rice, corn stover, maize, sugar cane, sugarcane bagasse, cassava, and jatropha. Among these, sugarcane and sugarcane bagasse are categorized as first- and second-generation biofuels, respectively (Santoyo-Castelazo et al., 2023).

12.6.10 Biobutanol

The management of the sugarcane waste into biobutanol starts with the pre-treatment process. During this process, the impurities that are attached to the sugarcane will be removed. There are three major types of pre-treatment processes that can be applied to sugarcane waste: chemical treatment, physical treatment, and biological treatment (Cardona et al., 2010). The aim of having pre-treatment in the biobutanol process is to release the fermentable sugar to biocatalysts. This is because biobutanol is made up of lignocellulosic feedstock, which can be gained from sugarcane bagasse (Cardona et al., 2010).

Then, once the initial stage of biobutanol production is completed, the fermentation process ensues, wherein careful attention is paid to optimizing various factors such as temperature, pH, and several other essential conditions to achieve the highest possible yield of biobutanol. This meticulous optimization enables the process to naturally generate by-products such as acetone and ethanol

(Shangdiar et al., 2022). Consequently, to enhance the purity and quality of the primary products, it becomes imperative to undertake an additional separation and purification process subsequent to the fermentation stage.

The purpose of this supplementary procedure is to eliminate any impurities or contaminants and thereby elevate the overall purity levels of the desired end products. Moreover, it is worth noting that the aforementioned by-products obtained during this process are not left to waste; rather, they are judiciously repurposed and utilized in various other production processes, ensuring maximum resource efficiency and minimizing any potential waste (Jannah et al., 2019). The application of biobutanol in the industry is as a fuel additive. This biobutanol produced from the sugarcane bagasse can be used as a fuel additive in gasoline, similar to bioethanol. It can enhance the octane rating and improve the overall performance of gasoline, making it a suitable alternative to traditional fossil fuel-based additives (Maheshwari & Shrivastava, 2023).

12.6.11 Natural Acids

The extraction of natural acids from sugarcane bagasse involves a number of procedures that take advantage of the organic elements that are already present in this agricultural by-product. As shown in Figure 12.9, one of the pre-treatment processes for the sugarcane bagasse is using citric acid solutions (Gomes et al., 2022). A feasible feedstock for the manufacture of several natural acids is sugarcane bagasse, the fibrous by-product of sugar extraction. Collecting and preprocessing the bagasse is the first stage in turning sugarcane waste into natural acids. To simplify the bagasse's complicated structure and aid future extraction procedures, it is often subjected to physical or chemical treatments.

The needed natural acids can be obtained from bagasse once it has been processed using a variety of extraction techniques. For instance, solvent extraction entails dissolving the target acids from the bagasse using a suitable solvent in a targeted manner. The bagasse's organic constituents can also be transformed into natural acids using other procedures including hydrolysis, fermentation, or enzymatic ones (Narayanasamy et al., 2020). Following the extraction or conversion procedure,

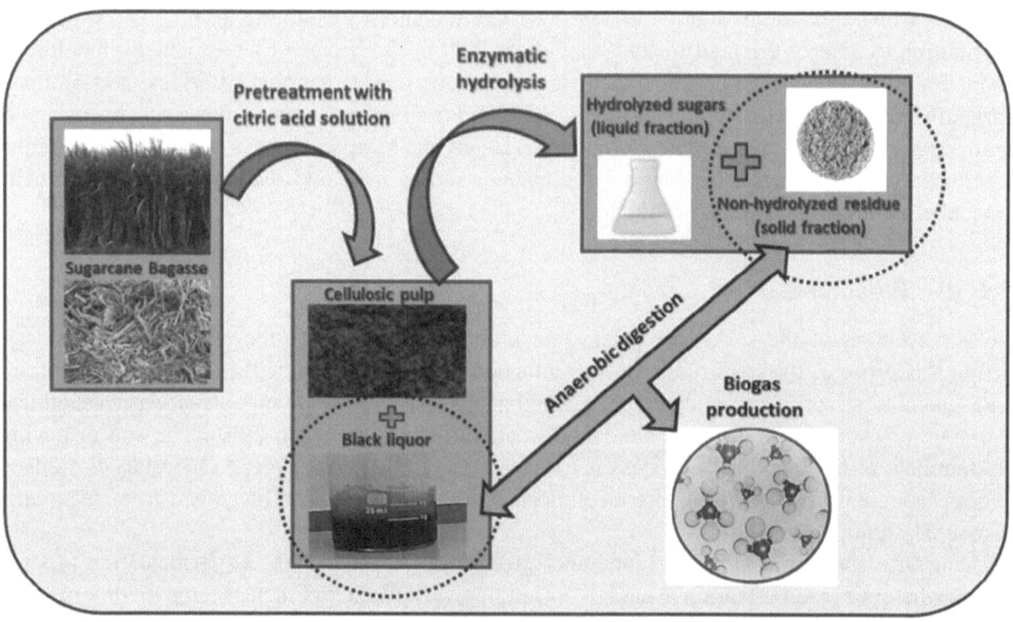

FIGURE 12.9 Pre-treatment of sugarcane bagasse (Gomes et al., 2022).

the resulting mixture or solution containing natural acids goes through a purification phase. This purification is necessary to get rid of contaminants and concentrate the target acids to a desired purity level. To achieve the desired purification, methods like chromatography, distillation, or filtration can be used.

The exact composition of sugarcane bagasse and the chosen extraction or conversion techniques determine the kinds of natural acids that can be made from it. Acetic acid, lactic acid, citric acid, succinic acid, and malic acid are a few examples of common natural acids that can be obtained. These organic acids have a wide range of uses in the food and beverage, pharmaceutical, cosmetic, and agricultural industries (Shabbirahmed et al., 2022). There are various benefits to using sugarcane bagasse to produce natural acids. As a result, there is less reliance on non-renewable resources because it uses a widely accessible and renewable feedstock. Additionally, it offers a synthetic acid substitute that is less harmful to the environment because natural acids are made from biomass and have less of an effect on the environment (Narayanasamy et al., 2020). Furthermore, by transforming a waste product into a useful resource, the valorization of sugarcane bagasse through the production of natural acid supports the circular economy.

12.6.12 Bioelectricity

Bagasse, the fibrous residue that remains after the sugarcane processing activities, possesses remarkable potential as a valuable resource for sustainable energy production. When efficiently harnessed and put to productive use, this by-product of sugarcane processing exhibits the capability to generate an impressive proportion of approximately 3.6% of electricity, contributing significantly to the overall energy mix and fostering greener and more eco-conscious power generation practices (Savla et al., 2021).

While endeavoring to produce electricity from diverse biomass materials through the use of microbial fuel cells (MFCs), researchers discovered that lignocellulosic biomass (LCB) cannot be directly utilized in its raw state by various microbial consortia to generate electricity, as illustrated in Figure 12.10. Consequently, to overcome this limitation, different pre-treatment approaches encompassing physical, physico-chemical, and microbial methods are employed. These pre-treatment techniques aim to convert the LCB into fermentable sugars, thereby making it more suitable and

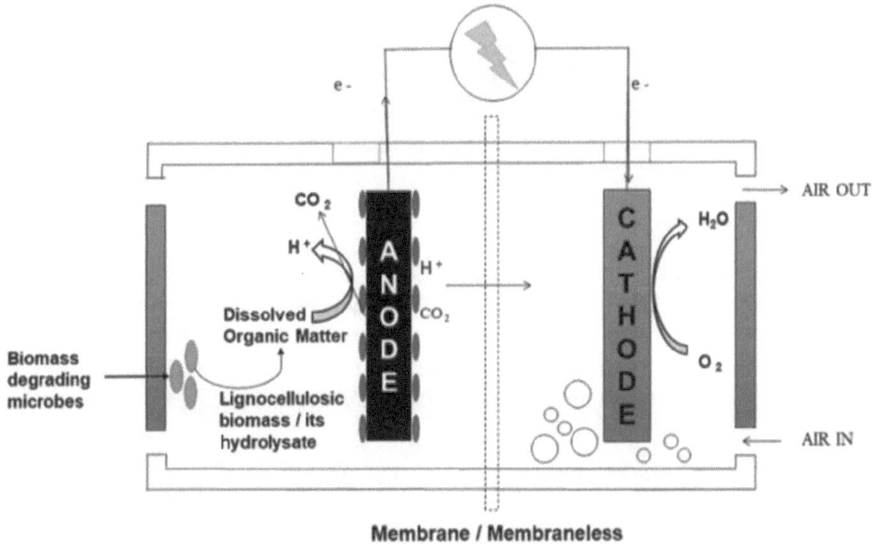

FIGURE 12.10 Biological treatment to produce lignocellulosic biomass from sugarcane bagasse (Savla et al., 2021).

conducive for electricity production within the MFC system (ElMekawy et al., 2013). The production of electricity and heat generation from sugarcane bagasse is reliant on the lower heating value of bagasse, which is estimated to be approximately 15.4 MJ/kg of dry matter, indicating the energy potential of this biomass residue (Kumar et al., 2021). By utilizing sugarcane bagasse as a renewable fuel source, the emission of greenhouse gases such as carbon dioxide (CO_2) and carbon monoxide (CO) is directly related to the carbon content released during the combustion process, as bagasse primarily consists of carbon-rich organic matter (Carvalho et al., 2019). Consequently, the use of bagasse as a feedstock for electricity and heat generation not only harnesses its significant energy value but also offers the potential to minimize the environmental impact by mitigating CO_2 and CO emissions traditionally associated with fossil fuel-based energy sources.

12.7 POTENTIAL APPLICATIONS

Bagasse, the fibrous residue remaining after the extraction of sugar juice from sugarcane, accounts for approximately 30% to 40% of the sugarcane's weight. It is commonly utilized in the production of various commercial goods for sale, including biodegradable packaging, paper, and composite materials.. Around 280 kg of bagasse is produced from each ton of sugarcane. The degree of crushing and milling operations carried out on the sugarcane at the sugar mill, as well as harvest time, agronomic practices, soil quality, and other factors, are all mentioned by Gunawan et al. (2018). Bagasse, the fibrous residue left after sugarcane or agave stalks are crushed to extract their juice, has numerous potential applications across various industries. One of the key uses is as a biofuel and energy source. Bagasse can be burned in boilers to generate steam, which is then used to produce electricity and heat, reducing reliance on fossil fuels. Additionally, bagasse fibers can be processed for the production of paper and pulp products, offering a sustainable alternative to wood-based pulp. It is particularly suitable for specialty papers and eco-friendly packaging materials.

Bagasse can also be transformed into animal feed and livestock bedding, providing a renewable and cost-effective option. Moreover, bagasse can be molded into biodegradable packaging items such as plates, bowls, and food containers, contributing to the reduction of plastic waste. It can also be composted and used as a soil amendment, enhancing soil fertility and water retention. Furthermore, bagasse can be utilized in the manufacturing of construction materials, acting as a reinforcing agent in particleboards and fiberboards. It can be a feedstock for the production of biochemicals and bioplastics, contributing to the development of sustainable alternatives to conventional materials. Lastly, bagasse can be used as a substrate for mushroom cultivation, creating additional income streams for farmers and utilizing agricultural waste effectively. With ongoing research and technological advancements, the potential applications of bagasse continue to expand, providing environmentally friendly solutions across industries.

12.7.1 Bioelectricity from Bagasse and Sugarcane Tops

Arshad and Ahmed (2016) reviewed power cogeneration using bagasse. According to their research, 2.2 tons of steam are produced at a pressure of 20–22 bars and a temperature of 350°C using the thermal energy from the combustion of 1 ton of bagasse in the grate boiler. As residues can be put into the boiler, together with bagasse for the creation of steam, cane waste is typically left in the field. The backpressure steam is then operated using the generated steam.

12.7.2 Vinasse and Filter Mud-Derived Methane Biogas

The residue left over after the sugarcane juice has been clarified and filtered is known as filter mud or press mud. It is both fibrous and solid. Filter mud contains a lot of soil nutrients, making it a good fertilizer. However, the expense of transporting the filter mud mounds from the sugar mill is high. As a result, these leftovers are frequently ignored or dumped into water bodies, polluting and contaminating the water (Christofoletti et al. 2013).

12.8 CHALLENGES ASSOCIATED WITH PROCESSING AND COMMERCIALIZING

The challenge associated with processing sugarcane waste is the availability and quality of sugarcane waste. This is because the amount and quality of sugarcane waste are influenced by a number of variables, including sugar production processes, harvesting techniques, and weather conditions. Locating a regular and dependable feedstock for processing can be difficult because of the variable supply and quality of sugarcane waste, which could impact the final product quality and performance (Menyuka et al., 2020). The technology challenges represent the second set of challenges that arise in the context of converting sugarcane waste into value-added goods, such as biocomposites or biofuels. This intricate process often necessitates the utilization of specialized tools and technologies, which may still be undergoing development and refinement. Consequently, there can be notable technical obstacles associated with deploying and scaling up these emerging technologies (Ukanwa et al., 2019).

Factors such as high capital costs, the presence of complex processing steps, and the absence of standardized processing techniques can impede the smooth integration and widespread adoption of these technologies in the sugarcane waste conversion process. These challenges underscore the need for continued research, innovation, and collaborative efforts to overcome the existing barriers and drive the advancement and commercialization of efficient and cost-effective technologies in this field (Chilakamarry et al., 2022). From a sustainability perspective, the use of waste materials in industrial production is considered a sustainable practice.. However, to ensure that sugarcane waste is processed and commercialized in a way that takes into consideration factors such as land use, water use, energy consumption, and social implications can be difficult (Koul et al., 2022). Planning, monitoring, and implementation must be done with great care to achieve a balance between economic viability and environmental and social sustainability concerns.

12.9 CONCLUSIONS

In conclusion, the abundant and versatile resource of sugarcane bagasse holds great promise due to its advantageous characteristics, including high cellulose content and low ash content. These properties make it an attractive raw material for a range of applications. Biofuel production stands out as a significant opportunity, as the cellulose and hemicellulose present in bagasse can be converted into biofuels like ethanol, reducing our dependence on fossil fuels and mitigating environmental impact. Furthermore, the high cellulose content of bagasse makes it suitable for paper manufacturing, offering a sustainable alternative to traditional wood pulp and contributing to forest conservation. Additionally, the extraction and processing of cellulose fibers from bagasse can be utilized in the production of biodegradable plastics, thus addressing the issue of plastic waste. However, the commercialization of sugarcane bagasse faces challenges, including high processing costs and limited infrastructure. To overcome these obstacles, it is crucial to invest in research and development to devise cost-effective and energy-efficient processes. Innovation plays a pivotal role in exploring new applications and value chains for bagasse, fostering its utilization in various industries. Collaborative efforts between researchers, industry players, and policymakers are essential to exchange knowledge, share best practices, and drive investment in infrastructure development. By addressing these challenges through research, innovation, and collaboration, we can fully unlock the potential of sugarcane bagasse, leading to sustainable and diverse applications across industries and contributing to a more environmentally conscious future.

ACKNOWLEDGMENT

This chapter was written with Nur Fatin Nabilah Mohd Rasdan, Jonathan Wong Qin Xiang, and Lavintheran Shanmukam.

REFERENCES

Adeniyi, A. G., Abdulkareem, S. A., Ighalo, J. O., Onifade, D. V., & Sanusi, S. K. (2021). Thermochemical co-conversion of sugarcane bagasse-LDPE hybrid waste into biochar. *Arabian Journal for Science and Engineering, 46*(7), 6391–6397. https://doi.org/10.1007/s13369-020-05119-9

Agarwal, N. K., Kumar, M., Ghosh, P., Kumar, S. S., Singh, L., Vijay, V. K., & Kumar, V. (2022). Anaerobic digestion of sugarcane bagasse for biogas production and digestate valorization. *Chemosphere, 295*, 133893.

Ajala, E. O., Ighalo, J. O., Ajala, M. A., Adeniyi, A. G., & Ayanshola, A. M. (2021). Sugarcane bagasse: a biomass sufficiently applied for improving global energy, environment and economic sustainability. *Bioresources and Bioprocessing, 8*(1), 87. https://doi.org/10.1186/s40643-021-00440-z

Akinkuade, K. A., Olasesan, I. P., Fakayode, O. A., Odesanmi, A. A., & Ajibade, O. A. (2022). Thermocatalytic conversion of sugarcane bagasse (SCB) for bioethanol production: a review. *World News of Natural Sciences, 45*, 117–142.

Anitha, K., & Senthilselvan, S. (2022). Agricultural waste materials applications in building industry–an overview. *ECS Transactions, 107*(1), 2371.

Ansari, M., Zafar, U., Ejaz, U., Sohail, M., Pirzada, A., & Aman, A. (2021). Comparison of composting of chemically pretreated and fermented sugarcane bagasse for zero-waste biorefinery. *Journal of Material Cycles and Waste Management, 23*(3), 911–921. https://doi.org/10.1007/s10163-021-01176-w

Arshad, M., & Ahmed, S. (2016). Cogeneration through bagasse: a renewable strategy to meet the future energy needs. *Renewable and Sustainable Energy Reviews, 54*, 732–737. Elsevier BV. https://doi.org/10.1016/j.rser.2015.10.145

Arun Solomon, A., & Hemalatha, G. (2017). Inspection of properties of expanded polystyrene (EPS), compressive behaviour, bond and analytical examination of insulated concrete form (ICF) blocks using different densities of EPS. *International Journal of Civil Enginering and Technology, 8*(1), 209–221.

Boontima, B., Noomhorm, A., Puttanlek, C., Uttapap, D., & Rungsardthong, V. (2015). Mechanical properties of sugarcane bagasse fiber-reinforced soy based biocomposites. *Journal of Polymers and the Environment, 23*(1), 97–106. https://doi.org/10.1007/s10924-014-0679-2

Bos, A., & Hamelinck, C. (2014). Greenhouse gas impact of marginal fossil fuel use. Project number: BIENL14773 2014.

Bundhoo, Z. M. (2019). Potential of bio-hydrogen production from dark fermentation of crop residues: a review. *International Journal of Hydrogen Energy, 44*(32), 17346–17362.

Cao, Y., Shibata, S., & Fukumoto, I. (2006). Mechanical properties of biodegradable composites reinforced with bagasse fibre before and after alkali treatments. *Composites Part A: Applied Science and Manufacturing, 37*(3), 423–429. https://doi.org/10.1016/j.compositesa.2005.05.045

Cardona, C. A., Quintero, J. A., & Paz, I. C. (2010). Production of bioethanol from sugarcane bagasse: status and perspectives. *Bioresource Technology*, *101*(13), 4754–4766.

Carvalho, M., Da Silva Segundo, V. B., De Medeiros, M. G., Dos Santos, N. A., & Junior, L. M. C. (2019). Carbon footprint of the generation of bioelectricity from sugarcane bagasse in a sugar and ethanol industry. *International Journal of Global Warming, 17*(3), 235–251. https://doi.org/10.1504/IJGW.2019.098495

Chilakamarry, C. R., Sakinah, A. M., Zularisam, A. W., Sirohi, R., Khilji, I. A., Ahmad, N., & Pandey, A. (2022). Advances in solid-state fermentation for bioconversion of agricultural wastes to value-added products: opportunities and challenges. *Bioresource Technology, 343*, 126065.

Chong, M. L., Sabaratnam, V., Shirai, Y., & Hassan, M. A. (2009). Biohydrogen production from biomass and industrial wastes by dark fermentation. *International Journal of Hydrogen Energy, 34*(8), 3277–3287.

Christofoletti, C. A., Escher, J. P., Correia, J. E., Marinho, J. F. U., & Fontanetti, C. S. (2013). Sugarcane vinasse: environmental implications of its use. *Waste Management, 33*(12), 2752–2761). Elsevier BV. https://doi.org/10.1016/j.wasman.2013.09.005

Danso, H., Martinson, D., Ali, M., & Williams, J. (2015). Effect of sugarcane bagasse fiber on the strength properties of soil blocks. *Academic Journal of Civil Engineering, 33*, 251–256.

Doroudiani, S., & Kortschot, M. T. (2003). Polystyrene foams. III. Structure e tensile properties relationships. *Journal of Applied Polymer Science, 90*, 1427–34.

ElMekawy, A., Diels, L., De Wever, H., & Pant, D. (2013). Valorization of cereal based biorefinery byproducts: reality and expectations. *Environmental Science & Technology, 47*(16), 9014–9027. https://doi.org/10.1021/es402395g

El-Ramady, H., Brevik, E. C., Bayoumi, Y., Shalaby, T. A., El-Mahrouk, M. E., Taha, N., … & Ling, W. (2022). An overview of agro-waste management in light of the water-energy-waste nexus. *Sustainability, 14*(23). https://doi.org/10.3390/su142315717

Getu, M., Birhanu, A., Jida, M., Genet, T., Endale, W., & Fedi, G. B. (2021). Briquette production from sugar cane bagasse and its potential as clean source of energy. *African Journal of Environmental Science and Technology, 15*, 339–348. https://doi.org/10.5897/AJEST2021.3006

Ghaleb, A. A. S., Kutty, S. R. M., Salih, G. H. A., Jagaba, A. H., Noor, A., Kumar, V., … & Saleh Al-dhawi, B. N. (2021). Sugarcane bagasse as a co-substrate with oil-refinery biological sludge for biogas production using batch mesophilic anaerobic co-digestion technology: effect of carbon/nitrogen ratio. *Water, 13*(5), 590.

Gomes, M. G., de Oliveira Paranhos, A. G., Camargos, A. B., Baêta, B. E. L., Baffi, M. A., Gurgel, L. V. A., & Pasquini, D. (2022). Pretreatment of sugarcane bagasse with dilute citric acid and enzymatic hydrolysis: use of black liquor and solid fraction for biogas production. *Renewable Energy, 191*, 428–438.

Graminha, E. B. N., Goncalves, A. Z. L., Pirota, R. D. P. B., Balsalobre, M. A. A., Silva, R., & Gomes, E. (2008). Enzyme production by solid-state fermentation: application to animal nutrition. *Animal Feed Science and Technology, 144*, 1–22.

Gunawan, B. T., Romli, M., & Noor, E. (2018). Biomass by-product from crystal sugar production: a comparative study between Ngadirejo and Mauritius sugar mill. *IOP Conference Series: Earth and Environmental Science, 141*, 012009. IOP Publishing. https://doi.org/10.1088/1755-1315/141/1/012009

Gupta, J., Kumari, M., Mishra, A., Akram, M., & Thakur, I. S. (2022). Agro-forestry waste management-a review. *Chemosphere, 287*, 132321.

Hidayat, A., Kurniawan, W., & Hinode, H. (2021). Sugarcane bagasse biochar as a solid catalyst: from literature review of heterogeneous catalysts for esterifications to the experiments for biodiesel synthesis from palm oil industry waste residue. *Indonesian Journal of Science and Technology, 6*(2), 337–352.

Jalalah, M., Khaliq, Z., Malik, Z., Ahmed, A., Qadir, M. B., Afzal, A., … & Harraz, F. (2022). Preliminary studies on conversion of sugarcane bagasse into sustainable fibers for apparel textiles. *Sustainability, 14*, 16450. https://doi.org/10.3390/su142416450

Jannah, A. M., Amalia, D. U., & Hana, N. S. (2019, June). Biobutanol production from bagasse using ammonia pre-treatment and acid hydrolysis method. *IOP Conference Series: Materials Science and Engineering, 543*(1), 012053. IOP Publishing.

Koul, B., Yakoob, M., & Shah, M. P. (2022). Agricultural waste management strategies for environmental sustainability. *Environmental Research, 206*, 112285.

Kulkarni, N. G., & Rao, A. B. (2016). Carbon footprint of solid clay bricks fired in clamps of India. *Journal of Cleaner Production, 135*, 1396–1406. https://doi.org/10.1016/j.jclepro.2016.06.152

Kumar, A., Kumar, V., & Singh, B. (2021). Cellulosic and hemicellulosic fractions of sugarcane bagasse: potential, challenges and future perspective. *International Journal of Biological Macromolecules, 169*, 564–582

Lee, S. C., & Mariatti, M. (2008). The effect of bagasse fibers obtained (from rind and pith component) on the properties of unsaturated polyester composites. *Materials Letters, 62*(15), 2253–2256. https://doi.org/10.1016/j.matlet.2007.11.097

Maheshwari, V., & Shrivastava, S. (2023). Agricultural waste: a potential renewable energy source for biobutanol production. *Indian Journal of Agricultural Biochemistry, 36*(1), 1–9.

Mahmud, M. A., & Anannya, F. R. (2021). Sugarcane bagasse – a source of cellulosic fiber for diverse applications. *Heliyon, 7*(8), e07771. https://doi.org/10.1016/j.heliyon.2021.e07771

Mahmud, M. A., Belal, S. A., & Gafur, M. A. (2023). Development of a biocomposite material using sugarcane bagasse and modified starch for packaging purposes. *Journal of Materials Research and Technology, 24*, 1856–1874.

Manu, T., Nazmi, A. R., Shahri, B., Emerson, N., & Huber, T. (2022). Biocomposites: a review of materials and perception. *Materials Today Communications, 31*, 103308.

Manyuchi, M. M., Mbohwa, C., & Muzenda, E. (2019). Evaluating the usability of bio coal from sugar cane bagasse as a solid fuel. *Procedia Manufacturing, 33*, 516–521.

Mariana, M., Alfatah, T., Abdul Khalil, H. P. S., Yahya, E. B., Olaiya, N. G., Nuryawan, A., … & Ismail, H. (2021). A current advancement on the role of lignin as sustainable reinforcement material in biopolymeric blends. *Journal of Materials Research and Technology, 15*, 2287–2316. https://doi.org/10.1016/j.jmrt.2021.08.139

Mehrzad, S., Taban, E., Soltani, P., Samaei, S. E., & Khavanin, A. (2022). Sugarcane bagasse waste fibers as novel thermal insulation and sound-absorbing materials for application in sustainable buildings. *Building and Environment, 211*, 108753. https://doi.org/10.1016/j.buildenv.2022.108753

Mendes, R. F., Vilela, A. P., Farrapo, C. L., Mendes, J. F., Denzin Tonoli, G. H., & Mendes, L. M. (2017). 1- Lignocellulosic residues in cement-bonded panels. In: H. Savastano Junior, J. Fiorelli, & S. F. dos Santos (Eds.), *Sustainable and Nonconventional Construction Materials using Inorganic Bonded Fiber Composites* (pp. 3–16): Woodhead Publishing.

Menyuka, N. N., Sibanda, M., & Bob, U. (2020). Perceptions of the challenges and opportunities of utilising organic waste through urban agriculture in the Durban south basin. *International Journal of Environmental Research and Public Health, 17*(4). https://doi.org/10.3390/ijerph17041158

Morais, L. C. D., Maia, A. A., Resende, P. R., Rosa, A. H., & Nunes, L. J. R. (2022). Thermochemical conversion of sugarcane bagasse: A comprehensive analysis of ignition and burnout temperatures. *Clean Technologies, 4*(4), 1127–1137. Retrieved from https://www.mdpi.com/2571-8797/4/4/68

Nadaleti, W. C., & Lourenco, V. A. (2021). A mathematical, economic and energetic appraisal of biomethane and biohydrogen production from Brazilian ethanol plants' waste: towards a circular and renewable energy development. *International Journal of Hydrogen Energy, 46*(54), 27268–27281.

Narayanasamy, S., Chan, K. L., Cai, H., Abdul Razak, A. H. B., Tay, B. K., & Miao, H. (2020). Biobutanol production from sugarcane bagasse by Clostridium beijerinckii strains. *Biotechnology and Applied Biochemistry, 67*(5), 732–737. https://doi.org/10.1002/bab.1865

Okonko, I. O., Adeola, O. T., Aloysius, F. E., Damilola, A. O., & Adewale, O. A. (2009). Utilization of food wastes for sustainable development. *Electronic Journal of Environmental, Agricultural and Food Chemistry, 8*(4), 263–286.

Pallavi, H. B., Srikantaswamy, S., Kiran, M. B., Vyshnavi, D. R., & Ashwin, C. A. (2013). Briquetting agricultural waste as an energy source. *Journal of Environmental Science, Computer Science and Engineering and Technology, 2*(1), 160–172.

Pérez, N. P., Pedroso, D. T., Machin, E. B., Antunes, J. S., Tuna, C. E., & Silveira, J. L. (2019). Geometrical characteristics of sugarcane bagasse for being used as fuel in fluidized bed technologies. *Renewable Energy, 143*, 1210–1224. https://doi.org/10.1016/j.renene.2019.05.082

Rainey, T. J., & Covey, G. (2016). Pulp and paper production from sugarcane bagasse. In: O'Hara, I. M., & Mundree, S. G. (Eds.), *Sugarcane-based Biofuels and Bioproducts* (pp. 259–280). Wiley Blackwell. https://doi.org/10.1002/9781118719862.ch10

Rasul, M. G., Rudolph, V., & Carsky, M. (1999). Physical properties of bagasse. *Fuel, 78*(8), 905–910. https://doi.org/10.1016/S0016-2361(99)00011-3

Rautray, P., Roy, A., Mathew, D. J., & Eisenbart, B. (2019). Bio-brick - development of sustainable and cost effective building material. *Proceedings of the International Conference on Engineering Design, ICED, 2019-August, 1*, 3171–3180. https://doi.org/10.1017/dsi.2019.324

Rautray, P., Roy, A., Mathew, D. J., & Eisenbart, B. (2021). Bio-bricks: circular economy and new products. In: Chakrabarti, A., Poovaiah, R., Bokil, P., & Kant, V. (Eds.), *Design for Tomorrow—Volume 1: Proceedings of ICoRD 2021* (pp. 845–857). Springer Singapore.

Rodríguez-Couto, S. (2008). Exploitation of biological wastes for the production of value-added products under solid-state fermentation conditions. *Biotechnology Journal, 3*, 859–870.

Salman, M., Inamullah, Jamal, A., Mihoub, A., Saeed, M. F., Radicetti, E., Ahmad, I., Naeem, A., Ullah, J., & Pampana, S. (2023). Composting sugarcane filter mud with different sources differently benefits sweet maize. *Agronomy, 13*(3), 748. https://doi.org/10.3390/agronomy13030748

Santoyo-Castelazo, E., Santoyo, E., Zurita-García, L., Luengas, D. C., & Solano-Olivares, K. (2023). Life cycle assessment of bioethanol production from sugarcane bagasse using a gasification conversion process: bibliometric analysis, systematic literature review and a case study. *Applied Thermal Engineering, 219*, 119414.

Savla, N., Pandit, S., Mathuriya, A. S., Gupta, P. K., Khanna, N., Babu, R. P., & Kumar, S. (2021). Recent advances in bioelectricity generation through the simultaneous valorization of lignocellulosic biomass and wastewater treatment in microbial fuel cell. *Sustainable Energy Technologies and Assessments, 48*, 101572.

Shabbirahmed, A. M., Haldar, D., Dey, P., Patel, A. K., Singhania, R. R., Dong, C. D., & Purkait, M. K. (2022). Sugarcane bagasse into value-added products: a review. *Environmental Science and Pollution Research, 29*(42), 62785–62806.

Shangdiar, S., Lin, Y. C., Ponnusamy, V. K., & Wu, T. Y. (2022). Pretreatment of lignocellulosic biomass from sugar bagasse under microwave assisted dilute acid hydrolysis for biobutanol production. *Bioresource Technology, 361*, 127724.

Soleymani Angili, T., Grzesik, K., Rödl, A., & Kaltschmitt, M. (2021). Life cycle assessment of bioethanol production: a review of feedstock, technology and methodology. *Energies, 14*(10), 2939.

Song, C., Zhang, C., Zhang, S., Lin, H., Kim, Y., Ramakrishnan, M., Du, Y., Zhang, Y., Zheng, H., & Barceló, D. (2020). Thermochemical liquefaction of agricultural and forestry wastes into biofuels and chemicals from circular economy perspectives. *Science of the Total Environment, 749*, 141972.

Sosa-Arnao, J., & Nebra, S. A. (2009). Bagasse dryer role in the energy recovery of water tube boilers. *Drying Technology, 27*, 587–594. https://doi.org/10.1080/07373930802716326

Srivastav, A. L., & Kumar, A. (2021). An endeavor to achieve sustainable development goals through floral waste management: a short review. *Journal of Cleaner Production, 283*, 124669.

Tanigawa, S. (2017). Fact sheet: biogas: converting waste to energy, EESI. Available at: https://www.eesi.org/papers/view/fact-sheet-biogasconverting-waste-to-energy (Accessed: 02 July 2023).

Thommes, M., Kaneko, K., Neimark, A. V., Olivier, J. P., Rodriguez-Reinoso, F., Rouquerol, J., & Sing, K. S. W. (2015). Physisorption of gases, with special reference to the evaluation of surface area and pore size distribution (IUPAC Technical Report). *Pure and Applied Chemistry, 87*(9–10), 1051–1069. https://doi.org/10.1515/pac-2014-1117

Thyavihalli Girijappa, Y. G., Mavinkere Rangappa, S., Parameswaranpillai, J., & Siengchin, S. (2019, September 3). Natural fibers as sustainable and renewable resource for development of eco-friendly composites: a comprehensive review. *Frontiers in Materials, 6*. https://doi.org/10.3389/fmats.2019.00226

Toda, M., Takagaki, A., Okamura, M., Kondo, J. N., Hayashi, S., Domen, K., & Hara, M. (2005). Biodiesel made with sugar catalyst. *Nature, 438*(2005), 178.

Ukanwa, K. S., Patchigolla, K., Sakrabani, R., Anthony, E., & Mandavgane, S. (2019). A review of chemicals to produce activated carbon from agricultural waste biomass. *Sustainability, 11*(22), 6204.

Ungureanu, N., Vlăduţ, V., & Biriş, S.-Ş. (2022, September 19). Sustainable valorization of sugarcane waste from sugarcane processing. Encyclopedia. Available at: https://encyclopedia.pub/entry/27272#:~:text=It%20is%20also%20estimated%20that,of%20bagasse%20are%20generated%20annually (Accessed: 01 March 2025).

Wang, L., Lurina, M., Hyytiäinen, J., & Mikkonen, E. (2014). Bio-coal market study: macro and micro-environment of the bio-coal business in Finland. *Biomass and Bioenergy, 63*, 198–209.

Yamane, T. (2023, May 12). Sugarcane. Encyclopædia Britannica. Available at: https://www.britannica.com/plant/sugarcane (Accessed: 01 March 2025).

Yaqoob, H., Teoh, Y. H., Din, Z. U., Sabah, N. U., Jamil, M. A., Mujtaba, M. A., & Abid, A. (2021). The potential of sustainable biogas production from biomass waste for power generation in Pakistan. *Journal of Cleaner Production, 307*, 127250.

Zanatta, E. R., Reinehr, T. O., Awadallak, J. A., Kleinübing, S. J., dos Santos, J. B. O., Bariccatti, R. A., … & da Silva, E. A. (2016). Kinetic studies of thermal decomposition of sugarcane bagasse and cassava bagasse. *Journal of Thermal Analysis and Calorimetry, 125*(1), 437–445. https://doi.org/10.1007/s10973-016-5378-x

Zhu, M. J., & Lin, H. N. (2016). Biohydrogen production from waste biomass. *Trends in Renewable Energy, 2*(2), 54–55.

13 Corn Waste Management

13.1 INTRODUCTION

The agriculture industry currently has a considerable issue controlling organic waste created throughout various production stages. Corn waste stands out as a significant contributor among these agricultural leftovers due to its widespread cultivation and high annual production (Choi et al., 2022). Corn waste, also known as corn stover or corn residue, includes the stalks, leaves, husks, cobs, and other non-edible elements of the corn plant that remain after the kernels have been harvested. The residue left over from harvesting has enormous potential that is unexplored (Serna-Saldivar, 2019). Corn waste has traditionally been used as animal bedding or left to decay in the field. Still, recent advances have emphasized its significance as a renewable resource that may be successfully exploited to address pressing concerns (Tang et al., 2018).

Furthermore, corn waste contains a high concentration of cellulose, hemicellulose, lignin, and other organic components, making it an attractive feedstock for a variety of industrial applications (Song et al., 2020). Its properties make it suitable for conversion into value products such as biofuels, biogas, biochemicals, bioplastics, animal feed, and soil amendments. Corn waste can also be turned into sustainable alternatives to fossil fuels, chemicals, and materials using innovative processes such as biochemical conversion, thermochemical conversion, and microbial fermentation (Narisetty et al., 2022).

In addition, efficient corn waste management has numerous benefits. First, it reduces greenhouse gas (GHG) emissions related to decomposition and prevents soil erosion, which helps lessen the environmental impact of agricultural activities (Galindo et al., 2022). Second, it minimizes the demand for traditional fossil-based resources, resulting in a lower carbon impact. Furthermore, the use of corn waste creates new markets and job opportunities in the bioeconomy sector while also providing farmers with an extra income stream.

The purpose of this chapter is to study the efficient management of corn waste and its potential industrial applications. It strives to provide insights into developing sustainable waste management techniques and maximizing the value of this abundant agricultural by-product by analyzing the composition, environmental impact, and economic potential of corn waste.

Corn waste, which includes stalks, cobs, husks, and leaves, is a by-product of corn production that must be managed properly to reduce pollution and GHG emissions. Current disposal options, such as burning or landfilling, are unsustainable and have negative environmental and human health consequences. There is, however, a great possibility of using corn waste as a valuable resource for the manufacture of biofuels, bioproducts, and other industrial applications that can support a more circular economy and minimize waste. It is crucial to note that the content and qualities of corn waste biomass and natural fiber differ depending on the part of the plant and the time of harvest, determining their possible applications and processing methods. Therefore, the study of the current corn waste management, the properties of corn waste fibers, and their potential industrial applications were conducted in this research to explore the potential industrial applications of corn waste and the sustainable methods for their management to provide new opportunities for farmers and the bioeconomy industry while reducing negative impacts on the environment. Thus, the objectives of this chapter are to study the properties of corn waste; characterize its physical, chemical, and biological attributes; investigate the current methods of corn waste management in the agricultural industry; and study the potential application of corn waste for various industries.

DOI: 10.1201/9781003472209-18

13.2 NATURAL FIBER FROM CORN WASTE

Natural fiber derived from corn waste is produced by extracting the fibrous material from the stalks and leaves of the corn plant after the kernels have been harvested. Typically discarded or used as animal feed, these agricultural residues contain valuable components: cellulose (~35% w/w), hemicellulose (~20% w/w), and lignin (~12% w/w) (Pan et al., 2019). To obtain corn fiber, the corn waste undergoes a processing stage to remove contaminants and isolate the fibrous material. This is achieved through mechanical methods such as grinding or chopping. Subsequently, non-fibrous components are broken down, and cellulose fibers are isolated using chemical or enzymatic treatments (Bousfield et al., 2018).

The fiber, generated from corn waste, is used in a variety of sectors to manufacture bio-based products. Corn fiber is processed into pulp in the paper and cardboard sector, providing strength and stability to paper products such as packaging materials, stationery, and specialty papers (Castrillón et al., 2021). Corn fiber is spun into yarn and woven into garments in the textile industry, where its qualities can be altered through treatments such as dyeing and finishing. It can also be combined with other natural or synthetic fibers to produce fabrics with specialized properties such as durability, softness, or moisture-wicking (Hazrol et al., 2022). Textiles made from corn fibers are used in clothing, household textiles, and industrial fabrics.

Corn fiber can also be used as a reinforcement in composites. Corn fiber can improve the mechanical qualities of a composite material when mixed with a bio-based matrix, such as a resin generated from plant oils (Chun et al., 2020). This qualifies it for use in the automotive, construction, and aerospace industries, where lightweight and sustainable materials are desired. Corn fiber's adaptability allows for tweaking its qualities to suit various uses. Chemical treatments, like bleaching or crosslinking, can further improve its strength, flexibility, and flame retardancy (Chong et al., 2021). Mechanical treatments, such as refining or stretching, can affect the fineness and tensile strength of the fiber (Sequeira et al., 2012).

Using corn waste as a source of natural fiber is an important step toward sustainability. By making use of agricultural by-products, we can reduce waste and promote eco-friendly practices. Corn fiber comes from a crop that is easy to grow, making it a renewable resource and a great alternative to traditional materials. Because it is biodegradable, it breaks down naturally and doesn't harm the environment over time. Additionally, producing corn fiber uses less energy and generates fewer GHG emissions compared to synthetic fibers, which helps lessen its environmental impact. Recognizing corn waste as a valuable resource highlights the need for innovative solutions to support sustainability and address environmental challenges across different industries and applications.

13.2.1 Environmental Impact from Biomass

Corn is a versatile and widely farmed crop that can be used for many different purposes, such as food, fuel, animal feed, and industrial products. However, inappropriate corn waste removal and management can have far-reaching environmental repercussions. Some of the environmental implications of corn waste were investigated, focusing on air pollution, soil degradation, water contamination, GHG emissions, and waste accumulation. Hence, by researching each of these key points, we can fully grasp the challenges involved with corn waste management and the critical need for long-term solutions (Lambert, 2019).

First, improper disposal practices, such as burning, considerably contribute to the air pollution generated by corn waste. Whenever corn waste is burned, hazardous pollutants like particulate matter, volatile organic compounds, and nitrogen oxides are released into the atmosphere Shafer (2020). These contaminants are hazardous to both human health and the environment. The particulate matter, made up of tiny particles and soot, can be breathed and cause respiratory issues like asthma and bronchitis Russell and Brunekreef (2009). Volatile organic compounds and nitrogen oxides contribute to the development of ground-level ozone and smog, which degrade air quality and negatively affect human respiratory systems (David & Niculescu, 2021; Salonen et al., 2019). Furthermore, the production of GHGs during the burning of corn waste, such as carbon dioxide (CO_2) and methane (CH_4), contributes to climate change, leading to increasing global warming catastrophes (Tiammee & Likasiri, 2020).

In addition, corn waste accumulation on agricultural land also can cause soil deterioration, posing severe hazards to soil fertility and erosion (Galindo et al., 2022). When corn waste is not properly managed and is left on the soil surface, it produces a coating that inhibits sunlight from reaching the underlying soil. Lack of sunlight restricts the growth of essential plants and microbes, reducing organic matter content in the soil. Low organic matter content causes nutrient depletion, lower water-holding capacity, and increased soil compaction Bot and Benites (2005). As a result, the soil's ability to support healthy crop development deteriorates, resulting in lower agricultural yield. Furthermore, the accumulation of corn waste can cause soil erosion because strong rainfall or wind can shift the loose residue, further deteriorating the soil's integrity (Wilson et al., 2008).

Lastly, improper corn waste management also increases the danger of water pollution, deteriorating water quality and harming aquatic life (Garcia et al., 2017). For example, excess nutrients, such as nitrogen and phosphorus, can reach neighboring water bodies when corn waste is improperly disposed of, such as runoff or leaching Long and Sun (2012). This nutrient pollution causes dangerous algal blooms to proliferate, depleting oxygen levels in the water and resulting in hypoxic or dead zones. These oxygen-depleted zones destabilize aquatic ecosystems, killing fish and other marine animals (Mishra et al., 2022). Furthermore, pesticides and herbicides used in the cultivation of corn can leak into groundwater, causing dangers to both human and aquatic health.

Thus, corn waste creates a slew of environmental issues, including air pollution, soil deterioration, and water contamination. To effectively address these concerns, it is critical to encourage sustainable waste management practices. Composting, which allows for the recycling of organic matter back into the soil, and anaerobic digestion, which allows for the controlled breakdown of corn waste to generate energy and reduce methane emissions, are two practices. Furthermore, a change toward circular economy concepts, in which corn waste is considered a helpful resource rather than a burden, can help reduce trash accumulation and promote long-term use of this agricultural residue.

13.2.2 Availability of Corn Waste

Corn production levels heavily influence the availability of corn waste in a specific location. Corn is one of the widely used farmed crops, with large-scale production in nations such as the United States, China, Brazil, and Argentina accounting for around 15%, 14%, and 7% of the global corn production, respectively (Shahbandeh, 2023). In addition to their extensive corn production and harvesting practices, these countries considerably contribute to the global supply of corn waste. According to Shell Global, under the article topic Reimagine Corn Husk, about 4.4 million tons of corn waste was discarded and thrown away from corn production only in the Philippines (University of San Carlos, 2023).

In Malaysia, the exact value of corn waste generated was hard to obtain, but as of 2021, 74,718.65 tons of corn were produced, and only the corn grains and stem are consumed by both humans and animals; a large percentage of the waste is created (FAO, 2022). Plus, according to previous research, data from Jabatan Pertanian Malaysia show that there are around 6,390 ha of corn planting, which generated approximately 36,420 metric tons where the corn husk component of corn waste which was left over from the plantation activities was thrown away without recycling or renewing to new material (Ahmad et al., 2014).

The availability of corn waste is influenced by several factors, including agricultural practices, harvesting techniques, and the demand for corn grain. Modern farming practices, such as mechanized harvesting, have made it easier to efficiently collect corn waste. Furthermore, advances in

FIGURE 13.1 Part of the corn plant and corn husk fiber.

equipment and technology have permitted more efficient separation of grain from plant material, making corn waste collection and utilization easier.

13.3 PROPERTIES OF CORN WASTE FIBERS

Every plant species is made up of cells. A fiber is a cell that is exceptionally lengthy in comparison to its breadth. The fiber is similar to a microscopic tube (a wall enclosing a central void known as the lumen). Furthermore, when the cell wall is composed primarily (85% or more) of cellulose, hemicellulose, and lignin, we refer to lignocellulosic fibers, including woody species, scrubs, and most agricultural crops. Straws, flax, hemp, jute, and sisal are typical agricultural lignocellulosic fibers. Non-lignocellulosic fibers present in potatoes, beets, and cotton, among other crops, do not contain lignin. Different raw materials are separated using methods to create fibers appropriate for particular end products Olesen and Plackett (1999). Figure 13.1 shows an example of corn husk fiber. Often discarded corn cobs, husks, stalks, and stovers make up corn waste fibers, which are lignocellulosic fibers (Chong et al., 2021). Understanding the properties of corn waste fiber is vital as its properties are the deciding factor on which particular fiber should be used to suit a particular requirement of a product.

13.3.1 Physical Properties

Table 13.1 lists the physical properties of different parts of corn fibers. For fiber length, the corn sheath recorded the longest among other corn fibers, which is 1,487 μm. Meanwhile, corn stalk fiber and corn husk fiber have a similar length of around 218 μm. The fibers of corn stalk, corn husk, and corn sheath had around the same value of fiber diameter, around 17–18 μm. The density of corn husk is the highest, with a value of 1.4913 g/cm^3, while the corn hull recorded a lower value at about 1.3 g/cm^3. Densities for corn stalks fiber and corn cob fiber are 0.21–0.38 and 0.8–1.2 g/cm^3, respectively. Fibers of higher density were reported to be stiffer and stronger as the mechanical properties rely on the overall fiber cross-section (Chong et al., 2021). The low density of biomaterials made it more attractive for manufacturing biocomposites compared to artificial composite materials, such as fiberglass (2.5 g/m^3) (Mendes et al., 2015). Fibers of higher density were reportedly stiffer and stronger as the mechanical properties rely on the overall fiber cross-section (Chong et al., 2021). Hence, corn cob and corn stalk are suitable raw materials for biocomposite.

TABLE 13.1
Physical Properties of Corn Waste

Fibers	Density (g/cm^3)	Fiber Length (μm)	Fiber Diameter (μm)	References
Corn stalks	0.21–0.38	218.45±184.05	18.32±9.07	Chong et al. (2021) and Mohammed et al. (2022)
Corn cob	0.8–1.2	—	—	Mohammed et al. (2022)
Corn husk	1.4913	218.59±165.36	17.78±15.59	Ibrahim et al. (2019a) and Kambli et al. (2016)
Corn hull	1.3231	—	—	Ibrahim et al. (2019)
Corn sheath	—	1,487±344	18.049±6.66	Alagbe et al. (2019a)

TABLE 13.2
Chemical Composition of Corn Waste

Fibers	Cellulose (%)	Hemicellulose (%)	Lignin (%)	References
Corn cobs	45	35	15	Abdul Khalil et al. (2012)
Corn stalks	35	25	35	
Corn husk	50–55	39.39	7.5	Kambli et al. (2016)
Corn hull	15.3	40.4	2.87	Ibrahim et al. (2019a)

13.3.2 Chemical Composition

Table 13.2 lists the chemical properties of different parts of corn fibers. Among all the corn fibers, corn husk has the highest cellulose content with the range of 50–55%, followed by corn cobs at 45%, corn stalks at 35%, and corn hulls at 15.3%. For hemicellulose, the fiber with the highest percentage is corn hull, followed by corn husk, corn cobs, and, lastly, corn stalks with values of 40.4%, 39.39%, 35%, and 25%, respectively. The lignin content in corn husks is the lowest at 2.87%, followed by corn husks at 7.5%. Meanwhile, corn cob fibers and corn stalk fibers have moderately high lignin contents of 15% and 35%, respectively. Fibers with high cellulose content have a relatively stable structure due to the cellulose OH bonds and a linear shape for hydrogen bonds between consecutive sequences. Furthermore, the series cellulose and hemicellulose combine to form a variety of fibril chains via van der Waals, and hydrogen bonds have reinforced the cellulose chains. Thus, celluloses are accountable for a structure's mechanical strength (Chong et al., 2021). The paper also stated that fibers with higher cellulose content had higher tensile strength and elastic modulus. Because of the orientation of crystalline cellulose, the fibers indicated above have a high mechanical property.

Table 13.3 provides the functional group and its wavenumber discovered in corn stem stalk fiber. Figure 13.2 compares the FT-IR spectra of corn stalk fibers treated with various chemical solutions to those of untreated fiber. The distinctive peaks of cellulose, hemicellulose, pectin, and lignin are apparent in the FT-IR spectrum of untreated corn stalk fiber. According to Sepe et al. (2018), the peak around 1,723 cm^{-1} after alkali treatment can be attributed to the C=O linkages starting with the raw cellulose molecules and collagen-corn stem fibers. Additionally, in the corn stem fibers treated with alkali solution and NaOH-silane solution, the peak seen at around 1,243 cm^{-1} (C–O bond stretching) was no longer present. This suggested that some lignin and hemicellulose had been removed from the treated fibers (Shanmugasundaram et al., 2018).

Figure 13.3 shows the fiber from the corn stem and husk from the FT-IR analysis. As can be observed, the patterns were similar for both fibers. The significant variations between the corn

TABLE 13.3
Functional Group and Wavenumber of Corn Stem Stalk (Liu et al., 2019)

Functional Group	Wavenumber (cm^{-1})
O–H	3,349
C–H	2,917
C=O	1,723
C=C	1,513
C–O	1,243
C1–H	898

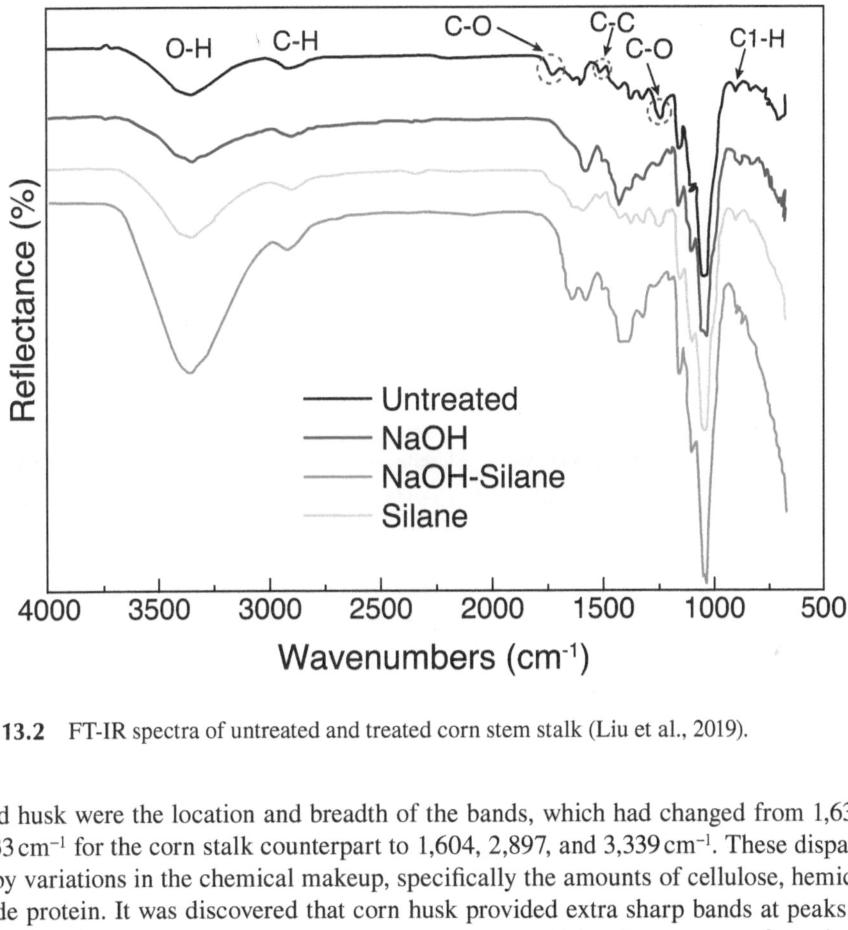

FIGURE 13.2 FT-IR spectra of untreated and treated corn stem stalk (Liu et al., 2019).

stalk and husk were the location and breadth of the bands, which had changed from 1,635, 2,917, and 3,333 cm^{-1} for the corn stalk counterpart to 1,604, 2,897, and 3,339 cm^{-1}. These disparities are caused by variations in the chemical makeup, specifically the amounts of cellulose, hemicellulose, and crude protein. It was discovered that corn husk provided extra sharp bands at peaks of 2,921 and 2,852 cm^{-1}. The prominence of these peaks was attributable to the presence of starch remnants from the extraction process connected to hull fiber (Ibrahim et al., 2019a,b).

13.3.3 Mechanical Properties

Table 13.4 lists the mechanical properties of different pulping methods on corn stalks and husks. For corn stalks, the tensile strength for untreated fibers was 112.96 MPa and increased to 113.87, 152.90, and 223.33 MPa as it was treated with NaOH-silane, NaOH, and silane, respectively. Young's

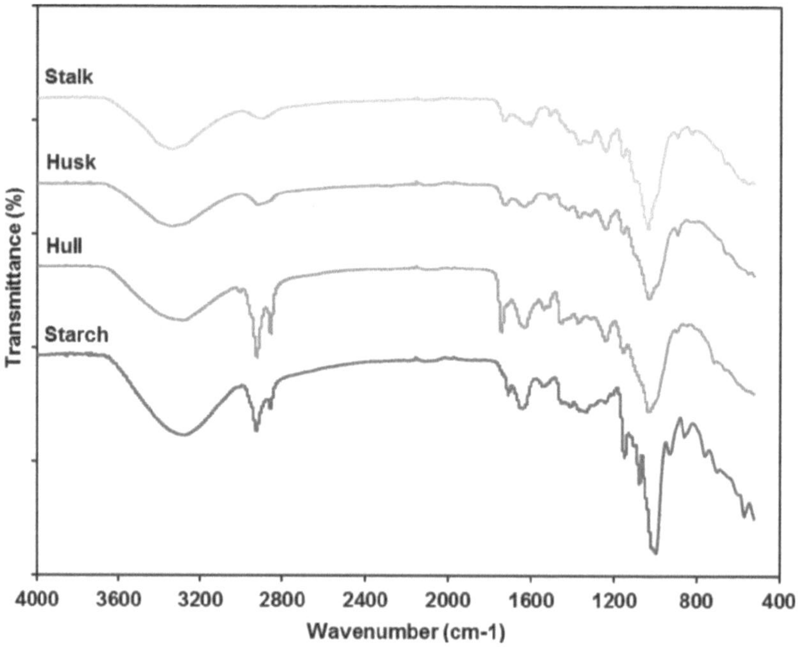

FIGURE 13.3 FT-IR spectra of corn stalk, husk, hull, and starch (Ibrahim et al., 2019a).

TABLE 13.4
Mechanical Properties of Corn Stalk and Husk

Fiber	Treatment Process	Tensile Strength (MPa)	Young's Modulus (GPa)	Elongation at Break (%)	References
Corn stalk	Untreated	112.95±24.15	3.69±0.96	3.64±0.52	Liu et al. (2019)
	NaOH	152.90±32.51	5.13±1.18	3.67±1.05	
	NaOH-silane	113.87±19.36	4.63±0.81	2.74±0.36	
	Silane	223.33±41.22	7.05±1.07	4.02±1.24	
Corn husk	Untreated	160.49±17.12	4.57±0.54	21.08±2.860	Herlina Sri et al. (2017)
	NaOH (0.5%)	224.05±22.14	9.95±1.34	8.66±4.28	
	NaOH (1%)	230.90±41.85	11.16±4.12	6.85±2.60	
	NaOH (2%)	176.59±12.01	7.09±0.52	10.58±3.64	
	NaOH (5%)	237.43±104.82	9.19±3.68	11.35±4.48	
	NaOH (8%)	368.25±78.97	15.87±1.87	7.41±0.61	

modulus exhibited the same increasing pattern in corn stalks where untreated fiber was 3.69 GPa, NaOH-silane-treated fiber was 4.63 GPa, NaOH was 5.13 GPa, and silane was 7.05 GPa. Meanwhile, the percentage of elongation at break was the lowest at corn stalk fiber treated with NaOH-silane, with a percentage of 2.74. Silane-treated corn stalk fiber had the highest elongation at a break of 4.02%, untreated fiber with 3.64%, and corn stalk fiber treated with NaOH at 3.67%. For corn husks that underwent NaOH treatment at different concentrations, 8% of NaOH has the highest Young's modulus and tensile strength with values of 368.25 MPa and 15.87 GPa, respectively. However, corn husk fiber treated with 8% NaOH has a relatively low percentage of elongation at a break of 7.41%.

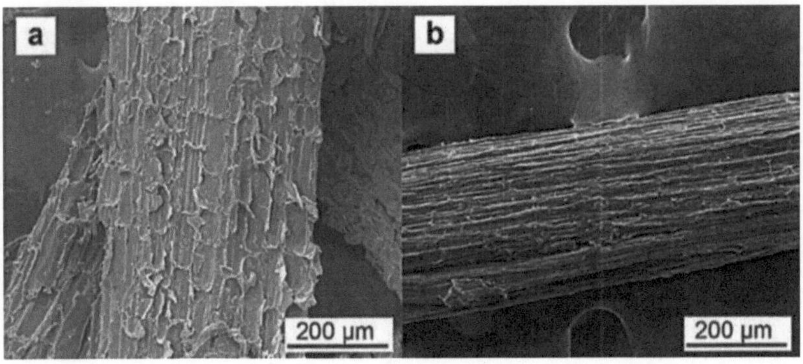

FIGURE 13.4 (a) Untreated corn fiber and (b) alkali-treated fiber (Mohammed et al., 2022).

Untreated corn husk has the lowest tensile strength, highest Young's modulus, and longest elongation at break with values of 160.49 MPa, 4.57 GPa, and 21.08%, respectively.

In the pulp and paper industry, the tensile strength of the paper from pulp is highly dependent on the quality and quantity of cellulose and hemicellulose present in the fiber, which are carriers of the –OH group responsible for the hydrogen bonding in the paper (Alagbe et al., 2019). Thus, additional cellulose in the fiber is anticipated to have a favorable effect on physical properties such as tensile strength. Generally, corn husk has higher tensile strength and Young's modulus compared to corn stalk as it has higher cellulose content. Based on the review, it clearly shows that treated corn fiber shows greater mechanical properties due to the loss of lignin and hemicellulose in the fibers after treatment with alkali.

13.3.4 Morphological Properties

Figure 13.4 shows untreated corn fiber and alkali-treated corn fiber. SEM findings demonstrate that, when treated with alkali, hemicellulose and lignin levels significantly decreased, enhancing the corn composite product's structure. However, Figure 13.5a shows the morphology of untreated corn stem fiber, and (b–d) present treated corn stem with different chemicals. It was found that the natural fiber's surface was even, smooth, and unbroken. According to Nadlene et al. (2018), wax, pectin, lignin, hemicelluloses, and cellulose are typically present on the fiber surface. The treated fibers in Figure 13.5b and c were jagged and had rough surfaces in comparison to the untreated corn stem fiber. The removal of hemicellulose, lignin, and ash from fibers after chemical treatments is one explanation that may be offered. However, the most desired natural fibers would have a rough surface (Ramamoorthy et al., 2015).

13.3.5 Thermal Properties

Thermogravimetric analysis (TGA) examines a material's thermal characteristics, such as weight loss, breakdown, and thermal stability. TGA may be used to investigate the thermal behavior of corn fiber, including the temperature ranges at which it decomposes or loses weight as a result of thermal impacts. At temperatures ranging from 40°C to 130°C, raw and silane-treated corn stem fibers lost 12.3% and 7.4% of their mass, respectively. The alterations were linked to moisture vaporization in natural fibers. Thermal degradation occurred at roughly 180–300°C in the second phase, with a decrease in weight of 42.4% for raw corn stem fiber and 36.3% for silane-treated corn stem fiber due to the destruction of hemicellulose and glycosidic connections of cellulose (Liu et al., 2019).

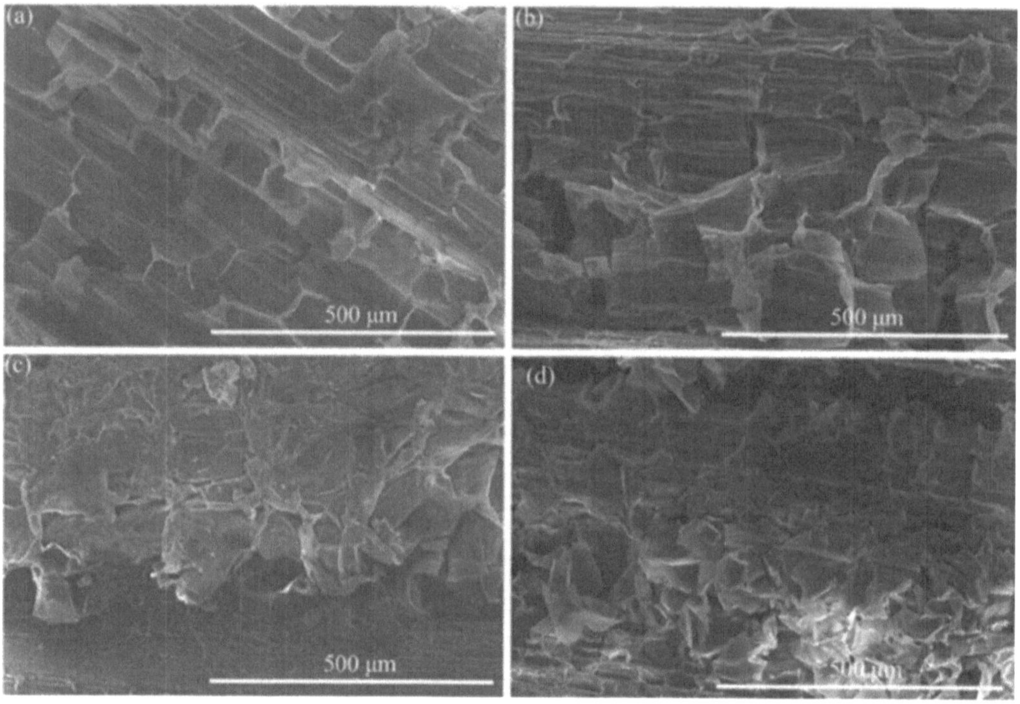

FIGURE 13.5 SEM images of lateral surfaces of corn stem fibers. (a) Untreated fiber, (b) alkali-treated fiber, (c) NaOH-silane-treated fiber, and (d) silane-treated fiber (Liu et al., 2019).

13.4 CORN WASTE MANAGEMENT

The practice of processing and properly discarding corn-producing leftovers such as stalks, cobs, husks, and leaves is known as corn waste management. Corn waste management is significant in addressing agricultural sustainability and resource efficiency. There are several ways to make use of these residues to increase value and decrease waste. Solid wastes must be managed properly to turn uncontrollable landfills into clean energy, environment, and healthy communities. Various corn waste recreations include compost, biofuel, and biocomposite material. The final line of products gives numerous advantages to the communities and environment, especially for sustainability development. For instance, resource optimization, maximizing the potential value of corn crops, and efficient management of corn waste are essential.

13.4.1 Compost

Soil incorporation of composted municipal solid waste usually positively affects the growth and yield of a wide variety of crops and the restoration of ecologic and economic functions of land (Shiralipour et al., 1992). Corn wastes such as corn cobs and rotten corn are highly suitable for compost piles. Both are highly composites that lead to a faster decomposition process. In addition, the decomposition process leaves no adverse effect on the environment. Corn husks can also be used for compost piles. Higher levels of soil organic matter may arise from spreading a lot of organic waste on a specific area of land, such as a landfill. Since there are excessive corn husks, especially after the harvest season, corn husks can be collected and composed. Corn husks are a source of natural fertilizer for soil and plants. Corn husk can easily break down compared to corn cobs and rotten corn. Thus, corn waste can become a green product that does not harm the environment.

The compost must be appropriately made to achieve higher composition efficiency. The quality of the compost determines the growth and development of plants (Lima et al., 2004).

Compost has many advantages for consumers and the environment. First, it is cost-effective. Composting can result in significant financial savings, particularly in areas with limited landfill or burningcapacity.Next,thecompostisrichinnutrients,whicharenitrogen,phosphorus,andpotassium. It helps to supply nutrients to the soil and regulates the pH of the soil. In addition, compost is environmentally friendly. It helps reduce methane gas emissions due to the proper decomposition process. By retaining carbon in the soil through composting, the carbon footprint left by decomposing matter is reduced.

13.4.2 Biofuel

The demand for fossil fuels can be overcome by enhancing biofuel usage from biomass such as corn waste. The main qualities of biomass include abundant availability, wide-spreading, sustainability, and renewability (Khan et al., 2023). Figure 13.6 shows the conversion process of biofuel. Biofuel is produced by using biomass as its primary source. Hydrothermal and biochemical methods are proposed for biofuel production from biomass. Both methods can be integrated to produce high-quality biofuel. Thermochemical processes are classified into low temperature, below 300°C, and high temperature, higher than 300°C, i.e., gasification, combustion, and pyrolysis Osman et al. (2021).

Biofuel offers many advantages to consumers and the environment. Biofuel offers high energy efficiency compared to fossil fuels. In addition, it helps to reduce foreign oil dependency. For example, experts say that, using biogas, enacting tax incentives for hybrids and fuel-cell vehicles, and raising fuel economy standards for motor vehicles will help wean the United States off the need for foreign oil (DXP Marketing, 2021). Because of their affordability and ecological sustainability, biofuels are an excellent replacement for fossil fuels (Akram et al., 2023). This is because the lower emissions produced from biofuel make it safer than fossil fuels. Other than that, biofuels can be manufactured from a wide range of materials (Amer et al., 2021). In this case, corn cobs and corn husks can be used as the main sources of biofuel.

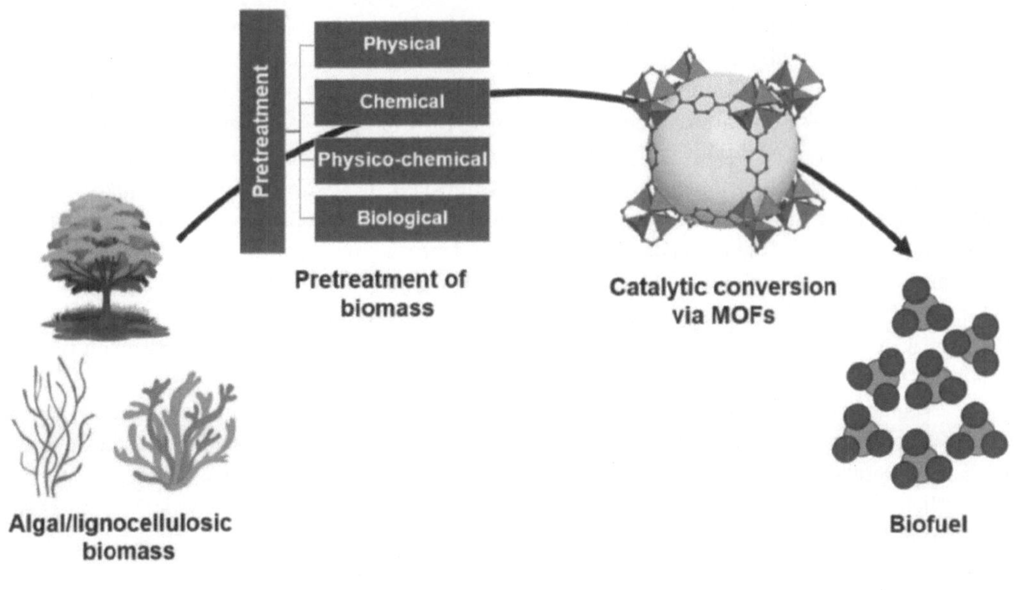

FIGURE 13.6 The conversion of biofuel by using catalytic conversion via metal-organic frameworks (MOFs) (Akram et al., 2023).

13.4.3 Biocomposite Materials

A biocomposite is a substance made of two or more different constituent materials, at least one of which is naturally produced, combined to create a new substance that performs better than the sum of its parts (Rudin & Choi, 2013). The shortage of petroleum sources for composite material is one of the major concerns in the industry. Thus, biocomposite material has significant potential to replace composite material. The weight-related mechanical properties enable biocomposites to be employed in applications still dominated by glass fiber-reinforced plastics (Riedel, 2012). Other than that, bio-filler can be used in the biocomposite to promote sustainability. Corn cob can be the ideal biocomposite application. Corn cob utilization as reinforcement in practical industrial applications can add value to a significant crop residue and provide macroeconomic benefits to rural communities (Gairola et al., 2022).

Nowadays, biocomposite is widely used in many industries. The automotive industry has a high demand for biocomposite materials. This is because of its lightweight properties, making it suitable for improving vehicle efficiency. Compared to composites of non-renewable origin, biocomposites have excellent acoustic and thermal properties, which make them ideal for vehicle interior parts (Rius, 2019). It is a fact that the automotive sector can benefit from adopting these materials in relation to weight reduction, not only due to the depletion of oil reserves but also due to the high capability and significance of these materials in automobiles (Ramli et al., 2018).

Using grafting modification enhanced the mechanical properties of corn composite whereby the addition of maleic anhydride polyethylene to the composite of low-density polyethylene (LDPE)/corn stalk (CS) improves the interaction between the LDPE/CS composites; at 10% of filler loading, tensile strength of the biocomposite was 8.1 MPa without grafting, and it increased to 9.5 MPa. However, the addition of MAOE decreased the elongation and young modulus (Husseinsyah et al., 2015a). Reinforcing corn starch with polypropylene coupled with maleic anhydride increased the tensile strength and reduced tensile modulus while there was no significant change in elongation (Lenz et al., 2018). Generally, mechanical properties can be improved by fiber surface treatments and the addition of plasticizers and crosslinking agents. Most of the researchers focus on mechanical properties as part of their investigation. Table 13.5 lists the mechanical properties behavior of biocomposite-based corn.

The water vapor barrier and mechanical and optical properties of *Kappaphycus alvarezii* seaweed and corn starch blend biocomposite films were investigated (Khalil et al. [115]). The researchers stated that the concentration of seaweed had the biggest impact on the mechanical and optical behavior of the seaweed/starch biocomposite films. However, the water vapor permeability of the composite films improved with increasing amounts of starch and seaweed. In addition, the FT-IR measurement showed intermolecular interactions between the two compounds, which is in line with the high miscibility degree between the seaweed and the starch in the mixture. The primary peak shift in FT-IR to higher wavenumbers and the decrease in crystallinity in XRD are evidence that starch molecules and chitosan may interact via hydrogen bonding [116]. In addition, Ren et al. [116] stated that the film's solubility, tensile properties, elasticity, and water vapor permeability of the film were increased with the inclusion of natural fiber. Wang et al. [117] researched on water solubility index and mechanical characteristics of collagen-corn starch biocomposite films, as well as the evaluation of the effects of concentration and type of starch. The authors found that the addition of starch to collagen enhanced the composite films' thickness and opacity. Furthermore, the surface of starch–collagen films was rougher than that of pure collagen films, which smoothed out during heating, according to images taken using scanning electron microscopy. SEM images were reported in Figure 13.7.

TABLE 13.5
Mechanical Properties of Biocomposite-Based Corn

Fiber-Reinforced Material	Matrix	Plasticizer	Plasticizer %	Reinforcing Material %	Tensile Strength (MPa)	Young's Modulus (MPa)	Flexural Strength (MPa)	Elongation (MPa)	References
—	Corn starch	Glycerol	10%	—	60	62	—	10	Mansour (2019)
			20%		30	32		25	
			30%		17	20		45	
			40%		9	12		55	
Graphene oxide (GO)	Corn starch+soy protein isolate (SPI)	Glycerol	30 wt%	GO 0	6.65±0.9	252±5.2	—	7.25±43	Mansoori et al. (2019)
			30 wt%	GO 0.5	2.84±0.48	236±48		19.73±3.38	
			30 wt%	GO 1	2.32±0.35	332±26		17.9±3.4	
			30 wt%	GO 1.5	3.60±0.6	359±25		10±2.02	
			30 wt%	GO 2	5.9±0.27	449±28		11.83±2.08	
Graphene (G)	Corn starch+soy protein isolate (SPI)	Glycerol	30 wt%	G 0	6.65±0.9	252±502	—	7.25±0.43	Mansoori et al. (2019)
			30 wt%	G 0.5	16.53±0.26	578±105		2.7±0.12	
			30 wt%	G 1.0	11.38±1.4	380±48		1.2±0.21	
			30 wt%	G 1.5	8.09±1.4	329±34		2.83±1.83	
			30 wt%	G 2.0	6.66±1.3	132±55		6.58±0.98	
Corn cob	Low-density polyethylene	—	—	0	14.9	130	—	11.3	Obasi (2013)
				40	13.8	133		10.2	
				45	13.0	138		9.2	
				50	15.8	141		7.8	
				55	11.3	148		7.1	
				60	9.4	163		5.8	

(Continued)

TABLE 13.5 (*Continued*)
Mechanical Properties of Biocomposite-Based Corn

Fiber-Reinforced Material	Matrix	Plasticizer	Plasticizer %	Reinforcing Material %	Tensile Strength (MPa)	Young's Modulus (MPa)	Flexural Strength (MPa)	Elongation (MPa)	References
Corn husk	Epoxy	—	—	0	34	—	48	—	Verma (2014)
				5	29.2		40.8		
				10	26.2		33.7		
				15	22.5		29.3		
				20	19.4		25.6		
				25	16.2		23.5		
				30	14.8		21.8		
Corn flour	Poly(butylene succinate-co-butylene adipate) blends	—	—	70	3.93±0.58	189±15	—	16±2	Jbilou et al. (2013)
				50	7.33±0.96	235±11		38±14	
				30	11.9±0.6	279±10		64±36	
				0	50 ± 0.2	265±5		800±10	
Corn cob, epichlorohydrin (EP), adipic acid (ADP), glutaraldehyde (GLA)	Chitosan (CS)	—	—	0	52.8±2.3	2,269±186	—	13.8±0.4	Yeng et al. (2015)
				20	33.1±1.7	2,571±121		8.7±0.7	
				20+EP	46.9±0.9	2,703±177		8.2±0.4	
				20+ADP	33.71±2.01	2,616±210		12.18±0.3	
				20+GLA	42.0±1.6	3,178±174		7.7±0.2	
Corn cob treated by acrylic acid	Soy protein isolated	—	—	0	2.6	35	—	49	Husseinsyah et al. (2018)
				10	3.6	65		34	
				20	4.5	130		19	
				30	4.8	140		12	
				40	5.7	195		8.5	

(*Continued*)

TABLE 13.5 (*Continued*)
Mechanical Properties of Biocomposite-Based Corn

Fiber-Reinforced Material	Matrix	Plasticizer	Plasticizer %	Reinforcing Material %	Tensile Strength (MPa)	Young's Modulus (MPa)	Flexural Strength (MPa)	Elongation (MPa)	References
Corn husk fiber, alkali (A)	Polylactic acid	—	—	0	47	—	80.5	—	Luo et al. (2016)
				5 A	56		89		
Alkali and silane (AS) treatment				10 A	64		109		
				15 A	61		102		
				10 A-1S	60		103		
				10 A-2S	69		113		
				10 A-3S	67.5		114		
				10 A 10–4S	66		106		
Corn husk	Corn starch	Fructose	25%	2	8.8	320	—	—	Ibrahim et al. (2019b)
			25%	4	7.8	190			
			25%	6	10.9	510			
			25%	8	12.9	615			
Corn husk (CH), sugar palm fiber (SPF)	Corn starch	Fructose	25%	8%CH+	17	1,050	—	1.3	Ibrahim et al. (2020)
			25%	2%SPF	17.3	1,100		1.15	
			25%	8%CH+	19	1,160		0.9	
			25%	4%SPF	18	1,110		1.1	
				8%CH+					
				6%SPF					
				8%CH+					
				8%SPF					

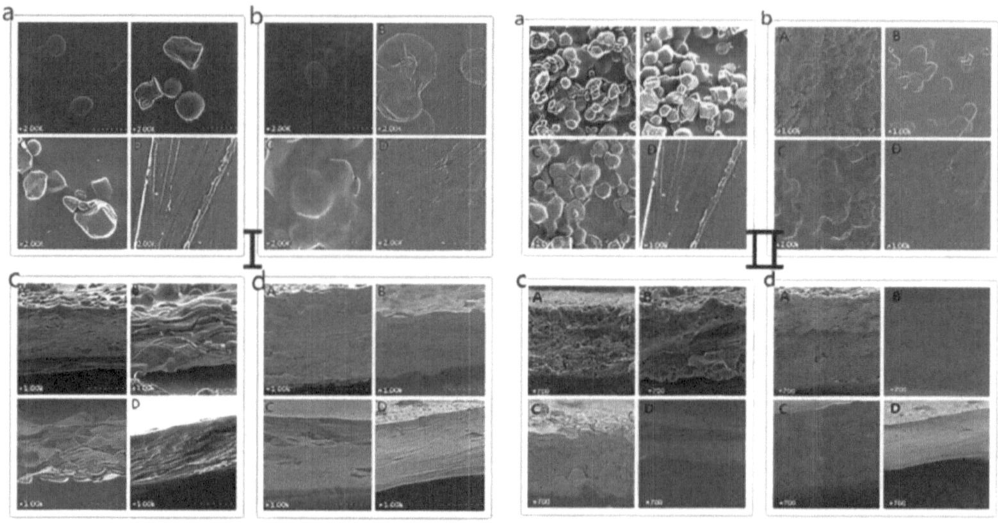

FIGURE 13.7 SEM micrographs of collagen and collagen-based biocomposite film surface (a) before heating, (b) after heating at a scale of 20 μm; (c) cross-section of biocomposite film before heating, and (d) after heating at a scale of 50 μm. Collagen film blended with corn starch (I) 10 wt%, (II) 50 wt% and hybridized with (A) high amylose content, (B) waxy corn starch content, (C) without starch, (D) neat collagen composites.

13.4.4 Bioethanol

Bioethanol production from lignocellulosic biomass involves three major unit operations: pre-treatment, enzymatic saccharification, and fermentation (Sindhu et al., 2019). Bioethanol is a renewable biofuel that is also oxygenated (35% oxygen), thus providing the potential to reduce automobile emissions (Verma et al., 2019). The increase in ethanol demand in the industry causes a shortage of ethanol. Hence, bioethanol can be synthesized to tackle the problem. Bioethanol from agricultural waste could be a promising technology, though the process has several challenges and limitations, such as biomass transport and handling and efficient pre-treatment methods for total delignification of lignocellulosic (Sarkar et al., 2012).

The bioethanol manufacturing process is started with a pre-treatment process. Some feedstocks require pre-treatment conditions to release fermentable sugars into the media. Without pre-treatment, fermentation progress can be slowed due to the limited availability of fermentable sugars for metabolism (Tse et al., 2021). During hydrolysis, the water breaks down the polysaccharides, such as cellulose, present in the corn. Lastly, it will undergo microbial fermentation. It is a biological method that uses sugars to release energy chemically without the presence of oxygen.

13.4.5 Pulp and Paper

Corn husk has a high cellulose fiber content, so it can be used as a raw material for making paper (Rahmayanti & Asma, 2022). Corn husk is used due to its high accessibility, low cost, and excellent biodegradability properties. The production of pulp from corn husk gives high tensile strength and tear strength, indicating strong mechanical properties of pulp. The use of corn husk material can reduce newsprint production costs because it uses cheap and waste materials and can help overcome the problem of corn husk waste Rahmayanti and Asma (2022). Producing paper pulp involves a number of steps, including drying corn husks, chopping them up into little pieces, boiling them with water and NaOH, and filtering and cleaning them before blending them. Another chemical method of pulp production is the nitrate-alkali process, in which the bond between cellulose and lignin is hydrolytically released during cooking in nitric acid (Hájková et al., 2023). Nitrolignin is created when lignin reacts with nitrate and oxidizes at the same time. Thus, it can be completely dissolved in the base.

Increased demand for biomass increases pressure on the availability of this resource, which might strengthen the need for paper recycling (Laurijssen et al., 2010). Hence, making paper from biomass such as corn waste helps to reduce CO_2 emissions. This helps to reduce GHG emissions. Reducing our carbon emissions helps reverse the impact of global warming overall, but more specifically, it benefits the overall air quality. In addition, it can help reduce the energy consumption of paper production.

13.4.6 Bioplastic

Plastic is now found in chemicals, food, medicine, electronics, automobiles, construction materials, and every other aspect of industrial production and daily life. A bioplastic can be defined as a polymer that is manufactured into a commercial product from a natural source or renewable resource Rudin and Choi (2013). Lignocellulose in corn can be transformed into bioplastic. Using straightforward methods, cellulose and lignin were reduced to micro or nano-sizing in molecules, resulting in viscous and homogenous sludge (Zhou et al., 2023). Other than that, bioplastic has advanced mechanical properties compared to other plastics.

Bioplastic has many advantages compared to conventional plastic. Microorganisms, industrial or home composting as an end-of-life option, as well as anaerobic digestion, may also decompose bioplastics, encouraging a more sustainable circular economy (Boey et al., 2023). Bioplastic has better energy efficiency. On the other hand, plastics are made from about 4% of the oil that the world uses every year (Ezgi & Havva, 2015). Oil shortage increases the exposure of the production of plastics to economic fluctuations. In addition, bioplastic is less harmful as compared to conventional plastic. This is because the bioplastic uses natural ingredients such as corn waste. Corn compost of sugar can be changed into a carbon-neutral product. Thus, it shows that bioplastic is environmentally friendly.

13.5 POTENTIAL APPLICATIONS

13.5.1 Biodegradable Textile

A new development in the use of corn husk for textile manufacture incorporates an environmentally friendly method that includes alkali retting, pressing, cutting, and twisting. This method efficiently converts corn husk into corn husk textile fibers (CHTFs) while retaining lignocellulosic fibers and removing flaws and microchannels from the material (Chen et al., 2020). The produced CHTFs outperform raw corn husk and wood-based textile fibers in terms of mechanical strength (Zheng et al., 2022). This approach is scalable, low cost, and has a much lower carbon footprint, making it an appealing and sustainable textile production solution.

The effective use of corn waste in textile production has enormous potential for addressing environmental issues and increasing sustainability. In China alone, the predicted yearly yield of 14.6 million tons of CHTFs reveals a significant reduction in land usage, agricultural waste recycling, and GHG emissions (Zheng et al., 2022). Furthermore, the great mechanical strength of CHTFs opens up new possibilities for their use in textile products, packaging materials, and wearable electronics. More research and development can be conducted to investigate alternate biomass sources and optimize production processes for increased efficiency and affordability, opening the way for a more sustainable future in the textile sector.

Hence, the use of corn husk as a raw material for textile production is a long-term solution to the environmental challenges faced by synthetic fiber-based textiles. Corn husk-based textile bio-fibers with high mechanical strength and a low carbon footprint can be generated using a green and efficient approach (Youssef et al., 2015). This method minimizes land consumption and GHG emissions and allows for the recycling and reuse of agricultural waste. Corn waste-based textile products have enormous promise for developing a more sustainable and environmentally friendly textile industry with further advances and wider use.

13.5.2 Animal Feed

Manufacturing animal feed from corn waste has received much attention as a long-term solution to environmental and economic issues. Corn, a significant agricultural product, provides a rich feed source. Traditional ways of producing animal feed from corn necessitate significant resources such as water, land, and energy (Dou et al., 2018). However, using corn waste as a feed source reduces food rivalry between humans and animals and delivers economic savings (Ferguson, 2020; zu Ermgassen et al., 2016). This economic advantage is due to lower expenses associated with corn waste versus the preparation of animal feed from agricultural products such as corn. By diverting corn waste from the waste stream and transforming it into valuable feed, food waste is reduced, and livestock production becomes more economically viable.

According to studies, the annual production and supply of feed reached a mind-boggling 1 billion tons in 2018, putting a strain on critical resources such as water, land, and energy (Alltech, 2018). The demand for these resources can be relieved using a method that transforms corn waste into animal feed. Furthermore, this strategy contributes to addressing the issue of food waste, which accounts for one-third of all food produced annually (FAO, 2013). The feed industry can help reduce food waste while satisfying animal nutritional needs by repurposing corn waste, high in proteins and essential nutrients (Godfray et al., 2010).

To summarize, using corn waste to produce animal feed is a sustainable and cost-effective solution to environmental and economic concerns. The demand for resources such as water, land, and energy can be minimized by transforming corn waste into feed. Furthermore, this technique addresses the issue of food waste while also providing economic benefits to the livestock business (zu Ermgassen et al., 2016), FAO (2013). In the drive for a more sustainable and efficient food system, the conversion of corn waste into animal feed represents a viable option (Georganas et al., 2020; United Nations, 2019).

13.5.3 Raw Building Material

Despite a decline in cultivation, corn cobs pose an environmental concern since they are frequently burned, adding to pollution. However, there is potential for addressing this issue and stimulating social and economic growth in rural areas by employing corn cobs as a sustainable raw building material (Santolini et al., 2021). It is able to determine its suitability for construction applications by evaluating qualities such as density, water absorption, fire resistance, and thermal insulation capability (Pinto et al., 2012).

The construction industry wants environmentally favorable raw materials, and corn cob, a worldwide abundant agricultural waste product, provides a solution. Its macrostructure has three unique layers and displays alveolar characteristics (Pinto et al., 2012). A comparison with traditional construction materials reveals microstructure and chemical composition similarities, implying possible thermal insulation capabilities (Choi et al., 2022). Hence, corn cob is emerging as a promising raw material for green building.

Examining the fundamental features of corn cob is critical for determining its potential as a building material. It has the same density as cork and outperforms extruded polystyrene (XPS) and expanded polystyrene (EPS) (Pinto et al., 2012). Despite its microstructure's expressive water absorption, corn cob keeps material integrity even after prolonged contact with water. It has equal fire resistance to cork and greater thermal transmission coefficients and thermal conductivity values than traditional materials (Pinto et al., 2012). These preliminary findings suggest that corn cob could be used in thermal insulation products, partition walls, ceiling coatings, internal doors, and furniture, thereby improving sustainability in the building industry.

13.5.4 Mushroom Cultivation

Long-term production and environmental protection are both enhanced by mushroom culture on corn waste. The oyster mushroom (*Pleurotus* sp.) is particularly popular in the global market for edible mushrooms due to its ease of cultivation and low production costs (Zárate-Salazar et al., 2020). *Pleurotus* sp. can be produced on a wide range of organic substrates and temperatures, making them adaptable to a wide range of conditions.

The disposal of agricultural by-products and waste presents considerable issues and has the potential to pollute the environment. However, using lignocellulosic biomasses as substrates for mushroom production, such as corn straw, corn cob, and corn husk, offers a more sustainable approach (Adjapong et al., 2015). These wastes are high in lignocellulosic content and provide an abundance of carbohydrates for fungus (Badu et al., 2011). Trash disposal can be efficiently addressed while improving mushroom productivity and environmental conservation by utilizing corn husks with high cellulose content.

Wood sawdust has long been used as a substrate for mushroom culture, but its availability is limited due to forest preservation considerations (Rambey et al., 2019). Furthermore, wood sawdust may include harmful compounds or fungicide treatments that interfere with mushroom growth. As a result, investigating alternate substrates such as corn husks becomes critical for increasing sustainable mushroom production. Corn husks provide a cellulose-rich, readily available substrate that can gradually replace wood sawdust in mushroom production, reducing reliance on limited resources while minimizing environmental effects.

Finally, mushroom production using corn waste, specifically corn husks, is sustainable and environmentally benign. Oyster mushroom growth on corn husk substrates has the potential to contribute to the worldwide edible mushroom market while also solving waste disposal difficulties and conserving forest resources. Mushroom growers can boost production, reduce environmental pollution, and promote a more sustainable future for mushroom agriculture by gradually replacing wood sawdust with corn husks.

13.6 CHALLENGES ASSOCIATED WITH PROCESSING AND COMMERCIALIZING

During the processing of the bioproduct, there are many challenges that need to be overcome. First is storage and transportation. Transferring wet biomass from the farm to the manufacturing site becomes more expensive and energy-unfavorable with distance due to moisture present in the collected biomass Raychaudhuri and Ghosh (2016). Next is the high capital cost. The annual global investment requirements for renewable energy, bio-based chemicals and fuels, and ecosystem services are estimated at USD 1–2 trillion over the next three decades (Bähner et al., 2021). The existing plant for bio-based production is on a small scale. Thus, a significant investment is required for a larger scale production in the future. In addition, the inefficiency of the storing and processing plant can be a drawback of the project. Technical challenges are brought because of the lack of standards for bioenergy facilities and equipment, mainly when energy sources are so broad that proper pre-treatment is necessary to avoid biological degradation and the loss of thermal value, increasing equipment investment costs as well as production costs (Raychaudhuri & Ghosh, 2016). For instance, there might be a lack of proper logistics and handling facilities at the farm level. The farm area can be limited, and no proper storage is available to store the waste.

As for commercializing, the lack of awareness of the bioproduct can be a challenge to further promoting sustainable products. The lack of standardization and quality of the final product might not meet the consumer standard since it is from waste. Consumers are too comfortable with conventional products in the market. It is indeed one of the most complex challenges to expose the bioproduct to the consumer and, at the same time, educate them regarding the bioproduct. Thus, the processing and commercializing challenges need to be overcome to ensure the longevity of the project. Next is the price comparison, such as for bioplastic. As there has become some competition in the industry and low production amounts, the cost of producing bioplastics and the high cost of input materials make the switch to bioplastics unattractive (Neves et al., 2020). As for biofuel, the price is higher compared to fossil fuel. This is due to the fact that biofuels are derived from renewable resources, such as crops and waste materials, which are more expensive to grow and process than fossil fuels.

13.7 CONCLUSIONS

In conclusion, sustainability and sustainable development have become integral concepts in contemporary society, emphasizing the need to enhance processes, activities, and innovations that promote these ideals across all human endeavors. The utilization of corn waste as a novel resource underscores the considerable potential to reintegrate raw materials into processes with reduced environmental impact, thereby mitigating the carbon footprint associated with the use of virgin materials. This chapter underscores the remarkable properties of corn waste fiber, including its mechanical strength and chemical composition, highlighting its versatility in various applications. From composting to biofuel production, serving as a filler in composite materials, and contributing to the pulp and paper industry, corn waste offers numerous avenues for repurposing and value creation.

Despite the evident benefits, there are notable challenges and risks associated with the utilization of corn waste. However, as advancements in corn waste management expand and its superior properties gain recognition across different sectors, the potential for creating new business opportunities becomes increasingly tangible. As industries continue to explore innovative ways to harness the potential of corn waste, there exists a genuine opportunity to drive economic growth while simultaneously promoting sustainability. Moving forward, concerted efforts to address challenges, foster collaboration, and leverage the inherent properties of corn waste will be essential in realizing its full potential as a sustainable resource and catalyst for positive change in various industries and sectors.

ACKNOWLEDGMENT

This chapter was written with Jessica Lim Joe Yee, Aina Natasha Suhaimi, and Muhammad Naim Mohd Hairi.

REFERENCES

Abdul Khalil, H. P. S., Bhat, A. H., & Ireana Yusra, A. F. (2012). Green composites from sustainable cellulose nanofibrils: A review. *Carbohydrate Polymers, 87*(2), 963–979. https://doi.org/10.1016/J.CARBPOL.2011.08.078

Adjapong, A. O., Ansah, K. D., Angfaarabung, F., & Sintim, H. O. (2015). Maize residue as a viable substrate for farm scale cultivation of oyster mushroom (Pleurotus ostreatus). *Advances in Agriculture,* 2015, 1–6. https://doi.org/10.1155/2015/213251

Ahmad, N. H., Md Nawawi, N., & Mohamad, S. (2014). Innovation of corn husk fiber into songket weaving towards sustainability of cottage industry. Paper presented at the International Colloquium of Art & Design Education Research (i-CADER 2014). Universiti Teknologi MARA (UiTM), Malaysia.

Akram, H. A., Imran, M., Javaid, A., Latif, S., Rizvi, N. B., Jesionowski, T., & Bilal, M. (2023). Pre-treatment and catalytic conversion of lignocellulosic and algal biomass into biofuels by metal organic frameworks. *Molecular Catalysis, 539*, 112893. Elsevier BV. https://doi.org/10.1016/j.mcat.2022.112893

Alagbe, E. E., Bassey, E. S., Daniel, O. E., Shongwe, M. B., Ojewumi, M. E., & Igwe, C. C. (2019). Physical, chemical and mechanical properties of corn sheath as pulp and paper raw material. *Journal of Physics: Conference Series, 1378*(3), 032083. https://doi.org/10.1088/1742-6596/1378/3/032083

Alltech. (2018). 2018 Alltech Global Feed Survey estimates world feed production in excess of 1 billion metric tons for second consecutive year. [Press release]. https://www.alltech.com/press-release/2018-alltech-global-feed-survey-estimates-world-feed-production-excess-1-billion

Amer, M. W., Aljariri Alhesan, J. S., Ibrahim, S., Qussay, G., Marshall, M., & Al-Ayed, O. S. (2021). Potential use of corn leaf waste for biofuel production in Jordan (physio-chemical study). *Energy, 214*, 118863. https://doi.org/10.1016/j.energy.2020.118863

Arikan, E. B., & Ozsoy, H. D. (2015). A review: Investigation of bioplastics. *Journal of Civil Engineering and Architecture, 9*(2). David Publishing Company. https://www.researchgate.net/publication/277593613_A_Review_Investigation_of_Bioplastics

Badu, M., Twumasi, S. K., & Boadi, N. O. (2011). Effects of lignocellulosic in wood used as substrate on the quality and yield of mushrooms. *Food & Nutrition Sciences, 2*(7), 780–784. https://doi.org/10.4236/fns.2011.27107

Bähner, F. D., Prado-Rubio, O. A., & Huusom, J. K. (2021). Challenges in optimization and control of biobased process systems: An industrial-academic perspective. *Industrial & Engineering Chemistry Research, 60*(42), 14985–15003. https://doi.org/10.1021/acs.iecr.1c01792

Boey, J. Y., Lee, C. K., & Tay, G. S. (2022). Factors affecting mechanical properties of reinforced bioplastics: A review. *Polymers (Basel), 14*(18), 3737. https://doi.org/10.3390/polym14183737

Bot, A., & Benites, J. (2005). The importance of soil organic matter. https://www.fao.org/3/a0100e/a0100e09.htm

Bousfield, G., Morin, S., Jacquet, N., & Richel, A. (2018). Extraction and refinement of agricultural plant fibers for composites manufacturing. *Comptes Rendus Chimie, 21*(9), 897–906. https://doi.org/10.1016/j.crci.2018.07.001

Bušić, A., Marđetko, N., Kundas, S., Morzak, G., Belskaya, H., Ivančić Šantek, M., Komes, D., Novak, S., & Šantek, B. (2018). Bioethanol production from renewable raw materials and its separation and purification: A review. *Food Technology and Biotechnology, 56*(3). Faculty of Food Technology and Biotechnology - University of Zagreb. https://doi.org/10.17113/ftb.56.03.18.5546

Castrillón, H. D. C., Aguilar, C. M. G., & Álvarez, B. E. A. (2021). Circular economy strategies: Use of corn waste to develop biomaterials. *Sustainability, 13*(15), 8356. https://doi.org/10.3390/su13158356

Chen, Q., Xiong, J., Chen, G., & Tan, T. (2020). Preparation and characterization of highly transparent hydrophobic nanocellulose film using corn husks as main material. *International Journal of Biological Macromolecules, 158*, 781–789. https://doi.org/10.1016/j.ijbiomac.2020.04.250

Choi, J. Y., Nam, J., Yun, B. Y., Kim, Y. U., & Kim, S. (2022). Utilization of corn cob, an essential agricultural residue difficult to disposal: Composite board manufactured improved thermal performance using microencapsulated PCM. *Industrial Crops and Products, 183*, 114931. https://doi.org/10.1016/j.indcrop.2022.114931

Chong, T. Y., Law, M. C., & Chan, Y. S. (2021). The potentials of corn waste lignocellulosic fiber as an improved reinforced bioplastic composites. *Journal of Polymers and the Environment, 29*(2), 363–381. https://doi.org/10.1007/s10924-020-01888-4

Chun, K. S., Maimunah, T., Yeng, C. M., Yeow, T. K., & Kiat, O. T. (2020). Properties of corn husk fiber reinforced epoxy composites fabricated using vacuum-assisted resin infusion. *Journal of Physical Science, 31*(3), 17–31. https://doi.org/10.21315/jps2020.31.3.2

David, E., & Niculescu, V.-C. (2021). Volatile organic compounds (VOCs) as environmental pollutants: Occurrence and mitigation using nanomaterials. *International Journal of Environmental Research and Public Health, 18*(24), 13147. https://doi.org/10.3390/ijerph182413147

wwDou, Z., Toth, J. D., & Westendorf, M. L. (2018). Food waste for livestock feeding: Feasibility, safety, and sustainability implications. *Global Food Security, 17*, 154–161. https://doi.org/10.1016/j.gfs.2017.12.003

DXP Marketing. (2021). *7 Hidden Benefits of Using Biofuel*. The Industrial Distribution Experts.

FAO. (2013). *Food Wastage Footprint: Impacts on Natural Resources*. FAO.

FAO. (2022). *FAO Publications Catalogue 2022*. https://doi.org/10.4060/CC2323EN

Ferguson, J. D. (2020). Food residue, loss and waste as animal feed. In: Hashmi, S., & Imtiaz Choudhury, A. (Eds.), *Encyclopedia of Renewable and Sustainable Materials* (pp. 395–407). Elsevier. https://doi.org/10.1016/b978-0-12-803581-8.11169-5

Gairola, S., Naik, T. P., Sinha, S., & Singh, I. (2022). Corncob waste as a potential filler in biocomposites: A decision towards sustainability. *Composites Part C: Open Access, 9*, 100317. https://doi.org/10.1016/j.jcomc.2022.100317

Galindo, F. S., Strock, J. S., & Pagliari, P. H. (2022). Impacts of corn stover management and fertilizer application on soil nutrient availability and enzymatic activity. *Scientific Reports, 12*(1), 1985. https://doi.org/10.1038/s41598-022-06042-9

Garcia, V., Cooter, E., Crooks, J., Hinckley, B., Murphy, M., & Xing, X. (2017). Examining the impacts of increased corn production on groundwater quality using a coupled modeling system. *Science of the Total Environment, 586*, 16–24. https://doi.org/10.1016/j.scitotenv.2017.02.009

Georganas, A., Giamouri, E., Pappas, A. C., Papadomichelakis, G., Galliou, F., Manios, T., Tsiplakou, E., Fegeros, K., & Zervas, G. (2020). Bioactive compounds in food waste: A review on the transformation of food waste to animal feed. *Foods, 9*(3), 291. MDPI AG. https://doi.org/10.3390/foods9030291

Godfray, H. C. J., Beddington, J. R., Crute, I. R., Haddad, L., Lawrence, D., Muir, J. F., Pretty, J., Robinson, S., Thomas, S. M., & Toulmin, C. (2010). Food security: The challenge of feeding 9 billion people. *Science, 327*(5967), 812–818. American Association for the Advancement of Science (AAAS). https://doi.org/10.1126/science.1185383

Hájková, K., Jurczyková, T., Filipi, M., & Bouček, J. (2023). Chemical pulp from corn stalks. *Biotechnology Reports, 37*, e00786. https://doi.org/10.1016/j.btre.2023.e00786

Hazrol, M., Sapuan, S., Zainudin, E., Wahab, N., & Ilyas, R. (2022). Effect of kenaf fiber as reinforcing fillers in corn starch-based biocomposite film. *Polymers (Basel), 14*(8), 1590. https://doi.org/10.3390/polym14081590

Herlina Sari, N., Wardana, I. N. G., Irawan, Y. S., & Siswanto, E. (2017). The effect of sodium hydroxide on chemical and mechanical properties of corn husk fiber. *Oriental Journal of Chemistry, 33*(6), 3037–3042. https://doi.org/10.13005/ojc/330642

Husseinsyah, S., Ismail, A., & Osman, H. (2015a). Corn stalk filled low density polyethylene bio composites: Influence maleic anhydride grafted polyetylene. *Applied Mechanics and Materials, 754–755*, 187–191. https://doi.org/10.4028/www.scientific.net/amm.754-755.187

Husseinsyah, S., Yeng, C.M., & Mann, T.K., (2018). Chemical treatment on corn cob by acrylic acid and its reinforced soy protein isolated/corn cob biocomposite films. *Journal of Vinyl and Additive Technology, 24*(1), pp.68–74. https://doi.org/10.1002/vnl.21528

Ibrahim, M. I. J., Sapuan, S. M., Zainudin, E. S., & Zuhri, M. Y. M. (2019a). Extraction, chemical composition, and characterization of potential lignocellulosic biomasses and polymers from corn plant parts. *Bioresources, 14*(3), 6485–6500.

Ibrahim, M. I. J., Sapuan, S. M., Zainudin, E. S., & Zuhri, M. Y. M. (2019b). Potential of using multiscale corn husk fiber as reinforcing filler in cornstarch-based biocomposites. *International Journal of Biological Macromolecules, 139*, 596–604. https://doi.org/10.1016/j.ijbiomac.2019.08.015

Ibrahim, M. I. J., Sapuan, S. M., Zainudin, E. S., & Zuhri, M. Y. M. (2020). Preparation and characterization of cornhusk/sugar palm fiber reinforced cornstarch-based hybrid composites. *Journal of Materials Research and Technology, 9*(1), 200–211. https://doi.org/10.1016/j.jmrt.2019.10.045

Jbilou, F., Joly, C., Galland, S., Belard, L., Desjardin, V., Bayard, R., Dole, P., & Degraeve, P. (2013). Biodegradation study of plasticised corn flour/poly(butylene succinate-co-butylene adipate) blends. *Polymer Testing, 32*(8), 1565–1575. https://doi.org/10.1016/j.polymertesting.2013.10.006

Kambli, N., Basak, S., Samanta, K. K., & Deshmukh, R. R. (2016). Extraction of natural cellulosic fibers from cornhusk and its physico-chemical properties. *Fibers and Polymers, 17*(5), 687–694. https://doi.org/10.1007/s12221-016-5416-0

Khalil, H. P. S. A., Yap, S. W., Tye, Y. Y., Tahir, P. M., Rizal, S., & Nurul Fazita, M. R. (2018). Effects of Corn Starch and Kappaphycus alvarezii Seaweed Blend Concentration on the Optical, Mechanical, and Water Vapor Barrier Properties of Composite Films. *BioResources, 13*(1), 1157–1173.

Khan, A., Niazi, M. B. K., Ansar, R., Jahan, Z., Javaid, F., Ahmad, R., Anjum, H., Ibrahim, M., & Bokhari, A. (2023). Thermochemical conversion of agricultural waste to hydrogen, methane, and biofuels: A review. *Fuel, 351*, 128947. Elsevier BV. https://doi.org/10.1016/j.fuel.2023.128947

Lambert, J. (2019). Growing corn is a major contributor to air pollution, study finds. NPR. https://www.npr.org/sections/thesalt/2019/04/01/708818581/growing-corn-is-a-major-contributor-to-air-pollution-study-finds

Laurijssen, J., Marsidi, M., Westenbroek, A., Worrell, E., & Faaij, A. (2010). Paper and biomass for energy? *Resources, Conservation and Recycling, 54*(12), 1208–1218. https://doi.org/10.1016/j.resconrec.2010.03.016

Lenz, D. M., Tedesco, D. M., Camani, P. H., & dos Santos Rosa, D. (2018). Multiple reprocessing cycles of corn starch-based biocomposites reinforced with curauá fiber. *Journal of Polymers and the Environment, 26*(7), 3005–3016. https://doi.org/10.1007/s10924-018-1179-6

Lima, J. S., de Queiroz, J. E. G., & Freitas, H. B. (2004). Effect of selected and non-selected urban waste compost on the initial growth of corn. *Resources, Conservation and Recycling, 42*(4), 309–315. https://doi.org/10.1016/j.resconrec.2004.02.006

Liu, Y., Xie, J., Wu, N., Ma, Y., Menon, C., & Tong, J. (2019). Characterization of natural cellulose fiber from corn stalk waste subjected to different surface treatments. *Cellulose, 26*(8), 4707–4719. https://doi.org/10.1007/s10570-019-02429-6

Long, G.-Q., & Sun, B. (2012). Nitrogen leaching under corn cultivation stabilized after four years application of pig manure to red soil in subtropical China. *Agriculture, Ecosystems & Environment, 146*(1), 73–80. https://doi.org/10.1016/j.agee.2011.10.013

Luo, H., Zhang, C., Xiong, G., & Wan, Y., (2016). Effects of alkali and alkali/silane treatments of corn fibers on mechanical and thermal properties of its composites with polylactic acid. *Polymer composites*, *37*(12), pp.3499-3507. https://doi.org/10.1002/pc.23549

Mansoori, E., Reza, A., & Esfahani, S. (2019). Preparation and characterization of corn starch/soy protein biocomposite film reinforced with graphene and graphene oxide nanoplatelets. *Polymers Advanced Technologies, 30*, 2301–2312. https://doi.org/10.1002/pat.4657

Mansour, G., et al. (2019). Investigation on the effects of glycerol and clay contents on the structure and mechanical properties of maize starch nanocomposite films. 1–37. https://doi.org/10.1002/star.201900166.

Mendes, C. A., Adnet, F., Leite, M. C. A. M., Furtado, C., & Maria, F. (2015). Chemical, physical, mechanical, thermal and morphological characterization of corn husk residue. *Cellulose Chemistry and Technology, 49*, 727–735.

Mishra, P., Naik, S., Babu, P. V., Pradhan, U., Begum, M., Kaviarasan, T., Vashi, A., Bandyopadhyay, D., Ezhilarasan, P., Panda, U. S., & Murthy, M. V. R. (2022). Algal bloom, hypoxia, and mass fish kill events in the backwaters of Puducherry, Southeast coast of India. *Oceanologia, 64*(2), 396–403. Elsevier BV. https://doi.org/10.1016/j.oceano.2021.11.003

Mohammed, A. A. B. A., Hasan, Z., Omran, A. A. B., Kumar, V. V., Elfaghi, A. M., Ilyas, R. A., & Sapuan, S. M. (2022). Corn: Its structure, polymer, fiber, composite, properties, and applications. *Polymers, 14*(20), 4396. MDPI AG. https://doi.org/10.3390/polym14204396

Nadlene, R., Sapuan, S.M., Jawaid, M., Ishak, M.R., & Yusriah, L. (2018). The effects of chemical treatment on the structural and thermal, physical, and mechanical and morphological properties of roselle fiber-reinforced vinyl ester composites. *Polymer Composites*, *39*(1), 274–287. https://doi.org/10.1002/pc.23927

Narisetty, V., Reshmy, R., Maitra, S., Tarafdar, A., Alphy, M. P., Kumar, A. N., Madhavan, A., Sirohi, R., Awasthi, M. K., Sindhu, R., Varjani, S., & Binod, P. (2022). Waste-derived fuels and renewable chemicals for bioeconomy promotion: A sustainable approach. *BioEnergy Research, 16*(1), 16–32. Springer Science and Business Media LLC. https://doi.org/10.1007/s12155-022-10428-y

Neves, A. C., Moyne, M. M., Eyre, C., & Casey, B. P. (2020). Acceptability and societal impact of the introduction of bioplastics as novel environmentally friendly packaging materials in Ireland. *Clean Technologies, 2*(1), 127–143. https://doi.org/10.3390/cleantechnol2010009

Obasi, H. C., & Onuegbu, G.C. (2013). Biodegradability and mechanical properties of low density polyethylene/waste maize cob flour blends. *International Journal of Applied Science and Engineering Research, 2*(3), 233–240.

Olesen, P., & Plackett, D. (1999). Perspectives on the performance of natural plant fibers. Natural Fibres Performance Forum, Copenhagen, Denmark, May 27–28.

Osman, A. I., Mehta, N., Elgarahy, A. M., Al-Hinai, A., Al-Muhtaseb, A. H., & Rooney, D. W. (2021). Conversion of biomass to biofuels and life cycle assessment: A review. *Environmental Chemistry Letters, 19*(6), 4075–4118. https://doi.org/10.1007/s10311-021-01273-0

Pan, Z., Zhang, R., & Zicari, S. (Eds.). (2019). *Integrated Processing Technologies for Food and Agricultural By-Products*. Elsevier. https://doi.org/10.1016/C2017-0-00901-1

Pinto, J., Cruz, D., Paiva, A., Pereira, S., Tavares, P., Fernandes, L., & Varum, H. (2012). Characterization of corn cob as a possible raw building material. *Construction and Building Materials, 34*, 28–33. Elsevier BV. https://doi.org/10.1016/j.conbuildmat.2012.02.014

Rahmayanti, H., & Asma, A. (2022). Development and characterization of paper from corn husks as a new material for newsprint. *Proceedings of the First Jakarta International Conference on Multidisciplinary Studies Towards Creative Industries, JICOMS 2022*, 16 November 2022, Jakarta, Indonesia. EAI. https://doi.org/10.4108/eai.16-11-2022.2326122

Ramamoorthy, S. K., Skrifvars, M., & Rissanen, M. (2015). Effect of alkali and silane surface treatments on regenerated cellulose fiber type (Lyocell) intended for composites. *Cellulose, 22*(1), 637–654. https://doi.org/10.1007/s10570-014-0526-6

Rambey, R., Sitepu, I. D. B., & Siregar, E. B. M. (2019). Productivity of oyster mushrooms (Pleurotus ostreatus) on media corncobs mixed with sawdust. *IOP Conference Series: Earth and Environmental Science, 260*(1), 012076. https://doi.org/10.1088/1755-1315/260/1/012076

Ramli, N., Mazlan, N., Ando, Y., Leman, Z., Abdan, K., Aziz, A. A., & Sairy, N. A. (2018). Natural fiber for green technology in automotive industry: A brief review. *IOP Conference Series: Materials Science and Engineering, 368*, 012012. IOP Publishing. https://doi.org/10.1088/1757-899x/368/1/012012

Raychaudhuri, A., & Ghosh, S. K. (2016). *Challenges Related to Biomass*. European Biomass Industry Association.

Ren, L., Yan, X., Zhou, J., et al. (2017). Influence of chitosan concentration on mechanical and barrier properties of corn starch/chitosan films. *International Journal of Biological Macromolecules, 98*, 1–8.

Riedel, U. (2012). Biocomposites. In: Matyjaszewski, K., & Möller, M. (Eds.), *Polymer Science: A Comprehensive Reference* (pp. 295–315). Elsevier. https://doi.org/10.1016/B978-0-444-53349-4.00268-5

Rius, B. L. (2019). Biocomposites in the automotive industry: potential applications and benefits. AIMPLAS. https://www.aimplas.net/blog/biocomposites-in-the-automotive-industry-potential-applications-and-benefits/

Rudin, A., & Choi, P. (Eds.). (2013). Biopolymers. In: *The Elements of Polymer Science & Engineering* (pp. 521–535). Elsevier. https://doi.org/10.1016/B978-0-12-382178-2.00013-4

Russell, A. G., & Brunekreef, B. (2009). A focus on particulate matter and health. *Environmental Science & Technology, 43*(13), 4620–4625. https://doi.org/10.1021/es9005459

Salonen, H., Salthammer, T., & Morawska, L. (2019). Human exposure to NO_2 in school and office indoor environments. *Environment International, 130*, 104887. https://doi.org/10.1016/j.envint.2019.05.081

Santolini, E., Bovo, M., Barbaresi, A., Torreggiani, D., & Tassinari, P. (2021). Turning agricultural wastes into biomaterials: Assessing the sustainability of scenarios of circular valorization of corn cob in a life-cycle perspective. *Applied Sciences, 11*(14), 6281. https://doi.org/10.3390/app11146281

Sarkar, N., Ghosh, S. K., Bannerjee, S., & Aikat, K. (2012). Bioethanol production from agricultural wastes: An overview. *Renewable Energy, 37*(1), 19–27. https://doi.org/10.1016/j.renene.2011.06.045

Sepe, R., Bollino, F., Boccarusso, L., & Caputo, F. (2018). Influence of chemical treatments on mechanical properties of hemp fiber reinforced composites. *Composites Part B: Engineering, 133*, 210–217. https://doi.org/10.1016/j.compositesb.2017.09.030

Sequeira, S., Cabrita, E. J., & Macedo, M. F. (2012). Antifungals on paper conservation: An overview. *International Biodeterioration & Biodegradation, 74*, 67–86. https://doi.org/10.1016/j.ibiod.2012.07.011

Serna-Saldivar, S. O. (Ed.) (2019). *Corn: Chemistry and Technology*. Elsevier. https://doi.org/10.1016/C2016-0-01986-1

Shafer, M. (2020). Global crop waste burning – micro-biochar; how a small community development organization learned experientially to address a huge problem one tiny field at a time. *Sustainable Earth, 3*(1), 18. https://doi.org/10.1186/s42055-020-00037-y

Shahbandeh, M. (2023). Top corn producing countries worldwide 2022/23. Statista. https://www.statista.com/statistics/254292/global-corn-production-by-country/

Shanmugasundaram, N., Rajendran, I., & Ramkumar, T. (2018). Characterization of untreated and alkali treated new cellulosic fiber from an Areca palm leaf stalk as potential reinforcement in polymer composites. *Carbohydrate Polymers, 195*, 566–575. https://doi.org/10.1016/j.carbpol.2018.04.127

Shiralipour, A., McConnell, D. B., & Smith, W. H. (1992). Uses and benefits of MSW compost: A review and an assessment. *Biomass and Bioenergy, 3*(3–4), 267–279. https://doi.org/10.1016/0961-9534(92)90031-K

Sindhu, R., Binod, P., Pandey, A., Ankaram, S., Duan, Y., & Awasthi, M. K. (2019). Biofuel production from biomass. In: Pandey, A., Negi, S., & Soccol, C. R. (Eds.), *Current Developments in Biotechnology and Bioengineering* (pp. 79–92). Elsevier. https://doi.org/10.1016/B978-0-444-64083-3.00005-1

Song, H., Jia, H., Wang, Q., Zhao, X., Yang, G., Zhang, M., Zhou, H., Xu, S., Zang, Y., Wang, Y., & Ma, L. (2020). A new environmentally-friendly system for extracting cellulose from corn straw: The low temperature laccase system. *Materials, 13*(2), 437. MDPI AG. https://doi.org/10.3390/ma13020437

Tang, X., Zhang, X., Zhang, H., Zhuang, X., & Yan, X. (2018). Corn husk for noise reduction: Robust acoustic absorption and reduced thickness. *Applied Acoustics, 134*, 60–68. https://doi.org/10.1016/j.apacoust.2018.01.012

Tiammee, S., & Likasiri, C. (2020). Sustainability in corn production management: A multi-objective approach. *Journal of Cleaner Production, 257*, 120855. https://doi.org/10.1016/j.jclepro.2020.120855

Tse, T. J., Wiens, D. J., & Reaney, M. J. T. (2021). Production of bioethanol—A review of factors affecting ethanol yield. *Fermentation, 7*(4), 268. https://doi.org/10.3390/fermentation7040268

United Nations. (2019). World population prospects 2019: Highlights. Statistical Papers - United Nations (Ser. A), Population and Vital Statistics Report. UN. https://doi.org/10.18356/13bf5476-en

University of San Carlos. (2023). Reimagine corn husks as eco-friendly packaging. https://nxplorers.com/en/case-studies/philippines

Verma, D., Fortunati, E, Jain, S., & Zhang, X. (2019). *Biomass, Biopolymer-Based Materials, and Bioenergy*. Elsevier. https://doi.org/10.1016/c2017-0-00839-x

Verma, M. (2014). Taguchi analysis of erosion wear maize husk based polymer composite. *IJMER*, *4*(3), 130–138.

Wang, K., Wang, W., Ye, R., Liu, A., Xiao, J., Liu, Y., & Zhao, Y. (2017). Mechanical properties and solubility in water of corn starch–collagen composite films: Effect of starch type and concentrations. *Food Chemistry*, *216*, 209–216.

Wilson, G. V., McGregor, K. C., & Boykin, D. (2008). Residue impacts on runoff and soil erosion for different corn plant populations. *Soil & Tillage Research, 99*(2), 300–307. https://doi.org/10.1016/j.still.2008.04.001

Yeng, C. M., Husseinsyah, S., & Ting, S. S. (2015). A comparative study of different crosslinking agent-modified chitosan/corn cob biocomposite films. *Polymer Bulletin*, *72*(4), 791–808. https://doi.org/10.1007/s00289-015-1305-8

Youssef, A. M., El-Gendy, A., & Kamel, S. (2015). Evaluation of corn husk fibers reinforced recycled low density polyethylene composites. *Materials Chemistry and Physics, 152*, 26–33. https://doi.org/10.1016/j.matchemphys.2014.12.004

Zárate-Salazar, J. R., Santos, M. N., Caballero, E. N. M., Martins, O. G., & Herrera, Á. A. P. (2020). Use of lignocellulosic corn and rice wastes as substrates for oyster mushroom (Pleurotus ostreatus Jacq.) cultivation. *SN Applied Sciences, 2*(11), 1904. https://doi.org/10.1007/s42452-020-03720-z

Zheng, M., Zhang, K., Zhang, J., Zhu, L., Du, G., & Zheng, R. (2022). Cheap, high yield, and strong corn husk-based textile bio-fibers with low carbon footprint via green alkali retting-splicing-twisting strategy. *Industrial Crops and Products, 188*, 115699. Elsevier BV. https://doi.org/10.1016/j.indcrop.2022.115699

Zhou, H., Guan, Y., Yan, X., Pan, Z., Xu, J., Dai, L., Zhang, M., & Si, C. (2023). All-lignocellulose-based hard bioplastic. *Industrial Crops and Products, 193*, 116164. Elsevier BV. https://doi.org/10.1016/j.indcrop.2022.116164

zu Ermgassen, E. K. H. J., Phalan, B., Green, R. E., & Balmford, A. (2016). Reducing the land use of EU pork production: Where there's swill, there's a way. *Food Policy, 58*, 35–48. Elsevier BV. https://doi.org/10.1016/j.foodpol.2015.11.001

14 Rice Waste Management

14.1 INTRODUCTION

Rice holds tremendous significance as a primary cereal crop globally, catering to the dietary needs of more than half of the world's population. As reported in the "World Food and Agriculture—Statistical Yearbook 2021," the total production of paddy rice in 2020 reached approximately 500 million tons globally. Significantly, rice is a staple food in many countries, particularly in Asia, where it forms an essential part of daily meals (Ma, 2015). Rice cultivation also produces substantial quantities of materials, which could harm both the natural world and the well-being of humans. Rice trash consists of the husks of rice, straws, and grains, among other things. Worldwide waste production from rice cultivation is predicted to be 731 million tons, with Asia accounting for 667.7 million tons. Rice straw, the stem and leaf that stay after the grain is harvested, is one of the most prevalent types of rice waste. Rice husk, the hard outer coating of the rice grain, is another important type of rice trash. Rice bran, cracked rice, and rice shells are other types of rice waste (Figures 14.1 and 14.2).

Rice husk is another rice waste produced during rice crop harvesting. On an annual basis, approximately 4 billion tons of rice straw is generated globally (Nguyen et al., 2020). Rice straw can be used to make biofuels like biogas and bioethanol, as well as charcoal and plastics (Mofijur et al., 2019).

FIGURE 14.1 Rice.

DOI: 10.1201/9781003472209-19

FIGURE 14.2 Rice straw (Moraes et al., 2014).

FIGURE 14.3 Rice husk (Moraes et al., 2014).

These husks are the most common rice waste, making up around 20% of total harvested rice (Anggraeni et al., 2022). They are typically burned or thrown away, producing pollution and wasting a valuable resource. Rice husks, on the contrary, can be used to make renewables like ethanol from biomass and biogas, as well as other goods like activated carbon and silica (Mofijur et al., 2019) (Figure 14.3).

Rice grain, a by-product of the cultivation of rice, is high in nutrients and bioactive substances. Each year, 63–76 million tons of rice husk are generated worldwide (Chiou et al., 2013). Rice bran can be processed into high-value goods such as oil for cooking, livestock feed, and beneficial food ingredients (Tan et al., 2023) (Figure 14.4).

Because of its harmful effects on the environment and human health, rice waste has become an increasingly relevant concern. According to the "World Food and Agriculture—Statistical Yearbook 2021" (2021), an estimated 150 million tons of rice are produced globally each year, resulting in a large accumulation of rice waste. This research aims to look into the technical, biological, chemical, mechanical, morphological, and thermal aspects of rice waste; analyze the environmental implications of rice biomass; and determine innovative prospective uses for rice waste. This discarded rice is susceptible to carbon dioxide emissions, pollution of the environment, and human wellness problems, such as breathing disorders caused by burning rice straws.

FIGURE 14.4 Rice bran (Moraes et al., 2014).

TABLE 14.1
Proximate Composition of Different Varieties of Rice (Ghanghas et al., 2022)

Rice Variety	Moisture	Ash	Fat	Protein	Dietary Fiber	Carbohydrate	Energy (kcal)
Red	11.55±0.04	1.33±0.01	2.86±0.02	9.21±0.13	3.63±1.30	75.04±0.15	362.78±0.24
Bahng Gawk	12.38±0.03	1.40±0.01	2.91±0.05	7.40±0.11	4.18±0.14	75.92±0.11	359.43±0.30
Haek yah	11.45±0.03	1.5±0.07	2.37±0.06	7.16±0.01	3.17±0.84	77.53±0.01	360.06±0.49
Niaw Look Pueng	9.28±0.06	1.42±0.12	2.67±0.06	10.36±0.04	4.51±1.60	76.27±0.13	370.53±0.95
Sung YodPhatthalung	12.01±0.01	1.45±0.06	3.19±0.06	9.62±0.16	3.75±0.79	73.73±0.10	362.10±0.26
NiawDawk Yong	13.12±0.16	1.26±0.05	3.08±0.08	7.35±0.06	3.27±1.20	75.20±0.16	357.86±0.37
Niaw Lan Tan	12.94±0.03	0.82±0.14	1.15±0.03	9.63±0.04	2.82±0.55	75.45±0.09	350.72±0.42
Sri Lanka Red Rice 1	11.12±0.06	1.12±0.44	2.19±0.10	9.52±0.12	2.87±0.59	76.05±0.27	362.01±2.03
Sri Lanka Red Rice 2	9.85±0.15	0.98±0.19	1.17±0.05	8.72±0.06	2.88±0.55	79.27±0.26	362.51±0.65
Sri Lanka Red Rice 3	11.90±0.20	1.37±0.02	2.35±0.03	9.72±0.04	2.52±0.24	76.66±0.17	358.72±0.82
China Red Rice Black	12.59±0.16	1.42±0.01	3.72±0.06	8.17±0.41	4.01±0.58	74.09±0.48	362.55±0.85
Niaw Dam Pleuak Khao	12.03±0.13	1.48±0.02	3.65±0.05	10.85±0.09	3.41±0.24	71.99±0.08	364.22±0.79
China Black Rice	11.26±0.28	1.74±0.02	2.85±0.09	8.44±0.02	4.08±0.54	75.71±0.37	362.25±0.96

Nonetheless, rice waste has a lot of promise as an essential asset. Rice straw, for example, can be used as feed for livestock or as an input in the manufacturing of bioenergy. Rice husk, on the reverse side, can generate power or manufacture construction materials and insulation. Rice waste management gives a chance to lessen its adverse environmental impact while also offering potential revenue.

14.2 BIOMASS FROM NATURAL RICE FIBERS

In nations where rice is consumed daily, it provides a key source of protein and food for a huge section of the global population. Extensive studies have shown that rice grains are a good source of energy, giving 348.79–370.53 kcal/100 g (Verma & Srivastav, 2017). According to Singh et al. (2014a), the transplantation date of paddy can affect the physical qualities and composition of rice. The development of fat, protein, and carbohydrates in rice seeds can be affected by prolonged transplant, which results in lower evening hours weather. Sompong et al. (2011) identified the subsequent comparable measurements of the closest example of rice seeds per 100g on the assumption of dry matter in a study evaluating several kinds of rice from different countries: Protein is 7.16–10.85g, moisture is 9.28–13.12g, ash is 0.82–1.74g, fat is 1.15–3.72g, total dietary fiber is 2.52–4.51g, and total carbs are 71.99–79.27g. The closest representation of various rice types is provided in Table 14.1 (Ghanghas et al., 2022).

14.3 ENVIRONMENTAL IMPACT OF BIOMASS

14.3.1 Biomass is Not Nutrient Neutral

During the gasification and pyrolysis of the biomass raw materials, ash, wastewater, ash, indifferent petroleum products, traces of gases such as carbon dioxide (CO_2), ammonia (NH_3), hydrogen sulfide (H_2S), hydrogen cyanide (HCN), carbonyl sulfide (COS), and carbon disulfide (CS_2), along with organic substances such as phenols and POMs, are generated. Environmental circumstances, control measures, and input composition all play a role in the physical features of the air emissions generated by these conversion processes (Trimble et al., 1984).

It comprises more than simply carbon; it also includes ammonium and a few additional crucial elements. Aside from absorbing carbon from the atmosphere, excessive production of biomass harms the ecosystem due to horticulture's significance in global nitrogen cycles (Wang et al., 2023). Agriculture-related activities are responsible for more than 75% of the reactive nitrogen compounds that are now emitted. Whereas the level of nitrogen dioxide in the earth's atmosphere has increased substantially by 15% since 1750, carbon dioxide has risen by more than a third. When you consider that each nitrogen dioxide molecule has 300 times the capacity to contribute to climate change as a single unit of carbon dioxide (Abbasi & Abbasi, 2010), the number takes on scary aspects.

Due to industrial ecosystem modifications, the conversion of inactive nitrogen in the atmosphere into reactive nitrogen-containing substances has risen by approximately 1,100%. Nitrogen remains a reactive response in surroundings as sodium hydroxide, nitrogen dioxide, nitrogen oxides, and nitrogen oxide have an array of adverse impacts on the natural environment. This includes the development of ozone at the ground level, algal blooms, hypoxia, ozone depletion in the stratosphere, and global warming. The degradation of the surface waters and contamination of subsurface aquifers are two of these consequences, which are also the most extensive. The degradation is caused by disrupted phosphoric cycles (Abbasi & Abbasi, 2010).

It is still entirely unknown what happens to the waste produced during post-harvest processes, particularly damaged fruit. It is assumed that a sizeable percentage of this residue is just thrown away, which could have a negative impact on the environment. Post-harvest agricultural losses certainly have an impact on environmental problems in addition to financial and food security difficulties. Due in part to the lack of leachate protection measures at Indian dumps, the leaching from these sites has the potential to affect groundwater. When estimating greenhouse gas emissions, it is assumed that all post-harvest waste is dumped or landfilled uncontrolled, resulting in an estimated annual emission of about 2.8 million tons of CH_4, or almost 60 million tons of CO_2 equivalent. In contrast, it was calculated that India's total CO_2 emissions in 2006 were 1,510 million tons (United Nations Statistics Division—Environment Statistics, 2007). In 1999, landfill emissions from municipal solid waste were about 0.4 million tons of CH_4, comparable to 8.4 million tons of CO_2-eq. However, it is important to note that this assumption probably overestimates the emissions, as post-harvest losses may include weight reductions due to moisture content reductions and consumption by insects and rodents in addition to spoiled food that must be disposed of (Cardoen et al., 2015).

14.3.2 Land and Water Resources

A significant biomass energy generation program necessitates abundant land and water resources. Horticulture, in particular, demands a tremendous amount of water, requiring several orders of magnitude more water per hectare compared to domestic and industrial needs. In Asia, the Indian subcontinent, characterized by the fertile Gangetic Delta and Punjab Plains, holds vast expanses of rice-cropping land. States like West Bengal, Uttar Pradesh, and Punjab in India, for instance, collectively produced around 124 million metric tons of rice in the year 2021–2022 (Rathour et al., 2023). The extensive use of pesticides and fertilizers in intensive farming also contributes significantly to water contamination. Additionally, as biomass production for energy increases, it leads to competition for the same land among crops, forests, and urban development (Shashidara et al., 2022).

The difference between the amount of cropland needed to feed one person and the amount needed to fuel one car for a year may be used to demonstrate this competition. Additionally, the demand for agricultural and forestry goods is expanding quickly over time, widening the already significant gap between supply and demand and escalating competition for land and water resources. While jatropha cultivation has already influenced food costs in some locations due to farmers switching from food crops to jatropha cultivation, jatropha biofuel extraction has yet to significantly impact India's fossil fuel bill. In comparison to the usage of fossil fuels, the net consequence of this change in land use is an increase in CO_2 emissions lasting decades or millennia (Abbasi & Abbasi, 2010).

The elimination of rice residue from the environment for energy generation initiatives causes flooding, nitrogen loss, and soil and water damage. The natural ecosystems and wildlife are also impacted. The manufacturing of biomass poses these and other environmental risks.

14.3.3 Soil Erosion and Water Runoff

Projects that produce electricity from rice biomass will likely worsen soil erosion issues. Although modern technology can reduce erosion, doing so is expensive and time-consuming. They might not even be adopted in emerging nations like India and Brazil. To produce energy crops like maize for ethanol, more agricultural acreage is needed. This would require cultivating maize on marginal cropland that is very sensitive to soil erosion (Swain, 2009). Groundwater recharge is delayed as a result of soil erosion, which also contributes considerably to water runoff. Nutrient-rich runoff can degrade the quality of receiving rivers, lakes, or estuaries by contributing to eutrophication (Abbasi & Abbasi, 2010).

14.3.4 Nutrient Removal and Losses

The collection of agricultural wastes for rice biomass energy results in substantial nutritional loss. With a 7,840 kg/ha maize yield, the nutrients are as follows: 224 kg N, 37 kg P, 140 kg K, and 6 kg Ca, with the residues containing about half of the nutrients. At least 460 L of oil per hectare in energy would be required to restore the nutrients lost when the biofuel plants are harvested (Pimentel et al., 1984).

14.3.5 Loss of Natural Biota, Habitats, and Wildlife

The habitat and food supplies for animals and other biota will alter as a result of the conversion of natural ecosystems into plantations for energy crops. Numerous favored habitats and breeding grounds for various animals, birds, and other biota will be reduced as a result of changes to forests and wetlands (Ghani et al., 2023). Fast-growing tree monoculture plantations lessen the vegetation's diversity and the importance of the regions as habitats for several animal species. These monocultures require more energy inputs in the form of pesticides and fertilizers to sustain production since they are less stable than climatic forests. Trees in commercial plantations are two to three times more densely packed than those in wild forests, which might lead to more insect issues (Abbasi & Abbasi, 2010).

14.4 AVAILABILITY OF RICE WASTE

Agro-waste materials, including rice waste, are widely available in many parts of the world. According to a study by Kumar et al. (2021), rice straw, rice husks, and rice bran are abundant agricultural by-products in rice-growing countries such as India and China. These materials are often considered waste and are either burnt or disposed of in landfills, causing environmental problems.

Increasing interest in utilizing agro-waste materials for various applications, including energy production, biofuels, and value-added products. For example, a study by Hossain et al. (2021) explored the potential of rice husks for producing biofuels and other products, such as activated carbon and silica. In addition to rice waste, other types of agro-waste materials are available for various applications. For example, a study by Emdadi et al. (2021a) explored the use of agro-waste fibers, such as rice straw and wheat straw, for reinforcing composite materials. Another study by Jawaid et al. (2021a) investigated the potential of coconut coir and jute fibers for reinforcing polymer composites.

Hence, agro-waste materials, including rice waste, are abundant in many parts of the world and can be utilized for various applications. However, more research is needed to explore the full potential of these materials and to develop sustainable and cost-effective methods for their utilization.

14.5 PROPERTIES OF RICE WASTE

14.5.1 Physical Properties

Korotkova et al. (2016) studied on physical properties of rice waste (husk) using clay as a medium. The control sample had a one-cycle thermal resistance, but the sample containing husk had a thermal resistance of ten cycles at 1,200°C. The silica concentration ranges between 50% and 70%. For point metals with low melting points, it is recommended to employ materials such as the lining of thermal and melting furnaces. Elevated quantities of iron oxide suggest that the material is suitable for manufacturing ceramics. Table 14.2 physical properties of rice husk and dust. Laboratory experiments revealed that the grain dust surface area was 0.6–0.9 m^2/g, the bulk density of rice dust was 0.221 g/cu·cm, and the density of rice dust particles was 1.46 g/cu·cm. Table 14.3 lists the elemental composition of different parts of the grain dust (Korotkova et al., 2016).

TABLE 14.2
Physical Properties of Rice Husk and Dust (Korotkova et al., 2016)

Sample	Shrinkage (%)	Apparent Porosity (%)	Bulk Density (g/cu·cm)	Thermal Resistance, Cycles	Fire Resistance, °C
Clay	3.89	27.15	1.98	1	1,300
Clay mixed with husk	3.00	36.74	1.52	10	1,200

TABLE 14.3
The Elemental Composition of Different Parts of the Grain Dust (Korotkova et al., 2016)

Sample	Shrinkage (%)	Apparent Porosity (%)	Bulk Density (g/cu·cm)	Thermal Resistance, Cycles	Fire Resistance, °C
Clay	3.89	27.15	1.98	1	1,300
Clay mixed with husk	3.00	36.74	1.52	10	1,200

TABLE 14.4
Trace Element Values in Various Types of Rice Samples (You et al., 2017)

Constructed Rice	Chem Level (mg/kg)							
	Mg	Ca	Zn	Cu	Cr	Cd	Ni	Pb
SUMO Calrose	287.0	3.0	12.0	3.0	Nd	Nd	Nd	Nd
Jasmine Nutririce	42.0	14.0	18.0	14	Nd	Nd	Nd	Nd
Floral Glutinous	33.0	1.0	14.0	1.0	Nd	Nd	Nd	Nd
Sun Rise Australian Fragrant	31.0	6.0	14.0	6.0	Nd	Nd	Nd	Nd
Bird of Paradise Thai Fragrant	22.0	Nd	9.0	Nd	Nd	Nd	Nd	Nd
Phkarkhnei Cambodia Organi	35.0	13.0	13.0	13.0	Nd	Nd	Nd	Nd
Sakura Super Thai Brown Rice	29.0	10.0	11.0	10.0	Nd	Nd	Nd	Nd
Sakura Super Basmathi Pakistan	34.0	11.0	11.0	11.0	Nd	Nd	Nd	Nd
Giant Super Special White	91.0	8.0	11.0	8.0	Nd	Nd	Nd	Nd
Average value n.d.	62.7	8.3	12.3	7.5	Nd	Nd	Nd	Nd

Nd, not detected.

14.5.2 Chemical Properties

A number of variables influence the chemical characteristics of agricultural waste fibers, including fiber category, fiber length, and fiber orientation. Table 14.4 lists the results of the trace element content analysis in a factor of distinct types of rice samples. In these rice samples, the average amount of copper (Cu), magnesium (Mg), calcium (Ca), and zinc (Zn) were 7.5, 62.7, 8.3, and 12.3 mg/kg, respectively. However, none of the rice samples contained any chromium (Cr), cadmium (Cd), nickel (Ni), or lead (Pb). The average concentration of Cu ranged from 0 to 14 mg/kg, Ca ranged from 0 to 14 mg/kg, Zn ranged from 9 to 18 mg/kg, and Mg had a higher range from 22 to 287 mg/kg across all the rice samples. When the data in Table 14.4 were compared to the concentrations of heavy metals in rice specimens from various countries, there was little difference in levels of heavy metals among the two different sources. According to You et al. (2017), the average Zn, Pb, Ni, Cu, Cr, and Cd contents in Taiwanese rice samples were 14.6, 0.01, 0.26, 2.20, 0.07, and 0.02 mg/kg, respectively.

14.5.3 Mechanical Properties

A number of variables influence the physical characteristics of agricultural waste fibers, including fiber category, fiber length, and fiber orientation. Longer fibers offer stronger tensile strength and stiffness than shorter fibers, according to a study by Liu et al. (2021). Furthermore, fibers positioned in the opposite way of the applied force have better stiffness and strength than fibers positioned perpendicular to the load.

Agro-waste fibers also have high interfacial interaction with substrate substances, which may enhance composite products' overall durability. It was discovered that using agro-waste fibers like reinforcing materials in polyethylene materials made of composites such as coconut coir and jute significantly increased flexural strength, tensile strength, and impact strength. However, moisture content, fiber treatment, and aging may impact the tensile strength of agro-waste materials. Finally, agro-waste fibers have desired mechanical qualities such as high tensile strength and stiffness, making them appropriate for reinforcement components in composite substances. Several parameters influence the tensile strength of agro-waste fibers, including fiber length, orientation, and interfacial bonding. However, moisture content, fiber treatment, and aging can all have an impact on the tensile strength of agro-waste materials.

14.5.4 Morphological Properties

According to a study conducted by Fávaro et al. (2010), through scanning electron microscopy (SEM) images of rice husks in Figure 14.5, the outside region is formed by aligned and symmetrical bumps composed mainly of silica onto the cellulose, while the inside region presents a smooth cellulose surface. The energy-dispersive spectrograms of both rice husk particle areas are shown in Figure 14.6. A silica layer predominates on the exterior, providing a protective barrier against microbial assault. Figure 14.7 depicts micrographs of both the exterior (Figure 14.7a and b) and

FIGURE 14.5 SEM micrographs of two rice husk particles (Fávaro et al., 2010).

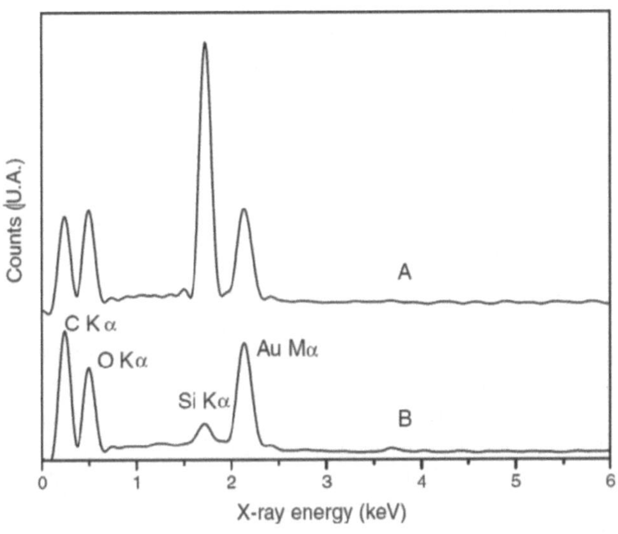

FIGURE 14.6 EDX spectra of rice husk: (A) outside and (B) inside regions (Fávaro et al., 2010).

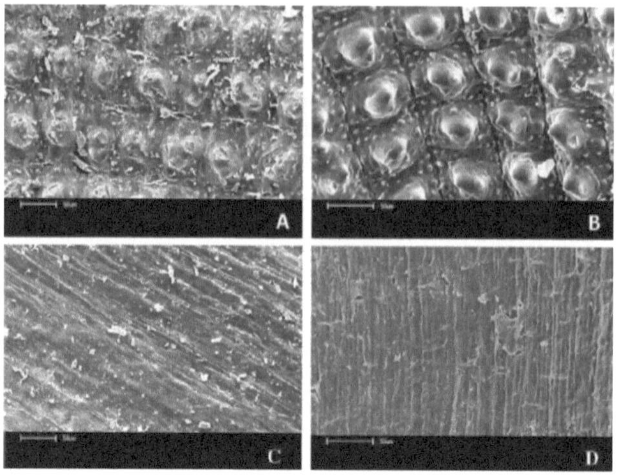

FIGURE 14.7 Rice husk SEM micrographs: (a) unmodified outside region, (b) mercerized outside region, (c) unmodified inside region, and (d) mercerized inside region (Fávaro et al., 2010).

interior (Figure 14.7c and d) of rice husk particles before (Figure 14.7a and c) and after mercerization (Figure 14.7b–d). Mercerization was used to remove the microscopic particles on both the interior and exterior surfaces of the rice husk. No further impact was detected on the exterior. However, the protective covering of waxes, hemicelluloses, and lignin were stripped from the inside, revealing the cellulose microfibers.

14.5.5 Thermal Properties

The contribution of buildings to climate change as a result of energy consumption and greenhouse gas emissions necessitates the use of agro-industrial by-products as alternative construction materials. In all instances, the curves exhibit remarkable similarity, with the initial decline attributed to moisture elimination, followed by a significant decrease resulting from the removal of volatiles and, ultimately, the degradation of fixed carbon. At elevated temperatures, the curves indicate that a certain percentage remains unaltered, which corresponds to the residual ash from Figure 14.5 and the combined addition of ash and fixed carbon in Figure 14.6. Additionally, it is noticeable that higher heating rates lead to slightly elevated thermal decomposition temperatures, a trend commonly observed in the thermal gravimetric (TG) curves of various substances (Arranz et al., 2021).

The thermal characteristics of rice can alter when it is stored at various temperatures. Rice aged at 37°C may have developed a better-organized grain structure since it has a greater breaking point temperature than rice aged at 4°C. In contrast to storage at lower temperatures, the differential thermal analysis revealed that storage at the higher temperature increased the peak temperature (T_p) and conclusion temperature (T_c) of rice, resulting in a wider differential scanning calorimetry (DSC) thermogram. Additional research was carried out employing cellulase and protease treatments to comprehend the mechanism underlying these modifications in the thermal characteristics of the rice. The amount of cell wall remains along the rice grain significantly increased when rice was stored at 37°C, according to the study of the remnants' amount. This implies that the structure of the aged rice may have become more organized. These results suggest that storage temperature and aging can impact rice's thermal properties, which may impact the quality and shelf life of rice-based products (Zhou et al., 2010) (Table 14.5).

TABLE 14.5
DSC Data for Rice Flour Ground from Rice Grains Following Storage for 12 Months at 4°C and 37°C (Zhou et al., 2010)

Variety	Storage Temperature	T_o (°C)	T_p (°C)	T_c (°C)	Peak Width (°C)	ΔH (J/g Rice Flour)[b]
Koshihikari	Fresh	61.2	67.7	73.1	7.1	7.4
	4	61.0	67.7	73.3	7.1	7.5
	37	61.6	68.4*	75.9**	8.3**	8.5
Kyeema	Fresh	73.5	77.9	81.9	4.9	10.3
	4	73.6	78.0	81.9	4.8	10.3
	37	74.1	78.9	85.4*	6.4**	11.1
Doongara	Fresh	69.4	74.2	79.1	5.5	8.7
	4	69.4	74.3	79.2	5.6	8.8
	37	69.9	75.0*	82.1**	7.0**	9.4

*$p<0.01$, **$p<0.001$, ***$p<0.05$.
T_o=onset of gelatinization
T_p=peak temperature
T_c=conclusion temperature
ΔH=enthalpy of gelatinization
[b]Standard error 0.3 J/g.

14.5.6 Calorific Value

The quantity of energy that can be released through the combustion or oxidation of rice residues is referred to as the caloric or calorific value of rice waste. This number represents the amount of heat energy rice waste produces when it is burned or utilized as fuel per unit mass or volume. It is essential to remember that the calorific content of rice waste varies based on a number of variables. The calorific value of rice waste is significantly influenced by its moisture content. Due to the energy needed to evaporate the extra water during burning, materials with a higher moisture content typically have lower energy content.

Rice straw, one of the main components of rice waste, has been the subject of numerous research examining its calorific value. Research in the *Journal of the Energy Institute* stated that rice straw normally has a calorific value of 14–15 megajoules per kilogram (MJ/kg). Rice straw has a comparatively high energy level due to its high carbon content. In addition, rice straw contains lignocellulosic substances, including cellulose and hemicellulose, that raise its calorific value (Huang et al., 2008). The calorific value of rice waste can also be impacted by impurities and pollutants like soil particles or inorganic minerals. As a result, these aspects must be taken into account when estimating the potential energy value of rice waste for certain uses. The calorific value of rice waste is of particular interest for its utilization as a bioenergy feedstock. Bioenergy conversion technologies, such as combustion, gasification, or pyrolysis, can utilize rice waste as a renewable energy source. By harnessing the energy stored in rice waste, it can be transformed into useful forms such as heat, electricity, or biofuels, thereby contributing to energy sustainability and reducing reliance on fossil fuels.

It is crucial to keep in mind that the calorific value of rice waste might change based on the kind of rice, the amount of moisture, and the presence of contaminants. Therefore, additional investigation and analysis may be required to pinpoint the precise calorific content of particular forms of rice waste under various circumstances. Through cutting-edge conversion technology and process optimization, continual efforts are also being made to increase the effectiveness of energy recovery from rice waste.

14.6 AGRO-WASTE MANAGEMENT

14.6.1 Compost

Rice straw, which is abundant organic waste, can be transformed into fertilizer through composting. As stated by Tiquia and Tam (2002), composting is a cost-effective biological treatment for various types of organic waste. This concept was initially introduced to utilize effective microorganisms (EM) in environmental management. Rice waste is a promising candidate for composting due to its high cellulose and lignin content. A high-quality compost can be produced by mixing rice straw with goat manure, green waste from fruits and vegetables, and EM. However, achieving optimal microbial degradation requires specific conditions.

In the research involving rice husk utilizing EMs done by Jusoh et al. (2013), all evaluated parameters revealed the occurrence of a breakdown process over the 90-day treatment period. Temperature and pH levels both suggested organic matter breakdown. The drop in the total organic carbon or TOC levels and the carbon/nitrogen (C/N) ratio revealed that bacteria were consuming organic substances. Furthermore, the use of EM in compost boosted both macro and micronutrient composition.

14.6.2 Pulp and Paper

Pulp and paper manufacturing has become a component of worldwide industrial output. Paper consumption, as well as demand, has grown significantly in tandem with population growth. The global demand for cellulose, documents, and board exceeds 402 million metric tons yearly, and this projected demand has only been realized in the preceding 2 years. It is possible that there could be a 521 million ton annual demand for paper by 2021. Paper consumption increased from 316 to 351 million tons between 1999 and 2005, and it is expected to reach 500 million tons (Mt) by 2025, representing a 1.6% yearly increase (Kaur et al., 2017).

In developing nations, primarily India, rice straw is a plentiful and convenient non-wood raw material. These nations generate a lot of rice straw, which causes significant issues for farmers. Around the world, an excessive amount of paddy (755 million metric tons) is produced each year, and an equivalent amount of straw is also generated. Turning this trash into a useful raw material for making paper is an environmentally responsible method of waste management (Maeno & Sano, 1950). It could be a viable choice for protecting our natural resources. Burning the leftovers allows farmers to get rid of them while lowering their environmental pollution load. Furthermore, it decreases the seriousness of medical issues associated with contaminants in the air caused by smoldering fiber. Because of its chemical composition, which consists primarily of hemicelluloses, cellulose, a substance called and silicon dioxide, rice straw is receiving an increasing amount of interest in the paper and pulp industry to extract additional usefulness from farm waste. Rice straw is used to create paper in Egypt and Southern and Eastern Asia (China, India, and Sri Lanka). Many paper and pulping facilities combine straw from rice with other agriculture-produced components in variable ratios to create paper for a variety of tasks such as writing, printing, low-grade wrapping sheets, greaseproof papers, and so on (Kaur et al., 2017).

The benefits of using rice straw for paper and board manufacture may be summed up as follows: Rice straw is available in large quantities and has been available in a consistent supply for many years (Dulay et al., 2012). Additionally, it is simple to gather, bale, and transport rice straw to the factory. Rice straw is inexpensive and has no further useful or lucrative uses. Without deteriorating, rice straw can be kept for at least a year. Storage facilities should have inexpensive ways of protection. It should be noted that rice straw is easily pulpable and does not require strong pulping conditions due to its open structure. Additionally, rice straw is substantially less expensive than other agricultural waste. Even without the recovery of chemicals, the production of chemical pulp from rice straw was economically feasible (Elhelece, 2020).

Agricultural leftovers are higher in cellulose and pentosans than raw forest material. Additionally, these residues have a lower lignin concentration, facilitating pulping and bleaching using fewer chemicals (Jagaba et al., 2022). Hemp, bagasse, and kenaf are three non-wood raw materials that can generate high-quality pulp but are not commercially feasible. Due to the enormous economic potential, there is also significant pressure on agricultural wastes, particularly bagasse and cereal straws for pulp and paper manufacture (Kaur et al., 2017).

Rice straw can be quickly converted into pulp and paper using semichemical pulp. Semichemical pulping can be used to produce coarse or fine-quality pulp. As was previously demonstrated, the low lignin concentration of these residues—much lower than that of wood—enables delignification or pulping. Agricultural leftovers are also easier for chemical reagents used in cooking to reach due to their open structures. The reagent molecules diffuse and enter the fine structure of the cell wall fast and profoundly. So, compared to wood, agricultural leftovers are more pulpable (Elhelece, 2020).

14.6.3 Biocomposite Materials

The utilization of rice waste in the development of biocomposite materials has gained attention as an innovative approach. Biocomposites are composite materials made by combining a matrix material with natural fibers or particles derived from renewable resources. Rice husk, a major component of rice waste, has been studied as a reinforcing filler in biocomposites. In a study by Hardinnawirda and SitiRabiatull Aisha (2012), the researchers investigated the effect of incorporating rice husk particles into a polymer matrix. The study demonstrated that the inclusion of rice husk led to improved mechanical properties, such as increased tensile strength and flexural modulus, in the resulting biocomposites (Figure 14.8).

Similarly, rice straw, another abundant rice waste, has been explored for its potential in biocomposite materials. In a study conducted by Aladejana et al. (2020), they examined the potential of rice straw fibers as a reinforcement material in biodegradable composites. The results of their research demonstrated that the inclusion of rice straw fibers led to enhancements in both the tensile strength and modulus of the biocomposites, all while maintaining their ability to biodegrade naturally. A

FIGURE 14.8 Biocomposite material from rice husk (Johnson & Nordin, 2009).

study by Dutta et al. (2018) explored the utilization of rice bran, a by-product of rice milling, as a filler in polymer composites. The findings of their investigation revealed that incorporating rice bran into the composites had a positive impact on their tensile and flexural properties, further enhancing the overall performance of the biocomposites. These studies collectively highlight the potential of utilizing different types of rice waste, such as rice husk, rice straw, and rice bran, in the development of biocomposite materials. Incorporating rice waste into biocomposites not only provides a sustainable solution for agro-waste management but also enhances the mechanical properties of the resulting materials, making them suitable for various applications in industries such as construction, packaging, and automotive.

These studies demonstrate the potential of rice waste, such as rice husk, rice straw, and rice bran, in the development of environmentally friendly biocomposite materials, contributing to sustainable agro-waste management and the transition toward more sustainable and greener material solutions. By converting rice waste into valuable biocomposite materials, the agricultural residues that would otherwise be discarded can be effectively utilized, reducing environmental impact and promoting a circular economy approach. Additionally, biocomposites offer advantages such as renewability, biodegradability, and potential reductions in greenhouse gas emissions compared to conventional synthetic composites.

14.6.4 Bio-Coal

Excessive mining and consumption of fossil fuels will not only steadily deplete the earth's storage capacity but will also bring serious climate change and pollution concerns. Biofuels, on the contrary, can be mass-produced and are acknowledged as a viable future energy source. The current strategy of producing biofuel from post-harvest rice crop leftovers is a potential step toward producing renewable energy sources to ameliorate energy demand. The environmental deterioration and energy crises caused by increasing daily consumption levels are major worldwide issues. The production of biofuel from post-harvest rice crop wastes effectively utilizes these residues and may enhance the long-term energy reserve (Das et al., 2019).

In Figure 14.9A, the process for preparing bio-coal is illustrated. In brief, renewable biomass, such as rice husk, sawdust, and maize stalk, undergoes rapid pyrolysis at 500°C under anaerobic conditions, producing bio-oil and biochar. Subsequently, the bio-oil is distilled in the presence of air, ranging from room temperature to approximately 240°C, to extract liquid chemicals and bio-coal. Bio-coal production follows standard chemical engineering methods and does not necessitate complex or expensive refinery equipment or operations, making it easily scalable (Cheng et al., 2020).

The biomass oil obtained from the hulls of rice was initially condensed to produce bio-coal, and its properties were investigated. According to the element analysis, C and H were substantially greater in bio-coal. At the same time, O remained decreased in renewable energy following extraction, implying that the groups containing oxygen were eliminated. As-prepared bio-coal was a solid mass with a glossy black base that resembled commercialized coal. SEM pictures of the bio-coal show that it was a transparent or imporous material (Figure 14.9B and C) (Cheng et al., 2020).

14.6.5 Biochar

Since rice is a primary crop cultivated in Asia, adopting sustainable agricultural practices in rice fields is of utmost importance, especially concerning soil fertility and handling agricultural waste. Like many other plantations, rice farmers often rely on excessive use of chemical fertilizers, particularly nitrogen-based ones, leading to increased operational costs and negative environmental consequences. The leftover material formed during the cultivation of rice is an important source of by-products in the whole rice manufacturing method (Zeidabadi et al., 2018). Rice straw cultivation is vast, with approximately 600–800 million metric tons processed annually in Asia and between 800 and 1,000 million tons produced globally each year. Furthermore, rice husks and bran generate significant amounts of

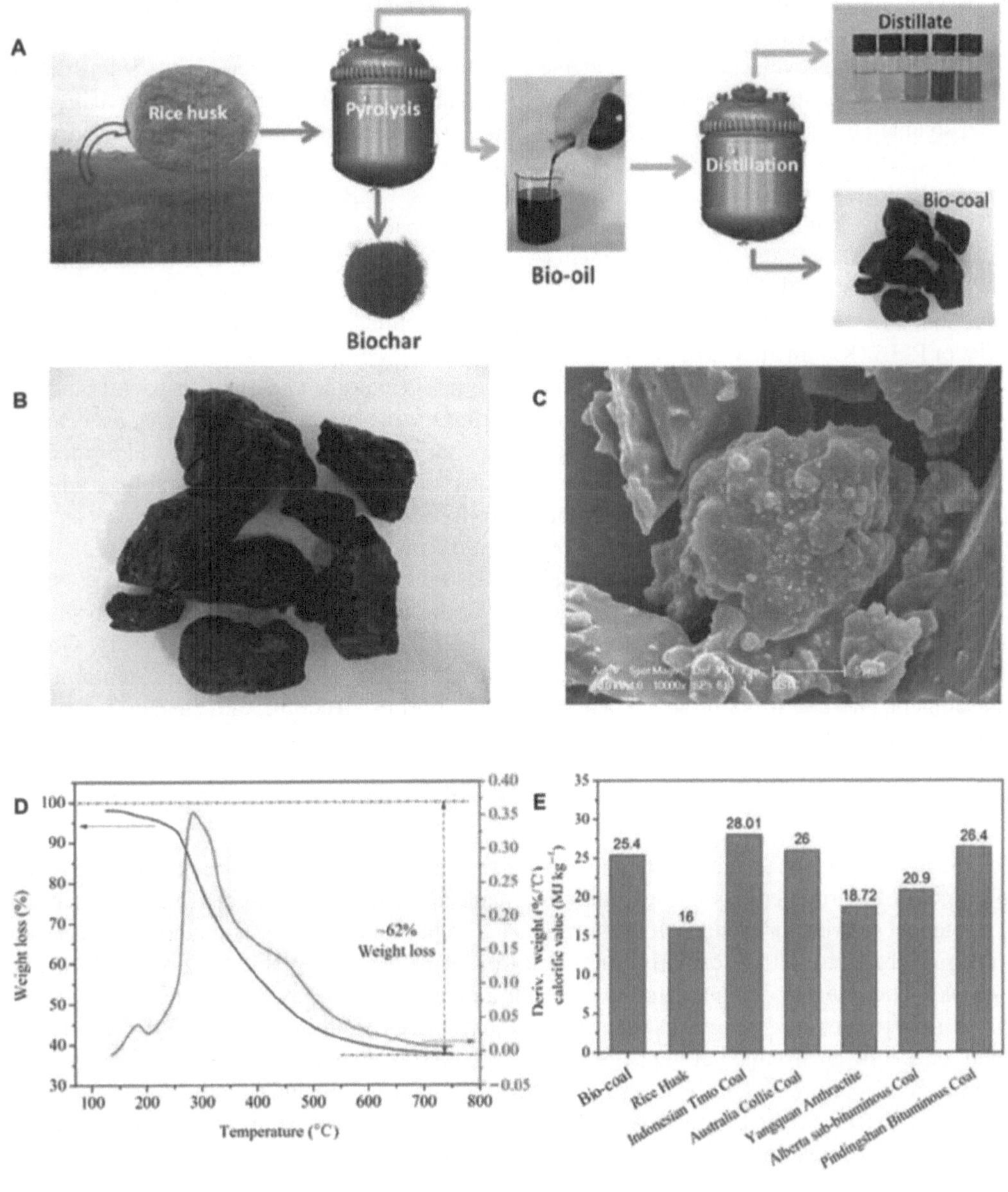

FIGURE 14.9 Preparation route and characteristics of lignocellulosic biomass–derived bio-coal. (A) Bio-coal formation from biomass that contains lignin is depicted schematically. (B and C) Bio-coal photograph and SEM picture. (D) Bio-coal analysis using thermogravimetric analysis (TGA) and difference thermo gravitation (DTG) spectrum. (E) Coal and bio-coal mass-energy densities (Cheng et al., 2020).

garbage, with 120 and 76 million metric tons produced globally, respectively. Although some rice husk is used as energy in mills that grind rice to generate heat during the drying procedure, consuming it in agriculture may contaminate the surrounding environment and increase greenhouse gas emissions. As a result, there will be an immediate necessity for developing environmentally friendly practices in rice farming as well as the residue of rice processing (Karam et al., 2022).

Common biochar raw materials include waste from agriculture, manures, organic waste from industry, and forest by-products. Biochar has a lot to offer in terms of potential solutions because it can both manage agricultural leftovers in an ecologically appropriate way and maybe improve soil quality without causing harm. According to studies, using rice husk bagasse (RHB) has been proven to boost crop output, enhance water retention, promote carbon sequestration, boost cation exchange capacity, lower toxicity in polluted soils, and reduce nitrogen leaching (Karam et al., 2022).

RHB has a high pH value, making it a viable substitute for soil liming chemicals. Raising the pH of the soil will increase CEC, especially in acidic soils, and make it simpler to reduce exchangeable Al and soluble Fe levels. Furthermore, RHB's high Si content benefits crops in a number of ways. Different volumes of RHB were administered at various rates depending on the type of crops, the type of soil, and the temperature at which RHB pyrolyzes. RHB reduces the bulk density of the soil, but applying it at different rates has no appreciable impact on the biomass of lentils above and below ground. Additionally, the stem size and leaf length of water spinach were both boosted by the administration of RHB at rates of 3.0–4.0 kg/m^3. The panicle length, tiller number, rice grain production, and straw yield were all boosted by applying commercial bio-fertilizer combined with 10 t/ha of RHB (Karam et al., 2022).

The major applications for rice husk biochar are in agriculture and environmental cleanup. Utilizing biochar in soil reduces CO_2 emissions from burning biomass while acting as a carbon sink (Asadi et al., 2021). Additionally, the use of biochar raises the soil's quality and production by boosting its nutrient content and structure. Water retention, soil fertility, cation exchange capacity, soil pH, and enhanced soil microbial activity all increased when biochar was applied to the soil (Pham et al., 2022). A higher temperature of 500–600°C was used to make rice husk biochar, which resulted in high availability of nutrients like phosphate and potassium. As the temperature increased from 350°C to 700°C, the nutritional availability of phosphorus and potassium improved from 0.22 to 1.8 mg/g and 0.16 to 0.68 mg/g, respectively. On the contrary, biochar has been used as an adsorbent for a variety of contaminants from aqueous or gaseous phases due to its large surface area, microporous structure, and surface functional groups (Bushra & Remya, 2020).

14.6.6 Bio-Bricks

One innovative approach to agro-waste management is the production of bio-bricks using rice waste. Bio-bricks are environmentally friendly and sustainable alternatives to traditional clay bricks, offering a potential solution for reducing the environmental impact of the construction industry. A review by Ologunagba et al. (2015) explored the feasibility of incorporating rice husk ash (RHA) into bio-bricks. The research demonstrated that the addition of RHA improved the mechanical properties, such as compressive strength and density, of the bio-bricks. Moreover, using RHA in bio-bricks helped reduce the environmental impact associated with brick manufacturing.

Similarly, rice straw has been investigated as a potential ingredient for bio-brick production. Munshi and Sharma (2019) conducted a study on the utilization of rice straw ash (RSA) as a partial replacement for cement in bio-bricks. The research revealed that incorporating RSA improved the compressive strength and durability of the bio-bricks, making them suitable for construction applications. Several benefits can be achieved by utilizing rice waste in the production of bio-bricks. First, it offers a sustainable solution for managing rice waste, reducing its environmental impact. Second, the incorporation of RHA or RSA enhances the mechanical properties of the bio-bricks, making them suitable for structural applications. Additionally, the use of bio-bricks reduces the demand for traditional clay bricks, which typically involve energy-intensive manufacturing processes and contribute to environmental degradation.

The development of bio-bricks from rice waste aligns with the principles of a circular economy, where waste materials are transformed into valuable resources. The utilization of rice waste in bio-bricks reduces waste accumulation and promotes the efficient use of agricultural by-products. These studies collectively demonstrate the potential of rice waste, such as RHA and RSA, in the production of environmentally friendly bio-bricks. By harnessing the benefits of rice waste, agro-waste management can be improved, leading to more sustainable construction practices and reduced environmental impact.

14.6.7 Biohydrogen

A significant task that will be documented in the coming years is improving sustainable energy. The world's energy needs can be satisfied by using fossil fuels. Fossil fuel consumption, however, has negative effects on the environment, such as greenhouse gas emissions and climate change. Therefore, the best way to lessen reliance on fossil fuels is through the production of clean energy from alternative renewable sources. Agricultural biomass has aroused a lot of interest in energy generation due to its abundance. An agricultural waste product called rice straw can be used as a starting point for the production of bioenergy (Singh et al., 2015). Hydrogen is regarded as clean energy because it has a better potential for energy production than conventional energy sources (fossil fuels) and burns to produce just water. For instance, one may get an energy production of about 122 kJ/g H_2 (Yukesh Kannah et al., 2019).

Numerous biological activities, such as direct or indirect photolysis, dark fermentation, and photo-fermentation, can produce hydrogen. It has been found that biohydrogen production from agricultural waste biomass—particularly rice straw—is more advantageous than other biofuels. Because of its higher sugar content, many studies, for instance, have suggested that agricultural waste, particularly rice straw, could be exploited as a potential source of feedstock for biohydrogen (Fang et al., 2006). Furthermore, cellulose, hemicellulose, and lignin are all included in the complex makeup of rice straw. Cellulosic compounds are not easily available for hydrogen generation because of the intricacy of rice straw (Yukesh Kannah et al., 2019).

14.6.8 Biomethane

Enhancing volumetric methane yields and process stability to produce synergistic effects is the primary goal of anaerobic co-digestion of rice straw. However, the high carbon-to-nitrogen ratio of rice straw results in poor mono-digestion process stability. Co-digestion of RS and de-oiled rice bran (DORB) yields results that are quite encouraging in terms of the amount of biogas and methane generated. The high C/N ratio, which is the main limiting factor in the AD of RS, may be reduced to a more suitable range by adding nitrogen-rich feed, and DORB is an excellent and sensible option for this (Grisolia et al., 2022). The mixing ratios of RS and DORB were the major factors that determined how effective co-digestion systems were. In comparison to digesters maintained at greater total solids concentrations, the amount of biogas generated was greatest at 5% TS concentration of the substrate while maintaining the ratio of 3:2 (DORB: RS) (Jha et al., 2021).

Each year, there are often two to three harvests on rice fields, yielding enormous amounts of RS in a short length of time. Given this operational reality, it is remarkable that so little research has been done on the effects of erratic RS production patterns on AD, a bioprocess that typically requires a steady and predictable feedstock. Bombardiere et al. (2007) evaluated the impact of 1–12 RS feeds per day and discovered that operations were steadier at lower feeding rates. However, their feed frequencies were quite brief compared to seasonal cycles in rice fields, which the investigation here was intended to evaluate. Overall, the findings indicate that low OLRs and occasional feeding can produce substantially higher specific quantities of biogas and CH_4 at a 1/21 day feed rate (148 mL CH_4/g VS/day), which was greater than the reactors that were fed more often. RS AD field applications should find this encouraging. Higher OLRs and more frequent feeding, however, can result in higher biogas volume production. As a result, RS AD has two distinct operational options: feeding less frequently at low OLRs or feeding more frequently at higher OLRs. However, there are additional factors to take into account; the optimal solution will rely on the space available for RS storage prior to use, the quality of the biogas for direct burning, and related expenses (Zealand et al., 2017).

The study focused on the environmental performance of using biomethane as a replacement for liquefied petroleum gas (LPG) in ovens and for light-duty vehicles (Otto-cycle engines) and heavy-duty vehicles (Diesel-cycle engines). The system boundary encompassed the production of biomethane through a desulfurization and enrichment process and its delivery through pipelines to households as well as to light-duty and heavy-duty vehicles. Comparatively, LPG was assumed to be produced at the Replan-Paulnia refinery in Brazil and transported in trucks in 28 kg cylinders, with round trips of 250 km full for delivery and another 250 km (empty) on the return route to the distribution center. For the ovens' case, it was assumed that biomethane emitted the same amount of greenhouse gas emissions as natural gas, except for non-methane volatile organic compounds, which were considered zero for biomethane. The inventory data was collected from literature, considering that 60% of the energy produced from biomethane was utilized as useful heat, while the remaining 40% was lost to the environment.

For the vehicles' case, which included light-duty and heavy-duty vehicles, the study evaluated the emissions of nitrous oxide (N_2O) and carbon dioxide from biomethane, taking into account the guidelines established by the Intergovernmental Panel on Climate Change. Biomethane was injected into Otto-cycle engines at a pressure of 200 bar, and the compression of biomethane to this pressure was considered, taking into account the gas energy balance and assuming 90% compressor efficiency. Overall, the research examined the entire lifecycle of biomethane usage, from production to delivery, and its substitution for LPG and diesel oil, with a focus on environmental performance (Ferreira et al., 2019; Kavitha et al., 2020).

14.6.9 Bioethanol

The creation of bioethanol from rice waste is one promising strategy for managing agro-waste. In place of fossil fuels, bioethanol is a renewable and environmentally beneficial fuel that can lower emissions of greenhouse gases and reliance on resources that are not renewable. It has been investigated to use rice straw, a significant portion of rice trash, as a feedstock for manufacturing bioethanol. In a 2019 study, Julianto et al. (2019) looked into the fermentation-based conversion of rice straw to bioethanol. According to the study, rice straw can be a useful source for bioethanol manufacturing, with high ethanol yields attained through effective enzymatic hydrolysis and fermentation processes.

In addition to rice straw, rice bran has been studied as a potential source for bioethanol production. A study by Yuliani et al. (2018) investigated the enzymatic hydrolysis and fermentation of rice bran for bioethanol production. The research revealed that rice bran contains a significant amount of fermentable sugars, making it a valuable feedstock for bioethanol production with the potential to contribute to the bioenergy sector. Bioethanol production from rice waste offers several advantages. First, it provides a sustainable solution for managing rice waste, reducing its environmental impact and promoting a circular economy approach. Second, bioethanol is a renewable and cleaner-burning fuel compared to fossil fuels, contributing to the reduction of greenhouse gas emissions and air pollution. Furthermore, using rice waste for bioethanol production can help reduce reliance on non-renewable energy sources and mitigate the impacts of climate change. It also offers economic opportunities for rural communities by creating a value-added market for agricultural residues. The studies highlighted the potential of rice waste, such as rice straw and rice bran, as valuable feedstocks for bioethanol production. By converting rice waste into bioethanol, agro-waste management can be improved, offering both environmental and economic benefits. Thus, bioethanol production from rice waste represents a promising avenue for sustainable energy production and agro-waste management, contributing to the transition to a greener, more sustainable energy future.

14.6.10 Biobutanol

Bio-butanol has emerged as a promising alternative to petroleum and natural gas, offering significant potential to reduce carbon dioxide emissions. It is a major product of the acetone–butanol–ethanol (ABE) fermentation process and holds various industrial applications, including use as an engine fuel, a biochemical intermediate, an extractant, and a potential biofuel. Chemically, bio-butanol is a colorless, straight-chain four-carbon alcohol with superior combustion properties compared to other alcohol-based biofuels. These advantages include a higher octane number, greater heating value, improved thermal and oxidative stability, better miscibility with gasoline, reduced ignition issues, and enhanced clarity.(Vivek et al., 2019).

Rice straw has a normal structure that includes 37 wt% cellulose, 24 wt% cellulose fibers, and 14 wt% lignin. Despite its high-calorie content, the direct use of newly harvested rice husks in bio-conversions is problematic due to its high lignin level. The preliminary treatment of rice straw is thus required for the breakdown of simple sugars (glucose, xylose, mannose, etc.). It can then be employed as a possible material for the fermentation-based synthesis of alcohol renewables (Ranjan et al., 2013). For biobutanol fermentation, carbohydrates must be delivered in some form of the best chemical, physical, or consecutive biological hydrolysis of straw made of rice. Later, multiple solventogenic anaerobic organisms will use the sugars to carry out ABE process of fermentation (Vivek et al., 2019).

The elevated temperatures of hydrolysis in acids of lignin-based biomass generate hydrolyzate, which can be utilized as a feedstock in fermentation. The use of increased stress due to shear and elevated tension throughout hydrolysis, as performed in the present investigation, has led to the rapid release of considerable amounts of sugar. Yet it primarily generates sugar (from cellulose), with trace amounts of other sugars such as xylose, arabinose, and others. Fermentation consumed around 63% of the total decreasing sugar and 46% of the glucose content, producing approximately 15 g of L1, 13 g of L1, and 10 g of L1 of the remaining sugar content, reducing glucose and sugar (Ranjan et al., 2013).

14.6.11 Natural Acids

Organic acids like citric acid are widely employed in industrial products. Since Indonesia's citric acid production cannot keep up with demand, there is a need to find an inexpensive and simple way to increase raw material supply (Menzel et al., 2020). *Aspergillus niger* submerged fermentation is a promising technique for producing citric acid. In this process, citric acid is synthesized by microorganisms using a carbon-rich substrate. Rice husks, due to their high cellulose content (28.6%–43.3%), can serve as an effective substrate for citric acid production, providing a sustainable and low-cost carbon source.(Ritonga et al., 2022).

The capacity of the fungus *A. niger* to make vinegar may be affected by the optimal dosage of substrate, sugar level, alcohol level, and agitating velocity. The following were the ideal circumstances: the percentage by volume of rice husk base is 22.42% w/v, sugar is 10% w/v, methanol is 3% v/v, and the rate of agitation is 400 rpm. This was determined by the Minitab program evaluation of the recorded level of pH and the citric acid information from the 19 ferment modifications. The largest amount of acidity produced on day 5 of the cultivation was 19.12 g/L (Ritonga et al., 2022).

The solubilization of insoluble phosphates is significantly influenced by the type of organic acids that are generated. Following citric acid in the order of P solubilizing effectiveness were oxalic, maleic, and formic acids. Each acid's location and the kind of functional group appear to be the main determinants of the release of P. Sulfuric acid produced less P than organic acids (Kumari et al., 2008).

Levulinic acid, also known as 4-oxopentanoic acid, plays a crucial role as a platform chemical in the production of alternative energy sources. It is utilized in the creation of various compounds such as valeric acid, 2-methyl-THF, 5-bromolevulinic acid, methyl pyrrolidone, and Y-valerolactone (GVL). Additionally, levulinic acid serves as a renewable resource for polymer synthesis, particularly its derivative GVL, which acts as a fundamental building block for polymer precursors. Researchers like Kon et al. have combined levulinic and valeric biofuels through selective hydrogenation with a Pt/HMFI zeolite catalyst. However, the use of microporous zeolite (pore diameter >2 nm) as support for heterogeneous catalysis has certain limitations when it comes to converting or adsorbing larger substrates, as noted by Mazhar et al. (2017).

The production of levulinic acid is hindered in many industries due to the high cost of its synthetic raw ingredients. This underscores the importance of exploring affordable and easily producible biomass raw materials. Despite being the third-largest rice producer globally, Indonesia has yet to fully utilize rice husks. These rice husks can serve as organic (biomass) alternatives for synthetic raw materials in levulinic acid synthesis, utilizing the cellulose present in biomass to overcome limitations and make effective use of abundant resources. To maximize cellulose extraction from lignocellulosic biomass, the delignification pre-treatment process must be optimized, as highlighted by Pratama et al. (2020).

14.6.12 Bioelectricity

Bioelectricity generation from rice waste involves harnessing the electrical energy produced through the microbial degradation of organic matter in rice residues. Rice waste, including rice straw, husks, and bran, is a rich source of organic materials that can be utilized for bioenergy production. Microorganisms, such as bacteria and fungi, naturally present in the environment or introduced via anaerobic digestion, can decompose the complex organic compounds in rice waste, releasing electrons. These released electrons can be captured and converted into electricity through various bioelectrochemical systems.

Rice straw, a significant component of rice waste, has been investigated for its potential in bioelectricity production. A study by Gu et al. (2014) focused on the anaerobic digestion of rice straw for biogas production. The researchers demonstrated that rice straw could be efficiently converted into biogas, which can be further utilized for the generation of bioelectricity. Moreover, rice husk, another significant by-product of rice processing, has been explored for bioelectricity production. A study by Pandit et al. (2021) investigated the use of rice husk in microbial fuel cells for electricity generation. The research demonstrated that rice husk could serve as an effective substrate in microbial fuel cells, providing a sustainable source of bioelectricity.

The conversion of rice waste into bioelectricity offers several benefits. First, it provides a renewable and sustainable source of energy, reducing dependence on fossil fuels. Second, it addresses the issue of rice waste management by utilizing it as a valuable resource for energy generation. Additionally, bioelectricity production from rice waste contributes to greenhouse gas mitigation, as it avoids the release of methane, a potent greenhouse gas, during natural decomposition (Hunter, 2008). Bioelectricity generation from rice waste aligns with the principles of a circular economy, promoting the efficient utilization of agricultural by-products. By harnessing the energy potential of rice waste, it contributes to sustainable agro-waste management and supports the transition toward cleaner energy alternatives. All in all, using rice waste for bioelectricity generation presents a viable and environmentally friendly solution for agro-waste management. It offers a sustainable energy source and helps reduce the environmental impact associated with the disposal of rice waste.

14.7 POTENTIAL APPLICATIONS

14.7.1 Bioenergy Production

A useful material that is readily accessible and renewable is rice straw. It complies with the Sustainable Development Goals of the United Nations by producing methanol, ethanol, heat, biodiesel, steam, coal, and raw materials. These objectives stress the prudent and efficient use of resources within the framework of a circular economy. A possible strategy to control the rising output of rice straw is to recycle agricultural waste like rice straw and turn it into other commercial sources. Biofuels like biodiesel and ethanol, affordable, renewable, biodegradable, and environmentally benign fuels that meet future energy and environmental needs, can be made from rice straw (Harun et al., 2022).

Figure 14.10 shows the many innovations that can be used to transform rice husks into bioenergy. Biology transformation (fermentation and anaerobic digestion, also known as anaerobic) and thermal processing (combustion, gas production, and degradation) are the two basic types of conversion technologies. Because it is simple to use, the combustion approach is extensively used. Burning can turn the chemical compounds in waste into thermally or electrically mechanical energy

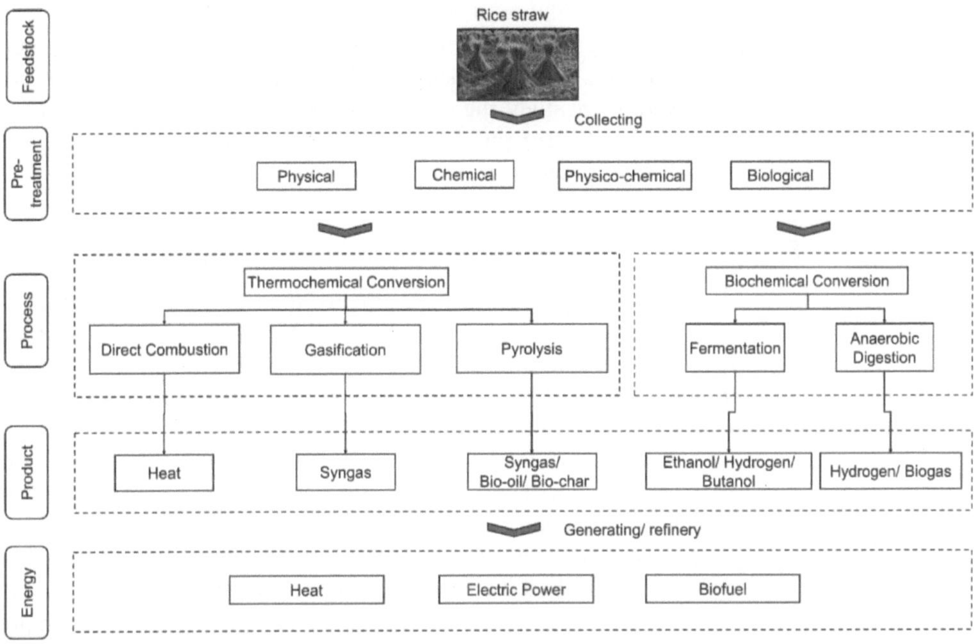

FIGURE 14.10 Technologies for the conversion of rice straw to bioenergy (Harun et al., 2022).

using boilers, generators, furnaces, and other devices. The combustion is the regulated combustion of biomass at degrees ranging from 800°C to 1,100°C, resulting in a combination of ignited gases (Luo & Zhou, 2022). The process of drying, pyrolysis, combustion, and elimination are all intermediate steps in the combustion process. Carbon monoxide, nitrogen, methane, and hydrogen are just some of the various gases generated. They are able to be utilized immediately in thermal applications or as gasoline in combustion engines to generate mechanical or electrical power. Pyrolysis is the method of exposing biomass without air to degrees ranging from 300°C to 800°C and converting it into various non-condensable liquids, gases, and solids. Pyrolysis generates biochar, biological oil, and gas such as hydrogen, carbon monoxide, carbon monoxide, carbon dioxide, nitrogen, and more (Singh et al., 2014b).

14.7.2 Construction Materials

In the Spanish city of Valencia, the paddy fields surrounding Albufera, a natural park, yield an abundance of rice straw, with over 75,000 tons produced annually. Traditionally, some harvested rice straw is either burned or mixed with the soil, leading to greenhouse gas emissions and negative environmental impacts. To mitigate these emissions and promote sustainable practices, researchers propose using rice straw as a raw material for construction materials, specifically for building façade panels as an alternative to the common double-layered brick façades in the region.

The study conducted by Quintana-Gallardo et al. (2021) compared the life cycle assessment (LCA) of the rice straw façade to that of the standard brick façade. Employing the Environmental Footprint approach, the LCA revealed that the rice straw façade had significantly lower environmental impacts in all investigated categories. Specifically, it emitted 78% less CO_2e compared to the standard brick façade. Each square meter of the rice straw façade saved 18.85 kg of CO_2e when compared to burning the straw, and 52.64 kg of CO_2e when compared to mixing the straw with the soil.

The acoustic insulation of both façades is comparable, meeting local building codes, with the brick façade having a sound reduction index of 49 dB, and the rice straw façade with a sound reduction index of 47.93 dB. Regarding thermal insulation, the rice straw façade outperforms the benchmark type, with a thermal transmittance of 0.29–0.08 W/m^{2}°C, double that of the standard brick façade (0.629 W/m^{2}°C). The hygrothermal behavior of the rice straw façade also complies with local building requirements, with no signs of surface condensation and no risk of interstitial condensation in any month. Overall, the study suggested that rice straw panels offer a sustainable and environmentally friendly option for the Spanish building sector, providing a viable solution to address the environmental challenges posed by rice straw mismanagement.

14.7.3 Animal Feed

Rice straw is a convenient, practical, and inexpensive source of food for ruminants such as buffaloes, cattle, goats, and sheep. During lean months or when good-quality roughages are limited, livestock farmers typically collect and stack rice straw from their rice farm, which then creates conserved feed for their animals. Pure rice straw feeding to ruminants during fast development and early lactation has been proven to impact both body condition score and animal performance. This is due to the straw's decreased dry matter intake and protein content (which has increased from 4.0% to 4.7% crude protein). The straw's high silica and lignin concentration also contributes to its low nutritional (dry matter and protein) digestibility (50%).

As a result, the processing of straw is required to maximize its contribution to enhancing meat and milk production. Farm techniques based on science and technology to enhance the nutritional and feeding properties of rice straw have been established, with considerable improvements in intake, nutrient digestibility, and animal performance. These technologies have also been shown to be useful in providing additional money to livestock producers through the sale of milk or live animals. This chapter describes and investigates current advances and innovative methods for increasing the nutritional (nutrient composition and fiber fraction) and feeding aspects of rice straw.

14.8 CHALLENGES ASSOCIATED WITH PROCESSING AND COMMERCIALIZING

14.8.1 Introduction

If straw is not used and utilized to its fullest potential, it will be greatly lost in the industrial development and alternative energy sectors. A smart economy should be efficiently implemented to reduce straw waste and repurpose straw as a resource. From an economic, agronomic, and environmental standpoint, moving from open-burning grain and conventional grain management to agricultural expansion can be challenging. The most successful agricultural company development appears to be to expand the industrial chain with other industries. The "waste-to-wealth" idea might offer a more eco-efficient way to dispose of straws while also being more ecologically friendly. As a result, an expanding body of study aims to assess the obstacles and difficulties faced by company owners when using and developing rice straws in their agricultural operations (Rosmiza et al., 2015).

Businesses often encounter challenges in meeting the demand for green products. Consumers may resist using these products due to their beliefs in technology, consumption patterns, and lifestyle choices, making it essential for green products to demonstrate reasonable pricing, superior quality, uniqueness, and practicality to persuade consumers. Moreover, legal restrictions, a lack of regulations, insufficient incentives, and limited consumer awareness pose additional obstacles. To address these challenges, businesses can modify their marketing strategies, showcasing creativity and innovation to attract potential customers while highlighting the financial benefits of green products. Employing cultural symbols in packaging design and leveraging the influence of socio-religious movements can also play a role in driving systemic change and promoting sustainable entrepreneurship (Rosmiza et al., 2015).

14.8.2 Less Support from Institutions

Institutions are essential for enforcing laws, advancing technology, attracting investment, building infrastructure, and advertising and selling products. The present norms and regulations in various institutional systems might be changed by government policies, which could substantially impact the growth of entrepreneurship. This calls for thorough documentation and methodical approaches. Strong support services could increase demand and supply for straw by-products. Most company owners firmly concur that the fundamental factor contributing to the stall in rice straw agribusiness is the lack of adequate support services from competent authorities or government organizations.

About 83.3% of business owners in the MADA region and 70.0% of business owners in Sekinchan, Selangor, Malaysia, agreed with this. The development of straw is today viewed by institutions as a modest effort to reduce open burning of straw, air pollution, and field management costs. The major agricultural by-products strategy of palm oil and rubber is the only thing that directly controls how straw is managed and used. Consequently, the government only offers a small amount of assistance for straw growth. Additionally, inadequate institutional promotion and regularly giving business owners advice on improving the quality of their products have a significant impact on the market's supply of straw by-products (Rosmiza et al., 2015).

14.8.3 Lack of Community Support

Most farmers are unaware of the possibilities and benefits of using straw compost instead of artificial fertilizer, claim compost entrepreneurs. For instance, despite straw compost's superior performance in stabilizing soil structure, storing and recycling nutrients, encouraging microorganism activity, and improving agronomic productivity, particularly over the long term, many farmers continue to use chemical fertilizer in their agricultural operations. Additionally, chemical fertilizer is far more accessible on the market than composted straw. All artisans firmly agree that straw crafts appeal only to metropolitan rather than rural societies. Because straw crafts are made locally from agricultural leftovers, rural populations think that they are not unique (Rosmiza et al., 2015).

14.8.4 Lack of Research and Access to Technology Facilities

Craft items made from straw cannot compete locally due to a lack of study on straw quality. When wet, straw paper items like paper bags, gift wraps, and key chains are readily damaged, business owners see a decline in demand for their products. The demand for straw for use as cow feed is likewise not much better. Due to its high fiber content, straw is provided as a supplement; nevertheless, its mineral value for the development of cattle is not as important. More work must be done to study and develop straw as a substrate for mushroom production. According to an interviewed mushroom business owner, imported Thai straw yielded higher-quality mushrooms than local straw. This could happen because of chemical pesticides and fertilizers used during rice farming. The toxins in rice straw impacted the growth of mushrooms (Rosmiza et al., 2015).

14.8.5 Lack of Skills and Knowledge

According to interviews, every entrepreneur had sought help from these organizations and taken their courses and seminars. Additionally, a number of business owners had acquired equipment, sheds, and workshops. Entrepreneurs claimed they still lacked the confidence to compete in the market. They continue to require ongoing guidance and better instruction from institutions. The institutes also provide trips and exhibitions for craft business owners in regional and international destinations, including the Philippines, Hong Kong, and Indonesia so that they may develop their talents and creativity. Craft business owners claim that participating in this training is helpful in increasing knowledge, fostering creativity, and developing new innovations in the methods used to manufacture straw craft goods (Rosmiza et al., 2015).

14.8.6 Entrepreneurial Attitude

According to observation, entrepreneurs are less likely to venture into new markets, especially domestically in Malaysia. They like to keep doing the same thing. As a result, it appears that the straw craft market is restricted to a single small community. When massively expanding the market, entrepreneurs do not appear to be searching for variety. Their company's profitability suffers as a result. One craft business owner claimed that being older also makes business owners less innovative in their contributions to product ideas and design. They don't take contemporary demands into consideration when designing. The concepts put forth by established entrepreneurs, such as key chains and catchphrases, are conventional and readily accessible in the market. On the contrary, young entrepreneurs are more adept at developing fresh ideas and innovative product designs. Their knowledge of the latest trends and innovations is also significantly influenced by their access to the internet and reading (Rosmiza et al., 2015).

14.8.7 Limited Product in the Market

According to the assessment, there is a lack of originality in straw craft goods, and they tend to be more community- and religion-specific. The same product is produced by every craft business owner, including keychains, Islamic calligraphy, and boxes for Malay wedding supplies. As a result, the market will be small and unpopular among Malaysia's many races. Additionally, neither the business sector nor rural communities show any interest in the products. This can then have an impact on how the market performs (Rosmiza et al., 2015).

14.8.8 Continuous Supplying of Straw Raw Material

Straw has flooded the states of the Kedah and Sekinchan. It is distributed two times throughout the year, and this organization keeps a steady supply in its straw warehouse. For example, most MADA company owners get their straw from the District Farmer's Organizations more commonly referred to as Pertubuhan Peladang Kawasan (PPK). The average straw barrel weighing 145 kg costs MYR 80 (Malaysian Ringgit) with delivery included and RM70 without delivery. However, some business owners find it difficult to obtain an appropriate quantity of straw to maintain their operations. This is due to a lack of nearby storage space to meet the need for organic matter, livestock feed, and other needs from the neighborhood and surrounding areas. As a result, PPK is looking to obtain the hay in different locations that may be far apart. This condition would have an impact on the growth of the straw agribusiness. However, the majority of livestock entrepreneurs are also farmers, obtaining straw for cattle feed from their own agricultural land (Rosmiza et al., 2015).

14.8.9 Lack of Capital

The market's potential for too sluggish demand prevents short-term capital investment and recovery. Additionally, there is sporadic demand for the craft, which is not a daily consumer good. They employ their own money since it is harder for them to obtain loans, which makes it take longer for them to become profitable. This circumstance may cause their business to stall. According to one business owner in the craft industry, innovation calls for new technologies and materials. The use of novel techniques like marbling and "batik canting" raised the cost of manufacture. These advancements make excellent use of colors and laser technology. Packaging materials must be more appealing to increase the value of the straw goods. To avoid damage, straw craft goods sold on the international market need to be packed to the highest caliber. According to the results of interviews, entrepreneurs' capacity to compete on the world market is constrained by a lack of funding. Craft business owners' experiences indicate that damaged products were caused by shipment in subpar packing. Customers lost faith, and business owners faced losses. Additionally, he mentioned that the demand for exports often required a huge amount of capacity, needing at least 5,000–10,000 straw crafts every month during a 3-year period (Rosmiza et al., 2015).

14.8.10 Lack of Skilled Workers

Typically, homemakers and secondary school graduates provide the local labor force. Craft business owners reported having trouble finding skilled labor. Most temporary employees are housewives looking to supplement their income. They lack ideas and skills, produce poor-quality work, are inefficient producers, and show little initiative to improve themselves by enrolling in skill-building classes. The final product seems to be of low quality as a result—unattractive, messy, and unable to satisfy the requests. Customers first lack confidence in the items, which costs entrepreneurs money when orders are cancelled (Rosmiza et al., 2015).

14.9 CONCLUSION

The production and processing of rice have significant implications for both food security and waste generation. Rice waste, including husks, straws, and bran, poses environmental challenges, such as greenhouse gas emissions, water pollution, soil degradation, and biodiversity loss. However, this waste also holds potential for sustainable solutions. Given its organic nature, rice waste can be effectively utilized as compost material. Additionally, it can serve as an alternative raw material for the pulp and papermaking industries, reducing the reliance on wood resources. Despite these potential applications, the utilization of rice waste faces challenges. Limited support from institutions and communities and gaps in research and knowledge hinder the development and adoption of sustainable practices. Overcoming these obstacles is crucial to fully harness the potential benefits of rice waste management. Addressing these challenges can enhance waste management strategies, mitigate environmental impacts, and promote the sustainable use of rice waste, contributing to a more environmentally friendly and resource-efficient agricultural sector.

ACKNOWLEDGMENT

This chapter was written with Azriena Haninah Binti Abd Aziz, K. Darshan A.L. N. Kumaran, and Noor Ain Syafiqah Binti Badrol Hizam.

REFERENCES

Abbasi, T., & Abbasi, S. A. (2010). Biomass energy and the environmental impacts associated with its production and utilization. *Renewable and Sustainable Energy Reviews*, *14*(3), 919–937. https://doi.org/10.1016/j.rser.2009.11.006

Aladejana, J. T., Wu, Z., Fan, M., & Xie, Y. (2020). *Key advances in development of straw fibre bio-composite boards: An overview. Materials Research Express*, 7(1), 012005. https://doi.org/10.1088/2053-1591/ab66ec

Anggraeni, S., Bayu Dani Nandiyanto, A., Restu Pribadi, A., & Ghiyats Al-kadzim, M. (2022). The effect of rice husk composition on porous concrete performance. *Journal of Engineering Science and Technology, 17*(2), 1346–1355.

Arranz, J. I., Miranda, M. T., Montero, I., & Sepúlveda, F. J. (2021). Thermal study and emission characteristics of rice husk using TG-MS. *Materials (*Basel, Switzerland), *14*(20), 6203. https://doi.org/10.3390/ma14206203

Asadi, H., Ghorbani, M., Rezaei-Rashti, M., Abrishamkesh, S., Amirahmadi, E., Chengrong, C. H. E. N., & Gorji, M. (2021). Application of rice husk biochar for achieving sustainable agriculture and environment. *Rice Science*, *28*(4), 325–343.

Bombardiere, J., Espinosa-Solares, T., Domaschko, M., & Chatfield, M. (2007). Thermophilic anaerobic digester performance under different feed-loading frequency. In: Mielenz, J. R., Klasson, K. T., Adney, W. S., & McMillan, J. D. (Eds.), *Applied Biochemistry and Biotecnology* (pp. 765–775). Humana Press. https://doi.org/10.1007/978-1-60327-181-3_63

Bushra, B., & Remya, N. (2020). Biochar from pyrolysis of rice husk biomass—Characteristics, modification and environmental application. *Biomass Conversion and Biorefinery, 14*. https://doi.org/10.1007/s13399-020-01092-3

Cardoen, D., Joshi, P., Diels, L., Sarma, P. M., & Pant, D. (2015). Agriculture biomass in India: Part 2. Post-harvest losses, cost and environmental impacts. *Resources, Conservation and Recycling, 101*, 143–153. https://doi.org/10.1016/j.resconrec.2015.06.002

Cheng, B.-H., Huang, B.-C., Zhang, R., Chen, Y.-L., Jiang, S.-F., Lu, Y., Zhang, X.-S., Jiang, H., & Yu, H.-Q. (2020). Bio-coal: A renewable and massively producible fuel from lignocellulosic biomass. *Science Advances, 6*, eaay0748. https://advances.sciencemag.org/

Chiou, T.-Y., Ogino, A., Kobayashi, T., & Adachi, S. (2013). Characteristics and antioxidative ability of defatted rice bran extracts obtained using several extractants under subcritical conditions. *Journal of Oleo Science, 62*(1), 1–8. https://doi.org/10.5650/jos.62.1

Das, P. K., Das, B. P., & Dash, P. (2019). Potentials of post-harvest rice crop residues as a source of biofuel. In: Praveen Kumar, R., Gnansounou, E., Raman, J. K., & Baskar, G. (Eds.), *Refining Biomass Residues for Sustainable Energy and Bioproducts: Technology, Advances, Life Cycle Assessment, and Economics* (Vol. 2014, pp. 275–301). Elsevier Inc. https://doi.org/10.1016/B978-0-12-818996-2.00013-2

Dulay, R. M. R., Parungao, A. G., Kalaw, S. P., & Reyes, R. G. (2012). Aseptic cultivation of Coprinus comatus (OF Mull.) Gray on various pulp and paper wastes. *Mycosphere, 3*(3), 392–397.

Dutta, D., Parvej, M. S., & Afsar, A. M. (2018). Valuation of mechanical properties of de-oiled rice bran reinforced polypropylene composites. *AIP Conference Proceedings, 1980*, 030003. https://doi.org/10.1063/1.5044282

Elhelece, W. A. (2020). Rice straw as a raw material for pulp and paper production. In: Hashmi, S., & Choudhury, I. A. (Eds.), *Encyclopedia of Renewable and Sustainable Materials* (pp. 296–304). Elsevier. https://doi.org/10.1016/b978-0-12-803581-8.10596-x

Fang, H. H. P., Li, C., & Zhang, T. (2006). Acidophilic biohydrogen production from rice slurry. *International Journal of Hydrogen Energy, 31*(6), 683–692. https://doi.org/10.1016/j.ijhydene.2005.07.005

Fávaro, S. L., Lopes, M. S., Vieira de Carvalho Neto, A. G., Rogério de Santana, R., & Radovanovic, E. (2010). Chemical, morphological, and mechanical analysis of rice husk/post-consumer polyethylene composites. *Composites Part A: Applied Science and Manufacturing, 41*(1), 154–160. https://doi.org/10.1016/j.compositesa.2009.09.021

Ferreira, S. F., Buller, L. S., Berni, M., & Forster-Carneiro, T. (2019). Environmental impact assessment of end-uses of biomethane. *Journal of Cleaner Production, 230*, 613–621. https://doi.org/10.1016/j.jclepro.2019.05.034

Ghanghas, N., Mukilan, M. T., Sharma, S., & Prabhakar, P. K. (2022). Classification, composition, extraction, functional modification and application of rice (Oryza sativa) seed protein: A comprehensive review. *Food Reviews International, 38*(4), 354–383. https://doi.org/10.1080/87559129.2020.1733596

Ghani, H. U., Mahmood, A., Finkbeiner, M., Kaltschmitt, M., & Gheewala, S. H. (2023). Evaluating the absolute eco-efficiency of food products: A case study of rice in Pakistan. *Environmental Impact Assessment Review, 101*, 107119. https://doi.org/10.1016/j.eiar.2023.107119

Grisolia, G., Fino, D., & Lucia, U. (2022). Biomethanation of rice straw: A sustainable perspective for the valorization of a field residue in the energy sector. *Sustainability, 14*(9), 5679.

Gu, Y., Chen, X., Liu, Z., Zhou, X., & Zhang, Y. (2014). Effect of inoculum sources on the anaerobic digestion of rice straw. *Bioresource Technology, 158*, 149–155. https://doi.org/10.1016/j.biortech.2014.02.011

Hardinnawirda, K., & SitiRabiatull Aisha, I. (2012). Effect of rice husks as filler in polymer matrix composites. *Journal of Mechanical Engineering and Sciences, 2*, 181–186. https://doi.org/10.15282/jmes.2.2012.5.0016

Harun, S. N., Hanafiah, M. M., & Noor, N. M. (2022). Rice straw utilization for bioenergy production: A brief overview. *Energies, 15*(15), 5542. https://doi.org/10.3390/en15155542

Hossain, M. S., Iqbal, M. T., Islam, M. S., & Al-Mamun, M. R. (2021). Production of biofuels from rice husk: An overview. *Renewable Energy, 168*, 1136–1151.

Huang, C., Han, L., Liu, L., & Yang, Z. (2008). Proximate analysis and calorific value estimation of rice straw by near infrared reflectance spectroscopy. *Journal of the Energy Institute, 81*, 153–157. https://doi.org/10.1179/174602208X330284

Hunter, P. (2008). Natural solutions to climate change. *EMBO Reports, 9*(6), 508–511. https://doi.org/10.1038/embor.2008.87

Jagaba, A. H., Kutty, S. R. M., Baloo, L., Birniwa, A. H., Lawal, I. M., Aliyu, M. K., … & Usman, A. K. (2022). Combined treatment of domestic and pulp and paper industry wastewater in a rice straw embedded activated sludge bioreactor to achieve sustainable development goals. *Case Studies in Chemical and Environmental Engineering, 6*, 100261.

Jawaid, M., Abdul Khalil, H. P. S., Bakar, A. A., & Mohd Ishak, Z. A. (2021a). Mechanical properties of biodegradable polymer composite reinforced with natural fibers: A review. *Polymer Composites, 42*(5), 2045–2067. https://doi.org/10.1002/pc.25862

Jha, B., Isha, A., Trivedi, A., Chandra, R., & Subbarao, P. M. V. (2021). Anaerobic co-digestion of rice straw and de-oiled rice bran for biomethane production. *Energy Reports*, *7*, 704–710. https://doi.org/10.1016/j.egyr.2021.01.032

Johnson, A. C., & Nordin, Y. (2009). Particleboards from rice husk: A brief introduction to renewable materials of construction. https://myiem.org.my/assets/download/Feature-Particleboard0609.pdf

Julianto, T. S., Kurniawan, S. M. A., & Arifin, I. (2019). Production of bioethanol from rice straw assisted by cellulosic enzyme oyster mushroom stem using simultaneous saccharification and fermentation (SSF). *EKSAKTA: Journal of Sciences and Data Analysis, 19*, 193–199. https://www.researchgate.net/publication/336100907_Production_of_Bioethanol_from_Rice_Straw_Assisted_by_Cellulosic_Enzyme_Oyster_Mushroom_Stem_Using_Simultaneous_Saccharification_and_Fermentation_SSF

Jusoh, M. L. C., Manaf, L. A., & Latiff, P. A. (2013). Composting of rice straw with effective microorganisms (EM) and its influence on compost quality. *Journal of Environmental Health Science and Engineering*, *10*(1), 1–9.

Karam, D. S., Nagabovanalli, P., Sundara Rajoo, K., Fauziah Ishak, C., Abdu, A., Rosli, Z., Melissa Muharam, F., & Zulperi, D. (2022). An overview on the preparation of rice husk biochar, factors affecting its properties, and its agriculture application. *Journal of the Saudi Society of Agricultural Sciences*, *21*(3), 149–159. https://doi.org/10.1016/j.jssas.2021.07.005

Kaur, D., Bhardwaj, N. K., & Lohchab, R. K. (2017). Prospects of rice straw as a raw material for paper making. *Waste Management*, *60*, 127–139. https://doi.org/10.1016/j.wasman.2016.08.001

Kavitha, S., Kannah, R. Y., Kasthuri, S., Gunasekaran, M., Pugazhendi, A., Rene, E. R., … & Banu, J. R. (2020). Profitable biomethane production from delignified rice straw biomass: The effect of lignin, energy and economic analysis. *Green Chemistry*, *22*(22), 8024–8035.

Korotkova, T. G., Ksandopulo, S. J., Donenko, A. P., Bushumov, S. A., & Danilchenko, A. S. (2016). Physical properties and chemical composition of the rice husk and dust. *Oriental Journal of Chemistry*, *32*(6), 3213–3219. https://doi.org/10.13005/ojc/320644

Kumar, A., Chauhan, A. S., Bains, R., & Das, P. (2021). Rice straw (Oryza sativa L.) biomass conversion to furfural, 5-hydroxymethylfurfural, lignin and bio-char: A comprehensive solution. *Journal of Industrial and Engineering Chemistry*, *104*, 286–294. https://doi.org/10.1016/j.jiec.2021.08.025

Kumari, A., Kapoor, K.K., Kundu, B.S., & Kumari Mehta, R. (2008). Identification of organic acids produced during rice straw decomposition and their role in rock phosphate solubilization. *Plant Soil and Environment*, *54*(2), 72–77. https://doi.org/10.17221/2783-PSE

Liu, J., Song, J., Ma, H., Li, J., & Liu, C. (2021). The mechanical properties of natural fiber reinforced polymer composites: A review. *Polymer Composites, 42*(S1), E820–E836. https://doi.org/10.1002/pc.25813

Luo, Z., & Zhou, J. (2022). Thermal conversion of biomass. In: Lackner, M., Sajjadi, B., & Chen, W.-Y. (Eds.), *Handbook of Climate Change Mitigation and Adaptation* (pp. 965–1021). Cham: Springer International Publishing.

Ma, G. (2015). Food, eating behavior, and culture in Chinese society. *Journal of Ethnic Foods*, *2*(4), 195–199. https://doi.org/10.1016/j.jef.2015.11.004

Maeno, S., & Sano, K. (1950). On the manufacture of rice straw pulp. *Journal of the Japanese Technical Association of the Pulp and Paper Industry*, *4*(4), 2–10, 87. https://doi.org/10.2524/jtappij1947.4.4_2

Mazhar, B., Jahan, N., Ali, N. M., Andleeb, S., & Ali, S. (2017). Production of vanillin by a novel bacterium from waste residues of rice bran oil. *Punjab University Journal of Zoology*, *32*(1), 137–142.

Menzel, C., González-Martínez, C., Vilaplana, F., Diretto, G., & Chiralt, A. (2020). Incorporation of natural antioxidants from rice straw into renewable starch films. *International Journal of Biological Macromolecules*, *146*, 976–986.

Mofijur, M., Mahlia, T. M. I., Logeswaran, J., Anwar, M., Silitonga, A. S., Ashrafur Rahman, S. M., & Shamsuddin, A. H. (2019). Potential of rice industry biomass as a renewable energy source. *Energies*, *12*, 4116. https://doi.org/10.3390/en12214116

Moraes, C.A., Fernandes, I.J., Calheiro, D., Kieling, A.G., Brehm, F.A., Rigon, M.R., Berwanger Filho, J.A., Schneider, I.A. & Osorio, E. (2014). Review of the rice production cycle: by-products and the main applications focusing on rice husk combustion and ash recycling. *Waste Management & Research*, *32*(11), 1034–1048. https://doi.org/10.1177/0734242X14557379

Munshi, S., & Sharma, R. P. (2019). Utilization of rice straw ash as a mineral admixture in construction work. *Materials Today: Proceedings*, *11*, 637–644. https://doi.org/10.1016/j.matpr.2019.03.021

Nguyen, V. H., Detras, M. C., Migo, M., Quilloy, R., Balingbing, C., Chivenge, P., & Gummert, M. (2020). *Rice Straw Overview: Availability, Properties, and Management Practices* (pp. 1–13). https://doi.org/10.1007/978-3-030-32373-8_1

Ologunagba, I. B., Daramola, A. S., & Aliu, A. O. (2015). Feasibility of using rice husk ash as partial replacement for concrete. *International Journal of Engineering Trends and Technology (IJETT), 30*(5), 267–269.

Pandit, S., Savla, N., Sonawane, J. M., Sani, A. M., Gupta, P. K., Mathuriya, A. S., Rai, A. K., Jadhav, D. A., Jung, S. P., & Prasad, R. (2021). Agricultural waste and wastewater as feedstock for bioelectricity generation using microbial fuel cells: Recent advances. *Fermentation, 7*(3), 169. https://doi.org/10.3390/fermentation7030169

Pham, T. H., Chu, T. T. H., Nguyen, D. K., Le, T. K. O., Al Obaid, S., Alharbi, S. A., … & Nguyen, M. V. (2022). Alginate-modified biochar derived from rice husk waste for improvement uptake performance of lead in wastewater. *Chemosphere, 307*, 135956.

Pimentel, D., Berardi, G., & Fast, S. (1984). Energy efficiencies of farming wheat, corn, and potatoes organically. In: Bezdicek, D. F., Power, J. F., Keeney, D. R., & Wright, M. J. (Eds.), *Organic Farming* (pp. 151–161). https://doi.org/10.2134/asaspecpub46.c12

Pratama, A. P., Rahayu, D. U., & Krisnandi, Y. K. (2020). Levulinic acid production from delignified rice husk waste over manganese catalysts: Heterogeneous versus homogeneous. *Catalysts, 10*(3), 327. https://doi.org/10.3390/catal10030327

Quintana-Gallardo, M., et al. (2021). *Waste valorization of rice straw as a building material in Valencia and its implications for local and global ecosystems. Journal of Cleaner Production, 316*, 128–137. https://doi.org/10.1016/j.jclepro.2021.128137

Ranjan, A., Khanna, S., & Moholkar, V. S. (2013). Feasibility of rice straw as alternate substrate for biobutanol production. *Applied Energy, 103*, 32–38. https://doi.org/10.1016/j.apenergy.2012.10.035

Rathour, R. K., Devi, M., Dahiya, P., Sharma, N., Kaushik, N., Kumari, D., Kumar, P., Baadhe, R. R., Walia, A., Bhatt, A. K., & Bhatia, R. K. (2023). Recent trends, opportunities and challenges in sustainable management of rice straw waste biomass for green biorefinery. *Energies, 16*(3), 1429. https://doi.org/10.3390/en16031429

Ritonga, E. R., Manurung, A., & Martgrita, M. M. (2022). Optimization of citric acid production by utilizing rice husk waste as a substrate using submerged fermentation. *Proceedings of the 4th International Conference on Life Sciences and Biotechnology (ICOLIB 2021), 2022*, 88–97. https://doi.org/10.2991/978-94-6463-062-6_10

Rosmiza, M., Davies, W., Rosniza, A. C., Jabil, M., Mazdi, M., Wan Toren, W., & Che Rosmawati, C. (2015). Stagnation of rice straw agribusiness development in Malaysia: The entrepreneurs' perspectives. *Mediterranean Journal of Social Sciences, 6*, 523–530. https://doi.org/10.5901/mjss.2015.v6n4p523

Shashidara, P. S., Nagesha, N., & Raghavendrachari, B. (2022). Estimation of environmental impact from rice mill cluster. *International Journal of Darshan Institute on Engineering Research & Emerging Technology, 11*(1), 19–23. https://doi.org/10.32692/ijdi-eret/11.1.2022.2204

Singh, A., Sevda, S., Abu Reesh, I. M., Vanbroekhoven, K., Rathore, D., & Pant, D. (2015). Biohydrogen production from lignocellulosic biomass: Technology and sustainability. *Energies, 8*(11), 13062–13080. https://doi.org/10.3390/en81112357

Singh, N., Paul, P., Virdi, A. S., Kaur, P., & Mahajan, G. (2014a). Influence of early and delayed transplantation of paddy on physicochemical, pasting, cooking, textural, and protein characteristics of milled rice. *Cereal Chemistry, 91*(4), 389–397. https://doi.org/10.1094/CCHEM-09-13-0193-R

Singh, V. K., Soni, A. B., Kumar, S., & Singh, R. K. (2014b). Pyrolysis of sal seed to liquid product. *Bioresource Technology, 151*, 432–435. https://doi.org/10.1016/j.biortech.2013.10.087

Sompong, R., Siebenhandl-Ehn, S., Linsberger-Martin, G., & Berghofer, E. (2011). Physicochemical and antioxidative properties of red and black rice varieties from Thailand, China and Sri Lanka. *Food Chemistry, 124*(1), 132–140. https://doi.org/10.1016/j.foodchem.2010.05.115

Swain, D. (2009). Simulating the impact of climate change on rice yield using ceres-rice model. *Journal of Environmental Informatics, 13*(2), 104–110. https://doi.org/10.3808/jei.200900145

Tan, B. L., Norhaizan, M. E., & Chan, L. C. (2023). Rice bran: From waste to nutritious food ingredients. *Nutrients, 15*(11), 2503. https://doi.org/10.3390/nu15112503

Tiquia, S. M., & Tam, N. F. Y. (2002). Characterization and composting of poultry litter in forced-aeration piles. *Process Biochemistry, 37*(8), 869–880. https://doi.org/10.1016/S0032-9592(01)00274-6

Trimble, J. L., Van Hook, R. I., & Gray Folger, A. (1984). Biomass for energy: The environmental issues. *Biomass, 6*(1–2), 3–13. https://doi.org/10.1016/0144-4565(84)90003-9

United Nations Statistics Division—Environment Statistics. (2007). Environmental indicators: Greenhouse gas emissions. [Online]. Retrieved from https://unstats.un.org/unsd/environment/air_co2_emissions.htm

Verma, D. K., & Srivastav, P. P. (2017). Proximate composition, mineral content and fatty acids analyses of aromatic and non-aromatic Indian rice. *Rice Science, 24*(1), 21–31. https://doi.org/10.1016/j.rsci.2016.05.005

Vivek, N., Nair, L. M., Mohan, B., Nair, S. C., Sindhu, R., Pandey, A., Shurpali, N., & Binod, P. (2019). Bio-butanol production from rice straw – Recent trends, possibilities, and challenges. *Bioresource Technology Reports, 7*, 100224. https://doi.org/10.1016/j.biteb.2019.100224

Wang, X., Chang, X., Ma, L., Bai, J., Liang, M., & Yan, S. (2023). Global and regional trends in greenhouse gas emissions from rice production, trade, and consumption. *Environmental Impact Assessment Review, 101*, 107141. https://doi.org/10.1016/j.eiar.2023.107141

World Food and Agriculture – Statistical Yearbook 2021. (2021). *World Food and Agriculture – Statistical Yearbook 2021*. FAO. https://doi.org/10.4060/cb4477en

You, K. Y., You, L. L., Yue, C. S., Mun, H. K., & Lee, C. Y. (2017). Physical and chemical characterization of rice using microwave and laboratory methods. In: Amanullah, & Fahad, S. (Eds.), *Rice—Technology and Production*. Intech. https://doi.org/10.5772/66001

Yukesh Kannah, R., Kavitha, S., Sivashanmugham, P., Kumar, G., Nguyen, D. D., Chang, S. W., & Rajesh Banu, J. (2019). Biohydrogen production from rice straw: Effect of combinative pre-treatment, modelling assessment and energy balance consideration. *International Journal of Hydrogen Energy, 44*, 2203–2215. https://doi.org/10.1016/j.ijhydene.2018.07.201

Yuliani, D., Julianto, K. A., & Jannah, A. (2018). Acid and enzymatic hydrolyses of rice bran for bio-ethanol production. *ALCHEMY Journal of Chemistry*, *6*(1), 24. https://d1wqtxts1xzle7.cloudfront.net/99100321/7190-libre.pdf?1677314270=&response-content-disposition=inline%3B+filename%3DAcid_and_Enzymatic_Hydrolyses_of_Rice_Br.pdf&Expires=1752226630&Signature=c5uPy5z~7JRDpw~Ih4j58tqG1kwbzdK4pywtCf7rjFo4xcdfE~G9jzNEhQR4Q7fCnkrDMemi-hItU5mzd48p5eyDqoZP2smCvPSHt31oUt2aLWgLqrTZO5O-irN4KFSngup8dH~w406HC5yU63pUkXiRcnK0s21nJzbQAylcpG~RKy1Ox6Dtsvf28xDT4VlXeOt7AViXGuEk4w8oG-Lfj8Lz5QeaM75hhDyUxZUzSUbQ6wLPgf2V3eFfj7~mEoVmGcLzWEtGxrAy~88b8DypbxXVXd30fQ~onwD6LDLi09xuQhL19pVZxi-8AKKTDUV~S~PWERarXc1R43SMc-He7Q__&Key-Pair-Id=APKAJLOHF5GGSLRBV4ZA

Zealand, A. M., Roskilly, A. P., & Graham, D. W. (2017). Effect of feeding frequency and organic loading rate on biomethane production in the anaerobic digestion of rice straw. *Applied Energy*, *207*, 156–165. https://doi.org/10.1016/j.apenergy.2017.05.170

Zeidabadi, Z. A., Bakhtiari, S., Abbaslou, H., & Ghanizadeh, A. R. (2018). Synthesis, characterization and evaluation of biochar from agricultural waste biomass for use in building materials. *Construction and Building Materials*, *181*, 301–308.

Zhou, Z., Robards, K., Helliwell, S., & Blanchard, C. (2010). Effect of storage temperature on rice thermal properties. *Food Research International*, *43*(3), 709–715. https://doi.org/10.1016/j.foodres.2009.11.002

15 Roselle Waste Management

15.1 INTRODUCTION

Agriculture is one of the most comprehensive words indicating that crop plants and domestic animals can sustain the human population by providing food and other products. It has been incorporated into a wide range of activities, including cultivation, domestication, and vegeculture, along with livestock management (Harris & Fuller, 2014). Aligning with the soaring growth in the human population, it has boosted agricultural production threefold more in the past decades. Within agriculture production, each step of its process produces agricultural solid waste based on its type of agricultural product, procedure, and purpose of use. Therefore, this sector particularly is one of the main sectors that has been generating massive amounts of agricultural solid wastes and allowing for it to accumulate aimlessly and constitute inconvenience toward the global health and using raw materials for bioeconomy or threat to food security (Oluseun & Adebukola, 2021).

Agriculture waste or agro-waste is described as the waste material produced from any chemical, pesticide, or fertilizer used during farming practices. These are typically hazardous materials; therefore, they must be used in small amounts. However, the materials are needed in large amounts to produce crop products at the optimum stage (Mohamed et al., 2022). Suppose no actions are taken to prevent these materials from being littered in the environment. In that case, there will be a potential habitat for pathogenic microorganisms, therefore nudging nuisance toward the environmental, economic, and social sectors. Aggravatedly, the major problem of agricultural production is the illegitimate agro-waste management system, where most of the waste is dumped in landfills, outdoor open fires, or improper disposal methods (Fadzil & Othman, 2021).

Poor waste management will cause the deterioration of soils and farmland, destroying the existing natural landscape and inducing more greenhouse gas (GHG) emissions, thus leading to economic loss. Approximately 998 million tons of agro-waste is generated globally annually, and Malaysia has been disposing of a total of 1.2 million tons of agro-waste in landfills. Hence, essential waste management, which applies the concept of circular economy, should be obligatory for more sustainable agricultural management. The circular economy operates on the extract-manufacture-disposal approach, where the waste is reused, reduced, and recycled to be derived from renewable resources. This will increase their economic value and also optimize the waste produced by exploiting all the residues and creating a large range of valuable materials and energy for a profit-oriented bioproducts technologies development (Fadzil & Othman, 2021).

Therefore, this study focuses on identifying the properties of roselle (*Hibiscus sabdariffa* L.) and the type of waste management that has been developed to fully optimize the waste generated. This study discussed the application of roselle in pulp and paper, biocomposite materials, bio-coal, biochar, bio-bricks, bio-hydrogen, biomethanol, bioethanol, biobutanol, and bioelectricity.

15.2 BIOMASS FROM NATURAL FIBERS

Approximately 140 gigatons of biomass are discarded internationally as waste every year. This has significantly become an environmental burden as it contributes to soil-water quality deterioration, the greenhouse effect, and major gas emissions. As the population increases, there is a surge in the

DOI: 10.1201/9781003472209-20

amount of food to fuel debates as more than half of the biomass (approximately 60%) is used for energy generation. This has greatly pressured the agriculture sectors as they are expected to produce more biomass (Jayaprakash et al., 2022).

Not only is it needed for energy generation, but more residues from biomass disposal are used in numerous fields, such as wastewater treatment as absorbents and possibly ceramic synthesis processes. The disposal-utilization pattern of disposed biomass has shifted from the production of fertilizers, animal feeds, or burning toward biogas–biofuel production or value-added bioproduct generation. Considering that, in modern days, it has been largely established to apply the concept of biorefining and circular bioeconomy, now, options such as recycling to produce fibers are overshadowed as techno-economics of process production become the main challenge to face (Jayaprakash et al., 2022).

Biomass that comes from natural fibers has different properties, and, due to its favorable properties, such as its sustainable and eco-friendly nature, it has been in considerable demand in these neoteric years. Natural fiber manages to attract society as it has shown several advantages for its utility, especially when it is reinforced into polymer-based composites. One of the advantages that it has been preferred over synthetic fibers is its low density. Other than that, reinforced natural fibers are more lightweight, have a higher strength-to-weight ratio, are biodegradable, have better thermal and insulation properties, and are of low cost (Abdul et al., 2012; Wong et al., 2010; Bledzki et al., 2006). However, despite having tons of advantages, this reinforced natural fiber has its downfall, as it has a higher tendency to moisture absorption, poor adhesion with synthetic counterparts, and low thermal stability during processing (Abdul et al., 2012; Demir et al., 2006; Son et al., 2001).

15.3 ENVIRONMENTAL IMPACT FROM BIOMASS

To fulfill the rising need for energy, it is strategically important to produce bioenergy from forests. Due to the great potential exploitation of wood leftovers, the utilization of biomass from forests and wood processing industries for energy has significantly expanded recently (Paletto et al., 2019). The strain on agriculture to produce biomass is increased by the rising demand for biofuels and bio-based products. Increased land usage can result in more frequent adverse outcomes such as erosion, nitrogen leakage, soil carbon loss, and floods (Berndes & Börjesson, 2022).

The most significant factors affecting climate change and other environmental implications are the scale of the biomass energy plant and the feedstock employed. A thorough understanding of these factors is crucial for creating an effective communication and information plan that highlights the potential effects on human health, ecosystems, and resources. The distinctions between the environmental effects of renewable energy sources (such as woody biomass) and fossil fuels should be highlighted in the definition of a communication and information plan clearly and concisely (Paletto et al., 2019).

The effects on the environment of generating biodiesel in Vietnam from a mixture of roselle seeds and trau to replace traditional diesel in cruise ships were studied. However, the authors noted that, depending on the fertilizers used in the manufacture of the agricultural feedstock, the bio-alternative might perform worse in the category of ecosystem quality. In general, biodiesel fared better in most areas. Two life cycle assessment (LCA) studies, but none as a food colorant, were published to assess the environmental impact of roselle-based products. In Monteiro's study, they used LCA to evaluate a natural coagulant produced by *H. sabdariffa* through freeze-drying and a commercial coagulant for treating wastewater from a 25,000-person town. The bio-based coagulant made from roselle was found to function better environmentally than the chemical one, although having a substantially greater production cost (Monteiro & Moura, 2022).

15.4 AVAILABILITY OF ROSELLE WASTE

Malvaceae is a family of plants that includes the bast fiber known botanically as roselle, or *H. sabdariffa* L., and Figure 15.1 shows a close-up picture of roselle calyces. This plant's native habitat

FIGURE 15.1 Roselle calyces (Blackwood, 2022).

is thought to be either in Asia or tropical Africa. The plant is known by several names in different parts of India. The plant has a dense taproot system and is around 3.5 m tall. It grows only once a year. These plants are edible and grown primarily for their calyces, fruits, and stem fiber. The calyces are picked after 15–20 days of flowering, while the other parts of the plant are left in the field for threshing. The agriculture waste's long, branchless steam is used to create fiber. The rosella is naturally lignocellulosic. As a result, its ability to absorb moisture is weak, preventing hydroxyl groups from interacting with polar matrices in the process. Therefore, it cannot be used as reinforcement while employing an interlock (Dharmendra et al., 2021).

The bleaching effects of sulfites on anthocyanins have long been recognized. Plants contain anthocyanin, particularly in their flowers and fruits. An excellent source of anthocyanin is roselle (*H. sabdariffa* L.), which is frequently used as a food, herbal beverage, hot beverage, cold beverage, or flavoring agent in the food sector. Instant dried roselle powder is readily available in local stores and is frequently used in Thailand to make tea or juice (Reanpang et al., 2021).

As a reagent in an acceptor solution of the flow injection–gas diffusion (FI-GD) system, spray-dried roselle powder crude extract can be utilized. In an acidic donor stream, sulfite is transformed into sulfur dioxide, which diffuses throughout a polytetrafluoroethylene (PTFE) membrane of the GD unit and dissolves in an acceptor stream. It causes the natural reagent's anthocyanin to react, bleaching the reagent's color, which is observed using a homemade photometer. Applying a common PTFE plumber tape as a gas-permeable membrane in the GD unit offered an environmentally friendly and green analytical approach. With the advantages of simplicity, cheap cost, environmental friendliness, and a reduction in chemicals, it could detect sulfite in wine samples (Reanpang et al., 2021).

For the current year, 100 tons of roselle output is expected to be processed by a single firm. After a thorough washing to remove any remaining acidity, the leftover biomass produced is occasionally used as animal feed; otherwise, it is dumped as garbage. Agro-industrial by-products are treated to recover or manufacture valuable products due to environmental and economic factors. The residual biomass of *H. sabdariffa* has not yet been considered with the intention of raising its value. The pectin-free biomass of *H. sabdariffa* cultivars, their overall monosaccharide content, and their lignin content were also assessed. Enzymes and the production of mono- and oligosaccharides can be used to cure this residue. The entire thing represents an unusual strategy for potentially fully utilizing this agro-waste (Cheikh et al., 2015).

15.5 PROPERTIES OF ROSELLE WASTE FIBERS

15.5.1 Physical Properties

The physical properties of roselle were determined by following the standardization of the American Society of Agricultural Engineers (ASAE) (Bamgboye & Adejumo, 2009), whereas the water value, water content, solubility, bulk density, cohesiveness, particle density, porosity, hygroscopicity, and flowability are listed in Table 15.1 (Archaina et al., 2019).

The bulk density of the roselle was determined using the water displacement method. In the preliminary stage, a bundle of roselle seeds was coated with a thin layer of Araldite, known as an epoxy resin adhesive, to prevent the water from being absorbed during the experiment. The bundle was hung in a measuring cylinder, and the ratio of the amount of the displaced water to the weight of the roselle seed was to obtain the true bulk density of the roselle, which is 0.42 g/mL (Archaina et al., 2019).

15.5.2 Chemical Composition

As for the chemical properties of roselle (*H. sabdariffa* L.), it was found that roselle is rich in malic acid, anthocyanins, ascorbic acid, and minerals, mainly calcium and ferum. However, roselle is low in glucose, and more than 18 volatile compounds were identified by gas chromatography and gas chromatography–mass spectrometry. Roselle fruit is also rich in phenolic compounds and is a good source of natural antioxidants (Jung et al., 2013). Figure 15.2 shows a few organic structures that were identified within roselle, such as phenolic compounds, particularly protocatechuic acid and eugenol, flavonoid-type polyphenol compounds, mainly 3-glucoside anthocyanins, anthocyanidins, and flavonols quercetin and vitamin C, B1, B2, along with carotenoid (Carvajal-Zarrabal et al., 2012).

15.5.3 Mechanical Properties

Natural fibers, including those derived from plants such as roselle, can exhibit a range of mechanical properties depending on factors such as fiber diameter, length, orientation, and the inherent

TABLE 15.1
Physical Properties of Roselle Seed

Properties	Value
Water value	0.2606±0.0003
Water content (g H_2O/100 g d.m)	1.67±0.06
Solubility (%)	99.35±0.01
Bulk density (g/mL)	0.42±0.01
Cohesiveness	1.06±1.03
Particle density (g/mL)	1.18±0.15
Porosity	61.3±5.7
Hygroscopicity (g a.w/100 g d.m)	11.1±0.2
Flowability	4.3±0.1

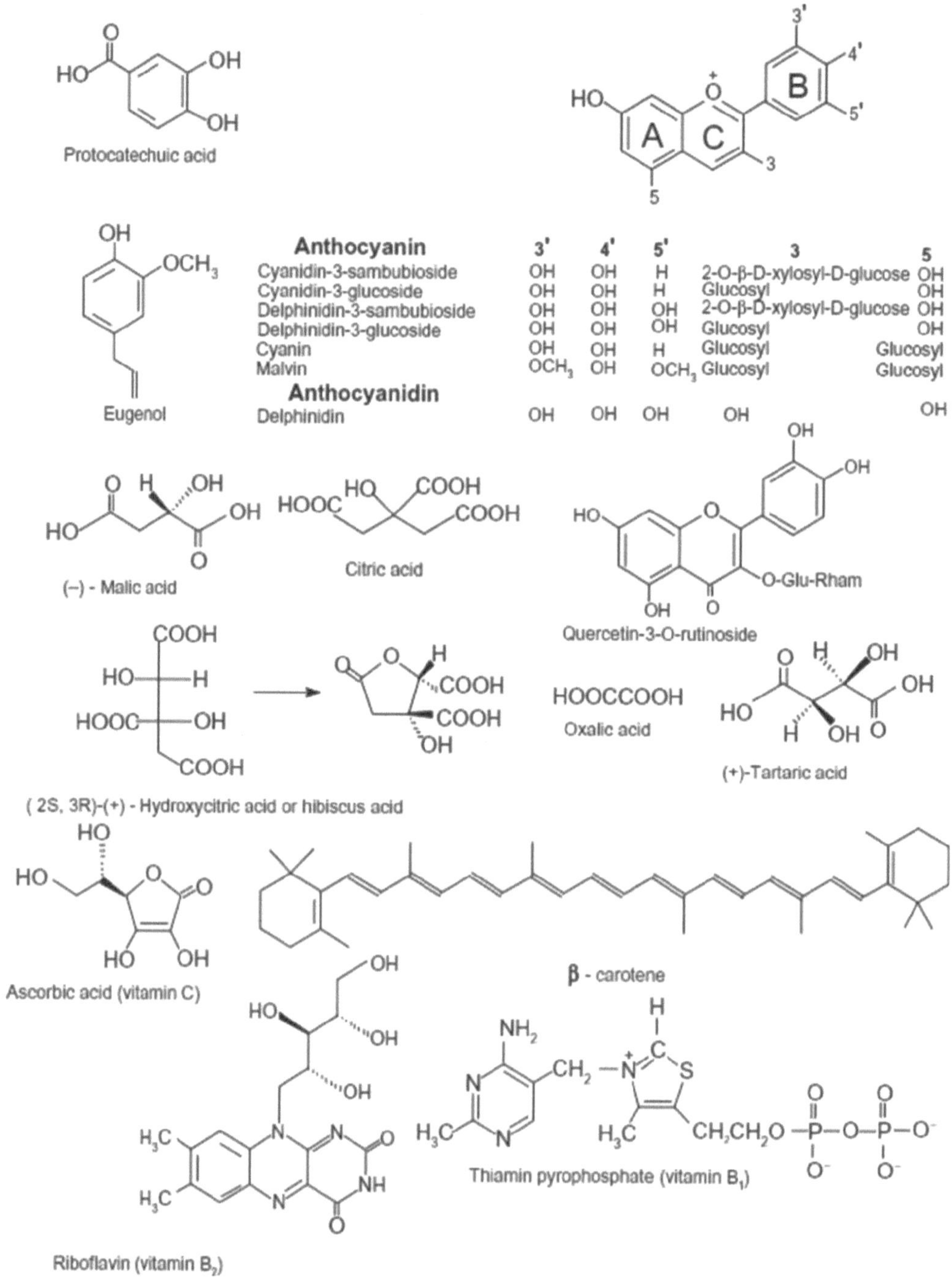

FIGURE 15.2 Organic structure found in roselle (*Hibiscus sabdariffa* L.) (Carvajal-Zarrabal et al., 2012).

characteristics of the material. Natural fibers typically have lower tensile strength and modulus than synthetic fibers like glass or carbon fibers. However, they are biodegradable, are renewable, and have lower density, making them environmentally friendly alternatives for certain applications. A tensile test can measure several important mechanical properties, including Young's modulus, tensile stress, maximum elongation, tensile strain, and yield stress. Table 15.2 shows the tensile

TABLE 15.2
Tensile Strength of Roselle Fiber

Age	Tensile Strength (MPa)
3 months	414.72
6 months	252.64
9 months	228.5738

strength of roselle fiber. The maximum tensile stress and elongation at the break of roselle fiber, in terms of mechanical qualities, can give the composite product good mechanical capabilities (Nadlene et al., 2015). Roselle fiber exhibits tensile strength values ranging from 130 to 562 MPa, placing it within the typical range for natural fibers and making it a suitable candidate for use as a reinforcement material in composite applications. When subjected to a sudden load, roselle fiber tends to fracture in a brittle manner. This highlights the importance of its tensile properties, as the mechanical load in a composite is primarily transferred to the reinforcing fiber. An increase in the fiber's cellulose content enhances both its tensile strength and Young's modulus. This is attributed to the crystalline structure of cellulose, which contributes to the mechanical stability of the plant stem. Moreover, the alignment of microfibrils relative to the fiber axis plays a key role in determining the fiber's ductility; better alignment generally results in improved tensile performance. It is also noted that the tensile properties of roselle fiber tend to decrease with younger plant age, as younger fibers typically possess less mature and less crystalline cellulose structures.

15.5.4 Morphological Properties

The morphological properties of roselle fiber, as presented in Table 15.3 (Nadlene et al., 2015; Tamta & Kalita, 2020), offer important insights into its structural characteristics and potential applications. The fiber's length ranges between 2 and 2.5 m (7–8 ft), making it suitable for use in various industrial applications, especially in textiles and biocomposites where longer fibers are preferred for enhancing mechanical strength. The fiber diameter was measured at three different points: the bottom (2.90 cm), middle (2.35 cm), and top (2.23 cm). This gradual decrease in diameter from the base to the top is typical of natural fibers and may affect how the material performs in terms of tensile strength and flexibility, making the lower part of the fiber potentially stronger due

TABLE 15.3
Morphological Properties of the Roselle Fiber (Nadlene et al., 2015; Tamta & Kalita, 2020)

Parameter	Measurements
Length (m)	2–2.5 (7–8 ft)
Diameter (cm)	
1. Bottom	1. 2.90
2. Middle	2. 2.35
3. Top	3. 2.23
Weight (g)	
1. Core	1. 300
2. Bark	2. 220
Moisture (%)	73.1

to its larger diameter. In terms of weight, the core of the roselle fiber weighs 300 g, while the bark weighs 220 g. The core likely contributes more to the fiber's structural integrity, whereas the bark might offer additional properties, such as flexibility or durability, depending on its application. These differences in core and bark weight suggest that different parts of the fiber could be used for various specialized purposes. The moisture content is notably high at 73.1%, which is common for natural fibers but also indicates the need for drying treatments during processing to ensure stability and prevent degradation. This high moisture content can impact the fiber's mechanical properties and necessitates careful handling in applications where moisture sensitivity is a concern.

Due to the extremely fine nature of roselle fiber, it is difficult to determine its cross-sectional shape using a scanning electron microscope (SEM). To overcome this challenge, researchers utilized a specific approach. First, the roselle fibers were treated by immersing them in liquid nitrogen to increase their hardness. Subsequently, a gold coating was applied to enhance the quality of the results (Nadlene et al., 2015; Tamta & Kalita, 2020). To examine the morphology of the roselle fiber, a Hitachi S-3400N SEM equipped with energy-dispersive X-ray (EDX) using a 15 kV accelerating voltage was employed. To prevent any charging effects, the roselle fiber samples were coated with gold before observation. Additionally, EDX diffraction was utilized to determine the elemental composition of the roselle fiber. Furthermore, the particle sizes of the roselle fiber were analyzed using a Malvern Master Sizer 2000 particle size analyzer (Kian et al., 2017). Morphology is an element that affects the mechanical and physical properties of roselle fiber as a reinforcement material for composite materials. Figure 15.3 shows the cross-section of roselle fiber at 3 months, 6 months,

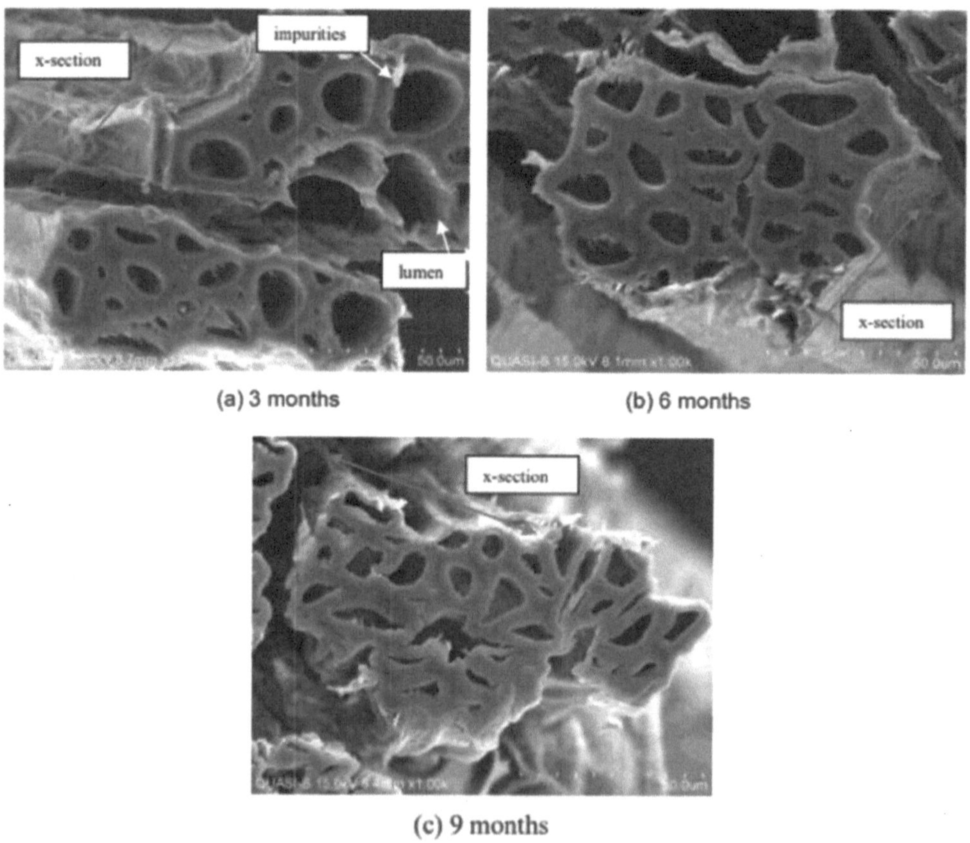

FIGURE 15.3 Cross-section of roselle fiber at (a) 3 months, (b) 6 months, and (c) 9 months (Munawar et al., 2007).

and 9 months. The cross-sectional view of roselle fiber reveals a distinct central lumen structure. As the fiber matures, the lumen size decreases due to the thickening of the cell walls. Fibers with larger lumen diameters tend to exhibit higher moisture content and increased water absorption capacity. This is because a larger lumen enhances the fiber's ability to retain water, making younger roselle fibers more hydrophilic compared to older, more lignified ones.

The roselle fiber bundle is overlapped. The diameter of the bundle fiber is increased when the thickness of the cell wall is increased. In composites, the surface structure of a plant that is 9 months old is preferable (Munawar et al., 2007).

15.5.5 Thermal Properties

In a study conducted by Bamgboye and Adejumo (2010), the thermal properties of roselle seeds were investigated in relation to their moisture content. The ASAE standard method was used to determine the initial moisture content of the seeds, as outlined in the work of Azizan and Zalani (1998). The researchers tested the seeds at five different moisture contents, ranging from 8.8% to 19%. To achieve the desired moisture content levels within this range, the samples were conditioned using the technique described in the study by Bamgboye and Adejumo (2010).

The method of combination described by Hsu et al. (1991) was employed to measure the specific heat capacity of the roselle seeds. In this method, a sample weighing 80 g with known moisture content and at an ambient temperature of 28°C was added to water within a calorimeter that had been preheated to 80°C. The temperature readings of the grain were taken at 10-min intervals until a steady measurement, known as the equilibrium temperature, was achieved. The enthalpy change of the roselle seed was determined based on the heat exchange between the water, calorimeter, and seed mixture, using the equilibrium temperature for the water and seed mixture (Bamgboye & Adejumo, 2010).

$$C_{ps} = \frac{C_{pw}(M_w + E)(T_{oc} - T_{em})}{M_s(T_{os} - T_{em})} \tag{15.1}$$

where

C_{pw} is the specific heat of water in kJ/kg/K.
M_w is the mass of water in kg.
M_s is the mass of the sample in kg.
E is the water equivalent of the system in kg.
T_{oc} is the initial temperature of the calorimeter in °C.
T_{os} is the initial temperature of the sample in °C.
T_{em} is the equilibrium temperature for the mixture in °C.

In the steady-state heat of vaporization method used to assess thermal conductivity, the sample was placed between two copper plates. One copper plate was in contact with water boiling at 100°C, while the other was in contact with ethanol with a lower boiling point of 78°C. The researchers measured the time it took for 1 ml of ethanol to vaporize under these conditions. The experiment was conducted at a specific moisture content, and the thickness of the entire sample was maintained as the minor axial dimension of the seed (Camelo-Méndez et al., 2013).

$$K = \frac{QL}{\theta(t_A - t_B)A} \tag{15.2}$$

where

L is the latent heat of vaporization of ethanol in J/kg.
A is the surface area of the sample in m^2.
t is the time taken for 1 mm of ethanol to vaporize in seconds.

Through the utilization of the empirical equation provided below, the thermal diffusivity of the roselle seed was determined using the data on thermal conductivity, specific heat capacity, and bulk density:

$$\propto = \frac{k}{\rho_b C_{ps}} \tag{15.3}$$

where

k is the thermal conductivity (W/mK).
ρ is the bulk density of the seed (kg/m^3).
C_{ps} is the specific heat capacity of the sample (kJ/kg K).

Table 15.4 provides the temperature characteristics of roselle seeds. With an increase in moisture content, the specific heat capacity of the roselle seeds consistently decreased in a linear fashion, dropping from 5.63 to 4.04 kJ/kg/K. Although these values fell below those of cooked meals like beef and pork rolls, they closely resembled the specific heat capacities of hydrated soybean hull and oil bean seed (Raheem, 2013). This trend is similar to what has been observed in lighter seeds like faba bean and oil bean seeds, which generally exhibit higher specific heat capacities compared to heavier seeds. The effect of moisture content on specific heat capacity was statistically significant ($p<0.05$).

Likewise, the thermal conductivity of the seeds also exhibited a linear decrease as moisture content increased, with mean values ranging from 1.56 to 1.22 W/m/K. The higher moisture levels led to more porous samples with increased intergranular void spaces, resulting in looser packing of individual seeds and reduced heat transfer within the seeds. This behavior resembled observations made in tortilla chips, maize, and wheat (Moreira' et al., 1995). Furthermore, with increased moisture content, the mean values of thermal diffusivity showed a linear growth from 4.274 to 4.477×10^{-4} m^2/s. This increasing trend could be attributed to the influence of bulk density on seed porosity, affecting the heat transport and storage capabilities of the seeds (Bamgboye & Adejumo, 2010).

15.6 ROSELLE WASTE MANAGEMENT

15.6.1 Compost

With a complex organic structure, roselle (*H. sabdariffa* L.) cannot easily disintegrate naturally, yet the amount of nutrients within roselle is high. Considering the chemical and physical properties of roselle, roselle can be one of the raw materials for compost substrate (Seyedbagheri, 2010). An organic substrate—a surface on which organisms grow—is broken down by aerobic (oxygen-requiring) bacteria under regulated circumstances to produce compost, a soil additive. The agricultural practices of composting, which date to biblical times, have significant contemporary agricultural applications.

TABLE 15.4
Thermal Properties of Roselle Seeds

Property	Unit of Measurement	Moisture Content (% d.b.)				
		8.8	10.0	12.5	16.0	19.0
Specific heat capacity	kJ/kg/K	5.63 (±0.53)	5.49 (±0.43)	5.06 (±0.44)	4.57 (±0.35)	4.04 (±0.45)
Thermal conductivity	W/m/K	1.56 (±0.22)	1.44 (±0.21)	1.37 (±0.23)	1.23 (±0.22)	1.22 (±0.23)
Thermal diffusivity	m^2/s	4.274×10^{-4}	4.297×10^{-4}	4.335×10^{-4}	4.440×10^{-4}	4.477×10^{-4}

Compost for contemporary farming systems has witnessed a resurgence in interest in recent years. It reserves as the crop's main source of nutrients in organic cropping systems and contributes additional nitrogen to fertilizer nitrogen in conventional cropping systems, resulting in a more sustainable farming system (Seyedbagheri, 2010).

The nutrients required for microbial development must be provided by organic resources. The carbon to nitrogen (C:N) ratio is one of the most crucial elements. Microbes involved in the composting process require both carbon and nitrogen. A high C:N ratio (too much carbon) indicates that there is insufficient nitrogen present to meet the needs of the microorganisms. A high amount of nitrogen allows more bacteria to break down and produce an unpleasant smell from the compost pile. Composting has been done in a range of 20:1 to 40:1, while a C:N ratio of 25:1 to 30:1 is ideal. Low C:N ratios increase the risk of loss into the atmosphere or soil because the available carbon is used up before the nitrogen stabilizes (Seyedbagheri, 2010). The physical properties of a compost substrate—such as porosity, structure, texture, and particle size, play a crucial role in influencing the composting process. These characteristics directly affect oxygen availability and the surface area accessible to microbial activity. Adequate oxygen supply is essential, with a minimum of 5% oxygen content required in the pore spaces of the compost pile to sustain aerobic microbial activity. Additionally, the moisture content should be maintained between 40% and 65%. This range ensures sufficient water for microbial metabolism while preventing excessive saturation that could hinder air circulation and lead to anaerobic conditions. (Seyedbagheri, 2010).

15.6.2 Pulp and Paper

The pulp and paper manufacturing process using wood as the primary raw material is shown in Figure 15.4. The process consists of two main steps: the extraction of fibrous material into pulp and the conversion of this pulp into paper. Both mechanical and chemical methods are employed to produce the pulp, which can be further refined depending on the required type and grade of paper. The pulp can be derived from wood or other lignocellulosic materials, which are either mechanically or chemically separated. These fibers are then suspended in water and transformed into a web of woven pulp. Mechanical pulping, a method used in this process, breaks fiber connections by applying mechanical energy to wood chips, with electricity serving as the main power source (Afiqah et al., 2021).

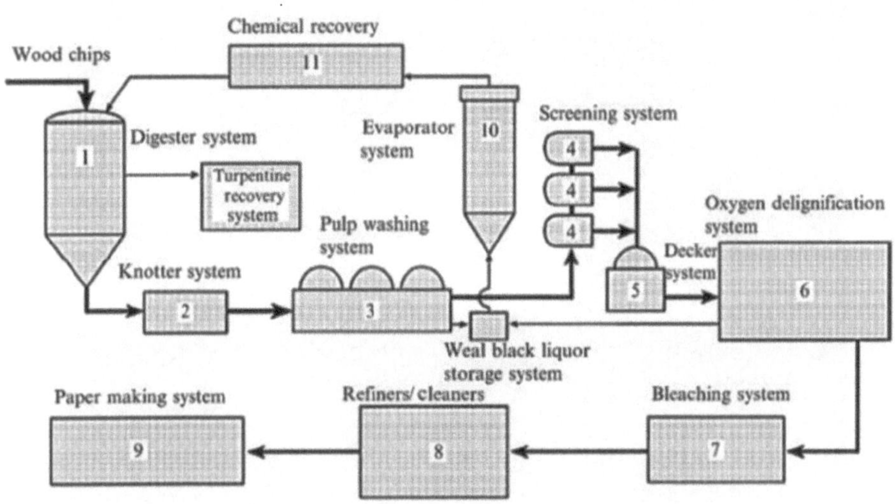

FIGURE 15.4 Pulp and paper manufacturing process using wood (Bajpai, 2015).

The main goal of mechanical pulping is to keep the lignin, essential for achieving the high potential of pulp with adequate strength and luminous properties intact. Chemical pulping breaks down the wood pulp using chemicals with enough chemical heat and pressure. During this procedure, lignin in the fiber walls, as well as other elements of the inter-fiber matrix material, disintegrates. Creating hydrogen bonds between the cellulose surfaces of the fibers enables the fibers to bind together during the papermaking process. In this instance, raw materials are cooked to make chemical pulps using the kraft and sulfite techniques (Omotoso & Folahan, 2015; Hart, 2011).

Wood is the primary source of fiber materials for paper companies worldwide. However, the rising price of pulpwood and its depletion in many nations have drawn attention to using non-wood materials, such as solid-waste plant fibers, to create various grades of pulp. However, there are not enough studies regarding producing pulp and paper from roselle waste.

15.6.3 Biocomposite Materials

Cellulose is a renewable, plentiful, and naturally occurring biopolymer generated from biomass. Roselle is a common crop in tropical regions like Malaysia, Borneo, Indonesia, India, Tanzania, Thailand, Sri Lanka, Sudan, Tanzania, Togo, and Guyana. The roselle plant's fiber composition is made up of 58.63–64.50% cellulose, 16.27–20.82% hemicellulose, and 6.21–10.26% lignin. The roselle fruit is typically used in the culinary and medicinal industries, but its underutilized stem could be used to make bio-based composites. Soaking the plant in water enables the separation of the fiber from the roselle (Kian et al., 2017). The past 10 years have seen incredible advancements in green technology, particularly in the field of materials science, through the creation of biocomposites. These advancements are due to environmental and sustainability challenges. The creation of biocomposites utilizing leftover agricultural leftovers, such as the stalks of most cereal crops, rice husks, and coconut fibers, is currently a topic of interest among potential alternatives like wood and plastic-based materials. Figure 15.5 shows the bio-reinforced roselle fiber mug pad product. There are a few biocomposites that were made of roselle fiber waste, such as bio-reinforced roselle fiber mug pad products where the mug pad was made on a laboratory scale. Despite being able to be transformed into a bio composite, using roselle waste or fiber is most unlikely in aerospace, transportation, or even heavy furniture.

FIGURE 15.5 Bio-reinforced roselle fiber mug pad product (Ilyas et al., 2021).

Most developing nations produce a lot of agricultural fiber and use a lot of their agricultural waste as fuel. More than 400 million tons of agricultural waste, including bagasse, maize cobs, peanut shells, and other wastes, are produced in India each year. Due to their accessibility, agricultural leftovers are good prospective alternative waste materials to replace plastic items (Prithivirajan et al., 2015). Aside from their abundance and renewability, agricultural residues have several advantages over thermoplastic polymer composites reinforced with inorganic fillers, including their low density, low manufacturing energy requirement, low CO_2 emission, and high level of biodegradability. The efficient use of corn straw, soy stalk, wheat straw, and other fibers with biodegradable resin improves the strength of the agricultural leftovers. The prospect of hybrid particle-reinforced composites has been seen in a recent study on the use of agricultural fibers as reinforcement in composites. Due to the appealing physical and mechanical characteristics of bioparticulates, the current inquiry concentrates on creating an epoxy composite reinforced by coir pith and rice husk hybrid particulates to manufacture customer value-added goods (Prithivirajan et al., 2015).

15.6.4 Bio-Coal

Bio-coal made from biomass is a promising material for replacing fossil coal in reaching net-zero GHG emissions from the industrial sector. High-temperature pyrolysis at 500°C can produce bio-coal with quality comparable to that of fossil coal, but the production efficiency is very low due to the low bio-coal yield at high pyrolysis temperatures. The viability of producing bio-coal economically is impacted by this trade-off (Aekjuthon Phounglamcheik, 2021; Baba et al., 2022a, 2022b). Figure 15.6 shows an image of bio-coal briquettes' final form of its production. During manufacture, storage, and transportation, bio-coal frequently spontaneously ignites and is prone to self-heating, which can have an adverse effect on worker health and production. This work used numerical models to describe the changes in local characteristics inside various bio-coal containers, such as closed metal containers and woven plastic bags, to evaluate self-heating at low temperatures. Measurements were made, and the bio-coal's kinetic characteristics were incorporated into the model. The heat produced by O_2 chemisorption was seen to cause the bio-coal temperature to rise gradually from its starting point (Baba et al., 2022a, 2022b; Aekjuthon Phounglamcheik, 2021).

The biggest obstacles to using bio-coal as a substitute for fossil coal are its characteristics. The bio-coal created in this study has an elemental composition and heating value that are identical to those of fossil coal, but it has a comparatively high level of reactivity. It is recommended that bio-coal's reactivity be comparable to that of fossil coal to replace fossil coal in already-existing

FIGURE 15.6 Bio-coal briquettes or pellets can be made from organic waste like food scraps (Merrifield, 2020).

FIGURE 15.7 Biochar composite (Spears, 2018).

industrial processes and prevent significant process modifications (Baba et al., 2022b; Aekjuthon Phounglamcheik, 2021). Bio-coal from roselle waste is unlikely to be produced. Due to the amount of moisture content in roselle, it will take longer and incur more cost to generate bio-coal from roselle fiber waste.

15.6.5 Biochar

Due to its great value and easy access to international markets, the roselle (*H. sabdariffa* L.) plant, a native of Africa, is one of the most significant medicinal and nutritional plants. Anthocyanins, organic acids, pectin, phenolic compounds, and vitamins are abundant in the roselle plant. Biochar, a carbon-rich substance produced by thermochemically converting plant biomass in an oxygen-deficient environment, is the subject of a lot of research. Utilizing biochar improves soil structure, diversity of microorganisms, nutrient content, cation exchange capacity (CEC), and environmental sustainability (Figure 15.7). Moreover, this biomaterial has been shown to be a great alternative to conventional materials because it uses less energy and produces fewer carbon emissions while still providing thermal comfort for buildings. The addition of biochar increased plant photosynthetic rate, chlorophyll content, and stomatal conductance (Al-Sayed et al., 2022).

While biochar derived from roselle waste offers several advantages, it does present some challenges, particularly its high acidity and salt content, which can negatively impact alkaline soils. These characteristics may alter soil chemistry, reducing its suitability for certain agricultural applications. Additionally, biochar produced at temperatures between 600°C and 700°C tends to have a limited CEC. This limitation is primarily due to the high proportion of stable aromatic carbon formed at these temperatures, which results in fewer hydrogen and oxygen functional groups. These functional groups are essential for enhancing CEC, a key property that influences nutrient retention and availability in soils (Al-Sayed et al., 2022). Using biochar and applying *Azolla* to roselle plants enhanced soil qualities, increased nutrient availability and uptake, and markedly increased roselle growth, yield, and quality. Depending on the amount of biochar or *Azolla* used, these increases changed. The addition of charcoal and *Azolla* improved the photosynthetic process, which is evident in the yield, yield components, and yield quality. These improvements also enhanced nutrient availability, metabolite production, and chlorophyll synthesis (Al-Sayed et al., 2022).

15.6.6 Bio-Bricks

In India, burning agricultural waste is a substantial cause of air pollution, particularly following harvest. A sustainable alternative building material called bio-brick was created using agricultural waste. Additionally, it will simultaneously decrease air pollution and create new local jobs (Gupta et al., 2022; Rautray et al., 2021). Building construction is one of India's fastest-growing businesses, which significantly strains the country's meager natural resources. One of the main building materials is fired clay brick, which generates significant GHGs. One such material that has the potential to be a viable and affordable option is bio-brick or agro-waste-based brick. It has a low overall carbon footprint and serves as an excellent insulator against heat and sound. It also discourages the practice of stubble burning, which is common in northern India and contributes significantly to air pollution. Due to its low density, RCC construction is more cost-effective since it decreases dead load in high-rise structures (Rautray et al., 2019). However, there was barely any research that suggested roselle waste could be used to make bio-bricks.

15.6.7 Bio-Hydrogen

One type of sustainable energy source is bio-hydrogen. In contrast to typical fossil fuels, hydrogen gas produces water after combustion and emits no carbon dioxide, making it a clean energy source. Hydrogen has an energy yield nearly 2.75 times higher than that of other hydrocarbon fuels, which has an energy yield of 122 kJ/g (Dutta & Kumar, 2021). Anaerobic digestion (AD) is a highly valuable technology within the circular economy framework, as it enables the treatment of organic-rich waste while simultaneously generating energy vectors (EV) such as biogas and biohydrogen (Aboudi et al., 2021). In comparison to other techniques of producing hydrogen, the biological production of hydrogen is regarded as one of the most practical alternatives. The creation of bio-hydrogen from organic waste materials also lays the way for the creation of renewable energy sources and a sustainable waste management method. Numerous bacteria, including *Clostridium*, *Citrobacter*, and *Enterobacter*, have been employed in literature to produce bio-hydrogen (Dutta & Kumar, 2021). Feedstocks with a lot of carbohydrates are the most suitable candidates for bio-hydrogen production (Kulkarni & Ghanegaonkar, 2019).

15.6.8 Biomethane

The waste from roses cannot be used to make biomethane. AD, a process that can transform municipal solid waste, wastewater sludge, and/or the chain of agro-industrial waste into sustainable energy, can produce biogas, which can then be converted into biomethane. Roselle waste has the potential to produce biogas, bio-hydrogen, bioethanol, and biocharcoal. Animal, industrial, food processing, plant residues, and other wastes have been used to produce biogas. Research on numerous additional trashes as a potential feedstock for biogas production is ongoing. Vegetable scraps and withered flowers are being examined as potential feedstocks. In gardens, on the streets, and at flower marketplaces or even temples, withered flowers are readily accessible (Ranjitha et al., 2014). The biogas can be utilized in cogeneration systems to produce biomethane, which can then be delivered to the national natural gas network, or as a biocombustible fuel in the car industry (Baba et al., 2022a, 2022b; Molino et al., 2013).

Additionally, food waste is the main source of biomethane production. The amount of food waste accessible for AD and the particular amount of methane produced from that waste determine the resource's availability. This is significant because it exceeds the resources needed to electrify 10% of Ireland's private car fleet, which is currently the country's preferred option for renewable energy in transport. Biomethane from food waste is shown to have the potential to replace 2.8% of the energy used in transport. However, source segregation of food waste must be implemented for this resource to be utilized within the European Union. Digestate from source-segregated food waste may be used on agricultural land after AD, according to the Animal By-Products Regulations. Food waste-derived soil from a mixed waste source is not allowed to be used on agricultural land. So, source segregation of food waste is a prerequisite for producing biomethane (Browne & Murphy, 2013).

Energy from biomass meets the main categories of energy requirements, offering electricity, heat, and transportation fuel, with the benefit of being able to be stored and used as needed. This production cannot be used as a source of nutrition since they are used to produce biofuels, and their usage as biofuels may have a negative effect on soil fertility and biodiversity (Barbot et al., 2016).

15.6.9 Biobutanol

Microbial fermentation has the potential to create butanol from roselle (*H. sabdariffa* L.). Butanol is a form of alcohol that can be generated by microbes from a variety of feedstocks, including plant biomass. In the case of roselle, butanol manufacturing would entail extracting sugars from the plant material and then fermenting them with certain strains of bacteria such as *Clostridium acetobutylicum* (Arowosafe et al., 2022). Through a metabolic mechanism, these bacteria can transform carbohydrates into butanol.

However, it is crucial to highlight that large-scale butanol production from roselle or any other feedstock is currently not economically competitive compared to other butanol production techniques. Traditional butanol production methods, such as petrochemical synthesis or maize as a feedstock, are more efficient and cost-effective. Nonetheless, researchers are constantly investigating and developing more ecological and cost-effective methods of manufacturing butanol from renewable resources such as plant biomass. While it is theoretically conceivable to make butanol from roselle, it is not currently commercially viable.

15.6.10 Bioethanol

Roselle can be used as a feedstock to produce bioethanol. By fermenting the sugars and starches from diverse plant sources, bioethanol is created as a type of biofuel. A sizable portion of the carbohydrates in roselle calyces can be transformed into sugars, which can subsequently be fermented to produce bioethanol. Due to their high sugar content and low lignin concentration, roselle calyces have been found to offer great potential for the generation of bioethanol (Arowosafe et al., 2022).

Pretreatment, hydrolysis, fermentation, and distillation are some of the phases in the process of producing bioethanol from roselle (Arowosafe et al., 2022). The calyces' complicated carbohydrates undergo pretreatment to be converted into simple sugars that can be fermented. The simple sugars are hydrolyzed to create glucose, which is fermented by yeast to create ethanol. Distillation, which separates the ethanol from the fermentation broth, is the last step. Overall, roselle has enormous promise as a source of biomass to produce bioethanol, and continuing studies are being conducted to improve the method and raise the bioethanol output from this plant material.

15.6.11 Natural Acids

The limited research available on roselle extracts indicates that the main compounds found in them are organic acids, with the prominent ones being citric acid, hydroxycitric acid, hibiscus acid, malic acid, and tartaric acid. Although traces of oxalic and ascorbic acids have also been detected, the overall percentage of these organic acids in the *H. sabdariffa* L. species is quite similar, with hibiscus acid being the most abundant, ranging from 13% to 24%. The remaining organic acids are present in the following percentages: (1) citric acid comprises 12–20%, (2) malic acid constitutes 2–9%, (3) tartaric acid is about 8%, and (4) ascorbic acid falls within the range of 0.02–0.05% (Riaz & Chopra, 2018).

For instance, ascorbic acid can be produced via roselle waste. The aqueous extract of roselle has been found to contain ascorbic acid, but the concentration of this compound varies depending on factors such as the variety, season, harvest conditions, genetics, and origin of the roselle. For Nigerian white, red, and deep red roselle, the respective concentrations of ascorbic acid were 86.5 mg/100 g, 63 mg/100 g, and 54 mg/100 g. The equivalent concentrations in Sudanese white and red roselle were lower, measuring 11.0 mg/100 g and 15.5 mg/100 g, respectively (Tahir et al., 2017). Comparatively, the ascorbic acid levels in Malaysian red roselle were notably higher than those reported for samples from Sudan and Nigeria (Ismail et al., 2008). In the case of 35 genotypes of roselle collected from Bangladesh, the levels of ascorbic acid ranged from 2.62 to 42.42 mg/100 g. These findings highlight the variability in ascorbic acid content among different types of roselle from various regions.

Hydroxycitric acid is citric acid with an additional hydroxyl group at carbon number two. It exists in lactone forms and as stereoisomers, including (2*S*, 3*S*), (2*R*, 3*R*), (2*S*, 3*R*), and (2*R*, 3*S*). Among these, the predominant organic acid found in roselle calyces is believed to be (2*S*, 3*R*)-hydroxycitric acid. It's important to note that this (2*S*, 3*R*)-hydroxy acid in roselle differs from the more commonly known (2*S*, 3*S*)-hydroxycitric acid found in other sources like *Garcinia* sp. Due to this difference, it becomes essential to understand whether the pharmacological profiles of these two diastereomers, (2*S*, 3*R*) and (2*S*, 3*S*), are entirely the same or exhibit some variations (Hida et al., 2007).

Hibiscus acid is a prominent organic acid found in the extracts of roselle calyces and leaves, as stated in the research by Herranz-López et al. (2012). This acid possesses two chiral centers, resulting in two diastereomers and a citric acid fraction with an additional hydroxyl group at the second carbon in its structure. Flowers obtained from roselle plants typically contain hibiscus acid in levels ranging from 13% to 24%. The variation in the proportions of hibiscus acid in the flowers could be attributed to different strains, genetic variations, and various environmental factors. These factors play a role in influencing the organic acid content in the roselle plant.

15.6.12 Bioelectricity

Although certain plants are known to create electricity, there is no proof that roselle produces bioelectricity in any appreciable quantity. By converting chemical energy into electrical energy, living things produce bioelectricity. During photosynthesis, plants produce a tiny amount of bioelectricity, but this energy is required for internal metabolic activities and is not used by other organisms. Roselle is, therefore, unlikely to be a substantial source of bioelectricity (Rizwan et al., 2018). Other plants and animals, on the contrary, have the ability to produce bioelectricity and are being researched for possible application in the production of alternative energy.

15.7 POTENTIAL APPLICATIONS

15.7.1 *H. sabdariffa* L. as a Natural Coagulant to Treat Congo Red in Wastewater

Coagulation is a widely used physicochemical treatment in water and wastewater treatment due to its simplicity and effectiveness in reducing dissolved chemical levels, turbidity, and other factors. This process involves using a chemical coagulant to neutralize negatively charged colloids and dissolved solids in wastewater, which are responsible for turbidity and color. The neutralized particles then combine to form larger particles known as flocs, which can be easily separated and removed through sedimentation (Gao et al., 2007). As a result, wastewater experiences reduced turbidity, color, and contamination.

Aluminum sulfate (alum) has been the most commonly used coagulant; however, concerns about residual aluminum in drinking water and sludge may contribute to Alzheimer's disease and environmental issues. To address these concerns, researchers have developed natural coagulants that are non-toxic and environmentally friendly (Yong & Ismail, 2015).

The use of natural coagulants derived from plants and animals has garnered significant attention as potential alternatives to traditional coagulants. These natural coagulants are highly biodegradable and safe for human consumption and the environment. Moreover, locally grown plants offer economic advantages over chemical coagulants. Several natural coagulants, such as *Moringa oleifera*, Nirmali seed, and *Jatropha curcas*, have been explored for water and wastewater treatment due to their water-soluble proteins that aid in the coagulation process (Muthuraman & Sasikala, 2014). Additionally, *H. sabdariffa* L. has been investigated as a potential natural coagulant for removing turbidity in water treatment (Yani et al., 2019). However, further research is needed to assess its effectiveness in treating dye wastewater (Yong & Ismail, 2015).

In a study by Yong and Ismail (2015), *H. sabdariffa* L. was examined as a natural coagulant to treat dye wastewater containing Congo red. The coagulation agent was extracted from the seeds using various solvents, including distilled water, 0.5 M NaCl and 0.05 M NaOH. Response surface methodology was employed to optimize the working parameters, and the most effective solvent was found to be 0.5 M NaCl, with 95.1% color removal. *H. sabdariffa* L. seed was found to be a potent coagulant, achieving 91.2% color removal at pH 2, 190 mg/L coagulant dosage, and 400 ppm dye concentration. Moreover, natural coagulants demonstrated comparable performance to conventional coagulants like aluminum sulfate, with color removal rates of 91.2% and 92.3%, respectively (Yong & Ismail, 2015).

15.7.2 Roselle Uses in Traditional Medicine

Roselle is widely consumed as a beverage in various regions, including West Africa, the United States, India, Iran, Mexico, Thailand, Taiwan, and Egypt, as mentioned by Osei-Kwarteng et al. (2021). Additionally, different parts of the roselle plant have been utilized in traditional medicine for various purposes (Figure 15.8). Folk medicine practices have employed roselle to treat conditions such as urinary tract infections, toothaches, hangovers, and colds (Maganha et al., 2010). In specific regions like Africa, India, and Mexico, ethnomedicinal studies have revealed the use of roselle calyces or leaf infusions for addressing various health issues. These traditional uses include treating hypotensive (lowering blood pressure), choleretic (promoting bile production and secretion), and diuretic effects (increasing urine production). Roselle has also been utilized for managing hyperlipidemia (high levels of lipids or fats in the blood), stimulating intestinal peristalsis (movement of the intestines), and reducing blood viscosity (Riaz & Chopra, 2018). The widespread use of roselle in traditional medicine in these regions indicates its significance as a medicinal plant with various potential health benefits.

In traditional Thai medicine, roselle treats urinary bladder and kidney stones (Maganha et al., 2010). When combined with common salt, the dried powder and extract of the calyces are used to alleviate conditions like diarrhea, flatulence, waist pain, dysentery, and gynecological disorders, particularly in post-delivery cases for humans and animals (Singh et al., 2006). The infusion of the calyces, known as Sudan tea, is a useful remedy for addressing issues such as biliousness, coughing, and elevated body temperature (Osei-Kwarteng et al., 2021).

In Egypt, the calyces are utilized to treat nerve and cardiac diseases and promote urine production. In Guatemala, the infusion of the calyces is employed as a remedy for drunkenness, while, in North Africa, it is used to treat coughs, sore throats, and genital problems. Moreover, in North Africa, the roselle plant leaves are used to treat abscesses and external wounds (Neuwinger, 2000). These diverse traditional uses of roselle across different regions demonstrate its versatility as a medicinal plant and its role in addressing various health conditions in traditional and folk medicine practices.

FIGURE 15.8 Roselle (*Hibiscus sabdariffa* L.) as medicine tea (James & Morton, 2017).

15.8 CHALLENGES ASSOCIATED WITH PROCESSING AND COMMERCIALIZATION

Roselle is a widely cultivated plant used in various industries, including food, cosmetics, and pharmaceuticals. It is primarily grown in Africa, Asia, and South America. However, there are difficulties involved in processing and commercializing roselle. The lack of suitable harvesting and processing methods is one of the major obstacles to the processing and commercialization of roselle. Roselle has a brief shelf life, and improper harvesting and processing methods can cause the product to spoil quickly and lose quality. Additionally, the crop is vulnerable to pest and disease infestations, which can drastically lower yields and quality (Khan, 2017).

Lack of production and quality control standardization is another issue with roselle processing and commercialization. Since there are no established guidelines for roselle production, there may be wide variations in the quality of the finished product. It may be difficult to market and sell roselle products to consumers due to the lack of standardization. The lack of understanding and awareness of the crop's potential benefits is a significant obstacle to roselle processing and commercialization. Despite having numerous health advantages, including strong anti-inflammatory and antioxidant properties, roselle is still largely unheard of by many consumers. It may be difficult to market and sell roselle products effectively due to this lack of knowledge (Shruthi & Ramachandra, 2019).

Another difficulty with processing and commercializing roselle is the high cost of infrastructure and processing machinery. Small-scale farmers and processors may find it challenging to enter the market due to the high cost of processing equipment, which could result in little competition and higher prices for consumers (Khan, 2017). Another significant obstacle is the limited availability of credit and financing options for producers and processors. Small-scale farmers may find it challenging to invest in the infrastructure and machinery needed for processing and commercialization due to limited access to credit (Shruthi & Ramachandra, 2019).

Hence, various obstacles to the processing and commercialization of roselle, such as poor harvesting and processing methods, a lack of production and quality control standards, a lack of understanding of the crop's potential advantages, expensive processing infrastructure and equipment, and restricted access to credit and financing options for farmers and processors. The establishment of appropriate policies and regulations, increased stakeholder cooperation, and increased investment in research and development are all necessary to address these issues.

15.9 CONCLUSION

In conclusion, the rising demand for sustainable materials has brought natural fibers like roselle (*H. sabdariffa* L.) into focus as a promising alternative to synthetic fibers. The plant's agro-waste, rich in valuable properties, offers numerous advantages such as biodegradability, lightweight, low cost, and strong mechanical characteristics, making it an attractive resource for industries concerned with environmental sustainability. This study emphasizes the potential of utilizing roselle waste in the production of biocomposites, biochar, bioethanol, bioelectricity, and natural coagulants for wastewater treatment, as well as its applications in traditional medicine. Despite progress in managing roselle waste, there is still potential for further optimization. Beyond conventional composting methods, roselle waste can be converted into more advanced materials, including biochar and biocomposites. Research into innovative waste management strategies could make these processes more economical and sustainable, contributing to the bioeconomy. Additionally, addressing challenges such as poor harvesting and processing methods, inconsistent quality control, and the high costs associated with processing equipment is crucial for unlocking the full commercial potential of roselle. Limited awareness of its health and industrial benefits, combined with restricted access to credit for small-scale farmers, also hampers commercialization efforts. Overcoming these challenges through policy support, stakeholder collaboration, and increased investment in research and development will help maximize the economic and environmental benefits of roselle. Expanding research on hybrid products and developing new applications can increase demand for roselle and foster more innovative waste management solutions, contributing to a circular economy and enhanced sustainability.

ACKNOWLEDGMENT

The chapter was written with Muhammad Harraz Bin Azmi, Patricksya, and Nur Afiqah Binti Abd Razak.

REFERENCES

Abdul, H. P. S., Jawaid, M., Hassan, A., Paridah, M. T., & Zaido, A. (2012). *Oil Palm Biomass Fibers and Recent Advancement in Oil Palm Biomass Fibers Based Hybrid Biocomposites*. InTech. https://doi.org/10.5772/48235

Aboudi, K., Fernández-Güelfo, L.A., Álvarez-Gallego, C.J., & Romero-García, L.I., (2021). Biogas, biohydrogen, and polyhydroxyalkanoates production from organic waste in the circular economy context. In *Sustainable biofuels*, 305–343. Academic Press. https://doi.org/10.1016/B978-0-12-820297-5.00002-5

Afiqah, A.N., Sapuan, S.M., & Ilyas, R.A. (2021). Pulp and paper production: A review. In *Seminar on Advanced Bio-and Mineral based Natural Fibre Composites*, 10–13.

Aekjuthon Phounglamcheik. (2021). Bio-coal for the sustainable industry a scientific approach to optimizing production, storage, and usage. Doctoral Thesis, Luleå University of Technology.

Al-Sayed, H. M., Ali, A. M., Mohamed, M. A., & Ibrahim, M. F. (2022). Combined effect of prickly pear waste biochar and azolla on soil fertility, growth, and yield of roselle (*Hibiscus sabdariffa L.*) plants. *Journal of Soil Science and Plant Nutrition, 22*(3), 3541–3552. https://doi.org/10.1007/s42729-022-00908-7

Archaina, D., Vasile, F., Jiménez-Guzmán, J., Alamilla-Beltrán, L., & Schebor, C. (2019). Physical and functional properties of roselle (Hibiscus sabdariffa L.) extract spray dried with maltodextrin-gum Arabic mixtures. *Journal of Food Processing and Preservation, 43*(9), e14065.

Arowosafe, K.O., Bamgboye, A.I., Adesanya, A.A., Odugbose, B.D., Olannye, U.D., & Fatoye, O.V. (2022). Biofuel production from locally sourced roselle (*Hibiscus Sabdariffa* L.) seed oil using transesterification process. *Nigerian Journal of Technology*, *41*(1), 191–196. https://doi.org/10.4314/njt.v41i1.24

Baba, N., Agmour, I., El Foutayeni, Y., & Achtaich, N. (2022b). The tide effects on bioeconomic model of Sardina pilchardus, Engraulis encrasicolus and Xiphias gladius in Atlantic Moroccan zone. *Earth Systems and Environment, 6*(1), 295–305.

Baba, N., Bashir, A., & Jika, A. (2022a). Production of briquettes from agro: Wastes/Coal blends and its effect on water boiling studies. *International Journal of Multidisciplinary Research and Growth Evaluation, 3*, 381–385.

Bajpai, P. (Ed.). (2015). Basic overview of pulp and paper manufacturing process. In: *Green Chemistry and Sustainability in Pulp and Paper Industry*. Springer, Cham. https://doi.org/10.1007/978-3-319-18744-0_2

Bamgboye, A. I., & Adejumo, O. I. (2009). Physical properties of roselle (*Hibiscus sabdariffa L.*) seed. *Agricultural Engineering International: The CIGR Ejournal*. Manuscript 1154. Vol. XI, 1–13.

Bamgboye, A. I., & Adejumo, O. I. (2010). Thermal properties of roselle seeds. *International Agrophysics, 24*(1), 85–87.

Barbot, Y. N., Al-Ghaili, H., & Benz, R. (2016). A review on the valorization of macroalgal wastes for biomethane production. *Marine Drugs* 2016, *14*(6), 120. https://doi.org/10.3390/MD14060120

Berndes, G., & Börjesson, P. (2022). More biomass and less negative environmental impact – Press Release by Task 45. IEA Bioenergy. Retrieved from https://www.ieabioenergy.com/blog/publications/press-release-more-biomass-and-less-negative-environmental-impact/

Blackwood, M. (2022). Roselle (*Hibiscus Sabdariffa*). Retrieved from: https://healthiersteps.com/roselle-hibiscus-sabdariffa/

Bledzki, A. K., Faruk, O., & Sperber, V. E. (2006). Cars from bio fibers. *Macromolecular Materials and Engineering, 291*, 449–457.

Browne, J. D., & Murphy, J. D. (2013). Assessment of the resource associated with biomethane from food waste. *Applied Energy, 104*, 170–177.

Camelo-Méndez, G. A., Ragazzo-Sánchez, J. A., Jiménez-Aparicio, A. R., Vanegas-Espinoza, P. E., Paredes-López, O., & Del Villar-Martínez, A. A. (2013). Comparative study of anthocyanin and volatile compounds content of four varieties of Mexican roselle (*Hibiscus sabdariffa L.*) by multivariable analysis. *Plant Foods for Human Nutrition, 68*(3), 229–234. https://doi.org/10.1007/s11130-013-0360-2

Carvajal-Zarrabal, O., Barradas-Dermitz, D. M., Orta-Flores, Z., Hayward-Jones, P. M., Nolasco-Hipólito, C., Aguilar-Uscanga, M. G., Miranda-Medina, A., & Bujang, K. B. (2012). *Hibiscus sabdariffa L.*, roselle calyx, from ethnobotany to pharmacology. *Journal of Experimental Pharmacology*, *4*, 25–39. https://doi.org/10.2147/JEP.S27974

Cheikh, B., Mario, A., Serge, H., Lat, S. T., Aurore, R., & Philippe, T. (2015). Preliminary characterization of residual biomass from *Hibiscus sabdariffa* calyces. *African Journal of Biotechnology*, *14*(36), 2683–2692. https://doi.org/10.5897/ajb2015.14795

Demir, H., Atikler, U., Balköse, D., & Tihm, F. (2006). The effect of fiber surface treatments on the tensile and water sorption properties of polypropylene-luffa fiber composites. *Composites Part A: Applied Science and Manufacturing, 373*, 447–456.

Dharmendra, B. V., Vivek, S., Ramu, P., Srinivasan, T., Suresh, G., Meenakshi, C. M., & Lavanya, R. (2021). Static investigation of roselle waste powder reinforced bio polymer composite. *Journal of Physics: Conference Series*, *2054*(1). https://doi.org/10.1088/1742-6596/2054/1/012058

Dutta, S., & Kumar, M. S. (2021). Potential of value-added chemicals extracted from floral waste: A review. *Journal of Cleaner Production*, *294*, 126280. https://doi.org/10.1016/J.JCLEPRO.2021.126280

Fadzil, N. F., & Othman, S. A. (2021). The growing biorefinery of agricultural wastes: a short review. *Journal of Sustainable Natural Resources*, *2*(2), 46–51. https://penerbit.uthm.edu.my/ojs/index.php/jsunr/article/view/9238

Gao, B. Y., Wang, Y., Yue, Q. Y., Wei, J. C., & Li, Q. (2007). Colour removal from simulated dye water and actual textile wastewater using a composite coagulant prepared by polyferric chloride and poly dimethyldiallylammonium chloride. *Separation and Purification Technology*, *54*(2), 157–163. https://doi.org/10.1016/j.seppur.2006.08.026

Gupta, N., Mahur, B. K., Izrayeel, A. M. D., Ahuja, A., & Rastogi, V. K. (2022). Biomass conversion of agricultural waste residues for different applications: A comprehensive review. *Environmental Science and Pollution Research, 29*(49), 73622–73647. https://doi.org/10.1007/S11356-022-22802-6

Harris, D. R., & Fuller, D. Q. (2014). Agriculture: Definition and overview. In: Smith, C. (Ed.), *Encyclopedia of Global Archaeology* (pp. 104–113). Springer, New York.

Hart, P. W. (2011). Production of high yield bleached hardwood kraft pulp: Breaking the kraft pulp yield barrier. *Tappi Journal, 10*(9), 37–41.

Herranz-López, M., Fernández-Arroyo, S., Pérez-Sanchez, A., Barrajón-Catalán, E., Beltrán-Debón, R., Menéndez, J. A., Alonso-Villaverde, C., Segura-Carretero, A., Joven, J., & Micol, V. (2012). Synergism of plant-derived polyphenols in adipogenesis: Perspectives and implications. *Phytomedicine*, *19*(3–4), 253–261. https://doi.org/10.1016/j.phymed.2011.12.001

Hida, H., Yamada, T., & Yamada, Y. (2007). Genome shuffling of Streptomyces sp. U121 for improved production of hydroxycitric acid. *Applied Microbiology and Biotechnology*, *73*(6), 1387–1393. https://doi.org/10.1007/s00253-006-0613-1

Hsu, M.-H., Mannapperuma, J. D., & Singh, R. P. (1991). Physical and thermal properties of pistachios. *Journal of Agricultural Engineering Research, 49*, 311–321.

Ilyas, R. A., Asyraf, M. R. M., Sapuan, S., Afiq, T. M. N., Suhrisman, A., Atikah, M. S. N., & Ibrahim, R. (2021). Development of roselle fiber- reinforced polymer biocomposite mug pad using the hybrid design for sustainability and pugh method. In: Sapuan, S. M., Nadlene, R., Radzi, A. M., & Ilyas, R. A. (Eds.), *Roselle: Production, Processing, Products and Biocomposites*. Academic Press. https://doi.org/10.1016/B978-0-323-85213-5.00002-0

Ismail, A., Ikram, E. H. K., & Nazri, H. S. M. (2008). Roselle (Hibiscus sabdariffa L.) seeds nutritional composition protein quality and health benefits. *Food, 2*(1), 1–16.

James, C., & Morton Jr. (2017). *Hibiscus sabdariffa* L. tea (Tisane) reduces blood pressure in pre-and mildly hypertensive patients. Retrieved from https://www.naturalhealthresearch.org/61616-2/

Jayaprakash, K., Osama, A., Rajagopal, R., Goyette, B., & Karthikeyan, O. P. (2022). Agriculture waste biomass repurposed into natural fibers: A circular bioeconomy perspective. *Bioengineering (*Basel, Switzerland), *9*(7), 296. https://doi.org/10.3390/bioengineering9070296

Jung, E., Kim, Y., & Joo, N. (2013). Physicochemical properties and antimicrobial activity of roselle (*Hibiscus sabdariffa* L.). *Journal of the Science of Food and Agriculture, 93*, 3769–3776. https://doi.org/10.1002/jsfa.6256

Khan, M. (2017). Nutritional and health importance of *Hibiscus Sabdariffa*: A review and indication for research needs. *Journal of Nutritional Health & Food Engineering*, *6*(5). https://doi.org/10.15406/jnhfe.2017.06.00212

Kian, L. K., Jawaid, M., Ariffin, H., & Alothman, O. Y. (2017). Isolation and characterization of microcrystalline cellulose from roselle fibers. *International Journal of Biological Macromolecules*, *103*, 931–940. https://doi.org/10.1016/j.ijbiomac.2017.05.135

Kulkarni, M. B., & Ghanegaonkar, P. M. (2019). Biogas generation from floral waste using different techniques. *Global Journal of Environmental Science and Management, 5*(1), 17–30. https://doi.org/10.22034/gjesm.2019.01.02

Maganha, E. G., Halmenschlager, R. da C., Rosa, R. M., Henriques, J. A. P., Ramos, A. L. L. de P., & Saffi, J. (2010). Pharmacological evidence for the extracts and secondary metabolites from plants of the genus Hibiscus. *Food Chemistry, 118*(1), 1–10. https://doi.org/10.1016/j.foodchem.2009.04.005

Merrifield, R. (2020). Making coal from food waste, garden cuttings – and even human sewage. Horizon: The EU Research & Innovation Magazine. Retrieved from: https://ec.europa.eu/research-and-innovation/en/horizon-magazine/making-coal-food-waste-garden-cuttings-and-even-human-sewage

Mohamed, M. G., El-Mahdy, A. F., Kotp, M. G., & Kuo, S. W. (2022). Advances in porous organic polymers: Syntheses, structures, and diverse applications. *Materials Advances, 3*(2), 707–733.

Molino, A., Nanna, F., Ding, Y., Bikson, B., & Braccio, G. (2013). Biomethane production by anaerobic digestion of organic waste. *Fuel, 103*, 1003–1009. https://doi.org/10.1016/J.FUEL.2012.07.070

Monteiro, H., & Moura, B. (2022). Life cycle energy and carbon emissions of colorants extraction from *Hibiscus sabdariffa. Energy Reports, 8*, 277–283.

Moreira, R. G., Palau, J., Sweat, V. E., & Sun, X. (1995). Thermal and physical properties of tortilla chips as a function of frying time. *Journal of Food Processing and Preservation, 19*(3), 175–189.

Munawar, S. S., Umemura, K., & Kawai, S. (2007). Characterization of the morphological, physical, and mechanical properties of seven nonwood plant fiber bundles. *Journal of Wood Science, 53*(2), 108–113. https://doi.org/10.1007/s10086-006-0836-x

Muthuraman, G., & Sasikala, S. (2014). Removal of turbidity from drinking water using natural coagulants. *Journal of Industrial and Engineering Chemistry, 20*(4), 1727–1731. https://doi.org/10.1016/j.jiec.2013.08.023

Nadlene, R., Sapuan, S. M., Jawaid, M., Ishak, M. R., & Yusriah, L. (2015). Material characterization of roselle fiber (*Hibiscus sabdariffa L.)* as potential reinforcement material for polymer composites. *Fibers and Textiles in Eastern Europe, 23*(6), 23–30. https://www.researchgate.net/profile/Mohammad-Jawaid/publication/280560885_Material_characterization_of_Roselle_fibre_as_potential_reinforcement_material_for_polymer_composites/links/5649308408ae451880aecb84/Material-characterization-of-Roselle-fibre-as-potential-reinforcement-material-for-polymer-composites.pdf

Neuwinger, H. D. (2000). *African Traditional Medicine: A Dictionary of Plant Use and Applications. With Supplement: Search System for Diseases*. Medpharm.

Oluseun Adejumo, I., & Adebukola Adebiyi, O. (2021). *Agricultural Solid Wastes: Causes, Effects, and Effective Management*. IntechOpen. https://doi.org/10.5772/intechopen.93601

Omotoso, M. A., & Owolabi, A. W. (2015). Pulp and paper evaluation of solid wastes from agricultural produce. *International Journal of Chemistry, 7*, 113–121. https://doi.org/10.5539/ijc.v7n2p113

Osei-Kwarteng, M., Gweyi-Onyango, J. P., & Mahunu, G. K. (2021). Harvesting, storage, postharvest management, and marketing of Hibiscus sabdariffa. In V. R. Thakur, M. R. Islam, & S. Thomas (Eds.), *Roselle (Hibiscus sabdariffa)* (pp. 15–31). Academic Press.

Paletto, A., Bernardi, S., Pieratti, E., Teston, F., & Romagnoli, M. (2019). Assessment of environmental impact of biomass power plants to increase the social acceptance of renewable energy technologies. *Heliyon, 5*(7). https://doi.org/10.1016/j.heliyon.2019.e02070

Prithivirajan, R., Jayabal, S., & Bharathiraja, G. (2015). Bio-based composites from waste agricultural residues: Mechanical and morphological properties. *Cellulose Chemistry and Technology, 49*, 65–68.

Raheem, A. A. (2013). Revamping the industrialization of the Nigerian economy: The role of science, engineering and technology. In Proceedings of the *2nd International Conference on Engineering & Technology Research.* Faculty of Engineering and Technology, Ladoke Akintola University of Technology, Ogbomoso, Nigeria.

Ranjitha, J., Vijayalakshmi, S., Kumar, V., & Ralph, N. (2014). Production of bio-gas from flowers and vegetable wastes using anaerobic digestion. *International Journal of Research in Engineering and Technology, 3*(8), 279–283.

Rautray, P., Roy, A., Mathew, D. J., & Eisenbart, B. (2019). Bio-brick - Development of sustainable and cost effective building material. *Proceedings of the Design Society: International Conference on Engineering Design, 1*(1), 3171–3180. https://doi.org/10.1017/DSI.2019.324

Rautray, P., Roy, A., Mathew, D. J., & Eisenbart, B. (2021). Bio-bricks: Circular economy and new products. In B. Pradhan, M. Bui, & A. Singh (Eds.), *Smart Innovation, Systems and Technologies, 221*, 845–857. Springer. https://link.springer.com/chapter/10.1007/978-981-16-0041-8_69

Reanpang, P., Pun-Uam, T., Jakmunee, J., & Khonyoung, S. (2021). An environmentally friendly flow injection-gas diffusion system using roselle (*Hibiscus sabdariffa L.*) extract as natural reagent for the photometric determination of sulfite in wines. *Journal of Analytical Methods in Chemistry, 2021*. https://doi.org/10.1155/2021/6665848

Riaz, G., & Chopra, R. (2018). A review on phytochemistry and therapeutic uses of *Hibiscus sabdariffa L. Biomedicine and Pharmacotherapy, 102*, 575–586. Elsevier Masson SAS. https://doi.org/10.1016/j.biopha.2018.03.023

Rizwan, M., Tse, J. W., Nori, A., Leong, K. W., & Yim, E. K. F. (2018). Cell–Substrate interactions. In: Atala, A., Lanza, R., Mikos, A. G., Nerem, R. M. (Eds.), *Principles of Regenerative Medicine* (pp. 437–468). Elsevier. https://doi.org/10.1016/B978-0-12-809880-6.00027-8

Seyedbagheri, M. M. (2010). Compost: Production, quality, and use in commercial agriculture. University of Idaho Extension, CIS 1171. Retrieved from https://www.extension.uidaho.edu/publishing/pdf/CIS/CIS1171.pdf

Shruthi, V. H., & Ramachandra, C. T. (2019). Roselle (Hibiscus sabdariffa L.) calyces: A potential source of natural color and its health benefits. In: Deka, S. C., Seth, D., & Hulle, N. R. S. (Eds.), *Food Bioactives* (pp. 169–190). Apple Academic Press.

Singh, R. K., Sureja, A. K., & Singh, D. (2006). Amta and amti (Hibiscus sabdariffa L.)-cultural and agricultural dynamics of agrobiodiversity conservation. *Indian Journal of Traditional Knowledge, 5*, 151–157.

Son, J., Kim, H. J., & Lee, P. W. (2001). Role of paper sludge particle size and extrusion temperature on performance of paper sludge–thermoplastic polymer composites. *Journal of Applied Polymer Science, 82*(11), 2709–2718.

Spears, S. (2018). What is biochar? Regeneration International. Retrieved from: https://regenerationinternational.org/2018/05/16/what-is-biochar/

Tahir, H. E., Xiaobo, Z., Mariod, A. A., Mahunu, G. K., Abdualrahman, M. A. Y., & Tchabo, W. (2017). Assessment of antioxidant properties, instrumental and sensory aroma profile of red and white karkade/roselle (*Hibiscus sabdariffa L.*). *Journal of Food Measurement and Characterization, 11*(4), 1559–1568. https://doi.org/10.1007/s11694-017-9535-0

Tamta, M., & Kalita, B. B. (2020). Properties of roselle and its blends. *International Journal of Current Microbiology and Applied Sciences, 9*(10), 3616–3621. https://doi.org/10.20546/ijcmas.2020.910.418

Wong, K. J., Nirmal, U., & Lim, B. K. (2010). Impact behavior of short and continuous fiber-reinforced polyester composites. *Journal of Reinforced Plastics and Composites,* 29(23), 3463–3474.

Yani, N. F. A., Ismail, N., & Oh, K. S. (2019). Potential of using *Hibiscus Sabdariffa* in treating greywater. *AIP Conference Proceedings, 2137*. https://doi.org/10.1063/1.5120993

Yong, M. Y., & Ismail, N. (2015). Optimisation of *Hibiscus sabdariffa* as a natural coagulant to treat Congo red in wastewater. *Journal of Engineering Science and Technology, 4*, 153–165.

Section VI

Leaf-Based Plant Waste

16 Pineapple Waste Management

16.1 INTRODUCTION

Pineapple is a tropical fruit native to South America and grown in many tropical places worldwide. Because of its extensive fruit growing, India is now the world's second-largest producer of fruits after Brazil. Growing pineapples contributes $9 billion annually to the global economy (Sarangi et al., 2022a). The other main producers are Thailand, the Philippines, China, Mexico, Brazil, Nigeria, Indonesia, and Colombia. It is a delicious, juicy fruit with a prickly outer peel and sweet, fragrant flesh. Pineapples are high in vitamin C and manganese and contain a variety of other important minerals, such as bromelain, an enzyme that aids digestion. Pineapple can be eaten fresh, cooked, or canned and is widely used in sweets, smoothies, and salads.

About 60% of pineapple fruits are edible, leaving processing residues ranging from 40% to 60%. Researchers found that pineapple processing results in 9.13% of the middle portion, 13.49% from the peels, 14.5% from the pulp, 14.86% from the top part, and 48.05% from the finished products (Hikal et al., 2021). Pineapple waste refers to the parts of the pineapple plant left over after the edible fruit has been harvested. This can include the skin, core, and leaves. Pineapple waste is often discarded as agricultural waste, which can contribute to environmental pollution and waste management issues. However, pineapple waste can be repurposed for various purposes.

One potential use for pineapple waste is in animal feed. Pineapple waste contains a significant amount of fibers, which can be beneficial for animals such as cattle and pigs. Additionally, pineapple waste can be used in composting or as a natural fertilizer for crops (Sukri et al., 2023). Another potential use for pineapple waste is in the production of biofuels. The sugars and carbohydrates in pineapple waste can be converted into biofuels, such as ethanol, through a process called fermentation.

Figure 16.1 shows types of pineapple waste. It can also be processed to extract valuable compounds such as bromelain, an anti-inflammatory and digestive-beneficial enzyme. It can produce textiles, paper products, and building materials (Tran et al., 2023). Overall, pineapple waste has the potential to be a valuable resource rather than a waste product. Finding ways to repurpose pineapple waste can reduce environmental pollution and waste while creating new economic opportunities.

Growing pineapple production leads to a consistent generation of pineapple waste. With the yearly rise in the production of processed pineapple items, a significant amount of waste is produced. Moreover, due to physical handling and exposure to unfavorable conditions, up to 50% of the pineapples can be lost or damaged, contributing to further waste. The utilization of pineapple waste also poses a significant challenge, as it can lead to environmental pollution (Salve & Ray, 2020). The effective and sustainable use of pineapple waste is necessary for their value addition. The vast supply of phytochemicals, antioxidants, biofuels, and phenolic compounds makes it possible to apply biotechnological and strong methods to sustain pineapple waste.

16.2 PINEAPPLE WASTE FROM NATURAL FIBERS

Pineapple leaves are considered natural wastes that can be used to make paper and textile fibers (Asim et al., 2015). After being harvested, fruit leaves are typically returned to the ground to be used as fertilizer. Intake of pineapple leaf fiber normally occurs when plants are between 1 and

DOI: 10.1201/9781003472209-22

FIGURE 16.1 Type of pineapple waste (Tran et al., 2023).

5 years old (Pandit et al., 2020). The fiber extracted from young pineapple leaves is often short and weaker. Plants that grow in open areas with excessive sunlight without protection may produce moderate fiber that is too old and will result in short, brittle fibers that are harsh and brittle (Widowati & Amalia, 2021). Therefore, it is necessary to select pineapple leaves that are relatively ripe so that the growth is partially shielded from sunlight to obtain a strong, smooth, and soft fiber.

Pineapple fibers are lightweight, soft, and breathable, making them an excellent choice for clothing and textile production. They are also strong and durable, making them a sustainable alternative to synthetic materials. In addition to their use in fashion and textile production, pineapple fibers are used in producing other products, such as paper and packaging materials (Rasmi, 2023). They are also being explored as a potential material for use in the automotive and construction industries.

One of the major benefits of pineapple fibers shown in Figure 16.2 is that they are sustainable and environmentally friendly. The production of pineapple fibers requires minimal water and energy compared to other fibers, and the waste from the production process can be used as fertilizer for the next crop of pineapple plants (Aili Hamzah et al., 2021). Overall, pineapple fibers offer a promising alternative to traditional synthetic fibers, and their use could help to reduce the environmental impact of the textile industry.

16.3 ENVIRONMENTAL IMPACT FROM BIOMASS (PINEAPPLE WASTE)

Overall, pineapple biomass has very little of a negative environmental impact. Its extensive use of agrochemicals, monoculture farming, consumption of plastic packaging, and low composting rates are contributing factors (Castillo-González et al., 2020). When it comes to their effects on the environment, pineapple biomass benefits the planet in a variety of ways (Castillo-González et al., 2020). It conducts intensive farming, leaves little carbon impact, and doesn't need a lot of

FIGURE 16.2 Pineapple leaf fiber (Aili Hamzah et al., 2021).

irrigation. Despite that, they still have a lot of potential to harm the ecosystem (Ruslan et al., 2017). The monocultures used in the pineapple agricultural sector have a terrible environmental effect. Since the pineapple industry in Costa Rica has proliferated, more and more land has been cleared (Amores-Monge et al., 2022). Biodiversity suffers greatly from monocultures as they restrict the variety of food available to local fauna (Amores-Monge et al., 2022).

Other than that, one of the methods for gauging the effects of human-caused global climate change is the carbon footprint. Today, the production of food is responsible for 26% of the world's greenhouse gas emissions. The carbon footprint of pineapple biomass is relatively small, at 0.09 kg (0.20 lb) per pound of pineapple (Castillo-González et al., 2020). The majority of these emissions are caused by their heavy reliance on pesticides, transportation emissions such as those involved in moving them from Costa Rica to the United States, and low composting rates (Amores-Monge et al., 2022). The amount of pesticides used to grow pineapples is very large, and these chemicals can harm groundwater, wildlife, insect populations, and soil microbes. This indicates that their usage of pesticides has a highly harmful influence on the environment (Ghazi et al., 2023).

The growth of the biomass industry seems to be a top priority and is crucial for promoting the circular economy and sustainability worldwide (Ruslan et al., 2017). It is because, through its conversion into more environmentally friendly valuable bio-based products, biomass can help create a more balanced carbon cycle (Castillo-González et al., 2020). Direct combustion to generate heat is one of the most common procedures for turning biomass into energy. Using thermochemistry, fuels can be converted into solid, gaseous, and liquid forms, while liquid fuels are produced chemically (Ghazi et al., 2023). As agricultural waste was readily available, it was not foreign for Malaysia to use biomass conversion into high-value biofuels and biochemicals. Research accomplishments showed that biomass seems to be a promising renewable feedstock for synthesizing chemicals and energy (Williams et al., 2017). The Malaysian government took advantage of this opportunity by implementing numerous green policies that helped the biomass industry advance and brought more biomass-capable industrial waste, such as pineapple, to light.

Pineapple wastes are a classic example of lignocellulosic biomass, which has enormous potential for energy generation but is reportedly underutilized in terms of clean, green, and sustainable sources (Amores-Monge et al., 2022). The production of fuels and chemicals from this biomass is appealing because it is one of the planet's renewable resources. Various microbes can break down pineapple waste biomass, yielding valuable compounds (Ruslan et al., 2017). According to the report, pineapple wastes are made up, by dry weight, of peel cellulose (19.8%), peel hemicellulose (11.7%), pulp cellulose (14.3%), pulp lignin (2.3%), and peel hemicellulose (22.1%) (Amores-Monge et al., 2022). These substances have enormous potential for producing fermentable sugars (Franklyn da Cruz et al., 2018). The biomass from pineapple can also be utilized in the production of second-generation biofuels such as bioethanol, biobutanol, biomethanol, biomethane, biohydrogen, and bioethane through a one or two-stage fermentation processes, which are considered as the breakthrough for environmentally friendly processing product (Sarangi et al., 2022b).

Before being replanted, most pineapple leaves are discarded in the open and burned. However, these days, pineapple product manufacturers have found a more profitable and environmentally friendly approach to utilize pineapple waste, transforming it into useful products like animal feed because it is high in sugars and organic acids (Amores-Monge et al., 2022). Ensilaging could improve feed quality and make it more appealing to animals (Ghazi et al., 2023). Pineapple waste is an excellent digestive feed since it is high in fiber and can be utilized as a source of energy for animals like chickens and cows (Williams et al., 2017). Additionally, giving pineapple waste to dairy cows might speed up their digestion, resulting in increased milk output (Williams et al., 2017).

Positive environmental impact aside, however, pineapple waste still often shares the same fate as other biomass products in which dried pineapple waste products are introduced to combustion technology for energy generation (Williams et al., 2017). Burning of common biomass in a solid, liquid, or gaseous state can release additional pollutants and particulate matter into the air, such as carbon monoxide, volatile organic compounds, and nitrogen oxides, in addition to contributing to carbon dioxide emissions (Ghazi et al., 2023). In some cases, burning biomass can produce more pollution than burning fossil fuels. In contrast to carbon dioxide emissions, many of these contaminants cannot be sequestered by young plants (Williams et al., 2017). If these substances are not properly contained, they may cause a variety of problems for both the environment and human health.

16.4 AVAILABILITY OF PINEAPPLE WASTE

Pineapple waste is a by-product of the pineapple industry, and its availability varies according to certain factors, such as pineapple-producing region, season, and specific processing methods utilized. Pineapple waste comprises the fruit's stem, leaves, and rind, often forming during the preparation of the pineapple for market or processing. Pineapple waste may be widely available in areas where pineapple is farmed commercially, such as Southeast Asia, South America, and Africa. It may be collected and utilized as animal feed or compost in some instances. However, in other areas where pineapple is not extensively farmed, obtaining pineapple waste may be more challenging. Malaysia is one of Southeast Asia's largest producers, with substantial pineapple growing in areas such as Johor, Negeri Sembilan, and Selangor (Ruslan, 2017).

In addition, these leftover components have potential as building blocks for future products. Malaysia has progressively increased its acreage from a 2010 low of 10,000 ha to a projected 2020 high of 17,222 ha, or 9.5% of the world's total fruit farming. Pineapple waste accumulates at a rate of 3 kg/kilo of fruit produced. This fibrous and voluminous garbage is burned and composted as usual. The majority of waste can be obtained from Johor, which produces 56% of pineapple waste in Malaysia, followed by Sarawak, which produces 19%, and Sabah, which produces 14% (Ruslan, 2017).

16.5 PROPERTIES OF PINEAPPLE WASTE FIBERS

16.5.1 Physical Properties

The texture of pineapple leaf fiber is more delicate than that of any other vegetable fiber. By limiting soil erosion, it aids in climate restoration and soil quality. It has the same white, creamy, and shiny appearance as silk fiber and is ten times as coarse as cotton, despite being fine (as fine quality jute) with well-separated filaments. Colors are easily retained by the fiber. The bundle strength of pineapple leaf fiber decreases when wet, but the yarn strength improves (Asim et al., 2015).

Pineapple fibers possess a significant number of unique qualities. They exhibit exceptional hygroscopic properties, meaning they efficiently absorb moisture. Colored in a pristine white hue with a sleek, shiny appearance akin to silk, these fibers boast a low microfibrillar angle and moderate length. However, their fiber bundle strength decreases significantly by 50% when wet, contrasting with the remarkable 50% increase in yarn strength under similar conditions. Despite their absorbent nature, pineapple fibers present challenges in dye penetration due to their high coarseness. Their hardness prevents wrinkling while showcasing impressive antibacterial and deodorization capabilities, rendering them a versatile and valuable material in various applications.

16.5.2 Chemical Composition

Pineapple waste contains lignin (19.4%), cellulose (32.4%), and hemicellulose (23.2%), according to Dahunsi et al. (2021). Potassium, calcium, and magnesium are present in untreated peels. Pineapple peels have a crude fiber content of 23.71%, a crude fat content of 0.46%, and a crude protein content of 0.33% (Chu et al., 2020). The lignin content of pineapple leaf waste is 13.05%, the hemicellulose content is 21.02%, and the cellulose content is 41.15% (Aili Hamzah et al., 2021). Pineapple waste is an excellent feedstock for the formation of fermentable sugar for energy production because of its high quantities of lignin, cellulose, and hemicellulose (Pilap et al., 2022). Aside from that, the carboxyl, carbonyl, and hydroxyl groups present in pineapple waste can be harnessed as precursors in the production of activated carbon (Foo & Hameed, 2012). The authors of the study found that the pineapple leaf has the lowest quantity of protein among the analyzed parts, but it has the highest amount of ash. Additionally, they assert that the peel of the pineapple contains the highest fiber content compared to the pulp, leaf, and core. Furthermore, when compared to other plants, the pineapple leaf contains higher levels of lignin and cellulose.

Pineapple leaf is composed of numerous chemical elements. The pineapple leaf is a complex multicellular lignocellulosic fiber, abundant in polysaccharides and lignin, along with various other compounds like fat, wax, pectin, uranic acid, anhydride, pentosan, color pigment, and inorganic substances. The fibers have a thread-like structure, consisting of thin and tiny multicellular fibers that are firmly bound together by pectin. The composition of the leaves includes cellulose, comprising approximately 70–82% of its content, with a fiber arrangement similar to that of cotton (82.7%).

16.5.3 Mechanical Properties

As fiber-reinforced composite constructions become more and more prevalent in practically every area of material science, lignocellulosic natural fibers, such as pineapple fibers, offer affordable and plentiful alternatives to the pricey and non-renewable synthetic fibers. The mechanical characteristics of the polymer matrix were enhanced by those with high specific strengths (Abdullah et al., 2020).

Pineapple leaf fibers, among other natural fibers, have exceptional mechanical qualities besides being lignocellulosic and multicellular (Abdullah et al., 2020). The high cellulose content of pineapple leaf fibers is linked to their outstanding mechanical qualities (George et al., 1995). The examined pineapple fiber-reinforced polyethylene composites' stress relaxation characteristics (Nordin et al., 2020). Due to a stronger reinforcing impact, they discovered that increased fiber content lowered the ability to relax under stress (Cheung et al., 2009). It is determined that several variables, including the adhesion of the fiber to the matrix and the volume fraction of the fiber, affect the attributes of fiber-reinforced composites, the aspect ratio of the fiber, the orientation of the fiber, and the effectiveness of the interface's ability to transfer stress (Mansor et al., 2019).

The results of the inquiry into the mechanical properties of composites made of pineapple leaf fiber–polypropylene (PALF–PP) were described in the current study (Henrique et al., 2022). Analysis was done on how volume fraction affected the tensile and flexural characteristics of composite materials (Henrique et al., 2022). This study focuses on examining the tensile and flexural properties of laminated composites made from pineapple leaf fiber–polypropylene. They are treated with isocyanates to enhance the durability of the raw, short pineapple leaf fibers. This treatment induces a reaction between the hydroxyl groups present in cellulose and lignin within the fibers, resulting in improved properties for the composites (George et al., 1996). It strips the cellulose fibers of their lignin and pentosan. It improves the integrity of fiber-to-matrix adhesion more effectively. Pineapple leaf fiber/low-density polythene was claimed to have enhanced melt rheological performance and durability (Cheung et al., 2009).

16.5.4 Morphological Properties

Fresh pineapple peel displayed erratic morphologies with various particle sizes, huge wrinkled, and depressed blocks (Nordin et al., 2020). Non-cellulosic elements like pectin, lignin, and hemicellulose were present in their intact state, forming a dense structure (Henrique et al., 2022). The morphology property of the pineapple waste was determined by extracting fibers from pineapple leaves carried out by mechanical processes (Henrique et al., 2022). The waxy layers of pineapple leaves are removed to obtain the PALF. The PALF is a green leaf that is long, thin, waxy, pointed, and covered with sharp spines. It is also bent inward toward the cross-section to keep the leaf firm (Mansor et al., 2019). Aside from being as white, creamy, and shiny as silk, PALF is said to have a softer surface than other vegetable fibers and the capacity to absorb and hold color well. As a result, paper with exceptional features, including thinness, smoothness, and pliability, is created (Cheung et al., 2009). Pineapple leaves contain a medium-length, fine, silky fiber with high tensile strength. PALF is a significant natural fiber with high specific strength and stiffness that is also quite hygroscopic and reasonably priced (Wan Nadirah et al., 2012) (Figure 16.3).

The cell wall structure of PALF, which comprises a central lamella, primary wall, and secondary wall, was validated by transmission electron microscopy (Nordin et al., 2020). This structure resembles the cell walls seen in wood and other lignocellulosic fibers like flax and jute. The thickness of the cell wall makes it simple to discern between the variations in cell wall layers (Mansor et al., 2019). A thin layer called the primary wall is created as stem xylem cells divide and continue to expand (Wan Nadirah et al., 2012).

16.5.5 Thermal Properties

Thermal properties are the characteristics of a material or substance that relate to its ability to transport, store, and conduct heat. These qualities include specific heat capacity, thermal conductivity, thermal stability, thermal absorptivity, thermal deterioration, thermal diffusivity, and thermal

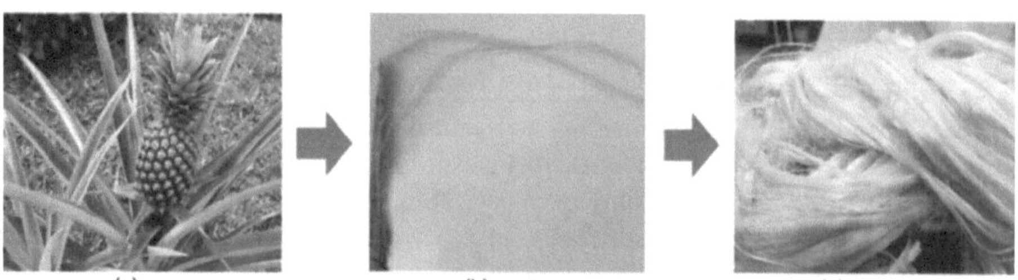

(a) (b) (c)

FIGURE 16.3 Manufacture of PFs with (a) pineapple cultivation, (b) extraction of fibers from pineapple leaves, and (c) pineapple fiber (Asim et al., 2015).

expansion. Thermal conductivity best describes a material's ability to conduct heat, whether it is a good conductor or a good insulator of heat. In contrast, thermal absorptivity describes the amount of heat a material can absorb, which differs from thermal diffusivity, defined as the promptness of heat diffusion through that specific material. On the contrary, thermal expansion describes a material's ability to expand or shrink in response to temperature changes. However, suppose the material has great thermal stability. In that case, it clearly shows that it is stable at elevated temperatures until thermal degradation occurs, which explains how the material breaks down after being subjected to feverish temperatures. A substance's specific heat capacity is the amount of energy required to raise the temperature of a unit mass of the material by 1°C, which is also a crucial factor to consider in the thermal properties of a material.

Based on Table 16.1, it has been seen that pineapple waste fiber acts as a good insulator due to its poor heat conductivity with low thermal conductivity at a range of 0.0791–0.0741 W/m·K. This quality makes it an excellent choice for uses requiring heat insulation. In Malaysia, the weather is hot sometimes, and it is highly recommended to be used as a cement admixture. First, pineapple waste is an eco-friendly material easily obtainable as a by-product of the pineapple industry. In addition, adding pineapple leaf fibers to concrete can make the resulting cement brick stronger and more thermally efficient, which justifies sustainable development in the built environment by using pineapple leaf fibers.

Moreover, it has a high specific heat capacity, indicating that it requires more energy to increase its temperature. On the contrary, the thermal resistivity of pineapple waste proves it has a very high thermal resistivity material that creates an enormous difficulty or resistance for heat to flow through it. However, as the waste's temperature changes throughout heat absorption, it could store large amounts of heat for a more extended period due to its high thermal absorptivity. The thermal diffusivity of *Ananas comosus* is relatively high, which clearly explains heat spreads through a material at a good speed. There will also be no thermal degradation due to its high thermal stability, which avoids degradation (Mangal et al., 2003).

16.5.6 Calorific Value

The calorific value of a substance is the quantity of energy generated by completely burning a unit mass of the material. The calorific value of pineapple waste fiber has been reported to range from 16.38 to 17.51 MJ/kg, depending on the exact type and analytical method utilized (Abdullah et al., 2020). The nature of pineapple waste fiber, which contains both lignocellulosic and non-lignocellulosic components, influences its calorific value. Because they are constituted of carbon, hydrogen, and oxygen, lignocellulosic components such as cellulose and hemicellulose have significant calorific values. Non-lignocellulosic components, including proteins and lipids, contribute to the calorific value of pineapple waste fiber but to a lesser amount than lignocellulosic components. The calorific content of pineapple waste fiber is significant in terms of its prospective usage as a biofuel or feedstock for energy generation. This shows that it has the potential to be a great biomass source. According to the provided data, pineapple waste leaf fiber

TABLE 16.1
Thermal Properties of Pineapple Waste Fibers (Mangal et al., 2003)

Property	Value
Specific heat capacity (J/kg·K)	1,948.95–1,949.37
Thermal conductivity (W/m·K)	0.0719–0.0741
Thermal resistivity (m·K/W)	3.091
Thermal absorptivity (J/g·°C)	27.9–28.3
Thermal diffusivity (m^2/s)	8.6–9.1
Thermal stability (°C)	0.0122

has a high calorific value and can be incinerated or converted into another form of useful energy (Abdullah et al., 2020). The high calorific value of pineapple waste leaf fiber makes it a promising biomass resource for energy generation or conversion technologies. It can be used to produce renewable energy in the form of bioenergy or biofuel, which can then be consumed in power plants or other combustion systems. The high calorific value of the waste fiber indicates that it contains a great deal of energy and could be used to generate electricity sustainably.

16.6 PINEAPPLE WASTE MANAGEMENT

16.6.1 Compost

Due to the growing agricultural production globally, notably in Malaysia, there are significant concerns regarding the management of plant waste. The country's rapid development and expanding agricultural activities lead to the disposal of around 1.2 million tons of agricultural waste each year in Malaysia alone. Tackling these waste management issues has become a crucial aspect, given the continuous growth of the agricultural sector (Eeda, 2020). In Malaysia, various types of waste, including pineapple leaves, are often disposed of through burning or decomposition, leading to increased environmental issues. However, there is a growing consciousness of the environment and the importance of adopting sustainable practices. As a result, there is an immediate requirement to tackle the long-term consequences of plant waste, notably the burning of pineapple leaves, and find more environmentally friendly solutions (Rabiu et al., 2017). Recently, researchers have worked to utilize pineapple waste as a valuable resource for economic development with a zero-waste concept.

Pineapple (*A. comosus*) is explored as a biomass derived from waste and utilized to produce valuable goods. It represents the most significant productive system within the agro-industrial domain and has demonstrated considerable potential as a useful source for composting purposes (Santulli et al., 2022). Pineapples are naturally acidic, which can interfere with the composting process. They contain citric acid, which retains rather than allows breakdown. To compost pineapple waste, some deacidifying items such as hydrated white lime should be added, which will react with the acidity of the pineapple and eventually diminish it.

There are many benefits of putting pineapple into compost. First, pineapple compost helps improve soil texture. When pineapple compost is applied to the soil, it improves its texture. Pineapple includes calcium, which aids in water penetration and loosens compacted soil. Improved soil results in increased agricultural production and favorable environmental consequences. The second benefit of using pineapple in composting is that it contains a lot of moisture. Composts must be hydrated or watered to produce an effective end result. Fortunately, pineapple has such a high moisture content that it minimizes the frequency needed to water the entire pile. As a result, the microorganisms continue to proliferate, and the pile decomposes (Sarangi et al., 2023). Other than that, composting pineapple waste leads to a much more agriculturally productive and ecological.

Composting is the process of decomposing organic matter into nutrient-rich compost that can be used to improve soil health and enhance plant growth. Pineapple waste can be incorporated into compost by following a few steps. First, the peel and leaves must be thoroughly rinsed to remove any pesticide residues that may be present. Then, the peel, leaves, and core must be chopped into 1-inch pieces. The green-to-brown ratio or the moisture level in the compost must be complied with when adding the pineapple waste. Adding a lot of water-dense pineapple flesh to your compost will overly wet the compost. Adding absorbent browns (such as sawdust or shredded paper) is necessary to adjust the moisture level. Turning the compost regularly allows the composting microorganisms to breathe.

16.6.2 Pulp and Paper

A pressing concern in the agriculture industry today is the substantial amount of waste generated from agricultural activities, with tons of discarded materials being deposited into the environment. To address this issue, individuals and businesses are increasingly exploring ways to upcycle various agricultural wastes. Pineapple leaves, for instance, which are commonly discarded, have shown potential in treating flu and nosebleeds when boiled and consumed as tea. Additionally, these leaves possess lignocellulosic fibers that make them suitable for paper production. The cellulose found in pineapple leaves is crucial for paper-making due to its high tensile strength, suppleness, and flexibility. Moreover, the low lignin content in these leaves helps enhance the efficiency of the pulping process by reducing energy consumption and chemical usage (Laftah & Wan Abdul Rahman, 2015a).

Employing non-wood resources offers several fascinating benefits, foremost among them being the preservation of raw wood materials for more advantageous uses, leading to decreased deforestation and the need for replanting. Additionally, it can potentially reduce wood and cellulose fiber imports in countries facing limited wood resources. Moreover, a growing demand for papers produced through eco-friendly methods or utilizing recycled and non-wood fibers is growing.

Transforming pineapple leaves into paper sheets is a challenging and time-consuming yet viable process. It commences with gathering discarded pineapple leaves, which are then sun-dried. Once adequately dried, the leaves are cut into approximately one-inch pieces to ensure uniformity. The next step involves subjecting the cut leaves to chemical pulping, which extracts cellulose from the raw material, releases fibers, and eliminates impurities that could cause discoloration.

Chemical pulping involves cooking the leaves for approximately 2–3 h. During this process, a pulping agent dissolves most of the lignin and separates the cellulose, forming pulp. A sizing agent is applied to impart water resistance to the resulting pineapple sheets and prevent them from smudging when exposed to water or ink. Subsequently, the pulp is beaten and mixed with a reinforcing solution to enhance the quality of the paper. The mixture is then molded into sheets, and excess water is removed by pressing them with a screw press before the drying phase. Lastly, the dried sheets are pressed again using a heat press to achieve a smoother surface for the final paper product (Mahatme et al., 2018) (Figure 16.4).

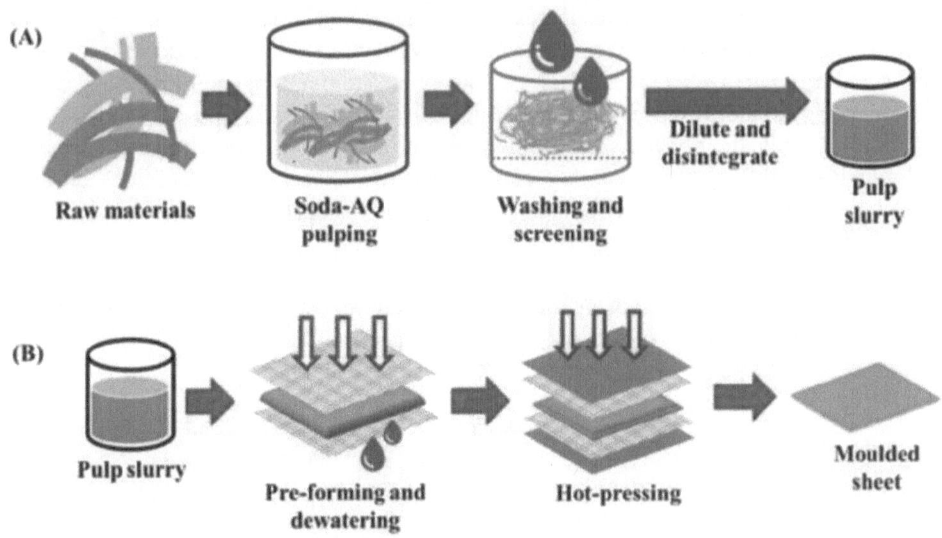

FIGURE 16.4 (A) Pulp and (B) paper preparation process.

16.6.3 Biocomposite Materials

Packaging waste contributes significantly to municipal solid waste and is disposed of in landfills. Landfill disposal can result in the release of greenhouse gases and waste. It could also contaminate land that may be utilized in the future. Because of the high volume-to-weight ratio and resistance to degradation of plastic waste, the rapid increase in production and use of plastics has resulted in a severe plastic waste problem, in addition to landfill depletion. Natural fiber-reinforced composites will create a new class of materials with significantly improved qualities without sacrificing desirable properties (Ali et al., 2017). The use of biocomposites, which are produced from local and renewable resources, represents a profoundly sustainable approach. By embracing principles of industrial ecology, eco-efficiency, and green chemistry, we are fostering the development of innovative materials, products, and processes for the next generation, all focused on environmental responsibility and conservation.

Pineapple waste biocomposite materials are a type of sustainable material made from by-products of the pineapple plant, such as pineapple leaves, fruit peels, and other agricultural waste, to create a sustainable and environmentally friendly material (Upadhyay et al., 2013). Biocomposites are materials that blend a natural polymer matrix obtained from pineapple waste with reinforcing fibers or particles to improve mechanical characteristics (Kumar et al., 2021). Pineapple waste bio composite materials utilize agricultural waste that would otherwise be discarded or incinerated, providing an eco-friendly alternative to conventional materials. The natural polymer matrix in pineapple waste biocomposites is typically derived from cellulose and lignin found in pineapple leaves and fruit peels. These natural polymers act as the matrix that binds the reinforcing fibers or particles together (Eixenberger et al., 2022). To enhance the mechanical properties of the biocomposite materials, reinforcing fibers or particles are added to the natural polymer matrix. Common reinforcing materials include natural fibers like pineapple leaf, jute, and hemp or synthetic fibers like glass or carbon fibers. Pineapple waste biocomposite materials have various applications, including packaging, automotive parts, construction, furniture, and consumer goods. Their mechanical properties can be tailored to meet specific requirements by selecting reinforcing fibers and manufacturing processes.

The use of pineapple waste biocomposite materials offers several benefits. They are renewable, reduce dependence on non-renewable resources, have lower carbon footprints (Upadhyay et al., 2013) compared to synthetic materials, and can contribute to waste reduction and sustainable agricultural practices. Another advantage of pineapple waste biocomposite materials is their biodegradability. They can decompose naturally over time, reducing their environmental impact compared to synthetic materials. Despite their benefits, pineapple waste bio composites face some challenges. These include limited availability and scalability of pineapple waste feedstock, variations in material properties due to agricultural and environmental factors, and the need for optimized manufacturing processes to achieve consistent and reliable material performance.

Overall, pineapple waste biocomposite materials represent a novel and long-term solution to reusing agricultural waste and decreasing the environmental impact of many businesses. Continued research and development in this subject may result in broader use and greater performance of these materials in the future.

16.6.4 Bio-Coal

Bio-coal, also known as biomass briquettes or white coal, is a solid fuel produced from biomass materials. While it is more commonly made from agricultural residues such as rice husks, wheat straw, or sawdust, it is also possible to produce bio-coal from pineapple waste. Torrefaction can be used to make bio-coal from pineapple waste, such as pineapple leaves, peels, and stems. Torrefaction is the process of heating biomass in the absence of oxygen to high temperatures, eliminating moisture, volatile chemicals, and other contaminants (Jeng et al., 2018). Briquettes are made from the residual solid material.

There are various advantages of manufacturing bio-coal from pineapple waste. For starters, it promotes the use of agricultural wastes that would otherwise be thrown away, thus decreasing waste and potential environmental repercussions (Sanchez et al., 2022). Second, bio-coal can be used as a renewable energy source, replacing fossil fuels and lowering greenhouse gas emissions. Bio-coal has a higher energy density than raw biomass, making it better suited for combustion (Cheng et al., 2020). It can be used as a fuel source in industrial boilers, heating systems, and power plants to generate heat and electricity. Furthermore, bio-coal has less moisture, more energy, and superior combustion properties than raw biomass, making it a more efficient and cleaner-burning fuel (Bhatia et al., 2023). However, it should be noted that synthesizing bio-coal from pineapple waste or other biomass feedstocks may necessitate the use of specialized equipment and expertise. This process's availability and economic viability can vary based on factors such as the local pineapple industry, trash collecting infrastructure, and regional demand for bio-coal.

Overall, utilizing pineapple waste to produce bio-coal can be a promising way to convert agricultural residues into a valuable energy source, contributing to sustainability and waste reduction efforts.

16.6.5 Biochar

Biochar is referred to as the product created from the natural biological organism that undergoes the carbonization process. With this being said, biochar can be prepared by utilizing the carbon content originating from the pineapple waste and other biomass materials converted to aromatic carbon compounds via the pyrolysis process (Bohari et al., 2020). Two common preparational conditions are used nowadays, and the first condition is that the biochar is burned at relatively low temperatures at 450–550°C, which can produce high carbon content that capable of undergoing a long-lasting bind of groundwater and dissolving ions (Chávez-García et al., 2020). The second condition is the material is pyrolyzed at temperatures of 600–650°C or higher, with the addition of plentiful calcium phosphate and apatite minerals can produce significantly lower carbon content (Chávez-García et al., 2020). However, the second condition is usually only applied if the biomass is animal bone-based (Kocsis et al., 2022). The physical form of biochar is commonly uneven, bulky, and grainy composite charred with a blackish color.

Biochar offers numerous benefits, especially in enriching the yield of crops due to its natural capability of enriching and enhancing the soil (Williams et al., 2017). The natural abundance of the porous structure of biochar can improve the physicochemical and biological properties of supplemented soil. First and foremost, biochar can improve the water retention capacity of the soil. In contrast, when mixed with soil, the composite can have a new changed texture, pore size distribution, and bulk density (Williams et al., 2017). These traits can intensify the water-holding capacity of the soil, enhancing the mineral distribution, while the diverse porous structure can positively influence soil aeration (Kocsis et al., 2022) (Figure 16.5).

Other than that, biochar is also capable of enriching the organic matter content of the soil, where the microorganisms may be indirectly protected by the biochar's enormous porous surface area, and the growth of bacteria and fungi may be aided by the adsorbed organic molecules (Fu et al., 2016). In addition, biochar can also increase the pH of the soil to a certain degree via the liming effect that increases the soil cation exchange capacity, affecting the abundance of nutrients and avoiding their leaching (Fu et al., 2016). The preferred nitrogen source for plants, ammonium, is scarcely available when the pH is raised, which enhances microbial nitrification and decreases nitrate availability while increasing phosphorus and nitrogen availability (Mndzebele et al., 2020). In addition, depending on the raw material used and the pyrolysis temperature, biochar may have a high concentration of macro- and micronutrients with varying degrees of availability (Mndzebele et al., 2020). As a

FIGURE 16.5 Biochar (Kocsis et al., 2022).

result, by increasing the potential uptake of meso- and micronutrients, biochar can play a direct role in the replenishment of nutrients for plants (Kocsis et al., 2022).

16.6.6 Bio-Bricks

India and China are among the fastest developing countries in terms of economic expansion, specifically in the investment of construction field. In India, over 590 million people are projected to reside in cities by 2030. Large housing investments will be required for this level of urbanization, and the raw materials needed are anticipated to reach approximately 15 billion tons by 2030 and approximately 25 billion tons by 2050 (Rautray et al., 2019). Therefore, such demand will only push the production of around 140,000 bricks, which will produce approximately 66 million tons of carbon dioxide, carbon monoxide, sulfur dioxide, nitrous oxide, and other fine particulate emissions, which can cause a total of 9% greenhouse emission (Rautray et al., 2019). To cater to the matter at hand, bio-brick is one of the solutions that can be approached since most of these countries, including Malaysia, are some of the significant agricultural-economic-based countries.

Because hempcrete blocks (bio-brick) have a negative carbon footprint and function as efficient carbon sinks, the total emission balance is advantageous (Krishan et al., 2017). This is because hempcrete blocks absorb carbon dioxide through carbonation and biogenic carbon dioxide uptake during hemp growth (Krishan et al., 2017). This is the dual advantage of such an inventive building material, as it also serves as a superb sustainable building material in addition to its great heat insulation and sound absorption capabilities (Mndzebele et al., 2020). Bio-bricks are generally more environmentally friendly than burned clay bricks because they fix more carbon dioxide during their lifetime than they produce (Arrigoni et al., 2017).

To create the bio-brick, the dried pineapple waste and other agro-waste are carefully chosen and then cut to the proper size. Slake lime, binder, stone dust, and water are combined to create a slurry with a lime foundation (Fu et al., 2016). The chopped agro-waste is added to the slurry and thoroughly mixed by hand or a mechanical mixer to make a uniform mixture. This mixture is poured into molds and rammed with wooden blocks to remove excess water or cavities and create a compact brick (Mndzebele et al., 2020). These molds are given 1 or 2 days to dry before having their sides

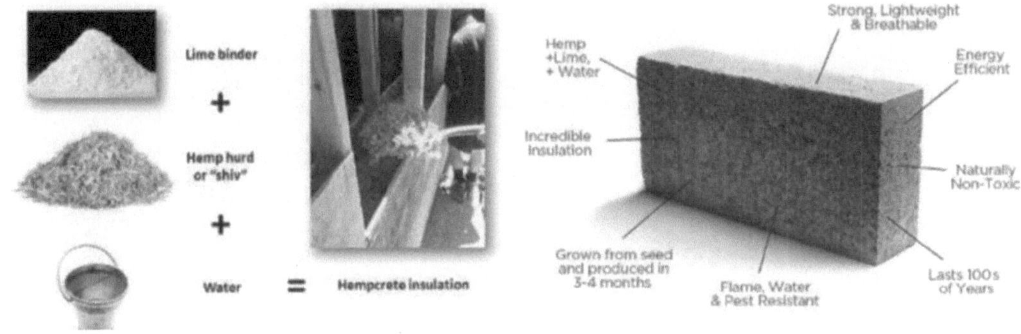

FIGURE 16.6 Bio-brick (Arrigoni et al., 2017).

removed, and the brick is allowed to dry for 15–20 days. These bio-bricks take about a month to develop their operating strength by air drying fully (Rautray et al., 2019) (Figure 16.6).

These bio-bricks are lighter in weight (1.43 kg per block), which is almost one-eighth of fired clay bricks and one-tenth of concrete blocks of the same volume, but have less compressive strength than fired clay bricks or concrete blocks (Rautray et al., 2019). Because of their excellent heat and sound insulation and low dead load requirements, they can be used efficiently as non-load-bearing walls in framed structures. These bio-bricks can be employed in low-cost housing with a mix of wooden or metal structural frameworks even though they cannot be used to build load-bearing structures directly, with a relatively low thermal conductivity of 0.27 W/mK (Son et al., 2017). Due to their superior sound and heat insulation, these bio-bricks can be utilized in walls. These bio-bricks' porosity and low-density aid in controlling the buildings' humidity levels, making these homes suited for hot, humid climates.

16.6.7 Biohydrogen

The production of biohydrogen from pineapple waste is a breakthrough in solving the renewable energy production crisis while simultaneously resolving the waste management of pineapple industrial residue (McCance et al., 2021). When it comes to fossil fuel substitution, hydrogen is one of the best clean energy carriers for a better environment. It produces fewer greenhouse pollutants, has a fuel cell platform that converts energy more efficiently (up to 49.7%), has better energy densities, and is renewable (Cahyari et al., 2018). Hydrogen is the gas with the most energy per unit weight. Since hydrogen gas combustion yields just water, it is regarded as a clean and carbon-neutral energy source (Ahmad et al., 2023). However, since fossil fuels make up the majority of its current production, its position as a green fuel is dubious (Cahyari et al., 2018).

Water electrolysis is the most practical and frequent way to produce hydrogen but only if there is enough free excess electricity produced by solar, hydropower, wind, geothermal energy, or another source (Ghazi et al., 2023). Recent research indicates that the electrolysis process still has a high energy input-to-energy-output ratio, indicating a negative energy balance for this technology (Khedkar et al., 2017). Therefore, dark fermentation as a substitute alternative is a promising technique to generate hydrogen from organic waste with the least energy input. This dark fermentation method of producing hydrogen uses the microbial community to break down organic material into volatile fatty acids (VFAs) and electron transfer to hydrogenase to turn a proton into hydrogen (McCance et al., 2021). Numerous variables, including the type and concentration of the substrate, temperature, pressure, pH, locally available chemical inhibitors, organic loading rate, hydraulic retention duration, and reactor structure, have an impact on this process (Reungsang & Sreela-or, 2013).

Due to the rising demand for hydrogen and the encouragement provided by the Malaysian government for industrial investment through a variety of incentives, it is thought that this process has a promising future both in Malaysia and on the international market (Ahmad et al., 2023). Japan and India are the top two economies gaining from increased biohydrogen yield efficiency.

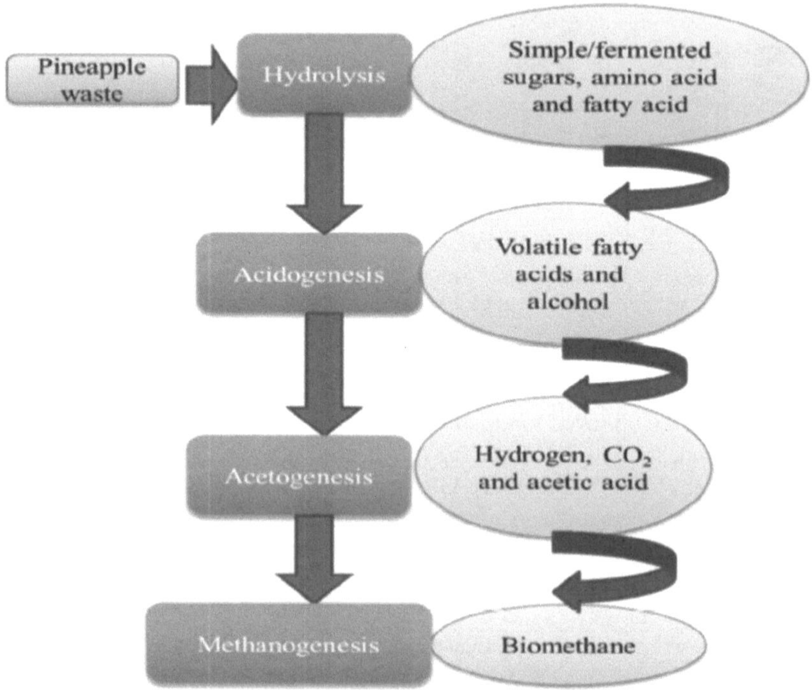

FIGURE 16.7 Biogas production from pineapple waste (Sarangi et al., 2022b).

Enhancing the yield efficiency of biohydrogen can boost macroeconomic growth and promote the biohydrogen and renewable hydrogen industries (Sepúlveda et al., 2018).

16.6.8 Biomethane

A bioconversion technique is used, which includes transforming pineapple waste into biomethane. This transformation is accomplished by the utilization of a particular kind of fermentation known as solid-state anaerobic digestion. To get the process started, the waste from the pineapple is subjected to treatment with sulfuric acid and alkaline hydrogen peroxide. This is done to decompose a component known as lignin found in the waste. Lignin is made more soluble and manageable when applied in this procedure. After the pre-treatment, the digestive process is helped, and biochar, a charcoal form, is added. Within 20 days when digestion has taken place, approximately 48% of the waste's biomethane can be recovered. Methanogens and acetogens are two examples of the types of microorganisms involved in creating biomethane. These microorganisms transform organic material into methane and other by-products. When making these forms of biofuels, it is necessary to take a number of considerations into account, including the levels of pH, the kinds of microorganisms that are utilized, the influence that the quantity of the organic material has on its performance, and the presence of VFAs, which can have an impact on the process as a whole (Chulalaksananukul, 2017) (Figure 16.7).

16.6.9 Bioethanol

Pineapple waste products such as peels contain a significant amount of carbohydrates that may be turned into fermentable sugar and can be used as a biomass source for the manufacture of bioethanol (Ahmad et al., 2023). Total structural carbohydrates, which comprise 37% of the dry pineapple waste and are therefore regarded as the primary component of dry pineapple waste, are represented as carbohydrates in the biomass (McCance et al., 2021). The portion of the pineapple peelings that can be converted into ethanol is called ethanol extractive. At 22.2%, ethanol extract covers a significant portion of dry pineapple waste (Hossain & Fazliny, 2016). Because of its alternative green energy sources that reduce greenhouse gas emissions and, ultimately, serve to mitigate the rise in fuel prices, bioethanol has long been a subject of significant attention (Khedkar et al., 2017).

Compared to fossil fuels, bioethanol is non-toxic, is biodegradable, and does not pollute the environment (Sepúlveda et al., 2018). After fermentation technology, carbon-based feedstock is converted to create bioethanol. Because they use photosynthesis to obtain sun energy, agricultural waste feedstocks are regarded as renewable components (Hossain & Fazliny, 2011). Concern over climate change and the potential for biofuels to lower greenhouse gas emissions are two main factors driving bioethanol development globally (Maynard et al., n.d.). Fermentation was used to produce bioethanol from pineapple trash using *Saccharomyces cerevisiae* (a type of yeast) and *Aspergillus niger* together with various ratios of pineapple waste (as substrate) and microbial inoculum (Dl & Ss, 2022). The optimum bioethanol yield could be produced at 32°C, with pH 4, using 3 g/L yeast.

Other than that, bioethanol from pineapple waste is also suitable for running a petrol engine and may make bioethanol from rotten pineapple waste (Sepúlveda et al., 2018). In terms of parameters like pH, temperatures, fermentation time, and substrate concentrations, it was discovered that a yield of 8.7% with a pH of 4 at 30°C and 3 g/L of substrate was ideal for running the petrol engine. The American Society for Testing Materials (ASTM) standard was used to get the viscosity results for various yeast concentrations (Sepúlveda et al., 2018). Analyses of the anhydrous ethanol revealed no potential harmful components, making it acceptable for use as a transportation fuel under ASTM standards (Hossain & Fazliny, 2016). Therefore, it is less harmful to the environment than other fuels. The construction of renewable bioenergy, specifically bioethanol production, utilizing the waste from rotten pineapples could assist in clearing the environment of waste and solve the issue of fossil fuel depletion (McCance et al., 2021). The creation of bioethanol from pineapple trash can lessen the financial burden on society that comes with waste management and the rising cost of renewable energy sources (Dl & Ss, 2022).

16.6.10 Biobutanol

This section will explain the process of converting hemicellulose, which can be found in waste pineapple peel, into a useful biofuel called butanol. At the beginning of the process, the hemicellulose is heated to a temperature of 180°C in a high-pressure reactor that contains phenol and sulfuric acid (pyruvic acid). This combination is used to solubilize the hemicellulose. The hemicellulose is converted into soluble components as a result of this treatment, which acts as an enhancer. This method is successful in solubilizing approximately 81.2% of the total amount of hemicellulose based on its weight and volume (Sarangi et al., 2022a).

The process of fermentation comes next, after the acid treatment and the enzymatic reaction that came before it. During this stage of the fermentation process, a particular microbial strain known as *Clostridium acetobutylicum*, NRRL-B527, is utilized. This particular microbe strain can break down sugar into butanol as a by-product of its metabolic process. The microbial fermentation process produces 0.13 g of butanol for every gram of sugar that is used. Butanol is a useful biofuel, and this series of steps reveals a method for efficiently converting the hemicellulose found in discarded pineapple peels into butanol. Overall, the method is quite effective. This method contributes to the development of a source of energy that is both renewable and sustainable. It does so by efficiently using waste material and leveraging microbial fermentation (Sarangi et al., 2022a) (Figure 16.8).

16.6.11 Natural Acids

Both the rind and the core of pineapple have a high concentration of fermentable sugars that can be used to make organic acids. The food and beverage, pharmaceutical, and bioplastics industries are

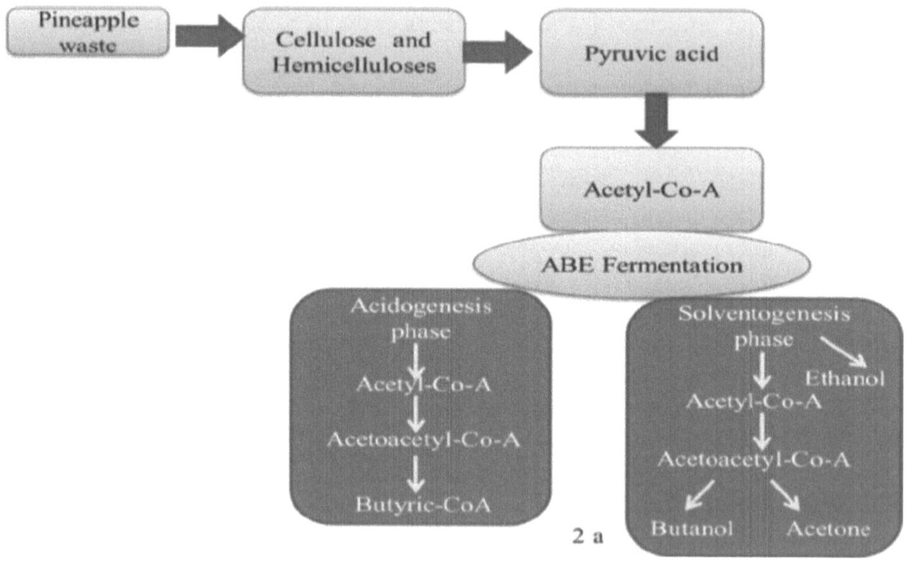

FIGURE 16.8 Biobutanol production from pineapple waste (Sarangi et al., 2022a).

TABLE 16.2
Production of Various Acids Using Different Microorganisms on the Waste (Aili Hamzah et al., 2021)

Type of Organic Acid	Microorganisms	Type of Pineapple Waste	Maximum Acid Production (mg/g)
Lactic acid	*Rhizopus oryzae*	Solid	103.69
	R. oryzae	Solid	23.6
	Lactobacillus delbrueckii	Liquid	54.97
Citric acid	*Aspergillus niger*	Solid (peels)	60.61
	A. niger	Solid (peels)	15.51
	Yarrowia lipolytica	Solid	202.35
Succinic acid	*Escherichia coli*	Liquid	6.26

just a few of the many commercial uses for these organic acids. Various bacteria are tested to find the one that maximizes the production of organic acids from pineapple waste. The primary features of the microorganisms are crucial, as they can affect the formation of organic acids. Factors including humidity, incubation period, temperature, pH, and inoculum (microbial culture) quantity may all have a role (Aili Hamzah et al., 2021).

First, the best circumstances for organic acid production are screened for, typically with the help of a two-level factorial design. This screening investigates the impact of varied factors on organic acid output. The optimized parameters and organic acid yield from solid-state fermentation studies are commonly tabulated and reported in Table 16.2. This table shows how various experimental factors affect the yield of organic acids from pineapple rinds, providing important context for the experiments (Aili Hamzah et al., 2021).

16.6.12 Bioelectricity

When fossil fuels are used to create power, greenhouse gases are released, and a lot of wasted energy is produced. In contrast, microbial fuel cells (MFCs) can convert pineapple waste,

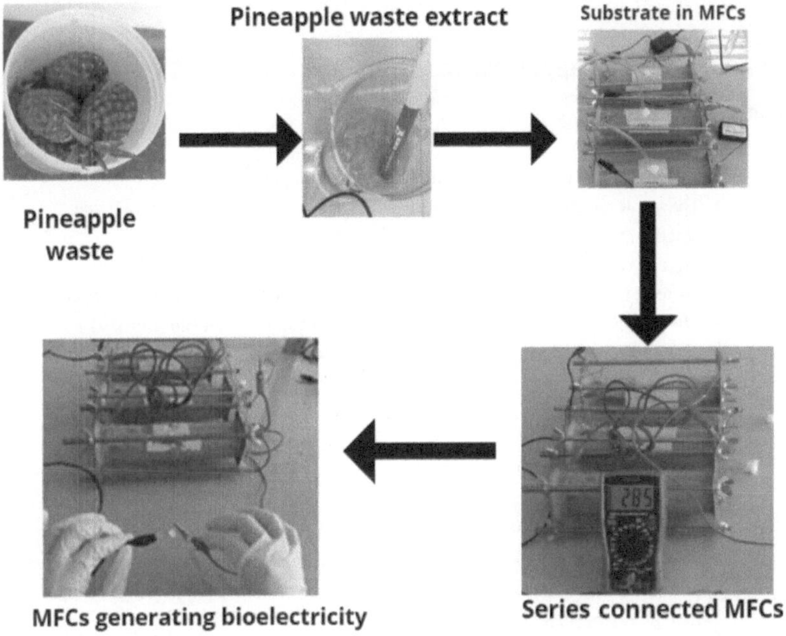

FIGURE 16.9 Bioelectricity production from pineapple waste (Rojas-Flores et al., 2022).

including a lot of substrates, into bioelectricity. MFCs are electronic devices that, like electrolysis, transform chemical energy into electrical energy. A proton exchange membrane and an external circuit connect the cationic and anionic chambers of an MFC (Rojas-Flores et al., 2022). Pineapple waste contains bacteria essential to the MFCs' ability to generate power. As particles are in motion, substrates aid in the vital electrochemical reactions to produce energy. Based on Figure 16.9, electricity is generated by extracting the waste from pineapple waste using some extraction technique so that the substrate obtained can be used to produce electricity where the voltage is observed on the voltmeter.

Bioelectricity can be generated as a renewable and sustainable alternative to conventional fossil fuel-based electricity generation by tapping into the energy potential of pineapple waste through MFCs. Both environmental and energy issues are effectively addressed by this method since greenhouse gas emissions are reduced and the copious substrate present in pineapple waste is put to good use (Rojas-Flores et al., 2022).

16.7 POTENTIAL APPLICATIONS

Pineapple wastes, often derived from processing pineapples for consumption or juice extraction, have found diverse applications across several industries. In the realm of antioxidants, pineapple wastes contain compounds such as phenolic acids and flavonoids, which exhibit antioxidant properties. These properties make them valuable in various health and wellness products, including dietary supplements and skincare formulations. For instance, in the pharmaceutical industry, pineapple wastes are utilized for their medicinal properties. They contain enzymes like bromelain, known for its anti-inflammatory and digestive benefits. Bromelain is often extracted from pineapple stems and cores and used in pharmaceutical formulations for conditions such as osteoarthritis and sinusitis. Furthermore, pineapple wastes contribute to the food industry in multiple ways. They can be processed into dietary fiber supplements, natural food additives, or utilized in animal feed formulations. Additionally, pineapple wastes can serve as raw material for vinegar production, where they undergo

TABLE 16.3
Common Techniques for Pineapple Waste Extractions

Types of Extraction Techniques	Description	References
Solvent extraction	• Using solvents to extract antioxidants from pineapple waste. • The waste material (peels or cores) is typically dried and ground into a fine powder. • Mixed with a suitable solvent (ethanol or water) and subjected to extraction through techniques like maceration, reflux, or ultrasound-assisted extraction. • The solvent helps dissolve the antioxidants, and subsequent filtration and evaporation processes are used to obtain the concentrated extract.	Lourenço et al. (2020)
Solid–liquid extraction	• Pineapple waste is directly immersed in a suitable solvent (ethanol or water). • Subjected to heat or agitation to facilitate the release of antioxidants into the liquid phase. • The mixture is then filtered to separate the solid waste from the antioxidant-rich extract.	
Supercritical fluid extraction	• Antioxidants are extracted using a supercritical fluid, such as carbon dioxide. • Carbon dioxide reaches a supercritical state under specific temperature and pressure conditions, where it exhibits properties of both liquids and gases, making it an effective solvent for extraction. • Producing high-quality extracts with minimal solvent residue.	
Enzyme-assisted extraction	• Enzymes, such as cellulases or pectinases, can be used to facilitate the release of antioxidants from pineapple waste. • The waste material is treated with specific enzymes that break down the cell walls and release the antioxidants. • This method is often combined with other extraction techniques to enhance the overall extraction efficiency.	

fermentation processes to yield vinegar with distinct flavor profiles. Hence, the utilization of pineapple wastes in various applications underscores their potential to add value to different industries while promoting sustainable practices through waste reduction and resource optimization.

16.7.1 Antioxidants

Antioxidants are beneficial compounds that safeguard cells from harm caused by unstable molecules known as free radicals, which can lead to various health problems. Pineapple peels contain bioactive elements like vitamin C, carotene, phenolic compounds, and flavonoids, all possessing antioxidant properties and other biological benefits. Studies have indicated that fruit remnants, particularly those from pineapples, exhibit the same antioxidant activity as the fruit's edible pulp (Saraswaty et al., 2017). Even though these by-products are discarded, they could be exploited as an alternative source of polyphenols and natural antioxidants. Extracting antioxidants from pineapple waste can be done through various methods. Table 16.3 lists common techniques for pineapple waste extraction.

Pineapple waste, particularly the core and peels, has high antioxidant compounds such as phenolic acids, flavonoids, and vitamin C levels. These chemicals can be extracted and used as natural additives in functional meals and beverages to boost their antioxidant effects (Hajar et al., 2012). Pineapple waste extracts can be processed into nutraceutical supplements in the form of capsules, powders, or liquid extracts. These supplements can be taken orally to provide antioxidant benefits and support overall health. Other than that, pineapple waste extracts, such as bromelain and vitamin C, are known for their skin rejuvenating and antioxidant properties. They can be incorporated into skincare products like creams, serums, and masks to provide antioxidant protection, improve skin tone, and reduce

FIGURE 16.10 Pineapple waste as a skincare product (Hajar et al., 2012).

the signs of aging. The extracted material is often concentrated and purified following procedures such as evaporation, filtration, or chromatography. Depending on the intended application, the resulting antioxidant-rich extract can be included in various products or formulations (Figure 16.10).

16.7.2 Pharmaceutical and Food Industries

Researchers have investigated the utilization of pineapple waste in the pharmaceutical and food industries and successfully extracted a promising protease called bromelain (Nor et al., 2016). The commercial use of bromelain requires absolute purity owing to its extensive range of applications. In the food industry, bromelain derived from pineapple waste finds applications in meat tenderization, brewing, baking, apple juice processing, prevention of browning, and beer clarification and as a nutritional supplement (Chaurasiya & Umesh Hebbar, 2013). Bromelain facilitates meat softening by breaking down tough fibers during the tenderization process. Bromelain enhances dough relaxation in baking, ensuring even rising and producing hypoallergenic flour suitable for individuals allergic to wheat. Due to its favorable temperature range of 50–70°C, bromelain finds practical application in various food processing endeavors.

Bromelain serves multiple purposes, being utilized as an active ingredient in tooth-whitening toothpaste and skincare products for treating acne, wrinkles, and dry skin. Additionally, it is employed to reduce bruising and swelling after injections. The enzyme acts as a cleansing agent and gently exfoliates the skin. In the pharmaceutical industry, bromelain finds extensive application as a medication for inflammatory conditions, intestinal disorders, blood-coagulation-related diseases, and improving antibiotic absorption. It is also used as a wound debridement agent, as a fibrinolytic agent, and for the oral treatment of third-degree burns. Furthermore, it has therapeutic applications for antibodies and displays mucolytic action (Ramli et al., 2016). Bromelain has gained popularity as a natural remedy due to its oral effectiveness and lack of side effects, and it is available in both pill and capsule forms. According to Tropea et al. (2014), bromelain is only poorly absorbed when taken orally, resulting in plasma levels of less than 10 ng/mL in humans given 4 g/day.

The extraction of bromelain from pineapple waste can be achieved using various methods (Manzoor et al., 2016). The pineapple waste, particularly the core and stem, is collected as it contains higher concentrations of bromelain. Any outer skin or rough parts are removed from the waste and chopped into smaller pieces. This will increase the surface area and facilitate extraction. The chopped pineapple waste was placed into a blender or grinder and processed until a homogeneous pulp or paste was formed. Ensure that the waste is finely ground to maximize the surface area for extraction. Then, an extraction solution is prepared. A common choice is a buffer solution containing Tris–HCl or phosphate buffer, adjusted to a suitable pH range for bromelain stability (typically pH 6–8). This buffer helps maintain the enzyme's activity during the extraction process. The extraction solution is applied to the ground pineapple waste and mixed thoroughly. The mixture should have a sufficient volume of buffer solution to fully immerse the waste material. The ratio of extraction solution to waste can vary, but typically a ratio of 1:5 to 1:10 (w/v) is used. The mixture is allowed to stand at a controlled temperature for a specific duration. The extraction time can range from several hours to overnight, depending on the desired yield and bromelain concentration. After the extraction period, the mixture is filtered using a fine mesh or cheesecloth to separate the solid waste residue from the liquid extract by squeezing. To further clarify the extract and remove any remaining solid particles, the liquid extract is centrifuged at a low speed for a short duration. This step helps separate the clear from any sediment or particulate matter. Bromelain can be further purified using various methods, such as chromatography (gel filtration or ion exchange chromatography), precipitation, or membrane filtration, depending on the desired purity and application.

It is important to remember that the extraction process can change based on the particular needs and equipment that are available. Additionally, to maximize bromelain yield and activity, extraction parameters like temperature, pH, extraction duration, and solvent-to-solid ratio must be optimized. To guarantee the security and effectiveness of the extracted bromelain, quality control procedures, and regulatory requirements must also be followed.

16.7.3 Production of Vinegar

According to Joy (2010), a comparison of the phytochemical screening and antioxidant capacity of vinegar revealed that vinegar made from pineapple peel had higher antioxidant activity (2,077 mg acetate equivalent/100 mL) than vinegar made from other fruit wastes. Due to its beneficial properties and environmental friendliness, this product can be manufactured and sold in huge quantities, in contrast to synthetically created chemical vinegar. Pineapple vinegar may aid in digestion. The acetic acid in vinegar has been shown to promote the production of digestive enzymes, helping to break down food and improve nutrient absorption. It may also help to regulate stomach acid levels and reduce symptoms of indigestion. Pineapple vinegar's acetic acid may help to regulate blood sugar levels. According to studies, consuming vinegar—including pineapple vinegar—can enhance insulin sensitivity and lessen blood sugar rises after meals. However, before adding pineapple vinegar to their diet, those with diabetes should closely monitor their blood sugar levels and speak with a healthcare provider (Shu et al., 2019). Other than that, pineapple vinegar adds a tangy and fruity flavor to various dishes, making it a versatile ingredient in culinary preparations. It can be used in salad dressings, marinades, sauces, or even in beverages. Adding pineapple vinegar to your recipes can enhance the taste profile of your dishes (Figure 16.11).

Pineapple waste can be utilized directly in the production of vinegar without extracting the juice. This method is known as the "waste-to-vinegar" or "vinegar fermentation on pineapple waste" process (Roda et al., 2014). First, the by-product peels were removed from the edible pulp and the crown after washing the pineapples. The peels were manually chopped into small pieces with a knife and blended

FIGURE 16.11 Pineapple vinegar.

into a homogeneous mixture. The samples were then placed in a freezer at –18°C before being used. Then, chopped pineapple waste was transferred into the fermentation vessel. Pack the waste tightly, but not too tightly, to allow air circulation during fermentation. It is important to fill the vessel while leaving some headspace to accommodate the formation of a vinegar mother and the production of carbon dioxide during fermentation. A tiny amount of vinegar mother or a commercial starter culture containing acetic acid bacteria (*S. cerevisiae*) is added to the pineapple waste. During fermentation, the acetic acid bacteria convert the alcohol in the waste into acetic acid. Follow the directions included with the vinegar starter culture to determine the optimum amount to use. The fermentation process can take several weeks to months, depending on factors such as temperature and desired acidity.

16.8 CHALLENGES ASSOCIATED WITH PROCESSING AND COMMERCIALIZATION

Various sources have discovered various challenges associated with processing and commercializing pineapple waste. Four main challenges contribute to the growth of pineapple waste management industries. In this instance, the cost of managing pineapple waste is very high. The management of pineapple waste can incur fees for collection, transportation, storage, and treatment (Matius et al., 2022). These costs can be substantial and have a negative impact on the overall commercial viability of pineapple waste utilization. Cost-effective waste management strategies, such as optimizing transportation routes, exploring value-added product options, and seeking partnerships for resource sharing, can help reduce costs and make the processing and commercialization of pineapple waste economically viable.

Furthermore, time is required to produce valuable products from pineapple waste (Matius et al., 2022). Certain processing techniques, such as extraction, dehydrating, or fermentation, may necessitate specific durations for optimal outcomes. In addition, increasing production to meet market demand may contribute to time constraints. To surmount this obstacle and ensure the timely production of valuable products, efficient processing techniques and streamlined production workflows are necessary.

In addition, a sufficient supply of competent laborers is essential for monitoring and production processes (Matius et al., 2022). Frequently, skilled personnel are required to operate the apparatus, monitor fermentation or extraction processes, and ensure quality control in processing pineapple waste. In some regions or industries, however, there may be a dearth of skilled laborers. Training programs and workforce development initiatives can assist in bridging the gap and guaranteeing a sufficient workforce for the efficient management and commercialization of pineapple waste.

Lastly, pineapple waste can have an offensive odor and attract insects, making it difficult to maintain hygiene during processing and storage (Matius et al., 2022). The decomposition of waste materials can result in offensive odors and an environment hospitable to insect infestations. Hygiene concerns associated with pineapple waste processing can be alleviated by implementing proper waste management practices, such as timely removal, storage in suitable containers, and odor control measures.

16.9 CONCLUSION

wwThere has been a significant increase in trash output in recent years due to the increased production of pineapple around the globe due to the food and drink industry's economic push. Therefore, if not managed and utilized properly, these wastes may lead to environmental pollution issues. One of the concerns is that, if left unattended, these pineapple residuals will start to rot, emit an unpleasant odor, and, in some cases, catalyze pest infestation that can negatively affect Malaysian food hygiene quality. To reduce wastage and create a product that can be commercialized as an environmentally friendly alternative to using carbon sources, it is necessary to sustainably employ pineapple waste into a value-added product. Consequently, the finished product that is clean, readily marketable, and sustainable will result from well-managed waste. The perk is that it will help Malaysia reduce environmental pollution and introduce the country to renewable energy and the knowledge and technological advancement that come with it. Additionally, these new market opportunities would create a sustainable economy in line with global tendencies toward a circular economy because of the repurposing of pineapple residuals.

ACKNOWLEDGMENT

This chapter is written with Ismahani Rosazhar, Karisma Kaur Paremjeet Singh, and Wan Mohamad Sharul Nizam Wan Rostan.

REFERENCES

Abdullah, N., Jamiluddin, J., Hagos, F. Y., & Azmi, N. S. N. (2020). Experimental investigation on pineapple leaf fiber as biomass source for renewable energy application. *IOP Conference Series: Materials Science and Engineering*, 788(1), 012059. https://doi.org/10.1088/1757-899X/788/1/012059

Ahmad, S. I., Rashid, R., Hashim, Z., Meng, C. C., Lun, C. K., Jumaatuden, D. M. H., Yasin, N. A., Jati, A., & Hassim, M. H. (2023). Economic study on biohydrogen production from liquid pineapple waste. *Clean Technologies and Environmental Policy*, 25(2), 703–716. https://doi.org/10.1007/s10098-022-02282-5

Aili Hamzah, A. F., Hamzah, M. H., Che Man, H., Jamali, N. S., Siajam, S. I., & Ismail, M. H. (2021). Recent updates on the conversion of pineapple waste (Ananas comosus) to value-added products, future perspectives and challenges. *Agronomy*, 11(11), 2221. https://doi.org/10.3390/agronomy11112221

Ali, R., Aizan, W., Rahman, W., Kasmaini, R., Ibrahim, N., Hasbullah, H., Sadikin, A., Asli, U., & Abouzari, E. (2017). Pineapple peel fiber biocomposite: Characterisation and biodegradation studies. *Chemical Engineering Transactions*, 56. https://doi.org/10.3303/CET1756223

Amores-Monge, V., Goyanes, S., Ribba, L., Lopretti, M., Sandoval-Barrantes, M., Camacho, M., Corrales-Ureña, Y., Roberto, J., & Baudrit, V. (2022). Pineapple agro-industrial biomass to produce biomedical ap-plications in a circular economy context in Costa Rica. *Polymers*, 14(22), 4864. https://doi.org/10.20944/preprints202209.0365.v1

Arrigoni, A., Pelosato, R., Melià, P., Ruggieri, G., Sabbadini, S., & Dotelli, G. (2017). Life cycle assessment of natural building materials: The role of carbonation, mixture components and transport in the environmental impacts of Hempcrete blocks. *Journal of Cleaner Production*, 149, 1051–1061. https://doi.org/10.1016/j.jclepro.2017.02.161

Asim, M., Abdan, K., Jawaid, M., Nasir, M., Dashtizadeh, Z., Ishak, M. R., & Hoque, M. E. (2015). A review on pineapple leaves fiber and its composites. *International Journal of Polymer Science*, 2015, 1–16. https://doi.org/10.1155/2015/950567

Bhatia, L., Jha, H., Sarkar, T., & Sarangi, P. K. (2023). Food waste utilization for reducing carbon footprints towards sustainable and cleaner environment: A review. *International Journal of Environmental Research and Public Health*, 20(3), 2318. https://doi.org/10.3390/ijerph20032318

Bohari, N., Mohidin, H., Idris, J., Andou, Y., Man, S., Saidan, H., & Mahdian, S. (2020). Nutritional characteristics of biochar from pineapple leaf residue and sago waste. *Pertanika Journal of Science and Technology*, 28(Special Issue 2), 273–286. https://doi.org/10.47836/pjst.28.S2.21

Cahyari, K., Putri, A. M., Oktaviani, E. D., Hidayat, M. A., & Norajsha, J. D. (2018). Biohydrogen production from pineapple waste: Effect of substrate concentration and acid pre-treatment. *IOP Conference Series: Materials Science and Engineering*, 358(1). https://doi.org/10.1088/1757-899X/358/1/012001

Castillo-González, E., Giraldi-Díaz, M. R., De Medina-Salas, L., & la Cruz, R. V. D. (2020). Environmental impacts associated to different stages spanning from harvesting to industrialization of pineapple through life cycle assessment. *Applied Sciences (Switzerland)*, 10(19), 1–21. https://doi.org/10.3390/app10197007

Chaurasiya, R. S., & Umesh Hebbar, H. (2013). Extraction of bromelain from pineapple core and purification by RME and precipitation methods. *Separation and Purification Technology*, 111, 90–97. https://doi.org/10.1016/j.seppur.2013.03.029

Chávez-García, E., Aguillón-Martínez, J., Sánchez-González, A., & Siebe, C. (2020). Characterization of untreated and composted biochar derived from orange and pineapple peels. *Revista Internacional de Contaminacion Ambiental*, 36(2), 413–427. https://doi.org/10.20937/RICA.53591

Cheng, B.-H., Huang, B.-C., Zhang, R., Chen, Y.-L., Jiang, S.-F., Lu, Y., Zhang, X.-S., Jiang, H., & Yu, H.-Q. (2020). Bio-coal: A renewable and massively producible fuel from lignocellulosic biomass. *Science Advances*, 6(1), eaay0748. https://doi.org/10.1126/sciadv.aay0748

Cheung, H., Ho, M., Lau, K., Cardona, F., & Hui, D. (2009). Natural fiber reinforced composites for bioengineering and environmental engineering applications. *Composites Part B*, 40(7), 655–663. https://doi.org/10.1016/j.compositesb.2009.04.014

Chu, C.-Y., Vo, T.-P., & Chen, T.-H. (2020). A novel of biohythane gaseous fuel production from pineapple peel waste juice in two-stage of continuously stirred anaerobic bioreactors. *Fuel*, 279, 118526. https://doi.org/10.1016/j.fuel.2020.118526

Chulalaksananukul, S., Sinbuathong, N., & Chulalaksananukul, W. (2017). Bioconversion of pineapple solid waste under anaerobic condition through biogas production. *Asia-Pacific Journal of Science and Technology*, 17(5), 734–742. https://so01.tci-thaijo.org/index.php/APST/article/view/83282

Dahunsi, S. O., Ogunwole, J. O., Owoseni, A. A., Olutona, G. O., Nejo, Y. T., & Atobatele, O. E. (2021). Valorization of pineapple peel and poultry manure for clean energy generation. *Food and Energy Security*, 11(1). https://doi.org/10.1002/fes3.228

Dl, B., & Ss, I. (2022). Issue 8 www.jetir.org (ISSN-2349–5162). *JETIR2208304 Journal of Emerging Technologies and Innovative Research*, 9. www.jetir.org

Eeda, N. (2020). Agricultural waste management system [AWMS] in Malaysian. *Journal of Waste Management & Xenobiotics*, 3, 000140. https://doi.org/10.23880/oajwx-16000140

Eixenberger, D., Carballo-Arce, A.-F., Vega-Baudrit, J.-R., Trimino-Vazquez, H., Villegas-Peñaranda, L. R., Stöbener, A., Aguilar, F., Mora-Villalobos, J.-A., Sandoval-Barrantes, M., Bubenheim, P., & Liese, A. (2022). Tropical agroindustrial biowaste revalorization through integrative biorefineries—Review part II: Pineapple, sugarcane and banana by-products in Costa Rica. *Biomass Conversion and Biorefinery*. https://doi.org/10.1007/s13399-022-02721-9

Foo, K. Y., & Hameed, B. H. (2012). Porous structure and adsorptive properties of pineapple peel based activated carbons prepared via microwave assisted KOH and K_2CO_3 activation. *Microporous and Mesoporous Materials*, 148(1), 191–195. https://doi.org/10.1016/j.micromeso.2011.08.005

Franklyn da Cruz, L., Simões, A. J. A., Vieira, I. M. M., Silva, D. P., & Ruzene, D. S. (2018). An overview of applications in pineapple agro-industrial residues. *Acta Agriculturae Slovenica*, 111, 445–462. https://doi.org/10.14720/aas.2018.111.2.18

Fu, B., Ge, C., Yue, L., Luo, J., Feng, D., Deng, H., & Yu, H. (2016). Biochar oxytetracycline sorption. *BioResources*, 11(4), 9017–9035.

George, J., Bhagawan, S. S., & Prabhakaran, T. S. (1995). Short pineapple-leaf-fiber-reinforced low-density polyethylene composites. *Journal of Applied Polymer Science*, 57, 843–54.

George, K., Janardhan, R., Anad, J. S., Bhagawan, S. S., & Thomas, S. (1996) Melt rheological behavior of short pineapple fiber reinforced low density polyethylene composites. *Polymer*, 37, 5421e31

Ghazi, N. A. M., Khairuddin, F., & Noor, W. N. W. M. (2023). The factors that influence operation risk on pineapple production: A case study in Muar, Johor. *IOP Conference Series: Earth and Environmental Science*, 1182(1), 012027. https://doi.org/10.1088/1755-1315/1182/1/012027

Hajar, N., Zainal, S., Nadzirah, K. Z., Roha, A. M. S., Atikah, O., & Elida, T. Z. M. T. (2012). Physicochemical properties analysis of three indexes pineapple (Ananas Comosus) peel extract variety N36. *APCBEE Procedia*, 4, 115–121. https://doi.org/10.1016/j.apcbee.2012.11.020

Henrique, P., Pereira, F., Magalhães De Oliveira, D., & Pereira, B. (2022). *Effect of Chemical Treatment Sequence on Pineapple Peel Fiber: Chemical Composition, Thermal Stability and Thermal Degradation Kinetics*. Heitor Luiz Ornaghi Universidade Federal da Integracao Latino-Americana. https://doi.org/10.21203/rs.3.rs-1121648/v1

Hikal, W. M., Mahmoud, A. A., Said-Al Ahl, H. A. H., Bratovcic, A., Tkachenko, K. G., Kačániová, M., & Rodriguez, R. M. (2021). Pineapple (Ananas comosus L. Merr.), waste streams, characterisation and valorisation: An overview. *Open Journal of Ecology*, 11(09), 610–634. https://doi.org/10.4236/oje.2021.119039

Hossain, A. B. M. S., & Fazliny, A. R. (2016). Creation of alternative energy by bio-ethanol production from pineapple waste and the usage of its properties for engine. *Advanced Journal of Microbiology Research*, 2016, 10(8), 1–7. https://internationalscholarsjournals.org/journal/ajmr

Jeng, S., Lim, W., Ho, J., Klemeš, Maisyarah Mansor, A., Lim, J., Nasir Ani, F., Hashim, H., Ho, W., & Bahru, J. (2018). Ultimate and proximate analysis of Malaysia pineapple biomass from MD2 cultivar for biofuel application. *Chemical Engineering Transactions*, 63. https://doi.org/10.3303/CET1863022

Joy, P. P. (2010). Benefits and uses of pineapple. https://doi.org/10.13140/RG.2.1.2782.4888

Khedkar, M. A., Nimbalkar, P. R., Gaikwad, S. G., Chavan, P. V., & Bankar, S. B. (2017). Sustainable biobutanol production from pineapple waste by using clostridium acetobutylicum B 527: Drying kinetics study. *Bioresource Technology*, 225, 359–366. https://doi.org/10.1016/j.biortech.2016.11.058

Kocsis, T., Ringer, M., & Biró, B. (2022). Characteristics and applications of biochar in soil–plant systems: A short review of benefits and potential drawbacks. *Applied Sciences (Switzerland)*, 12(8). MDPI. https://doi.org/10.3390/app12084051

Krishan, R., Tiwari, O. P., Rana, Y. S., Mishra, C. M., & Sharma, A. K. (2017). Changes in levels of soil carbon and forest floor carbon stocks in the different temperate forests of Garhwal Himalaya. *International Journal of Current Research and Review*, 9(7), 5–10. https://www.ijcrr.com/uploads/279_pdf.pdf

Kumar, S., Saha, A., & Bhowmik, S. (2021). Accelerated weathering effects on mechanical, thermal and viscoelastic properties of kenaf/pineapple biocomposite laminates for load bearing structural applications. *Journal of Applied Polymer Science*, 139(2), 51465. https://doi.org/10.1002/app.51465

Lourenço, S. C., Moldão-Martins, M., & Alves, V. D. (2020). Microencapsulation of pineapple peel extract by spray drying using maltodextrin, inulin, and arabic gum as wall matrices. *Foods*, 9(6), 718. https://doi.org/10.3390/foods9060718

Mahatme, S., Kanse, N., Bandsode, A., Student, B., & Professor, A. (2018). Pulp and paper production from pineapple leaves as a substitute to wood source: A review. *International Journal of Creative Research Thoughts*, 6(2), 2320–2882. https://ijcrt.org/papers/IJCRT1813604.pdf

Mangal, R., Saxena, N. S., Sreekala, M. S., Thomas, S., & Singh, K. (2003). Thermal properties of pineapple leaf fiber reinforced composites. *Materials Science and Engineering: A*, 339(1–2), 281–285. https://doi.org/10.1016/s0921-5093(02)00166-1

Mansor, A. M., Lim, J. S., Ani, F. N., Hashim, H., & Ho, W. S. (2019). Characteristics of cellulose, hemicellulose and lignin of MD2 pineapple biomass. *Chemical Engineering Transactions*, 72, 79–84. https://doi.org/10.3303/CET1972014

Manzoor, Z., Nawaz, A., Mukhtar, H., & Haq, I. (2016). Bromelain: Methods of extraction, purification and therapeutic applications. *Brazilian Archives of Biology and Technology*, 59(0). https://doi.org/10.1590/1678-4324-2016150010

Matius, A., Saili, A. R., Hamzah, N. M., Abdul Fatah, F., Che Ramli, N. K., Nur Hazwani, M. A., & Fatimah, S. (2022). Current practices and challenges of pineapple smallholder growers in managing the pineapple waste in Samarahan, Sarawak Malaysia. *Food Research,* 6(S1), p. 6. https://doi.org/10.26656/fr.2017.6(S1).001

Maynard, R., Antonio, D. R., Angelica, A., Dela Cruz, C., Quinto, A. S., Cordero, P. R., Natalia, M., & Dimaano, R. (n.d.). Bioethanol production from pineapple (Ananas comosus) peelings using Saccharomyces cerevisiae as fermenting yeast with focus on fermentation pH. *International Journal of Engineering Research & Technology*, 4, 356–362. www.ijert.org

McCance, K. R., Suarez, A., McAlexander, S. L., Davis, G., Blanchard, M. R., & Venditti, R. A. (2021). Modeling a biorefinery: Converting pineapple waste to bioproducts and biofuel. *Journal of Chemical Education*, 98(6), 2047–2054. https://doi.org/10.1021/acs.jchemed.1c00020

Mndzebele, B., Ncube, B., Fessehazion, M., Mabhaudhi, T., Amoo, S., du Plooy, C., Venter, S., & Modi, A. (2020). Effects of cowpea-amaranth intercropping and fertilizer application on soil phosphatase activities, available soil phosphorus, and crop growth response. *Agronomy*, 10(1), 79. https://doi.org/10.3390/agronomy10010079

Nor, M. Z. M., Ramchandran, L., Duke, M., & Vasiljevic, T. (2016). Separation of bromelain from crude pineapple waste mixture by a two-stage ceramic ultrafiltration process. *Food and Bioproducts Processing*, 98, 142–150. https://doi.org/10.1016/j.fbp.2016.01.001

Nordin, N., Md Illias, R., Hasmaliana Abdul Manas, N., Nor Mazila Ramli, A., & Izyan Wan Azlee, N. (2020). *Efficient delignification of pineapple waste by low-pressure steam heating pre-treatment.* In *Proceedings of the Third International Conference on Separation Technology 2020 (ICoST 2020)* (pp. 10–16). Atlantis Press.

Pandit, P., Pandey, R., Singha, K., Shrivastava, S., Gupta, V., & Jose, S. (2020). Pineapple leaf fiber: Cultivation and production. In *Pineapple Leaf Fibers,* 1–20. https://doi.org/10.1007/978-981-15-1416-6_1

Pilap, W., Thanonkeo, S., Klanrit, P., & Thanonkeo, P. (2022). The potential of multistress tolerant yeast, Saccharomycodes ludwigii, for second-generation bioethanol production. *Scientific Reports*, 12(1). https://doi.org/10.1038/s41598-022-26686-x

Rabiu, Z., Maigari, F. U., Lawan, U., & Mukhtar, Z. G. (2017). Pineapple waste utilization as a sustainable means of waste management. In *Sustainable Technologies for the Management of Agricultural Wastes*, 143–154. https://doi.org/10.1007/978-981-10-5062-6_11

Ramli, A. N. M., Aznan, T. N. T., & Illias, R. M. (2016). Bromelain: From production to commercialization. *Journal of the Science of Food and Agriculture*, 97(5), 1386–1395. https://doi.org/10.1002/jsfa.8122

Rasmi. (2023, July 13). Pineapple fiber: A sustainable option for your future fashion. Ecomaniac.org. https://ecomaniac.org/pineapple-fiber/

Rautray, P., Roy, A., Mathew, D. J., & Eisenbart, B. (2019). Bio-brick - Development of sustainable and cost effective building material. *Proceedings of the International Conference on Engineering Design*, 1, 3171–3180. https://doi.org/10.1017/dsi.2019.324

Reungsang, A., & Sreela-or, C. (2013). Bio-hydrogen production from pineapple waste extract by anaerobic mixed cultures. *Energies*, 6(4), 2175–2190. https://doi.org/10.3390/en6042175

Roda, A., Marco, D., Dordoni, R., & Lambri, M. (2014). Vinegar production from pineapple wastes - Preliminary saccharification trials. *Chemical Engineering Transactions*, 37, 607–612. https://doi.org/10.3303/cet1437102

Rojas-Flores, S., Nazario-Naveda, R., Benites, S. M., Gallozzo-Cardenas, M., Delfín-Narciso, D., & Díaz, F. (2022). Use of pineapple waste as fuel in microbial fuel cell for the generation of bioelectricity. *Molecules*, 27(21). https://doi.org/10.3390/molecules27217389

Ruslan, N. A., Aris, N. F. M., Othman, N., Saili, A. R., Muhamad, M. Z., & Aziz, N. N. H. (2017). A preliminary study on sustainable management of pineapple waste: Perspective of smallholders. *International Journal of Academic Research in Business and Social Sciences*, 7(6). https://doi.org/10.6007/ijarbss/v7-i6/2937

Salve, R. R., & Ray, S. (2020). Comprehensive study of different extraction methods of extracting bioactive compounds from pineapple waste—A review. *The Pharma Innovation International Journal*, 9, 327–340.

Sanchez, P. D., Aspe, M. M., & Sindol, K. (2022). An overview on the production of bio-briquettes from agricultural wastes: Methods, processes, and quality. *Journal of Agricultural and Food Engineering*, 3(1), 1–17.

Santulli, C., Palanisamy, S., & Kalimuthu, M. (2022). Pineapple fibers, their composites and applications. In S. M. Rangappa, S. Siengchin, & M. Jawaid (eds.) *Plant Fibers, their Composites, and Applications*, 323–346. https://doi.org/10.1016/b978-0-12-824528-6.00007-2

Sarangi, P. K., Anand Singh, T., Joykumar Singh, N., Prasad Shadangi, K., Srivastava, R. K., Singh, A. K., Chandel, A. K., Pareek, N., & Vivekanand, V. (2022). Sustainable utilization of pineapple wastes for production of bioenergy, biochemicals and value-added products: A review. *Bioresource Technology*, 351, 127085. https://doi.org/10.1016/j.biortech.2022.127085

Sarangi, P. K., Singh, A. K., Srivastava, R. K., & Gupta, V. K. (2023). Recent progress and future perspectives for zero agriculture waste technologies: Pineapple waste as a case study. *Sustainability*, 15(4), 3575. https://doi.org/10.3390/su15043575

Saraswaty, V., Risdian, C., Primadona, I., Andriyani, R., Andayani, D. G. S., & Mozef, T. (2017). Pineapple peel wastes as a potential source of antioxidant compounds. *IOP Conference Series: Earth and Environmental Science*, 60, 012013. https://doi.org/10.1088/1755-1315/60/1/012013

Sepúlveda, L., Romaní, A., Aguilar, C. N., & Teixeira, J. (2018). Valorization of pineapple waste for the extraction of bioactive compounds and glycosides using autohydrolysis. *Innovative Food Science & Emerging Technologies*, 47, 38–45. https://doi.org/10.1016/j.ifset.2018.01.012

Shu, H., Sun, W., Xu, G., Zhan, R., & Chang, S. (2019). The situation and challenges of pineapple industry in China. *Agricultural Sciences*, 10(5), 683–688. https://doi.org/10.4236/as.2019.105053

Son, J. M., Lee, C., Hong, S. K., Kang, J. J., & Cho, Y. H. (2017). Fast thermal response of silicon nanowire-heater for heat shock generation. *International Journal of Precision Engineering and Manufacturing-Green Technology*, 4(1), 45–52. https://doi.org/10.1007/s40684-017-0006-1

Sukri, S. A. M., Andu, Y., Sarijan, S., Khalid, H.-N. M., Kari, Z. A., Harun, H. C., Rusli, N. D., Mat, K., Khalif, R. I. A. R., Wei, L. S., Rahman, M. M., Hakim, A. H., Lokman, N. H. N., Hamid, N. K. A., Khoo, M. I., & Van Doan, H. (2023). Pineapple waste in animal feed: A review of nutritional potential, impact and prospects. *Annals of Animal Science*, 23(2). https://doi.org/10.2478/aoas-2022-0080

Tran, T. V., Nguyen, D. T. C., Nguyen, T. T. T., Nguyen, D. H., Alhassan, M., Jalil, A. A., Nabgan, W., & Lee, T. (2023). A critical review on pineapple (Ananas comosus) wastes for water treatment, challenges and future prospects towards circular economy. *Science of the Total Environment*, 856, 158817. https://doi.org/10.1016/j.scitotenv.2022.158817

Tropea, A., Wilson, D., Torre, L. G. L., Curto, R. B. L., Saugman, P., Troy-Davies, P., Dugo, G., & Waldron, K. W. (2014). Bioethanol production from pineapple wastes. *Journal of Food Research*, 3(4), 60. https://doi.org/10.5539/jfr.v3n4p60

Upadhyay, A., Lama, J. P., & Tawata, S. (2013). Utilization of pineapple waste: A review. *Journal of Food Science and Technology Nepal*, 6, 10–18. https://doi.org/10.3126/jfstn.v6i0.8255

Wan Nadirah, W. O., Jawaid, M., Al Masri, A. A., Abdul Khalil, H. P. S., Suhaily, S. S., & Mohamed, A. R. (2012). Cell wall morphology, chemical and thermal analysis of cultivated pineapple leaf fibers for industrial applications. *Journal of Polymers and the Environment*, 20(2), 404–411. https://doi.org/10.1007/s10924-011-0380-7

Widowati, T., & Amalia, N. S. A. (2021). Utilization of pineapple leaf fiber (Ananas comosus) as material for false eyelashes production. *IOP Conference Series: Earth and Environmental Science*, 700(1), 012034. https://doi.org/10.1088/1755-1315/700/1/012034

Williams, P. A., Crespo, O., & Essegbey, G. O. (2017). Economic implications of a changing climate on smallholder pineapple production in Ghana. *Journal of Economics and Sustainable Development*, 8(18), 34–43. Online. www.iiste.org

Section VII

Fruit-Based Plant Waste

17 Durian Waste Management

17.1 INTRODUCTION

For thousands of years, agriculture has been essential to human life. To facilitate growing populations, global food production urgently needs to improve. Significant resources are needed for food production, including water and energy. Millions of tons of solid waste are produced each year because of the rapid urbanization and population growth, which has major consequences for the ecosystem on a global scale. A waste of energy and water resources follows any losses of agricultural products or food. A substantial amount of these losses occurs after harvest, during processing and storage. Agro wastes make up close to 30% of all agricultural products generated worldwide. Raw agricultural goods can consist of both plant and animal leftovers. Agricultural wastes, referred to as agro wastes, are by-products of farming and the initial processing of agricultural goods such as vegetables, fruits, dairy products, meat, poultry, and other products (Affandi et al., 2019).

Both process and field residues are further categories for agricultural waste. Field residues are residues from the crop harvesting process that is still taking place in the field. The field residues are made up of stems, seed capsules, leaves, and stalks. Crop residues, waste from the agro-industry's processing, animal waste, produce waste, dairy waste, meat waste, poultry waste, and other garbage are examples of these wastes. While the crop is transformed into another useful resource, the process residues are generated, such as husks, bagasse, seeds, molasses, straw, stalks, shells, and other materials that make up these wastes used during processes. Then, industrial by-products are typically made by food processing sectors, including the chips, juice, meat, confectionery, and fruit industries, which produce a sizable amount of organic waste and associated effluents each year. The example of the different types of agro-industrial waste is shown in Figure 17.1. Different energy sources can be made from these organic wastes. To protect the environment and effectively utilize resources, agro-waste must be managed properly due to its commercial viability, renewability, and biodegradability. Methods like composting, producing bioactive nanomaterials, or using biorefinery technologies manage agricultural waste to achieve maximum benefits (El-Ramady et al., 2022).

In addition, examples of crop waste can be classified into field-based and process-based, as given in Table 17.1. Such classification is crucial, especially when it comes to the application of energy, since the accessibility and availability of these sources heavily depend on this quality. Since collecting field residues for recourse is challenging, the availability of field-based residues for energy application could be seasonal and limited.

The durian, also known as *Durio zibethinus*, is the source of agricultural waste discussed in this research. In Asian countries, numerous fruit trees thrive in the tropical climates. The climate's consistency facilitates the production of tropical fruits with minimal variation that can meet the demands of the world market. One type of tropical fruit identified as the "king of fruits" is the durian (*Durio zibenthinus*). The Malay word "Duri," which translates to "thorn of the fruits," is where the name durian originated. Durian's morphology has spiky thorns on the surface. The dicotyledonous species of the fruit known as durian is a genus *Durio* and the family Bombacaceae. There are numerous varieties of durian; however, there are about 30 Durio species that are widely known. There are at least nine species that produce edible durian fruit, which include *Durio testudinarum*, *Durio oxleyanus*, *Durio lowianus*, *Durio kutejensis*, *Durio dulcis*, and *Durio grandifloras* (Tabatabaee

FIGURE 17.1 Agro-industrial wastes and their types (Riseh et al., 2024).

TABLE 17.1
Examples of Frequently Available Crops and Residues (Dyjakon, 2019)

Crop	Category	Residue
Wheat	Field-based	Straw
	Process-based	Husk
Rape	Field-based	Straw
	Process-based	Rapeseed cake
Maize	Field-based	Straw
	Process-based	Cobs
Tea	Field-based	Prune branches
	Process-based	Refuse tea
Coffee	Field-based	Prune branches
	Process-based	Refuse coffee
Vegetables	Field-based	Leaves
	Process-based	Spoiled vegetables
Animals	Field-based	Manure
	Process-based	Wastewater
Apple	Field-based	Prune branches/leaves
	Process-based	Spoiled apples/pomace
Cherry	Field-based	Prune branches/leaves
	Process-based	Spoiled cherries/seeds
Vineyard	Field-based	Prune branches/leaves
	Process-based	Spoiled fruit/pomace

Amid et al., 2012). It is one of the priciest fruits in the region. A variety has been suggested for commercial planting in Malaysia, including D24 (also known as Bukit Merah), D99 (also known as Kop Kecil), and D145 (also known as Beserah). Durian species registered in Thailand used regional designations, including Puang Manee, Kradum, and Monthong. Some species are similar in Malaysia and Thailand but have different names, such as Chanee D123, Kan Yao D158, and Monthong D169. Indonesia registers durian species based on their regional designations, which is similar to Thailand. Examples include Pelangi Atururi, Salisun, Nangan, Matahari, and Sitokong (Aziz & Jalil, 2019). In Southeast Asia, including Thailand, Malaysia, Indonesia, and the Philippines, this tropical fruit is particularly well-known because of the pulp's richness, sweetness, creaminess, and unique scent (Lubis et al., 2018). Although, due to its incredibly potent and distinctive aroma, consumer choice is subject to individual acceptance and tolerance, the durian's popularity is restricted to certain geographic areas (Cheok et al., 2018).

Environmental standards and laws managing pollution from agricultural waste streams have recently become more demanding and constraining the durian operation. These have encouraged research into developing efficient, long-term alternatives for utilizing the severely polluted durian leftovers. The utilization of tropical fruit wastes is the main topic of this review. The fruit and its by-products have recently been studied as possible food additives for a variety of functions, including preservatives, thickeners, antibacterial agents, and potential medicinal applications. Researchers are now interested in the potential of converting the durian's high percentage of inedible sections, which ranges from 60% to 81%, into highly valued biomass products (Cheok et al., 2018). Agricultural waste production is anticipated to rise over the following 20 years, resulting in the formation of durian residues (Foo & Hameed, 2011). The current chapter investigates the potential application of durian residues as a component of alternative sustainable biomass products. The objectives of this research are to research the structural composition and properties of durian waste and investigate the utilization of durian waste to produce biomass or any industrial applications.

The usage of agricultural waste, specifically durian waste, is still playing a crucial role in generating biomass. Due to the growth in the population, natural resource consumption has increased. The abundance of durian waste and the limitations of research for durian waste-generating biomass should be tackled. Therefore, the transformation of durian waste is needed to produce goods such as pulp and paper with higher economic value and market demand. New and cost-effective production processes are essential to enable the commercial manufacturer of durian biomass products. This can help to deal with a large amount of durian waste in the long term. Most of the current research focused on the structural composition and characteristics of the fruit. Limited research was conducted to investigate durian waste producing everyday goods or using durian waste in any industrial application. Finally, this research will encourage further investigation and discovery in new areas.

17.2 BIOMASS FROM NATURAL FIBERS

Biomass refers to any organic matter, such as plants, agricultural residues, or forestry by-products, that can be used as a renewable energy source or precursor for various bio-based products (Morgan et al., 2018). Durian consists of 5–15% seeds, 20–35% flesh (aril), and 55–66% husk (Nordin et al., 2017). Along with its great flavor, durian is a nutritious food high in vitamins A, β-carotene, and C, as well as potassium, magnesium, salt, and calcium (Cheok et al., 2018). Due to the high nutritional and organic content, durian waste can be utilized to produce biomass products. Recycling durian shell waste has produced culinary products, fertilizer, and biopesticides (Mahathaninwong et al., 2022). Another example is a discarded durian shell, which has been shown to be a useful resource for creating activated carbon adsorbent for acid or basic dye removal from aqueous solutions. It may also be processed into biochar, desiccant for air conditioning systems, and a variety of other materials (Zhao et al., 2017).

With more than 250,000 ha planted in durians in 2008, Malaysia, Thailand, and Indonesia are the top three suppliers of the fruit (Teh et al., 2017). This region, including the Philippines, annually produces more than 1,600,000 tons of durian (Zhao et al., 2017). Each year, Malaysia produces a large amount of durian fruit, mostly for use in commerce. According to estimates, Malaysia produced 320,164 MT of durian in 2013 (Payus et al., 2019). According to data from the Indonesian Statistic Institute, Indonesia produced up to 735,423 tons of durian in 2016 and an additional 514,796 tons of by-products (Lubis et al., 2018). The Agriculture Economics Office of Thailand stated that 1,111,928

TABLE 17.2
Comparison of Durian Production (MT) in Several Southeast Asian Countries (Safari et al., n.d.)

	Thailand	Malaysia	Indonesia
2012	524,000.00	347,704.40	888.13
2013	569,000.00	373,082.90	759.06
2014	632,000.00	351,517.90	859.12
2015	602,000.00	360,270.70	995.73
2016	518,000.00	302,645.80	735.42

FIGURE 17.2 Composting of durian waste (Year of the Durian, 2015).

tons of durian fruit were produced in Thailand in 2020 (Mahathaninwong et al., 2022). Table 17.2 provides the data on durian production (MT) in Malaysia, Thailand, and Indonesia from 2012 until 2016.

Furthermore, in comparison with different types of tropical fruits like pineapple, papaya, mango, and jackfruit, durian has the highest energy content (Aziz & Jalil, 2019). Durian waste contains high energy due to the sugars, fat, potassium, dietary fiber, and bioactive and volatile chemicals found in abundance in durian, along with other macro- and micronutrients (Aziz & Jalil, 2019). The high energy content of durian waste allows the production of bioethanol using durian seed waste (Purnomo et al., 2016) and biogas from durian shell waste (Shen et al., 2019).

It may be feasible to use durian waste to benefit biomass as in the agribusiness of horticulture products, durian has a significant economic worth (Affandi et al., 2019). Durian residues are not considered to be "wastes," given the fruit's high nutritional content. Rather, they are viewed as raw materials for developing and manufacturing additional goods. These nutrients are present in raw materials, which gives suitable conditions for developing microbes. The bacteria in durian waste have the capacity to utilize fermentation techniques to repurpose the source ingredients. SSF or solid-state fermentation advancements utilize agro-industrial waste as a solid foundation to produce a range of beneficial products. Agro-industrial wastes are converted into antibiotics, animal feed, biofuels, enzymes, vitamins, antioxidants, and other compounds (Sadh et al., 2018). For example,

Composting is a microbial process that accelerates the biodegradation and conversion of agricultural residue from complex materials into organic or inorganic forms that may be more quickly broken down (El-Ramady et al., 2022). The process of composting can be seen in Figure 17.2, as the durian waste can be composted with soil or mixed with other agro-waste.

17.3 ENVIRONMENTAL IMPACT FROM BIOMASS

The great majority of agricultural residues created annually by industries are not properly disposed of, which has a negative impact on the environment. Most agro-industrial wastes are not well taken advantage of. Therefore, they are usually burned, dumped, or unintentionally landfilled (Chiemchaisri & Visvanathan, 2008). This increase in global production has increased environmental pressure to the point where it has a detrimental effect on soil, air, and water resources, negatively affecting human health and the environment. Typically, people may only eat half of the durian from the entire bulk while the parts, such as the shell, are completely disposed of. Therefore, durian husk is discarded in massive quantities, which endangers the environment, particularly given the present shortage of space at waste disposal facilities for the waste to degrade.

Due to the high moisture content and number of biodegradable organic compounds found in this fruit waste, as decomposition starts, an intolerable odor is released. These untreated wastes are dumped for an extended period, which causes more greenhouse gas (GHG) emissions and climate change-related problems (Sadh et al., 2018). Research shows that 21% of the emissions of GHGs result from agriculture (Duque-Acevedo et al., 2020). During the growing season from May to June, problems brought on by fruit peels and shells are evident. For example, durian shells are disposed of in vast quantities in landfills, while some are discarded (Mahathaninwong et al., 2022). Durian skin adds significantly to agricultural wastes when burned or discarded in a landfill, creating an environmental concern that includes the spread of mice, pests, and bacteria to breed, which transmits the plague. Across Malaysia, the average durian fruit weighs 255,353 MT, and landfill loading rates are increasing, particularly due to the massive volume and size of durian garbage, which requires an even larger landfill area (Payus et al., 2019).

Durian garbage is enormous and demands a larger location to dispose of. An increase in durian waste will have detrimental effects toward the environment, especially when disposed of improperly, as shown in Figure 17.3. According to its definition, climate change refers to long-term, worldwide changes in the weather that are typified by increases in atmospheric temperatures and alterations in the amounts and patterns of precipitation. Ecosystems may be negatively impacted by climate change, especially agroecosystems, through increased temperature, precipitation, and a rise in the

FIGURE 17.3 Big-size durian waste (Charoenvai, 2014).

frequency and intensity of extreme weather. Agricultural yields will be disrupted by these changes, with yield losses for most crops in most places, which will impact human society and, hence, the global agricultural economy, which will impact the globe's capacity to feed an expanding human population. The consequences of climate change on organisms that interact with crops have the potential to impact production from agriculture indirectly. For instance, many insects pollinating crops have already been harmed by climate change, and agricultural viruses and pests are expected to grow higher as temperatures rise. Consequently, climate change poses an imminent hazard and is a severe social and environmental concern (El-Ramady et al., 2022). Additionally, an increase in this waste may result in several issues with human health, such as disease outbreaks or air pollution and pose an adverse impact on the environment.

Therefore, agro-waste management is essential since it can result in environmental degradation and health risks. Agro-waste can, nevertheless, have a huge range of industrial applications if handled properly. Utilizing agro-industrial wastes as raw materials can reduce production costs and environmental concerns. Adopting innovative technical methods to prevent negative environmental effects on agriculture makes the transition to sustainable agriculture of the utmost importance.

17.4 AVAILABILITY OF DURIAN WASTE

The expansion of the durian market demonstrates the fruit's potential for the global market. Consumption of durian has actively increased and further established in a number of food production businesses, including ice cream flavors, jams, rolls, tarts, pies, and milkshakes (Foo & Hameed, 2011). Durian market increase for the global market is about 1.4 MT. Indonesia (265 kt), Malaysia (376 kt), and Thailand (781 kt) are the top three exporters in this market, followed by Florida, Hawaii, Papua New Guinea, Madagascar, Australia, Cambodia, Laos, Vietnam, Myanmar, India, Sri Lanka, and the Philippines (Davao). On the contrary, with a ratio of 65% to the total quantity of durian exports, importer countries include Taiwan (5 kt), Singapore (40 kt), and China (65 kt) (Lubis et al., 2018). The biggest importers of durian fruits are Hong Kong, Taiwan, and Singapore, while 90% of all exports originate in Thailand, Malaysia, and Indonesia (Husin et al., 2018). The increased waste from durians is an outcome of the increased durian production.

Typically, 70–85% of the fruit is wasted if not properly disposed of, which would result in pollution (Payus et al., 2019). Durian seeds, which account for 20–25% of the fruit's total weight, are typically discarded as agricultural trash due to the mucus that covers them and gives the tongue an unpleasant sensation (Siti Faridah et al., 2021). With an estimated 480,000 metric tons of durian skin waste produced each year, durian waste is considered the main source of agricultural trash with rising volumes (Payus et al., 2019).

17.5 PROPERTIES OF DURIAN WASTE FIBERS

17.5.1 Physical Properties

There are over 30 different species of durian, and they range in size from 2 to 4.5 kg on average, having a diameter between 14 and 18 cm and a length between 19 and 32 cm (Husin et al., 2018). There are large varieties of durian, referred to as durian Kepala Gajah 3, which weigh from 4.5 to 7 kg (Brown, 1997). The fruit is ovoid and nearly spherical in shape. The fruit tends to open at full ripening and has 3–5 longitudinal sutures covering it from the apical to the basal end. A durian fruit covers an area of around 400 cm^2 when all its dehiscence zones are open. Durian seeds resemble chestnuts in form and size, measuring 2–4 cm wide and 2–6 cm long, respectively. They are also light brownish (Husin et al., 2018). Typically, the durian's outer layer of skin weighs more than half of the entire fruit, 20–25% of the durian's weight can be attributed to its seeds, and 75–80% to its rinds (Lubis et al., 2018).

The fruit's edible part ranges in color from yellowish-brown to green and is extremely thick and slightly woody with pyramidal thorns that are sharply pointed. Only 15–30% of the fruit's total mass can be consumed (Payus et al., 2019). The flesh of the durian ranges in hue from pink to red and from yellow to a darker orange or yellow tone (Siti Faridah et al., 2021). In research conducted by Hamzah et al. (2023), durian husk is treated as a biosorbent or feasible adsorbent for enhancing the retting process of soaking water of pepper berries. Although only a small portion of the fruit is edible, the discarded durian seed and shell offer enormous potential for the development of higher-worth products.

17.5.2 Chemical Composition

The chemical composition of durian can be analyzed by examining its mineral content, protein fraction, carbohydrate profile, fatty acid composition, and amino acid content. Understanding the chemical composition of durian is essential for various reasons. First, it can provide valuable insights into the flavor and aroma profile of durian. Durian skins comprise lignin, hemicellulose, and cellulose. The cellulose concentration of durian rinds is considerable at 57%, followed by 22% hemicellulose, 13% lignin, and 3.3% moisture (Lubis et al., 2018). Durian skin is hydrophilic because it is made of lignocellulose, a natural fiber that contains a hydroxyl group (Manshor et al., 2014). It has a high nutritional value for its seeds and is a rich source of lipids, proteins, and carbs. Additionally, it contains a lot of dietary fiber, potassium, sugar, vitamin C, minerals, and trace metals (Nordin et al., 2017). It tastes sweet-bitter and has a creamy texture. Different durian varieties have different nutritional compositions in Table 17.3 and sugar in Table 17.4 (in 100 g fresh weight [FW]). Table 17.5 lists the mineral and vitamin contents of different types of durian (in grams, based on 100 g of FW).

Alcohols, volatile compounds comprising esters and sulfur-containing substances, which include thioesters, thioacetals, and thiolanes, as well as sugar, fat, and volatile compounds which include esters, are all the causes of the distinctive flavor of durian (Aziz & Jalil, 2019). The aroma of durian fruits is distinctive and intense in general. The unpleasant odor, which is comparable to the smell of rotten eggs or onions, is set on by volatile sulfur compounds, which are inhibited by methionine gamma lyases. Due to its overbearing aroma, durian fruit is prohibited in hotels and airports, and there are limits on its transportation and storage (Husin et al., 2018). The sugar analysis also showed that the chemical composition of durian seed gum contained only very small amounts of arbinose and xylose (Mirhosseini & Amid, 2013). Previous studies have shown that the durian seed contains 76.8% carbohydrate, 7.6% protein, 6.6% moisture, 4.8% crude fiber, 3.8% ash, and 0.4% fat, and 52.9% of the recommended daily intake of fiber (Siti Faridah et al., 2021).

The carbohydrate content of durian seed gum contains the most prevalent monosaccharides were xylose (0.3–3.21%), arabinose (0.58–3.41%), glucose (37.1–45.1%), and galactose (48.6–59.9%). The majority of the lipid fraction derived from durian seed gum was composed of the fatty acids oleic acid (C18:1), linoleic acid (C18:2), linolenic acid (C18:2), stearic acid (C18:0), palmitic acid (C16:0), and palmitoleic acid (C16:1). The chemical composition of gum from durian seeds contains necessary amino acids, which increases its nutritional value. The most prevalent amino acids in durian seed gum included valine (4.5–5.50%), proline (3.87–4.81%), serine (4.39–5.18%), threonine (3.44–6.50%), isoleucine (3.30–4.07%), lysine (6.04–8.36%), aspartic acid (6.10–7.19%), glycine (6.07–7.42%), leucine (30.9–37.3%), glutamic acid (5.57–7.09%), and alanine (5.24–6.14%) (Tabatabaee Amid et al., 2012). A polysaccharide–protein structure can be found in the durian seed-based biopolymer. In the durian seed's carbohydrate composition, glucose, and D-galactose were the two most abundant monosaccharides (Mirhosseini & Amid, 2013). These factors suggest that durian seeds provide a number of health benefits, including enhancing good skin, regulating blood sugar levels, fending off depression, avoiding constipation, and maintaining healthy bone structure (Kai et al., 2022).

There have been numerous volatile compounds discovered in Malaysian, Thailand, and Indonesian durian types, which include ethanol, thioacetals (1-(methylthio)-propane), thioesters (1-(methylthio)-ethane), thiolates (3,5-dimethyl-1,2,4-trithiolane isomers), sulfur compounds (diethyl disulfide, diethyl trisulfide, and ethanethiol), and esters (ethyl propanoate, methyl-2-methylbutanoate, and propyl propanoate) (Aziz & Jalil, 2019). Then, durian fruit ripening is a complex process with molecular and biochemical changes. Durian undergoes postharvest ripening, increasing respiration and ethylene production, breaking down carbohydrates, and reducing acidity (Sangpong et al., 2021).

TABLE 17.3
Nutritional Composition of Durian Aril (Flesh) of Different Durian Varieties (g per 100 g Fresh Weight) (Aziz & Jalil, 2019)

	Indonesian Variety				Thailand Variety						
Durian Variety	**Ajimah**	**Hejo**	**Matahari**	**Sukarno**	**Mothong**	**Chance**	**Kradum**	**Kobtakam**	**Unknown Variety**	**Unknown Variety**	**Unknown Variety**
Nutrients Energy (kcal)	151	84	163	134	134–162	145	185	145	147	153	134
Carbohydrate (g)	28.90	15.65	34.65	27.30	21.70–27.10	20.13	29.15	21.15	27.09	27.90	28.00
Protein (g)	2.36	1.76	2.33	2.13	1.40–2.33	3.10	3.50	2.86	1.47	2.70	2.50
Fat (g)	2.92	1.59	1.69	1.86	3.10–5.39	4.48	4.67	4.40	5.33	3.40	3.00

TABLE 17.4
Sugar Content of Various Durian Varieties (g per 100 g Fresh Weight) (Aziz & Jalil, 2019)

Sugars	Fructose	Glucose	Sucrose	Maltose	Total Sugar
		Malaysian Variety			
Durian Kampung	1.60	2.21	12.58	0.51	16.90
D2	1.66	2.51	7.70	NA	11.87
D24	0.76	0.73	6.03	NA	7.52
MDUR78	1.82	2.77	8.02	NA	12.61
D101	1.29	1.97	5.57	NA	8.83
Chuk	1.28	1.87	10.65	NA	13.80
		Thailand Variety			
Monthong	0.15	0.74	13.69	0.25	14.83
Chance	0.26	0.58	15.71	0.00	16.55
Kradum	0.33	0.71	17.89	1.04	19.97
Kobtakam	0.10	0.45	17.30	0.26	18.11
		Indonesian Variety			
Ajimah	NA	NA	NA	NA	14.05
Hejo	NA	NA	NA	NA	3.10
Matahari	NA	NA	NA	NA	8.10
Sukarno	NA	NA	NA	NA	8.12

Durian seed starch has been determined to provide a high amount of dietary fiber for its pasting capabilities that can be applied to future pharmaceutical, food, and cosmetics applications. Water-in-oil-in-water emulsion stabilization has been attempted using durian seed gum (Cheok et al., 2018). Earlier research found that the natural seed gum polymer possessed the required emulsifying properties for an oil-in-water emulsion (O/W). The small concentration of the proteinaceous component (4%) in durian seed gum may be the cause of this interfacial action. The durian fruit seeds' heteropolysaccharide–protein polymer also showed only moderately effective thickening characteristics in water. This may have been based on the structure's comparatively low molecular weight (Mirhosseini & Amid, 2013). By studying these various aspects of the chemical composition, researchers can gain a comprehensive understanding of durian's molecular structure and its implications for flavor, nutrition, and potential health benefits.

17.5.3 Mechanical Properties

Durian is a tropical fruit that is known for its distinct flavor and distinctive physical properties. Its outer shell is covered in thorny spikes, making it difficult to handle without protective gloves. The durian fruit is typically large, weighing up to several kilograms, and has a characteristic oblong shape. The mechanical properties of durian are an important aspect to consider when studying this fruit. Durian has a relatively tough and rigid outer shell, which protects the soft and delicate flesh inside. This outer shell is composed of a thick and fibrous husk that acts as a barrier against external forces, such as physical impact or compression.

Durian exhibits several mechanical properties that make it an interesting material to study. One notable mechanical property of durian is its high stiffness. This means that durian is resistant to deformation and can withstand high loads without undergoing significant changes in shape. Durian's stiffness can be attributed to its structural characteristics, specifically the peel layer and spines that provide

TABLE 17.5
Vitamin and Minerals Contents of Various Durian Types (Aziz & Jalil, 2019)

Durian Variety	Thailand Variety				Malaysian Variety	Unknown Variety	Unknown Variety	Unknown Variety
	Monthong	Chance	Kradum	Kobkatam	Unknown			
Macrominerals (mg per 100 g Fresh Weight)								
Calcium	4.298–6.134	5.44	3.75	3.21	NA	6.00	40.00	7.00
Phosphorus	25.79–33.99	32.96	36.70	37.56	NA	39.00	44.00	44.00
Sodium	6.14–15.66	11.84	19.60	21.51	NA	2.00	40.00	1.00
Potassium	377.00–489.42	539.20	439.52	438.17	NA	436.00	70.00	601.00
Magnesium	19.28–24.87	23.36	23.35	22.79	NA	30.00	NA	NA
Microminerals (mg per 100 g Weight)								
Iron	0.18–0.23	0.45	0.23	0.36	NA	0.43	0.1.90	1.30
Copper	0.13–0.15	0.27	0.23	0.17	NA	NA	NA	0.12
Manganese	0.23–0.26	NA	NA	NA	NA	NA	NA	NA
Zinc	0.15–0.21	0.45	0.37	0.32	NA	0.28	NA	0.30
Vitamins (μg per 100 g Fresh Weight)								
A (RAE)	NA	NA	NA	NA	NA	2.00	NA	NA
B_1/Thiamine	NA	NA	NA	NA	NA	374.00	100.00	100.00
B_2/Riboflavin	NA	NA	NA	NA	NA	200.00	100.00	100.00
B_3/Niacin	NA	NA	NA	NA	NA	1,074.00	NA	13,650.00
B_6/Pyridoxine	NA	NA	NA	NA	NA	316.00	NA	NA
E/Tocopherol or Tocotrienol (μg per 100 g Fresh Weight)								
α-Tocopherol	NA	NA	NA	NA	3,774.00	NA	NA	NA
γ-Tocopherol	NA	NA	NA	NA	1,013.00	NA	NA	NA
δ-Tocopherol	NA	NA	NA	NA	11.00	NA	NA	NA
δ-Tocopherol	NA	NA	NA	NA	1.00	NA	NA	NA

energy absorption and protection for the pulp during impact events such as durian falling to the ground (Ha et al., 2020). This is due to the durian shell comprising several porous, irregular fibers. The durian shell's intricate mesocarp and exocarp structure is made up of bundles of fibers. Furthermore, durian possesses good impact resistance due to its unique structural composition. The peel layer and spines of durian act as a protective shield, absorbing and dispersing impact energy. In a study conducted by researchers, it was determined that particular energy absorption in the durian shell's mesocarp layer was higher under lateral loading than under axial loading. This means that the durian shell's mesocarp layer has a higher ability to absorb energy when subjected to forces applied in a sideways direction rather than forces applied in a straight line. These findings suggested that the durian shell does not exhibit a prominent fiber structure, similar to that of wood, which may be a factor in its reduced strength and dampening capability when loads are applied axially as opposed to laterally (Ha et al., 2020).

Mechanical properties of durian fiber are affected by the cell wall structure and chemical content, particularly the amount of cellulose, hemicellulose, and lignin. The fibers in durian rinds are stiff and inelastic. The mechanical characteristics of fiber were significantly influenced by the high cellulose content, as about 22% of the hemicellulose in durian rinds is cellulose. Cellulose is a component of a structure that can give the wall cell of a plant or fiber robustness and stability. Cellulose has several characteristics, including high stiffness, high crystallinity, and an inability to dissolve in organic solvents. However, the durian rinds hemicellulose might weaken the fiber and make it easier to decompose (Lubis et al., 2018). As a result of its macro-porous structure, acceptance of pH, mechanical stress, and prolonged immersion in water, it can absorb water contaminants (Payus et al., 2019). A behavior study conducted with durian seed gums indicated that the seed gums showed pseudoplastic flow behavior in which viscosity decreased under shear strain (Amid & Mirhosseini, 2012). Additionally, studies were done on the use of durian shell-derived activated carbon for the adsorption of methylene blue in the removal of colors from water effluent (Chandra et al., 2007). Understanding the mechanical properties of durian is crucial to aid in developing appropriate handling and processing techniques to minimize damage and maximize shelf life.

17.5.4 Morphological Properties

Different durian fruit varieties tend to have a range of characteristics, including differences in shape and fruit quality. One of these differences was that fruit and plants each have distinctive qualities. Genetics and environment also have an impact on physical differences among durian trees. For plants to fully express their genetic makeup, they require a specific environmental condition that is ideal (Selvy Handayani & Ismadi, 2018). Durian flowers have peltate scales covering their ovary, but as they expand, spines start to form. The flowers are often pentamerous, and the calyx, which is united, typically has five lobes. Durian leaves have an acuminate tip, are oval-oblong in shape, and are covered in clear spots on both their top and lower surfaces. The pollen grains are spherical. The pollen grains have an exine that is smooth and sticky and range in size from 80 to 150 μm. When durian pollen is shed, it is binucleated, and if it is stored at room temperature, it is viable for around 48 h.

The funicular end of the ovule is where the aril grows. Most wild species, including *D. zibethinus*, have an aril that surrounds the seed. The mature durian fruit's fruit stalk is made up of a parenchyma, tanning-containing-cell, and scleroid-filled cortex surrounded by an outer periderm. The core of the stalk is where the vascular tissue is located. The placement of seeds affects the form of durian fruits. Unfertilized ovule-containing locules frequently fail to mature, which causes the fruits to have irregular shapes. The development of durian fruit and the fruit shape significantly influences marketability (Brown, 1997). Durian genome is about twice as large as humans, with 46,000 genes (Husin et al., 2018). Depending on the species, durians range in weight from 2 to 4.5 kg on average and have an oval to nearly spherical shape, as shown in Figure 17.4. After the germination of their seeds, mature durian trees are known to yield fruit just 5–7 years later. Each durian tree yields between 15 and 800 fruits during the fruiting season (Husin et al., 2018).

Durian trees, which range from 25 to 50 m in height, flourish in a warm, humid climate with balanced temperatures between 25°C and 30°C and rainfall rates between 1.5 and 2 my^{-1} (Foo & Hameed, 2011). The peak durian season in Malaysia occurs in the middle of the year, from June to August. The durian fruit begins to ripen during the rainy seasons, from April to September for the southwest monsoon and from October to March for the north-east monsoon. The fruit takes 96–130 days to fully grow into a durian (Lubis et al., 2018). Different durian variation sometimes has different morphological properties of their trees, as noted by Affandi et al. (2019). Table 17.6 provides the morphological characters of several durian varieties based on the Ministry of Agriculture Decision Letter in Indonesia.

As depicted in Figure 17.5, the durian consists of a thick bottom white inner husk paired with a green thorny outer husk, a tiny neck section of white inner husk, and a thick bottom part of the white inner husk (Nordin et al., 2017). Decoctions of durian leaves and roots have been reported to be febrifuges and anti-malarial (Husin et al., 2018).

17.5.5 Thermal Properties

According to research conducted by Aziz and Jalil (2019), durian has an energy content per 100 g of FW that ranges from 84 to 185 kcal, which is the same as the food composition datasets from Malaysia, Indonesia, and the U.S. Department of Agriculture. In comparison, the Thai Kradum durian aril had the highest calorie content at 185 kcal, whereas the caloric level of the Indonesian type of Hejo was the lowest, with 84 kcal per 100 g FW of durian aril. The variation in carbohydrate

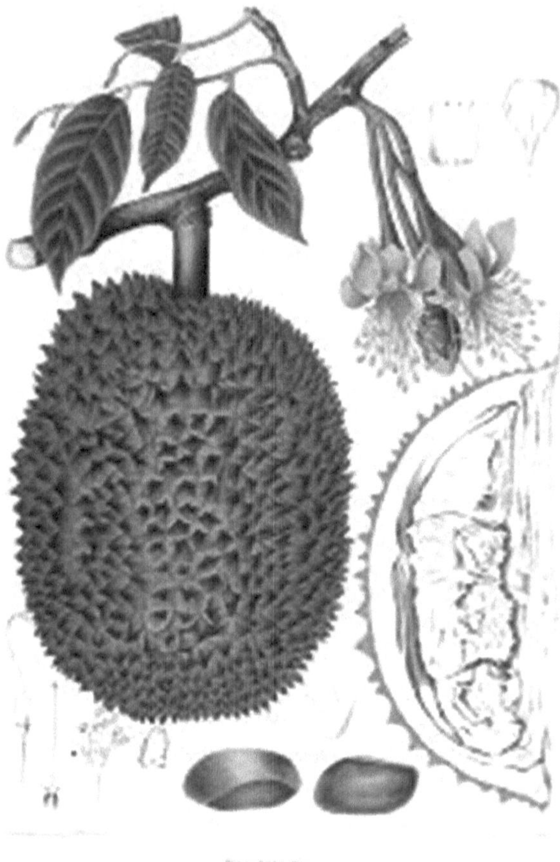

FIGURE 17.4 Typical shape of a durian plant.

TABLE 17.6
Morphological Characters of Several Durian Varieties (Affandi et al., 2019)

		Morphological Characters					
No.	Clones	Crown Shape	Tree Growth Habit	Branching Density	Leaf Density	Leaves Width (cm)	Leaves Length (cm)
1	Kalumpang	Elliptical	Intermediate	Medium	Medium	4.0–6.3	12.0–17.0
2	Sukun	Irregular	Intermediate	Medium	Medium	5.7–6.5	14.0–16.3
3	Matahari	Semi-circular	Intermediate	Dense	Dense	3.0–4.1	9.8–14.2
4	Sunan	Semi-circular	Intermediate	Dense	Dense	4.4–4.6	11.3–12.0
5	Sijantung Kampar	Semi-circular	Intermediate	Medium	Medium	4.4–5.1	12.1–12.4
6	Selat	Semi-circular	Spreading	Dense	Dense	4.0–5.0	14.0–16.0
7	Sitokong	Semi-circular	Spreading	Dense	Dense	4.1–4.9	12.5–16.7
8	Petruk	Semi-circular	Intermediate	Dense	Dense	3.2–4.2	11.7–13.0
9	Sikapal	Irregular	Intermediate	Medium	Medium	4.5–5.0	14.0–15.0
10	Ginting	Pyramidal	Spreading	Medium	Medium	5.0–6.0	12.0–16.0
11	Bintana	Spherical	Spreading	Dense	Dense	5.5–6.5	15.0–18.0
12	Kani	Elliptical	Spreading	Dense	Dense	4.5–5.4	12.0–15.5

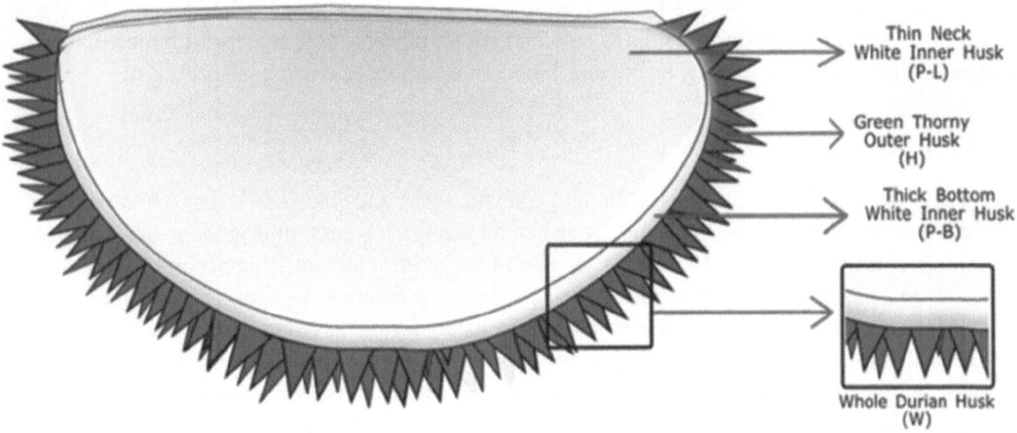

FIGURE 17.5 Durian husk's locule and its differential segments (Nordin et al., 2017).

content causes higher and lower energy levels (Aziz & Jalil, 2019). The durian husk fibers, including cellulose, hemicellulose, and lignin, give it the capacity to withstand heat (Payus et al., 2019). The fruit is known to have a high water content, affecting its thermal properties. During a pyrolysis experiment conducted in research, the removal of moisture occurred at 100°C, and lignin and hemicelluloses were then removed during pyrolysis from pre-treated durian skin fiber (DSF). Hemicelluloses withstood temperatures up to 220°C, while cellulose and lignin decomposed at roughly 330°C. Char residue was created as cellulose decomposed at higher temperatures. High thermal stability was achieved due to strong bonding in the durian skin (Manshor et al., 2014). In a study, the side chains in the durian seed gum's molecular structure thermally degraded at a high temperature of 105°C (Amid & Mirhosseini, 2012).

17.5.6 Calorific Value

In charcoal briquettes, durian peel can be used as an alternative fuel as it contains combustible materials, cellulose (50–60%), lignin (5%), and starch (5%). According to Sitti Rahmawati et al. (2023), the results of evaluating the calorific value, moisture content, ash content, volatile matter value, and bounds carbon content of briquettes made from durian peels and starch from sago were 5,189.13 cal/g, 3.60 g/cm^3 of density, 0.48 g/cm^3 of ash, and 0.22 g/s of combustion rate.

17.6 DURIAN WASTE MANAGEMENT

Malaysia produces a lot of durian fruit each year, mostly for export. Malaysia was anticipated to have produced 320,164 MT of durian in 2013 (Payus et al., 2019). The average weight of durian fruit in the country is about 255,353 MT, and the rate at which landfills are filling up continues to increase in light of the durian garbage's massive volume and size, which necessitates an even larger area to be disposed of. There are about 30 different varieties of durians. The outer skin of the durian typically weighs more than half of the whole fruit. The fruit's edible portion is exceedingly thick and semi-woody with sharply pointed pyramidal thorns. Its color ranges from green to yellowish-brown. Just 15–30% of the fruit's mass can be eaten. If not properly disposed of, from 70% to 85% of the fruit is wasted as trash and would harm the ecosystem (Mahathaninwong et al., 2022).

As a result, a lot of durian garbage is generated each year. There will be a greater need for disposal space, particularly in landfills, and there are currently few areas where rubbish can decompose. Typically, landfills or even burning were used to dispose of this agricultural trash. As a result, it is unquestionably crucial to reuse and utilize them to eliminate this agricultural waste (Cheok et al., 2018). Few applications of durian waste can manage and reduce the volume of durian husks produced each year. The durian waste can be applied to produce compost, pulp and paper, biocomposite materials, bio-coal, biochar, concrete, biohydrogen, biomethane, bioethanol, biobutanol, natural acids, and bioelectricity.

17.6.1 Compost

Durian is a tropical fruit well-known for its powerful scent and unusual flavor. After the fruit is harvested, the leftover durian husks and seeds can be used as a source of organic material for composting. Compost is a process of decaying organic material into rich, dark, and crumbly soil amendment with the presence of oxygen, moisture, and the right balance between carbon and nitrogen. Compost also functioned as a provider of nutrient elements for plants (Sidauruk et al., 2017).

Durian waste contains a high percentage of organic matter, making it an excellent carbon source for composting. Carbon-rich materials, also known as brown materials, include dried leaves, twigs, and paper. The composting process can start when green or nitrogen-rich items, including food waste and grass clippings, are added.

Durian waste can be composted successfully with other organic elements to yield superior compost (Mahathaninwong et al., 2022). According to this study, compost quality could be improved by mixing durian shells with citrus peel scraps. Prior to being transferred to decomposers for the subsequent procedure, the durian shell and the citrus peels will be cut and chipped into smaller sizes, as shown in Figure 17.6. Decomposers made the composting process more efficient. The remaining processes in the composting process used in this study include aerobic composting or vermicomposting. The study also discovered that adding durian trash to the composting process raised the level of carbon and enhanced the compost's total nutrient content. Comparing the amount of carbon released into the atmosphere by composting versus disposing of the trash in a landfill or burning it helps (Mahathaninwong et al., 2022).

Durian compost also has wide applications in the agricultural industry. Durian shell compost acts as a fertilizer and has effective functions in enhancing the growth and yield of green cabbage as it contains 14.6% C organic matter, 1.69% N, 0.16% P_2O_5, and 1:20% K_2O (Fauzi & Puspitawati, 2017). Durian waste composting can also aid in reducing trash and advancing sustainable practices. Instead of disposing of this trash in landfills, it can be composted and turned into a useful resource for horticulture and agriculture. Utilizing durian waste in composting can enhance soil health,

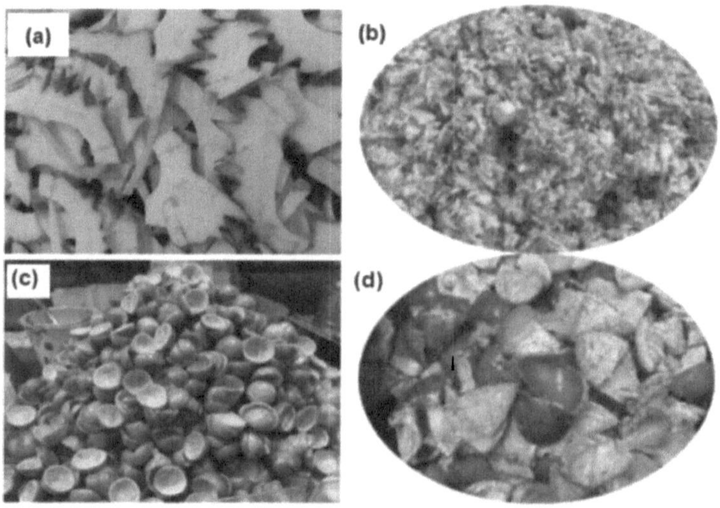

FIGURE 17.6 (a) Durian shell chips, (b) milled durian shells, (c) as-received citrus peels, and (d) citrus peels cut to a smaller size (Mahathaninwong et al., 2022).

cut waste, and encourage sustainable practices. High-quality compost may be created and used to encourage healthy plant growth by using proper composting methods and adding durian waste to the process (Tudsanaton et al., 2021).

17.6.2 Pulp and Paper

In the pulp and paper industry, durian waste, including the husks and seeds, can be used as a raw material for papermaking. Durian trash contains significant amounts of cellulose and lignin, the primary components of paper (Masrol et al., 2015). The lignin and hemicelluloses contained in durian peel and fiber have altered the characteristics of natural fibers used in papermaking in durian waste. Peels from durians have a high concentration of hemicellulose. Hemicellulose is the difference between cellulose and holocellulose. The moisture, ash, carbon, and oxygen contents of the durian shell are 11.27%, 4.84%, 39.3%, and 53.74%, respectively (Chowdhury et al., 2016).

Significant amounts of wood fiber are consumed by the pulp and paper sector, taking a toll on natural forests and accelerating deforestation. Utilizing alternate raw material sources, such as durian trash, might lessen the industry's environmental impact.

The potential of durian waste as a raw material for pulp and papermaking has been investigated in various research. According to a study by Masrol et al. (2015), chemi-mechanical pulping can be used to turn durian husk into pulp. Before moving on to the chemo-mechanical pulping stage, the durian rinds went through a washing, slicing, and drying procedure. Then, as illustrated in Figures 17.7–17.10, the durian rinds underwent a chemo-mechanical pulping process that includes pulping, moisture content analysis, sodium hydroxide treatment, refiner mechanical pulping, screening test, spin-drying, and paper sheet fabrication. Next, a typical test was conducted to examine the paper's structural, mechanical, and optical characteristics. The research revealed that durian husk pulp's

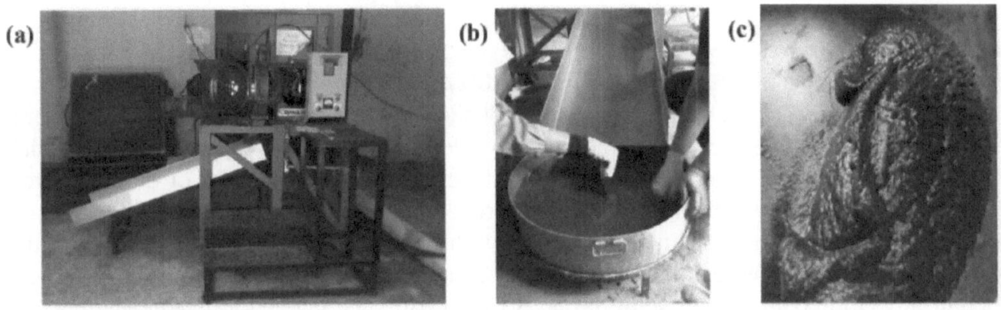

FIGURE 17.7 Refiner mechanical pulping process: (a) refiner mechanical pulping machine (RMP), (b) refined durian rinds screening, and (c) refined durian rinds (Masrol et al., 2015).

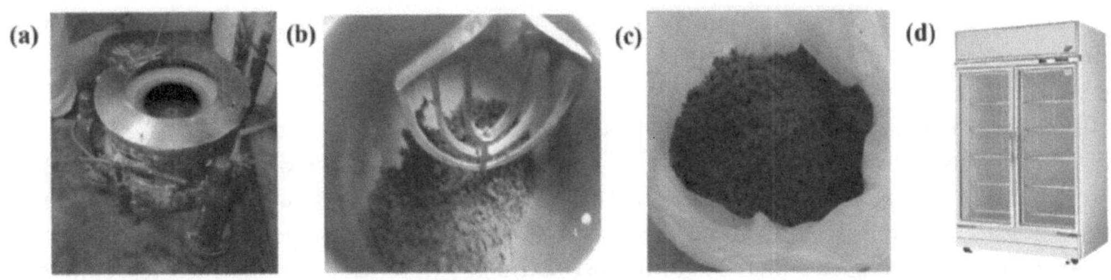

FIGURE 17.8 Spin-drying process: (a) spin dryer, (b) Hobart mixer, (c) durian rinds Chemico-Mechanical Pulping (CMP) pulp, and (d) storage chiller (Masrol et al., 2015).

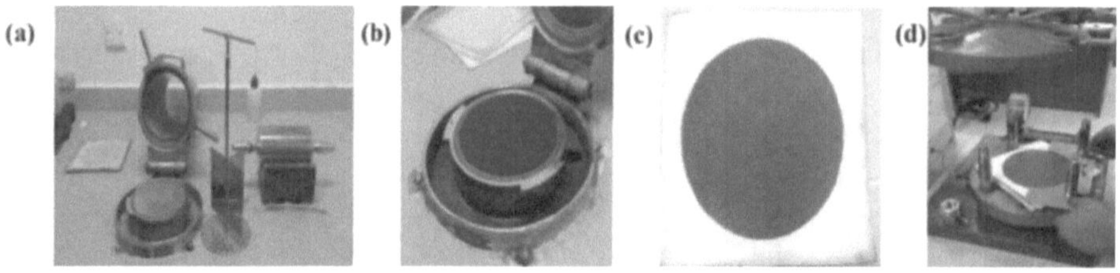

FIGURE 17.9 Paper sheet preparation process: (a) paper sheet former, (b) fiber sheet, (c) paper sheet on blotter paper, and (d) press machine (Masrol et al., 2015).

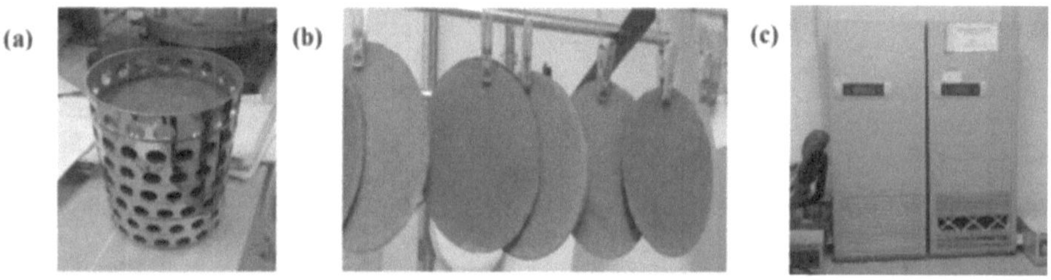

FIGURE 17.10 Paper sheet preparation (a) dry ring, (b) durian rind CMP paper, and (c) control room conditioning (Masrol et al., 2015).

characteristics were comparable to those of wood pulp, making it a suitable replacement for wood fiber in papermaking (Masrol et al., 2015).

Another study by Masrol et al. (2018) evaluated the feasibility of using durian seed as a source of lignocellulosic material for papermaking by using soda-anthraquinone pulping of the durian rinds. The study found that the durian rind could be used to produce high-quality and high-elasticity paper with properties comparable to wood pulp because the bleaching and beating process enhanced the fiber-to-fiber bonding and strength (Masrol et al., 2018).

Utilizing durian waste in pulp and paper production might assist communities that generate a lot of durian waste economically in addition to offering an alternate supply of raw materials. In Malaysia, for instance, durian is a crucial agricultural product, and a lot of durian waste is generated each year. When employed in the papermaking process, durian waste can be a useful raw material source for the pulp and paper industry. Utilizing this garbage can decrease the industry's reliance on wood fiber and benefit the local communities that generate a lot of durian waste economically. To maximize the use of durian waste in papermaking and to handle any potential technical or financial hurdles, more research and development efforts are required.

17.6.3 Biocomposite Materials

The husks and seeds of durians can be used as raw material to make biocomposite products. Biocomposites are composite materials that can replace synthetic composites that are detrimental to the environment since they are made from natural fibers and a matrix ingredient (Nagalakshmaiah et al., 2019). An example of a biocomposite can be seen in Figure 17.11.

In recent years, there has been a lot of research done on the utilization of durian waste in the creation of biocomposites. Cellulose and lignin, two vital natural fibers utilized in the creation of biocomposites, are abundant in the husk and seed of the durian. A biodegradable polymer or other matrix material can be used with the natural fibers extracted from durian trash to create a composite material that is both strong and environmentally beneficial (Chee Lee et al., 2020).

Durian husk fiber's potential as a reinforcement material for biocomposite manufacturing has been examined in a study (Manshor et al., 2014). The research revealed that the fiber from durian skin had exceptional mechanical qualities and could be used to create a biocomposite material that

FIGURE 17.11 Biocomposites (Centre of Expertise Biobased Economy, 2020).

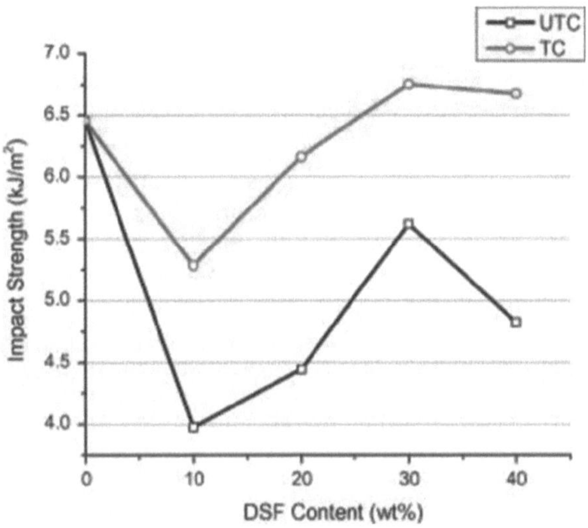

FIGURE 17.12 The effect of DSF content and pre-treatment on the impact strength of PLA–DSF biocomposite (Manshor et al., 2014).

was both strong and light. The impact and thermal properties of polylactic acid (PLA) biocomposites have been greatly improved by the invention of PLA biocomposites with natural fiber, specifically DSF. The graph in Figure 17.12 depicts how pre-treatment and DSF content affect the impact strength of PLA–DSF biocomposites. The graph demonstrates that treated biocomposites, which contain 30% of DCF as DSF, have a higher impact strength than untreated biocomposites (Manshor et al., 2014).

Another study by Anuar et al. (2017) investigated the possibilities of a plasticized PLA biocomposite film reinforced with DSF and cinnamon essential oil (CEO) for use as active food packaging. It is translucent, and the CEO's gifts to the biocomposite enhanced its barrier qualities by lowering its water vapor permeability. The antibacterial effectiveness of CEO-containing biocomposites has been demonstrated against both gram-positive and gram-negative bacteria. The translucent plasticized PLA/DSF biocomposite with CEO indicates its potential as an active food packaging material.

The use of durian waste in biocomposite production can provide several benefits, including reduced environmental impact, improved sustainability, and economic opportunities for communities that produce large quantities of durian waste. For example, durian is a significant agricultural product in Malaysia, and large amounts of durian waste are generated each year. Utilizing this waste in biocomposite production can be converted into a value-added product, creating opportunities for economic development and job creation. The use of durian waste in biocomposite production can provide several benefits, including reduced environmental impact due to its biodegradability, improved sustainability as biocomposite is recyclable, and economic opportunities for communities that produce large quantities of durian waste (Mohd Nordin et al., 2021).

17.6.4 Bio-Coal

Durian waste, such as durian husks and shells, can be used as a potential feedstock for bio-coal production. Bio-coal, or bio-briquette, also known as torrefied biomass, is a solid fuel produced from biomass feedstock through a process called torrefaction. Torrefaction is a thermal treatment process that involves heating biomass feedstock to a high temperature (200–300°C) in the absence of oxygen, which results in the removal of volatile compounds and the production of a solid, energy-dense fuel (Haryati et al., 2018). The experimental equipment scheme for the torrefaction process can be seen in Figure 17.13.

Durian waste, including the husks and shells, has the potential to be used as a feedstock for the manufacturing of bio-coal. Torrefaction is a technique used to create bio-coal, also known as bio-briquette or torrefied biomass, a solid fuel made from biomass feedstock. In the thermal treatment process of torrefaction, biomass feedstock is heated to a high temperature (200–300°C) without oxygen, removing volatile chemicals and producing a solid, energy-dense fuel (Haryati et al., 2018). Figure 17.13 depicts the experimental setup for the torrefaction procedure.

The use of durian waste in bio-coal production has been investigated in several studies. Durian husks and shells are rich in lignocellulosic materials (with 60% cellulose content), the main components of biomass feedstock used in production. The high lignin content of durian waste makes it an ideal candidate for torrefaction, as lignin is highly resistant to thermal degradation and can produce a stable, high-quality bio-coal product. The torrefaction process of durian waste is also safe for the environment as durian waste will have zero emissions during combustion and also contains a high calorific value (Crnogaca, 2017).

A study by Haryati et al. (2018) evaluated the potential of durian shells as a feedstock for bio-coal production. The study found that the torrefaction of durian shells resulted in a high-quality bio-coal product with a high energy density and low moisture content, which makes it a suitable alternative to traditional fossil fuels. Torrefaction also makes biomass hydrophobic, which means it almost never reabsorbs moisture accumulation during storage (Li et al., 2012).

Therefore, by utilizing this waste in bio-coal production, the biomass can be converted into a value-added product, creating opportunities for economic development and job creation. The massive production of bio-coal by utilizing durian waste can reduce the amount of coal needed in the industry as renewable energy fuels are now available.

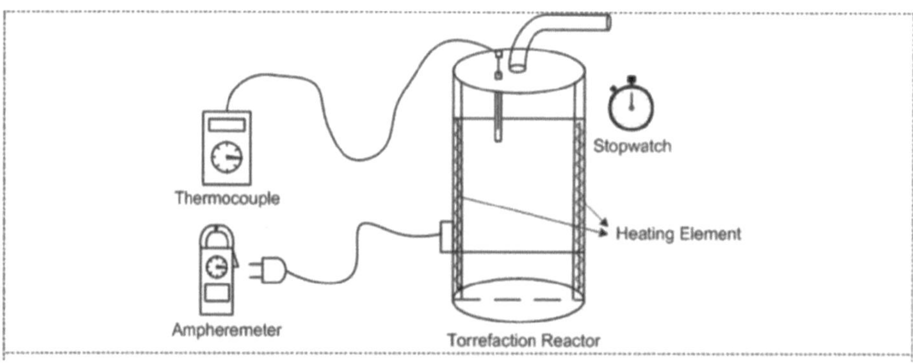

FIGURE 17.13 Experimental equipment scheme for the torrefaction process (Haryati et al., 2018).

17.6.5 Biochar

Durian waste, including the husks and shells, has the potential to be utilized as a feedstock to make biochar. Biochar is a material rich in carbon that is created by the thermal decomposition of biomass at low oxygen levels and temperatures under 700°C via pyrolysis (Nartey & Zhao, 2014). Biochar is a highly stable and carbon-rich substance with numerous uses in agriculture, environmental cleanup, and energy production (Hertsgaard, 2014).

Durian waste is an excellent source of lignocellulosic material for making biochar because of its high composition. The durian rinds will be pyrolyzed at a temperature of 325°C for 2h to make biochar. With a heating value of 28.7 MJ/kg, the biochar yield at the optimal temperature was 40.8wt%. The outcomes of the pyrolysis, therefore, demonstrated the durian rind's ability to be transformed into solid biofuel for energy applications. The biochar yield at the optimum temperature was 40.8wt%, with a heating value of 28.7 MJ/kg. The energy business can then employ solid biofuel (Selvarajoo, 2021). The experimental setup for pyrolyzing durian trash to create biochar is shown in Figure 17.14.

One of the numerous potential uses for biochar is in fertilizer or soil improvement. Numerous studies have been done on the application of biochar soil amendment to support plant development in the agriculture industry (Chua et al., 2023). Particularly, biochar made from durian shells can be used as fertilizer (Prakongkep et al., 2014; Takolpuckdee, 2014). The N, P, and K contents of durian shell biochar are outstanding, and they can significantly enhance the growth of stem length and leaf number (Takolpuckdee, 2014). Adding durian shell biochar with a ratio of durian shell biochar and soil of 3:4 offered the maximum improvement on the development of the morning glory plant when compared to adding banana peel, mangosteen, corncob, and standard soil. Figure 17.15 shows a biochar example that served as a soil improvement.

Biochar formed from durian shells can be used as fertilizer because of its chemical qualities, as is well known. Durian shells can both reduce GHG emissions and boost crop output when added to the soil. In the pores of biochar, there are crystalline minerals that can serve as plant nutrition. The major water-soluble minerals found in durian shell charcoal included sylvite (KCl), kalicinite ($KHCO_3$), archerite (KH_2PO_4), chlorocalcite ($KCaCl_3$), and struvite ($KMgPO_4 \cdot 6H_2O$). Biochar also serves as a fertilizer and an alkaline substance that balances acidic soil (Mutolib et al., 2023; Prakongkep et al., 2014). To conclude, the application of durian waste to produce biochar has significant advantages, especially in agriculture. By having this biochar in the industry, the food chain supply can also be improved globally.

17.6.6 Concrete

In recent years, the utilization of agricultural waste to produce concrete materials has gained significant attention in the construction industry as it gives the advantage of lowering construction costs and reducing environmental impact (Itam et al., 2023). Durian waste has many potential parts that can be used in cement production, including the husk, peel, and other by-products generated during durian fruit processing. In concrete production, durian shell needs to be converted to durian shell ash and mixed with the concrete mixture to function as a partial cement replacement. This helps

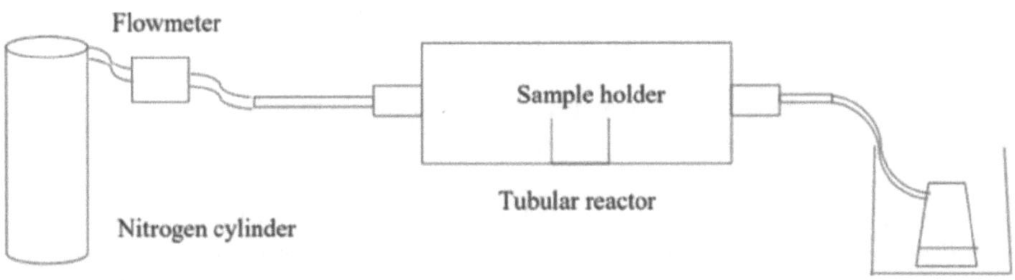

FIGURE 17.14 Experimental setup for the pyrolysis of durian waste (Selvarajoo, 2021).

FIGURE 17.15 Example of biochar functions in soil amendment (Are & Are, 2019).

minimize the environmental issue due to durian waste and the use of cement in making concrete (Itam et al., 2023). Concrete and cement bonded board from durian waste as the concrete comprises of durian waste has good compressive strength with more than 30 MPa, good fire resistance capacity, and high flexural strengths (Gonzales & Tabanguil, 2006; Itam et al., 2023).

A study also found that a mixture of durian pectin and rice husk ash (RHA) can temporarily substitute cement in concrete (Chia et al., 2020). Durian pectin and RHA act as supplementary cementitious materials. Durian pectin was used in cement as it has the same characteristics, which are that it is able to hydrate and form a paste when water is added to it. RHA has the same chemical properties as cement, which they have calcium oxide (CaO), silica dioxide (SiO_2), aluminum oxide (Al_2O_3), magnesium oxide (MgO), iron oxide (Fe_2O_3), and sulfur trioxide (SO_3). It is also able to increase the strength and helps reduce the voids in concrete (Chia et al., 2020).

The process of converting durian waste into concrete involves several key steps. Initially, the waste materials are collected and sorted to remove impurities. Subsequently, the durian husks and paddy are mechanically processed to extract the durian pectin and RHA, which contribute to the strength and stability of the concrete. Extracted durian pectin and RHA are then mixed with a suitable binder material, such as cement, sand, and coarse aggregate, to hold them together. This mixture is molded into concrete cubes using molds or presses and left to cure and dry, allowing the binder to harden and create structurally sound bricks (Chia et al., 2020).

The utilization of durian waste for concrete production offers several noteworthy benefits. First, it provides an eco-friendly solution to manage a significant agricultural waste stream, reducing environmental pollution and landfill accumulation. Second, the incorporation of durian fibers enhances the strength and durability of the concrete, making them suitable for construction purposes. Additionally, concrete derived from durian waste can contribute to a reduction in carbon emissions compared to traditional brick production methods, thus mitigating climate change impacts (Damasco et al., 2022).

17.6.7 Biohydrogen

Durian waste, such as durian husks and shells, can be a potential feedstock to produce biohydrogen. Biohydrogen is a renewable and clean energy source produced through the fermentation of organic materials, such as agricultural waste, forestry residues, and other organic waste materials.

Durian waste is a suitable feedstock for biohydrogen production due to its high content of carbohydrates and lignocellulosic materials. Carbohydrates are easily fermentable and can be converted to hydrogen gas through microbial processes. At the same time, lignocellulosic materials provide structural support and contribute to the stability of the biohydrogen production process (Cheok et al., 2018).

Several studies have investigated the potential of durian waste as a feedstock for biohydrogen production. A study evaluated the potential of durian shell waste for biohydrogen production using a mixed culture of bacteria. The study found that durian shell waste is a suitable feedstock for biohydrogen production, with a high yield of hydrogen gas and a low production of methane (Zhao et al., 2017).

Another study investigated the use of durian peel waste as a feedstock for biohydrogen production. The study found that durian peel waste had high contents of carbohydrates and lignocellulosic materials, resulting in a high hydrogen gas yield during the fermentation process (Ab Rasid et al., 2020). Ozone pre-treatment functions in enhancing the production of biohydrogen. Durian lignin was effectively degraded, and ozonolysis helped reduce sugar yield (Ab Rasid et al., 2020).

The use of durian waste in biohydrogen production can provide several benefits, including reducing GHG emissions and providing a renewable and sustainable energy source. In addition, using durian waste for biohydrogen production can provide an economic opportunity for communities that produce large amounts of durian waste (Osman et al., 2020).

17.6.8 Biomethane

Biomethane production can be done from anaerobic co-digestion of durian shells with chicken, dairy, and pig manures (animal manures). The main technique for producing biomethane from organic waste is anaerobic digestion (AD). In this method, durian waste is put in a digester, a closed tank where microorganisms decompose the waste without oxygen. As a result, methane (CH_4) and carbon dioxide (CO_2), which make up the majority of biogas, are produced. AD has been proven to be an economical method that converts organic wastes (durian waste) into biogas/biomethane (mainly methane and carbon dioxide) by the action of specific microorganisms in the absence of atmospheric oxygen. AD often produces biogas that need to be improved to add more methane since it contains contaminants. Purified biomethane can be created via upgrading techniques, including pressure swing adsorption, water washing, or membrane separation by removing impurities, typically CO_2 (Werkneh, 2022). The purified biomethane is compressed to the required pressure and stored in tanks or pipelines to be distributed and used as a renewable fuel.

The residue obtained from the AD process serves as a stable organic material suitable for use as a biofertilizer. Despite its efficiency, AD has limitations, such as the accumulation of volatile fatty acids, which can disrupt methanogenic activity and affect process stability. Durian wastes (durian shells) have been recognized as viable feedstock for AD with positive methane potential, but their high lignocellulose content can create a nutrient imbalance for microorganisms in the AD process. Similarly, the AD of animal manures faces challenges due to their high nitrogen content (Wang et al., 2023).

To address these issues, the co-digestion of durian wastes and animal manures has been found to be effective in improving nutrient balance, providing synergistic effects, and enhancing the specific methane yield of substrates. The microbial community structure in the AD process directly influences its performance and efficiency. However, the specific microbial community structure of co-digestion systems involving durian shells and animal manures is poorly understood. To explore this further, high-throughput sequencing (HTS) has been widely employed to analyze microbial community structures in various settings, including anaerobic digesters, food, groundwater, and activated sludge. HTS employs the principle of the pyrophosphate cascade reaction, enabling the identification of microorganisms at a deeper level with high throughput and efficiency (Wang et al., 2023).

Natural gas can be replaced by biomethane in a variety of applications. It can be used to generate energy and heat, be pumped into the natural gas system, or used as a fuel for automobiles (like compressed natural gas vehicles) (Ammenberg et al., 2021). Reducing trash and environmental effects, producing renewable energy, and promoting a circular economy are all advantages of the conversion of durian waste to biomethane. Waste from durians is converted to biomethane to assist in reducing the build-up of organic waste and decrease open burning or landfilling, both of which can increase environmental pollution and GHG emissions. A sustainable and renewable energy source, biomethane created from durian waste lessens reliance on fossil fuels and aids in the shift to a low-carbon economy. Through the use of garbage as a valuable resource, the reduction of waste generation, and the creation of a clean energy source, the conversion of durian waste into biomethane advances the idea of a circular economy (Sharma et al., 2021).

It should be noted that the precise method and process utilized to convert durian waste to biomethane may differ based on the size of the plant, the resources available, and local laws. To determine whether adopting such a waste management system is feasible, feasibility studies and local expertise should be used.

17.6.9 Bioethanol

Bioethanol, an eco-friendly alternative energy source, serves as a substitute for petroleum. The main ingredients for bioethanol production include sucrose, starch, and fibrous cellulose (lignocellulose). It proves cost-effective and environmentally beneficial, making it a viable replacement for fossil fuels in transportation, with lower carbon emissions per liter compared to high bioethanol content (Balat & Balat, 2009). Bioethanol offers several advantages, as it can be produced from various plant sources containing sugar and starch. During the bioethanol production process, the carbon dioxide released is neutralized by the same amount of carbon the plants absorbed during photosynthesis, contributing to the overall reduction of ozone depletion in the atmosphere, which is of environmental significance. Ethanol's combustion emissions are less reactive with sunlight than gasoline, reducing the risk of ozone damage (Sun et al., 2019). Additionally, bioethanol-blended gasoline minimizes GHG emissions, and fuel spills are more biodegradable or diluted to non-toxic levels. Ethanol's exhaust gases are notably cleaner since it is derived from renewable agricultural waste rather than finite resources. Blending ethanol with petrol extends the life of depleting oil supplies and enhances global fuel security. Moreover, bioethanol is less harmful and biodegradable than fossil fuels, making it a more sustainable and environmentally friendly choice (Nichodemus, 2017).

The resource is readily available and abundant in agricultural crop waste. In Malaysia, durian seeds represent one such agricultural waste that has not been fully utilized. Despite their high carbohydrate content, these durian seeds are currently discarded as waste and remain untapped for commercial purposes. However, they hold the potential to serve as a promising new source for bioethanol production (Masturi et al., 2020). The carbohydrate content of durian seeds ranges from 50% to 70%. However, these seeds have not yet been exploited as a viable carbohydrate source for food or starch production. Due to its significant carbohydrate content, specifically an amylose content of 20.8%, durian seeds possess the potential to serve as a viable raw material for the production of bioethanol (Purnomo et al., 2016).

The general outline of the procedure for turning durian waste into bioethanol is as follows. First, waste must be collected and prepared, like when making biomethane. Durian waste, including peels, seeds, and other discarded parts, must be gathered and properly sorted for processing. The complex carbohydrates in durian waste must then undergo pre-treatment to be converted into simpler sugars for ethanol fermentation (Aditiya et al., 2016). Pre-treatment techniques may include mechanical actions like shredding or grinding to enhance the release of sugars, as well as chemical or enzymatic treatments. Enzymatic hydrolysis is utilized to transform complex carbohydrates, like cellulose and hemicellulose, into fermentable sugars, primarily glucose, from the pre-treated durian waste (Obeng et al., 2021). The fermentable sugars obtained from the enzymatic hydrolysis are then fermented using specific strains of yeast, which convert the sugars into ethanol (Obeng et al., 2018). The fermentation process typically occurs in large vessels under controlled temperature and pH conditions.

After fermentation, a small amount of ethanol, water, and other by-products are present in the mixture. The mixture is distilled to separate and concentrate the ethanol from the water. To attain a higher ethanol concentration, distillation is often carried out utilizing numerous distillation columns. Typically, a little amount of water is included in the ethanol produced by distillation. To get anhydrous (pure) ethanol, water is removed using dehydration procedures like molecular sieves or azeotropic distillation (Kumar et al., 2010). Depending on the bioethanol's intended application, it might be denatured by a small amount of petrol or other chemicals. The ethanol can be used for industrial or fuel purposes after denaturing it to render it unfit for human consumption. The produced bioethanol can be combined with petrol to make ethanol–gasoline blends for use in cars (Chakraborty & Mukhopadhyay, 2020). It can also be further processed to create other compounds produced from ethanol or utilized as a feedstock for a variety of industrial purposes.

17.6.10 Biobutanol

The same feedstocks as ethanol are used to make butanol, which are 4-carbon alcohol (butyl alcohol), durian waste, and other biomass. Biobutanol is the name given to butanol produced using biomass as a feedstock. Biobutanol is a renewable fuel that can be blended into fuel formulations or used as an alternative to petrol. The complex carbohydrates in durian waste often must be broken down into simpler sugars before fermentation can produce biobutanol. Pre-treatment techniques may include physical procedures like shredding or grinding to enhance the release of sugars, as well as chemical or enzymatic treatments. The pre-treated durian waste is subjected to enzymatic hydrolysis, where specific enzymes convert complex sugars, such as cellulose and hemicellulose, into fermentable sugars, primarily glucose (Torres-Mayanga et al., 2019).

Enzymatic hydrolysis yields fermentable sugars, which are then fermented by particular bacterial strains, such as *Clostridium acetobutylicum*, to produce biobutanol. Large fermentation tanks with regulated pH and temperature levels are commonly used for the fermentation process. Following fermentation, biobutanol, water, and other by-products are produced. The mixture is distilled to concentrate and separate the biobutanol from the remaining components. To achieve a greater biobutanol concentration, distillation is frequently carried out utilizing several distillation columns (Green, 2011). Additional purification processes may be applied to the concentrated biobutanol obtained from distillation to remove contaminants and reach the appropriate purity level. Adsorption or solvent extraction are two possible purification methods. There is often a minor amount of water in the purified biobutanol. The remaining water is eliminated using dehydration techniques like molecular sieves or azeotropic distillation to produce anhydrous (pure) biobutanol. When combined with petrol to form biobutanol–gasoline blends, the resulting biobutanol can be used as a fuel additive; it can also be used to make other butanol-derived compounds as a precursor.

By effectively utilizing a by-product that would otherwise be thrown away and converting durian waste to biobutanol, waste can be reduced, and a circular economy can be fostered. A renewable fuel made from durian waste, biobutanol, can aid in the shift to a more sustainable energy system by lowering GHG emissions. Particularly in durian-producing regions, converting durian waste to biobutanol can open up economic opportunities, including creating jobs and developing bioenergy enterprises. One advantage of biobutanol is that it has more energy. Among petrol substitutes, biobutanol has a comparatively high energy content. However, the energy density of biobutanol is 10–20% lower than that of petrol. In comparison to ethanol, biobutanol has a lower Reid vapor pressure, which results in less volatility and evaporative emissions (Obergruber et al., 2021). Additionally, it has improved energy security because biobutanol can be produced locally from a range of feedstocks and produces fewer emissions when used in place of petroleum fuels. Balancing the carbon dioxide emitted during biobutanol combustion with the carbon dioxide absorbed during feedstock growth decreases overall GHG emissions. Because biobutanol and water are insoluble, they could be transported through pipes to save money on transportation (Wu et al., 2008).

17.6.11 Natural Acids

Converting durian waste to natural acids is another potential avenue for waste management and value addition. Durian waste, which consists of peels, seeds, and other discarded parts, can be processed to extract and produce natural acids. Waste from durian processing plants or marketplaces is collected and properly sorted for further processing. It is necessary to clean and prepare the garbage to eliminate any impurities or toxins. The natural acids found in durian waste can be isolated using a variety of extraction techniques, such as solvent extraction or enzymatic extraction. These procedures entail removing the specific acids from the waste material using the proper solvents or enzymes. After extraction, the natural acid-containing solution is put through purifying procedures to eliminate impurities and undesired substances. A pure natural acid solution can be made using methods like chromatography, distillation, or filtration (Mellinas et al., 2020).

To increase the concentration of the desired acids, the purified natural acid solution is concentrated. Techniques like vacuum distillation or evaporation can be used to get rid of extra water and reach the appropriate concentration level. Analysis and characterization are performed to identify the specific types and amounts of acids present in the concentrated natural acid solution. Quality control measures are implemented to guarantee the uniformity and purity of the natural acid product. Diverse industries use natural acids made from durian waste, including food and beverage, cosmetics, medicines, and agriculture. They can be utilized in a variety of products as active compounds, flavoring agents, preservatives, or pH regulators (Ben-Othman et al., 2020).

As an illustration, lactic acid serves multiple functions as a food preservative, curing agent, and flavoring agent. It is commonly utilized in processed foods as a decontaminant during meat processing. Commercial production of lactic acid involves the fermentation of carbohydrates like glucose, sucrose, or lactose or by chemical synthesis. Lactic acid can also be synthesized from durian waste, as ripe durian seeds contain 51.1% water, 46.2% carbohydrates, 2.5% protein, and 0.2% fat, making their high carbohydrate content suitable for lactic acid fermentation (Abdullah et al., 2021). Lactic acid holds significant importance due to its broad applications, particularly in the food, chemical, cosmetic, and pharmaceutical industries. Microbial fermentation offers advantages in lactic acid production, resulting in high purity levels (90–95%). *Lactobacillus plantarum*, a homofermentative lactic acid bacteria, is expected to yield abundant lactic acid as it is the primary metabolite. The fermentation process is notably influenced by various operating conditions, such as pH, temperature, substrate concentration, inoculum concentration, and the addition of other nutrients (Abdullah et al., 2021).

Furthermore, lactic acid possesses significant promise in the production of PLA, a biodegradable and biocompatible polymer. PLA, a type of bioplastic, is synthesized from natural materials like corn starch, tapioca root, and sugarcane (Lee et al., 2018). Notably, PLA, being derived from renewable resources, exhibited biodegradable properties, excellent mechanical strength, and is non-toxic. The versatility of PLA allows for its application in a wide range of products, including packaging, fibers, and foams, making it a sustainable choice for various industries. The utilization of a by-product that would otherwise be thrown away, waste reduction, and the advancement of a circular economy are all advantages of converting durian waste to natural acids. Natural acids are precious commodities that can be marketed as valuable goods. Opportunities for commercial expansion and revenue generation are created by the extraction and manufacturing of natural acids from durian trash (Ahmad Khorairi et al., 2021). A sustainable method that lessens reliance on synthetic or chemically generated acids is using durian waste to manufacture natural acids, which helps create a more sustainable and renewable resource base. Diverse goods supporting health and wellness can employ natural acids from durian waste.

17.6.12 Bioelectricity

Waste from durians may be effectively managed and used to produce renewable energy by turning it into bioelectricity. Electricity produced by the burning or gasification of organic materials is known as bioelectricity, biomass electricity, or biopower (Lee et al., 2019). The following gives a general description of the steps necessary to turn durian waste into bioelectricity. To proceed with processing, durian trash, including peels, seeds, and other leftover pieces, is first gathered and correctly sorted. To reach the highest combustion or gasification efficiency level, the waste must be cleaned and dried.

Combustion and gasification are the two main processes for turning durian waste into bioelectricity. Durian waste is burned in a boiler or furnace during combustion to produce heat. The steam created by the heat is then utilized to power a steam turbine connected to a generator to provide electricity. In a controlled setting with little oxygen, gasification partially burns durian waste. Syngas is the result of this process, and it mostly contains carbon monoxide (CO), hydrogen (H_2), and methane (CH_4). After that, a generator, gas turbine, or internal combustion engine is connected to the syngas to create power. The generated heat or syngas is utilized to power a turbine or engine, coupled to a generator, in both the combustion and gasification processes. Electricity is created by the generator using mechanical energy (Alves et al., 2020). The generated electricity is often used locally to power adjacent facilities, such as durian processing plants, agricultural operations, or towns, or fed into the local electrical system.

Benefits of turning durian waste into bioelectricity include keeping organic waste out of landfills, lessening the environmental impact of trash decomposition, and perhaps reducing methane emissions. A renewable energy source that lessens reliance on fossil fuels and aids in the decarbonization of the electrical sector is bioelectricity produced from durian waste (Kpalo et al., 2020). Locally produced bioelectricity from durian waste can act as a decentralized energy source, fostering resilience and energy self-sufficiency in durian-producing regions. In some instances, the waste heat produced during bioelectricity production can be used for heating or other industrial activities, increasing the system's efficiency through the formation of combined heat and power (CHP). It is important to consider the environmental impacts and emissions associated with combustion or gasification processes. Proper emission control measures, such as flue gas cleaning systems, should be implemented to minimize air pollutants and ensure compliance with local regulations.

17.7 POTENTIAL APPLICATIONS

The synthesis of valuable products from durian waste presents numerous potential applications. Composts derived from durian waste can find utility in agronomic and horticultural practices as a soil amendment, seed starter, mulch, container mix ingredient, or natural fertilizer, depending on their specific characteristics (Dam, 2022). Composting also effectively reduces or eliminates weed seeds and plant pathogens in organic residues. Biocomposites made from local and renewable resources offer significant sustainability benefits. The next generation of materials, products, and processes is guided by principles of industrial ecology, eco-efficiency, and green chemistry (Ortega et al., 2022). The use of biocomposites has notably grown in the domestic sector, building materials, aerospace industry, circuit boards, and automotive applications over the past decade. However, their application in other sectors has been limited until now.

Bio-coal derived from biomass like durian waste shows promise as a substitute for fossil coal, aiding in achieving net-zero GHG emissions in the industrial sector. To produce bio-coal with a quality comparable to fossil coal, high-temperature pyrolysis at temperatures above 500°C is required. However, this process faces challenges as the production efficiency is relatively low due to the limited bio-coal yield at high pyrolysis temperatures (Chua et al., 2023). This trade-off raises concerns about the economic viability of bio-coal production. Initially, biochar was primarily created for soil applications, focusing on carbon storage or sequestration, improving soil performance by enhancing nutrient availability, reducing soil compaction, and enhancing soil pH. It was also employed for soil water filtration (Armah et al., 2023). In recent times, biochar has found various applications, including energy production, enhancing biochemical process stability, mitigating climate change, and serving as a construction additive (Figure 17.16).

Using durian waste to create bio-bricks for construction purposes can lead to a reduction in the consumption of natural resources and energy. Additionally, this practice can provide an extra source of income for farmers who can sell the leftover durian stubbles instead of burning them, thus mitigating air pollution. Biomethane production from durian waste has emerged as a significant energy source, supported by private and public organizations seeking sustainable alternatives (Cryospain, 2022). This biomethane is converted into electricity and heat, proving to be a more efficient source than biogas. Consequently, it opens up various possibilities for industrial applications, including serving as a transportation fuel to replace compressed natural gas, diesel, and liquid natural gas, generating steam and heat, facilitating CHP production, and injecting into the natural gas grid (Bujang et al., n.d.) (Figure 17.17).

In the industrial sector, bioethanol has the potential to serve as a viable alternative to petrol for transportation purposes. Additionally, it can be utilized as a fuel source for power generation through thermal combustion, as well as for fuel cells via thermochemical reactions. Furthermore, bioethanol can be employed as a fuel in cogeneration systems and as a chemical industry feedstock (Pirolini, 2020). Pure bioethanol and ethanol–gasoline blends are widely used as alternative fuels in automotive vehicles. When used as a substitute for gasoline in spark ignition engines, the brake thermal efficiency can be improved by increasing the compression ratio without pre-ignition to

FIGURE 17.16 Bio-coal (Yirka, 2020).

FIGURE 17.17 Bio-brick (Deccan Herald, 2019).

take advantage of the ethanol's higher octane rating. Also, greater latent heat of vaporization provides a higher charge density than petroleum gasoline. However, it is true that ethanol fuel still has limitations in applying ethanol to engines and is widely used for automobile applications because of a cold-start problem resulting from modifications of the engine design and fuel supply system in the engines. Moreover, the relatively low Reid vapor pressure compared with gasoline poses a challenge in spite of many advantages of ethanol (Yoon et al., 2009). However, since the energy yield of ethanol is lower than that of petrol, it is recommended to utilize bioethanol in producing biodiesel rather than replacing conventional petrol (Alam & Tanveer, 2020). The utilization of durian waste has been found to have the potential to produce biobutanol, which is considered a viable alternative to conventional petroleum-derived fuels due to its increased energy density and molecular resemblance to petrol. The current fuel engines can more easily utilize it, either by blending it with diesel due to its favorable intersolubility or by using it without requiring any modifications (Nawkarkar et al., 2021).

Lactic acid, a naturally occurring acid, can be derived through the processing of durian waste materials. Lactic acid finds application in the cosmetic industry for producing hygiene and aesthetic products due to its skin moisturizing, antimicrobial, and rejuvenating properties. Additionally, it is utilized in the manufacturing of oral hygiene products. Lactate esters, which are lactic acid derivatives, are extensively employed due to their hygroscopic and emulsifying characteristics. Within the pharmaceutical sector, it serves as a supplementary component in the synthesis of dermatological medications and is employed in the treatment of osteoporosis (Castillo Martinez et al., 2013).

Around 70% of the total lactic acid production is allocated for utilization within the food industry, primarily owing to its significant involvement in the manufacturing processes of yogurt and cheese. The co-fermentation of *Streptococcus thermophilus* and *Lactobacillus bulgaricus* is the primary process involved in the production of yogurts. The process of cheese production involves a decrease in pH due to the release of lactic acid, which subsequently leads to the aggregation of casein micelles. In certain cases, the utilization of direct acidification with lactic acid is employed to mitigate the potential growth of undesirable microorganisms based on the desired sensory attributes of the end product. In grain production, lactic acid is generated spontaneously due to the activity of microorganisms that engage in the process of lactic acid fermentation of the raw material, such as the wet processing of corn. This fermentation process results in alterations in the aroma and taste of the preparations while concurrently lowering the pH level, thereby inhibiting the proliferation of pathogenic bacteria (Castillo Martinez et al., 2013).

17.8 CHALLENGES ASSOCIATED WITH PROCESSING AND COMMERCIALIZATION

Durian waste processing and commercialization can be difficult in some ways. Waste collection and handling, waste characterization and variability, technology selection and scale-up, economic viability and market demand, environmental considerations, as well as public perception and social acceptance are some of the major challenges connected with the management and utilization of durian waste. Durian trash is made up of a variety of pieces, including peels, seeds, and other left-overs that must be collected skillfully and successfully. Infrastructure, logistics, and coordination with durian processing plants, marketplaces, or households are necessary for the gathering procedure. It can be difficult to guarantee a steady trash supply, particularly during the off-season or in areas with little durian production (Safari et al., 2021).

The content of durian waste can change based on the durian variety, ripeness, and processing methods. The variable content of durian trash may impact the efficiency and effectiveness of waste conversion operations. For processing technology optimization and assuring consistent quality and yield of finished products, thorough waste classification and understanding of its variability are essential (Mohd Ali et al., 2020). It is important to carefully think while choosing the right methods for turning durian waste into products with added value. Technologies for conversion include AD, fermentation, combustion, and gasification. Various criteria, including waste composition, desired end products (such as biomethane, bioethanol, and bioelectricity), scalability, cost-effectiveness, and local infrastructure, must be considered when selecting a technique. There may be technical and financial difficulties when transferring these technologies from pilot or laboratory to commercial operations.

The cost of waste collection, processing methods, the value of the finished product, and market demand are some variables that affect how economically viable durian waste utilization is. Establishing a sustainable business model requires evaluating the future market demand and identifying workable uses for the final products (such as biomethane, bioethanol, and bioelectricity). Market conditions, laws, and subsidies for renewable energy or waste management can impact the commercial viability of durian waste processing initiatives.

Processing durian waste requires careful environmental control. Some conversion techniques, including combustion or gasification, may emit air pollutants or call for proper ash disposal. To reduce environmental consequences, emission control methods and adherence to environmental rules must be considered (Kusumaningtyas & Syah, 2020). The ecology must also be protected by making sure that the waste management procedure does not contaminate the land or the water. The public view of novel waste management and utilization techniques may be problematic, particularly when they involve odoriferous or possibly divisive trash, such as durian waste. Gaining community support and acceptance requires addressing issues with odor, noise, visual impact, and potential health risks through good communication.

17.9 CONCLUSIONS

In response to the growing need for food sources that alleviate hunger and support overall human health, there is an increasing demand worldwide. Despite containing valuable bioactive compounds, fruit waste is unfit for direct consumption due to its undesirable bitter taste. Consequently, converting fruit waste into an alternative food source poses a significant challenge for researchers and food manufacturers. Nevertheless, the search for alternative food sources, particularly from fruit waste, has become essential due to the current imbalance between global population growth and available food resources. The management of durian waste offers both opportunities and challenges. Durian waste, which consists of peels, seeds, and other leftover pieces, can be properly managed and used through a variety of procedures like the creation of biomethane, the conversion of bioethanol, or the creation of bioelectricity. Several advantages can be realized by putting good waste management practices into practice by keeping organic waste out of landfills, reducing methane emissions, and having a minimal negative impact on soil and water resources, durian waste management aids in the reduction of environmental pollution. By turning garbage into useful resources, it encourages a circular economy. Renewable energy is produced by converting durian waste into biomethane, bioethanol, or bioelectricity. This lessens the need for fossil fuels, lowers GHG emissions, and aids in the shift to a low-carbon, more sustainable energy system. Durian waste management makes it possible to effectively use a by-product that would otherwise be thrown away. This offers regions that produce durians the chance to prosper economically, create jobs, and start bioenergy companies. Despite the potential benefits, durian waste management faces challenges such as waste collection and characterization, technology selection, economic viability, and social acceptance. Addressing these challenges requires collaboration among stakeholders, feasibility studies, and the adoption of appropriate technologies. The durian fruit contains only 20.52% edible flesh, leaving approximately 79.48% as unused parts like the shell and seeds, resulting in around 556,360 tons of waste annually. The difficulty of degrading durian waste, particularly the shell, highlights the importance of effective waste management. Although durian peels are currently used for biosorbent, activated carbon, and biocomposite purposes, it is suggested that their potential as a source of health-promoting bioactive compounds like dietary fiber and pectin be further investigated. Overall, managing durian waste can help make resources more sustainable and effective, lessen the environmental impact, and open the door to the creation of renewable energy. To maximize the efficiency and viability of durian waste management, local factors, laws, and market demand must be taken into account.

ACKNOWLEDGMENT

This chapter is written with Nur Ariena Izzaty Sulaiman, Sanjay Devan, and Faisal Ardianshah Adinegara.

REFERENCES

Ab Rasid, N. S., Zainol, M. M., & Amin, N. A. S. (2020). Pretreatment of agroindustry waste by ozonolysis for synthesis of biorefinery products. In: *Refining Biomass Residues for Sustainable Energy and Bioproducts: Technology, Advances, Life Cycle Assessment, and Economics*. pp. 303–336. https://doi.org/10.1016/B978-0-12-818996-2.00014-4

Abdullah, A., Winaningsih, I., & Hadiyarto, A. (2021). Lactic acid fermentation from durian seeds (Durio zibethinus Murr.) using *Lactobacillus plantarum*. *IOP Conference Series: Materials Science and Engineering*, *1053*(1), 012032. https://doi.org/10.1088/1757-899x/1053/1/012032

Aditiya, H. B., Mahlia, T. M. I., Chong, W. T., Nur, H., & Sebayang, A. H. (2016). Second generation bioethanol production: A critical review. *Renewable and Sustainable Energy Reviews*, *66*, 631–653. Elsevier Ltd. https://doi.org/10.1016/j.rser.2016.07.015

Affandi, Lististio, D., Santoso, P. J., Mansyah, E., & Sartiami, D. (2019). Identification, distribution and abundance of scale insect associate with several clones of durian (Durio zibethinus merr.). *Agrivita*, *41*(2), 206–212. https://doi.org/10.17503/agrivita.v41i2.2162

Ahmad Khorairi, A. N. S., Sofian-Seng, N. S., Othaman, R., Abdul Rahman, H., Mohd Razali, N. S., Lim, S. J., & Wan Mustapha, W. A. (2021). A review on agro-industrial waste as cellulose and nanocellulose source and their potentials in food applications. *Food Reviews International*, *39*, 663–688. https://doi.org/10.1080/87559129.2021.1926478

Alam, M. S., & Tanveer, M. S. (2020). Conversion of biomass into biofuel: A cutting-edge technology. In: *Bioreactors*. pp. 55–74. https://doi.org/10.1016/b978-0-12-821264-6.00005-x

Alves, J. L. F., da Silva, J. C. G., Mumbach, G. D., Domenico, M. Di, da Silva Filho, V. F., de Sena, R. F., Machado, R. A. F., & Marangoni, C. (2020). Insights into the bioenergy potential of jackfruit wastes considering their physicochemical properties, bioenergy indicators, combustion behaviors, and emission characteristics. *Renewable Energy*, *155*, 1328–1338. https://doi.org/10.1016/j.renene.2020.04.025

Amid, B. T., & Mirhosseini, H. (2012). Influence of different purification and drying methods on rheological properties and viscoelastic behaviour of durian seed gum. *Carbohydrate Polymers*, *90*(1), 452–461. https://doi.org/10.1016/j.carbpol.2012.05.065

Ammenberg, C., Gustafsson, J., O'shea, M., Gray, R., Lyng, N., Eklund, K.-A., & Murphy, M. D. (2021). Perspectives on biomethane as a transport fuel within a circular economy, energy, and environmental system. In J. Ammenberg & J. D. Murphy (Eds.), *IEA Bioenergy Task* (Vol. 37) Report 2021:12. IEA Bioenergy, Paris, France.

Anuar, H., Nur Fatin Izzati, A. B., Sharifah Nurul Inani, S. M., Siti Nur E'zzati, M. A., Siti Munirah Salimah, A. B., Ali, F. B., & Manshor, M. R. (2017). Impregnation of cinnamon essential oil into plasticised polylactic acid biocomposite film for active food packaging. *Journal of Packaging Technology and Research 2017 1:3, 1*(3), 149–156. https://doi.org/10.1007/S41783-017-0022-1

Are, K. S., & Are, K. S. (2019). Biochar and soil physical health. In: *Biochar - An Imperative Amendment for Soil and the Environment*. https://doi.org/10.5772/INTECHOPEN.83706

Armah, E. K., Chetty, M., Adebisi Adedeji, J., Erwin Estrice, D., Mutsvene, B., Singh, N., & Tshemese, Z. (2023). Biochar: Production, application and the future. In: *Biochar - Productive Technologies, Properties and Applications*. https://doi.org/10.5772/intechopen.105070

Aziz, N. A. A., & Jalil, A. M. M. (2019). Bioactive compounds, nutritional value, and potential health benefits of indigenous durian (Durio zibethinus Murr.): A review. *Foods, 8*(3), 96. MDPI Multidisciplinary Digital Publishing Institute. https://doi.org/10.3390/foods8030096

Balat, M., & Balat, H. (2009). Recent trends in global production and utilization of bio-ethanol fuel. *Applied Energy, 86*(11), 2273–2282. Elsevier Ltd. https://doi.org/10.1016/j.apenergy.2009.03.015

Ben-Othman, S., Jõudu, I., & Bhat, R. (2020). Bioactives from agri-food wastes: Present insights and future challenges. *Molecules, 25*(3). MDPI AG. https://doi.org/10.3390/molecules25030510

Brown, M. J. (1997). *Durio, A Bibliographic Review*. International Plant Genetic Resources Institute, Office for South Asia.

Bujang, A. S., Bern, C. J., Brumm, T. J., Rosentrater, K. A., Richard, M. M.-W., & Gladon, J. (n.d.). *Utilization of durian biomass for biorenewable applications*.

Castillo Martinez, F. A., Balciunas, E. M., Salgado, J. M., Domínguez González, J. M., Converti, A., & Oliveira, R. P. (2013). Lactic acid properties, applications and production: A review. *Trends in Food Science & Technology, 30*(1), 70–83. https://doi.org/10.1016/j.tifs.2012.11.007

Centre of Expertise Biobased Economy. (2020). *Biocomposites - MaterialDistrict*. Retrieved June 10, 2023, from https://materialdistrict.com/material/biocomposites/

Chakraborty, R., & Mukhopadhyay, P., (2020). Green fuel blending: A pollution reduction approach. In: Hashmi, S., Choudhury, I. A. (Eds.), *Encyclopedia of Renewable and Sustainable Materials*. Elsevier, pp. 487–500. https://doi.org/10.1016/B978-0-12-803581-8.11019-7.

Chandra, T. C., Mirna, M. M., Sudaryanto, Y., & Ismadji, S. (2007). Adsorption of basic dye onto activated carbon prepared from durian shell: Studies of adsorption equilibrium and kinetics. *Chemical Engineering Journal*, *127*(1–3), 121–129. https://doi.org/10.1016/j.cej.2006.09.011

Charoenvai, S. (2014). Durian peels fiber and recycled HDPE composites obtained by extrusion. *Energy Procedia*, *56*(C), 539–546. https://doi.org/10.1016/j.egypro.2014.07.190

Chee Lee, M., Chun Koay, S., Yeng Chan, M., Leng Choo, H., Meng Pang, M., May Chou, P., & Yeow Tshai, K. (2020). Properties of poly(lactic acid)/durian husk fiber biocomposites: Effects of fiber content and processing aid. *Journal of Thermoplastic Composite Materials, 33*(1). https://doi.org/10.1177/0892705719831734

Cheok, C. Y., Mohd Adzahan, N., Abdul Rahman, R., Zainal Abedin, N. H., Hussain, N., Sulaiman, R., & Chong, G. H. (2018). Current trends of tropical fruit waste utilization. *Critical Reviews in Food Science and Nutrition*, *58*(3), 335–361. https://doi.org/10.1080/10408398.2016.1176009

Chia, S. L., Zamri, Z., Muniandy, A., Joseph, E., & Sathasivam, J. (2020). Durian pectin and rice husk ash (RHA) as partial replacement of cement in concrete. *IOP Conference Series: Earth and Environmental Science*, *476*(1). https://doi.org/10.1088/1755-1315/476/1/012022

Chiemchaisri, C., & Visvanathan, C. (2008). Greenhouse gas emission potential of the municipal solid waste disposal sites in Thailand. *Journal of the Air and Waste Management Association*, *58*(5), 629–635. https://doi.org/10.3155/1047-3289.58.5.629

Chowdhury, Z. Z., Ziaul Karim, M., Ashraf, M. A., & Khalid, K. (2016). Influence of carbonization temperature on physicochemical properties of biochar derived from slow pyrolysis of durian wood (Durio zibethinus) sawdust. *BioResources*, *11*(2), 3356–3372. https://doi.org/10.15376/biores.11.2.3356-3372

Chua, J. Y., Pen, K. M., Poi, J. V., Ooi, K. M., & Yee, K. F. (2023). Upcycling of biomass waste from durian industry for green and sustainable applications: An analysis review in the Malaysia context. *Energy Nexus, 10*, 100203. Elsevier Ltd. https://doi.org/10.1016/j.nexus.2023.100203

Crnogaca, B. (2017). Torrefaction as a process for biomass conversion into bio-coal. *Tehnika*, *72*(3), 323–327. https://doi.org/10.5937/TEHNIKA1703323C

Cryospain. (2022). Biomethane: Industrial uses and particularities. https://cryospain.com/biomethane-industrial-uses-particularities

Dam, M. (2022, June 1). Reducing waste by making compost out of durian husks. Vietnamagriculture. https://vietnamagriculture.nongnghiep.vn/reducing-waste-by-making-compost-out-of-durian-husks-d324346.html

Damasco, I. Y., Gallawan, J. M. M., & Lawagon, C. P. (2022). Synthesis of nanocellulose from durian rinds for the preparation of a self-healing smart concrete with augmented mechanical properties. *Chemical Engineering Transactions*, *92*, 667–672. https://doi.org/10.3303/CET2292112

Deccan Herald. (2019). Researchers develop bio-bricks from agricultural waste. https://www.deccanherald.com/science/researchers-develop-bio-bricks-from-agricultural-waste-769036.html

Duque-Acevedo, M., Belmonte-Ureña, L. J., Cortés-García, F. J., & Camacho-Ferre, F. (2020). Agricultural waste: Review of the evolution, approaches and perspectives on alternative uses. *Global Ecology and Conservation, 22*, e00902. Elsevier B.V. https://doi.org/10.1016/j.gecco.2020.e00902

Dyjakon, A. (2019). The influence of apple orchard management on energy performance and pruned biomass harvesting for energetic applications. *Energies*, *12*(4). https://doi.org/10.3390/en12040632

El-Ramady, H., Brevik, E. C., Bayoumi, Y., Shalaby, T. A., El-Mahrouk, M. E., Taha, N., Elbasiouny, H., Elbehiry, F., Amer, M., Abdalla, N., Prokisch, J., Solberg, S., & Ling, W. (2022). An overview of agro-waste management in light of the water-energy-waste nexus. *Sustainability (Switzerland), 14*(23). MDPI. https://doi.org/10.3390/su142315717

Fauzi, A. R., & Puspitawati, M. D. (2017). Pemanfaatan kompos kulit durian untuk mengurangi dosis pupuk N anorganik pada produksi tanaman sawi hijau (Brassica juncea). *Agrotrop*, *7*(1), 22–30. Retrieved from https://ojs.unud.ac.id/index.php/agrotrop/article/view/29177

Foo, K. Y., & Hameed, B. H. (2011). Transformation of durian biomass into a highly valuable end commodity: Trends and opportunities. *Biomass and Bioenergy, 35*(7), 2470–2478. https://doi.org/10.1016/j.biombioe.2011.04.004

Gonzales, A. R., & Tabanguil, E. E. (2006). Cement-bonded board from durian waste. *UM Research Journal*, *3*(1), 1–1. https://ejournals.ph/article.php?id=2013

Green, E. M. (2011). Fermentative production of butanol-the industrial perspective. *Current Opinion in Biotechnology, 22*(3), 337–343. https://doi.org/10.1016/j.copbio.2011.02.004

Ha, N. S., Lu, G., Shu, D. W., & Yu, T. X. (2020). Mechanical properties and energy absorption characteristics of tropical fruit durian (Durio zibethinus). *Journal of the Mechanical Behavior of Biomedical Materials*, *104*. https://doi.org/10.1016/j.jmbbm.2019.103603

Hamzah, M. H., Amran, A. S., Aili Hamzah, A. F., Mohd Nor, M. Z., Shamsudin, R., Che Man, H., & Wan Razali, W. A. (2023). Durian waste husks as an adsorbent in improving soaking water during the retting process of Piper nigrum L. (Pepper Berries). *Separations*, *10*(2). https://doi.org/10.3390/separations10020096

Haryati, S., Rahmatullah, R., & Putri, R. W. (2018). Torrefaction of durian peel and bagasse for bio-briquette as an alternative solid fuel. *IOP Conference Series: Materials Science and Engineering*, *334*(1), 012008. https://doi.org/10.1088/1757-899X/334/1/012008

Hertsgaard, M. (2014). As uses of biochar expand, climate benefits still uncertain - Yale E360. https://e360.yale.edu/features/as_uses_of_biochar_expand_climate_benefits_still_uncertain

Husin, N. A., Rahman, S., Karunakaran, R., & Bhore, S. J. (2018). A review on the nutritional, medicinal, molecular and genome attributes of Durian (Durio zibethinus L.), the King of fruits in Malaysia. *Bioinformation*, *14*(06), 265–270. https://doi.org/10.6026/97320630014265

Itam, Z., Syamsir, A., Ismail, I., Fatah, A. F. A., Barzani, A. R. M., Razeman, N. A., Itam, Z., Syamsir, A., Ismail, I., Fatah, A. F. A., Barzani, A. R. M., & Razeman, N. A. (2023). The efficiency of durian shell ash as partial replacement of cement in concrete for mechanical properties improvement. *AIPC*, *2544*(1), 020001. https://doi.org/10.1063/5.0116620

Kai, K. Y., Zubairi, S. I., Shaiful Hisham, I. H., Ying, M. Y., Lynn, G. V, Lani, N. N., Fazial, F. F., & Nurzahim, Z. (2022). Production of gummy confection using seeds gum from Durio zibethinus seeds. *IOP Conference Series: Materials Science and Engineering*, *1232*(1), 012005. https://doi.org/10.1088/1757-899x/1232/1/012005

Kpalo, S. Y., Zainuddin, M. F., Manaf, L. A., & Roslan, A. M. (2020). A review of technical and economic aspects of biomass briquetting. *Sustainability (Switzerland), 12*(11). MDPI. https://doi.org/10.3390/su12114609

Kumar, S., Singh, N., & Prasad, R. (2010). Anhydrous ethanol: A renewable source of energy. *Renewable and Sustainable Energy Reviews*, *14*(7), 1830–1844. https://doi.org/10.1016/j.rser.2010.03.015

Kusumaningtyas, R. D., & Syah, A. F. A. (2020). Conversion of durian shell agroindustrial waste into various valuable products to support the food security during the COVID-19 new normal era: Review. *Jurnal Teknologi Hasil Pertanian*, *13*(2), 111. https://doi.org/10.20961/jthp.v13i2.43599

Lee, M. C., Koay, S. C., Chan, M. Y., Pang, M. M., Chou, P. M., & Tsai, K. Y. (2018). Preparation and characterization of durian husk fiber filled polylactic acid biocomposites. *MATEC Web of Conferences, 152*. https://doi.org/10.1051/matecconf/201815202007

Lee, S. Y., Sankaran, R., Chew, K. W., Tan, C. H., Krishnamoorthy, R., Chu, D.-T., & Show, P.-L. (2019). Waste to bioenergy: A review on the recent conversion technologies. *BMC Energy, 1*(1). https://doi.org/10.1186/s42500-019-0004-7

Li, J.-X., Schieberle, P., & Steinhaus, M. (2012). Characterization of the major odor-active compounds in Thai durian (Durio Zibethinus L. 'Monthong') by aroma extract dilution analysis and headspace gas chromatography–olfactometry. *Journal of Agricultural and Food Chemistry, 60*(45), 11253–11262. https://doi.org/10.1021/jf303881k

Lubis, R., Saragih, S. W., Wirjosentono, B., & Eddyanto, E. (2018). Characterization of durian rinds fiber (Durio zubinthinus, murr) from North Sumatera. *AIP Conference Proceedings, 2049*. https://doi.org/10.1063/1.5082474

Mahathaninwong, N., Wandee, S., Totwaree, N., & Romyen, P. (2022). Aerobic composting and vermicomposting of durian shell and citrus peel wastes. *BioResources, 17*(1), 1144–1160. https://doi.org/10.15376/biores.17.1.1144-1160

Manshor, M. R., Anuar, H., Nur Aimi, M. N., Ahmad Fitrie, M. I., Wan Nazri, W. B., Sapuan, S. M., El-Shekeil, Y. A., & Wahit, M. U. (2014). Mechanical, thermal and morphological properties of durian skin fiber reinforced PLA biocomposites. *Materials and Design, 59*, 279–286. https://doi.org/10.1016/j.matdes.2014.02.062

Masrol, S. R., Ibrahim, M. H. I., & Adnan, S. (2015). Chemi-mechanical pulping of durian rinds. *Procedia Manufacturing, 2*, 171–180. https://doi.org/10.1016/j.promfg.2015.07.030

Masrol, S. R., Ibrahim, M. H. I., Adnan, S., Abdul Raub, R., Sa'adon, A. M., Sukarno, K. I., & Yusoff, M. F. H. (2018). Durian rind soda-anthraquinone pulp and paper: Effects of elemental chlorine-free bleaching and beating. *Journal of Tropical Forest Science, 30*(1), 106–116. https://doi.org/10.26525/jtfs2018.30.1.106116

Masturi, M., Alighiri, D., Dwijananti, P., Widodo, R. D., Budiyanto, S. P., & Drastisianti, A. (2020). Optimization of bioethanol synthesis from durian seeds using Saccharomyces cerevisiae in fermentation process. *Journal Bahan Alam Terbarukan, 9*(1), 36–46. https://doi.org/10.15294/jbat.v9i1.23574

Mellinas, C., Ramos, M., Jiménez, A., & Garrigós, M. C. (2020). Recent trends in the use of pectin from agro-waste residues as a natural-based biopolymer for food packaging applications. *Materials, 13*(3). MDPI AG. https://doi.org/10.3390/ma13030673

Mirhosseini, H., & Amid, B. T. (2013). Effect of different drying techniques on flowability characteristics and chemical properties of natural carbohydrate-protein gum from durian fruit seed. *Chemistry Central Journal, 7*(1). https://doi.org/10.1186/1752-153X-7-1

Mohd Ali, M., Hashim, N., Aziz, S. A., & Lasekan, O. (2020). Exploring the chemical composition, emerging applications, potential uses, and health benefits of durian: A review. *Food Control, 113*, 107189. Elsevier Ltd. https://doi.org/10.1016/j.foodcont.2020.107189

Mohd Nordin, N., Anuar, H., Buys, Y. F., Ali, F., Thomas, S., & Mohd Nasir, N. A. (2021). Effect of freeze-dried durian skin nanofiber on the physical properties of poly(lactic acid) biocomposites. *Polymer Composites, 42*(2), 842–848. https://doi.org/10.1002/PC.25869

Morgan, H. M., Xie, W., Liang, J., Mao, H., Lei, H., Ruan, R., & Bu, Q. (2018). A techno-economic evaluation of anaerobic biogas producing systems in developing countries. *Bioresource Technology, 250*, 910–921. Elsevier Ltd. https://doi.org/10.1016/j.biortech.2017.12.013

Mutolib, A., Rahmat, A., Triwisesa, E., Hidayat, H., Hariadi, H., Kurniawan, K., Sutiharni, S., & Sukamto, S. (2023). Biochar from agricultural waste for soil amendment candidate under different pyrolysis temperatures. *Indonesian Journal of Science and Technology, 8*(2), 243–258. https://doi.org/10.17509/IJOST.V8I2.55193

Nagalakshmaiah, M., Afrin, S., Malladi, R. P., Elkoun, S., Robert, M., Ansari, M. A., Svedberg, A., & Karim, Z. (2019). Biocomposites: Present trends and challenges for the future. In: Koronis, G., Silva, A. (Eds.), *Green Composites for Automotive Applications*. Woodhead Publishing is an imprint of Elsevier, pp. 197–215. https://doi.org/10.1016/B978-0-08-102177-4.00009-4

Nartey, O. D., & Zhao, B. (2014). Biochar preparation, characterization, and adsorptive capacity and its effect on bioavailability of contaminants: An overview. *Advances in Materials Science and Engineering, 2014*. https://doi.org/10.1155/2014/715398

Nawkarkar, P., Singh, A. K., Abdin, M. Z., & Kumar, S. (2019). Life cycle assessment of chlorella species producing biodiesel and remediating wastewater. *Journal of Biosciences, 44*(4). https://doi.org/10.1007/s12038-019-9896-0

Nichodemus, C. O. (2017). Bioethanol: Production, advantages, disadvantages and environmental impacts. https://www.biotecharticles.com/Environmental-Biotechnology-Article/Bioethanol-Production-Advantages-Disadvantages-And-Environmental-Impacts-3816.html

Nordin, N., Shamsudin, R., Azlan, A., & Ya'acob, M. E. (2017). Dry matter, moisture, ash and crude fiber content in distinct segments of 'durian kampung' husk. *International Scholarly and Scientific Research & Innovation*, *11*(12), 762–766.

Obeng, A. K., Premjet, D., & Premjet, S. (2018). Fermentable sugar production from the peels of two durian (Durio zibethinus Murr.) cultivars by phosphoric acid pretreatment. *Resources*, *7*(4). https://doi.org/10.3390/resources7040060

Obeng, A. K., Premjet, D., & Premjet, S. (2021). Improved glucose recovery from durian peel by alkaline-catalyzed steam pretreatment. *PeerJ, 9*. https://doi.org/10.7717/peerj.12026

Obergruber, M., Hönig, V., Procházka, P., Kučerová, V., Kotek, M., Bouček, J., & Mařík, J. (2021). Physicochemical properties of biobutanol as an advanced biofuel. *Materials*, *14*(4), 1–21. https://doi.org/10.3390/ma14040914

Ortega, F., Versino, F., López, O. V., & García, M. A. (2022). Biobased composites from agro-industrial wastes and by-products. *Emergent Materials, 5*(3), 873–921. Springer Nature. https://doi.org/10.1007/s42247-021-00319-x

Osman, A. I., Deka, T. J., Baruah, D. C., & Rooney, D. W. (2020). Critical challenges in biohydrogen production processes from organic feedstocks. *Biomass Conversion and Biorefinery*, *12*, 159–177. https://link.springer.com/article/10.1007/s13399-020-00965-x

Payus, C. M., Refdin, M. A., Zahari, N. Z., Rimba, A. B., Geetha, M., Saroj, C., Gasparatos, A., Fukushi, K., & Alvin Oliver, P. (2019). Durian husk wastes as low-cost adsorbent for physical pollutants removal: Groundwater supply. *Materials Today: Proceedings*, *42*, 80–87. https://doi.org/10.1016/j.matpr.2020.10.006

Pirolini, A. (2020). What is bioethanol?. AZoCleantech. https://www.azocleantech.com/article.aspx?ArticleID=403

Prakongkep, N., Gilkes, R. J., & Wiriyakitnateekul, W. (2014). Agronomic benefits of durian shell biochar Fabrication of porous alumina ceramics from alumina powder with sol-gel view project the use of soil quality index in site capability assessment for extensive grazing in a semi-arid alpine rangeland in Iran view project Nattaporn Prakongkep Land Development Department Agronomic benefits of durian shell biochar. *Journal of Metals, Materials and Minerals*, *24*(1), 7–11. https://doi.org/10.14456/jmmm.2014.2

Purnomo, A., Yudiantoro, Y. A. W., Putro, J. N., Nugraha, A. T., Irawaty, W., & Ismadji, S. (2016). Subcritical water hydrolysis of durian seeds waste for bioethanol production. *International Journal of Industrial Chemistry*, *7*(1), 29–37. https://doi.org/10.1007/s40090-015-0059-3

Riseh, R. S., Vazvani, M. G., Hassanisaadi, M., & Thakur, V. K. (2024). Agricultural wastes: A practical and potential source for the isolation and preparation of cellulose and application in agriculture and different industries. *Industrial Crops and Products*, *208*, 117904. https://doi.org/10.1016/j.indcrop.2023.117904

Sadh, P. K., Duhan, S., & Duhan, J. S. (2018). Agro-industrial wastes and their utilization using solid state fermentation: A review. *Bioresources and Bioprocessing, 5*(1). Springer Science and Business Media Deutschland GmbH. https://doi.org/10.1186/s40643-017-0187-z

Safari, S., Fathiah Athirah, N., Jamaludin, J. A., Sunthralingam, C., & Mustaffa, R. (n.d.). *Durian as New Source of Malaysia's Agricultural Wealth.*

Safari, S., Razali, A., Mahfuzah, W., Ibrahim, W., Sukri, M., & Rahim, A. (2021). From farm to China: A case study of Malaysian frozen whole durian export supply chain (Dari ladang ke China: Kajian kes rantaian bekalan eksport durian sejuk beku Malaysia dan bentuk biji). *Economic and Technology Management Review, 16*, 1–20.

Sangpong, L., Khaksar, G., Pinsorn, P., Oikawa, A., Sasaki, R., Erban, A., Watanabe, M., Wangpaiboon, K., Tohge, T., Kopka, J., Hoefgen, R., Saito, K., & Sirikantaramas, S. (2021). Assessing dynamic changes of taste-related primary metabolism during ripening of durian pulp using metabolomic and transcriptomic analyses. *Frontiers in Plant Science*, *12*. https://doi.org/10.3389/fpls.2021.687799

Selvarajoo, A. (2021). Slow pyrolysis of Durio zibethinus rind and the influence of carbonization temperature on biochar properties. *IOP Conference Series: Materials Science and Engineering*, *1092*(1), 012042. https://doi.org/10.1088/1757-899x/1092/1/012042

Selvy Handayani, R., & Ismadi. (2018). Inventory and morphological characterization of durian (Durio zibethinus) in Langkahan and Sawang sub-district of north Aceh Indonesia. *Emerald Reach Proceedings Series, 1*, 601–608. Emerald Group Holdings Ltd. https://doi.org/10.1108/978-1-78756-793-1-00027

Sharma, P., Gaur, V. K., Sirohi, R., Varjani, S., Hyoun Kim, S., & Wong, J. W. C. (2021). Sustainable processing of food waste for production of bio-based products for circular bioeconomy. *Bioresource Technology, 325*. Elsevier Ltd. https://doi.org/10.1016/j.biortech.2021.124684

Shen, J., Zhao, C., Liu, Y., Zhang, R., Liu, G., & Chen, C. (2019). Biogas production from anaerobic co-digestion of durian shell with chicken, dairy, and pig manures. *Energy Conversion and Management, 198*. https://doi.org/10.1016/j.enconman.2018.06.099

Sidaruk, I., Rohanah, A., & Daulay, S. B. (2017). Uji jenis dekomposer pada pembuatan kompos dari limbah kulit durian terhadap mutu kompos yang dihasilkan. *Journal Rekayasa Pangan dan Pertanian*, 5(1), 166–170. Retrieved from https://jurnal.usu.ac.id/index.php/jrpp/article/view/16642

Siti Faridah, M. A., Nur Atikah, M. R. L., Jau-Shya, L., Hasmadi, M., Mohd Rosni, S., Wolyna, P., & Noorakmar, A. W. (2021). Physicochemical and thermal properties of durian seed flour from three varieties of durian native of Sabah. *Food Research, 5*(4), 374–381. https://www.myfoodresearch.com/uploads/8/4/8/5/84855864/_46__fr-2020-659_siti_faridah_2.pdf

Sitti Rahmawati, Rabasia, Afadil, Suherman, Tri Santoso, Sangkota, V. D., & Abram, P. H. (2023). The utilization of durian peels (Durio zibethinus) for the manufacturing of charcoal briquettes as alternative fuel. *Jurnal Pengelolaan Sumberdaya Alam Dan Lingkungan (Journal of Natural Resources and Environmental Management), 13*(1), 76–87. https://doi.org/10.29244/jpsl.13.1.76-87

Sun, W., Gao, X., Wu, B., & Ombrello, T. (2019). The effect of ozone addition on combustion: Kinetics and dynamics. *Progress in Energy and Combustion Science, 73*, 1–25. Elsevier Ltd. https://doi.org/10.1016/j.pecs.2019.02.002

Tabatabaee Amid, B., Mirhosseini, H., & Kostadinovic, S. (2012). Chemical composition and molecular structure of polysaccharide-protein biopolymer from Durio zibethinus seed: Extraction and purification process. *Chemistry Central Journal, 6*, 117. https://link.springer.com/article/10.1186/1752-153X-6-117

Takolpuckdee, P. (2014). Transformation of agricultural market waste disposal to biochar soil amendments. *Procedia Environmental Sciences, 20*, 64–70. https://doi.org/10.1016/j.proenv.2014.03.010

Teh, B. T., Lim, K., Yong, C. H., Ng, C. C. Y., Rao, S. R., Rajasegaran, V., Lim, W. K., Ong, C. K., Chan, K., Cheng, V. K. Y., Soh, P. S., Swarup, S., Rozen, S. G., Nagarajan, N., & Tan, P. (2017). The draft genome of tropical fruit durian (Durio zibethinus). *Nature Genetics, 49*(11), 1633–1641. https://doi.org/10.1038/ng.3972

Torres-Mayanga, P. C., Lachos-Perez, D., Mudhoo, A., Kumar, S., Brown, A. B., Tyufekchiev, M., Dragone, G., Mussatto, S. I., Rostagno, M. A., Timko, M., & Forster-Carneiro, T. (2019). Production of biofuel precursors and value-added chemicals from hydrolysates resulting from hydrothermal processing of biomass: A review. *Biomass and Bioenergy, 130*. https://doi.org/10.1016/j.biombioe.2019.105397

Tudsanaton, C., Pattamapitoon, T., Phewnil, O., Semvimol, N., Wararam, W., Chanthasoon, C., & Thaipakdee, S. (2021). Limiting factors of durian rind composting by natural technology during the wet period. *Ecology Environment and Conservation, 27*(2), 590–598.

Wang, Y., Dai, Z., Zha, J., & Wei, W. (2023). Development of durian rind/polypyrrole composite and its application in removing of anionic dyes. *Adsorption Science & Technology, 2023*. https://doi.org/10.1155/2023/6985657

Werkneh, A. A. (2022). Biogas impurities: Environmental and health implications, removal technologies and future perspectives. *Heliyon, 8*(10). Elsevier Ltd. https://doi.org/10.1016/j.heliyon.2022.e10929

Wu, M., Wang, M., Liu, J., & Huo, H. (2008). Review: Bioseparations and downstream processing assessment of potential life-cycle energy and greenhouse gas emission effects from using corn-based butanol as a transportation fuel. *Biotechnology Progress, 24*, 1204–1214. https://doi.org/10.1002/btpr.71

Year of the Durian. (2015, December 18). Karak Organic Durian Farm, A Licensed Organic Farm | Karak, Malaysia. https://www.yearofthedurian.com/2015/12/karak-organic-durian-farm-licensed.html

Yirka, B. (2020). Making bio-coal from plant waste to mitigate climate change. Phys.org. https://phys.org/news/2020-01-bio-coal-mitigate-climate.html

Yoon, S. H., Ha, S. Y., Roh, H. G., & Lee, C. S. (2009). Effect of bioethanol as an alternative fuel on the emissions reduction characteristics and combustion stability in a spark ignition engine. *Proceedings of the Institution of Mechanical Engineers, Part D: Journal of Automobile Engineering, 223*(7), 941–951. https://doi.org/10.1243/09544070jauto1016

Zhao, C., Zhang, R., He, Y., Wang, W., Chen, C., Liu, G., Cui, X., Liu, Y., & Zhang, R. (2017). Maximization of the methane production from durian shell during anaerobic digestion. *Bioresource Technology, 238*, 433–438. https://doi.org/10.1016/j.biortech.2017.03.184

18 Tomato Waste Management

18.1 INTRODUCTION

The agricultural sector produces huge amounts of biomass as a by-product of crop cultivation and processing, often leading to environmental challenges when not properly managed. Therefore, strategies are required to overcome the environmental problems (Fritsch et al., 2017). An example is the tomato (*Lycopersicon Esculentum*), which is known as the world's second-most important vegetable and a crucial ingredient for dietitians, especially in sauce production. Its intake is linked to lowering the risk of inflammation, various chronic diseases, and cancer development, and it helps reduce blood cholesterol levels. According to the World Processing Tomato Council, over 130 million tons of tomatoes are processed annually, with nearly 8 million tons becoming waste, tomatoes with undesirable color, ripeness, or shape, which can lead to economic losses for producers and negative environmental impact (Trombino et al., 2021). This biomass waste can lead to ecological problems such as methane emissions, soil degradation, and water contamination as well as depletes the ozone layer when burned, thereby increasing the impact of climate change. Therefore, these wastes can potentially be upgraded in other production processes and may be converted into biodegradable products.

However, tomato biomass holds significant potential for sustainable solutions, such as bioenergy production, organic fertilizers, and even bioplastics. The efficient management of this waste not only addresses environmental concerns but also presents economic opportunities through the valorization of biomass. This chapter will further explore the current issues surrounding tomato biomass waste management and discuss innovative strategies for harnessing its potential, emphasizing sustainability and resource efficiency.

18.2 AGRO-BIOMASS

18.2.1 Classification of Agro-Biomass

Agricultural biomass is the renewable organic matter derived from biological organisms that can be used to produce energy. It is produced by a variety of agricultural processes and activities as raw materials, by-products, or final products. The production and consumption of food to feed a growing population and supply raw materials for the industrial sector have led to a substantial increase in the amount of waste generated by the agriculture sector (Awogbemi & Kallon, 2022).

Agricultural waste is categorized as lignocellulosic biomass, which is made up of complex molecular structures like cellulose, hemicellulose, lignin, ash, and some extractives (Imraan et al., 2023; Norfarhana et al., 2022). The majority of agricultural wastes include a higher percentage of cellulose than hemicellulose or lignin combined. To preserve the stability of a plant's structure, cellulose—a rigid, fibrous, impenetrable polysaccharide—is structured in chains with packs of microfibrils. Its cellulose properties dictate the mechanical stability, strength, and chemical fingerprint of biomass. Furthermore, hemicellulose is more vulnerable to chemical attacks and alterations and has a lower mechanical strength than cellulose. Biofuel production from biomass is influenced by the amount of hemicellulose present (Williams et al., 2016).

Agricultural wastes are classified into crop residues, industrial processing wastes, livestock wastes, and fruit wastes. Figure 18.1 shows the classification and major examples of agricultural

DOI: 10.1201/9781003472209-25

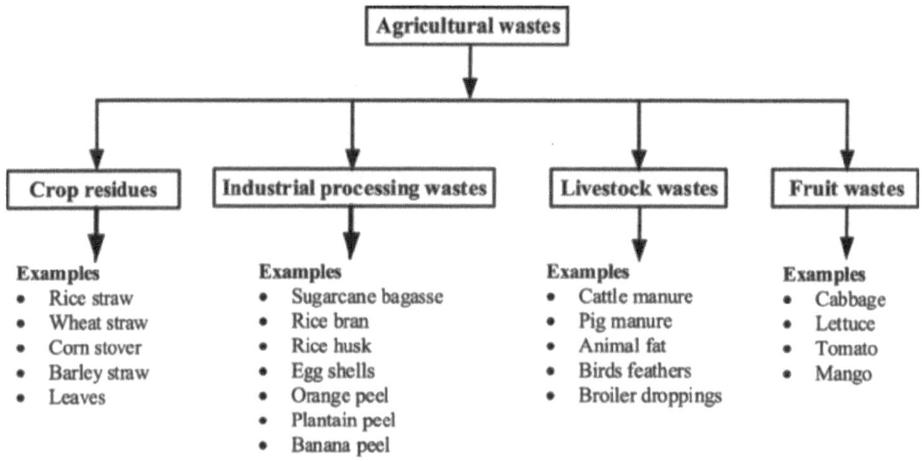

FIGURE 18.1 Classification and major examples of agricultural wastes.

waste. Crop residues are the portion of a plant in the field after harvest of the crop that is not used domestically or sold commercially such as rice straw. Moreover, industrial processing wastes are the wastes produced by industrial activity, which includes any material that is rendered useless during a manufacturing process of factories such as rice husks. Livestock wastes are the animal urine and fecal matter, including absorbent bedding such as cattle manure. Lastly, fruit wastes are solid wastes derived from fruits and vegetables such as tomatoes (El Mashad et al., 2019).

18.2.2 Availability of Tomato Biomass

Fruits and vegetables account for 45% of all food waste worldwide, making them the food category with the highest rate of waste. Fruit and vegetable processing by-products, which are mainly made up of pomace, which includes peel, pulp, and seeds, are excellent sources of bioactive compounds, natural colors, dietary fiber, proteins, peptides, and polysaccharides. The food and beverage industry is currently dealing with significant difficulty in managing and getting rid of processed wastes and by-products while trying to meet the increasing demand from customers for natural ingredients that are beneficial to their health. Food waste valorization and its conversion into useful products are being extensively researched globally to address these issues (Teigiserova et al., 2019).

Tomatoes are one of the most widely grown fruits and vegetables in the world, with an estimated yearly production of over 180 million tons in 2021. In Figure 18.2, the top ten tomato-producing nations are shown. The tomato processing sector produces large amounts of residue every year and decides on disposal concerns related to both the rise in significant environmental burdens and the waste of priceless resources (Eslami et al., 2023).

Tomatoes are seasonal fruits that can be eaten raw or processed into sauces, ketchup, purees, juices, and pastes. Peel and seeds are considered by-products, and only pulp is used in processed tomato-based goods. Tomato by-product disposal is currently seen as a social, economic, and environmental issue. Tomato by-products are often disposed of in landfills or used as animal feed. They can be applied to food in a variety of ways, such as by using the fresh by-product directly in the formulation, incorporating it with other ingredients in a dry powder form, or substituting some of the flour used during manufacture. The primary goal of adding tomato by-products to food items is to enhance their nutritional value by turning them into a source of fiber, protein, or lycopene. Tomato by-products are used to improve food items' performance and replace artificial additives,

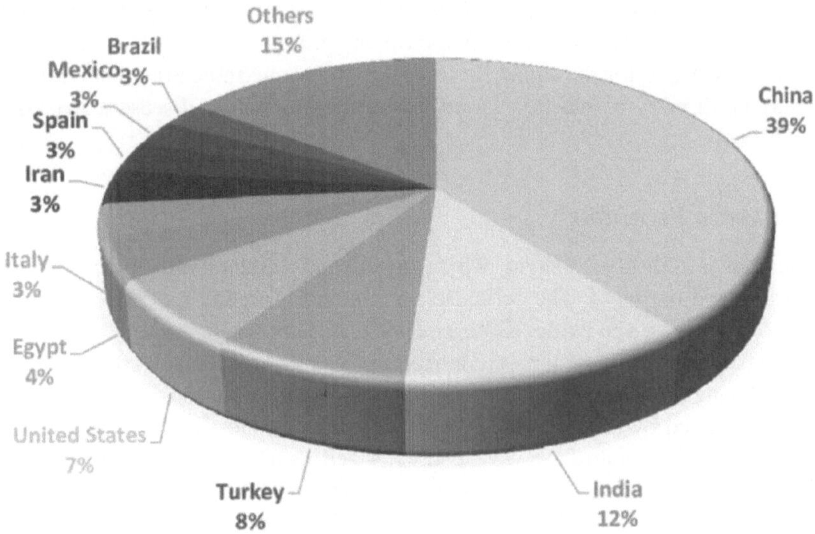

FIGURE 18.2 Top ten nations producing tomatoes

such as natural coloring and fat alternatives in meat products, or as a source of hydrocolloids in bakery products (Silva et al., 2023).

Furthermore, tomato pomace, which makes up 4% of the fruit weight, is a well-liked by-product that is high in tomato skin and water. It is primarily composed of fibers, but it also includes minerals, proteins, lipids, pectins, and the carotenoid lycopene, among other beneficial antioxidants. Tomato waste materials have biological activity as antioxidants and, in vitro, also control cell development and confer antimutagenic qualities. Furthermore, fermenting tomato waste rich in protein to produce amino acids and peptides with anti-inflammatory and antioxidant properties is an intriguing use for this waste product. Research has shown that the anti-inflammatory qualities originate from the tomato skin's naringenin chalcone, which successfully lowers edema in animal models (Giorgio, 2017). Tomato pomace oil can therefore be used as a protective component for skin care, enhancing the skin's condition on the face, particularly by lowering inflammation and skin sensitivity and balancing the skin tone (Rajkowska et al., 2024).

18.3 CHARACTERISTICS OF TOMATO BIOMASS

18.3.1 Physical Properties

The high moisture content of tomato wastes will lead to an unpleasant stench and encourage the growth of germs and fungi. Tomato waste is capable of being broken down by microbes due to its high initial moisture content and high nutrient content. Furthermore, the bulk density of tomato waste makes it difficult to dispose of in landfills. Tomato wastes are red because they contain the pigment lycopene, yet while a tomato is young, it is green in color (Fărcaş et al., 2019).

18.3.2 Chemical Composition

The approximate chemical composition of tomato pomace at different stages of industrial processing was found to be 25.73% total sugars, 19.27% protein, 7.55% pectin, 5.85% lipids, and 3.92% mineral content on a dry weight basis (Szabo et al., 2018). It has also been noted that tomato pomace is a great source of antioxidants, such as ascorbic acid, phenolic compounds, and carotenoids, which are well known for their potent ability to scavenge free radicals. Lycopene is the main carotenoid in tomatoes, accounting for 80–90% of the total quantity; however, other carotenoids such as lutein, beta, gamma, and alpha-carotene are also present. Numerous studies have provided scientific evidence for the significance of this molecule, indicating that eating foods high in lycopene may be directly linked to a lower risk of cardiovascular disease and prostate cancer (Fărcaş et al., 2019).

18.3.3 Mechanical Properties

Tomato pomace has a high fiber content, which significantly alters technological food properties including cohesion and hardness. The cellulose, lignin, and lycopene-rich tomato peel makes the tomato waste more cohesive and harder (Silva et al., 2023). The tomato waste's thin, fibrous texture, which makes it easily broken under stress, contributes to its low tensile and compressive strengths. Because of this, during processing, tomato waste can be readily distorted.

In the study conducted by Li et al. (2012), the mechanical properties of the exocarp, mesocarp, and locular gel tissues of tomato fruit were characterized using various tests including compression, tension, shear, and bending. The tension failure stress and elastic modulus of exocarp tissue were found to range from 0.421 to 0.582 and 4.601 to 9.59 MPa, respectively, while the mesocarp exhibited compression failure stress and elastic modulus values between 0.122 and 0.229 MPa, and 0.726–0.846 MPa, respectively. Locular gel tissues showed lower values, with compression failure stress ranging from 0.012 to 0.016 MPa and an elastic modulus between 0.048 and 0.124 MPa. Notably, the outer layer tissues of the tomato fruit exhibited higher failure stress and elastic modulus compared to the inner layers. These mechanical properties varied significantly between the two tomato varieties tested. The study further modeled the tomato fruit as a multibody system, with the mechanical parameters used in developing a finite element model to predict internal stress distribution and damage regions when subjected to external forces. This data holds the potential for improving quality assessment processes in tomato handling and processing.

18.4 CURRENT ISSUES AND CHALLENGES

18.4.1 Volume of Waste Produced

The need for a steady availability of fresh tomatoes for consumption for processing frequently results in excessive production, leading to significant waste. Overproduction is also a common practice in the industry to offset potential losses during processing, such as when tomatoes are rejected because of quality defects (Laranjeira et al., 2022). In 2010, global waste from tomato processing ranged from 4.3 to 10.2 million metric tons. The processing of tomatoes into various fresh and preserved products generates large amounts of solid waste, primarily made up of lignin, cellulose, and hemicellulose (Lin et al., 2012). Each year, approximately 31% of fresh tomatoes are wasted, meaning the average person in the country discards around 21 tomatoes annually (Howarth, 2024). Shockingly, up to 86.7% of tomatoes are discarded based on their appearance before they even reach the market.

18.4.2 Types of Waste Produced

18.4.2.1 Tomato Pomace

The primary by-product in the tomato processing industry is tomato pomace, mainly composed of peels and seeds. This waste accounts for nearly 3% of all processed fresh tomatoes, amounting to roughly 0.8 million tons annually. Managing tomato pomace is particularly challenging due to its seasonal production. The processing window lasts 2–3 months, during which a large quantity of waste is produced daily, complicating proper waste management efforts (Mangut et al., 2006). Additionally, tomato seed waste contains various bioactive compounds that may provide wellness advantages. Research using both in vitro and in vivo models has shown that extracts from tomato seeds exhibit several positive effects, such as antiplatelet, antioxidant, anticancer, antimutagenic, antibacterial, and neuroprotective properties (Sorrenti et al., 2023). The concentrated nature of the processing season poses significant difficulties for efficient waste disposal. Tomato waste, rich in antioxidants like lycopene (510.6 mg/kg), β-carotene (95.6 mg/kg), and total phenols (1,229.5 mg GAE/kg), presents potential for potential future uses. This by-product initially has a moisture content of about 66–67% (wet basis), with tomato peels (61.5%) and seeds (38.5%) making up most of its composition (Nour et al., 2018).

18.4.2.2 Tomato Peel

The structure of tomato peel is crucial because it contains numerous valuable health and nutritional elements. These include lycopene, dietary fiber, pectin, proteins, oils, and antioxidants, all of which contribute to its health benefits (Lu et al., 2019). Cutin, a major component of tomato peel, accounts for 45–80% of its structure. This amorphous polyester is composed of long-chain fatty polyhydroxy acids, containing 16 or 18 carbon atoms, linked by cross-ester bonds, with these fatty acids forming the primary structure of cutin (Koçak et al., 2023). The tomato processing industry produces a significant amount of waste; for example, from 10 kg of tomatoes, only 1 kg of pulp is extracted, leaving 9 kg as waste. This waste, mainly consisting of tomato peel, serves as a secondary raw material that can be utilized to extract bioactive compounds through various techniques (Prasad & Murugavelh, 2020).

18.4.3 Biomass Utilization

Tomato fruits are very perishable, having a limited shelf life of just 12–72 h. Due to this perishability, many tomatoes spoil before they can be transported across the country (Irokanulo et al., 2015). Their high water content makes them susceptible to microbial deterioration, complicating storage and transportation. Tomatoes are rich in carbohydrates, lipids, organic acids, water, minerals, vitamins, and pigments. Ripe tomatoes are estimated to contain approximately 94% water, 4.3% carbohydrates, 1% protein, 0.1% fat, 0.6% fiber, and various vitamins (Wogu & Ofuase, 2014). These nutrients foster the growth of microorganisms such as bacteria and fungi, which produce enzymes that break down the nutrients (Maddox, 1998). As a result, microorganisms not only diminish the nutritional quality but also reduce the economic value of tomato fruits (Wogu & Ofuase, 2014).

18.4.4 Environmental Impacts

Tomato (*Lycopersicon esculentum*) cultivation, despite its global significance and economic value, presents various environmental challenges. Tomatoes are the second-most important vegetable worldwide and are heavily consumed in both raw and processed forms. However, the production process generates substantial environmental impacts due to factors such as high land usage, agrochemical reliance, and water consumption. Table 18.1 provides the environmental impact of each factor of tomatoes. Dried tomato waste, which contains 53% organic carbon, poses additional difficulties due to its high carbon-to-nitrogen (C:N) ratio of 39.1, making it resistant to decomposition in soil (Grenier et al., 2022). Tomato cultivation requires significant land, contributing to soil erosion, biodiversity loss, and habitat reduction, particularly in regions like Mexico and Canada, where sandy soils increase the risk of erosion (Howarth, 2024). This degradation not only affects soil health but also exacerbates flooding and water contamination. Furthermore, agriculture occupies 44% of the world's habitable land, with crop cultivation accounting for 14.7%. Given that tomatoes are land-intensive compared to other plant-based foods, their production contributes to the reduction of habitable land and adversely impacts wildlife and biodiversity. Pesticide use during tomato farming often leads to contamination of ecosystems, posing risks to small animals and pollinators. Water pollution is another critical issue, especially in regions like California, where intensive irrigation

TABLE 18.1
Environmental Impact of Each Factor of Tomatoes (Howarth, 2024)

Environmental Impact	Explanation
Soil fertility and erosion	Tomatoes require a significant amount of land for cultivation, resulting in a considerable environmental impact. As one of the plant-based foods that require the most land, they are frequently sourced from Mexico and Canada, leading to issues like soil erosion, biodiversity loss, and carbon emissions from transportation. Grown in sandy soils, tomatoes face a higher risk of erosion compared to other crops such as cereals or oilseeds. This erosion negatively affects the ecological impact of tomato production, as it diminishes soil health and may result in flooding and water contamination.
Reduction of livable land	Agriculture utilizes 44% of the world's habitable land, with livestock grazing representing two-thirds of that area. Crop cultivation accounts for 14.7% of all usable land. Since tomatoes require more land in relation to other plant-based foods, the tomato industry can contribute to the reduction of habitable land.
Impact wildlife and biodiversity	The tomato industry, similar to other agricultural sectors, adversely affects wildlife. During harvesting, small animals like mice, rabbits, birds, and insects are often displaced or killed. Additionally, the use of pesticides in tomato farming can contaminate ecosystems, leading to declines in populations of amphibians, birds, and pollinators.
Water and ocean pollution	In California, tomatoes require irrigation methods to fulfill crop water consumption or evapotranspiration needs for processing. Intensive irrigation can lead to freshwater and ocean pollution by causing erosion and transporting pesticides and heavy metals into water sources. Additionally, it can result in the accumulation of harmful selenium, which diminishes the natural flow of water in streams. Consequently, tomato irrigation contributes to increased water contamination.
Agrochemical usage	Tomatoes rank 12th in pesticide usage among all fruits and vegetables in the United States. On average, each tomato has four different pesticides applied, with some containing as many as 15. The production, shipping, and application of these pesticides release carbon dioxide, making tomatoes a high-impact crop in terms of emissions. Vertical farming could help reduce chemical use. To minimize the environmental impact of tomatoes, opting for organic produce is advisable. Tomatoes generally require more fertilizers than other vegetables, especially nitrogen, phosphate, and potassium. However, nitrogen fertilizer is used cautiously, as it can negatively affect growth. Most commercially available tomato varieties are highly susceptible to disease and crop loss, leading to the use of various agrochemicals to enhance yields. These toxic chemical residues, along with water and air pollution, can harm both ecosystems and human health.

practices for tomato crops contribute to the leaching of pesticides and heavy metals into freshwater and ocean ecosystems. The use of nitrogen, phosphate, and potassium fertilizers in tomato farming exacerbates these environmental impacts, while pesticide application contributes to carbon emissions and ecological damage. Given these challenges, sustainable agricultural practices, such as organic farming and vertical farming, are being explored to reduce the environmental footprint of tomato production.

18.4.5 Logistical and Storage Constraints

Tomatoes are more sensitive than other fruits and vegetables (Mai et al., 2021) and are extremely perishable because of vibrations and impact during shipping, market handling, and storage temperatures (Kabir et al., 2020). Vibrations in transit cause tomatoes to rub and rotate against other products and packaging, leading to softening and other mechanical issues (Idah et al., 2012). Additionally, prolonged shipping times can accelerate enzymatic and metabolic activities, increasing the risk of mechanical damage and diminishing market value (Cherono & Workneh, 2018). Proper storage of fresh produce is crucial in the postharvest supply chain (Arah et al., 2015). Many microbiological, physiological, and metabolic processes that lead to quality decline are temperature-dependent (Majidi et al., 2014). A significant rise in storage temperature increases respiration, transpiration, and ethylene production rates. Conversely, a drastic drop in temperature can cause chilling injuries and reduce quality (Atanda et al., 2011). Maintaining storage conditions is particularly challenging for high-moisture fruits like tomatoes (Arah et al., 2015).

18.4.6 Diseases and Pests

Tomato production faces significant challenges from various diseases and pests, which can lead to substantial crop losses and increase the biomass that requires management due to the disposal of infected plant parts. Table 18.2 lists the diseases in tomato production. Diseases such as anthracnose, caused by *Colletotrichum coccodes*, lead to sunken lesions with fungal fruiting bodies on fruits, stems, and leaves, particularly under warm spring conditions. Black mold, caused by *Alternaria alternata*, forms black or brown lesions on ripe fruits, especially when mature fruit remains on the vine. Early blight, triggered by *Alternaria solani*, produces concentric leaf spots and can result in severe blighting and reduced yield. Table 18.3 lists examples of pests in tomato production.

TABLE 18.2
Diseases in Tomato Production

Diseases	Symptoms	Causes	Comments
Anthracnose (*Colletotrichum coccodes*) Examples:  *Anthracnose* on tomato fruit  *Anthracnose* symptoms on tomato leaf	• Fruit symptoms are predominant, yet infections can also affect the stem, leaves, and roots. • Disease results in the fruit's typical sunken, round imperfection. • There may be obvious concentric bands of brown and yellow in the indentations. • Lesion centers evolve to a brown color and develop minute black fungal fruiting bodies, or microsclerotia, scattered throughout. • Lesions have the potential to enlarge greatly.	• Fungus	• Warm weather is favorable for disease onset. • Occurs early in the spring.
Black mold (*Alternaria alternata*) Examples:  Black mold-infected tomato leaves 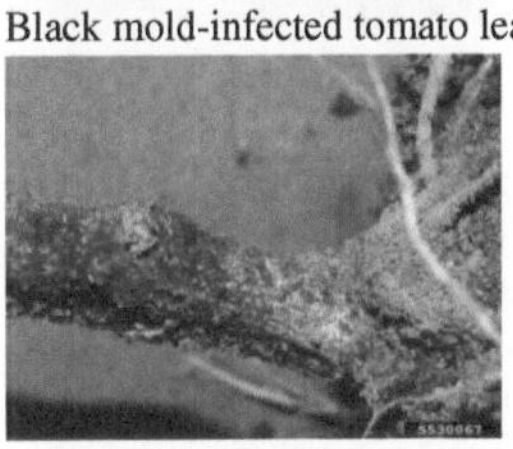 Canker at the base of the plant, blackish growth of *Alternaria conidia*	• On the surface of ripe fruit, the disease develops lesions that appear black or brown. • Lesions may be microscopic specks or huge regions of rotting tissue. • In humid situations, the lesions could have velvety black spore masses covering them.	• Fungus	• Despite where tomatoes are cultivated, disease exists. • The longer mature fruit remains on the vine after ripening, the more vulnerable it is to the fungus.

(*Continued*)

TABLE 18.2 (*Continued*)
Diseases in Tomato Production

Diseases	Symptoms	Causes	Comments
Early blight (*Alternaria solani*) Examples:  Early blight symptoms on tomato stem  Early blight symptoms on tomato fruit	• Oval-shaped lesions with a yellow chlorotic area spanning the lesion are the initial signs of early blight. • On diseased leaves, concentric leaf sores may be visible. • There is a complete collapse of the leaf due to acute infections that damage the tissue between the veins. • As the illness worsens, leaves become badly blighted, which lowers yield. • Alternaria stem canker can occur when a fungus infects tomato stems. • The appearance of dark brown patches on the stem is one of the early signs of stem canker. • The plant as a whole may die if stem cankers grow to the point where they encircle the entire stem. • The vascular tissue both above and below the canker area has brown streaks. • Small black v-shaped lesions located at the fruit's shoulders are one of the symptoms (the condition is also known as black shoulder). • On the fruit, lesions can sometimes show up as dark specks with a concentric ring pattern. • Fruit lesions may appear in the field or may appear while the fruit is being transported to the market. • A velvety appearance may result from the fungus's sporulation in the lesions.	• Fungus	• As soon as plants begin to bear fruit, disease can spread quickly. • The illness spreads through contact with infected soil and the movement of airborne spores.

Additionally, pests like aphids and the curly top virus, transmitted by beet leafhoppers, can stunt growth, deform leaves, and inhibit fruiting, adding to the burden of managing affected biomass in tomato production.

18.5 INNOVATIVE SOLUTIONS AND POTENTIAL APPLICATIONS

Tomato biomass is a major by-product of the global tomato industry, one of the world's largest agricultural sectors. With a wide variety of tomato cultivars grown for both fresh market and food processing applications such as sauces, juices, concentrates, paste, canning, and others, the production of tomatoes results in considerable organic waste. This waste includes unsold fruits, plant residues, and by-products from processing plants like skins, seeds, and pulp. Together with the growing demand for tomatoes all over the world, a quantity of biomass waste from tomato production is accordingly produced. If this biomass waste is not managed properly, it can lead to several environmental issues, such as the emission of greenhouse gases from rotting organic material, water pollution through runoff, and the depletion of nutrients in the soil. However, modern technological

TABLE 18.3
Pests on Tomato Production

Diseases	Symptoms	Cause	Comments
Category: Insects Aphids (peach aphid, potato aphid) *Myzus persicae Macrosiphon euphorbiae* **Examples:** 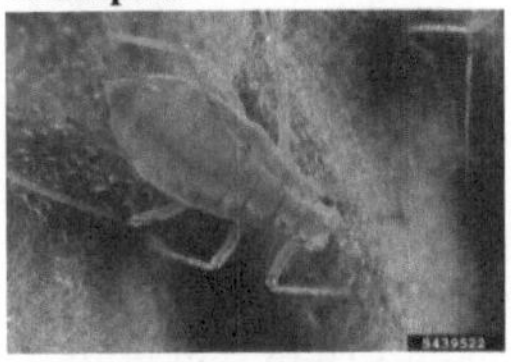 Potato aphid on tomato leaf	• Small soft-bodied insects found on the underside of plant leaves and/or stems • Typically green or yellow but can be pink, brown, red, or black depending on the species and host plant • A heavy aphid infestation may cause leaves to yellow and/or become deformed, necrotic patches on leaves, and/or stunted shoots • Aphids excrete a sticky, sugary fluid called honeydew, which promotes the formation of sooty mold on plants	• Insect	• Aphids are distinguished by the presence of cornicles (tubular structures) that extend backward from the insect's body. • Aphids may not move fast when disturbed. • They may potentially transmit plant viruses to the plant when they eat.
Category: Others **Curly top** Beet curly top virus (BCTV)  BCTV symptoms on tomato	• The virus normally kills young plants • Older plants become stunted and become yellow to bronze, with purple-tinged leaves • Leaves thicken and roll upward • Petioles on leaves roll downhill • Regardless of age, green fruit becomes red, becoming wrinkled and drab in color. • After infection, plants do not recover from the disease and do not blossom or bear fruit.	• Virus	• The virus is transmitted by beetroot leafhoppers. • Insects spread the illness among approximately 300 plant species, including beets, tomato, squash, Swiss chard, cucumber, and melon.

advancements present opportunities to treat this biomass as a valuable resource. This section explores various innovative solutions for addressing the challenges associated with the volume and types of tomato biomass waste, environmental impacts, logistical and storage constraints, and risks posed by diseases and pests, stressing potential applications such as bioenergy production, material recovery, and environmental remediation, all of which can contribute to more sustainable and efficient resource use in the agricultural industry.

18.5.1 Volume of Waste Produced

One of the biggest issues in the tomato industry is the enormous volume of waste generated throughout the production cycle. From the field to the processing plants, tomato biomass waste accumulates rapidly, particularly during peak harvesting seasons. This not only burdens local waste management systems but often leads to improper handling, increasing environmental impacts. A promising solution to this high volume of waste is the application of biorefinery technologies. Biorefineries operate similarly to oil refineries but instead convert biomass into a wide array of bio-based products. In the case of tomato biomass, the organic matter can be separated into fractions such as fibers, sugars, and organic acids, which are processed into value-added products. For example, cellulose and lignin from tomato plant stems and leaves can be used to produce biofuels, while sugars and organic acids from fruit waste can be processed into bioplastics and biochemicals (Zuorro et al., 2014).

By converting waste into valuable products, biorefineries not only reduce the volume of waste but also contribute to the circular economy by returning these products to the market (Almeida et al., 2024). Moreover, biorefineries are capable of operating continuously, ensuring that even during peak harvesting periods, the biomass is processed efficiently, preventing local waste management systems from being overwhelmed. This also supports the global shift toward renewable energy and bio-based materials, reducing dependence on fossil fuels. Another option for managing organic waste in biorefinery systems is anaerobic digestion, which transforms biomass into biogas, a renewable source of energy. Anaerobic digestion is a microbial process that decomposes organic material in the absence of oxygen, producing biogas which consists of methane and carbon dioxide, along with digestate, a nutrient-rich slurry that can be used as fertilizer (Tallou et al., 2022). The resulting biogas can be used to produce electricity, heat, or as a vehicle fuel, and the digestate can be directly used on fields for agricultural production, hence closing the nutrient loop and further reducing waste.

18.5.2 Types of Waste Produced

Tomato biomass waste is a heterogeneous mixture consisting of various types of organic material that differ in composition and potential applications. The most common types of waste include skins, seeds, stems, leaves, and unsellable or overripe fruits. Each component offers potential for resource recovery, provided the appropriate processing technologies are applied. For instance, tomato skins are rich in lycopene, a powerful antioxidant with numerous health benefits, including protection against heart disease and certain types of cancer (Vella et al., 2017). The extraction of lycopene from tomato skins transforms what would otherwise be waste into a valuable nutraceutical or functional food ingredient. Extraction techniques such as supercritical fluid extraction or solvent extraction can yield pure lycopene, which isolates lycopene in its pure form while the remaining biomass can be used for energy generation or composting.

Tomato seeds, another by-product, can be processed into oil, which is rich in essential fatty acids and antioxidants. This oil has applications in cosmetics, skincare products, and even health supplements. Seeds can be cold-pressed to extract oil, and the leftover seed cake can be further processed into animal feed or used as raw material for bio-based products.

In addition, the stems and leaves of tomato plants contain high levels of fibers and lignocellulose, which can be used to produce biocomposites. These materials, made by reinforcing natural fibers with bioplastics, provide a biodegradable alternative to conventional plastic products. For instance, biocomposites made from tomato fibers can be used in packaging, automotive parts, and construction, offering eco-friendly solutions for industries that are committed to reducing their environmental impact. By adopting a component-specific approach to waste utilization, the tomato industry can ensure that each type of biomass is processed in the most efficient and sustainable way possible. This not only maximizes the value of the waste but also reduces the environmental impact of traditional waste disposal methods, such as landfilling and incineration.

18.5.3 Biomass Utilization

The fundamental principle of biomass utilization is to transform waste into resources by using innovative technologies to recover valuable products from organic material. In the case of tomato biomass, the applications range from energy production to the creation of materials. A particularly promising application of tomato biomass is in the production of biocomposites and biodegradable packaging materials. The fibers from tomato plants, which remain after harvest, can serve as reinforcement in biocomposites. These composites are formed by blending natural fibers with biodegradable polymers, creating materials that are both strong and lightweight, as well as environmentally friendly. Their uses span industries such as packaging, automotive manufacturing, and construction, where there is a growing demand for biodegradable substitutes for plastic (Morone et al., 2015).

Tomato biomass also holds potential for the production of biodegradable coatings and films for food packaging. Polysaccharides and proteins from tomato residues can be processed into films that could replace petroleum-based plastic films. These films not only reduce plastic waste but also extend the shelf life of fresh produce and possess antimicrobial properties, making them ideal for food preservation. Biocomposites and biodegradable films produced from tomato biomass contribute to solving two problems at once, the accumulation of waste from agriculture and the growing pollution problem caused by plastic waste. As petroleum-based plastics can take centuries to decompose, biocomposites and biodegradable films offer an environmentally friendly alternative that degrades naturally over time, particularly suited for single-use products.

18.5.4 Environmental Impacts

Tomato biomass waste can cause severe environmental impacts if not managed properly. Organic waste decomposing in landfills emits methane, a potent greenhouse gas that contributes to global warming. Additionally, as biomass breaks down, it produces leachate, which can contaminate water sources. Improper disposal also leads to the loss of nutrients that could otherwise be used to enrich soils. Composting is one of the most effective ways to reduce these environmental impacts. Tomato biomass, including plant residues and fruit waste, can be composted either on farms or at processing plants (Durmuş & Kızılkaya, 2022). Composting reduces the need for transportation to landfills and provides a nutrient-rich soil amendment that can be applied to agricultural fields, helping reduce reliance on chemical fertilizers.

In addition to composting, the production of biochar from tomato biomass offers a promising solution to both waste management and carbon sequestration. Biochar is a stable form of carbon produced through pyrolysis, which involves heating organic material in an oxygen-free environment. When applied to soil, biochar enhances its structure, increases nutrient and water retention, and promotes the growth of beneficial microorganisms. Additionally, biochar sequesters carbon in a form that can persist in the soil for hundreds to thousands of years, making it a valuable tool for climate change mitigation (Lehmann & Joseph, 2015). Incorporating biochar production into tomato waste management systems allows farmers to enhance soil health while supporting global initiatives to lower atmospheric carbon dioxide levels (El-Bassi et al., 2021; Rasool et al., 2021). This method not only mitigates the environmental effects of tomato waste but also boosts the sustainability of farming practices by fostering soil preservation and carbon capture.

18.5.5 Logistical and Storage Constraints

Due to its high moisture content, tomato biomass is highly perishable and must be processed quickly after harvesting to avoid spoilage. This presents significant logistical challenges, particularly in transporting and storing large quantities of wet biomass, which can be costly and inefficient. Wet biomass can also decompose during storage, releasing methane and other harmful gases that contribute to environmental degradation. Mobile biomass processing units offer a potential solution to these logistical challenges. These units can be installed directly to farms or processing plants, allowing biomass to be processed at the source. By reducing the need for transportation, these mobile units offer an efficient way to handle biomass quickly and cost-effectively. For example, small-scale anaerobic digesters can convert organic waste into biogas and compost, or drying units can reduce the moisture content of biomass, making it easier to transport and store (Sharma et al., 2020). Processing biomass directly at the source with mobile units helps to alleviate the logistical difficulties of transporting wet biomass while also offering immediate advantages to farmers, such as renewable energy and soil improvements. This decentralized waste management approach is especially beneficial in rural regions where access to centralized waste processing facilities is often limited.

18.5.6 Disease and Pest Challenges

The presence of diseases and pests in the crops is one of the biggest challenges regarding the effective management of tomato biomass. Poor management of infected plant materials may result in the harboring of pathogens that can be spread to healthy crops, causing further losses. Such infected biomass might not be appropriate for composting or direct application into agricultural fields since it would introduce harmful organisms to the soil. Innovative solutions, such as controlled environment agriculture (CEA) systems like greenhouses and vertical farms, offer ways to reduce disease and pest outbreaks (Ragaveena et al., 2021). These systems allow farmers to control environmental conditions, such as temperature, humidity, and light, reducing the likelihood of pests and diseases affecting crops. Under these controlled environment conditions, tomato plants are less likely to be affected by pests and pathogens, thus yielding more with less biomass wastage (Ravindran et al., 2020).

Additionally, infected biomass can be repurposed for value-added applications, such as the production of biopesticides. These biopesticides, made from fermented plant material, can be used to protect crops from pests while minimizing the need for chemical interventions, contributing to more sustainable farming practices. The challenges posed by tomato biomass waste, including its volume, types, environmental impacts, logistical constraints, and disease and pest issues, require innovative solutions. Approaches such as biorefinery technologies, component-specific waste utilization, biocomposites, composting, biochar production, and mobile processing units offer promising ways to address these challenges. Moreover, the risk of diseases and pests will be minimized, because the system allows for the integration of CEA and biopesticide production methods for more viable farming.

By transforming waste into valuable resources, these solutions can help reduce environmental impacts, improve resource efficiency, and promote sustainability within the agricultural industry. The potential applications of tomato biomass are very broad, ranging from the production of bioenergy and recovery of materials to environmental remediation and pest management. As these technological advancements continue, the tomato industry has the potential to serve as a model for waste reduction, resource recovery, and environmental management.

18.6 CONCLUSIONS AND FUTURE PERSPECTIVES

The future of tomato biomass utilization is promising, with numerous opportunities for innovation and sustainability year by year. Biorefineries will most likely continue to evolve, giving more efficient ways to convert tomato biomass into high-value products such as biofuels, bioplastics, and biochemicals. These advancements will eventually contribute to the global shift toward renewable energy sources and bio-based materials, thus reducing dependence on fossil fuels, which are non-renewable energy and addressing the growing problem of plastic pollution. The use of biodegradable packaging materials derived from tomato biomass could become more widespread, particularly in industries in a way to reduce the environmental impact. As the demand for sustainable alternatives to traditional plastics grows, tomato-based biocomposites and biodegradable films could play a significant role in reducing plastic waste and improving food preservation. In addition, mobile processing units will likely become more advanced and widely used, particularly in rural regions. These units could provide farmers with immediate access to renewable energy and soil amendments, reducing transportation costs and enhancing the sustainability of their operations by processing biomass on-site. The use of biochar and composting will further contribute to sustainable farming practices by improving soil health and sequestering carbon, supporting global climate change mitigation efforts. Moreover, the integration of tomato biomass waste management into a broader circular economy model presents opportunities for industries, farmers, and policymakers to collaborate on sustainable agricultural practices. Policy frameworks and incentives will be critical to encouraging the adoption of these technologies at scale. Finally, further exploration of partnerships between the food processing industry, waste management companies, and research institutions could drive innovation and implementation of practical, sustainable solutions. As the tomato industry continues to adopt innovative waste management practices, it could become a model for sustainability within the broader agricultural sector. By transforming waste into valuable resources, the industry has the potential to significantly reduce its environmental impact while promoting resource efficiency and sustainability. The continuous development of bio-based products, renewable energy, and CEA will not only enhance the value of tomato biomass but also contribute to more resilient and sustainable agricultural systems worldwide. In a nutshell, tomato biomass waste presents both a challenge and an opportunity for the agricultural industry. As the demand for tomatoes and their processed forms continues to grow, so does the need for effective waste management strategies and planning. This chapter has highlighted the current issues associated with tomato biomass, including environmental degradation and resource inefficiency, while also exploring its untapped potential. Various pathways, innovative technologies, and sustainable practices in tomato biomass management exist for turning waste into valuable resources. Not only will doing so mitigate the environmental impacts of agricultural waste, but also the agricultural sector can benefit economically by embracing resource-efficient and circular economy principles. Ultimately, the future of agro-biomass management lies in integrating sustainability into every aspect of agricultural production, with tomato biomass being a prime example of how waste can be converted into opportunity.

ACKNOWLEDGMENT

This chapter was written by Adrienne Chan Yi Ning, Lim Chun Xi, Mohamad Zahin Bin Mohamad Zaidi, and Tan Zi Ying.

REFERENCES

Aba, I. P., Gana, Y. M., Ogbonnaya, C., & Morenikeji, O. O. (2012). Simulated transport damage study on fresh tomato (Lycopersicon esculentum) fruits. *Agricultural Engineering International: CIGR Journal*, 14, 119–126.

Almeida, P. V., Gando-Ferreira, L. M., & Quina, M. J. (2024). Tomato residue management from a biorefinery perspective and towards a circular economy. *Foods*, 13(12), 1873. https://doi.org/10.3390/foods13121873

Arah, I. K., Kumah, E. K., Anku, E. K., & Amaglo, H. (2015). An overview of post-harvest losses in tomato production in Africa: Causes and possible prevention strategies. *Journal of Biology, Agriculture and Healthcare*, 5(16), 78–88.

Atanda, S. A., Pessu P. O., Agoda S., Isong I. U., & Ikotun I. (2011). The concepts and problems of post–harvest food losses in perishable crops. *African Journal of Food Science,* 5(11): 603–613.

Awogbemi, O., & Kallon, D. V. V. (2022). Valorization of agricultural wastes for biofuel applications. *Heliyon*, 8(10), e11117. https://doi.org/10.1016/j.heliyon.2022.e11117

Cherono, K., & Workneh, T. S. (2018). A review of the role of transportation on the quality changes of fresh tomatoes and their management in South Africa and other emerging markets. *International Food Research Journal,* 25(6), 2211–2228.

Durmuş, M., & Kızılkaya, R. (2022). The effect of tomato waste compost on yield of tomato and some biological properties of soil. *Agronomy*, 12(6), 1253. https://doi.org/10.3390/agronomy12061253

El Mashad, H. M., Zhao, L., Zhang, R., & Pan, Z. (2019). Tomato. In: Pan, Z., Zhang, R., & Zicari, S. (Eds.), *Integrated Processing Technologies for Food and Agricultural By-Products* (pp. 107–131). Elsevier. https://doi.org/10.1016/B978-0-12-814138-0.00005-8

El-Bassi, L., Azzaz, A. A., Jellali, S., Akrout, H., Marks, E. A. N., Ghimbeu, C. M., & Jeguirim, M. (2021). Application of olive mill waste-based biochars in agriculture: Impact on soil properties, enzymatic activities and tomato growth. *Science of the Total Environment*, 755, 142531. https://doi.org/10.1016/j.scitotenv.2020.142531

Eslami, E., Carpentieri, S., Pataro, G., & Ferrari, G. (2023). A comprehensive overview of tomato processing by-product valorization by conventional methods versus emerging technologies. *Foods*, 12(1), 166. https://doi.org/10.3390/foods12010166

Fărcaş, A. C., Socaci, S. A., Michiu, D., Biriş, S., & Tofană, M. (2019). Tomato waste as a source of biologically active compounds. Bulletin of University of Agricultural Sciences and Veterinary Medicine Cluj-Napoca. *Food Science and Technology*, 76(1). https://www.cabidigitallibrary.org/doi/full/10.5555/20193361560

Fritsch, C., Staebler, A., Happel, A., Márquez, M. A. C., Aguiló-Aguayo, I., Abadias, M., Gallur, M., Cigognini, I. M., Montanari, A., López, M. J., Suárez-Estrella, F., Brunton, N., Luengo, E., Sisti, L., Ferri, M., & Belotti, G. (2017). Processing, valorization and application of bio-waste derived compounds from potato, tomato, olive and cereals: A review. *Sustainability (Switzerland)*, 9(8). MDPI. https://doi.org/10.3390/su9081492

Giorgio, D. A. (2017). Garbage to glamour: Recycling food byproducts for skin care. https://www.cosmeticsandtoiletries.com/research/methods-tools/article/21836720/garbage-to-glamour-recycling-food-byproducts-for-skin-care

Grenier, V., Moingt, M., Lucotte, M., & Pitre, F. E. (2022). Dissipation and effect of glyphosate during composting of organic wastes. *Journal of Environmental Quality, 51*(3), 300–410.

Howarth, G. (2024, March 29). The environmental impact of tomatoes: From farm to table. *Impactful Ninja*. https://impactful.ninja/the-environmental-impact-of-tomatoes/

Imraan, M., Ilyas, R. A., Norfarhana, A. S., Bangar, S. P., Knight, V. F., & Norrrahim, M. N. F. (2023). Sugar palm (Arenga pinnata) fibers: New emerging natural fibre and its relevant properties, treatments and potential applications. *Journal of Materials Research and Technology*, 24, 4551–4572. https://doi.org/10.1016/j.jmrt.2023.04.056

Irokanulo, E. O., Egbezien, I. L., & Owa, S. O. (2015). Use of Moringa oleifera in the preservation of fresh tomatoes. *IOSR Journal of Agriculture and Veterinary Science*, 8, 127–132.

Kabir, M. S. N., Rasool, K., Lee, W.-H., Cho, S.-I., & Chung, S.-O. (2020). Influence of delayed cooling on the quality of tomatoes (Solanum lycopersicum L.) stored in a controlled chamber. *AIMS Agriculture & Food,* 5(2), 272–285.

Koçak, E., Esmer, O. K., & Sahiner, A. (2023). Optimization of the conditions of alkaline extraction of tomato peels and characterization of tomato peel extracts obtained under atmospheric and oxygen free conditions. *Anais da Academia Brasileira de Ciências,* 95(suppl 1), e20220077.

Laranjeira, T., Costa, A., Faria-Silva, C., Ribeiro, D., de Oliveira, J. M. P. F., Simões, S., & Ascenso, A. (2022). Sustainable valorization of tomato by-products to obtain bioactive compounds: Their potential in inflammation and cancer management. *Molecules,* 27(5), 1701.

Lehmann, J., & Joseph, S. (Eds.). (2015). *Biochar for Environmental Management: Science, Technology, and Implementation*. Earthscan/Routledge. https://doi.org/10.4324/9780203762264

Li, Z., Li, P., Yang, H., Liu, J., & Xu, Y. (2012). Mechanical properties of tomato exocarp, mesocarp and locular gel tissues. *Journal of Food Engineering*, 111(1), 82–91. https://doi.org/10.1016/j.jfoodeng.2012.01.023

Lin, Y., Munroe, P., Joseph, S., Henderson, R., & Artur Ziolkowski, A. (2012). Water extractable organic carbon in untreated and chemical treated biochars. *Chemosphere*, 87, 151–157.

Lu, Z., Wang, J., Gao, R., Ye, F., & Zhao, G. (2019). Sustainable valorisation of tomato pomace: A comprehensive review. *Trends in Food Science & Technology,* 86, 172–187.

Maddox, D. A. (1998). Implications of new technologies for seed health testing and the worldwide movement of seed. *Seed Science Research,* 8(2), 277–284.

Mai, A. L.-D., & Pathare, P. B. (2021). Kinetic modeling of quality changes of tomato during storage. *Agricultural Engineering International: CIGR Journal,* 23(1), 183–193.

Majidi, H., Minaei, S., Almassi, M., & Mostofi, Y. (2014). Tomato quality in controlled atmosphere storage, modified atmosphere packaging and cold storage. *Journal of Food Science and Technology,* 51, 2155–2161.

Mangut, V., Sabio, E., Gañán, J., González, J. F., Ramiro, A., González, C. M., Román, S., & Al-Kassir, A. (2006). Thermogravimetric study of the pyrolysis of biomass residues from the tomato processing industry. *Fuel Processing Technology,* 87(2), 109–115.

Morone, P., Tartiu, V. E., & Falcone, P. M. (2015). Assessing the potential of bioplastics made of tomato peel waste. *Journal of Cleaner Production, 93*, 42–54. https://doi.org/10.1016/j.jclepro.2015.01.011

Norfarhana, A. S., Ilyas, R. A., & Ngadi, N. (2022). A review of nanocellulose adsorptive membrane as multifunctional wastewater treatment. *Carbohydrate Polymers*, 291, 119563. https://doi.org/10.1016/j.carbpol.2022.119563

Nour, V., Panaite, T. D., Ropota, M., Turcu, R., Trandafir, I., & Corbu, A. R. (2018). Nutritional and bioactive compounds in dried tomato processing waste. *CyTA-Journal of Food,* 16(1), 222–229.

Prasad, K. M., & Murugavelh, S. (2020). Experimental investigation and kinetics of tomato peel pyrolysis: Performance, combustion and emission characteristics of bio-oil blends in diesel engines. *Journal of Cleaner Production,* 254, 120115.

Ragaveena, S., Shirly Edward, A., & Surendran, U. (2021). Smart controlled environment agriculture methods: A holistic review. *Reviews in Environmental Science and Bio/Technology*, 20(4), 887–913. https://doi.org/10.1007/s11157-021-09591-z

Rajkowska, K., Otlewska, A., Raczyk, A., Maciejczyk, E., & Krajewska, A. (2024).Valorisation of tomato pomace in anti-pollution and microbiome-balance face cream. *Scientific Reports,* 14, 20516. https://doi.org/10.1038/s41598-024-71323-4

Rasool, M., Akhter, A., Soja, G., & Haider, M. S. (2021). Role of biochar, compost and plant growth promoting rhizobacteria in the management of tomato early blight disease. *Scientific Reports*, 11(1), 6092. https://doi.org/10.1038/s41598-021-85633-4

Ravindran, B., Nguyen, D. D., & Bharagava, R. N. (2020). Challenges and opportunities in bioconversion of agricultural waste into high value-added materials: Emerging technologies and future trends. *Bioresource Technology, 300*, 122–134. https://doi.org/10.1016/j.biortech.2019.122134

Sharma, H. B., Vanapalli, K. R., & Imani, M. (2020). Challenges and solutions to the sustainable management of food waste in developing countries. *Waste Management, 105*, 227–238. https://doi.org/10.1016/j.wasman.2020.02.013

Silva, F. D., Garcia, V. A. D. S., Vanin, F. M., Yoshida, C. M. P., & de Carvalho, R. A. (2023). Application of tomato byproduct in food products: A review. *Food Science and Engineering*, 4(1), 103–15. https://ojs.wiserpub.com/index.php/FSE/article/view/1566

Sorrenti, V., Burò, I., Consoli, V., & Vanella, L. (2023). Recent advances in health benefits of bioactive compounds from food wastes and by-products: Biochemical aspects. *International Journal of Molecular Sciences* 24(3), 2019.

Szabo, K., Cătoi, A.-F., & Vodnar, D. C. (2018). Bioactive compounds extracted from tomato processing by-products as a source of valuable nutrients. *Plant Foods for Human Nutrition*, 73(4), 268–277. https://doi.org/10.1007/s11130-018-0691-0

Tallou, A., Aziz, F., Garcia, A. J., Salcedo, F. P., El Minaoui, F. E., & Amir, S. (2022). Bio-fertilizers issued from anaerobic digestion for growing tomatoes under irrigation by treated wastewater: targeting circular economy concept. *International Journal of Environmental Science and Technology*, 19(4), 2379–2388. https://doi.org/10.1007/s13762-021-03265-7

Teigiserova, D. A., Hamelin, L., & Thomsen, M. (2019). Review of high-value food waste and food residues biorefineries with focus on unavoidable wastes from processing. *Resources, Conservation and Recycling*, 149, 413–426. https://doi.org/10.1016/j.resconrec.2019.05.003

Trombino, S., Cassano, R., Procopio, D., Di Gioia, M. L., & Barone, E. (2021). Valorization of tomato waste as a source of carotenoids. *Molecules*, 26(16). MDPI AG. https://doi.org/10.3390/molecules26165062

Vella, F. M., Laratta, B., & Calabrese, G. (2017). Lycopene from tomato by-products: A biochemical and molecular approach for waste valorization. *Journal of Food Processing and Preservation*, 41(2), e12943. https://doi.org/10.1111/jfpp.12943

Williams, C. L., Westover, T. L., Emerson, R. M., Tumuluru, J. S., & Li, C. (2016). Sources of biomass feedstock variability and the potential impact on biofuels production. *BioEnergy Research*, 9(1), 1–14. https://doi.org/10.1007/s12155-015-9694-y

Wogu, M. D., & Ofuase, O. (2014). Microorganisms responsible for the spoilage of tomato fruits, Lycopersicum esculentum, sold in markets in Benin City, Southern Nigeria. *School of Academics and Journal of Biosciences,* 2(7), 459–466.

Zuorro, A., Lavecchia, R., & Medici, F. (2014). Valorization of tomato processing residues for the production of bioplastic materials. *Journal of Cleaner Production*, 67, 195–202. https://doi.org/10.1016/j.jclepro.2014.01.035

19 Olive Waste Management

19.1 INTRODUCTION

Agricultural waste is the organic or inorganic materials discarded from agricultural operations like farming, poultry, crop residues, or runoff fertilizer. The waste can include processing or post-process waste like stalks, rice straws, dead leaves, seeds, peels, or plastic packaging. When all these unused yields are utilized as a valuable source, environmental pollution issues can be controlled. The importance of managing agro-biomass waste is the continuous production of bioenergy to conquer the high demand for useful energy worldwide. Biomass from agricultural fields consists of plants, crops, fisheries, livestock, forests, municipal waste, and many more are utilized as valuable feedstock to produce bioenergy. Bioenergy as renewable energy substitutes conventional fossil fuels like oil, natural gas, and coal, which in turn decreases the high emission of carbon dioxide into the atmosphere. In this way, air pollution and greenhouse effect are controlled. Moreover, reusing agricultural solid waste to be transformed into other useful products also helps save a lot of raw resources to make the products. For example, in livestock operations, large quantities of feathers are generated during poultry processing, mainly the feathers of mature chickens. These feathers are rich in keratinous protein, which is good for nerve generation, tissue repair, and healing wounds. Therefore, feathers are suitable to be made as livestock animal feed, which will enhance the quality of meat (Nagal & Jain, 2010). Furthermore, agricultural solid waste like cocoa pod husk can be used to make potash and black soap traditionally for cleaning and bathing (Bannor et al., 2024).

Olive biomass refers to the organic material discarded from olive trees, including branches, pits, and leftover pulp during olive oil production. Olive trees play a huge role in sustaining the clean and green energy used in agriculture and industry. In the wide scale of olive tree plantation, the unproductive branches from the olive trees will be cut down every 2 years, leading to a large amount of olive tree pruning (OTP), which yields a massive available resource to create bioenergy or use as biofuel (Maggiotto et al., 2023). A study shows Greece is one of the ideal countries for generating massive bioenergy to be used in electricity generation because Greece can generate a high amount of OTP (Sagani & Hagidimitriou, 2014). Table 19.1 shows that a very

TABLE 19.1
Olive Tree Pruning Production and Relevant Thermal Energy Capacity Per Prefecture (Sagani & Hagidimitriou, 2014)

Prefecture	OTP Production (Dry Tons/Year)	Thermal Energy Capacity (GJ/Year)
Messenia	55,467	914,092
Laconia	44,783	738,023
Chalkidiki	20,475	337,428
Phthiotis	22,645	373,195
Euboca	20,909	344,581
Lesvos	29,979	494,062
Heraklion	54,783	902,822
Chania	28,313	466,595

DOI: 10.1201/9781003472209-26

huge amount of OTP can be produced every year in each prefecture in Greece, alongside the thermal energy generated annually. This thermal energy can be used as a heating source rather than using fossil fuels.

In the construction industry, ash from burnt olive can be used as a cement improver. A study shows that olive biomass concrete, which contains olive ash, has similar or higher compressive strength than normal concrete. When producing mortars, Al-Akhras and Abdulwahid (2010) mixed olive waste ash (OWA) into the mortars to replace Portland cement and mortar sand. The innovative mortars with appropriate portions of replacement reveal significant effects on the reduction of workability, an increase in compressive and flexural strength when replacing sand, and a decrease when replacing Portland cement. However, when more OWA is replaced for the sand, the compressive and flexural strength will significantly decrease. It seems that, although OWA is not perfect, the excessive biowaste lets the industry have more material choices to be manipulated so that the other resources can be preserved (Mohamed et al., 2023).

Uncollected waste could have a massive impact on the environment. Municipal solid waste (MSW) is the organic and inorganic waste produced from domestics, commercials, and industrials. Poor management of MSW due to lack of regulatory policies, financial ability, and infrastructure like in India leads to an increase in unprocessed street garbage, which will jeopardize environmental quality. Agriculture waste like livestock manure, ashes, dead plants, or rotten crops can pollute the soil and water when left uncleaned (Sajid et al., 2023). Throwing agricultural wastes like pesticides, which are chemically toxic, into the rivers or ponds will result in water pollution and food poisoning. In addition, agricultural solid waste that is burnt will also result in massive emissions of harmful gases like carbon dioxide and other toxic gases. This causes the rise of the greenhouse effect, and the temperature of Earth will gradually increase. Consequently, climate change will become more unpredictable, and more disasters will occur (Oluseun & Adebukola, 2021). Therefore, waste is like a double-edged sword; it only brings a positive impact if humankind can brightly utilize waste to improve the environmental situation.

Agricultural waste, including olive biomass, holds significant potential for generating bioenergy and reducing environmental pollution. By utilizing unused yields from olive trees, we can not only produce renewable energy but also reduce reliance on fossil fuels and minimize carbon emissions. Moreover, reusing agricultural waste for valuable products, such as animal feed and natural cleaning materials, contributes to resource conservation and environmental sustainability. Thus, efficient management of agro-biomass waste is essential for a cleaner and more sustainable future. Thus, this chapter aims to explore the potential of various agricultural wastes, particularly olive biomass, in bioenergy production and resource conservation.

19.2 CURRENT ISSUES AND CHALLENGES ASSOCIATED WITH OLIVE BIOMASS

19.2.1 Type of Waste Produced

Twenty percent of olive oil is produced during the extraction process, and, as by-products, there is also 30% of wet solid waste with 65% moisture (known as pomace or jift) (Alouiz et al., 2025) and 50% of liquid waste known as olive mill wastewater (OMW) or zibar.

19.2.1.1 Olive Oil Pomace

Olive pomace, or crushed olive stones, constitutes the majority of the residual by weight from the olive oil extraction process, together with vegetative water, process water, and all fruit by-products other than olive oil. The most common way of producing olive oil in Spain, which is the primary producer of olive oil worldwide (Erses Yay et al., 2012; Millán et al., 2014), is the two-phase separation mode, as shown in Figure 19.1, which yields pomace with higher moisture content (55–70%)

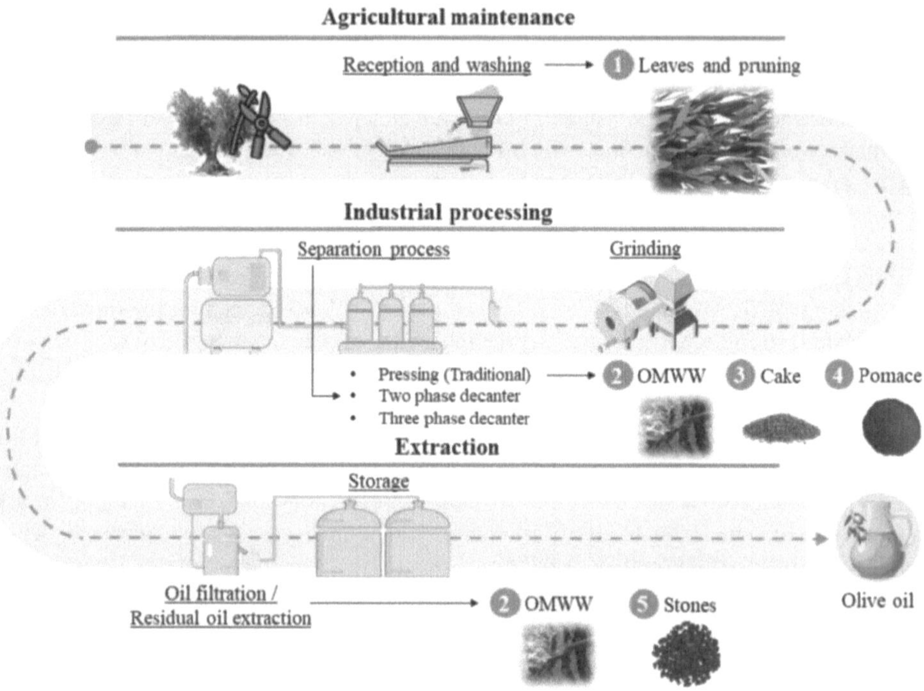

FIGURE 19.1 Simplified process of olive processing for olive oil production. By-products of the olive oil industry from agricultural maintenance (leaves and pruning biomass) and industrial processes (aqueous olive mill wastewater, cake, pomace, and stones) (Otero et al., 2021).

than the three-phase operation (40–45%) (Borja et al., 2006). Olive pomace is mostly composed of water, although it also contains lipids (remaining oil), polyphenols, carbohydrates, and many metals and salts with an average pH of 4.8–5.2. Its large organic charge also contributes to its high chemical and biological oxygen consumption (Romero-García et al., 2014).

19.2.1.2 Olive Mill Wastewater

OMW is olive fruit vegetative water mixed with soft tissues, which is used in several stages of olive oil processing (Ayoub, 2017; Oreopoulou & Russ, 2007). Three distinct processes result in OMW generation which is cleaning the olive fruits, washing the virgin olive oil from the secondary centrifuge, and the horizontal centrifuge (decanter) during the three-phase separation step. OMW contains sugars (1.6–4%) and phenolic compounds responsible for the difficulty of proper treatment. Since OMW contains a range of toxins, it becomes an environmental burden if improperly controlled and treated before disposal.

19.2.2 Volume of Waste Produced

When extracting olive oil, several procedures are carried out, such as leaf removal, washing the olives, grinding, beating, and oil separation. The extraction process selected determines the properties and quantity of wastewater generated. The press extraction system, which uses pressure to extract material, is the oldest known technique (Ribalta et al., 2021). However, continuous centrifugation systems, such as the two- and three-phase systems, are being used because of technological improvement. Water is supplied in a three-phase system to speed up the extraction process, up to 130 L for every 100 kg of olives (Foti et al., 2021). OMW is generated more than with other methods. Water is only injected at the final vertical centrifugation step in the two-phase continuous

TABLE 19.2
Input–Output Material and Energy in the Olive Mill Extraction Systems (Medouni-Haroune et al., 2018; Mercy Corps, 2017)

System	Input	Quantity	Output	Quantity
Traditional press extraction	Olive	1 ton	Oil	200 kg
	Rinsing water	100–200 L	Olive pomace (35% humidity)	400–600 kg
	Energy	40–60 kWh	OMW	400–600 L
Three-phase extraction	Olive	1 ton	Oil	200 kg
	Rinsing water	100–120 L	Olive pomace (50% water content)	500–600 kg
	Water added	700–1,000 L	OMW	1,000–1,200 L
	Energy	90–117 kWh		
Two-phase extraction	Olive	1 ton	Oil	200 kg
	Rinsing water	100–120 L	Olive wet pomace (65% humidity)	800 kg
	Energy	<90–117 kWh	Wastewater	100–150 L

centrifugation system (Ochando-Pulido et al., 2020), which uses less energy and water than the other systems. As a result, the volume of wastewater produced is reduced by an average of one-third compared to the three-phase system, and it mostly comes from washing water (Azbar et al., 2004). The two-phase method is therefore thought to be more ecologically beneficial (Khdair & Abu-Rumman, 2020). In Spain, Italy, Portugal, Greece, and other Mediterranean nations, the two-phase approach is becoming increasingly popular (Sygouni et al., 2019). Hence, the different methods will induce different amounts of water being used and bring the consequences of the different volumes of wastewater and oil produced. The input and output of the process are summarized in Table 19.2.

19.2.3 Challenges Posed by Olive Pomace and Oil Production Waste

19.2.3.1 Environmental Concerns

The primary issues facing the olive oil business are related to the disposal and management of this liquid and solid waste because of its high organic load (Cardinali et al., 2010) and phenolic compounds that are phytotoxic and antibacterial yet challenge biological breakdown. To conduct an environmental audit and analyze the chemical characteristics of their waste, 15 olive mills in the governorates of Irbid and Ajloun in northern Jordan were chosen. Table 19.3 lists the pollution loads per 1 ton of processed olives as a function of extraction techniques. These mills used a variety of oil extraction techniques, including traditional, two-phase, and three-phase continuous techniques. Several parameters in the OMW's chemical characteristics are directly influenced by the oil extraction system including residual oil, phenols, biological oxygen demand (BOD), and chemical oxygen demand (COD). OMWs contain high concentrations of total suspended solids (TSSs), phenols, and other organic matter as shown in this study. The organic load is characterized by high levels of BOD (20–22 kg/ton) and COD (52–56 kg/ton), in addition to having a very high concentration of residual olive oil of 7 kg/ton from traditional mills.

For every ton of processed olives, the conventional mills generate the most pollutants. Table 19.3 shows that, by using the traditional method, they produced 21.94 kg/ton of BOD5, 56.18 kg/ton of COD, 7.02 kg/ton of residual olive oil, 20.33 g/ton of phenol, and 27.79 kg/ton of TSSs. In terms of BOD_5, COD, residual olive oil, phenol, and TSSs, the three-phase mill yields the second-highest pollution load of 20.29 kg/ton, 52.63 kg/ton, and 0.32 g/ton (Khdair et al., 2019). Conversely, the lowest pollutant load per ton of processed olives was generated by the two-phase mills. They produced 23.5 g/ton of BOD5, 142.2 g/ton of COD, 5.7 g/ton residual olive oil, 0.31 g/ton phenol, and 68.1 g/ton TSSs, which may be classified as an ecologically friendly machine in terms of generated

TABLE 19.3
The Pollution Loads Per 1 Ton of Processed Olives as a Function of Extraction Techniques (Khdair et al., 2019)

Type of Mill	BOD_5 (kg/ton)	COD (kg/ton)	Residual Olive Oil (kg/ton)	Phenols (g/ton)	TSS (Total Suspended Solid) (kg/ton)	Phosphorus (g/ton)
Traditional	21.94	56.18	7.02	20.33	27.79	0.13
Two-phase	0.0235	0.1422	0.0057	0.3007	0.0681	0.0008
Three-phase	20.29	52.63	0.32	0.32	24.68	0.15

wastewater quantity and quality. In a nutshell, the disposal of olive waste has brought a lot of pollutants to the environment, and the ways to dispose of all that waste are very important to conserve the environment.

Water utilization in the mills is influenced by pressing methods and labor practices (three-phase mills use a lot of water). Spreading odorous and potentially infectious elements throughout the environment is a drawback of the unregulated discharge of untreated OMWs on land and in water. OMWs, as they are known locally, are composed of the olive plant's vegetative waters, often combined with additional water. They have a strong, harmful impact and a large capacity for pollution. Due to its COD content of up to 220 g/L, OMWs are one of the "strongest" industrial effluents (Salmoral et al., 2010).

OMW is characterized by a high organic load, COD, and BOD. If this type of effluent is released into water supplies, the organic materials will be consumed by ambient microorganisms, which will lower the oxygen content of the water and have an impact on aquatic life (Azzam et al., 2004). Additionally, it has a high concentration of phenol, which is toxic and harmful to plants. If it meets chlorinated water, it can create chlorinated phenols, which are the source of the disagreeable odor (Azzam et al., 2004). OMW is difficult to treat biologically because of its low acidity pH (Stoller et al., 2016). OMW contains a lot of fat and solid matter. A lot of solid matter might encourage sediments to accumulate in the wastewater network, which can cause blockage and corrosion in the sewer systems. Fat will float to the top of the water's surface and block the depth at which sunlight may penetrate the water. OMW contains a lot of phosphorus, which, if it reacts with water bodies, may speed up the growth of algae and cause eutrophication.

19.2.3.2 Poor Waste Management

The production of odors was noticed in the vicinity of mills, which are mostly caused by poor housekeeping, unsanitary conditions, inappropriate waste storage locations, extended pomace or OMW storage periods, and meteorological factors. Due to gas emissions from the fermentation process that occurs when the OMWs are dumped into a land or water body and/or kept in open ponds, there have been significant odor complaints reported around the mills. There are a number of volatile organic acids and other low-boiling organic materials that give off offensive odors. Methane and other strong gases, such as hydrogen sulfide, are also emitted as a result of the ponds evaporating, polluting the air, water streams, and soil. Since traditional mills depend heavily on manual labor, they are less hygienic and maintain poor housekeeping because most of their machinery and production processes are open to the air. This means that spills and leaks occur more frequently in traditional mill operations than in modern ones. Once more, inappropriate and extended storage directly affects the total amount of odor emissions.

According to the results of the field audits, most contemporary mills (those using two and three phases of technology) have improved hygienic conditions and better housekeeping practices, which reduce odor emissions. This feature, however, might be countered by long-term on-site storage of generated pomace (olive cake) or OMWs or by poor storage (Khdair et al., 2019). The waste management practices of the olive oil sectors in Jordan are nonexistent (Ayoub, 2017); thus, it is necessary to introduce cleaner production options and to put preventative, control, and treatment methods in place. To motivate stakeholders to uphold the green processing standards for olive oil, it is imperative that mills implement an appropriate system for managing environmental waste, training their personnel, and providing technical support. Utilizing oil/water separators at the mills to recover leftover olive oil and lower pollution loads to the surrounding environment should be promoted. This will recover 150 tons of olive oil annually and minimize water consumption during production by using ecological decanters (two-phase mills).

To avoid and eliminate pollution from olive oil extraction processes, it is essential to launch a clear framework or guidelines that fill the existing legislative gap to regulate and control the safe disposal of olive mill wastes (Ayoub, 2017; Halalsheh et al., 2021). This study explores the availability of international legislation and policies implemented to guarantee a safe olive mill waste disposal and its applicability in a hot climate using Jordan as a case study.

19.3 INNOVATIVE SOLUTIONS AND POTENTIAL APPLICATIONS

19.3.1 Waste Management Hierarchy

The waste management hierarchy is a systematic structure for prioritizing eco-friendly trash reduction, pollution control, and by-product management solutions. At the top of this hierarchy is waste prevention, which emphasizes the importance of reducing waste generation at the source through efficient design and consumption behaviors. The emphasis then changes to reducing waste through reuse and repair, which promotes product longevity and reduces resource extraction. Following that, recycling and composting are critical for turning garbage into useful resources, diverting it from landfills, and cutting greenhouse gas emissions. Energy recovery, albeit lower on the hierarchy, still contributes by converting trash into energy and lowering dependency on fossil fuels. Finally, landfilling and incineration are regarded as the least desirable solutions, reserved for garbage that cannot be managed in other ways. Adhering to this hierarchy allows communities and organizations to drastically reduce pollution, encourage resource conservation, and develop a circular economy that values sustainability.

19.3.2 Energetic Valorization of Olive Biomass

Converting olive-derived biomass into energy is a highly effective and sustainable method for managing waste from olive oil production. Biomass, including olive pomace, OMW, and OTPs, serves as a renewable energy source through various processes such as direct combustion, biogas production, and torrefaction. These methods convert organic matter into heat, electricity, and biofuels like bioethanol and methane (Najafi et al., 2021). Utilizing olive biomass for energy not only reduces the environmental burden of waste disposal but also generates valuable energy that can support agricultural operations or supply local communities.

19.3.2.1 Incineration and Combustion

Incineration of olive biomass is one of the practical methods of energy recovery, particularly in regions where olive processing generates significant organic waste, such as olive stones, husks, and pomace (Christoforou & Fokaides, 2016). The energy produced from burning these biomasses can be used in small-scale energy applications, such as heating homes and greenhouses or powering small industries. This approach is particularly beneficial in rural or agricultural areas where access to traditional energy sources may be limited or expensive. By using olive biomass as a fuel, waste from olive production is reduced, while energy is generated in an eco-friendly and sustainable manner. However, this energy source typically has limitations in terms of capacity and scale. Olive biomass incineration might not be sufficient to power large industries or supply extensive energy grids, but it can significantly contribute to reducing reliance on fossil fuels at a local level. Moreover, the incineration of olive biomasses also leads to by-products like ash, which can be further used in agriculture as a soil amendment, offering a secondary benefit of waste valorization. For instance, ashes produced from combusting olive tree pruning biomass (OTPB) can be used as a source of phosphorus and as a component in the construction of red clay bricks (Lo Giudice et al., 2021). However, emissions of greenhouse gases and ensuring clean combustion processes are difficult to control to minimize environmental impacts, such as air pollution or the release of particulate matter.

19.3.2.2 Bioethanol Production

Bioethanol production from olive biomass entails converting organic resources such as olive pomace, leaves, and pruning into ethanol via a series of biochemical processes (Contreras et al., 2020; Fernandes et al., 2016). The process starts with the pretreatment of olive biomass, which breaks down its complex lignocellulosic structure and makes the sugars more available for fermentation. Mechanical or chemical processes, such as a steam explosion or acid hydrolysis, are commonly employed to extract fermentable sugars, chiefly glucose, and xylose. These sugars are subsequently fermented by yeast, such as *Saccharomyces cerevisiae*, yielding ethanol and carbon dioxide. Following fermentation, ethanol is separated by distillation, and the residual residue, known as stillage, can be used for various purposes including energy generation. This approach provides a renewable alternative fuel source, lowering reliance on fossil fuels while utilizing the plentiful waste from olive oil manufacturing.

19.3.2.3 Methane Production

Methane synthesis from olive biomass is primarily accomplished by anaerobic digestion, a biological process in which microbes break down organic material in the absence of oxygen (Messineo et al., 2020). Olive biomass, which includes olive pomace, leaves, and prunings, is crushed and occasionally pre-treated to aid in the breakdown of its lignocellulosic structure. During anaerobic digestion, bacteria break down complex chemical components like carbohydrates and lipids into simpler ones. The biomass undergoes hydrolysis, acidogenesis, acetogenesis, and methanogenesis, resulting in methane (CH_4) as the main biogas component. This methane-rich biogas may be harvested and used as a renewable energy source, such as power generation, heating, or automobile fuel, helping to reduce greenhouse gas emissions and promote sustainable energy practices.

19.3.2.4 Gasification

Gasification of olive biomass is an efficient method of energy utilization that turns organic waste into a useful fuel by partial oxidation (Fernández-Lobato et al., 2022). In this process, olive biomass such as olive pomace, branches, and pruning is heated to temperatures exceeding 700°C in a controlled atmosphere with little oxygen or steam. Unlike combustion, which completely burns the biomass, gasification happens in an oxygen-depleted environment, resulting in partial breakdown of the substance. This process produces syngas, a combination of hydrogen (H_2), carbon monoxide (CO), and other gases that may be utilized as fuel for power production or industrial purposes. Furthermore, gasification generates condensable organic molecules that can be converted into bio-oil or other useful by-products. This process is useful because it optimizes the energy content of the biomass while minimizing hazardous emissions, resulting in a sustainable method of converting olive biomass into clean energy.

19.3.2.5 Pyrolysis

Pyrolysis is a thermal conversion process that takes place in the lack or near absence of oxygen, usually at temperatures ranging from 300°C to 900°C (Alcazar-Ruiz et al., 2021; Dorado et al., 2021). Organic materials, such as olive biomass, are heated to a high temperature, causing thermal breakdown to produce three major products: bio-oil, syngas, and solid biochar. Unlike burning, which completely oxidizes the biomass, pyrolysis decomposes it into volatile gases and a solid carbon-rich residue called biochar. The volatile gases, which mostly consist of hydrogen, carbon monoxide, methane, and other organic molecules, can be utilized as fuel for energy production, while bio-oil can be converted into biofuels or chemicals. Biochar, a solid by-product, finds applications in agriculture as a soil supplement and as a soil additive to sequester carbon. Pyrolysis is a sustainable approach to extract energy from olive biomass while also producing valuable by-products that may be utilized in a variety of industries.

19.3.2.6 Torrefaction

Torrefaction is a moderate thermal treatment technique that takes place in a non-oxidizing environment at temperatures up to roughly 300°C. During this process, biomass, such as olive waste materials, undergoes partial decomposition, which removes moisture and volatile organic components while increasing its energy density (Alshareef et al., 2021; Trubetskaya et al., 2020). Torrefaction, unlike pyrolysis or gasification, takes place at lower temperatures, causing the biomass to become more hydrophobic and simpler to ground, making it ideal for use as a solid fuel in a variety of combustion systems. Torrefied biomass, the resultant product, has enhanced qualities such as higher energy content, lower oxygen concentration, and increased carbon content, making it a better fuel source than untreated biomass. Torrefied biomass is also easy to handle and transport, with energy properties like coal, making it an attractive alternative fuel for renewable energy applications.

19.3.3 Chemical Valorization of Olive Biomass

Olive biomass is usually thought to possess a high concentration of antioxidants and lignin, making it an important resource for a variety of businesses. These molecules have a wide range of uses, including food preservation, health supplements, and bioplastics. The high antioxidant concentration of olive biomass, notably polyphenols such as hydroxytyrosol and oleuropein, has the potential to increase product shelf life while boosting health outcomes (Gagour et al., 2024). Furthermore, the existence of lignin, a complex organic polymer, opens up prospects for manufacturing sustainable building materials and producing biodegradable goods. By harnessing these qualities, the chemical valorization of olive biomass can help to reduce waste and promote a circular economy in which agricultural by-products are converted into useful items.

19.3.3.1 Extraction of Antioxidants

Olive biomass has been found to be abundant in antioxidants, particularly phenolic compounds, which contribute to its high antioxidant activity. These antioxidants include hydroxytyrosol, oleuropein, verbascoside, and quercetin derivatives, commonly found in olive leaves, pomace, and stones. Bioactive compounds vary in concentration depending on the type of olive biomass and the extraction methods used. For example, olive leaves are particularly rich in oleuropein, while olive pomace contains higher levels of hydroxytyrosol. Table 19.4 refers to the components of antioxidants found in various olive of biomass (Ruiz et al., 2017). According to Ruiz et al. (2017), olive-derived biomass is a significant source of antioxidant chemicals that may be used in a variety of industries. OTPB includes hydroxytyrosol, oleuropein, and other phenolics (47–139 mg/g), which have antioxidant activity similar to synthetic antioxidants. Chemlali olive stones have a high concentration of nuezhenide (up to 325.78 mg/100 g), which has good antioxidant potential. Olive leaves include chemicals such as hydroxytyrosol, verbascoside, and oleuropein, which have antibacterial, antiviral, and cosmetic properties, with Picual olive leaves having the greatest phenolic concentration. Olive pomace, which contains a total phenolic concentration of 59.5 mg/g, has been shown to be effective in lowering fat and oxidative stress, while pomace-derived squalane has important use in cosmetic and pharmaceutical industries. Another effective source of antioxidants is olive mill effluent, which contains hydroxytyrosol and its derivatives. It has the potential to enhance olive oil or be employed as a stable bio-antioxidant for medicinal purposes.

TABLE 19.4
Component of Antioxidants Found in Various Olive of Biomass

Olive Biomass	Component of Antioxidants Found
Olive tree pruning biomass (OTPB)	Hydroxytyrosol, homovanillyl alcohol, oleuropein
Olive stones	Hydroxytyrosol, tyrosol, nuezhenide, methoxy derivative of nuezhenide, nuezhenide11-methyloleoside, and oleuropein
Olive leaves	Hydroxytyrosol, verbascoside, luteolin-*O*-glucoside, apigenin7-*O*-glucoside, oleuropein ligstroside, oluropein-aglycone, caffeic acid
Olive pomace	Tyrosol, apigenin, oleuropein, squalene
Olive mill wastewater	Hydroxytyrosol
	3,4-DHPEA-EDA
	p-HPEA-EDA
	3,4-DHPEA-EA
	Triacetylhydroxytyrosol

19.3.3.2 Extraction of Polysaccharide

Olive-derived biomass is a valuable source of polysaccharides, including pectin, which can be utilized as stabilizers, gelling agents, and emulsifiers in various industries (Gómez-Cruz et al., 2024). Pectin fractions have been successfully extracted from both OMW and two-phase olive pomace, demonstrating functional properties comparable to commercial citrus pectin. This makes pectin derived from olive biomass a promising alternative for applications in food processing, pharmaceuticals, and cosmetics due to its bio-functional characteristics. Additionally, olive-derived biomass has shown potential in nanocellulose production. Specifically, OTP waste has been utilized to create cellulose nanofibers (Kian et al., 2020). These nanocellulose materials have applications in fields such as pulp and paper reinforcement, improving the barrier properties of coatings, and even drug delivery systems. The efficient extraction and utilization of polysaccharides from olive biomass align with the concept of a biorefinery, supporting the sustainable use of agricultural by-products for high-value applications.

19.3.4 Olive Biomass Products for Animal Feed

Olive oil by-products have gained recognition as a valuable addition to animal feeds, including those used in aquaculture. Numerous studies have highlighted their positive effects on farmed fish, particularly in terms of carcass composition, growth, intestinal health, microbial diversity, and immunity. Olive-derived biomass, rich in antioxidants, polyphenols, and other bioactive compounds, has proven to be a beneficial dietary component. By carefully designing feeds that incorporate specific nutrients and minerals, farmers can enhance carcass composition to meet consumer preferences. Research on species like African catfish (*Clarias gariepinus*) (Yilmaz et al., 2004), rainbow trout (*Oncorhynchus mykiss*) (Sicuro et al., 2010a), red sea bream (*Pagrus major*) (Arsyad et al., 2018), and gilthead sea bream (*Sparus aurata*) (Nasopoulou et al., 2011; Sicuro et al., 2010c; Sioriki et al., 2016) has revealed the impact of olive oil by-products on their growth and carcass quality. Most studies have focused on fatty acid profiles rather than overall body composition. For example, feeding juvenile African catfish diets containing 9% and 3% olive pomace oil (OPO) with and without L-carnitine increased muscle lipid content compared to controls (Yilmaz et al., 2004). However, no significant changes were noted in rainbow trout (Sicuro et al., 2010a) or gilthead sea bream (Sicuro et al., 2010c) diets enriched with OMW. OPO also appeared to reduce fatty acids in African catfish muscles, although the statistical significance of these results was not clearly determined. Additionally, Arsyad et al. (2018) found that olive leaf powder increased collagen and myofibril proteins in red sea bream muscles but did not affect moisture content.

Fish growth is often associated with improvements in immune function, such as humoral and mucosal immunity, and heightened antioxidant enzyme activity (Hoseinifar et al., 2020b). Baba et al. (2018b) found that incorporating olive leaves into *O. mykiss* diets improved survival rates and serum biochemical markers. Fish fed with 0.1% olive leaf showed the lowest mortality rate, suggesting that this concentration is an optimal choice for aquaculture. Growth improvements are likely due to immune-nutritional components like polysaccharides (Zahran et al., 2014), though studies on *Cyprinus carpio* and *O. mykiss* have reported no significant differences in growth characteristics when using varying amounts of olive leaf extract, indicating that olive oil by-products offer health benefits without impairing growth.

In terms of intestinal health, olive oil by-products enhance microbial diversity. Fish fed with 0.17% and 0.42% olive oil bioactive extract showed an increase in goblet cells in the intestinal lining, boosting mucosal immunity. The upregulation of goblet cell differentiation factors, like the zinc-finger transcription factor (Klf4), indicated a strong defense against harmful gut bacteria (Gisbert et al., 2017). Compounds in olive oil by-products, such as hydroxytyrosol, tyrosol, and oleuropein, have anti-inflammatory effects and are often found in the fish gut (Centrone et al., 2021; Corona et al., 2009). Replacing fish oil with olive oil and rapeseed oil in aquafeed also led to lower cholesterol levels in fish intestinal tissue (Liland et al., 2018).

Olive wastewater extracts have been shown to have antibacterial properties, particularly against pathogenic gram-positive bacteria like *Staphylococcus aureus* (Leouifoudi et al., 2015; Rebollada-Merino et al., 2019). Olive oil by-products hold promise as immunomodulators in aquaculture, enhancing immune responses in farmed fish. Fatty acids like oleic, linoleic, stearic, palmitic, and palmitoleic acids, found in refined OPO, provide additional immunological benefits that can be used for the prevention of infection. Table 19.5 collectively provides the result of giving and incorporating olive biomass in aquafeed. In summary, research highlights the antioxidant, antibacterial, and antifungal properties of olive by-products, which improve fish gut health and immune responses without negatively affecting growth. This makes olive oil by-products a promising addition to aquafeed formulations, supporting both fish health and sustainable aquaculture practices.

19.4 FUTURE PERSPECTIVES OF MANAGEMENT OLIVE BIOMASS

The olive oil industry is one of the important contributors to both agricultural landscapes and local economies, particularly in the Mediterranean Region (Donner et al., 2022). However, it could lead to environmental concerns if the biomass waste generated is not managed properly. Thus, to improve the management of olive biomass waste, some recommendations involving policy changes, technological advancements, and industry partnerships are suggested. One of the effective ways to improve the management of olive biomass waste is through strategic policy changes.

Government can play an important role by incentivizing waste-to-energy projects. The literature suggests that a variety of incentives can speed up the deployment of renewable energy (Qadir et al., 2021). This is because implementing incentives like tax credits or subsidies by the government can help encourage businesses to invest in technologies that transform olive biomass into valuable products such as compost, animal feed, and biofuels (Yuan et al., 2023). Incentives like these can also encourage innovation and the development of a circular economy within the industry. Additionally, building a strong regulatory framework is also crucial. A clear guideline for sustainable management of biomass waste is needed to help ensure compliance among producers (Wang & Tester, 2023). Regulations that require proper recycling and disposal of biomass waste can reduce negative environmental effects and encourage ethical waste management practices. Mandating olive oil producers to provide information on the generation of biomass waste can also improve transparency and encourage the use of waste reduction solutions. Furthermore, the government can promote research and innovation in the industry of olive biomass waste management (Francioso, 2024). New and innovative uses for olive biomass can be explored through the allocation of research funds and the enhancement of

TABLE 19.5
Recommended Inclusion Level of Olive Plant Part in Different Species of Fish

Type of Olive Plant Part	Recommended Inclusion Level of Olive Oil	Duration	Species	Effects	Source
Olive oil	5%	2 months	Persian sturgeon (*Acipenser persicus*)	Lowest Feed Conversion Ratio (FCR) and the highest Specific Growth Rate (SGR), Body Weight Increase (BWI), and Growth Rate (GR)	Hosseinnia et al. (2021)
Olive oil bioactive extract (OBE)	0.17% and 0.42%	90 days	Gilthead sea bream (*Sparus aurata*)	OBE meliorated fish excrescency primarily by perfecting the condition and protective part of the intestine.	Gisbert et al. (2017)
c	200 mg/kg of olive leaf	75 days	Common carp (*Cyprinus carpio*)	Weight Gain (WG) SGR, and Protein Efficiency Ratio (PER) values were significantly higher in the groups fed the diet containing 200 mg/kg of olive leaf extract compared with the control group.	Sokooti et al. (2021)
Olive leaf	1%	2 months	Nile tilapia (*Oreochromis niloticus*)	Olive leaves extract can be used to enhance the immunity, growth, and health of fish.	Fazio et al. (2021)
Olive leaf	1 g/kg of olive leaf extract	60 days	Common carp (*C. carpio*)	Enhance fish immune function and increase the survival rate.	Zemheri-Navruz et al. (2019)
Olive leaf powder	8%	40 days	Red sea bream (*Pagrus major*)	Boosted the collagen levels and myofibril protein content in the dorsal muscles.	Arsyad et al. (2018)
Olive leaf extract	0.1%	60 days	Rainbow trout (*Oncorhynchus mykiss*)	Ameliorate serum biochemical labels, boost survival rate, and boost vulnerable gene expression.	Baba et al. (2018a)
Olive oil extract	0.1–0.5%	8 weeks	Common carp (*C. carpio*)	Enhance the intestinal health of fish.	Rajabiesterabadi et al. (2020)
Olive waste	2.5 g of olive waste kg^{-1}	6 weeks	Rainbow trout (*O. mykiss*)	Ameliorated excrescency rate and feed effectiveness.	Hoseinifar et al. (2020a)
Olive oil	Partial or total dietary replacement	40 days	Young yellowtail (*Seriola quingqueradiata*)	Stops abrasion of tenebrous brawn while not influencing the excrescency of young yellowtail.	Seno-o et al. (2008)
Olive oil	Low level of olive oil inclusion	10 weeks	Large yellow croaker (*Larimichthys crocea*)	Elevated situations of salutary olive oil painting downgraded excrescency interpretation and antioxidant capacity.	Li et al. (2019)

the collaboration between educational institutions and industry stakeholders. This can lead to the development of new products which utilize the economic value of olive biomass waste.

Next, another effective way to improve the management of olive biomass waste is through strategic technological advancements. Figure 19.2 shows the working of olive machines and treatment of OMW. One promising direction is the use of technologies for biomass conversion. For example, through processes like pyrolysis and anaerobic digestion, olive waste can be converted into

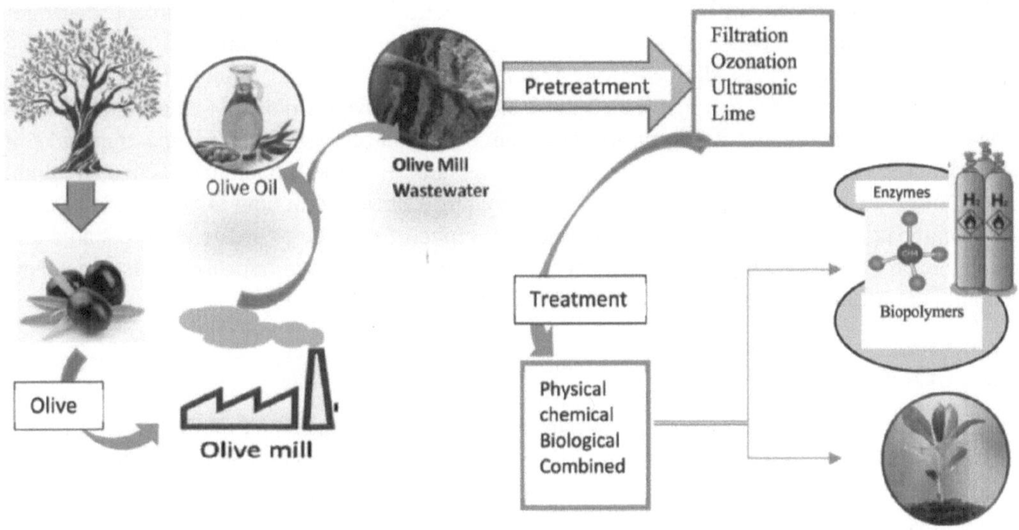

FIGURE 19.2 The working of the olive machine and treatment of olive mill wastewater (Shabir et al., 2023).

renewable energy sources such as biogas and bioethanol (Ali et al., 2024). These technologies not only reduce the environmental impact of the olive oil industry but also produce useful by-products that can be utilized in other industries. Technology advancements are especially beneficial for the treatment of wastewater from olive mills. With the use of modern treatment methods such as membrane bioreactors, evaporation systems, and advanced filtration, valuable compounds can be recovered from OMW. Mills can supply raw materials for industries like food production, pharmaceuticals, and cosmetics that create a new revenue stream from waste by extracting polyphenols, antioxidants, and other bioactive substances from OMW.

Another opportunity lies in composting and soil enrichment technologies. The solid waste from olive oil production, called olive pomace, can be composted and utilized as a natural fertilizer (Ameziane et al., 2020). Better monitoring and management of important variables such as temperature and moisture enhance the composting techniques, which accelerate the process, producing high-quality organic fertilizer that enriches soil fertility. Furthermore, the production of biochar from biomass derived from olives is another promising technological solution. Biochar is a carbon-rich by-product of biomass pyrolysis that can be added to soil as a soil amendment to increase soil health and water retention (Singh Yadav et al., 2023).

Moreover, the effective way to improve the management of olive biomass waste is through strategic industry partnerships. To optimize waste management of olive biomass, collaboration between different industries is crucial. One significant partnership is between the olive oil sector and the energy industry. These industries can produce integrated solutions to convert olive biomass into bioenergy. For example, co-locating biomass-to-energy facilities close to olive mills can improve energy recovery and reduce transportation costs, making the process more economically viable (Donner et al., 2022). In addition, the agricultural sector can work with olive mills to recycle olive biomass into organic fertilizer. After the olive pomace and pruned branches are processed, they can be returned to the agricultural cycle, thereby lowering the need for artificial fertilizers (Alaoui et al., 2024). By promoting sustainable agricultural practices, a closed-loop system that benefits both the olive and farming sectors is established. Furthermore, raising awareness among stakeholders is crucial for the successful adoption of sustainable waste management practices (Debrah et al., 2021). A more sustainable method can be promoted by educating olive growers, mill owners, and local communities about the environmental and economic benefits of proper waste management. Workshops, training sessions, and public campaigns are effective ways to share the best practices.

The effective management of olive biomass waste requires a multi-faceted approach that integrates policy changes, technological advancements, and industry partnerships. By incentivizing waste-to-energy projects, building a strong regulatory framework, and promoting research and innovation, the government can facilitate a more sustainable olive oil industry. Additionally, technological advancements such as biomass conversion, treatment of wastewater, composting and soil enrichment technologies, and production of biochar can maximize the economic and environmental value of olive waste. Finally, collaboration between energy, agricultural, and olive oil industries as well as campaigns to raise awareness and educate the public are able to create a more stable and sustainable olive industry. This effort will not only reduce environmental impacts but also open new economic opportunities for the valuation of olive biomass.

19.5 CONCLUSION

Extracting olive oil produces significant waste including olive pomace and OMW, which poses a high risk of polluting the environment due to the high organic load and phenolic compounds contained in the produced waste if no management is made. The major waste produced during olive oil extraction is wastewater OMW. Poor storage and treatment of OMW could contaminate the water due to OMW being a highly chemical container and toxic if the OMW is thrown into the water source. OMW will also release unpleasant odors that could jeopardize the air quality. Therefore, the two-phase extraction system is preferable to the traditional three-phase extraction system as two-phase extraction produces the least OMW (100–120L) with the most olive wet pomace humidity (65%). Countries with high yields of olive oil should start to practice the two-phase extraction system in extracting olive oil so that OMW can be reduced. Methods are required to further reduce the environmental pollution from olive waste. Olive waste like OTP, olive stones, pomace, and husks are commonly transformed into sources for biofuels through incineration, bioethanol production, methane production, gasification, pyrolysis, and torrefaction. The olive waste is reused as a fuel for industry combustion so that the usage of fossil fuels can be reduced. Olive oil residues like OPO, OMW, olive leaf powder, and olive pomace are reused as feed for aquaculture, which significantly shows an increase in muscle, fat content, and the nutritional value of fish meat. All these available strategies help reduce the production of olive waste and reuse the olive waste for other benefits. When the olive waste like mill wastewater is reduced, the water pollution can be controlled, keeping the habitats for aquatic living things safe. Moreover, industries can use clean water for operation, which in turn maintains hygiene. Olive waste like OTP or olive pomace can be a good source of fuel when reusing it for industrial purposes. This greatly helps minimize the use of fossil fuels and hence in their preservation. Countries like Greece which can produce a lot of OTP resources should make good use of this OTP to generate clean energy.

ACKNOWLEDGMENT

The chapter was written with Soh Fui Yuan, Lee Siao Yan, Nafiz Waqiuddin Bin Mohamad Nazri, and Micheal Hii Sieng Ann.

REFERENCES

Alaoui, I., El Ghadraoui, O., Tanji, K., Harrach, A., & Farah, A. (2024). The olive mill pomace: a sustainable biofertilizer to improve soil proprieties and plant nutrient uptake. *Waste and Biomass Valorization, 15*(5), 2575–2590. https://doi.org/10.1007/s12649-023-02324-z

Alouiz, I., Benhadj, M., Sennoune, M., Amarouch, M. Y., & Mazouzi, D. (2025). Elimination of olive mill wastewater polyphenols through an olive pomace-derived activated charcoal. *Moroccan Journal of Chemistry*, 13(1), 1–10.

Al-Akhras, N. M., & Abdulwahid, M. Y. (2010). Utilisation of olive waste ash in mortar mixes. *Structural Concrete*, 11(4), 221–228.

Alcazar-Ruiz, A., Garcia-Carpintero, R., Dorado, F., & Sanchez- Silva, L. (2021). Valorization of olive oil industry subproducts: ash and olive pomace fast pyrolysis. *Food and Bioproducts Processing, 125*, 37–45. https://doi.org/10.1016/j.fbp.2020.10.011

Ali, F., Dawood, A., Hussain, A., Alnasir, M. H., Khan, M. A., Butt, T. M., Janjua, N. K., & Hamid, A. (2024). Fueling the future: biomass applications for green and sustainable energy. *Discover Sustainability, 5*(1), 156. https://doi.org/10.1007/s43621-024-00309-z

Alshareef, S. A., Otero, M., Alanazi, H. S., Siddiqui, M. R., Khan, M. A., & Alothman, Z. A. (2021). Upcycling olive oil cake through wet torrefaction to produce hydrochar for water decontamination. *Chemical Engineering Research and Design, 170*, 13–22. https://doi.org/10.1016/j.cherd.2021.03.031

Ameziane, H., Nounah, A., Khamar, M., & Zouahri, A. (2020). Composting olive pomace: evolution of organic matter and compost quality. *Agronomy Research, 18*(1), 5–17.

Arsyad, M. A., Akazawa, T., Nozaki, C., Yoshida, M., Oyama, K., Mukai, T., & Ogawa, M. (2018). Effects of olive leaf powder supplemented to fish feed on muscle protein of red sea bream. *Fish Physiology and Biochemistry*, 44(5), 1299–1308. https://doi.org/10.1007/s10695-018-0521-1

Ayoub, S. (2017). Management of olive by-products in Jordan. *Ensuring Sustainability of Jordanian Olive Sector Conference*, Amman, Jordan, pp. 24–25.

Azbar, N., Bayram, A., Filibeli, A., Muezzinoglu, A., Sengul, F., & Ozer, A. (2004). A review of waste management options in olive oil production. *Critical Reviews in Environmental Science and Technology*. https://doi.org/10.1080/10643380490279932

Azzam, M. O. J., Al-Malah, K. I., & Abu-Lail, N. I. (2004). Dynamic post-treatment response of olive mill effluent wastewater using activated carbon. *J. Environ. Sci. Health A Tox. Hazard. Subst. Environ. Eng.* 39, 269–280. https://doi.org/10.1081/ESE-120027383

Baba, E., Acar, Ü., Yılmaz, S., Zemheri, F., & Ergün, S. (2018a). Dietary olive leaf (*Olea europaea* L.) extract alters some immune gene expression levels and disease resistance to *Yersinia ruckeri* infection in rainbow trout (*Oncorhynchus mykiss*). *Fish & Shellfish Immunology*, 79, 28–33. https://doi.org/10.1016/j.fsi.2018.04.063

Baba, E., Acar, Ü., Öntaş, C., Kesbiç, O. S., Yilmaz, S., & Gürbüz, M. (2018b). The effects of dietary olive leaf extract on growth performance, immune response and disease resistance of rainbow trout (*Oncorhynchus mykiss*). *Aquaculture Research*, 49(1), 345–352.

Bannor, R. K., Amoako, D., & Kwabena Chaa Kyire, S. (2024). Sustaining beyond the production of cocoa for beans: the preference and impact of products from the cocoa husk. *Environmental and Sustainability Indicators, 21*. https://doi.org/10.1016/j.indic.2024.100333

Borja, R., Raposo, F., & Rincón, B. (2006). Treatment technologies of liquid and solid wastes from two-phase olive oil mills. *Grasas y Aceites, 57*(1), 32–46. https://doi.org/10.3989/gya.2006.v57.i1.20

Calduch-Giner, J. A., Ipharraguerre, I. R., & Pérez-Sánchez, J. (2017). Olive oil bioactive compounds increase body weight, and improve gut health and integrity in gilthead sea bream (*Sparus aurata*). *British Journal of Nutrition*, 117(3), 351–363. https://doi.org/10.1017/S0007114517000228

Cardinali, A., Cicco, N., Linsalata, V., Minervini, F., Pati, S., Pieralice, M., Tursi, N., & Lattanzio, V. (2010). Biological activity of high molecular weight phenolics from olive mill wastewater. *Journal of Agricultural and Food Chemistry, 58*(15), 8585–8590.

Centrone, M., Luzzana, U., Antonini, M., & Messina, M. (2021). The olive oil by-product in 'rainbow trout Oncorhynchus mykiss (Walbaum)' farming: Productive results and quality of the product. *Aquaculture Research, 41*(10), e1–e9.

Christoforou, E., & Fokaides, P. A. (2016). A review of olive mill solid wastes to energy utilization techniques. *Waste Management, 49*, 346–363. https://doi.org/10.1016/j.wasman.2016.01.012

Contreras, M. del M., Romero, I., Moya, M., & Castro, E. (2020). Olive-derived biomass as a renewable source of value-added products. *Process Biochemistry, 97*, 43–56. Elsevier Ltd. https://doi.org/10.1016/j.procbio.2020.06.013

Corona, G., Tzounis, X., Assunta Dessi, M., Deiana, M., Debnam, E. S., Visioli, F., & Spencer, J. P. E. (2009). The fate of olive oil polyphenols in the gastrointestinal tract: Implications of gastric and colonic microflora-dependent biotransformation. *Free Radical Research*, 43(6), 581–591.

Debrah, J. K., Vidal, D. G., & Dinis, M. A. P. (2021). Raising awareness on solid waste management through formal education for sustainability: a developing countries evidence review. *Recycling, 6*(1), 6. https://doi.org/10.3390/recycling6010006

Donner, M., Erraach, Y., López-i-Gelats, F., Manuel-i-Martin, J., Yatribi, T., Radić, I., & El Hadad-Gauthier, F. (2022). Circular bioeconomy for olive oil waste and by-product valorisation: Actors' strategies and conditions in the Mediterranean area. *Journal of Environmental Management, 321*, 115836. https://doi.org/10.1016/J.JENVMAN.2022.115836

Dorado, F., Sanchez, P., Alcazar-Ruiz, A., & Sanchez-Silva, L. (2021). Fast pyrolysis as an alternative to the valorization of olive mill wastes. *Journal of the Science of Food and Agriculture, 101*(7), 2650–2658. https://doi.org/10.1002/jsfa.10856

Erses Yay, A. S., Oral, H. V., Onay, T. T., & Yenigün, O. (2012). A study on olive oil mill wastewater management in Turkey: a questionnaire and experimental approach. *Resources, Conservation and Recycling, 60*, 64–71. https://doi.org/10.1016/j.Resconrec.2011.11.009

Fazio, F., Habib, S. S., Naz, S., Filiciotto, F., Cicero, N., Rehman, H. U., Sad Calduch-Giner, J. A., Ipharraguerre, I. R., & Pére Calduch-Giner, J. A., Ipharraguerre, I. R., & Pérez-Sánchez, J. (2017). Olive oil bioactive compounds increase body weight, and improve gut health and integrity in gilthead sea bream (*Sparus aurata*). *British Journal of Nutrition*, 117(3), 351–363. https://doi.org/10.1017/S0007114517000228

Fazio, F., Habib, S. S., Naz, S., Fillicioto, F., cicero, N., ur Rehman, H., Saddozai, S., Rind, K. H., Rind, N. A., & Shar, A. H. (2021). Effect of fortified feed with olive leaves extract on the haematological and biochemical parameters of *Oreochromis niloticus* (Nile tilapia). *Natural Product Research*, 36(6), 1575–1580. https://doi.org/10.1080/14786419.2021.1883606

Fernandes, M. C., Torrado, I., Carvalheiro, F., Dores, V., Guerra, V., Lourenço, P. M. L., & Duarte, L. C. (2016). Bioethanol production from extracted olive pomace: dilute acid hydrolysis. *Bioethanol, 2*(1). https://doi.org/10.1515/bioeth-2016-0007

Fernández-Lobato, L., Aguado, R., Jurado, F., & Vera, D. (2022). Biomass gasification as a key technology to reduce the environmental impact of virgin olive oil production: a life cycle assessment approach. *Biomass and Bioenergy, 165*, 106585. https://doi.org/10.1016/j.biombioe.2022.106585

Foti, P., Romeo, F. V, Russo, N., Pino, A., Vaccalluzzo, A., Caggia, C., & Randazzo, C. L. (2021). Applied sciences olive mill wastewater as renewable raw materials to generate high added-value ingredients for agro-food industries. *Applied Sciences, 11*(16), 7511.

Francioso, O. (2024). Current and future perspectives for biomass waste management and utilization. *Scientific Reports, 14*(1), 9635. https://doi.org/10.1038/s41598-024-59623-1

Gagour, J., Hallouch, O., Asbbane, A., Bijla, L., Laknifli, A., Lee, L., Zengin, G., Bouyahya, A., Sakar, E. H., & Gharby, S. (2024). A review of recent progresses on olive oil chemical profiling, extraction technology, shelf-life, and quality control. *Chemistry & Biodiversity, 21*(4). https://doi.org/10.1002/cbdv.202301697

Gisbert, E., Andree, K. B., Quintela, J. C., Calduch-Giner, J. A., Ipharraguerre, I. R., & Pérez-Sánchez, J. (2017). Olive oil bioactive compounds increase body weight, and improve gut health and integrity in gilthead sea bream (*Sparus aurata*). *British Journal of Nutrition*, 117(3), 351–363. https://doi.org/10.1017/S0007114517000228

Gómez-Cruz, I., del Mar Contreras, M., Romero, I., & Castro, E. (2024). Towards the integral valorization of olive pomace-derived biomasses through biorefinery strategies. *ChemBioEng Reviews, 11*(2), 253–277. https://doi.org/10.1002/cben.202300045

Halalsheh, M., Kassab, G., & Shatanawi, K. (2021). Impact of legislation on olive mill wastewater management: Jordan as a case study. *Water Policy, 23*(2), 343–357. https://doi.org/10.2166/Wp.2021.171.

Hoseinifar, S. H., Khalili, M., & Esteban, M. Á. (2020b). Humoral and skin mucosal immune parameters, intestinal immune-related genes, and antioxidant defense in rainbow trout (*Oncorhynchus mykiss*) fed dietary olive (*Olea europaea* L.) waste. *Fish & Shellfish Immunology*, 98, 574–580.

Hoseinifar, S. H., Yousefi, S., Van Doan, H., & Ashouri, G. (2020a). Humoral and skin mucosal immune parameters, intestinal immune-related genes, and antioxidant defense in rainbow trout (Oncorhynchus mykiss) fed dietary olive (Olea europaea L.) waste. *Fish & Shellfish Immunology*, 97, 135–142. https://doi.org/10.1016/j.fsi.2019.12.073

Hosseinnia, E., Khara, H., Farokhroz, M., Yousefi Jourdehi, A., & Kazemi, R. (2021). Effects of dietary olive oil and butylated hydroxytoluene (BHT) on growth, blood, and immunity indices in juvenile Persian sturgeon (*Acipenser persicus*). *Iranian Journal of Fisheries Sciences*, 20(3), 810–827.

Khdair, A., & Abu-Rumman, G. (2020). Sustainable environmental management and valorization options for olive mill byproducts in the Middle East and North Africa (MENA) region. *Processes*. https://doi.org/10.3390/PR8060671.

Khdair, A. I., Abu-Rumman, G., & Khdair, S. I. (2019). Pollution estimation from olive mills wastewater in Jordan. *Heliyon, 5*(8). https://doi.org/10.1016/j.heliyon.2019.e02386

Kian, L. K., Saba, N., Jawaid, M., Alothman, O. Y., & Fouad, H. (2020). Properties and characteristics of nanocrystalline cellulose isolated from olive fiber. *Carbohydrate Polymers, 241*, 116423. https://doi.org/10.1016/j.carbpol.2020.116423

Li, X., Cui, K., Fang, W., Chen, Q., Xu, D., Mai, K., Zhang, Y., & Ai, Q. (2019). High level of dietary olive oil decreased growth, increased liver lipid deposition, and induced inflammation by activating the p38 MAPK and JNK pathways in large yellow croaker (*Larimichthys crocea*). *Fish & Shellfish Immunology*, 94, 157–165. https://doi.org/10.1016/j.fsi.2019.09.001

Liland, N. S., Rosenlund, G., Espe, M., Waagbø, R., Mørkøre, T., & Torstensen, B. E. (2018). High levels of dietary plant proteins and vegetable oil affect the intestinal morphology and plasma cholesterol levels in Atlantic salmon (*Salmo salar* L.). *British Journal of Nutrition*, 120(6), 569–582.

Lo Giudice, V., Faraone, I., Bruno, M. R., Ponticelli, M., Labanca, F., Bisaccia, D., Massarelli, C., Milella, L., & Todaro, L. (2021). Olive trees by-products as sources of bioactive and other industrially useful compounds: a systematic review. *Molecules, 26*(16), 5081. https://doi.org/10.3390/molecules26165081

Maggiotto, G., Colangelo, G., Milanese, M., & de Risi, A. (2023). Thermochemical technologies for the optimization of olive wood biomass energy exploitation: a review. *Energies, 16*(19). https://doi.org/10.3390/en16196772

Medouni-Haroune, L., Zaidi, F., Medouni-Adrar, S., & Kecha, M. (2018). Olive pomace: from an olive mill waste to a resource, an overview of the new treatments. *Journal of Critical Review*. https://doi.org/10.22159/jcr.2018v5i5.28840

Medouni-Haroune, L., Zaidi, F., Medouni-Adrar, S., & Kecha, M. (2018). *Olive pomace: From an olive mill waste to a resource—An overview of the new treatments. Journal of Critical Reviews*, 5(5), 1–9.

Messineo, A., Maniscalco, M. P., & Volpe, R. (2020). Biomethane recovery from olive mill residues through anaerobic digestion: a review of the state of the art technology. *Science of the Total Environment, 703*, 135508. https://doi.org/10.1016/j.scitotenv.2019.135508

Millán, G., Arjona, J. M., & Amador, L. (2014). A new market segment for olive oil: olive oil tourism in the south of Spain. *Agricultural Sciences, 5*, 179–185. https://doi.org/10.4236/as.2014.53022.

Mohamed, A. M., Tayeh, B. A., Abu Aisheh, Y. I., & Salih, M. N. A. (2023). Utilising olive-stone biomass ash and examining its effect on green concrete: a review paper. *Journal of Materials Research and Technology, 24*. https://doi.org/10.1016/j.jmrt.2023.05.039

Nagal, S., & Jain, P. C. (2010). Feather degradation by strains of bacillus isolated from decomposing feathers. *Brazilian Journal of Microbiology, 41*, 196–200.

Najafi, E., Castro, E., & Karimi, K. (2021). Biorefining for olive wastes management and efficient bioenergy production. *Energy Conversion and Management, 244*, 114467. https://doi.org/10.1016/j.enconman.2021.114467

Nasopoulou, C., Stamatakis, G., Demopoulos, C. A., & Zabetakis, I. (2011). Effects of olive pomace and olive pomace oil on growth performance, fatty acid composition and cardioprotective properties of gilthead sea bream (*Sparus aurata*) and sea bass (*Dicentrarchus labrax*). *Food Chemistry*, 129(3), 1108–1113.

Ochando-Pulido, J. M., Vellido-Pérez, J. A., González-Hernández, R., & Martínez-Férez, A. (2020). Optimization and modeling of two-phase olive-oil washing wastewater integral treatment and phenolic compounds recovery by novel weak-base ion exchange resins. *Separation and Purification Technology, 249*. https://doi.org/10.1016/j.seppur.2020.117084 117084

Oluseun Adejumo, I., & Adebukola Adebiyi, O. (2021). *Agricultural Solid Wastes: Causes, Effects, and Effective Management*. IntechOpen. https://doi.org/10.5772/intechopen.93601. https://www.intechopen.com/chapters/73517#

Oreopoulou, V., & Russ, W. (Eds.). (2007). *Utilization of By-Products and Treatment of Waste in the Food Industry*. Integrating Food Science and Engineering Knowledge into the Food Chain, Springer. https://doi.org/10.1007/978-0-387-35766-9

Otero, P., Garcia-Oliveira, P., Carpena, M., Barral-Martinez, M., Chamorro, F., Echave, J., Garcia-Perez, P., Cao, H., Xiao, J., Simal-Gandara, J., & Prieto, M. A. (2021). Applications of by-products from the olive oil processing: revalorization strategies based on target molecules and green extraction technologies. *Trends in Food Science & Technology, 116*, 1084–1104. https://doi.org/10.1016/j.tifs.2021.09.007

Qadir, S. A., Al-Motairi, H., Tahir, F., & Al-Fagih, L. (2021). Incentives and strategies for financing the renewable energy transition: a review. *Energy Reports, 7*, 3590–3606. https://doi.org/10.1016/J.EGYR.2021.06.041

Rajabiesterabadi, H., Ghelichi, A., Jorjani, S., Hoseini, S. M., & Akrami, R. (2020). Effects of dietary olive leaf extract on intestinal immune-related gene expressions in common carp, *Cyprinus carpio. International Journal of Aquatic Biology*, 8(2), 91–97. https://doi.org/10.22034/ijab.v8i2.839

Ribalta, M., Mateu, C., Bejar, R., Rubión, E., Echeverria, L., Varela Alegre, F. J., & Corominas, L. (2021). Sediment level prediction of a combined sewer system using spatial features. *Sustainability, 13*. https://doi.org/10.3390/su13074013

Romero-García, J. M., Niño, L., Martínez-Patiño, C., Álvarez, C., Castro, E., & Negro, M. J. (2014). Biorefinery based on olive biomass. State of the art and future trends. *Bioresource Technology, 159*, 421–432. Elsevier Ltd. https://doi.org/10.1016/j.biortech.2014.03.062

Ruiz, E., Romero-García, J. M., Romero, I., Manzanares, P., Negro, M. J., & Castro, E. (2017). Olive-derived biomass as a source of energy and chemicals. *Biofuels, Bioproducts and Biorefining, 11*(6), 1077–1094. John Wiley & Sons Ltd. https://doi.org/10.1002/bbb.1812

Sagani, A., & Hagidimitriou, M. (2014). A study of burning olive tree pruning biomass for electricity generation. https://www.researchgate.net/publication/303864726

Sajid, M., Akram, A., Fatima Sajjad, S., Siddique, T., & Arshad, M. (2023). *Biological Waste Management*. IntechOpen. https://doi.org/10.5772/intechopen.1003266

Salmoral, G., Aldaya, M. M., Chico, D., Garrido, A., & Llamas, M. R. (2010). *The Water Footprint of Olive Oil in Spain*. Fundación Marcelino Botín.

Seno-o, A., Takakuwa, F., Hashiguchi, T., Morioka, K., Masumoto, T., & Fukada, H. (2008). Replacement of dietary fish oil with olive oil in young yellowtail *Seriola quinqueradiata*: Effects on growth, muscular fatty acid composition, and prevention of dark muscle discoloration during refrigerated storage. *Fisheries Science*, 74(6), 1297–1306. https://doi.org/10.1111/j.1444-2906.2008.01645.x

Shabir, S., Ilyas, N., Saeed, M., Bibi, F., Sayyed, R. Z., & Almalki, W. H. (2023). Treatment technologies for olive mill wastewater with impacts on plants. *Environmental Research, 216*, 114399. https://doi.org/10.1016/J.ENVRES.2022.114399

Sicuro, B., Luzzana, U., Antonini, M., & Messina, M. (2010). The olive oil by-product in 'rainbow trout *Oncorhynchus mykiss* (Walbaum)' farming: Productive results and quality of the product. *Aquaculture Research*, 41(10), e1–e9.

Singh Yadav, S. P., Bhandari, S., Bhatta, D., Poudel, A., Bhattarai, S., Yadav, P., Ghimire, N., Paudel, P., Paudel, P., Shrestha, J., & Oli, B. (2023). Biochar application: a sustainable approach to improve soil health. *Journal of Agriculture and Food Research, 11*, 100498. https://doi.org/10.1016/J.JAFR.2023.100498

Sioriki, E., Smith, T. K., Demopoulos, C. A., & Zabetakis, I. (2016). The effect of dietary olive pomace on the growth performance and health status of gilthead sea bream (*Sparus aurata*). *Aquaculture International*, 24(5), 1299–1311.

Sokooti, R., Chelemal Dezfoulnejad, M., & Javaheri Baboli, M. (2021). Effects of olive leaf extract (*Olea europaea* Leccino) on growth, haematological parameters, immune system and carcass composition in common carp (*Cyprinus carpio*). *Aquaculture Research*, 52(6), 2415–2423. https://doi.org/10.1111/are.15091

Stoller, M., Azizova, G., Mammadova, A., Vilardi, G., Di Palma, L., & Chianese, A. (2016). Treatment of olive oil processing wastewater by ultrafiltration, nanofiltration, reverse osmosis and biofiltration. *Chemical Engineering Transactions, 47*. https://doi.org/10.3303/CET1647069

Sygouni, V., Pantziaros, A. G., Iakovides, I. C., Sfetsa, E., Bogdou, P. I., Christoforou, E. A., & Paraskeva, C. A. (2019). Treatment of two-phase olive mill wastewater and recovery of phenolic compounds using membrane technology. *Membranes, 2019*(9), 27. https://doi.org/10.3390/MEMBRANES9020027

Trubetskaya, A., Grams, J., Leahy, J. J., Johnson, R., Gallagher, P., Monaghan, R. F. D., & Kwapinska, M. (2020). The effect of particle size, temperature and residence time on the yields and reactivity of olive stones from torrefaction. *Renewable Energy, 160*, 998–1011. https://doi.org/10.1016/j.renene.2020.06.136

Wang, K., & Tester, J. W. (2023). Sustainable management of unavoidable biomass wastes. *Green Energy and Resources, 1*(1), 100005. https://doi.org/10.1016/J.GERR.2023.100005

Yilmaz, E., Ergün, S., & Yigit, M. (2004). Effect of dietary olive pomace oil and L-carnitine on growth and chemical composition of African catfish (*Clarias gariepinus* Burchell, 1822). *The Israeli Journal of Aquaculture – Bamidgeh*, 56(3), 169–178.

Yuan, S., Li, C., Wang, M., Wu, H., & Chang, L. (2023). A way toward green economic growth: role of energy efficiency and fiscal incentive in China. *Economic Analysis and Policy, 79*, 599–609. https://doi.org/10.1016/J.EAP.2023.06.004

Zahran, E., Risha, E., & Mahgoub, H. (2014). Effects of olive leaf extract on the growth performance, immune response and resistance of common carp (*Cyprinus carpio*) to *Aeromonas hydrophila* infection. *Research Journal of Pharmaceutical, Biological and Chemical Sciences*, 5(4), 1190–1200.

Zemheri-Navruz, F., Acar, Ü., & Yılmaz, S. (2019). Dietary supplementation of olive leaf extract increases haematological, serum biochemical parameters and immune-related genes expression level in common carp (*Cyprinus carpio*) juveniles. *Fish & Shellfish Immunology*, 89, 672–676. https://doi.org/10.1016/j.fsi.2019.04.037

20 Citrus Waste Management

20.1 INTRODUCTION

Farming practices, along with alterations in land utilization, currently account for around 25% of the overall human-caused greenhouse gas (GHG) emissions (Bennetzen et al., 2016). These releases play a major role in driving global climate change, underscoring the pressing demand for increased sustainable farming methods. With this in mind, the main aim of this task is to extensively research citrus waste (CW) management, looking at present methods, the difficulties faced, and the possible applications of CW. Citrus biomass, ranging from peelings to seeds, is a plentiful by-product in areas where citrus farming is prevalent. This biomass has great potential for numerous applications in both the agriculture and industry sectors. In Italy, the annual production of citrus fruits like lemons, oranges, bergamot, and clementines is about 3 million tons each year. Improper handling of citrus biomass can result in significant environmental problems, especially pollution, despite its potential for reuse.

If CW is not properly handled, it will break down without oxygen in landfills, emitting methane gas a strong contributor to climate change into the air. This garbage can also lure pests and create bad smells, becoming a bother to neighboring communities. Furthermore, the accumulation of organic waste in landfills can lead to soil deterioration, water contamination due to runoff, and air pollution caused by the emission of harmful gases. In spite of these difficulties, citrus biomass has the potential to be a valuable asset with proper management. Repurposing CW offers various advantages for both agriculture and industry. In the field of agriculture, CW can serve various functions, such as:

- CW can be composted and utilized as an organic fertilizer, enhancing the soil with vital nutrients and decreasing the need for artificial fertilizers.
- CW pulp and peels can be turned into livestock feed. This not only offers an economical food source but also helps decrease the volume of garbage sent to dumps.
- Materials for mulching: Citrus peels and other organic waste can be utilized to retain soil moisture, manage weed growth, and enhance soil quality.

Within the industrial sector, there are numerous potential uses for CW. Bioenergy can be generated from CW by converting its organic material into biogas using anaerobic digestion, which can then be utilized as a sustainable energy source. This assists in decreasing reliance on fossil fuels and tackling the problem of waste management. Citrus peels contain essential oils (EOs) like D-limonene that are widely used in the food, pharmaceutical, and cosmetic industries. Obtaining these oils from discarded materials not only decreases the amount of waste but also increases their economic worth. Bioplastics: The use of CW in making biodegradable plastics provides an eco-friendly option, decreasing plastic pollution and serving as a sustainable alternative to traditional plastics.

Effective management of CW is essential to reduce environmental pollution, foster sustainability, and enhance resource recovery. Inadequate handling of citrus by-products can lead to numerous adverse impacts on the environment. For instance, growing volumes of CW in landfills not only decreases space for other types of waste but also accelerates land degradation. Additionally, runoff from improperly managed waste sites can pollute nearby water bodies with organic materials, nutrients, and chemicals, potentially causing eutrophication, algal blooms, and damage to aquatic ecosystems. The decomposition of CW in landfills releases methane and other harmful

DOI: 10.1201/9781003472209-27

gases, contributing to air pollution and climate change. Moreover, when CW is left to decompose inappropriately, it can disrupt soil nutrient balance, diminishing fertility and causing long-term soil degradation. This chapter aims to provide an in-depth exploration of CW management practices, with a focus on addressing the environmental impacts and evaluating the potential applications of citrus biomass. Ultimately, this chapter highlights the need for enhanced management strategies that not only mitigate pollution but also harness the resource potential of CW, supporting both environmental sustainability and circular economy initiatives.

20.2 CURRENT ISSUES AND CHALLENGES

Over the years, the citrus processing industry has become a crucial player in the agro-industrial sector, contributing significantly to global production and economic growth. Oranges, in particular, account for approximately 50–60% of the total citrus output, while other citrus species such as lemons, limes, and mandarins also hold considerable industrial importance (Satari & Karimi, 2018). However, the production of citrus fruits also generates significant waste in the form of CW, which can reach up to 50% of the fruit's total weight. Traditionally, this waste is either sent to landfills or left to decompose naturally. Unfortunately, such practices lead to the emission of GHGs, including methane, carbon dioxide, and nitrous oxide, which contribute to climate change. Additionally, CW is characterized by its low pH, high levels of organic compounds, and the presence of EOs, particularly D-limonene, which inhibits microbial activity and makes traditional biological treatments challenging (Zema et al., 2018).

From the processing of citrus fruits, two primary by-products are generated: a solid or semi-solid fraction composed of peels and fruit residues, and citrus wastewater (CWW) (Lucia et al., 2022). A large body of research has focused on solutions for valorizing the solid portion of these by-products. Studies indicate that the solid fraction is often reused as a source of valuable bioactive compounds with various health benefits, making it applicable in industries such as pharmaceuticals, cosmetics, and food production. In some cases, this fraction is repurposed for dietary fibers (Zema et al., 2018) or as absorbent materials (Calabrò et al., 2016). Despite these options, the solid waste from citrus processing still presents environmental challenges due to its high organic matter content, which complicates disposal and raises concerns over potential soil degradation and eutrophication of water bodies (Lucia et al., 2022).

CWW comprises both the water inherent in the fruit and process water from activities such as fruit washing, equipment cleaning, cooling, EO extraction, and peel drying (Lucia et al., 2022). CWW is known for its low pH and high electrical conductivity (Rojas Flores et al., 2020). Its chemical composition can fluctuate daily, making it difficult to manage. Currently, CWW management follows a "take-make-use-dispose" model, where wastewater undergoes treatment before being discharged into the environment. However, the volume of CWW produced is substantial, ranging from 1 to 17 m^3/ton of processed fruit, depending on the specific technology used in the processing plant (Zema et al., 2018). Handling such large volumes presents significant environmental and economic challenges, particularly due to the low pH, high organic content, and EO concentration. Improper disposal of this wastewater could result in severe environmental impacts, and increasingly stringent environmental regulations require advanced, costly treatments to mitigate these effects. In contrast, developing countries are often permitted to use simplified treatment systems due to different environmental standards and economic conditions (Zema et al., 2019). Technologies such as membrane filtration, advanced oxidation, or bioenergy recovery systems may be necessary to meet these regulatory requirements, increasing the financial burden on processing facilities.

Much of the current literature on CWW management has focused on optimizing treatment methods to minimize the environmental impact of wastewater disposal. However, global environmental policies are now pushing for a shift beyond mere pollution reduction to the utilization of wastewater through a biorefinery approach. This approach aims to transform waste into valuable resources, promoting the reuse of wastewater in various industries. The biorefinery model has recently gained traction due to its potential for extracting valuable compounds from CW, offering an avenue for the creation of high-value products (Kim et al., 2022).

One of the most promising aspects of this biorefinery approach is microbial fermentation, which has garnered significant attention. This process involves several stages: preparing the substrate by breaking down organic materials like carbohydrates, inoculating the substrate with specific microorganisms (such as yeast for alcoholic fermentation), maintaining low-oxygen conditions to promote microbial growth, and allowing the microorganisms to metabolize the sugars, producing ethanol, carbon dioxide, or other by-products depending on the desired outcome. The final products can then be harvested, purified, or further processed based on their intended use.

In recent years, studies have demonstrated that innovative technologies can be used to harness CWW for the extraction of value-added compounds and the recovery of chemicals (Zema et al., 2019). Despite these promising developments, there is still limited knowledge about the advantages and disadvantages of these technologies. The scientific community and stakeholders must work together to better understand and implement these approaches, as the current understanding remains fragmented. Developing a more comprehensive view of these technologies will be crucial for fully realizing their potential and ensuring that they are applied effectively to support both environmental sustainability and economic growth.

The current issues surrounding CWW biomass management require an innovative engineering approach that can address the high volume, environmental risks, and even technical difficulties associated with waste treatment. Effective strategies might include integrating biorefinery approaches, enhancing treatment technologies for organic matter and EO separation, and promoting sustainable practices like bioenergy recovery and the production of value-added products. Developing cost-effective, scalable solutions is crucial for reducing the environmental impacts of citrus biomass while aligning with global Sustainable Development Goals.

20.3 INNOVATIVE SOLUTIONS AND POTENTIAL APPLICATIONS

20.3.1 Compositions of CWs

More than 100 million tons of processed CW are produced each year from the industrial processes which are mainly for juice production, producing one of the most popular juices all over the world. Unfortunately, from its total fruit weight, only 45% is used in the production, whereas the disposal remnants from the remaining weight include the seeds (2%), pulp (26%), and peel (27%). These numbers are further increased when the fruits do not meet the minimum quality standard, and they were disposed of. Simultaneously, pruning is a common practice used to improve the quality of citrus fruits by removing tree branches, thus producing large amounts of leaves that greatly increase the amount of waste throughout the entire process (Russo et al., 2021).

Some of these wastes are frequently thrown into rivers or landfills, especially in developing countries, causing environmental pollution and reducing levels of dissolved oxygen in the water. CW is abundant, complex chemically, and biodegradable, causing it to have a high tendency to ferment, which requires high amounts of oxygen for it to break down. One of the alternatives for disposing of CW is to burn it to generate thermal energy. However, the process is risky and not a satisfactory method to dispose of CW as it releases high levels of sulfur oxides, carbon, and nitrogen into the atmosphere (Russo et al., 2021). Hence, in this report, several alternatives are suggested as a better method to manage CW.

The by-products of citrus can be divided into a few categories, namely the peels, seeds, and pulp residue. The peel on its own has two different parts, which are the flavedo and albedo. The flavedo, which contains the oil sacs, is the outermost colorful part, while the interior white part is the albedo and is rich in pectin. The peels have high amounts of *S*-limonene, a powerful antimicrobial compound, and are also rich in sugars. The seeds are mostly made of nitrogen-free extract, lipids, crude protein, and fiber. The pulp residue is composed of segment membranes that contain the juice of the fruit.

Water makes up 85% of the chemical makeup of citrus fruits, along with sugars, fibers, fats, vitamins, organic acids, pectin, minerals, proteins, and secondary metabolites, such as carotenoids, phenolic compounds, and alkaloids among others (Lu et al., 2023). Citrus fruits are rich in dietary fiber and bioactive substances such as citric acid, polyphenols, vitamins, minerals, and EOs, which are primarily responsible for their nutraceutical and therapeutic properties. They have great potential for both illness prevention and treatment due to their diverse phytochemical profile. Citrus fruits have differing proportions of soluble and insoluble fibers, including cellulose, hemicellulose, and lignin, as well as pectin, fructans, and psyllium. Soluble fibers can withstand digestion in the small intestine and be fermented into a gooey material in the large intestine by the microflora, while the insoluble fibers maintain their structure as they pass through the digestive tract and soften the stool, reducing the occurrence of constipation (Ben Hsouna et al., 2023).

The active compounds with powerful bioactive activities in these fruits with their antimicrobial and antioxidant properties have a crucial role in the food industry. Citrus fruits are also known for the amount of vitamin C and carotenoids with pro-vitamin A. Carotenoids, natural compounds that show the red, orange, and yellow colors in plants, assist photosynthesis in plants and defend against light oxidation. These compounds are essential for human health as they help prevent degenerative and cardiovascular diseases (Andrade et al., 2022). With appropriate solutions and innovations, these by-products can be turned into invaluable resources in many industries.

20.3.2 Innovations on Bioethanol

As the demand for energy increases as the years go by, the search for renewable and sustainable sources of energy has increased. With the current advancements in renewable technologies, renewable sources of energy can be produced. These energies produce minimal amounts of waste, lessen the impact on the environment, and are sustainable while considering the needs of current and future generations. Among the available sources used nowadays, ethanol is a rising source of energy. Ethanol is considered a substitute for gasoline fuel, where fewer pollutants and carcinogens are produced than the usual gasoline or methyl tertiary butyl ether, an additive to petroleum fuel (Mahato et al., 2021).

One promising alternative is the production of bioethanol from waste materials. It is found that ethanol can be produced from citrus peel waste (Sarkar et al., 2022; Teke et al., 2023; Vadalà et al., 2023). The main industrial production of ethanol these days is by fermenting agricultural feedstock, a biochemical method, and through catalytic hydration of ethylene, which is a chemical method. In the case of using CW, which is composed largely of carbohydrates, fermentation can be done to form ethanol. When starchy and cellulosic materials are pretreated with chemicals or enzymes to hydrolyze the polymers into sugars, ethanol is formed from the sugars. In the process, the peels are used as a cellulosic material, which has a substantial amount of soluble sugars, which are glucose, fructose, and sucrose, that could be applied for the production of a hydrolysate rich in carbon sources for ethanol fermentations (Patsalou et al., 2019). Usually, the starch used is sourced from other plants such as potatoes, corn, sugarcane, and molasses. However, the usage of citrus peel in this process helps to reuse a large amount of the waste rather than being discarded into landfills or the environment.

The process begins with pre-treatment to break down the fibrous structure of CW. This process helps to release the sugars. Enzymatic hydrolysis or acid hydrolysis are some of the more frequently used methods to break down cellulose and hemicellulose into fermentable sugars. Fermentation starts when the sugars are ready to be used. *Saccharomyces cerevisiae*, a commonly used species of yeast for ethanol production, ferments the sugars into ethanol. When fermentation is done, the bioethanol is distilled and purified for use as a fuel.

20.3.3 Innovations on EO

Citrus EOs are liquids that hold the volatile aroma compounds in a citrus plant. The flavedo, or the upper shell, and exocarp of citrus fruits contain tiny vesicles that hold the EOs. Citrus EOs have long been utilized in traditional medicine and are known to have antimicrobial qualities

(Köse & Bayraktar, 2018). The EOs can be used in natural cleaning products and disinfectants. This is due to the presence of limonene, citral, and linalool compounds in the EOs, which exhibit antibacterial and antimicrobial properties. In addition, citrus oils can be utilized as a fragrance and natural flavoring agent. The scent of citrus is widely known to enhance mood and promote relaxation. Some even use it for aromatherapy for its stress-relieving properties. Hence, the EOs can be added to personal care products, such as lotions, bathing products, and perfumes for their refreshing scent.

The most industrially established technologies are cold-press and hydrodistillation. For cold-press, the EO yield at the lab scale was lower, but it gave a high yield at the industrial scale. On the contrary, hydrodistillation provides higher yields but is an energy-intensive process and takes a relatively long time (Teigiserova et al., 2021). Cold pressing is widely used as it preserves the volatile compounds that contribute to their aroma and therapeutic properties. In this method, the peels are punctured mechanically and pressed to release the EOs. The oil is then separated from the other components such as water and plant matter. This method does not require heat or chemicals, which makes it an eco-friendly and efficient process, yielding high-quality oil. Hydrodistillation uses the principle of applying simultaneous internal and external heat and mass transfer to the peeks.

20.3.4 Innovations on Bioplastic

Plastic is a versatile material for its durability and affordability. The creation of plastic has brought many conveniences from packaging foods to sealing apparatus for the medical field, ensuring sterile packaging for ease of use. Its wide range of applications and ability to be molded into countless forms have changed the manufacturing world and made products much more accessible. Unfortunately, the same qualities that make plastic a wonderful material are also causing devastating effects on the environment. Almost all plastics are non-biodegradable. This causes serious pollution in oceans, rivers, and landfills where they stay unchanged for years, contributing to a climate crisis. Not only that, but wildlife has also been found harmed by these plastic wastes, disrupting the ecosystem.

The pectin extracted from citrus peels can be transformed into a bioplastic with the addition of some components such as citric acid, glycerol, carboxymethylcellulose, and water. To extract the pectin, an aqueous extraction method was used where distilled water was added to the biomass of citrus peels and heated for a specific amount of time. After filtration, the aqueous part was separated, and alcohol was added to precipitate the pectin. After centrifugation to separate the alcohol from the pectin, the pectin is dried by baking. It is then added to the other components and cast into the desired form. After testing, it was found that the bioplastic formed has mechanical physical characteristics similar to synthetic bioplastics such as its water absorption, elongation, and tensile strength. It also complies with the regulated percentage of biodegradability (Jara Ramirez et al., 2023). Hence, the production of bioplastics from citrus peel waste is a plausible method to recycle waste and reduce pollution in the environment.

20.3.5 Innovations for Food Ingredients

Microbial growth in food is a main concern in the food industry. Manufacturers usually utilize antimicrobial compounds such as benzoic acids, nitrates, and nitrites to inhibit microorganism growth and extend the shelf life of food. However, there have been concerns about the safety of these compounds for human health and the search for a safer substitute is a priority. Hence the application of citrus by-products has been studied for their antimicrobial properties (Andrade et al., 2022). The antimicrobial properties contained in the natural extracts help increase food safety. In some studies, the addition of citrus to bread showed positive effects on its shelf life. This is due to the high pectin and fiber content of the citrus. The research showed that the food has a better ability to bind larger volumes of water when a portion of the regular wheat flour was replaced with dried fruit peel powder. This is due to the water and fat-binding properties found in the dietary fiber (Ben Hsouna et al., 2023).

The focus of consumers has shifted in recent years toward eating foods that are enhanced with dietary fiber. Citrus fruits are known as a great source of dietary fiber and antioxidants. Adding them to daily consumed foods such as meat, dairy, and baked goods has been a topic of study for scientists. Citrus powder or flour has been suggested by a number of writers as a useful ingredient for baked goods or confectionary products. The addition of the dietary-rich extract reduced the predicted glycemic index after consuming the fiber-enriched food. There was also no difference in mouthfeel for fiber-rich confectionary goods compared to a regular one (Ben Hsouna et al., 2023).

Other studies in relation to the usage of citrus extracts as food additives are also observed. The studies looked into the possibility of utilizing citrus peel extracts to preserve the oxidative stability of meatballs for storage as frozen food (Nishad et al., 2018). The authors showed that by blocking the enzymatic processes that cause oxidative damage, the naturally occurring antioxidant extracts from citrus peels can regulate the oxidation of lipids in meat products. There was also a positive effect on the appearance and mouthfeel of the food, suggesting that, for foods rich in fatty acid, the extract from citrus can be used as a natural preservative (Ben Hsouna et al., 2023).

Citrus fruit peels are widely known for being a good source of carotenoids. The food industry has been interested in these compounds as a potential replacement for toxic synthetic colorants because they not only give fruits color but also bring benefits to health. In a research, beta-carotene (β-carotene) is extracted from orange peel waste and utilized to create a stable nanoemulsion that could be used as a natural food colorant in the future. According to the authors, adding the nano-emulsion to fruit juice considerably improves its color, offering a substitute for artificial colorants. The methodology of extracting total carotenoids from the citrus, Mandarin epicarp with ultrasound assistance was improved and showed that this method might potentially decrease the amount of tartrazine in baked goods, such as breads and cakes (Ben Hsouna et al., 2023). This has greatly improved the potential of it being used as a natural coloring additive.

In the food industry, synthetic flavors are still often utilized, although there is growing interest in citrus EOs as flavoring agents as a replacement. EOs, derived primarily from citrus peel, are notable sources of terpenoids, which have extensive application in food flavoring and possess antibacterial, antifungal, and insecticidal characteristics. There are not many citrus applications in the food business as flavoring agents as most studies have concentrated on identifying the volatile profiles of various citrus species. In certain recipes for flour-based confections, bergamot oils, a rich source of linalool and linalyl acetate with promising flavor characteristics, have been used in place of bergamot peels (Ben Hsouna et al., 2023).

A common dietary fiber made from citrus peel is pectin. The main applications of pectin are as an emulsifier, thickener, and stabilizer (Ben Hsouna et al., 2023). It is often used to make fruit juice, confections, marmalade, jams, and bakery fillings. Additionally, yogurts and other acidified milk beverages are stabilized with it. In a research conducted by Khan et al. (2014), jellies are prepared using two varieties of grapefruit peel pectin. Both varieties produced jelly that set in 10–25 min, proving that they were rapid-set pectin and showed promising results as a thickener or stabilizer for culinary applications. This research is promising as it shows the potential of extracted pectin from citrus to act as a gelling agent in place of commercial pectin in many foods.

As per the definition provided in EC Regulation No 450/2009, "active materials and articles" refer to materials and articles that are intentionally designed to incorporate components that would release or absorb substances into or from the packaged food or the surrounding environment, to extend its shelf life or maintain or improving its condition. Other than adding active compounds directly to the food, active packaging technology offers a number of benefits. A smaller amount of active substances is needed than the conventional method, localization of activity at the food's surface, migration of the active compounds from the film into the food matrix, controlled release systems, and the removal of steps that were previously part of a standard process meant to introduce the active compounds at the industrial processing level, such as mixing, immersion, or spraying. Controlled-release systems are crucial because they can avoid sensory or toxicological issues or system inefficiencies brought on by excessively low or high concentrations of the delivered drug (Gómez-Estaca et al., 2014).

As technology developed, new ideas and materials started to appear. From a traditional standpoint, the primary function of packaging is to shield food from the surrounding influences without interfering with the food's matrix. These days, smart and active food packaging has emerged as an attempt to address the inadequacies of traditional food packaging. The goal of active packaging solutions is to prolong food shelf life by interacting directly with the packaged food. "Changes to the composition or the organoleptic properties of the food only if the changes comply with the Community provisions applicable to food, such as the provisions of Directive 89/107/EEC (4) on food additives," states the European Legislation regarding active packages (Andrade et al., 2022).

Active packaging comes in two varieties: absorbing packaging and releasing packaging. The first kind works by interacting with food by absorbing substances from the food itself or from the package's headspace. There is no migration of active ingredients into the food. Within this category, moisture and oxygen absorbents are the most widely available on the market. When it comes to the releasing packages, the polymeric matrix is packed full of active ingredients that will slowly seep into the food to extend its shelf life by delaying the processes that cause food to deteriorate, like the inhibition of bacteria and/or lipid oxidation. The organoleptic properties of food can also be preserved, enhanced, or improved with the usage of these packages (Andrade et al., 2022).

Citrus by-products are used to produce polyhydroxyalkanoate and bacterial cellulose because of their high carbohydrate content. First described in 1988, BC is a β-1,4-linked glucose polymer generated by aerobic bacteria. High crystallinity, high cellulose purity, high tensile strength, and water-holding capacity are just a few of BC's intriguing characteristics. Fruit cocktails and jellies are made using it in the food industry. Since coconuts only grow in tropical regions, the bacteria (usually *Gluconacetobacter xylinum*) used to generate the BC are mostly grown in coconut water, which makes it a limited source. Citrus pulp water is used in place to overcome this issue and the results are compared with the conventional method of using coconut water. It was concluded that the levels of BC grown in citrus pulp water were almost the same as industrial levels, making it a viable alternative (Andrade et al., 2022).

20.4 CHALLENGES AND FUTURE PERSPECTIVE

Effective management of citrus biomass waste is crucial for reducing environmental impacts, promoting sustainable practices, and unlocking the potential of waste-generated biogas, bio-based products, and EOs. To improve the management of citrus biomass waste, particularly orange peels and pulp, several key strategies can be implemented to optimize waste utilization and reduce environmental impact. These recommendations focus on policy improvements, technological advancements, and industry partnerships that can enhance both sustainability and economic value in CW management.

20.4.1 Policy Enhancements

Governments and regulatory bodies should establish incentives for the adoption of circular economy principles in the citrus processing industry. Offering tax reductions, subsidies, or grants for companies investing in biomass conversion technologies can accelerate waste-to-energy projects, such as biogas production from citrus peels. Additionally, creating mandates for companies to track and report their waste management efforts can encourage the industry to explore alternative waste streams, such as EO extraction and the production of bio-based products. Moreover, implementing extended producer responsibility policies, which hold citrus fruit processing industries accountable for the waste generated throughout their supply chain. This can promote waste reduction, recycling, and proper disposal practices (Cai & Choi, 2021).

20.4.2 Technological Advancements

Investing in new technologies for CW processing is critical. For instance, advancements in anaerobic digestion can increase the efficiency of biogas production from citrus peels, which contain high levels of pectin that can impede microbial digestion. Research into pre-treatment processes, such as thermal or enzymatic methods, could enhance digestibility and boost biogas yields. This can be achieved through the development of more efficient bioreactor designs, improved microbial communities, and optimized operating conditions (Guerrero-Martin et al., 2024). Additionally, the integration of green extraction technologies, such as supercritical CO_2 extraction, for obtaining EOs and biorefinery concepts that integrate multiple conversion technologies to produce a range of bio-based products can reduce the use of harmful solvents and improve product purity (Kim et al., 2022).

20.4.3 Industry Partnerships

Collaborative efforts between citrus processors, bio-refineries, and energy companies can create a robust supply chain for utilizing citrus biomass. These partnerships could streamline the collection, transportation, and processing of biomass waste, ensuring that valuable by-products, such as biofuels, pectin, or EOs, are recovered efficiently. For example, citrus processors could collaborate with bio-based chemical companies to develop bioplastics or with cosmetic industries interested in using citrus oils for fragrance and skincare products (Taghizadeh-Alisaraei et al., 2017). Furthermore, collaboration along the citrus fruit supply chain to reduce waste generation, promote waste reuse, and develop closed-loop systems can be done (Liao et al., 2020).

20.4.4 Public–Private Research Initiatives

Universities, research institutions, and industry players should establish joint research programs to explore innovative uses of citrus biomass. Such collaborations could focus on optimizing the extraction of high-value compounds, such as flavonoids and polyphenols, from CW, which have applications in nutraceuticals and pharmaceuticals. Moreover, further research into biotechnological approaches for converting citrus residues into bio-based chemicals, such as limonene or bioethanol, could contribute to greener energy solutions and chemical production (Vilariño et al., 2017). By addressing these areas, citrus biomass waste can be better managed and transformed from a disposal problem into an opportunity for sustainable growth and product innovation.

20.5 CONCLUSION

Citrus biomass waste, if mismanaged, poses significant environmental challenges, contributing to methane emissions, air and water pollution, and landfill overload. These problems are compounded by the complex chemical nature of CW, particularly its low pH and high content of EOs, which hinder traditional waste treatment methods. However, the potential of citrus biomass as a valuable resource is immense. Through the adoption of innovative waste management strategies, such as biorefinery models and advanced technological solutions, this waste can be transformed into valuable by-products. These include biogas, EOs, bioethanol, bioplastics, and food ingredients, all of which contribute to both environmental sustainability and economic growth. The integration of biorefinery technologies, anaerobic digestion, and microbial fermentation can optimize the valorization of CW, turning it into biofuels, biodegradable plastics, and nutraceutical products. Implementing these innovative approaches can reduce environmental pollution, decrease reliance on fossil fuels, and offer sustainable alternatives for the food, pharmaceutical, and cosmetic industries. Collaboration between industries, governments, and research institutions is essential to realizing the full potential of citrus biomass. Ultimately, by managing CW efficiently, industries can mitigate environmental harm, contribute to global sustainability goals, and unlock new economic opportunities.

ACKNOWLEDGMENT

The chapter was written with Fiorena A.P. Santhana Dass, Muhammad Hazim Azatul Amery, and Ruth Tay Soong Yi.

REFERENCES

Andrade, M. A., Barbosa, C. H., Shah, M. A., Ahmad, N., Vilarinho, F., Khwaldia, K., Silva, A. S., & Ramos, F. (2022). Citrus by-products: Valuable source of bioactive compounds for food applications. *Antioxidants (Basel, Switzerland)*, *12*(1). https://doi.org/10.3390/antiox12010038

Ben Hsouna, A., Sadaka, C., Generalić Mekinić, I., Garzoli, S., Švarc-Gajić, J., Rodrigues, F., Morais, S., Moreira, M. M., Ferreira, E., Spigno, G., Brezo-Borjan, T., Akacha, B. B., Saad, R. B., Delerue-Matos, C., & Mnif, W. (2023). The chemical variability, nutraceutical value, and food-industry and cosmetic applications of citrus plants: A critical review. *Antioxidants*, *12*(2), 481. https://doi.org/10.3390/antiox12020481

Bennetzen, E. H., Smith, P., & Porter, J. R. (2016). Agricultural production and greenhouse gas emissions from world regions—The major trends over 40 years. *Global Environmental Change, 37*, 43–55. https://doi.org/10.1016/J.GLOENVCHA.2015.12.004

Cai, Y.-J., & Choi, T.-M. (2021). Extended producer responsibility: A systematic review and innovative proposals for improving sustainability. *IEEE Transactions on Engineering Management*, *68*(1), 272–288. https://doi.org/10.1109/TEM.2019.2914341

Calabrò, P. S., Pontoni, L., Porqueddu, I., Greco, R., Pirozzi, F., & Malpei, F. (2016). Effect of the concentration of essential oil on orange peel waste biomethanization: Preliminary batch results. *Waste Management, 48*, 440–447. https://doi.org/10.1016/j.wasman.2015.10.032

Gómez-Estaca, J., López-de-Dicastillo, C., Hernández-Muñoz, P., Catalá, R., & Gavara, R. (2014). Advances in antioxidant active food packaging. *Trends in Food Science & Technology*, *35*(1), 42–51. https://doi.org/10.1016/j.tifs.2013.10.008

Guerrero-Martin, C. A., Rojas-Sanchez, A. N., Cruz-Pinzón, D. F., Milquez-Sanabria, H. A., Sotelo-Tobon, D. L., da Cunha, A. L. R., Salinas-Silva, R., Camacho-Galindo, S., Costa Gomes, V. J., & Cunha Malagueta, D. (2024). The advantage of citrus residues as feedstock for biogas production: A two-stage anaerobic digestion system. *Energies*, *17*(6), 1315. https://doi.org/10.3390/en17061315

Jara Ramirez, V. N., Ocaña Gonzales, Z. I., Lizarzaburu-Aguinaga, D., Ccuro, F. M., Pérez, H. R., & Benites-Alfaro, E. (2023). Circular economy: Use of fruit waste to obtain bioplastics. *Chemical Engineering Transactions*, *100*, 103–108. https://doi.org/10.3303/CET23100018

Khan, A. A., Randhawa, M. A., Karim, R., & Ahmed, W. (2014). Extraction and characterization of pectin from grapefruit (Duncan cultivar) and its utilization as gelling agent. *International Food Research Journal, 21*(6), 2195–2199.

Kim, I. J., Jeong, D., & Kim, S. R. (2022). Upstream processes of citrus fruit waste biorefinery for complete valorization. *Bioresource Technology*, *362*, 127776. https://doi.org/10.1016/J.BIORTECH.2022.127776

Köse, M. D., & Bayraktar, O. (2018). Valorization of citrus peel waste. *Volatiles & Essent. Oils, 5*(2), 10–18.

Liao, Y., Kaviyani-Charati, M., Hajiaghaei-Keshteli, M., & Diabat, A. (2020). Designing a closed-loop supply chain network for citrus fruits crates considering environmental and economic issues. *Journal of Manufacturing Systems*, *55*, 199–220. https://doi.org/10.1016/J.JMSY.2020.02.001

Lu, X., Zhao, C., Shi, H., Liao, Y., Xu, F., Du, H., Xiao, H., & Zheng, J. (2023). Nutrients and bioactives in citrus fruits: Different citrus varieties, fruit parts, and growth stages. *Critical Reviews in Food Science and Nutrition*, *63*(14), 2018–2041. https://doi.org/10.1080/10408398.2021.1969891

Lucia, C., Laudicina, V. A., Badalucco, L., Galati, A., Palazzolo, E., Torregrossa, M., Viviani, G., & Corsino, S. F. (2022). Challenges and opportunities for citrus wastewater management and valorisation: A review. *Journal of Environmental Management, 321*, 115924. https://doi.org/10.1016/j.jenvman.2022.115924

Mahato, N., Sharma, K., Sinha, M., Dhyani, A., Pathak, B., Jang, H., Park, S., Pashikanti, S., & Cho, S. (2021). Biotransformation of citrus waste-I: Production of biofuel and valuable compounds by fermentation. *Processes*, *9*(2), 220. https://doi.org/10.3390/pr9020220

Nishad, J., Koley, T. K., Varghese, E., & Kaur, C. (2018). Synergistic effects of nutmeg and citrus peel extracts in imparting oxidative stability in meat balls. *Food Research International*, *106*, 1026–1036. https://doi.org/10.1016/j.foodres.2018.01.075

Patsalou, M., Samanides, C. G., Protopapa, E., Stavrinou, S., Vyrides, I., & Koutinas, M. (2019). A citrus peel waste biorefinery for ethanol and methane production. *Molecules (Basel, Switzerland)*, *24*(13). https://doi.org/10.3390/molecules24132451

Rojas Flores, S. J., Cruz-Noriega, M. D. L., Chirinos, L. C., Benites, S. M., Palacios, F. S., Altamirano-León Rosa, A.-L. R., Altamirano-León Zoilita, A.-L. Z., & Luis, A.-S. L. (2020). Using lime (citrus×aurantiifolia), orange (citrus×sinensis), and tangerine (citrus reticulata) waste as a substrate for generating bioelectricity. *Environmental Research, Engineering and Management, 76*(3), 24–34. https://doi.org/10.5755/j01.erem.76.3.24785

Russo, C., Maugeri, A., Lombardo, G. E., Musumeci, L., Barreca, D., Rapisarda, A., Cirmi, S., & Navarra, M. (2021). The second life of citrus fruit waste: A valuable source of bioactive compounds. *Molecules, 26*(19), 5991. https://doi.org/10.3390/molecules26195991

Sarkar, R., Nain, L., Dutta, A., Kundu, A., & Saha, S. (2022). Unraveling the utilization feasibility of citrus peel solid distillation waste as potential source for antioxidant as well as bioethanol. *Biomass Conversion and Biorefinery, 14*, 27379–27391. https://doi.org/10.1007/s13399-022-03367-3

Satari, B., & Karimi, K. (2018). Citrus processing wastes: Environmental impacts, recent advances, and future perspectives in total valorization. *Resources, Conservation and Recycling, 129*, 153–167. https://doi.org/10.1016/j.resconrec.2017.10.032

Taghizadeh-Alisaraei, A., Hosseini, S. H., Ghobadian, B., & Motevali, A. (2017). Biofuel production from citrus wastes: A feasibility study in Iran. *Renewable and Sustainable Energy Reviews, 69*, 1100–1112. https://doi.org/10.1016/j.rser.2016.09.102

Teigiserova, D. A., Tiruta-Barna, L., Ahmadi, A., Hamelin, L., & Thomsen, M. (2021). A step closer to circular bioeconomy for citrus peel waste: A review of yields and technologies for sustainable management of essential oils. *Journal of Environmental Management, 280*, 111832. https://doi.org/10.1016/j.jenvman.2020.111832

Teke, G. M., De Vos, L., Smith, I., Kleyn, T., & Mapholi, Z. (2023). Development of an ultrasound-assisted pre-treatment strategy for the extraction of D-limonene toward the production of bioethanol from citrus peel waste (CPW). *Bioprocess and Biosystems Engineering, 46*(11), 1627–1637. https://doi.org/10.1007/s00449-023-02924-y

Vadalà, R., Lo Vecchio, G., Rando, R., Leonardi, M., Cicero, N., & Costa, R. (2023). A sustainable strategy for the conversion of industrial citrus fruit waste into bioethanol. *Sustainability, 15*(12), 9647. https://doi.org/10.3390/su15129647

Vilariño, M. V., Franco, C., & Quarrington, C. (2017). Food loss and waste reduction as an integral part of a circular economy. *Frontiers in Environmental Science, 5*. https://doi.org/10.3389/fenvs.2017.00021

Zema, D. A., Calabrò, P. S., Folino, A., Tamburino, V., Zappia, G., & Zimbone, S. M. (2018). Valorisation of citrus processing waste: A review. *Waste Management, 80*, 252–273. https://doi.org/10.1016/j.wasman.2018.09.024

Zema, D. A., Calabro, P. S., Folino, A., Tamburino, V., Zappia, G., & Zimbone, S. M. (2019). Wastewater management in citrus processing industries: An overview of advantages and limits. *Water, 11*(12), 2481. https://doi.org/10.3390/w11122481

21 Rambutan Waste Management

21.1 INTRODUCTION

One of the most significant tropical commercial fruits is the rambutan (*Nephelium lappaceum L.*), which is grown extensively in Southeast Asia, South America, Australia, and Africa. It has a close connection to the Sapindaceae family of subtropical fruits, which includes the lychee (*Litchi chinensis* Sonn.), longan (*Euphoria longan* Steud.), and pulasan (*Nephelium mutabile* Blume). Thailand, Malaysia, and Indonesia are the top countries for rambutan fruit production and exportation (Ahmad & Chua, 2013). Rambutan has become widely consumed due to its refreshing and sweet flavor, appealing appearance, and exotic aroma (Li et al., 2018; Zhuang et al., 2017). Normally, rambutan is consumed by people freshly, but it is also consumed in the form of jelly, jam, juice, marmalades, chips, and spreads, mainly produced in Malaysia and Thailand (Chai et al., 2019) (Figure 21.1).

The issue after rambutan consumption, whether directly or industrially processed, is the seeds and peels considered residues or wastes (Solís-Fuentes et al., 2010). In Thailand, an annual average of 1,900 tons of rambutan seeds is disposed of Mahisanunt et al. (2017). This large amount of rambutan industrial co-product could harm the environment if not managed properly. In fact, rambutan seeds have some nutrient sources that are helpful for the human diet, such as a huge presentation of carbohydrates (46–48.1%), crude fat (33.4–39.13%), protein (7.8–12.4%), and fiber (11.6%) (Chai et al., 2019; Nauli Harahap et al., 2012). Not only a great source of fat, rambutan seeds also contain engaging fatty acids such as arachidic, oleic, and stearic acids (Chai et al., 2018). Additionally, rambutan seeds have the ability to function as an antibacterial agent antioxidant by combating gram-negative and gram-positive bacteria (Bhat & Al-daihan, 2014). Interestingly, Evaristus et al. (2018) found that in Malaysians the seed converted into powder can manage the blood sugar levels.

Another co-product of rambutan consumption is the peel. The fact that the peel constitutes the highest weight (13.2%) of rambutan makes it a major contributor to the rambutan waste

FIGURE 21.1 Rambutan fruit. (Source: Healthline.com.)

DOI: 10.1201/9781003472209-28

(Hernández-Hernández et al., 2019). In fact, rambutan peels have certain biological properties that could positively impact human health. A study conducted by Hernández-Hernández et al. (2019) explained that rambutan peels can have antioxidant, antibacterial, antidiabetic, antiproliferative, anti-inflammatory, antiviral, and anticancer properties. Even though rambutan seed and peel have many potential sources in the human diet, most of them are considered industrial waste and will become pollutants if not managed properly (Evaristus et al., 2018). The co-products will harm the environment and become an economic burden due to their disposal costs (Jahurul et al., 2015). The right utilization of rambutan seed will have an excellent economic impact, nurturing health, decreasing disposal costs, and, most importantly, minimizing environmental impacts.

21.2 RAMBUTAN SEED AND PEEL AND THEIR VALUATION

Depending on the variety, rambutan's single oval seed—which makes up 4–9% of the entire fruit—is encircled by a white basal scar, as shown in Figure 21.2 (Akhtar et al., 2017). The seed is 1–1.3 cm long and has a brown tint. The rather bitter and narcotic rambutan seed is edible (Nauli Harahap et al., 2012). It contains a lot of protein, carbohydrates, and fat. This seed is handled as a by-product of rambutan fruit that is industrially processed into cans, juice, jam, jelly, marmalades, and spreads in Malaysia and Thailand (Chai et al., 2019). The abundance of these industrial by-products, which are both highly nutritive and include bioactive substances, is causing significant worry (Torres-León et al., 2016). Rambutan seeds contain a number of phenolic compounds, including corilagin, geraniin, and ellagic acid, which may have positive effects on one's health (Akhtar et al., 2017). Minerals, including calcium, zinc, iron, magnesium, and manganese, are also abundant in rambutan seeds. Rambutan seed can be consumed by humans or applied to food products from a nutritional and functional standpoint.

On the contrary, the fruit rind, often known as the peel, is a natural barrier protecting the edible flesh and preserving the fruit's structural integrity. Rambutan fruit skin can make up to 50% of the weight of the fruit, depending on the cultivar and level of maturity (Solís-Fuentes et al., 2010). Along with nutritious components, several antinutritional substances are found. These substances include saponins, alkaloids, tannins, phytates, and oxalates, with relative concentrations of 0.53 g^{-1}, 2.17 g^{-1}, 1.35 g^{-1}, 0.17 g^{-1}, and 0.12 g 100 g^{-1}, respectively (Eteng, 2012). However, because of their antioxidant properties, these antinutrients are also credited with a variety of bioactivities. Pectin is also found in the cell walls of rambutans, but it is much less abundant than it is in citrus peel. Pectin ranges from 1.05 to 1.9 wt% in fresh rambutan peel. In addition to being dark in color and having a

FIGURE 21.2 Rambutan components composed of flesh, seed, and peel (Akhtar et al., 2017).

low methoxy content (11.0–11.7%), the extracted pectin contains 8.73 wt% ash and 43 wt% moisture. Rambutan peel pectin has a lower water solubility than pomelo pectin (Chai et al., 2019).

21.3 ENVIRONMENTAL IMPACT FROM RAMBUTAN WASTE

Rambutan is a popular fruit grown in many tropical countries. While the fruit is enjoyed for its sweet taste and nutritional value, the waste produced from rambutan can have negative environmental impacts. One of the most significant environmental impacts of rambutan waste is the production of methane gas (Hadeel et al., 2011). Methane is a potent greenhouse gas contributing to climate change and global warming. When rambutan waste is disposed of in landfills, it undergoes anaerobic decomposition, which produces methane gas. This is because the landfill environment is oxygen-deprived, which causes the organic waste to break down in a way that produces methane gas (Ooi et al., 2012). According to the U.S. Environmental Protection Agency, methane has a global warming potential that is 28 times greater than carbon dioxide over a 100-year timescale. Therefore, the emissions of methane from rambutan waste can significantly contribute to climate change.

Furthermore, the disposal of rambutan waste can contribute to water pollution. When the waste is not properly disposed of, it can end up in water bodies and contaminate them. This can have negative impacts on aquatic life and the health of individuals who rely on these water sources. The contaminants from the waste can cause the water to become toxic, which can harm fish and other aquatic organisms. Furthermore, people who consume contaminated fish or water can become ill. This can lead to a host of health problems, including nausea, vomiting, diarrhea, and even death (Solangi et al., 2021).

In addition to water pollution, rambutan waste disposal can lead to soil quality degradation. When the waste is not properly disposed of, it can release harmful chemicals and pathogens that can negatively impact soil health and fertility. The chemicals in the waste can be absorbed by the soil, which can make it toxic and unsuitable for growing crops. This can lead to a reduction in agricultural yields and can have negative impacts on the livelihoods of farmers. Soil degradation can also make it difficult to restore the ecosystem, leading to a loss of biodiversity (Yap, 2013).

To mitigate the negative environmental impacts of rambutan waste, it is essential to properly dispose of the waste. Composting rambutan waste can be an effective way to reduce its negative environmental impacts. Composting is the process of breaking down organic matter into a nutrient-rich soil amendment. By composting the waste, it can be converted into a fertilizer that can be used to improve soil quality and promote plant growth. Composting can also help to reduce methane emissions by creating an environment that promotes aerobic decomposition instead of anaerobic decomposition. This can significantly reduce the amount of methane that is produced.

In summary, the disposal of rambutan waste can have significant negative environmental impacts. Properly disposing of the waste can reduce these impacts and promote a cleaner and healthier environment. Therefore, it is important for individuals and businesses to take responsibility for the proper disposal of rambutan waste to mitigate its negative environmental impacts. Governments and organizations can also be critical in creating policies and systems that encourage proper waste disposal methods and reduction practices.

21.4 AVAILABILITY

Rambutan is mainly found in Southeast Asian countries like Malaysia, Thailand, and Indonesia and minimally in India. Additionally, rambutan is found in humid-tropical weather countries in the Americas like Mexico, Ecuador, Colombia, Honduras, Trinidad and Tobago, and Cuba (Hernández-Hernández et al., 2019). Annually, the cultivation of rambutan yields 1.27 million metric tons, but substantial quantities of peels and seeds are lost during processing as by-products. Although certain portions of the waste are utilized as fertilizers or animal feed, the majority remain unused, posing a threat to the ecosystem (Eteng, 2012). Additionally, it has been reported that the rambutan peel constitutes 45% of the fruit's total weight, and studies have shown that rambutan peel extract (RPE) contains higher levels of phenolic compounds and exhibits stronger antioxidant properties compared to the fruit's leaves or seeds (Phuong et al., 2020b). Large quantities of antioxidant molecules, including geraniin, ellagic acid, and corilagin, are thought to be responsible for RPE's antioxidant activity (Hernández-Hernández et al., 2019). Among them, geraniin is recognized as the primary component of RPE and has an impact on RPE's antioxidant properties (Phuong et al., 2020a,b).

21.5 PROPERTIES OF RAMBUTAN WASTE FIBERS

Rambutan waste refers to the by-products of the rambutan fruit, including its peel, seeds, and pulp, that are typically discarded after consumption. These waste materials, often overlooked, possess valuable properties that make them suitable for various industrial applications. The fibers derived from rambutan waste, particularly from the peel and seeds, are rich in bioactive compounds and natural polymers such as cellulose, hemicellulose, and lignin. These components give rambutan waste fibers unique mechanical, thermal, and chemical properties, which can be utilized for applications in biocomposites, packaging, and others. Understanding the properties of rambutan waste fibers is essential for unlocking their potential in sustainable technologies.

21.5.1 Physical Properties

Fruit rind, often known as the peel, is a naturally occurring layer of protection that stabilizes the fruit's structural integrity. Rambutan fruit skin can make up to 50% of the weight of the fruit, depending on the cultivar and maturity (Augustin & Chua, 1988; Solís-Fuentes et al., 2010). Depending on the specific type of waste under consideration, the physical characteristics of rambutan waste can change. Nevertheless, rambutan waste typically exhibits the following physical characteristics.

The color of rambutan waste varies depending on which part of the fruit it originates from. The exterior peel is often bright red or yellow, while the seeds and pulp remains are white. Strong fibers in the peel and seeds give rambutan waste a fibrous feel. The remaining pulp is soft and pulpy. Rambutan waste includes a high concentration of moisture, particularly in the pulp residues. The density of rambutan waste varies according to the section of the fruit. The peel and seeds are heavier than the remaining pulp. Due to the presence of organic acids in the fruit, rambutan waste has a slightly acidic pH. Rambutan waste may have a pleasant, fruity odor, but this varies according to the maturity stage and the waste type being evaluated. In general, rambutan waste is a complex collection of organic elements with distinct physical qualities that might differ based on the fruit section and other factors. Table 21.1 provides the weight and percentage of constituent portions proposed by Solís-Fuentes et al. (2010) in rambutan fruit.

TABLE 21.1
Weight and Percentage of the Constituent Portions of Rambutan (*Nephelium lappaceum L*) Fruit (Solís-Fuentes et al., 2010)

Fruit Portion	Weight (g)	%
Whole fruit	27.4	100
Peel	13.2	45.7
Pulp	11.7	44.8
Seed	2.53	9.5
Embryo	1.60	6.1

21.5.2 Chemical Composition

Rambutan has several parts, which are the peel, seed, and antioxidants with different chemical compositions from one to another.

21.5.2.1 Rambutan Fruit Peels

Table 21.2 provides a summary of the nutritional composition of the peel (Mahmood et al., 2018). Apart from essential nutrients, it contains saponins, alkaloids, tannins, phytates, and oxalates, with quantities of 0.53, 2.17, 1.35, 0.17, and 0.12 g, respectively (Fila et al., 2012). Nevertheless, due to their antioxidant capabilities, these antinutrients are credited with a variety of bioactivities. Moreover, akin to citrus peel, the rambutan peel also contains pectin within its cell wall composition, albeit in a smaller proportion. The fresh rambutan peel has a pectin content ranging from 1.05 to 1.9 wt%. The extracted pectin from rambutan peel exhibits a dark hue, with a low methoxy concentration (11.0–11.7%), comprising 43 wt% moisture and 8.73 wt% ash. Nonetheless, it is essential to note that rambutan peel pectin demonstrates less water solubility compared to pomelo pectin (Suhaila & Zaharah, 1995; Normah & Hasnah, 2000).

21.5.2.2 Rambutan Peel Antioxidants

Notably, certain fruit peels exhibit greater bioactivity than their edible portions, attributed to their higher antioxidant content (Jayaprakasha et al., 2001). Rambutan fruit is no exception to this phenomenon. The primary antioxidants responsible for the beneficial properties of rambutan peel are phenolic acids and ellagitannins (Thitilertdecha et al., 2010). Palanisamy et al. (2008) conducted research on the peel's antioxidant activity through total phenolics assessment, revealing that the peel is abundant in natural antioxidants. Okonogi et al. (2007) also validated the peel ethanol extract's free radical scavenging efficacy. Figure 21.3 shows the main bioactive compounds in RPE.

According to Palanisamy et al. (2008), the total phenolic content of the ethanolic extract from the peel (762 mg gallic acid equivalent (GAE)/g extract) is comparable to that of commercially available grape seed extract. To assess the antioxidants in red and yellow rambutan cultivars, Gusman and Tsai (2015) conducted a study contrasting traditional extraction (12 h) versus ultrasonic extraction (2 min). The sonicated extract showed improved recovery with greater total phenolics and improved ferric reduction activity. Total phenolics were increased by 20% as a result of sonication, while FRAP increased by 8%. Similarly, Thitilertdecha et al. (2010) extracted phenolics from the

TABLE 21.2
Rambutan Peel Chemical Composition

Ingredient	Amount	References
Moisture (g $100 g^{-1}$)	72.05	Fila et al. (2013)
Lipids (g $100 g^{-1}$)	0.23	
Carbohydrates (g $100 g^{-1}$)	23.78	
Protein (g $100 g^{-1}$)	2.04	
Fiber (g $100 g^{-1}$)	0.7	
Ash (g $100 g^{-1}$)	1.2	
Vitamins		Johnson et al. (2013)
Carotene (μg $100 g^{-1}$)	10.60	
Thiamine (mg $100 g^{-1}$)	0.04	
Riboflavin (mg $100 g^{-1}$)	0.06	
Niacin (mg $100 g^{-1}$)	0.31	
Ascorbic acid (mg $100 g^{-1}$)	7.43	

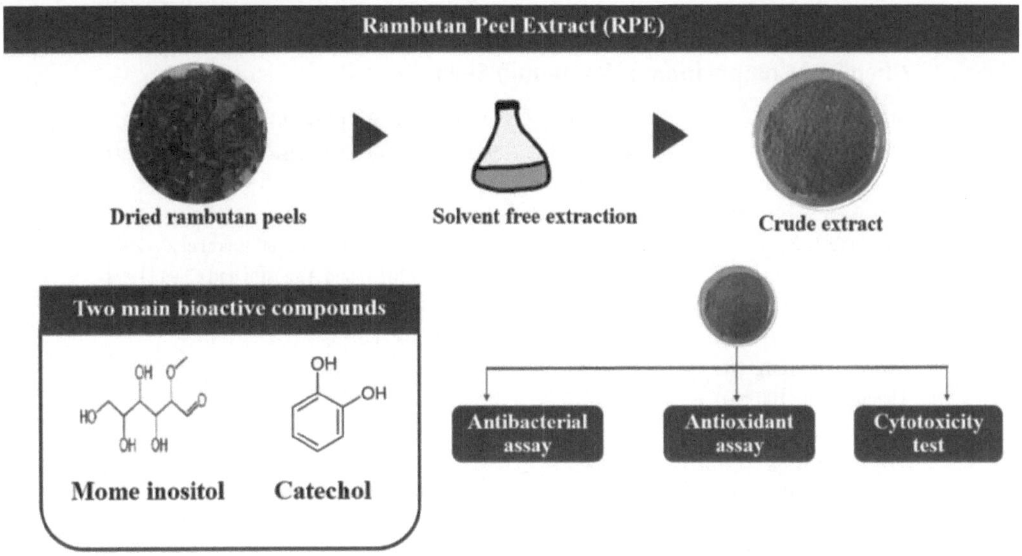

FIGURE 21.3 Main bioactive compounds in rambutan peel extract.

peel, and their ability to scavenge lipid peroxidation was assessed. A 100 g methanolic peel extract was tested, and the results showed that coraligin (71.9 mg), ellagic acid (53.5 mg), and geraniin (568.0 mg) were all present.

Additionally, the rambutan extract demonstrated noticeably better activity against lipid peroxidation than the synthetic antioxidant butylated hydroxytoluene (BHA), with a range of 77–186 times greater effectiveness. According to Palanisamy et al. (2011), the dried RPE contains 37.9 mg GAE/g. Corilagin and ellagic acids were later discovered in the peel at amounts of 0.47 and 0.14 mg/g, respectively, in further investigations (Samuagam et al., 2014). Muhtadi et al. (2014) investigated the antioxidant activity of the peel in four different solvents (ethyl acetate, hexane, chloroform, and methanol), and they found that ethyl acetate had the best bioactivity. With a lower IC50 value (concentration at which bioactivity is reduced by half) of 7.74 g/mL compared to vitamin E's 8.48 g/mL, peel extract revealed greater bioactivity in comparison to vitamin E.

Ellagic acid, a secondary metabolite that is found in plant cell vacuoles and has a number of positive effects, including antioxidant, anti-atherogenic, antiproliferative, and chemopreventive qualities (Atkinson et al., 2006), is categorized as a secondary metabolite. Due to its antioxidant, nitrous oxide scavenging, and anticancer properties, geraniin, a crystalline ellagitannin, has been the focus of study (Okabe et al., 2001). Successfully isolating pure geraniin from rambutan peel in 2012, Perera et al. (2012) produced varying yields of 10–20%.

21.5.2.3 Seed Chemical Composition

Rambutan seed is viewed as a waste product in the fruit processing business. On the contrary, rambutan seeds that have been roasted are edible and popular across Asia (Solís-Fuentes et al., 2010). According to Sirisompong et al. (2011), the flattened oblong seed makes up around 4.1% of the weight of the entire rambutan fruit. The rambutan seed's chemical composition is given in Table 21.3 (Mahmood et al., 2018). The dried seeds, on the contrary, have higher concentrations of protein, fat, and carbohydrates (Fila et al., 2013). According to the amino acid profile of defatted seed flour, the protein is of a high caliber and contains both necessary and non-essential amino acids in large amounts (Augustin & Chua, 1988). Lysine has the highest relative percentage of the essential amino acids, with leucine, valine, isoleucine, phenylalanine, methionine, and histidine following in decreasing order. Contrarily, the proportional percentages of the non-essential amino

TABLE 21.3
Chemical Composition of Rambutan Seed

Ingredient	Amount	References
Moisture (g $100 g^{-1}$)	34.28–34.6	Fila et al. (2013) and Augustin and Chua (1998)
Lipids (g $100 g^{-1}$)	18.19–38.9	
Carbohydrates (g $100 g^{-1}$)	33.65–61.6	
Protein (g $100g^{-1}$)	11.06–14.1	Fila et al. (2013) and Maisuthisakul et al. (2008)
Fiber (g $100 g^{-1}$)	0.62–6.6	Fila et al. (2013) and Augustin and Chua (1998)
Ash (g $100 g^{-1}$)	2.20–2.9	
Vitamins		Johnson et al. (2013)
Carotene (μg $100 g^{-1}$)	3.42	
Thiamine (mg $100 g^{-1}$)	0.02	
Riboflavin (mg $100 g^{-1}$)	0.09	
Niacin (mg $100 g^{-1}$)	0.08	
Calcium (mg $100 g^{-1}$)	30.4	Maisuthisakul et al. (2008)
Iron (mg $100 g^{-1}$)	2.8	

TABLE 21.4
Fatty Acid Composition of Rambutan Seed Lipids

Composition (%)	*a*	*b*	*c*	*d*
Saturated				
Myristic acid ($C_{14:0}$)	0.02	0.01	—	0.13
Palmitic acid ($C_{16:0}$)	4.69	7.39	6.1	4.60
Stearic acid ($C_{18:0}$)	7.03	16.58	7.1	7.88
Arachidic acid ($C_{20:0}$)	34.32	12.34	34.5	31.53
Heneicosanoic acid ($C_{21:0}$)	0.05	—	—	—
Behenic acid ($C_{22:0}$)	3.10	8.91	2.9	2.1
Tricosanoic acid ($C_{23:0}$)	0.03	—	—	—
Lignoceric acid ($C_{24:0}$)	0.33	—	—	—
Mono-Unsaturated				
Palmitolic acid ($C_{16:1}$)	0.49	0.49	1.5	0.72
Elaidic acid ($C_{18:1}$)	0.03	0.02	—	—
Oleic acid ($C_{18:1}$)	36.79	52.18	40.3	43.09
Gadoleic acid ($C_{20:1}$)	—	—	—	5.89
Erucic acid ($C_{22:1}$)	0.66	—	—	0.10
Poly-Unsaturated				
Linolenic acid ($C_{18:3}$)	6.48	2.02	—	0.74
Linoleic acid ($C_{18:3}$)	1.37	—	—	3.22
Eicosadienoic acid ($C_{20:2}$)	0.04	—	—	—

acids are as follows: glutamic acid>glycine>aspartic acid>arginine>serine>threonine>alanine>tyrosine>proline>cysteine (Augustin & Chua, 1988). According to Manaf et al. (2013), the seeds contain the following primary fatty acids: palmitic acid (4.36–4.86%), stearic acid (5.93–7.49%), oleic acid (37.91–40.15%), and arachidic acid (36.14–36.77%).

Table 21.4 summarizes the saturated and unsaturated fatty acid profiles of seeds (Mahmood et al., 2018). In 2015, Chimplee and Klinkesorn conducted a study to investigate how the moisture content of seeds affects the production of extracted fat. It was reported that a drying temperature of 65°C was optimum for decreasing seed moisture and greatly enhancing extraction yield. The difference in seed moisture, on the contrary, had no effect on fatty acid concentrations. However, fatty acid analysis revealed the presence of palmitic, palmitoleic, stearic, oleic, linoleic, arachidic, eicosanoic, behenic, and erucic acid. Previous research by Chimplee and Klinkesorn (2015) indicated that oleic acid (38%) and arachidic acid (36%) levels were higher across all moisture levels investigated (4%, 11%, and 24%). Moreover, fresh seeds were found to contain antinutritional chemicals (mg/100 g) such as saponins (0.98), alkaloids (0.82), tannins (0.15), phytates (0.40), and oxalates (0.26). However, after ten days of fermentation, tannins and saponins were reduced by approximately 50% and 60%, respectively (Olaniyi & Mehdizadeh, 2013). Additionally, a discovery by Fang and Ng (2015) revealed the presence of an atrypsin inhibitor in fresh seeds.

21.5.2.4 Seed Antioxidants

Different bioactive chemicals, including phenols and carotenoids, are extracted from the seed using a variety of solvents, including water, butanol, hexane, ethanol, ether, ethyl acetate, and methanol (Fidrianny et al., 2015). According to Maisuthisakul et al. (2008), the ethanolic extract of the seed contains 13.3 mg/g rutin equivalent of flavonoids and 43.5 mg/g GAE of total phenols. Similar to this, 124.14 and 3.05 mg GAE/g of total phenolics, respectively, were present in the methanolic and ethanolic extracts of the dried seed (Fidrianny et al., 2015). Using the DPPH and ABTS (Trolox equivalent antioxidant) assays, the IC50 values for the methanolic and ethanolic extracts were 7.0 and 7.34 g/mL, respectively. Research by Thitilertdecha et al. (2008) used a variety of assays, including reducing power, free radical scavenging, linoleic peroxidation, and carotene bleaching to assess the antioxidant capacity of different extracts (water, methanol, and ethanol). Similar to this, Soeng (2015) used the superoxide-dismutase assay to investigate the antioxidant properties of the ethanolic extract. The extract was separated into fractions by the researchers into water, butanol, ethyl acetate, and hexane. Ethyl acetate, which had concentrations of 3.37 and 3.03 g/mL, respectively, was the most active of these fractions, followed by water.

21.5.3 Morphological Properties

Numerous studies have shown that rambutan seeds have a high lipid content, ranging from 14% to 41% (Augustin & Chua, 1988; Sirisompong et al., 2011; Manaf et al., 2013). Rambutan seed lipids' physicochemical and thermal properties have been studied, showing prospective uses for them in the food industry (Solís-Fuentes et al., 2010). The extracted fat appears white and has a brightness value of 86.87, a yellowness value of 3.55, and a greenness value of –2.06 at room temperature. According to Sonwai and Ponprachanuvut (2012), the refractive index of fat is 1.469, and its slide melting point is 38.47°C. The acid value is 0.77 (% oleic acid), and the iodine value is 32.31 (g I2/100 g of fat). The fat has a saponification value of 199.38 (mg KOH/g of fat) and includes a minimal quantity of unsaponifiable material (0.19 g/100 g). Tocopherol (0.10 mg/g), stigmasterol (0.32 mg/g), and β-sitosterol (0.61 mg/g) are among the phytosterols that are found (Sirisompong et al., 2011).

21.5.3.1 Fatty Acids and Triglycerides Composition of Seed Lipids

The structural makeup of lipids affects their functional properties (Reddy & Jeyarani, 2001). Sirisompong et al. (2011) determined that the seed fat contains 45.9 g of unsaturated fatty acids/kg and 49.6 g of saturated fatty acids/kg. Arachidic acid (C20:0), with 34.3 g/100 g, and oleic acid (C18:1), at 36.8 g/100 g, are the main fatty acids found. Oleic acid (40.45%) and arachidic acid (36.36%) were also shown to be the two main fatty acids in a different research (Harahap et al., 2012), along with the presence of palmitic, gondoic, behenic, stearic, and palmitoleic acids. Similar to this, Sonwai and Ponprachanuvut (2012) discovered that the two fatty acids that were most prevalent in seed fats were oleic acid (C18:1) at 33.13% and arachidic acid (C20:0) at 42.5%. About 8.70% and 3.38% of the total came from stearic acid (C18:0) and palmitic acid (C16:0), respectively. Similar ratios of the main fatty acids were seen by Azam et al. (2005) in their previous investigation as well. It is an

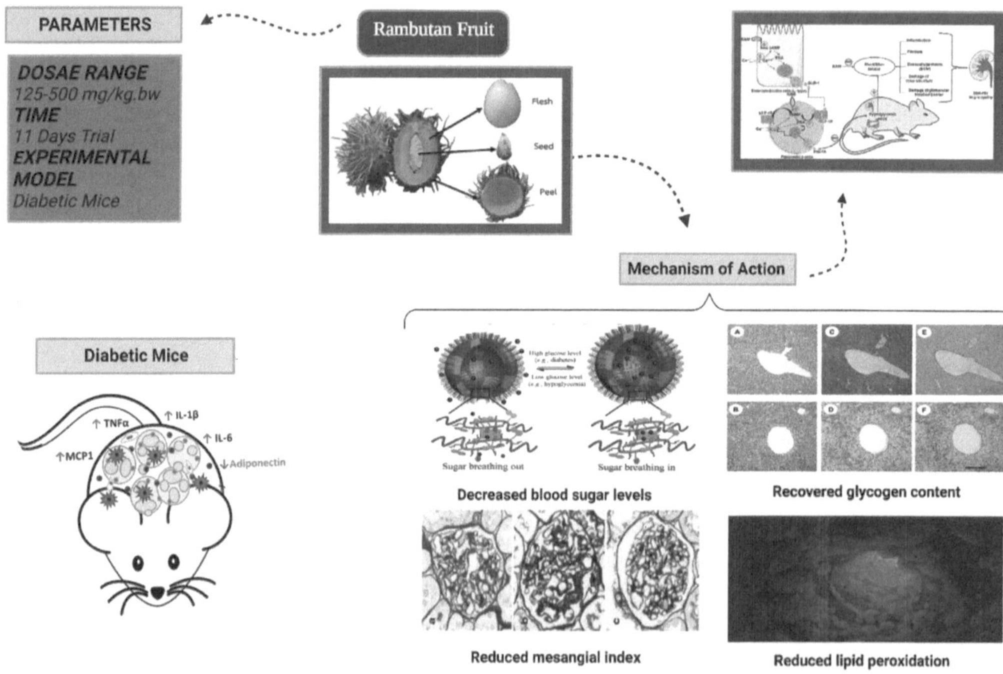

FIGURE 21.4 Therapeutic potential of rambutan fruit in diabetic mice (Afzaal et al., 2023).

intriguing discovery that the larger amount of arachidic acid makes it appropriate for food applications without requiring further hydrogenation, especially in cases where autoxidation is a problem. Three fatty acids and the sugar alcohol glycerol join chemically to form a triglyceride, which is an ester. Saturated and unsaturated fatty acids can both have the same length carbon chain or a different one. The majority of plant lipids are made up of triglycerides, and the fat extracted from rambutan seeds has the following triglyceride pattern: 1-arachidoyl-2,3-dioleoylglycerol (AOO) > 1-arachidoyl-2-steareoyl-3-oleoylglycerol (ASO) > 1-arachydoyl-2-oleoyl-3-palmitoglycerol (AOP) > 1-arachidyl-2-stearoyl-3-palmitoylglycerol (ASP) > 1-arachidyl-2-linoleoyl-3-palmitoylglycerol (ALP) > 1-arachidyl-2-linolenyl-oleoylglycerol(ALnO) > 1-arachidyl-2-linolenyl-3-stearoylglycerol(ALnS) > triolein (OOO) > tri-archidin (AAA) > 1-arachidyl2-linoleoyl-3-oleoylglycerol (ALO) (as shown in Figure 21.4). AOO, ASO, and AOP, the first three triglycerides in this series, comprise 77.74% of the total triglycerides.

Additionally, previous studies (Harahap et al., 2012; Sirisompong et al., 2011; Sonwai & Ponprachanuvut, 2012) demonstrated that seed triglycerides consist of two significant fatty acid types, namely, arachidic acid and oleic acid. The seed triglyceride composition was also investigated by Manaf et al. (2013), who verified their high molecular weights. Due to the absence of some triglyceride standards, reverse-phase HPLC was used to identify the substance based on its carbon number. Similar to previous studies, the predominant glycerides with carbon chain lengths of C56, C58, and C60 were identified, and the combination of arachidic and oleic acids was suggested as the cause of these long-chain glycerides. Kheiri and Mohd Som (1987) investigated the triglycerides in eight clones of rambutan fruit and found the presence of C50, C52, C54, C56, C58, and C60 with respective concentrations of 0.1%, 0.7%, 12.5–21.6%, 34.6–39.1%, 28.8–47.2%, and 1.5–6.6%.

21.5.3.2 Seed Solid Fat Content

Extracted lipids are frequently classed as either oils or fats, depending on their solid fat content (SFC). Differential scanning calorimetry (DSC) was used to examine the SFC of rambutan seed fat (RSF) (Solís-Fuentes et al., 2010). Figure 21.5 shows the SFC (%) of rambutan seed oil of varieties R4, R7,

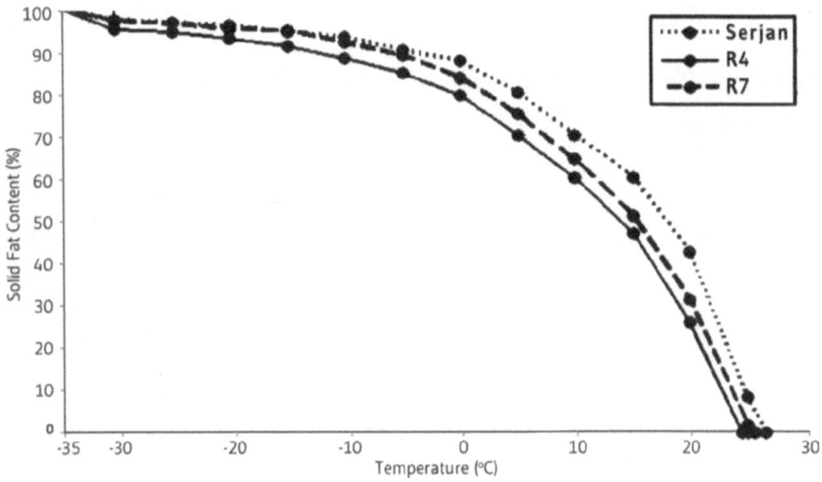

FIGURE 21.5 Solid fat content (%) of rambutan seed oil of varieties R4, R7, and Serjan (Ghobakhlou et al., 2019).

and Serjan (Ghobakhlou et al., 2019). According to the thermal profile, the SFC was 47.8 wt% at 10°C, dropped to 25 wt% with a 10°C rise, and attained full SFC at 51.8°C. Due to the presence of high molecular weight triglycerides and long-chain saturated fatty acids, this substance has a higher melting point. In contrast, Sonwai and Ponprachanuvut (2012) showed that solid fat was no longer detectable at 40°C, refuting this conclusion. Extracted lipids are frequently classed as either oils or fats, depending on their SFC. DSC was used to examine the SFC of RSF (Solís-Fuentes et al., 2010). According to the thermal profile, the SFC was 47.8 wt% at 10°C, dropped to 25 wt% with a 10°C rise, and attained full SFC at 51.8°C. Due to the presence of high molecular weight triglycerides and long-chain saturated fatty acids, this substance has a higher melting point. Sonwai and Ponprachanuvut (2012), in contrast, showed that solid fat was no longer detectable at 40°C, refuting this conclusion.

21.5.3.3 Fat Crystallization Behavior

Fat crystal polymorphism, a critical characteristic used to assess the functional properties of fats, indicates structural similarities among fats derived from different sources. Researchers discovered seed fat begins to harden before reaching the temperature necessary for complete crystallization using the capillary X-ray diffractometer's static isothermal mode (Sonwai & Ponprachanuvut, 2012). While the peak at 4.15°A corresponds to the pseudo-β′ structure, the occurrence of peaks at 3.85°A and 4.24°A proves the existence of the β-structure. The β-structure of fat crystals is also linked to the diffraction peak at 4.55°A. However, after 7 days of storage, a dominant single peak emerged at 4.60°A, indicating the typical β-structure. An in-depth analysis of crystal morphology suggests that the crystals take the form of loosely packed spherulites that appear to be practically aggregated. As reported by Sirisompong et al. (2011), the fat crystals in RSF consist of 15.3% of the β-form and 84.7% of the β-form. RSF's crystalline structure resembles cocoa butter, indicating that it can replace cocoa butter in processed goods. A recent research by Luma and Yang (2016) showed that, up to a certain proportion, RSF may be utilized in place of cocoa butter (as a filler) to obtain functionality similar to cocoa butter.

21.5.4 Thermal Properties

The thermal stability of fats intended for food processing is a crucial factor. In their study, Solís-Fuentes et al. (2010) employed thermogravimetric analysis to investigate the thermal stability of rambutan fat under both inert (N_2) and oxidizing conditions. The analysis involved calculating

TABLE 21.5
Melting and Crystallization Temperatures of Rambutan Seed Fats

Melting			Crystallization			
T_{onset} (°C)	T_{offset} (°C)	*dH* (J/g)	T_{onset} (°C)	T_{offset} (°C)	*dH* (J/g)	References
−14.5	51.8	124.3	33.6	−45.6	89	Solís-Fuentes et al. (2010)
−27.6 to −31.3	24.8–50.6	71.2–141.7	23.9–39.4	−43.7 to −45.6	60.4–88.9	Chai et al. (2018)
−4.5	58.9	85.4	37.8	−14.8	92.1	Sirisompong et al. (2011)
−0.9	58.2	87.5	52.4	−10.2	76.3	Chai et al. (2019)
−3.2 to −0.7	58.2–59.7	85.7–90.2	47.4–57.2	−12.6 to −16.3	72.5–78.6	Chai et al.[a] (2019)
−31.2	28.6	89.3	25.6	−41.3	71.3	Chai et al. (2019)
−5.3 to −1.9	48.5–58.6	84.5–92.1	29.8–53.8	−12.2	76.2–82.6	Chai et al.[b] (2019)
—	>50.0	—	50.0	−10.0	—	Febrianto et al.[b] (2016)
9.42	41.55	97.98	28.13	0.95	83.27	Sonwai and Ponprachanuvut (2012)
—	40.0	—	—	−43.8	—	Manaf et al. (2013)

[a] Roasted rambutan seed fat.
[b] Fermented rambutan seed fat.

the mass change with temperature, starting at 25°C and increasing at a rate of 10°C/min. The decomposition process and mass loss commenced at 240°C in a nitrogen atmosphere. Two specified temperatures—281°C, with a loss rate of 1.84% min/°C, and 409°C, with a loss rate of 2.11% min/°C—saw the most mass loss. At 475°C, the whole breakdown of the fat was seen.

However, compared to an inert atmosphere, the first breakdown temperature in an oxidizing (air) environment was lower (237.3°C). However, it was noted that the ultimate stage of decomposition was hotter, reaching 529°C. A three-stage breakdown with maxima at 278.8°C, 399.6°C, and 512.6°C was seen when oxygen was present. Nevertheless, the temperature at which mass change was determined to be the largest (1.57% min/°C) was 399.6°C. The wide temperature degradation range in an oxygen environment suggested that the fat was reactive with oxygen. This runs counter to the prevalent view that quality decline is mostly caused by fat oxidation. Instead, in this instance, the prolonged process of fatty acid mass loss under heat was compensated for by the absorption and interaction of the fat with ambient oxygen.

To investigate the possibility of employing rambutan seeds as a source of edible fat with possible industrial uses, several studies have looked at the melting and crystallization patterns of RSF, as given in Table 21.5 (Jahurul et al., 2020). However, as given in Table 21.5, there was considerable variability in the melting and crystallization starting and offset temperatures in the same fruit co-product across several research studies.

21.5.5 Calorific Properties

According to Manaf et al. (2013), rambutan seeds have a high lipids content ranging from 14% to 41%. Studies have looked at the rambutan seed lipids' physicochemical and thermal properties, which have shown to be appealing for certain uses in the food sector (Solís-Fuentes et al., 2010). The extracted fat has a lightness (*L**) value of 86.87, a yellowness (3.55), and a greenness (−2.06) value at room temperature. The slip melting point is calculated to be 38.47°C, and its refractive index is 1.469 (Sonwai & Ponprachanuvut, 2012). Similar to this, the iodine value is 32.31 g/100 g of fat, and the acid value is 0.77 (% oleic acid). The saponification value is 199.38 (mg KOH/g fat), with only a minor presence of unsaponifiable materials (0.19 g/100 g). Furthermore, the fat contains phytosterols such as -sitosterol (0.61 mg/g), stigmasterol (0.32 mg/g), and -tocopherol (0.10 mg/g) (Sirisompong et al., 2011).

Lipids' functional characteristics are determined by their structural makeup (Reddy & Jeyarani, 2001). The fat from rambutan seeds includes 45.9 g of unsaturated fatty acids and 49.6 g of saturated fatty acids per 100 g, according to Sirisompong et al. (2011). Arachidic acid (C20:0) (34.3 g/100 g) and oleic acid (C18:1) (36.8 g/100 g) were the two primary fatty acids found. Oleic acid (40.45%) and arachidic acid (36.36%) were the main fatty acids, according to a different research by Harahap et al. (2012). Palmitic, gondoic, behenic, stearic, and palmitoleic acids are some other fatty acids that have been identified. Arachidic acid (C20:0) and oleic acid (C18:1) were discovered by Sonwai and Ponprachanuvut (2012) to be the main constituents of rambutan seed lipids, making up 42.5% and 33.13%, respectively. Palmitic acid (C16:0) and stearic acid (C18:0) made up 8.70% and 3.38%, respectively, of the total. Major fatty acid levels were observed to be roughly equal in earlier studies by Azam et al. (2005). Surprisingly, the greater arachidic acid concentration in RSF renders it appropriate for culinary applications without further hydrogenation, especially where autoxidation is a problem. Triglycerides are chemical esters created when three fatty acids and glycerol, a sugar alcohol, combine. These fatty acids can be saturated or unsaturated, and their carbon chains can be the same length or various lengths.

The SFC of extracted lipids determines whether they should be classified as oil or fat. Researchers used DSC to analyze the SFC in RSF (Solís-Fuentes et al., 2010). The thermal profile shows that the SFC was 47.8 wt% at 10°C, dropped to 25 wt% with each every 10°C increase, and reached total fusion at 51.8°C. The longer melting point is a result of the existence of long-chain saturated fatty acids and high molecular weight triglycerides. This assertion is refuted by Sonwai and Ponprachanuvut's (2012) discovery that solid fat was no longer present at 40°C. At 0°C, 53.5 wt% solid fat was present, however, Manaf et al. (2013) found that at 40°C, all solid fat had changed to oil. Furthermore, the solid fat behaved like palm oil, mee fat, and lard between 25°C and 30°C (Manaf et al., 2012; Manaf et al., 2011). This result suggested that at room temperature, RSF may demonstrate improved stability and resistance to exudation, making it a useful choice to increase product stability.

21.6 RAMBUTAN WASTE MANAGEMENT

Currently, many of the by-products generated during fruit processing are being thrown away, raising processing costs and environmental issues. However, there is increased interest in using these by-products more efficiently because of the expanding demand for natural supplements and bioactive substances. Numerous research studies have been done to investigate the possibility of creating prebiotics, antioxidants, and other useful compounds from fruit waste. As a result, rambutan waste by-products are becoming more and more valued as raw materials or components for a variety of uses in both culinary and non-food goods.

21.6.1 Peel Extract as Antioxidant in Oil

Oxidative rancidity is a primary cause of quality degradation in edible oils, leading to the development of undesirable off-flavors that render the oil unsuitable for consumption. Figure 21.6 shows rambutan phenolic compounds (Phuong et al., 2020b). Using 80% methanol significantly improved the extraction yield of phenolic compounds from rambutan peel. Ultrasonic-assisted extraction further enhanced the phenolic yield, increasing it by 15% compared to conventional methanol extraction without ultrasound. The predominant phenolic compounds identified in RPEs include geraniin, corilagin, rutin, ellagic acid, and quercetin. The optimized extraction process yielded 55–64 g of lyophilized phenolic powder from 1 kg of fresh rambutan peel, with a total phenolic content of 250–300 mg GAE/g. Notably, the antioxidant activity of the extracted phenolic powder (at 100 μg

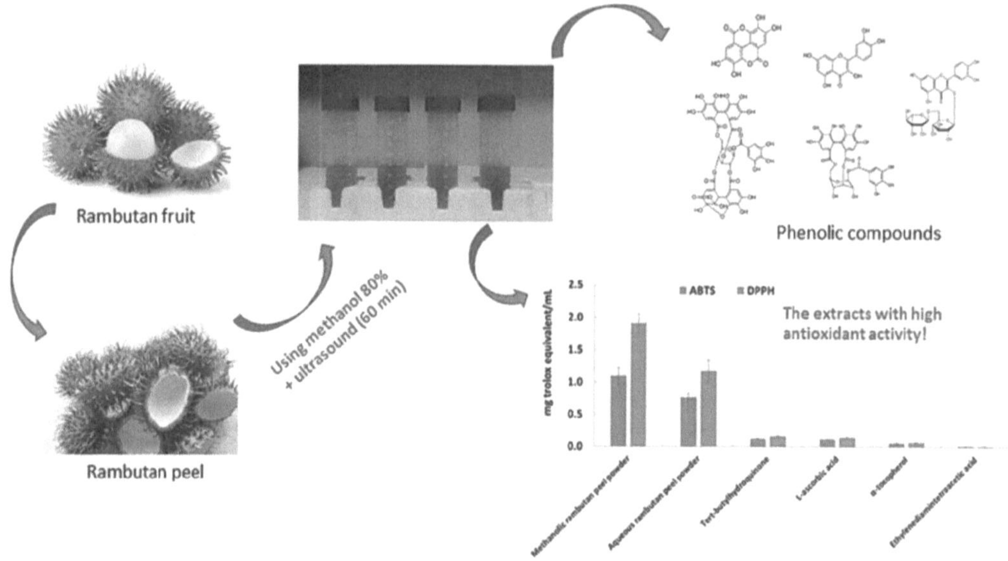

FIGURE 21.6 Rambutan phenolic compounds (Phuong et al., 2020b).

GAE/mL) exceeded that of common antioxidants like tert-butylhydroquinone, L-ascorbic acid, α-tocopherol, and EDTA at equivalent concentrations, highlighting its potent bioactive properties (Phuong et al., 2020b). Plant oils containing higher levels of unsaturated fatty acids are more susceptible to this issue, prompting the addition of synthetic antioxidants to extend shelf life and preserve flavor. However, growing concerns about chemical additives in food products have led to a search for natural antioxidant sources. In a study, researchers compared the efficacy of rambutan peel ethanolic extract to tocopherols and BHA in sunflower oil. During an accelerated storage experiment (60°C for 24 days), the rambutan extract revealed considerably greater oxidative stability for the oil than the control at all tested doses (100, 200, and 300 ppm). Another experiment involved storing oil enhanced with different extract fractions (300 ppm) for two years at room temperature (25°C). The extract outperformed tocopherol while displaying equal efficacy to BHA (Mei et al., 2014).

21.6.2 Seed Lipids as Cocoa Butter Substitute

The importance of cocoa butter as a key ingredient in chocolate and confectionery goods is seen in Figure 21.7. The search for a more affordable substitute for this in-demand component is still underway. According to Issara et al. (2014), rambutan fat has gained substantial interest as a potential alternative to cocoa butter in a variety of confectionery products. Zzaman et al. (2014) investigated the possibility of using RSF instead of cocoa butter in a separate investigation. Unexpectedly, it was discovered that the oleic acid concentration of cocoa butter and RSF was similar. To what degree cocoa butter might be substituted without losing its properties, various ratios of RSF and cocoa butter were examined. They discovered that a blend of 70% cocoa butter and 30% rambutan fat might deliver properties resembling pure cocoa butter. It should also be noted that at a certain shear rate, rambutan fat had a viscosity greater than cocoa butter's.

In a different research, Febrianto et al. (2016) proposed a compatibility study between fermented-roasted RSF and cocoa butter. Melted rambutan seed was added to cocoa butter at a rate of 10–90% wt%. Both the unstabilized and the stabilized phases of the mixes were evaluated. Next, they underwent two weeks of conditioning at 5°C after spending 24 h at room temperature. However, the unstabilized mixture merely underwent melt mixing, followed by solidification. This result showed that, at a mixing ratio of 30% or less, seed fat could take the place of cocoa butter in

FIGURE 21.7 Cocoa butter (Britannica).

confectionery goods. In an effort to mimic the properties of cocoa butter, Luma and Yang (2016) experimented with combining RSF at various ratios (20%, 40%, 60%, and 80%). The RSF:cocoa butter 80:20 combination showed cocoa butter-like X-ray diffraction peak patterns, and polarized light microscopy revealed crystals with similar sizes. Food processors might be able to mimic the benefits of cocoa butter and utilize a more affordable substitute without sacrificing the required texture in chocolate goods by properly blending rambutan fat and cocoa butter.

21.6.3 Seed Lipids as Vitamin Carrier

Micro- and nano-encapsulation techniques have become increasingly popular for delivering delicate nutrients, with a focus on achieving targeted distribution and controlled release of the active ingredients. Vitamin E is a fat-soluble vitamin, and Uraiwan and Satirapipathkul (2016) developed nanostructured lipid carriers for it using the melt emulsification technique. Tween-20 was employed as a surfactant, rambutan seed oil as a liquid lipid, stearic acid (C18:0) as a solid lipid, and vitamin E was enclosed within the carriers in their investigation. Smaller and more stable nanoparticles were created by changing the surfactant concentration, and a concentration of 5 wt% generated nanoparticles that were 139.43 nm in size. According to the study, rambutan seed oil may be used to transport other bioactive compounds, including fat-soluble vitamins.

21.6.4 Seed Flour as Stabilizer and Thickener

The defatted seed flour that is produced when the fat is removed from rambutan seeds has the potential to be a valuable source of carbs (87.04%) and proteins (10.07%) (Eiamwat et al., 2016). The amount of amylose in the defatted flour is also higher (32.16%). With a maximum viscosity of 1,300 cP (centipoise) and acceptable viscometric characteristics, it is useful as a thickening agent in confectionery products. According to recent studies, defatted wheat treated with diluted alkali (0.075N NaOH) showed enhanced bulk density, water absorption, swelling power, emulsion capacity, and stability while exhibiting lower solubility, turbidity, and oil absorption. By reducing the least gelation concentration (the quantity needed to produce a gel), it displayed good thickening abilities. It worked as an unusual supply of protein and starch (Eiamwat et al., 2016). In earlier research, the possibility of using rambutan seed flour (2%), which has fewer calories than egg yolk or vegetable oil, to make "Thousand Island Dressing" was investigated. The resulting dressing had four times fewer calories than the standard recipe and was well-received by 63% of the panelists in terms of sensory attributes (Phanthanapratet et al., 2012).

21.6.5 Seed Protein Concentrate

Human dietary supplements high in proteins, or protein concentrates, can be obtained from a variety of animal and plant sources. However, efforts are ongoing to find more affordable substitutes. Rambutan seeds have been suggested as a novel source of protein due to their high protein content, which ranges from 10.07% to 16.21% (Augustin & Chua, 1988; Maisuthisakul et al., 2008; Fila et al., 2013; Eiamwat et al., 2016). Additionally, a high-quality protein profile is indicated by the presence of both essential and non-essential amino acids in appreciable amounts (Augustin & Chua, 1988).

Water-soluble albumin proteins, which have a number of functions in the human body, including controlling colloidal osmotic pressure in the blood, are denatured by heat. While albumin is typically obtained from egg whites, Vuong et al. (2016) recently made an effort to extract an albumin concentration (80.8%) from rambutan seeds. Further research revealed that the maximum solubility was attained at a pH of 4, with capacities for absorbing water and oil being 6.13 and 0.79 mL/g, respectively. On the contrary, a pH of 12 was shown to have the maximum emulsification and foaming capacity.

21.6.6 Rambutan Seed Mucilage

The globular, water-soluble albumin proteins get denaturized when heated. The regulation of colloidal osmotic pressure in the blood is one of the many roles these proteins play in the human body. Albumin is frequently sourced from egg whites. On the contrary, a recent study by Vuong et al. (2016) examined the extraction of seed-albumin concentrate (80.8%) from rambutan seeds. Further research showed that a pH of 4 was where the greatest solubility was found. The capacity for absorbing water at this pH was 6.13 mL/g, while the capacity for absorbing oil was 0.79 mL/g. At a pH of 12, the highest emulsification and foaming capabilities were attained.

21.6.7 Seed Lipids in Personal Care Products

In cosmetics and personal care products, natural ingredients are frequently used since they have the fewest adverse effects. Similarly, Lourith et al. (2016) created solid bars and liquid soaps containing RSF. According to the authors, the process for producing fat was reportedly calculated for potential scaling up for industrial demands. It was also contrasted with other popular commercial sources of vegetable oils. The fat displayed higher compatibility with the other formulation elements. Thus, it was suggested that this unusual fat might serve as a viable raw material for personal care products, similar to other vegetable oils.

21.6.8 Peel Extract as Ligand in Nanoparticles Preparation

Metal nanoparticles have drawn a lot of interest because of their numerous uses and frequent use in pharmaceutical and personal care products. Nevertheless, the creation of nanoparticles frequently necessitates using chemicals that may be hazardous to the environment. Researchers are looking toward environmentally friendly strategies that lessen the need for chemical agents to address this issue. Using RPE as a natural ligand, Yuvakkumar et al. (2014) suggested an eco-friendly technique for manufacturing zinc oxide nanocrystals. The zinc-ellagate complexation process was used to effectively complete the synthesis. Agar disc diffusion was used to investigate the antibacterial efficacy of these nanocrystals against *Escherichia coli* and *Staphylococcus aureus* after they had been coated onto cotton fabric. The antibacterial characteristics of the Zn-nanocrystals were fascinating. According to the scientists, this novel green synthesis method has the potential for use in biomedical nanotechnology. In a different research, zinc oxide nanochains were made using rambutan peel, and the effectiveness of the method was checked using scanning electron microscopy. Human liver cancer cells (HepG2) were used to assess the anticancer abilities of the generated Zn-nanochains. At different incubation days, the results revealed decreased cell viability (40–60%) and alterations in cell shape (10–50%), indicating probable anticancer action (Yuvakkumar et al., 2015).

21.6.9 Peel Flour as Filler in Packaging

Petroleum-based commercial plastics are widely used for producing sheets and films, with polyethylene being a major component that contributes to the vast production of shopping bags and bottles annually. The mechanical properties of polyethylene can be influenced by the addition of fillers, which are materials combined with the parent component to reduce manufacturing costs or enhance functionality. In an attempt to explore this, rambutan peel flour was melt-blended with low-density polyethylene, and the tensile characteristics were examined. The study revealed that a maximum of 15% peel flour could be blended, which decreased polyethylene tensile strength. However, an interesting finding was that Young's modulus increased with higher flour content and larger particle size (63,250 μm) compared to plain polyethylene films (Nadhirah et al., 2014).

Another study examined the impact of adding adipic acid (0.25%) to polyethylene/rambutan peel flour blends. The incorporation of adipic acid led to an improvement in the mechanical strength of polyethylene films. Surprisingly, adipic acid counteracted the effects of increased rambutan flour on Young's modulus (Nadhirah et al., 2015). In a previous investigation, Ooi et al. (2012) investigated the effects of adding plasticizers (glycerol or sorbitol) to polyvinyl alcohol (PVA) and rambutan peel flour mixes. Glycerol was added, and as a result, there was more elongation-at-break than with sorbitol. On the contrary, when sorbitol was added, the films' tensile strength and Young's modulus increased. Both plasticizers increased water vapor transfer rates, while sorbitol had a lower transmission. In addition, films plasticized with sorbitol or glycerol lost more weight in a biodegradability test than films without any plasticizer.

21.6.10 Seed Oil as Biodiesel

Previous research (Harrington, 1986; Klopfenstein, 1988; Osman et al., 2023) has explored the use of plant oil and its derivatives as diesel engine fuel (Figure 21.8). However, several factors, such as availability, cost, and technical challenges, including reduced stability and increased viscosity, need to be considered. One potential way to modify viscosity is by substituting the glycerol component

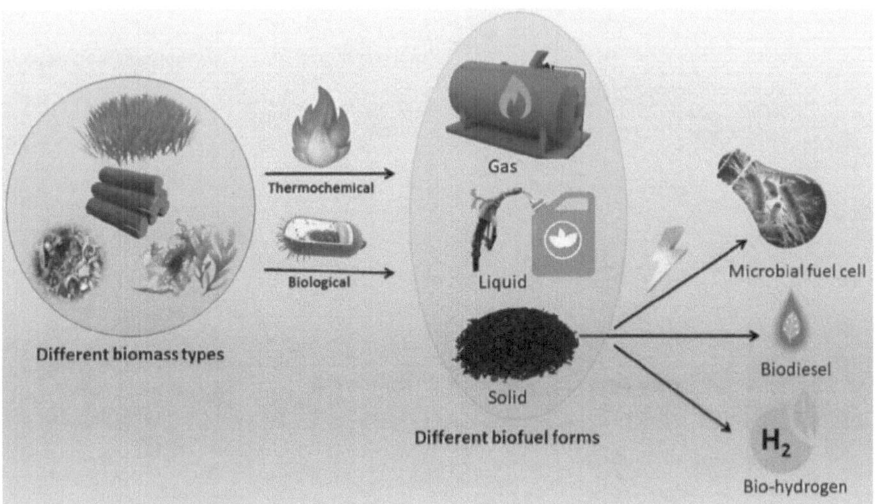

FIGURE 21.8 Biofuel (Osman et al., 2023).

with various simple monohydric alcohols. Nonetheless, oils with higher iodine content are more susceptible to oxidation and may lose effectiveness. A study by Kalayasiri et al. (1996) found that rambutan seed oil possesses a high cetane index of 67.1. The cetane number indicates the combustion rate of diesel fuel and the compression required for ignition. The significant presence of C20 chain lengths in rambutan oil contributes to its elevated cetane index. Moreover, the heat of combustion for extracted oils and esters falls within the range of 40.0–40.4 J/g. Due to these promising attributes, efforts are being made to explore rambutan processing by-products as potential alternative sources for biofuels.

21.6.11 Seed as Bio-Sorbent

Many bio-based sorbents are being developed to reduce the environmental load of heavy metals, and they are employed in the filtering process. Njoku et al. (2014) used potassium hydroxide (KOH) to chemically activate rambutan peel. However, activation time was lowered when microwave heating was combined with KOH. A common dye, acid yellow 17, was utilized as a test adsorbate. Contact time, pH, and dye concentration were all changed to track the process's efficiency. Even with a greater starting dye concentration, more adsorption was seen after 4 h of equilibrium. According to kinetic experiments utilizing the Langmuir model, dye adsorption occurred by producing a monolayer with a maximum adsorption capacity of 215.05 mg/g. Wattanakornsiri et al. (2022)conducted research focuses on the adsorption of Pb(II) and Cd(II) ions from wastewater using chemically modified natural biosorbents derived from three local fruit peels: dragon fruit peel (DFP), rambutan peel (RP), and passion fruit peel (PFP) (Figure 21.9). Among the modified biosorbents, RP exhibited the highest removal efficiency, achieving 97.867% for Pb(II) and 97.100% for Cd(II). The adsorption process followed a pseudo-second-order kinetic model and aligned well with both Langmuir and Freundlich isotherm models, though the Langmuir model provided the best fit. The Langmuir monolayer adsorption capacities were 97.087 mg/g (Pb^{2+}) and

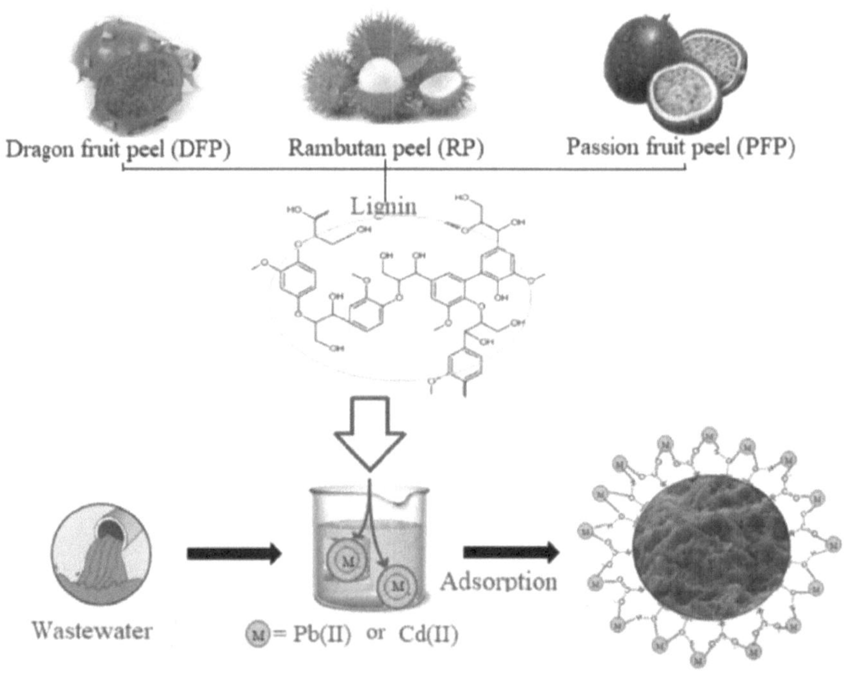

FIGURE 21.9 Rambutan peel used in the adsorption process (Wattanakornsiri et al., 2022).

86.207 mg/g (Cd^{2+}) for DFP, 114.943 mg/g (Pb^{2+}) and 102.041 mg/g (Cd^{2+}) for RP, and 103.093 mg/g (Pb^{2+}) and 89.286 mg/g (Cd^{2+}) for PFP. These results demonstrate that chemically modified fruit peels, particularly RP, are highly effective and economical biosorbents for the removal of heavy metals from contaminated wastewater.

Amid growing environmental degradation, the pressing need to mitigate pollutants and safeguard the ecosystem has become paramount. Consequently, various strategies and processes are being investigated to reduce environmental contaminants. One such approach involves the conversion of fruit and vegetable waste into raw or activated powders or transforming this waste into membranes for pollutant removal through adsorption or filtration. The technique of "activation" utilizes heat or chemicals to generate natural carbon particles with large surface areas. In industrialized countries, activated carbon has been found to be an application in filtering effluent wastewater within the food and beverage sectors, as shown in Figure 21.9. Past studies have already explored the use of agricultural waste by-products such as almond shells (Nabais et al., 2010), avocado kernels (Elizalde-Gonzalez et al., 2007), bamboo debris (Wang et al., 2010), and plum stones (Nowicki et al., 2010) for the production of activated carbon. In line with this, Ibrahim et al. (2013) investigated rambutan activated carbon as a potential sorbent. They successfully created porous carbon particles in a nitrogen atmosphere, employing two temperatures (450°C and 650°C). It was observed that higher temperatures led to increased surface area and total pore volume, albeit with decreased yield.

21.6.12 Seed as Biocoagulant

Coagulation refers to the process of neutralizing electrostatic charges and stabilizing particles through the attraction of opposite charges to form flocs. Alum, known for its wide availability and cost-effectiveness, is a commonly used water treatment agent to remove turbidity through coagulation. In a study by Abidin et al. (2014), rambutan seeds were evaluated for their efficacy in reducing water turbidity through coagulation. The researchers optimized the extraction solvent, coagulant content (rambutan seed), and pH for the study, using commercial kaolin as the model material to be extracted. The results revealed that the best extraction solvent was 1.0 N sodium chloride (NaCl), which achieved over 99% turbidity removal. At a pH of 3, the optimal dosage of rambutan seed was found to be 100 g/L. Additionally, a mixture of alum and rambutan seed in a 1:1 ratio demonstrated superior performance compared to using alum or rambutan seed alone. Notably, using rambutan seed resulted in larger flocs with lower sludge volume, indicating its effectiveness as a biocoagulant in reducing turbidity.

21.7 POTENTIAL APPLICATIONS

Rambutan is a tropical fruit widely consumed in many parts of the world. The fruit produces significant waste, including the peel, seed, and stem. Researchers have been investigating the potential of rambutan waste as a source of bioactive compounds, natural dyes, and other valuable products. This concept stresses the widespread application of rambutan peel across a variety of industries to maximize its value extraction and utilization. The phrase "multisectoral" indicates the engagement of numerous industries and sectors, including energy, agriculture, biotechnology, and a number of other businesses and sectors. The term "bioresource valorization" refers to the procedure of transforming rambutan peel into useful products such as bioenergy, biocomposites, charcoal, and animal feed. The term "framework" refers to a method organized to direct the effective and environmentally responsible usage of rambutan peel as a bioresource in a variety of different applications.

21.7.1 Food Industry

In recent years, there has been an increasing interest in both environmentally responsible methods of food production and methods that reduce the amount of wasted food. Consequently, academics and experts working in the food business are looking into innovative ways to utilize by-products and waste materials, including the peels of fruits. The peel of the rambutan, a tropical fruit that is most known for the luscious and tasty flesh it contains, is also a rich resource that is frequently overlooked. Rambutan peel in the food industry investigates the possibilities of rambutan peel in the food business. It illustrates the different ways in which it may be incorporated into culinary creations, resulting in benefits for both the environment and the consumers.

21.7.1.1 Nutritional Value and Health Benefits

The peel of a rambutan has a high fiber concentration and a wide variety of useful elements, including antioxidants, polyphenols, and vitamins. These components contribute to its possible health advantages, which may include better digestion, increased immunological function, and protection against oxidative stress. As mentioned by Jayasinghe (2013), food producers may offer their customers more nutritious options while lowering their overall waste output if they use rambutan peel to produce their goods.

21.7.1.2 Culinary Application

a. **Tea and Infusion**: The peel of the rambutan fruit may be dried and then used to produce teas and infusions that are fragrant and tasty. As mentioned by Kusuma et al. (2017), these drinks may be enjoyed either hot or cold, and provide a reviving alternative to the more often used choices. Infusions made from rambutan peel may also be mixed with a variety of other herbs and spices to produce distinctive taste profiles.
b. **Natural Food Colorant**: A natural food colorant that may be made from rambutan peel is a brilliant red pigment that can be removed from the peel and utilized. The color of the peel may be used in place of artificial color additives in a range of different culinary items, including drinks, confectioneries, and desserts.
c. **Extracts and Flavorings**: Extracts of rambutan peel have the potential to be used in a variety of applications as a natural flavoring ingredient. The extract has a flavor that is both sweet and somewhat acidic, and its addition to ice cream, yogurt, sauces, or baked products may make such flavors more prominent. It is also possible to use it as a component in marinades and sauces, providing a tropical flavor to the dishes.
d. **Dietary Fiber Enrichment**: Food items can benefit from adding rambutan peel, which is rich in dietary fiber and may be used as a source of natural fiber. Bread, cereals, and snacks may all have their fiber content increased and their digestive health improved if the producers use rambutan peel powder or fiber as an ingredient.

21.7.2 Health Industry

Peel from rambutans, which is often disregarded and thrown away, has in recent years acquired notice for the crucial role it plays in the health business. The rambutan peel is loaded with critical elements, such as dietary fiber, vitamins (such as vitamin C and vitamin A), and minerals (such as potassium and iron), making it an excellent source of nourishment. Rambutan peel is bursting with nutritional content. This nutrient-dense composition helps improve overall health and supports various processes throughout the body (Jayasinghe, 2013).

The amazing antioxidant capabilities of rambutan peel are one of the primary reasons why it has garnered the attention of those working in the health business. The peel has a wide variety of powerful antioxidants, including phenolic chemicals and flavonoids. These may be found in the peel. Oxidative stress is known to contribute to the development of chronic illnesses such as heart disease, cancer, and neurological disorders. These antioxidants serve a critical role in protecting the body against oxidative stress, which is known to contribute to the development of these diseases. The antioxidants in rambutan peel help minimize the risk of cellular damage and inflammation, which in turn promotes a healthy internal environment by neutralizing the potentially damaging free radicals.

It has been discovered that the peel of rambutans has amazing anti-inflammatory effects and powerful antioxidant capabilities. It has been established that the bioactive substances that are found in the peel have the power to reduce inflammation throughout the body. This activity has an anti-inflammatory effect, which is important since chronic inflammation has been linked to various health issues, including arthritis, cardiovascular disorders, and some forms of cancer. The inflammatory response may be modulated by rambutan peel, which may give prospective advantages in lowering the risk of various inflammatory disorders and the severity of their symptoms (Pereira et al., 2017).

The anticancer benefits of rambutan peel are only one of the possible therapeutic uses of rambutan peel. Preliminary research indicates that extracts and chemicals obtained from rambutan peel may have the ability to inhibit the growth of cancer cells. These bioactive components have shown the capacity to prevent the development of cancer cells, promote apoptosis (the planned death of cells), and limit the spread of malignant cells to other parts of the body. Even while further study is required to completely understand the processes that are responsible for these results, there is reason to be optimistic about the use of rambutan peel in the prevention and treatment of cancer (Uy et al., 2020).

In addition, the peel of rambutans has a high concentration of dietary fiber, which is beneficial to the health of the digestive system. By increasing the volume of food consumed and making it easier to have consistent bowel motions, fiber contributes to the maintenance of good digestion. There is a correlation between eating enough fiber and having a lower risk of digestive diseases such as constipation, diverticulosis, and colon cancer. Rambutan peel is a natural supplement to the diet that is high in fiber and helps promote digestive health. This, in turn, contributes to the maintenance of a healthy digestive system.

In addition to its beneficial effects on digestion, the peel of rambutans also helps the immune system function. The nutritious composition of the peel, notably its high vitamin C concentration, is an essential component in the process of improving immune function. It is well known that vitamin C has the ability to improve the immune system, which in turn helps the body better defend itself against common illnesses and infections. Individuals may take advantage of the possible immune-enhancing properties of rambutan peel and encourage the development of a strong immune system by including rambutan peel in their diet.

In addition to these well-established functions, the peel of rambutans may also have benefits for the health of the cardiovascular system. According to research findings, extracts obtained from the rambutan peel may have cardioprotective properties. These extracts have shown the capability of lowering total cholesterol levels, inhibiting the oxidation of LDL cholesterol (the "bad" cholesterol), and improving the function of blood vessels. These activities reduce the risk of cardiovascular disorders such as heart disease and stroke, highlighting the potential of rambutan peel as a natural supplement for maintaining healthy heart function. Research is still being done, but preliminary findings show that the rambutan peel may have additional health advantages. Antidiabetic, antimicrobial, and possibly anti-allergic activity are among the possible benefits of this compound. However, further research is necessary to completely comprehend and confirm these possible uses (Jayasinghe, 2014).

21.7.3 Textile Industry

Rambutan peel, the outer skin of the rambutan fruit, contains natural tannins that have potential applications in the textile industry. Tannins are a class of naturally occurring polyphenols that bind to proteins, making them useful in tanning and dyeing textiles. Tannins from rambutan peel can be used in tanning leather due to their ability to bind to the collagen fibers in the leather. This results in a durable and water-resistant leather that is also resistant to fungal and bacterial growth. RPE has been found to have antimicrobial properties due to the presence of tannins. This property can be useful in finishing textiles to make them resistant to microbial growth and improve their durability with UV-absorbing properties due to the presence of flavonoids. This property can be useful in finishing textiles to provide UV protection to the wearer. Finishing rambutan peels gives antioxidants that can protect textiles from damage caused by free radicals. This property can be useful in finishing textiles to improve their longevity and durability. Overall, the natural tannins found in rambutan peel make it a promising ingredient in the textile industry for tanning leather, dyeing textiles, and improving their durability and functionality. However, further research is needed to explore its full potential and ensure its safety for use in textile products (Saleh et al., 2020). The following is the fiber used and processed in rambutan peel.

Peel from rambutans, which is generally discarded as trash in the fruit business, contains fiber components that can be used in the manufacture of textiles. Cellulose and lignin, two essential elements of plant fibers, may be found in the skin of the fruit. Removing these fibers and subsequent processing can result in an alternative to synthetic textiles that is both sustainable and kind to the environment. There are a few different phases involved in the process of extracting fibers from rambutan peel. In the beginning, the peel is meticulously removed from the fruit, and then it is washed many times to get rid of any contaminants. After that, the peel goes through a mechanical process to separate the fibers. It is possible to acquire fibers with acceptable qualities for use in textile applications via several different procedures. Some examples of these methods include retting, mechanical extraction, and chemical treatments (Ahmed & Mondal, 2021; Olhan et al., 2021). The fibers from rambutan peels can be colored using natural or environmentally friendly dyes, decreasing the need for synthetic dyes. Different finishing methods can further increase the qualities of the fibers, such as the fibers' ability to repel water, resist flames, and prevent the growth of bacteria.

21.7.4 Cosmetics Industry

The peel of rambutans is an excellent source of many types of bioactive chemicals, including antioxidants, flavonoids, and phenolic compounds. Because of their capacity to protect the skin from the damage caused by free radicals, increase skin hydration, and reduce inflammation, these chemicals have the potential to find applications in the cosmetics sector. Because it includes antioxidants like ascorbic acid and phenolic compounds, it can shield the skin from the damage caused by free radicals, which is one of the primary factors in the aging process. Because of this, it is suitable for incorporation into anti-aging treatments such as face creams, serums, and lotions. Flavonoids have the ability to suppress melanin formation, which is the pigment responsible for giving skin its color. Because of its ability, it can be utilized in a variety of skin-lightening products, including creams and serums (Naqvi et al., 2014). Because of the presence of phenolic chemicals, extracts of rambutan peel have been discovered to have qualities similar to those of a moisturizer and a sunscreen. Sunscreen products that contain it help shield the skin from the damaging effects of ultraviolet radiation. Overall, the bioactive compounds in rambutan peel make it a promising ingredient in the cosmetics industry. However, further research is needed to explore its full potential and ensure its safety in cosmetic products.

21.7.5 Biomass Energy Industry

In addition to cellulose, hemicellulose, and lignin, the peel of a rambutan also contains lignin, an essential component of lignocellulosic biomass. Because it has a high percentage of cellulose and a relatively low amount of lignin, it is excellent for bioconversion. In the past, waste from rambutans was exploited for the production of energy by using it as a source of biomass. According to Rukmana et al. (2020), the trash was converted into biogas through anaerobic digestion. This biogas had the potential to be utilized as a source of renewable energy.

21.8 CHALLENGES

Although various companies might potentially reap many benefits from the utilization of rambutan waste, the processing and commercialization of this waste present a number of difficulties. One of the challenges is the difficulty and expense involved in processing rambutan waste. The extraction of the necessary bioactive chemicals or natural pigments from the peel and seed can be challenging, time-consuming, and expensive. It is possible that the procedure calls for specific equipment and a certain level of knowledge, both of which can drive up production costs. The variable nature of the content of rambutan waste presents still another obstacle. The type of rambutan, the time of year, and the region of the world in which it was grown all impact the amount of nutrients and bioactive compounds it contains. Because of this heterogeneity, designing standardized goods of consistently high-quality and effective doses might be challenging. In addition, there is a possibility that there is a limited demand in the market for waste products derived from rambutan, particularly in areas where rambutan is not extensively farmed or consumed. This can be a hurdle when attempting to commercialize waste products made from rambutans and make such items economically viable.

Regulatory agencies may need evidence of safety and efficacy before approving novel goods for commercialization. This can add to the expense and complexity of bringing rambutan waste products to market because of the regulatory challenges linked to the safety and efficacy of rambutan waste products. In general, the waste from rambutans has a significant amount of potential for use in a variety of different industries; nevertheless, there are a number of obstacles that need to be overcome before it can be considered a viable commercial commodity (Abas et al., 2020). There are a few obstacles to overcome to successfully include rambutan peel into environmental management across a variety of companies and sectors. The effective collection and disposal of discarded rambutan peels is one of the most significant challenges facing the industry of waste management today. The vast amount of peel produced during the processing and consumption of rambutans requires careful management to avoid causing pollution to the environment and creating potential health hazards. For effective environmental management, it is essential to establish adequate infrastructure for composting and other treatment processes, as well as efficient waste collecting systems, to raise knowledge among consumers about appropriate disposal techniques, and to raise awareness among consumers about proper disposal methods.

21.8.1 Energy Sectors

In the field of energy production, one of the primary problems is maximizing the efficiency with which rambutan peel may be converted into usable energy. The composition of the peel and the amount of moisture it contains can change, which can impact how well processes like combustion, gasification, or anaerobic digestion work. To achieve constant and high energy yields, technological breakthroughs in preprocessing techniques, storage technology, and energy conversion technology are required. In addition, the economic viability and scalability of producing energy from rambutan peel provide issues that need to be addressed to entice investment and encourage widespread usage.

21.8.2 Agricultural Sectors

The use of rambutan peel as a resource presents difficulties for the agricultural industry in terms of nutrient management and maintaining the health of the soil. Although the peel has the potential to be utilized as an organic fertilizer or as an addition to soil, further research is required to fully comprehend the composition of its nutrients and the rate at which they are released. Careful agricultural methods and recommendations based on research are required to minimize over-application, maintain the correct nutrient balance, and mitigate any potential negative effects on the surrounding ecosystem, such as runoff from nutrients.

21.8.3 Food and Beverage Sectors

The management of waste rambutan peel presents another obstacle for businesses in the food and beverage industry. Important factors to take into account include the successful removal of peels during processing and the reduction of food waste. Creating cutting-edge solutions, such as goods with additional value derived from rambutan peel, can help reduce waste and open up new avenues for commercialization. Nevertheless, there are additional challenges that need to be overcome, such as the demand of the market, the acceptance of the product by consumers, and the development of viable business models.

21.8.4 Manufacturing Sectors

In addition, the manufacturing industry faces difficulties in the exploitation of rambutan peel as a feedstock for biocomposite materials or other value-added products. This problem is caused by the fact that rambutan peel is fibrous. For successful integration, it is essential to maintain a consistent quality and set of qualities in the peel-derived products, utilize economical extraction and processing methods, and set up supply networks that can be relied upon. In addition, for there to be an awareness of and a market demand for environmentally friendly products generated from rambutan peel, there needs to be efficient communication and coordination between customers, merchants, and manufacturers. To overcome the issues that are linked with rambutan peel in environmental management, interdisciplinary collaboration, technical improvements, support from policies, and participation from stakeholders are required. Research and development efforts are necessary to overcome technological obstacles, and governmental incentives can help promote sustainable waste management practices and encourage investment in novel solutions. To fully realize the potential of rambutan peel as an environmental management tool across a variety of sectors, it is necessary to take a comprehensive approach that includes the participation of consumers, businesses, academic institutions, and government organizations.

21.9 CONCLUSION

In summary, after the consumption of rambutan, the seeds and peels are considered residues or wastes, and their large amounts could harm the environment if not managed properly. Nevertheless, rambutan seeds and peels have valuable nutrients and biological properties that could positively impact human health. The right utilization of rambutan seed and peel will make a good economic impact, nurturing health, decreasing disposal costs, and, most importantly, minimizing the environmental impacts. The seed is often handled as a by-product of rambutan fruit and can be processed into various food products, including cans, juice, jam, jelly, marmalades, and spreads. The seed is also rich in phenolic compounds and minerals, which may have positive effects on one's health. In addition, the fruit rind, or peel, is a natural barrier that protects the edible flesh and preserves the fruit's structural integrity. The peel contains nutritious components and several antinutritional substances, including saponins, alkaloids, tannins, phytates, and oxalates, which may have positive and negative effects on one's health. Furthermore, rambutan waste has potential applications in various industries, including food, health, textile, cosmetics, biomass energy, and environmental management. However, there are several challenges associated with processing and commercializing rambutan waste, including the cost and complexity of processing, variability of composition, and limited market demand.

ACKNOWLEDGMENT

The chapter was written with Muhammad Azwan Suid, Muhammad Rohman Al Hasan, and Abizard Baskara.

REFERENCES

Abas, M., Jarona, M. M., Mehue, B., Mulyono, S., Mulyani, W., & Mofu, R. M. (2020). Risk factor level of knowledge about the use of long lasting insectiside nets (LLIN) to malaria incidence (A Study in Genyem Primary Health Care Unit Areas, Jayapura Regency, Papua Province Indonesia). *International Journal of Innovative Science and Research Technology, 7*. https://doi.org/10.5281/zenodo.6893271

Abidin, Z. Z., Abdullah, A. G. L., & Mohamed, M. F. (2014). Preliminary study of rambutan (*Nephelium lappaceum*) seed as potential biocoagulant for turbidity removal. *Advanced Materials Research, 917*, 96–105. https://doi.org/10.4028/www.scientific.net/AMR.917.96

Afzaal, M., Saeed, F., Bibi, M., Ejaz, A., Shah, Y. A., Faisal, Z., Ateeq, H., Akram, N., Asghar, A., & Shah, M. A. (2023). Nutritional, pharmaceutical, and functional aspects of rambutan in industrial perspective: An updated review. *Food Science & Nutrition*, *11*(7), 3675–3685. https://doi.org/10.1002/fsn3.3379

Ahmad, I., & Chua, P. C. (2013). Trends in production and trade of tropical fruits in ASEAN countries. *Acta Horticulturae*, *975*, 559–580. https://doi.org/10.17660/ActaHortic.2013.975.73

Ahmed, F., & Mondal, M. I. H. (2021). Introduction to natural fibres and textiles. In: Mondal, M. I. H. (Ed.), *Fundamentals of Natural Fibres and Textiles* (pp. 1–32). Elsevier. https://doi.org/10.1016/B978-0-12-821483-1.00001-2

Akhtar, M. T., Ismail, S. N., & Shaari, K. (2017). Rambutan (*Nephelium lappaceum* L.). In: Yahia, E. M. (Ed.), *Fruit and Vegetable Phytochemicals: Chemistry and Human Health: Second Edition*, *2* (pp. 1227–1234). John Wiley & Sons, Ltd. https://doi.org/10.1002/9781119158042.CH64

Atkinson, C. J., Dodds, P. A. A., Ford, Y. Y., Le Miere, J., Taylor, J. M., Blake, P. S., & Paul, N. (2006). Effects of cultivar, fruit number and reflected photosynthetically active radiation on Fragaria× ananassa productivity and fruit ellagic acid and ascorbic acid concentrations. *Annals of Botany*, *97*(3), 429–441. https://doi.org/10.1093/aob/mcj046

Augustin, M. A., & Chua, B. C. (1988). Composition of rambutan seeds. *Pertanika*, *11*(2), 211–215.

Azam, M. M., Waris, A., & Nahar, N. M. (2005). Prospects and potential of fatty acid methyl esters of some non-traditional seed oils for use as biodiesel in India. *Biomass and Bioenergy*, *29*(4), 293–302. https://doi.org/10.1016/j.biombioe.2005.05.001

Bhat, R. S., & Al-Daihan, S. (2014). Antimicrobial activity of Litchi chinensis and Nephelium lappaceum aqueous seed extracts against some pathogenic bacterial strains. *Journal of King Saud University - Science*, *26*(1), 79–82. https://doi.org/10.1016/J.JKSUS.2013.05.007

Chai, K. F., Adzahan, N. M., Karim, R., Rukayadi, Y., & Ghazali, H. M. (2018). Characteristics of fat, and saponin and tannin contents of 11 varieties of rambutan (Nephelium lappaceum L.) seed. *International Journal of Food Properties*, *21*(1), 1091–1106. https://doi.org/10.1080/10942912.2018.1479857

Chai, K. F., Adzahan, N. M., Karim, R., Rukayadi, Y., & Ghazali, H. M. (2019). Fat properties and antinutrient content of rambutan (Nephelium lappaceum L.) seed during solid-state fermentation of rambutan fruit. *Food Chemistry*, *274*, 808–815. https://doi.org/10.1016/J.FOODCHEM.2018.09.065

Chimplee, S., & Klinkesorn, U. (2015). Thin layer drying model of rambutan (Nepheliumlappaceum L.) kernel and its application in fat extraction process. *International Journal of Food Engineering, 11*(2), 243–253. https://doi.org/10.1515/ijfe-2014-0209

Eiamwat, J., Wanlapa, S., & Kampruengdet, S. (2016). Physicochemical properties of defatted rambutan (Nephelium lappaceum) seed flour after alkaline treatment. *Molecules, 21*, 364. https://doi.org/10.3390/molecules21040364

Elizalde-Gonzalez, M. P., Mattusch, J., Pelaez-Cid, A. A., & Wennrich, R. (2007). Characterization of adsorbent materials prepared from avocado kernel seeds: Natural, activated and carbonized forms. *Journal of Analytical and Applied Pyrolysis, 78*, 185–193.

Eteng, O. E. (2012). Comparative antinutrients assessment of pulp, seed and rind of rambutan (Nephelium Lappaceum). *Annals of Biological Research*, *3*(11), 5151–5156. https://www.researchgate.net/profile/Odey-Oko/publication/259176521_Comparative_anti-nutrients_assessment_of_pulp_seed_and_rind_of_rambutan_Nephelium_Lappaceum/links/0c96052a2000bafccb000000/Comparative-anti-nutrients-assessment-of-pulp-seed-and-rind-of-r

Evaristus, N. A., Wan Abdullah, W. N., & Gan, C. Y. (2018). Extraction and identification of α-amylase inhibitor peptides from Nephelium lappacheum and Nephelium mutabile seed protein using gastro-digestive enzymes. *Peptides*, *102*, 61–67. https://doi.org/10.1016/J.PEPTIDES.2018.03.001

Fang, E. F., & Ng, T. B. (2015). A trypsin inhibitor from rambutan seeds with antitumor, anti-HIV-1 reverse transcriptase, and nitric oxide-inducing properties. *Applied Biochemistry and Biotechnology*, *175*, 3828–3839. https://doi.org/10.1007/s12010-015-1550-1

Febrianto, N. A., Yang, T. A., & Wan Abdullah, W. A. (2016). Cocoa-like flavor compound development of rambutan seed fat as the effect of fermentation and roasting. *International Food Research Journal, 23*(5), 2179–2187.

Fidrianny, I., Fikayuniar, L., & Insanu, M. (2015). Antioxidant activities of various seed extracts from four varieties of rambutan (Nelpheliumlappaceum) using 2, 2-diphenyl-1-picrylhydrazyl and 2,2′-azinobis (3-ethyl-benzothiazoline-6-sulfonic acid) assays. *Asian Journal of Pharmaceutical and Clinical Research, 8*(5), 215–219.

Fila, W., Ifam, E. H., Johnson, J., Oko, O., Effiong, E. E., Dasofunjo, K., & Ambo, E. E. (2013). Comparative proximate compositions of watermelon Citrullus lanatus, squash Cucurbita pepo'l and rambutan Nephelium lappaceum. *International Journal of Science and Technology, 2*, 81–88.

Fila, W. O., Johnson, J. T., Edem, P. N., Odey, M. O., Ekam, V. S., Ujong, U. P., & Eteng, O. E. (2012). Comparative antinutrients assessment of pulp, seed and rind of rambutan (Nephelium Lappaceum). *Annals of Biological Research, 3*(11), 5151–5156.

Ghobakhlou, F., Mohd Ghazali, H., Karim, R., & Sabo Mohammed, A. (2019). Characterization of three varieties of Malaysian rambutan seed oil. *Revista Española de Nutrición Humana y Dietética, 23*(3), 183–193. https://doi.org/10.14306/renhyd.23.3.756

Gusman, J. A., & Tsai, P. J. (2015). Extraction of antioxidant compounds from rambutan (Nephelium lappaceum L.) peel as agricultural waste in Taiwan. *Journal of Tropical Crop Science, 2*(2), 10–16.

Hadeel, A., Hossain, A. B. M. S., Latifa, K., Alnaqeb, H., Abear, J., & Norah, A. (2011). Bioethanol fuel production from rambutan fruit biomass as reducing agent of global warming and greenhouse gases. *African Journal of Biotechnology, 10*(50), 10157–10165. https://doi.org/10.5897/AJB11.790

Harahap, S. N.,Ramli, N., Vafaei, N., & Said, M. (2012). Physicochemical and nutritional composition of rambutan anak sekolah (Nephelium lappaceum L.) seed and seed oil. *Pakistan Journal of Nutrition : PJN, 11*(11), 1073–1077. https://doi.org/10.3923/pjn.2012.1073.1077

Harrington, K. J. (1986). Chemical and physical properties of vegetable oils esters and their effect on diesel fuel performance. *Biomass, 9*, 1–17.

Hernández-Hernández, C., Aguilar, C. N., Rodríguez-Herrera, R., Flores-Gallegos, A. C., Morlett-Chávez, J., Govea-Salas, M., & Ascacio-Valdés, J. A. (2019). Rambutan (Nephelium lappaceum L.): Nutritional and functional properties. *Trends in Food Science & Technology, 85*, 201–210. https://doi.org/10.1016/J.TIFS.2019.01.018

Ibrahim, N. M., MohdSarif, S. F. Z., CheAmat, R., Salehuddin, S., & Rahim, N. L. (2013). Characterization of rambutan seed (NepheliumLappaceum) as natural adsorbent for wastewater treatment. *Advanced Materials Research, 701*, 408–411.

Issara, U., Zzaman, W., & Yang, T. A. (2014). Rambutan seed fat as a potential source of cocoa butter substitute in confectionary product. *International Food Research Journal, 21*(1), 25–31.

Jahurul, M. H. A., Azzatul, F. S., Sharifudin, M. S., Norliza, M. J., Hasmadi, M., Lee, J. S., Patricia, M., Jinap, S., Ramlah George, M. R., Firoz Khan, M., & Zaidul, I. S. M. (2020). Functional and nutritional properties of rambutan (Nephelium lappaceum L.) seed and its industrial application: A review. *Trends in Food Science & Technology, 99*, 367–374.

Jahurul, M. H. A., Zaidul, I. S. M., Ghafoor, K., Al-Juhaimi, F. Y., Nyam, K. L., Norulaini, N. A. N., Sahena, F., & Mohd Omar, A. K. (2015). Mango (Mangifera indica L.) by-products and their valuable components: A review. *Food Chemistry, 183*, 173–180. https://doi.org/10.1016/J.FOODCHEM.2015.03.046

Jayaprakasha, G. K., Singh, R. P., & Sakariah, K. K. (2001). Antioxidant activity of grape seed (Vitis vinifera) extracts on peroxidation models in vitro. *Food Chemistry, 73*(3), 285–290. https://doi.org/10.1016/S0308-8146(00)00298-3

Jayasinghe, L. (2013). Nutritional and therapeutic potential of rambutan (Nephelium lappaceum L.): An overview. *European Journal of Integrative Medicine, 5*(6), 512–520.

Jayasinghe, L. (2014). Potential health benefits of rambutan. In: Nelson, D. L. (Ed.), *Functional Foods: Sources, Health Effects and Future Perspectives* (pp. 331–343). Nova Science Publishers.

Johnson, O. O., Ayoola, G. A., & Adenipekun, T. (2013). Antimicrobial activity and the chemical composition of the volatile oil blend from Allium sativum (Garlic Clove) and Citrus reticulata (Tangerine Fruit). *International Journal of Pharmaceutical Sciences and Drug Research, 5*(4), 187–193.

Kalayasiri, P., Jeyashoke, N., & Krisnagkura, K. (1996). Survey of seed oils for use as diesel fuels. *Journal of the American Oil Chemists' Society, 73*(4), 471–474.

Kheiri, M. S. A., & Mohd Som, M. N. (1987). Physicochemical characteristic of rambutan kernel fat. MARDI report. No.186. Mardi Publication. Serdang.

Klopfenstein, W. E. (1988). Effect of moleuclar weights of fatty acid esters on cetane numbers as deiesel fules. *Journal of American Oil Chemistry Society, 65*, 1029–1031.

Kusuma, H. S. W., Mahfud, M., & Wijaya, K.. (2017). Characteristics of rambutan (Nephelium lappaceum) peel extracts as potential food bio-colorant. *International Food Research Journal, 24*(2), 598–605.

Li, W., Zeng, J., & Shao, Y. (2018). Rambutãn—Nephelium lappaceum. In: Rodrigues, S., de Oliveira Silva, E., & de Brito, E. S. (Eds.), *Exotic Fruits Reference Guide* (pp. 369–375). Academic Press. https://doi.org/10.1016/B978-0-12-803138-4.00048-4

Lourith, N., Kanlayavattanakul, M., Mongkonpaibool, K., Butsaratrakool, T., & Chinmuang, T. (2016). Rambutan seed as a new promising unconventional source of specialty fat for cosmetics. *Industrial Crops and Products, 83*, 149–154.

Luma, K. H., & Yang, T. A. (2016). Comparing the structural parameters of rambutan seed fat incorporated wholly and partially with cocoa butter. *International Journal of Scientific and Engineering Research, 7*(5), 921–928.

Mahisanunt, B., Na Jom, K., Matsukawa, S., & Klinkesorn, U. (2017). Solvent fractionation of rambutan (Nephelium lappaceum L.) kernel fat for production of non-hydrogenated solid fat: Influence of time and solvent type. *Journal of King Saud University – Science*, *29*(1), 32–46. https://doi.org/10.1016/J.JKSUS.2016.08.004

Mahmood, K., Fazilah, A., Yang, T. A., Sulaiman, S., & Kamilah, H. (2018). Valorization of rambutan (Nephelium lappaceum) by-products: Food and non-food perspectives. *International Food Research Journal*, *25*(3), 892–902.

Maisuthisakul, P., Pasuk, S., & Ritthiruangdej, P. (2008). Relationship between antioxidant properties and chemical composition of some Thai plants. *Journal of Food Composition and Analysis, 21*, 229–240. https://doi.org/10.1016/j.jfca.2007.11.005

Manaf, Y. N., Marikkar, J. M. N., Che Man, Y. B., & Long, K. (2011). Composition and thermal analysis of lard stearin and lard olein. *Journal of Oleo Science, 60*, 333–338.

Manaf, Y. N., Marikkar, J. M. N., Long, K., & Ghazali, H. M. (2013). Physicochemical characterization of the fat from red-skin rambutan (Nepheliumlappaceum) seed. *Journal of Oleo Science, 62,* 335–343. https://doi.org/10.5650/jos.62.335

Manaf, Y. N., Marikkar, J. M. N., & Miskandar, M. S. (2012). Comparing the thermos-physical characteristics of lard and selected plant fats. *Grasas y Aceites, 63*, 328–334.

Mei, W. S. C., Ismail, A., Esa, M. N., Akowuah, G. A., Wai, H. C., & Seng, Y. H. (2014). The effectiveness of rambutan (Nephelium lappaceum L.) extract in stabilization of sunflower oil under accelerated conditions. *Antioxidants, 3*(2), 371–386.

Muhtadi, M., Haryoto, H., Sujono, T. A., Indrayudha, A., Suhendi, A., & Yen, K. H. (2014). Antioxidant activity and chemical constituents of some Indonesian fruit peels. *Medicinal Plants- International Journal of Phytomedicines and Related Industries, 6*(1), 1–4. https://doi.org/10.5958/j.0975-6892.6.1.006

Nabais, J. M. V., Laginhas, C. E. C., & Carrot, M. M. L. R. (2010). Production of activated carbon from almond shell. *Fuel Processing Technology, 92*(2), 234–240.

Nadhirah, A., Sam, S. T., Zulkepli, N. N., Voon, C., & Samera, S. S. (2014). Tensile properties of linear low density polyethylene/rambutan peels (Nephelium c hryseum Blum.) flour blends. *Materials Science Forum, 803*, 348–352. https://doi.org/10.4028/www.scientific.net/MSF.803.348

Nadhirah, A. A., Sam, S. T., Noriman, N. Z., Ragunathan, S., & Ismail, H. (2015, July). Influence of adipic acid on tensile and morphology properties of linear low density polyethylene/rambutan peels flour blends. *AIP Conference Proceedings, 1669*(1), 020031. AIP Publishing. https://doi.org/10.1063/1.4919169

Naqvi, S. A. R., Shah, S. A., & Riaz, A. (2014). Phytochemical and antioxidant analysis of Nephelium lappaceum L: fruit pulp and peel. *Journal of Food and Nutrition Research, 2*(10), 673–677.

Nauli Harahap, S., Ramli, N., Vafaei, N., & Said, M. (2012). Physicochemical and nutritional composition of rambutan anak sekolah (Nephelium lappaceum L.) seed and seed oil. *Pakistan Journal of Nutrition, 11*(11), 1073–1077. https://doi.org/10.3923/PJN.2012.1073.1077

Njoku, V. O., Foo, K. Y., Asif, M., & Hameed, B. H. (2014). Preparation of activated carbons from rambutan (Nepheliumlappaceum) peel by microwave-induced KOH activation for acid yellow 17 dye adsorption. *Chemical Engineering Journal, 250*, 198–204.

Normah, O., & Ku Hasnah, K. A. (2000). Pectin content of selected local fruit by-products (Kandunganpektindalamhasilansampinganbuah-buahantempatanterpilih). *Journal of Tropical Agriculture and Food Science, 28*(2), 195–201.

Nowicki, P., Wachowska, H., & Pietrzak, R. (2010). Active carbons prepared by chemical activation of plum stones and their application in removal of NO_2. *Journal of Hazardous Material, 181*, 1088–1094.

Okabe, S., Suganuma, M., Imayoshi, Y., Taniguchi, S., Yoshida, T., & Fujiki, H. (2001). New TNF-a releasing inhibitors, geraniin and corilagin, in leaves of Acer nikoense,Megusurino-ki. *Biological and Pharmaceutical Bulletin, 24*, S1145–S1148. https://doi.org/10.1248/bpb.24.1145

Okonogi, S., Duangrat, C., Anuchpreeda, S., Tachakittirungrod, S., & Chowwanapoonpohn, S. (2007). Comparison of antioxidant capacities and cytotoxicities of certain fruit peels. *Food Chemistry, 103*(3), 839–846. https://doi.org/10.1016/j.foodchem.2006.09.034

Olaniyi, L. O., & Mehdizadeh, S. (2013). Effect of traditional fermentation as a pretreatment to decrease the antinutritional properties of rambutan seed (Nepheliumlappaceum L.). *International Conference on Food and Agricultural Sciences*, Melaka, Malaysia. IACSIT Press Singapore, pp. 67–72.

Olhan, S., Khatkar, V., & Behera, B. K. (2021). Review: Textile-based natural fibre-reinforced polymeric composites in automotive lightweighting. *Journal of Materials Science*, *56*(34), 18867–18910. https://doi.org/10.1007/s10853-021-06509-6

Ooi, Z. X., Ismail, H., Bakar, A. A., & Aziz, N. A. A. (2012). The comparison effect of sorbitol and glycerol as plasticizing agents on the properties of biodegradable polyvinyl alcohol/rambutan skin waste flour blends. *Polymer – Plastics Technology and Engineering*, *51*(4), 432–437. https://doi.org/10.1080/03602559.2011.639827

Osman, A. I., Lai, Z. Y., Farghali, M., Yiin, C. L., Elgarahy, A. M., Hammad, A., Ihara, I., Al-Fatesh, A. S., Rooney, D. W., & Yap, P.-S. (2023). Optimizing biomass pathways to bioenergy and biochar application in electricity generation, biodiesel production, and biohydrogen production. *Environmental Chemistry Letters*, *21*(5), 2639–2705. https://doi.org/10.1007/s10311-023-01613-2

Palanisamy, U., Cheng, H. M., Masilamani, T., Subramaniam, T., Ling, L. T., & Radhakrishnan, A. K. (2008). Rind of the rambutan, Nephelium lappaceum, a potential source of natural antioxidants. *Food Chemistry*, *109*(1), 54–63. https://doi.org/10.1016/j.foodchem.2007.12.018

Palanisamy, U., Manaharan, T., Teng, L. L., Radhakrishnan, A. K., Subramaniam, T., & Masilamani, T. (2011). Rambutan rind in the management of hyperglycemia. *Food Research International*, *44*(7), 2278–2282.

Pereira, L. F. R., et al. (2017). Comparative study of nutritional composition, antioxidant activity, and phenolic compounds of three exotic fruits: Pitanga, jambo, and rambutan peel. *Food Science and Technology*, *37*(1), 139–145. https://doi.org/10.1590/1678-457x.09316

Perera, A., Appleton, D., Ying, L. H., Elendran, S., & Palanisamy, U. D. (2012). Large scale purification of geraniin from Nephelium lappaceum rind waste using reverse-phase chromatography. *Separation and Purification Technology*, *98*, 145–149. https://doi.org/10.1016/j.seppur.2012.06.019

Phanthanapratet, W., Trangwacharakul, S., Wanlapa, S., Eiamwat, J., Boonkosum, P., & Lauhakul, K. (2012). การ ประยุกต์ ใช้ แป้ง เมล็ด เงาะ ใน ผลิตภัณฑ์ น้ำ สลัด แค ล อ รี ต่ำ [Application of rambutan seed starch in low-calorie salad dressing products]. *Agricultural Science Journal*, *43*(3), 1–10.

Reddy, S. Y., & Jeyarani, T. (2001). Trans-free bakery shortenings from mango kernel and mahua fats by fractionation and blending. *Journal of the American Oil Chemists' Society*, *78*(6), 635–640.

Rukmana, R., Barlian, A., & Arifin, H. S. (2020). Potential of rambutan (Nephelium lappaceum L.) waste as raw material for biogas production. *Journal of Physics: Conference Series*, *1468*(1), 012069.

Saleh, T. A., Ali, A., Hamid, A. A., & Amir, N. H. (2020). Rambutan (Nephelium lappaceum L.) peel waste: A potential source of natural dye for textile industry. *Sustainable Chemistry and Pharmacy*, *18*, 100329.

Samuagam, L., Khoo, H. E., Akowuah, G. A., Okechukwu, P. N., & Yim, H. S. (2014). HPLC analysis of antioxidant compounds in some selected tropical fruits peel Innovative Romanian. *Food Biotechnology*, *14*, 61–68.

Sirisompong, W., Jirapakkul, W., & Klinkesorn, U. (2011). Response surface optimization and characteristics of rambutan (Nephelium lappaceum L.) kernel fat by hexane extraction. *LWT-Food Science and Technology*, *44*(9), 1946–1951. https://doi.org/10.1016/j.lwt.2011.04.011

Soeng, S. (2015). Antioxidant and hypoglycemic activities of extract and fractions of rambutan seeds (Nephelium lappaceum L.). *Biomedical Engineering*, *1*(1), 6.

Solangi, N. H., Kumar, J., Mazari, S. A., Ahmed, S., Fatima, N., & Mubarak, N. M. (2021). Development of fruit waste derived bio-adsorbents for wastewater treatment: A review. *Journal of Hazardous Materials*, *416*. https://doi.org/10.1016/j.jhazmat.2021.125848

Solís-Fuentes, J. A., Camey-Ortíz, G., Hernández-Medel, M. del R., Pérez-Mendoza, F., & Durán-de-Bazúa, C. (2010). Composition, phase behavior and thermal stability of natural edible fat from rambutan (Nephelium lappaceum L.) seed. *Bioresource Technology*, *101*(2), 799–803. https://doi.org/10.1016/J.BIORTECH.2009.08.031

Sonwai, S., & Ponprachanuvut, P. (2012). Characterization of physicochemical and thermal properties and crystallization behavior of krabok (Irvingia Malayana) and rambutan seed fats. *Journal of Oleo Science*, *61*(12), 671–679.

Suhaila, M., & Zaharah, H. (1995). Extraction and characterization of pectin from various tropical agrowastes. *Asean Food Journal*, *10*(2), 43–50.

Thitilertdecha, N., Teerawutgulrag, A., Kilburn, J. D., & Rakariyatham, N. (2010). Identification of major phenolic compounds from Nepheliumlappaceum L. and their antioxidant activities. *Molecules*, *15*(3), 1453–1465. https://doi.org/10.3390/molecules15031453

Torres-León, C., Rojas, R., Contreras-Esquivel, J. C., Serna-Cock, L., Belmares-Cerda, R. E., & Aguilar, C. N. (2016). Mango seed: Functional and nutritional properties. *Trends in Food Science & Technology*, *55*, 109–117. https://doi.org/10.1016/J.TIFS.2016.06.009

Uraiwan, K., & Satirapipathkul, C. (2016). The entrapment of vitamin E in nanostructured lipid carriers of rambutan seed fat for cosmeceutical uses. *Key Engineering Materials*, *675*, 77–80.

Uy, M. M., et al. (2020). Rambutan peel extract suppresses cell proliferation and induces apoptosis in human breast cancer cells. *Journal of Medicinal Food*, *23*(2), 130–140. https://doi.org/10.1089/jmf.2019.0126

Vuong, H. T. H., Tran, N. M. C., Tran, T. T. T., Ton, N. M. N., & Le, V. V. M. (2016). Effects of pH and salt concentration on functional properties of rambutan (Nephelium lappaceum L.) seed albumin concentrate. *International Journal of Food Science & Technology*, *51*(5), 1212–1219.

Wang, F. Y., Wang, H., & Ma, J. M. (2010). Adsorption of cadmium (II) ions from aqueous solution by a new low-cost adsorbent – Bamboo charcoal. *Journal of Hazardous Material, 177*, 300–306.

Wattanakornsiri, A., Rattanawan, P., Sanmueng, T., Satchawan, S., Jamnongkan, T., & Phuengphai, P. (2022). Local fruit peel biosorbents for lead(II) and cadmium(II) ion removal from waste aqueous solution: A kinetic and equilibrium study. *South African Journal of Chemical Engineering, 42*, 306–317. https://doi.org/10.1016/j.sajce.2022.09.008

Yap, L. L. (2013). Performance of rambutan seed extracts as iron and manganese removal in groundwater. *Journal of Tropical Resources and Sustainable Science (JTRSS), 5*(2), 79–82.

Yuvakkumar, R., Suresh, J., Nathanael, A. J., Sundrarajan, M., & Hong, S. I. (2014). Novel green synthetic strategy to prepare ZnO nanocrystals using rambutan (Nephelium lappaceum L.) peel extract and its antibacterial applications. *Materials Science and Engineering: C, 41*, 17–27.

Yuvakkumar, R., Suresh, J., Saravanakumar, B., Nathanael, A. J., Hong, S. I., & Rajendran, V. (2015). Rambutan peels promoted biomimetic synthesis of bioinspired zinc oxide nanochains for biomedical applications. *Spectrochimica Acta Part A: Molecular and Biomolecular Spectroscopy, 137*, 250–258.

Zhuang, Y., Ma, Q., Guo, Y., & Sun, L. (2017). Protective effects of rambutan (Nephelium lappaceum) peel phenolics on H_2O_2-induced oxidative damages in HepG2 cells and D-galactose-induced aging mice. *Food and Chemical Toxicology, 108*, 554–562. https://doi.org/10.1016/J.FCT.2017.01.022

Zzaman, W., Issara, U., Febrianto, N. F., & Yang, T. A. (2014). Fatty acid composition, rheological properties and crystal formation of rambutan fat and cocoa butter. *International Food Research Journal, 21*(3), 983–987.

22 Apple Waste Management

22.1 INTRODUCTION

22.1.1 Importance of Managing Agro-Biomass Waste in Pollution Control Engineering

Management of agro-biomass waste is important in pollution control engineering. Examples of agro-biomass waste are crop stalks, husks, and fruit pomace. If this agro-biomass waste is handled improperly, there will be environmental pollution. The most important aspect of managing agro-biomass well is the reduction of greenhouse gas (GHG) emissions. When biomass decomposes without adequate management, a powerful GHG called methane (CH_3) is released. Advanced biomass management methods such as anaerobic digestion, pyrolysis, and biomass gasification can trap methane for use as energy and keep it from escaping into the atmosphere.

Moreover, the reduction of landfill waste is the importance of well-managed agro-biomass. Agricultural waste takes up significant space in landfills, contributing to land contamination and the production of leachate, which can contaminate groundwater. By converting agro-biomass into bioplastics, energy, or fertilizer, we lessen the load on landfills and the pollution hazards that go along with it.

The reduction of air pollution is the importance of well-managed agro-biomass. Crop stalks and other agricultural wastes are mostly burned outdoors, which contributes to air pollution. It degrades air quality and contributes to climate change by releasing harmful pollutants such as particulate matter, carbon monoxide (CO), volatile organic compounds, and GHGs (CO_2, CH_4). Emissions can be greatly decreased by proper management, which includes energy conversion techniques like composting and the creation of biofuels.

22.1.2 Brief Overview from Apple Pomace

Apples are among the most popular and well-known fruits in the Rosaceae family, specifically in the genus *Malus* (Ginni et al., 2020). Figure 22.1 shows the basic information on apple consumption and apple pomace (AP) composition. When it comes to daily food intake, apples and apple-derived products are key sources of phytochemicals and antioxidants. A residue known as AP is produced in large quantities when apple products, including juice, sweet cider, jelly, and vinegar, are processed. About 95% of AP is made up of white flesh and skin, 2–4% are seeds, and 1% are stems (Perussello et al., 2017). AP and its solids are abundant in pectin, polyphenols, carbs, and other vital elements. AP is the term for the fruit leftover after apple juice is extracted. It is typically disposed of in a way that could be hazardous to the environment and the general public's health. AP is an excellent source of carbohydrates, phenolic compounds, dietary fiber, and minerals (Lyu et al., 2020). It is a by-product of the apple processing industry that is considered a valuable resource due to its high nutritional content, making its use very desirable.

22.1.3 Current Role in Agriculture Field

From the view of agriculture, AP currently plays an important role. In conventional ways of AP utilization, as it is a material obtained from plants, it can be utilized as fertilizer or animal feed.

DOI: 10.1201/9781003472209-29

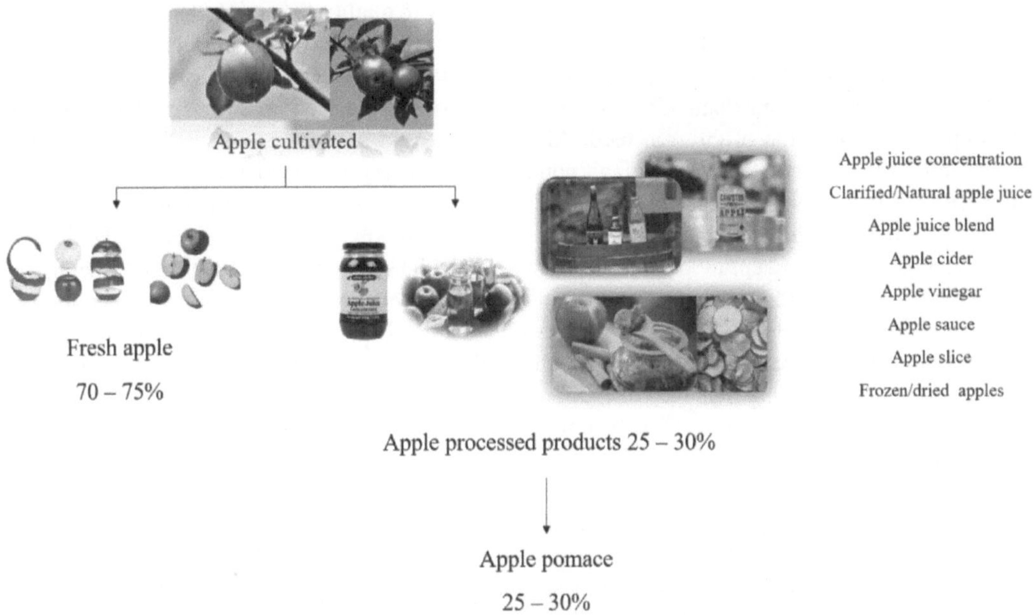

FIGURE 22.1 Basic information on apple consumption and apple pomace composition.

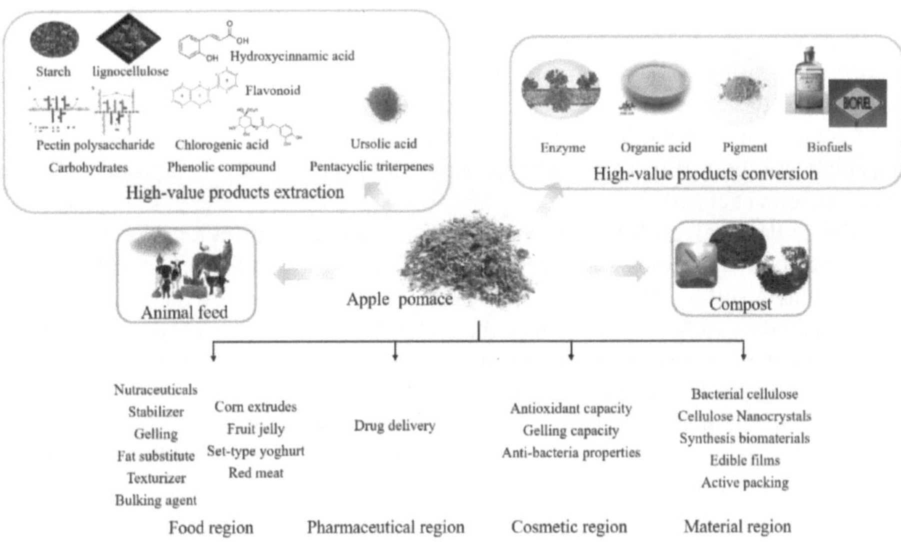

FIGURE 22.2 The specific application of apple pomace.

Figure 22.2 shows the specific application of AP. The two suggested components of AP utilization were the primary emphasis of early conventional methods: direct use in animal feed and basic fermentation conversion, depending on whether or not fermentation required separation and purification. In direct utilization in animal feeding, the main way that AP is used as a nutritional supplement is added to the diets of cattle and poultry (Guil-Guerrero et al., 2016). The growing economy has led to a shift in the desire for meat consumption from quantity to quality (Zhang et al., 2021). Customers are aware of the misuse of antibiotics while feeding animals, and natural materials are increasingly suggested as feed supplements to enhance the quality of the meat (Rushton, 2015).

Since AP is a cheap, nutrient-rich fruit residue, it is used extensively in a variety of animal diets. To improve the nutrition of chicken feed, AP has been used as a feedstock. Following a dietary intervention with four different dried fruit pomaces (strawberry, seedless strawberry, blackcurrant, and AP), the turkey breast muscles with the highest levels of vitamin A and α-linolenic acid were found to be treated with strawberry pomace. A meal high in polyphenols decreased lipid oxidation, and polyphenols were essential for the growth of turkeys (Juskiewicz et al., 2017). Consequently, AP is a useful component for boosting the nutrition in chicken feeds since it is high in polyphenols, vitamins, and other elements. Second, finishing pigs' diets are additionally supplemented with AP. The study conducted by Fang et al. (2016) aimed to investigate the meat quality and growth performance of castrated male finishing pigs fed fresh AP at varying levels (2%, 4%, 6%, and 8%). The findings indicated that the ideal ratio of AP supplementation was 2%, resulting in the best performance of pig growth and meat quality.

Aquatic animals also play a significant role in the breeding sector. Probiotics are added to the feed to partially or completely replace the antibiotics to address the issue of antibiotic overuse. Silver catfish (*Rhamdia quelen*) were fed AP and passion fruit peel as supplements, and 2.5 g/kg of AP effectively improved butyric acid synthesis and muscle layer thickness (Mombach et al., 2019). The other view from agriculture is that AP is utilized in composting. Biodegradable materials are broken down by microorganisms during the intricate aerobic process of composting, producing both organic and inorganic by-products (Toledo et al., 2018). It is one of the most well-known processes for getting rid of organic waste and making fertilizers, and it has a promising future given how quickly organic agriculture has grown in recent years. More significantly, wet AP might be added straight to compost, saving money and energy during the drying process. According to Jiang et al. (2014), decreasing the starting pH, adding physical adsorption materials, and adding metal salts are thought to be the most efficient approaches to decrease nitrogen loss and enhance the quality of composting during the process. Consequently, AP is appropriate for use in composting due to its high level of organic acid, carbohydrates, and other materials that promote the growth of microbes (Barreira et al., 2019).

According to Pan et al. (2018), AP was the perfect acidic supplement for the composting of sewage sludge and showed a comparable impact when combined with other acidic additions, including citric acid, element sulfur, phosphoric acid, magnesium hydrogen phosphate, and calcium superphosphate. According to Mao et al. (2017), the addition of 20% AP reduced the emission of NH_3 and N_2O during the composting of pig manure by 57% and 24%, respectively. This resulted in a 13% increase in the total nitrogen in the finished products. The C/N ratio and pH level are crucial variables that must be managed during the composting process since they have a big impact on the proliferation of microorganisms (Ghinea et al., 2019). Consequently, adding AP serves as both a substrate supplement and a possible means of modifying the pH and C/N ratio to quicken the composting process. In addition to efficiently resolving the AP processing issue, using AP as an acidic supplement to animal manure also enhanced compost quality and addressed soil fertility issues. This is an excellent way to increase composting quality and preserve nitrogen. However, even though adding AP is beneficial, the quantity added should be determined by a concrete experiment because adding too much AP will prevent the improvement of compost quality.

22.1.4 Current Role in the Industry

From the view of industry, AP also plays an important role in bioactive compound extraction. Figure 22.3 shows the flowchart of full-component utilization in AP. Apples' antioxidant components and dietary fiber are principally responsible for their health advantages (Koutsos et al., 2015; Tu et al., 2017). As previously described, AP has a significantly lower water content than fresh

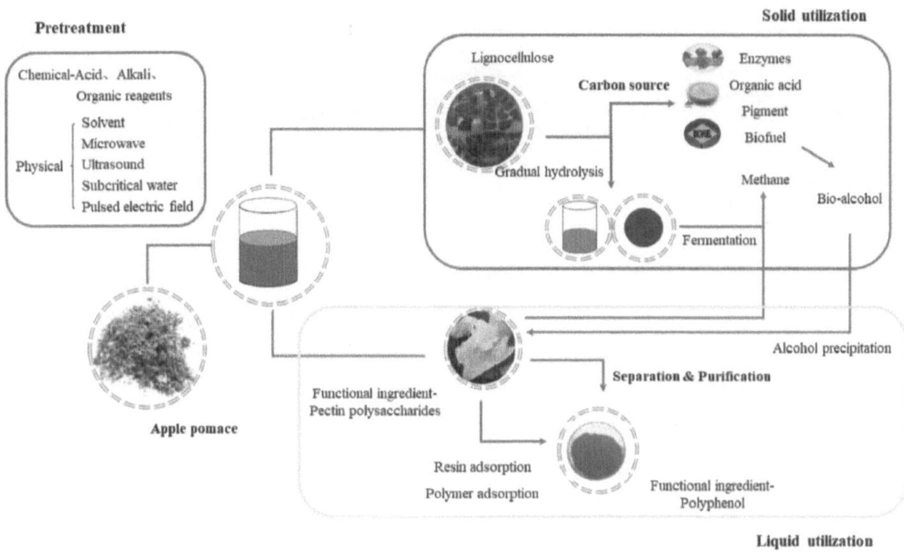

FIGURE 22.3 The flowchart of full-component utilization in apple pomace.

fruit. This, combined with the apple peel's naturally higher percentage of phenolic compounds (despite their partial transference to the juice), results in a by-product that has a higher phenolic concentration than the fresh fruit itself (Krawitzky et al., 2014). AP is a cheap source of bioactive substances and phytochemicals, including pectin, dietary fiber, polyphenols, triterpenoids, and volatiles. However, given its significance in the food and pharmaceutical industries, this by-product is not being utilized to its full potential (Waldbauer et al., 2017). AP has a high level of phenolic compounds, dietary fiber, and other nutrients, making it a good functional ingredient to use in a variety of food products. Nonetheless, there are reports that adding AP results in a drop in a few food product quality metrics. Pomace is therefore added at a somewhat low level as a functional element (Gómez et al., 2018), and these amounts need to be closely watched. Various bakery products, including bread, cakes, and cookies, have been consumed by humans for hundreds of years and are widely accepted (Gómez et al., 2018). The application of AP in bakery foods is considered to improve the dietary fiber content and health benefits (Masoodi et al., 1998; Wang et al., 1989). However, pomace addition results in a decrease in quality and sensory properties of baked products in the majority of cases (Jung et al., 2015; Sudha et al., 2007; Sudha et al., 2016). Therefore, efforts have been undertaken to evaluate the effects of AP on each individual bakery product and optimize the utilization level to incorporate (Lyu et al., 2020).

AP also utilized in agro-industrial wastes as feedstocks for bioenergy production. An intriguing field of research is the use of agro-industrial wastes in the creation of biofuels. It has been discovered that wastes from farms, cattle, animal husbandry, and other agricultural processing sectors are effective sources of biofuels (Nair et al., 2022). Carbohydrate polymers such as starch, cellulose, and hemicellulose, proteins, lipids, fibers, and other organic materials make up their principal component (Chen et al., 2019; Sindhu et al., 2019). As these wastes have a significant organic content, improper disposal could have a detrimental effect on the ecosystem. Improper waste management will contribute to environmental pollution. Environmental pollution includes air pollution, water pollution, soil contamination, global warming, and climate change. Addressing these issues requires proper waste treatment methods. Hence, well-managed agro-biomass waste keeps providing reductions of environmental pollution in aspects of pollution control engineering.

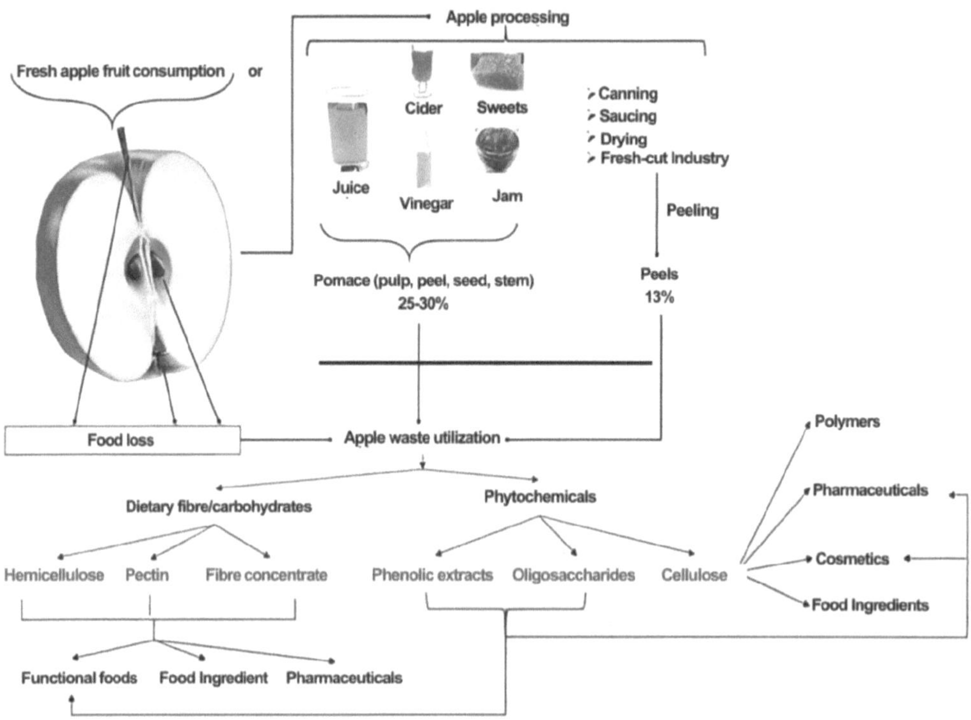

FIGURE 22.4 The types of waste produced from apples.

22.2 CURRENT ISSUES AND CHALLENGES IN MANAGING APPLE AGRO-BIOMASS WASTE

A number of difficulties and problems must be resolved in the management of apple biomass, especially the waste generated during apple processing, such as AP. Figure 22.4 shows the types of waste produced from apples. Environmental, logistical, and financial issues are among the technical difficulties in waste treatment and the utilization of these by-products. The management of apple biomass, particularly the waste generated during apple processing is a pressing challenge for the agricultural and food industries worldwide. About 25–30% of the fruit's bulk is made up of AP, the main by-product of apple processing. Large amounts of this trash are created, particularly in nations with extensive apple-producing sectors. In 2019, more than over 80 million tons of apples were produced worldwide, with major producers including China, the United States, and Turkey accounting for a large portion of this total (Zhang et al., 2021). This amounts to an astounding 5–7 million tons of AP produced each year, a quantity that presents serious logistical, environmental, and financial difficulties.

22.2.1 Volume and Types of Waste Produced

AP consists of a variety of organic materials, including the peel, pulp, seeds, stems, and core that are left over after juice extraction or processing for products like cider, vinegar, or applesauce. Its high moisture content, which ranges between 75% and 80%, makes it highly perishable and difficult to store or transport without further processing. This water-rich composition causes AP to spoil quickly, generating foul odors and potentially harmful by-products if not managed properly. The perishability of AP limits its utility as a raw material in further processing unless it is quickly dried or treated, which introduces additional costs and environmental concerns related to energy consumption (Lyu et al., 2020; Zhang et al., 2021).

Managing the large amount of AP generated in areas with a high level of apple processing activity is a significant concern. Producers must consider not only the costs of disposal but also the environmental implications of such large-scale waste. The weight and space needed for AP's movement are greatly increased by its mass and water-heavy composition, which makes transportation challenging and costly. Moreover, untreated pomace cannot be stored for extended periods without undergoing rapid decomposition, which can lead to the formation of harmful microorganisms and gases. These issues underscore the necessity for efficient, sustainable, and cost-effective waste management practices to handle the vast amounts of AP generated annually (Zhang et al., 2021).

22.2.2 Environmental Impacts: Pollution and Emissions

The consequences of improper disposing of AP are extensive, significantly causing both localized pollution and more general climate change issues. If AP is left untreated or sent to landfills, its natural decomposition process can result in the release of methane, a potent GHG with a global warming potential many times higher than that of carbon dioxide. AP's high organic content speeds up the decomposition of organic waste, which releases methane emissions that are a primary cause of climate change. Therefore, without proper management, apple processing industries can unintentionally contribute to a significant increase in GHG emissions, exacerbating the global challenge of climate change.

In addition to GHG emissions, the improper disposal of AP can lead to water pollution. Leachates produced by the pomace's breakdown have the potential to leak into the soil and contaminate nearby water bodies or groundwater. High amounts of organic material in these leachates can cause eutrophication, a condition in which an abundance of nutrients in water bodies encourages the fast growth of algae. By reducing the amount of oxygen in the water, this algal bloom damages aquatic life and affects regional ecosystems (Zhang et al., 2021). Water pollution is a serious environmental issue that needs to be handled with appropriate waste treatment techniques in areas with a high volume of apple processing.

Furthermore, even though composting AP is a commonly advised waste management technique, there are some negative environmental effects. Ammonia (NH_3) and nitrous oxide (N_2O), two nitrogen-based gases that are major pollutants, can be released when large amounts of wet AP are composted. While nitrous oxide is another strong GHG with a long atmospheric lifetime and ammonia is a contributing factor to acid rain and the deterioration of air quality (Pan et al., 2018; Mao et al., 2017). The environmental advantages of composting as a sustainable waste management method may be compromised by these emissions, particularly if the procedure is not properly controlled.

The environmental impact is further complicated by the potential for odor pollution. Particularly in the early phases of decomposition, the organic components in AP can break down and release disagreeable aromas. Tensions between the industry and local communities may arise as a result of this, which impacts not just the immediate area around apple processing facilities but also the standard of living for those who live nearby (Urbina et al., 2019). Therefore, although AP may be a valuable resource, its management may have negative and permanent impacts on the ecosystem.

22.2.3 Difficulties in Waste Treatment

Due to its short lifespan and complicated composition, AP presents several technological challenges for value-adding and treatment. One of the more popular ways to deal with organic waste, such as AP, is composting, although it has disadvantages. The carbon-to-nitrogen (C/N) ratio, moisture content, pH, and aeration must all be carefully controlled for composting to be successful. Due to its high sugar and carbohydrate content, AP frequently has insufficient nitrogen to maintain an ideal C/N ratio, which can cause incomplete decomposition and slow down the composting process (Ghinea et al., 2019) (Figures 22.5 and 22.6).

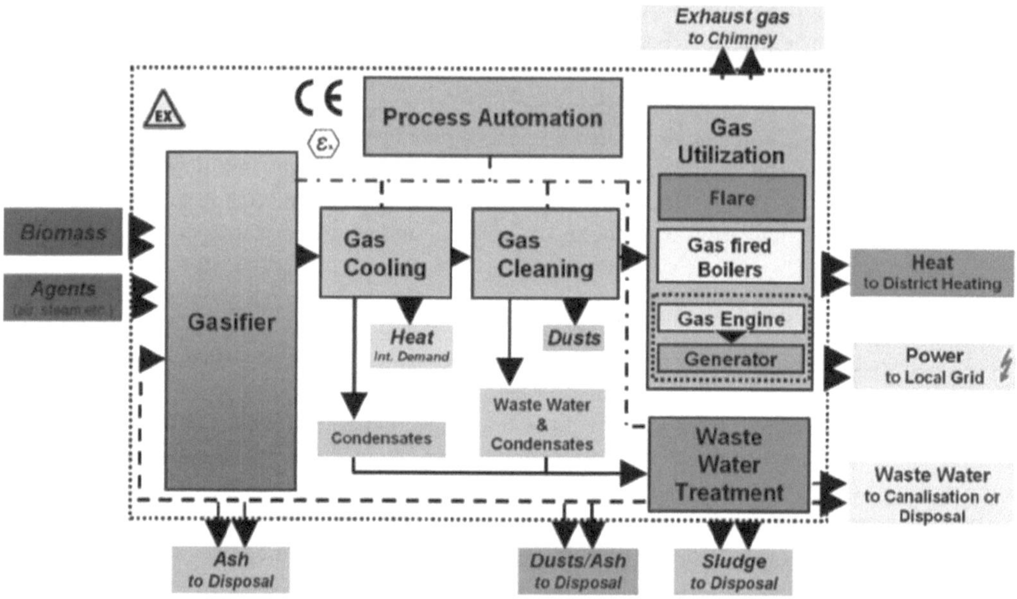

FIGURE 22.5 Biomass treatment process flow.

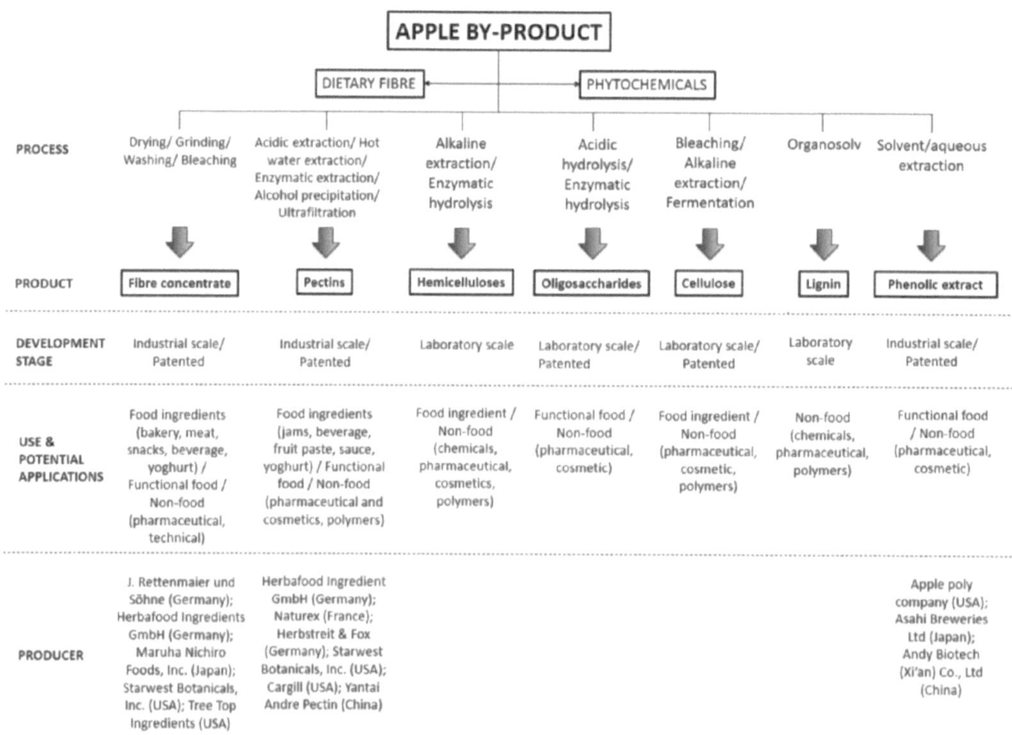

FIGURE 22.6 Use and potential applications of apple by-products and their stages of development.

Furthermore, composting AP on a large scale takes a lot of time, space, and work. Because of its high moisture content, AP also needs to be dried out or combined with drier materials for composting to work effectively. Anaerobic conditions brought on by too much moisture in the compost pile can produce methane and other unwanted gases, further harming the ecosystem (Zhang et al., 2021). Additionally, the composting process may cause organic waste and nutrients to seep into nearby soils and water bodies, escalating the previously mentioned environmental issues.

Fermenting AP is another potential treatment option, especially for high-value products including organic acids, biofuels, and enzymes. The fermenting process is not simple. It needs precise environmental parameters, such as regulated pH, temperature, and microbial activity. To maximize the fermentation process and generate the intended by-products, particular strains of microorganisms must be carefully chosen (Marín et al., 2019; Zhu et al., 2018). Any deviation from these guidelines may result in low yields or the creation of undesirable or hazardous by-products, which would make the treatment procedure more difficult and raise waste management expenses.

22.2.4 Cost Associated with Waste Management and Processing

One of the biggest issues facing the apple processing business is the cost of controlling apple biomass. Because of its high moisture content, AP is costly to ship and store, particularly when it is unprocessed and uncooked. Because of the water-heavy nature of AP, which makes it less cost-effective to ship over long distances, transportation expenses are especially expensive. Additionally, due to its high preservation, AP needs to be treated or disposed of quickly, which restricts businesses' ability to manage their waste streams (Scherhaufer et al., 2020). A potential solution is to dry AP to lower its moisture content and increase its shelf life, but this is a costly procedure. Significant amounts of heat are needed for the drying process, which can raise GHG emissions and operating expenses if renewable energy sources are not used. Furthermore, the market value of the dried product or any by-products that can be removed during the process greatly influences the economic viability of drying AP. Businesses might be hesitant to use this technique extensively if the expenses of drying outweigh the financial gains (Scherhaufer et al., 2020).

Other valorization techniques, such as the extraction of pectin, polyphenols, or other valuable components from AP, involve intricate procedures that call for costly machinery and substantial energy inputs, in addition to drying. For instance, pectin extraction usually uses heat and acidic solvents to break down the pomace's cell walls and liberate the pectin. The labor and energy expenses of this process are increased by the need for precise supervision and extra processes for separation and purification (Alongi et al., 2019). AP processing costs are further increased by the complexity of polyphenol extraction, which frequently calls for organic solvents or sophisticated methods like ultrasound-assisted extraction (Zhang et al., 2021). A further economic difficulty is the variation in AP's composition. It is challenging to standardize the waste treatment procedures since the chemical composition of AP might differ greatly based on the apple variety, growing environment, and processing techniques. This unpredictability might make it more difficult to forecast the financial rewards from valorization operations and result in inefficiencies in the extraction of high-value compounds. To guarantee that their waste management and processing techniques are successful and economical, businesses must thus make investments in advanced testing and quality control procedures (Bhushan et al., 2008).

22.3 INNOVATIVE SOLUTION AND POTENTIAL APPLICATION

AP, the by-product of juice and cider production, represents a significant challenge and opportunity for waste management. Typically, AP consists of the apple skin, seeds, and pulp left over after juice extraction. Instead of being discarded, this biomass has immense potential to be transformed into valuable resources through innovative technologies and sustainable processes. AP is often treated as waste, leading to environmental concerns, such as landfilling and the production of GHGs.

However, this by-product contains valuable bioactive compounds, including polyphenols, pectin, and dietary fibers, which can be harnessed for various industrial applications.

22.3.1 Bioactive Compound Extraction

AP, the primary by-product of juice and cider production, holds immense potential in the realm of sustainable waste management and resource utilization. Instead of being discarded, this by-product can be a rich source of bioactive compounds, particularly antioxidants and phenolic compounds, which have significant applications in the nutraceutical, pharmaceutical, and food industries. The extraction of these compounds can be achieved through various techniques such as Soxhlet extraction and subcritical water extraction, enabling the recovery of essential nutrients from what was once considered waste. Among the key compounds are polyphenols, flavonoids, and dietary fibers, all of which exhibit strong antioxidant properties. These compounds play a pivotal role in preventing oxidative stress, making them ideal ingredients for dietary supplements and functional foods. For instance, polyphenols can help reduce the risk of chronic diseases such as cardiovascular disorders and certain types of cancer (Fernandes et al., 2019). In the pharmaceutical industry, the anti-inflammatory properties of AP-derived bioactives, including triterpenoids, can be leveraged to create new treatments for metabolic and inflammatory diseases. In food processing, AP serves as a natural preservative and a valuable functional ingredient, enriching the nutritional profile of foods like snacks, beverages, and bakery products (Barreira et al., 2019). By transforming AP into valuable bioactive compounds, industries can not only reduce waste but also contribute to sustainable production practices (Tables 22.1 and 22.2).

The first and second values for the DPPH• and ABTS•+ are expressed in terms of EC50 (μg of extract/mL) and as relative antioxidant capacity with reference to total polyphenols (μg GAE of extract/mL), respectively. OH• scavenging was expressed as mannitol equivalents (mmol/g of extract), and for the NO• method as a percentage of inhibition at 130 μg/mL. Values are compared to ascorbic acid (AA).

22.3.2 Renewable Energy

AP, a by-product of juice and cider manufacturing, represents an opportunity to transform waste into renewable energy. One of the most promising methods to achieve this is anaerobic digestion, a biochemical process in which microorganisms break down organic matter in the absence of oxygen. Through this process, AP can be converted into valuable biogas and bioethanol, aligning with the principles of circular economy and sustainable resource management. Anaerobic digestion of AP produces biogas, primarily composed of methane, which can be used as a clean, renewable fuel source. In addition to energy generation, bioethanol, a liquid fuel, can also be extracted, providing further use as an alternative to fossil fuels. This process not only reduces the environmental burden of waste disposal but also mitigates reliance on non-renewable energy sources. Studies have shown that AP contains sufficient free sugars and structural carbohydrates, making it an ideal feedstock for biofuel production (Molinuevo-Salces et al., 2020).

22.3.2.1 Bioethanol Production Process

This biorefinery approach demonstrated that AP is a valuable feedstock for bioethanol production, particularly when using appropriate yeast strains. In addition to the well-known *Saccharomyces cerevisiae*, other yeasts like *Kluyveromyces marxianus*, *Kluyveromyces lactis*, and *Lachancea thermotolerans* were also effective. The alcoholic fermentation process enhanced methane production, showing that the exhausted broth could also be utilized for methane generation. This dual-process approach maximizes resource efficiency, offering a sustainable method for biofuel production from agricultural waste. The potential benefits extend beyond energy generation. AP valorization helps close the loop in industrial processes, converting waste that would otherwise contribute to pollution into energy that powers industries and homes. This shift exemplifies the role of biorefineries in creating a sustainable energy landscape (Vaez et al., 2023) (Figure 22.7).

TABLE 22.1
Yield (%), Monosaccharide (Molar%), Carbohydrate (%), Protein (%), and Polyphenolic Composition (g Gallic Acid Equivalents (GAE)/kg) of Industrial Apple Pomace, Hot Water Extract (HWE), High Molecular Weight Material (HMWM), Low Molecular Weight Material (LMWM), and Extraction Residue (Fernandes et al., 2019)

Fractions	Yield (%)	Yield of Carbohydrate (%)		Carbohydrate (Mol%)									Total Carbohydrate (%)		Total Protein (%)	Total PC (g GAE/kg)
				Rha	Fuc	Ara	Xyl	Man	Gal	Glc	Fru	GalA				
Apple pomace			Polysaccharides	1	1	12	10	5	9	41		19	53.4	71.7	5.2	MD
			Free sugars							23	77		18.3			
HWE	29.6	39.2	Polysaccharides	3	1	25	3		10	9		50	42.9	94.9	1.3	11
			Free sugars							18	82		52.0			
HMWM	6.9	7.5	Polysaccharides	1	t	35	6	1	10	9		36	77.7	77.7	ND	5
LMWM	22.3	27.7	Polysaccharides	1	t	15	2		6	34		42	33.5	89.0	ND	9
			Free sugars							6	94		55.5			
Residue	67.4	55.9	Polysaccharides	1	1	8	14	4	8	51		13	59.5	59.5	7.1	

t, trace; ND, not determined.

TABLE 22.2
Total Polyphenolic Content (TPC) and Antioxidant (DPPH•, ABTS•+, OH•, NO•) Activity of the Hot Water Extract before (HWE) and after the Purification (pHWE)

Extract	TPC	DPPH•	ABTS•+	OH•	NO•
HWE	10.7 ± 0.2	1,339 ± 211 (14.2 ± 1.7)	532 ± 11.5 (5.69 ± 0.12)	—	—
pHWE	149 ± 1.87	82.4 ± 11.2 (12.3 ± 1.7)	35.2 ± 5.9 (5.23 ± 0.44)	6.75 ± 0.45	35.2 ± 5.9
AA	—	2.70 ± 0.30	2.68 ± 0.03	—	57.3 ± 2.3

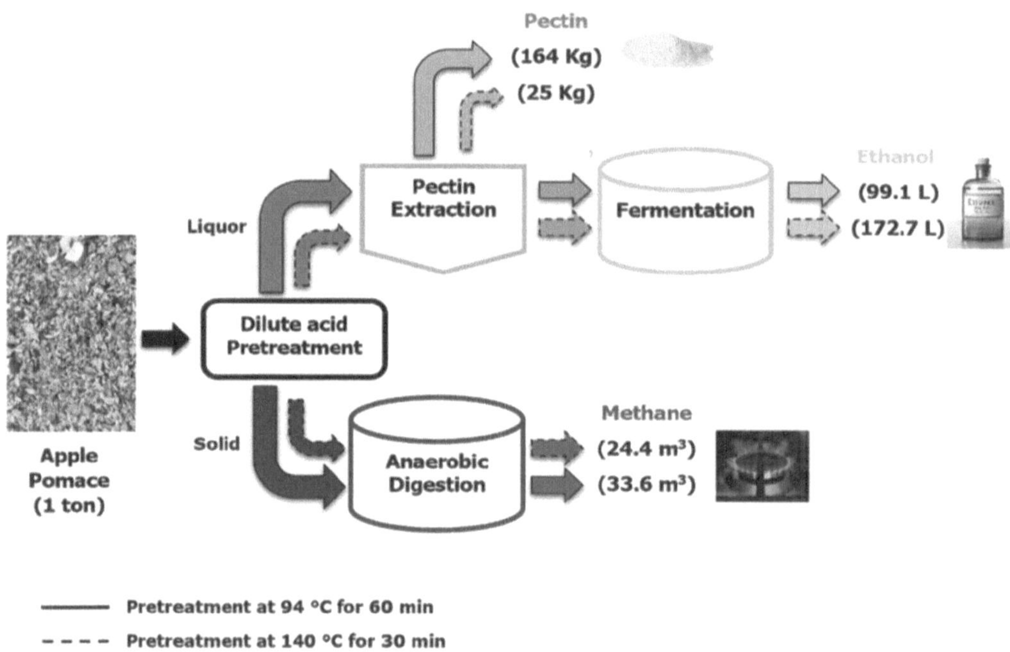

FIGURE 22.7 Biorefinery Process for the Biochemical Transformation of Apple Pomace into Biofuels (Ethanol and Methane) and Pectin (Vaez et al., 2023)

22.3.3 Biodegradable Packaging

By combining AP with biopolymers, such as potato starch or other biodegradable materials, researchers have developed innovative biodegradable films and biocomposites. These films serve as environmentally friendly alternatives to conventional plastic packaging. AP, which is the solid residue left after the extraction of juice from apples, is rich in fibers, vitamins, and organic matter. This makes it an ideal candidate for creating biocomposites with structural integrity, flexibility, and durability (Figure 22.8).

These biodegradable materials help tackle two major environmental concerns: plastic pollution and agricultural waste. Conventional plastic packaging has contributed significantly to global waste, much of which ends up in landfills or oceans. Biocomposites from AP decompose naturally, reducing the environmental footprint. Furthermore, utilizing agricultural by-products like AP aligns with

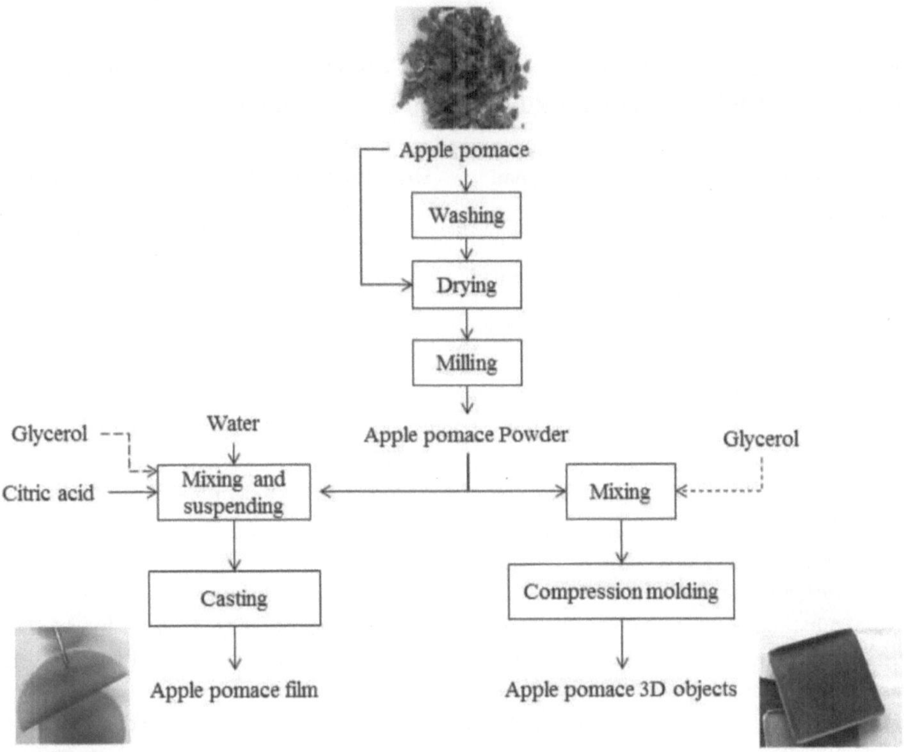

FIGURE 22.8 Flowchart of the Methods Used for Production of Bio-Based Films and 3D Objects from Apple Pomace (Fernandes et al., 2019)

the principles of the circular economy by converting waste into valuable resources. In addition to reducing plastic waste, these biocomposites can be engineered to be water-resistant, heat-resistant, and mechanically strong, making them viable alternatives in industries that require sustainable packaging solutions (Gustafsson et al., 2019) (Tables 22.3 and 22.4).

22.4 CHALLENGES AND FUTURE PERSPECTIVE

AP, a by-product of the manufacture of juice and cider, has many creative and sustainable uses, ranging from the extraction of bioactive compounds to the creation of renewable energy and biodegradable packaging. There are many recommendations for maximizing the potential of AP through its transformation into valuable resources. First, adopting advanced extraction

TABLE 22.3
Different Preparation Conditions and Main Properties of Bio-Based Films Prepared from Apple Pomace, Using 1% (w/v) Citric Acid Solution at 70°C (Fernandes et al., 2019)

Washing Step	Glycerol (%)	Particle Size[a] (mm)	Thickness (mm)	Tensile Strength (MPa)	Elongation (%)
Yes	7	~0.08	0.11 ± 0.01	16.49 ± 2.54	10.77 ± 3.19
No	7	~0.08	0.11 ± 0.01	3.27 ± 0.31	55.41 ± 5.38
No	0	~0.08	0.09 ± 0.00	4.20 ± 0.70	37.39 ± 10.38

[a] Apple pomace particle size used for the preparation of the films.

TABLE 22.4
Different Preparation Conditions and Main Features of Fiberboards Prepared from Apple Pomace Applying 8 MPa Pressure for 20 min at 100°C (Fernandes et al., 2019)

Washing Step	Plasticizer (%)	Particle Size[a] (mm)	Thickness (mm)	Tensile Strength (Mpa)	Elongation (%)	Young's Modulus (Mpa)
Yes	Glycerol 30%	0.2	3.18 ± 0.07	5.79 ± 0.79	1.54 ± 0.09	633.4 ± 65.6
No	Free sugars[b]	0.2	2.17 ± 0.51	3.71 ± 0.80	1.56 ± 0.13	367.1 ± 82.6
No	Free sugars[b]	1	2.91 ± 0.02	3.02 ± 0.65	0.93 ± 0.21	485.7 ± 94.7

[a] Apple pomace particle size used for the preparation of fiberboards.
[b] From non-washed apple pomace.

technologies for bioactive compounds. Modern, environmentally friendly extraction methods including Soxhlet, supercritical, and subcritical fluid extraction should be given top priority by industries. With the use of these techniques, AP's polyphenols, flavonoids, and dietary fibers can be effectively recovered. These substances have potent antioxidant qualities and are used in medicines, nutraceuticals, and functional foods. Use subcritical water extraction to increase the output of bioactive substances while reducing the impact on the environment. In addition to reducing waste, this would encourage the creation of functional foods, dietary supplements, and natural preservatives (Fernandes et al., 2019).

In addition, scaling up the production of bioethanol from AP is recommended, especially considering the successful results achieved using yeast strains such as Kluyveromyces marxianus and Lachancea thermotolerans. In line with the objectivesof renewable energy, the synthesis of bioethanol from AP can offer a sustainable substitute for fossil fuels. Create biorefineries that combine anaerobic digestion with other processes to produce biogas and bioethanol. This two-pronged strategy optimizes the use of resources, minimizes waste, and helps produce renewable energy. Work together with academic institutions to improve fermentation techniques and maximize bioethanol production by utilizing a variety of yeast strains (Vaez et al., 2023).

Moreover, promoting the development of biodegradable packaging is essential. AP serves as an excellent alternative to conventional plastic packaging when used as a raw material for producing biodegradable films and biocomposites. This strategy promotes the development of the circular economy, waste minimization, and the decrease of plastic pollution. Invest in the study and creation of biopolymers made from AP for the packaging sector. AP can be combined with other biodegradable materials, like potato starch, to create robust, naturally decomposing packaging materials that are resistant to water. To assess the feasibility of AP-based biodegradable films in practical applications, particularly in the food industry, carry out pilot studies with packaging businesses (Gustafsson et al., 2019). Furthermore, supporting circular economy practices. Circular economy concepts are aligned with the conversion of AP into valuable products like bioactive chemicals, biofuels, and biodegradable materials. Businesses should include several valorization processes in their waste management strategy to handle trash holistically. Encourage collaborations between waste management firms, biorefineries, and the food processing industry to create a seamless waste-to-resource pipeline. Through this partnership, it will be possible to guarantee that AP is effectively recycled and used in a variety of industries, including food, energy, and packaging (Molinuevo-Salces et al., 2020).

Lastly, one possible way to solve resource constraints and environmental issues is through the value-adding of AP. Industries may transform this waste ago-biomass into a useful resource, lowering dependency on non-renewable resources and promoting the development of a circular economy, by implementing sustainable technologies and practices.

22.5 CONCLUSION

Innovative uses of AP offer revolutionary possibilities for industrial applications, environmental sustainability, and waste management. By repurposing this usually wasted agricultural by-product, enterprises may solve multiple global issues such as plastic pollution, waste reduction, and the creation of renewable energy. The culinary, pharmaceutical, and nutraceutical industries have great potential for the bioactive compounds found in AP, which include dietary fibers, polyphenols, and flavonoids. The production of health-promoting goods like functional foods and nutritional supplements is made possible by the extraction of these chemicals using techniques like Soxhlet and subcritical water extraction, which also helps with sustainable waste management. The anti-inflammatory and antioxidant qualities of these chemicals are particularly interesting for improving food products and creating remedies for chronic illnesses. AP exhibits great potential as a feedstock for the production of biofuels, particularly biogas and bioethanol. AP can be transformed into useful energy sources through anaerobic digestion and fermentation processes, assisting in the switch from fossil fuels to greener, renewable alternatives. Waste is turned into energy in these operations, which is an example of a circular economy strategy that lowers GHG emissions and dependency on non-renewable resources. AP has been effectively utilized in the field of material sciences to create biodegradable packaging options. Through the use of biopolymers and AP, researchers have developed sustainable substitutes for conventional plastics. These biocomposites are a sustainable answer to the world's plastic waste problem because they break down naturally. These biocomposites' mechanical qualities, like their resistance to heat and water, also make them suitable for use in a range of industrial packaging applications. In conclusion, the value-adding of AP indicates a movement in several industries toward environmentally friendly manufacturing methods. The creative application of this waste, which illustrates the circular economy's tenets, lowers environmental effects while simultaneously creating economic value. Examples include bioactive substances, renewable energy, and biodegradable packaging. AP has the potential to be used in these domains in ways that promote resource efficiency and sustainability, providing a model for other agricultural wastes.

ACKNOWLEDGMENT

The chapter was written with Khor Jia Jin, Ahmad Arsyad Muhammad Gani, Tam Yee Bong, and Teh Xiong Siang.

REFERENCES

Alongi, M., Melchior, S., & Anese, M. (2019). Reducing the glycemic index of short dough biscuits by using apple pomace as a functional ingredient. *LWT – Food Science and Technology, 100*, 300–305.

Barreira, J. C. M., Arraibi, A. A., & Ferreira, I. C. F. R. (2019). Bioactive and functional compounds in apple pomace from juice and cider manufacturing: Potential use in dermal formulations. *Trends in Food Science & Technology*, *90*, 76–87. https://doi.org/10.1016/J.TIFS.2019.05.014

Bhushan, S., Kalia, K., Sharma, M., Singh, B., & Ahuja, P. S. (2008). Processing of apple pomace for bioactive molecules. *Critical Reviews in Biotechnology, 28*(4), 285–296.

Chen, J., Fan, Y., Jiaqiang, E., Cao, W., Zhang, F., Gong, J., … & Xu, W. (2019). Effects analysis on the gasification kinetic characteristics of food waste in supercritical water. *Fuel*, *241*, 94–104. https://www.sciencedirect.com/science/article/pii/S001623611832060X

Fang, J., Cao, Y., Matsuzaki, M., Suzuki, H., & Kimura, H. (2016). Effects of apple pomace-mixed silage on growth performance and meat quality in finishing pigs. *Animal Science Journal*, *87*(12), 1516–1521. https://onlinelibrary.wiley.com/doi/abs/10.1111/asj.12601

Fernandes, P. A. R., Ferreira, S. S., Bastos, R., Ferreira, I., Cruz, M. T., Pinto, A., Coelho, E., Passos, C. P., Coimbra, M. A., Cardoso, S. M., & Wessel, D. F. (2019). Apple pomace extract as a sustainable food ingredient. *Antioxidants*, *8*(6), 189. https://doi.org/10.3390/antiox8060189

Ghinea, C., Apostol, L. C., Prisacaru, A. E., & Leahu, A. (2019). Development of a model for food waste composting. *Environmental Science and Pollution Research*, *26*, 4056–4069. https://link.springer.com/article/10.1007/s11356-018-3939-1

Ginni, G., Kumar, S. A., Usman, T. M. M., Pakonyi, P., & Banu, J. R. (2020). Integrated biorefineries of food waste. In: J. Rajesh Banu, M. Gunasekaran, G. Kumar, & S. Kavitha (Eds.), *Food Waste to Valuable Resources: Applications and Management* (pp. 275–298). Academic Press. https://doi.org/10.1016/B978-0-12-818353-3.00013-4

Gómez, M., & Martinez, M. M. (2018). Fruit and vegetable by-products as novel ingredients to improve the nutritional quality of baked goods. *Critical Reviews in Food Science and Nutrition, 58*, 2119–2135. https://www.tandfonline.com/doi/full/10.1080/10408398.2017.1305946

Guil-Guerrero, J. L., Ramos, L., Moreno, C., Zúñiga-Paredes, J. C., Carlosama-Yépez, M., & Ruales, P. (2016). Plant-food by-products to improve farm-animal health. *Animal Feed Science and Technology*, *220*, 121–135. https://www.sciencedirect.com/science/article/pii/S0377840116304370

Gustafsson, J., Landberg, M., Bátori, V., Åkesson, D., Taherzadeh, M. J., & Zamani, A. (2019). Development of bio-based films and 3D objects from apple pomace. *Polymers*, *11*(2). https://doi.org/10.3390/polym11020289

Jiang, J., Huang, Y., Liu, X., & Huang, H. (2014). The effects of apple pomace, bentonite and calcium superphosphate on swine manure aerobic composting. *Waste Management*, *34*(9), 1595–1602. https://www.sciencedirect.com/science/article/pii/S0956053X14001822

Jung, J., Cavender, G., & Zhao, Y. (2015). Impingement drying for preparing dried apple pomace flour and its fortification in bakery and meat products. *Journal of Food Science and Technology, 52*, 5568–5578. https://link.springer.com/article/10.1007/s13197-014-1680-4

Juskiewicz, J., Jankowski, J., Zielinski, H., Zdunczyk, Z., Mikulski, D., Antoszkiewicz, Z., Kosmala, M., & Zdunczyk, P. (2017). The fatty acid profile and oxidative stability of meat from turkeys fed diets enriched with n-3 polyunsaturated fatty acids and dried fruit pomaces as a source of polyphenols. *PLoS One*, *12*(1), e0170074. https://journals.plos.org/plosone/article?id=10.1371/journal.pone.0170074

Koutsos, A., Tuohy, K. M., & Lovegrove, J. A. (2015). Apples and cardiovascular health—Is the gut microbiota a core consideration? *Nutrients*, *7*(6), 3959–3998. https://www.mdpi.com/2072-6643/7/6/3959

Krawitzky, M., Arias, E., Peiro, J. M., Negueruela, A. I., Val, J., & Oria, R. (2014). Determination of color, antioxidant activity, and phenolic profile of different fruit tissue of Spanish 'Verde Doncella' apple cultivar. *International Journal of Food Properties*, *17*(10), 2298–2311. https://www.tandfonline.com/doi/abs/10.1080/10942912.2013.792829

Lyu, F., Luiz, S. F., Azeredo, D. R. P., Cruz, A. G., Ajlouni, S., & Ranadheera, C. S. (2020). Apple pomace as a functional and healthy ingredient in food products: A review. *Processes*, *8*(3), 319. https://www.mdpi.com/2227-9717/8/3/319

Mao, H., Zhang, T., Li, R., Zhai, B., Wang, Z., Wang, Q., & Zhang, Z. (2017). Apple pomace improves the quality of pig manure aerobic compost by reducing emissions of NH_3 and N_2O. *Scientific Reports*, *7*(1), 870. https://www.nature.com/articles/s41598-017-00987-y

Marín, D., Romero, I., & Manzanares, P. (2019). Production and recovery of cellulases through solid-state fermentation of selected lignocellulosic wastes. *Journal of Cleaner Production, 211*, 844–852.

Masoodi, F.A., & Chauhan, G. S. (1998). Use of apple pomace as a source of dietary fiber in wheat bread. *Journal of Food Processing and Preservation, 22*, 255–263. https://onlinelibrary.wiley.com/doi/10.1111/j.1745-4549.1998.tb00349.x

Molinuevo-Salces, B., Riaño, B., Hijosa-Valsero, M., González-García, I., Paniagua-García, A. I., Hernández, D., Garita-Cambronero, J., Díez-Antolínez, R., & García-González, M. C. (2020). Valorization of apple pomaces for biofuel production: A biorefinery approach. *Biomass and Bioenergy*, *142*. https://doi.org/10.1016/j.biombioe.2020.105785

Mombach, P. I., Adorian, T. J., Pianesso, D., dos Santos, T. S., de Lima, J. S., Robalo, S. S., & da Silva, L. P. (2019). Pectic hydrolysates in the diet of silver catfish (Rhamdia quelen): Growth performance, blood and liver biochemistry, histological parameters and intestinal contents. *Aquaculture Nutrition*, *25*(6), 1378–1387. https://onlinelibrary.wiley.com/doi/abs/10.1111/anu.12958

Nair, L. G., Agrawal, K., & Verma, P. (2022). An overview of sustainable approaches for bioenergy production from agro-industrial wastes. *Energy Nexus*, *6*, 100086. https://doi.org/10.1016/J.NEXUS.2022.100086

Pan, J., Cai, H., Zhang, Z., Liu, H., Li, R., Mao, H., Awasthi, M. K., Wang, Q., & Zhai, L. (2018). Comparative evaluation of the use of acidic additives on sewage sludge composting quality improvement, nitrogen conservation, and greenhouse gas reduction. *Bioresource Technology*, *270*, 467–475. https://www.sciencedirect.com/science/article/pii/S0960852418313038

Perussello, C. A., Zhang, Z., Marzocchella, A., & Tiwari, B. K. (2017). Valorization of apple pomace by extraction of valuable compounds. *Comprehensive Reviews in Food Science and Food Safety*, *16*(5), 776–796. https://ift.onlinelibrary.wiley.com/doi/abs/10.1111/1541-4337.12290

Rushton, J. (2015). Anti-microbial use in animals: How to assess the trade-offs. *Zoonoses and Public Health*, *62*, 10–21. https://onlinelibrary.wiley.com/doi/abs/10.1111/zph.12193

Sindhu, R., Gnansounou, E., Rebello, S., Binod, P., Varjani, S., Thakur, I. S., Nair, R. B., & Pandey, A. (2019). Conversion of food and kitchen waste to value-added products. *Journal of Environmental Management*, *241*, 619–630. https://www.sciencedirect.com/science/article/pii/S030147971930204X

Sudha, M. L., Baskaran, V., & Leelavathi, K. (2007). Apple pomace as a source of dietary fiber and polyphenols and its effect on the rheological characteristics and cake making. *Food Chemistry, 104*, 686–692. https://www.sciencedirect.com/science/article/pii/S0308814606009903?via%3Dihub

Sudha, M. L., Dharmesh, S. M., Pynam, H., Bhimangouder, S. V., Eipson, S. W., Somasundaram, R., & Nanjarajurs, S. M. (2016). Antioxidant and cyto/DNA protective properties of apple pomace enriched bakery products. *Journal of Food Science and Technology, 53*, 1909–1918. https://link.springer.com/article/10.1007/s13197-015-2151-2

Toledo, M., Siles, J. A., Gutiérrez, M. C., & Martín, M. A. (2018). Monitoring of the composting process of different agroindustrial waste: Influence of the operational variables on the odorous impact. *Waste Management*, *76*, 266–274. https://www.sciencedirect.com/science/article/pii/S0956053X18301843

Tu, S. H., Chen, L. C., & Ho, Y. S. (2017). An apple a day to prevent cancer formation: Reducing cancer risk with flavonoids. *Journal of Food and Drug Analysis*, *25*(1), 119–124. https://www.sciencedirect.com/science/article/pii/S1021949816301788

Vaez, S., Karimi, K., Denayer, J. F. M., & Kumar, R. (2023). Evaluation of apple pomace biochemical transformation to biofuels and pectin through a sustainable biorefinery. *Biomass and Bioenergy*, *172*, 106757. https://doi.org/10.1016/j.biombioe.2023.106757

Waldbauer, K., McKinnon, R., & Kopp, B. (2017). Apple pomace as potential source of natural active compounds. *Planta Medica*, *83*(12/13), 994–1010. https://www.thieme-connect.com/products/ejournals/html/10.1055/s-0043-111898

Wang, H. J., & Thomas, R. L. (1989). Direct use of apple pomace in bakery products. *Journal of Food Science, 54*, 618–620. https://ift.onlinelibrary.wiley.com/doi/10.1111/j.1365-2621.1989.tb04665.x

Zhang, F., Wang, T., Wang, X., & Lü, X. (2021). Apple pomace as a potential valuable resource for full-components utilization: A review. *Journal of Cleaner Production*, *329*, 129676. https://doi.org/10.1016/J.JCLEPRO.2021.129676

Zhu, J., Li, F., Wang, X., & Wang, Y. (2018). Bioconversion of apple pomace into microbial protein feed based on extrusion pretreatment. *Bioresource Technology, 253*, 1–7.

Section VIII

Tuber-Based Plant Waste

23 Potato Waste Management

23.1 INTRODUCTION

The potato is one of the most widely consumed, reasonably priced crops in the world. The demand for agro-food products based on potatoes is growing as a result of the worsening food crisis (Chauhan et al., 2023). Since food waste (FW) increases carbon emissions, it is currently viewed as a major cause for concern and a pertinent issue to address (Khanal et al., 2023). It is also observed that managing this ecological pollution pathway from large-scale wastes is difficult. Waste from the food industry can be better managed through the use of biotechnological techniques (Chauhan et al., 2023). The potato industry is one of the industries that produce large amounts of waste that harm the environment. By-products such as potato peels and starch are disposed of as waste, which eventually can be used in biotechnological processing due to their bioactive ingredients. This chapter addresses the potential management strategies for potato waste and highlights its significance in industrial applications, renewable energy generation, and other sustainable practices. The discussion explores how effective utilization of potato waste can contribute to various sectors, emphasizing its role as a renewable resource with applications that support sustainability initiatives and environmental management.

As a crop for food security, potatoes rank fourth in importance among all agricultural vegetables worldwide. At over one-third of the world's potato production, China is the leading producer in the globe (Khanal et al., 2023). For many populations in developing nations, growing potatoes is crucial for their economic and nutritional well-being. Potatoes are usually peeled during processing, consequently producing large quantities of peels which are seen as waste to the industry. Potato peels contain polyphenols and glycoalkaloids, which may be recovered and used as natural antioxidants and steroid hormone precursors, respectively (Schieber et al., 2009). Therefore, one of the uses of potato peels (raw, steamed, or dried) is as a feed ingredient for animals. Numerous articles investigate the use of potato peel as a productive animal feed source. It needs to be processed before being used in animal feed since it contains too much fiber to be digested for feed intended for non-ruminants. Conversely, the same has shown effectiveness in several experiments when fed to animals with multi-gastric (Javed et al., 2019).

Potatoes have a high food production value per unit area and can thrive in a range of situations, making them a gratifying nutritional choice. As mentioned above, they are a fantastic source of protein, carbs, vitamins, minerals, and dietary fiber, and they are among the best providers of antioxidants (Wang et al., 2022). Their tubers are rich in niacin, vitamin B6, and vitamin C. The potato plant nevertheless has one of the largest input needs for fertilizer, the price of which has been rising over time. Fresh and chilled potatoes are the two main segments of the global potato market. Fresh potatoes are utilized for everyday dining, whereas cold potatoes are delivered or imported and used in the food processing industry. Even in countries where potatoes have not been traditionally consumed, potatoes have become more and more popular due to their lengthy shelf life. In the processed food industry, where processed foods account for most potato consumption, there is a global demand for fresh potatoes that is extraordinary (Fritsch et al., 2017).

Potato starch has generated a large amount of waste residue during processing. The processing of potatoes into starch results in waste items including effluents and leftover potatoes that have a negative environmental impact. For 100 days, a factory producing 10 kilotons of potato starch

DOI: 10.1201/9781003472209-31

annually releases 720 tons of effluents and 192 tons of residual potatoes daily. The research found that Chinese enterprises that produce potato starch pollute rivers significantly by leaking wastewater and potato leftovers into them. The fresh waste residue is not suitable for long-term storage or transportation because it is about 80% water. As such, it is often disposed of surroundings, where the organic materials might decompose and harbor a variety of bacteria, so contaminating the surrounding ecosystem and emitting an unpleasant stench. However, if this leftover potato waste was dried and made into dry feed, the cost would climb. But even if it's buried nearby or utilized as feed right away, it might contaminate the ground and subsurface water (Chauhan et al., 2023).

Most studies conclude that processing wastes range from 16% to 25% of the potatoes' starting weight: "359,071,403 metric tonnes of potatoes were expected to be produced worldwide in 2020." If 60% of all potatoes are processed, this leads to a waste stream that is expected to reach at least 44 million metric tons globally (Hong et al., 2020). Most of the wastes produced by the potato processing industry are inert organic materials that are easily destroyed by bacteria and hence harm the environment when dumped on land. When wastes with a high moisture content are dumped in public sanitary landfills, leaching issues might also arise (Jacob et al., 2016). Depending on the kind of potatoes used, several waste streams are produced as a result of potato processing. Potato may contain foreign materials such as fat, caustic, cleaning agents, dirt, and preservative chemicals. Usually, an effluent is created by combining multiple waste streams and then released from the potato processing plant. It is difficult to generalize these numbers since different operations generate varying amounts of wastewater using different processing strategies. There are significant differences in water use, peeling losses, and waste flow reporting methods, as evidenced by numerous studies and references in this area (Hung et al., 2006).

Therefore, poor handling of potato waste can seriously contaminate the environment. Potato waste breaks down anaerobically in landfills and produces methane, a powerful greenhouse gas if it is not treated. Additionally, by contaminating soil and water sources, the leachate from decaying waste can affect aquatic habitats and cause eutrophication (Kiran et al., 2015). Furthermore, the incorrect handling of organic waste leads to the release of unpleasant smells and the growth of disease-carrying organisms. To reduce these environmental concerns, pollution control engineering emphasizes the necessity of appropriate waste disposal methods, such as composting, anaerobic digestion (AD), and waste-to-energy systems.

23.2 POTATOES: PROPERTIES AND CHARACTERISTICS

Potatoes are one of the world's most widely cultivated root vegetables, known for their nutritional value, versatility in cooking, and utility in industrial applications. This tuber crop primarily consists of carbohydrates, fiber, protein, and essential vitamins like vitamin C, B-complex, and minerals such as potassium, making it a valuable food source globally. In addition to their nutritional significance, potatoes exhibit unique properties that have drawn attention in fields ranging from food science to material engineering, particularly in the development of bio-based materials. Crop productivity is fundamentally determined by the genes in plant genomes. However, specific enzymes and proteins often act as bottlenecks in essential processes like photosynthesis, photosynthate translocation, and carbon storage, thereby limiting overall yield potential (Figure 23.1) (Katoh et al., 2015).

Potato starch consists of two primary polysaccharides: amylose and amylopectin (Figure 23.2). Amylose is mostly linear with about 99% α-(1,4) linkages and few branches, while amylopectin is highly branched, containing 95% α-(1,4) and 5% α-(1,6) linkages that form complex clusters. On a fresh weight basis, potatoes have a starch content of 13.5–15%, which reaches 75–80% when dried. Typically, potato starch contains 20–33% amylose, with the rest being amylopectin. High-amylopectin "waxy" potato starch, produced traditionally and through genetic modification, exhibits low amylose (0–3%). Recently, waxy starch varieties have been developed naturally, and high-amylose varieties with up to 87% amylose have also been created. Properties like digestibility, gelatinization, and viscosity are influenced by the amylose-to-amylopectin ratio and branching characteristics. Textural differences among potato types, such as waxy versus mealy, stem from

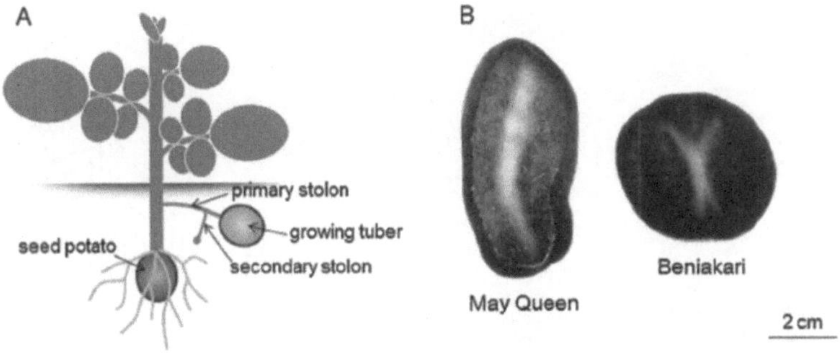

FIGURE 23.1 Morphology of whole potato plant (A) and iodine-stained longitudinal section of potato tubers (B). See text for details for (A). In (B), tubers were cut into 1-mm-thick slices, which were stained with diluted iodine solution (10 mM) for 1 min. Slices were washed in water to reveal differential staining of starch accumulation (Katoh et al., 2015).

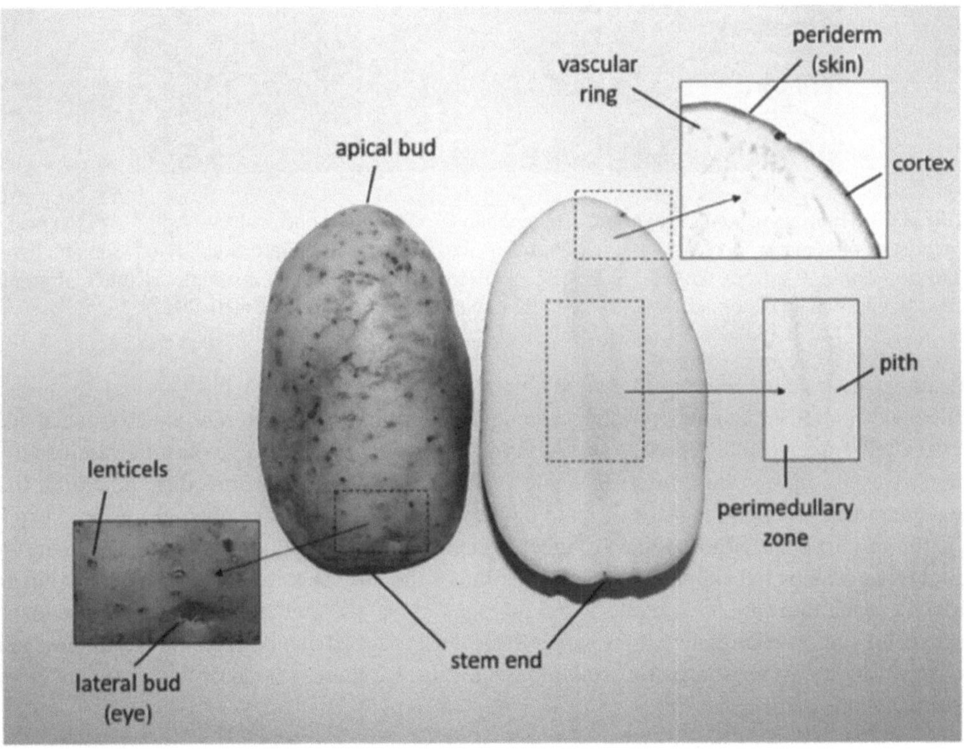

FIGURE 23.2 Structure of a potato tuber of the Bernina variety (Miller et al., 2022).

cellular structures, which affect cooking qualities. Potato starch granules range from 10 to 110 μm and have alternating crystalline and amorphous layers that yield unique patterns under polarized light microscopy. Genetic factors play a key role in starch composition, and amylose may form double helices with amylopectin in these granules (Dupuis & Liu, 2019).

Sedlák et al. (2022) discuss the use of somatic hybridization to bridge the sexual incompatibility between *Solanum tuberosum* (Stub), a primary cultivated potato species, and *Solanum bulbocastanum* (Sblb), a wild relative known for its robust resistance to late blight (Figure 23.3). Through electrofusion of mesophyll protoplasts, researchers generated 23 somatic hybrids to assess potential

FIGURE 23.3 Flowers of somatic hybrids (a, b) compared to Sblb66 (c) and Stub cv. "Apta" (d), intermediate characteristics of somatic hybrids visible in habitus of plants (e—from the left: StubDH165, somatic hybrid, Sblb66) and leaves (f—from the left: Sblb66, StubDH165, somatic hybrid), example of tuber of somatic hybrids (g), entire plant of somatic hybrid (h), entire plant of Sblb66 (i) (Sedlák et al. 2022).

applications in potato breeding. Detailed phenotypic, molecular, and biochemical evaluations revealed substantial somaclonal variation among hybrids, particularly in morphology, tuber yield, glycoalkaloid content, and genetic stability across 30 nuclear and 27 cytoplasmic microsatellite markers. Despite the hybrids' infertility, which limits their immediate breeding potential, these variants demonstrate high resistance to late blight and offer valuable insights into genetic expression patterns, particularly those linked to tuber dormancy and glycoalkaloid synthesis. The findings highlight that somatic hybridization not only facilitates the introgression of desirable traits but also broadens genetic diversity by generating unique phenotypic and genetic variations. Consequently, these hybrids can serve as essential "isogenic lines" for advanced comparative transcriptomic analyses, providing a novel resource for studying gene expression and regulatory mechanisms within critical metabolic pathways.

23.3 CURRENT ISSUES AND CHALLENGES

23.3.1 Volume and Types of Waste Produced

Potato production levels, processing techniques, and consumption habits are some of the variables that affect the amount of garbage produced from potato peels. After wheat, rice, and maize, potatoes are the fourth most eaten food crop in the world, with an annual production of over 368 million metric tons (FAO, 2020). Up to 20–30% of potatoes are wasted during the processing of potatoes into goods including chips, fries, starch, and dehydrated flakes. Peels make up a sizable amount of this loss (Pathak et al., 2018).

An estimated 5.8 million tons of potato peel waste (PPW) are produced annually in the European Union alone, mostly by the food processing industries (Galanakis, 2020). Comparably, the potato processing industries in the United States produce millions of tons of peel waste annually, especially in areas where French fry and chip production is done on a significant scale. Comparatively speaking, industrial potato peeling which uses steam treatment or mechanical abrasion to remove skins usually produces more waste than home or agricultural peeling (Ramasamy et al., 2020). Due to variations in consumption patterns and processing methods, the quantity of garbage might also differ between nations.

PPW is a heterogeneous material categorized according to its chemical, biological, and physical properties. Solid waste is the most obvious and abundant sort of waste generated after potato peeling. This includes the peels, mostly made up of cellulose, starch, lignin, and fibrous material. The amount of skin on a potato can vary from 5% to 15% of its weight, depending on the kind of potato and how it is peeled. After industrial processing, these peels are often thrown away and can build up significantly in landfills or other disposal locations, creating environmental issues (Zhang et al., 2017).

Potato peels are a rich source of organic substances, including proteins, carbohydrates, and polyphenols, that, if improperly managed, may contribute to environmental damage when it comes to organic waste. AD occurs when PPW is allowed to break down in open spaces or landfills, releasing carbon dioxide and methane into the environment. Strong greenhouse gas methane has the potential to cause 28 times more global warming than carbon dioxide over a 100-year timeframe (Kiran et al., 2015). Moreover, contaminated water bodies from potato peels' organic leachates can cause pollution and oxygen depletion in aquatic environments.

PPW contains compounds, like phenols and unsaturated fatty acids but also harmful glycoalkaloids such as α solanine and α solanidine that can be toxic to humans and animals as well as microorganisms. Nevertheless, PPW has started to receive recognition as a source of digestion (AD), an essential method for producing sustainable energy. At times anaerobic bacteria break down waste in environments without oxygen to create biogas containing 55–75% methane and 24–45% carbon dioxide along, with traces of other gases in small quantities. The quantity of waste plays a role, in the amount of biogas produced since its rich content of carbohydrates and starches greatly impacts the energy generated. Following the production of biogas, materials like waste products and leftovers rich in nutrients such, as nitrogen and phosphates can be repurposed as a fertilizer. This process contributes to enhancing energy efficiency and promoting sustainability (Awogbemi et al., 2022).

23.3.2 Environmental Impacts

The biomass of *S. tuberosum* contributes to the pollution caused by the synergies of nutrient practices, soil quality, and water resources. Potatoes require high levels of nitrogen, phosphorus, and potassium, which are supplied as fertilizers. Conversely, excessive application of fertilizer can lead to contamination of aquifers with nitrogen while unleached phosphorus ends up polluting surface water. Leaching of nutrients can lead to high levels of nitrate in dangerous drinking water. In addition, the cultivation of potatoes may result in runoff, which may cause water pollution in water bodies nearby leading to eutrophication, resulting in the depletion of oxygen and the death of fish. In addition, some of the negative effects can be mitigated by adopting better practices in nutrient management, such as precision application of nutrients and application at the right time. It also helps to reduce the loss of nutrients and soil when covering crops, and crop rotation with potatoes is employed. It's important to note that strategies to curb pollution from potato cultivation should seek to find a balance between the agronomic target of potato yield and the environmental impacts of potato cultivation.

The environmental ramifications of potato biomass and nitrogen (N) management practices are considerable due to nitrate leaching and nitrous oxide (N_2O) emissions. Potato cultivation, on the contrary, has increased N fertilizer use rates over the years, with yields having increased from 22 to 40 tons per hectare in the past 40 years. Nevertheless, enhanced N evidenced poor efficiency of N utilization because of the shallow-root nature of potatoes, which made it possible for nitrates to leach through the soil into the aquifer, more so for sandy soils. About 80% of nitrogen absorption occurs between 20 and 60 days after emergence: hence wrong application timing made the problem worse. It has also been noted that more N_2O emissions are produced in the production of potatoes than in other crop production. Improved nitrogen use efficiency, lessening nitrate leaching and N_2O emission, is achieved using controlled-release fertilizers and split-N applications. The effectiveness of N application relative to crop uptake must be maintained with the introduction of improved irrigation systems to minimize environmental impact (Munoz et al., 2005).

Potato biomass serves a beneficial purpose concerning pollution through the application of alterations on the activity of soil microbes, and, consequently, the levels of toxicity are associated with the use of pesticides, which further enhances pollution (Figure 23.4). In potato cultivation, herbicides, such as metribuzin and linuron, and a fungicide known as fluazinam, are frequently applied to manage weeds and control late blight. However, these chemicals have adverse effects on the soils; for instance, although fluazinam is used only in very small quantities, it is very crippling as it can inhibit the growth of bacteria luminescence for about 50% and it has a very long half-life in the soil, thus causing a reduction in enzyme activities that promote nutrient cycling like phosphomonoesters and cellobioses responsible for the degradation of organic materials. Appropriate use of these agrochemicals goes a long way in preserving the microbial populations and enhancing soil

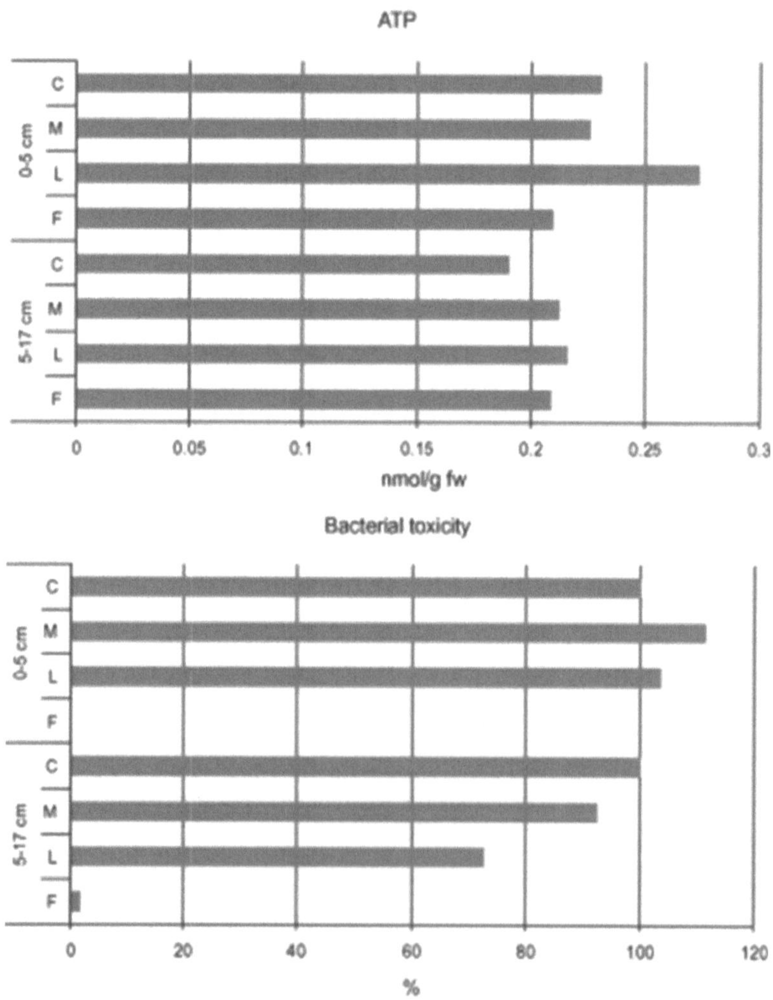

FIGURE 23.4 Soil ATP content and soil toxicity in the luminescent bacteria toxicity test in the mesocosm with separate treatments with different pesticides at two soil depths at harvesting potatoes (Niemi et al., 2009).

fertility; however, misuse leads to the situation of low microbial diversity. Enzymes like metribuzin may stimulate some enzyme levels after use, but other pesticides like fluazinam are toxic and do not change with repeated use. This inefficiency in the activities of soil enzymes and soil microbial toxicity leads to the deterioration of soil quality over time an issue that threatens the long-term sustainability of potato production (Niemi et al., 2009).

23.3.3 Difficulties in Waste Treatment or Disposal

23.3.3.1 High Moisture Content of Potato Waste Leads to Environment Concern

Following an extensive discussion on the primary challenges encountered in managing potato biomass, particularly the substantial waste volumes and associated environmental impacts, such as air and water pollution, the subsequent section provides an overview of how these disposal difficulties contribute to rising challenges in potato waste handling. It is evident that managing product-specific waste presents significant issues due to factors such as insufficient biological stability, pathogenic potential, high water content, susceptibility to rapid autoxidation, and elevated enzymatic activity (Russ & Schnappinger, 2007). Previous sections have established that improper disposal of potato biomass contributes to environmental degradation, primarily through the release of harmful substances, including methane, carbon dioxide, and organic leachates from potato peel, which collectively worsen both air and water pollution.

Various factors may contribute to the challenging disposal of potato waste, one of them is the high moisture content and high organic load of potato. Potato biomass, especially peels and residual potatoes, often contains a high amount of water. This makes it bulky and difficult to handle, transport, and store. Admittedly, potatoes have established themselves as a substantial source of various goods with value addition and as a great initial feedstock for many biorefinery processes (Khanal et al., 2024). Regarding this, the skin generated after peeling is not 100% a zero-value by-product that will be thrown away immediately as domestic waste. These two substances found in PPW including polyphenols and glycoalkaloids are both nutritionally and pharmacologically useful. These substances may be extracted and employed as natural antioxidants and steroid hormone precursors, respectively (Khanal et al., 2024).

Considering the potential products that could be derived from PPW, it would undoubtedly serve as a favorable starting material for the long-term growth of circular bio-economy. However, such wastes, due to the high moisture content (85–30%) and the rate of decomposition of the compounds, may lead to bacterial contamination of land and surface water (Rubanka et al., 2021). Because of its high moisture-holding capacity, drying pulp poses a significant challenge due to the substantial amount of energy it necessitates. Additionally, dried pulp is difficult to digest with the enzymes of domestic animals. The results of a study that examined the effects of moisture on greenhouse gases, including methane, nitrous oxide, and carbon dioxide, indicate that large-scale composting can reduce greenhouse gas emissions by adjusting the moisture content at different stages, thus lowering its overall environmental impact (Ermolaev et al., 2019).

In addition, potatoes are rich in steroidal alkaloids, a class of secondary metabolites commonly present in the plants of the Solanaceae family. Steroidal alkaloids play a crucial role in defense against bacterial, fungal, and insect attacks. These compounds are toxic and their negative effects on human health are typically manifested through symptoms such as colic pain, diarrhea, gastroenteritis, vomiting, fever, low blood pressure, rapid pulse, and neurological disorders (Fritsch et al., 2017). The extraction of this toxic compound from potato residues is necessary to enhance the valorization possibilities of potato peels as technical limitations like alkaloid content will prevent their use in more valuable applications (Fritsch et al., 2017).

The problem of recycling and the costs of this process cause a decrease in the efficient level of waste management that's why in most cases, existing logistics systems used in the food industry are not able to cope with the obstacles that arise on the way of their disposal (Ravindran & Jaiswal, 2016). In conclusion, waste disposal poses challenges due to the high moisture content in its peel thus hindering the efficiency of conventional disposal methods like incineration or composting without prior dehydration or treatment, as well as the toxic alkaloids content in the peels must be removed before disposal.

23.3.3.2 Energy-Intensive Treatment Processes and Poor Economic Feasibility

Beyond the challenges previously outlined, it is essential to consider that advanced waste treatment methods, such as AD for biogas production and other bioenergy processes, demand substantial investment in infrastructure and energy inputs. Economic feasibility remains a fundamental factor in the pursuit of sustainable waste disposal, as both policy implementation and technological advancements rely heavily on adequate financial support. Traditional FW disposal methods, including landfill, incineration, and aerobic composting, have faced limitations; for example, landfilling FW is now largely restricted in many countries, and incineration proves energy-intensive due to high moisture content, often resulting in air pollution (Zhang et al., 2014). Utilizing FW as animal feed also presents risks, as direct use in shorter food chains can increase the propagation of diseases (Zhang et al., 2014).

Some of the more environmentally friendly approaches are being discovered to replace these legally right, yet ethically wrong disposal methods, especially in those third-world countries that are poor and suffering from high rates of poverty. Their central government didn't allocate enough funds for the local governments to carry out the advanced waste management operations. Nonetheless, due to the increasing public awareness of agro-biomass waste management globally, more and more countries are starting to practice alternative methods that are more sustainable to tackle the waste crisis. Potato biomass can be treated through various processes like AD, bioenergy production, or composting. These methods can convert waste into useful products such as biogas, bioethanol, or organic fertilizers. However, these processes often require significant amounts of energy and investment in infrastructure, making them energy-intensive and sometimes costly.

According to the definition, AD is a process through which bacteria break down organic matter such as animal manure, wastewater biosolids, and FWs in the absence of oxygen (US EPA, 2024). It has been proposed as a relatively cost-effective technology for renewable energy production and waste treatment of this high-moisture and energy-rich material (Romero-Güiza et al., 2016; Xu et al., 2018). AD of FW is promising but has not yet been widely applied. This is because although AD is already a relatively mature technology in the treatment of high-strength wastewater, sewage sludge, and animal manure, adopting AD for FW management like potato waste still faces several challenges, such as volatile fatty acid accumulation and process instability, foaming, low buffer capacity, and high cost of transportation and operation (Xu et al., 2018). Regarding this, issues are generally classified as economic, technical, social, or regulatory issues. Some of the critical challenges to the economics of digester operations are the slow rate of biogas generation, the low energy content of the biogas, and the costs of upgrading the biogas (Shen et al., 2015).

Moving to the technical problem, one important technical challenge is that when lacking concise process control and optimization, harmful intermediate compounds can be easily produced, reducing system stability, or causing low methane yield or foaming (Grimberg et al., 2015). To elaborate in detail, AD requires constant temperature control, particularly in cold climates, to maintain optimal conditions for microbial activity. Temperature is indeed one of the most significant parameters influencing AD because it not only influences the activity of enzymes and co-enzymes but also influences the methane yield and digestate (effluent) quality (Zhang et al., 2014). Heating and maintaining the anaerobic digester's temperature consumes energy, making the process resource-intensive. Hence, not only does the installation of anaerobic digesters involve high upfront capital expenses, but maintenance and operational energy needs also add to the overall costs.

Potato waste generation is seasonal, depending on the harvest and processing periods. There are two main peaks in global potato distribution by latitude, the major peak is between 45°N and 57°N, where potato is a summer crop while the other is between 23°N and 34°N, representing production zones in the subtropical lowlands, where potato is a winter crop (Hijmans, 2001). This creates challenges for consistent and continuous waste management, with periods of high waste generation followed by off-seasons where waste treatment facilities may be underutilized. For small or medium-sized potato processors, the investment in such technology may not always be economically feasible. Additionally, achieving optimal efficiency in energy recovery depends on various factors, such as the consistency and quantity of waste, as well as regional energy costs and policies supporting renewable energy. For instance, potato waste from processing facilities or farms in colder climates will require even more energy for AD to maintain temperatures conducive to microbial activity. In conclusion, the economic return might not be sufficient to justify the initial and ongoing energy expenditures, which can dissuade producers from adopting energy-intensive solutions.

23.3.4 Costs

The preceding discussion explored the functionality of composting facilities utilizing advanced agro-biomass disposal technologies, such as AD, highlighting the significant management costs that pose a primary challenge in waste management. This section now examines additional cost factors, including hidden expenditures associated with transportation, labor, and supplementary expenses, which further complicate effective waste management. These accumulated costs, arising at various stages of the waste-handling process, considerably impact the overall financial requirements of managing potato biomass, underscoring the need for economically viable waste disposal solutions.

Firstly, transporting potato biomass in the form of potato peels, spoiled potatoes, or processing residues, to treatment facilities or composting sites can be a significant financial challenge. This is because potato biomass is high in volume and water content. This makes transportation less efficient because trucks quickly reach their weight limits due to moisture, leading to higher fuel and transport costs. Most of the potato farms or processing plants are located in rural areas, far from waste treatment facilities like composting sites or AD plants. The further the distance, the higher the transportation costs, especially when those areas are limited in basic infrastructure and suffer from poor road conditions. Transporting large amounts of potato biomass also has environmental costs. The increased fuel consumption will lead to more greenhouse gas emissions, resulting in further pollution thus reducing the sustainability of this waste management operation.

Secondly, managing potato biomass requires skilled labor for handling, sorting, and processing waste material. This apparently will increase the overall expense of waste management. For example, waste treatment methods such as AD, require skilled and specialized labor with the knowledge of advanced technology to operate and maintain the equipment, which drives up the labor costs. Seasonal fluctuations in potato production lead to substantial waste generation during peak harvest and processing periods, necessitating temporary labor that further elevates operational costs. This seasonal surge in labor requirements contributes to heightened expenses, rendering large-scale, sustainable disposal practices financially challenging, particularly for smaller producers. Consequently, the economic burden of waste management during these high-output periods underscores the need for more cost-effective and scalable disposal solutions within the sector.

Thirdly, potato biomass often requires further processing before it can be safely disposed of, adding layers of cost to waste treatment. To begin with, there are several mandatory pretreatment requirements such as drying the waste to reduce moisture content. Composting may seem like an economical and environmentally friendly option, but it also has hidden costs such as the cost to build up large-scale infrastructures such as facilities to carry out aeration to accomplish polymerization and humification of compost (Zhang et al., 2021). Not to mention many regions have strict regulations regarding waste disposal, especially for household hazardous waste that can contaminate water sources, including Malaysia (Zainu, 2021). Potato processors might need to invest in conversion technologies such as combustion or gasification (Zainu, 2021), to comply with environmental laws. These compliance costs further escalate the overall financial burden. Besides, some waste treatment methods such as AD, will generate by-products that need further treatment or disposal. For example, digestate from AD may require further treatment before it can be used as fertilizer, which can add to the cost of the process (Tampio et al., 2015).

To sum up, everything that has been stated so far, these costs together create substantial barriers to effective waste management for potato biomass. Without proper economic incentives, and subsidies by the government, potato waste disposal and treatment will remain costly and complex, leading many producers to default to less sustainable methods like landfilling or open dumping, which have negative environmental impacts that need to be avoided.

23.4 INNOVATIVE SOLUTIONS AND POTENTIAL APPLICATIONS

23.4.1 Bioenergy

The substantial annual production of potato biomass across various regions has driven research into innovative, high-value applications for transforming this waste. Utilizing this carbon-rich resource has the potential to prevent resource wastage, reduce reliance on raw materials, minimize landfill contributions, and yield environmental and societal benefits. The global biogas market, valued at USD 6.9 billion in 2021, is projected to reach USD 60 billion, underscoring an increasing focus on renewable feedstocks like potato biomass for biogas production (Market Analysis Report, 2021). Despite limited data on the exact contribution of potato biomass to global biogas output, its growing use as a feedstock highlights its potential. Post-biogas production, the nutrient-rich residue, containing essential elements like nitrogen, phosphate, and potash, can be effectively repurposed as an organic fertilizer, offering a sustainable and valuable resource for agricultural applications (Kreidenweis et al., 2021).

The primary feedstock, potato biomass, is first completely cleaned under running water to get rid of any debris stuck to the peel's body. After draining and sun-drying, the potato biomass is dried once more for around 60 min in an oven set to a temperature between 80°C and 100°C. To enhance the surface area for reaction and boost product output, the dried potato biomass is ground into a powder (Awogbemi et al., 2022). In a previous study, Suntikunaporn examined the use of potato biomass in a 20 L floating drum digester to produce biogas (Suntikunaporn, 2014). They credited the high glucose content of potato biomass for the remarkable 3,500 mL yield each day. Looking ahead, it is undeniable that potato biomass is a suitable substrate for producing biogas due to its chemical composition, oxidation state, and pH; however, different pretreatment and co-digestion methods are required to increase the concentration of CH_4 and increase the yield of biogas (Bożym et al., 2015). More research should be done on the statistical parametric optimization studies of process parameters such as reactor type, operating temperature, pH, mixing ratio, and pretreatment techniques. More research must be done on the use of modeling tools in methane generation to improve cumulative biogas synthesis.

Over the previous ten years, there has been a steady increase in the annual global output of ethanol. Global ethanol production increased from 13,123 million gallons (MG) in 2007 to 21,812, 26,583, and 29,026 MG in 2012, 2016, and 2019, respectively, based on the data that is currently available. The lower production volume of 26,059 MG in 2020 can be ascribed to the limitations and lockdowns brought on by the COVID-19 outbreak. The global ethanol market size, measured in U.S. dollars, increased from USD 89.1 billion in 2019 to USD 93.7 billion in 2020, with a projected 5.2% annual growth rate from 2021 to 2030 (Ethanol Market Size, 2021). This represents a significant increase in market share. The rising use of ethanol as fuel is driving the predicted growth. At least 73% of the world's ethanol production is thought to be used as fuel, with the remainder going toward drinks and industrial ethanol (Arapoglou, 2010).

Potato biomass was used in a chemical investigation to produce bioethanol using termamyl and amyloglucosidase, alpha-amylase and cellulase, and *Clostridium acetobutylicum* MTCC 11274, respectively. The results of their research supported the viability of turning PPW into bioethanol. To increase the conversion of potato biomass to acetone butanol and ethanol by fermentation using *C. acetobutylicum* MTCC 11274, Kamboj and Ms added powdered orange peel (*Citrus sinensis*) extract (Kamboj, 2021). According to reports, adding orange peel extract boosted the amount of butanol produced. Potato biomass was successfully employed in both cases as an inexpensive, environmentally friendly feedstock with a gentle production technique.

For a variety of uses, potato biomass has been successfully pyrolyzed into syngas, biochar, and bio-oil. For instance, Liang reported a yield of 30.5% biochar, 22.7% bio-oil, and 46% syngas from pyrolyzed potato biomass, whereas Daimary pyrolyzed potato biomass at 500°C for 30 min to obtain biochar (29.50%) and bio-oil (23.60%) for catalytic biodiesel production (Liang, 2015). The scientists stated that biofuel was produced from potato biomass without the need for a catalyst or enzyme. Based on the collective views of many researchers, potato biomass is an efficient, affordable, and easily accessible feedstock for the manufacture of bioethanol.

23.4.2 Composting

Organic debris that has broken down and been repurposed as fertilizer for soil amendments is called compost. It is an essential component of organic farming. Good soil structure is provided by the aggregates of soil particles known as aggregates, which are bound together by compost. Such soil is full of microscopic holes and air channels that retain moisture, air, and compost, which in turn brings and nourishes a variety of soil life (bacteria, fungi, insects, worms, and more) and further supports the growth of healthy plants. Maintaining healthy soil is essential to safeguarding our waterways. Compost improves the soil's water-retention capacity and lowers runoff. By bringing dirt, fertilizers, and pesticide residues to neighboring streams, runoff contaminates water (Camire et al., 1993). In Pakistan, compost is marketed commercially to replace fertilizers. One important component of the lignocellulose waste used in composting is potato peels. Numerous research has revealed that potato peels can be used to make compost.

An enormous amount of organic waste is generated annually as a by-product of the food and agro-industrial sectors. The environment's preservation and the value-adding of these by-products depend on the management of these wastes. The agro-food sectors in Castilla y León, Spain, produce a lot of wastewater and solid waste, both of which have a negative impact on the environment. Castilla y León is the leading producer and consumer of vegetables in Spain, with high levels of potato and cereal consumption (8.5 million tons of grain produced annually) and cultivation (900,000 tons). Additionally, the region is home to a sizable alimentary industry centered on these crops (Gullón and Siro Co.), which generates a sizable amount of residues like straw and potato peelings. The United Nations has proclaimed 2008 to be the International Year of the Potato to increase awareness of the crucial role that potatoes play in ensuring food security and reducing poverty. Experts estimate that, in developing nations, its use will quadruple over the next two decades. The current yearly production of potatoes is close to 300 million tons.

Temperature has shown to be the primary control and monitoring parameter for certain wastes, as will be discussed below. It has been seen nine times a day at the following locations: (1) four locations 10–15 cm above the reactor base; (2) four locations at a comparable distance from the upper surface; and (3) one location in the center of the material being composted. Temperature was recorded outside simultaneously for the following reasons: (1) a rise in temperature signals the start of the process; (2) a drop in temperature, once the process has begun, indicates a lack of oxygen or moisture, necessitating rolling and/or irrigation; and (3) the active stage of composition ends when, following cautious turnovers and irrigations, the temperature inside the reactor does not exceed the outer one.

23.4.3 Bioplastic Production

Because of their durability, affordability, and lightweight, plastics are more beneficial than metals, papers, and other materials. As a result, they are employed in practically all industrial sectors. In 2015, around 300 million tons of plastic were used globally. Plastic garbage is found throughout the world, including in the oceans. Furthermore, there are several drawbacks to the plastics business that are associated with both economic and environmental issues (Zarate-Ramirez, 2014).

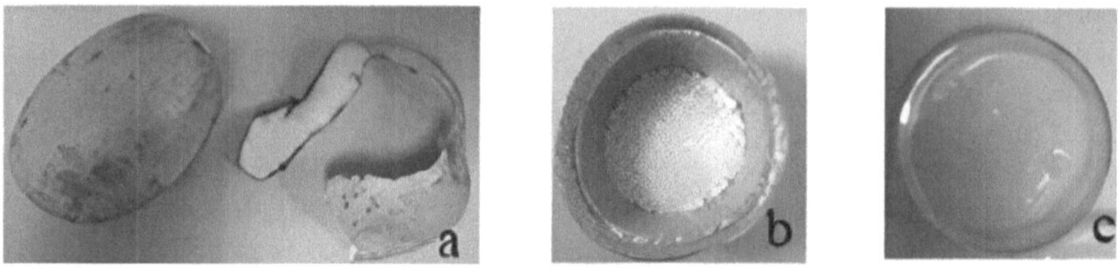

FIGURE 23.5 The process from potato peel to bioplastics product called PPB. (a) Raw potato peel used as the feedstock for bioplastic production. (b) Processed potato peel pulp after treatment, forming the base material for PPB. (c) Finished potato peel bioplastic (PPB) product after solution casting and drying.

Investigating the use of food industry wastes for the creation of bioplastics was the overarching goal. The process of turning leftover potato peels into bioplastics was looked into to accomplish this goal. Furthermore, certain characteristics of the generated bioplastics, such as their ability to absorb water and their biodegradability, were examined.

The potatoes were washed and peeled in preparation for more research. After pulverizing potato peels, they were centrifuged for 20 min at 15,000 rpm. After filtering the supernatant, starch was extracted. About 330 g of wet potato peels were converted into 13.5 g of dried starch. Following filtration, the starch was dried for 2 h at 50°C and stored in an airtight zip-locked container until processing (Arıkan & Bilgen 2019). Following the starch's extraction, 13.5 g of it was weighed and mixed with 135 mL of tap water, 16.2 mL of vinegar, and 10.8 mL of glycerin. A hot plate was used to heat this combination to 100°C, and it was left there for 20 min. The mixture was allowed to dry by air for roughly 48 h, and the bioplastics were produced in a sheet form.

In various controlled settings, the biodegradability of potato peel bioplastic (PPB) and Commercial Bioplastic (CB) samples was examined (Figure 23.5) (Ashok, 2018). Two separate PPB and CB samples were weighed and then buried in Petri dishes with 50 g of moist soil and 50 g of vermicompost, respectively. Weekly weigh-ins were conducted on bioplastic samples that had known beginning masses before being buried. Every experiment was run via three replications. The European Committee for Standardization defines biodegradation as a deterioration that results in a substantial alteration in a material's chemical structure and is brought on by biological activity, particularly by enzyme action. Additionally, a common technique for measuring the biodegradation of polymers is weight loss. Biodegradability studies, according to International Advanced Research, revealed that the PPB biodegraded at a rate of roughly 71% in moist soil and 100% in vermicompost in just 4 weeks (Figure 23.6).

23.5 CHALLENGES AND FUTURE PERSPECTIVE

23.5.1 Policy Changes

Governments need to implement laws and guidelines that promote the nutritional value of leftover potato peels. This involves creating policies and guidelines that support the production of bioenergy, bio-based goods, and organic fertilizers using potato peels. The Circular Economy Action Plan of the European Union, for instance, places a strong emphasis on waste-to-resource techniques that might be modified especially for agricultural wastes like potato peels (European Commission, 2020). Adopting such guidelines on a worldwide scale can encourage businesses to invest in technology that turns leftover potato peels into profitable goods. Governments can offer tax reductions and financial aid to businesses that use composting and AD to handle PPW. These kinds of rewards could make the initial investment less daunting and motivate businesses to incorporate these technologies into their daily operations. As an example, the UK's Renewable Heat Incentive program has been effective in encouraging the use of biogas plants to turn organic waste into electricity

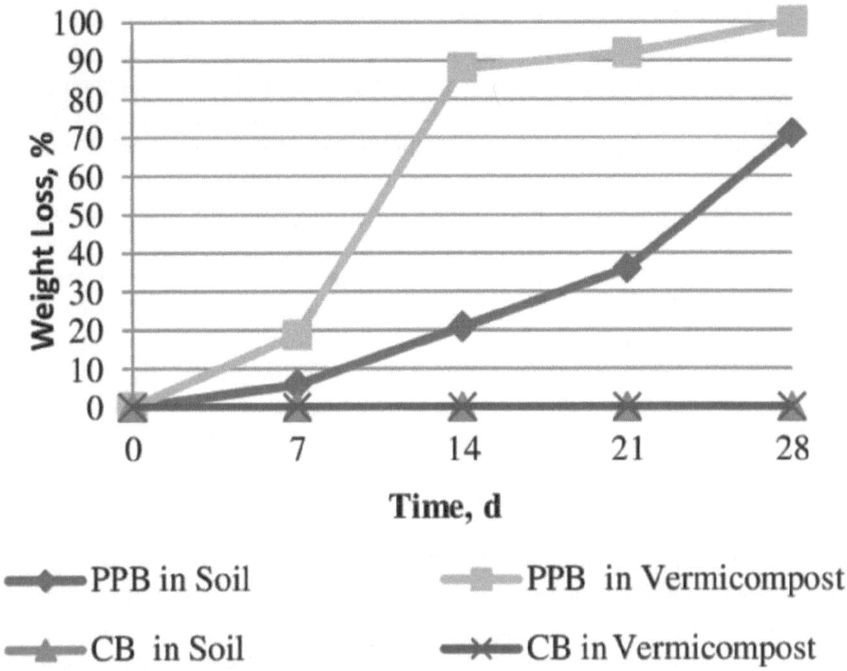

FIGURE 23.6 The biodegradability analysis of PBB.

(DEFRA, 2019). Methane emissions from inappropriate waste disposal can be decreased, and biogas generation can be scaled up by modifying similar models to target PPW.

At the same time, the effectiveness of managing PPW can be greatly increased by requiring waste separation procedures for potato processing facilities. To facilitate the utilization of organic waste in composting units or biorefineries, this policy would require big agricultural enterprises and food processing industries to separate organic waste from non-organic materials at the point of production. Cities such as San Francisco have established comparable initiatives requiring FW to be separated and directed toward composting facilities (CalRecycle, 2021). Such regulations can be modified for agricultural areas to expedite the gathering and recycling of leftover potato peels. Governments need to encourage the agriculture and food processing sectors to adopt circular economy concepts. This entails turning waste back into resources to close the loop. Policies that give priority to the generation of biofuel, bioplastics, or soil amendments using organic waste can assist industries in aligning with sustainable practices. Achieving long-term environmental goals, including those stated in the UN Sustainable Development Goals (SDGs), especially SDG 12 (Responsible Consumption and Production), depends on the implementation of circular economy practices (UN, 2020).

23.5.2 Technological Advancements

Potato peels and other organic waste can be effectively converted into digestate rich in nutrients and biogas through the process of AD. To increase the productivity and efficiency of biogas from PPW, though, innovations are required. Methane yield during digestion can be increased by investigating the best ways to pretreat potato peels, such as via enzymatic hydrolysis or mechanical grinding (Kiran et al., 2015). Additionally, integrating co-digestion methods in which potato peels are mixed with other organic waste streams can balance the carbon-to-nitrogen ratio and improve the efficiency of biogas generation (Awasthi et al., 2019).

PPW can be effectively managed using biorefineries, which transform it into a variety of products with additional value. Potato peels can be processed using advanced biorefineries to extract proteins, starch, and phenolic chemicals that are utilized in food, medicine, and cosmetic products (Pathak et al., 2018). The economic viability of waste management activities can be increased by industries processing potato peels into various high-value goods in an efficient manner through investment in research and development for biorefinery technology. Governments and academic institutions should work together to develop pilot programs that show these kinds of biorefinery models are feasible.

The most important way to recycle PPW is still through composting, yet conventional composting techniques can be laborious and sluggish. Microbial inoculants and in-vessel composting may accelerate the decomposition process and enhance the caliber of the final compost. Composting in-vessel entails breaking down organic waste in a confined, regulated space, like a tunnel, silo, or drum. By precisely controlling variables like temperature, moisture content, and air circulation, this approach greatly speeds up the composting process. In-vessel composting can decompose PPW faster than conventional methods by upholding ideal circumstances; this can frequently result in a composting time reduction from months to a few weeks. This system is appropriate for usage in urban or highly populated regions where odor control is crucial since it reduces odor emissions, a major problem with open-air composting. Microbial inoculants are composed of certain strains of actinomycetes, fungi, or bacteria that are well-known for their capacity to decompose organic materials. These inoculants increase microbial activity in compost piles, accelerating the breakdown of complex organic substances like cellulose and starch that are present in potato peels (Singh et al., 2020). This accelerates the composting process in general and produces a compost product that is higher in nutrients that are good for the health of the soil.

Additionally, adding aeration systems to composting installations can improve microbial activity and lower odor emissions, which will improve the breakdown of organic matter. Composting may become a more appealing alternative for handling PPW at the industrial and community levels as a result of these technological advancements. Aeration systems, such as perforated pipes or blowers, help keep the compost pile's oxygen levels stable, which encourages aerobic microbial activity, a critical component of effective decomposition. Anaerobic conditions, which can cause unpleasant odors and lower the quality of compost, are avoided by proper aeration. Research has demonstrated that enhanced aeration during the composting process can raise the pile's temperature to levels that encourage thermophilic bacterial activity, which is necessary for the breakdown of more resilient organic components (Pathak et al., 2018). By lowering the possibility of pathogens in the finished compost product and ensuring a more uniform decomposition process, this method makes compost safer for use in agriculture.

Another creative method is to use pyrolysis to turn potato peels into charcoal. According to Kumar et al. (2019), biochar is a carbon-rich substance that can be applied to improve soil moisture retention, soil fertility, and carbon sequestration. By using leftover potato peels to make biochar, organic waste can be reduced sustainably and a useful product for agriculture can be produced. This practice can be expanded with investments in pyrolysis technology and research into the best conditions for producing biochar from potato peels.

23.5.3 Industry Partnerships

Collecting and processing PPW can be centralized with the support of agricultural cooperatives. Smaller farmers can now utilize waste processing facilities that would be too expensive for them to run on their own, thanks to this strategy. To provide a consistent supply of raw materials for these operations, agricultural cooperatives can collect PPW from multiple farms by partnering with composting or biogas plants (Ramasamy et al., 2020). Through this partnership, waste management systems may be more scalable, and the logistical expenses of waste transportation can be decreased. Creating partnerships between biorefineries and potato processing industries can help maximize the utilization of PPW. Potato peels, for instance, can be sent straight from processors to biorefineries, where they can be processed into valuable goods like bioethanol and bioplastics. The relationship between agricultural processors and biofuel facilities in countries such as Brazil, where agricultural waste is transformed into bioethanol, is a successful example of this type of cooperation (Goldemberg, 2018). To optimize the financial gains from waste conversion, such collaborations should be promoted in areas where potatoes are produced in large quantities.

Universities and research centers are essential for developing the science underlying waste management technology. Working with these organizations can assist the potato business in creating innovative uses for leftover potato peels, such as advanced biodegradable packaging materials or fresh techniques for removing bioactive ingredients. Research supported by businesses may promote the creation of solutions that are especially suited to local waste management requirements and difficulties (Galanakis, 2020). Patents, knowledge transfers, and the founding of spin-off businesses with a waste valorization focus are all possible outcomes of these kinds of partnerships. Last but not least, public–private partnerships, or PPPs, can make it easier to build composting facilities, waste-to-energy facilities, and biorefinery plants. Governments and commercial businesses can work together to split the expenses and gains of building waste management infrastructure. These kinds of partnerships can be especially beneficial in emerging areas with limited infrastructure. PPPs can facilitate the establishment of modern facilities for handling PPW by combining resources and skills, which will help to create jobs and boost the local economy in rural regions (Pathak et al., 2018).

23.6 CONCLUSION

This chapter aims to examine the challenges associated with developing sustainable disposal methods for potato waste, a by-product of one of the world's most widely consumed agro-food products. Through a comprehensive analysis, the chapter seeks to address the current obstacles in potato waste management, beginning with an exploration of the environmental degradation caused by the volume and nature of potato biomass. Following this, the discussion explores the significant environmental impacts, illustrating how organic potato waste contributes to pollution and intensifies the complexities of waste treatment. An emphasis on treatment challenges reveals the multifaceted nature of the problem, including the financial burdens of various disposal methods. In light of these challenges, the chapter introduces innovative approaches and potential applications to enhance waste management strategies, including bioenergy production, composting, and bioplastic development. Each method is evaluated for its feasibility and potential benefits, with a focus on aligning environmental sustainability with economic practicality. Furthermore, this chapter presents strategies for optimizing agricultural biomass waste management systems in both international contexts and Malaysia. These strategies include policy adjustments, technological innovations, and collaborative efforts within industry, aimed at promoting the reuse of potato biomass as a valuable resource. The suggestions encourage cooperation among stakeholders to bolster the practical implementation of these solutions in real-world settings. By highlighting these innovative strategies, this chapter underscores the potential of potato waste as a resource and contributes valuable insights toward advancing sustainable and resilient waste management practices for the future.

ACKNOWLEDGMENT

The chapter was written with Cheah Pin Chian, Manoaharan A.L. Saravanan, So Jia Xin, and Mohamad Syarif Luqman Ibrahim.

REFERENCES

Arapoglou, D., Varzakas, T., Vlyssides, A., & Israilides, C. (2010). Ethanol production from potato peel waste (PPW). *Waste Management, 30*(10), 1898–1902. https://doi.org/10.1016/j.wasman.2010.04.017

Arıkan, B., & Bilgen, G. (2019). Production of bioplastic from potato peel waste and investigation of its biodegradability. *International Advanced Researches and Engineering Journal, 3*(2), 93–97. Retrieved from https://dergipark.org.tr/tr/download/article-file/767735

Ashok, R. B., Srinivasa, C. V., & Basavaraj, B. (2018). A review on the mechanical properties of areca fiber reinforced composites. *Science and Technology of Materials, 30*(1), 1–11. https://doi.org/10.1016/j.stmat.2018.05.004

Awasthi, M. K., Sarsaiya, S., & Awasthi, S. K. (2019). Anaerobic digestion of food waste: Benefits and challenges. *Waste Management, 91*, 29–41.

Awogbemi, O., Kallon, D. V. V., & Owoputi, A. O. (2022). Biofuel generation from potato peel waste: current state and prospects. *Recycling, 7*(2), 23. https://doi.org/10.3390/recycling7020023

Bożym, M., Florczak, I., Zdanowska, P., Wojdalski, J., & Klimkiewicz, M. (2015). An analysis of metal concentrations in food wastes for biogas production. *Renewable Energy, 77*, 467–472.

CalRecycle. (2021). San Francisco's mandatory recycling and composting ordinance. https://www.calrecycle.ca.gov (Accessed on December 1, 2024)

Camire, M. E., Zhao, J., & Violette, D. A. (1993). In vitro bind ing of bile acids by extruded potato peels. *Journal of Ag ricultural and Food Chemistry, 41*(12), 2391–2394.

Chauhan, A., Islam, F., Imran, A., Ikram, A., Zahoor, T., Khurshid, S., & Shah, M. A. (2023). A review on waste valorization, biotechnological utilization, and management of potatoes. *Food Science & Nutrition, 11*(10), 5773–5785. https://doi.org/10.1002/fsn3.3546

DEFRA. (2019). *Renewable Heat Incentive: Evaluation and Outcomes*. UK Department for Environment, Food & Rural Affairs.

Dupuis, J. H., & Liu, Q. (2019). Potato starch: A review of physicochemical, functional and nutritional properties. *American Journal of Potato Research, 96*(2), 127–138. https://doi.org/10.1007/s12230-018-09696-2

Ermolaev, E., Sundberg, C., Pell, M., Smårs, S., & Jönsson, H. (2019). Effects of moisture on emissions of methane, nitrous oxide, and carbon dioxide from food and garden waste composting. *Journal of Cleaner Production, 240*, 118165. https://doi.org/10.1016/j.jclepro.2019.118165

European Commission. (2020). Circular economy action plan. https://ec.europa.eu/environment/circular-economy/

Food and Agriculture Organization (FAO). (2020). FAO statistical database. https://www.fao.org/faostat

Fritsch, C., Staebler, A., Happel, A., Cubero Márquez, M. A., Aguiló-Aguayo, I., Abadias, M., Gallur, M., Cigognini, I. M., Montanari, A., López, M. J., Suárez-Estrella, F., Brunton, N., Luengo, E., Sisti, L., Ferri, M., & Belotti, G. (2017). Processing, valorization, and application of bio-waste derived compounds from potato, tomato, olive, and cereals: A review. *Sustainability, 9*(8), 1492. https://doi.org/10.3390/su9081492

Galanakis, C. M. (2020). *Food Waste Recovery: Processing Technologies and Industrial Techniques*. Academic Press.

Goldemberg, J. (2018). Ethanol for a sustainable energy future. *Science, 315*(5813), 808–810.

GlobeNewswire. (2021, January 18). *Ethanol market size worth around USD 155.6 billion by 2030.* https://www.globenewswire.com/news-release/2021/01/18/2160198/0/en/Ethanol-Market-Size-Worth-Around-USD-155-6-Billion-by-2030.html (Accessed on September 9, 2021)

Grimberg, S. J., Hilderbrandt, D., Kinnunen, M., & Rogers, S. (2015). Anaerobic digestion of food waste through the operation of a mesophilic two-phase pilot scale digester – Assessment of variable loadings on system performance. *Bioresource Technology, 178*, 226–229. ISSN 0960–8524. https://doi.org/10.1016/j.biortech.2014.09.001

Hijmans, R. J. (2001). Global distribution of the potato crop. *American Journal of Potato Research, 78*, 403–412. https://doi.org/10.1007/BF02896371

Hong, J., Mu, T., Sun, H., Richel, A., & Blecker, C. (2020). Valorization of the green waste parts from sweet potato (Impoea batatas L.): Nutritional, phytochemical composition, and bioactivity evaluation. *Food Science & Nutrition, 8*(8), 4086–4097.

Hung, Y.-T., Awad, A., & Salman, H. (2006). Potato wastewater treatment. In: Lo, H. H. (Ed.), *Potato Wastewater Treatment* (pp. 193–254). Taylor & Francis Group, LLC.

Jacob, S., Chintagunta, A. D., & Banerjee, R. (2016). Selective digestion of industrial potato wastes for efficient bio methanation: A sustainable solution for safe environmental disposal. *International Journal of Environmental Science and Technology, 13*(10), 2363–2374. https://doi.org/10.1007/s13762-016-1051-y

Javed, A., Ahmad, A., Tahir, A., Shabbir, U., Nouman, M., & Hameed, A. (2019). Potato peel waste- its nutraceutical, industrial and biotechnological applications. *AIMS Agriculture and Food, 4*(3), 807–823. https://doi.org/10.3934/agrfood.2019.3.807

Kamboj, R., & Ms, D. (2021). Orange peel extract enhanced sugar recovery and butanol production from potato peel by Clostridium acetobutylicum. International Journal of Green Energy, 18(2), 1–11. https://doi.org/10.1080/15435075.2021.1890083

Katoh, A., Ashida, H., Kasajima, I., Shigeoka, S., & Yokota, A. (2015). Potato yield enhancement through intensification of sink and source performances. *Breeding Science*, *65*(1), 77–84. https://doi.org/10.1270/jsbbs.65.77

Khanal, S., Karimi, K., Majumdar, S., Kumar, V., Verma, R., Bhatia, S. K., Kuca, K., Esteban, J., & Kumar, D. (2023). Sustainable utilization and valorization of potato waste: State of the art, challenges, and perspectives. *Biomass Conversion and Biorefinery, 14*, 23335–23360. https://doi.org/10.1007/s13399-023-04521-1

Kiran, E. U., Trzcinski, A. P., & Ng, W. J. (2015). Bioconversion of food waste to energy: A review. *Current Opinion in Biotechnology, 33*, 280–285.

Kreidenweis, U., Breier, J., Herrmann, C., Libra, J., & Prochnow, A. (2021). Greenhouse gas emissions from broiler manure treatment options are lowest in well-managed biogas production. *Journal of Cleaner Production, 280*, 124969.

Kumar, R., Sarsaiya, S., & Awasthi, S. K. (2019). *Management of Wastewater in the Food Processing Industry*. Springer.

Liang, S., Han, Y., Wei, L., & McDonald, A. G. (2015). Production and characterization of bio-oil and bio-char from pyrolysis of potato peel wastes. *Biomass Conversion and Biorefinery, 5*(3), 237–246. https://doi.org/10.1007/s13399-014-0130-x

Market Analysis Report. https://www.grandviewresearch.com/industry-analysis/biogas-market/toc (accessed on 31 August 2021).

Miller, K., Reichert, C. L., Schmid, M., & Loeffler, M. (2022). Physical, chemical and biochemical modification approaches of potato (peel) constituents for bio-based food packaging concepts: A review. *Foods, 11*(18), 2927. https://doi.org/10.3390/foods11182927

Munoz, F., Mylavarapu, R. S., & Hutchinson, C. M. (2005). Environmentally responsible potato production systems: A review. *Journal of plant nutrition, 28*(8), 1287–1309. https://doi.org/10.1081/PLN-200067434

Niemi, R. M., Heiskanen, I., Ahtiainen, J. H., Rahkonen, A., Mäntykoski, K., Welling, L., Laitinen, P., & Ruuttunen, P. (2009). Microbial toxicity and impacts on soil enzyme activities of pesticides used in potato cultivation. *Applied Soil Ecology, 41*(3), 293–304. https://doi.org/10.1016/j.apsoil.2008.12.002

Pathak, P. D., Mandavgane, S. A., & Kulkarni, B. D. (2018). Valorization of potato peel: A review. *Waste and Biomass Valorization, 9*(1), 73–82.

Ramasamy, D., Yu, M., Xu, W., & Kim, H. (2020). Potato peel waste and its valorization: A comprehensive review. *Waste Management, 118*, 242–256.

Ravindran, R., & Jaiswal, A. K. (2016). Exploitation of food industry. Waste for high-value products. *Trends in Biotechnology, 34*(1), 58–69.

Romero-Güiza, M. S., Vila, J., Mata-Alvarez, J., Chimenos, J. M., & Astals, S. (2016). The role of additives on anaerobic digestion: A review. *Renewable and Sustainable Energy Reviews, 58*, 1486–1499. ISSN 1364-0321. https://doi.org/10.1016/j.rser.2015.12.094

Rubanka, K., Terletska, V., Pysarev, M., & Abramova, A. (2021). Prospects of potato pulp disposal: A review. *Journal of Hygienic Engineering and Design, 35*, 83–95. https://dspace.nuft.edu.ua/handle/123456789/38011

Russ, W., & Schnappinger, M. (2007). Waste related to the food industry: A challenge in material loops. In: Oreopoulou, V., & Russ, W. (Eds.), *Utilization of By-products and Treatment of Waste in the Food Industry* (pp. 1–13). Springer.

Schieber, A., Marleny, B., & Aranda Saldaña, M. D. (2009) Potato peels: A source of nutritionally and pharmacologically interesting compounds–A review. *Food, 3*. https://doi.org/10.7939/R33T9DM0H

Sedlák, P., Sedláková, V., Vašek, J., Zeka, D., Čílová, D., Melounová, M., Orsák, M., Domkářová, J., Doležal, P., & Vejl, P. (2022). Phenotypic, molecular and biochemical evaluation of somatic hybrids between Solanum tuberosum and S. bulbocastanum. *Scientific Reports*, *12*(1), 4484. https://doi.org/10.1038/s41598-022-08424-5

Shen, Y., Linville, J. L., Urgun-Demirtas, M., Mintz, M. M., & Snyder, S. W. (2015). An overview of biogas production and utilization at full-scale wastewater treatment plants (WWTPs) in the United States: Challenges and opportunities towards energy-neutral WWTPs. *Renewable and Sustainable Energy Reviews, 50*, 346–362. https://doi.org/10.1016/j.rser.2015.04.129

Singh, A., Yadav, S., & Singh, R. P. (2020). Potential of potato peel waste for bioethanol production: A review. *Biofuels, 11*(5), 641–650.

Suntikunaporn, M., Echaroj, S., & Rimpikul, W. (2014). Evaluation of agricultural wastes for biogas production. Science & Technology Asia, 19(1), 1–8. Retrieved from https://ph02.tci-thaijo.org/index.php/SciTechAsia/article/view/17054

Tampio, E., Ervasti, S., & Rintala, J. (2015). Characteristics and agronomic usability of digestates from laboratory digesters treating food waste and autoclaved food waste. *Journal of Cleaner Production, 94*, 86–92. https://doi.org/10.1016/j.jclepro.2015.01.086

United Nations (UN). (2020). Sustainable development goals. https://sdgs.un.org

United States Environmental Protection Agency. (2024). How does anaerobic digestion work? https://www.epa.gov/agstar/how-does-anaerobic-digestion-work

Wang, Z. J., Liu, H., Zeng, F. K., Yang, Y. C., Xu, D., Zhao, Y. C., Liu, X. F., Kaur, L., Liu, G., & Singh, J. (2022). Potato processing industry in China: Current scenario, future trends and global impact. *Potato Research, 66*, 543–562. https://doi.org/10.1007/s11540-022-09588-3

Xu, F., Li, Y., Ge, X., Yang, L., & Li, Y. (2018). Anaerobic digestion of food waste – Challenges and opportunities. *Bioresource Technology, 247*, 1047–1058. ISSN 0960–8524. https://doi.org/10.1016/j.biortech.2017.09.020

Zainu, Z. A. (2021). Development of policy and regulations for hazardous waste management in Malaysia. *Journal of Science, Technology and Innovation Policy, 5*(2), 34–42. https://jostip.utm.my/index.php/jostip/article/view/45

Zarate-Ramirez, A. (2014). *Plastic pollution: Economic and environmental impacts. Journal of Environmental Studies, 22*(3), 45–58.

Zhang, C., Su, H., Baeyens, J., & Tan, T. (2014). Reviewing the anaerobic digestion of food waste for biogas production. *Renewable and Sustainable Energy Reviews, 38*, 383–392. ISSN 1364-0321. https://doi.org/10.1016/j.rser.2014.05.038

Zhang, L., Xu, C., & Champagne, P. (2017). Overview of recent advances in thermo-chemical conversion of biomass. *Energy Conversion and Management, 142*, 524–548.

Zhang, S., Wang, J., Chen, X., Gui, J., Sun, Y., & Wu, S. (2021). Industrial-scale food waste composting: Effects of aeration frequencies on oxygen consumption, enzymatic activities, and bacterial community succession. *Bioresource Technology, 320*(Part A), 124357. ISSN 0960–8524. https://doi.org/10.1016/j.biortech.2020.124357

24 Cassava Waste Management

24.1 INTRODUCTION

Cassava (*Manihot esculenta* Crantz) originated in tropical America, where the indigenous Indian population found it more than 4,000 years ago (Waisundara, 2018). Cassava is a major crop farmed in many countries of the world, including Africa, Asia, and Latin America. It provides sustenance for millions of people and is used in animal feed and industrial purposes. However, cassava processing creates a huge quantity of trash, which, if not adequately handled, can cause environmental difficulties. In the past few years, there has been a rise in interest in exploring new and imaginative solutions to manage cassava waste and convert it into valuable products for a variety of industrial purposes (Mo & Omojola, 2013). Cassava waste is classified into two types: solid waste and liquid waste. Cassava peels, stems, and leaves are considered solid trash, whereas wastewater created during cassava processing is considered liquid waste. Cassava processing creates significant amounts of both solid and liquid waste. Cassava wastes include leaves, stems, peels, wastewater, and starch bagasse from production and processing. Cassava waste, which includes high quantities of organic matter, nitrogen, and phosphorus, may be a substantial cause of environmental contamination if not adequately managed. Furthermore, cassava waste can attract pests and emit unpleasant odors, affecting the health and well-being of local people (Oghenejoboh, 2015).

Today, worldwide cassava starch output has increased quickly and is expected to reach 8.8 million tons by 2020 (Wang et al., 2022). Cassava yields around 40% more carbs than rice and 25% more carbohydrates than maize. Cassava is also high in micronutrients such as iron, zinc, and vitamins A, B, and C (Waisundara, 2018). Cassava has a global production of 276 million tons and is largely produced in the Asia Pacific area, accounting for over three-fourths of the global market share, according to the Global Cassava Starch Market (2021–2026). Thailand is the global leader in cassava starch production, followed by Vietnam and Cambodia; nevertheless, their cassava output level was not as high as Nigeria's. Cassava starch is in high demand due to its distinct features, which include being allergy-free, having a bland flavor, characterized by a high paste viscosity, transparency, shear, and freeze–thaw stability, and being extremely appealing to various food and non-food businesses. Because of these advantages, cassava starch production in the world has increased in recent years, reaching 8.8 million tons in 2020 (Wang et al., 2022).

Cassava waste management is an urgent issue that must be addressed by legislators, researchers, and industry players. Cassava waste has several industrial uses that might provide a sustainable solution for trash management while producing value-added goods. Cassava trash might be transformed into a lucrative resource that helps both the environment and the economy with ongoing study and investment. Figure 24.1 shows cassava production worldwide, and Table 24.1 shows global cassava production in 2017.

24.2 BIOMASS FROM NATURAL FIBERS

Cassava, a starchy root crop grown extensively in tropical and subtropical countries, may also be utilized to generate biomass from natural fiber. Cassava biomass may be used for a variety of applications, including energy generation and the creation of bio-based goods.

DOI: 10.1201/9781003472209-32

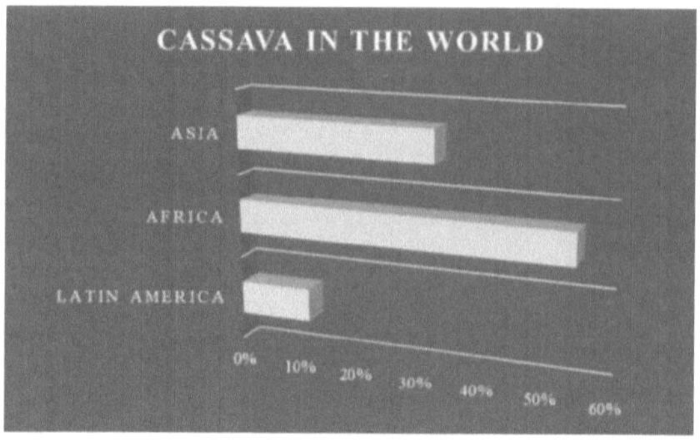

FIGURE 24.1 The cassava production worldwide (GCP21, 2021).

TABLE 24.1
Global Cassava Production in 2017(FAOSTA, 2019)

Producing Countries	Quantity Produced (MMT)	Percentage Contribution	Position
Nigeria	59.83	20.4	1st
Thailand	31.70	10.8	2nd
Congo Democratic Republic	30.00	10.6	3rd
Indonesia	19.05	6.5	4th
Brazil	18.88	6.5	5th
Ghana	18.47	6.3	6th
Angola	11.75	4.0	7th
Cambodia	10.58	3.6	8th
Vietnam	10.28	3.5	9th
Rest of the World	80.95	27.7	

Cassava processing by-products such as peels, stems, and leaves can be utilized as feedstock for bioenergy production. These wastes can be treated using processes such as anaerobic digestion or gasification to generate biogas or syngas, which can be used to generate power or as a cooking fuel. Cassava fiber may be compressed into briquettes or pellets, which can then be utilized as a sustainable fuel source. These densified biomass products are suitable for direct combustion in stoves, boilers, or biomass power plants (Wu et al., 2023).

Cassava fiber, particularly the peels, can be used as a component in animal feed compositions. It is a source of nutritional fiber and can be used to enhance cattle diets, notably for ruminants, or as a poultry feed element. Compostable cassava leftovers, including leaves and stems, can be utilized as organic fertilizers or soil additives. Cassava biomass composting can increase soil structure, nutritional content, and water-holding capacity (Fanelli et al., 2023).

Biocomposite materials may be created by combining cassava fibers with polymers. These materials may be utilized in a variety of applications, including packaging, building materials, and automobile parts, and provide a more sustainable alternative to traditional materials. It is vital to emphasize that cassava is grown primarily for its edible roots, and utilizing the fiber or waste for biomass should not jeopardize food security or degrade the environment. Sustainable farming practices, resource efficiency, and environmental effect mitigation should all be considered in the management of cassava fiber biomass production.

24.3 ENVIRONMENTAL IMPACT FROM BIOMASS

Cassava provides carbohydrates and nutrients to millions of people. However, cassava processing creates enormous amounts of waste materials, which can have negative environmental consequences if not properly handled. Figure 24.2 shows the typical cassava peel dumps in Nigeria. This section will look at the environmental impact of cassava waste and propose strategies to lessen its consequences. Cassava waste disposal can harm the ecosystem. Inadequate disposal can result in air, soil, and water pollution. Methane, carbon dioxide, and nitrous oxide are all emitted by cassava waste when disposed of in landfills or burnt, contributing to climate change. Furthermore, when cassava waste is dumped into rivers or lakes, it includes significant quantities of organic matter and nitrogen, which can promote the eutrophication of water bodies. Excessive algal blooms, oxygen deprivation, and fish fatalities can all be caused by eutrophication caused by high levels of nutrients in cassava waste (Wu et al., 2023).

Farmers in Nigeria typically dispose of generated wastes thrown into pools of water, unfinished buildings, vacant spaces, and any available open area along major roadways and streets, despite the negative influence on both the environment and people's general health (Nweke, 1994; Omotioma et al., 2013; Oghenejoboh, 2015). The absence of strong and enforceable cassava waste management legislation makes understanding the relationship between the financial efficacy of waste treatment and the value-added benefit of cassava waste management difficult for smaller farmers. Another key factor is that proper waste management is much beyond the capabilities of smallholder farmers, who typically grow and prepare cassava for food and minimal family income (Ekop et al., 2019; Omilani et al., 2019). Cassava solid waste comprises the cassava plant's wastes (leaves, stems, and peels), making up over 18% of the cassava plant (Ezekiel et al., 2012). Peels are the most plentiful sort of cassava leftovers, and they are the most harmful to the environment. Nigeria produces from 20% to 35% of the total weight of processed roots annually (Adesanya et al., 2007; Ukwuru & Egbonu, 2013). Nigeria eliminates more than 10 tons of cassava peels from 59.5 million metric tons of the annual cassava crop with little consideration for the environment (Coker et al., 2015; Sivamani et al., 2018).

Cassava tuber in its raw form comprises around 60–70% water, with the majority of it converting into wastewater during processing. Aside from cassava root's high moisture content, a large amount

FIGURE 24.2 Typical cassava peel dumps in Nigeria (Olukanmi & Adetunji, 2018).

of water is consumed during the processing of the roots into food items. As a result, the effluent burden of cassava processing is quite large. A substantial amount of water is used in washing the peeled cassava tubers, and this water, along with the water contained in the tuber released during processing (mechanical squeezing or settling depending on the desired final product), is freely discharged into the environment (Ketiku & Oyenuga, 1992; Omosuli et al., 2017). Cassava wastewater, in the meantime causing harm to soil, streams, and groundwater, serves as a haven for mosquitoes and other parasite creatures, as well as emitting an irritating odor when left in open, sluggish channels (Oghenejoboh, 2015; Onifade et al., 2015). No surprise, daily, significant rates of malaria and other insect-borne ailments are recorded in several cassava-producing villages in Nigeria.

24.4 AVAILABILITY OF AGRO-WASTE

Cassava is an important source of food and income for millions of people worldwide since it is one of the most extensively produced crops in tropical and subtropical climates. Cassava consumption is predicted to rise in the next few years, perhaps leading to a considerable rise in the quantity of cassava waste created. Many factors, including cassava production volume, processing technology, and cassava product demand determine the availability of cassava waste. The accumulation of enormous amounts of cassava trash occurs in many places where cassava is farmed and processed due to a lack of proper waste management methods and infrastructure. In Nigeria, for example, cassava waste is estimated to account for 30–50% of total cassava output, with the majority of this trash being allowed to rot in fields or thrown into rivers, creating environmental contamination (Ogunsina et al., 2010).

Cassava waste, on the contrary, is utilized for a variety of reasons in some areas, including animal feed, bioenergy generation, and composting. Cassava waste is utilized as a substrate for bioethanol manufacturing in Thailand, a significant renewable energy source (Phitsuwan et al., 2019). Similarly, cassava waste is utilized as a feedstock for biogas generation in Vietnam (Hoang et al., 2019). According to the Cassava Market Outlook (2022–2026), Cassava consumption is expected to reach 110 million metric tons by 2026, representing a 0.3% reduction from 2021 levels. Nigeria was the largest cassava consumer in 2021, with 25.2 million metric tons. Second, third, and fourth place went to Indonesia, Ghana, and Mozambique, respectively. Cassava production is predicted to reach 335 million metric tons by 2026, representing a 1.2% rise over 2021 levels. Nigeria was the biggest producer of cassava in 2021, with 60.8 million metric tons. Chaudary (2022) projected that the cassava global market will increase from 183.25 billion USD in 2023 to 254.28 billion USD in 2032.

The availability of cassava waste and its potential for industrial use is inextricably linked to developing adequate cassava processing and waste management technologies. Composting, biogas generation, and bioenergy production can lessen the environmental effect of cassava waste while also providing economic advantages to farmers and other stakeholders in the cassava value chain.

24.5 PROPERTIES OF AGRO-WASTE FIBERS

24.5.1 Physical Properties

Cassava, *M. esculenta*, commonly known as tapioca, yucca, and cassada, belongs to the Euphorbiaceae family, and the one being mass cultivated is *M. esculenta* Crantz, mainly in Latin America, Africa, and Asia. Figure 24.3 shows the physical appearance of the cassava plant. The cassava root is the main organ of cassava, and it is edible and has a very high starch content. The cassava cannot be used for vegetative propagation when the roots mature. The edible root that occurs 85% of the total weight is called parenchyma, which consists of xylem vessels in a matrix of starch-containing cells (Wheatley & Chuzel, 1993). Cassava roots have long brown peel and firm, homogeneous white or yellowish flesh. The size varies from 5 to 10 in diameter and is 50–80 cm long. Starch is the major component of cassava, which has small quantities of protein and lipids.

FIGURE 24.3 Physical appearance of the cassava plant (Abotbina et al., 2022).

Cassava has sympodial branching, where the main stems can produce secondary branches. These braces are induced by flowering, which has been called reproductive branchings. The leaves are formed by simple lamina and petiole and lobed with palmate veins. The flowers can be found in the leaf axils on the upper part of the plant. The plant produces both male and female flowers on the same plant, where the female is located on the lower part of the inflorescence and is fewer in number compared to the male flowers. The fruit is a trilocular capsule with sizes from 1 to 1.5 cm in diameter. Each locule has a single carunculate seed with a dark brown coating that usually germinates soon after collection and 16 days for germination. Worldwide, cassava is classified as sweet and bitter.

24.5.2 Chemical Composition

The root of the bitter variety is poisonous when it is still raw since it contains hydrocyanic acid, which is reported by the USDA NRCS National Plant Data Center. The moisture content of local cassava was lower than that of the improved varieties. The shows the protein values of sweet cassava are higher compared to bitter ones, with 2.69% for sweet and 3.37% for bitter varieties, as reported by Sarkiyayi and Agar (2010). The differences in protein can be affected by genotype rather than environment. Lipid content remains low, with only 3.92% for sweet and 3.82% for the bitter varieties. Zinc, copper, magnesium, and calcium are the minerals distinct from cassava and vary depending on variety, as indicated in the chart above. Cassava cultivars produced in Thailand were observed to have substantially greater amounts of calcium (136–369 mg/100 g) and magnesium (31–43 mg/100 g). The copper concentrations in this investigation ranged from 0.002 to 0.14 mg/100 g, which is within the Ugandan Standards (US 235) tolerances. Higher concentrations of copper (between 0.037 and 0.057 mg/100 g), zinc (between 13 and 19 mg/100 g), and iron (between 29 and 40 mg/100 g) were found by Charles et al. (2005). Thus, Thai cassava varieties are more mineral-rich than Ugandan ones.

24.5.3 Mechanical Properties

The mechanical properties of an agricultural material are those that influence its response to an applied force. Agricultural production is affected by mechanical properties such as hardness, compressive strength, impact, and shear resistance. The information on these properties is beneficial for designing equipment for handling, milling, storing, transporting, and food processing. When the moisture content is reduced, the tuber becomes firmer and increases its resistance to abrasion and cutting. This implies that mechanical properties are dependent on the relative humidity (Kolawole et al., 2007). Oupathum et al. (2019) reported that the shearing stress and specific shearing energy increase from 20° to 40° at the knife bevel angle. Due to the increase in tuber density and starch content, Lomchangkum et al. (2020) found that the maximal shear stress and cutting force increased with increasing tuber age.

According to a study by Parra et al. (2004) on the tensile strength and elongation percentage, the additives glycerol and glutaraldehyde have a statistically significant effect on tensile strength, with a 95.0% level of confidence. Additionally, it was observed that all three additives significantly affected the films' elongation. According to the same journal, as glutaraldehyde concentration increased, film tensile strength increased while mechanical properties decreased. One gram of glycerol was then added. The tensile strength of films containing greater than 1 g of glycerol was diminished. Therefore, at concentrations greater than 1 g of glycerol, this additive functions as a plasticizer, altering the mechanical behavior of the film; this restricts the use of these materials in moist environments. Due to the crosslinking effect, the water permeability of films containing up to 0.5 g of glutaraldehyde was reduced. Films formulated from cassava starch and the composites investigated in this paper exhibited excellent flexibility, indicating that they could be used as edible coatings (Parra & Tadini, 2004).

A separate study discovered that cassava scratch foams could influence the mechanical properties of fibers by applying a portion of dough to the bottom surface of a 200°C mold (Parra, 2004). Heating generates vapor, which has a significant effect on the flow of dough. The steam is then expelled through steam exhaust vents, and the heating promotes solidification and porosity in the dough. According to statistical analysis (ANOVA), only fiber quantities have statistical significance (P 0.05) in foam strength. The type of fiber lacked statistical significance ($P>0.05$). The foam containing 1% yucca fiber has the greatest tensile strength. Adding up to 1% cassava fiber diminishes compression strength. The strength values of foam produced with 1% and 3% wheat fiber were identical, with 2% yielding the lowest result (Parra, 2004).

Low concentrations of short or powdered fibers (cassava or wheat fibers) perform well as reinforcing additives in starch cassava foams. The compression resistance of foam containing 1% cassava fiber was greater, increasing the percentage decreased compression resistance. In 2%, wheat fiber-based foam produced a poorer result. In terms of foam tensile strength, pliability, and density, there was no statistical significance between fiber type and these variables. Only the amount of fiber was significant. During SEM microscopy analysis of samples containing 1% fiber, adhesion of the fiber-matrix interface was observed. Foams containing 1% fiber had more regular structures, and cassava fiber foam contained fewer internal exposed cells than wheat fiber foam. Both fibers had limited dimensions and could only enhance the open-cell structure of the starch foams by 1% (Carr et al., 2006).

24.5.4 Morphological Properties

The highly variable morphological characteristics of cassava indicate a high degree of interspecific hybridization. Numerous cassava cultivars are stored in the germplasm repositories of international and domestic research institutions. The largest germplasm bank is located in Colombia at the Centro Internacional de Agricultura Tropical, with approximately 4,700 accessions (Bonierbale et al., 1997), followed by EMBRAPA's collection in Cruz das Almas, Bahia, with approximately 1,700 accessions (Fukuda et al., 1997). Given the large number of commercially cultivated cassava genotypes and the diversity of ecosystems in which cassava is grown, it is challenging to precisely describe the morphological descriptors due to the interaction between genotype and environmental conditions. In addition to morphological characterization, molecular characterization, based primarily on DNA molecular markers, has proven to be extremely useful for assessing the genetic diversity of germplasm (Beeching et al., 1993; Fregene et al., 1994).

24.5.4.1 Roots

Cassava's primary storage organs are its roots. Similar to dicot species, a conventional primary tap root system develops in plants grown from true seeds. The radicle of a germinating seed grows vertically downward and becomes a taproot, from which adventitious roots develop. Eventually, the taproot and a few adventitious roots develop into storage roots. Cassava is a tuberous root but is not used for vegetative propagation. The mature cassava storage root comprises three distinct tissues: epidermis (periderm), peel (or cortex), and parenchyma. The parenchyma, which is the digestible portion of the fresh root, consists of xylem vessels radially distributed in a matrix of starch-containing cells and accounts for approximately 85% of the total weight (Wheatley & Chuzel, 1993).

A study is being conducted to ascertain the range of variation for carotene content in a subset of the global cassava germplasm collection, as well as the inheritance of carotenoids and their stability in response to various processing techniques (Safo-Katanga et al., 1984). Although root color is highly correlated with carotene content, quantitative evaluation of genotypes selected on the basis of root color is necessary to increase selection efficiency. A limited number of major genes determine the accumulation of carotene in the roots of cassava. Carotenoid stability in response to various processing methods is genotype-dependent, indicating the need to evaluate a trait after selecting high carotene concentration in fresh roots. This study laid the groundwork for the development of future strategies to improve the nutritional quality of cassava. Based on the work of Safo-Katanga et al. (1984), an extraction procedure for cassava root parenchyma was developed. To obtain root parenchyma, petroleum ether was used. Before being injected into the HPLC column, the organic phase is dried with sodium sulfate and concentrated through rotary distillation.

Table 24.2 lists the cassava genotypes with the highest concentration of carotene (fresh weight basis) in the roots. Within the range of variability for carotene concentration in the roots, a few genotypes with greater than 2 mg/100 g of roots may be chosen. These high levels should be combined with other agronomic and root quality traits to produce as productive and as high-quality cassava as the most popular cultivars in each region, but also has a high carotene content (Iglesias et al., 1997).

24.5.4.2 Leaves

Cassava leaves are composed of a lamina and a petiole. The veins of the leaf are palmately lobed. On average, the number of lobes ranges from three to nine (occasionally 11). Rogers and Fleming (1973) report that only a few cultivars have mature leaves with three lobes, which may represent the primitive ancestral form. The leaf closest to the base of the inflorescence is frequently simple

TABLE 24.2
Cassava Genotypes with the Highest Concentration of Carotene (Fresh Weight Basis) in the Roots

Genotype	Common Name	Origin	Root Color	Carotene (mg/100 g)
MBRA 516	Olho Verde	Manaus—Brazil	Deep yellow	2.55
MCOL 2285	Furetsikae	Guainia—Col.	Orange	2.40
MCOL 2109	Manaca	Amazonas—Col.	Orange	2.14
MCOL 2295	Bitsurikae	Guainia—Col.	Orange	2.12
MBRA 481	Baixota	Manaus—Brazil	Orange	2.04

Source: Brice et al. (2014).

and unlobed. Generally, the leaves near the inflorescence are smaller and have fewer lobes (most often three), but the leaf closest to the inflorescence is frequently simple and unlobed. The upper leaf surface is covered by an oily, reflective epidermis. A few stomata are located along the main vein on the upper (adaxial) surface of the leaf (Cerqueira, 1989). Only 2% of the 1,500 cultivars examined had stomata on the adaxial surface (El-Sharkawy & Cock, 1987a). Moreover, the stomata on the upper surface are larger and functional than those on the underside. Both possess two small sentinel cells surrounded by two subsidiary cells (Cerqueira, 1989). The number of stomata per leaf area varies between 278 and 700 mm^{-2}, and stomatal apertures can occupy between 1.4% and 3.1% of the total leaf area.

24.5.4.3 Flowers

Cassava is a monoecious species that produces both male and female blossoms on the same plant. In general, the inflorescence is formed at the site of insertion of the reproductive branchings; however, inflorescences can also be found in the upper leaf axils. The female flowers, which are located at the bottom of the inflorescence, are fewer in number than the male flowers, which are abundant at the top. Female flowers on the same inflorescence open 1–2 weeks before male flowers (protogyny).

24.5.4.4 Fruit and Seeds

The fruit is a trilocular capsule that is ovoid or globose, 1–1.5 cm in diameter, and has six prominent, straight longitudinal ridges, or aristae. Each locule contains a solitary carunculate seed. The fruit has bicidal dehiscence, a combination of septicidal and loculicidal dehiscences, with apertures along the parallel plane of the dissepiments and along the mid veins of the carpels, respectively. With this combination of dehiscences, the fruit opens into six valves, resulting in an explosive dehiscence that propels the seeds a considerable distance (Rogers, 1965). Typically, fruit maturity occurs between 75 and 90 days after pollination (Ghosh et al., 1988). The seed is approximately 100 mm long, 6 mm wide, and 4 mm thick. The weight per seed ranges from 95 to 136 mg (Ghosh et al., 1988). The silky seed coat is mottled dark brown and grey. Typically, the seeds germinate within 16 days of being collected.

24.5.5 Thermal Properties

In most tropical countries, cassava, plantain fruit, and tubers are important crops with the potential for industrial starch production and are principally made up of carbohydrates. In scratch-making, thermal properties are crucial to controlling the sun drying process as well as other post-harvest processing operations, including thermal conductivity (k), specific heat capacity (C_p), and thermal diffusivity (d). Tables 24.3 and 24.4 provide the thermal conductivity and specific heat capacity of cassava.

TABLE 24.3
Thermal Conductivity of Cassava (Njie et al., 1998)

	M (%)	k ($W{\cdot}m^{-1}{\cdot}°C^{-1}$) k ($Wm^{1}°C^{1}$)	Number of Replicates	Standard Deviation
Cassava	7060	0.57	9	0.013
	55	0.55	9	0.015
	47	0.53	10	0.028
	31	0.50	10	0.017
	18	0.38	8	0.054
		0.16	7	0.022

TABLE 24.4
Specific Heat Capacity of Cassava (Njie et al., 1998)

	Moisture Content, *M* (%)			
	10	20	45	68
Temperature (°C)	Specific Heat Capacity, c_{p*} (kJ/kg/°C)			
36	1.636	2.013	2.736	3.074
39	1.728	2.102	2.860	3.196
42	1.791	2.143	2.895	3.229
45	1.827	2.151	2.918	3.245
48	1.858	2.174	2.938	3.260
51	1.866	2.195	2.958	3.275

24.6 AGRO-WASTE MANAGEMENT

Over 260 million tons of cassava was produced globally in 2012, with Nigeria being the largest producer, accounting for over 20% of the total (Otekunrin & Sawicka, 2019). Cassava is primarily grown and processed on a small scale in Nigeria by family or village farmers. For people in sub-Saharan Africa, especially in Nigeria, where its production, processing, and consumption are most prevalent and significant on a global scale, cassava offers a dependable and affordable source of carbohydrates. From the beginning of production until the end, it offers various job opportunities to both men and women.

The versatile root vegetable cassava is processed using a variety of techniques to create a variety of goods. The cassava root's outer skin is removed after harvesting, either manually or mechanically (Bellotti et al., 2012). After being peeled, the roots are grated into a fine pulp using a grater or, more recently, powered machines. The grated pulp is pressed to remove extra liquid using a hydraulic press or manual squeezing. An essential step is fermentation, where the pressed pulp ages 24–72 h to improve flavor and lower cyanide levels (Cardoso et al., 2005). The cassava pulp is spread out and dried after fermentation, either using mechanical dryers or the sun, to extend its shelf life and make storage easier. Wet milling is another method for separating cassava starch from other ingredients and extracting it from dried flour (Mbanjo et al., 2021). The resulting starch or flour can be processed further to make things like chips, flakes, tapioca pearls, starch noodles, and chips. These techniques may change based on the desired product and regional cultural customs.

The technology used and the processing method determine the type and composition of the waste (Mbanjo et al., 2021). Several tons of cassava peels are produced as a waste product from processing in most cassava processing communities, and they are generally considered a significant source of environmental pollution (Ubalua, 2007). The generation of waste is anticipated to rise along with the anticipated increase in cassava production. Cassava peels can be used as livestock feed. Still, because of the volume produced and the isolation of many communities where processing occurs, a lot of waste is left behind. It is burned or allowed to rot, which has a negative impact on the environment (Nhassico et al., 2008).

24.6.1 Compost

Compost made from decomposing cassava plant residues and other organic matter is referred to as cassava compost. It is created through composting, which involves the carefully regulated microbial breakdown of organic materials (Adeyeye et al., 2017). Compost made from cassava is a valuable fertilizer and soil amendment because it is full of nutrients and organic matter. It can be applied to enhance soil structure, increase soil fertility, and encourage strong plant growth. Gathering cassava leaves, stems, peels, and other plant parts, combining them with other organic materials, making compost bins or piles, and supplying the right levels of moisture and aeration for microbial activity are all part of the composting process. The organic materials eventually decompose into compost that is rich in nutrients and can be used in gardens, fields, or potted plants.

One hectare of cassava would provide 140 kg of protein based on the average protein content of cassava leaves, and the estimated litter left by cassava would compensate for the nitrogen needed for the succeeding maize crop (Hawlader et al., 2020). Cassava leaf compost is a potential source of organic matter and plant nutrients, just like many other organic compost materials. By enhancing soil's physical, chemical, and biological qualities, the practice also raises crop yield and nutritional value. By supplying significant nitrogen and other essential nutrients, cassava leaf compost reduces the need for expensive inorganic agrochemicals like urea and treats soil nutrient deficiencies. Because of this, using cassava leaves as a source of nitrogen in rice farming is very sustainable.

In some areas, the term "cassava compost" may refer to the process of composting leftover cassava peels and other unused cassava plant parts. This method can be used to improve the soil in gardens or agricultural fields and reduce food waste. Composting with cassava is a sustainable agricultural practice that encourages the recycling of organic waste. To avoid any potential negative effects, such as the production of harmful by-products or offensive odors, it is crucial to make sure the composting process is done correctly. In addition, to optimize the decomposition process in composting, it is essential to maintain the correct balance of green materials, which are rich in nitrogen, and brown materials, which provide carbon. Adequate aeration and moisture levels are also crucial for supporting microbial activity and ensuring efficient breakdown of organic matter.

24.6.2 Pulp and Paper

When the starchy component (cassava starch) has been removed from the cassava roots through processing, the fibrous residue is referred to as cassava pulp. It is the chunk of solid cassava that remains after being grated and pressed to remove extra liquid. The pulp from cassava is typically made up of a mixture of cassava fibers, leftover starch, and moisture (Kosugi et al., 2009). It frequently has a tan or off-white color. Although it is generally regarded as less desirable than the extracted cassava starch, the pulp still has some nutritional value and can be used in a variety of applications. The pulp is frequently further processed in traditional cassava processing techniques through fermentation, drying, and milling to produce goods like gari (fermented and roasted cassava granules) or fufu (a starchy food that resembles dough). However, cassava pulp can be used in various ways depending on the culture and region.

Cassava pulp, a solid, moist by-product of the production of cassava starch, makes up roughly 10–15% of the weight of the original root (Khempaka et al., 2009). The quantity of waste by-products produced rises along with cassava starch production. At least 1 million tons of pulp is produced annually in Thailand. After drying, some of the by-products are incorporated into ruminant and swine diets or used to make fertilizer. There are still a lot of by-products, though. However, dried cassava pulp can be used as a feed ingredient to supply part of the dietary energy requirement of broilers, but it should be limited to 8% or less in diets. Not only is dried cassava pulp useful as an energy source, but it may also improve gut health and reduce abdominal fat deposition in broilers.

The following steps are usually involved in making paper from cassava pulp. First, the cassava pulp is gathered after starch extraction and prepared for papermaking by removing any remaining starch or impurities. The cassava pulp may then go through mechanical or chemical processes to enhance the quality of its fiber and remove impurities. To separate the individual fibers, the cassava pulp is further broken down through mechanical or chemical pulping processes. To make a pulp slurry, the fibers from the processed cassava pulp are combined with water. Then, using different papermaking methods like Fourdrinier or cylinder molding, this slurry is formed into paper sheets. To create the finished paper product, the freshly formed paper sheets are dried, pressed, and finished.

24.6.3 Biocomposite Materials

Cassava fibers or particles combined with a matrix material, usually a polymer, create a composite material known as a cassava biocomposite. It is a kind of biocomposite in which a reinforcement component made of cassava biomass is used (Abotbina et al., 2022). The stalks or peels of cassava plants are used to make the fibers, which are prized for their tensile and tenacious properties. They can be processed and incorporated into different matrix materials like thermoplastics or bio-based polymers to improve their mechanical properties.

The development of biodegradable materials has received more attention recently due to growing environmental concerns. To investigate this, composites made of cassava starch and Brazilian green coconut fibers were prepared (Lomelí-Ramírez et al., 2014). Thermal molding is used to create these composites with varying amounts of coir fibers, with glycerol acting as a plasticizer for the starch. Thermal treatment is applied to the matrix and composites. Both the addition of fibers and thermal processing enhanced the tensile properties of cassava starch. Scanning electron microscopy fractographic studies were used to explain the observed strength properties. Adding fibers decreases the water uptake, swelling, and moisture absorption attributable to improved interfacial bonding between the matrix and fibers, as well as the fibers' ability to prevent absorption.

Cassava biocomposites have a wide range of potential uses, including in the construction industry, consumer goods, packaging, and the automotive sector. The success of cassava biocomposites depends, as with any material development, on a number of variables, including cost, raw material accessibility, producibility, scalability, and the capacity to meet particular performance specifications for various applications. The use of cassava and other natural resources in sustainable materials is still being explored and improved upon through ongoing research and innovation in the field of biocomposites.

24.6.4 Bio-Coal

Cassava bio-coal, also referred to as cassava charcoal or cassava biomass briquettes, is a solid fuel made from cassava biomass using a process known as carbonization. It entails turning leftover cassava plant material, like leaves, stems, and peels, into a high-energy fuel that can take the place of conventional coal or charcoal (Asmara et al., 2021). Cassava bio-coal production typically involves a number of steps. The biomass from cassava is first gathered and dried to lower its moisture content. The dried biomass is then carbonized, which entails heating the substance without oxygen. As a result, volatile compounds are released, and the biomass is changed into more carbon- and energy-dense forms.

To enhance its handling and combustion qualities, the resulting cassava char is typically ground or crushed after carbonization. These particles are then compressed under pressure to create solid, uniform fuel units called briquettes or pellets. Comparing cassava bio-coal to conventional coal or charcoal, there are a number of benefits. It is a renewable energy source that lessens reliance on fossil fuels using agricultural waste (Citrasari et al., 2017). Because it makes use of biomass that would otherwise be wasted, it has a lower carbon footprint. Additionally, because it produces fewer pollutants and greenhouse gases when it burns, cassava bio-coal has the potential to be a cleaner-burning fuel. Cassava bio-coal, or any other bio-coal made from renewable biomass sources, has the potential to reduce greenhouse gas emissions, promote sustainable energy practices, and make use of agricultural waste or residues that would otherwise be thrown away.

24.6.5 Biochar

Cassava biomass is heated in a low-oxygen environment during the pyrolysis process to create a type of charcoal-like substance known as cassava biochar. It is a type of biochar made specifically from the peels, stems, and leaves of the cassava plant (Hasnan et al., 2018). The first step in making cassava biochar is gathering and drying the cassava biomass to lower its moisture content. After that, the dried biomass is pyrolyzed, usually in a specialized kiln or retort. During pyrolysis, the biomass goes through thermal decomposition, releasing volatile gases and transforming the solid biomass into biochar, a carbon-rich substance.

Because of its high carbon content, porous design, and potential for long-term carbon sequestration in the soil, cassava biochar is prized (Lubwama et al., 2020). Biochar can increase soil fertility, increase water retention, and encourage microbial activity in agricultural soils. Additionally, it can increase the availability of nutrients to plants and decrease nutrient leaching. Cassava biochar's porous structure allows it to hold onto moisture and nutrients, which promotes the growth of plants. In a variety of agricultural and horticultural applications, it can be added to soil to improve its physical, chemical, and biological qualities. In addition to helping with agriculture, carbon sequestration by cassava biochar has the potential to slow down global warming.

The use of cassava biochar as a soil amendment to increase soil fertility, water retention, and nutrient-holding capacity has a number of potential uses and advantages. As a long-term carbon sink, it helps to reduce greenhouse gas emissions by storing carbon in the soil. The carbon that was once a component of the cassava plant is stabilized and stored in a form that resists decomposition by being converted into biochar, effectively sequestering carbon for a long time. It offers a useful method of putting leftover cassava plant material to use, cutting down on agricultural waste, and promoting environmentally friendly waste disposal techniques.

24.6.6 Bio-Bricks

Cassava starch is used as a binder along with other sustainable ingredients to create the innovative building material known as cassava bio-bricks. As they use renewable resources and have less of an impact on the environment than conventional clay bricks or concrete blocks, these bio-bricks serve as an environmentally friendly substitute (Bency et al., 2023). Cassava bio-bricks are typically produced in a multi-step process. First, the starch from the cassava roots is extracted frequently by grinding the roots and separating the starch from the fibrous parts. After extraction, the starch is combined with other ingredients like water, natural fibers, and occasionally a stabilizing substance like lime or cement.

To achieve a consistent composition, the mixture is thoroughly mixed before being poured into brick molds or formed into the desired shapes. The starch hardens and binds the other components to form a solid and long-lasting structure as the bio-bricks dry and cure. Cassava bio-bricks have a number of benefits for the environment and sustainability. They lessen the need for conventional brick-making materials like concrete or clay, which involve energy-intensive processes and raise carbon dioxide emissions. Because it is a renewable resource, cassava starch aids in reducing reliance on non-renewable resources.

By transforming agricultural waste into a useful energy source, the use of cassava plant residues for the production of bio-bricks encourages sustainable waste management. Similar to other biomass briquettes, cassava bio-bricks can be used in boilers, cookstoves, and other heating applications. They are especially helpful when cassava is grown as a staple crop and when sustainable energy options are necessary for both socioeconomic and environmental reasons.

24.6.7 Biohydrogen

Cassava biohydrogen is produced using cassava biomass as a feedstock through a procedure known as dark fermentation or biohydrogen production. To produce hydrogen gas as a by-product, specific microorganisms must anaerobically ferment cassava biomass (Balbin et al., 2022). The collection and pretreatment of cassava biomass, which can include cassava peels, leaves, stems, or other plant residues, is the usual first step in the production of cassava biohydrogen. Under anaerobic conditions, the biomass is subjected to a fermentation process, frequently in specialized bioreactors.

Cassava biomass can be used to create hydrogen gas, which can be used as a clean and sustainable energy source (Zong et al., 2009). Power generation, transportation, and industrial processes are just a few of the industries in which it is used. Fuel cells, automobiles, and chemical production all use hydrogen gas, which can also be used to produce electricity. However, it is important to remember that research and development are still ongoing for biohydrogen production from cassava biomass. For cassava biohydrogen to be economically and technically feasible, issues like low hydrogen yields, the need for effective microbial strains, and process optimization must be dealt with.

Using cassava biomass to produce hydrogen lessens reliance on fossil fuels, which is one of the potential benefits of cassava biohydrogen production. Since the carbon dioxide captured during cassava growth balances out the carbon dioxide released during hydrogen combustion, the production of biohydrogen from cassava biomass is regarded as a carbon-neutral process. In addition, turning agricultural waste into a useful energy source helps to promote sustainable waste management by using cassava biomass for biohydrogen production.

24.6.8 Biomethane

Cassava biomethane is the term for biomethane gas that is produced using cassava as the feedstock. Anaerobic digestion is a process that creates biomethane, a renewable natural gas. When there is no oxygen present, microorganisms break down organic materials during anaerobic digestion, creating biogas as a by-product. Methane (CH_4) and carbon dioxide (CO_2) make up the majority of biogas, with smaller amounts of other gases (Mari et al., 2020). The harvested cassava roots are typically processed and shredded to increase their surface area to produce cassava biomethane. In an anaerobic digester, where bacteria break down the cassava's organic material, biogas is created. Through a procedure known as upgrading, impurities like carbon dioxide, water vapor, and other trace gases can be taken out of the biogas. As a result, biomethane is produced, which is more pure than raw biogas and consists primarily of the gas methane.

There are many uses for the biomethane that cassava produces. It can be used for heating, cooking, and the production of electricity after being injected into natural gas pipelines and blended with regular natural gas (Varongchayakul et al., 2021). Natural gas vehicles can also compress biomethane and use it as fuel, making it a renewable and low-carbon substitute for traditional fuels. There are several advantages to cassava biomethane production. In addition to providing a renewable energy source, it uses agricultural waste, lowers greenhouse gas emissions compared to fossil fuels, and promotes rural development and energy independence. When using cassava for energy, it's crucial to use sustainable farming techniques and to take into account the possible effects on food security and land use.

The biogas plant or sealed anaerobic digester receives the cassava biomass. The organic material is broken down by microorganisms in the digester through a series of biochemical processes that produce biogas, a mixture of methane and carbon dioxide. The raw biogas's typical composition is methane, which makes up between 50% and 70% of it, with the remainder being carbon dioxide and minor amounts of other gases. A procedure called "biogas upgrading" is carried out to boost the biogas's methane content and raise its quality. This involves filtering out impurities like CO_2 and other trace gases to produce purified biomethane. The biomethane can be used as a renewable energy source for many purposes, such as the production of electricity, heating, cooking, and fuel for vehicles.

24.6.9 Bioethanol

Cassava, a starchy root vegetable also known as manioc or yuca, cassava is used to make bioethanol, a type of ethanol fuel. In place of fossil fuels, bioethanol is a renewable and sustainable alternative that is frequently blended with petrol or used as a fuel for transportation (Sriroth et al., 2010). Cassava bioethanol production is a multi-step process. When the cassava roots are fully mature, they are first harvested and cleaned to get rid of contaminants (Lyu et al., 2020). After cleaning, the outer layer is typically peeled off to remove the majority of the bitter compounds from the roots. The cassava roots are now ready for processing. The next step involves chopping, grating, or mashing the peeled cassava roots.

By increasing the cassava's surface area, this process makes it simpler to extract the starch. A slurry is frequently made by combining the cassava mash and water, which helps the extraction procedure.

The starch extraction stage starts after the slurry is created. Starch can be extracted using a variety of techniques, including enzymatic hydrolysis or wet milling (Sriroth et al., 2010). Wet milling is the mechanical processing of the slurry to isolate the starch from other constituents, such as the fibrous material. On the contrary, enzymatic hydrolysis involves the addition of particular enzymes to convert the starch into fermentable sugars. After being obtained, the starch or sugar solution goes through fermentation. The solution contains particular yeast strains, such as *Saccharomyces cerevisiae*. The sugars are consumed by the yeast, which then ferments them into ethanol. To maximize the conversion of sugars to ethanol, this process takes place in a controlled environment with set temperatures (Han et al., 2010).

The mixture undergoes distillation after fermentation. By taking advantage of the variances in their boiling points, distillation separates the ethanol from water and other impurities (Han et al., 2010). To create a more concentrated ethanol solution, the fermented mixture is heated, the vapor is captured, and then condensed. Dehydration, a subsequent step in the processing process, can be used if necessary (Sriroth et al., 2010). Dehydration improves the ethanol's purity and energy content by removing the remaining water. For this, methods like molecular sieves or additional distillation are used.

24.6.10 Bioelectricity

When cassava biomass is used as a renewable energy source, it is referred to as cassava bioelectricity. It entails using various processes to capture the energy potential found in cassava and transform it into electrical power (Agarry et al., 2016). Here is a summary of how cassava can be used to produce bioelectricity. First is cassava biomass, which can be used as a feedstock for the creation of bioelectricity and includes the stems, leaves, and other plant parts. The plant matter that remains after harvesting cassava roots for use in food production or other industrial processes can be used to produce this biomass.

Additionally, the biomass from cassava is typically transformed into a more useful form for producing electricity (Aduba et al., 2023). This may entail procedures like anaerobic digestion, gasification, or combustion. Next, electricity is produced using the energy released during the combustion, gasification, or anaerobic digestion of cassava biomass. This is accomplished by integrating suitable power generation technologies, such as steam turbines, gas turbines, or gas engines, with the energy conversion processes. Electrical generators that turn mechanical energy into electrical energy are connected to these devices.

The benefit of using a locally accessible and renewable biomass resource for power generation is provided by cassava bioelectricity. It helps to lessen reliance on fossil fuels, reduce greenhouse gas emissions, and promote sustainable energy production, making sure that the production of cassava for bioelectricity does not conflict with food production or negatively impact the environment. Sustainable practices and careful land management should be used to lessen any potential adverse effects.

24.7 POTENTIAL APPLICATIONS

Cassava peels are a solid waste that has recently attracted increasing study interest. This is because it contributes significantly more to environmental harm than other cassava waste products combined. Another advantage is that it has a significant potential for biomass production for the treatment of wastewater. Table 24.5 provides the research in transforming cassava solid waste into value-added goods to reduce pollution. Figure 24.4 shows the sustainable use of cassava solid waste the leaves are used as animal feed, and the stems and rhizome are sun-dried and utilized as fuel, posing a

TABLE 24.5
Research in Transforming Cassava Solid Waste into Value-Added Goods to Reduce Pollution (Oghenejoboh et al., 2021)

Cassava Solid Waste	Value-Added Product	Area of Application and Effectiveness	References
Cassava peels	Biomass—activated carbon	Used to treat wastewater from refineries. When phenol, COD, BOD, and trace metals were removed from the target effluent, the activated carbon produced was more effective than commercial-grade carbon.	Oghenejoboh et al. (2016)
	Activated carbon	Used to remove Cr(VI) from aqueous effluent. Generated activated carbon removed Cr(VI) from treated wastewater with 99% effectiveness.	Vasudevan et al. (2016)
	Activated carbon	Treatment of various types of industrial wastewater using various adsorption isotherms. Activated carbon was shown to be highly effective from coefficient r^2 of the experimental results.	Ilaboya et al. (2013)
	Protein	The protein content of the biomass produced was 56.7% for cassava peel hydrolysates, 47.5% for glucose media, and 49.1% for acid media. As a result, they recommended cassava peels as a source of carbs for protein production.	Ezekiel et al. (2012)
Cassava leaves	Livestock feed	According to the report, well-processed cassava leaves contain up to 37% calories, 3.7% protein, vitamins, and minerals. Cassava leaves should be prepared before being consumed by people.	Haider (2017)
Cassava stem	Ethanol	Biological conversion of cassava stems into ethanol. The goal of the stem hydrolysis was to convert the cellulose to saccharine, which was then fermented to convert the monomers to ethanol via distillation. The ethanol output was 32%, and the alcohol content was 1.622 g/L. Cassava stem could be a promising substrate for ethanol synthesis.	Sivamani and Baskar (2018)

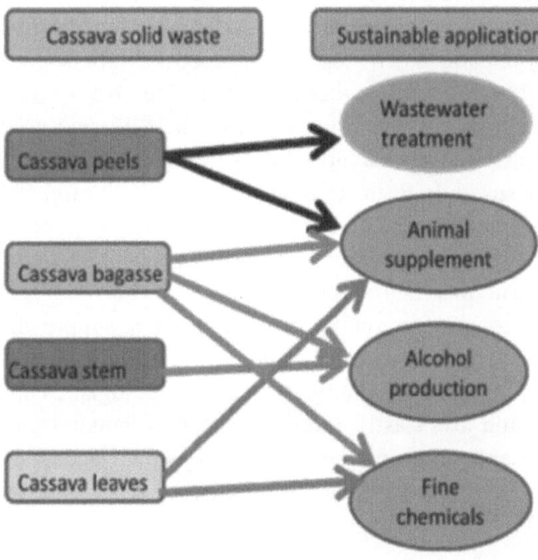

FIGURE 24.4 Sustainable use of cassava solid waste (Oghenejoboh et al., 2021).

pollution risk to the environment by farmers. Cassava waste, including peels, stems, and leaves, may be used in a variety of applications, helping to reduce waste and optimize resources. Cassava waste can be used as an ingredient in animal feed compositions. The peels and stems, which are high in fiber and minerals, can be used as feed additives for animals, offering a low-cost and long-term source of nutrition (Smah et al., 2020).

Cassava waste may be used to create biogas in anaerobic digestion systems. In the absence of oxygen, anaerobic digestion degrades organic materials, producing methane-rich biogas that may be utilized for cooking, heating, or energy generation (Aisien & Aisien, 2020). Cassava trash may be composted or processed to make organic fertilizers. Composting cassava waste aids the decomposition process, resulting in nutrient-rich compost that can increase soil fertility, structure, and plant development. The anaerobic fermentation of cassava peels produces biofertilizers (Ogbo, 2010; Okolie et al., 2018). Cassava waste, such as peels and other types of residual biomass, can be used as feed for bioethanol manufacturing. Waste may be treated to extract starch or fermentable sugars, which are subsequently transformed into bioethanol, a renewable and sustainable fuel source. Cassava waste, in the form of biopolymers derived from starch or cellulose, can be utilized to manufacture biodegradable packaging materials. These materials provide an environmentally friendly alternative to traditional plastics, lowering environmental impact and boosting sustainability (Oghenejoboh et al., 2021).

24.8 CHALLENGES ASSOCIATED WITH PROCESSING AND COMMERCIALIZING

While there are potential applications for cassava waste, there are also challenges involved with its processing and commercialization. Among the major challenges are composition variability, high moisture content, fiber and starch separation, enzyme and microbial degradation, cost-effectiveness, infrastructure and supply chain, and market acceptance (Alfred, 2007).

Firstly, the composition of cassava waste can vary depending on factors such as cassava variety, processing methods, and storage conditions. This diversity makes creating standardized processing procedures and guaranteeing the consistent quality of cassava waste end products difficult. Cassava waste, particularly peels and stems, frequently has a high moisture content. This can reduce the efficiency of storage, transit, and processing. To minimize moisture content and avoid spoiling, drying technology or proper storage procedures are necessary (Alfred, 2007).

Next, cassava waste comprises a mixture of fiber and starch, which must be successfully separated to be used in various applications. The separation process might be complicated, necessitating the use of specialized equipment or processing processes. Cassava waste is susceptible to enzymatic and microbiological degradation, resulting in lower shelf life and product quality. To reduce deterioration difficulties, appropriate preservation measures such as adequate storage conditions or enzymatic treatments must be used. In terms of cost-effectiveness, it might be difficult to determine the economic feasibility of processing and commercializing cassava waste products. Cassava waste utilization projects' profitability might be impacted by high processing costs, limited economies of scale, and market demand variations (Sriroth et al., 2000).

Furthermore, building a strong infrastructure and an effective supply chain for collecting, processing, and transporting cassava trash can be difficult, particularly in distant or rural locations where cassava growing is common. Adequate infrastructure and logistics are required to guarantee efficient waste management and use. Lastly, increasing market demand for cassava waste-derived goods might be difficult. To encourage market uptake, consumer acceptability, marketing methods, and education on the environmental and economic benefits of these goods are required (Hierro-Iglesias et al., 2022).

24.9 CONCLUSION

Cassava waste management, which involves properly handling and utilizing the by-products and residues produced during various processing stages, is an essential component of cassava production. Effective waste management techniques provide economic advantages and potential opportunities for value addition in addition to supporting environmental sustainability. Composting, anaerobic digestion, and conversion into useful products like animal feed, biofertilizers, and bioenergy are just a few ways that cassava waste can be managed. By composting, organic matter can break down and become nutrient-rich soil amendments, improving soil fertility and lowering the need for synthetic fertilizers. Cassava waste can be converted through anaerobic digestion into biogas, a renewable energy source that can be used to generate heat and electricity. Effective waste management techniques not only reduce environmental pollution but also open up new economic opportunities and revenue streams. For instance, the commercialization of the production of biofertilizers and biogas from cassava waste can result in income and job opportunities. Additionally, by maximizing resource utilization, integrating waste management systems into cassava processing facilities can result in increased efficiency and cost savings. However, it is crucial to put in place waste management procedures that prioritize sustainability and take into account potential effects on local communities and ecosystems. To minimize any negative effects on water quality, air emissions, and soil contamination, waste by-products should be handled and disposed of properly in accordance with environmental regulations and guidelines. In conclusion, efficient cassava waste management is crucial for fostering sustainability, minimizing environmental effects, and maximizing the financial benefits of cassava processing. The cassava industry can maximize resource utilization, reduce waste generation, and support a more sustainable and circular agricultural system by implementing cutting-edge and environmentally responsible waste management practices.

ACKNOWLEDGMENT

This chapter was written with Almira Salsabila, Muhammad Shafiq Bin Musa, and Yirthick A.L. Kathiravan.

REFERENCES

Abotbina, W., Sapuan, S. M., Ilyas, R. A., Sultan, M. T., Alkbir, M. F., Sulaiman, S., Harussani, M. M., & Bayraktar, E. (2022). Recent developments in cassava (Manihot esculenta) based biocomposites and their potential industrial applications: A comprehensive review. *Materials*, *15*(19), 6992. https://doi.org/10.3390/ma15196992

Adesanya, O., Oluyemi, K., Josiah, S., Adesanya, R., Shittu, L., Ofusori, D., Bankole, M., & Babalola, G. (2007). Ethanol production by saccharomyces cerevisiae from cassava peel hydrolysate. *The Internet Journal of Microbiology*, *5*(1). https://www.researchgate.net/profile/Lukeman-Shittu/publication/259104809_Ethanol_Production_By_Saccharomyces_Cerevisiae_From_Cassava_Peel_Hydrolysate/links/567f78aa08ae1e63f1e7fc8a/Ethanol-Production-By-Saccharomyces-Cerevisiae-From-Cassava-Peel-Hydrolysate.pdf

Adeyeye, A., Akanbi, W., Sobola, O., Lamidi, W., & Olalekan, K. (2017). The growth and tuber yield of cassava as affected by cassava peel compost application. *Journal of Experimental Agriculture International*, *16*(1), 1–5. https://doi.org/10.9734/jeai/2017/32048

Aduba, C. C., Ndukwe, J. K., Onyejiaka, C. K., Onyeiwu, S. C., & Moneke, A. N. (2023). Integrated valorization of cassava wastes for production of bioelectricity, biogas and biofertilizer. *Waste and Biomass Valorization, 14*, 4003–4019. https://doi.org/10.1007/s12649-023-02126-3

Agarry, S., Oghenejobor, K., & Solomon, B. (2016). Bioelectricity production from cassava mill effluents using microbial fuel cell technology. *Nigerian Journal of Technology*, *35*(2), 329. https://doi.org/10.4314/njt.v35i2.13

Aisien, F.A., & Aisien, E.T. (2020). Biogas from cassava peels waste. *Detritus Multidisciplinary Journal Waste Resources and Residues, 10*, 100–108. https://pdfs.semanticscholar.org/fea8/f3174dd105eb1c1858d3837b05a7ec0d08ff.pdf

Alfred, O. U. (2007). Cassava wastes: Treatment options and value addition alternatives. *African Journal of Biotechnology*, *6*(18), 2065–2073. https://doi.org/10.5897/ajb2007.000-2319

Asmara, S., Lanya, B., Tamrin, Putri, A. N., Rahmat, A., & Mutolib, A. (2021). The effect of various varieties of cassava stems waste and tapioca adhesive concentrations on the quality of bio-coal Briquette. *IOP Conference Series: Earth and Environmental Science*, *739*(1), 012084. https://doi.org/10.1088/1755-1315/739/1/012084

Balbin, A. J., Gonzales, H. A., Bornillo, L. A., Tagupa, J. H., Jaro, R. H., Ido, A. L., & Arazo, R. O. (2022). High biohydrogen yield from fresh cassava waste pulps via gas dissolution in a dark fermentative environment. *Cleaner Waste Systems*, *3*, 100039. https://doi.org/10.1016/j.clwas.2022.100039

Beeching, J. R., Marmey, P., Gavalda, M.-C., Noirot, M., Hayson, H. R., Hughes, M. A., & Charrier, A. (1993). An assessment of genetic diversity within a collection of cassava (Manihot esculenta Crantz) germplasm using molecular markers. *Annals of Botany, 72*(6), 515–520.

Bellotti, A., Herrera Campo, B. V., & Hyman, G. (2012). Cassava production and pest management: Present and potential threats in a changing environment. *Tropical Plant Biology*, *5*(1), 39–72. https://doi.org/10.1007/s12042-011-9091-4

Bency, P., Anish, M., Jayaprakash, V., Jayaprabakar, J., Yanmaz, E., Joy, N., & Ramulu, P. J. (2023). Comparative assessment of compression strength of solid biobriquette using different binding materials. *Journal of Nanomaterials*, 2023, 1–7. https://doi.org/10.1155/2023/3685782

Bonierbale, M., Guevara, C., Dixon, A. G. O., Ng, N. Q., Asiedu, R., & Ng, S. Y. C. (1997). *Biodiversity in Trust: Conservation and Use of Plant Genetic Resources in CGIAR Centres.* Cambridge University Press; pp. 29–40.

Brice, D. K. E., Pierre, Z. G., Sidoine, E. B., & Alexandre, D. A. (2014). Morphological characterization of cassava (Manihot esculenta Crantz) accessions collected in the centre-west, south-west and west of Cote d'Ivoire. *Greener Journal of Agricultural Sciences, 4*(6), 220–231.

Cardoso, A. P., Mirione, E., Ernesto, M., Massaza, F., Cliff, J., Rezaul Haque, M., & Bradbury, J. H. (2005). Processing of cassava roots to remove cyanogens. *Journal of Food Composition and Analysis*, *18*(5), 451–460. https://doi.org/10.1016/j.jfca.2004.04.002

Carr, L. G., Parra, D. F., Ponce, P., Lugão, A. B., & Buchler, P. M. (2006). Influence of fibers on the mechanical properties of cassava starch foams. *Journal of Polymers and the Environment, 14*, 179–183. https://doi.org/10.1007/s10924-006-0008-5

Cerqueira, Y. M. (1989) Efeito da deficiência de água na anatomia foliar de cultivares de mandioca (Manihot esculenta Crantz). MSc thesis, Universidade Federal da Bahia, Cruz das Almas, Brazil.

Charles, A. L., Sriroth, K., & Huang, T. C. (2005). Proximate composition, mineral contents, hydrogen cyanide and phytic acid of 5 cassava genotypes. Food Chemistry, 92(4), 615–620. https://doi.org/10.1016/j.foodchem.2004.08.024

Chaudary, T. (2022). Cassava market research report information by category (conventional and organic), by application (food & beverages, animal feed, and others), by form (solid and liquid) and by region (North America, Europe, Asia-Pacific, and Rest of the World) - Market forecast till 2032. Market Research Future. www.marketresearchfuture.com

Citrasari, N., Pinatih, T. A., Kuncoro, E. P., Soegianto, A., Salamun, & Irawan, B. (2017). Potency of biocharcoal briquette from leather cassava tubers and industrial sludge. *AIP Conference Proceedings, 1854*(1), 020007. https://doi.org/10.1063/1.4985398

Coker, A. O., Achi, C. G., & Sridhar, M. C. K. (2015). Utilization of cassava processing waste as a viable and sustainable strategy for meeting cassava processing energy needs: Case study from Ibadan City, Nigeria. *The Journal of Solid Waste Technology and Management, 41*(4), 628–639.

Ekop, I. E., Simonyan, K. J., & Evwierhoma, E. T. (2019). Utilization of cassava wastes for value added products: An overview. *International Journal of Engineering Science, 3*(1), 31–39. https://doi.org/10.5281/zenodo.2556465

El-Sharkawy, M. A., & Cock, J. H. (1987). Cassava Botany and Physiology. In R. J. Hillocks, J. M. Thresh, & A. C. Bellotti (Eds.), Cassava: Biology, Production and Utilization (pp. 39–59). CABI Publishing.

Ezekiel, O. O., Aworh, O. C., Du Preez, J. C., & Steyn, L. (2012). Cultivation of candida utilis on cassava peel hydrolysates for single-cell protein production. *Journal of Food Engineering, 2*, 452–461. https://www.davidpublisher.com/Public/uploads/Contribute/5681e2764d4e8.pdf

Fanelli, N. S., Torres-Mendoza, L. J., Abelilla, J. J., & Stein, H. H. (2023). Chemical composition of cassava-based feed ingredients from south-East Asia. *Animal Bioscience, 36*(6), 908–919. https://doi.org/10.5713/ab.22.0360

FAOSTAT. (2019). Food and Agriculture Organization of the United Nations. Retrieved from https://www.fao.org/faostat/en/#data

Fregene, M. A., Vargas, J., Angel, F., Tohme, J., Asiedu, R. A., Akorada, M. O., & Roca, W. M. (1994) Chloroplast DNA and nuclear ribosomal DNA variability in cassava (Manihot esculenta Crantz) and its wild relatives. *Theoretical and Applied Genetics, 89*, 719—727.

Fukuda, W. M. G., Silva, S. O., & Porto, M. C. M. (1997). *Caracterização e Avaliação de Germoplasma de Mandioca (Manihot esculenta Crantz). Catalog*. EMBRAPA/CNPMF, Cruz das Almas BA, Brazil.

GCP21. (2021). Gcp21.org. https://gcp21.org/

Ghosh, S. P., Ramanujam, T., Jos, J. S., Moorthy, S. N., & Nair, R. G. (1988). *Tuber Crops*. Oxford & IBH Publishing Co., New Delhi; pp. 3–146.

Haider, P. (2017). Cassava leaves: Its health benefits.

Han, M., Kim, Y., Kim, Y., Chung, B., & Choi, G.-W. (2010). Bioethanol production from optimized pretreatment of cassava stem. *Korean Journal of Chemical Engineering, 28*(1), 119–125. https://doi.org/10.1007/s11814-010-0330-4

Hasnan, F. I., Iamail, K. N., Musa, M., Jaapar, J., Alwi, H., & Hamid, K. K. (2018). Characterization of bio char derived from Tapioca skin. *IOP Conference Series: Materials Science and Engineering, 334*, 012016. https://doi.org/10.1088/1757-899x/334/1/012016

Hawlader, N. H., Ali Fakir, M. S., Ahmad, M., Nesa, H., Rahman, M. M., Hasan, I., Islam, M. M., & Majumder, M. S. (2020). Cassava leaf compost influences growth, yield and nutrient uptake of rice. *Annual Research & Review in Biology, 35*, 23–33. https://doi.org/10.9734/arrb/2020/v35i930268

Hierro-Iglesias, C., Chimphango, A., Thornley, P., & Fernández-Castané, A. (2022). Opportunities for the development of cassava waste biorefineries for the production of polyhydroxyalkanoates in Sub-Saharan Africa. *Biomass and Bioenergy, 166*, 106600. https://doi.org/10.1016/j.biombioe.2022.106600

Hoang, T. M. L., Le, Q. T., Le, V. M., & Hoang, T. T. H. (2019). Biogas production from cassava waste: Effect of inoculum to substrate ratio and hydraulic retention time. *Sustainable Environment Research, 29*(1), 1–7.

Iglesias, C., Mayer, J., Chavez, L., & Calle, F. (1997). Genetic potential and stability of carotene content in cassava roots. *Euphytica, 94*, 367–373. https://doi.org/10.1023/A:1002962108315

Ilaboya, I. R., Oti, E. O., Ekoh, G. O., & Umukoro, L. O. (2013). Performance of activated carbon from cassava peels for the treatment of effluent wastewater. *Iranica Journal of Energy & Environment, 4*(4), 361–375.

Ketiku, A. O., & Oyenuga, V. A. (1992). Preliminary report on the carbohydrate constituents of cassava roots and yam tubers. *Nigerian Journal of Science, 4*, 25–30.

Khempaka, S., Molee, W., & Guillaume, M. (2009). Dried cassava pulp as an alternative feedstuff for broilers: Effect on growth performance, carcass traits, digestive organs, and nutrient digestibility. *Journal of Applied Poultry Research, 18*(3), 487–493. https://doi.org/10.3382/japr.2008-00124

Kolawole, O. P., Agbetoye, L. A. S., & Ogunlowo, A. S. (2007). Cassava mash dewatering parameters. *International Journal of Food Engineering, 3*. https://doi.org/10.2202/1556-3758.1088

Kosugi, A., Kondo, A., Ueda, M., Murata, Y., Vaithanomsat, P., Thanapase, W., Arai, T., & Mori, Y. (2009). Production of ethanol from cassava pulp via fermentation with a surface-engineered yeast strain displaying glucoamylase. *Renewable Energy, 34*(5), 1354–1358. https://doi.org/10.1016/j.renene.2008.09.002

Lomchangkum, C., Junsiri, C., Sudajan, S., & Laloon, K. (2020). A study on the mechanical characteristics of cassava tuber cutter. *International Journal of Agricultural Technology, 16*(1), 63–76.

Lomelí-Ramírez, M. G., Kestur, S. G., Manríquez-González, R., Iwakiri, S., de Muniz, G. B., & Flores-Sahagun, T. S. (2014). Bio-composites of cassava starch-green coconut fiber: Part II—structure and properties. *Carbohydrate Polymers, 102*, 576–583. https://doi.org/10.1016/j.carbpol.2013.11.020

Lubwama, M., Yiga, V. A., Muhairwe, F., & Kihedu, J. (2020). Physical and combustion properties of agricultural residue bio-char bio-composite briquettes as sustainable domestic energy sources. *Renewable Energy, 148*, 1002–1016. https://doi.org/10.1016/j.renene.2019.10.085

Lyu, H., Zhang, J., Zhai, Z., Feng, Y., & Geng, Z. (2020). Life cycle assessment for bioethanol production from whole plant cassava by integrated process. *Journal of Cleaner Production, 269*, 121902. https://doi.org/10.1016/j.jclepro.2020.121902

Mari, A. G., Andreani, C. L., Tonello, T. U., Leite, L. C. C., Fernandes, J. R., Lopes, D. D., Rodrigues, J. A. D., & Gomes, S. D. (2020). Biohydrogen and biomethane production from cassava wastewater in a two-stage anaerobic sequencing batch biofilm reactor. *International Journal of Hydrogen Energy, 45*(8), 5165–5174. https://doi.org/10.1016/j.ijhydene.2019.07.054

Mbanjo, E. G., Rabbi, I. Y., Ferguson, M. E., Kayondo, S. I., Eng, N. H., Tripathi, L., Kulakow, P., & Egesi, C. (2021). Technological innovations for improving cassava production in Sub-Saharan Africa. *Frontiers in Genetics, 11*. https://doi.org/10.3389/fgene.2020.623736

Mo, O., & Omojola, M. (2013). Tacca starch: A review of its production, physicochemical properties, modification and industrial uses. *African Journal of Food, Agriculture, Nutrition and Development, 13*(4). https://refhub.elsevier.com/S0268-005X(22)00062-5/sref56

Nhassico, D., Muquingue, H., Cliff, J., Cumbana, A., & Bradbury, J. H. (2008). Rising African cassava production, diseases due to high cyanide intake and control measures. *Journal of the Science of Food and Agriculture, 88*(12), 2043–2049. https://doi.org/10.1002/jsfa.3337

Njie, D. N., Rumsey, T. R., & Singh, R. P. (1988). Thermal properties of cassava, yam and platain. *Journal of Food Engineering, 37*(1), 63–76, 0260–8774. https://doi.org/10.1016/S0260-8774(98)00068-5

Nweke, F. I. (1994). Cassava processing in sub-Saharan Africa: The implications for expanding cassava production. *Outlook on Agriculture, 23*(3), 197–205. https://doi.org/10.1177/003072709402300307

Ogbo, F. (2010). Conversion of cassava wastes for biofertilizer production using phosphate solubilizing fungi. *Bioresource Technology, 101*(11), 4120–4124. https://doi.org/10.1016/j.biortech.2009.12.057

Oghenejoboh, K. M. (2015). Effects of cassava wastewater on the quality of receiving water body intended for fish farming. *Current Journal of Applied Science and Technology, 6*(2), 164–171. https://doi.org/10.9734/BJAST/2015/14356

Oghenejoboh, K. M., Orugba, H. O., Oghenejoboh, U. M., & Agarry, S. E. (2021). Value added cassava waste management and environmental sustainability in Nigeria: A review. *Environmental Challenges, 4*, 100127. https://doi.org/10.1016/j.envc.2021.100127

Oghenejoboh, K. M., Otuagoma, S. O., & Ohimor, E. O. (2016). Application of cassava peels activated carbon in the treatment of oil refinery wastewater—A comparative analysis. *Journal of Ecological Engineering, 17*(2), 52–58. https://doi.org/10.12911/22998993/62290

Ogunsina, B. S., Ogunade, O. M., & Ogunade, J. O. (2010). Utilization of cassava peels in the diet of pigs: Effects on performance, nutrient digestibility and nitrogen balance. *Tropical Animal Health and Production, 42*(5), 865–871.

Okolie, O. C., Stanley, H. O., Frank-Peterside, N., & Ugboma, C. J. (2018). The production of liquid biofertilizer from cassava peels and spent mushroom substrates using microbial inoculants. *Asian Journal of Biotechnology and Bioresource Technology, 4*(3), 1–13. https://doi.org/10.9734/AJB2T/2018/45642

Olukanni, D. O., & Olatunji, T. O. (2018). Cassava Waste Management and Biogas Generation Potential in Selected Local Government Areas in Ogun State, Nigeria. *Recycling*, 3(4), 58. https://doi.org/10.3390/recycling3040058

Omilani, O., Abass, A.B., Okoruwa, V.O., 2019. Smallholder agroprocessors' willingness to pay for value-added solid-waste management solutions. *Sustainability, 11*, 1759. https://doi.org/10.3390/su11061759

Omosuli, S.V., Ikujenlola, A.V., & Abisuwa, A.T. (2017). Quality assessment of stored fresh cassava roots and 'fufu' flour produced from stored roots. *Journal of Food Science & Nutrition, 3*(1), 009–013. https://doi.org/10.17352/JFSNT.000008

Omotioma, M., Mbah, G. O., Akpan, I. J., & Ibezim, O. B. (2013). Impact assessment of cassava effluents on Barika Stream in Ibadan, Nigeria. *International Journal of Environmental Science, Management, Engineering and Research, 2*(2), 50–52. https://www.researchgate.net/profile/M-Omotioma/publication/337883232_IMPACT_ASSESSMENT_OF_CASSAVA_EFFLUENTS_ON_BARIKA_STREAM_IN_IBADAN_NIGERIA/links/5df049424585159aa473f5e7/IMPACT-ASSESSMENT-OF-CASSAVA-EFFLUENTS-ON-BARIKA-STREAM-IN-IBADAN-NIGERIA.pdf

Onifade, T., Adeniran, K., & Ojo, O. (2015). Assessment of cassava waste effluent effects on some water sources in Ilorin, Nigeria. *African Journal of Engineering Research, 3*, 56–68 https://www.netjournals.org/pdf/AJER/2015/3/15-025.pdf

Otekunrin, O. A., & Sawicka, B. (2019). Cassava, a 21st century staple crop: How can Nigeria harness its enormous trade potentials? *Acta Scientific Agriculture, 3*(8), 194–202. https://doi.org/10.31080/asag.2019.03.0586

Oupathum, C., Charee, S., Sudajan, S., & Thivavarnvongs, T. (2019). Effects of moisturecontent and knife bevel angle on shearing properties of cassava stems. *IOP Conference Series: Earth and Environmental Science, 301*(1). https://doi.org/10.1088/1755-1315/301/1/012010

Parra, D. F., Tadini, C. C., Ponce, P., & Lugão, A. B. (2004). Mechanical properties and water vapor transmission in some blends of cassava starch edible films. *Carbohydrate Polymers*, 58(4), 475–481. https://doi.org/10.1016/j.carbpol.2004.08.021

Phitsuwan, P., Luengnaruemitchai, A., & Chompupoung, C. (2019). Cassava waste as a feedstock for bioethanol production: A review. *Sustainable Energy Technologies and Assessments, 35*, 107–118.

Rogers, D. J. (1965). Some botanical and ethnological considerations of Manihot esculenta. *Economic Botany, 19*(4), 369–377. https://doi.org/10.1007/bf02904807

Rogers, D. J., & Fleming, H. S. (1973). A monograph of Manihot esculenta with an explanation of the taximetrics methods used. *Economic Botany, 27*(1), 1–113. https://doi.org/10.1007/bf02862222

Safo-Katanga, O., Aboagye, P., Amartey, S. A., & Olaham, J. H. (1984). Studies on the content of yellow-pigmented cassava. In: Terry, E. R., Doku, E. V., Arene, O. B., & Mahungu, N. M. (Eds.), *Tropical Roots Crops Production and Uses in Africa*. IDRC, Ottawa, Canada; pp. 103–104.

Sarkiyayi, S., & Agar, T. M. (2010). Comparative analysis on the nutritional and anti-nutritional contents of the sweet and bitter cassava varieties. *Advance Journal of Food Science and Technology, 2*(6), 328–334.

Sivamani, S., & Baskar, R. (2018). Bioconversion of cassava stem to ethanol: Oxalic acid pretreatment and co-culture fermentation. *Biofuels*, *9*(5), 559–566. https://doi.org/10.1080/17597269.2018.1432271

Sivamani, S., Chandrasekaran, A. P., Balajii, M., Shanmugaprakash, M., HosseiniBandegharaei, A., & Baskar, R. (2018). Evaluation of the potential of cassava-based residues for biofuels production. *Reviews in Environmental Science and Biotechnology, 17*, 553–570. https://doi.org/10.1007/s11157-018-9475-0

Smah, A. C., Garba, B. S., Salam, S. B., & Josephine, S. (2020). The effects of cyanide concentration on the environment and the consumption of varieties of cassava. *Journal of Pollution Effects & Control, 8*, 1–7. https://www.researchgate.net/profile/Adamu-Cornelius-2/publication/351037860_The_Effects_of_Cyanide_Concentration_on_the_Environment_and_the_Consumption_of_Varieties_of_Cassava/links/60808645907dcf667bb5aaea/The-Effects-of-Cyanide-Concentration-on-the-Environment-and-the-Consumption-of-Varieties-of-Cassava.pdf

Sriroth, K., Piyachomkwan, K., Wanlapatit, S., & Nivitchanyong, S. (2010). The promise of a technology revolution in cassava bioethanol: From Thai practice to the World Practice. *Fuel*, *89*(7), 1333–1338. https://doi.org/10.1016/j.fuel.2009.12.008

Sriroth, K., Piyachomkwan, K., Wanlapatit, S., & Oates, C. G. (2000). Cassava starch technology: The Thai experience. *Starch - Stärke, 52*(12), 439–449. https://doi.org/10.1002/1521-379x(200012)52:12<439::aid-star439>3.0.co;2–e

Ubalua, A. O. (2007). Cassava wastes: Treatment options and value addition alternatives. *African Journal of Biotechnology*, *6*(18), 2065–2073. https://doi.org/10.5897/ajb2007.000-2319

Ukwuru, M. U., & Egbonu, S. E. (2013). Recent development in cassava-based products research. *Academia Journal of Food Research, 1*(1), 001–013. https://www.new.academiapublishing.org/journals/ajfr/pdf/2013/Feb/Ukwuru%20and%20Egbonu.pdf

Varongchayakul, S., Songkasiri, W., & Chaiprasert, P. (2021). Optimization of cassava pulp pretreatment by liquid hot water for biomethane production. *BioEnergy Research*, *14*(4), 1312–1327. https://doi.org/10.1007/s12155-020-10238-0

Vasudevan, M., Ajithkumar, P. S., Singh, R. P., & Natarajan, N. (2016). Mass transfer kinetics using two-site interface model for removal of Cr(VI) from aqueous solution with cassava peel and rubber tree bark as adsorbents. *Environmental Engineering Research, 21*(2), 152–163. https://doi.org/10.4491/eer.2015.152

Waisundara, V. Y. (Ed.). (2018). Introductory chapter: Cassava as a staple food. In: *Cassava*. IntechOpen. https://doi.org/10.5772/intechopen.70324

Wang, Z., Mhaske, P., Farahnaky, A., Kasapis, S., & Majzoobi, M. (2022). Cassava starch: Chemical modification and its impact on functional properties and digestibility, a review. *Food Hydrocolloids*, *129*, 107542. https://doi.org/10.1016/j.foodhyd.2022.107542

Wheatley, C., & Chuzel, G. (1993). Cassava: The nature of the tuber and the use as a raw material. In: Macrae, R., Robinson, R., & Sadler, M. (Eds.), *Encyclopedia of Food Science,* and Food Technology and Nutrition. Academic Press, San Diego; pp. 734–743. https://doi.org/10.1016/b0-12-227055-x/00181-4

Wu, Y., Lu, H., Thanh, N. C., Al Obaid, S., Alfarraj, S., Jhanani, G. K., & Xia, C. (2023). Mixed pollutants adsorption potential of Eichhornia crassipes biochar on Manihot esculenta processing industry effluents. *Environmental Research, 231*, 116074. https://doi.org/10.1016/j.envres.2023.116074

Zong, W., Yu, R., Zhang, P., Fan, M., & Zhou, Z. (2009). Efficient hydrogen gas production from cassava and food waste by a two-step process of dark fermentation and photo-fermentation. *Biomass and Bioenergy, 33*(10), 1458–1463. https://doi.org/10.1016/j.biombioe.2009.06.008

Section IX

Challenges and Future Perspectives

25 Waste Management Challenges Associated with Processing of Agricultural Biomass in Various Industries

25.1 INTRODUCTION

The improper management of agro-biomass waste, including crop residues, animal manure, and organic by-products, poses significant environmental, economic, and public health challenges. Open burning, uncontrolled decomposition, and insufficient waste management infrastructure contribute to air pollution, greenhouse gas (GHG) emissions, soil degradation, and water contamination. Hence, managing agro-biomass waste is becoming increasingly important in pollution control engineering. This focus is not just about tackling environmental issues but also fostering sustainability and enhancing rural economies. Agricultural practices generate substantial amounts of organic waste. If this waste is not managed effectively, it can lead to significant environmental challenges, such as air pollution, GHG emissions, and water contamination. Implementing effective waste management strategies, including methods like composting, anaerobic digestion, and pyrolysis, alongside advanced applications like using agro-biomass for dye removal from wastewater, plays a critical role in reducing environmental harm and transforming waste into valuable resources (Mishra et al., 2021).

One of the most compelling environmental benefits of managing agro-biomass waste is its ability to convert this waste into biofuels, biogas, and biochar. This conversion process is essential for decreasing our dependence on fossil fuels, a significant step toward combating climate change. Moreover, these processes produce beneficial by-products, such as biofertilizers, which not only enrich soil health but also help close nutrient cycles within agricultural ecosystems (Kiran et al., 2020). By integrating these innovative technologies, the agricultural sector can effectively reduce its environmental footprint while simultaneously progressing toward a circular bioeconomy, an economic model where waste is minimized, and resources are continuously reused (Kumar Sarangi et al., 2023; Bux et al., 2023).

The economic implications of agro-biomass waste management are equally significant. When farmers convert agricultural residues into biofuels, biofertilizers, or other value-added products, they can significantly reduce waste disposal costs and create new revenue streams. This not only improves resource efficiency but also fosters job creation in rural areas, providing opportunities in waste processing, bioenergy production, and product development (Bux et al., 2023; Kumar Sarangi et al., 2023). These developments promote sustainable agricultural practices, which ultimately enhance productivity and resilience within the sector, leading to a more robust rural economy (Duque-Acevedo et al., 2020).

DOI: 10.1201/9781003472209-34

From a public health perspective, effective management of agro-biomass waste is vital. Proper handling of this waste minimizes the risks associated with improper disposal, which can lead to air pollution from combustion and water contamination from runoff. By implementing sound waste management practices, the agricultural sector can help protect both the environment and public health by reducing exposure to harmful pollutants. This not only promotes cleaner air and water but also contributes to overall community well-being (Duque-Acevedo et al., 2020).

In the end, effective management of agro-biomass waste is a key element in pollution control engineering. It plays an essential role in preventing environmental degradation while simultaneously supporting economic growth and protecting public health. By embracing these management practices, the agricultural sector can pave the way for a future that is not only more sustainable but also healthier and more resilient for everyone involved. The interconnectedness of environmental health, economic vitality, and public well-being makes agro-biomass waste management a crucial focus for the future of agriculture.

25.1.1 Agro-Biomass Waste Management Challenges Across Industries and Its Current Role in Agriculture and Industry

The management of agro-biomass waste is increasingly recognized as a critical issue, particularly in agriculture and manufacturing. This waste, which includes crop residues, animal manure, and organic by-products from food processing, holds significant potential for conversion into valuable resources like bioenergy, fertilizers, and bio-based materials. However, inefficient management practices and inadequate infrastructure often lead to environmental degradation, as large quantities of biomass waste end up in landfills (Pandey et al., 2020). Without proper systems for collection and processing, valuable organic materials are wasted, and opportunities to reduce pollution and improve sustainability are missed.

In the agricultural sector, managing biomass waste poses several challenges. The seasonal nature of crop residues and their uneven regional availability complicate both the collection and processing of this waste. A lack of efficient recycling systems and composting infrastructure contributes to unsustainable practices such as open burning, which is a significant source of air pollution and GHG emissions (Kumar et al., 2021). Additionally, high operational costs and complex regulatory frameworks act as barriers to the development of sustainable waste management systems. In many developing countries, agricultural regions struggle with insufficient technical expertise and inadequate logistical infrastructure, making it difficult to implement effective recycling and reuse strategies (Gupta et al., 2020).

In the industrial sector, agro-biomass waste presents significant opportunities for integration into a circular economy. Industries have the potential to transform this waste into bioenergy or utilize it as raw material for bio-based products, reducing reliance on fossil fuels. However, these efforts are hindered by the variability in types of agro-biomass waste generated, as well as its often low energy density. Such variability complicates the scalability of bioenergy projects, particularly in regions where consistent biomass supplies are not guaranteed (Chauhan et al., 2019). Aligning industrial production processes with circular economy principles requires overcoming these challenges and ensuring that the conversion of waste into energy and materials is both economically viable and environmentally sustainable.

25.1.2 How Improper Agro-Biomass Waste Management Contributes to Environmental Pollution?

Improper management of agro-biomass waste presents significant environmental challenges, impacting soil, water, air quality, and food safety. One of the primary concerns is soil pollution, where the improper disposal of agricultural residues, such as crop leftovers and animal manure, leads to contamination with harmful chemicals and pathogens. This not only degrades soil quality but also disrupts essential microbial communities crucial for nutrient cycling and overall soil health (Gontard et al., 2018; Joshi et al., 2020). The accumulation of heavy metals and pesticides in soils can also have long-term detrimental effects on agricultural productivity and ecosystem balance (Mishra et al., 2021).

Water pollution is another critical issue associated with poor agro-biomass management. Nutrient runoff from fertilizers and untreated organic waste often finds its way into nearby water bodies, triggering eutrophication. This process results in algal blooms that deplete oxygen levels in water, harming aquatic ecosystems and leading to fish kills. Moreover, contaminated runoff poses risks to human health by compromising drinking water supplies with pathogens and toxic substances (Kumar et al., 2021; Misra & Pandey, 2005).

Air quality is adversely affected by practices such as the open burning of crop residues, which releases particulate matter and GHGs into the atmosphere. This contributes to respiratory issues in humans and animals while exacerbating climate change. Additionally, volatile organic compounds released during the decomposition of organic waste can form ground-level ozone, further deteriorating air quality (Kumar et al., 2015; Giller et al., 2021). The improper disposal of agro-biomass also raises food safety concerns. Contaminants from polluted soil and water can be absorbed by crops, entering the food chain and posing health risks to consumers (Jain et al., 2020; Gontard et al., 2018). This cycle of contamination illustrates how ineffective waste management practices threaten both environmental health and food security.

Hence, the implications of improper agro-biomass management are extensive, leading to soil degradation, water contamination, air quality issues, and food safety risks. Addressing these challenges requires implementing sustainable agricultural practices that prioritize effective waste management strategies. By viewing agro-biomass waste as a resource rather than a liability, stakeholders can mitigate its negative environmental impacts while promoting sustainability (Sarkar et al., 2020; Haider et al., 2022).

25.2 AGRO-BIOMASS WASTE MANAGEMENT

25.2.1 What Is Agro-Biomass?

Agro-biomass refers to organic materials that originate from agricultural activities, including plant residues, animal waste, and by-products of agro-industrial processes. These materials are generated because of farming, forestry, and food production. Agro-biomass includes a wide variety of resources such as crop residues (e.g., straw, husks, stalks), animal manure, and agro-industrial by-products like sugarcane bagasse, fruit peels, and palm oil waste (Kumar Sarangi et al., 2023). These materials are typically classified as waste, but they hold significant potential if utilized properly.

The importance of agro-biomass lies in its dual nature as both a waste product and a valuable resource. On the one hand, if not managed properly, agro-biomass can cause serious environmental issues. The accumulation of agricultural waste can lead to soil degradation, water pollution through nutrient runoff, and emissions of methane, a potent GHG, when organic matter decomposes anaerobically. In many regions, open-field burning of crop residues further contributes to air pollution and environmental degradation (Subhash Babu, 2022).

Agro-biomass plays a critical role in agriculture and the economy. It has great potential for sustainable use. It is an important feedstock for renewable energy generation, particularly in bioenergy production. Agricultural residues like wood chips, straw, and bagasse can be converted into biofuels or used directly for heat and power generation in biomass plants (Kumar Sarangi et al., 2023). In addition, agro-biomass can be used as organic fertilizers, contributing to soil health by adding essential nutrients and reducing the dependency on chemical fertilizers (El-Ramady et al., 2022).

Agro-biomass can also be used in innovative industrial applications. For instance, agro-industrial by-products are being explored for use in the production of bioplastics, biocomposites, and biochar, which provide environmentally friendly alternatives to traditional fossil-fuel-based materials (Zhou & Wang, 2020).

25.2.2 Agro-Biomass Waste: Background and Its Types

Agro-biomass waste refers to organic materials generated as by-products from agricultural activities, encompassing any biodegradable matter produced during the cultivation, processing, and consumption of agricultural products. The significance of effective management of agro-biomass waste cannot be overstated. It is essential for minimizing environmental pollution, reducing GHG emissions, and fostering sustainable practices within agriculture and related industries. By converting these materials into valuable resources, we can mitigate waste while enhancing resource efficiency, which is critical in our quest for sustainability (Francioso, 2024). Agro-biomass waste, a by-product of agricultural and forestry activities, presents both challenges and opportunities for sustainable resource management. Generated from a variety of sources such as crop residues, animal waste, agro-industrial by-products, forestry residues, and food waste, this material can contribute to environmental degradation if not properly managed. However, with effective strategies, agro-biomass waste can be transformed into valuable resources that support a circular economy. Crop residues like straw, stalks, and husks, for instance, can be repurposed for bioenergy production or soil enrichment, while animal waste and agro-industrial by-products offer potential for renewable energy generation and organic fertilization. Forestry residues and food waste further enhance sustainability efforts through biofuel production and nutrient recycling. Understanding the different types of agro-biomass waste is crucial for developing effective waste reduction and resource recovery strategies, which can simultaneously mitigate environmental impacts and generate economic benefits.

1. **Crop Residues:**

   Crop residues constitute a substantial proportion of agro-biomass waste, consisting of the leftover plant material remaining in fields post-harvest. The main categories include:
   - **Straw**: The dry stalks of cereal crops, such as wheat and rice, are frequently left in the fields after harvest or burned, contributing to air pollution.
   - **Stalks and Leaves**: Residual materials from crops like corn and soybeans can serve as animal feed or be composted to enrich soil.
   - **Husks and Shells**: By-products from grains and legumes, like rice husks, can be processed into biofuels or organic fertilizers.

     When properly managed, crop residues can provide numerous benefits, such as improving soil fertility and generating bioenergy through anaerobic digestion or biomass combustion. These practices not only enhance agricultural productivity but also contribute to a circular economy within the agricultural sector (Kumar Sarangi et al., 2023).
2. **Animal Waste:**

   This category includes manure and other by-products from livestock and poultry, which are rich in nutrients and beneficial for soil amendment. The main types are:
   - **Manure**: Comprising solid and liquid waste from livestock, which can be processed into organic fertilizers, improving soil quality and reducing dependency on chemical inputs.
   - **Slaughterhouse Waste**: By-products from meat processing, such as bones and offal, can be converted into bioenergy or compost.

     The effective treatment of animal waste can result in the production of organic fertilizers and the generation of renewable energy through biogas production, providing a dual benefit of waste reduction and resource recovery (Subhash Babu, 2022).

3. **Agro-Industrial By-Products:**
   Agro-industrial processes generate a variety of waste materials that can be repurposed effectively. Notable examples include:
   - **Sugarcane Bagasse**: The fibrous residue left after sugar extraction, which can be used for bioenergy production or as a raw material for biocomposites, contributing to sustainable manufacturing practices.
   - **Fruit and Vegetable Peels**: Organic waste from food processing can be composted or converted into biogas, significantly reducing landfill contributions and enhancing nutrient recycling.
   - **Palm Oil Waste**: Including empty fruit bunches (EFBs) and palm kernel cake, which are valuable for energy recovery and soil enhancement.

     Transforming these by-products into usable materials helps industries minimize waste while generating economic benefits and promoting sustainability (Kumar Sarangi et al., 2023).
4. **Forestry Residues**: Residues from forestry operations, such as branches, leaves, and sawdust, also classified as agro-biomass waste. Their common applications include:
   - **Biofuel Production**: Utilizing wood chips or sawdust in biomass power generation facilities, providing a renewable energy source that can reduce reliance on fossil fuels.
   - **Composite Materials**: Transforming forestry residues into particleboard or other sustainable products, supporting the circular economy in the forest products industry.

     Proper utilization of forestry residues not only enhances carbon sequestration efforts but also contributes to the production of renewable energy and sustainable materials (Kumar Sarangi et al., 2023).
5. **Food Waste:**
   While often categorized separately, food waste generated from agricultural production also plays a significant role in agro-biomass waste. Key types include:
   - **Spoiled Produce**: Fruits and vegetables that become unsellable due to spoilage, which can be redirected to composting or biogas production to minimize environmental impacts.
   - **Processing Waste**: Leftover materials from food processing, such as seeds and skins, which can be converted into feed or bioenergy.

Transforming food waste into compost or biogas can significantly mitigate environmental impacts while enhancing nutrient recycling within ecosystems, thus closing the loop in agricultural production (Francioso, 2024).

In conclusion, the effective management of agro-biomass waste is essential for advancing sustainable agricultural practices and contributing to a circular economy. By recognizing the diverse types of agro-biomass waste—ranging from crop residues and animal waste to agro-industrial by-products, forestry residues, and food waste—stakeholders can implement strategies that transform these materials into valuable resources. Practices such as bioenergy production, organic fertilization, and composting not only reduce environmental impact but also enhance agricultural productivity and resource recovery. By fostering these waste-to-resource pathways, the agricultural sector can significantly mitigate waste, lower carbon emissions, and promote a more sustainable and resilient future. The management of agro-biomass waste is critical for promoting sustainable agricultural practices and protecting environmental health. Understanding the various types of agro-biomass waste enables stakeholders to develop effective waste reduction strategies and recovery processes, transforming waste into valuable resources. By fostering a circular economy in agriculture, where waste management and resource recovery work in tandem, we can ultimately benefit both the environment and the economy.

25.2.3 Environmental and Economic Impacts Due to Agro-Biomass Waste Mismanagement

The mismanagement of agricultural biomass waste presents significant environmental and economic challenges. Improper disposal methods, such as open burning, landfilling, or incineration, release harmful GHGs into the atmosphere, particularly methane and carbon dioxide. These emissions contribute to climate change and degrade air quality, leading to serious health implications for nearby populations (Subhash Babu, 2022). Additionally, the environmental pollution resulting from mismanaged agricultural waste can contaminate soil and water sources, threatening agricultural productivity and biodiversity (Francioso, 2024). Pollutants can leach into groundwater, leading to long-term damage to local ecosystems and impacting human health through the food chain (Phiri et al., 2023). Moreover, the degradation of soil quality due to the accumulation of toxic substances can reduce crop yields, ultimately affecting food security and economic stability in farming communities.

In terms of economic impacts, the mismanagement of agro-biomass waste leads to substantial financial losses. Farmers often incur additional costs for waste disposal or face fines due to non-compliance with environmental regulations (Francioso, 2024). The loss of potential economic benefits arises from failing to convert agricultural waste into value-added products such as biofuels, organic fertilizers, or other materials. Proper management practices could enable the recovery of valuable resources, thus promoting sustainable agriculture. However, many farmers lack the necessary knowledge or resources to implement effective waste management strategies, perpetuating a cycle of waste and loss (Kamarudin et al., 2022).

Moreover, the health-related costs associated with the environmental degradation caused by mismanaged agricultural waste can be significant. Communities affected by pollution may experience increased healthcare expenses due to illnesses linked to poor air and water quality (Subhash Babu, 2022). Respiratory diseases, gastrointestinal disorders, and other health issues can emerge from exposure to pollutants, resulting in a heavy burden on public health systems (Jayaprakash et al., 2022). This economic strain on healthcare resources further exacerbates the challenges faced by communities already struggling with the impacts of agricultural mismanagement.

The social implications of poor agricultural waste management also deserve attention. Communities may experience tensions arising from environmental degradation and health issues, leading to conflicts over resources and land use (Kamarudin et al., 2022). Additionally, the aesthetic impacts of improperly managed waste, such as unpleasant odors and litter, can decrease property values and affect the quality of life for residents. As a result, communities may become less attractive to potential residents or investors, stalling economic growth and development (Jayaprakash et al., 2022).

In contrast, effective management of agricultural biomass waste can yield numerous benefits. Implementing sustainable practices, such as composting, anaerobic digestion, and bioenergy production, can transform waste into valuable resources. These methods not only mitigate environmental impacts but also create economic opportunities by fostering green jobs and sustainable agricultural practices. The valorization of agricultural waste through innovative technologies can lead to a circular economy where waste is minimized, and resources are reused effectively (Subhash Babu, 2022).

Innovative approaches to waste management can help farmers adopt more sustainable practices and improve their economic resilience. For instance, community-based initiatives can provide training and resources to farmers, enabling them to implement better waste management strategies (Phiri et al., 2023). By enhancing their knowledge of sustainable practices, farmers can reduce waste, increase their profitability, and contribute positively to the environment (Francioso, 2024).

25.3 CURRENT ISSUES AND CHALLENGES ASSOCIATED WITH AGRO-BIOMASS WASTE MANAGEMENT ACROSS INDUSTRIES

According to the Ministry of Plantation and Commodities, the agro-biomass sector is considered one of the eight strategic and high-impact industries that regenerate economic growth in Malaysia

(KPK, Ministry of Plantation and Commodities, 2023). Based on this action plan, the Malaysian government has recognized the important role that biomass resources play in helping the country achieve its Sustainable Development Goals (SDGs). Several agricultural sectors significantly contribute to Malaysia's economy and generate biomass waste, including the palm oil sector, paddy cultivation, rubber plantations, and sugarcane farming. These industries not only provide essential products for local and international markets but also result in substantial biomass waste, which presents both challenges and opportunities for sustainable management and utilization.

25.3.1 Inadequate Collection and Processing Facilities

One significant challenge in agro-biomass waste management is the inadequate collection and processing facilities for these materials. The logistical difficulties arise from the dispersed locations of agricultural sites across the country, which complicate the collection and transport of waste to processing facilities. For example, the palm oil industry, one of the largest sectors in agriculture and the second-largest palm oil producer in the world, Malaysia encompasses an estimated 5.67 million hectares of palm oil plantations in Malaysia, producing more than 90 million tons of dried palm biomass, while many regions still lack sufficient processing infrastructure. Figure 25.1 shows the location of palm oil plantations in Malaysia. The limited number of processing facilities cannot accommodate this large number of production sites. Currently, only 60% of palm waste, including palm fibers and husks, is utilized for energy conversion, leaving the remaining 40% unexploited and considered agro-biomass waste transported to the landfill site (Pandey et al., 2020). Beyond the palm oil sector, other industries, such as sugarcane plantations and paddy fields, contribute millions of tons to production. The northern regions of Malaysia, often referred to as the country's rice bowl, are home to extensive areas dedicated to both rice cultivation and sugarcane farming. This issue is worsened by the seasonal variability of biomass waste production, as some biomass waste fluctuates with the seasons. Consequently, the seasonal production of this biomass waste leads to inconsistent supply and insufficient storage facilities during those times. For instance, paddy cultivation is classified as a seasonal crop, typically planted twice a year (Shafie et al., 2012). This seasonal nature results in the generation of large amounts of paddy biomass waste left to be wasted at once.

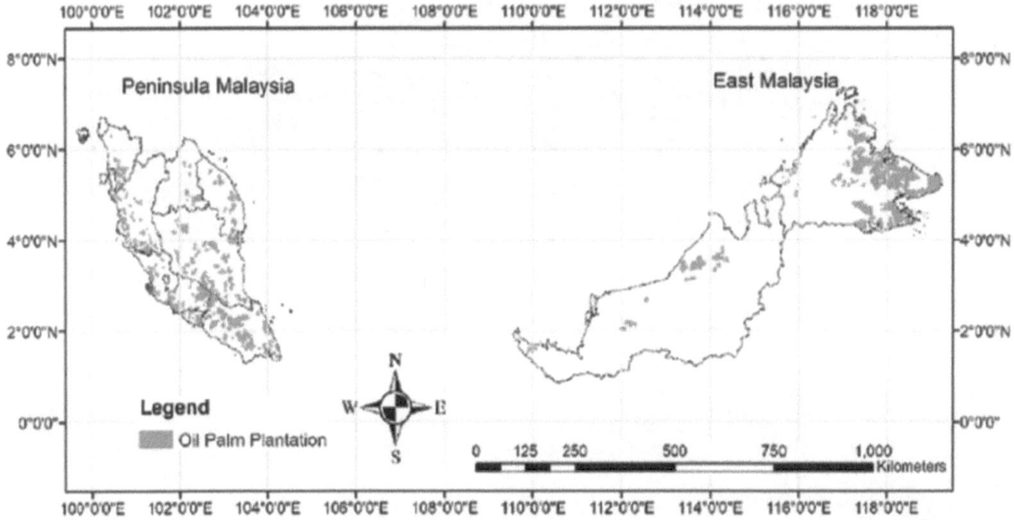

FIGURE 25.1 Location of palm oil plantations in Malaysia.

25.3.2 Establishing the Technical and Economic Viability of Effective Waste Management

An additional significant challenge in agro-biomass waste management is establishing the technical and economic viability of effective waste management solutions. Figure 25.2 shows the agro-biomass processing plant located in Teluk Intan, Perak, Malaysia. This entails developing technologies that effectively transform biomass waste into usable products while ensuring that these processes are affordable for producers. The diverse nature of agro-biomass such as sugarcane bagasse, palm oil residues, and paddy straw requires tailored processing and utilization techniques to maximize efficiency. For instance, sugarcane bagasse can be converted into bioethanol or used as a substrate for mushroom cultivation, while palm oil waste can be transformed into biodiesel or biomass pellets for energy generation (Shafie et al., 2012). Similarly, paddy straw can be processed into biodegradable materials or used for composting to enrich soil quality. Customizing these approaches not only enhances the effectiveness of waste management but also aligns with sustainable agricultural practices by reducing environmental impact. Moreover, investing in research and development can facilitate the adoption of innovative technologies, ensuring that these processes remain economically viable for producers. By maximizing the value derived from agro-biomass, Malaysia can strengthen its agricultural sector, promote a circular economy, and contribute to environmental sustainability (KPK, Ministry of Plantation and Commodities, 2023). However, the potential uses for agro-biomass waste are hindered by insufficient technology and restricted development. Consequently, a significant amount of this surplus biomass ends up in landfills, leading to environmental problems and squandering valuable resources that could be utilized for beneficial products. Maximizing the value of waste and minimizing environmental impact requires tailored processing techniques, but without

FIGURE 25.2 Agro-biomass processing plant located in Teluk Intan, Perak, Malaysia.

substantial investment in research and development, these prospects are largely unexplored. Bridging this technological gap is vital to improve the economic feasibility of agro-biomass waste management in Malaysia, encourage sustainable agricultural practices, and promote a circular economy.

25.3.3 Lack of Exposure among Agricultural Stakeholders

Other than that, the challenges arising in agro-biomass waste management are exposure to this issue. There is a lack of awareness among agricultural stakeholders about the potential uses of biomass waste, posing a significant challenge in agro-biomass waste management. This lack of understanding can lead to missed opportunities for innovation and economic growth in the sector, as many farmers, producers, and policymakers may not fully comprehend how biomass can be effectively utilized to create value-added products or contribute to sustainability goals. The regulatory landscape surrounding biomass waste management is often complex and lacks clarity, which can create uncertainty for stakeholders. Inconsistent regulations and policies make it difficult to navigate compliance requirements, potentially deterring investment in new technologies and practices. Producers may be hesitant to engage in initiatives that are not clearly supported by existing regulations due to this situation. Effective biomass waste management strategies face hindrances from these factors, impeding progress in enhancing sustainability and resource efficiency in the agricultural sector (KPK, Ministry of Plantation and Commodities, 2023). Addressing this challenge requires the implementation of targeted educational programs to raise awareness among stakeholders and the establishment of transparent, supportive regulatory frameworks that encourage innovation and the adoption of sustainable practices.

25.3.4 Lack of Effective Waste Segregation Practices

The lack of effective waste segregation practices is a major issue in managing agro-biomass waste in Malaysia's diverse agricultural environment. Efficient waste management relies on proper segregation at the source, but many agricultural operations do not have systematic methods for separating biomass from non-biomass materials. For example, in Malaysia's palm oil industry, which is crucial to the country's economy, palm fronds, EFBs, and palm kernel shells (PKSs) are often mixed with non-biomass waste, leading to contamination problems that affect the quality of biomass. Contaminated biomass is often unsuitable for processing into valuable products like biofuels, bioplastics, or organic fertilizers, resulting in reduced economic benefits. Furthermore, the presence of contaminants complicates the processing of biomass, necessitating additional steps to remove impurities, which can prolong processing times and increase operational expenses. Facilities that lack the necessary equipment to handle mixed waste may face challenges in operating profitably. This difficulty is exacerbated by reduced marketability; products derived from biomass that do not meet regulatory standards or consumer expectations due to contamination may struggle to attract buyers, negatively impacting the economic feasibility of biomass utilization projects. The mixing of waste can lead to environmental hazards if pollutants seep into the soil or water sources during disposal, making effective segregation crucial. Agricultural workers' lack of knowledge and training in proper waste management practices worsens this problem, as they may not give priority to segregation. Failure to comply with waste segregation regulations can lead to penalties, adding complexity to operations. To tackle these difficulties, various strategies can be put in place. Educational and training programs targeted at agricultural workers can increase awareness about the significance of waste segregation and provide practical implementation methods. Investing in suitable infrastructure, such as separate bins for different types of waste, can facilitate more efficient segregation in agricultural environments. Government policy support, including monetary incentives for establishing segregation infrastructure and penalties for non-compliance, can also encourage better practices (KPK, Ministry of Plantation and Commodities, 2023). Involving local communities in awareness campaigns can cultivate a sustainability mindset and promote collective action. By enhancing waste segregation practices in industries like palm oil, sugarcane, and rubber, stakeholders in the Malaysian agro-biomass sector can improve the quality and value of biomass waste, ultimately contributing to more sustainable and economically feasible waste management.

25.4 INNOVATIVE SOLUTIONS AND POTENTIAL APPLICATIONS

25.4.1 Innovative Solutions for Agro-Biomass Waste Management

In today's continuously progressing world, new and revolutionary technologies are constantly being developed to solve problems that we face today or in the future. This is especially relevant in the waste management industry as it serves a valuable function in maintaining health and safety. Due to this, various parties have proposed and researched numerous solutions for agro-biomass waste management that are renewable, inexpensive, and, most importantly, realistic.

One of the solutions proposed is regarding palm oil waste in Malaysia. As we all know, palm oil plays a crucial part in the country's economy and serves as a lifeline for a large chunk of the population. Unfortunately, the pursuit of palm oil as a renewable resource is not without its downsides as the large amount of waste it produces in the form of EFBs, palm oil mill effluent (POME), and PKSs, all contribute to growing environmental and management challenges. Early efforts seek to convert biomass into renewable sources of energy or incorporate it into building materials to strengthen while reducing the cost of said materials. The strategy proposed by Mah Jun Kit, the CEO of BioLoop Sdn Bhd, involves the usage of biotechnology, as well as black soldier fly larvae, to siphon a good amount of organic matter from landfills into usable products with economic value (Figure 25.3). The solution projects drastically faster and cleaner waste management with regard to GHGs (Aufa Mardhiah, 2024).

An alternative approach to enhance the sustainability of palm oil production involves making the process more environmentally friendly by reusing by-products in a sustainable manner. One example is the utilization of palm kernel, which was traditionally discarded, composted, or sold as low-cost livestock feed. Instead, it can be repurposed as a sustainable feed for larvae, which in turn can produce high-quality protein and fertilizers. This transformation of waste into valuable products and reintroduction into the market contributes to improving the sustainability of palm oil production and supports the development of a circular economy. To optimize this process, collaboration between palm oil producers and waste management companies is crucial. It is important

FIGURE 25.3 BioLoop's egg trap which replicates the crevices black soldier fly (BSF) prefers to lay eggs in.

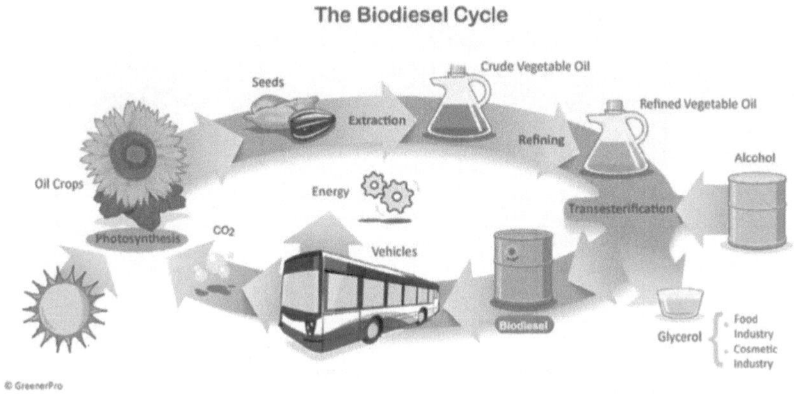

FIGURE 25.4 The biodiesel cycle.

to avoid the fragmentation of palm oil plantations, particularly in countries like Indonesia, as this could undermine waste management efforts. Additionally, expanding logistical networks between plantations and waste management facilities could enhance the overall efficiency of waste management systems, further promoting sustainable practices in the palm oil industry.

Another critical aspect that requires greater attention is the biogas industry, as it plays a significant role in the waste management of palm oil. In Malaysia, currently, only 30% of palm oil mills are capturing biogas generated from POME (Aufa Mardhiah, 2024). By harnessing this biogas, a renewable energy source can be produced, allowing the mills to become self-sufficient in energy generation. This not only enhances the sustainability of the production process but also helps maintain operational costs within a reasonable range. Reducing reliance on external energy sources can significantly lower the overall environmental impact, with an estimated reduction of approximately 5.5 million tons of CO_2 equivalent through this process (Ministry of Plantation and Commodities, 2024).

Through the conversion of waste oil, biodiesel and biojet fuel can be produced using various biochemical processes, such as transesterification and hydrotreated vegetable oil (HVO) synthesis (Aufa Mardhiah, 2024). This system is both practical and cost-effective, as it can leverage existing refinery and petrochemical infrastructure to produce HVO. The raw materials used for this process, including waste cooking oil, palm fatty acid distillates, sludge palm oil, and microalgae oil, are inexpensive and abundant. However, the system is dependent on a consistent supply of feedstock to maintain the production of both fuel and chemicals. Figure 25.4 illustrates the biodiesel cycle, highlighting the process from feedstock conversion to fuel production. One often overlooked product in the palm oil industry is biodiesel. Compared to other fuels, biodiesel is both renewable and environmentally friendly, particularly in terms of reducing GHG emissions (Ministry of Plantation and Commodities, 2024). With cost-effective policies, biodiesel production could be integrated into the palm oil industry, reducing operational costs and enhancing industry growth. Its renewable nature makes biodiesel a more sustainable alternative for long-term energy consumption.

25.4.2 Potential Application for Agro-Biomass Waste Management

Agro-waste requires appropriate pretreatment to enhance its functional properties, making its benefits more pronounced post-modification (Dey et al., 2021). Typically, agro-waste undergoes a series of physical, chemical, and biological treatments, which help to simplify the material and make it more accessible for further processing (de Souza & Shivakumar, 2021). These treatments enable the extraction of bioactive compounds, naturally occurring substances that can interact with organic matter in various ways. Incorporating bioactive compounds into food products significantly improves food quality, extends shelf life, and enhances health benefits. The growing consumer

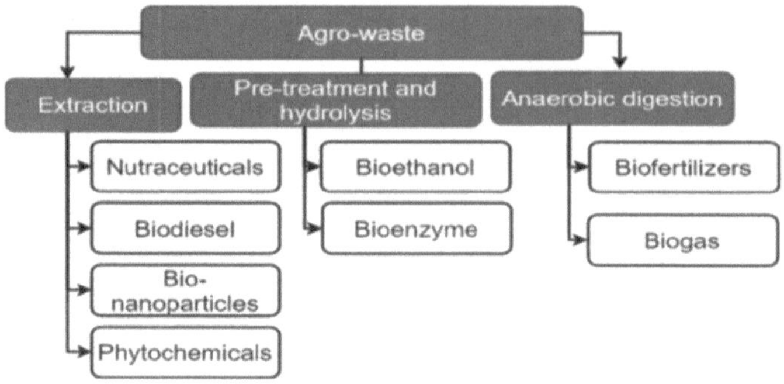

FIGURE 25.5 Agro-waste-derived bioproducts and bioactive compounds.

demand for such functional foods has led to a substantial expansion of the market for these "superior foods." To obtain these bioactive compounds, they must be extracted from agro-waste. Various bioconversion processes, spread across multiple industry sectors, are employed to produce a range of valuable bioproducts. Figure 25.5 illustrates the agro-waste-derived bioproducts and bioactive compounds, demonstrating the potential of agro-waste as a resource for developing high-value products that promote sustainability and innovation across industries.

25.4.2.1 Agriculture Sector

Organic household waste that is not repurposed can be efficiently decomposed into bioactive compounds, offering a cost-effective and sustainable solution for managing organic waste. One of the most effective methods for achieving this is vermicomposting, a process that utilizes microorganisms and macroorganisms, such as earthworms, to break down organic matter. Vermicomposting not only stabilizes organic waste but also accelerates decomposition, making it a valuable method for converting household and agricultural waste into high-quality biofertilizers. Figure 25.6 illustrates the vermicomposting process, showing how organic waste is transformed into nutrient-rich compost (Thirunavukkarasu et al., 2022).

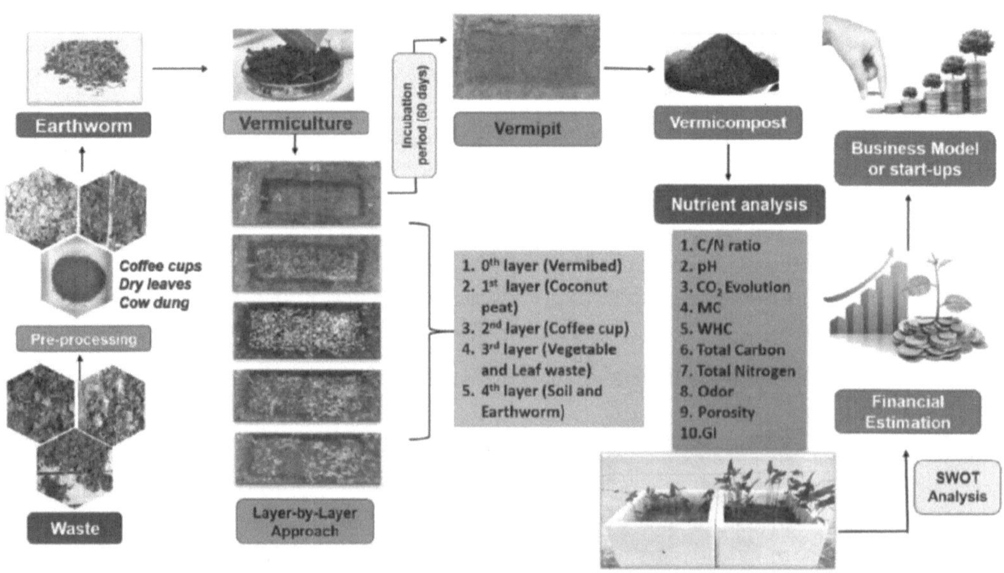

FIGURE 25.6 Vermicomposting process.

In the agriculture industry, fertilizers are essential for reintroducing nutrients into the soil to support plant growth. Recent studies have shown that biofertilizers derived from bio-waste are particularly effective, as the vermicomposting process naturally eliminates harmful elements, leaving behind a nutrient-rich product. This bio-fertilizer enhances soil health, improves crop yields, and promotes sustainable agricultural practices. Furthermore, vermicomposting reduces the environmental impact associated with traditional waste disposal, such as methane emissions from landfills, while offering a scalable solution for both household and industrial waste management. By encouraging the use of biofertilizers created through vermicomposting, the agriculture industry can adopt more eco-friendly practices, reduce dependency on synthetic fertilizers, and contribute to a more sustainable food production system.

25.5 FUTURE PERSPECTIVE

The utilization of agro-biomass in biocomposite applications and the production of advanced materials holds significant promise for sustainable development. Agro-biomass, derived from agricultural residues such as crop waste, fibers, and husks, can be processed into biocomposites, which are environmentally friendly alternatives to conventional synthetic materials. These biocomposites find applications in various industries, including automotive, construction, and packaging, offering enhanced properties such as biodegradability, reduced carbon footprint, and improved mechanical strength. For instance, agro-waste such as rice husks, coconut shells, and sugarcane bagasse have been successfully integrated into polymer matrices to produce lightweight, durable, and eco-friendly materials. In addition to biocomposites, agro-biomass is being explored for producing other advanced materials like biochar, bio-based polymers, and nanocellulose, which can be used in filtration, packaging, and medical applications. These materials contribute to a circular bioeconomy by transforming agricultural waste into high-value products, reducing landfill waste, and decreasing reliance on fossil fuel-based resources. With increasing global interest in sustainable and renewable materials, the integration of agro-biomass in the production of biocomposites and advanced materials represents a critical step toward mitigating environmental impacts, enhancing resource efficiency, and supporting green innovation in various sectors.

Sustainable management of agro-biomass waste is critical in addressing environmental degradation and ensuring the long-term viability of agricultural and industrial processes. One of the most promising approaches to managing this waste is anaerobic digestion, where organic waste is broken down in oxygen-free environments, producing biogas and nutrient-rich digestate. The biogas, composed mainly of methane and carbon dioxide, can be harnessed as a renewable energy source, reducing dependency on fossil fuels and mitigating GHG emissions. Digestate, the by-product of this process, serves as a valuable fertilizer, enriching the soil with essential nutrients and reducing the need for synthetic fertilizers, which are linked to soil degradation and water contamination. This closed-loop system improves both environmental and economic sustainability. Numerous studies support anaerobic digestion as one of the most efficient methods to manage agricultural waste, helping reduce methane emissions while promoting renewable energy use (Pandey et al., 2020; Kumar Sarangi et al., 2023; Chauhan et al., 2019).

Another critical approach to agro-biomass waste management is pyrolysis, a thermochemical process that transforms organic materials into biochar, bio-oil, and syngas. Biochar, a carbon-rich solid, is a highly effective soil amendment that enhances soil's water retention and nutrient availability, promoting sustainable agriculture. Additionally, biochar sequesters carbon, playing a crucial role in climate change mitigation by storing carbon dioxide that would otherwise be released into the atmosphere. The production of bio-oil and syngas through pyrolysis also contributes to energy generation, offering an alternative to fossil fuels. This multipurpose strategy not only addresses waste management issues but also supports energy needs and soil fertility. Research shows that biochar's soil-enhancing properties make it a vital tool for improving agricultural resilience, particularly in arid regions (Bux et al., 2023; Francioso, 2024).

Biomass conversion technologies, including combustion, fermentation, and gasification, offer additional opportunities for sustainable energy production. Combustion of biomass waste can generate heat and electricity, providing a more sustainable alternative to burning fossil fuels. Fermentation, particularly of crops like sugarcane, produces ethanol, a cleaner biofuel used in the transportation sector. Gasification, on the contrary, converts biomass into syngas, a mixture of hydrogen, carbon monoxide, and carbon dioxide, which can be used for power generation or processed into biofuels. These processes contribute to a circular economy by transforming waste into valuable energy products, reducing landfill waste, and decreasing environmental pollution. Governments and industries are increasingly adopting these technologies to align with global sustainability goals and reduce the carbon footprint of agricultural operations (Chauhan et al., 2019; Mishra & Pandey, 2005; Kumar et al., 2021).

Another environmentally beneficial method is the use of agro-biomass waste for mulching and composting. Mulching with crop residues, leaves, or straw helps conserve soil moisture, suppress weed growth, and prevent soil erosion. This method improves crop health by stabilizing soil temperatures and providing essential nutrients. Additionally, composting agricultural waste transforms organic matter into nutrient-rich compost that enhances soil fertility and structure. Composting also reduces methane emissions from organic waste that would otherwise decompose anaerobically in landfills, making it an environmentally friendly solution. These practices help maintain soil health, which is crucial for long-term agricultural productivity and reducing reliance on chemical fertilizers, which can degrade soil quality over time (Francioso, 2024; Subhash Babu, 2022).

The potential for agro-biomass waste to contribute to a circular bioeconomy is vast, especially through its use in producing bio-based products such as bioplastics, biocomposites, and biochemicals. By integrating agricultural waste into industrial processes, industries can reduce their reliance on fossil fuels and promote sustainability. For example, crop residues like rice husks and sugarcane bagasse can be processed into materials that replace petroleum-based plastics. Such bioproducts are not only biodegradable but also significantly reduce the environmental footprint of manufacturing industries. Countries such as India and Malaysia are leading in this area, with numerous research initiatives focused on maximizing the value of agro-biomass waste in industrial applications. The use of agro-waste in manufacturing not only addresses environmental concerns but also creates new economic opportunities, contributing to job creation and rural development (Kumar Sarangi et al., 2023).

In addition to technological solutions, policy incentives and educational programs play an essential role in promoting sustainable agro-biomass waste management. Governments can provide subsidies, tax reductions, and grants to encourage farmers and industries to adopt sustainable practices. Educational initiatives are also crucial in raising awareness about the benefits of proper waste management and training farmers in sustainable techniques. By integrating these programs into national and regional strategies, countries can drive the adoption of waste-to-energy technologies and support the growth of the bioeconomy. Research highlights the importance of aligning waste management practices with broader sustainability goals, such as those outlined in the U.N. SDGs, to achieve long-term environmental and economic resilience (Ghosh, 2016; Kiran et al., 2020).

In the end, sustainable agro-biomass waste management offers a range of benefits that extend beyond environmental protection. Technologies such as anaerobic digestion, pyrolysis, and biomass conversion enable the transformation of waste into valuable resources, contributing to renewable energy generation, soil health, and reduced GHG emissions. Moreover, integrating agro-waste into industrial processes helps create bio-based products that support a circular economy. By adopting these strategies, coupled with robust policy frameworks and educational initiatives, agricultural systems can become more sustainable, economically viable, and resilient to the challenges posed by climate change and resource depletion. Addressing agro-biomass waste management is thus essential not only for environmental sustainability but also for achieving long-term agricultural productivity and economic development (Subhash Babu, 2022; Kumar Sarangi et al., 2023).

25.6 CONCLUSION

In a nutshell, the mismanagement of agricultural biomass waste poses significant environmental and economic challenges, contributing to climate change, water pollution, soil degradation, and adverse health impacts on surrounding communities. Improper disposal methods such as burning or allowing waste to decay release harmful GHGs and pollutants, while also degrading the quality of agricultural land, further threatening food security. Economically, ineffective waste management leads to financial losses for farmers and missed opportunities for generating revenue through the conversion of biomass into biofuels, compost, or other value-added products. Sustainable management strategies like composting or bioenergy production can reduce disposal costs, offer new income streams, and promote a circular economy. Policy interventions are critical to driving the adoption of eco-friendly practices, with governments playing a pivotal role in offering incentives and enforcing regulations. Ultimately, transforming agro-waste into a resource through responsible management can mitigate environmental harm, enhance economic viability, and foster a more sustainable and resilient agricultural system.

ACKNOWLEDGMENT

The chapter was written with Firaas Nur Saheran, Ivy Ling Tsae Hwa, Lim Chee Jie, and Muhammad Alif Syukri Rosmadi.

REFERENCES

Aufa Mardhiah. (9 May 2024). *Innovative solutions address Malaysia's palm oil waste challenge.* The Malaysian Reserve Thursday, May 9th, 2024. https://themalaysianreserve.com/2024/05/09/innovative-solutions-address-malaysias-palm-oil-waste-challenge/

Bux, C., Cangialosi, F., & Amicarelli, V. (2023). Biomethane and compost production by anaerobic digestion of organic waste: Suggestions for rural communities in Southern Italy. *Sustainability* 2023, *15*(21), 15644. https://doi.org/10.3390/SU152115644

Chauhan, A., Tyagi, V. V., & Pandey, A. (2019). Biomass resources and conversion technologies for bioenergy: Challenges and opportunities. *Renewable and Sustainable Energy Reviews, 102*, 759–774.

de Souza L., & Shivakumar, S. (2021). Conversion of agro-industrial wastes for the manufacture of bio-based plastics. In: Kuddus, M., & Roohi (Eds.), *Bioplastics for Sustainable Development* (pp. 177–204). Springer: Singapore.

Dey, T., Bhattacharjee, T., Nag, P., Ghati, A., & Kuila, A. (2021). Valorization of agro-waste into value added products for sustainable development. *Bioresource Technology Reports, 16*, 100834. https://doi.org/10.1016/j.biteb.2021.100834

Duque-Acevedo, M., Belmonte-Ureña, L. J., Plaza-Úbeda, J. A., & Camacho-Ferre, F. (2020). The management of agricultural waste biomass in the framework of circular economy and bioeconomy: An opportunity for greenhouse agriculture in Southeast Spain. *Agronomy* 2020, *10*(4), 489. https://doi.org/10.3390/AGRONOMY10040489

El-Ramady, H., Brevik, E. C., Bayoumi, Y., Shalaby, T. A., El-Mahrouk, M. E., Taha, N., Elbasiouny, H., Elbehiry, F., Amer, M., Abdalla, N., Prokisch, J., Solberg, S. Ø., & Ling, W. (2022b). An overview of agro-waste management in light of the water-energy-waste nexus. *Sustainability, 14*(23), 15717. https://doi.org/10.3390/su142315717

Francioso, O. (2024). Current and future perspectives for biomass waste management and utilization. *Scientific Reports, 14*(1). https://doi.org/10.1038/s41598-024-59623-1

Giller, K. E., Witter, E., Corbeels, M., & Tittonell, P. (2021). Characterization of ambient volatile organic compounds, source apportionment, and health risk assessment in Zhengzhou, China. *Atmospheric Chemistry and Physics, 21*(15), 15239–15258. https://doi.org/10.5194/acp-21-15239-2021

Gontard, N., Sonesson, U., Birkved, M., Majone, M., Bolzonella, D., Celli, A., Angellier-Coussy, H., Jang, G. W., Verniquet, A., Broeze, J., Schaer, B., Batista, A. P., & Sebok, A. (2018). A research challenge vision regarding management of agricultural waste in a circular bio-based economy. *Critical Reviews in Environmental Science and Technology, 48*(6), 614–654. https://doi.org/10.1080/10643389.2018.1471957

Gupta, R., Sharma, S., & Yadav, S. (2020). Agro-residue biomass management: Issues and challenges in sustainable bioenergy production. *Renewable and Sustainable Energy Reviews, 130*, 109946.

Haider, F. U., Coulter, J. A., Cai, L., Hussain, S., Cheema, S. A., Wu, J., & Zhang, R. (2022). An overview on biochar production, its implications, and mechanisms of biochar-induced amelioration of soil and plant characteristics. *Pedosphere*, *32*(1), 107–130. https://doi.org/10.1016/S1002-0160(20)60094-7

Jain, R., Singh, B., & Sharma, R. (2020). Impacts of soil and water pollution on food safety and health risks in China. *International Journal of Civil Engineering and Technology, 6*(11), 45–58. https://iaeme.com/Home/article_id/IJCIET_06_11_004

Jayaprakash, K., Osama, A., Rajagopal, R., Goyette, B., & Karthikeyan, O. P. (2022). Agriculture waste biomass repurposed into natural fibers: A circular bioeconomy perspective. *Bioengineering*, *9*(7), 296. https://doi.org/10.3390/bioengineering9070296

Joshi, A., Kanthaliya, B., & Arora, J. (2020). Current scenario of potential renewable energy sources for sustainable development in India. *Journal of Plant Sciences Research, 35*, 205–214.

Kamarudin, S. H., Basri, M. S. M., Rayung, M., Abu, F., Ahmad, S., Norizan, M. N., Osman, S., Sarifuddin, N., Desa, M. S. Z. M., Abdullah, U. H., Tawakkal, I. S. M. A., & Abdullah, L. C. (2022). A review on natural fiber reinforced polymer composites (NFRPC) for sustainable industrial applications. *Polymers*, *14*(17), 3698. https://doi.org/10.3390/polym14173698

Kiran, S., Ahmad, T., Gulzar, T., Ashraf, A., Naqvi, S. A. R., & Naz, S. (2020). Agro-waste applications for bioremediation of textile effluents. In: Pandit, P., Ahmed, S., Singha, K., & Shrivastava, S. (Eds.), *Recycling from Waste in Fashion and Textiles* (pp. 391–421). Wiley-Scrivener: Hoboken, NJ. https://doi.org/10.1002/9781119620532.CH18

Kumar, A., Samadder, S. R., & Kumar, N. (2021). Assessment of the potential of agricultural biomass for bioenergy in India. *Energy Reports, 7*, 795–801.

Kumar, P., Kumar, S., & Joshi, L. (2015). *Socioeconomic and Environmental Implications of Agricultural Residue Burning—A Case Study of Punjab, India.* New Delhi, India: Springer Briefs in Environmental Science. https://library.oapen.org/bitstream/handle/20.500.12657/27961/1/1002038.pdf

Kumar Sarangi, P., Subudhi, S., Bhatia, L., Saha, K., Mudgil, D., Prasad Shadangi, K., Srivastava, R. K., Pattnaik, B., & Arya, R. K. (2023). Utilization of agricultural waste biomass and recycling toward circular bioeconomy. *Environmental Science and Pollution Research*, *30*(4), 8526–8539. https://doi.org/10.1007/s11356-022-20669-1

Ministry of Plantation and Commodities. (2023). National Biomass Action Plan 2023–2030. Putrajaya, Malaysia. Retrieved from https://www.kpk.gov.my/kpk/images/mpi_biomass/27122023-National_Biomass_Action_Plan.pdf

Ministry of Plantation and Commodities. (2024). National Biomass Action Plan 2023–2030. Putrajaya, Malaysia. Retrieved from https://www.kpk.gov.my/kpk/images/mpi_biomass/27122023-National_Biomass_Action_Plan.pdf

Mishra, R., & Pandey, S. D. (2005). Hazardous waste, impact on health and environment for development of better waste management strategies in future in India. *Environmental International, 31*(3), 417–431.

Mishra, S., Cheng, L., & Maiti, A. (2021). The utilization of agro-biomass/byproducts for effective bio-removal of dyes from dyeing wastewater: A comprehensive review. *Journal of Environmental Chemical Engineering*, *9*(1), 104901. https://doi.org/10.1016/J.JECE.2020.104901

Pandey, A. K., Chauhan, R. S., Kumar, S., & Rani, P. (2020). Biomass management for energy generation and biochar production: Challenges and opportunities. *Bioresource Technology Reports, 12*, 100573.

Phiri, R., Rangappa, S. M., & Siengchin, S. (2023). Agro-waste for renewable and sustainable green production: A review. *Journal of Cleaner Production*, *434*, 139989. https://doi.org/10.1016/j.jclepro.2023.139989

Sarkar, B., Meena, R. S., & Meena, V. S. (2020). Advancements and challenges in agriculture waste management: A comprehensive review. *KUEY, 6*(11), 45–58. https://kuey.net/index.php/kuey/article/view/4150

Shafie, S. M., Mahlia, T. M. I., Masjuki, H. H., & Ahmad-Yazid, A. (2012). A review on electricity generation based on biomass residue in Malaysia. *Renewable and Sustainable Energy Reviews*, 16(8), 5879–5889. https://doi.org/10.1016/j.rser.2012.06.031

Subhash Babu, A. (2022, September). *Exploring agricultural waste biomass for energy, food and feed production and pollution mitigation: A review*. ScienceDirect. Retrieved October 13, 2024, from https://www.sciencedirect.com/science/article/abs/pii/S0960852422008951

Thirunavukkarasu, A., Nithya, R., Kumar, S. M., Priyadharshini, V., Kumar, B. P., Premnath, P., Sivashankar, R., & Sathya, A. B. (2022). A business canvas model on vermicomposting process: Key insights onto technological and economical aspects. *Bioresource Technology Reports*, *18*, 101119. https://doi.org/10.1016/j.biteb.2022.101119

Zhou, C., & Wang, Y. (2020). Recent progress in the conversion of biomass wastes into functional materials for value-added applications. *Science and Technology of Advanced Materials, 21*(1), 787–804. https://doi.org/10.1080/14686996.2020.1848213

Index

Note: **Bold** page numbers refer to tables and *italic* page numbers refer to figures.

D

E

F

G

H

I

K

L

M

N

O

P

R

S

T